Molecular Structure and Statistical Thermodynamics

World Scientific Series in 20th Century Chemistry – Vol. 1

MOLECULAR STRUCTURE AND STATISTICAL THERMODYNAMICS

Selected Papers of Kenneth S. Pitzer

Editor

Kenneth S. Pitzer

University of California, Berkeley
USA

World Scientific
Singapore • New Jersey • London • Hong Kong

Published by

World Scientific Publishing Co. Pte. Ltd.
P O Box 128, Farrer Road, Singapore 9128
USA office: Suite 1B, 1060 Main Street, River Edge, NJ 07661
UK office: 73 Lynton Mead, Totteridge, London N20 8DH

Library of Congress Cataloging-in-Publication Data

Pitzer, Kenneth S. (Kenneth Santon), 1914–
Molecular structure and statistical thermodynamics: selected papers of Kenneth S. Pitzer / edited by Kenneth S. Pitzer.
p. cm. -- (World Scientific series in 20th century chemistry : vol. 1)
Includes bibliographical references.
ISBN 9810214391
1. Molecular structure. 2. Statistical thermodynamics.
I. Title. II. Series.
QD461.P57 1993
541.3'69--dc20 93-40841
CIP

We are grateful to the following publishers for their permission to reproduce the articles found in this volume:

Academic Press (*J. Chem. Thermodyn.*)
American Association for the Advancement of Science (*Science*)
American Chemical Society (*Accts. Chem. Res., ACS Symp. Series, Ind. Eng. Chem. and J. Am. Chem. Soc.*)
American Institute of Physics (*J. Chem. Phys. and J. Phys. Chem.*)
Elsevier Science Publishers (*Chem. Phys. Lett.*)
John Wiley & Sons Ltd (*Adv. Chem. Phys.*)
National Academy of Sciences (*Proc. Natl. Acad. Sci.*)
Pergamon Press Ltd (*Geochim. Cosmochim. Acta*)
Plenum Publishing Corporation (*J. Sol. Chem.*)
Royal Society of Chemistry (*Faraday Soc. Disc. and J. Chem. Soc. Chem. Comm.*)

Printed in Singapore by Utopia Press.

Kenneth S Pitzer

In the course of a distinguished career, spanning more than 50 years, Kenneth S Pitzer made the key discovery or advance in several areas. First was his pioneering research on molecules with internal rotation which was recognized by the ACS Award in Pure Chemistry in 1943. Later discoveries included the acentric factor theory for fluids, relativistic quantum chemistry for molecules with heavy atoms and the "Pitzer" equations for complex aqueous solutions. His many honors and awards include the National Medal of Science, the Priestley Medal of the ACS, the Robert A Welch Award, several honorary degrees and most recently the Clark Kerr Award in 1991. In 1949–1951, he served as the Director of Research of the U.S. Atomic Energy Commission and later as Chairman of its General Advisory Committee. In the 60's, he was President of Rice University and later of Stanford University. Today, he is Professor of Chemistry at the University of California, Berkeley, a position he has been holding since 1971

FOREWORD

This is a book of selected papers which I hope will be useful for several purposes. The cohesive aspect of certain advances is shown by bringing together papers published in different journals including, in a few cases, journals of limited circulation. This should aid historical studies. And I hope that it will give to a young reader a feeling for the general nature of research. When the territory is just being opened up, initial approaches are only partially successful and must be abandoned or altered, but a clear map eventually unfolds.

As suggested by the title, molecular structural information, statistical methods, and thermodynamic measurements are all involved throughout. But the relative role of each differs greatly. In some of the earlier work there was no spectroscopic or other molecular level source for important parameters. These could only be obtained by inference from quantum statistical mechanics combined with experimental thermodynamic measurements. But this process was not straightforward; other molecular parameters were required which should have been measurable spectroscopically, but the spectral measurements or assignments were uncertain. In later years, the molecular information was reasonably certain, but the system of interest, a liquid or dense gas, was so complex that a deductive solution was impractical. Then a molecular model and the statistical relationships were used to suggest a form of equation and, possibly, to evaluate one or two terms, but the remaining parameters were evaluated from the thermodynamic measurements.

For certain groups of papers I present a brief introduction; in two cases this calls attention to a short review paper written about a decade after that area of research was started. The reader may wish to read it first. In other cases "comments" are appended which sometimes include a reference to a subsequent review or to one or more papers extending the methods already developed. Extensive lists of subsequent papers are not given, however, and the "comments" primarily are just my current view of some aspect of the paper which is reprinted.

CONTENTS

D. Other Papers on Molecules

Part II. Dense Phases

A. Extended Corresponding States and the Acentric Factor

PART I
MOLECULES

A. Barriers to Internal Rotation

Introduction

In 1936 it was the widely accepted view that the internal rotation of methyl groups within a simple molecule such as ethane was essentially "free". The barrier (the change of potential energy) for such rotation was though to be of the order of kT at 300 K or less and the effect of such a small barrier would be trivial for macroscopic properties. But inconsistencies had been noted which demanded attention. These are cited in the early papers which show their resolution. The next-to-last paper concerns formic acid where the potential barrier has an additional cause and is considerably higher.

At the end of this group of papers is a review written in 1951 which some readers may wish to consult earlier out of sequence. The final, best equations and tables for the contribution of internal rotation to thermodynamic properties were given in a series of four papers, of which only the first and most useful is included as the second from last paper in this group. The other three papers are cited in a "comment" to that paper.

Reprinted from THE JOURNAL OF CHEMICAL PHYSICS, Vol. 4, No. 11, 749, November, 1936
Printed in U. S. A.

Hindered Rotation of the Methyl Groups in Ethane

A reliable value of the entropy of ethane from calorimetric measurements extending to 15°K has been obtained by Witt and Kemp.[1] Several investigators[2, 3] have given the treatment for obtaining the entropy of ethane from statistical-mechanical considerations, assuming a model with freely rotating methyl groups. The value of the entropy of ethane obtained from such a treatment, however, differs from the experimental value by 1.57 E.U. The reason for this disagreement could be either the assumption of an incorrect molecular model or an incorrect experimental entropy due to the persistence of rotation in the crystal at low temperatures, as suggested by Aston and Messerly[4] and Kassel[3] for disagreements in the values for other hydrocarbons.

A further discrepancy has been found between an accurate determination of the heat of hydrogenation of ethylene by Kistiakowsky and co-workers,[5, 6] and a large number of concordant experiments on the free energy of hydrogenation of ethylene Using a model of ethane requiring free rotation of the methyl groups, Teller and Topley[7] and Smith and Vaughan,[8] have been unable to correlate these data by a statistical-mechanical calculation. The assumption that ethane rotates in the crystal at low temperatures would not remove this discrepancy.

Confronted with these disagreements, we felt that the trouble lay perhaps in the assumption of completely free rotation of the methyl groups. Accordingly, we have made statistical-mechanical calculations[9] assuming various potential barriers of the form $\frac{1}{2}V_0(1-\cos 3\phi)$. Table I gives a preliminary survey of the results of these calculations, together with the experimental values of the entropy of ethane and the heat of hydrogenation of ethylene.

The values of $\Delta H_{355°}$ calculated for the reaction $C_2H_4+H_2=C_2H_6$ were obtained using the best value of the free energy of reaction selected by Teller and Topley.[7] It is seen that agreement between the various data is obtained only for a potential barrier of about 3150 cal., which, with suitable selection of the uncertain vibration frequency, we have found to be in accord with the heat capacity of ethane gas determined by Eucken and co-workers.[10, 11]

TABLE I. *The heat of hydrogenation of ethylene and the entropy of ethane.*

Calc. with	$\Delta H_{355°}$ (cal.)	$S_{184.1°K}$ (1 atmos.) cal./deg./mole
$V_0=0$ cal.	$-31{,}500\pm300$	51.21 ± 0.2
$V_0=315$ cal.	$-31{,}560\pm300$	51.10 ± 0.2
$V_0=3150$ cal.	$-32{,}940\pm300$	49.44 ± 0.2
experimental	$-32{,}824\pm50$[5]	49.64 ± 0.15[1]

In view of these considerations, we feel that the entropies of ethane and similar hydrocarbons obtained with the aid of the third law of thermodynamics from calorimetric data extending to low temperatures are the correct ones for use in thermodynamic calculations. Thus the disagreements found by Aston and Messerly[4] for tetramethylmethane and by Kassel[3] for *n*-butane are due probably to an incorrect assumption of completely free rotation of the methyl groups. A potential barrier of 3150 cal. in no way conflicts with the existing knowledge of organic chemistry concerning isomerism.

A more detailed discussion of both the experimental data and statistical-mechanical calculation will be published later.

J. D. KEMP
KENNETH S. PITZER

Department of Chemistry,
University of California,
Berkeley, California,
October 2, 1936.

[1] To be published.
[2] Mayer, Brunauer and Mayer, J. Am. Chem. Soc. **55**, 37 (1933).
[3] Kassel, J. Chem. Phys. **4**, 276 (1936).
[4] Aston and Messerly, J. Chem. Phys. **4**, 391 (1936).
[5] Kistiakowsky, Romeyn, Ruhoff, Smith and Vaughn, J. Am. Chem. Soc. **57**, 65 (1935).
[6] Kistiakowsky, Ruhoff, Smith and Vaughn, J. Am. Chem. Soc. **58**, 137 (1936).
[7] Teller and Topley, J. Chem. Soc. 876 (1935).
[8] Smith and Vaughan, J. Chem. Phys. **3**, 341 (1935).
[9] A quantum-mechanical treatment of this model has been given by Nielsen, Phys. Rev. **40**, 445 (1932).
[10] Eucken and Parts, Zeits. f. physik. Chemie **B20**, 184 (1933).
[11] Eucken and Weigert, Zeits. f. physik. Chemie **B23**, 265 (1933).

[Reprint from the Journal of the American Chemical Society, 59, 276 (1937).]

[CONTRIBUTION FROM THE CHEMICAL LABORATORY OF THE UNIVERSITY OF CALIFORNIA]

The Entropy of Ethane and the Third Law of Thermodynamics. Hindered Rotation of Methyl Groups

BY J. D. KEMP AND KENNETH S. PITZER[1]

Since the development of methods of calculating entropy from spectroscopic and molecular structure data, it has become possible to investigate the practical applicability of the third law of thermodynamics in a larger number of cases than previously. For many molecules, there is no question about the correctness of directly applying the third law to low temperature heat capacity measurements in obtaining entropies to be used in thermodynamic calculations. However, in certain cases, false equilibrium, encountered in obtaining low temperature calorimetric measurements, has complicated the use of the third law. The reasons for the complications in the several known cases are understood, and the immediate problem is to gather more information as to which molecules will be affected in order to increase the general usefulness of the third law.

The discrepancy found in hydrogen was first explained correctly by Giauque and Johnston,[2] who suggested the possibility of similar discrepancies in other molecules containing symmetrically placed hydrogens. Later investigations of the entropies of ammonia,[3] hydrogen sulfide[4] and methane[5] have shown that there is no persistence of rotation in the crystal lattices of these molecules which contain symmetrically placed hydrogens. A discrepancy has been found to exist in the case of water, but it has been explained by Pauling[6] as due to the formation of hydrogen bonds in the crystal. In all of these cases, there is no reason to doubt that the calculation of the entropies from molecular structure data is correct.

As attempts are made to deal with larger and more complicated molecules, it becomes more difficult to obtain sufficient molecular data to make even approximately correct calculations of thermodynamic quantities. An increased knowledge as to which molecules the third law may be applied unambiguously will thus be very valuable.

It is of interest therefore to consider the correlation between the third law value of the entropy of ethane obtained by Witt and Kemp[7] and the various possible values of the entropy which may be calculated assuming different molecular models.

In this paper, a comparison of the third law value for the entropy of ethane with one derived

(1) Shell Research Fellow, academic year 1936–1937.

(2) Giauque and Johnston, THIS JOURNAL, **50**, 3221 (1928). See also Giauque, *ibid.*, **52**, 4816 (1930).

(3) Overstreet and Giauque, *ibid.*, **59**, 254 (1937).

(4) Giauque and Blue, *ibid.*, **58**, 831 (1936).

(5) Giauque, Blue and Overstreet, *Phys. Rev.*, **38**, 196 (1931).

(6) Pauling, THIS JOURNAL, **57**, 2680 (1935).

(7) Witt and Kemp, *ibid.*, **59**, 273 (1937).

from reliable data on the heat and free energy of hydrogenation of ethylene will be found to prove definitely that no false equilibrium exists in ethane. This, of course, means that it is correct to use the third law value in thermodynamic calculations.

In order to make a statistical mechanical calculation of the entropy of ethane, it will be necessary to consider the energy levels of the restricted rotator.

The Energy Levels of Ethane.—Since the spectroscopic interpretation of ethane is uncertain, it was necessary to make certain assumptions with regard to the potential barrier hindering the internal rotation of the methyl groups. A potential function, $^1/_2\ V_0\ (1 - \cos 3\phi)$, which gives three humps of height V_0, was assumed to modify the relative rotation of the methyl groups. Although this form of the potential function is entirely arbitrary, it is a reasonable one and is the only form for which solutions of the wave equation are available.

The quantum mechanical problem as set up in the preceding paragraph has been solved by Nielsen[8] and by Teller and Weigert,[9] who found that the wave equation separates into two parts. One of these is the usual equation for the rotation of a symmetric top with energies

$$W_{J,K} = \left(\frac{h^2}{8\pi^2}\right)\left[\frac{J(J+1)}{I_z} + K^2\left(\frac{1}{I_z} - \frac{1}{I_x}\right)\right]$$

while the second may be transformed easily into Mathieu's differential equation. Of the acceptable solutions for the latter, some are the ordinary Mathieu functions for which the characteristic values and thereby the energies are given by Ince.[10] The energies for the remaining solutions were obtained easily with sufficient accuracy for the present purpose by approximate means.

Nuclear spin was neglected in all calculations, and the molecules were assumed to be equally divided between states with odd and even values of the quantum number K. A symmetry number of 18 was used.

For vibrational frequencies we have used the assignment of Teller and Topley[11] except for one frequency $^2\nu_\alpha{}^M$ which they fixed only from specific heat data assuming free internal rotation. We have fixed $^2\nu_\alpha{}^M$ by the same method, obtaining different values depending on the restricting potential, V_0. The order of magnitude has been restricted to about 1000 cm.$^{-1}$ as indicated by the analogous frequencies in ethylene.

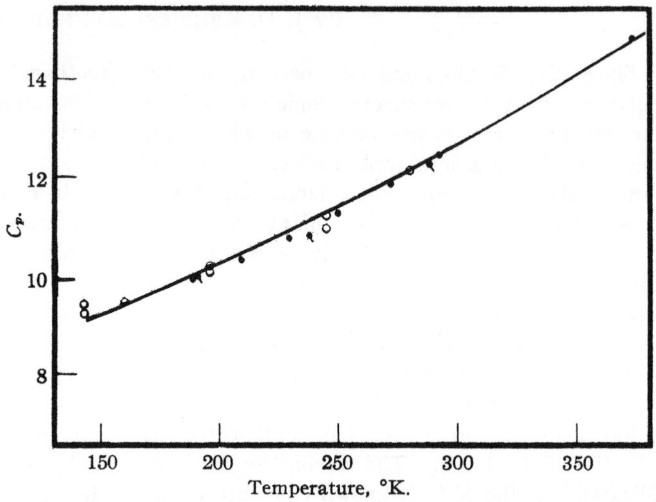

Fig. 1.—Heat capacity of ethane gas in calories per mole per degree: ● Eucken and Parts; ● Heuse; ○ Eucken and Weigert; the smooth curve represents the theoretical calculation.

It has been assumed that each methyl group has the same configuration as in methane and that the carbon–carbon distance is 1.52Å. One thus obtains $I_z = 40.1 \times 10^{-40}$ g. cm.^{2} and $I_{CH_3} = I_x/2 = 5.4 \times 10^{-40}$ g. cm.^{2}.

The Heat Capacity of Ethane Gas.—The heat capacity of ethane gas has been measured recently by Eucken and Parts[12] and by Eucken and Weigert.[13] Their results agree very well with the earlier measurements of Heuse.[14] Eucken and Parts, whose measurements extend down to the boiling point, find that their data may be fitted either by the assumption of free rotation with the unknown frequency $^2\nu_\alpha{}^M = 712$ cm.$^{-1}$ or with a torsional vibration of frequency 298 cm.$^{-1}$ and $^2\nu_\alpha{}^M = 920$ cm.$^{-1}$. Eucken and Weigert compared the thermal conductivity of ethane and ethylene and, assuming the same accommodation coefficient for both, calculated the heat capacity for ethane down to 140°K. Using the calculations of Teller and Weigert,[9] they found

(8) Nielsen, *Phys. Rev.*, **40**, 445 (1932).

(9) Teller and Weigert, *Nachr. Ges. Wiss. Göttingen. Math. physik. Klasse*, 218 (1933).

(10) Ince, *Proc. Roy. Soc. Edinburgh*, **46**, 316 (1925–26).

(11) Teller and Topley, *J. Chem. Soc.*, 876 (1935).

(12) Eucken and Parts, *Z. physik. Chem.*, **B20**, 184 (1933).

(13) Eucken and Weigert, *ibid.*, **B23**, 265 (1933).

(14) Heuse, *Ann. Physik*, **59**, 86 (1919).

that the best fit is obtained with a potential barrier of 315 cal.

We have also calculated the heat capacity of ethane in the range 140 to 400°K. and find that in addition to the combination V_0 = 315 cal., $^2\nu_\alpha{}^M$ = 750 cm.$^{-1}$,[15] a fit may also be obtained for V_0 near 3150 cal. and $^2\nu_\alpha{}^M$ = 1160 cm.$^{-1}$ as is seen in Fig. 1. The latter is, of course, essentially the second alternative of Eucken and Parts, but is based on a restricted rotator model rather than an oscillator. Since no other reasonable combinations of V_0 and $^2\nu_\alpha{}^M$ even approximate the specific heat curve, we need consider only these two in later calculations.

The Entropy of Ethane.—We have calculated the entropy of ethane from molecular structure data using the familiar expression for the rotational entropy of a rigid symmetric top and adding to it the entropy due to the hindered internal rotation of the methyl groups and the ordinary vibrational and translational contributions. A symmetry number of 18 was used.

It has been shown that in order to explain the observed heat capacity of ethane gas, a potential barrier of about 3150 cal. or one near 315 cal. must be assumed. Table I gives a comparison of the entropy of ethane, calculated from molecular structure data, assuming various potential barriers, with the calorimetric entropy obtained by Witt and Kemp.[7]

TABLE I

THE ENTROPY OF ETHANE CALCULATED FROM MOLECULAR STRUCTURE DATA COMPARED WITH THE ENTROPY OBTAINED CALORIMETRICALLY FROM THE THIRD LAW[16]

T, °K.	Potential barrier V_0, cal./mole	Statistical mechanical entropy, cal./deg./mole	Third law entropy, (exptl.) cal./deg./mole
184.1	0	51.09	
184.1	315	50.98	
184.1	3150	49.30	49.54 ± 0.15
298.1	0	56.36	
298.1	315	56.26	
298.1	3150	54.62	54.85 ± 0.2

It may be seen from Table I that there is agreement between the calorimetric and calculated entropies, within the limits of uncertainty, when a potential barrier of 3150 cal. is assumed.

A further comparison may be obtained between the third law entropy and entropies calculated from the heat and free energy of hydrogenation of ethylene. The heat of hydrogenation of ethylene has been measured accurately by Kistiakowsky and co-workers.[17] A preliminary value calculated from heats of combustion by Rossini[18] is in very good agreement with the direct determination of Kistiakowsky. The equilibrium data have been summarized by Teller and Topley[11] and the value they select as the most reliable ($\log K_p$ = 1.613 at 863°K.) was used in our calculations. The entropy change in the reaction $C_2H_4 + H_2 = C_2H_6$ may then be obtained at 355°K., provided that the heat capacities of the various gases are known from 355 to 863°K. It is given by the expression

$$\Delta S^\circ_{355} = \frac{\Delta H^\circ_{355} - \Delta F^\circ_{863}}{863} - \int_{355}^{863} \Delta C_p \left(\frac{1}{T} - \frac{1}{863}\right) dT$$

Since no experimental data are available on the heat capacity of ethane above 400°K., it was necessary to calculate that quantity, using the various molecular models found to give satisfactory heat capacities below 400°K. Fortunately, as can be seen in the expression above, the effect of the heat capacity is small, particularly at the higher temperatures. In calculating the entropy of ethane from the heat and free energy of the hydrogenation of ethylene, the assignment of frequencies of ethylene given by Bonner[19] and the moments of inertia given by Badger[20] were used. The entropy of ethylene at 355°K. obtained by using these values is 54.43 e.u. The thermodynamic constants for hydrogen were taken from the tables given by Davis and Johnston.[21] The values of the entropy of ethane so obtained depend chiefly upon the ΔF and ΔH of hydrogenation of ethylene and only to a small extent upon the assumption made regarding the potential barrier.

Table II gives a comparison between the third law value of the entropy of ethane and the value obtained from the ΔH and ΔF of hydrogenation of ethylene. For further comparison, values of the entropy obtained from a statistical mechanical calculation are also included.

The comparison of the entropies obtained from ΔF and ΔH with the entropy obtained from the third law shows that, regardless of the

(15) The difference between Teller and Topley's assignment of the remaining vibration frequencies and that of Eucken and co-workers leads to the difference in $^2\nu_\alpha{}^M$.

(16) The values in this table differ slightly from the preliminary survey by Kemp and Pitzer, *J. Chem. Phys.*, **4**, 749 (1936).

(17) Kistiakowsky, Romeyn, Ruhoff, Smith and Vaughan, THIS JOURNAL, **57**, 65 (1935). See also Kistiakowsky, Ruhoff, Smith and Vaughan, *ibid.*, **58**, 137 (1936).

(18) Kassel, *J. Chem. Phys.*, **4**, 435 (1936). A personal communication from F. D. Rossini.

(19) Bonner, THIS JOURNAL, **58**, 34 (1936).

(20) Badger, *Phys. Rev.*, **45**, 648 (1934).

(21) Davis and Johnston, THIS JOURNAL, **56**, 1045 (1934).

TABLE II

A SUMMARY OF THE ENTROPY OF ETHANE AT 355°K.

Potential barrier, V_0, cal./mole	0	315	3150
S_{355} from molecular structure data, cal./deg./mole	58.74	58.64	57.06
S_{355} from ΔH°_{355} and ΔF°_{863}, cal./deg./mole (exptl.)	56.61	56.61	57.34
S_{355} from the third law, cal./deg./mole (exptl.)		57.20 ± 0.2	

height of the potential barrier assumed, no residual entropy, such as would be left if there were false equilibrium in the crystal at very low temperatures, remains after the ordinary extrapolation of the heat capacity data to absolute zero. Indeed, for barriers lower than 3150 cal., the impossible situation of obtaining too much entropy from the third law arises. As the correctness of the third law value for the entropy of ethane is now established, the values calculated assuming potential barriers of 315 cal. or less must be in error. It is strong evidence for the existence of a potential barrier of the order of 3000 cal. per mole, which has heretofore been considered improbable.[10,22]

Unfortunately, numerous statistical mechanical calculations have been published with respect to molecules allowing similar rotations for which no potential barrier was assumed. The results for ethane suggest that the entropies so calculated may be in error by as much as 1.5 cal./deg./mole for each internal rotation.[23] Several authors have obtained such disagreements. Kassel[18] was unable to correlate the ΔF and ΔH for the hydrogenation of some higher olefins, finding entropy differences of from 2 to 3 cal./mole/deg. Since one single bond has been formed and the restricting potentials for others probably have been changed by the reaction, the apparent disagreement may well be due to the assumption of completely free rotation in the calculations. Aston and Messerly[24] in the case of tetramethylmethane obtained a disagreement of about 8 cal./deg./mole between the third law entropy and a calculated value of the entropy, while Kassel[25] obtained a similar difference of 3.76 cal./deg./mole for *n*-butane. They suggested that the calorimetric rather than the calculated entropies (assuming free rotation) are in error and ascribe the disagreement to a lack of equilibrium in the crystal at very low temperatures. The work on ethane indicates that this view is probably erroneous and that the entropies of simple hydrocarbons obtained from the application of the third law of thermodynamics to calorimetric data extending to hydrogen temperatures are the correct entropies to use in thermodynamic calculations. As we now know that the calorimetric entropies of ammonia, hydrogen sulfide, methane and ethane are the correct ones to use in thermodynamic calculations, an even more general statement may be made. That is: very probably no molecules other than hydrogen and deuterium retain any entropy due to the persistence of rotation in the crystal below 10°K.

(22) (a) Eyring, THIS JOURNAL, **54**, 3191 (1932).

(23) Where six rather than three potential humps would be present it is quite possible that free rotation may be good approximation.

(24) Aston and Messerly, *J. Chem. Phys.*, **4**, 391 (1936).

(25) Kassel, *ibid.*, **4**, 276 (1936).

Summary

The entropy of ethane has been calculated from the heat and free energy of hydrogenation of ethylene considering ethane as a restricted internal rotator. A comparison of the values so obtained, with the entropy obtained from calorimetric measurements extending to low temperatures by use of the third law of thermodynamics, proves the correctness of the third law value.

Very strong evidence has been presented for the existence of a potential barrier of about 3150 cal./mole restricting the internal rotation of the methyl groups in ethane by the comparison of the experimental values of the entropy of ethane with values obtained from statistical mechanical calculations. It is believed that the discrepancies noted by some authors in the cases of other hydrocarbons are due to the assumption of incorrect molecular models involving free internal rotation of the methyl groups. From a knowledge of the correctness of the third law entropies in the cases of ammonia, hydrogen sulfide, methane and ethane, it may be stated that very probably no molecules other than deuterium and hydrogen will have any residual entropy due to the persistence of rotation in the crystal at temperatures below 10°K.

BERKELEY, CALIF. RECEIVED NOVEMBER 16, 1936

JUNE, 1937 JOURNAL OF CHEMICAL PHYSICS VOLUME 5
Printed in U. S. A.

Thermodynamic Functions for Molecules Having Restricted Internal Rotations

KENNETH S. PITZER[1]
Department of Chemistry, University of California, Berkeley, California

(Received March 3, 1937)

The partition function is considered for molecules having restricted internal rotations, and expressions are obtained which are reasonably accurate for most actual molecules. Tables are presented which list the contributions of a single restricted rotational degree of freedom to the entropy, energy, free energy, and heat capacity.

THE presence of a potential barrier of about 3000 cal./mole restricting rotation about the single bond in ethane, which was shown to exist by Kemp and Pitzer[2] and confirmed by Howard,[3] suggests that the internal rotations of a large number of molecules may be appreciably hindered. It is therefore desirable that as general and simple a procedure as possible be developed for calculating the thermodynamic functions of these molecules. Assuming a reasonable shape of restricting potential, the pertinent theory will be briefly discussed and tables presented, which make possible easy and quite accurate calculations of the complete rotational contribution to the entropy, energy, and heat capacity of almost any molecule. The moments of inertia and potential barriers must, of course, be known, unless the calculation is to be reversed, and one or more of these quantities derived from thermodynamic data. In a second paper the method here developed will be applied to a number of molecules.

Little is known concerning the exact shape of the restricting potentials, except that they must be the same qualitatively as that expressed by $\frac{1}{2}v(1-\cos n\phi)$ where v is the height and n the number of potential maxima. This, moreover, is the only form for which the mathematics has been even partially developed, and must, consequently, be adopted. For low barriers the exact shape should make little difference, while for higher barriers the shape near the minimum is the important element so that, although a somewhat incorrect height may need to be used, the thermodynamic functions should be quite accurate.

It is, of course, necessary to use approximate methods if results of general applicability are desired. In the case of high potential barriers, when the restricted rotation becomes essentially a torsional oscillation, the method commonly used with vibrating molecules is applicable. The entropy (or other thermodynamic quantity) is

[1] Shell Research Fellow, Academic Year 1936–37.
[2] Kemp and Pitzer, J. Chem. Phys. **4**, 749 (1936); J. Am. Chem. Soc. **59**, 276 (1937).
[3] Howard, Phys. Rev. **51**, 53 (1937).

TABLE I. $(S_f - S)$ *cal./deg. mole.*

V/RT	$10^{-36}n^2/IV$ 0	1	2	4	8	16	32	64	128
0.0	0.00	0.00	0.00	0.00	0.00	0.00	0.00	0.00	0.00
0.5	0.03	0.03	0.03	0.03	0.03	0.03	0.03	0.03	0.03
1.0	0.12	0.12	0.12	0.12	0.12	0.11	0.11	0.10	0.09
2.0	0.42	0.41	0.41	0.41	0.40	0.38	0.36	0.29	0.21
3.0	0.79	0.77	0.76	0.75	0.74	0.68	0.62	0.47	
4.0	1.13	1.11	1.10	1.08	1.05	0.97	0.84	0.61	
5.0	1.42	1.39	1.37	1.35	1.29	1.18	0.98		
6.0	1.66	1.62	1.60	1.57	1.49	1.34	1.08		
7.0	1.87	1.82	1.79	1.74	1.63	1.44	1.12		
8.0	2.04	1.98	1.95	1.88	1.74	1.50	1.12		
9.0	2.19	2.11	2.07	1.98	1.81	1.54			
10.0	2.32	2.21	2.16	2.06	1.86	1.55			
12.0	2.53	2.37	2.32	2.17	1.91	1.53			
14.0	2.69	2.51	2.42	2.23	1.91	1.46			
16.0	2.83	2.63	2.50	2.27	1.89	1.36			

TABLE II. $(F - F_f)/T$.

V/RT	$10^{-36}n^2/IV$ 0	1	2	4	8	16	32	64	128
0.0	0.00	0.00	0.00	0.00	0.00	0.00	0.00	0.00	0.00
0.5	0.47	0.40	0.38	0.36	0.32	0.29	0.23	0.16	0.08
1.0	0.87	0.77	0.73	0.68	0.61	0.53	0.42	0.28	0.12
2.0	1.52	1.35	1.28	1.18	1.04	0.86	0.63	0.35	0.11
3.0	1.98	1.73	1.63	1.49	1.29	1.03	0.70	0.32	
4.0	2.33	1.99	1.85	1.67	1.43	1.10	0.70	0.26	
5.0	2.60	2.18	2.00	1.78	1.50	1.11	0.65		
6.0	2.81	2.30	2.10	1.85	1.52	1.08	0.58		
7.0	2.98	2.39	2.16	1.88	1.51	1.03	0.49		
8.0	3.13	2.45	2.20	1.89	1.48	0.96	0.38		
9.0	3.26	2.50	2.22	1.89	1.45	0.88			
10.0	3.38	2.53	2.24	1.88	1.40	0.82			
12.0	3.57	2.57	2.23	1.83	1.31	0.69			
14.0	3.72	2.59	2.21	1.76	1.20	0.57			
16.0	3.86	2.59	2.18	1.68	1.08	0.43			

taken as the sum of the translational and rotational entropies of the rigid molecule, and the entropies due to vibration, torsional or otherwise. With very high potentials one may use the harmonic oscillator or Einstein functions in which:

$$h\nu/kT = \theta/T = (nh/2\pi kT)(v/2I)^{\frac{1}{2}} = 4.463(n/T)(V/10^{40}I)^{\frac{1}{2}},$$

where I is the reduced moment of inertia (I_1I_2/I_1+I_2), n the number of potential maxima, v the height of the potential maxima in ergs/molecule, and V, the same in cal./mole. For moderately high potentials it is preferable to use the values obtained with the above potential function itself.

The quantum mechanics of the one dimensional restricted rotator has been discussed by Nielsen.[4] The wave equation in this case is easily transformed into Mathieu's differential equation:

$$M'' + [\alpha + 16q \cos 2x]M = 0, \qquad (1)$$

where $q = \pi^2 Iv/n^2h^2$ and $E - v/2 = n^2h^2\alpha/32\pi^2 I$.

The single value condition requires, in general, solutions for which the characteristic values α are not known; however, it can be shown that the characteristic values for these solutions must lie in regions which are closely enough bounded so as to allow a reasonably accurate partition function to be calculated. It is convenient to take the ratio of this function to the partition function of the free rotator, since this ratio depends on only two variables (V/RT) and (n^2/IV). By means of numerical sums the above partition function ratio and its first two temperature derivatives were calculated for a range of values of each of the two variables. (These calculations were actually made for the case $n=2$ and $I_1=\infty$, $I_2=I$, which involves only the ordinary Mathieu functions, for which some characteristic values have been tabulated by Ince[5] and Goldstein[6] and others were calculated as needed. The results, however, are quite accurately applicable to the general case, as stated above.) The results thus obtained for the quantities $(S_f - S)$, $(F - F_f)/T$, H/T and C are listed in Tables I, II, III and IV. The absolute values of S and F/T are given by the equations:

$$S = R(-0.767 + \tfrac{1}{2}\ln T + \tfrac{1}{2}\ln I \times 10^{40} - \ln n) - (S_f - S), \qquad (2)$$

$$-F/T = R(-1.267 + \tfrac{1}{2}\ln T + \tfrac{1}{2}\ln I \times 10^{40} - \ln n) - (F - F_f)/T, \qquad (3)$$

where n is the symmetry number of the free rotator which will ordinarily be the same as the n in the potential function. If there are n potential minima, at least approximately equal, yet the molecule is not identical in the various equilibrium positions, an approximate treatment may be obtained with the n's not the same. A

[4] Nielsen, Phys. Rev. **40**, 445 (1932). See also Pauling, Phys. Rev. **36**, 430 (1930); Teller and Wiegert, *Nachr. Ges. Wiss. Göttingen. Math. physik. Klasse.* (1933) p. 218.

[5] Ince, Proc. Roy. Soc. Edinburgh, **46**, 316 (1925–26).

[6] Goldstein, Trans. Camb. Phil. Soc. **23**, 303 (1927).

case of this sort, n-butane, will be discussed in the following paper.

As the potential barrier becomes lower the uncertainty in the above treatment becomes greater, since the molecule spends an increasing fraction of the time in configurations appreciably different from the equilibrium one. In order to investigate this situation a study will be made of the limiting high temperature (or large moment of inertia) form of the partition function:

$$Q=\frac{1}{h^m}\int_{-\infty}^{+\infty}\cdots\int_{-\infty}^{+\infty}\int_{\alpha_m}^{\beta_m}\cdots\int_{\alpha_1}^{\beta_1}e^{-1/kT(K.E.+\Phi)}\times dq_1\cdots dq_m dp_1\cdots dp_m, \quad (4)$$

where $K.E.$ and Φ are the kinetic and potential energies, respectively, m the number of rotational degrees of freedom, and α_1, β_1 the range of the ith angular coordinate. Edinoff and Aston[7] have shown that this form gives a very satisfactory approximation in the case of free rotation, while for small barriers the degree of approximation may be estimated by noting the variation of the functions in Tables I to IV with (n^2/IV), holding (V/RT) constant. The first column $(n^2/IV=0)$ was calculated with the aid of this limiting function as is explained below. Using the results of Edinoff and Aston[7] who discussed the integration with respect to the momenta, one obtains:

$$Qr=\frac{(2\pi kT)^{m/2}}{h^m}\int_{\alpha_m}^{\beta_m}\cdots\int_{\alpha_1}^{\beta_1}[A]^{\frac{1}{2}}\cdot\times e^{-\Phi/kT}dq_1\cdots dq_m, \quad (5)$$

where $[A]$ is the determinant of

$$2K.E.=\sum_{i,\,j=1}^{m}A_{ij}\dot{q}_i\dot{q}_j$$

and Φ may be expressed:

$$\Phi=\sum_{i=4}^{m}\tfrac{1}{2}v_i(1-\cos n_iq_i).$$

It may be noted at this point that the limits of integration of all internal coordinates will be 0 and 2π and that by substituting $q_1'=n_1q_1$ the n's may be removed except possibly for the factor $[A]^{\frac{1}{2}}$. For many molecules one may take

$$[A]=\sin^2 q_1\prod_{i=1}^{m}I_i$$

in which the I's are all constants, moments of inertia. This gives:

$$Qr=\frac{(2^9\pi^7k^3T^3I_1I_2I_3)^{\frac{1}{2}}}{h^3}\prod_{i=4}^{m}\left(\frac{(2\pi kTI_i)^{\frac{1}{2}}}{h}\times\int_0^{2\pi}e^{-(v_i/kT)(1-\cos x)}dx\right). \quad (6)$$

TABLE III. H/T.

V/RT	$10^{-36}n^2/IV$								
	0	1	2	4	8	16	32	64	128
0.0	0.99	0.99	0.99	0.99	0.99	0.99	0.99	0.99	0.99
0.5	1.43	1.36	1.34	1.32	1.28	1.25	1.19	1.12	1.04
1.0	1.74	1.64	1.60	1.55	1.48	1.41	1.30	1.17	1.02
2.0	2.09	1.93	1.86	1.76	1.63	1.47	1.26	1.05	0.89
3.0	2.18	1.95	1.86	1.73	1.54	1.34	1.07	0.84	
4.0	2.19	1.87	1.74	1.58	1.37	1.12	0.85	0.64	
5.0	2.17	1.78	1.62	1.42	1.20	0.92	0.66		
6.0	2.14	1.67	1.49	1.27	1.02	0.73	0.49		
7.0	2.10	1.56	1.36	1.13	0.87	0.58	0.36		
8.0	2.08	1.46	1.24	1.00	0.73	0.45	0.25		
9.0	2.06	1.38	1.14	0.90	0.63	0.33			
10.0	2.05	1.31	1.07	0.81	0.53	0.26			
12.0	2.03	1.19	0.90	0.65	0.39	0.15			
14.0	2.02	1.07	0.78	0.52	0.28	0.10			
16.0	2.02	0.95	0.67	0.40	0.18	0.06			

TABLE IV. C.

V/RT	$10^{-36}n^2/IV$								
	0	1	2	4	8	16	32	64	128
0.0	0.99	0.99	0.99	0.99	0.99	0.99	0.99	0.99	0.99
0.5	1.06	1.06	1.06	1.06	1.06	1.06	1.06	1.05	1.04
1.0	1.22	1.22	1.21	1.21	1.21	1.20	1.19	1.17	1.14
2.0	1.69	1.68	1.68	1.67	1.65	1.61	1.54	1.43	1.22
3.0	2.10	2.08	2.07	2.05	2.01	1.91	1.73	1.47	
4.0	2.29	2.27	2.25	2.21	2.13	1.98	1.73	1.34	
5.0	2.33	2.30	2.27	2.23	2.12	1.89	1.56		
6.0	2.32	2.26	2.22	2.16	2.01	1.70	1.31		
7.0	2.27	2.19	2.12	2.05	1.85	1.52	1.09		
8.0	2.22	2.11	2.04	1.92	1.68	1.34	0.91		
9.0	2.18	2.05	1.94	1.80	1.53	1.19			
10.0	2.15	1.99	1.86	1.68	1.39	1.03			
12.0	2.11	1.89	1.73	1.45	1.15	0.77			
14.0	2.08	1.81	1.61	1.24	0.93	0.55			
16.0	2.06	1.75	1.51	1.06	0.72	0.34			

[7] Edinoff and Aston, J. Chem. Phys. 3, 379 (1935).

In (6) the first factor pertains to the free rotation of the molecule as a whole while each of the other factors pertain to an internal rotation with restricting potential v_i. The integral in the latter factors has been evaluated by suitable expansions for various values of (v/kT), the results being used to obtain the contribution of a single degree of freedom to the functions $(S_f - S)$, $(F - F_f)/T$, H/T and C as listed in the first columns of Tables I, II, III and IV, respectively. It is apparent from Eq. (6) that for this group of molecules, the treatment of the entropy as a sum of independent contributions is valid for all magnitudes of restricting potential.

Where this treatment is not exact, it is, nevertheless, a good approximation, particularly for high potential barriers. This can be seen from Eq. (5) in which the principal contribution to the integral will be from configurations for which Φ is small. Therefore, if $[A]$ varies only slowly with the coordinates, it can be removed from the integral and given the value corresponding to the configuration of minimum potential. The partition function will be most in error under this procedure in the case of free rotation, when, however, an exact value for the integral in Eq. (5) can often be obtained.[7, 8] Thus an upper limit can be set to this error for all values of the potential. For the temperature derivatives of the partition function no such simple relationship holds; however, an analysis of their behavior with a continuous and slowly varying $[A]$ indicates that the limiting error in $-F/T$ will apply in order of magnitude at least to the other thermodynamic functions. Fortunately in most actual molecules the limiting error in $-F/T$ will probably be small, of the order of 0.1 cal./deg. mole. If the uncertainty of this method is too great the proper procedure appears to be the further analysis of Eq. (5) for the individual molecule or type of molecules at hand.

The accuracy of the data in Tables I to IV is about 0.02 cal./deg. mole except for certain portions of Table IV which may contain values erroneous by 0.05 cal./deg. mole. However, considering the arbitrary form of the potential function and the approximate methods involved in application to actual molecules, the results obtained with the aid of these tables cannot ordinarily be considered significant beyond 0.1 cal./deg. mole. It should also be emphasized that this treatment will fail at very low temperatures.

As a summary, the methods of calculating thermodynamic quantities for molecules having restricted internal rotations will be briefly reviewed.

1. If (V/RT) is of the order of one or greater, the desired function is calculated for the rigid molecule and to this is added contributions from the various oscillational degrees of freedom. Unless (V/RT) is very large the contributions from torsional oscillations should be calculated from Eqs. (2) and (3) and Tables I to IV.

2. If (V/RT) is fairly large (greater than about three) this result may be considered accurate; however, if (V/RT) is smaller the result may be somewhat in error. An upper limit to the error in $-F/T$ may be found by comparing the result obtained by this method for a hypothetical molecule having free rotation, with the accurate value for that molecule.[7, 8] The limiting error thus obtained for $-F/T$ applies approximately for S, H/T and C.

3. For very small values of (V/RT) it is preferable to obtain S or $-F/T$ by applying a small correction taken from Table I or II to the value calculated for the freely rotating molecule.

[8] Kassel, J. Chem. Phys. **4**, 276 (1936); **4**, 435 (1936).

JUNE, 1937 JOURNAL OF CHEMICAL PHYSICS VOLUME 5
Printed in U. S. A.

Thermodynamics of Gaseous Hydrocarbons: Ethane, Ethylene, Propane, Propylene, *n*-Butane, Isobutane, 1-Butene, *Cis* and *Trans* 2-Butenes, Isobutene, and Neopentane (Tetramethylmethane)

KENNETH S. PITZER[1]
Department of Chemistry, University of California, Berkeley, California

(Received March 3, 1937)

It is pointed out that the assumption of completely free internal rotation in the simpler hydrocarbon molecules is probably responsible for the discrepancies between the results of previous statistical mechanical calculations and the experimental data. Using the formulas and tables of the preceding paper, calculations are presented which show that, for reasonable values of rotation restricting potentials, complete agreement can be obtained with all experimental results. The uncertainty as to the exact height and shape of these potential barriers, together with the possible errors in estimated vibration frequencies make highly precise calculations of thermodynamic functions out of the question at present. Nevertheless the general agreement with experiment indicates that the potentials and frequencies selected must be approximately correct. These molecular structure data together with the available values of heats of combustion and hydrogenation are then employed in calculations which yield thermodynamic constants and the free energy of formation as a function of the temperature in the range from 300 to 1500°K. The various calculations have been made for all of the hydrocarbons listed in the title.

THE rotational partition functions for several of the simpler hydrocarbons have been calculated with the assumption of completely free rotation about single carbon—carbon bonds.[2, 3] The results of these calculations are not, however, in accord with experimental data, and in the case of ethane the error has been shown to be the assumption of free internal rotation.[4, 5] These results strongly suggest that a situation similar to that in ethane exists in many other molecules. In the preceding paper[6] the procedure is given for treating molecules having restricted internal rotations; however, moments of inertia, potential barriers, and vibration frequencies need to be known. The moments of inertia for the simpler hydrocarbons can be calculated with sufficient accuracy by using the reasonably well established values of the interatomic distances, while the vibration frequencies are known or can be satisfactorily estimated in many cases. The potential barriers are not known independently for any case, but if they are obtained from thermodynamic data for a few molecules, values can sometimes be interpolated for others. It is thus apparent that purely theoretical-spectroscopic calculations of thermodynamic functions are not possible at present; nevertheless in many cases the procedure and formulas for such calculations are useful as a framework for the correlation of all experimental data available, whether thermodynamic or spectroscopic in nature. If all of the various molecular constants can be determined by this procedure, then accurate calculations of thermodynamic functions can be made just as if the spectroscopic data were complete. In the present paper the various paraffins and olefins having less than four carbon atom and tetramethylmethane are discussed from this point of view. It is found that for reasonable values of potential barriers and vibration frequencies complete agreement with all experimental data can be obtained. While it is not possible at present to make precise calculations of the free energy or other functions, reasonably accurate values of thermodynamic constants are given and approximate values are tabulated for the free energy in the region from 300 to 1500°K.

MOLECULAR STRUCTURE DATA AND CALCULATED ENTROPIES

In the following calculations the C—C distance was taken as 1.53A, the C=C distance as 1.36A and the C—H distance as 1.11A. The angle between single bonds from a given carbon atom was assumed to be the normal tetrahedral angle. These data suffice for the calculation of

[1] Shell Research Fellow, academic year 1936–1937.
[2] Edinoff and Aston, J. Chem. Phys. **3**, 379 (1935).
[3] Kassel, J. Chem. Phys. **4**, 276 (1936); **4**, 435 (1936).
[4] Kemp and Pitzer, J. Chem. Phys. **4**, 749 (1936); J. Am. Chem. Soc. **59**, 276 (1937).
[5] Howard, Phys. Rev. **51**, 53 (1937).
[6] Pitzer, J. Chem. Phys. **5**, 469 (1937).

TABLE I. *Translation-rotation entropies for rigid molecules.*

SUBSTANCE	S_{t+r}	S_{t+r} at 298.1°K
Ethane	$7.21+4R \ln T$	52.50
Propane	$13.33+4R \ln T$	58.62
Isobutane	$15.22+4R \ln T$	60.51
Neopentane	$14.24+4R \ln T$	59.53
(St)*n*-butane	$16.33+4R \ln T$	61.62
(Be)*n*-butane	$17.40+4R \ln T$	62.69
Isobutene	$15.76+4R \ln T$	61.05

moments of inertia and thereby translation-rotation partition functions for either a rigid molecule or a freely rotating one. The resulting entropies for the latter case have been presented by Kassel[3] while those for the former appear in Table I.

The vibration frequencies of the simpler hydrocarbons have been discussed by Kassel[3, 7] and the treatment used here will be very similar, including, however, more recent experimental results. The vibrations of the hydrogens are grouped into three classes: those involving primarily C−H bond distances, H−C−H bond angles, and H−C−C bond angles. The frequencies of vibrations of the first type are always very near 3000 cm^{-1} and may be taken as such without appreciable error. Each hydrogen contributes one such frequency. The H−C−H bending frequencies are 1306 (triple) and 1536 (double) in methane;[8] 1460, 1370, 1480 (double), and 1460 (double) in ethane;[5] 1342 and 1443 in ethylene.[9] For these vibrations an averaged frequency of 1440 cm^{-1} will be used. Each methyl group contributes three such frequencies, while each methylene group contributes one. The third type of vibrations, which depend essentially on the H−C−C angle, have the lowest frequencies and are therefore the most important. Ethane has two such frequencies, each double. The value 827 cm^{-1} can be definitely assigned to one of these. The other has been estimated as 1005 cm^{-1} by Howard[5] from a force constant calculation, and as 1160 cm^{-1} by Kemp and Pitzer[4] from a consideration of gas specific heat data. In ethylene[9] the frequencies of this group have the values 1100, 950, 950, and 940 cm^{-1}. Although the frequencies for ethylene appear to be slightly higher than those for ethane, the difference is insignificant for the present purpose and the value 950 cm^{-1} will be used throughout. Each methyl group has two such frequencies, each $-CH_2-$ group three, each $=CH_2$ group two, and each CH group two. The total number of degrees of freedom in each class is shown in Table II for a number of molecules.

The vibrations of the carbon skeleton are easily classified as bending or stretching modes. From both the available spectral data and the calculations of Kassel,[7] it is apparent that for each C−C bond there is a frequency near 1000 cm^{-1}. The data on ethylene[9] suggest that

[7] Kassel, J. Chem. Phys. **3**, 326 (1935).
[8] Childs, Proc. Roy. Soc. **A153**, 555 (1936).
[9] Bonner, J. Am. Chem. Soc. **58**, 34 (1936).

TABLE II. *Data concerning internal degrees of freedom.*

SUBSTANCE	CARBON SKELETON VIBRATIONS	HYDROGEN VIBRATIONS 950	1440	3000	TORS. VIB.	S AT 298.1°K	RESTRICTED ROT. V, $I\times10^{40}$	S_{rr} AT 298°K
Ethane	1000 (1)	4	6	6		0.62	3000, 2.7 (1)	1.70
Propane	380 (1), 1000 (2)	7	7	8		2.12	3400, 4.7 (2)	4.0
Isobutane	380 (3), 1000 (3)	8	9	10		4.44	3800, 5.2 (3)	5.9
Neopentane	380 (5), 1000 (4)	8	12	12		6.65	4200, 5.3 (4)	7.5
n-Butane	320 (2), 1000 (3)	10	8	10		4.14	{ 3400, 5.0 (2) 3800, 18.6 (1) }	7.3
Propylene	380 (1), 1000 (1), 1600 (1)	6	4	6	570 (1)	2.38	<800, — (1)	—
1-Butene	320 (2), 1000 (2), 1600 (1)	9	5	8	570 (1)	4.4	{ 3400, 5.0 (1) <800, — (1) }	—
Cis 2-butene	320 (2), 1000 (2), 1600 (1)	8	6	8	270 (1)	5.4	<800, — (2)	—
Trans 2-butene	320 (2), 1000 (2), 1600 (1)	8	6	8	270 (1)	5.4	<800, — (2)	—
Isobutene	380 (3), 1000 (2), 1600 (1)	6	7	8	570 (1)	4.6	1800, 5.2 (2)	5.7

The numbers in parentheses or in the columns under "Hydrogen Vibrations" indicate the number of degrees of freedom. The contributions to thermodynamic quantities may be calculated using tables of Einstein functions for the vibrational degrees of freedom, and using the equations and tables of the preceding paper[4] for the restricted rotational degrees of freedom. The symmetry number for the restricted rotations is three, except for the central bond of *n*-butane and 1-butene. These two rotations are discussed in the text.

for the C=C bond the value should be about 1600 cm^{-1}. The bending frequencies are much lower and consequently much more important. The values from Raman spectra[10] are for propane 374; for *n*-butane 222, 320 and 428; for isobutane 370, 438; and for tetramethylmethane 330, 416 cm^{-1}. The data for the series propane, isobutane, neopentane are thus reasonably well known and the experimental frequencies may be used; or an averaged value, 380 cm^{-1}, may be taken for each of the one, three, and five degrees of freedom, respectively, which are present in these molecules. The case of *n*-butane is less certain, since two isomers are probably present in equilibrium. This situation will be discussed below. In any event no gross error would be made if the two expected frequencies are assigned the value 320 cm^{-1}. Unfortunately data are not available for propylene or the various butenes; consequently it was necessary to make some assumption. Comparison of the bending frequencies of ethane and ethylene, as mentioned above, indicates that the carbon skeleton bending trequencies of the olefins are probably but slightly higher than those of the corresponding paraffins, and consequently the same values will be used. The selected vibration frequencies are summarized in Table II.

The only remaining data needed are those concerning the internal rotational oscillational degrees of freedom. Unfortunately these frequencies are active as fundamentals in neither the Raman nor the infrared spectra. Thus their determination by these methods is difficult at best. In the case of the double carbon — carbon bond the rotation restricting barriers are so high that the harmonic oscillator treatment is satisfactory. Bonner, in discussing ethylene, gives 825 cm^{-1} as the most probable, though not completely certain value of this frequency. Since better data are not available this value will be used to determine the force constant for all the olefins. The frequency then becomes about 570 cm^{-1} when heavy groups are substituted at one end as in propylene, 1-butene and isobutene, and about 270 cm^{-1} for both of the 2-butenes. Fortunately this frequency appears but once, so that any moderate error would not be serious. The frequency cannot be much smaller than the value given in the case of ethylene, for the total vibrational heat capacity of the gas is only 0.08 cal./deg. at 143°K.[11]

The heights of potential barriers restricting rotation about single bonds have, as yet, been determined only from thermodynamic data, although Howard[5] has employed spectral data to set a lower limit of 2000 cal./mole for the barrier in ethane. Previous calculations[4] have shown that the various thermodynamic data yield a value of 3000 cal./mole for ethane, while an application of the formulas and tables of the preceding paper[6] to the accurate third law entropy of tetramethylmethane, which was obtained by Aston and Messerly,[12] gives 4200 cal./mole for that case. Both of these values are probably correct within about 300 cal. The interpolation of 3400 cal. for propane and 3800 cal. for isobutane should be a safe procedure. The situation of the end methyl groups in *n*-butane and 1-butene is so similar to that in propane as to allow considerable confidence in the use of the same height of barrier. The central bond in *n*-butane does not have exact threefold symmetry, although there are presumably three maxima and minima. About all that can be said is that the potential maxima are probably about 3800 cal. high and that the minimum corresponding so the straightened position is most likely lowest.

The single bonds adjacent to a double bond will have a plane group on one end and a group with threefold symmetry (at least approximately) on the other end. As the two sides of the plane group become more equal, the three potential barriers will tend to be canceled out, eventually leaving six much lower ones. Consequently one would expect in propylene, 1-butene, and the 2-butenes a much lower barrier than in ethane. Actually heat and free energy of hydrogenation data indicate quite free rotation, the potential maxima being not more than 800 cal. high. Isobutane, however, has two methyl groups at the same end of the double bond, and their mutual restriction should predominate over any from the other end of the molecule, giving a higher barrier in this case. The experimental

[10] Landolt-Bornstein, *Tabellen*.

[11] Eucken and Parts, Zeits. f. physik. Chemie **B20**, 184 (1933).

[12] Aston and Messerly, J. Am. Chem. Soc. **58**, 2354 (1936).

TABLE III. *Comparison of calculated and observed free energies for hydrogenation reactions.*

OLEFIN	$\Delta F/T_{\text{exp}}$	$\Delta F/T_{\text{calc}}$ ($V=0$)	$\Delta F/T_{\text{calc}}$ ($V=800$ cal.)
Propylene	12.16	12.26	12.37
1-Butene	12.16	12.53	12.64
Cis 2-butene	11.90	12.06	12.28
Trans 2-butene	11.02	11.24	11.46
Isobutene	9.15	(8.54)	9.38 ($V=1800$ cal.)

TABLE IV. *Comparison of calculated and third law entropies for olefin hydrocarbons at their respective boiling points.*

SUBSTANCE	b.p.	S_{exp}	S_{calc} ($V=0$)	S_{calc} ($V=800$ cal.)
Propylene	225.3	60.4	61.8	61.5
Cis 2-butene	276.8	72.7	72.4	71.9
Trans 2-butene	274.1	70.5	71.7	71.2
Isobutene	266.0	67.8	(70.5)	68.8 ($V=1800$ cal.)

In Tables III and IV, V is the potential barrier restricting rotation about bonds adjacent to the double bond.

value is 1800 cal./mole, obtained chiefly from the heat and free energy of hydrogenation, but giving some weight to the less accurate third law entropy. The resulting data regarding potential barriers are listed in Table II, while the calculations involving heat and free energy of hydrogenation data are summarized in Table III and the comparison with third law entropies in Table IV.

Before considering in detail these comparisons with experiment, the calculations for n-butane and 1-butene should be further discussed. All the other molecules have internal rotations involving only symmetrical methyl groups; consequently the simple treatment given in the preceding paper[6] is at once applicable for any height of potential barrier. The torsional oscillations about the double bonds are not included as rotations in the preceding statement; however, their treatment as harmonic oscillators is satisfactory.

Rotation about the middle bond in 1-butene seems to be so slightly restricted that the free rotation calculation of Kassel[3] should be quite satisfactory. A small correction can be made for a low barrier if desired.

In n-butane, on the other hand, the potential is large compared to RT at ordinary temperatures. Consequently, the rotational partition function should be set up for the rigid molecule in the equilibrium positions. Of these one may be described as a configuration in which the molecule is straightened out and the other two as "bent" configurations which are mirror images of each other. The translation-rotation entropies for these molecules are listed in Table I, wherein an amount $R \ln 2$ has been added to the entropy of (Be) n-butane because of the two optically isomeric forms present. Strictly, the internal entropy contributions would be different for the (St) and (Be) forms, however, this difference will be neglected since it is probably small but not readily ascertained. We may then write for the reaction:

$$\text{(St) } n\text{-butane} = \text{(Be) } n\text{-butane}$$

$$R \ln K = -\Delta F^\circ/T = 1.07 - \Delta H_0^\circ/T,$$

where ΔH_0° is probably positive and not more than a few hundred calories. Assuming for the moment that ΔH_0° is known, then the mole fractions N_{St} and N_{Be} can be obtained and the entropy of the mixed gas written:

$$S = N_{\text{St}}S_{\text{St}} + N_{\text{Be}}S_{\text{Be}} - R(N_{\text{St}} \ln N_{\text{St}} + N_{\text{Be}} \ln N_{\text{Be}}).$$

Although the magnitude of ΔH_0° is not known, for temperatures near 300°K, values of ΔH_0° up to 300 cal. have a negligible effect on the entropy while a value of 600 cal. reduces the entropy by less than 0.3 cal./deg. mole from the result for ΔH_0° equal to zero. Since the temperatures of interest are near or above 300°K and it seems unlikely that ΔH_0° is more than 600 cal. one may use the result for a zero value with considerable confidence. Note should be made of one additional element concerning the calculation of the entropy contribution from the torsional oscillation about the central C−C bond. While the symmetry number for a rotator would be one, nevertheless a term $(-R \ln 3)$ is introduced because only the energy levels corresponding to oscillation about one of the three minima are desired. For calculation of the quantity $(10^{-36}n^2/IV)$ or the equivalent harmonic oscillator frequency, n should be given the value three. If the existence of the different forms of n-butane had been neglected, and averaged values of the moments of inertia used, then the $(-R \ln 3)$ term would be omitted as would the entropy of mixing terms which have as a maximum value, ($+R$ ln 3). This treatment, however, would be less accurate.

Comparison with Experiment

The experimental data with which comparison can be made are entropies obtained from low temperature heat capacities and combinations of heat and free energy data for hydrogenation reactions. The entropies of all of the paraffins here considered except propane are known experimentally, however the more accurate values which are for ethane and tetramethylmethane have been used to fix the heights of potential barriers. The assumptions made in the calculations for propane and isobutane were reasonably safe and very similar; consequently a satisfactory agreement in the case of isobutane will serve as confirmation of the values for both molecules. Parks, Shomate, Kennedy, and Crawford[13] obtained 67.2±1.1 cal./deg. mole for the entropy of isobutane gas at its boiling point (260.9°K) while the calculated value is 67.7 cal./deg. mole.

Huffman, Parks, and Barmore[14] obtained 72.5 cal./deg. mole for the entropy of *n*-butane at its boiling point. Recently the extropolation below liquid-air temperatures has been reconsidered by Parks and co-workers,[13] who give a revised value of 73.7 cal./deg. mole. The calculated result 72.8 cal./deg. mole lies between those obtained from the experimental data. This is a very satisfactory confirmation of the calculated value and indicates that the rather uncertain assumptions were at least approximately correct.

The heats of hydrogenation at 355°K of ethylene, propylene, and the various butenes

[13] Parks, Shomate, Kennedy and Crawford, J. Chem. Phys. **5**, 359 (1937).

[14] Huffman, Parks and Barmore, J. Am. Chem. Soc. **53**, 3876 (1931).

Table V. *Thermodynamic constants for twelve hydrocarbons in the gaseous state.*

Substances	$S°_{298.1}$ cal./deg. mole	Constants of formation from $H_{2(g)}$ and $C_{(\beta\text{-graphite})}$: $\Delta S°_{298.1}$ cal./deg. mole	$\Delta H°_{298.1}$ kcal./mole	$\Delta F°_{298.1}$ kcal./mole	$\Delta H_0°$ kcal./mole
Methane	44.60±0.0	−19.22	−18.07±0.12	−12.34	−16.15
Ethane	54.82±0.2	−41.59	−20.60±0.23	− 8.20	−16.86
Propane	64.8 ±0.5	−64.2	−25.39±0.32	− 6.25	−20.14
n-Butane	75.1 ±1	−86.5	−30.57±0.43	− 4.78	−23.90
Isobutane	70.8 ±0.8	−90.8	−32.20±0.43	− 5.13	−25.45
Neopentane	73.7 ±0.5	−120.5	−40.7 ±1.2	−4.8	−32.4
Ethylene	52.75±0.1	−12.43	+12.09±0.25	+15.80	+14.06
Propylene	65.1 ±0.7	−32.7	+4.48±0.38	+14.23	+8.12
1-butene	75.4 ±1	−55.0	−0.41±0.5	+15.99	+4.72
Cis 2-butene	73.6 ±1	−56.0	−2.14±0.5	+14.79	+2.92
Trans 2-butene	73.0 ±1	−57.4	−3.09±0.5	+14.02	+1.97
Isobutene	71.3 ±1	−59.5	−4.02±0.5	+13.72	+0.85

The uncertainty in $S°_{298.1}$ applies also to $\Delta S°_{298.1}$ and the uncertainty in $\Delta H°_{298.1}$ to $\Delta F°_{298.1}$ and $\Delta H_0°$.

Table VI. *Values of $\Delta F°/T$ of formation for twelve hydrocarbons in the gaseous state.* (*Units: cal./deg. mole*).

T, °K	Methane*	Ethane	Propane	*n*-Butane	Isobutane	Neo-pentane	Ethyl-ene	Propyl-ene	1-Bu-tene	*Cis* 2-Butene	*Trans* 2-Butene	Iso-butene
300	−40.9	−27.0	−20.3	−15.2	−16.4	−15.1	52.8	47.7	53.7	49.7	47.2	46.1
350	−32.2	−17.0	− 8.1	− 0.5	− 0.9	+ 4.5	47.2	45.8	54.1	51.0	48.8	48.0
400	−25.5	− 9.4	+ 1.4	+10.9	+11.2	19.8	43.0	44.5	54.6	52.1	50.4	49.8
450	−20.2	− 3.2	8.9	20.0	20.7	31.7	39.9	43.6	55.2	53.2	51.7	51.3
500	−16.0	+ 1.8	15.1	27.4	28.4	41.5	37.6	43.1	55.8	54.2	53.0	52.6
600	− 9.4	9.7	24.5	38.7	40.2	56.4	34.0	42.4	56.9	56.0	55.1	54.9
700	− 4.5	15.4	31.5	47.1	49.0	67.5	31.7	42.0	57.9	57.6	56.8	56.7
800	− 0.8	19.7	37.0	53.6	55.9	76.0	30.0	41.9	58.9	59.0	58.3	58.3
900	+ 2.1	23.1	41.2	58.8	61.2	82.6	28.7	41.9	59.7	60.1	59.6	59.6
1000	4.5	25.7	44.6	62.6	65.1	87.5	27.6	41.8	60.1	60.7	60.4	60.5
1200	8.1	29.1	48.7	67.2	69.8	93.4	25.9	41.3	60.0	61.1	61.0	61.3
1500	11.8	31.5	50.9	68.8	71.8	96.2	22.8	39.0	58.1	59.7	59.6	59.9

* The values for methane are taken from Kassel, J. Am. Chem. Soc. **55**, 135 (1933).

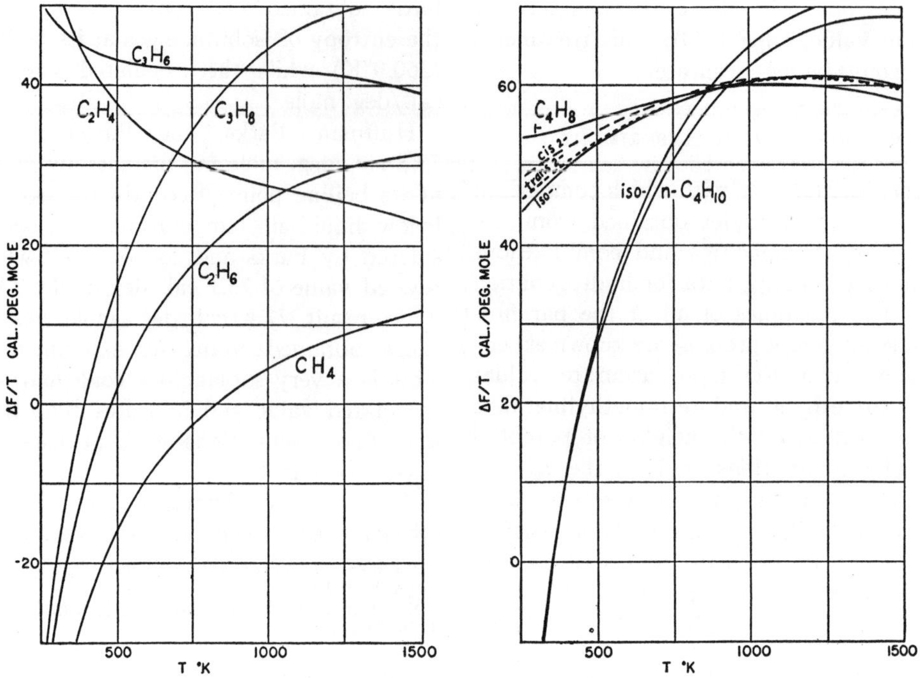

FIG. 1. The free energy of formation of gaseous hydrocarbons.

have been accurately measured by Kistiakowsky and co-workers.[15] The equilibrium constants of the same reactions have been obtained by Frey and Huppke[16] for several temperatures, including 673°K. Table III gives a comparison of the experimental values of $(-R \ln K = \Delta F/T)$ for 673°K with those calculated with the aid of the heat data. In all the olefins except isobutene, the potentials restricting rotation should be about the same, and rather small. Table III shows that the agreement is satisfactory for values up to about 800 cal., since the experimental errors may be taken as at least 0.4 cal./deg. mole. With isobutane a higher potential is expected, and a value of 1800 cal. gives a good agreement.

The experimental entropies of the olefins are all from the work of Parks and co-workers[14, 17] and have been increased over the originally published values by 10 percent of the extropolated portion as has been recently recommended.[13] The comparison with calculated values in Table IV shows complete agreement within the experimental errors.

Thermodynamic Functions for Twelve Hydrocarbons

While the partition functions set up in the preceding paragraphs are not very precise, the general agreement of thermodynamic quantities calculated from them with experimental values indicates that they must be approximately correct. These partition functions appear to be, in fact, accurate enough for use with the best heat of formation data now available without significantly increasing the uncertainties except at high temperatures. Table V contains the values of several thermodynamic functions at 298.1°K and the values of the heat of formation at the absolute zero for twelve hydrocarbons. In Table VI the free energies of formation are listed for various temperatures in the range from 300 to 1500°K. The data of Table VI are presented graphically in Fig. 1.

In the various calculations the heats of formation of the paraffins are taken from the work of

[15] Kistiakowsky, Romeyn, Ruhoff, Smith, and Vaughan, J. Am. Chem. Soc. 57, 65 (1935); Kistiakowsky, Ruhoff, Smith, and Vaughan, ibid. 57, 876 (1936).
[16] Frey and Huppke, Ind. Eng. Chem. 25, 54 (1933).
[17] Todd and Parks, J. Am. Chem. Soc. 58, 134 (1936).

Rossini,[18] while the values for the olefins are obtained by combining the heat of hydrogenation[14] with the heat of formation of the resulting paraffin. In the case of ethylene, precise heat of combustion data[19] are available and the value used is a compromise one. The remaining calculations were made using the data and methods described in this and the preceding paper.[6] All values of the constants of formation are based on H_2 gas and β-graphite. The contribution due to nuclear spin is omitted from all entropies. In obtaining the values listed in Tables V and VI the following constants were used either in the present work or that of Rossini.

For $H_{2(g)}+1/2O_{2(g)}=H_2O_{(l)}$,

$$\Delta H°_{298.1}=-63.313\pm0.01 \text{ kcal./mole},$$

$C_{(\beta\text{-graphite})}+O_{2(g)}=CO_{2(g)}$,

$$\Delta H°_{298.1}=-94.24\ \pm0.10 \text{ kcal./mole},$$

$C_{(\text{diamond})}=C_{(\beta\text{-graphite})}$,

$$\Delta H°_{298.1}=-\ 0.22\ \pm0.07 \text{ kcal./mole},$$

[18] Rossini, Nat. Bur. Stand. J. Research 13, 21 (1934); 15, 357 (1935); J. Chem. Phys., 3 438 (1935).
[19] Rossini, Nat. Bur. Stand. J. Research 17, 629 (1936).

1 kcal. = 1000 cal. = 4183.3 int. joules;

for H_2, $S°_{298.1}=31.23$ cal./deg. mole

and $(H°_{298.1}-H_0°)=2.028$ kcal./mole;

for $C_{(\beta\text{-graphite})}S°_{298.1}=1.36$ cal./deg. mole

and $(H°_{298.1}-H_0°)=0.25$ kcal./mole.

In Table VI some of the values have been obtained by graphical interpolation. The values for 450, 1000 and 1500°K were directly calculated while those calculated for 298.1 and 763°K (in Tables III and V) were also used in making the interpolations. The slope of the curve at 298.1°K was made to correspond to the known $\Delta H°$ at that temperature. It should be noted that the values in Table VI are not sufficiently precise to allow accurate differences to be taken. Thus heat contents and heat capacities must be calculated separately if other than very rough estimates are desired.

The data in Table VI should be useful in obtaining equilibrium constants for a large number of hydrocarbon reactions.

Errata: Thermodynamics of Gaseous Hydrocarbons: Ethane, Ethylene, Propane, Propylene, *n*-Butane, Isobutane, 1-Butene, *Cis* and *Trans* 2-Butenes, Isobutene, and Neopentane (Tetramethylmethane)

(J. Chem. Phys. 5, 473 (1937))

The calculated values of the free energy of formation of the various hydrocarbons at 1500°K as listed in Table VI are in error. Certain of the interpolated values are also slightly affected. The corrected values are given below:

TABLE VI. *Values of $\Delta F°/T$ of formation for hydrocarbons in the gaseous state. (Units: cal./deg. mole.)*

T, °K	800	900	1200	1500
Ethane	19.7	23.1	29.9	34.5
Propane	37.0	41.2	49.5	54.9
n-Butane	53.5	58.6	68.5	73.8
Isobutane	55.7	61.0	71.0	76.8
Neopentane	75.8	82.3	94.9	112.2
Ethylene	30.0	28.7	26.1	24.8
Propylene	41.8	41.8	41.9	42.0
1-Butene	58.8	59.5	61.1	62.1
Cis 2-Butene	58.8	59.9	62.1	63.7
Trans 2-Butene	58.2	59.5	61.8	63.6
Isobutene	58.3	59.6	62.1	63.9

KENNETH S. PITZER

Department of Chemistry,
University of California,
Berkeley, California.

Reprinted from THE JOURNAL OF CHEMICAL PHYSICS, Vol. 8, No. 9, 711–720, September, 1940
Printed in U. S. A.

The Vibration Frequencies and Thermodynamic Functions of Long Chain Hydrocarbons

KENNETH S. PITZER
Department of Chemistry, University of California, Berkeley, California

(Received February 28, 1940)

A method is developed for calculating thermodynamic functions for long chain molecules with particular attention to normal paraffins. In this connection the infinite chain approximation method for calculating vibration frequencies is considered. Starting with the results of Kirkwood, a modification is made which improves the agreement with the exact values for the simpler cases. In addition this method of attack is extended to out of plane motions. This vibrational analysis shows that all skeletal frequencies for molecules of the normal paraffin type can be put into two groups, one fairly narrow band near 1000 cm⁻¹, and a broader band extending from 0 to 460 cm⁻¹. The partition function is then set up on the assumption that motions in the low frequency group can be treated classically, and that the high frequency band can be replaced by a suitable number of 1000 cm⁻¹ frequencies. Contributions from hydrogen atom vibrations are added on later. A formula is finally obtained which is quite simple, considering the complexity of the molecules. The calculated entropies can be brought into agreement with experimental values on the basis of very reasonable internal rotation restricting barriers.

THE writer has recently measured with considerable accuracy the entropies of *n*-heptane and of 2,2,4 trimethyl pentane.[1] In calculating statistical entropies for comparison, it became apparent that the existing methods are not very satisfactory for such molecules.[2] By making extensive use of the integral form of the partition function, a somewhat different method is presented below which has certain advantages. Considerable use also is made of the interesting scheme for calculating vibration frequencies recently described by Kirkwood[3] and in somewhat different form by Thomas and Whitcomb[4] and by Barriol.[5] Kirkwood's formulation will be used with, however, one important modification and with an extension to include out of plane vibrations.

[1] Pitzer, J. Am. Chem. Soc. **62**, 1224 (1940).
[2] Most recently Huggins (J. Chem. Phys. **8**, 181 (1940)) has published calculations of the entropies of the normal paraffins using a method quite different from that given below. The molecular picture assumed by Huggins is, in the writer's opinion, not as detailed and specific as that assumed below. Consequently the latter method is likely to give more molecular structure information. From an empirical point of view the formula given below is linear in the number of carbons while Huggins' contains also a logarithm term. While the available experimental values are probably not accurate enough to decide this point, they seem to be better fitted by the simpler linear formula which was also extensively used by Parks and Huffman (reference 13).
[3] Kirkwood, J. Chem. Phys. **7**, 506 (1939).
[4] Thomas and Whitcomb, Phys. Rev. **56**, 383 (1939).
[5] Barriol, J. de phys. et rad. **10**, 215 (1939).

Let us consider the vibration frequency problem first. With Kirkwood we shall separate the motions of the hydrogen atoms from those of the carbon skeleton, and shall consider the latter to lie at equilibrium in a planar, zigzag chain. The interesting step is treatment of the problem as one of an infinitely long chain after which those modes of vibration which are approximately correct for the molecule in question are selected. We shall consider first the motions within the plane, i.e., bond stretching and bending but not internal rotation, and shall take Kirkwood's nomenclature. He finds in his Eq. (7):

$$\begin{aligned}\omega^2 &= \omega_0^2 \pm [\omega_0^4 - \omega_1^4]^{\frac{1}{2}},\\ \omega_0^2 &= \alpha(1+\cos\gamma\cos\lambda)\\ &\quad +2\beta(1+\cos\lambda)(1-\cos\gamma\cos\lambda),\\ \omega_1^4 &= 8\alpha\beta(1+\cos\lambda)\sin^2\lambda,\end{aligned}\tag{1}$$

where α and β are the ratios k_1/m and k_2/m, of the bond stretching and bending force constants to the effective mass m of a CH_2 group, γ is the C–C–C bond angle, and ω is 2π times the frequency. The motion of the nth particle is given in terms of ω and λ as follows:

$$\xi_n = Ae^{i(\omega t+n\lambda)}, \quad \eta_n = (-1)^n Be^{i(\omega t+n\lambda)}, \tag{2}$$

where ξ is the displacement from equilibrium along the length of the chain and η is the displacement perpendicular to the chain. The wave-

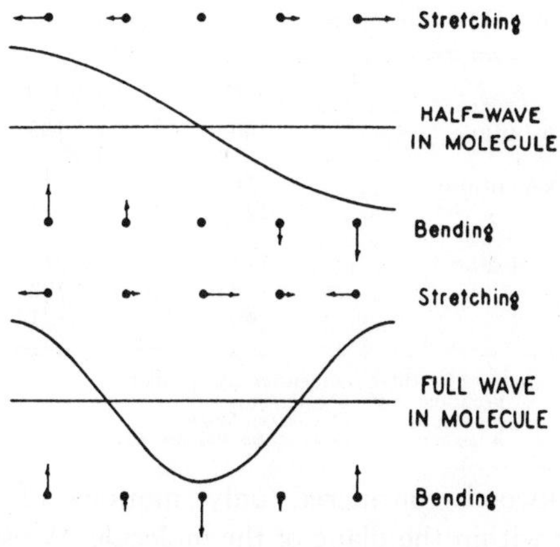

FIG. 1. Illustration of the longitudinal and transverse (stretching and bending) modes of motion for a linear pentatomic molecule with half-wave and full wave in the molecule length.

length of the longitudinal motion is thus $2\pi/\lambda$ in terms of a chain link as a unit. The transverse motion has the wave-length $2\pi/(\pi-\lambda)$.

The next step is to select values of λ so as to approximate the motion of a finite chain of N elements. Kirkwood followed Born and Karman in using the condition:

$$\lambda = 2\pi l/N, \tag{3}$$

which the writer does not believe to be quite correct.* In Eq. (3) l is to have integral values from 0 or 1 to $N-1$. In this form the maximum wave-length is N or the length of the molecule for the fundamental longitudinal mode. However, in the true fundamental stretching mode of vibration, the half-wave-length corresponds to the chain length. The better requirement is that an integral number of half-wave-lengths must be completed within the molecule, or:

$$\lambda = \pi l/N \qquad l = 1, 2, 3, \cdots N-1. \tag{4}$$

On this basis the bending modes also fit more satisfactorily. The mode of longest wave-length closely approximates a rotation of the molecule in the plane and is to be discarded on that basis, as was stated by Kirkwood. The next mode closely approximates the actual fundamental bending mode of vibration of the molecule. Some of the higher bending modes include slight rotations of the molecule. However, the error from this source is very small and seems to be inherent in this approximation method. Certainly this error is less serious than taking modes of vibration that fail to qualitatively correspond to the true ones. Figure 1 illustrates for a straight chain the motions just discussed.

It is interesting to note that this correction removes the peculiar degeneracy found by Kirkwood. The correspondence with the true frequencies is greatly improved, especially for the smaller molecules. This is shown in Table I. Here we have compared for ethane, propane, butane, and pentane, the frequencies calculated by Kirkwood's method, those calculated by the modification of it just discussed, those calculated with the accurate equations of Kassel,[6] and the experimental values. Kassel's equations approximate the CH_2 group as a unit and assume the same forces and equilibrium configuration, but are exact solutions for the molecules in question. Thus a comparison of the columns headed "Kirkwood" and "This Research" with that headed "Kassel" offers a good test of the approximations of the two infinite chain methods. Comparison of the latter with the "observed" tests the remaining approximations. The force constants are the same for each method of

TABLE I. *The skeletal vibration frequencies of some simple hydrocarbons* (cm^{-1}).

	Calculated by Method of:			
	Kirkwood	This Research	Kassel	Observed
Ethane	1120	970	970	993
	1034(?)	(408)		
Propane	1057	1057	1050	1053
	1057	934	915	867
	208	447	375	375
	(208)	(208)		
n-butane	1120	1087	1082	1058
	1034	970	970	954
	970	960	945	835
	970	408	359	326
	408	380	288	?
	408(?)	(142)		
n-pentane	1102	1102	1090	1073
	1102	1025	1020	1033
	933	983	977	900
	933	933	937	837
	460	460	394	334
	460	317	320	?
	90	304	190	?
	(90)	(90)		

* This error is entirely trivial for the applications to crystals which interested Born and Karman.

[6] Kassel, J. Chem. Phys. **3**, 326 (1935).

calculation but differ slightly from those used by Kirkwood and Kassel.*

The higher or stretching frequencies are surprisingly well approximated by the infinite chain method. The more serious approximations appear to be rather in the separation of the hydrogen vibrations and possibly in an oversimplified expression of forces.

The infinite chain method is less satisfactory for the lower or bending modes. However, the results are always of the correct general magnitude. It is pleasing that the spurious frequency which is a rotation of the individual molecule, takes a value much smaller than the other bending frequencies.

The assignment of observed frequencies[7] for butane and pentane is quite an arbitrary procedure at present, since the planar configuration here assumed is only one of many possible ones.

We may give as equations for calculation the following:

$$\begin{aligned}
\nu_{\pm}(l) &= \bar{\nu}[\tfrac{1}{2}\eta(l) \pm \tfrac{1}{2}\zeta(l)]^{\frac{1}{2}}, \quad l=1, 2, \cdots, N-1 \\
\bar{\nu} &= (\tfrac{1}{2}\pi)(2k_1/m)^{\frac{1}{2}}, \quad \epsilon = k_2/k_1 \\
\eta &= 1+\cos\gamma \cos(\pi l/N) \\
&\quad +2\epsilon[1+\cos(\pi l/N)] \\
&\quad \times[1-\cos\gamma\cos(\pi l/N)], \\
\zeta^2 &= \eta^2 - 8\epsilon[1+\cos(\pi l/N)]\sin^2(\pi l/N).
\end{aligned} \tag{5}$$

For the values in Table I, γ was given the tetrahedral angle, $\bar{\nu}$ was given the value 970 cm^{-1} and ϵ, the bending to stretching force constant ratio, was given the value 0.088. These seemed to give the best all-around agreement.

* The writer wishes to express his appreciation to Professor Kirkwood for his comments after reading this section. There is no absolutely correct criterion for selecting those modes most closely approximating the true ones in a finite molecule. Kirkwood suggests the requirement of zero acceleration for the center of mass, and states that either the Born-Karman condition or the one suggested above is sufficient. Thus further considerations are necessary. Besides the wave-length criterion for qualitative similarity to the true modes suggested in the text, one may also mention that a requirement of the proper number of different modes of vibration would lead definitely to the new condition, for molecules with N odd. For N even the latter argument does not lead to a definite result, however, the same condition might be selected by analogy. In any event the results in Table I justify a claim of superiority for the new condition; and Professor Kirkwood states definitely his agreement with this conclusion.

[7] Kohlrausch and Köppl, Zeits. f. physik. Chemie **B26**, 209 (1934).

TABLE II. *Out of plane vibration frequencies* (cm^{-1}).

	EXACT*	INFINITE CHAIN
n-butane	140	198
n-pentane	156	214
	121	157
n-hexane	172	210
	148	198
	86	121

* The equations for these frequencies are as follows:
n-butane $\omega^2 = 4k_3/m$
n-pentane $\omega^2 = 3k_3/m,\ 5k_3/m$
n-hexane $\omega^2 = 3k_3/2m,\ 9k_3/2m,\ 6k_3/m$.

Kirkwood considered only motions of the atoms within the plane of the molecule. We shall find some interest in the modes of motion where the atoms move perpendicular to this plane and will treat this problem in an analagous manner. These motions are the beginnings of internal rotations, but if of small enough amplitude they may be treated as harmonic oscillations. If the potential energy is connected with rotation about bonds, we may take an increase of energy of $\frac{1}{2}k_3(\Delta Z)^2$ when of four successive atoms one is a distance ΔZ from the plane of the other three. In these terms the equation of motion of the nth atom is:

$$m\ddot{Z}_n + k_3(4Z_n - Z_{n-1} - Z_{n+1} - 2Z_{n-2} - 2Z_{n+2} + Z_{n-3} + Z_{n+3}) = 0 \tag{6}$$

where the Z's are perpendicular distances from the plane of the infinite chain at equilibrium. Then making the usual substitution:

$$Z_n = Ae^{i(\omega t + n\lambda)}, \tag{7}$$

we find:

$$\begin{aligned}
\omega^2 &= (k_3/m)(4-2\cos\lambda - 4\cos 2\lambda + 2\cos 3\lambda). \\
\lambda &= \pi l/N, \quad l=1, 2, \cdots, N-1, \quad \nu = \omega/2\pi.
\end{aligned} \tag{8}$$

Of these solutions the ones with $l=1$ and $l=N-1$ should be discarded as corresponding to rotations. The latter approximates rotation about an axis along the chain, the former about an axis perpendicular to the chain.

In Table II the frequencies calculated by this method are compared with exact values calculated for each of the molecules using the same forces. For the simpler molecules the exact equations are easily obtained by the usual methods.

It is seen in Table II that the infinite chain method is even less satisfactory for the out of plane vibrations. However, it may still be of value in a statistical sense for very long molecules.

It is interesting to note that the success of the infinite chain method is closely connected with the range of the principal forces involved. Thus the bond stretching depends on two atoms, the bond bending on three, and the out of plane bond rotations on four.

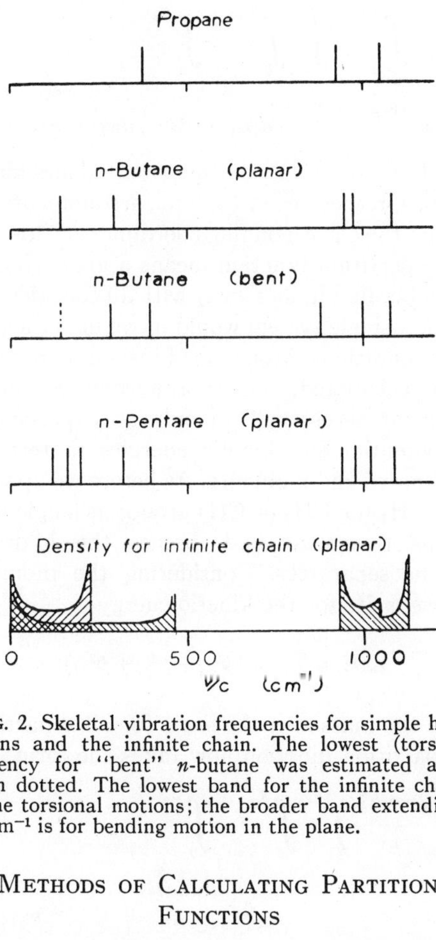

FIG. 2. Skeletal vibration frequencies for simple hydrocarbons and the infinite chain. The lowest (torsional) frequency for "bent" n-butane was estimated and is shown dotted. The lowest band for the infinite chain is for the torsional motions; the broader band extending to 460 cm^{-1} is for bending motion in the plane.

Methods of Calculating Partition Functions

Before actually setting up the partition function for the chain molecules, it seems desirable to discuss general methods briefly. It would be very desirable, if possible, to obtain a general formula for the n atom chain. The usual methods do not permit this generalization, except possibly on an empirical basis after calculating a number of specific examples. Approximations which would be undesirable in simple cases are worth while here if they simplify the complex problem, particularly since only relatively rough molecular data are available for these molecules. This work aims at an accuracy of a few tenths cal. per degree.

If we consider the distribution of skeletal vibration frequencies in these molecules, as summarized in Fig. 2, the possibility immediately suggests itself of replacing the high frequency band by a suitable number of 1000 cm^{-1} frequencies. This approximation will be reasonably satisfactory for all lengths of chain from propane up. Also if we may generalize from n-butane, it will be satisfactory for various bent configurations in addition to the planar one. On the other hand similar approximations for the low frequency group would not be satisfactory, particularly when one considers that these bands extend to zero frequency.

Another possibility which arises is that the low frequency group may be treated by the classical or integral partition function. Although results obtained thereby will necessarily fail at low enough temperatures, this is not necessarily a serious limitation for practical purposes. It will be shown below that the classical formula allows the separation of a factor for each additional atom in the chain. This, combined with the approximation of a number of 1000 cm^{-1} frequencies permits the desired type of formula.

Our fundamental equation can then be taken as:

$$Q = Q^{\text{classical}}_{(\text{trans, rot, tors, bend})} Q^{n-1}_{(1000\ \text{cm}^{-1}\ \text{vib})}, \qquad (9)$$

where the first term is a classical partition function for translation, rotation, torsion, and bending and the second term accounts for $(n-1)$ vibrations of frequency 1000 cm^{-1}. For convenience in calculation we make the transformation:

$$Q = \{ Q^{\text{classical}}_{(\text{trans, rot, tors, bend})} [Q^{\text{classical}}_{(10{,}000\ \text{cm}^{-1}\ \text{vib})}]^{n-1} \} \times [Q(1000\ \text{cm}^{-1}\ \text{vib.}) / Q^{\text{classical}}_{(1000\ \text{cm}^{-1}\ \text{vib})}]^{n-1}. \qquad (10)$$

Now the factors in the braces comprise a complete classical partition function which can be set up more easily than one wherein only the low frequency skeletal motions are considered. This interpretation depends upon the fact established above, that the molecules will have $(n-1)$ motions of approximately 1000 cm^{-1} frequency.

The form of Eq. (10) suggests the possibility of including similar factors for the low frequency bands to correct for small deviations from the classical formula in these cases. In order to investigate this possibility, we make the following expansions which are valid when $(h\nu/kT)$ is less than about 1.7. This includes the low frequency bands at or above room temperature.

$$Q_{(\nu)}=1/(1-e^{-h\nu/kT}),\quad Q_{(\nu)}{}^{\text{class}}=kT/h\nu,$$

$$[Q(\nu)/Q_{(\nu)}{}^{\text{class}}]=(h\nu/kT)/(1-e^{-(h\nu/kT)})$$
$$=1+\tfrac{1}{2}(h\nu/kT)+1/12(h\nu/kT)^2+\cdots,$$

$$[(F/T)-(F/T)^{\text{class}}]$$
$$=-R[\tfrac{1}{2}(h\nu/kT)-1/24(h\nu/kT)^2+\cdots],$$

$$[(H/T)-(H/T)^{\text{class}}]$$
$$=-R[\tfrac{1}{2}(h\nu/kT)-1/12(h\nu/kT)^2+\cdots],$$

$$[S-S^{\text{class}}]=R[1/24(h\nu/kT)^2+\cdots],$$

$$[C-C^{\text{class}}]=-R[1/12(h\nu/kT)^2+\cdots].$$

Under these conditions, the equations show that the use of the arithmetic mean for even a broad band of frequencies would give the correct leading terms in (F/T) and (H/T). These leading terms may be regarded as corrections for zero point energy in $(H_0{}^0)$ and therefore of little practical value, as is indicated by their disappearance in S and C. For the square terms which lead in S and C, a root mean square frequency should be used in approximating a band. This will not differ grossly from the ordinary mean if the band has a definite upper limit even though it continues to zero. On the other hand the computation of the partition function itself requires, in the classical case, the geometric mean frequency which would be grossly affected by the continuance of a band to zero. Thus other methods must and will be used to calculate the classical function itself. If $(h\nu/kT)$ is about one, the second-order corrections are about 0.1 cal. per degree or less and are consequently almost negligible.

We may summarize this section as follows. A partition function may be calculated as a product of the complete classical function and correcting factors for the higher frequencies provided one of the following conditions is fulfilled: (1) the exact frequencies are used in the corrections: (2) mean frequencies are used for narrow bands: (3) mean frequencies are used when $(h\nu/kT)$ is less than about 1.7. In any case force constants must be used in the classical function which are consistent with the vibration frequencies.

The Partition Function for the General Normal Paraffin

Following the method outlined above we first calculate the complete classical partition function.[8]

$$Q=\frac{1}{h^{3n}}\int_{-\infty}^{\infty}\cdots\int_{-\infty}^{\infty}\int_0^{X_0}\cdots\int_0^{Z_0}$$
$$\times e^{-(\text{P.E.}+\text{K.E.})/kT}dp_{x1}\cdots dp_{zn},\, dx_1\cdots dz_n. \quad (11)$$

Here P. E. and K. E. are the potential and kinetic energies, respectively, p_{x1}, etc., are momenta and x_1, etc., the corresponding coordinates. This form of the partition function means a great simplification because it does away with all consideration of how a single system would move under a given set of conditions. Moments of inertia do not need to be calculated, nor is a normal coordinate treatment necessary, but only an expression of the potential and kinetic energies in terms of coordinates and momenta. As before we approximate CH_2 (or CH_3 or CH) groups as single units in this treatment and discuss the hydrogen motions separately. Considering the momenta first we write for the kinetic energy:

$$\text{K.E.}=\sum_{i=1}^{n}\tfrac{1}{2}m(p^2{}_{x_i}+p^2{}_{y_i}+p^2{}_{z_i}),$$

where the mass m is the same for all n particles. The partition function now takes the form:

$$Q=\left(\frac{2\pi mkT}{h^2}\right)^{3n/2}\int_0^{X_0}\cdots\int_0^{Z_0}$$
$$\times e^{-(\text{P.E.})/kT}dx_1\cdots dz_n. \quad (12)$$

For the potential energy the expression will be the same as in the preceding section for bond stretching and bending motions. For internal rotations, however, a three-peaked sinusoidal barrier will be used: $\frac{1}{2}V_0(1-\cos 3\phi)$. It is also necessary to assign a high energy to those configurations which involve serious steric repulsions.

[8] Fowler, *Statistical Mechanics* (Cambridge University Press, second edition, 1936), p. 56 ff.

Since we assume a perfect gas state, the first atom of a molecule may be placed anywhere in the container, and the second may be oriented in any direction with respect to the first. A potential energy term $[\frac{1}{2}k_1(r-r_0)^2]$ must be included for the distance between these atoms. Integrating for these atoms we have:

$$Q=\left(\frac{2\pi mkT}{h^2}\right)^{3n/2} V(4\pi r_0^2)\left(\frac{2\pi kT}{k_1}\right)^{\frac{1}{2}}\int_0^{X_0}\cdots\int_0^{Z_0} \times e^{-(\mathrm{P.E.})/kT} dx_3\cdots dz_n, \quad (13)$$

where r_0 is the C−C distance, V the volume, and k_1 the stretching force constant.

The motion of the next atom involves another bond stretching term, a bond bending term $[\frac{1}{2}k_2\delta^2]$, where δ is the distance from equilibrium position, and a third rotation. The circumference of the allowed circle is $2\pi r_0 \sin\gamma$ where γ is the C−C−C angle. We then have:

$$Q=\left(\frac{2\pi mkT}{h^2}\right)^{3n/2} V(8\pi^2 r_0^3 \sin\gamma)\frac{(2\pi kT)^{\frac{3}{2}}}{(k_1^2k_2)^{\frac{1}{2}}} \int_0^{X_0}\cdots\int_0^{Z_0} e^{-(\mathrm{P.E.})/kT} dx_4\cdots dz_n. \quad (14)$$

All atoms along a normal chain after the third have bond stretching and bending motions and a restricted internal rotation. If one assumes three equal maxima and minima and the usual sinusoidal shape, the factor in the partition function is:

$$r_0 \sin\gamma \int_0^{2\pi} e^{-(V_0/2kT)(1-\cos 3\varphi)} d\varphi.$$

With long chain molecules it seems certain that some configurations will involve considerable steric repulsion and that as a result the assumption of three equal minima for each rotation will not be satisfactory. In order to account for these steric effects without too great complexity, the following procedure was adopted. By some means the energy for each configuration corresponding to all internal rotations at minimum points was decided upon. This special "steric" energy was then assumed to remain a constant addition to the internal rotation potential energy for oscillations about a given position as far as the potential peaks. Thereafter energies appropriate to new minimal positions apply. Mathematically this procedure involves changing the upper limit of integration in each [I. Rot.] factor from 2π to $2\pi/3$, and the introduction of a factor:

$$F_{\mathrm{steric}}=\sum e^{-E_i/kT}, \quad (15)$$

where E_i is the energy of the ith configuration. The partition function is not sensitive to changes in the shape of the potential peaks, where this assumption would err most.

We shall assume that the most stable configuration is the planar, zig-zag one discussed above and observed in solids. Rotation about one bond by $(2\pi/3)$ always involves the same interaction which probably increases the energy. Let us call this energy a. By the use of models it was found that rotation about several bonds led either to that number of these same interactions, or else to an impossible configuration from the steric point of view. By assigning a numerical value to a, and counting the number of each kind of position, the sum in Eq. (15) can be calculated easily.

We may now write the complete classical partition function as follows:

$$\begin{aligned} Q^{\mathrm{class}} &= \tfrac{1}{2}F_{0(T)}[C-C_{\mathrm{str}}]^{n-1}[C-C_{\mathrm{bend}}]^{n-2} \times[\mathrm{I.\ Rot.}]^{n-3}F(\mathrm{steric}),\\ F_{0(T)} &= (2\pi mkT/h^2)^3 V(8\pi^2 r_0^3 \sin\gamma),\\ [C-C_{\mathrm{str}}] &= (2\pi kT/h)(m/k_1)^{\frac{1}{2}}, \qquad (16)\\ [C-C_{\mathrm{bend}}] &= (2\pi kT/h)(m/k_2)^{\frac{1}{2}},\\ [\mathrm{I.\ Rot.}] &= (2\pi mkTr_0^2 \sin^2\gamma/h^2)^{\frac{1}{2}} \times\int_0^{2\pi/3} e^{-(V_0/2kT)(1-\cos 3\varphi)} d\varphi,\\ F_{(\mathrm{steric})} &= \sum e^{-E_i/kT}. \end{aligned}$$

The molecule symmetry number 2 (which appears as the factor $\frac{1}{2}$ above) has been introduced since both ends are the same. For N_0 identical molecules the partition function is raised to that power and divided by $(N_0/e)^{N_0}$ which approximates $(N_0!)$.

Recalling now the results of the preceding section, we wish to add factors of the type $[Q_{(\nu)}/Q_{(\nu)}^{\mathrm{class.}}]$ for $(n-1)$ 1000 cm^{-1} frequencies and similar factors for the low frequency bands. For these we take $(n-2)$ 290 cm^{-1} frequencies

TABLE III. *Hydrogen vibration frequencies* (cm^{-1}).[9]

(NUMBERS IN PARENTHESES ARE DEGENERACIES)

	(1)	(2)	(1)	(2)	(2)	
CH_3F	2900	2990	1460	1476	1200	
CH_3Cl	2920	3050	1355	1460	1020	
CH_3Br	2910	3060	1305	1450	957	
CH_3 adopted	3000 (3),		1400 (3),		1000 (2)	
CH_2Cl_2	2985	3045	1266	1420	1060	900
CH_2O	2900	2800	1460	—	1039	917
$(CH_2)_2$	3019	3069	1342	—	950	940
	2988	3107	1444	—	950	1100
CH_2 adopted	3000 (2),		1400 (2),		1000 (2)	
$CHCl_3$	3018			1216 (2)		
$CHBr_3$	3021			1144 (2)		
CH adopted	3000 (1)			1200 (2)		

and $(n-3)$ 120 cm^{-1} frequencies. The justification of using mean frequencies for these broad bands was explained above. It was also pointed out that these latter corrections are very small so that they might even be omitted without very serious error.

It is interesting to note that the factor $[C-C_{str}]$ in Eq. (16) is not just the classical factor for a 1000-cm^{-1} vibration but is somewhat larger. The difference may be thought of as a contribution to mass and moment of inertia terms in the usual formulas because of the additional atom in the molecule. This effect can be easily traced through in the simpler cases by comparing with the usual type of calculation. The situation is, of course, similar with $[C-C_{bend}]$ and [I. Rot.].

Finally the contributions of the hydrogen vibrations must be considered. This will be done by selecting frequencies characteristic of the CH_3 and CH_2 (and for branch chain compounds CH)

TABLE IV. *Formula for calculating thermodynamic functions of paraffin hydrocarbons.*

$$\text{Th. F.} = [F_0(T)] + n_1[C-C_{str}] + n_2[C-C_{bend}] + n_3[\text{I. Rot.}] + F_{(steric)} + F_{(\sigma)} + n_4[CH_3] + n_5[CH_2] + n_6[CH]$$

	$-F/T$	H/T	C
$F_0(T)$	$R\ln\left(\frac{64\pi^5 m^3 r_0^3 k^3 \sin\gamma R}{h^6 N_0}\right)+4R\ln T-R\ln P=4.274+4R\ln T-R\ln P$	$4R$	$4R$
$[C-C_{str}]$	$R\ln(2\pi\nu_1(m/k_1)^{\frac{1}{2}})+\text{Ein.}(h\nu_1/kT)=0.75+\text{Ein.}\left(\frac{hc1000}{kT}\right)$	$\text{Ein.}(h\nu_1/kT)$	$\text{Ein.}(h\nu_1/kT)$ ($\nu_1/c=1000\ cm^{-1}$)
$[C-C_{bend}]$	$R\ln(2\pi km^{\frac{1}{2}}/hk_2^{\frac{1}{2}})+R\ln T+(R/2)(h\nu_2/kT)-(R/24)(h\nu_2/kT)^2=-11.275+R\ln T+413/T-14200/T^2$	$R-(R/2)(h\nu_2/kT)+(R/12)(h\nu_2/kT)^2=R-413/T+28400/T^2$	$R-(R/12)(h\nu_2/kT)^2=R-28400/T^2$ ($T>250$)
[I.Rot.]	$R[-1.267+\frac{1}{2}\ln T+\frac{1}{2}\ln(mr_0^2\sin^2\gamma\times10^{40})-\ln 3]-\frac{(F-F_f)}{T}\left(V_0/RT,\frac{n^2}{IV}=0\right)+\frac{h\nu_3}{2T}=\frac{1}{2}R\ln T-0.55+\frac{170}{T}-\frac{(F-F_f)}{T}\left(\frac{3600}{RT}\right)$	$\frac{H}{T}\left(\frac{V_0}{RT},\frac{n^2}{IV}=0\right)-\frac{h\nu_3}{2T}=\frac{H}{T}\left(\frac{3600}{RT}\right)-\frac{170}{T}$	$C\left(\frac{V_0}{RT},\frac{n^2}{IV}=0\right)=C\left(\frac{3600}{RT}\right)$
$F_{(steric)}$	$R\ln Q$ $Q=\Sigma e^{-Ei/kT}$	$R(Q'/Q)$ $Q'=\Sigma(Ei/kT)e^{-Ei/kT}$	$R[Q''/Q-(Q'/Q)^2]$ $Q''=\Sigma(Ei/kT)^2e^{-Ei/kT}$
$F_{(\sigma)}$	$-R\ln\sigma$	0	0
$[CH_3]$	$\Sigma\ \text{Ein.}(h\nu/kT)+R(-1.267+\frac{1}{2}\ln T+\frac{1}{2}\ln I_R\times10^{40}-\ln 3)-(F-F_f)/T_{(V_0/RT)}$ $\nu/c=3000$ (3), 1400 (3), 1000 (2)	$\Sigma\ \text{Ein.}(h\nu/kT)+H/T_{(V_0/RT)}$ $V_0=3600$ cal.	$\Sigma\ \text{Ein.}(h\nu/kT)+C_{(V_0/RT)}$ $I_R=5\times10^{-40}$
$[CH_2]$, [CH]	$\Sigma\ \text{Ein.}(h\nu/kT)$ $CH_2-\nu/c=3000$ (2), 1400 (2), 1000 (2) $CH-\nu/c=3000$ (1), 1200 (2)	$\Sigma\ \text{Ein.}(h\nu/kT)$	$\Sigma\ \text{Ein.}(h\nu/kT)$

For a normal paraffin of N carbon atoms: $n_1=N-1$, $n_2=N-2$, $n_3=N-3$, $n_4=2$, $n_5=N-2$, $n_6=0$, $\sigma=2$, summations in $F_{(steric)}$ are over 3^{N-3} terms.

Ein. refers to an Einstein function for one degree of freedom and of the proper type, i.e., $-F/T$, H/T, or C. $(F-F_f)/T$, H/T, C in [I. Rot.] and $[CH_3]$ are the restricted rotation functions calculated by the writer[10] in 1937.

[9] All data are from Sponer, *Molekulspektren* (Julius Springer, Berlin, 1935) except for CH_2Cl_2 from Corin and Sutherland, Proc. Roy. Soc. London **A165**, 43 (1938), and $(CH_2)_2$ from Bonner, J. Am. Chem. Soc. **58**, 34 (1936).

[10] Pitzer, J. Chem. Phys. **5**, 469 (1937).

TABLE V. *Numerical values for some of the terms in Table IV (units: cal. per degree).*

T	$F_0(T)$	[C−Cstr]	[C−Cbend]	[I. Rot.]	[CH_3]	[CH_2]	[CH]	ISOBUTANE* CORR.	NEO-PENTANE* CORR.
				CONTRIBUTIONS TO $(H_0^0-F_T^0)/T$					
298.1	49.55	0.76	1.27	2.86	0.89	0.04	0.01	0.00	−1.39
400.	51.89	.80	1.57	3.35	1.39	.14	.05	− .19	−1.80
500.	53.67	.86	1.84	3.76	1.91	.30	.13	− .32	−1.95
600.	55.11	.93	2.07	4.10	2.47	.52	.23	− .36	−2.05
800.	57.40	1.11	2.50	4.64	3.60	1.08	.50	− .45	−2.20
1000.	59.17	1.29	2.85	5.07	4.79	1.71	.81	− .60	−2.30
1500.	62.40	1.71	3.51	5.81	7.54	3.37	1.63	− .59	−2.16
				CONTRIBUTIONS TO $(H_T^0-H_0^0)/T$					
298.1	7.95	0.08	0.91	1.57	1.38	0.19	0.07	−0.73	−1.39
400.	7.95	.20	1.13	1.76	2.04	.54	.23	− .60	−1.16
500.	7.95	.34	1.27	1.85	2.73	.99	.46	− .50	− .90
600.	7.95	.47	1.37	1.89	3.39	1.47	.72	− .37	− .55
800.	7.95	.71	1.51	1.91	4.62	2.40	1.18	− .27	− .11
1000.	7.95	.89	1.60	1.86	5.73	3.27	1.61	− .22	+ .11
1500.	7.95	1.19	1.72	1.70	7.82	4.94	2.46	+ .07	+ .41

* The empirical corrections for isobutane and neopentane are to be added (algebraically) whenever a carbon atom is bonded to three or four other carbons, respectively. See text discussion.

groups. These are given in Table III, together with the data from simple molecules on which they are based.

The contributions from these vibration frequencies are easily calculated by the usual methods.

Before summarizing these data again a few additional numerical values will be adopted. An effective mass of a CH_2 group was obtained by calculating the product $[M^3I_1I_2I_3]$ for propane, both neglecting and considering the hydrogens. The effective mass of 18.6 which made both products equal will be used. This value is as high as it is because of the relatively large contribution of the hydrogens to the moments of inertia. The C−C bond distance will be taken as 1.54A and the tetrahedral angles assumed. The force constants have already been given; the potential barrier V_0 will be given the value 3600 cal. per mole. This may be regarded as estimated from simple molecules[11] or determined from the experimental data considered later. The steric interaction energy a will be taken as 800 cal. and was determined from the data considered below. In order to calculate the internal rotation contribution of a methyl group, the same 3600 cal. potential will be used and a reduced moment of inertia I_r equal to 5×10^{-40} g cm.[2]

Table IV gives the final formula. It is expressed as the sum characteristic of a thermodynamic function rather than the partition function product. The form of each term is given for the functions $(-F/T)$, (H/T), and C. The entropy S is the sum of the first two.

Numerical values for certain of the terms in Table IV are listed in Table V for a series of temperatures. In addition to their use in calculations for hydrocarbons, it seems likely that these data may be of value when the method is extended to alcohols and other molecules derivable from the hydrocarbons.

COMPARISON WITH OBSERVED NORMAL PARAFFIN ENTROPIES

We are now ready to calculate the entropies of the normal paraffins for comparison with experiment. Of all the constants appearing in Table IV only two can be regarded as adjustable to fit the data, V_0 and a. The former was actually assigned the value 3600 cal. per mole from comparison with simpler molecules, and then it was found that a value of 800 cal. for a gave a good fit with the experimental entropies of n-butane[12] and n-heptane.[1] The agreement might be slightly improved by lowering V_0 and raising a, however, this seems improbable on other grounds and the present constants give agreement within the

[11] Pitzer, Chem. Rev. **27**, August (1940).

[12] Aston, private communication, $S^\circ_{272.6}=72.0$ cal. per degree for gaseous n-butane.

experimental errors. For further comparison we have the entropies of the other normal paraffins as measured by Parks, Huffman, and coworkers,[13] which involve extrapolations below 90°K. The good agreement between Parks' extrapolated entropy at 90°K and the writer's experimental value for *n*-heptane suggests that the extrapolations for the other normal paraffins are probably about right.[14] For conversion to the gas state various data were used. The values of Parks, Shomate, Kennedy, and Crawford[15] were used for butane, the heat of vaporization of Griffiths and Awberry[16] for pentane, and the vapor pressure-temperature data[17] for the others.

Table VI summarizes these results, and includes the data on steric repulsions used in the computations.

The good agreement shown in Table VI certainly gives no reason to doubt the picture here set up, and suggests that general thermodynamic functions calculated by this method for various temperatures will be of considerable value. Such results will be presented elsewhere.[11] On the other hand, approximations have been made such that the values of the various molecular constants which give correct entropies may not be exactly the true ones. The treatment is an approximate derivation from statistical mechanics based on an admittedly oversimplified picture of the molecules. However, the errors will not change greatly from one molecule to a similar one; consequently these equations can be used empirically with expectation of considerable accuracy.

TABLE VI. *Molal entropies of the normal paraffin hydrocarbons in the gaseous state.*

SUBSTANCE	NUMBER OF POSITIONS WITH E_i (STERIC): 0	a	$2a$	$3a$	$4a$	$5a$	∞	$S^0_{298.1}$ (cal./deg.) calc.	obs.*
Butane	1	2	0	0	0	0	0	74.17	73.7
Pentane	1	4	2	0	0	0	2	83.27	82.2
Hexane	1	6	8	2	0	0	10	92.41	92.3
Heptane	1	8	18	12	2	0	40	101.54	101.3
Octane	1	10	32	38	16	2	144	110.67	110.0
Δ per CH_2	—	—	—	—	—	—	—	9.13	

* These observed entropies are probably accurate to within about 1.0 cal./deg. The more accurate values for butane[12] and heptane[1] at their boiling points were used to fix a parameter in the formula for calculation. ($S^0_{371.5}$ obs. 111.78±0.3, calc. 111.6 for heptane; $S^0_{272.6}$ obs. 72.0, calc. 72.1 cal. per degree for butane.)

ENTROPIES OF BRANCH CHAIN PARAFFINS

The present lack of knowledge of force constants and vibration frequencies for the branch chain paraffins prevents calculations of even the degree of approximation attained for the normal series. Even so, it seems that the formulas developed in the preceding section should serve to give rough estimates, of value until such time as calculations of greater certainty can be made. Only one accurate experimental entropy is available, that of 2,2,4-trimethylpentane.[1] The long extrapolation from 0 to 90°K needed in all other cases appears to be a doubtful procedure, except in a series of very similar compounds such as the straight chain paraffins.

In adapting the formulas of Table IV we note that a side chain carbon atom attached in the middle of a straight chain has the usual bond stretching degree of freedom. The other two degrees of freedom are of a bond bending nature. Thus one $[C-C_{str}]$ and two $[C-C_{bend}]$ terms should be added. The force constant in the latter is not necessarily the same, however. Vibration frequencies for the hydrogen of a CH group have been given. Steric factors must be considered as before, but are not necessarily as simple. The possibility of symmetry in the skeletal internal rotations, for example of tertiary butyl groups, must also be thought about.

If an entropy is calculated for 2,2,4-trimethyl pentane using the same force constants and potential barriers as for the normal paraffins, the result is somewhat too large. However, a potential barrier well over 4000 cal. was found for the simple highly branched molecule tetramethyl methane[11] as compared to 3300 for propane and less than 3000 for ethane. Also the vibration frequencies for isobutane and tetramethyl methane suggest higher bending force constants. As a simple way of correcting for these factors, the thermodynamic functions were calculated for isobutane and neopentane (tetramethyl methane) by these approximate methods and with the more exact method and constants available for

[13] Parks and Huffman, *Free Energies of Some Organic Compounds* (The Chemical Catalog Co., New York, 1932).

[14] This refers to Parks' original values, not to possible revised ones suggested by Parks, Shomate, Kennedy, and Crawford, reference 15.

[15] Parks, Shomate, Kennedy and Crawford, J. Chem. Phys. **5**, 359 (1937).

[16] Griffiths and Awberry, Proc. Phys. Soc. London **44**, 121 (1932).

[17] Egloff, *Physical Constants of Hydrocarbons*, Vol. 1 (Reinhold Publishing Company, New York, 1939).

TABLE VII. *Molal entropies of the branch chain paraffins together with data used in their calculation.*†

Substance	n_1	n_2	n_3	n_4	n_5	n_6	n_7	No. of Positions with Energy: 0	a	$2a$	$3a$	$4a$	∞	σ	$S^0_{298.1}$ (cal./deg.) calc. (gas)	exp. (liq.)
2-methyl butane	4	4	1	3	1	1	0	2	0	1	0	0	0	1	82.0	60.8
2,2-dimethyl butane	5	6	1	4	1	0	1	1	0	0	0	0	0	1	85.7	64.4
2,3-dimethyl butane	5	6	1	4	0	2	0	1	0	2	0	0	0	2	86.5	—
2-methyl pentane	5	5	2	3	2	1	0	2	0	3	0	0	4	1	90.1	69.9
3-methyl pentane	5	5	2	3	2	1	0	2	0	2	0	2	3	1	90.0	—
2,2,3-trimethyl butane	6	8	1	5	0	1	1	1	0	0	0	0	0	1	92.3	64.8
2,2-dimethyl pentane	6	7	2	4	2	0	1	1	0	0	0	0	2	1	93.4	68.1
2,3-dimethyl pentane	6	7	2	4	1	2	0	4	0	0	0	0	5	1	98.8	72.4*
2,4-dimethyl pentane	6	7	2	4	1	2	0	2	0	0	0	2	5	2	94.7	69.7
3,3-dimethyl pentane	6	7	2	4	2	0	1	5	0	2	0	0	2	2	95.4	70.1
3-ethyl pentane	6	6	3	3	3	1	0	11	0	0	0	0	16	3	98.3	74.6
2-methyl hexane	6	6	3	3	3	1	0	2	4	3	4	0	14	1	99.5	75.3
3-methyl hexane	6	6	3	3	3	1	0	4	4	4	0	0	15	1	101.3	74.0*
2,2,3,3-tetramethyl butane	7	10	1	6	0	0	2	1	0	0	0	0	0	6	94.1	61.4 (solid)
2,2,4-trimethyl pentane	7	9	2	5	1	1	1	2	0	0	0	0	1	1	101.4	75.2

† The experimental entropies are for the liquid except as noted. All are from Parks and Huffman (reference 13) except 2,2-dimethyl butane and 2-methyl pentane which are from Stull (J. Am. Chem. Soc. **59**, 2726 (1937)). Those values designated by * are especially uncertain; the others are probably self-consistent to within one or two cal. per degree and somewhat low. The substances marked by * are, interestingly, also the two cases where optical isomers occur. The calculated values include an $R \ln 2$ term to account for this fact.

such simple molecules.[11] The differences which are given in Table V, can then be applied as corrections whenever a carbon atom of a complex molecule is bonded to three or four other carbons respectively. On this basis the entropy of 2,2,4-trimethyl pentane is within about one cal. per degree of the observed value. Further refinements would not be justified at present.

The results for all branch chain paraffins through the heptanes and for two octanes are given in Table VII. Therein the n's refer to Table IV except n_7 which is the number of carbon atoms bonded to four other carbons, hence the number of "neopentane corrections." The number of carbons bonded to three others is n_6, also the number of *CH* groups.

The energies for the various internal rotation minima were assigned values in terms of the parameter a of the normal paraffins. This somewhat arbitrary procedure was carried out with the aid of the "Fisher-Hirshfelder" models which approximate proportional atomic sizes.

For rough comparison the available experimental entropies are listed, even though they involve uncertain extrapolations below 90°K. Values are given for the liquids since vaporization data are not available in most cases. On the whole the calculated and experimental values show remarkably similar variations and their differences are quite reasonable for entropies of vaporization.

Finally it may be mentioned that the heat capacities in the gas phase, calculated for n-heptane and 2,2,4-trimethylpentane, are in agreement with preliminary experimental values obtained recently by the writer.[1]

Reprinted from THE JOURNAL OF CHEMICAL PHYSICS, Vol. 10, No. 9, 605–606, September, 1942
Printed in U. S. A.

Internal Rotation in Molecules with Two or More Methyl Groups

KENNETH S. PITZER
Department of Chemistry, University of California, Berkeley, California

July 31, 1942

IN treating the internal rotation of molecules with two or more identical rotating groups, it has been commonly assumed, of necessity, that the potential energy for one rotation is independent of the position of the other groups. For some time the writer has been attempting to study this assumption but until recently the results have been disappointingly negative in character. However, one positive result may now be presented together with a brief discussion of the other work.

Kennedy, Sagenkahn, and Aston[1] have recently published measurements leading to the entropy of dimethyl ether. After subtracting contributions for the other modes of motion, the value 2.65 cal. per degree at 200°K was obtained for the entropy contribution of the two internal rotational degrees of freedom. This was interpreted on the basis of independent potential barriers for each methyl group of magnitude 3100 cal. per mole. However, Ananthakrishnan,[2] has reported Raman shifts of 160 cm^{-1} and 300 cm^{-1} for dimethyl ether. Application of the usual harmonic oscillator formula to these frequencies yields an entropy contribution of 2.63 cal. per degree at 200°K in essentially perfect agreement with the above calorimetric value. Furthermore, these Raman lines have never been explained on any other basis. Therefore it seems likely that these lines are the fundamental frequencies of the two modes of torsional oscillation.

However, this picture is by no means certain. Earlier investigators[3] failed to find one or both of the lines. Although the selection rules permit these torsional modes in the Raman effect, they predict depolarized lines whereas Ananthakrishnan reports both lines polarized. This latter difficulty is partially offset by the fact that depolarized lines are normally broad and both of these lines are reported as broad. Thus additional study is needed before this assignment can be regarded as proved or disproved.

So far as the writer is aware, if this assignment proves to be correct it will be the first set of spectroscopic frequencies for torsional oscillations which is consistent with reliable thermodynamic data. Assuming moderate anharmonicity, these frequencies are in agreement with the observed entropy at higher temperatures and with the observed heat capacity in the gaseous state.[4]

Figure 1 shows first the expected equilibrium configuration and then moderate displacements in each of the two normal modes of internal torsion. It is quite apparent that mode *A* rapidly brings hydrogen atoms very close together while mode *B* allows them to remain at considerable distance from one another. Thus it is not surprising to find the first frequency almost double the second. A normal coordinate calculation yields for the force constants: $\partial^2 V/\partial\varphi_1^2 = 8.0\times10^{-13}$, $\partial^2 V/\partial\varphi_1\partial\varphi_2 = -3.4\times10^{-13}$ erg per radian². This value of the first constant corresponds to a cosine potential barrier of about 2600 cal. per mole. It is quite evident that the cross derivative or interaction term, although somewhat smaller, is by no means zero.

The other two types of calculations concerning the interactions of two or more methyl groups will now be described briefly. The first concerns the possibility of using the entropy and heat capacity of a single gas to determine both the potential barrier for the rotation of one methyl group by itself and the magnitude of the interaction. Calculations for several typical cases showed that there was practically no chance of obtaining information by this method. The differences were too small to be detected experimentally. This means, inversely, that when potential barriers obtained from entropies are used only to calculate heat capacities, no serious error is likely to arise from the neglect of cross terms.

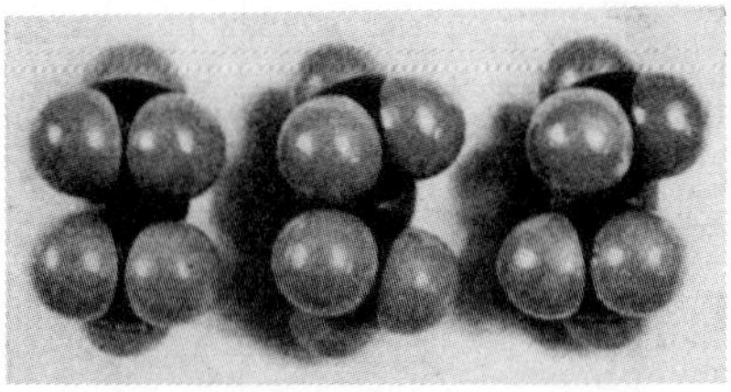

A *B*

FIG. 1. Left, the equilibrium position; center, moderate displacement in torsional mode *A*; right, moderate displacement in mode *B*.

The second calculation concerned the series propane, isobutane, and tetramethylmethane. A basic entropy was calculated for each molecule using the potential barrier from ethane (2800 cal. per mole) and the best available values for vibration frequencies, moments of inertia, etc. The amounts by which the observed entropies were smaller than these basic values might then be ascribed either to (1) interaction terms such as appear to be present in dimethyl ether, or to (2) a rise in the separate potential barriers upon substitution of methyl groups for the hydrogens in ethane. However, on the basis of reasonable approximations, both assumptions led to the same relative values, 1 to 3 to 6, for the entropy differences in propane, isobutane, and tetramethylmethane.[5] These ratios agree with the observed values within their respective experimental errors. However, this tells us nothing about the presence or absence of the interaction terms.

We may then draw the tentative conclusion that thermodynamic data alone will not give definite information concerning cross terms in the internal rotation potential energy, but that spectroscopic data, particularly when coupled with thermodynamic data, have interesting possibilities.

[1] R. M. Kennedy, M. Sagenkahn, and J. G. Aston, J. Am. Chem. Soc. **63**, 2267 (1941).
[2] R. Ananthakrishnan, Proc. Ind. Acad. Sci. **A5**, 285 (1937).
[3] S. C. Sikar, Ind. J. Phys. **7**, 257 (1932); A. Dadieu and K. W. F. Kohlrausch, Monatsh. f. Chemie **57**, 225 (1931); K. W. F. Kohlrausch, *ibid.*, **68**, 349 (1936); M. Wolkenstein and J. K. Syrkin, Nature **139**, 288 (1937).
[4] G. B. Kistiakowsky and W. W. Rice, J. Chem. Phys. **8**, 618 (1940).
[5] This calculation may be explained very roughly as follows: (1) The number of interaction terms will be the number of edges of the tetrahedron about the center carbon atom which connect two methyl groups. (2) The increments of potential barrier rise will be, respectively, 1, 2, and 3 for the series, but the number of degrees of freedom are, respectively, 2, 3, and 4. The products of these numbers are thus just double the numbers quoted in the text.

Reprinted from THE JOURNAL OF CHEMICAL PHYSICS, Vol. 10, No. 7, 428–440, July, 1942
Printed in U. S. A.

Energy Levels and Thermodynamic Functions for Molecules with Internal Rotation

I. Rigid Frame with Attached Tops

KENNETH S. PITZER AND WILLIAM D. GWINN
Department of Chemistry, University of California, Berkeley, California

(Received April 18, 1942)

A general treatment of internal rotation is given for molecules whose moments of inertia for over-all rotation are independent of internal rotational coordinates. Tables are presented for the various thermodynamic functions which are accurate for molecules with one internal rotation and for the potential energy $(V/2)(1-\cos n\phi)$. The tables are shown to be a good approximation for molecules with several internal rotational coordinates, provided the potential energy can be expressed as a sum of terms of this type. Methods are suggested for treating problems where cross terms involving more than one internal coordinate are present in the potential energy. The energy level expressions are developed for the more general case with the potential energy expressed by a Fourier series. Although a few specific cases were worked out with different shape potential barriers, it appears that the simple form assumed above will be satisfactory for many purposes.

ALTHOUGH a number of papers have already been published concerning the energy levels and thermodynamic functions of internal rotational degrees of freedom,[1–11] all of them have either treated highly specialized cases or have made approximations which need further examination. The treatment that follows applies to any molecule which can be regarded as a rigid frame with attached symmetrical tops; or in other words, any molecule whose moments of inertia for over-all rotation are independent of the coordinates of internal rotation.[12] (For example, *iso*butane and acetone fall within this class while *n*-butane and sym. dichloroethane do not.) Furthermore, in the present treatment all approximations that are made for convenience or necessity in calculation will be carefully examined and their range of validity defined.

In comparison with previous work it may be stated that the tables published by one of us[7] are probably a satisfactory approximation for the cases to which they have been applied. The formulas developed by Crawford[9] and recommended by Wilson[8b] as more accurate than these tables are actually less accurate insofar as the two differ significantly. The present results reduce exactly to those of Koehler and Dennison[10] who considered the special case of methyl alcohol.

The only formula which appears to be sufficiently general and useful to justify tabulation of the results for the full range of the variables leads to the same tables that were published previously. These have been recalculated, however, and now have an accuracy far beyond that of the present experimental thermodynamic data. (See Tables I–VI.)

THERMODYNAMIC RESULTS FOR THE SIMPLER CASES

We shall first present the thermodynamic results in the simple form which is accurate for all molecules with one symmetrical top and for the simple potential barrier $\frac{1}{2}V(1-\cos n\phi)$. In many cases this will be quite accurate for molecules with several tops. The proof, methods of computing the tables, and the more general formulas

[1] H. H. Nielsen, Phys. Rev. **40**, 445 (1932).
[2] J. E. Mayer, S. Brunauer, and M. G. Mayer, J. Am. Chem. Soc. **55**, 37 (1933).
[3] E. Teller and K. Weigert, Nachr. Ges. Wiss. Göttingen, Math. physik. Klasse, 218 (1933).
[4] L. J. B. La Coste, Phys. Rev. **46**, 718 (1934).
[5] M. L. Edinoff and J. G. Aston, J. Chem. Phys. **3**, 379 (1935).
[6] L. S. Kassel, J. Chem. Phys. (a) **3**, 115 (1935); (b) **3**, 326 (1935); (c) **4**, 276 (1936); (d) **4**, 435 (1936).
[7] K. S. Pitzer, J. Chem. Phys. **5**, 469 (1937).
[8] E. B. Wilson, Jr., (a) J. Chem. Phys. **6**, 740 (1938); (b) Chem. Rev. **27**, 17 (1940).
[9] B. L. Crawford, Jr., J. Chem. Phys. **8**, 273 (1940).
[10] J. S. Koehler and D. M. Dennison, Phys. Rev. **57**, 1006 (1940).
[11] D. Price, J. Chem. Phys. **9**, 807 (1941).
[12] It was originally planned to present at this time a second paper dealing with molecules not falling within this class. The pressure of other work makes this impossible. Two methods applicable to different types of molecules may be mentioned, however. For relatively heavy molecules the method used by one of us for the *n*-paraffin series is applicable (K. S. Pitzer, J. Chem. Phys. **8**, 711 (1940)). If the moments of inertia for over-all rotation change only slightly with internal rotation, the methods of this paper offer a satisfactory approximation.

TABLE I. Free energy, $-F/T$ (cal. per degree mole).

V/RT \ $1/Q_f$	0.25	0.30	0.35	0.40	0.45	0.50	0.55	0.60	0.65	0.70	0.75	0.80
0.0	2.754	2.392	2.086	1.821	1.587	1.377	1.188					
.2	2.710	2.359	2.061	1.803	1.574	1.368	1.182					
.4	2.623	2.296	2.014	1.765	1.543	1.342	1.160					
.6	2.518	2.208	1.944	1.708	1.498	1.309	1.134					
.8	2.406	2.106	1.856	1.636	1.442	1.266	1.100					
1.0	2.296	2.004	1.764	1.559	1.379	1.214	1.059					
1.5	2.040	1.770	1.548	1.370	1.210	1.069	.938					
2.0	1.819	1.563	1.360	1.193	1.052	.927	.817					
2.5	1.630	1.389	1.197	1.043	.912	.802	.709					
3.0	1.473	1.240	1.059	.914	.793	.695	.612	.53				
3.5	1.340	1.117	.943	.802	.694	.603	.528	.46				
4.0	1.225	1.013	.847	.713	.613	.527	.456	.40	.34			
4.5	1.133	.925	.764	.637	.543	.463	.397	.35	.29			
5.0	1.053	.849	.696	.577	.483	.408	.347	.30	.25	.22		
6	.919	.728	.586	.477	.393	.325	.272	.23	.19	.16	.14	
7	.819	.636	.503	.402	.325	.267	.218	.183	.15	.12	.10	.08
8	.735	.564	.440	.346	.275	.221	.178	.146	.119	.10	.08	.06
9	.667	.504	.388	.300	.235	.186	.149	.119	.096	.079	.06	.05
10	.610	.456	.345	.264	.203	.159	.125	.099	.079	.064	.051	.04
12	.521	.380	.280	.209	.157	.120	.092	.071	.055	.043	.033	.027
14	.452	.321	.232	.169	.124	.092	.069	.052	.038	.030	.022	.018
16	.396	.276	.195	.139	.100	.072	.053	.039	.028	.021	.016	.012
18	.351	.240	.166	.117	.082	.058	.042	.030	.022	.016	.012	.009
20	.315	.211	.144	.098	.068	.047	.033	.024	.017	.012	.009	.006

TABLE II. Free energy increase from free rotation, $(F-F_f)/T$ (cal. per mole degree)
$-F/T=-F_f/T-(F-F_f)/T=R\ln Q_f-(F-F_f)/T$.

V/RT \ $1/Q_f$	0	0.05	0.10	0.15	0.20	0.25	0.30	0.35	0.40	0.45	0.50	0.55
0.0	0.0000	0.000	0.000	0.000	0.000	0.000	0.000	0.000	0.000	0.000	0.000	0.000
0.2	.1937	.154	.117	.085	.061	.044	.033	.025	.018	.013	.009	.005
0.4	.3776	.326	.274	.225	.176	.131	.096	.072	.056	.044	.035	.026
0.6	.5516	.489	.424	.361	.298	.236	.184	.142	.113	.089	.068	.054
0.8	.7161	.640	.566	.493	.420	.348	.286	.230	.185	.145	.111	.088
1.0	.8711	.784	.699	.617	.537	.461	.389	.322	.262	.208	.163	.129
1.5	1.2200	1.114	1.010	.909	.809	.714	.622	.538	.451	.375	.308	.250
2.0	1.5182	1.395	1.276	1.159	1.045	.935	.829	.726	.628	.535	.450	.371
2.5	1.7724	1.635	1.501	1.371	1.246	1.124	1.004	.889	.778	.675	.575	.479
3.0	1.9893	1.839	1.693	1.552	1.415	1.282	1.152	1.027	.907	.794	.682	.576
3.5	2.1756	2.013	1.856	1.704	1.557	1.414	1.275	1.143	1.019	.893	.774	.660
4.0	2.3366	2.163	1.996	1.833	1.676	1.525	1.379	1.239	1.108	.974	.850	.732
4.5	2.4772	2.293	2.117	1.945	1.780	1.621	1.467	1.322	1.184	1.044	.914	.791
5.0	2.6012	2.408	2.221	2.042	1.868	1.703	1.543	1.392	1.244	1.104	.969	.841
6.0	2.8108	2.599	2.396	2.202	2.015	1.836	1.664	1.500	1.344	1.194	1.052	.916
7.0	2.9833	2.755	2.537	2.328	2.129	1.936	1.757	1.583	1.418	1.262	1.111	.971
8.0	3.1294	2.886	2.653	2.432	2.220	2.020	1.828	1.646	1.474	1.312	1.157	1.011
9.0	3.2563	2.998	2.753	2.520	2.298	2.087	1.888	1.698	1.520	1.351	1.192	1.039
10.0	3.3686	3.097	2.839	2.594	2.362	2.144	1.936	1.741	1.557	1.383	1.219	1.063
12.0	3.5602	3.263	2.982	2.718	2.468	2.233	2.013	1.806	1.612	1.429	1.258	1.096
14.0	3.7205	3.400	3.099	2.816	2.551	2.303	2.071	1.854	1.651	1.462	1.285	1.119
16.0	3.8584	3.517	3.197	2.897	2.618	2.358	2.116	1.891	1.682	1.486	1.305	1.135
18.0	3.9793	3.618	3.280	2.965	2.674	2.403	2.152	1.920	1.704	1.505	1.319	1.146
20.0	4.0872	3.707	3.353	3.024	2.720	2.440	2.181	1.942	1.722	1.519	1.331	1.155

for molecules with several tops or different potential barriers will be given in later sections.

The quantities entering in the formulas are:

I_1, I_2, I_3 the three principal moments of inertia of the rigid molecule (internal rotation frozen).

V_m the height of the potential barrier for the mth top.

n_m the number of potential minima per revolu-

TABLE III. Entropy, S (cal. per degree mole).

V/RT \ $1/Q_f$	0.25	0.30	0.35	0.40	0.45	0.50	0.55	0.60	0.65	0.70	0.75	0.80
0.0	3.748	3.386	3.079	2.814	2.580	2.371	2.181					
0.2	3.743	3.382	3.076	2.811	2.578	2.369	2.179					
0.4	3.730	3.370	3.065	2.801	2.568	2.359	2.171					
0.6	3.709	3.347	3.043	2.780	2.547	2.340	2.153					
0.8	3.679	3.318	3.013	2.750	2.519	2.315	2.128					
1.0	3.638	3.279	2.974	2.714	2.485	2.279	2.095					
1.5	3.512	3.156	2.854	2.600	2.376	2.173	1.992					
2.0	3.355	3.004	2.709	2.458	2.241	2.048	1.873					
2.5	3.180	2.836	2.548	2.303	2.091	1.907	1.741					
3.0	3.008	2.667	2.380	2.138	1.933	1.756	1.600	1.45				
3.5	2.838	2.500	2.218	1.978	1.782	1.610	1.459	1.34				
4.0	2.678	2.343	2.069	1.834	1.643	1.475	1.326	1.22	1.10			
4.5	2.528	2.199	1.926	1.698	1.511	1.348	1.204	1.10	1.00			
5.0	2.396	2.068	1.798	1.579	1.392	1.233	1.095	.97	.88	.81		
6.0	2.166	1.844	1.585	1.370	1.192	1.040	.913	.79	.70	.63	.56	
7.0	1.983	1.665	1.411	1.204	1.033	.891	.770	.665	.57	.50	.44	.39
8.0	1.830	1.519	1.272	1.071	.906	.770	.656	.564	.482	.41	.36	.31
9.0	1.703	1.397	1.156	.962	.804	.674	.569	.482	.411	.348	.29	.25
10.0	1.593	1.295	1.060	.872	.719	.596	.495	.418	.352	.295	.245	.21
12.0	1.417	1.125	.904	.728	.588	.476	.388	.315	.258	.212	.173	.143
14.0	1.275	.994	.783	.620	.492	.388	.309	.247	.196	.158	.126	.102
16.0	1.157	.890	.688	.533	.414	.322	.251	.196	.155	.120	.094	.075
18.0	1.058	.801	.609	.464	.353	.270	.205	.158	.121	.093	.072	.056
20.0	.975	.727	.542	.405	.303	.228	.170	.129	.097	.073	.056	.042

TABLE IV. Entropy decrease from free rotation, (S_f-S) (cal. per mole degree)
$S=S_f-(S_f-S)=R(\frac{1}{2}+\ln Q_f)-(S_f-S)$.

V/RT \ $1/Q_f$	0.0	0.05	0.10	0.15	0.20	0.25	0.30	0.35	0.40	0.45	0.50	0.55
0.0	0.0000	0.000	0.000	0.000	0.000	0.000	0.000	0.000	0.000	0.000	0.000	0.000
0.2	.0049	.005	.004	.004	.004	.004	.004	.003	.003	.002	.002	.002
0.4	.0198	.020	.018	.018	.018	.018	.016	.014	.013	.012	.012	.010
0.6	.0440	.044	.043	.043	.040	.039	.039	.036	.034	.033	.031	.028
0.8	.0771	.077	.077	.075	.072	.069	.068	.066	.064	.061	.056	.053
1.0	.1185	.118	.117	.115	.112	.110	.107	.105	.100	.095	.092	.086
1.5	.2527	.252	.250	.218	.212	.236	.230	.225	.211	.201	.198	.189
2.0	.4182	.417	.415	.410	.402	.393	.382	.370	.356	.339	.323	.308
2.5	.6001	.599	.594	.585	.577	.568	.550	.531	.511	.489	.464	.440
3.0	.7856	.783	.777	.768	.757	.740	.719	.699	.676	.647	.615	.581
3.5	.9660	.964	.957	.944	.929	.910	.886	.861	.836	.798	.761	.722
4.0	1.1356	1.133	1.126	1.111	1.094	1.070	1.043	1.011	.980	.937	.896	.855
4.5	1.2918	1.289	1.280	1.265	1.244	1.220	1.187	1.153	1.116	1.069	1.023	.977
5.0	1.4339	1.431	1.421	1.404	1.380	1.352	1.318	1.281	1.235	1.188	1.138	1.086
6.0	1.6781	1.674	1.662	1.643	1.616	1.582	1.542	1.494	1.444	1.388	1.331	1.268
7.0	1.8783	1.874	1.860	1.837	1.807	1.765	1.721	1.668	1.610	1.547	1.480	1.411
8.0	2.0447	2.040	2.024	1.998	1.962	1.918	1.867	1.807	1.743	1.674	1.601	1.525
9.0	2.1864	2.180	2.163	2.134	2.095	2.045	1.989	1.923	1.852	1.776	1.697	1.612
10.0	2.3095	2.303	2.284	2.252	2.208	2.155	2.091	2.019	1.942	1.861	1.775	1.686
12.0	2.5155	2.508	2.485	2.447	2.394	2.331	2.261	2.175	2.086	1.992	1.895	1.793
14.0	2.6847	2.676	2.650	2.607	2.547	2.473	2.392	2.296	2.194	2.088	1.983	1.872
16.0	2.8289	2.819	2.788	2.740	2.674	2.591	2.496	2.391	2.281	2.166	2.049	1.930
18.0	2.9545	2.943	2.910	2.855	2.781	2.690	2.585	2.470	2.350	2.227	2.101	1.976
20.0	3.0659	3.054	3.017	2.956	2.872	2.773	2.659	2.537	2.409	2.277	2.143	2.011

tion; also the symmetry number of the internal rotation.

I_m the reduced moment of inertia for the rotation of the mth top. Only I_m needs further discussion; it is given by the equation:

$$I_m=A_m\left(1-\sum_{i=1}^{3}A_m\lambda^2_{mi}/I_i\right), \tag{1a}$$

TABLE V. Heat content, $(H_T-H_0)/T$ (cal. per degree mole).

V/RT \ $1/Q_f$	0.0	0.05	0.10	0.15	0.20	0.25	0.30	0.35	0.40	0.45	0.50	0.55	0.60	0.65	0.70	0.75	0.80
0.0	0.9934	0.993	0.993	0.993	0.993	0.993	0.993	0.993	0.993	0.993	0.993	0.993					
0.2	1.1822	1.142	1.106	1.074	1.050	1.032	1.022	1.015	1.008	1.004	1.000	.997					
0.4	1.3513	1.300	1.249	1.200	1.151	1.106	1.073	1.051	1.036	1.025	1.015	1.009					
0.6	1.5011	1.437	1.374	1.311	1.251	1.190	1.138	1.099	1.072	1.049	1.030	1.019					
0.8	1.6324	1.556	1.482	1.411	1.340	1.272	1.211	1.157	1.114	1.077	1.048	1.028					
1.0	1.7460	1.660	1.576	1.495	1.418	1.344	1.275	1.211	1.155	1.106	1.065	1.037					
1.5	1.9607	1.856	1.753	1.654	1.561	1.472	1.385	1.306	1.230	1.164	1.103	1.054					
2.0	2.0934	1.971	1.854	1.742	1.636	1.536	1.440	1.350	1.265	1.190	1.120	1.056					
2.5	2.1657	2.031	1.900	1.779	1.662	1.550	1.448	1.351	1.260	1.179	1.104	1.032					
3.0	2.1971	2.049	1.909	1.777	1.651	1.535	1.426	1.321	1.224	1.140	1.060	.988	.92				
3.5	2.2030	2.043	1.893	1.753	1.621	1.497	1.382	1.275	1.176	1.088	1.006	.931	.88				
4.0	2.1944	2.024	1.864	1.715	1.577	1.448	1.329	1.221	1.121	1.030	.947	.870	.82	.77			
4.5	2.1788	1.998	1.829	1.673	1.529	1.394	1.273	1.162	1.061	.968	.884	.807	.75	.71			
5.0	2.1607	1.971	1.794	1.631	1.481	1.344	1.218	1.104	1.002	.909	.824	.748	.67	.63	.59		
6.0	2.1261	1.918	1.727	1.552	1.392	1.247	1.115	.999	.893	.799	.714	.641	.56	.51	.47	.42	
7.0	2.0984	1.875	1.670	1.484	1.315	1.164	1.029	.908	.802	.708	.624	.553	.482	.42	.38	.34	.31
8.0	2.0781	1.840	1.623	1.427	1.251	1.095	.955	.833	.725	.631	.549	.479	.418	.363	.31	.28	.25
9.0	2.0634	1.811	1.583	1.379	1.196	1.035	.892	.768	.661	.569	.488	.420	.363	.315	.269	.23	.20
10.0	2.0526	1.787	1.548	1.335	1.147	.982	.838	.715	.608	.515	.437	.371	.319	.273	.231	.194	.17
12.0	2.0382	1.749	1.492	1.264	1.067	.896	.745	.624	.519	.431	.356	.296	.244	.203	.169	.140	.116
14.0	2.0292	1.717	1.441	1.202	.997	.823	.672	.551	.450	.367	.295	.240	.195	.158	.128	.104	.084
16.0	2.0229	1.690	1.401	1.150	.937	.760	.613	.493	.394	.314	.249	.198	.157	.127	.099	.078	.063
18.0	2.0182	1.666	1.363	1.102	.886	.707	.561	.443	.347	.271	.211	.164	.128	.099	.077	.060	.047
20.0	2.0147	1.646	1.329	1.061	.841	.660	.515	.399	.307	.236	.181	.138	.105	.080	.061	.047	.036

TABLE VI. Heat capacity, C (cal. per degree mole).

V/RT \ $1/Q$	0.0	0.05	0.10	0.15	0.20	0:25	0.30	0.35	0.40	0.45	0.50	0.55	0.60	0.65	0.70	0.75	0.80
0.0	.9934	.993	.993	.993	.993	.993	.993	.993	.993	.993	.99	.99					
0.2	1.0033	1.003	1.003	1.002	1.001	1.000	.999	.998	.998	.998	1.00	1.00					
0.4	1.0326	1.033	1.032	1.030	1.028	1.025	1.024	1.021	1.019	1.017	1.02	1.01					
0.6	1.0799	1.080	1.079	1.076	1.073	1.068	1.065	1.060	1.056	1.051	1.05	1.04					
0.8	1.1433	1.143	1.141	1.138	1.133	1.128	1.121	1.114	1.106	1.099	1.09	1.08					
1.0	1.2201	1.219	1.217	1.212	1.206	1.199	1.190	1.180	1.169	1.157	1.14	1.13					
1.5	1.4506	1.449	1.444	1.435	1.423	1.408	1.391	1.370	1.348	1.324	1.30	1.27					
2.0	1.6975	1.695	1.687	1.673	1.655	1.632	1.606	1.574	1.541	1.505	1.469	1.43					
2.5	1.9211	1.917	1.908	1.888	1.866	1.840	1.801	1.756	1.717	1.670	1.623	1.58					
3.0	2.0986	2.095	2.082	2.062	2.033	1.996	1.952	1.900	1.846	1.794	1.738	1.68	1.7				
3.5	2.2223	2.218	2.204	2.180	2.146	2.106	2.054	1.995	1.934	1.869	1.803	1.74	1.7				
4.0	2.2986	2.294	2.276	2.249	2.213	2.168	2.110	2.048	1.980	1.907	1.834	1.76	1.69	1.6			
4.5	2.3354	2.330	2.312	2.280	2.238	2.190	2.129	2.062	1.990	1.911	1.832	1.75	1.67	1.6			
5.0	2.3443	2.338	2.318	2.285	2.241	2.186	2.120	2.056	1.972	1.890	1.808	1.719	1.62	1.55	1.4		
6.0	2.3155	2.307	2.283	2.245	2.192	2.130	2.059	1.979	1.893	1.803	1.711	1.616	1.51	1.43	1.33	1.2	
7.0	2.2647	2.256	2.228	2.185	2.126	2.055	1.973	1.883	1.787	1.688	1.588	1.491	1.394	1.30	1.21	1.12	1.0
8.0	2.2157	2.205	2.174	2.125	2.058	1.979	1.888	1.788	1.684	1.576	1.468	1.362	1.260	1.159	1.07	.99	.91
9.0	2.1759	2.164	2.130	2.074	1.999	1.909	1.808	1.699	1.587	1.474	1.362	1.252	1.149	1.049	.955	.86	.79
10.0	2.1454	2.133	2.094	2.033	1.951	1.854	1.745	1.630	1.507	1.382	1.262	1.151	1.047	.949	.853	.762	.68
12.0	2.1050	2.089	2.043	1.972	1.877	1.763	1.636	1.502	1.365	1.233	1.107	.989	.877	.774	.683	.600	.519
14.0	2.0810	2.063	2.009	1.923	1.814	1.686	1.546	1.400	1.254	1.112	.978	.855	.744	.644	.555	.476	.408
16.0	2.0654	2.044	1.983	1.887	1.764	1.622	1.468	1.311	1.156	1.009	.873	.749	.639	.542	.457	.384	.321
18.0	2.0544	2.031	1.961	1.853	1.717	1.562	1.397	1.232	1.070	.919	.780	.657	.549	.456	.378	.312	.256
20.0	2.0462	2.020	1.944	1.827	1.678	1.510	1.333	1.158	.991	.837	.701	.580	.477	.389	.316	.256	.207

which is exact for a single top and probably a satisfactory approximation for molecules with several tops. Here A_m is the moment of inertia of the mth top itself, λ_{mi} is the direction cosine between the axis of the mth top and the ith principal axis of the whole molecule, and I_i is the moment of inertia of the whole molecule about this axis.

For molecules with several tops the next approximation to I_m is:

$$I_m = I^0{}_m - (\tfrac{1}{2}) \sum_{m' \neq m}{}' \Lambda^2{}_{mm'}/I^0{}_{m'}, \qquad (1b)$$

where $I^0{}_m$ is the value given in Eq. (1a) and

$$\Lambda_{mm'} = A_m A_{m'} \sum_{i=1}^{3} \lambda_{mi}\lambda_{m'i}/I_i. \qquad (1c)$$

The essential criterion of accuracy of the simple formulas and tables for molecules with several tops is whether the potential can be expressed as:

$$V(\phi_1, \phi_2 \cdots \phi_N) = \sum_{m=1}^{N} \tfrac{1}{2} V_m(1-\cos n_m\phi_m). \qquad (2)$$

This expression should suffice where the rotating groups are far apart and are not coupled by long range forces. For example, Eq. (2) should suffice for *para*-xylene and *trans*-2-butene but possibly not for *ortho*-xylene, *cis*-2-butene, or propane.

The well-known formulas for translation and over-all rotation may be written as follows:

$$(H^0_0 - F^0_T)/T = R[\tfrac{3}{2}\ln M + 4\ln T - \ln P + \tfrac{1}{2}\ln(I_1I_2I_3\times 10^{117}) - \ln\sigma] - 10.237, \quad (3)$$

$$(H^0_T - H^0_0)/T = C_p = 4R = 7.948,$$

$$S^0_T = (H^0_0 - F^0_T)/T + (H^0_T - H^0_0)/T.$$

Here M is the molecular weight, P is the pressure in atmos., and σ is the symmetry number for over-all rotation. Physical constants have been taken from the *International Critical Tables* since these values are standard for this type of work. However, the only physical constant which enters the tables is the gas constant, $R = 1.9869$ cal. per degree mole. This constant is a factor in all entries and in the very improbable circumstance that it is appreciably in error an appropriate factor can be applied.

The two variables for the tables were selected as the dimensionless ratios (V/RT) and $(1/Q_f) = hn/(8\pi^3IkT)^{\frac{1}{2}}$. The latter is the reciprocal of the partition function for free rotation which also appears as $R\ln Q_f$. In terms of I in g cm² and T°K one obtains:

$$Q_f = 2.815(10^{38}IT)^{\frac{1}{2}}/n. \quad (4)$$

The use of the physical constants recently compiled by Birge[13] gives 2.7934 for the numerical constant in Eq. (4). These variables have the additional advantages of permitting a very compact form for the tables and in having the potential barrier in only one variable.

The accuracy of each entry in the tables is indicated by the number of significant figures given. Ordinarily linear interpolation will give values accurate to 0.01 cal. per degree. Graphical interpolations are recommended if the maximum accuracy is desired.

ENERGY LEVELS

Having stated the thermodynamic results for the case of greatest practical interest, let us proceed to the quantum mechanical analysis. In this section we will avoid all but one of the assumptions of the preceding section, discussing the still more general case below. The notation will be the same except for the use of T for kinetic energy instead of temperature.

Beginning with the kinetic energy we use the expression of Kassel[6c] and Crawford[9] as follows:

$$2T = \sum_{i=1}^{3} I_i(\omega'_i)^2 + 2\sum_{i=1}^{3}\sum_{m=1}^{N} A_m\lambda_{mi}\dot\phi_m\omega'_i + \sum_{m=1}^{N} A_m\dot\phi_m^2, \quad (5)$$

where ω'_i is the instantaneous angular velocity of the rigid frame about the ith principal axis, and $\dot\phi_m$ is the angular velocity of the mth top with respect to the rigid frame. Now introduce the transformation:

$$\omega_i = \omega'_i + \sum_{m=1}^{N} A_m\lambda_{mi}\dot\phi_m/I_i, \quad (6)$$

then

$$2T = \sum_i I_i\omega_i^2 + \sum_m A_m\dot\phi_m^2 - \sum_m\sum_{m'} \Lambda_{mm'}\dot\phi_m\dot\phi_{m'}. \quad (7)$$

The ω's are now weighted average angular velocities of the whole molecule. The Λ's were defined in Eq. (1c). This transformation has separated the terms for over-all rotation from those for internal rotation. It should also be noticed that since the A's will be usually much smaller than the I_i's, the Λ's will be much smaller than the A_m's.

Now introducing momenta:

$$P_i = \partial T/\partial\omega_i, \quad P_m = \partial T/\partial\dot\phi_m$$

one obtains

$$2T = \sum_i P_i^2/I_i + \sum_m\sum_{m'} \beta_{mm'}P_mP_{m'}. \quad (8)$$

Here the P_i's are the net angular momenta of the whole molecule, and $(\beta_{mm'})$ is the inverse matrix of the matrix $(A_m\delta_{mm'} - \Lambda_{m,m'})$ of Eq. (7). In order to separate variables it is desirable to remove all terms in the second sum with $m \neq m'$ and to regard them as a perturbation. This is justifiable because it is found that their coefficients $(\sim\Lambda_{mm'}/A_mA_{m'})$ are of a smaller order of magnitude than those with $m = m'$. While the exact expression for β_{mm} has been defined above in terms of matrix algebra, an approximate expression

[13] R. T. Birge, Rev. Mod. Phys. **13**, 233 (1941).

would seem to be much more useful. We then find for the kinetic energy the expression:

$$2T=\sum_i P_i^2/I_i+\sum_m P_m^2/I_m, \qquad (9)$$

where I_m was defined in Eq. (1a) or more accurately in (1b) omitting the factor ($\frac{1}{2}$) before the sum. This expression is exact for one top since no $P_mP_{m'}$ terms are possible there.

For the potential energy we assume a form more general than that of the preceding section, namely:

$$V=\sum_{m=1}^{N} V_m(\varphi_m), \qquad (10)$$

where for each value of m

$$V_{m(\varphi_m)}=\sum_{k=1}^{\infty} \tfrac{1}{2}V_{mk}(1-\cos kn_m\varphi_m). \qquad (11)$$

This form really involves two assumptions. The first is the absence of terms involving more than one φ_m which, as stated above, will probably be a good approximation in some cases and a poor one in others. The second assumption is the absence of sine terms in the Fourier series expansion of V_m. This may be justified by symmetry considerations in many cases. Both assumptions will be examined further.

Turning now to quantum mechanics, since the coefficients of the P's are constants in Eq. (9), one may write at once the separated wave equations:

$$(\textstyle\sum_i \mathbf{P}_i^2/2I_i)\psi_{OR}=W_{OR}\psi_{OR}, \qquad (12)$$

$$(\mathbf{P}_m^2/2I_m)\psi_m+\sum_k \tfrac{1}{2}V_{mk}(1-\cos kn_m\phi_m)\psi_m = W_m\psi_m \quad m=1,2,\cdots,N, \qquad (13)$$

where $\mathbf{P}_m^2=(h/2\pi i)^2\partial^2/\partial\phi_m^2$ and the $\mathbf{P}_i$'s are the operators studied by Klein.[14]

The transformation of Eq. (6) introduces some additional complexity into the representation of the $\mathbf{P}_i$ operators in terms of the Eulerian angles of the rigid frame but does not affect their essential relationships.

$$\begin{aligned}\mathbf{P}_x\mathbf{P}_y-\mathbf{P}_y\mathbf{P}_x&=(h/2\pi i)\mathbf{P}_z,\\ \mathbf{P}_y\mathbf{P}_z-\mathbf{P}_z\mathbf{P}_y&=(h/2\pi i)\mathbf{P}_x, \qquad (14)\\ \mathbf{P}_z\mathbf{P}_x-\mathbf{P}_x\mathbf{P}_z&=(h/2\pi i)\mathbf{P}_y.\end{aligned}$$

[14] O. Klein, Zeits. f. Physik **58**, 730 (1929).

The energy levels for over-all rotation, W_{OR}, are therefore just those of the rigid molecule of moments of inertia I_1, I_2, I_3 as given by Klein and others.

Each wave Eq. (13) for internal rotation may be converted into the Hill differential equation by the following substitutions:

$$2x=n\phi,\quad \theta_k=8\pi^2IV_k/n^2h^2,$$
$$W-\tfrac{1}{2}\sum_k V_k=n^2h^2\theta_0/32\pi^2I, \qquad (15)$$

$$M''(x)+(\theta_0+2\sum_{k=1}^{\infty}\theta_k\cos 2kx)M(x)=0,$$

where the subscript m has been omitted throughout. In general, the boundary conditions on $M(x)$ are very complicated since the requirement is

$$\psi_{OR}(\varphi_m)M_m(n\varphi_m/2) =\psi_{OR}(\varphi_m+2\pi)M_m((n/2)(\varphi_m+2\pi)), \qquad (16)$$

where the wave function for over-all rotation ψ_{OR} depends on ϕ_m in addition to the Eulerian angles of the rigid frame. The dependence of ψ_{OR} on ϕ_m cannot be stated simply. Nevertheless, it is apparent that Eq. (16) cannot be satisfied unless $M_m(x)$ remains finite even when x approaches either positive or negative infinity. We shall proceed to consider all solutions of the Hill equation which satisfy this very general boundary condition and to show that for thermodynamic calculations further analysis is not necessary at this point.

We shall follow the solution offered by Hill.[15] A general solution is

$$M(x)=e^{\mu x}\sum_{k=-\infty}^{+\infty} b_k e^{2kix}, \qquad (17)$$

where the sum is evidently a purely periodic function. Ordinarily the second solution is given by $M(-x)$. In the special case $\mu=\pm 0$, $\pm i$ these two solutions become identical. If $M(x)$ is to remain finite for all x, it is seen that μ must be a pure imaginary number or zero. Otherwise $M(x)$ will contain a real exponential factor that will approach infinity either as x increases indefinitely in a positive direction or else as it increases

[15] G. W. Hill, Acta Math. **8**, (1886), or Whittaker and Watson, *Modern Analysis* (Cambridge University Press, 1935), pp. 413–417.

negatively. In the special case $\mu=\pm i$, 0 these two solutions become identical. By other methods a second, different solution may be obtained in this case, but it does not remain finite and therefore is not acceptable.

Now μ is given by the transcendental equation

$$\sin^2(\tfrac{1}{2}\pi i\mu)=\Delta(\theta_0\theta_1\cdots)\sin^2(\tfrac{1}{2}\pi\theta_0^{\frac{1}{2}})=F(\theta_0\theta_1\cdots), \quad (18)$$

where $\Delta(\theta)$ is the convergent, infinite determinant

$$\Delta(\theta_1\theta_1\cdots)=\begin{vmatrix}
\cdot & \cdot & \cdot & \cdot & \cdot & \cdot & \cdot & \cdot & \cdot \\
\cdots & 1 & \frac{\theta_1}{\theta_0-6^2} & \frac{\theta_2}{\theta_0-6^2} & \frac{\theta_3}{\theta_0-6^2} & \frac{\theta_4}{\theta_0-6^2} & \frac{\theta_5}{\theta_0-6^2} & \frac{\theta_6}{\theta_0-6^2} & \cdots \\
\cdots & \frac{\theta_1}{\theta_0-4^2} & 1 & \frac{\theta_1}{\theta_0-4^2} & \frac{\theta_2}{\theta_0-4^2} & \frac{\theta_3}{\theta_0-4^2} & \frac{\theta_4}{\theta_0-4^2} & \frac{\theta_5}{\theta_0-4^2} & \cdots \\
\cdots & \frac{\theta_2}{\theta_0-2^2} & \frac{\theta_1}{\theta_0-2^2} & 1 & \frac{\theta_1}{\theta_0-2^2} & \frac{\theta_2}{\theta_0-2^2} & \frac{\theta_3}{\theta_0-2^2} & \frac{\theta_4}{\theta_0-2^2} & \cdots \\
\cdots & \frac{\theta_3}{\theta_0} & \frac{\theta_2}{\theta_0} & \frac{\theta_1}{\theta_0} & 1 & \frac{\theta_1}{\theta_0} & \frac{\theta_2}{\theta_0} & \frac{\theta_3}{\theta_0} & \cdots \\
\cdots & \frac{\theta_4}{\theta_0-2^2} & \frac{\theta_3}{\theta_0-2^2} & \frac{\theta_2}{\theta_0-2^2} & \frac{\theta_1}{\theta_0-2^2} & 1 & \frac{\theta_1}{\theta_0-2^2} & \frac{\theta_2}{\theta_0-2^2} & \cdots \\
\cdots & \frac{\theta_5}{\theta_0-4^2} & \frac{\theta_4}{\theta_0-4^2} & \frac{\theta_3}{\theta_0-4^2} & \frac{\theta_2}{\theta_0-4^2} & \frac{\theta_1}{\theta_0-4^2} & 1 & \frac{\theta_1}{\theta_0-4^2} & \cdots \\
\cdots & \frac{\theta_6}{\theta_0-6^2} & \frac{\theta_5}{\theta_0-6^2} & \frac{\theta_4}{\theta_0-6^2} & \frac{\theta_3}{\theta_0-6^2} & \frac{\theta_2}{\theta_0-6^2} & \frac{\theta_1}{\theta_0-6^2} & 1 & \cdots \\
\cdot & \cdot & \cdot & \cdot & \cdot & \cdot & \cdot & \cdot & \cdot
\end{vmatrix}. \quad (19)$$

$\Delta(\theta)$ has singularities at $\theta_0=0, 2^2, 4^2\cdots$; however, $\sin^2(\frac{1}{2}\pi\theta_0^{\frac{1}{2}})$ approaches zero in an exactly compensating manner so that $F(\theta)$ is a continuous, single valued function for all θ_0, θ_1, etc. Figure 1 shows $F(\theta)$ as a function of θ_0 for several discrete values of θ_1 and θ_2, with $\theta_3=\theta_4\cdots=0$.

The region of acceptable solutions (μ to be a pure imaginary) is given by $0\leqslant F(\theta)\leqslant 1$ as can be easily seen from (18). In the special case $\theta_1=\theta_2=\cdots=0$, $\Delta=1$ all positive values and no negative values of θ_0 are in the acceptable region. Otherwise for θ_0 sufficiently negative Δ will always become positive and approach $+1$. Then F remains negative and no acceptable solutions are found. For positive or small negative values of θ_0, regions of acceptable solutions are obtained and between these are regions where no acceptable solutions exist.

We may now examine the completeness of our solution. For given values of θ_0, θ_1, etc., a linear combination of two independent solutions is a complete solution. By Eq. (18) the positive and negative values of μ correspond to the same values of θ_0, θ_1, etc., and, excepting the special case already mentioned, they give independent solutions. Thus we do have a complete solution.

We now return to the topic of boundary conditions in order to define the number of energy levels and so far as possible their exact energies. We shall not discuss the wave functions here; they are readily obtained in terms of the boundary conditions and the cited analysis of the Hill equation.[15] By symmetrical tops we mean tops whose moments of inertia perpendicular to the top axis are equal. Ordinarily this equality of moments of inertia will arise (as in CH_3) from identical atoms symmetrically placed. We shall call such tops "identically" symmetrical tops. It

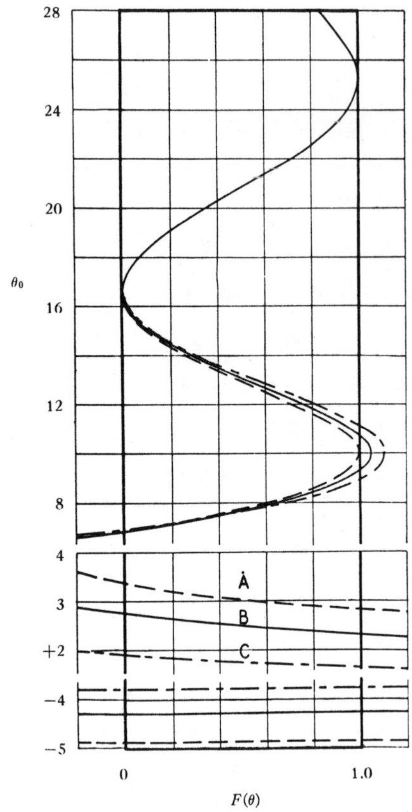

FIG. 1. $F(\theta_0, \theta_1, \theta_2 \cdots)$ as a function of θ_0 with $\theta_1=4$, $\theta_2=+0.8$ for A; $\theta_1=4$, $\theta_2=0$ for B; $\theta_1=4$, $\theta_2=-0.8$ for C; in all cases $\theta_3=\theta_4=\cdots=0$.

is possible, however, for a symmetrical top to consist of various different atoms so placed as to make the moments equal. We shall call these "accidentally" symmetrical tops.

Let us now consider a molecule made up of one n fold identically symmetrical top, and one accidentally symmetrical top. Group theoretical methods give the general boundary condition on the internal rotation to be:

$$\psi_{OR}\left(\varphi_m+\frac{2\pi}{n}\right)M_m\left(\frac{n}{2}\left(\varphi_m+\frac{2\pi}{n}\right)\right)$$

$$=e^{i(2\pi/n)k}\psi_{OR}(\phi_m)M_m\left(\frac{n}{2}\phi_m\right), \quad (20)$$

where the value of k, an integer, depends on the statistics obeyed by the identical atoms, their relative nuclear spin orientations, the symmetry of the vibrational state, and possibly other factors. For a molecule in a given state with respect to other factors one and only one value of k is allowed. Thus if the totally symmetric state for internal rotation is desired, k is zero.

Now using the results of Nielsen[1] for two coaxial tops we obtain

$$M[(n/2)(\phi+2\pi/n)]$$

$$=\exp\{i(2\pi/n)[k+K(A/I_3)]\}M[(n/2)\phi], \quad (21)$$

where A is the moment of inertia of the identically symmetrical top, I_3 is the moment of inertia of the whole molecule about the axis of the tops, and K is the quantum number for over-all rotation giving the angular momentum about the axis of the tops. This leads to an equation for μ (in Eq. (17)):

$$\mu=i(2/n)[k+K(A/I_3)]. \quad (22)$$

Thus for any given set of other conditions on this molecule, a single energy level for internal rotation is obtained in each of the acceptable regions discussed above.

We were unable to obtain any reasonably simple expression for μ in the more general case of a symmetric top attached to an unsymmetrical frame. However, since such a molecule must be related adiabatically (by a series of infinitesimal perturbations) to the case discussed above, the same number of energy levels must be present. Thus we may conclude for the general case of an identically symmetrical top attached to a rigid frame of any symmetry that there will be one energy level in each acceptable region for any given set of other quantum numbers or other specifying conditions.

Two special cases may be mentioned. If two coaxial tops have different symmetries the internal rotation may have an even higher symmetry. The molecules nitromethane and toluene exemplify this with two- and threefold symmetries combined in each case. The second case arises when the potential barrier has a higher symmetry, $n=ab$, (possibly only approximate) than the rotating groups, $n=b$. Then there will be a energy levels in each acceptable region if the value $n=ab$ is used in Eq. (15). This increases S and $-F/T$ by $R \ln a$ but does not change H or C. If $n=b$ is used in Eq. (15) then the case ceases to be a special one—only one energy level per region

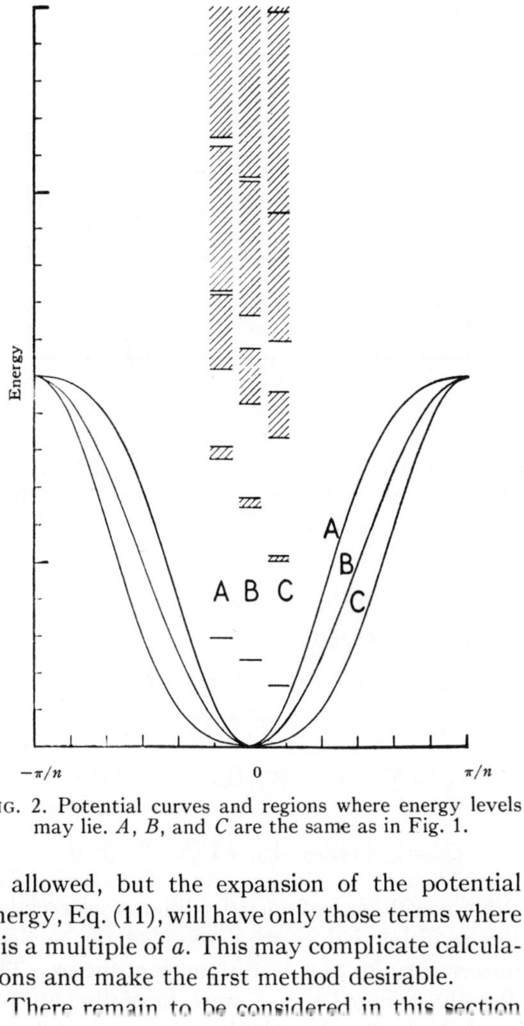

FIG. 2. Potential curves and regions where energy levels may lie. A, B, and C are the same as in Fig. 1.

is allowed, but the expansion of the potential energy, Eq. (11), will have only those terms where k is a multiple of a. This may complicate calculations and make the first method desirable.

There remain to be considered in this section the two energy terms which were dropped above and which we may now treat as perturbations. First are the cross terms in the kinetic energy of the type $\beta_{mm'}\mathbf{P}_m\mathbf{P}_{m'}$. The effect of these terms is somewhat similar to that of the coriolis perturbation or vibration-rotation angular momentum interaction in polyatomic molecules. The perturbation energies will be small but appreciable. However, since angular momenta are as likely to be positive as negative, as many energy levels will be raised as will be lowered. The effect on the partition function or thermodynamic properties will, therefore, be of a much smaller order of magnitude and probably negligible in all practical cases. For comparisons with spectroscopic data these terms will need to be considered, except of course in molecules with only one top where they are absent. However, in view of the lack of application at present and the extreme complexity of equations for the general case as compared to special cases, it does not seem worthwhile to pursue this subject further.

The second set of energy terms that was dropped is the portion of the Fourier series expansion of the potential energy involving sines:

$$V' = \sum_{k=1}^{\infty} S_k \sin kn\phi,$$

where in order that $\phi = 0$ shall be a minimum in the potential curve it is necessary that:

$$\left(\frac{\partial V'}{\partial \phi}\right)_{\phi=0} = n \sum_{k=1}^{\infty} kS_k = 0.$$

Thus the S_k's must vary the sign tending to cancel one another in effect. For many special cases (two coaxial tops, for example) it is possible to prove that the first-order perturbation energy will be exactly zero. In general this is not possible but then the situation is like that discussed above, i.e., some energy levels are raised and others lowered. Thus it seems certain that these terms will not affect the thermodynamic functions appreciably although they may perturb individual energy levels somewhat. In many cases the sine terms are eliminated completely by symmetry considerations.

PARTITION FUNCTION

We shall now calculate the partition function for the reasonably general case considered in the preceding section. Holding all other quantum

TABLE VII. The effect of μ on the thermodynamic functions.

$\sin^2 \frac{1}{2}i\pi\mu$	$-F/T$	H/T	C
	$(V/RT)=4$	$(1/Q)=0.5642$	
0.0	0.436	0.851	1.737
.2	.436	.851	1.735
.4	.436	.851	1.733
.6	.436	.852	1.728
.8	.436	.852	1.725
1.0	.436	.853	1.724
	$(V/RT)=6$	$(1/Q)=0.6910$	
0.0	0.167	0.475	1.378
.2	.166	.477	1.367
.4	.165	.481	1.358
.6	.165	.484	1.344
.8	.164	.488	1.340
1.0	.164	.492	1.334

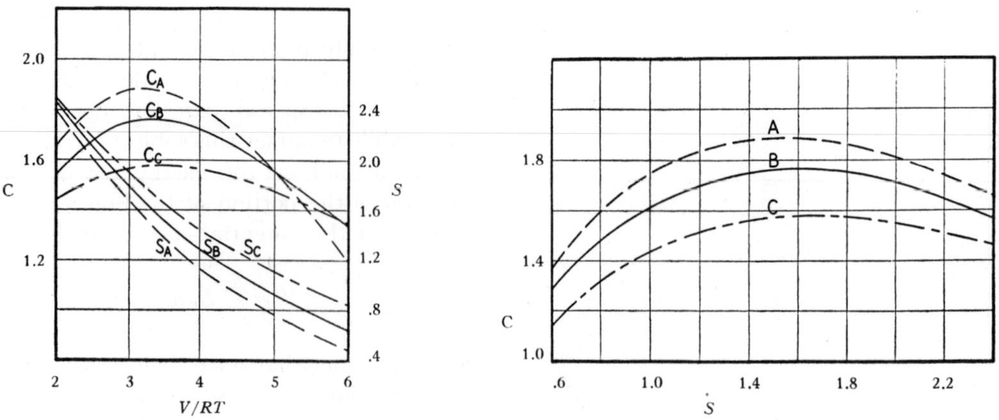

(a) (b)

FIG. 3. (a) Heat capacity and entropy as a function of temperature for the potential curves of Figs. 1 and 2. (b) Heat capacity as a function of entropy for the same potential curves. This shows the possibility of determining the shape of potential curve from thermodynamic data alone. It is apparent that the experimental error would have to be less than one-tenth cal. per degree to distinguish definitely A or C from B.

numbers or conditions constant, and therefore μ constant, one may write for the mth internal rotation

$$Q_m(\mu)=\sum_{l=0}^{\infty}\exp\left[-(W_{m,l}(\mu)-W_{m,0})/kT\right], \quad (23)$$

where l is simply an index numbering off the acceptable regions of energy levels, and W_0 is the energy of the lowest real state. Numerical evaluation of typical cases for various values of μ showed that the partition function was a smooth and nearly linear function of μ between its limits of 0 and i. Two examples are listed in Table VII. In other words, if the partition function is summed for the two cases $\mu=0,\ i(F(\theta)=0,\ 1)$ and the two values are essentially the same, then this value may be taken as a good approximation for all μ.

The tables given above were evaluated by this method. If only the first term is taken in the potential energy expression, $\theta_2=\theta_3\cdots=0$ in Eqs. (15) and (19), then the Hill Eq. (15) reduces to the Mathieu equation. Ince[16] has given the characteristic values, and hence the energy levels, for the cases $\mu=0,\ i$ (ordinary Mathieu functions) over a range of values of θ_1, the potential barrier height. Additional energy levels were calculated by Ince's methods as needed.

The equations for the thermodynamic functions are the usual ones:

$$\begin{aligned}
-F/T&=R\ln Q,\\
H/T&=R(Q'/Q),\\
S&=H/T-F/T=R(\ln Q+Q'/Q),\\
C&=R[Q''/Q-(Q'/Q)^2], \qquad (24)\\
Q'&=\sum_{l=0}^{\infty}[(W_l-W_0)/kT]e^{-(W_l-W_0)/kT},\\
Q''&=\sum_{l=0}^{\infty}[(W_l-W_0)/kT]^2e^{-(W_l-W_0)/kT}.
\end{aligned}$$

Wherever the sums for $\mu=0,\ i$ differed appreciably the resulting values were rounded off to limit uncertainty to the last significant figure. The directly summed values were carefully interpolated by graphical methods and the use of deviation functions to the even points desired in the tables. Use was made of the classical or limiting function:

$$Q=\frac{(2\pi IkT)^{\frac{1}{2}}}{h}\int_0^{2\pi/n}e^{-(V/2kT)(1-\cos n\phi)}d\phi, \quad (25)$$

which is closely related to certain Bessel functions. In these terms the thermodynamic functions take the following form:

$$\begin{aligned}
(F-F_f)/T&=R[y-\ln J_0(iy)],\\
H/T&=R[\tfrac{1}{2}+y+yiJ_1(iy)/J_0(iy)],\\
C&=R[\tfrac{1}{2}+y^2-yiJ_1(iy)/J_0(iy) \qquad (26)\\
&\quad-\{yiJ_1(iy)/J_0(iy)\}^2],\\
y&=\tfrac{1}{2}(V/RT).
\end{aligned}$$

[16] E. L. Ince, Proc. Roy. Soc. Edinburgh **52**, 355 (1931).

These results apply, of course, in the limit $(1/Q)$ zero.

In addition to these results which comprise the tables of the first section, a few calculations were made for different shapes of potential barrier by the direct evaluation of Hill's determinant. Figure 2 shows the potential curves and energy levels while Fig. 3 shows the thermodynamic functions. The entropy-heat capacity curves indicate that precise thermodynamic data would be needed to distinguish clearly even these extreme variations of shape. Spectroscopic data would seem more promising in this regard. Thus for most practical thermodynamic work the use of the simple cosine potential barrier should be satisfactory.

A USEFUL APPROXIMATION

Before proceeding to still more complex cases, a generally useful approximation will be described. Any potential function, $V(\phi_1\cdots\phi_N)$ can be expanded in a power series about a point of equilibrium.

$$V=\tfrac{1}{2}\sum_m\sum_{m'} b_{mm'}\phi_m\phi_{m'} + \frac{1}{3!}\sum_m\sum_{m'}\sum_{m''} c_{mm'm''}\phi_m\phi_{m'}\phi_{m''}+\cdots, \quad (27)$$

$$b_{mm'}=(\partial^2 V/\partial\phi_m\partial\phi_{m'})_e,$$

$$c_{mm'm''}=(\partial^3 V/\partial\phi_m\partial\phi_{m'}\partial\phi_{m''})_e, \quad \text{etc.}$$

The origins of the ϕ_m's are now of course at the equilibrium positions where the derivatives are evaluated. If only the first sum is used, the usual harmonic oscillator treatment is obtained. For internal rotation problems this is not a very good approximation by itself, because of the large anharmonicity (importance of the other sums) which appears even in the lower energy levels. However, the following combination appears to be quite accurate:

$$Q=Q_{\text{Class}}[Q_{\text{HO Quant}}/Q_{\text{HO Class}}], \quad (28)$$

where Q_{Class} is the complete classical partition function:

$$Q_{\text{Class}}=\frac{1}{h^N}\int_{-\infty}^{+\infty}\cdots\int_{-\infty}^{+\infty}\int_0^{2\pi/n_1}\cdots\int_0^{2\pi/n_N} \times e^{-(T+V)/kT}dp_1\cdots dp_N d\phi_1\cdots d\phi_N, \quad (29)$$

and Q_{HO} represents an harmonic oscillator partition function of either the quantum mechanical or classical form. Since the ratio in brackets in (28) approaches unity at high temperatures, the results must be correct there. Similarly, if the two lowest energy levels are given by the harmonic oscillator treatment with reasonable accuracy, the result is necessarily a good approximation at very low temperatures because Q_{Class} and $Q_{\text{HO Class}}$ will become equal. The best test is a comparison in a case where the exact answer is known. This is presented in Table VIII where a comparison is made for the case of one top and the simple cosine potential barrier. The exact answer comes of course from the earlier part of this paper.

TABLE VIII. Comparison of approximate (Eq. (28)) and exact partition functions in terms of heat capacity and entropy.

	V/RT \ $1/Q_f$	0.10	0.20	0.40
δC	0	0.000	0.000	0.000
approx.	1	− .002	− .007	− .030
−exact	4	+ .001	+ .004	+ .016
(cal./degree)	9	.000	− .000	− .017
	16	+ .001	− .001	− .020
δS	0	0.000	0.000	0.000
approx.	1	+ .001	+ .004	+ .022
−exact	4	.000	− .001	− .002
(cal./degree)	9	.000	.000	+ .002
	16	.000	+ .003	+ .007

FURTHER STUDY OF MOLECULES WITH SEVERAL TOPS

We shall now proceed to apply the approximate method just described to molecules with several attached tops, where the potential energy is not separable according to Eq. (10). By the procedure of Eqs. (5) to (7) the kinetic energy of external rotation was separated and we have for internal rotation the exact expression

$$2T=\sum_m A_m\dot{\phi}_m^2-\sum_m\sum_{m'}\Lambda_{mm'}\dot{\phi}_m\dot{\phi}_{m'}. \quad (30)$$

First let us consider the exact classical partition function (29). By the interesting theorem of Eidinoff and Aston[5] this becomes

$$Q_{\text{Class}}=(2\pi kT/h^2)^{N/2}[D]^{\frac{1}{2}} \times\int_0^{2\pi/n_1}\cdots\int_0^{2\pi/n_N} e^{-V/kT}d\varphi_1\cdots d\phi_N, \quad (31)$$

where

$$[D]=\begin{vmatrix} A_1-\Lambda_{11} & -\Lambda_{12} & \cdots & -\Lambda_{1N} \\ -\Lambda_{21} & A_2-\Lambda_{22} & \cdots & -\Lambda_{2N} \\ \cdot & \cdot & \cdots & \cdot \\ \cdot & \cdot & \cdots & \cdot \\ \cdot & \cdot & \cdots & \cdot \\ -\Lambda_{N1} & -\Lambda_{N2} & \cdots & A_N-\Lambda_{NN} \end{vmatrix}=\prod_{m=1}^{N} I_m. \tag{32}$$

I_m is given approximately by Eq. (1a) and more accurately by (1b). Using the latter values of I_m the second equality in (32) is accurate to terms of order Λ^2 which will be sufficient for all practical purposes.

Turning now to the harmonic oscillator approximation, we have for the potential energy from (27):

$$2V=\sum_m \sum_{m'} b_{mm'}\phi_m\phi_{m'}, \qquad b_{mm'}=(\partial^2 V/\partial\phi_m\partial\phi_{m'})_e. \tag{33}$$

We now follow the usual normal coordinate analysis which is given in greater detail in many books. In the lth normal mode of frequency ν_l the equation of motion will be

$$\phi_m=A_{ml}\sin(2\pi\nu_l t+\delta_l). \tag{34}$$

By the usual Lagrangian methods we find

$$-4\pi^2\nu_l^2(I_mA_{ml}-\sum_{m'\neq m}{}' \Lambda_{mm'}A_{m'l})+\sum_{m'} b_{mm'}A_{m'l}=0, \tag{35}$$

which leads to the determinantal equation

$$\begin{vmatrix} b_{11}-4\pi^2\nu^2 I_1 & b_{12}+4\pi^2\nu^2\Lambda_{12} & \cdots & b_{1N}+4\pi^2\nu^2\Lambda_{1N} \\ b_{21}+4\pi^2\nu^2\Lambda_{21} & b_{22}-4\pi^2\nu^2 I_2 & \cdots & b_{2N}+4\pi^2\nu^2\Lambda_{2N} \\ \cdot & \cdot & \cdots & \cdot \\ \cdot & \cdot & \cdots & \cdot \\ \cdot & \cdot & \cdots & \cdot \\ \cdot & \cdot & \cdots & \cdot \\ b_{N1}+4\pi^2\nu^2\Lambda_{N1} & b_{N2}+4\pi^2\nu^2\Lambda_{N2} & \cdots & b_{NN}-4\pi^2\nu^2 I_N \end{vmatrix}=0. \tag{36}$$

This equation has N roots $\nu_1\cdots\nu_N$ which may be obtained by suitable calculations.

The quantum mechanical energy levels corresponding to this approximation are

$$W=\sum_{l=1}^{N}(v_l+\tfrac{1}{2})h\nu_l, \tag{37}$$

and the partition function

$$Q_{\text{HO Quant}}=\prod_{l=1}^{N}[1/(1-e^{-h\nu_l/kT})]. \tag{38}$$

The classical or high temperature limit approached by this function is

$$Q_{\text{HO Class}}=(kT/h)^N/(\prod_l \nu_l). \tag{39}$$

We have now the factors necessary for the approximate partition function, Eq. (28).

It is possible to expand the exponentials in Eq. (38) to a form more convenient for most calculations. They will be satisfactory provided $(h\nu/kT)$ is not much greater than one.

Let $h\nu/kT=a$:

$$-F/T=R\{\ln Q_{\text{Class}}+\sum_{l=1}^{N}(a_l/2-a_l^2/24+a_l^4/2880\cdots)\},$$

$$\begin{aligned} H/T&=R\{N/2+P'/P+\sum_{l=1}^{N}(-a_l/2+a_l^2/12-a_l^4/720\cdots)\},\\ S&=R\{\ln Q_{\text{Class}}+N/2+P'/P+\sum_{l=1}^{N}(a_l^2/24-a_l^4/960\cdots)\},\\ C&=R\{N/2+P''/P-(P'/P)^2+\sum_{l=1}^{N}(-a_l^2/12+a_l^4/240\cdots)\}, \end{aligned} \tag{40}$$

$$Q_{\text{Class}}=P_{\text{Class}}(2\pi kT/h^2)^{N/2}\prod_{m=1}^{N} I_m^{\frac{1}{2}},$$

$$P_{\text{Class}}=\int_0^{2\pi/n_1}\cdots\int_0^{2\pi/n_N} e^{-V/kT}d\phi_1\cdots d\phi_N,$$

$$P'_{\text{Class}}=\int_0^{2\pi/n_1}\cdots\int_0^{2\pi/n_N}(V/kT)\times e^{-V/kT}d\phi_1\cdots d\phi_N,$$

$$P''_{\text{Class}}=\int_0^{2\pi/n_1}\cdots\int_0^{2\pi/n_N}(V/kT)^2\times e^{-V/kT}d\phi_1\cdots d\phi_N.$$

The potential energy, V, may or may not be expressible in simple form. If the various integrals (P, P', P'') cannot be handled directly, graphical or other numerical methods should be feasible. Satisfactory results may be obtained using Simpson's rule with points spaced at surprisingly large intervals.

This approximate method may even prove worthwhile in cases where the solution of the infinite determinant, Eq. (19), would lead to exact results but is too laborious. The two methods complement one another; that of this section is convenient and accurate for large moments of inertia and high temperatures while the exact quantum mechanical methods are necessary and not too difficult with small moments of inertia and moderate temperatures.

We have enjoyed discussions of certain phases of this work with Dr. Bryce L. Crawford.

Comment: Energy Levels and Thermodynamic Functions for Molecules with Internal Rotation I. Rigid Frame with Attached Tops

The "uscful approximation" described toward the end of this paper has been used successfully by others for several different problems in addition to restricted internal rotation.

The other papers in this series consider more complex structural situations or highly specialized cases where no new basic concepts are involved but the calculations become more complex or less accurate. They are: II. Unsymmetrical Tops Attached to a Rigid Frame, K. S. Pitzer, *J. Chem. Phys.* **14**, 239–243 (1946); III. Compound Rotation, J. E. Kilpatrick and K. S. Pitzer, *J. Chem. Phys.* **17**, 1064–1075 (1949); and IV. Extended Tables for Molecules with Small Moments of Inertia, J. C. M. Li and K. S. Pitzer, *J. Phys. Chem.* **60**, 466–474 (1956).

Reprinted from the JOURNAL OF CHEMICAL PHYSICS, Vol. 30, No. 4, 1076–1086, April, 1959
Printed in U. S. A.

Internal Rotation and Infrared Spectra of Formic Acid Monomer and Normal Coordinate Treatment of Out-of-Plane Vibrations of Monomer, Dimer, and Polymer*

TATSUO MIYAZAWA AND KENNETH S. PITZER
Department of Chemistry, University of California, Berkeley, California

(Received October 6, 1958)

The infrared spectra of the four isotopic species of formic acid were measured in the vapor phase as well as in the solid nitrogen matrix in the region 400–800 cm^{-1}. Absorption bands in the vapor phase were analyzed by considerations of band contours and comparisons with the frequencies distinctly observed in the matrix, and the previous assignment was revised. The torsional vibrations of the terminal OH groups of short chain polymers of HCOOH were observed in the matrix at 685 and 694 cm^{-1}. The OH torsional frequencies of the *trans* isomer were located in the vapor phase at 582 (HCOOH), 576 (DCOOH), 450 (HCOOD), and 448 cm^{-1} (DCOOD). The energy difference between the two isomers was determined to be 2.0±0.3 kcal/mole. The potential hindering internal rotation was calculated to be $2V\ (\text{kcal/mole}) = 2._1(1-\cos\theta)+9._9(1-\cos2\theta)-0._1(1-\cos3\theta)$ from the energy difference and the fundamental torsional frequencies. The entropy of mixing of the two isomers was found to be 0.3±0.1 cal deg^{-1} $mole^{-1}$ at 25°C. A normal coordinate treatment was made of the *cis* and *trans* isomers of monomer and of the dimer and polymer. The double bond characters calculated from the torsional potential constant agree with those calculated from bond lengths.

INTRODUCTION

THE stable configuration of formic acid monomer has been known to be *cis* by studies of the infrared vibration-rotation spectra,[1] microwave absorption,[2] electron diffraction,[3] and dielectric constant.[4] The latter study[4] has shown, however, that the effective moment of the monomer increases with temperature, suggesting an increase in mole fraction of the *trans* isomer at higher temperatures. Mariner and Bleakney[5] have made an electron impact study and explained pairs of appearance potentials by postulating the coexistence of the *cis* and *trans* isomers. An infrared band of acetic acid monomer observed at 538 cm^{-1} has been assigned[6] to the *trans* isomer; further work on acetic acid in this laboratory will be published soon.

The present study was undertaken to investigate the internal rotation of formic acid monomer. Infrared

* This research was assisted by the American Petroleum Institute through Research Project 50.

[1] V. Z. Williams, J. Chem. Phys. **15**, 232, 243 (1947).
[2] Lerner, Dailey, and Friend, J. Chem. Phys. **26**, 680 (1957).
[3] I. L. Karle and J. Karle, J. Chem. Phys. **22**, 43 (1954).
[4] Coop, Davidson, and Sutton, J. Chem. Phys. **6**, 905 (1938).
[5] T. Mariner and W. Bleakney, Phys. Rev. **72**, 792 (1947).
[6] W. Weltner, Jr., J. Am. Chem. Soc. **77**, 3941 (1955).

spectra of the four isotopic species of formic acid were measured in the vapor phase as well as in the solid nitrogen matrix in the region 400–800 cm^{-1}. A normal coordinate treatment was made of the out-of-plane vibrations of the monomer (*cis* and *trans*), dimer and polymer, and the potential constants were calculated from observed frequencies. The results will be presented in this paper.

EXPERIMENTAL

The preparation and purification of the samples of the four isotopic species of formic acid were carried out by the same method as given in a previous paper.[7] The contamination of other isotopic species was less than 1% in each case. The infrared spectra in the 400–800 cm^{-1} region were measured using a Perkin-Elmer Model 12C spectrometer with KBr optics. The calculated spectral slit widths were 2.8, 2.4, and 3.2 cm^{-1} at 400, 550, and 700 cm^{-1}, respectively. The absorption cell was predeuterated before admitting the samples of HCOOD or DCOOD in order to prevent conversion to HCOOH or DCOOH.

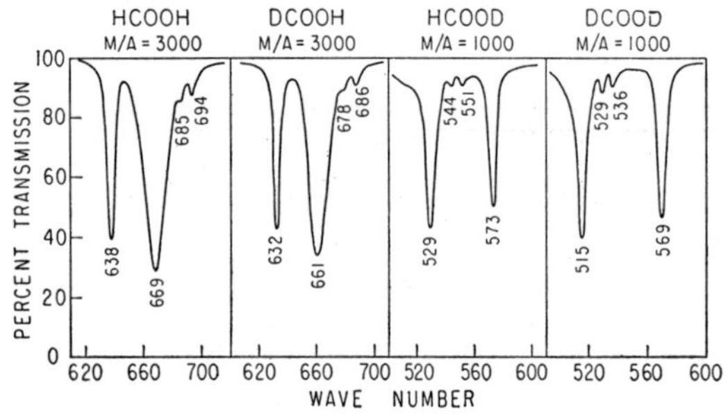

FIG. 1. Infrared spectra of formic acid in the solid nitrogen matrix.

The absorption spectra in the solid nitrogen matrix were measured by the experimental technique described before.[8] Twenty micromoles of formic acid vapor were mixed with enough nitrogen gas to give a mole ratio (M/A: moles of nitrogen divided by moles of formic acid) of 1000 or 3000 and then deposited on a cesium bromide window maintained at 20°K. The observed spectra are shown in Fig. 1.

The infrared spectra of monomeric species have been measured by Millikan and Pitzer[7] in the vapor phase at 150°, however, more accurate measurements were carried out with higher resolution than before, and thus many new bands (or lines) were found. The absorption curves are shown in Figs. 2 and 3. The optical path length of the cell with potassium bromide windows was 10 cm and the sample pressure (measured at room temperature) was 20 mm Hg.

[7] R. C. Millikan and K. S. Pitzer, J. Chem. Phys. **27**, 1305 (1957).

[8] E. D. Becker and G. C. Pimentel, J. Chem. Phys. **25**, 224 (1956).

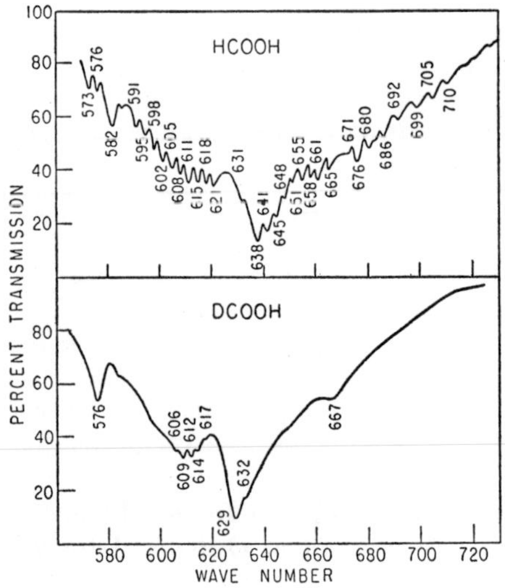

FIG. 2. Infrared spectra of HCOOH and DCOOH in the vapor phase at 150°C.

VIBRATIONAL FREQUENCIES IN THE SOLID NITROGEN MATRIX

It has been concluded by Millikan and Pitzer that in the potassium bromide region there are two vibrations for each isotopic species; i.e., the in-plane OCO bending vibration and the out-of-plane OH (or OD) torsional vibration. However only one strong peak is observed in the vapor spectrum for HCOOH and DCOOH whereas several peaks are observed for HCOOD and DCOOD. Thus the selection of the band centers has been somewhat uncertain. In the present study, therefore, the monomeric frequencies were measured in the solid nitrogen matrix in the hope of sharpening the bands and observing the two frequencies separately.

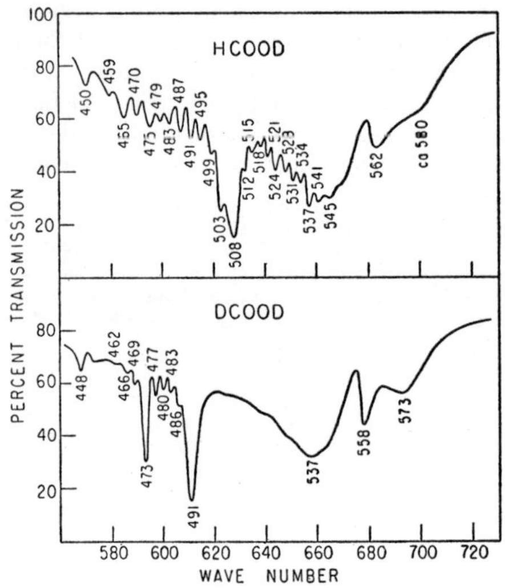

FIG. 3. Infrared spectra of HCOOD and DCOOD in the vapor phase at 150°C.

TABLE I. Infrared frequencies of formic acid observed in the solid nitrogen matrix at 20°K

	HCOOH	DCOOH	HCOOD	DCOOD	Assignment
Monomer	638	632	573	569	OCO bending (A')
	669	661	529	515	OH (or OD) torsion (A'')
Associated species	685 694	678 686	544 551	529 536	OH (or OD) torsion

As shown in Fig. 1, two strong bands of HCOOH were observed at 669 cm^{-1} and 638 cm^{-1} and two weak bands at 695 cm^{-1} and 686 cm^{-1}. The latter two are assigned to associated species since their intensities decrease on dilution, and the 669-cm^{-1} and 638-cm^{-1} bands are assigned to the monomeric species. Thus the two monomeric frequencies were observed distinctly in the matrix even though not in the vapor phase. The corresponding bands of HCOOD monomer are observed at 573 cm^{-1} and 529 cm^{-1} and the two weak bands of associated species at 551 and 544 cm^{-1}. The observed frequencies of DCOOH and DCOOD are also assigned to monomer or associated species as shown in Table I.

Monomeric Frequencies

As shown in the preceding, two monomeric frequencies were observed for each isotopic species corresponding to the OCO bending and OH (or OD) torsional vibrations. The vibrational assignment of these frequencies may be made by the use of the Teller-Redlich product rule. There are two out-of-plane vibrations for each isotopic species, i.e., the OH (or OD) torsion and the CH (or CD) bending. The latter frequencies have been observed in the solid nitrogen matrix[9] at 1041, 877, and 876 cm^{-1} for HCOOH, DCOOH, and DCOOD, respectively.† The theoretical and observed product rule ratios for the pair DCOOH/DCOOD do not agree with each other unless the 661-cm^{-1} band of DCOOH and the 515-cm^{-1} band of DCOOD are assigned to the OH and OD torsional vibrations (Table II). Therefore the band at 632 cm^{-1} of DCOOH and that at 569 cm^{-1} of DCOOD are assigned to the OCO bending vibration. It deserves mentioning that the frequency of this mode decreases by 10% on deuteration of the hydroxyl group. A normal coordinate treatment of the in-plane vibrations of formic acid[9] has indicated, however, that the fractional potential energy associated with the OH bending motion is only 5% for the 632-cm^{-1} vibration, whereas that associated with the OD bending motion amounts to 15% for the 569-cm^{-1} vibration. Therefore the frequency decrease on deuteration may be attributed to the interaction of the OCO and OD bending motions as postulated before.[7]

The frequency assignment of the monomer bands of HCOOH and HCOOD may be made by comparison with those of DCOOH and DCOOD (Table I). The product rule is, then, satisfied for the isotopic pair HCOOH/DCOOH as shown in Table II.

Frequencies of Associated Species

For each of the four isotopic species, two satellite bands were observed on the high-frequency side of the monomeric band due to the OH (or OD) torsional vibration. Since the frequency ratios of the satellite bands to the monomeric bands (1.027 and 1.041) are just the same for all the isotopic species, it is most reasonable to assign these satellite bands to the OH (or OD) torsional vibrations of associated species.

These satellite bands, however, may not be assigned to the hydroxyl groups which form hydrogen bond, since the OH torsional frequencies of polymer (in the crystalline state)[10] are much higher than the monomeric frequencies‡ (in the matrix). On the other hand, terminal hydroxyl groups of short chain polymers, such as dimer and trimer, are not involved in hydrogen bonding and their torsional frequencies will not be much different from the monomeric frequency. On hydrogen bond formation through the carbonyl group, the contribution of the polar electronic structure O^-—CH=O^+—H to the terminal monomer unit is expected to increase with accompanying increase in the potential constant for the torsional displacement. In fact, the frequencies of the satellite bands are higher than those of the monomer bands. On this basis, the satellite bands are assigned to the torsional vibrations of the terminal OH (or OD) groups of short chain polymers.

NORMAL COORDINATE TREATMENT OF OUT-OF-PLANE VIBRATIONS

Monomer in the Solid Nitrogen Matrix

A normal coordinate treatment was made of the out-of-plane normal vibrations in order to elucidate their nature as well as to calculate potential constants. The *GF* matrix method[11] was used for the calculation. The internal coordinates are the out-of-plane deformation of HCOO, Ω_0, and two torsional coordinates θ_{14} and θ_{24} (see Fig. 4 and Appendix I). The appropriate symmetry coordinates are

$$S_1=\Omega_0(8/9)^{1/2}+(-\theta_{14}+\theta_{24})/8^{1/2}$$

and

$$S_2=(\theta_{14}+\theta_{24})(3/8)^{1/2} \qquad (1)$$

[9] T. Miyazawa (to be published).

† It seems that the band due to the CH bending mode of HCOOD lies buried under a more intense band of an associated species at 1046 cm^{-1}.

[10] R. C. Millikan and K. S. Pitzer (to be published).

‡ The ratios are 1.44, 1.48, 1.35, and 1.37 for HCOOH, DCOOH, HCOOD, and DCOOD, respectively.

[11] E. B. Wilson, Jr., J. Chem. Phys. 7, 1047 (1939); 9, 76 (1941).

and $2\Omega_0/3+(\theta_{14}-\theta_{24})/2$ is the redundant coordinate. S_1 corresponds to the CH out-of-plane deformation and S_2 to the internal rotation about the C—O single bond. The elements of the inverse kinetic energy matrix G are

$$G_{11}=[\mu_0(\tau_{01}+\tau_{02}+\tau_{03})^2+\mu_1\tau_{01}^2+\mu_2\tau_{02}^2+\mu_3\tau_{03}^2](3/2),$$

$$G_{12}=[\mu_0(\tau_{01}+\tau_{02}+\tau_{03})(-\tau_{01}+\tau_{02}-\tau_{03})-\mu_1\tau_{01}^2+\mu_2\tau_{02}^2-\mu_3\tau_{03}(\tau_{03}+2\tau_{34})](3/4)^{\frac{1}{2}},$$

$$G_{22}=[\mu_0(-\tau_{01}+\tau_{02}-\tau_{03})^2+\mu_1\tau_{01}^2+\mu_2\tau_{02}^2+\mu_3(\tau_{03}+2\tau_{34})^2+4\mu_4\tau_{34}^2]/2, \quad (2)$$

where τ_{ij} is the reciprocal bond length ij. The numerical calculation of these elements was done using the bond lengths of $r(\text{C—O})=1.245$ A, $r(\text{C—O})=1.312$ A, $r(\text{C—H})=1.085$ A and $r(\text{O—H})=0.95$ A.

The potential function V used for the calculation was

$$2V=H_\Omega\Omega_0^2+2H'_{\Omega\theta}\Omega_0(\theta_{14}+\theta_{24})/2+H_\theta(\theta_{14}+\theta_{24})^2/4$$

where $(\theta_{14}+\theta_{24})/2$ is the angle of the internal rotation of the hydroxyl group about the C—O bond. The

TABLE II. Teller-Redlich product rule ratios of observed frequencies in the solid nitrogen matrix.

Isotopic pair	Calc[a]	Obs
DCOOD/DCOOH	0.769	0.778
DCOOH/HCOOH	0.827	0.832

[a] See reference 7.

elements of the potential energy matrix expressed in terms of the aforementioned symmetry coordinates are

$$F_{11}=H_\Omega/2,$$

$$F_{12}=H'_{\Omega\theta}/3^{\frac{1}{2}},$$

and

$$F_{22}=2H_\theta/3. \quad (3)$$

The potential constants H_Ω, $H_{\Omega\theta}'$, and H_θ were calculated from the observed frequencies of HCOOH (669 and 1041 cm^{-1}) and DCOOD (515 and 876 cm^{-1}) by the use of the method of least squares.[12] To begin with, the characteristic values ($\lambda_a=4\pi^2c^2\nu_a^2$) and vectors ($L_a$) of the secular determinant GF are calculated assuming an appropriate set of potential constants.

The Jacobian

$$J\equiv\partial(\ln\lambda_1,\ln\lambda_2\cdots\ln\lambda_a\cdots)/\partial(K_1,K_2\cdots K_h\cdots)$$

is also calculated using the equation[13]

$$J_{ah}\equiv\partial\ln\lambda_a/\partial K_h=\mathbf{L}_a\dagger\mathbf{A}_h\mathbf{L}_a/\lambda_a, \quad (4)$$

[12] T. Shimanouchi, *Zikken Kagaku Koza* (Maruzen, Tokyo, 1957) Vol. III, p. 236; Mann, Shimanouchi, Meal, and Fano, J. Chem. Phys. **27**, 43 (1957).

[13] T. Miyazawa, J. Chem. Soc. Japan **76**, 1132 (1955).

FIG. 4. Formic acid dimer.

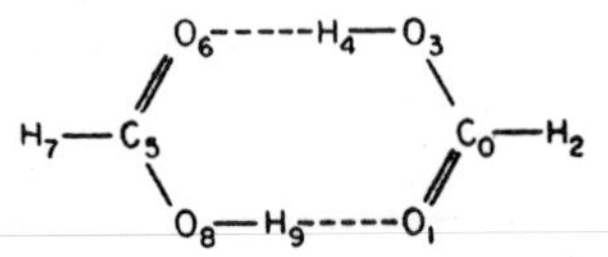

where the matrix $\mathbf{A}_h$ is defined as

$$\mathbf{F}=\sum_h K_h\mathbf{A}_h. \quad (5)$$

The first set of force constants K^I is now corrected so as to yield a better agreement between the observed and calculated characteristic values. The correction $\mathbf{\Delta K}$ is calculated using the equation[12]

$$\mathbf{J}\dagger\mathbf{J}\mathbf{\Delta K}+\mathbf{J}\dagger\mathbf{\Delta}\ln\boldsymbol{\lambda}=0, \quad (6)$$

where $(\mathbf{\Delta}\ln\boldsymbol{\lambda})_a$ is the difference between the calculated and observed values of $\ln\lambda_a$. The elements of the Jacobian depend on the ratios of potential constants and thus Eq. (6) applies only for small values of $\mathbf{\Delta K}$ and $\mathbf{\Delta}\ln\boldsymbol{\lambda}$. It is therefore necessary to repeat the procedure until $\mathbf{\Delta K}$ becomes negligible as compared with $\mathbf{K}$.

The potential constants (erg/radian²) thus calculated are

$$H_\Omega=0.58_7\times10^{-11},$$

$$H_{\Omega'\theta}=-0.00_3\times10^{-11},$$

and

$$H_\theta=0.14_5\times10^{-11}.$$

By the use of these constants, frequencies were calculated as shown in Table III. The calculated and observed frequencies agree within 1%. The nature of the out-of-plane vibrations may be elucidated in terms of the fractional potential energy associated with each internal coordinate (Table IV). Thus for all the isotopic species, the potential energy of the ν_1 vibration is associated almost exclusively with the CH (or CD) bending motion, and that of the ν_2 vibration with the OH (or OD) torsional motion. This allows

TABLE III. Calculated and observed frequencies in the solid nitrogen matrix.

		Obs	Calc	Deviation (%)
HCOOH	ν_1	1041	1047	+0.6
	ν_2	669	673	+0.6
DCOOH	ν_1	877	871	−0.7
	ν_2	661	665	+0.6
HCOOD	ν_1	...	1044	...
	ν_2	529	527	−0.4
DCOOD	ν_1	876	871	−0.6
	ν_2	515	512	−0.6

TABLE IV. Fractional potential energy distributions.

	HCOOH		DCOOD	
	ν_1	ν_2	ν_1	ν_2
$F_{11}L_{1a}^2/\lambda_a$	+0.98	+0.02	+1.00	+0.00
$2F_{12}L_{1a}L_{2a}/\lambda_a$	−0.01	+0.01	−0.00	+0.00
$F_{22}L_{2a}^2/\lambda_a$	+0.03	+0.97	+0.00	+1.00

the treatment of the OH torsional vibration separately from the CH (or CD) out-of-plane vibration.

Energy Levels of the OH Torsional Vibration

The potential energy V hindering the internal rotation of the OH group may be expressed as

$$V=V_2(1-\cos2\theta)/2=V_2\theta^2-V_2\theta^4/3+2V_2\theta^6/45, \quad (7)$$

where V_2 is the twofold potential barrier and θ the angle of internal rotation.

For infinitesimal amplitude, the harmonic oscillator approximation may be used. Then the potential constant H_θ is equal to $2V_2$ and the energy levels E (n= vibrational quantum number) and the transition frequencies ν (in wave number) are

$$E=(n+\tfrac{1}{2})hc\nu_0, \quad (8)$$

$$\nu(n, n+1)=\nu_0, \quad (9)$$

$$4\pi^2c^2\nu_0^2=2V_2/I_r, \quad (10)$$

where I_r is the reduced moment of inertia.

The vibrational amplitude, actually, is not infinitesimal, and it is necessary to take into account the second term $(-V_2\theta^4/3)$ in Eq. (7). The vibrational energy levels for reasonably high potential barrier may be calculated by the use of the first-order perturbation theory. Since the perturbation term H' is

$$H'=-V_2\theta^4/3, \quad (11)$$

the energy levels are calculated to be

$$E=(n+\tfrac{1}{2})hc\nu_0-(h^2/8\pi^2I_r)(2n^2+2n+1)/4 \quad (12)$$

and the transition frequency

$$\nu(n, n+1)=\nu_0-(n+1)h/8\pi^2cI_r. \quad (13)$$

It may be mentioned here that the second term on the right side of Eq. (13) depends only on the quantum number, the reduced moment of inertia, and the number of potential minima per complete revolution.

The accurate energy levels may be calculated.[14] Calculation was made of the *cis* isomer HCOOH for the twofold potential barrier of 9.044 kcal/mole. The transition frequencies are shown in Table V together with those calculated approximately by Eqs. (9), (10), and (13).

[14] U. S. Bureau of Standards, *Tables Relating to Mathieu Functions.* (Columbia University Press, New York, 1951).

As seen from Table V, the harmonic oscillator approximation is not good even for the fundamental frequency. On the other hand, if we take the first anharmonic term $(-V_2\theta^4/3)$ of the potential function, the calculated frequency of the fundamental and the first hot transition are accurate enough for most purposes, and therefore, this approximation will be used in the following sections of this paper.

The differences between the harmonic oscillator frequency ν_0 and the actual fundamental frequency may be calculated by the use of Eq. (13). They are found to be 32 (HCOOH), 30 (DCOOH), 19 (HCOOD) and 18 cm^{-1} (DCOOD). The twofold potential barrier (V_2) in the nitrogen matrix may now be calculated from ν_0(HCOOH:701, DCOOH:691, HCOOD:548, and DCOOD:533 cm^{-1}) and is found to be 11.$_2$ kcal/mole.

Dimer in the Vapor Phase

The formic acid dimer has four infrared active out-of-plane vibrations (a_u); two lying below 200 cm^{-1} and the other two in the region 700–1100 cm^{-1}. The lower frequency vibrations may be approximately described as the rotational vibrations of the two monomer units against each other and their frequencies depend primarily on the intermonomer potential function. On the other hand, the higher frequency vibrations may be correlated with the two out-of-plane vibrations of the monomer and their frequencies depend on the intramonomer potential as well as on the intermonomer potential.

The internal coordinates localized in each monomer unit are denoted as in the case of the monomer. The intermonomer coordinates (see Fig. 4) are the O—H··O bending coordinates, β_{346} and β_{891}; the O—C═O··H torsional coordinates, θ_{93} and θ_{48}; and the H—C═O··H torsional coordinates, θ_{92} and θ_{47}. The symmetry coordinates and the corresponding B matrix elements have been described elsewhere.[15] The bond distances adopted for the calculation of the inverse kinetic energy matrix were r(C═O)=1.25 A, r(C—O)=1.35 A, r(O—H)=0.95 A, r(C—H)=1.085 A, and r(O··O)=2.75 A and all the valence angles were taken to be 120° except for the O—H··O hydrogen bond.

TABLE V. Comparison of transition frequencies (in cm^{-1}) of *cis*-HCOOH for the twofold potential barrier 9.044 kcal/mole.

	$V=V_2(1-\cos2\theta)/2$	$V=V_2\theta^2-V_2\theta^4/3$	$V=V_2\theta^2$
$\nu(0\to1)$	599.2	601.1(+0.3%)	632.7(+5.6%)
$\nu(1\to2)$	562.9	569.5(+1.2%)	632.7(+12.4%)
$\nu(2\to3)$	521.9	537.9(+3.0%)	632.7(+21.2%)

[15] T. Miyazawa and K. S. Pitzer, J. Am. Chem. Soc. **81**, 74 (1959).

The potential function was chosen to be

$$2V=H_\Omega(\Omega_0^2+\Omega_5^2)+H_\theta[(\theta_{14}+\theta_{24})^2+(\theta_{69}+\theta_{79})^2]/4$$
$$+H_1''(\beta_{346}^2+\beta_{891}^2)+H_2''[(\theta_{93}+\theta_{92})^2+(\theta_{48}+\theta_{47})^2]/4. \quad (14)$$

The first two terms express the intramonomer potential where $H_{\Omega\theta}'$ has been taken to be zero, since its value for the monomer is negligible as compared with H_Ω and H_θ. H_1'' and H_2'' are the constants related to hydrogen bonding; they were found to be 0.027×10^{-11} and 0.024×10^{-11} erg/radian², respectively, in our previous study.[15] The intramonomer potential constants are now calculated so that the observed and calculated frequencies of $(HCOOH)_2$ and $(DCOOD)_2$ agree best with each other. The CH out-of-plane bending frequencies of $(HCOOH)_2$ and $(HCOOD)_2$ have not been observed distinctly, however, they were estimated on the assumption that the fractional change in the CH bending frequency from dimer to polymer is equal to that in the CD bending frequency. The potential constants thus calculated are

$$H_\Omega=0.59_9\times10^{-11}$$

and

$$H_\theta=0.21_4\times10^{-11}.$$

The out-of-plane frequencies calculated by the use of these constants are shown in Table VI together with the observed frequencies. The frequency agreement is within 2%.

It is remarkable to note that the OH torsional frequency increases by 39% on dimerization and the OD torsional frequency by 31%. This increase in frequency corresponds to an increase by 80% of the effective force constant. However the value of H_θ of the dimer ($0.21_4\times10^{-11}$) is only 50% larger than that of the monomer. Therefore an additional increase of 30% is due to the intermonomer potential energy associated with hydrogen bonding.

TABLE VI. Observed and calculated a_u frequencies of formic acid dimer.

	Obs	Calc	Deviation (%)
$(HCOOH)_2$	1073[a]	1089	+2
	917	897	−2
$(DCOOH)_2$	930	927	−0
	890	872	−2
$(HCOOD)_2$	1060[a]	1066	+1
	693	700	+1
$(DCOOD)_2$	890	877	−1
	678	693	+2

[a] Estimated from the frequencies observed in the crystalline state (see text).

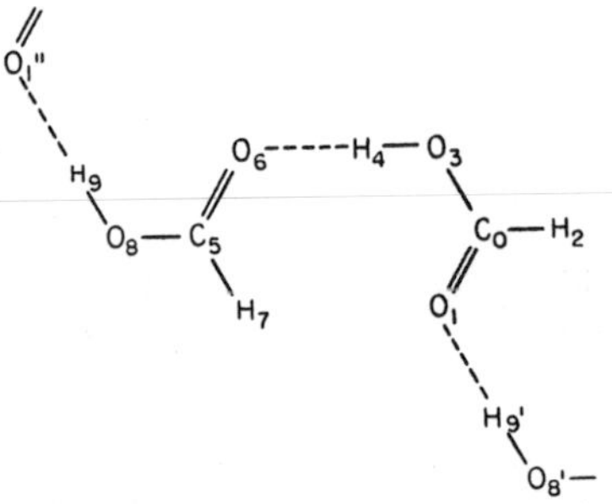

FIG. 5. Formic acid polymer.

Polymer in the Crystalline Phase

The x-ray diffraction study by Holtzberg, Post, and Fankuchen[16] has revealed that formic acid molecules form planar chain polymers in the crystal as shown in Fig. 5. The selection rules and infrared spectra of the crystal were discussed by Millikan and Pitzer[10] and a frequency assignment made. In the present normal coordinate treatment the interchain interaction will be neglected, since the interaction of the van der Waals type is generally much smaller than the interaction due to hydrogen bonding. The vibrations in which corresponding atoms in various repeating units move in phase will be treated because they are the only vibrations which are potentially active in absorption or in scattering. Since there are two molecules per repeating unit of the polymer chain, each intramonomer vibration will be split into two components, one in which all the intramonomer vibrations are in phase with each other and the other in which each vibration is 180° out-of-phase from the adjacent monomers in the chain. Of the out-of-plane vibrations, those belonging to the latter group§ are infrared active, and will now be treated.

The internal coordinates for the polymer are the same as those for the dimer except that β_{891}, θ_{93}, and θ_{92} are now replaced with $\beta_{8'9'1}$, $\theta_{9'3}$, and $\theta_{9'2}$, respectively (see Fig. 5). The symmetry coordinates and the corresponding B matrix elements are given in Appendix II. The inverse kinetic energy matrix G was calculated by the use of the following bond lengths[16] of r(C═O) = 1.23 A, r(C—O) = 1.26 A, r(O—H) = 0.95 A, r(C—H) = 1.085 A, and r(O··O) = 2.58 A. The potential energy per repeating unit was expressed in the same form as that for the dimer, and the corresponding potential energy matrix expressed in terms of the symmetry coordinates (Appendix II) was the same as that for the dimer.[15] The infrared active high frequencies have been observed by Millikan and Pitzer as shown in Table VII, however, the low-frequency out-of-plane vibrations have not been

[16] Holtzberg, Post, and Fankuchen, Acta Cryst. **6**, 127 (1953).

§ The interchain interaction is expected to split each of these vibrations into two components. Actually no such splitting has been observed,[10] thus rendering support to the neglect of interchain interaction in the present treatment.

TABLE VII. Observed and calculated infrared active out-of-plane frequencies of formic acid polymer.

	Obs	Calc	Deviation (%)	Calc low frequencies
$(HCOOH)_n$	1083 974	1095 945	+1 −3	184, 41
$(DCOOH)_n$	899 981	890 961	−1 −2	166, 39
$(HCOOD)_n$	1070 713	1077 732	+1 +3	174, 39
$(DCOOD)_n$	899 707	892 722	−1 +2	159, 37

observed and are not available for the determination of H_1'' and H_2''. The low-frequency in-plane vibrations have been observed, however, at 227 cm^{-1} for HCOOH polymer and at 218 cm^{-1} for DCOOD polymer.[9] The frequency ratio of 1.04 is in accord only with the in-plane vibration in which each monomer unit in the polymer chain carries out the rotational vibration|| and the frequency value indicates that all the rotational vibrations are in phase with each other throughout the chain.¶ Since the kinetic and potential energies of this vibration correspond to those of the low-frequency b_u vibration of the dimer, the close proximity of the polymer (227 cm^{-1}) and dimer (237 cm^{-1}) frequencies renders some support to the assumption of the same values of H_1'' and H_2'' to both cases.

The intramonomer potential constants were calculated so as to yield the best agreement between the observed and calculated high frequencies.

$$H_\Omega = 0.59_6 \times 10^{-11},$$
$$H_\theta = 0.23_6 \times 10^{-11}.$$

The results of frequency calculation for the four isotopic species are shown in Table VII. The frequency agreement within 3% is quite satisfactory for this kind of treatment.

Bond Order and Torsional Potential Constant

The torsional potential constant thus calculated for the monomer, dimer, and polymer will now be discussed in connection with the double bond character of the C—O single bond. The electronic structure of HCOOH may be expressed as a linear combination of the wave functions of the normal structure O═CH—OH and the polar structure

$$O^-—CH═OH^+.$$

|| For the low-frequency in-plane vibrations in which each monomer unit carries out the translational vibration, the ratio $\nu(HCOOH)/\nu(DCOOD) = 1.02$. For the out-of-plane vibrations, the ratios are 1.16 and 1.1 (see Table VII).

¶ There is another vibration of polymer in which each rotational vibration of monomer is 180° out-of-phase from those of adjacent units, however, this corresponds to the a_g vibration of the dimer[15] at 103 cm^{-1}.

The torsional constant, then, may be approximated to be $(1-a)H_N + aH_P$, where H_N and H_P are the potential constants for the hypothetical molecules with the nonpolar and polar structures, respectively, and a the fractional contribution of the polar structure, or in other words, the double bond character of the C—O bond in question. The constant H_P may be calculated to be $0.6_2 \times 10^{-11}$ erg/radian² from the value of the twofold potential barrier V_2 which is assumed to be equal to the bond energy difference (45 kcal/mole) between C═N and C—N bonds (note that —CH═O⁺— is isoelectronic with —CH═N—). On the other hand, the value of H_N (about $0.0_2 \times 10^{-11}$) calculated from the potential barrier of 1 kcal/mole of CH_3OH is negligible as compared with H_P. It follows, then, that $H_\theta = aH_P$, allowing the calculation of the double bond character from the torsional potential constant H_θ. The results are shown in Table VIII. In the same table are shown the values of a calculated from the C—O and C═O bond lengths by the use of the Pauling equation.[17] (The values of a have been adjusted so that the sum of a for the two CO bonds is equal to unity).

The values of double bond character a derived by the two methods agree with each other in the case of the monomer. In this case, the bond lengths determined by a microwave study[2] are highly accurate and the potential constant (monomer in the vapor phase, see later sections) is reliable, since it has been determined by the use of a complete harmonic potential function. Therefore the agreement brought about here justifies the calculation of a from H_θ.

For the polymer, the values of a derived by the two methods agree, indicating that the potential function adopted for the calculation has been reasonable.

On the other hand, in the case of the dimer, the values of a calculated by the two methods do not agree. The double bond character of the C—O bond of the dimer is calculated to be less than that of the monomer, which is highly unlikely. However, as has been stated by Karle and Brockway, "A decrease of 5% from the value of the ratio $r(C—O)/r(C═O)$ used in the best model produces an almost negligible change in the theoretical intensity curve..." A decrease by 5% (while retaining the O··O distance) yields $r(C—O) =$ 1.33 A and $r(C═O) = 1.28$ A. The double bond character of the C—O bond is now revised to be 0.3_1 which agrees well with that calculated from H_θ.

It may be mentioned that the values of a calculated from H_θ follow a reasonable sequence from monomer, dimer, to polymer without any adjustment. On the other hand, the calculation of a from bond lengths does not yield a reasonable sequence if not adjusted as mentioned before.**

[17] L. Pauling, *The Nature of the Chemical Bond* (Cornell University Press, Ithaca, New York, 1940), p. 175.

** If not adjusted, the sum of a for the two CO bonds amounts to 1.4 for the polymer, whereas it is only 0.7 for the dimer.

ABSORPTION BANDS IN THE VAPOR PHASE

OCO Bending and OH Torsional Frequencies of HCOOH and DCOOH

In a previous section we discussed monomeric frequencies in the solid nitrogen matrix, where absence of rotational fine structures made it possible to assign the observed frequencies in a straightforward manner. On the other hand, in the vapor phase a vibrational transition is accompanied by various rotational transitions, and even though the molecule of formic acid is only slightly asymmetric, the observed band contours are not too regular, possibly because of asymmetry and Coriolis interaction.†† The analysis of the band contours, however, was undertaken by a comparison with monomeric frequencies observed in the matrix.

The three principal moments of inertia (10^{-40} g cm^2) of HCOOH[2] are $I_A=10.9$, $I_B=69.6$, and $I_C=80.5$. Since the ratio $I_A/I_B=0.16$ is just equal to that of C_2H_4, the vibration-rotation spectra of the latter[18] are helpful in analyzing the spectra of formic acid. The transition moment of the OH torsional vibration is perpendicular to the molecular plane, and accordingly it should give rise to a type *C* band, which exhibits a strong *Q* branch and weak *P* and *R* branches just as the 949-cm^{-1} band (b_{1u}) of C_2H_4. On the other hand, the transition moment of the OCO bending vibrations is in the molecular plane and, in general, gives rise to a hybrid of the type *A* and *B* bands. However a normal coordinate treatment of the in-plane vibrations of HCOOH and DCOOH has indicated that the interaction of the OCO bending mode with other modes is almost negligible,[9] and the transition moment is expected to be nearly parallel to the axis of intermediate moment of inertia. Therefore the contours should be of *B* type with an intensity minimum near the band center as the 2943-cm^{-1} band of HCOOH[1] and the 3105-cm^{-1} band of C_2H_4.

As shown in Fig. 2, the most prominent peaks are observed at 638 cm^{-1} for HCOOH and at 629 cm^{-1} for DCOOH. These are most reasonably assigned to the *Q* branch of the OH torsional vibration. The most likely centers for the OCO bending vibration bands are the absorption minima at 625 cm^{-1} for HCOOH and at 620 cm^{-1} for DCOOH. The general band contours seem consistent with the superposition of these *B*- and *C*-type bands; the fine structure is complicated for reasons already stated.

The vapor monomeric frequencies may be compared with those observed in the solid nitrogen matrix. The OCO bending frequency increases by 2% from vapor to matrix (HCOOH, 625→638 cm^{-1}; DCOOH, 620→632 cm^{-1}) and the OH torsional frequency by 5% (HCOOH, 638→669 cm^{-1}; DCOOH, 629→661 cm^{-1}). It will be shown that exactly the same percentagewise changes are observed also for HCOOD and DCOOD.

†† Coriolis interaction is expected to be strong since the two frequencies, OCO bending and OH torsion, are not far apart.

[18] W. S. Gallaway and E. F. Barker, J. Chem. Phys. **10**, 88 (1942).

TABLE VIII. Double bond character of C—O bond of formic acid.

	From bond length	From potential constant
Monomer (vapor)	0.2_8[a]	0.2_2[e]
Dimer (vapor)	0.1_6[b]	
	0.3_1[c]	0.3_4
Polymer	0.3_8[d]	0.3_8

[a] r(C—O)=1.312 A and r(C=O)=1.245 A (see reference 2).
[b] r(C—O)=1.36 A and r(C=O)=1.25 A [J. Karle and L. O. Brockway, J. Am. Chem. Soc. **66**, 574 (1944)].
[c] r(C—O)=1.33 A and r(C=O)=1.28 A (see text).
[d] r(C—O)=1.26 A and r(C=O)=1.23 A (see reference 16).
[e] Calculated from the twofold potential barrier in the vapor phase (see later sections).

OCO Bending and OD Torsional Vibrations of HCOOD and DCOOD

In the cases of HCOOD and DCOOD, a normal coordinate treatment has indicated that the interaction of the OD bending and OCO bending modes is not negligible,[9] suggesting the observation of a hybrid of the type *A* and *B* bands which would exhibit a *Q* branch as well as *P* and *R* branches.‡‡ In fact such a band was observed for DCOOD with well-defined *P*, *Q*, and *R* branches at 537, 558, and 573 cm^{-1}, respectively. The separation (36 cm^{-1}) between *P* and *R* branches is equal to the separation observed for other in-plane vibration bands (at 2632, 2232, 1171, and 945 cm^{-1}).[1] In the case of HCOOD, the *P*, *Q*, and *R* branches were observed at 545, 562, and *ca* 580 cm^{-1}, and the separation (*ca* 35 cm^{-1}) between the *P* and *R* branches agrees with that observed for other bands (at 2948, 2632, and 1178 cm^{-1}). The frequencies of the band origin in the vapor phase are 2% lower than the corresponding ones for HCOOD or DCOOD observed in the matrix.

The bands due to the OD torsional vibrations of HCOOD or DCOOD are supposed to be of type *C*. In fact the *Q* branches of such bands were observed at 508 cm^{-1} for HCOOD and at 491 cm^{-1} for DCOOD. In both cases, the frequencies are 5% lower than the corresponding ones observed in the matrix just as in the cases of HCOOH and DCOOH.

In conclusion, the present assignment of band centers may be compared with the one made previously.[7] The present frequencies assigned to the OH or OD torsional vibrations agree to within 4 cm^{-1} with those measured previously. The earlier spectra

‡‡ The component of the transition moment along the axis of smallest moment of inertia will give rise to a type *A* band such as the 1890-cm^{-1} band of C_2H_4 which exhibits a fairly strong *Q* branch.

TABLE IX. Observed and calculated frequencies in the vapor phase.

	Cis			*Trans*		
	Obs	Calc	Deviation (%)	Obs	Calc	Deviation (%)
HCOOH	1033[a]	1040	+1	...	1080	...
	638	642	+1	582	584	+0
DCOOH	...	867	...	...	904	...
	629	633	+1	576	583	+1
HCOOD	...	1038	...	...	1072	...
	508	502	−1	450	442	−2
DCOOD	873[a]	867	−1	...	894	...
	491	488	−1	448	441	−2

[a] See reference 1.

did not include various minor features, consequently the previous selection of the band centers of the OCO bending vibrations was uncertain. The values (HCOOH, 638; DCOOH, 629; HCOOD, 541; and DCOOD, 537 cm^{-1}) differ by up to 21 cm^{-1} from the present ones (HCOOH, 625; DCOOH, 620; HCOOD, 562, and DCOOD, 558 cm^{-1}).

Normal Coordinate Treatment

The calculation of the monomeric out-of-plane frequencies in the vapor phase was carried out just in the case of monomer in the solid nitrogen matrix. The potential constants were calculated to be

$$H_{\Omega}=0.58_3\times10^{-11},$$

$$H_{\Omega\theta}'=-0.00_5\times10^{-11},$$

and

$$H_{\theta}=0.131\times10^{-11}.$$

The values of H_{Ω} and $H_{\Omega\theta}'$ in the vapor phase agree within $0.00_4\times10^{-11}$ with those in the matrix, and the value of H_{θ} in the matrix is about 10% larger than that in the vapor phase. The results of the frequency calculation are shown in Table IX. The frequency agreement is within 1%.

The out-of-plane frequencies of the monomer in the *trans* configuration were also calculated using the same potential function as that of the *cis* configuration. The results are shown in Table VIII.

INTERNAL ROTATION OF MONOMER IN THE VAPOR PHASE

Evidence for the *trans* Isomer

As shown in Fig. 2, prominent absorption peaks were observed at 582 cm^{-1} for HCOOH and at 576 cm^{-1} for DCOOH. These peaks are much stronger than neighboring rotational lines and their intensity increases with temperature, whereas other rotational lines do not change much in intensity. A quantitative study of the temperature behavior of the relative intensity of the 576-cm^{-1} band with respect to the 629-cm^{-1} band of DCOOH (the detail will be presented in the next section) has indicated that the species responsible for the 576-cm^{-1} band is 2.0 kcal $mole^{-1}$ higher in energy than the ground vibrational state of the *cis* configuration. Therefore, this species may be either (1) the *cis* isomer in its first excited state of the 629-cm^{-1} or 620-cm^{-1} vibration or (2) the *trans* isomer.

For the *cis* isomer in its first excited vibrational state, the transition to the second excited state may be considered. The observed frequency in question is 7% lower than the fundamental OCO bending frequency. This means, however, too much anharmonicity, which is highly unlikely, since the corresponding frequency decrease is less than 2% for the bending vibration of N_2O or HCN. On the other hand, the 1→2 transition frequency of the OH torsional vibration of DCOOH was found, as mentioned before, to be 30 cm^{-1} lower than the fundamental frequency, whereas the observed frequency in question is 53 cm^{-1} lower. Therefore the observed frequency cannot be assigned to hot transitions. Secondly the transition to the first excited state of another vibration may be discussed even though the observed intensity seems to be too high for this kind of transition. The similar behavior of the two bands in question strongly suggests that the same kind of transition be responsible for both cases. The only transition of the *cis* isomer which may account for the observed frequency of 582 cm^{-1} is 1229–638=591, however, the frequency of the corresponding one of DCOOH 1235§§ −629=606 is too much different from the observed frequency 576 cm^{-1}. Thus the intensity and frequency considerations rule out the *cis* isomer as the species which undergoes the observed transitions in question.

Therefore the observed frequencies in question may be explained to be due to the *trans* isomer. The OH torsional frequencies calculated for the *trans* isomers of HCOOH and HCOOD agree well with the observed frequencies (see Table IX). Such frequencies are also observed at 450 cm^{-1} for HCOOD and at 448 cm^{-1} for DCOOD, just as predicted by the normal coordinate treatment.

Energy Difference between the *trans* and *cis* Isomers

The energy difference between the two isomers of DCOOH was determined by measuring the relative intensity of the *trans* (576 cm^{-1}) and *cis* (629 cm^{-1}) bands at 25, 57, 99, and 149°C. The peak intensity of the latter band was measured from the 100% line,

§§ In the solid nitrogen matrix, the OH bending frequencies have been observed at 1264 cm^{-1} (HCOOH) and at 1270 cm^{-1} (DCOOH).[9] The missing frequency of DCOOH in the vapor phase may be estimated to be 1235 cm^{-1} from the observed frequency 1229 cm^{-1} of HCOOH[1] on the assumption that the frequency changes from matrix to vapor are the same for HCOOH and DCOOH.

and that of the former band was measured from a smooth curve drawn across it since it lies in the P branch of the latter band. The side arm of the vapor cell was immersed in a water bath, the temperature of which was adjusted so that the intensity of the 629-cm^{-1} band remained almost the same at various temperatures. In Fig. 6 are shown the plot of logarithm of the intensity ratio against $10^3/T$. The straight line was drawn by the method of least squares and from its slope the energy difference was calculated to be

$$\Delta E = 2.0 \pm 0.3 \text{ kcal/mole}$$

(probable error: 0.17 kcal/mole). This value is in accord with the value 3.0–1.2 kcal/mole estimated from the electron impact study[5] (from the curve of dissociation yield *versus* applied potential).

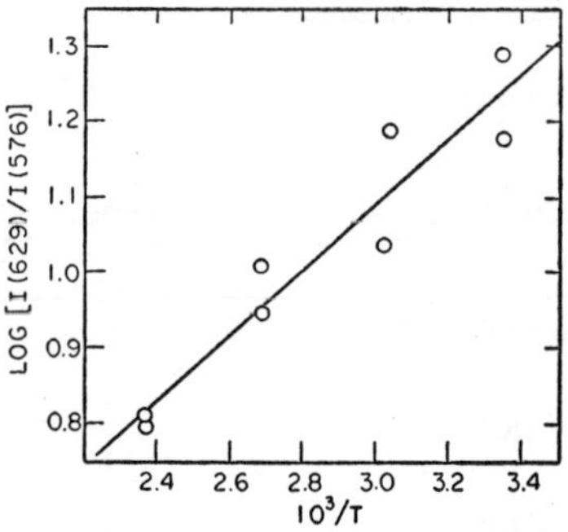

FIG. 6. Relative intensity of bands of DCOOH at 629 and 576 cm^{-1} *versus* temperature.

Potential Hindering Internal Rotation

The potential energy hindering internal rotation of formic acid monomer in the vapor phase may be expressed as

$$2V = V_1(1-\cos\theta) + V_2(1-\cos 2\theta) + V_3(1-\cos 3\theta) + \cdots, \tag{15}$$

where only the first three terms will be taken in the present treatment ($\theta = 0$ at the *cis* potential minimum). The three coefficients V_1, V_2, and V_3 may be determined from the energy difference (ΔE) and the fundamental OH (or OD) torsional frequencies of the *cis* (ν^c) and *trans* (ν^t) isomers of HCOOH and DCOOD after taking anharmonicity into account. The equations derived by the first-order perturbation are

$$\nu^c = \nu_0{}^c - (h/8\pi^2 c I_r{}^c) \times (V_1 + 16V_2 + 81V_3)/(4V_1 + 16V_2 + 36V_3), \tag{16}$$

$$(2\pi c\nu_0{}^c)^2 = (V_1 + 4V_2 + 9V_3)/2I_r{}^c, \tag{17}$$

$$\nu^t = \nu_0{}^t - (h/8\pi^2 c I_r{}^t) \times (-V_1 + 16V_2 - 81V_3)/(-4V_1 + 16V_2 - 36V_3), \tag{18}$$

$$(2\pi c\nu_0{}^t)^2 = (-V_1 + 4V_2 - 9V_3)/2I_r{}^t, \tag{19}$$

and

$$V_1 + V_3 = \Delta E. \tag{20}$$

The three coefficients are found to be (in kcal mole^{-1})

$$V_1 = +2._1,$$

$$V_2 = +9._9,$$

and

$$V_3 = -0._1.$$

The potential maximum is at 98° from the *cis* configuration and is $10._9$ kcal/mole above the *cis* minimum.

Abundance of *trans* Isomer

The equilibrium constant for the *cis* = *trans* isomerization reaction may be calculated from the energy difference of 2.0 ± 0.3 kcal/mole^{-1} together with the entropy difference. The data available do not suffice for a complete calculation of the entropy of the *trans* isomer but do indicate that the entropy difference between isomers is very small. Consideration of the spectroscopic data for nitrous acid[19] in its *cis* and *trans* forms supports this estimate. Consequently ΔS is assumed to be zero and the equilibrium constant is

$$K = \exp(-2000/RT).$$

This equation yields at 298°K, $K = 0.04$ and at 373°K, $K = 0.07$.

The entropy of the equilibrium vapor of monomeric formic acid exceeds that of the *cis* isomer by just the entropy of mixing which is 0.3 ± 0.1 cal/deg mole at 298°K. The reported experimental entropy[20] for the monomer agrees more closely with the value for the *cis* isomer alone but the discrepancy is small and presumably within the uncertainty of the experimental value. Indeed the independent value of the entropy of the monomer which is now available will allow an improved interpretation of the dissociation equilibrium data.

APPENDIX I

The internal coordinates and the corresponding s vectors have been given by Wilson *et al.*, however, the coordinates related to out-of-plane vibrations of planar molecules were modified in order to yield more convenient expressions for the corresponding B matrix elements. They will be described in this appendix.

Out-of-Plane Deformation Coordinate

For a planar structure this will be denoted as Ω_i.

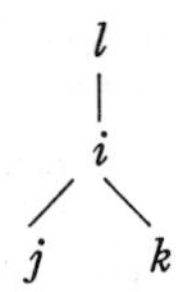

This is defined as $\sin\alpha_{kil}$ times the angle between the

[19] Jones, Badger, and Moore, J. Chem. Phys. **19**, 1599 (1951).

[20] J. O. Halford, J. Chem. Phys. **10**, 582 (1942); W. Waring, Chem. Revs. **51**, 171 (1952).

bond ij and the plane kil, which is equal to sin α_{lij} times the angle between the bond ik and the plane lij and also equal to $\sin\alpha_{jik}$ times the angle between the bond il and the plane jik. This coordinate may also be defined as $(\tau_{ij} \sin\alpha_{kil}+\tau_{ik} \sin\alpha_{lij}+\tau_{il} \sin\alpha_{jik})$ times the distance from the ith atom to the plane jkl, where the τ's are reciprocal bond lengths. The B matrix elements for this coordinate are

$$B_i=(\tau_{ij} \sin\alpha_{kil}+\tau_{ik} \sin\alpha_{lij}+\tau_{il} \sin\alpha_{jik}),$$

$$B_j=-\tau_{ij} \sin\alpha_{kil},$$

$$B_k=-\tau_{ik} \sin\alpha_{lij},$$

and

$$B_l=-\tau_{il} \sin\alpha_{jik}.$$

Torsional Coordinate

The torsional coordinate θ_{ad} for the sequence of atoms, a, b, c, and d, is defined as the dihedral angle between the plane of a, b, and cth atoms and that of the b, c, and dth atoms. It is positive when the bond cd is rotated counterclockwise with reference to the bond ab as viewed along the bond bc with bth atom closer to the eye. It will be seen that $\theta_{ad}=\theta_{da}$. The element of the corresponding B matrix are

$$B_a=-\tau_{ab}/\sin\alpha_{abc},$$

$$B_b=\tau_{ab}/\sin\alpha_{abc}-[\tau_{bc} \sin(\alpha_{abc}+\alpha_{bcd})]/\sin\alpha_{abc} \sin\alpha_{bcd},$$

$$B_c=-\tau_{cd}/\sin\alpha_{bcd}+[\tau_{bc} \sin(\alpha_{abc}+\alpha_{bcd})]/\sin\alpha_{abc} \sin\alpha_{bcd},$$

and

$$B_d=\tau_{cd}/\sin\alpha_{bcd},$$

where α_{abc} is the angle between the bonds ab and bc and it is positive when the bond bc is traced counterclockwise with reference to the bond ab.

APPENDIX II

The symmetry coordinates for the infrared active out-of-plane vibrations of formic acid polymer are

$$S_a=(\Omega_0+\Omega_5)/3^{\frac{1}{2}}+(-\theta_{14}+\theta_{69}+\theta_{24}-\theta_{79}+\theta_{47}-\theta_{9'2}-\theta_{48}+\theta_{9'3})(3/64)^{\frac{1}{2}},$$

$$S_b=(\theta_{14}-\theta_{69}+\theta_{24}-\theta_{79})(3/16)^{\frac{1}{2}},$$

$$S_1=(\theta_{47}-\theta_{9'2}+\theta_{48}-\theta_{9'3})(3/16)^{\frac{1}{2}},$$

and

$$S_2=(\beta_{346}+\beta_{8'9'1})/2^{\frac{1}{2}}.$$

The corresponding B matrix elements are

$$B_{a0}=\tau_{01}+\tau_{02}+\tau_{03}, \quad B_{b0}=(-\tau_{01}+\tau_{02}-\tau_{03})/2,$$

$$B_{a1}=-\tau_{01}, \quad B_{b1}=\tau_{01}/2,$$

$$B_{a2}=-\tau_{02}, \quad B_{b2}=-\tau_{02}/2,$$

$$B_{a3}=-\tau_{03}, \quad B_{b3}=\tau_{34}+\tau_{03}/2,$$

$$B_{a4}=0, \quad B_{b4}=-\tau_{34},$$

$$B_{10}=(\tau_{01}+\tau_{02}-\tau_{03})/2, \quad B_{20}=0,$$

$$B_{11}=-\tau_{01}/2-\tau_{46}, \quad B_{21}=-\tau_{46}/2^{\frac{1}{2}},$$

$$B_{12}=-\tau_{02}/2, \quad B_{22}=0,$$

$$B_{13}=\tau_{03}/2, \quad B_{23}=-\tau_{34}/2^{\frac{1}{2}},$$

$$B_{14}=+\tau_{46}, \quad B_{24}=(+\tau_{34}+\tau_{46})/2^{\frac{1}{2}},$$

where

$$B_{ij+5}=B_{ij}.$$

Reprinted from the *Faraday Society Discussion*, 1951, No. 10

POTENTIAL ENERGIES FOR ROTATION ABOUT SINGLE BONDS

BY KENNETH S. PITZER *

Received 18th January, 1951

The theoretical relationships between the potential energy function for internal rotation and the energy levels and thermodynamic properties are summarized briefly. The height and shape of the potential barrier in ethane are re-examined in relation to all pertinent data. It is concluded that the barrier is very close to the cosine function in shape and from 2750 to 3000 cal./mole in height. The barrier heights for a number of other hydrocarbons are summarized and a few comments are made concerning the cause of these barriers.

In the last fifteen years a great deal has been learned about the quantitative changes in potential energy associated with rotation about single covalent bonds within a molecule. Probably a majority of the examples investigated are hydrocarbons, and conversely all but the simplest hydrocarbon molecules allow internal rotations. Consequently a discussion of this phenomenon is appropriate to a symposium on hydrocarbons.

The present contribution will attempt to review principally the theoretical aspects of restricted internal rotation but will also summarize some present best values for potential barriers. The theoretical problems are of two general sorts. First are the phenomenological inter-relationships—the quantum energy levels and the statistical thermodynamic properties in terms of the potential energy function for internal rotation, the pertinent moments of inertia, etc. Second are theories as to the cause of potential barriers in terms of the behaviour of the valence electrons of the molecule.

The principal problem involved in relating other properties to potential barriers associated with internal rotation is the calculation of the quantum energy levels for the system. It is more convenient to assume a form of potential function and then calculate energy levels than the reverse. A simple and very useful form is

$$V = \tfrac{1}{2}V_0(1 - \cos n\phi), \qquad (1)$$

where V_0 is the height of the potential peaks above the valleys, ϕ is the rotation angle and n is the number of peaks and valleys, all identical, per revolution. Thus, if one or both of the rotating groups are symmetrical, such as a methyl group, the appropriate value of n is given by symmetry.

Nielson [1] first solved the Schrödinger equation with the potential of eqn. (1) and obtained the energy levels for certain cases. The chief complication is the interaction of the internal rotation with the rotation of the entire molecule. Koehler and Dennison [2] treated this matter in further detail and gave solutions for any molecule which can be described as a pair of coaxial symmetrical tops. Actually Koehler and Dennison were particularly interested in methyl alcohol for which their solution is only an approximation.

The Schrödinger equation is transformed by the separation of the over-all rotation co-ordinates and yields for a single internal rotation : [3]

$$M''(x) + (a + 2\theta \cos 2x)M(x) = 0, \qquad (2)$$

* On temporary leave from Department of Chemistry, University of California, Berkeley.

[1] Nielsen, *Physic. Rev.*, 1932, **40**, 445.

[2] Koehler and Dennison, *Physic. Rev.*, 1940, **57**, 1006.

[3] Pitzer and Gwinn, *J. Chem. Physics*, 1942, **10**, 428.

with
$$2x = n\phi,$$
$$\theta = 8\pi^2 I_r V_0/n^2h^2$$
$$W = \tfrac{1}{2}V_0 + n^2h^2a/32\pi^2 I_r,$$

wherein W is the energy of a quantum state (above the zero of the potential curve), I_r is the reduced moment of inertia for the internal rotation and the other symbols are conventional or defined in connection with eqn. (1). This is a relatively well-investigated differential equation, commonly known as the Mathieu equation. The boundary conditions for acceptable solutions depend on the relative moments of inertia of the two parts of the molecule and upon the quantum numbers of over-all rotation. Although these relationships can be fairly simple in special cases, they are so complex in the general case as to preclude presentation here. However, it is important to give the relationship for the reduced moment of inertia for the most general case of eqn. (2), namely, a symmetrical top attached to any rigid frame [3]

$$I_r = A(1 - \sum_{i=1}^{3} A\lambda_i^2/I_i), \quad . \quad . \quad . \quad . \quad (3)$$

where A is the moment of inertia (about the axis of internal rotation) of the symmetrical top, λ_i is the direction cosine between the axis of internal rotation and the ith principal axis of the whole molecule about which the overall moment of inertia is I_i. In the special case of two coaxial tops of moment of inertia A_1 and A_2 this reduces to the familiar expression [1]

$$I_r = A_1A_2/(A_1 + A_2). \quad . \quad (4)$$

The general character of the energy levels as a function of potential barrier height is shown in Fig. 1. The energy levels lie within the shaded regions, one within each region. The dashed line indicates the potential barrier height. Well below the barrier the levels are simply defined and essentially vibrational. Above the barrier the regions are wide, indicating large interaction with overall rotation, and the general spacing follows the pattern for rotation. Fig. 1 is strictly a plot of $(a + 2\theta)$ against θ. From eqn. (2) one then notes that the unit of energy on the figure is proportional to (n^2/I_r).

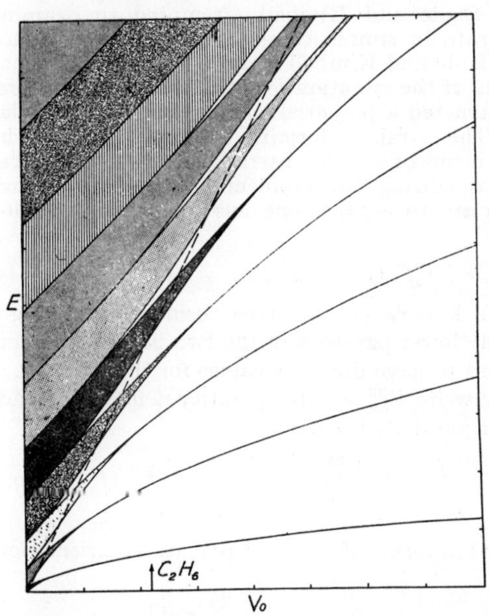

FIG. 1.—Energy levels for restricted internal rotation. The dashed line shows the top of the potential barrier.

The statistical mechanical formulae for thermodynamic properties involve sums over all the energy levels which, at sufficiently high temperatures, are not sensitive to small shifts in individual energy levels. Thus it is commonly sufficient to know that the energy levels for a given molecule are located one within each band shown in Fig. 1. The Tables prepared by Gwinn and the writer [3] for the thermodynamic functions are based on this assumption and are limited to the range of temperature, moment of inertia, etc., for which the assumption is valid. While this covers almost all molecules a few exceptions such as methyl alcohol have been found. Halford [4] has given an interesting method for these exceptional cases wherein a few low energy levels are calculated exactly by the method of Koehler and Dennison [2] and the higher levels are obtained by the much easier but approximate method of Halford [5] based on old quantum theory.

Treatments have also been given for less symmetrical [6] or more complex [7] classes of molecules than that of the rigid frame with attached symmetrical top. Although the classical mechanics can be carried out quite rigorously even for rotating groups attached to other rotating groups all unsymmetrical, etc., the quantum mechanics becomes very involved and has been handled only to a low approximation. However, actual molecules in these categories frequently approach classical behaviour so that the thermodynamic properties can be calculated relatively accurately.

Shape of Potential Barrier.—While most of the treatments of restricted internal rotation have been based on the simple cosine formula for potential energy, eqn. (1), it has been realized that this was an assumption, and occasional calculations for other shapes have been made.[3] Recently Halford [5, 8] and Blade and Kimball [9] have discussed general methods by which the energy levels can be calculated conveniently for other shapes of potential curves. Halford has considered primarily examples qualitatively different whereas Blade and Kimball attempted to examine whether eqn. (1) was accurate in simple cases such as ethane. Since some of the conclusions of Blade and Kimball seem to the present author to be misleading, an analysis of the situation for ethane is included here.

Blade and Kimball constructed a potential curve from two parabolas fitted smoothly together. The parabola forming the lower part of the potential curve is defined in terms of a cosine barrier with the same curvature at the minimum. Specializing our problem to a three-fold symmetrical barrier as appropriate to ethane, one has for the true cosine barrier equivalent to eqn. (1),

$$V = \tfrac{1}{2} V_0^{\cos} (1 - \cos 3\phi). \qquad (5)$$

It is important to distinguish between this true cosine barrier height $V_0^{\cos}$ and the constant for the lower parabola in the two parabola system because the two will be found to have different values for the same set of energy levels. Therefore we write $V_{\cos}^{BK}$ for the quantity defined by Blade and Kimball and their lower parabola becomes

$$V_1 = \frac{9}{4} V_{\cos}^{BK}\phi^2, \quad \left(-\frac{4}{3\pi}\frac{V_0^{BK}}{V_{\cos}^{BK}} < \phi < \frac{4}{3\pi}\frac{V_0^{BK}}{V_{\cos}^{BK}}\right). \qquad (6a)$$

The upper parabola is defined in terms of the total potential barrier height V_0^{BK} as follows,

$$V_2 = V_0^{BK}\left[1 - \frac{9}{\pi^2 - 4(V_0^{BK}/V_{\cos}^{BK})}\left(\phi - \frac{\pi}{3}\right)^2\right]$$
$$\left(\frac{4}{3\pi}\frac{V_0^{BK}}{V_{\cos}^{BK}} < \phi < \frac{2\pi}{3} - \frac{4}{3\pi}\frac{V_0^{BK}}{V_{\cos}^{BK}}\right). \qquad (6b)$$

[4] Halford, *J. Chem. Physics*, 1950, **18**, 444.
[5] Halford, *J. Chem. Physics*, 1947, **15**, 645, and 1948, **16**, 410.
[6] Pitzer, *J. Chem. Physics*, 1946, **14**, 239.
[7] Kilpatrick and Pitzer, *J. Chem. Physics*, 1949, **17**, 1064.
[8] Halford, *J. Chem. Physics*, 1948, **16**, 560.
[9] Blade and Kimball, *J. Chem. Physics*, 1950, **18**, 630.

Blade and Kimball derived the energy levels approximately by the phase integral or W.K.B. method and give graphs from which they may be calculated. In treating ethane, however, they considered only one set of heat capacity data on the gas from 92 to 134° K, ignoring other data in this region, all heat capacity data at higher temperatures, the value of the entropy, and the spectroscopic value for the separation of the first two energy levels. Thus their conclusion that barriers over the range 1550 to 2700 cal./mole were correct for ethane cannot be accepted without further examination.

In attempting to obtain the maximum information about the potential barrier in ethane, one should first consider the spectroscopic value recently obtained by Smith [10] for the separation between the lowest pair of energy levels. His value is based on two combination bands and in each case the rotational fine structure is verified as correct for the assignment given. Thus there can be little doubt but that his value of 290 cm.$^{-1}$ is essentially correct. Unfortunately this value is not as precise as might be desired. The two bands yielded 287·1 and 294·8 cm.$^{-1}$ and it is even conceivable that the band centres were incorrectly identified [11] by 10 or 20 cm $^{-1}$. Thus this energy level separation could be as low as the 275 cm.$^{-1}$ value derived by Kistiakowsky, Lacher and Stitt [12] from specific heat data.

The entropy of ethane was obtained by Witt and Kemp [13] from low temperature heat capacity measurements and is compared in Table I with calculated values using Smith's moments of inertia and vibration frequencies. One should first note that the entropy comparison at 184·1° K concerns primarily the lowest energy level separation. This is indicated by the fact that the value for a 3050 cal./mole cosine barrier, which corresponds to a 290 cm.$^{-1}$ separation of the lowest levels, is only 0·04 cal./deg. mole higher than the value for a 290 cm.$^{-1}$ harmonic oscillator, and similarly for the 2750 cal./mole barrier which has a 275 cm.$^{-1}$ separation of the lowest levels. Consequently it would be fruitless to compare a further variety of barrier shapes at this point.

TABLE I.—ENTROPY OF ETHANE AT 184·1° K
(Ideal gas at 1 atm. in cal./deg. mole)

Experimental	49·54 ± 0·15
Calc. for trans., rot., and vibr. . . .	48·69
Expl. for internal rotation . . .	0·85
Harmonic osc. $\nu =$ 290 cm.$^{-1}$. . .	0·74
Harmonic osc. 275 cm.$^{-1}$. . .	0·81
Cosine barrier $V_0 =$ 3050 cal./mole . .	0·78
Cosine barrier 2750 cal./mole . .	0·87

One can conclude that the separation of the lowest pair of energy levels is probably in the 275-290 cm.$^{-1}$ range, where the upper limit arises from the entropy value and the lower limit from the spectroscopic data.

The principal basis for discussion of the shape of the potential barrier is the heat capacity curve for the gas. Fig. 2 shows the experimental data [14] over the range 90-305° K, together with several calculated curves. The contributions of translation, rotation and vibration have been subtracted so that only the contribution of internal rotation is shown in Fig. 2. The two solid curves are for cosine barriers of the heights indicated, while

[10] Smith, *J. Chem. Physics*, 1949, **17**, 139.

[11] Dr. L. Smith mentioned this possibility in private communication.

[12] Kistiakowsky, Lacher and Stitt, *J. Chem. Physics*, 1939, **7**, 289.

[13] Witt and Kemp, *J. Amer. Chem. Soc.*, 1937, **59**, 273.

[14] Hunsmann, *Z. physik. Chem. B*, 1938, **39**, 23; Eucken and Parts, *Z. physik. Chem.*, *B* 1933, **20**, 184; Heuse, *Ann. Physik*, 1919, **59**, 86; Kistiakowsky and Nazmi, *J. Chem. Physics*, 1938, **6**, 18; Kistiakowsky and Rice, *J. Chem. Physics*, 1939, **7**, 281; also ref. (12).

the dashed curves are for a series of two parabola barriers of various heights. Each of the latter series has the lowest pair of energy levels spaced at 290 cm.$^{-1}$ which corresponds to $V^{\mathrm{BK}}_{\cos} = 2570$ cal./mole. One can see from the upper dashed curve, for $V^{\mathrm{BK}}_{0} = 2050$ cal./mole, how Blade and Kimball were led to false conclusions by considering only the data of Kistiakowsky, Lacher and Stitt. This curve fits this particular set of data fairly well but is totally unacceptable at higher temperatures. Also the difference between the quantities $V^{\cos}_{0}$ and $V^{\mathrm{BK}}_{\cos}$ is indicated by their respective values 3050 and 2570, each corresponding to the 290 cm.$^{-1}$ separation of lowest energy levels.

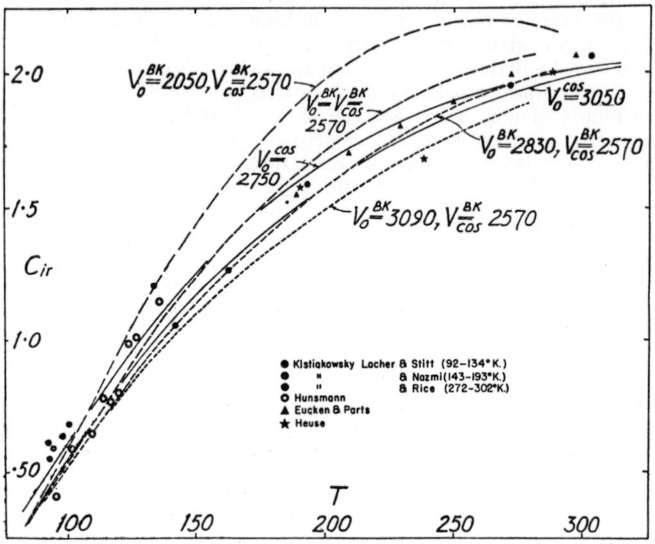

FIG. 2.—The heat capacity for internal rotation in ethane for various assumed potential barriers.

The 275-290 cm.$^{-1}$ range for the lowest energy level separation, which was found probable above, receives further support from the data near 100° K in Fig. 2. The cosine barriers fit the somewhat scattered data in the range around 150° K as well as is possible but yield slightly low values near 300° K. However, the difference at the higher temperatures is within the limit of the various errors involved. The best of the two parabola curves, that for $V^{\mathrm{BK}}_{0} = 2830$, is no net improvement being too low in the 150-200° K range while yielding slightly better fit near 300° K. Thus one can conclude that the potential barrier for ethane probably has a height in the range 2750-3000 cal./mole and is very close to the cosine function in shape. The deviation indicated is that of a slight lowering and flattening of the potential peaks as compared to the remainder of the curve.

Only recently has the work of Smith [10] provided strong evidence from data on ethane itself that the stable or minimum energy configuration is the staggered one (point group D_{3d}). However, the general pattern of evidence from substituted ethanes had long indicated this result.

Potential Barriers for Various Hydrocarbons.—There are assembled in Table II most of the available values for potential barriers to internal rotation in hydrocarbons. The cosine shape of barrier is ordinarily assumed. In cases of several equivalent rotations, the barriers are assumed independent and equal. Where this assumption seems doubtful the resulting value is indicated as an average, (av.). Ranges of uncertainty are given where the completeness of the data justified it—in other cases somewhat larger uncertainties must be allowed.

Most of the barriers in Table II are for methyl group rotations under circumstances in which the three-fold symmetry of the methyl group applies also to the potential function. In toluene and approximately in *m*- and *p*-xylene the combination of two-fold symmetry of the phenyl group with the three-fold symmetry of the methyl yields a six-fold symmetry for the potential barrier. Styrene has a two-fold symmetry for the vinyl against phenyl rotation while 1 : 3-butadiene has two equal potential peaks but the *cis* and *trans* minima are different.

TABLE II.—VALUES FOR POTENTIAL BARRIERS TO INTERNAL ROTATION IN VARIOUS HYDROCARBONS

Substance	V_0, cal./mole	Source
Ethane . .	2875 ± 125	Present paper
Propane . .	3400 (av.)	Pitzer,[15] also Kemp and Egan [16]
*iso*Butane . .	3620 (av.)	Pitzer and Kilpatrick,[17] also Aston *et al.*[18]
*neo*Pentane . .	4300 (av.)	Pitzer and Kilpatrick,[17] also Aston and Messerly [19]
Propylene . .	1950	Kilpatrick and Pitzer [20]
*iso*Butene . .	2350 (av.)	Kilpatrick and Pitzer,[20] also Aston *et al.*[21]
trans-2-Butene	1950	Kilpatrick and Pitzer [20]
cis-2-Butene	450 (av.)	Kilpatrick and Pitzer,[20] also Aston, *et al.*[21]
Toluene . .	500 ± 500	Pitzer and Scott [22]
m and *p*-Xylene	500 ± 500	Pitzer and Scott [22]
o-Xylene . .	2000 (av.)	Pitzer and Scott [22]
1 : 3-Butadiene .	{5000 *trans* to peak; 2575 *cis* to peak}	Aston, Szasz, Wooley and Brickwedde [21]
Styrene. . .	2200	Pitzer, Guttman and Westrum [23]
Dimethylacetylene	0	Kistiakowsky and Rice,[24] Osborne *et al.*[25]

Internal rotation about the middle bond in *n*-butane is related to the 1 : 3-butadiene case but somewhat more complex. Of the three potential minima for the *n*-butane, one is in the *trans* position while the other two are enantiomorphs called gauche or skew forms. The full details of this potential curve for *n*-butane are not known yet. Assuming that the potential barriers for the end methyl group rotations are similar to those in propane, one finds the general magnitude of the height of a potential peak above adjoining valleys to be in the 3000-4000 cal. range.[26] Also

[15] Pitzer, *J. Chem. Physics*, 1944, **12**, 310.
[16] Kemp and Egan, *J. Amer. Chem. Soc.*, 1938, **60**, 1521.
[17] Pitzer and Kilpatrick, *Chem. Rev.*, 1946, **39**, 435.
[18] Aston, Kennedy and Schumann, *J. Amer. Chem. Soc.*, 1940, **62**, 2059.
[19] Aston and Messerly, *J. Amer. Chem. Soc.*, 1936, **58**, 2354.
[20] Kilpatrick and Pitzer, *J. Res., Nat. Bur. Stand.*, 1946, **37**, 163; also 1947, **38**, 191.
[21] Aston, Szasz, Woolley and Brickwedde, *J. Chem. Physics*, 1946, **14**, 67.
[22] Pitzer and Scott, *J. Amer. Chem. Soc.*, 1943, **65**, 803.
[23] Pitzer, Guttman and Westrum, Jr., *J. Amer. Chem. Soc.*, 1946, **68**, 2209.
[24] Kistiakowsky and Rice, *J. Chem. Physics*, 1940, **8**, 618.
[25] Osborne, Garner and Yost, *J. Amer. Chem. Soc.*, 1941, **63**, 3492.
[26] Pitzer, *J. Chem. Physics*, 1940, **8**, 711; *Ind. Eng. Chem.*, 1944, **36**, 829.

the gauche or skew minima are known to be about 750-1000 cal./mole above the *trans* minimum. This latter value came first from the writer's statistical-thermodynamic treatment of a series of *n*-paraffins,[26] but has recently been measured by Szasz, Sheppard and Rank [27] from the shift of the relative intensity of Raman spectral lines for skew and *trans* forms with change of temperature. This spectroscopic method has been applied as yet only to the liquid state for *n*-butane, which leaves the uncertainty of the difference in heat of vaporization of the *trans* and skew forms. However, data in the gas phase have been obtained for some halogen derivatives using the infra-red instead of the Raman spectrum. Assuming that the necessary intensities can be had, gas phase data should be obtainable for hydrocarbons also.

The skew-*trans* energy difference in *n*-butane has found interesting application in the methyl substituted cyclohexanes where similar geometrical situations arise.[28] These results also tend to verify the value given above.

More complex hydrocarbons have, of course, additional possibilities of internal rotational minima corresponding to non-equivalent configurations with different energies. The writer has preferred the term *tautomers* (rather than isomers) in describing these configurations since they are in rapid reversible equilibrium with one another.

Cause of Potential Barriers.—There have been a number of investigations directed to the explanation of these potential barriers in terms of electron or valence theory. In the present degree of advance of quantum mechanical calculations an exact solution is impossible and attempts at approximation have as often as not yielded the wrong sign or a totally wrong order of magnitude.

However, one can say that there is every reason to believe that the actual single bond itself, i.e. the pair of electrons and the orbital describing their motion, does not resist rotation appreciably. The writer has made perturbation calculations for the ethane case and has found that the maximum potential energy differences associated with the carbon-carbon bond itself are several orders of magnitude smaller than the observed barrier. Thus the observed phenomena must be associated with the electron pairs attaching other groups, i.e. the C—H bond electrons in ethane. Of course, where the attached groups are large enough there may also be other interactions between them.

Kistiakowsky, Lacher and Ransom [29] stated in 1938 that the potentials were due to "an interaction of electron pairs forming single bonds on adjacent polyvalent atoms". Today we can add with certainty that the interaction is repulsive, i.e. the staggered configuration is the stable one, and that there is doubt as to the limitation to *single* bonds on the adjacent atoms. Otherwise this statement summarizes the current position on this question.

Of the more detailed calculations that of Lassettre and Dean [30] is probably the most interesting and plausible. They attempt to account for the barriers by an electrostatic model recognizing quadrupole as well as dipole moments for the bonds. In hydrocarbons the dipole moments are small but, if the bonding electrons are largely concentrated between the nuclei, the quadrupole moments could be large enough to account for the observed barriers. In extending their work to more complex molecules, Lassettre and Dean found peculiar unsymmetrical configurations to have minimum energy in *iso*butane and *neo*pentane. Thus these molecules would appear to deserve further study.

[27] Szasz, Sheppard and Rank, *J. Chem. Physics*, 1948, **16**, 704; see also *J. Chem. Physics*, 1949, **17**, 86.

[28] Beckett, Pitzer and Spitzer, *J. Amer. Chem. Soc.*, 1947, **69**, 2488.

[29] Kistiakowsky, Lacher and Ransom, *J. Chem. Physics*, 1938, **6**, 900.

[30] Lassettre and Dean Jr., *J. Chem. Physics*, 1949, **17**, 317.

It seems likely to the writer that the true explanation of these potential barriers may be more complex than any of the treatments offered as yet in that more than one effect is of importance. The same sort of " van der Waals " repulsive forces arise between pairs of electrons in different atoms or bonds within a molecule as between different molecules. It has been shown that this effect is too small to account for the full barrier in ethane if only the regions describable as the hydrogen atoms are considered. However, an additional contribution will arise from these same C—H bond electron pairs in the regions near the carbon nuclei and this term is very difficult to calculate. However, these effects can be shown to be of the right order of magnitude. Thus this " van der Waals " type of repulsive forces and the quadrupole electrostatic effects probably each contribute substantially to the net barrier.

U.S. Atomic Energy Commission,
Washington DC.

PRINTED IN GREAT BRITAIN AT
THE UNIVERSITY PRESS
ABERDEEN

B. Ring Molecules

Introduction

I realized before much of the work on ethane, propane and similar molecules had been finished that the resulting torsional potentials were applicable to cyclic molecules and would yield strain energies. But activities related to World War II occupies much of my time and only the short note to Science was published in 1945. Soon thereafter the more detailed studies on cyclopentane and substituted cyclohexanes were initiated. The first yielded the concept of pseudorotation which has since found wide application for several types of molecules. For the disubstituted cyclohexanes in the stable chair structure, Hassel had recognized in 1943 the pattern of axial and equatorial locations of the substituent groups. We calculated quantitative values of the entropies of the dimethylcyclohexanes with the interesting result that the cis and trans 1,3 isomers had been misassigned. This may or may not have been the first structural assignment based on calorimetric entropies. Different naming systems for the substituent locations came into use and a four-nation international discussion in 1953 led to a recommendation published in 1954 which has been accepted. The last paper concerns cyclobutane, which had long been thought to be planar but was found to depart significantly from planarity because the internal rotational strain was stronger than expected and the C-C-C bond angle strain weaker.

Reprinted from SCIENCE, June 29, 1945, Vol. 101, No. 2635, page 672.

STRAIN ENERGIES OF CYCLIC HYDROCARBONS[1]

THE strain energies of cyclic hydrocarbons have long been of interest to chemists. Since the discovery of appreciable energy barriers to rotation about single bonds, it has been apparent that this phenomenon may contribute a strain energy[2] in addition to the long-recognized deviation from tetrahedral bond angles. Conn, Kistiakowsky and Smith[2] discussed the relative energy contents of five- and six-membered cyclic hydrocarbons with zero, one and two double bonds without being able to obtain agreement between calorimetric values and theory. It is the purpose of this communication to point out that by recognizing the non-planarity of the cyclopentane ring[3] it is possible to obtain substantial agreement.

The calculations leading to the values in Table 1 are based on the well-established (though not very precise) values of 2,800 cal. per mole for the potential barrier[4] about an ethane-like single bond (staggered position having minimum energy) and 6×10^{-12} erg per radian[2] for the force-resisting[5] distortion of the bond angles about a carbon atom. Several amounts and types of puckering of cyclopentane were considered in order to estimate the energy of the

[1] The material in this communication was presented as a part of the American Chemical Society Award in Pure Chemistry address, Pittsburgh, 1943. Publication and further work have been delayed by war activities.

[2] J. B. Conn, G. B. Kistiakowsky and E. A. Smith, *Jour. Am. Chem. Soc.*, 61: 1868, 1939. These authors state that Dr. V. Schomaker first suggested this possibility to them.

[3] J. G. Aston, H. L. Fink and S. C. Schumann, *Jour. Am. Chem. Soc.*, 65: 341, 1943.

[4] K. S. Pitzer, *Chem. Rev.*, 27: 39, 1940.

[5] Rough average of values given by T. Y. Wu, "Vibration Spectra and Structure of Polyatomic Molecules," National University of Peking, Kun-ming (1939).

2

TABLE 1

STRAIN ENERGIES (KCAL. PER MOLE)

	Bond angles	Single bond twist	Total calc.	Exper.
Cyclopentane	2.6	6.6	9.2	7
Cyclopentane (planar)* ..	(0.1)	(14.0)	(14.1)*	(7)*
Cyclopentene	2.6	5.6	8.2	6
Cyclopentadiene—1, 3	7	0	7	6
Cyclohexane (chair form)	0	0	0	(0)
Cyclohexane (boat form)*	(0)	(5.6)	(5.6)*	(0)*
Cyclohexene	0	1.5	1.5	1
Cyclohexadiene—1, 3	0	4	4	4

* Hypothetical, unstable forms.

stable position. In the column headed "Bond Angles" are given the energies needed to bend the bond angles below their equilibrium values. To compute the twist energy, each single bond in the ring is considered separately. The energy needed to twist an ethane molecule into a similar configuration is taken as the contribution for that bond. In considering orientations adjacent to the double bond in the olefins, it is assumed that the equilibrium position is still the staggered one (if two single bonds were bent into a two-membered ring). This is confirmed by results for the simpler olefins. While the agreement in Table 1 is not perfect, it is within the various uncertainties present.

The experimental values in Table 1 are based on hydrogenation data[2] for the olefins and on combustion data[6] for the two naphthenes. The basic, no-strain heat of olefin hydrogenation is taken as that of trans 2-butene. For the diene hydrogenation, the value for 1,3 butadiene is adjusted for the effect of substitution on each end with the trans 2-butene-ethylene difference.

[6] H. M. Huffman, private communication.

3

It is planned, when circumstances permit, to refine these calculations by detailed consideration of the force constants and spectra of the molecules involved. At that time additional details of these calculations can be given.

KENNETH S. PITZER

DEPARTMENT OF CHEMISTRY,
UNIVERSITY OF CALIFORNIA,
BERKELEY

[Reprinted from the Journal of the American Chemical Society, **69**, 977 (1947).]

TAUTOMERISM IN CYCLOHEXANE DERIVATIVES; REASSIGNMENT OF CONFIGURATION OF THE 1,3-DIMETHYLCYCLOHEXANES

Sir:

It is now established that cyclohexane has predominantly the chair configuration.[1] Except for the work of Hassel,[2] however, the implications of this result with respect to cyclohexane derivatives have not been investigated. In the chair configuration, there are three hydrogen atoms above the carbon ring which together with the three equivalent hydrogens below the ring will be called polar (p). The other six hydrogens lie in an equatorial belt around the ring and will be called equatorial (e).

A monosubstituted cyclohexane can have the substituent either polar or equatorial. Since one form is converted to the other by mere distortion of bond angles (*i.e.*, passing through boat forms) they will be tautomers. For methylcyclohexane, the equatorial form is strain free, but the polar form involves steric interference. From comparison with similar interferences in the *n*-paraffins, this may be estimated to be about 2 kcal.

The possible configurational tautomers of the dimethylcyclohexanes are listed in Table I together with the estimated energy differences. Successive carbon atoms around the ring have their polar hydrogens alternately up and down. Thus *cis* 1,2, *trans* 1,3, and *cis* 1,4 have necessarily one polar and one equatorial methyl group. Correspondingly, *trans* 1,2, *cis* 1,3, and *trans* 1,4 can have both methyl groups equatorial or both polar, the former having much lower energy.

Skita and Schneck[3] assigned the configuration of the dimethylcyclohexanes on the basis of higher b.p., refractive index, etc., for the *cis* isomer. However, in this case there is no real geometrical similarity associated with the *cis* name. It seems more likely that the presence or absence of a polar methyl in the low energy tautomer would determine the physical properties. Thus if Skita's assignment is correct for 1,2- and 1,4-dimethylcyclohexanes probably it should be reversed for the 1,3-dimethylcyclohexanes. This last conclusion is strongly supported by the experimental entropies of these substances.[4] While detailed calculations will be presented later, the principal differences between isomers arise from tautomerism, optical isomerism and symmetry. These entropy contributions are listed in Table I, which shows that agreement is obtained only when the assignment of *cis* and *trans* configuration is reversed for the 1,3 dimethylcyclohexanes.

TABLE I

S^*, entropy contribution of tautomerism, optical isomerism and symmetry, cal./deg. mole.

Cmpd.	Tautomers Config.	Tautomers Description	S^* Calcd.	S^* Exptl.
cis 1,2	e, p p, e	Enantiomorphs	1.38[a]	0.83 ± 0.3
trans 1,2	e, e p, p	Energy diff. 3 kcal. *dl* isomers each with sym. no. 2	0.08	0.09 ± 0.3
cis 1,3	e, e p, p	Energy diff. 6 kcal.	0.00	0.06 ± 0.3[b]
trans 1,3	e, p p, e	Tautomers identical *dl* isomers	1.38	1.10 ± 0.3[c]
cis 1,4	e, p p, e	Identical	0.00	(.00)
trans 1,4	e, e p, p	Energy diff. 4 kcal. Sym. no. 2 each case	−1.36	−1.58 ± 0.3

[a] Greater interference to methyl rotation will reduce this value to agree with exptl. value. [b] For cmpd. previously labelled *trans*, b. p. 120.1°. [c] For cmpd. previously labelled *cis*, b. p. 124.5°.

(1) R. Rassmussen, *J. Chem. Phys.*, **11**, 249 (1943), and others there cited.

(2) O. Hassel, *Tids. Kjemi, Bergvesen Met.*, **3**, 32 (1943).

(3) A. Skita and A. Schneck, *Ber.*, **55B**, 144 (1922).

(4) Values for liquid from Oliver, Todd and Huffman, kindly communicated to us before publication. Vaporization data from Tables of Project 44 of the American Petroleum Institute.

PROJECT 44 OF THE AMERICAN PETROLEUM INSTITUTE
DEPARTMENT OF CHEMISTRY
UNIVERSITY OF CALIFORNIA
BERKELEY, CALIFORNIA

KENNETH S. PITZER
CHARLES W BECKETT

RECEIVED FEBRUARY 24, 1947

[Reprinted from the Journal of the American Chemical Society, **69**, 2483 (1947).]

[CONTRIBUTION FROM THE DEPARTMENT OF CHEMISTRY, UNIVERSITY OF CALIFORNIA]

The Thermodynamics and Molecular Structure of Cyclopentane[1]

BY JOHN E. KILPATRICK,[2] KENNETH S. PITZER[3] AND RALPH SPITZER[4]

The five-membered saturated ring is an important unit in organic chemistry. It is also particularly interesting because the torsional forces about the single bonds are in opposition to the forces tending to retain tetrahedral bond angles. Thus the latter forces tend to keep all five carbon atoms co-planar while the torsional forces tend to pucker the ring.

Aston, Schumann, Fink and Doty[5] have measured the entropy of gaseous cyclopentane and shown that it indicates a puckered ring, while one of us has found that calculations based on normal values of the force constants,[6] give lower energy to a puckered than to a planar ring. Douslin and Huffman[7] have also measured the entropy of cyclopentane and confirm the value of Aston and co-workers. Two of us[8] have recently measured the heat capacity of gaseous cyclopentane. It is the purpose of the present paper to analyze all of these and other data in order to give, if possible, a complete and reliable picture of the structure of the molecule and set of tables of thermodynamic properties.

Bond Distances and Angles.—Hassel and Viervoll[9] have recently reinvestigated cyclopentane by the electron diffraction method, using a rotating sector. They find C–C to be 1.54 Å. and C–H to be 1.09 Å. While they assumed a plane pentagonal model, the small puckering we shall find would have only a minor effect at this point. Indeed a study of their curves indicates that the fit at larger distances might be even more perfect if the slight changes due to puckering were taken into account. In any event this work gives us reliable distances; also it confirms approximate tetrahedral angles.

Vibration Frequencies.—The Raman spectrum of cyclopentane has been reported by Rosenbaum and Jacobson[10] who summarize earlier work. They remark about the unusual diffuseness of the lines. This we will find to be significant. The infrared spectrum has been reported by several laboratories.[11]

Since detailed and complete normal coördinate calculations would be excessively laborious for the puckered ring model, we first made calculations for the planar model which has the symmetry D_{5h}. At this point we are indebted to Dr. C. S. Lu and Dr. C. W. Beckett[12] who were studying certain features of the vibrational spectrum of the series cyclopropane through cyclohexane. Using force constants from this study, chiefly from cyclo-

(1) This work was supported by Research Project 44 of the American Petroleum Institute with headquarters at the National Bureau of Standards.

(2) Research Associate, Project 44 of the American Petroleum Institute. Now at Department of Chemistry, The Rice Institute, Houston, Texas.

(3) Associate Supervisor, Project 44 of the American Petroleum Institute, and Professor of Chemistry, University of California.

(4) National Research Fellow in Chemistry, 1945–46. Now at Department of Chemistry, Oregon State College, Corvallis, Oregon.

(5) J. G. Aston, S. C. Schumann, H. L. Fink and P. M. Doty, THIS JOURNAL, **63**, 2029 (1941); Aston, Fink and Schumann, *ibid.*, **65**, 341 (1943).

(6) K. S. Pitzer, *Science*, **101**, 672 (1945).

(7) D. R. Douslin and H. M. Huffman, THIS JOURNAL, **68**, 173 (1946).

(8) R. Spitzer and K. S. Pitzer, *ibid.*, **68**, 2537 (1946).

(9) O. Hassel and H. Viervoll, *Tids. Kjemi Bergvesen*, **3**, 31 (1946).

(10) E. J. Rosenbaum and H. F. Jacobson, THIS JOURNAL, **63**, 2841 (1941).

(11) Spectrograms 343 and 446 distributed by Project 44 of the American Petroleum Institute, National Bureau of Standards. 343 is from the Radiometry section of the Bureau of Standards, which also supplied data at wave lengths greater than 15μ. 446 is from the Shell Development Company, Emeryville, Calif.

(12) These results including these force constants will be published separately.

hexane, one may calculate the frequencies shown in Table I. The methods of Wilson[13] were employed and C–H stretching motions were reduced out immediately. The highest order matrix remaining is then the fourth order E_2'. Table I also shows the selection rules and describes the principal coördinates involved.

TABLE I

CALCULATED VIBRATION FREQUENCIES OF CYCLOPENTANE BASED ON PLANAR MODEL OF SYMMETRY D_{5h}

	Raman active			Infrared active		Inactive		
Motion	A_1'	E_2'	E_1''	A_2''	E_1'	A_2'	A_1''	E_2''
CH_2 sym. bend.	1450	1452	..	..	1449	..	..	..
CH_2 twist	..	..	1348	..	..	..	1295	1327
CH_2 wag in plane	..	1207	..	..	1298	1231	..	..
CH_2 rock out of plane	..	..	905	896	..	..	..	1020
C–C stretch	838	1049	..	..	951	..	..	..
C–C–C bend	..	549	..	..	..	..	..	[a]

[a] This motion is the ring puckering; hence the calculated value for this model is meaningless.

In Table II are listed the observed frequencies below 1470 cm.$^{-1}$ which appear to us to be reasonably well established, together with their intensities and polarizations where available. These are compared with the active fundamentals calculated for the planar model.

TABLE II

COMPARISON OF OBSERVED SPECTRAL LINES OF CYCLOPENTANE BELOW 1470 CM.$^{-1}$ WITH VALUES CALCULATED ON THE PLANAR MODEL (D_{5h})

Infrared		Raman	
Obs.	Calcd.	Obs.	Calcd.
1462 (7)	1449 cm.$^{-1}$	1448 (5, dp)	1452 (dp)
1376 (1)	..		1450 (p)
1317 (1)	1298	..	1348 (dp)
1250	..	1298 (1)	..
950 (0)	951	1016 (?)	1207 (dp)
895 (4)	896	1032 (4, dp)	1049 (dp)
827 (1)	..	..	905 (dp)
769 (1)	..	888 (10, p)	838 (p)
616	..	..	595 (dp)
545	..	288 (1)	..

It will be noted that Table II shows relatively good agreement at the higher frequencies, while below 800 cm.$^{-1}$ there is no agreement at all. This is about what would be expected if cyclopentane is actually somewhat puckered. The high frequency vibrations are not closely related to the puckering motion, nor are their frequencies greatly affected. Consequently we would expect to find approximately the calculated frequencies, although there might be some breakdown of selection rules. The biggest frequency difference is on the symmetrical C–C stretching motion which is calculated at 838 and observed at 888 cm.$^{-1}$. However, as Rosenbaum and Jacobson[10] remark, the diffuseness of these higher frequency lines is probably associated with small effects of puckering. Indeed we shall find later that this puckering is of an indefinite character so that this diffuseness is most reasonable.

There are four modes of vibration expected to have frequencies less than 700 cm.$^{-1}$. On the planar model these are two degenerate modes; the higher is calculated at 595 cm.$^{-1}$ while no reasonable calculation can be made of the lower mode because it depends on twisting about the single bonds. In this range no credence can be given the selection rules for the planar model. Indeed the strict symmetry is so low that there will be no selection rules. Consequently it seems reasonable to select 616 and 545 cm.$^{-1}$ as the frequencies arising when the degenerate 595 frequency splits due to ring puckering. This leaves 288 cm.$^{-1}$ for one of the twisting modes. The other twisting mode will be discussed in the next section.

In Table III we summarize the complete vibration assignment used for thermodynamic calculations. It will be noted by comparison with Tables I and II that all values for hydrogen motions are well supported by calculated or observed results, usually both. However, average values are used for the groups of frequencies. The 2950 cm.$^{-1}$ value for C–H stretching is the average of closely spaced and largely unresolved lines in both spectra. For C–C stretching, 1032, 950 and 888 cm.$^{-1}$ are well substantiated. The selection of the observed lines at 827 and 769 cm.$^{-1}$ for the other two is quite arbitrary and was based largely on the gaseous heat capacity. The three low frequencies have been discussed already.

TABLE III

VIBRATING FREQUENCIES OF CYCLOPENTANE USED FOR THERMODYNAMIC CALCULATIONS

Motion	Frequency, cm.$^{-1}$	Degeneracy
C–H stretching	2950	10
CH_2 sym. bend.	1462	5
CH_2 twist	1376	2
	1298	3
CH_2 wag in plane	1250	5
CH_2 rock out of plane	950	3
	895	2
C–C stretching	1032	1
	950	1
	888	1
	827	1
	769	1
C–C–C bending	616	1
	545	1
C–C twisting	288	1

Pseudo-Rotation of Ring Puckering.—In order to analyze further the ring puckering motions we first calculated the potential energy. There are two ways a five-membered ring can pucker and still retain some symmetry. As is shown in Fig. 1, one atom can move out of the plane of the other

(13) E. B. Wilson, Jr., *J. Chem. Phys.*, **9**, 76 (1941).

four giving the C_s mode or two atoms can twist relative to the others giving the C_2 mode. Calculations were made for these two modes in which the C–C bond lengths were kept at exactly their original value and the deviation of C–C–C bond angles from the tetrahedral value kept to a minimum. The potential energy of bond bending was evaluated by summing the squares of the deviations from tetrahedral angles together with the force constant $k = 0.8 \times 10^{-11}$ ergs/radian². The value of this constant is only approximate, but is consistent with the ring bending (616 and 545 cm.$^{-1}$) frequencies of cyclopentane, similar frequencies for cyclohexane and the skeletal bending frequency in propane. The potential energy associated with rotation or twist about the five C–C bonds was also included, using a cosine shape and the value from ethane,[14] 2800 cal. per mole, for the potential height, V_0.

The potential energy curve as a function of the deviation from planarity proved to be the same for the C_2 and C_s types of puckering as nearly as the two could be compared. This suggested that very probably intermediate types of puckering would also show the same potential curve. To investigate this we define the following coördinates. Let z_j be the displacement of jth carbon atom perpendicular to the plane of the unpuckered ring. Then write

$$z_j = \sqrt{\frac{2}{5}}\, q \cos 2\left(\frac{2\pi}{5} j + \varphi\right) \qquad (1)$$

where q is the amplitude of the puckering, φ is the phase angle of maximum amplitude (see Fig. 1) and j takes integral values from 1 to 5. Coördinates q and φ are orthogonal to each other and to translation and rotation of the whole molecule. The two symmetrical types of puckering are obtained by setting $\varphi = l\pi/10$ for C_s and $\varphi = (2l+1)\pi/20$ for C_2, where l is any integer. Figure 1 shows these two types of puckering, together with an intermediate position. This indicates how a rotation of the maximum amplitude of puckering gives successively the C_s and C_2 configurations.

We may now calculate the potential energy as a function of q and see if it is independent of φ. It proved too complicated mathematically to retain the condition of constant bond lengths at this stage since that introduces small motions in the x and y directions. Also we were unable to carry through the calculation with φ unspecified. However, it was possible to show that for various assumed values the C–C–C angle strain and the C–C twisting strain were each dependent on q and independent of φ. For example, calculations were made for $q = 0.308$ Å. and $\varphi = 0, \pi/40$ and $\pi/20$. The C–C–C angle strain was $1.445(k \times 10^{11})$ kcal./mole and the twisting strain $3.6036\ V_0$ kcal./mole, the same to four and five significant figures, respectively, for all three values of φ.

(14) K. S. Pitzer, *Chem. Rev.*, **27**, 39 (1940).

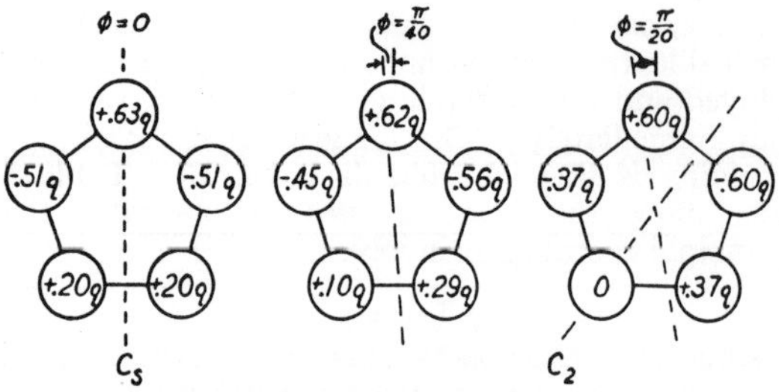

Fig. 1.—Pseudo-rotation of ring puckering in cyclopentane. Displacements perpendicular to the plane of the paper are given in the circles.

In the first method, where the bond distances were held constant, the following equations apply to the C_s mode. Let atom 1 be displaced from the plane of atoms 2, 3, 4 and 5. Ring angles $\omega_2 = \omega_5$ and $\omega_3 = \omega_4$. The following relations may be derived by the use of vector analysis

$$\cos \omega_3 = \frac{1}{2} - \sin \frac{1}{2}\omega_1$$

$$\cos \lambda = -\frac{\sin \frac{1}{2}\omega_1 \cos \omega_3 + \cos \omega_2}{\cos \frac{1}{2}\omega_1 \sin \omega_3}$$

$$\cos \theta_{12} = \frac{\cos \omega_2 (\cos \omega_1 - 1) - \sin \frac{1}{2}\omega_1 \left(1 - 2 \sin \frac{1}{2}\omega_1\right)}{\sin \omega_1 \sin \omega_2}$$

$$\cos \theta_{23} = \frac{\cos \omega_2 \cos \omega_3 + \sin \frac{1}{2}\omega_1}{\sin \omega_2 \sin \omega_3}$$

The angle θ_{mn} is the torsion of bond mn. By symmetry, $\cos \theta_{12} = \cos \theta_{51}$, and $\cos \theta_{23} = \cos \theta_{45}$. Torsion $\theta_{34} = 0$. The angle λ is the dihedral angle between the planes 251 and 2543. From λ the amplitude of the puckering can be calculated.

There are two independent variables in the above equations, which can be chosen as ω_1 and ω_2. It was found that the combination $\omega_1 = \omega_2$ gave substantially the lowest potential energy.

In the second method, with all displacements perpendicular to the ring, the following equations apply

$$\cos \omega_j = -\frac{\cos(2\pi/5) + (\Delta z_{j+1,j})(\Delta z_{j,j-1})}{(d_{j+1,j})(d_{j,j-1})}$$

$$\sin \theta_{j,j+1} = \frac{[\mathbf{r}_{j,j-1}\, \mathbf{r}_{j+1,j}\, \mathbf{r}_{j+2,j+1}]}{(d_{j,j-1})(d_{j+1,j})(d_{j+2,j+1}) \sin \omega_j \sin \omega_{j+1}}$$

$$d_{j+1,j} = (d_0^2 + \Delta z_{j+1,j}^2)^{1/2}$$

where Δz is the difference in the z coördinates indicated (from eq. 1), d is the absolute length of the bond indicated and in the second equation there is the triple vector product of the vectors, $\mathbf{r}$, from atoms $j-1$ to j, j to $j+1$ and $j+1$ to $j+2$.

In each method, after the values of all ω's and θ's are obtained, the potential energy is calculated from the standard equations.

While the two calculations mentioned above can not be expected to agree quantitatively, they indicate clearly that the potential energy is at maximum in the planar configuration and falls to a minimum about 4 kcal./mole lower at q approximately 0.3 Å. Near this minimum the curve is roughly parabolic with a curvature corresponding to a force constant of about 10^5 dynes per cm.

We may now introduce the Schrödinger equation and calculate the energy levels for this system.

$$\frac{-h^2}{8\pi^2 m}\sum_{j=1}^{5}\frac{\partial^2\psi}{\partial z_j^2} + V\psi = E\psi \qquad (2)$$

where m is the mass of a carbon atom, V is the potential energy and E the total energy for wave function ψ. In addition to the coördinates q and φ defined in equation (1) we have translation in the z direction and rotation about two axes. Equations for all of these coördinates may be written as

$$z_j = \sqrt{\frac{2}{5}}\left[q \cos 2\left(\frac{2\pi}{5} j + \varphi\right) + \alpha \cos \frac{2\pi}{5} j + \beta \sin \frac{2\pi}{5} j + \frac{t}{\sqrt{2}}\right] \qquad (3)$$

where α, β and t describe the two rotations and the translations, respectively. Following the usual procedure for the separation of variables, one obtains

$$-\frac{h^2}{8\pi^2 m}\frac{\partial^2\psi_t}{\partial t^2} = E_t\,\psi_t$$

$$-\frac{h^2}{8\pi^2 m}\frac{\partial^2\psi_\alpha}{\partial \alpha^2} = E_\alpha\,\psi_\alpha$$

$$-\frac{h^2}{8\psi^2 m}\frac{\partial^2\psi_\beta}{\partial \beta^2} = E_\beta\,\psi_\beta$$

$$-\frac{h^2}{32\pi^2 m q_0^2}\frac{\partial^2\psi_\varphi}{\partial \varphi^2} = E_\varphi\,\psi_\varphi$$

$$-\frac{h^2}{8\pi^2 m q}\frac{\partial}{\partial q}\left(q\frac{\partial\psi_q}{\partial q}\right) + V(q)\,\psi_q = E_q\,\psi_q \qquad (4)$$

$$E = E_t + E_\alpha + E_\beta + E_\varphi + E_q$$

$$\psi = \psi_t\,\psi_\alpha\,\psi_\beta\,\psi_\varphi\,\psi_q$$

The validity of the equations for α, β and q is limited to small values of these variables. The equations in α and β can be rearranged, when the x and y coördinates are also considered, to give the usual equations for rotation; however, the limitation of q to small values (as compared to the distance between atoms) remains. Since the puckering of the cyclopentane ring is small, this restriction is not serious. The separation of the q and φ equations is based on q remaining near an equilibrium value, q_0.

We may now proceed to solve the equations for q and φ. The latter yields the usual energy levels for a free, one dimensional rotation, however, the boundary condition is $\psi(\varphi) = \psi(\varphi + \pi)$ because there are two regions of positive amplitude around the ring (or because eq. 1 contains essentially $\cos 2\varphi$). Then

$$E_\varphi = n^2h^2/8\pi^2 m q_0^2 \qquad (5)$$

in which mq_0^2 is essentially a moment of inertia and n takes integral values both positive and negative including zero.

The effect of symmetry is to eliminate four-fifths of the values of the quantum number, n, for any given over-all rotational state. In particular if nuclear spins are such that symmetric states are allowed, then $K + n$ must be an integral multiple of five. Here K is the quantum number for the rotation about the five-fold axis.[15] Since the energy levels for successive multiples of five for n would be separated by approximately kT at the temperature of interest, it is important to note that all values of n are allowed for any nuclear spin species. For a given value of n, only one-fifth of the values of K are allowed, but since the energy level spacing even for increase of K by five is close compared to kT, the classical formulas with symmetry number are still valid for statistical calculations. In order to solve the equation in q, some assumption must be made about the nature of $V(q)$. As a first approximation we assume a harmonic oscillator potential about the equilibrium position q_0. Then the substitution $\psi_q = q^{-1/2}f(q)$ gives substantially the harmonic oscillator energy levels

$$E_q = \left(n' + \frac{1}{2}\right) h\nu_q - h^2/32\pi^2 m q_0^2 \qquad (6)$$

with

$$\nu_q = (1/2\pi)(k/m)^{1/2},\; V = (k/2)(q - q_0)^2$$

We now identify ν_q with the observed Raman line at 288 cm.$^{-1}$. The very small, constant second term in (6) is considered part of the vibrational zero point energy.

Consideration of the hydrogen atoms complicates the mathematics considerably but the only change in the essential results is to give the mass of the single atom, m, the value 28.1 on the atomic weight scale. This figure was calculated from the inverse kinetic energy matrix, G, for E_2'' (planar D_{5h} model) by making all of the coördinates except q rigid. This of course assumes that the CH_2 group is a rigid unit which should be a good approximation for such low frequency motion. Although we have assumed a puckered form for the equilibrium configuration of the ring, the reduced mass for the CH_2 group for the puckering motion should still be about this magnitude. In any event, a slight error in this reduced CH_2 mass is absorbed in the value of q_0 determined from the experimental entropy.

The essential result of this section has been to show that for the potential energy function expected for cyclopentane, the ring puckering motions are: first, an ordinary vibration in which the amount of puckering oscillates about a most stable value and second, a pseudo one-dimensional rotation in which the phase of the puckering rotates around the ring. This is not a real rotation since the actual motion of the atoms is perpendicular to the direction of rotation and there is no angular momentum about the axis of rotation. Since the higher vibration frequencies must be at least slightly affected by the coördinate φ, the observed diffuseness of the spectral lines is most reasonable. The 288 cm.$^{-1}$ frequency corresponds to a force constant of 1.4×10^5 dynes per cm. on the coördinate q. This agrees with our calculations within their accuracy, which is not very high because of uncertainties in the potential constants. Likewise our calculated value of the position of minimum potential energy, q_0, is not very accurate. Consequently we shall use the observed 288 cm.$^{-1}$ frequency and shall evaluate the exact value of q_0 later from the entropy value.

Experimental Entropy Value.—Before proceeding to the final calculations, a few words must be said about the selection of a best value of the entropy of cyclopentane from the experimental "Third Law" method. Although the heat capacity curves of Aston, Fink and Schumann[5] and of Douslin and Huffman[7] do not agree too well, their final values of the entropy of the liquid differ by only 0.08 cal. per degree. This is within the limit of error assigned by Douslin and Huffman and just outside the optimistically small limit assigned by Aston. However, the value given by Aston, Fink and Schumann for the heat of vaporization at 298.16° K. is unacceptable because it is in serious disagreement with their own vapor pressure curve and with that of Willingham, Taylor, Pignocco and Rossini.[16]

(15) L. Pauling and E. B. Wilson, Jr., "Introduction to Quantum Mechanics," McGraw-Hill Co., New York, N. Y., 1935, pp. 275–280.

(16) C. B. Willingham, W. J. Taylor, J. M. Pignocco and F. D. Rossini, *J. Research Natl. Bur. Standards*, **35**, 219 (1945).

The use of the vapor pressure data above to calculate the heat of vaporization at 298.16° K. is not very satisfactory since gas imperfection effects are considerable and may not be estimated accurately by available equations of state. We can avoid this difficulty by making the calculation at a lower temperature and pressure where the gas is nearly perfect. Table IV summarizes these calculations. We have used the vapor pressure equation selected by Project 44 of the American Petroleum Institute[17] from both sets of data mentioned above. In both cases the effect of gas imperfection was calculated with the Berthelot equation with critical constants[5] $T_c = 520°$ K. and $P_c = 44.2$ atm.

TABLE IV

THE ENTROPY OF CYCLOPENTANE

T, °K.	260°K.		298.16°K.	
Source	A., F., S.[5]	D. and H.[7]	A., F., S.[5]	D. and H[7]
Liquid at melting point	35.533	35.493	35.533	35.493
M. p. to T° K.	9.403	9.342	13.388	13.293
Vaporization to real gas	27.94	27.94	22.84	22.84
Gas imperfection	0.020	0.020	0.084	0.084
Correction to 1 atm.	−5.244	−5.244	−1.725	−1.725
Entropy of ideal gas, T° K. and 1 atm.	67.65	67.55	70.12	69.99
Calculated	67.54		70.00	

Thermodynamic Functions and Final Molecular Model.—We are now ready to make the final calculations of the entropy, heat capacity and other functions. The moments of inertia calculated on the planar model with the distances given above are $I_x = I_y = 13.028$ and $I_z = 23.405$ each $\times 10^{-39}$ g. cm². The vibration frequencies are given above. The effective symmetry number for over-all rotation of the puckered ring is one. (Aston, *et al.*[5] give $\sigma = 2$ for the C_2 model but they have neglected the fact that it has d and l forms which cancels the effect of the symmetry number.) When the pseudo rotation is added a symmetry number of five is introduced as explained above. In order to fit the entropy closely, the value of 0.236 Å. was selected for q_0. This is reasonably close to the rough value calculated above. Also in order to fit the heat capacity values more closely, an anharmonicity coefficient of 0.03 was assumed for the 288 cm.$^{-1}$ vibration. Since it is a double minimum vibration with the central peak only about 4 kcal. above the minima, this is quite reasonable. In any case the effect of the anharmonicity is quite small.

Table IV compares the calculated entropy of the gas with the experimental values while Table V similarly compares values for the heat capacity of gaseous cyclopentane. The experimental values of the gas heat capacity are from Spitzer and Pitzer.[8] The values of physical constants and working formulas for the calculations are given by Wagman, Kilpatrick, Taylor, Pitzer, and Rossini.[18]

(17) American Petroleum Institute Research Project 44: selected values of properties of hydrocarbons, Table 7k (part 2), 8/31/44.

(18) D. D. Wagman, J. E. Kilpatrick, W. J. Taylor, K. S. Pitzer and F. D. Rossini, *J. Research, Natl. Bur. Standards*, **34**, 143 (1945).

TABLE V

THE HEAT CAPACITY OF CYCLOPENTANE IN THE IDEAL GAS STATE

T, °K.	C_P^0 exptl.[8]	C_P^0 calcd.
353	24.38 ± .2	24.37
	24.45 ± .3	
372	25.85 ± .2	25.94
395	28.03 ± .3	27.83
424	30.19 ± .3	30.16
463	33.16 ± .3	33.16
503	36.02 ± .4	36.07
539	38.26 ± .4	38.52

Using the formulas and parameters already mentioned, values were calculated for the functions $-(F_T^0 - H_0^0)/T$, $(H_T^0 - H_0^0)/T$, S^0 and C_P^0 over a range of temperatures. These are given in Table VI.

TABLE VI

THE THERMODYNAMIC FUNCTIONS OF CYCLOPENTANE (CAL./DEG. MOLE FOR THE IDEAL GAS STATE)

T, °K.	$-(F_T^0 - H_0^0)/T$	$(H_T^0 - H_0^0)/T$	S^0	C_P^0
298.16	57.93	12.07	70.00	19.82
300	58.00	12.12	70.12	19.98
400	61.88	15.12	77.00	28.24
500	65.62	18.52	84.14	35.86
600	69.30	21.98	91.28	42.36
700	72.95	25.29	98.24	47.81
800	76.52	28.40	104.92	52.44
900	80.04	31.29	111.33	56.37
1000	83.48	33.97	117.45	59.75
1100	86.84	36.46	123.30	62.68
1200	90.13	38.75	128.88	65.18
1300	93.31	40.87	134.18	67.36
1400	96.40	42.83	139.23	69.24
1500	99.42	44.65	144.07	70.89

Discussion.—Initially we had hoped that it would be possible to correlate a number of quantities which depend on the bond twisting and bending forces in the cyclanes. These include the heats of formation per CH_2 group of cyclopentane, cyclohexane and cycloöctane,[19] the energy difference of chair and boat forms of cyclohexane and the detailed nature of the puckering of cyclopentane. While the use of the ethane potential for twisting gives rough agreement,[6] some of these relationships indicate a lower value while others indicate a higher value of this twisting potential. Numerous plausible assumptions are involved in these relationships at least one of which apparently must be in error but we are not prepared to say which it is. As an example, the intrinsic C–C bond energy might not be exactly the same for five- and six-membered rings although we are inclined to believe that it should be. Further study appears to be needed before these various relationships can be put on a quantitative basis. Nevertheless we believe the analysis given

(19) R. Spitzer and H. M. Huffman, THIS JOURNAL, **69**, 211 (1947).

in this paper for cyclopentane brings into agreement a number of previously unrelated facts, and that a final understanding of the broader problems will not essentially change this analysis.

Summary

The published data from electron diffraction, Raman and infrared spectra, entropy and specific heat measurements have been combined to develop as complete a picture of structure of the cyclopentane molecule as possible. Most striking is the result that the puckering of the ring is not of a definite type but that the angle of maximum puckering rotates around the ring. Other properties prove to be consistent with this conclusion. Among the results obtained are an assignment of vibration frequencies and tables of the thermodynamic functions, $-(F - H_0)/T$, S, $(H - H_0)/T$ and C_p for the range 298.16 to 1500° K.

BERKELEY, CALIF. RECEIVED FEBRUARY 28, 1947

1947, VOL. 69

John E. Kilpatrick, Kenneth S. Pitzer and Ralph Spitzer. The Thermodynamics and Molecular Structure of Cyclopentane.

Pages 2486–7. In the statistical calculations of this paper we failed to introduce a symmetry number factor of 2 for the identity of the two sides of the ring of the cyclopentane molecule. The peculiarly indefinite type of puckering of that molecule confuses the situation but we have now analyzed the problem in detail and find the additional factor to be required. Since the puckering parameter q_0 was determined to fit the entropy, the net effect of this change is to double q_0 and to leave all other numerical values unchanged. The new value of q_0 is 0.472 Å.—K. S. PITZER.

[Reprinted from the Journal of the American Chemical Society, **69**, 2488 (1947).]

[CONTRIBUTION FROM THE DEPARTMENT OF CHEMISTRY, UNIVERSITY OF CALIFORNIA]

The Thermodynamic Properties and Molecular Structure of Cyclohexane, Methylcyclohexane, Ethylcyclohexane and the Seven Dimethylcyclohexanes[1]

BY CHARLES W. BECKETT,[2] KENNETH S. PITZER[3] AND RALPH SPITZER[4]

This paper is a continuation of studies on the structural and thermodynamic properties of hydrocarbons. As compared to cyclopentane,[5] in which a large number of configurations with nearly equal stability result from the opposition of torsional forces about single bonds against forces tending to maintain tetrahedral angles, cyclohexane has one predominant stable configuration, the chair form, since it lies at the minimum in potential energy of each type. Other, higher energy tautomers exist in quite small amounts at room temperature. The energies of these, however, are low enough so that they may contribute to thermal properties. We[6] have recently measured the heat capacity of gaseous cyclohexane and methylcyclohexane and these data are compared with calculated values in the present paper.

In the course of our study very interesting effects of geometrical tautomerism in disubstituted cyclohexanes became apparent. These have been mentioned briefly in a communication to the editor,[6a] but further details are given in this paper.

These compounds are important as constituents of petroleum hence their properties should have considerable practical interest.

We shall first discuss the work on cyclohexane in some detail and then briefly describe the less detailed calculations on its derivatives.

(1) This research was supported by Project 44 of the American Petroleum Institute with headquarters at the National Bureau of Standards.

(2) Research Associate, Project 44 of the American Petroleum Institute.

(3) Associate Supervisor, Project 44 of the American Petroleum Institute and Professor of Chemistry at the University of California.

(4) National Research Council Fellow in Chemistry, 1945–1946. Present address: Department of Chemistry, Oregon State College, Corvallis, Oregon.

(5) J. E. Kilpatrick, K. S. Pitzer and Ralph Spitzer, THIS JOURNAL, **69**, 2483 (1947).

(6) R. Spitzer and K. S. Pitzer, THIS JOURNAL, **68**, 2537 (1946).

(6a) K. S. Pitzer and C. W. Beckett, *ibid.*, **69**, 977 (1947). See also F. D. Rossini and K. S. Pitzer, *Science*, **105**, 647 (1947).

Cyclohexane

Equilibrium Configuration.—Hassel and collaborators[7] have made an extensive study of cyclohexane by the electron diffraction method. They find that substantially all of it is in the more symmetrical chair (D_{3d}) form with a carbon–carbon distance of 1.54 Å. and approximately tetrahedral bond angles. They do not exclude a small percentage of the less symmetrical, boat form. If the staggered position is the potential minimum for the rotation about single bonds, then the chair form would be expected to have a lower energy since it maintains the staggered position throughout while the boat form twists two C–C bonds into an opposed configuration.[8]

Spectra and Vibration Frequencies.—Rasmussen[9] has very adequately summarized his own and previous spectroscopic work on cyclohexane and has shown that all lines can be explained on the chair (D_{3d}) model. This confirms the conclusion drawn from electron diffraction results. Since Rasmussen's work, the infrared spectrum has been investigated[10] further in the range 15–25μ. This later work shows that the frequency reported by Kettering and Sleator[11] at 673 cm.$^{-1}$ is spurious and that instead a band exists at 522 cm.$^{-1}$. Another weaker band at 459 cm.$^{-1}$ is also suggested.

A normal coördinate analysis has been carried out for the series of molecules cyclopropane, cyclobutane, cyclopentane and cyclohexane by Dr. C. S. Lu and certain of the present authors which will be published separately.[12] This analysis,

(7) O. Hassel, *Tids. Kjemi Bergvesen*, **3**, 32 (1943); O. Hassel and B. Ottar, *Arch. Math. Naturvidenskab*, **45**, No. 10, 1–7 (1942), and earlier papers there cited.

(8) K. S. Pitzer, *Science*, **101**, 672 (1945).

(9) R. S. Rasmussen, *J. Chem. Phys.*, **11**, 249 (1943).

(10) Dr. R. Halford, private communication; American Petroleum Institute Research Project 44, "Catalog of Infrared Spectrograms," Serial No. 466 contributed by the Radiometry Section, National Bureau of Standards.

(11) C. F. Kettering and W. W. Sleator, *Physics*, **4**, 39 (1933).

(12) C. S. Lu, C. W. Beckett and K. S. Pitzer, to be published.

while uncertain at some points, indicates that the vibration frequencies listed in Table I for cyclohexane are substantially correct. At least the remaining errors should not be serious for the thermodynamic calculations we have made. The frequency at 231 cm.$^{-1}$ is not observed since it is forbidden in the Raman effect and the infrared has not been investigated in this region. The value was selected to fit the experimental entropy of cyclohexane.

The A_{1g} frequency at 1443 and the E_g frequency at 802 cm.$^{-1}$ were calculated to be near these values which are those of stronger lines in different symmetry classes. Presumably these lines are buried under the stronger lines. There is a resonance between the E_u frequency at 1030 cm.$^{-1}$ and 231 + 802 = 1033 (also E_u) leading to the pair of bands observed.

TABLE I

THE VIBRATION FREQUENCIES OF CYCLOHEXANE

Symbol class	Type of motion	Frequency	Source
A_{1g}	C–C bending	382	R (p)
	C–C stretching	802	R (p)
	CH_2 wagging	1155	R (p)
	CH_2 bending	1443	Calcd.[a]
A_{1u}	C–C stretching	1030	Calcd.
	CH_2 wagging	1176	Calcd.
	CH_2 twisting	1375	Calcd.
A_{2g}	CH_2 wagging	1185	Calcd.
	CH_2 twisting	1308	Calcd.
A_{2u}	C–C bending	522	I. R.
	CH_2 rocking	903	I. R.
	CH_2 bending	1456	I. R.
E_g	C–C bending	426	R (d)
	C–C stretching	1028	R (d)
	CH_2 rocking	802	Calcd.[a]
	CH_2 wagging	1266	R (d)
	CH_2 twisting	1340	R (d)
	CH_2 bending	1443	R (d)
E_u	C–C bending	231	Calcd.
	C–C stretching	864	I. R.
	CH_2 rocking	1030	I. R.[a]
	CH_2 wagging	1261	I. R.
	CH_2 twisting	1348	I. R.
	CH_2 bending	1456	I. R.

[a] See text discussion. C–H stretching (12 degrees of freedom) not analyzed in detail, 2950 cm.$^{-1}$ taken as average value.

A study of the motions of the 231 and 382 cm.$^{-1}$ frequencies indicates that they may be rather anharmonic. The 382 cm.$^{-1}$ motion eventually moves the molecule through the planar (D_{6h}) configuration at which point the potential energy is at a maximum. This was estimated to be 31 kcal. above the minimum. A better estimate of the anharmonicity comes from the overtone observed by Langseth and Bak[13] which gives an anharmonicity coefficient, X = 3.0 cm.$^{-1}$. The 231 cm.$^{-1}$ motion passes over a lower potential maximum at about 14 kcal. and then down to the boat form. Thus greater anharmonicity may be expected here. Since we have no value from the spectrum, X = 4.0 cm.$^{-1}$ was selected primarily to give the best fit with the heat capacity of the gas.

The equation (3) of Wagman, Kilpatrick, Pitzer and Rossini[14] was employed in calculating the effect of anharmonicity on the thermodynamic properties. No corrections for interaction between vibrations, for anharmonicities of higher frequencies, for interaction between vibration and rotation, or for rotational stretching were included because the last two were estimated to be negligible and no information was available for the first two. However, the effect of anharmonicity upon thermal properties decreases rapidly with increasing frequency so that no serious error can be involved.

Geometrical Tautomerism.—The chair and boat forms have already been mentioned. We assume that a tautomeric equilibrium exists between these forms. While we shall find the fraction in the boat form at room temperature to be small, it increases rapidly with the temperature. Since the conversion absorbs considerable heat this contributes a very appreciable term to the heat capacity. Indeed it would be difficult to fit the observed data without this term. Velocity of sound measurements might determine the rate of the tautomerism reaction because of this effect.

We have assumed that the energy difference is twice the internal rotation potential barrier in ethane[15] or 5.6 kcal. because the change from chair to boat twists two carbon–carbons bonds from their potential minima to their maxima. Also, since the symmetry number of the boat form is only 2 as compared to 6 for the chair form, we assign an entropy $R \ln 3$ greater to the boat form. This assumes that apart from symmetry number, the sum of rotational and vibrational entropies of the two forms is the same. The usual thermodynamic equations then permit the calculations of the various properties desired.

Thermodynamic Properties.—The entropy and related properties of the ideal gas were calculated by the usual statistical methods. The physical constants and working formulas of Wagman, Kilpatrick, Taylor, Pitzer and Rossini[16] were used. Taking 1.54 and 1.09 Å., respectively, for the carbon–carbon and carbon–hydrogen bond distances and tetrahedral angles one obtains 12,583 × 10^{-117} g.^{3} cm.^{6} for the product of moments of inertia. Vibration frequencies and other parameters have already been discussed. Table II shows the magnitude of the various contributions to the heat capacity at 500°K. and the

(13) A. Langseth and B. Bak, *J. Chem. Phys.*, **8**, 403 (1940).

(14) D. D. Wagman, J. E. Kilpatrick, K. S. Pitzer and F. D. Rossini, *J. Research Natl. Bur. Standards*, **35**, 467 (1945).

(15) K. S. Pitzer, *Chem. Rev.*, **27**, 39 (1940).

(16) D. D. Wagman, J. E. Kilpatrick, W. J. Taylor, K. S. Pitzer and F. D. Rossini, *J. Research Natl. Bur. Standards*, **34**, 143 (1945).

entropy at 298.16°K. and also gives experimental values. The first experimental entropy (R. H.–A. P. I.) is obtained by adding the entropy of vaporization selected by Project 44 of the American Petroleum Institute[17] to the entropy of the liquid given by Ruehrwein and Huffman.[18]

TABLE II

CONTRIBUTIONS TO THERMAL PROPERTIES OF CYCLOHEXANE; COMPARISON OF CALCULATED AND OBSERVED VALUES; IDEAL GASEOUS STATE; CAL./DEG. MOLE

Calculated	$S^0_{(g)298.16}$	$C^0_{p500°K.}$
Vibration, harmonic	9.111	36.268
Vibration, anharmonic	0.187	0.578
Chair to boat tautomerism	.005	.676
Translation and rotation	61.973	7.949
Total	71.28	45.47

Experimental:

$C^0_{p500°K.}$ = 45.6 ± 0.5, Spitzer and Pitzer[6]
$S_{(liq.)298.16}$ = 48.84 ± 0.10, Ruehrwein and Huffman[18]
$\Delta S^0_{(vap.)298.16}$ = 22.43 A. P. I.[17]

$S^0_{(gas)298.16}$ = 71.27 (R. H.–A. P. I.)
$S^0_{(gas)298.16}$ = 71.42 ± 0.18, Aston, Szasz and Fink[19]

The value of the entropy of the gas given by Aston, Szasz and Fink[19] is also included.

Table III compares the calculated and experimental values for the heat capacity of the ideal gas over a range of temperature.[6,20]

Table IV gives the calculated values of S^0, $(F^0 - H^0_0)/T$, $(H^0 - H^0_0)/T$ and C^0 from 298.16 to 1500°K. Brickwedde, Moskow and Aston[21] have also presented values of these functions based on a crude vibrational assignment. Their calculated thermodynamic functions are, however, very similar to ours.

TABLE III

CALCULATED AND OBSERVED HEAT CAPACITIES OF CYCLOHEXANE IN THE IDEAL GASEOUS STATE, CAL./DEG. MOLE

Temp., °K.	C^0_p Calcd.	C^0_p Obs.	Reference
384	34.2	34.2 ± 0.3	Spitzer and Pitzer[6]
428	38.5	38.7 ± .4	Spitzer and Pitzer[6]
460	41.6	41.7 ± .4	Spitzer and Pitzer[6]
495	45.0	45.3 ± .5	Spitzer and Pitzer[6]
521	47.2	47.0 ± .5	Spitzer and Pitzer[6]
544	49.2	49.3 ± .5	Spitzer and Pitzer[6]
370	32.8	33.0	Montgomery and DeVries[20]
390	34.8	35.0	Montgomery and DeVries[20]
410	36.8	36.8	Montgomery and DeVries[20]

By employing the values in Table IV, a number of sources yield either the heat of hydrogenation of benzene to cyclohexane or the heat of formation of cyclohexane. Assuming the heat of formation of benzene[22] to be 24.000 kcal./mole at 0° K. one can calculate one quantity from the other. These values are listed in Table V. Comment is needed only on the values from the equilibrium studies of Burrows and Lucarini and of Zharkova and Frost. A value of ΔH^0_0 can be calculated from each equilibrium constant using the $(F_T - H_0)/T$ function for cyclohexane from Table IV and for hydrogen[16] and benzene.[22] These separate values of ΔH^0_0 were found to show no trend with temperature and were concordant within the average deviation shown in Table V.

TABLE IV

THERMODYNAMIC PROPERTIES OF CYCLOHEXANE IN THE IDEAL GASEOUS STATE, CAL./DEG. MOLE

Temp., °K.	C^0_p	$\frac{H^0 - H^0_0}{T}$	$-\frac{F^0 - H^0_0}{T}$	S^0
298.16	25.40	14.21	57.07	71.28
300	25.58	14.28	57.16	71.44
400	35.82	18.38	61.80	80.18
500	45.47	22.85	66.39	89.24
600	53.83	27.34	70.96	98.30
700	60.87	31.64	75.50	107.14
800	66.76	35.67	79.98	115.65
900	71.68	39.40	84.40	123.80
1000	75.80	42.85	88.74	131.59
1100	79.3	46.0	93.0	139.0
1200	82.2	48.9	97.1	146.0
1300	84.7	51.6	101.1	152.7
1400	86.8	54.0	105.0	159.0
1500	88.6	56.3	108.8	165.1

TABLE V

HEAT OF FORMATION AND OF DEHYDROGENATION OF CYCLOHEXANE AT 0°K.

(Dehydrogenation to benzene, all for ideal gas state in kcal./mole).

Ref.	Method	$-\Delta H^0_0$ formation	ΔH^0_0 dehydrogenation
23	Combustion, calorimetric	20.09 ± 0.2	44.09 ± 0.3
24	Combustion, calorimetric	20.02 ± .2	44.02 ± .3
25	Combustion, calorimetric	20.36 ± .4	44.36 ± .5
26	Hydrogenation, calorimetric	19.77 ± .2	43.77 ± .1
27	Hydrogenation, equilib.	20.3 ± .5	44.3 ± .4
28	Hydrogenation, equilib.	19.67 ± .3	43.67 ± .2

Alkyl Cyclohexanes

Tautomerism and Symmetry Effects.—As Hassel[7] has pointed out, the hydrogen atoms in cyclohexane may be divided into two types which

(17) American Petroleum Institute Research Project 44, "Selected Values of Properties of Hydrocarbons," Table 7g.

(18) R. A. Ruehrwein and H. M. Huffman, THIS JOURNAL, **65**, 1620 (1943).

(19) J. G. Aston, G. J. Szasz and H. L. Fink, *ibid.*, **65**, 1135 (1943).

(20) J. B. Montgomery and T. DeVries, *ibid.*, **64**, 2375 (1942).

(21) F. G. Brickwedde, M. Moskow and J. G. Aston, *J. Research Natl. Bur. Standards*, **37**, 263 (1946).

(22) W. J. Taylor, D. D. Wagman, M. G. Williams, K. S. Pitzer and F. D. Rossini, *ibid.*, **37**, 95 (1946).

(23) Recalculated from G. E. Moore, M. L. Renquist and G. S. Parks, THIS JOURNAL, **62**, 1505 (1940).

(24) E. J. Prosen, W. H. Johnson and F. D. Rossini, *J. Research, Natl. Bur. Standards*, **37**, 51 (1946).

(25) R. Spitzer and H. M. Huffman, THIS JOURNAL, **69**, 211 (1947).

(26) G. B. Kistiakowsky, J. R. Ruhoff, H. A. Smith and W. E. Vaughan, *ibid.*, **58**, 146 (1936).

(27) G. H. Burrows and C. Lucarini, *ibid.* ,**49**, 1157 (1927).

(28) V. R. Zharkova and A. V. Frost, *J. Gen. Chem., U. S. S. R.*, **2**, 534 (1932).

Fig. 1.—Cyclohexane (left), methylcyclohexane, equatorial (middle) and polar (right).

we shall call polar (p) and equatorial (e) (Hassel's ϵ and κ, respectively). If the carbon ring (in the chair configuration) be horizontal there are three polar hydrogens markedly above and three below the ring while six hydrogen atoms lie in an "equatorial" belt around the carbons. Of the two hydrogen atoms on each carbon, one is polar and one equatorial. If the six carbon atoms pass through a single plane over to the opposite chair configuration all hydrogens originally equatorial become polar and vice versa. So long as all are identical hydrogen atoms, nothing new is obtained, but with another group attached, tautomers become possible. Methylcyclohexane may have its methyl group in either a polar or an equatorial position. By studying a model of this molecule it is found that steric interactions arise which are similar to those of *n*-paraffins in their various configurations. One of us[29] has given a value of 0.8 kcal./mole for the strain energy, *a*, in *n*-butane twisted 120° about the central carbon–carbon bond as compared to the stable, planar configuration of that molecule. Polar methylcyclohexane has two similar interactions whereas the equatorial form has no strain of this type. Therefore, the energy of the former is greater than that of the latter by approximately twice the *n*-butane value. The assigned value, 1.8 kcal./mole, or $a = 0.9$ kcal./mole (together with the frequency assignment and potential barrier for the methyl rotation discussed below) appears to give the best agreement between calculated and observed entropies. However, other values from 1.6 to 2.4 kcal./mole will give agreement within experimental error.

Of the dimethylcyclohexanes, (1,1), (*cis*-1,2), (*trans*-1,3) and (*cis*-1,4) have one polar and one equatorial methyl and hence have only one tautomeric form of the type discussed above. The second and third, however, do have *d* and *l* configurations, which accounts for their double probability, since *d* and *l* forms have the same energy. In *cis*-1,2 the enantiomorphs are in tautomeric equilibrium while in *trans*-1,3 they are true isomers interconvertible only by inversion of two carbon atoms. The other dimethylcyclohexanes, (*trans*-1,2), (*cis*-1,3) and (*trans*-1,4) have two tautomers, one with both methyl groups polar and one with both methyl groups equatorial. The energy of the latter is lower in each case.

In the *trans*-1,4 compound the energy difference is clearly twice that in methylcyclohexane, 4*a* or 3.6 kcal./mole. In the *trans*-1,2 derivative there is one *n*-butane-like interaction in the equatorial form and four in the polar form giving an energy difference of 3*a* or 2.7 kcal./mole. For *cis*-1,3-dimethylcyclohexane the polar form encounters much worse steric interference because both methyl groups are on the same side of the plane. We have arbitrarily assigned an energy difference of 6*a* or 5.4 kcal./mole to this case. This value is so high that the polar form could be ignored at the lower temperatures.

The *trans*-1,2-dimethylcyclohexane has, further, *d* and *l* isomers with a symmetry number of 2 in each case so that no net change in the relative probability of this substance arises. It will be noted that in agreement with the simple theory, only *trans*-1,2 and *trans*-1,3 compounds can exist as long lived *d* or *l* isomers.

A previous publication[6a] has discussed these tautomers briefly and shown that, contrary to earlier suggestions, *trans*-1,3-dimethylcyclohexane is probably the higher boiling isomer.[29a]

In addition to the polar-equatorial tautomerism, the chair-boat tautomerism still exists in the

(29) K. S. Pitzer, *J. Chem. Phys.*, **8**, 711 (1940).

(29a) Added in proof: We have belatedly come upon the work of Mousseron and Granger (*Bull. soc. chim.*, **5**, 1618 (1938)) who prepared an optically active and therefore *trans*-1,3-dimethylcyclohexane. They found it to be the higher boiling isomer in agreement with our conclusion.

TABLE VI

TAUTOMERISM IN DIMETHYLCYCLOHEXANES

ΔS and ΔH are entropies and heats of isomerization from *cis* 1,4 for the gas state at 298.16°K. in cal./degree mole and kcal./mole, respectively.

Compd.	Config.	Tautomers description	Energy	Sum. no.	ΔS Calcd.	ΔS Obs.	ΔH Calcd.	ΔH Obs.
cis-1,2	e,p	Enantiomorphs	3*a*	1	0.97	0.83[a]	0.9	1.13[a]
	p,e							
trans-1,2	e,e	*d,l* isomers of each	a	2	.11	.09	− .9	−0.83
	p,p		4*a*	2				
cis-1,3	e,e		0	1	.00	.06[b]	−1.8	−1.94[b]
	p,p		6*a*	1				
trans-1,3	e,p	Tautomers identical *d,l* isomers	2*a*	1	1.38	1.10[c]	0.0	0.02[c]
	p,e							
cis-1,4	e,p	Identical	2*a*	1	(0.0)	(0.0)	(.0)	(.0)
	p,e							
trans-1,4	e,e		0	2	−1.35	−1.58	−1.8	−1.91
	p,p		4*a*	2				

[a] Estimated uncertainty: 0.2 in ΔS and 0.3 in ΔH for all compounds. [b] Isomer boiling at 120.1° formerly called *trans*. [c] Isomer boiling at 124.5° formerly called *cis*.

derivatives. Since the latter is less important, because of the greater energy difference, we have not attempted to study it in detail for each derivative. Rather we have assumed that the chair–boat tautomerism contributes the same amount to various properties of the derivatives as in cyclohexane itself.

Table VI summarizes various parameters arising from the tautomerism and symmetry of the dimethyl derivative. The entropies and heats of isomerism are given with *cis*-1,4 as a standard because that substance is free from tautomeric complications. The entropies are essentially those of symmetry and of mixing of tautomers, but they will be discussed in the next section. The calculated heats are merely the differences in energy, taking the lowest energy tautomer and giving *a* the value 0.9 kcal./mole mentioned above. (The use of the exact heat contents computed later would make a negligible difference.) The experimental heats of isomerization from Johnson, Prosen and Rossini[30] were received after all other calculations were complete. They agree remarkably well. While an *a* value of 0.97 kcal./mole would fit slightly better it did not appear worthwhile to repeat the calculations for this small change.

Indeed these heat of isomerization data afford a satisfying confirmation of the approximate value 0.8 kcal. originally assigned this *n*-butane interaction in the early and approximate work on the *n*-paraffins.[15]

In the earlier publication[6a] the present assignment of *cis* and *trans* configurations of 1,3-dimethylcyclohexanes was recommended on the basis of the entropy values and for other reasons given there. The excellent agreement of ΔH values in Table VI offers strong confirmation of that conclusion.

(30) W. H. Johnson, E. J. Prosen and F. D. Rossini, unpublished data.

Thermodynamic Properties.—There were available experimental values, from the Third Law, of the entropy of methyl, ethyl and several dimethylcyclohexanes at 298.16°K. The values for the liquid state are from the work of Huffman and collaborators,[31] while the data for the entropy of vaporization are from various measurements at the National Bureau of Standards.[32] These experimental entropies for the ideal gas at 298.16°K. are listed in Table VII. We have also experimental values of the heat capacity of gaseous methylcyclohexane[6] which are shown in Table VIII.

TABLE VII

ENTROPIES OF THE ALKYLCYCLOHEXANES THROUGH C_8 AT 298.16° K.

Units: cal. per degree mole

	$S_{liq.}$[31]	$\Delta S_{vap.}$[32]	S^0_{gas} exptl.	S^0_{gas} calcd.
Methylcyclohexane	59.25	22.80	82.05	82.06
Ethylcyclohexane	67.14	24.32	91.46	91.44
1,1-Dimethylcyclohexane	63.88	23.35	87.23	87.24
cis-1,2-	65.53	23.95	89.48	89.51
trans-1,2-	65.30	23.45	88.75	88.65
cis-1,3-	65.16	23.56	88.72	88.54
trans-1,3-	65.82	23.93	89.75	89.92
cis-1,4-	64.83	23.84	88.67	88.54
trans-1,4-	63.70	23.39	87.09	87.19

TABLE VIII

THE HEAT CAPACITY OF METHYLCYCLOHEXANE

Values for ideal gas state in cal./deg. mole

T, °K.	C_p exptl.[6]	C_p calcd.
398	44.1 ± 0.4	44.1
439	48.6 ± .5	48.6
480	53.2 ± .5	53.0
527	57.8 ± .6	57.7

(31) H. M. Huffman, private communication and H. M. Huffman, THIS JOURNAL, **68**, 173 (1946).

(32) F. D. Rossini, private communication.

A study of the infrared and Raman spectra of these molecules indicated that it was not feasible to make detailed vibrational assignments from spectral data alone. Consequently we have assumed that each methyl group contributes the same increment to the sum of translational, rotational and vibrational entropy (or other function). The internal rotational contribution was taken to be that for a reduced moment of inertia of 5.3×10^{-40} g. cm.^{2} and for a 3.6 kcal./mole potential barrier to methyl rotation in all cases except 1,1 and *cis*-1,2-dimethylcyclohexanes where the two methyl groups interfere with one another. In these two cases the barriers were adjusted to fit the experimental entropies yielding 6.6 and 4.3 kcal./mole, respectively. The 3.6 kcal./mole value is rounded from that for isobutane[33] which seems the best analogy.

TABLE IX

VIBRATION FREQUENCIES OF METHYLCYCLOHEXANE

Units: cm^{-1}; values in parentheses indicate multiple weights. 153, 220, 307, 335, 403, 445, 520, 543, 768, 784, 843(2), 872, 910(2), 969, 1032(2), 1059, 1090, 1109, 1164, 1200, 1247(2), 1263(2), 1305(3), 1345(3), 1363, 1376, 1450(7), 2950(14).

The anharmonicity contribution in cyclohexane was assumed to apply here.

ing full use of analogies, to similar molecules. However, certain features were still uncertain and these were arbitrarily adjusted to give agreement with the various experimental, thermodynamic data mentioned above. Thus we may regard this vibrational assignment, listed in Table IX, as essentially an empirical representation of the thermodynamic data, but with the great advantage that it must extrapolate to approximately correct values at higher temperatures because the correct number of degrees of freedom are included and because all high frequencies are reasonably certain from the known regularities between similar molecules. The product of moments of inertia was calculated from the same distances and angles mentioned above to be $39{,}294 \times 10^{-117}$ g.^{3} cm.^{6}.

Table VI summarizes various parameters in the calculations of the entropies of the methyl derivatives and Tables VII and VIII compare the calculated entropy and heat capacity of gaseous methylcyclohexane with the experimental data. It should be noted that the entropies of methyl, *trans*-1,2, *cis* and *trans*-1,3 and *cis* and *trans*-1,4-dimethylcyclohexanes, six in all, have been fitted with a single adjusted quantity, the entropy increment per methyl group, together with the value of a, which is also known from other sources. This serves to confirm our general picture of the tauto-

TABLE X

HEAT CAPACITIES OF METHYL CYCLOHEXANE, ETHYL CYCLOHEXANE AND THE SEVEN DIMETHYL CYCLOHEXANES IN THE IDEAL GASEOUS STATE

	C_p^0, cal./deg. mole								
			Dimethylcyclohexanes						
Temp., °K.	Methyl-cyclo-hexane	Ethyl-cyclo-hexane	1,1	*cis*-1,2	*trans*-1,2	*cis*-1,3	*trans*-1,3	*cis*-1,4	*trans*-1,4
298.16	32.27	37.9	36.9	37.4	38.0	37.6	37.6	37.6	37.7
300	32.51	38.2	37.2	37.7	38.3	37.9	37.9	37.9	38.0
400	44.35	51.6	50.7	51.1	51.9	51.2	51.1	51.1	51.6
500	55.21	63.8	63.3	63.5	64.2	63.6	63.4	63.4	64.0
000	04.40	74.1	74.1	74.0	71.0	71.0	73.8	73.8	74.6
700	72.23	82.8	83.2	82.8	83.3	83.1	82.5	82.5	83.3
800	78.74	90.1	90.7	90.1	90.5	90.5	89.8	89.8	90.6
900	84.20	96.2	97.0	96.3	96.6	96.7	95.9	95.9	96.8
1000	88.79	101.3	102.2	101.4	101.7	102.0	101.1	101.1	101.9
1100	92.7	105.7	106.6	105.8	106.0	106.4	105.5	105.5	106.2
1200	96.0	109.4	110.3	109.5	109.7	110.1	109.2	109.2	109.9
1300	98.8	112.5	113.5	112.7	112.8	113.3	112.4	112.4	113.0
1400	101.2	115.3	116.2	115.4	115.5	116.0	115.2	115.2	115.7
1500	103.2	117.6	118.4	117.7	117.8	118.3	117.5	117.5	118.0

This methyl increment was obtained by subtracting values for cyclohexane from those for methylcyclohexane without the polar-equatorial tautomerism. The latter were based on the best possible assignment of vibration frequencies for methylcyclohexane from the spectral data[34] mak-

(33) K. S. Pitzer and J. E. Kilpatrick, *Chem. Rev.*, **39**, 435 (1946).

(34) Raman effect: H. Okazu, *J. Chem. Soc. Japan*, **60**, 559 (1939); Kohlrausch and Stockmair, *Z. physik. Chem.*, **B31**, 382 (1936); J. Goubeau and E. Köhler, *Ber.*, **75A**, 65 (1942); Infrared: American Petroleum Institute Research Project 44, "Catalog of Infrared Spectrograms," Serial No. 258 and 369 contributed by the Universal Oil Products Co. and the Radiometry Section of the National Bureau of Standards, respectively.

meric equilibria and the assumption of a single increment for the methyl group. However, certain details within the methyl increment could be changed considerably without affecting its total value.

The more detailed calculations and experimental data for the methyl derivatives of benzene[35] indicate that the methyl increment method used here should be fairly satisfactory also for the gaseous heat capacity. However, there are no experimental values for the dimethylcyclohexanes to verify this.

(35) K. S. Pitzer and D. W. Scott, THIS JOURNAL, **65**, 803 (1943).

TABLE XI

HEAT CONTENT FUNCTIONS OF METHYL CYCLOHEXANE, ETHYL CYCLOHEXANE AND THE SEVEN DIMETHYL CYCLOHEXANES IN THE IDEAL GASEOUS STATE

Temp., °K.	$(H^0 - H_0^0)/T$, cal./deg. mole: Methyl-cyclo-hexane	Ethyl-cyclo-hexane	Dimethylcyclohexanes: 1,1	*cis*-1,2	*trans*-1,2	*cis*-1,3	*trans*-1,3	*cis*-1,4	*trans*-1,4
298.16	17.55	20.45	19.72	20.16	20.44	20.35	20.35	20.35	20.38
300	17.66	20.58	19.85	20.29	20.58	20.48	20.48	20.48	20.51
400	22.84	26.7	25.9	26.3	26.7	26.5	26.5	26.5	26.6
500	28.26	32.9	32.1	32.5	33.0	32.7	32.7	32.7	32.8
600	33.53	38.9	38.2	38.6	39.0	38.8	38.7	38.7	38.9
700	38.53	44.6	44.0	44.3	44.7	44.5	44.3	44.3	44.7
800	43.16	49.9	49.4	49.6	50.0	49.8	49.6	49.6	50.0
900	47.42	54.7	54.4	54.4	54.9	54.7	54.4	54.4	54.8
1000	51.33	59.1	58.9	58.9	59.3	59.1	58.8	58.8	59.3
1100	54.9	63.1	63.0	63.0	63.4	63.2	62.9	62.9	63.4
1200	58.2	66.8	66.8	66.7	67.1	67.0	66.6	66.6	67.1
1300	61.2	70.2	70.3	70.1	70.5	70.5	70.0	70.0	70.5
1400	64.0	73.3	73.5	73.3	73.6	73.6	73.1	73.1	73.6
1500	66.5	76.2	76.4	76.1	76.5	76.5	76.0	76.0	76.5

TABLE XII

ENTROPIES OF METHYL CYCLOHEXANE, ETHYL CYCLOHEXANE, AND THE SEVEN DIMETHYL CYCLOHEXANES IN THE IDEAL GASEOUS STATE

Temp., °K.	S^0, cal./deg. mole: Methyl-cyclo-hexane	Ethyl-cyclo-hexane	Dimethylcyclohexane: 1,1	*cis*-1,2	*trans*-1,2	*cis*-1,3	*trans*-1,3	*cis*-1,4	*trans*-1,4
298.16	82.06	91.44	87.24	89.51	88.65	88.54	89.92	88.54	87.19
300	82.28	91.70	87.52	89.79	88.94	88.82	90.20	88.82	87.47
400	93.22	104.5	100.0	102.4	101.7	101.5	102.8	101.5	100.2
500	104.32	117.3	112.7	115.2	114.7	114.3	115.6	114.2	113.1
600	115.21	129.9	125.2	127.7	127.3	126.8	128.1	126.7	125.7
700	125.77	142.0	137.4	139.8	139.5	139.0	140.2	138.8	137.9
800	135.86	153.6	149.0	151.4	151.1	150.6	151.7	150.3	149.5
900	145.46	164.5	160.1	162.4	162.2	161.6	162.6	161.2	160.6
1000	154.57	174.9	170.6	172.8	172.6	172.1	173.0	171.6	171.1
1100	163.2	184.8	180.5	182.7	182.5	182.0	182.9	181.5	181.0
1200	171.4	194.2	189.9	192.0	191.9	191.4	192.2	190.8	190.3
1300	179.2	203.0	198.8	200.8	200.8	200.3	201.0	199.6	199.2
1400	186.6	211.4	207.3	209.3	209.2	208.8	209.4	208.0	207.7
1500	193.7	219.5	215.5	217.4	217.3	216.9	217.5	216.1	215.8

TABLE XIII

FREE ENERGY FUNCTIONS OF METHYL CYCLOHEXANE, ETHYL CYCLOHEXANE, AND THE SEVEN DIMETHYL CYCLOHEXANES IN THE IDEAL GASEOUS STATE

Temp., °K.	$-(F^0 - H_0^0)/T$, cal./deg. mole: Methyl-cyclo-hexane	Ethyl-cyclo-hexane	Dimethylcyclohexanes: 1,1	*cis*-1,2	*trans*-1,2	*cis*-1,3	*trans*-1,3	*cis*-1,4	*trans*-1,4
298.16	64.51	70.99	67.52	69.35	68.21	68.19	69.57	68.19	66.81
300	64.62	71.12	67.67	69.49	68.36	68.34	69.72	68.34	66.96
400	70.38	77.8	74.1	76.1	75.1	75.0	76.4	75.0	73.6
500	76.06	84.5	80.6	82.6	81.7	81.6	83.0	81.6	80.2
600	81.68	91.0	87.0	89.1	88.3	88.1	89.4	88.1	86.7
700	87.24	97.4	93.3	95.5	94.7	94.5	95.8	94.5	93.2
800	92.70	103.7	99.6	101.8	101.1	100.8	102.1	100.8	99.6
900	98.04	109.9	105.7	107.9	107.3	107.0	108.3	106.9	105.7
1000	103.24	115.9	111.7	113.9	113.3	113.0	114.2	112.8	111.8
1100	108.3	121.7	117.5	119.7	119.1	118.8	120.0	118.6	117.6
1200	113.2	127.3	123.1	125.3	124.8	124.4	125.6	124.2	123.2
1300	118.0	132.8	128.5	130.7	130.2	129.9	131.0	129.6	128.7
1400	122.6	138.1	133.8	136.0	135.6	135.2	136.3	134.9	134.0
1500	127.1	143.3	139.1	141.3	140.8	140.4	141.5	140.1	139.3

The change from methyl to ethylcyclohexane is essentially similar to that from isobutane to isopentane, except for the symmetry number of three for isobutane. Consequently we have calculated the various thermodynamic functions for gaseous ethylcyclohexane from the equation

$$F(\text{ethylcyclohexane}) = F(\text{methylcyclohexane}) + F(\text{isopentane}) - F(\text{isobutane}) - R \ln 3$$

where F means C, S, $(H - H_0^0)/T$ or $-(F - H_0^0)/T$ and the $R \ln 3$ applies only to S and $-(F-H_0)/T$. This equation gives the value in Table VII.

By similar analogies one may offer equations for n-propyl and isopropylcyclohexane which may be of interest for making estimates.

$$F(n\text{-propylcyclohexane}) = F(\text{methylcyclohexane}) + F(\text{2-methylpentane}) - F(\text{isobutane}) - R \ln 3$$

$$F(\text{isopropylcyclohexane}) = F(\text{methylcyclohexane}) + F(\text{2,3-dimethylbutane}) - F(\text{isobutane}) - R \ln (3/2)$$

Again the $R \ln 3$ and $R \ln (3/2)$ terms apply only to the functions S and $-(F-H_0)/T$.

The values of several thermodynamic functions calculated by the various methods discussed above for a range of temperatures are listed in Tables X–XIII. Where values are given to the second decimal place, the error should not exceed 0.4, otherwise the error may be larger up to 1.0 or even more at the high temperatures.

These data of Tables X to XIII will be combined with the heat of formation values now being obtained by Rossini and collaborators in a later publication.

We wish to thank Dr. F. D. Rossini and Dr. H. M. Huffman for making available to us various data in advance of publication.

Summary

The data for cyclohexane from electron diffraction, Raman and infra-red spectra, and thermodynamic measurements were analyzed in various ways to give an assignment of vibration frequencies and calculated values of the heat capacity, entropy and heat and free energy functions for cyclohexane in the ideal gas state from 298.16 to 1500°K. It is concluded in agreement with others that at room temperature cyclohexane is predominantly in the chair form of symmetry D_{3d}. However, the heat capacity of gaseous cyclohexane definitely shows a contribution from conversion to the boat form which is approximately 5.6 kcal./mole higher in energy.

The possibilities of geometrical tautomerism in the methyl and dimethylcyclohexanes, mentioned in earlier publications, are discussed in greater detail. It is shown that the energy differences between these tautomers are related to the strain or steric energies in n-paraffins, and that there is good agreement between the corresponding energy quantity obtained from data on each series of compounds. The heats of isomerization confirm the assignment of the configuration *cis*-1,3-dimethylcyclohexane to the isomer boiling at 120.1° and *trans*-1,3 to the isomer boiling at 124.5°. The thermodynamic functions mentioned above were calculated also for methyl, ethyl and the seven dimethyl-cyclohexanes in the ideal gas state from 298.16 to 1500°K.

BERKELEY, CALIF. RECEIVED MARCH 12, 1947

Reprinted from SCIENCE, January 1, 1954, Vol. 119, No. 3079, page 49.

Nomenclature of Cyclohexane Bonds

IT was shown originally by x-ray and electron diffraction, and has been confirmed by other physical and by chemical means, that the most stable and permanent form of the cyclohexane ring is that particular strainless form which is sometimes likened to a chair or a staircase. Geometrically, its chief feature is a sixfold alternating axis of symmetry. Its 12 extracyclic bonds fall into two classes (*1*): 6 lie parallel to the axis, while 6 extend radially outward at angles of $\pm 109.5°$ to the axis. The stereochemical properties of substituents bound by these two classes of bond are so different that a need has been felt for verbal and symbolic means of distinguishing the classes.

The first suggestion (*2*) to this end designated the six parallel bonds as 'ε' and the others as 'κ'. However, although these symbols have been considerably used, their origins in the Greek language, and particularly their allocation between two classes of bond, have been found difficult to remember. A more obvious description was given (*3*) when the six parallel bonds were called "polar" and the others "equatorial" in analogy to the geographical terms. This nomenclature has had a wide use, but is unsatisfactory in that it employs the word "polar" for a stereochemical concept, thereby tending to confuse discussions which have to take account of the electropolar nature, side by side with the stereochemical character of cyclohexane substituents.

The purpose of this note is to suggest a simple change which avoids these difficulties. It is that the six bonds parallel to the main cyclohexane axis should be called "axial"[1] (and symbolized '*a*'), while the other six retain the name "equatorial" (and become symbolized '*e*'). The private discussions we have had with friends and colleagues lead us to hope that this suggestion may find favor.

D. H. R. BARTON
O. HASSEL
K. S. PITZER
V. PRELOG

Birkbeck College, London, England; Universitetes Kjemske Institutt, Oslo, Norway; University of California, Berkeley; and *Organischchemisches Laboratorium der Eidg. Technischen Hochschulen, Zurich, Switzerland*

References

1. KOHLRAUSCH, K. W. F., REITZ, A. W., and STOCKMAIR, W. *Z. physik. Chem.* **B32**, 229 (1936).
2. HASSEL, O. *Tidsskr. Kjemi, Bergvesen Met.* **3**, 32 (1943).
3. BECKETT, C. W., PITZER, K. S., and SPITZER, R. *J. Am. Chem. Soc.* **69**, 2488 (1947).

[1] The term "axial" was suggested to us by Professor C. K. Ingold, London.

Received December 7, 1953.

[Reprinted from the Journal of the American Chemical Society, **75**, 5634 (1953).]

[Contribution from the Department of Chemistry, Columbia University and the Department of Chemistry and Chemical Engineering, University of California]

Infrared Absorption Spectra, Structure and Thermodynamic Properties of Cyclobutane[1]

By G. W. Rathjens, Jr., N. K. Freeman, William D. Gwinn and Kenneth S. Pitzer

Received August 3, 1953

Infrared and Raman measurements on cyclobutane are reported. The spectroscopic data, when used in conjunction with electron diffraction measurements and the measured entropy, suggest that the equilibrium configuration for cyclobutane is one with D_{2d} symmetry, but that the barrier hindering inversion of the molecule is sufficiently low so that even at ordinary temperatures an appreciable number of the molecules obey D_{4h} selection rules. A normal coördinate analysis for cyclobutane has been performed, and force constants compared with those of related molecules. A possible potential function for the out-of-plane bending motion of the ring has been suggested, and some of the thermodynamic properties of cyclobutane calculated.

Introduction

Cyclobutane and its derivatives have been the subjects of a number of recent investigations in molecular structure.[2-5] Strained CCC bonds in the ring suggest that the equilibrium configuration of the ring should be planar; on the other hand, torsional forces about the C–C bonds tend to cause a non-planar configuration. Previously, spectroscopic results have been interpreted as most consistent with a planar configuration for the skeleton (D_{4h} symmetry).[2] However, the entropy determination reported in the accompanying paper is in disagreement with the previous analyses. The investigations here reported were undertaken in an effort to supply further information about cyclobutane.

Apparatus and Sample.—Some of the spectroscopic observations were made by (N.K.F.) at the University of California using a Perkin–Elmer spectrometer that was converted from Model 12A to 12C during the course of the work. Potassium bromide, sodium chloride and lithium fluoride optics were used. Most of these measurements were repeated at Columbia University (by G. W. R.) using a Savitsky–Halford modified Perkin–Elmer 12 C spectrometer.[6] In addition, the spectrum of the liquid and the temperature dependence of certain vapor phase intensities were obtained. Sodium chloride and calcium fluoride optics were used. The published data represent what we regard as the best combination of the two sets of results.

The Raman measurements were made by Dr. J. T. Neu to whom we are indebted for his unpublished results. His measurements were made on a liquid sample using a Steinheil type GH three prism spectrograph.

The sample used was the same as reported in the accompanying article.[5]

Infrared Spectrum.—The infrared spectrum at room temperature is shown in Fig. 1. The optics used were as indicated in the figure, except for the liquid spectrum. In that case, sodium chloride optics were used below and calcium fluoride optics above 1400 cm.$^{-1}$.

With increasing temperature certain of the bands diminished in both peak and integrated intensity. The most noticeable case is the 750 cm.$^{-1}$ band. We have estimated $\int \ln(I/I_0)d\nu$ at 459°K. to be 73 ± 5% of the value at 299°K. Otherwise, with the exception of the 1889 cm.$^{-1}$ combination band, it has generally not been possible to estimate quantitatively changes in intensity, and in some cases there is a question as to whether there actually was a change. Bands whose intensities were observed to decrease on heating are marked in Table II. In addition to the 750 cm.$^{-1}$ band, two other bands assigned as fundamentals, 1453 and 1261 cm.$^{-1}$, exhibited changes in structure on heating. These cases are discussed in the section on interpretation.

Comparison of the liquid spectrum with that of the gas has been of aid in making certain assignments; the single peak in the liquid at 750 cm.$^{-1}$ seems to indicate that the two observed peaks in the gas are part of the same band, and that the expected third branch does not appear; the pronounced shoulder on the 1261 cm.$^{-1}$ band seems to indicate definitely two assignments in that region; and the marked change in structure in the 2100 cm.$^{-1}$ region indicates that the very complex gas phase spectrum is in part probably due to rotational structure of the various combinations that occur in the region. In the assignment of combinations (Table II) this information has been used.

With minor exceptions, the spectra obtained by us are in agreement with that reported by Wilson.[2] Because of the accessibility to us of lithium fluoride optics, our spectrum shows better resolution in the higher frequency region, and it is in this region where our spectrum differs most significantly from that published by Wilson. We have also observed two weak bands in the gas phase, but not in the liquid, at 955 and 850 cm.$^{-1}$ which were not observed by Wilson. (These may be due to impurity.)

Raman Spectra.—Table II gives the Raman lines as observed by Neu, their estimated intensities, states of polarization and assignments. A comparison with Wilson's[2] data and that of Pajenkamp[3] shows certain differences. Wilson observed doubling of the lines at 1444, 1220 and 928 cm.$^{-1}$ while neither Pajenkamp's nor Neu's data show such doubling. The resolution available in the spectrum reported here seems to indicate that the doubling is not genuine. Three lines, 1148, 1515 and 634 cm.$^{-1}$ not reported by Wilson were observed by Neu.

Symmetry Consideration and Selection Rules.—As previously indicated, the planarity or nonplanarity of the carbon skeleton in cyclobutane is dependent on the balance of the forces tending to cause the two configurations. The most likely

(1) This work was assisted by the American Petroleum Institute through Research Project 50.

(2) T. P. Wilson, *J. Chem. Phys.*, **11**, 369 (1943).

(3) H. Pajenkamp, *Z. Elektrochem.*, **52**, 104 (1948).

(4) J. D. Dunitz and V. Schomaker, *J. Chem. Phys.*, **20**, 1703 (1952).

(5) G. W. Rathjens and W. D. Gwinn, This Journal, **75**, 5629 (1953). Other references are also given in this article.

(6) A. Savitsky and R. S. Halford, *Rev. Sci. Instruments*, **21**, 203 (1950).

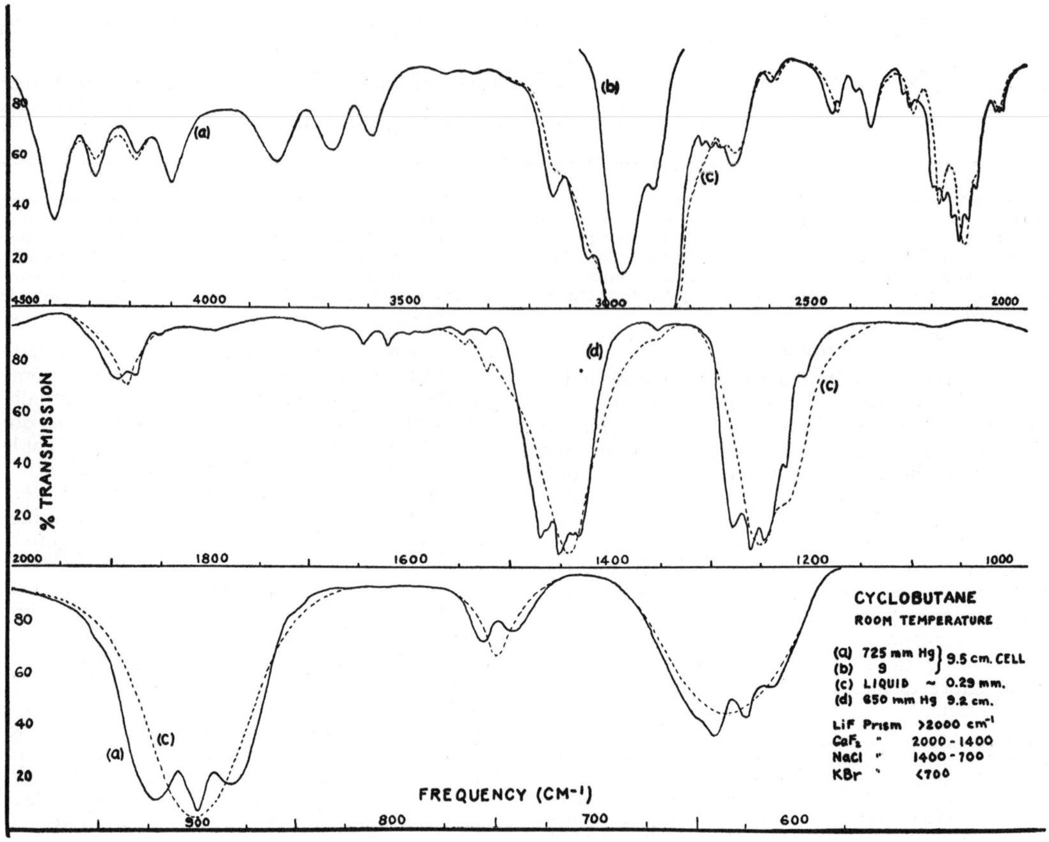

Fig. 1.—Infrared spectra of cyclobutane.

symmetries in the two cases are, respectively, D_{4h} and D_{2d}. Table I is a summary of the vibrations belonging to each class in the two symmetries, the selection rules, and the behavior of each class as the molecule goes from one symmetry to the other. From the table, it can be seen that B_{1u}, A_{1u}, A_{2u} and E_u vibrations become Raman active as the symmetry goes from D_{4h} to D_{2d}, and B_{2g} and E_g vibrations become infrared active as the symmetry goes to D_{2d}. It also can be seen that there are 23 fundamental vibrations and that for D_{4h} symmetry eleven fundamentals should be Raman active, and six infrared active with no coincidences. There are also six inactive fundamentals for this model. For D_{2d} symmetry nine vibrations are Raman active only, and twelve vibrations are active in both Raman and infrared. Two fundamentals remain inactive. Since the number of strong infrared bands and Raman bands actually observed is nearer the number predicted for D_{4h}, this is a factor in leading one tentatively to select D_{4h} symmetry. Wilson[2] interpreted his data on this model.

As a result of the discrepancy between the entropy measurement[5] and the value calculated from spectroscopic data with the assumption of a planar carbon skeleton, it has become necessary to consider more carefully the possibility of a non-planar skeleton, and to try to interpret the spectra in terms of D_{2d} symmetry. If this model is assumed, not all of the allowed bands or coincidences between Raman and infrared are observed. This may be due to intrinsic weakness of the bands, especially inasmuch as the deviation from skeletal planarity is expected to be small. There are, however, three observed coincidences (630, 750 and 1220 cm.$^{-1}$) in the low frequency range, where accidental coincidences are less likely. Since these are observed in the liquid state, and might otherwise be interpreted as violations of (D_{4h}) selection rules, conclusions therefrom regarding symmetry are equivocal. Nevertheless, it seems unlikely that all three coincidences would arise from a planar molecule. The following section will clarify the point of view from which neither symmetry is strictly correct, and from which *both* must be considered.

Interpretation

From a thermodynamic point of view the most serious problem in the interpretation of the cyclobutane spectra has been in the assignment of the B_{1u} out-of-plane bending vibration (assuming for the time being D_{4h} symmetry) since this vibration is expected to be of very low frequency and extremely anharmonic. Wilson assigned a value of 145 cm.$^{-1}$ to it, and Cottrell[7] has used this assign-

(7) T. L. Cottrell, *Trans. Faraday Soc.*, **44**, 716 (1948).

TABLE I

FUNDAMENTAL VIBRATIONS,[a] SYMMETRY CLASSES AND SELECTION RULES

Vibration type	Cm.$^{-1}$	Symmetry class and selection rules D_{4h}	D_{2d}
CH stretching	2870		
CH_2 deformation	1444*	A_{1g} R.(p.)	
CC stretching	1003*		A_1 R.(p.)
CH stretching	2921		
CH_2 rocking	(878)	B_{1u} inactive	
Ring bend. (out-of-plane for D_{4h})	[b]		
CH_2 wagging	(1299)	A_{2g} inactive	A_2 inactive
CH_2 twisting	1104 (1155)	B_{2u} inactive	
CH_2 wagging	(1286)		
CC stretching	928*	B_{1g} R.(dp.)	B_1 R.(dp.)
CH_2 twisting	1148 (1155)	A_{1u} inactive	
CH stretching	2981		
CH_2 deformation	1450*	B_{2g} R.(dp.)	
Ring bend. (in-plane for D_{4h})	750*		B_2 IR., R.(dp.)
CH stretching	2896		
CH_2 rocking	901*	A_{2u} IR.	
CH stretching	2960		
CH_2 twisting	1220*	E_g R.(dp.)	
CH_2 rocking	890[c]*		
CH stretching	2974		E IR., R.(dp.)
CH_2 deformation	1453 (1420)	E_u IR	
CH_2 wagging	1261*		
Ring deformation (stretch.-bend.)	625*		

[a] Frequencies used to evaluate force constants marked with an asterisk; calculated values in parentheses. [b] See text. [c] Assigned on the basis of combination bands.

ment to calculate thermodynamic properties for both harmonic and quartic potential functions. The measured entropy reported in the accompanying article is considerably greater than either of the values calculated. This suggests either that the frequency is even lower than 145 cm.$^{-1}$, or that the assumed shapes for potential functions are grossly in error, or both.

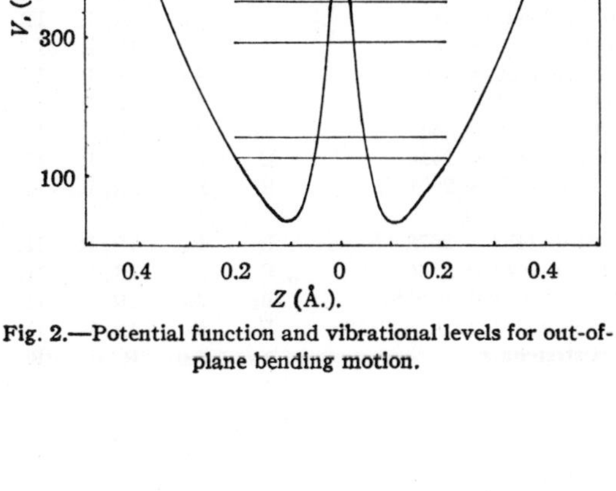

Fig. 2.—Potential function and vibrational levels for out-of-plane bending motion.

Our temperature dependence measurements were initiated with the hope that increase in intensities on heating would locate definitely the difference bands involving the B_{1u} vibration. No such increases have been observed, and we have been forced to conclude that no such differences are sufficiently intense to appear in our spectrum. (Possible exceptions are the 650 and 1355 cm.$^{-1}$ bands). However, a number of diminutions in intensity have been observed. Such behavior is consistent with a model in which the potential energy function for the puckering of the ring is of the form illustrated in Fig. 2. If the potential is of the illustrated shape and the barrier hindering inversion is sufficiently low, an appreciable fraction of the molecules will be in states whose energies lie above the barrier peak. These molecules and those slightly below the peak have effectively D_{4h} symmetry while molecules whose energies lie well below the barrier peak will have D_{2d} symmetry. On heating, the former population will be increased at the expense of the latter, and consequently there will be a decrease in intensity of all infrared bands that are allowed in D_{2d} symmetry but are forbidden in D_{4h} symmetry. This model is quite consistent with the observations, and also explains the weak Raman lines observed at 634 and 1148 cm.$^{-1}$.

The strength of the 750 cm.$^{-1}$ band, and the 1220 cm.$^{-1}$ shoulder in the liquid spectrum rather strongly suggest that an appreciable fraction of the molecules in the liquid obeys D_{2d} selection rules.

In the discussion that follows we shall generally prefer to use the D_{4h} nomenclature while keeping in mind the effects of D_{2d} symmetry considerations.

From a consideration of the paraffins and particularly the other saturated rings, the C–H stretching frequencies for cyclobutane may be expected to fall in the range from 2900 to 3100 cm.$^{-1}$. We have assigned the strongly polarized Raman band at 2870 to the A_{1g} vibration. No other polarized fundamentals should appear in this part of the spectrum for D_{4h} symmetry, but for D_{2d} symmetry the B_{1u} vibrations become active. Also there are overtones of vibrations in the region of 1450 cm.$^{-1}$ which are totally symmetric and could yield polarized bands in the vicinity of 2900 cm.$^{-1}$. We have rather arbitrarily chosen to assign the 2921 cm.$^{-1}$ band to the B_{1u} fundamental. All actual states in this region are presumably the resultants of the interaction of overtones and fundamentals, but we have no means to sort out the respective contributions.

Strong bands appear in the infrared at 2974 and 2896 cm.$^{-1}$. These have been assigned to vibrations of symmetry E_u and A_{2u}, respectively, since both are allowed even in D_{4h}.

CH_2 deformation vibrations are expected to lie at approximately 1450 cm.$^{-1}$. In the infrared spectrum there is a strong band in this region. The band appears to be of somewhat complex structure, and changed qualitatively with increasing temperature. We have interpreted this band as being the superposition of an E_u band at 1453 cm.$^{-1}$ and a B_{2g} band at 1450 cm.$^{-1}$. The latter

TABLE II

SPECTRA AND ASSIGNMENTS

Infrared Gas Cm.$^{-1}$	I	Infrared Liquid Cm.$^{-1}$	I	Raman Cm.$^{-1}$	I	Pol.	Assignment	Class D_{2d}	Class D_{4h}	Selection rules D_{2d}	Selection rules D_{4h}
611											
625	5	630	5	634 ± 5	0.5		Ring deformation	E	E_u	IR, R	IR
640											
650	0	648	0								
740											
758	2[a]	750	3	750 ± 5	1		Ring bending (in-plane for D_{4h})	B_2	B_{2g}	IR, R	R
850	1										
885											
901	10	901	10				CH_2 rocking	B_2	A_{2u}	IR, R	IR
922											
				928 ± 1	8	dp.	CC stretching	B_1	B_{1g}	R	R
955	1										
				1002.6 ± 0.5	10	p.	CC stretching	A_1	A_{1g}	R	R
				1104 ± 5	0.5		CH_2 twisting	A_2	B_{2u}	ia.	ia.
		1147	1	1148 ± 5	2	dp.	CH_2 twisting	B_1	A_{1u}	R	ia.
1210	2[a]										
1228	2	1220	3	1220 ± 1	5	dp.	CH_2 twisting	E	E_g	IR, R	R
1246											
1261	10	1253	10				CH_2 wagging	E	E_u	IR, R	IR
1277											
1355	1	1338	0?				625 + 750 = 1375	E	E_u	IR, R	IR
1433 }											
1450 }				1444 ± 1			CH_2 deformation	B_2	B_{2g}	IR, R	R
1471 }											
	10[b]	1445	10								
1433 }											
1453 }				1444 ± 1	7	p.?.	CH_2 deformation	E	E_u	IR, R	IR
1465 }											
				1444 ± 1			CH_2 deformation	A_1	A_{1g}	R	R
1530	0			1515 ± 5	1		625 + 890 = 1515	E	E_u	IR, R	IR
							625 + 901 = 1526	E	E_g	IR, R	R
1550	1	1549	0				625 + 928 = 1553	E	E_u	IR, R	IR
1605	0	1605	0								
1630	1[a]										
1650	1[a]	1650	0				750 + 890 = 1640	E	E_g	IR, R	R
1683	0?										
1817	0?	1817	0				890 + 928 = 1818	E	E_g	IR, R	R
1879							625 + 1261 = 1886	E	E_g	IR, R	R
	3[a]	1886	4								
1898							890 + 1003 = 1893	E	E_g	IR, R	R
2035	2[a]						928 + 1104 = 2032	B_2	A_{2u}	IR, R	IR
		2039	2								
2047	1[a]						890 + 1148 = 2038	E	E_u	IR, R	IR
2098	1	2098	2				878 + 1220 = 2098	E	E_u	IR, R	IR
2113	2										
2135	6	2123	7				901 + 1220 = 2121	E	E_u	IR, R	IR
2150	2						890 + 1261 = 2151	E	E_u	IR, R	IR
2176	1						928 + 1261 = 2189	E	E_u	IR, R	IR
2191	2[a]	2183	4				750 + 1444 = 2194	B_2	B_{2g}	IR, R	R
2202	0						750 + 1453 = 2203	E	E_u	IR, R	IR
2264	2[a]						1104 + 1148 = 2252	B_2	B_{2g}	IR, R	R
		2258	2								
2279	1[a]						1003 + 1261 = 2264	E	E_u	IR, R	IR
2357	3	2358	3				901 + 1444 = 2345	B_2	A_{2u}	IR, R	IR
							1148 + 1220 = 2368	E	E_u	IR, R	IR
2404	0	2397	1				1104 + 1286 = 2390	B_2	A_{2u}	IR, R	IR
							1148 + 1261 = 2409	E	E_g	IR, R	R
2437	1	2441	3				1148 + 1299 = 2447	B_2	A_{2u}	IR, R	IR
2455	2	2467	0				1003 + 1453 = 2456	E	E_u	IR, R	IR
2603	1	2577	1				1148 + 1453 = 2601	E	E_g	IR, R	R
2675	0?										
2694	4	2680	3				1220 + 1453 = 2673	E	E_u	IR, R	IR
		2715	2				1261 + 1444 = 2705	E	E_u	IR, R	IR
2755	1	2760	1				2 × 928 + 901 = 2757	B_2	A_{2u}	IR, R	IR
2775	1						625 + 928 + 1220 = 2773	E	E_u	IR, R	IR
				2870 ± 1	8	p.	CH stretching	A_1	A_{1g}	R	R

TABLE II (*Continued*)

Infrared Gas Cm.$^{-1}$	I	Infrared Liquid Cm.$^{-1}$	I	Raman Cm.$^{-1}$	I	Pol.	Assignment	Class D_{2d}	Class D_{4h}	Selection rules D_{2d}	Selection rules D_{4h}
2896	20	[c]					CH stretching	B_2	A_{2u}	IR, R	IR
				2921 ± 3	7	p.	CH stretching	A_1	B_{1u}	R	ia.
				2960 ± 3	7	dp.	CH stretching	E	E_g	IR, R	R
2974	20						CH stretching	E	E_u	IR, R	IR
				2981 ± 3	2	p.?	CH stretching	B_2	B_{2g}	IR, R	R
3058	3						901 + 928 + 1220 = 3049	E	E_u	IR, R	IR
3145	4[a]	3130	2				928 + 1003 + 1220 = 3151	E	E_g	IR, R	R
3238	0						750 + 1220 + 1261 = 3231	E	E_u	IR, R	IR
3341	0	[d]					2 × 1220 + 901 = 3341	E	E_u	IR, R	IR
3401	0						928 + 1220 + 1261 = 3409	E	E_u	IR, R	IR
							625 + 2960 = 3585	E	E_u	IR, R	IR
3585	3[a]?						625 + 2974 = 3599	E	E_g	IR, R	R
							625 + 2981 = 3606	E	E_u	IR, R	IR
3690	4						750 + 2921 = 3671	B_2	A_{2u}	IR, R	IR
3832	4[a]?						890 + 2921 = 3811	E	E_u	IR, R	IR
							878 + 2960 = 3838	E	E_u	IR, R	IR
4100	5						1148 + 2960 = 4108	E	E_u	IR, R	IR
4190	3[a]?						1220 + 2981 = 4201	E	E_g	IR, R	R
							1220 + 2974 = 4194	E	E_u	IR, R	IR
4290	3						1299 + 2974 = 4273	E	E_u	IR, R	IR
4390	6						1450 + 2921 = 4371	B_2	A_{2u}	IR, R	IR
							1453 + 2960 = 4413	E	E_u	IR, R	IR

[a] Diminishes in intensity on heating. [b] See text. [c] The CH stretching region was not investigated in the liquid. [d] No calibration was made on the liquid spectrum above 3300 cm.$^{-1}$. The 4190 cm.$^{-1}$ band appears slightly stronger in the liquid than in the gas, and the 4290 cm.$^{-1}$ band perhaps slightly weaker; otherwise, there is no apparent difference.

vibration should be infrared active in D_{2d} symmetry, but not for D_{4h}; therefore, a change in observed band structure on heating is to be expected though the intensity change might not be marked if the E_u absorption is much stronger than that of the B_{2g} vibration. The B_{2g} vibration should be Raman active for both symmetries, and we believe the 1444 cm.$^{-1}$ line observed in the Raman must be assigned B_{2g}. It is also necessary to assign 1444 cm.$^{-1}$ to the expected A_{1g} vibration since no other line appearing in the Raman spectra can reasonably be assigned to it. The 1515 cm.$^{-1}$ band is at a higher frequency than would be expected. The questionable state of polarization of the 1444 cm.$^{-1}$ line is consistent with the possibility that it is actually two lines.

CH_2 twisting and wagging frequencies may be expected to lie in the region 1100 to 1400 cm.$^{-1}$. The E_g twisting vibration should appear in both Raman and infrared if the symmetry is D_{2d}; for D_{4h} symmetry it is Raman active only. A Raman line is observed at 1220 cm.$^{-1}$ and a pronounced shoulder appears in the infrared spectrum of the liquid at the same value. In the gas phase two shoulders appear at 1228 and 1210 cm.$^{-1}$; because of the changing rotational structure of the 1261 cm.$^{-1}$ band, it is not possible to decide whether there is any variation in intensity of the 1228 cm.$^{-1}$ shoulder; however, the shoulder at 1210 cm.$^{-1}$ does appear to diminish in intensity on heating. We believe these two to be branches of the E_g twisting vibration. In view of the uncertainty in locating the peaks and the absence or obscuration of the third branch we have used the 1220 cm.$^{-1}$ observed in the liquid phase in our assignment. The 1261 cm.$^{-1}$ infrared band can only be the E_u wagging vibration (allowed in infrared, for D_4 symmetry). The B_l wagging vibration should be Raman active only, regardless of symmetry. The observed 1148 cm.$^{-1}$ Raman band may be that of this vibration; however, correlation with the 1261 cm.$^{-1}$ E_u wagging vibration is not very satisfactory. A large interaction constant is required in the normal coördinate analysis, and the resulting calculated value for the A_{2g} wagging vibration is well above the expected value. We, therefore, have assumed that the B_{1g} wagging vibration did not appear in Neu's spectrum, and have assigned the observed 1148 cm^{-1} line to the A_1 twisting vibration (allowed in the Raman for D_{2d} symmetry but not for D_{4h}.) The B_{2u} twisting vibration is forbidden in Raman and infrared for both symmetries. However, we have tentatively attributed the very weak Raman band at 1104 cm.$^{-1}$ to it assuming a selection rule violation in the liquid.

CH_2 rocking frequencies may be expected at from 900 to 1000 cm.$^{-1}$. The 901 cm.$^{-1}$ infrared band is almost certainly the A_{2u} vibration; it is quite clearly a parallel type band, and the P–R splitting is as predicted using the formula of Gerhard and Dennison.[8] Both the B_{1g} stretching and E_g rocking vibrations should appear in the same region of the Raman spectra as depolarized lines. Since only one line was observed, 928 cm.$^{-1}$, we have chosen to assign it to the stretching vibration, and to believe the rocking vibration was not observed. We have assigned a value of 890 cm.$^{-1}$ to the latter. Such an assignment is certainly a weak point in our analysis, but it is the only assignment we can advance consistent with the observed diminution in intensity of the 1630 and 1650 cm.$^{-1}$ combinations on heating.

The polarized Raman line at 1003 cm.$^{-1}$ is clearly the A_{1g} C–C stretching vibration. The E_u stretching–bending vibration should be active in infrared

(8) S. L. Gerhard and D. M. Dennison, *Phys. Rev.*, **43**, 197 (1933).

for D_{4h} symmetry, and both infrared and Raman active for D_{2d} symmetry. We believe that the infrared band at 625 cm.$^{-1}$ and the Raman line at 634 cm.$^{-1}$ correspond to it. The expected value of the frequency is much higher, but we can offer no other explanation for the observed infrared band and Raman line. Our interpretation requires a rather large interaction constant in the normal coördinate analysis. In hope of support for our assignment we have studied the published spectra of cyclobutane derivatives. Because of the removal of degeneracy, these should exhibit pairs of corresponding frequencies. Unfortunately, infrared investigations on derivatives seem to have been confined to the region above 700 cm.$^{-1}$. Kohlrausch and coworkers[9] have, however, obtained the Raman spectra of a large number of derivatives, and that of methylenecyclobutane is also available.[10] The corresponding lines are not found in cyclobutanone or methylenecyclobutane but, in general, lines that may correspond to this vibration are found for the derivatives of lower symmetry.

We have assigned the B_{2g} ring bending vibration as 750 cm.$^{-1}$. This vibration should be only Raman active for D_{4h} symmetry, but also infrared active for D_{2d}. This assignment (compared with Wilson's value of 595 cm.$^{-1}$)[2] is supported by the observance of the corresponding lines in the Raman spectra of methylenecyclobutane (657 cm.$^{-1}$)[9] and cyclobutanone (674 cm.$^{-1}$),[9] and also in other derivatives.[9] The observed diminution in intensity on heating of the infrared band at 750 cm.$^{-1}$ is perhaps the strongest single point supporting a low-barrier double minimum potential function for the out-of-plane bending.

The Assignment of Inactive Vibrations.—In order to calculate most of the frequencies of the inactive vibrations and also the force constants, we have resorted to a normal coördinate analysis, using Wilson's F and G matrices and the CH stretching frequencies have been factored out by his method.[11] The analysis has been made on the basis of D_4 symmetry. Since the deviation from planarity in going to D_2 symmetry is small, this is permissible. The fact that frequency shifts are not observed on heating may be taken as support for this procedure. The B_{1u} out-of-plane bending vibration has such a low frequency that interaction between this and other B_{1u} vibrations can be neglected. This has been done in our calculations, and the out-of-plane bending treated separately. The symmetry coördinates have been obtained in terms of the internal coördinates defined below

ξ = change in C–C bond length
α = change in CCC bond angle
β = change in HCH bond angle
χ = change in HCC bond angle
η = change in C–H bond length (above plane)
ρ = change in C–H bond length (below plane)
ψ = change in angle between Z and Y (see Fig. 3) where Z is fixed in the plane of the CH_2 group and Y in the plane of the carbon skeleton. This is the coördinate for CH_2 twisting motion.
ϕ = change in angle between Y and X (see Fig. 3), where Y is fixed in the plane of the carbon skeleton and X is the bisector of the CH_2 angle. This is the coördinate for the CH_2 wagging motion.
θ = change in angle between X and Z where Z is fixed perpendicular to the skeletal plane and X is the bisector of the CH_2 angle. The coördinate corresponds to the CH_2 rocking motion.

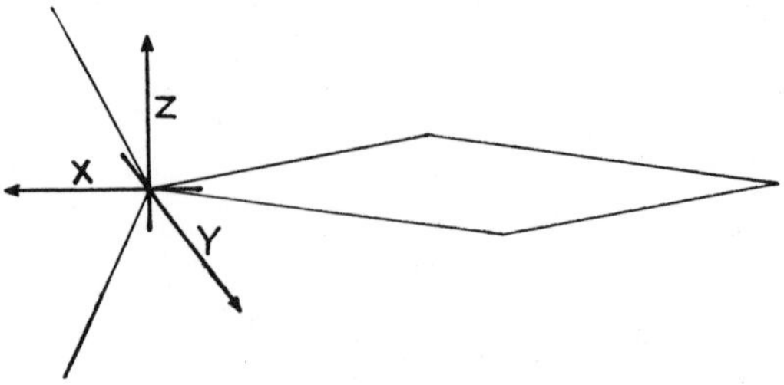

Fig. 3.—Vectors for describing CH_2 motions.

A redundancy relationship about each C atom exists between α, β and χ

$$\left(\sin\frac{\alpha_0}{2}\cos\frac{\beta_0}{2}\right)\alpha + \left(\cos\frac{\alpha_0}{2}\sin\frac{\beta_0}{2}\right)\beta + (\sin\chi_0)\chi_c = 0$$

where the zero subscripts refer to equilibrium angles, and χ_c is the normalized sum of the four HCC bond angles, $\chi_c = {}^1/_2(\chi_i + \chi_j + \chi_k + \chi_l)$. This redundancy relationship has been used to reduce the three coördinates α, β and χ to a new mutually orthogonal pair α' and β' where

$$\alpha' = \frac{\left(\sin^2\chi_0 + \cos^2\frac{\alpha_0}{2}\sin^2\frac{\beta_0}{2}\right)\alpha - \sin\frac{\alpha_0}{2}\cos\frac{\beta_0}{2}\left[\left(\cos\frac{\alpha_0}{2}\sin\frac{\beta_0}{2}\right)\beta + (\sin\chi_0)\chi\right]}{\sin^2\chi_0 + \cos^2\frac{\alpha_0}{2}\sin^2\frac{\beta_0}{2} + \sin^2\frac{\alpha_0}{2}\cos^2\frac{\beta_0}{2}}$$

$$\beta' = \frac{\sin\chi_0\left[(\sin\chi_0)\beta - \left(\cos\frac{\alpha_0}{2}\sin\frac{\beta_0}{2}\right)\chi\right]}{\sin^2\chi_0 + \cos^2\frac{\alpha_0}{2}\sin^2\frac{\beta_0}{2}}$$

The coördinates η and ρ have been transformed to a mutually orthogonal pair

$$\eta' = 1/\sqrt{2}\,(\eta + \rho) \text{ and } \rho' = 1/\sqrt{2}(\eta - \rho)$$

The definition of the $\bar{s}$ vectors is straightforward except in the case of that associated with the carbon atom for a β' vibration. In this case there is an α component as well as a β one; thus

$$\bar{S}_{\beta'_c} = \bar{S}_{\beta_c} + \frac{\sin\alpha_0\sin\beta_0}{4\left(\sin^2\chi_0 + \cos^2\frac{\alpha_0}{2}\sin^2\frac{\beta_0}{2}\right)}\bar{S}_{\alpha_c}$$

Symmetry coördinates are defined in Table III. The choice for the E_u skeletal deformation was particularly made so that a stretching–bending interaction term would occur in the potential energy function. In evaluating the elements of the G matrix, we have used the following parameters: C–C bond length 1.568 Å., C–H bond length 1.096 Å., HCH bond $<114°$.[4]

Force constants obtained are given in Table IV. The frequencies used for evaluating force constants, and the calculated frequencies are given in Table I. K's are principal force constants, L's are interac-

(9) K. W. F. Kohlrausch and R. Skrabel, *Z. Elektrochem.*, **43**, 282 (1937); *Monatsh.*, **70**, 44 (1937); A. W. Reitz and R. Skrabel, *ibid.*, **70**, 398 (1937); R. Skrabel, *ibid.*, **70**, 420 (1937).

(10) F. F. Cleveland, M. J. Murray and W. S. Gallaway, *J. Chem. Phys.*, **15**, 742 (1947).

(11) E. B. Wilson, Jr., *ibid.*, **9**, 76 (1941). We acknowledge also the aid of some preliminary calculation of C. S. Lu on cyclic molecules.

TABLE III

SYMMETRY COÖRDINATES

i, j, k, l refer to the four vertices of the carbon skeleton, with i opposite k.

A_{1g}	$R_1 = {}^1/_2(\eta'_i + \eta'_j + \eta'_k + \eta'_l)$	$R_2 = {}^1/_2(\beta'_i + \beta'_j + \beta'_k + \beta'_l)$
	$R_3 = {}^1/_2(\xi_i + \xi_j + \xi_k + \xi_l)$	
A_{1u}	$R = {}^1/_2(\psi_i + \psi_j + \psi_k + \psi_l)$	
A_{2g}	$R = {}^1/_2(\phi_i + \phi_j + \phi_k + \phi_l)$	
A_{2u}	$R_1 = {}^1/_2(\rho'_i + \rho'_j + \rho'_k + \rho'_l)$	$R_2 = {}^1/_2(\theta_i + \theta_j + \theta_k + \theta_l)$
B_{1g}	$R_1 = {}^1/_2(\phi_i - \phi_j + \phi_k - \phi_l)$	$R_2 = {}^1/_2(\xi_i - \xi_j + \xi_k - \xi_l)$
B_{1u}	$R_1 = {}^1/_2(\rho'_i - \rho'_j + \rho'_k - \rho'_l)$	$R_2 = {}^1/_2(\theta_i - \theta_j + \theta_k - \theta_l)$
	$R_3 = {}^1/_2(\gamma_i - \gamma_j + \gamma_k - \gamma_l)^a$	
B_{2g}	$R_1 = {}^1/_2(\eta'_i - \eta'_j + \eta'_k - \eta'_l)$	$R_2 = {}^1/_2(\beta'_i - \beta'_j + \beta'_k - \beta'_l)$
	$R_3 = {}^1/_2(\alpha'_i - \alpha'_j + \alpha'_k - \alpha'_l)$	
B_{2u}	$R = {}^1/_2(\psi_i - \psi_j + \psi_k - \psi_l)$	
E_g	$R_{1a} = {}^1/_2(\rho'_i + \rho'_j - \rho'_k - \rho'_l)$	$R_{1b} = {}^1/_2(\rho'_i - \rho'_j - \rho'_k + \rho'_l)$
	$R_{2a} = {}^1/_2(\theta_i + \theta_j - \theta_k - \theta_l)$	$R_{2b} = {}^1/_2(\theta_i - \theta_j - \theta_k + \theta_l)$
	$R_{3a} = {}^1/_2(\psi_i + \psi_j - \psi_k - \psi_l)$	$R_{3b} = {}^1/_2(\psi_i - \psi_j - \psi_k + \psi_l)$
E_u	$R_{1a} = {}^1/_2(\eta'_i + \eta'_j - \eta'_k - \gamma'_l)$	$R_{1b} = {}^1/_2(\eta'_i - \eta'_j - \eta'_k + \eta'_l)$
	$R_{2a} = {}^1/_2(\beta'_i + \beta'_j - \beta'_k - \beta'_l)$	$R_{2b} = {}^1/_2(\beta'_i - \beta'_j - \beta'_k + \beta'_l)$
	$R_{3a} = {}^1/_2(\phi_i + \phi_j - \phi_k - \phi_l)$	$R_{3b} = {}^1/_2(\phi_i - \phi_j - \phi_k + \phi_l)$
	$R_{4a} = 1/\sqrt{6}[(\xi_i - \xi_k) - \tau^{-1}(\alpha'_l + \alpha'_j - \alpha'_k - \alpha'_l)]^b$	
	$R_{4b} = 1/\sqrt{6}[(\xi_j - \xi_l) + \tau^{-1}(\alpha'_i - \alpha'_j - \alpha'_k + \alpha'_l)]$	

[a] Coördinate for the out-of-plane skeletal vibration. [b] τ = reciprocal of the C–C distance.

TABLE IV

FORCE CONSTANTS FOR CYCLIC HYDROCARBONS COMPARED[a]

	Cyclobutane		Cyclopropane[13]		Cyclohexane[14]	
	K	L	K	L	K	L
$\theta\theta$	1.418	0.232	1.145	0.286	1.502	0.166
$\phi\phi$	0.868	0	0.506	0	0.920	−.017
$\psi\psi$	2.273	0	1.778	−0.034	2.703	−.046
$\alpha'\alpha'$	2.457	0			1.966	.129
$\beta'\beta$	1.130	0	1.096	0	1.176	0
$\xi\xi$	7.141	0			6.791	0.170
	$(K + 2L)$		$(K + 2L)$		$(K + 2L)$	
$\xi\xi$	7.141		7.251		7.131	
	F		F		F	
$\xi\alpha'\xi\alpha'^b$	1.044		1.823		4.3	
	A					
$\xi\beta'$	−0.119					
$\zeta\phi$	0.107					
$\xi\alpha'$	2.495					
	K	L				
$\alpha\beta'$	0	0				
$\theta\psi$		0				
$\phi\alpha'$	0					
$\beta\phi$		0				

[a] In atomic weight, ångström units. For conversion to cgs. units divide by the factors below:[12]

Force constant	Factor
$K_{\theta\theta}$, $K_{\phi\phi}$, $K_{\psi\psi}$, $K_{\alpha'\alpha'}$, $K_{\beta'\beta'}$, $L_{\theta\theta}$	1.69765×10^{11}
$K_{\xi\xi}$, $K_{\xi\alpha'\xi\alpha'}$	1.69765×10^{-5}
$A_{\xi\beta'}$, $A_{\xi\theta}$, $A_{\xi\alpha'}$	1.69765×10^{3}

[b] This force constant is for the stretching–bending skeletal deformation; it includes $\xi\xi$, $\xi\alpha'$ and $\alpha'\alpha'$ terms. It is not possible to evaluate these terms separately for cyclopropane.

tion constants for adjacent vertices or bonds, and A's are interaction constants for bond stretching and the bending of an adjacent angle. Other interaction constants have been taken equal to zero.

The B_{1u} ring bending vibration is inactive except that it should be Raman active for D_{2d} symmetry. As was previously pointed out, it was hoped that an assignment could be made on the basis of difference bands. Since the line was neither observed in the Raman nor definitely assignable on the basis of combination and difference bands, we have had to assign it indirectly.

The calculation of a potential function for this vibration using the potential constants for the B_{2g} ring bending vibration and accepted values for the torsional interaction about C–C bonds is unsatisfactory because of uncertainties in the magnitude and the form of the two terms. The interaction of nonbonded carbon atoms introduces a further complication.

We have finally calculated a potential function that is in reasonable agreement with the measured entropy, the observed diminution in intensity of the 750 cm.$^{-1}$ band on heating, and the dihedral angle reported by Dunitz and Schomaker from their election diffraction work.[4] As all of this information, and particularly the last two pieces, is quite uncertain, our potential function must at best be regarded as a crude approximation. Because of the mathematical simplicity we have assumed a perturbed harmonic oscillator function of the form $V = 2800\,z^2 + 400e^{-500z^2}$ where z is the distance (in Å.) of alternate carbon atoms above and below the average plane, V is in cm.$^{-1}$, and an effective reduced mass of 20.7 mass units for the vibrations has been assumed.[15] We realize that probably

(12) J. E. Kilpatrick and K. S. Pitzer, *J. Research Natl. Bur. Standards*, **38**, 191 (1947).

(13) The force constants for cyclopropane were obtained from an analysis using the assignments given in Herzberg, "Infrared and Raman Spectra," D. Van Nostrand Co., Inc., New York, 1945, except that we have interchanged the E″ rocking and twisting assignments. The following parameters were used: C–C distance 1.525 Å., C–H distance 1.08 Å., and HCH angle 117°.

(14) From unpublished calculations by C. S. Lu, C. W. Beckett and K. S. Pitzer.

(15) The value 20.7 for μ, the effective reduced mass, is obtained using the molecular constants of Dunitz and Schomaker.

$$\mu = M_c + 2M_H\left(1 + \frac{2\sqrt{2}\epsilon}{\delta}\cos\beta/2\right)^2$$

where M_c = mass of C, M_H = mass of H, δ = C–C bond distance ϵ = C–H bond distance, β = HCH bond angle.

other terms, particularly a quartic term, should be added, but feel that refinement is not worthwhile in view of the several uncertainties and lack of more data. For the lower two states the effect of the perturbation is quite large so we have used simple variation functions $\psi = qe^{-az^2/2} + re^{-az^2}$ and $\psi = sze^{-az^2/2} + tze^{-az^2}$ to calculate the values 128 and 155 cm.$^{-1}$ for these two states. The next four levels we have calculated using the perturbation method: 292, 351, 464 and 542 cm.$^{-1}$. For still higher levels we have assumed the spacing to be constant and equal to 95 cm.$^{-1}$. Correlation with observed diminution in intensity is satisfactory if it is assumed that molecules in all of the above states except the two lowest have transition probabilities for D_{2d} allowed, D_{4h} forbidden transitions that are negligible when compared with those for these two states. Actual calculation of transition probabilities may indicate some modification of the potential function. Because of centrifugal effects, the barrier height will be reduced for molecules having high rotational quantum numbers. However, a simple calculation suggests that this effect will not be pronounced at temperatures at which we have worked, and while it may be a contributing cause in the diminution in intensity on heating, we believe the effect is principally the result of depopulation of the lowest two vibrational levels.

Combination Bands.—Observed infrared bands are listed in Table II with possible assignments for most of the combinations. Since no bands were observed to increase in intensity on heating, we have avoided any difference assignments. Most of the bands observed to decrease in intensity on heating are satisfactorily explained on the basis of our model. Several bands, notably those at 650, 850 and 955 cm.$^{-1}$ are not explained in Table II. We think it likely that these (and also probably those at 2755, 2775, 3058 and 3145 cm.$^{-1}$) may be combinations involving the B_{1u} out-of-plane bending motion. Because of uncertainties in the shape of the out-of-plane bending potential, we have not attempted to assign these bands. Those at 850 cm.$^{-1}$ and 650 cm.$^{-1}$ might well be sum and difference bands with the 750 cm.$^{-1}$ vibration. It is unfortunate that these bands are so weak since the behavior of their shapes and intensities with temperature could conceivably throw light on the out-of-plane bending potential function. In spite of the simplification that results in going to the liquid, the region around 2100 cm.$^{-1}$ is still fraught with uncertainties. A number of combinations besides those listed are allowed in this region; with this large number of combinations some cases of Fermi resonance are likely and the assignments given are most uncertain. Because of these complications, our failure to explain diminutions in intensity with increase in temperature for the bands at 2035 and 2047 cm.$^{-1}$ cannot be regarded as serious evidence against our analysis.

Force Constants.—Values of principal force constants for cyclobutane are compared with those for cyclohexane and cyclopropane in Table IV. The methylene constants for cyclobutane, as expected, generally have values intermediate between those for the other two compounds. The values obtained for $L_{\psi\psi}$ for cyclohexane and cyclopropane suggest that a small negative value be used also in the case of cyclobutane. The introduction of such a constant results in an increased value for the B_{2u} twisting vibration. It may be that the 1104 cm.$^{-1}$ Raman band has been erroneously assigned; a possible combination assignment could be skeletal B_{1u} + 1003 cm.$^{-1}$, A_{1g} = 1104 cm.$^{-1}$ (combination allowed for D_{2d} symmetry, but not for D_{4h}.)

Correlation of the skeletal force constants is necessarily less complete because of the limited information available from cyclopropane. That the value for $K_{\alpha'\alpha'}$ for cyclobutane is appreciably greater than for cyclohexane is not surprising. Because of the rather small diagonal distance across the ring, interaction between the opposite carbons may be very much greater in cyclobutane than in strain-free cases. (Dunitz and Schomaker[4] have estimated the magnitude of the effect from a consideration of lengthening of C–C bonds in cyclobutane.) Our failure to include an interaction term between opposite angles may well account for the large value for $K_{\alpha'\alpha'}$. A similar difficulty is implicit in our failure to include stretching interaction terms. The value obtained for $F_{\xi\alpha'\xi\alpha'}$ is unusually small, and we have assigned a large stretching–bending interaction term to explain this (see Table IV). The value of this interaction term will be reduced, but probably not greatly so, if the aforementioned α' and ξ interaction terms are included.

The observed strain energy[16] in cyclobutane, when taken in conjunction with our value for $K_{\alpha'\alpha'}$ requires that either the large value of $K_{\alpha'\alpha'}$ be in part due to an interaction term, or that C–C bonds be very much bent[17,18] or (more likely) both.[19]

Thermodynamic Functions

Thermodynamic functions based on our analysis are tabulated in Table V. The data cannot be regarded as being as reliable as is usual, particularly in view of the uncertainty in the potential function for the out-of-plane bending motion.

TABLE V

THERMODYNAMIC FUNCTIONS OF CYCLOBUTANE (IN CAL./DEG. MOLE)

T (°K.)	$\frac{H_0 - H_0^0}{T}$	$\frac{F_0 - H_0^0}{T}$	S^0	C_p^0
285.67	10.81	51.91	62.72	16.46
298.16	11.07	52.37	63.43	17.26
300	11.11	52.44	63.55	17.37
400	13.49	55.95	69.44	23.89
500	16.18	59.25	75.42	29.86
600	18.88	62.43	81.31	34.76
700	21.45	65.54	86.99	38.89
800	23.86	68.56	92.42	42.42
900	26.09	71.50	97.59	45.41
1000	28.15	74.36	102.51	47.96
1500	36.32	87.43	123.75	56.34

The entropy given in Table V may be combined with the measured heat of combustion for cyclobu-

(16) S. Kaarsemaker and J. Coops, *Rec. trav. chim.*, **71**, 261 (1952).

(17) J. E. Kilpatrick and R. Spitzer, *J. Chem. Phys.*, **14**, 463 (1946).

(18) C. A. Coulson and W. E. Moffitt, *Phil. Mag.*, **40**, 1 (1949).

(19) If our $K_{\alpha'\alpha'}$ is used, and the bonds are assumed to be distorted from 109° 28′ to 90°, the strain resulting from the distortion alone is much larger than is actually observed.

tane, 2774.1 kj./mole.[16] and other data[20] to obtain the free energy of formation. Free energy of isomerization to isobutene may also be calculated. These values together with the corresponding values for cyclopropane and cyclopentane are given in Table VI. The values given are consistent with the observation that synthesis of cyclobutane has been difficult.

Our analysis of cyclobutane suggests that some other strained cyclic compounds may also have double minimum potential functions with barriers sufficiently low so that similar effects might be observed. Spectroscopic investigation has confirmed this expectation in the cases of cyclopentene and cyclopentane. Preliminary spectroscopic investigation of octafluorocyclobutane has yielded negative results.

(20) "Properties of Hydrocarbons," Project #44, American Petroleum Institute.

TABLE VI

FREE ENERGY OF FORMATION ($\Delta F^0_{f\,298°}$) AND ISOMERIZATION ($\Delta F^0_{i\,298°}$) OF CYCLIC HYDROCARBONS

Hydrocarbon	$\Delta F^0_{f\,298°}$, kcal./mole	$\Delta F^0_{i\,298°}$ (to most stable alkene), kcal./mole
Cyclopropane[13]	24.98	− 9.99
Cyclobutane	26.31	−11.73
Cyclopentane	9.23	5.04

We are indebted to Professor R. S. Halford for helfpul suggestsions and advice in connection with this problem.

NEW YORK, N. Y.
BERKELEY, CALIF.

[Reprinted from the Journal of the American Chemical Society, **81**, 3213 (1959).]

[CONTRIBUTION FROM THE DEPARTMENT OF CHEMISTRY, UNIVERSITY OF CALIFORNIA, BERKELEY]

Conformations and Strain Energy of Cyclopentane and its Derivatives

BY KENNETH S. PITZER AND WILM E. DONATH

RECEIVED JANUARY 2, 1959

Earlier calculations of the conformation and strain energy of cyclopentane are refined by the addition of terms for the election correlation energy of non-bonded but nearby atoms and for the zero point vibrational energy. The agreement with observed properties of cyclopentane is substantially improved. It is shown that substitution of the cyclopentane will ordinarily yield a barrier to the pseudo-rotation of ring puckering and that the energy minimum will in some cases correspond to C_2 symmetry and in other cases to C_S symmetry. It is shown that certain of the parameters evaluated for cyclopentane may be transferred to derivatives and that strain energies may be calculated for the derivatives if potential barrier values are available for the appropriate open chain model compounds.

In 1941 Aston, Schumann, Fink and Doty[1] showed that the cyclopentane ring was puckered. The torsional forces associated with the single C–C bonds are undoubtedly responsible for this puckering and semi-quantitative calculations of the strain energy and of the detailed equilibrium conformation have been presented.[2–4] While the quantitative comparison of these results with the experimental heat of combustion value[5] was only partially satisfactory, a significant qualitative result emerged from the Kilpatrick, Pitzer and Spitzer[3] investigations, namely, "that the puckering of the ring is not of a definite type but that the angle of maximum puckering rotates around the ring." While certain authors[6–8] have been reluctant to accept this pseudo-rotation of the puckering and the consequent indefiniteness of the cyclopentane conformation, McCullough[9] has reported more precise heat capacity measurements for cyclopentane which definitely confirm the free pseudo-rotation.

It is the purpose of this paper to present: first, new and improved calculations of the total strain energy of cyclopentane and of the equilibrium degree of puckering, and second, a discussion of the expected equilibrium conformation of cyclopentane derivatives where, in general, the substitution will introduce a potential barrier restricting the pseudo-rotation.

Strain Energy of Cyclopentane.—Previous strain energy calculations for cyclopentane considered only the torsional energy associated with the rotation of C–C single bonds and the strain of bending C–C–C bond angles below the tetrahedral value. The force constant for the latter effect was taken from the normal vibration analysis of cyclohexane and open chain hydrocarbons, while the torsional potential was estimated from that of ethane and other simple hydrocarbons. These two terms do not yield any explanation of the difference in energy between such paraffin isomers as n-butane and isobutane or n-pentane and neopentane. Consequently the lack of quantitative agreement for the strain energy of cyclopentane was not too surprising.

Recently calculations for the paraffins were improved[10] by the addition of terms for the zero point vibration energy and for the interaction energy of nearby but non-bonded electron clouds. The correlation of the motion of these electrons yields an attractive force and an energy decrease which may be treated by the method used by F. London for attractive van der Waals forces between non-polar molecules. These calculations for the paraffin isomerization energies yielded quite good agreement and we now extend the same method to cyclopentane and cyclohexane.

The numerical constants of previous investigations were retained.[3,10] The London or correlation energy is given by the formula

$$\Delta E_C = -[22.6\Sigma R_{CC}^{-6} + 8.68\Sigma R_{CH}^{-6} + 3.42\Sigma R_{HH}^{-6}] \times 10^{-60} \text{ ergs} \quad (1)$$

where the sums are over all non-bonded interatomic distances of the types indicated. Bond distances are taken as C–C, 1.54 Å.; C–H, 1.10 Å. The C–C–C bending constant was taken as 0.8×10^{-11} erg/radian^2 and the torsional potential barrier as a cosine function of 2800 cal./mole height.

The puckering in cyclopentane is described by the formula

$$z_j = \sqrt{\frac{2}{5}}\, q \cos 2\left(\frac{2\pi}{5}j + \varphi\right) \quad (2)$$

where z_j is the displacement of the jth atom perpendicular to the plane of the unpuckered ring, q measures the amplitude, and φ the phase angle of the puckering.

The calculations follow the method of Kilpatrick, *et al.*[3]; additional details are given in the Appendix. The results for cyclopentane are given in Table I for six values of the parameter q. The bond energies, C–H, 102.4 and C–C, 80.2 kcal./mole, which were obtained in the corresponding treatment of the paraffins, together with the dissociation energies of the elements yield the term $-\Delta H_{\text{bond}} = 59.0$ kcal./mole. The zero point vibrational energy was calculated from the assignment of Miller and Inskeep.[6]

A plot (Fig. 1) of the heat of formation $-\Delta H_f°$

(1) J. G. Aston, S. C. Schumann, H. L. Fink and P. M. Doty, THIS JOURNAL, **63**, 2029 (1941); J. G. Aston, H. L. Fink and S. C. Schumann, *ibid.*, **65**, 341 (1943).

(2) K. S. Pitzer, *Science*, **101**, 672 (1945).

(3) J. E. Kilpatrick, K. S. Pitzer and R. Spitzer, THIS JOURNAL, **69**, 2483 (1947).

(4) R. Spitzer and H. M. Huffman, *ibid.*, **69**, 211 (1947).

(5) F. D. Rossini, *et al.*, "Selected Values of Physical and Thermodynamic Properties of Hydrocarbons and Related Compounds," Carnegie Press, Pittsburgh (1953 and subsequent supplements).

(6) F. A. Miller and R. G. Inskeep, *J. Chem. Phys.*, **18**, 1519 (1950).

(7) C. G. LeFevre and R. J. W. LeFevre, *J. Chem. Soc.*, 3549 (1956); *Chemistry and Industry*, 54 (1956).

(8) B. Curnutte and W. F. Shaffer, *J. Mol. Spectroscopy*, **1**, 239 (1957).

(9) J. P. McCullough, *J. Chem. Phys.*, **29**, 966 (1958).

(10) K. S. Pitzer and E. Catalano, THIS JOURNAL, **78**, 4844 (1956).

TABLE I

ENERGY OF CYCLOPENTANE AT 0°K.

q, Å.	0.00	0.236	0.308	0.40	0.50	0.60
$-E_C$, kcal./mole	44.49	44.57	44.67	44.94	45.72	47.03
$E_{torsion}$, kcal./mole	14.00	11.48	9.99	8.05	6.18	4.95
E_{C-C-C}, kcal./mole	0.08	0.62	1.10	2.19	4.26	7.72
$-\Delta H_{bond}$, kcal./mole	59.0	59.0	59.0	59.0	59.0	59.0
$\Sigma^1/_2 h\nu$, kcal./mole	85.26	85.26	85.26	85.26	85.26	85.26
$-\Delta H_f^0$, kcal./mole	4.2	6.2	7.3	8.4	9.0	8.1

vs. q yields a minimum of 9.0 kcal./mole at 0.50 Å. These values may be compared with the experimental $-\Delta H_f^\circ = 10.7$ kcal./mole[5] and the value of $q_0 = 0.47$ Å. which fits the experimental entropy.[11] The agreement for q_0 is excellent and for ΔH_f° reasonably satisfactory but not within experimental error.

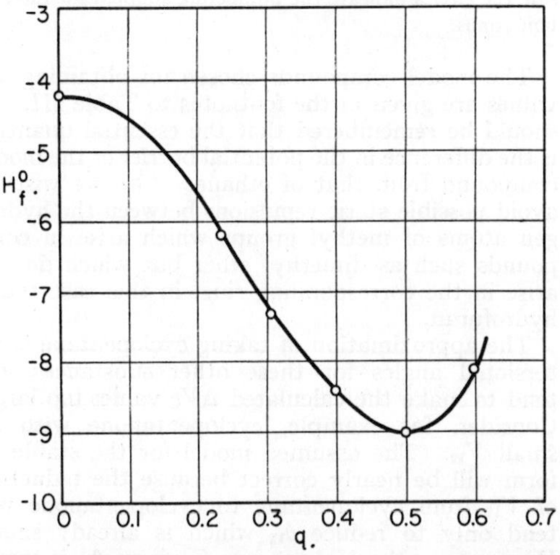

Fig. 1.—The calculated heat of formation (in kcal./mole) of cyclopentane as a function of the amplitude of puckering q in Å.

The calculations reported in Table I were made for cyclopentane in the conformation of C_S symmetry, *i.e.*, with $\varphi = 0$ or any integral multiple of $\pi/10$ in equation 2. The various terms were also calculated for the C_2 conformation which may be obtained on substitution of any odd integral multiple of $\pi/20$ for φ. The simplest case is $\varphi = -3\pi/20$ which yields $z_1 = 0$ and thus places atom one on the symmetry axis. A near equilibrium value 0.48 Å. was taken for q and the following values were obtained: $-E_C = 45.0$, $E_{torsion} = 6.5$, $E_{C-C-C} = 3.77$, $-\Delta H_f^\circ = 8.5$ kcal./mole. The torsional and bond bending strain energies are very close to the comparable values for $q = 0.48$ interpolated from Table I. The electron correlation energy and the net heat of formation differ by about one-half kilocalorie from the values for the C_S form; we are not sure whether this difference is significant or not (see Appendix). It may be concluded, nevertheless, that the calculated energy difference between C_2 and C_S forms is small and thus in practical agreement with experiment[9] and with the earlier calculation.[3]

The second derivative of the energy with respect to q gives the restoring force for the vibration (283 cm.$^{-1}$) in which the amplitude of puckering varies. The effective mass for this vibration was calculated only approximately and interaction with other vibrations was neglected. The resulting spectroscopic force constant, $k = 1.3 \times 10^5$ dynes/cm., agrees as well as could be expected with that obtained from the plot of the values of Table I, *ca.* 1.1×10^5.

While the 1.7 kcal./mole difference in ΔH_f° indicates that some further improvement in the theory of the energy is needed, the present results are qualitatively correct in all respects and yield as close a quantitative agreement as can be expected for a relatively simple theory.

A similar calculation for cyclohexane, where there is no torsional or C–C–C angle strain, yields $-\Delta H_f^\circ = 19.5$ kcal./mole at 0°K. in excellent agreement with the experimental value of 20.0 kcal./mole. In this calculation the zero point energy was obtained from the frequency assignment of Beckett, *et al.*[12]

Cyclopentane Derivatives and Related Five-membered Ring Compounds.—If cyclopentane is substituted in such a manner as to change significantly one or more of the torsional potential barriers, then the cancellation of forces that yield the free pseudo-rotation will no longer be as exact and a potential restricting the pseudo-rotation is to be expected. An example is thiacyclopentane where Hubbard, *et al.*,[13] found a potential barrier of 2800 cal./mole from their interpretation of spectroscopic and thermodynamic data.

An exact calculation of the potential barrier for the pseudo-rotation would require the readjustment of the puckering to minimize the energy for each phase angle between the puckering maximum and the substituted atom. However, we may obtain approximate results by assuming that the bond twist angles remain the same as in cyclopentane itself. These are shown for the two symmetrical conformations[14] in Fig. 2 and Table II for $q_0 = 0.48$ Å. The extreme difference in energy for a single substitution is obtained by placing the substituent in the unique position in each case (position one in Fig. 1). There are, of course, inter-

TABLE II

TORSIONAL ANGLES IN CYCLOPENTANE

	C_S	C_2
$\theta_{12} = \theta_{51}$	46.1°	15.15°
$\theta_{23} = \theta_{45}$	28.6°	39.45°
θ_{34}	0.0°	48.1°
$1 + \cos 3\theta_{12}$	0.25	1.70
$1 + \cos 3\theta_{23}$	1.07	0.52
$^1/_2(1 + \cos 3\theta_{34})$	1.00	0.09

(11) The original calculation of Kilpatrick, Pitzer and Spitzer[3] used an erroneous symmetry number of 5 instead of 10. Correction of this error doubles the value of q_0. See THIS JOURNAL, **80**, 6697 (1958).

(12) C. W. Beckett, K. S. Pitzer and R. Spitzer, *ibid.*, **69**, 2488 (1947).

(13) W. N. Hubbard, H. L. Finke, D. W. Scott, J. P. McCullough, C. Katz, M. E. Gross, J. F. Messerly, R. E. Pennington and Guy Waddington, *ibid.*, **74**, 6025 (1952).

(14) The names *envelope* and *half chair* have been suggested for the C_S and C_2 forms, respectively, by Brutcher (private communication). These terms seem to the writer to be appropriate if descriptive names are desired in addition to the symmetry symbols.

mediate conformations between C_2 and C_S which have the same energy in cyclopentane and smoothly varying energy in substituted molecules. Also conformations obtained by placing the substituent in an unsymmetrical position in the C_2 and C_S forms are also real but of intermediate energy.

The last three lines in Table II give the coefficients of the potential barriers $V_{12} = V_{51}$, $V_{23} = V_{45}$ and V_{34}, respectively, in the torsional strain energy. If we wish only the energy difference ΔV_C for the conformation change $C_S \rightarrow C_2$, we obtain

$$\Delta V_C = 1.45V_{12} - 0.55V_{23} - 0.91V_{34} \quad (3)$$

This torsional energy difference may be taken as a good approximation to the net difference in total energy unless some special complication such as a repulsive steric hindrance is present. If all potential barriers are the same, the result is zero within computational error as stated above.

Let us now consider in particular the case where substitution has occurred at position one in cyclopentane. The barriers are 2800 cal./mole except for $V_{12} = V_{51}$ which may be different. Then

$$\Delta V_C = 1.45(V_{12} - 2800) \text{ cal./mole} \quad (4)$$

where the change is $C_S \rightarrow C_2$ as before. Table III contains values of V_{12} taken from appropriate model molecules and the resulting ΔV_C values for several substituted cyclopentanes and related compounds containing saturated five-membered rings.

It should be emphasized that the ΔV_C values in Table III are potential barriers to the pseudo-rotation which is restricted in these cases. A positive value of ΔV_C indicates that the lowest energy phase angle of puckering is the symmetrical C_S form; a negative value indicates that the lowest energy phase angle is the symmetrical C_2 form. One may assume that a simple cosine function yields a good approximation for the energy of all intermediate values of the phase angle. Thus ΔV_C is *not* the energy difference between separate potential minima such as the equatorial and axial conformations of methylcyclohexane but is the height of the energy barrier restricting the rotation of the phase of puckering.

TABLE III

POTENTIAL BARRIERS TO PSEUDO-ROTATION IN CYCLOPENTANE DERIVATIVES

Substance	V_{12}	ΔV_C
Methylcyclopentane	3.40[a]	0.9
Fluorocyclopentane	3.30[b]	0.7
Methylenecyclopentane	1.98[c]	−1.2
Cyclopentanone	1.15[d]	−2.4
Pyrrolidine	1.90[e]	−1.3
Tetrahydrofuran	1.07[f]	−2.5
Thiacyclopentane	0.71[g]	−3.0

[a] Propane: K. S. Pitzer, *J. Chem. Phys.*, **12**, 310 (1944). [b] Ethyl fluoride: D. Herschbach, *ibid.*, **25**, 358 (1956). [c] Propylene: D. R. Lide, Jr., and D. E. Mann, *ibid.*, **27**, 868 (1957). [d] Acetaldehyde: R. W. Kilb, C. C. Lin and E. B. Wilson, Jr., *ibid.*, **26**, 1695 (1957). [e] Methylamine: D. R. Lide, Jr., *ibid.*, **22**, 1613 (1954); Shimoda, Nishikawa and Itoh, *J. Phys. Soc. Japan*, **9**, 947 (1954). [f] Methanol: E. V. Ivash and D. M. Dennison, *J. Chem. Phys.*, **21**, 1804 (1953); J. D. Swalen, *ibid.*, **23**, 1739 (1955). [g] Methanethiol: R. W. Kilb, *ibid.*, **23**, 1736 (1955).

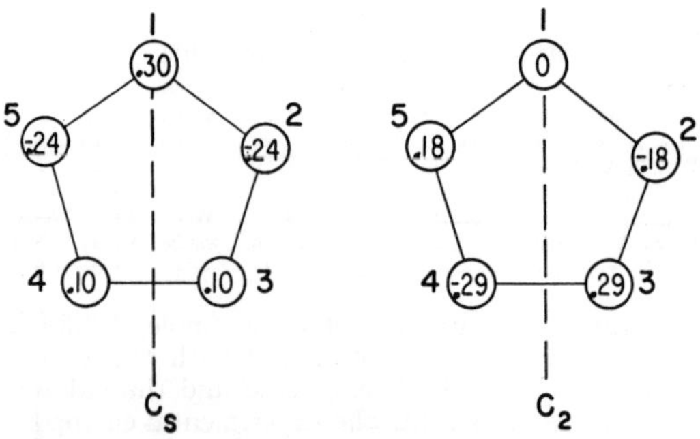

Fig. 2.—The two symmetrical conformations of cyclopentane. Figures inside the circles show equilibrium displacements perpendicular to plane of the paper in Å.; figures outside the circles indicate the numbering adopted for the carbon atoms.

The model compounds chosen for obtaining V_{12} values are given in the footnotes to Table III. It should be remembered that the essential quantity is the difference in the potential barrier of the model compound from that of ethane. Also we wish to avoid possible steric repulsions between the hydrogen atoms of methyl groups which arise in compounds such as dimethyl ether but which do not arise in the corresponding ring, in this case tetrahydrofuran.

The approximation of taking cyclopentane bond torsional angles for these other substances will tend to make the calculated ΔV_C values too large. Consider, for example, cyclopentanone with its small V_{12}. The assumed model for the stable C_2 form will be nearly correct because the reduction of V_{12} from cyclopentane to cyclopentanone will tend only to reduce θ_{12} which is already small. However, in the high energy C_S form θ_{12} is larger than optimum for the low V_{12} of cyclopentanone. Reduction in the puckering parameter reduces the C–C–C angle strain at moderate cost in torsional strain. A quantitative calculation for the C_S form of cyclopentanone (assuming still the E_C values of cyclopentane) indicates that the corrected potential is only 0.2 kcal./mole lower (at a value of q_0 about 0.04 Å. smaller) than was obtained on the approximate basis of Table III. Thus this effect is small and may be ignored in approximate calculations.

The calculated value of ΔV_C in Table III for thiacyclopentane of −3.0 kcal./mole is in good agreement with the experimental value ±2.8 kcal./mole obtained by Hubbard, *et al.*[13] The thermodynamic method used in their work determines the magnitude but not the sign of the potential barrier. McCullough[9] reports that the value for pyrrolidine is small without indicating an upper limit. The calculated value of −1.3 kcal./mole is certainly much smaller than that for thiacyclopentane and may be small enough to yield agreement with experiment.

Discussion

The data presented in Tables II and III provide the basis for estimates of strain energies for many five-membered ring systems. For maximum accuracy one should readjust all of the molecular parameters in order to minimize the strain energy for each individual case. A few exploratory calculations indicated, however, that useful estimates of strain energy differences may be obtained in many cases by retaining the geometry of cyclopentane but recognizing the difference in torsional potential barriers caused by the substitution. In making such calculations it is essential to select the conformation of cyclopentane which will be most stable in the substituted molecule. Thus a single substituent which raises the two adjacent barriers ($V_{12} = V_{51}$) stabilizes the symmetrical C_S form whereas if a single substituent lowers the adjacent barriers the symmetrical C_2 form becomes the most stable. One may then calculate torsional strain energy differences by multiplying the appropriate value of $(1 + \cos 3\theta_{12})$ in Table II by the difference in potential barrier height V_{12}. If two molecules are to be compared which have different stable conformations, then one must compare each with cyclopentane and sum the two differences. A case of this type will be illustrated below.

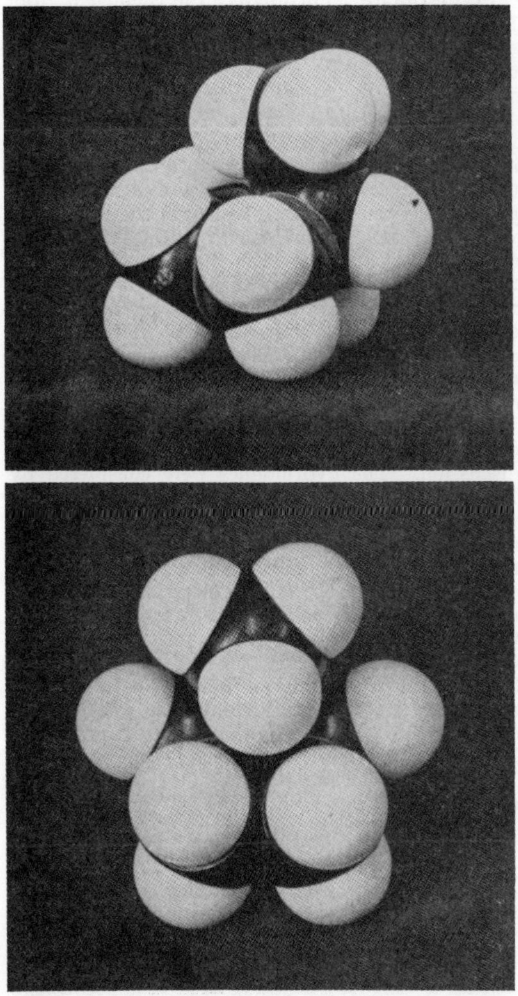

Fig. 3.—Two views of methylcyclopentane with the methyl group in the most axial conformation. The absence of steric repulsions analogous to axial methylcyclohexane is apparent.

The possibility of steric repulsions analogous to those of axial substituents in cyclohexane must also be considered. Examination of appropriate models (Fig. 3) shows no large interference in the cyclopentane case for the methyl group or any smaller group. An approximate quantitative value for the axial–equatorial energy difference can be obtained from the heat of formation values[5] found for *cis*- and *trans*-1,3-dimethylcyclohexanes which are −22.19 and −21.65 kcal./mole, respectively, at 0°K. The *cis* isomer has a di-equatorial conformation while the *trans* is axial–equatorial. Hence the energy difference of 0.54 kcal./mole may be ascribed to the extra strain of the axial methyl group. The corresponding value for the cyclohexane system is, of course, much larger (*ca.* 1.9 kcal./mole).

The terms axial and equatorial are descriptive only for that portion of the cyclopentane ring where the C–C bonds are in a nearly staggered orientation. We see from Table II that the 1,2- and 5,1-bonds in the C_S form are nearly staggered, hence the three successive positions 5, 1 and 2 qualify. In 1,3-dimethylcyclopentanes the methyl groups would be expected to be in positions 5 and 2 of the Table II numbering for the C_S form.

Since the e–a energy difference is small, the value $\Delta V_C = 0.9$ kcal./mole in Table III for methylcyclopentane may be interpreted as a good approximation to the net barrier to pseudo-rotation in that molecule. This agrees well with the value selected by Kilpatrick, *et al.*,[15] 0.75 kcal./mole which is consistent with the entropy of methylcyclopentane.[5] Conversely one may regard the agreement just cited as confirming the low e–a energy difference in methylcyclopentane.

Cases of axial conformation of groups larger than methyl or of conformations which are di-axial on the same side of the ring may have large strain but need individual consideration.

Let us now consider particular data of several types which relate to the strain of saturated five-membered rings. The heats of hydrogenation measured by Turner and Garner[16] for methylenecyclopentane (−26.9 kcal./mole) and methylenecyclohexane (−27.8 kcal./mole) may be interpreted to yield the difference in strain energy of methylcyclopentane and methylenecyclopentane. If all thermal energy terms are assumed to cancel and both cyclohexane compounds are strain free,[16a] then one assigns the difference of 0.9 kcal./mole to an increased amount of strain in methylcyclopentane over that in methylenecyclopentane.

(15) J. E. Kilpatrick, H. G. Werner, C. W. Beckett, K. S. Pitzer and F. D. Rossini, *J. Research Natl. Bur. Standards*, **39**, 523 (1947).

(16) R. B. Turner and R. H. Garner, THIS JOURNAL, **79**, 253 (1957); **80**, 1424 (1958).

(16a) H. C. Brown, J. H. Brewster, and H. Shechter, THIS JOURNAL, **76**, 467 (1954), assume that the introduction of an exo double bond into cyclohexane introduces strain. The structural explanation proposed was strain associated with the opposed orientation of the double bond relative to the C–H bonds of adjacent ring methylene groups. Recent microwave studies (ref. *d* of Table III and D. R. Herschbach and L. C. Krisher, *J. Chem. Phys.*, **28**, 728 (1958)) have shown that this opposed orientation is the stable orientation, hence there is no basis for expecting appreciable strain in methylenecyclohexane or cyclohexanone.

The equilibrium conformations will be C_S for methylcyclopentane and C_2 for methylenecyclopentane. The values of $(1 + \cos 3\theta_{12})$ in Table II then yield for the difference in torsional strain

$$\Delta E_{tors} = 0.25(3.40 - 2.80) + 1.70(2.80 - 1.98)$$
$$= 1.54 \text{ kcal./mole}$$

where the first term is the increase in torsional strain of methylcyclopentane over cyclopentane and the second term the decrease of methylenecyclopentane below cyclopentane.

The unstrained H–C–H bond angle in ethylene[17] is larger than tetrahedral (117°) although the force constant is somewhat smaller (0.7 instead of 0.8×10^{-11} erg/radian^2). The ring angles were readjusted approximately to minimize the strain under these new conditions and the result indicated an increase in C–C–C bond angle strain of about 0.9 kcal./mole in methylenecyclopentane above that in cyclopentane or in methylcyclopentane.

The net result (1.54 − 0.9) is the calculated strain energy difference of 0.64 kcal./mole which may be compared to Turner and Garner's experimental difference of 0.9 kcal./mole. This is as good agreement as can be expected in view of the approximations made and the omission of differences in electron correlation energies.

A somewhat similar argument may be applied to the data on the reaction of cyclic cyanohydrins to form the cyclic ketone and hydrogen cyanide. Prelog and Kobelt[18] report the standard free energy of dissociation to be 2.4 and 4.2 kcal./mole, respectively, for the cyclopentane and cyclohexane systems in 96% alcohol at 22–23°. The difference of 1.8 kcal./mole may be attributed to the appropriate difference in strain free energies.

Cyclohexanone is unstrained. The corresponding cyanohydrin may have either the OH or the CN group axial with the other in the equatorial position. Various investigations of cyclohexanol[19] indicate that the strain free energy of an axial OH group is 0.6 ± 0.2 kcal./mole. Such knowledge as we have of the van der Waals repulsive contour of the CN group indicates that it will be less hindered in the axial position than the OH group.

At this point we must note that it is strain free energy which we are estimating for the cyanohydrin, whereas it was strain energies (or enthalpies) in earlier examples. If the cyanohydrin were purely a-CN, e-OH and the a-CN were strain free, there would still be a strain free energy of $RT \ln 2$ (or 0.4 kcal./mole at 300°K.) because the elimination of the e-CN, a-OH conformation has reduced the entropy. In this case the 0.6 kcal./mole strain of a-OH will not eliminate the a-OH conformation; also the a-CN conformation may have a few tenths kcal. strain. Hence an exact calculation is not possible but the estimate of 0.5 ± 0.2 kcal./mole for the strain free energy of the cyclohexanone cyanohydrin seems reliable.

(17) L. S. Bartell and R. A. Bonham, *J. Chem. Phys.*, **27**, 1414 (1957); H. C. Allen and E. K. Plyler, THIS JOURNAL, **80**, 2673 (1958).

(18) V. Prelog and M. Kobelt, *Helv. Chim. Acta*, **32**, 1187 (1949).

(19) E. L. Eliel and C. A. Lukach, THIS JOURNAL, **79**, 5986 (1957); S. Winstein and N. J. Holmes, *ibid.*, **77**, 5562 (1955); R. A. Pickering and C. C. Price, *ibid.*, **80**, 4931 (1958).

The strain free energy in the cyclopentane system is, therefore, 2.3 kcal./mole greater in the cyanohydrin than in the ketone. The calculations follow closely the method explained above for methylenecyclopentane and methylcyclopentane. One readily calculates that the torsional strain energy of cyclopentanone is less than that of cyclopentane by 1.70(2.80 − 1.15) = 2.8 kcal./mole. The additional C–C–C bond angle strain in cyclopentanone must be very nearly the same as that in methylenecyclopentane, hence we take the same value of 0.9 kcal./mole.

There do not appear to be any serious steric hindrance effects in the cyclopentanone cyanohydrin, but the torsional potential barriers are presumably somewhat greater than in cyclopentane. Barrow[20] gives 3.3 kcal./mole for the barrier to methyl rotation in ethanol. The substitution of the CN group might be expected to increase this value somewhat but we have no basis for an accurate estimate. Fortunately the coefficient from Table II for the C_S form is only 0.25, hence a crude estimate will suffice. We take 4.0 kcal./mole for the barrier adjacent to the C(OH)CN group and obtain 0.25(4.0 − 2.8) = 0.3 kcal./mole for the excess torsional strain over that of cyclopentane.

Since the pseudo-rotation of the cyclopentane ring is substantially restricted for both the ketone and the cyanohydrin, the entropy effects should approximately cancel and we find from the terms previously discussed 0.3 + 2.8 − 0.9 = 1.2 kcal./mole for the excess strain in the cyanohydrin over that of the ketone. The agreement of this figure with the experimental value to 0.1 kcal./mole is, of course, fortuitous, since the calculation could easily err by 0.5 kcal./mole. Nevertheless, the results indicate that the important effects are adequately considered in these simple calculations.[21]

Fused rings offer a wide variety of situations; only one will be mentioned here in concluding this discussion. The hydrindanes were considered by Dauben and Pitzer[22] who emphasized the effects on the six-membered ring. We may now add that the ring fusion will also restrict the conformation of the five-membered ring. Since the shared C–C bond is staggered ($\theta = 60°$) in the normal chair conformation of cyclohexane, the five-membered ring will tend to place its most nearly staggered bond in that location. Table II shows that this is bond 3–4 in the C_2 conformation. Thus the five-membered ring in hydrindane will take the C_2 conformation with the unique CH_2 group (1 in Fig. 2) in the 8-position. We note, however, from Table II that $\theta_{12} = \theta_{51}$ for the C_S conformation is only 2° less than θ_{34} in C_2. The difference in energy

(20) G. M. Barrow, *J. Chem. Phys.*, **20**, 1739 (1952).

(21) The heat of hydrogenation values for the cyclic ketones may also be discussed in a similar fashion. Unfortunately the large difference in heat of hydrogenation of cyclohexanone (−15.42) from those of acetone (−13.41) and methyl ethyl ketone (−13.19) remains unexplained. The value for cyclopentanone is calculated to be about 2.0 kcal./mole smaller than that of an unstrained ketone, hence the observed value of −12.50 kcal./mole is reasonable but no more can be said. The measurements are by M. A. Dolliver, T. L. Gresham, G. B. Kistiakowsky, E. A. Smith and W. E. Vaughan, THIS JOURNAL, **60**, 440 (1938); J. B. Conn, G. B. Kistiakowsky and E. A. Smith, *ibid.*, **61**, 1868 (1939).

(22) "Steric Effects in Organic Chemistry," M. Newman, Ed., John Wiley and Sons, Inc., New York, N. Y., 1956, p. 37.

for this 2° change is very small, consequently substitution of hydrindane could easily introduce forces which would shift the conformation somewhat.

Appendix

The mathematical procedure in calculating atomic positions in a puckered cyclopentane (sym. Cs) is described briefly. First consider the carbon atoms; going clockwise around cyclopentane, the odd C-atom is given the number 1 and the position of each atom denoted by (x_i, y_i, z_i). z_i is given from equation 2 as a certain fraction of q, which is a measure of the puckering. All the x_i, y_i are unknowns except x_1 and y_1 which are set at zero.

$$(x_2 - x_1)^2 + (y_2 - y_1)^2 + (z_2 - z_1)^2 = (1.54)^2 \quad \text{(A-1)}$$

$$(x_3 - x_2)^2 + (y_3 - y_2)^2 + (z_3 - z_2)^2 = (1.54)^2 \quad \text{(A-2)}$$

$$x_3 = -x_4 = 0.77 \quad \text{(A-3)}$$

$$x_5 = -x_2 \quad \text{(A-4)}$$

$$y_3 = y_4 \quad \text{(A-5)}$$

$$y_2 = y_5 \quad \text{(A-6)}$$

$$y_1 = x_1 = 0 \quad \text{(A-7)}$$

$$z_1 = 0.63q \quad \text{(A-8)}$$

$$z_2 = z_5 = -0.51q \quad \text{(A-9)}$$

$$z_4 = z_3 = 0.20q \quad \text{(A-10)}$$

The scalar products of bond vectors yield

$$(1.54)^2 \cos \omega_1 = -x_2^2 + y_2^2 + (z_2 - z_1)^2 \quad \text{(A-11)}$$

$$(1.54)^2 \cos \omega_2 = x_2(x_2 - 0.77) + y_2(y_2 - y_3) + (z_2 - z_1)(z_2 - z_3) \quad \text{(A-12)}$$

Kilpatrick, Pitzer and Spitzer[3] found that the potential energy minimum lies very close to the configuration given by $\omega_1 = \omega_2$. We adopt this relation, which simplifies the calculations, and equate (A-11) and (A-12) which yields equation A-15. Equation A-1 may be reduced to A-13 and (A-2) to (A-14); thus we have three equations for x_2, y_2 and y_3 which must be solved simultaneously

$$x_2 = [(1.54)^2 - (1.14q)^2 - y_2^2]^{1/2} \quad \text{(A-13)}$$

$$y_3 - y_2 = [(1.54)^2 - (0.71q)^2 - (0.77 - x_2)^2]^{1/2} \quad \text{(A-14)}$$

$$y_2y_3 = 2x_2^2 - 0.77x_2 - (1.14q)(0.43q) \quad \text{(A-15)}$$

After several iterations have yielded a solution for x_2, y_2 and y_3 (which must be exact for A-13 and A-14 but may be only approximate for A-15), equations A-11 and A-12 may be solved for $\cos \omega_1$ and $\cos \omega_2$. $\cos \omega_3$ is given by

$$1.54 \cos \omega_3 = 0.77 - x_2 \quad \text{(A-16)}$$

The torsion angles are given by the equations

$$\cos \theta_{12} = \frac{\cos \omega_2(\cos \omega_1 - 1) - \sin {}^1/_2\omega_1(1 - 2 \sin {}^1/_2\omega_1)}{\sin \omega_1 \sin \omega_2} \quad \text{(A-17)}$$

$$\cos \theta_{23} = \frac{\cos \omega_2 \cos \omega_3 + \sin {}^1/_2\omega_1}{\sin \omega_2 \sin \omega_3} \quad \text{(A-18)}$$

$$\cos \theta_{34} = 1 \quad \text{(A-19)}$$

The above equations suffice for the calculation of torsional and C–C–C angle strain energies.

The hydrogen atom positions are determined in the following way: first define the vector

$$\mathbf{R}_i = 2\mathbf{r}_i - \mathbf{r}_{i-1} - \mathbf{r}_{i+1} \quad \text{(A-20)}$$

where the $\mathbf{r}_i$ are the vectors to the carbon atoms; then let

$$\mathbf{s}_i = \mathbf{R}_i/|\mathbf{R}_i| \quad \text{(A-21)}$$

Also let

$$\mathbf{T}_i = (\mathbf{r}_i - \mathbf{r}_{i+1}) \times (\mathbf{r}_i - \mathbf{r}_{i-1}) \quad \text{(A-22)}$$

and

$$\mathbf{t}_i = \mathbf{T}_i/|\mathbf{T}_i| \quad \text{(A-23)}$$

The two hydrogen atom positions on the ith CH_2 are now given by

$$\mathbf{h}_i = \mathbf{r}_i + 1.10\left[\sqrt{\frac{1}{3}}\,\mathbf{s}_i + \sqrt{\frac{2}{3}}\,\mathbf{t}_i\right] \quad \text{(A-24)}$$

$$\mathbf{h}'_i = + 1.10\left[\sqrt{\frac{1}{3}}\,\mathbf{s}_i - \sqrt{\frac{2}{3}}\,\mathbf{t}_i\right] \quad \text{(A-25)}$$

This treatment assumes a tetrahedral H–C–H angle and places the H atoms symmetrically with respect to the adjacent C–C bonds. It remains to calculate all of the interatomic distances and sum their inverse sixth powers in accordance with equation 1. This last calculation involves many steps and corresponding opportunities for error. Several calculations were made for Cs forms with increasing q values which gave an opportunity to check the trend of each interatomic distance and thereby identify errors. Since only one calculation was made for a C_2 form, it must be regarded as less reliable although it was carefully checked.

BERKELEY, CALIF.

C. Relativistic Effects on Molecular Properties

Introduction

For the calculation by quantum mechanics of most properties of chemical interest, relativistic effects are negligible for molecules comprised of light atoms. Thus, most quantum chemical calculations are based on the nonrelativistic Schrodinger equation. For elements of large atomic number, particularly for gold, mercury, and beyond, there are anomalies in periodic table trends which had long been recognized. These were usually ascribed to the effect of the 14-electrons in the $4f$ shell, but one wondered whether relativistic effects might also be important. By 1973, I though that calculational methods and approximations found to be valid for molecules with many-electron atoms in the middle range of atomic numbers had advanced to the point that exploration of relativistic effects in molecules was feasible. This sequence of papers includes an initial exploration by calculations for atoms, a first method for diatomic molecules which gave useful but not fully satisfactory results, and then a greatly improved method which has been used extensively for other molecules primarily by other scientists.

Also included as paper 27 is a review, written midway in this development, that is addressed to the nonspecialist reader.

Are elements 112, 114, and 118 relatively inert gases?

Kenneth S. Pitzer

Department of Chemistry and Inorganic Materials Research Division of the Lawrence Berkeley Laboratory, University of California, Berkeley, California 94720
(Received 14 April 1975)

Several recent papers present estimates of physical and chemical properties of super-heavy elements based on combinations of periodic table trends and relativistic Hartree–Fock (DHF) calculations for the atoms. The appropriateness of relativistic orbitals for covalent bonding must also be considered before reliable conclusions can be reached, and we are studying that aspect. Initial results lead to the striking conclusion that elements 112 and 114, with closed s^2 and $p_{1/2}^2$ shells, respectively, are volatile and relatively inert, as 118 is also expected to be. Since long nuclear lifetimes have been predicted for elements near 114, there is particular interest in this region. Earlier publications[1] offer no such striking conclusion for 112, but Keller *et al.*,[2] predict somewhat similar properties for 114 on the basis of rather different arguments.

In order to determine the effectiveness of a Dirac orbital for bonding in either a metal or a covalent compound, one may examine the angular function for each large component and its capacity to overlap the corresponding function on adjacent atoms. Alternatively, one may determine the appropriate functions for the very strong electric fields arising in bonding situations. Bethe and Salpeter[3] show that for strong electric fields m_l and m_s replace j and m as the good quantum numbers.

From the more detailed consideration of the angular functions,[4,5] one finds that relativistic s orbitals are appropriate for bonding. But a molecular orbital based on $p_{1/2}$ atomic orbitals is $\frac{1}{3}$ σ bonding and $\frac{2}{3}$ π antibonding on the basis of the large components. To obtain full σ-bonding character the combination $(\frac{1}{3} p_{1/2} + \frac{2}{3} p_{3/2})$ must be taken.

While our studies of this question are continuing, it appears that the best simple approximation for chemical bonding with p orbitals is to adopt valence states with weighted average $(\frac{1}{3} p_{1/2} + \frac{2}{3} p_{3/2})$ energies. Similarly for bonding with d orbitals the $(\frac{2}{5} d_{3/2} + \frac{3}{5} d_{5/2})$ average energies are appropriate while for s electrons the relativistic orbital energies may be used directly. These energies are best taken from the calculations of Desclaux[6] since these are for LS configuration averages and are otherwise the most accurate available.

In Table I are given the promotion energies and the heats of sublimation (in eV) for the group 2 and 4 elements. The valence states are sp and p^2, respectively; data for germanium and gray tin are omitted since they have tetravalent structures. The sums in the last column represent bonding energies, i.e., energies to dissociate the crystal to atoms in their valence states.

For group 2 elements the bonding energy is 5–6 eV which is insufficient to promote atoms of 112 from the s^2 closed shell to the sp valence state. Similarly for group 4 elements the bonding energy is about 4 eV which is insufficient to promote atoms of 114 from the $p_{1/2}^2$ closed shell to the p^2 valence state. Thus I estimate that both 112 and 114 in elementary form will be gases or very volatile liquids bound by dispersion (London) forces only.

Such a simple argument cannot preclude the possibility of some very weak metallic properties in the liquid state. Indeed extrapolation from Zn and Cd on this basis could suggest the same "prediction" for mercury as for element 112. But the excess of promotion energy over expected bond energy is much greater for either 112 or 114 than for Hg, and Hg is, indeed, a volatile liquid; hence our primary conclusion seems not too uncertain.

TABLE I. Energy values in eV.

	Promotion	Sublimation	Total
Zn	4.15	1.35	5.50
Cd	3.76	1.16	4.92
Hg	5.17	0.64	5.81
112	8.57	...	...
Sn	0.58	3.13	3.71
Pb	2.00	2.02	4.02
114	6.21	...	...

For element 112 the configuration $d_{5/2}^5 s^2 p$ is lower than $d_{5/2}^6 sp$. The former is not a good bonding state, however, because the large s^2 closed shell will generate strong repulsive forces before the d orbitals are close enough to interact and because the $d_{5/2}$ orbitals are not good bonding orbitals in any case.

For element 118 the conventional promotion energy np^6 to $np^5(n+1)s$ is 8.8 eV which is not much less than Rn (9.2) or Xe (9.9). One can also consider a valence state $np_{3/2}^3(n+1)s$ in which only the s electrons would be expected to bond effectively. Then for 118 the promotion energy is only 4.89 eV, but this far exceeds the bonding expected for a single s electron which, by analogy to the alkali metals, is about 1 eV. Hence 118 should also be a gas or very volatile liquid.

Estimates of the stability of primarily covalent compounds can be made on a similar basis. One assumes that the bonding energy of the compound (from the atom in its valence state) for 112 is the same as Hg and that for 114 is the same as Pb. If we assume elements 112 and 114 boil at room temperature, a heat of vaporization of about 6 kcal (0.26 eV) should be added to the promotion energies in Table I. Then 112 compounds will be less stable than the corresponding mercury compounds by 3.0_2 eV or 70 kcal/mole which is the difference in the energies to form atoms in their valence states. The corresponding difference of 114 from Pb is 2.4_5 eV or 56 kcal/mole. Addition of 70 kcal to the free energies of formation of mercury compounds[7] yields positive values (indicating instability) for most compounds including the neutral oxide, chloride, and bromide, although the $MCl_4^=$ and $MBr_4^=$ ions remain stable. There is no numerical value for HgF_2, but there is little doubt that the fluoride of 112 will be stable.

Earlier estimates[1] have not suggested any such extreme volatility and chemical inertness for element 112 as is here indicated.

The corresponding calculation for compounds of 114 yield positive free energies of formation for the oxide (+11) and the iodide (+15) but a small negative value for the bromide (−6). The chloride and fluoride are expected to be stable; however, with increasing ionic character the calculation on a covalent basis becomes less appropriate. Although our results indicate 114 to be somewhat more volatile and more noble than those of Keller *et al.*,[2] the difference is not large.

These properties of great volatility and ease of reduction to the element would appear to provide better separation methods than procedures based on uncertain quantitative similarities in solution chemistry of 112 to mercury and 114 to lead.

This research was sponsored by the Energy Research and Development Administration.

[1]B. Fricke and J. T. Waber, Actinides Rev. **1**, 433 (1971); B. Fricke, W. Greiner, and J. T. Waber, Theoret. Chim. Acta **21**, 235 (1971).

[2]O. L. Keller, Jr., J. L. Burnett, T. A. Carlson, and C. W. Nestor, Jr., J. Phys. Chem. **74**, 1127 (1970).

[3]H. A. Bethe and E. E. Salpeter, *Quantum Mechanics of One- and Two-Electron Atoms* (Academic, New York, 1957).

[4]R. E. Powell, J. Chem. Educ. **45**, 558 (1968).

[5]J. Oreg and G. Malli, J. Chem. Phys. **61**, 4349 (1974).

[6]J. P. Desclaux, At. Data Nucl. Data Tables **12**, 311 (1973).

[7]D. D. Wagman, W. H. Evans, V. B. Parker, I. Halow, S. M. Bailey, and R. H. Schumm, "Selected Values of Chemical Thermodynamic Properties," NBS Technical Notes 270-3 and 270-4.

760 J.C.S. Chem. Comm., 1975

Fluorides of Radon and Element 118

By Kenneth S. Pitzer

(*Department of Chemistry and Inorganic Materials Research Division of the Lawrence Berkeley Laboratory, University of California, Berkeley, California* 94720)

Summary Relativistic quantum mechanics as applied to radon (or element 118) fluoride structures indicates that ionic crystalline forms are probably more stable for the fluorides of these elements in contrast to the molecular form of xenon fluorides.

The formation and properties of radon fluoride have been studied by Stein[1] who found that a compound was readily formed but that its properties differed markedly from those of any xenon fluoride. Apparently radon fluoride does not vaporize as a molecular species but does decompose to the elements *in vacuo* above 250 °C.

The writer[2] has considered relativistic effects in relation to chemical bonding in certain heavy elements and their compounds. Briefly, it was found that relativistic s orbitals are suitable for bonding but that relativistic $p_{1/2}$ and $p_{3/2}$ orbitals are not. For example, a detailed consideration of a diatomic molecular orbital based on $p_{1/2}$ relativistic atomic orbitals indicates $1/3\ \sigma$-bonding and $2/3\ \pi$-antibonding character on the basis of the angular factors of the large components. To obtain full σ-bonding character the combination $1/3\ p_{1/2} + 2/3\ p_{3/2}$ must be taken. This restores the angular properties of familiar nonrelativistic orbitals. Relativistic differences in radial properties and in small components remain, of course.

As applied to the positive ion of a rare gas one concludes that the lowest energy state ($np^2_{1/2}\ np^3_{3/2},\ ^2P_{3/2}$) is not effective in bonding. The valence state for bonding is at the energy $1/3\ E(^2P_{1/2}) + 2/3\ E(^2P_{3/2})$, *i.e.*, higher by one third of the $p_{1/2}$–$p_{3/2}$ energy difference. This promotion energy is 0·44 eV for Xe, 1·42 eV for Rn, and 3·93 eV for element 118. The values for Rn and element 118 come from calculations of Desclaux.[3] While this promotion energy is moderate for Xe, it is comparable to the energy of an electron-pair bond in Rn and even larger for element 118. Thus it is plausible that Rn and element 118 fluorides are of an ionic rather than covalent character. While divalent ionic fluorides are possible, I consider in detail only the monovalent fluoride.

If molecular XeF_2 is stable in excess of xenon as compared to an ionic crystalline Xe^+F^-, the negative enthalpy of formation of XeF(s) cannot be greater than about half that of XeF_2, *i.e.*, *ca.* 0·61 eV. If this value, $\Delta H_f = -0{\cdot}61$ eV, is assumed for XeF, the ionic lattice energy is 10·76 eV, and a very approximate calculation indicates an Xe^+–F^- ionic distance of 2·34 Å. This is not unreasonable since Xe^+–F^- ionic distances range from 2·23 Å upward in various structures of solid XeF_6 and in complexes.[4]

For an estimate of the possible enthalpies of formation of ionic crystalline EF compounds of Rn and element 118,

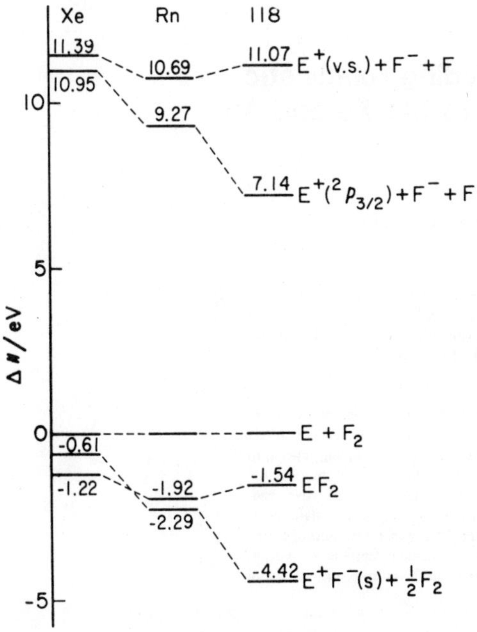

FIGURE. Energies or enthalpies of various states related to rare gas fluorides; v.s. means valence state as described in the text.

I assume the above ionic lattice energy and the appropriate ionization potentials.[3] Also one can assume the same bonding energy for a molecular EF_2 compound from $E^+ + F^- + F$ for radon and for element 118 as for xenon, and this yields values of the enthalpy of formation of EF_2. The results are shown in the Figure. In the presence of excess of fluorine, ionic XeF is unstable as compared to XeF_2 by 0·61 eV while ionic RnF is stable by 0·37 eV and ionic element 118–F by 2·88 eV. These are very rough estimates, since both ionic lattice energies and bond energies will certainly change somewhat from Xe to Rn to element 118. Nevertheless, this shows that radon fluoride might be a simple Rn^+F^- ionic crystal (although an ionic RnF_2 is not necessarily excluded) and that the ionic crystal is strongly indicated for element 118 fluoride.

An ionic crystalline form of RnF would be expected to be non-volatile, as observed, and to show the observed migration of Rn as a cation in electrolysis. This proposal of ionic crystalline character for radon fluoride appears to merit further consideration. Astatine fluoride shows a non-volatile character[5] similar to that of radon fluoride and in contrast to other halogen fluorides. Thus an ionic crystalline state should also be considered for astatine fluoride. Although no detailed calculations have been made for that case, the same relativistic effects will occur as for radon.

This research was sponsored by the Energy Research and Development Administration. Discussions with L. Stein and N. Bartlett are gratefully acknowledged.

(*Received*, *27th June* 1975; *Com.* 726.)

[1] L. Stein, *Science*, 1970, **168**, 362.
[2] K. S. Pitzer, *J. Chem. Phys.*, in the press.
[3] J. P. Desclaux, *Atomic Data and Nuclear Data Tables*, **1973**, **12**, **311**.
[4] N. Bartlett and F. O. Sladky, in 'Comprehensive Inorganic Chemistry,' J. C. Bailar, *et al.*, eds., Pergamon Press, 1973, Vol. I, pp. 213—330.
[5] E. H. Appelman, E. N. Sloth, and M. H. Studier, *Inorg. Chem.*, 1966, **5**, 766.

Reprinted from The Journal of CHEMICAL PHYSICS Vol. 67, No. 12, 15 December 1977

Ab initio effective core potentials including relativistic effects. I. Formalism and applications to the Xe and Au atoms

Yoon S. Lee, Walter C. Ermler, and Kenneth S. Pitzer

Department of Chemistry and Lawrence Berkeley Laboratory, University of California, Berkeley, California 94720
(Received 25 July 1977)

An effective core potential system has been developed for heavy atoms in which relativistic effects are included in the effective potentials (EP). The EP's are based on numerical Dirac–Hartree–Fock calculations for atoms and on the Phillips–Kleinman transformation with other aspects similar to the treatments of Goddard and Melius and Kahn, Baybutt, and Truhlar. The EP's may be written

$$U^{\mathrm{EP}} = \Sigma_l \, \Sigma_{j=|l-1/2|}^{l+1/2} \, \Sigma_{m=-j}^{j} \quad U_{lj}^{\mathrm{EP}}(r)|ljm\rangle\langle ljm|,$$

where $|ljm\rangle$ is a two-component angular basis function that is a product of a two-component Pauli spinor and spherical harmonics. The numerical functions $U_{lj}^{\mathrm{EP}}(r)$ are approximated as expansions in terms of Gaussian or exponential functions. The use of these EP's enables one to use the *jj*-coupling scheme for subsequent applications in all-valence-electron calculations on heavy atoms and their molecules. A standard atomic SCF program has been modified to accommodate these EP's and Gaussian and exponential basis sets having the proper *j* angular dependence. Energy levels for many atomic states of Xe and Au were calculated. The study of Xe excited states indicates that the spin–orbit splittings are reasonably approximated and that the numerical DHF calculations are adequately reproduced. Au has been treated as an atom with 1, 11, 17, 19, or 33 valence electrons to investigate the effects of redefinition of the core.

I. INTRODUCTION

The results of relativistic atomic structure calculations [Dirac–Hartree–Fock (DHF)[1] and Dirac–Hartree–Fock–Slater (DHFS)[2]] indicate that many properties of molecules containing heavy atoms may not be properly calculated without considering the relativistic effects.

Full self-consistent-field (SCF) calculations of those molecules, however, have not been practical even without relativistic effects due to the large number of two-electron integrals that arise. The H_2^+ molecular ion has been treated quite rigorously.[3] Relativistic effects on other molecules have been studied using such approximations as single-center expansions,[4,5] perturbation theory,[6] semiempirical models,[7] and a discrete-basis-set method using a local exchange approximation.[8] Other very recent treatments are discussed below.

One of the promising ways to handle these systems is to treat only the valence electrons explicitly using a frozen-core approximation. This usually involves the substitution of the effect of the core electrons with some form of pseudopotential. Many methods have been developed and we modify one of the *ab initio* approaches[9,10] to include the relativistic effects in the effective core potentials (EP's).

When the EP including relativistic effects is used for valence-electron calculations, the question of whether the relativistic or the nonrelativistic method should be used for valence electrons is still to be examined. The choice depends on the importance of relativistic effects on the valence electrons and this is discussed in the next section. The following sections describe the method to obtain EP's from the atomic DHF calculations and the application of these EP's to valence-electron atomic calculations.

It might also be possible to evaluate semiempirical model potentials to fit selected experimental properties of heavy atoms in a manner analogous to that which has been followed for lighter atoms where relativistic effects are small. Relativistic effects would be included, at least implicitly, for heavy atoms. We have chosen not to proceed in this way but rather to follow methods based upon *ab initio* calculations for atoms and the Phillips–Kleinman[11] transformation to effective core potentials.

II. RELATIVISTIC EFFECTS ON VALENCE ELECTRONS

Relativistic effects on valence electrons may be observed by comparing DHF with HF calculations. Although this topic has been already discussed by Desclaux and co-workers,[12] a slightly different approach is taken here.

The relativistic Hamiltonian used for atomic DHF calculations is given for n electrons by[13,14]

$$H^{\mathrm{rel}} = \sum_{i=1}^{n} h_D(i) + \sum_{i>j} \frac{1}{r_{ij}} \tag{1}$$

when many-electron relativistic interaction terms, which are usually approximated by Breit interaction terms,[14] are not included. In Eq. (1) h_D is the Dirac one-electron Hamiltonian which may be written as

$$h_D = c\boldsymbol{\alpha} \cdot \mathbf{p} + \beta c^2 - \frac{Z}{r} , \tag{2}$$

where r is the distance from the nucleus of charge Z and $\mathbf{p}$ is the momentum operator $(-i\nabla)$. Also,

$$\boldsymbol{\alpha} = \begin{pmatrix} 0 & \sigma^P \\ \sigma^P & 0 \end{pmatrix} ,$$

with σ^P being a Pauli matrix and

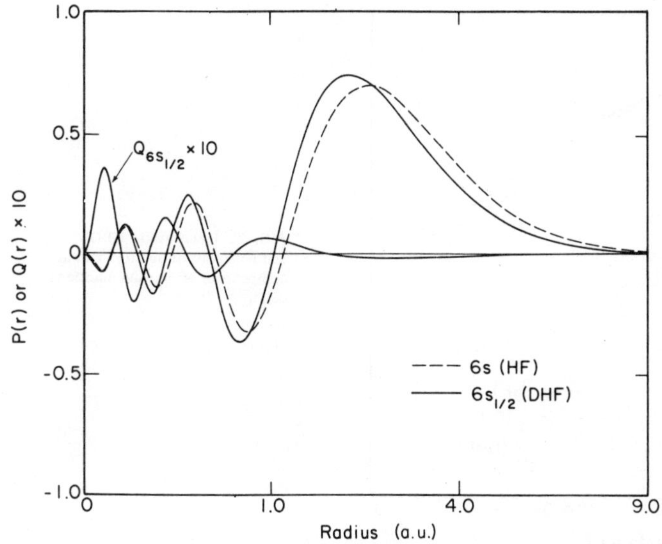

FIG. 1. Relativistic (DHF) and nonrelativistic (HF) radial components of $6s$ wavefunctions of Pb. Only the small component of DHF is labeled.

$$\beta=\begin{pmatrix} \mathrm{I} & 0 \\ 0 & -\mathrm{I} \end{pmatrix}$$

with I the 2×2 unit matrix. (In atomic units, which are used throughout this paper, c is equal to $1/\alpha$, where α is the hyperfine structure constant).

The eigenfunctions of the above Hamiltonian are four-component Dirac spinors which may be expressed as

$$\psi_{nkm}=\frac{1}{r}\begin{bmatrix} P_{nk}(r) & \chi_{km}(\theta,\varphi) \\ iQ_{nk}(r) & \chi_{-km}(\theta,\varphi) \end{bmatrix}\ , \tag{3}$$

where

$$\chi_{km}(\theta,\varphi)=\sum_{\sigma=\pm 1/2} C(l\tfrac{1}{2}j;\, m-\sigma,\sigma)\, Y_{\lambda}^{m-\sigma}(\theta,\varphi)\,\phi_{1/2}^{\sigma}\ ,$$

$Y_{\lambda}^{m-\sigma}$ is a spherical harmonic, $\phi_{1/2}^{1/2}=\binom{1}{0}$ and $\phi_{1/2}^{-1/2}=\binom{0}{1}$ are Pauli spinors, $C(l\frac{1}{2}j;\, m-\sigma,\sigma)$ are Clebsch–Gordan coefficients, and $\lambda=k-(j+\frac{1}{2})$ if $j=l-\frac{1}{2}$ and $\lambda=-(k+1)$ $=(j-\frac{1}{2})$ if $j=l+\frac{1}{2}$. In Eq. (3) k is the relativistic quantum number. P_{nk} and Q_{nk} satisfy the following coupled equations when V is a potential in the central field of the nucleus

$$\frac{dP_{nk}}{dr}+\frac{kP_{nk}}{r}-\left\{\frac{2}{\alpha}+\alpha[V(r)-\epsilon_{nk}]\right\}Q_{nk}=0\ , \tag{4}$$

$$\frac{dQ_{nk}}{dr}-\frac{kQ_{nk}}{r}+\alpha[V(r)-\epsilon_{nk}]P_{nk}=0\ . \tag{5}$$

In the nonrelativistic limit these two equations are reduced to the familiar Schrödinger equation after eliminating the small component Q_{nk}.[14] Therefore, by comparing the large component P_{nk} with the corresponding HF radial wavefunctions a reasonable qualitative picture of relativistic effects may be obtained. Also, the magnitude of the small component gives a measure of the relativistic effect.

Radial wavefunctions for several atoms have been calculated using the numerical DHF program of Desclaux.[13] The calculations have been performed with the assumption of point nuclei to facilitate comparison with the published HF calculations of Mann.[15] The effect of this approximation is small for the valence electrons although it may be substantial for $1s$ electrons.[16] Some valence orbitals of Pb and U are shown in Figs. 1–3, where the magnitude of the small component has been multiplied by ten for purposes of clarity. There is a relatively large contraction in the $6s$ orbital of Pb as shown in Fig. 1. This is due to the penetrating nature of $6s$ electrons and less effective screening by inner electrons with high angular quantum numbers (e.g., $5d$). The $6p_{1/2}$ orbital of Pb (Fig. 2) is noticeably different from the nonrelativistic $6p$, whereas the $6p_{3/2}$ is almost the same as the nonrelativistic $6p$. The $6d$ (Fig. 3) wavefunctions of U are more diffuse in the relativistic case due to the more effective shielding of the nuclear charge by electrons with smaller angular quantum numbers. This behavior is also found in the $5f$ wavefunctions.[16] For this reason the roles of $5f$ and $6d$ electrons in bonding are probably more important than would be estimated from nonrelativistic calculations.

Small components are relatively large near the nucleus and decrease rapidly as the radial distance increases. The electron densities represented by small components are negligible compared with those by large components for valence electrons of U (less than 0.03%) although 15% of the electron density is due to the small component for the $1s$ electrons.[16] Since small components become even less important in the outer region, it may be reasonable to assume that small components can be neglected in calculations that emphasize the description of valence electrons.[12(b)]

The total relativistic effect, or the difference between

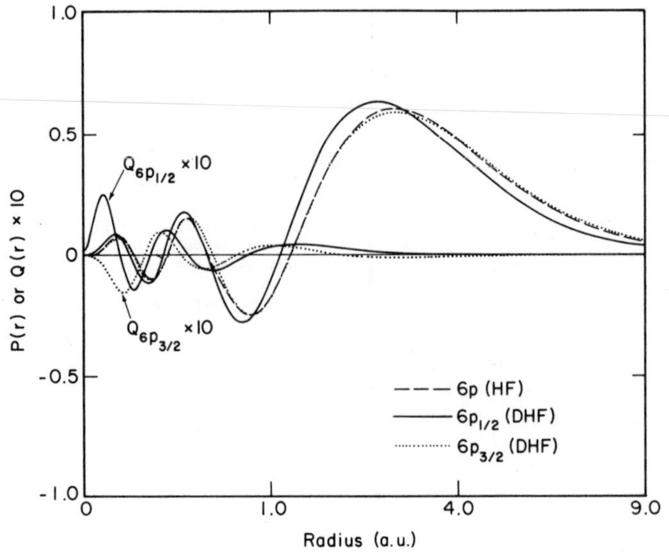

FIG. 2. Relativistic (DHF) and nonrelativistic (HF) radial components of $6p$ wavefunctions of Pb.

results of DHF and HF calculations in our case, may be divided into two kinds, one due to the use of different Hamiltonians (direct effects) and the other to rearrangement (indirect effects). The relativistic character of valence electrons apparently depends more on the magnitude of the former than that of the latter. Although the exact separation of the two may not be possible, the former can often be reasonably estimated from perturbation theory.[17]

From the Pauli approximation of the Dirac theory[17,18] the effective relativistic Hamiltonian, which is correct to the order of $1/c^2$, may be expressed as

$$H^{rel}_{eff}(r) = H_0(r) + H_m(r) + H_d(r) + H_{SO}(r) , \tag{6}$$

where H_0 is the nonrelativistic Hamiltonian, H_m the mass–velocity correction term, H_d the Darwin term, and H_{SO} the spin–orbit coupling term. The corresponding matrix elements, or energy expectation values, may be expressed as $E_0 \equiv E^0_{nl}$ (nonrelativistic energy eigenvalue of nl shell electrons) and

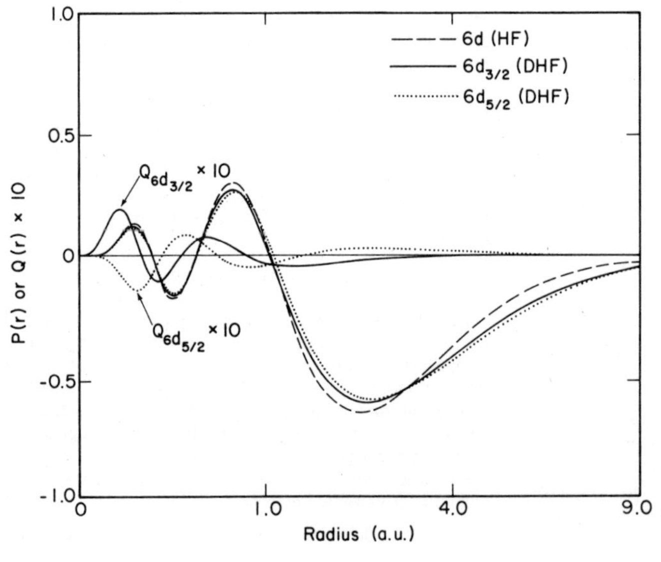

FIG. 3. Relativistic (DHF) and nonrelativistic (HF) radial components of $6d$ wavefunctions of U.

TABLE I. Orbital energies of Pb valence electrons (all are negative and in a.u.)

	DHF	HF[a]	DHFS[b]	HFS(I)[c]	HFS(II)[c,d]
$6s$	0.5670	0.4586	0.5259	0.4444	0.5883
$6p_{1/2}$	0.2750	0.2398	0.2376	0.2123	0.2847
$6p_{3/2}$	0.2198				0.2380

[a]From Ref. 15.
[b]From Ref. 2 (the electrons were considered to occupy only the $6p_{1/2}$ orbital).
[c]From Ref. 18.
[d]The relativistic energy calculated from the perturbation theory.

$$E_m = -\left(\frac{\alpha^2}{2}\right)\int_0^\infty dr\, r^2[R^0_{nl}(r)]^2[E^0_{nl} - V(r)]^2 , \tag{7}$$

$$E_d = -\left(\frac{\alpha^2}{4}\right)\int_0^\infty dr\, r^2 R^0_{nl}(r)\left[\frac{dV(r)}{dr}\right]\left[\frac{dR^0_{nl}(r)}{dr}\right] , \tag{8}$$

$$E_{SO} = -\left(\frac{\alpha^2}{4}\right)(k+1)\int_0^\infty dr\, r^2[R^0_{nl}(r)]^2\left[\frac{dV(r)}{dr}\right]\left(\frac{1}{r}\right) , \tag{9}$$

where R^0_{nl} is the nonrelativistic radial wavefunction.

Since $V(r)$ and P_{nk} decrease rapidly in the outer region, almost all contributions to the E_m terms are from the core region, as expected on physical grounds. Terms inside the integral decrease even more rapidly for Darwin and spin–orbit coupling effects. Therefore, valence electrons appear to be essentially nonrelativistic in the valence region, although there are substantial relativistic effects, direct or indirect, on valence electrons as indicated by the orbital energies shown in Table I. This evidence suggest that treating the valence electrons nonrelativistically, but in the field of an EP derived from a DHF calculation for the atom, may be a reasonable approximation.

III. EFFECTIVE CORE POTENTIALS FROM ATOMIC DHF CALCULATIONS

A. The DHF approximation

The general practice in atomic calculations is to express the many-electron wavefunctions as antisymmetrized products of one-electron wavefunctions. In the case of relativistic calculations each one-electron wavefunction is a four-component Dirac spinor as given by Eq. (3).

If we group the spinors into two classes, core orbitals and valence orbitals, the wavefunction for a many-electron atom may be expressed as

$$\Psi^R = \mathcal{A}[(\psi_1^c\psi_2^c \cdots \psi_m^c)(\psi_1^v\psi_2^v \cdots \psi_n^v)] \tag{10}$$

for a single configuration. In Eq. (10) $\mathcal{A}$ is an antisymmetrizer, m is the number of core electrons, and n is the number of valence electrons. The total relativistic energy of the atom corresponding to the wavefunction of Eq. (10) is

$$E_T^R = \langle \Psi^R | H^{rel} | \Psi^R \rangle / \langle \Psi^R | \Psi^R \rangle , \tag{11}$$

where H^{rel} is given in Eq. (1).

Since, in many respects, the relativistic formalism for the many-electron atom is essentially the same as the nonrelativistic one, when the Breit interaction and quantum electrodynamic effects are not included many methods used in the nonrelativistic theory are also valid here. The total energy may be expressed as the sum of core, valence, and core–valence interaction energies

$$E_T^R = E_c^R + E_v^R + E_{cv}^R . \tag{12}$$

As has been done in the nonrelativistic case[10] the last two terms may be combined by using a modified Hamiltonian and valence wavefunction

$$E_v + E_{cv} = \langle \Psi_v^R | H_v^{rel} | \Psi_v^R \rangle , \tag{13}$$

where

$$H_v^{rel} = \sum_v \left\{ h_D(v) + \sum_c [J_c(v) - K_c(v)] \right\} + \sum_{v>v'} \frac{1}{r_{vv'}} \tag{14}$$

and $J_c(v)$ and $K_c(v)$ are Coulomb and exchange operators, respectively. H_v^{rel} is a combination of the original Hamiltonian and additional terms which represent the effect of the core.

Our effective potentials are obtained from atomic DHF calculations using a method that is essentially parallel to that developed by Goddard[9] and Kahn[10] and their coworkers for the nonrelativistic case, although it was necessary to introduce some modifications and approximations to accommodate the relativistic characteristics.

B. Formal definition of effective core potentials

Effective potentials (EP's) have been most widely applied to one-valence-electron systems, and in these cases the physical meanings of EP's can be rather well understood. In this section EP's are derived first for the one-valence-electron atoms and later they are generalized to include many-valence-electron atoms.

When an orthonormal set is used for valence and core orbitals, the DHF equation for a single valence electron outside the closed-shell core becomes

$$\left[h_D + \sum_c (J_c - K_c)\right]\psi_v = \epsilon_v\psi_v + \sum_c \psi_c\epsilon_{cv} , \tag{15}$$

where J_c and K_c are Coulomb and exchange operators, respectively, and the ϵ_{cv}'s are off-diagonal Lagrange multipliers which may be written as

$$\epsilon_{cv} = \langle \psi_v | h_D + \sum_c (J_c - K_c) | \psi_c \rangle . \tag{16}$$

Equation (15) is essentially the same as that of the HF formulation[10] except that operators are 4×4 matrices operating on four-component wavefunctions which have the form given by Eq. (3).

A core projection operator may be defined as

$$\mathcal{P} = \sum_c |\psi_c\rangle\langle\psi_c| , \tag{17}$$

where the bra $\langle\psi_c|$ is a row vector with four components and the ket $|\psi_c\rangle$ is a column vector with four components. With Eqs. (15) and (17) it is possible to follow the nonrelativistic formalism to obtain the relativistic version of the generalized Phillips–Kleinman pseudopotential.[11] The complete derivation is omitted here since it may be found in the nonrelativistic work.[10] Several essential parts, however, will be reproduced in the following to clarify the development. Using Eq. (17), Eq. (15) may be rewritten as

$$(1-\mathcal{P})H_v^{rel}\psi_v=\epsilon_v\psi_v\ , \tag{18}$$

where

$$H_v^{rel}=h_D+\sum_c (J_c-K_c)\ . \tag{19}$$

If we define the pseudo-orbital χ_v^R as

$$\chi_v^R\equiv\psi_v^R+\sum_c a_c\psi_c^R \tag{20}$$

such that

$$\psi_v^R=(1-\mathcal{P})\chi_v^R\ , \tag{21}$$

then Eq. (18) becomes

$$(1-\mathcal{P})H_v^{rel}(1-\mathcal{P})\chi_v^R=\epsilon_v(1-\mathcal{P})\chi_v^R\ . \tag{22}$$

From Eq. (22) one can define the relativistic analog of the generalized Phillips–Kleinman pseudopotential

$$V^{RGPK}=-\mathcal{P}H_v^{rel}-H_v^{rel}\mathcal{P}+\mathcal{P}H_v^{rel}\mathcal{P}+\epsilon_v\mathcal{P}\ , \tag{23}$$

where

$$(H_v^{rel}+V^{RGPK})\chi_v^R=\epsilon_v\chi_v\ , \tag{24}$$

or

$$(h_D+U^{core})\chi_v^R=\epsilon_v\chi_v\ , \tag{25}$$

if

$$U^{core}=\sum_c (J_c-K_c)+V^{RGPK}\ . \tag{26}$$

It may be noted that U^{core} in Eq. (25) is again a 4×4 matrix operating on four-component spinors χ_v^R's.

To investigate the characteristics of the EP desired we assume that U^{core} is expanded in terms of an angular basis set of Dirac spinors. The following equations are then obtained for the radial pseudowavefunction:

$$\frac{d}{dr}\begin{bmatrix}P_v^{ps}(r)\\ Q_v^{ps}(r)\end{bmatrix}=\begin{Bmatrix}-\dfrac{k_v}{r} & \left(\dfrac{2}{\alpha}\right)+\alpha[\epsilon_v-U_v^Q(r)]\\ -\alpha[\epsilon_v-U_v^P(r)] & \dfrac{k_v}{r}\end{Bmatrix}\begin{bmatrix}P_v^{ps}(r)\\ Q_v^{ps}(r)\end{bmatrix}\ , \tag{27}$$

where $U_v^Q(r)$ and $U_v^P(r)$ are radial components of U^{core}.

In theory one may derive $U_v^P(r)$ and $U_v^Q(r)$ in Eq. (27) from any $P_v^{ps}(r)$ and $Q_v^{ps}(r)$ satisfying Eq. (22). Relativistic effective core potentials obtained from these pseudo-orbitals would not be continuous at points where nodes of $P_v^{ps}(r)$ or $Q_v^{ps}(r)$ occur. This difficulty may be avoided if the transformation of Eq. (20) can be carried out in a manner to yield nodeless $P_v^{ps}(r)$ and $Q_v^{ps}(r)$ functions. However, nodes may not, in general, be removed simultaneously for the large and small components.

Since the role of small components and direct relativistic effects are assumed to be negligible in the valance region, as discussed in the previous section, we develop a method which does not include manipulations with small components. The Dirac Hamiltonian can be expanded after substitution with the wavefunctions containing only large components. The radial equation obtained this way may be given as[17]

$$\{H_0(r)+H_m(r)+H_d(r)+H_{SO}(r)+H[O(\alpha^4)]\}P_v(r)=\epsilon_vP_v(r)\ , \tag{28}$$

where all terms have the same meaning as in Eq. (6) and $H[O(\alpha^4)]$ denotes the higher order terms which are not included in the Pauli approximation.

If higher order terms are neglected in Eq. (28), a $U_v^{core}(r)$ which is correct within the Pauli approximation can be obtained from a nodeless $P_v^{ps}(r)$. This involves solving a differential equation [Eq. (28)] for $U_v^{core}(r)$. Furthermore, it is correct only to order α^2 and all terms of the Hamiltonian have to be included when the EP obtained is applied in molecular calculations. It may be noted that the EP from Eq. (28) still contains higher-order relativistic corrections, since ϵ_v is the solution of the DHF equation.

However, for our purposes it is of little importance to have an accurate explicit representation of relativistic effects arising from the core region, where most of these effects exist as shown in the previous section, since we are mainly concerned with a reasonable representation of valence orbitals in the valence region. Thus, all relativistic effects may be incorporated into the EP.

With these approximations the radial equation becomes

$$[H_0'(r)+U_v^{EP}(r)]P_v(r)=\epsilon_vP_v(r)\ , \tag{29}$$

where $H_0'(r)$ denotes the absence of core potential terms from the nonrelativistic Hamiltonian. The EP in Eq. (29) contains all one-electron relativistic effects of the atom being considered. When this EP is used for calculations on molecules or excited states of atoms, the validity of the results will depend on the difference between relativistic effects of the molecule and its constituent atoms or between that in ground and excited electronic states. This approximation appears to be reasonable since the most pronounced relativistic effects are present near the nucleus, as was discussed above.

The original form of Eq. (29), before the reduction to the radial equation, is

$$\left(-\tfrac{1}{2}\nabla^2-\frac{Z}{r}+U^{EP}\right)\chi_v'=\epsilon_v\chi_v'\ , \tag{30}$$

where χ_v' is a two-component pseudowavefunction containing only the large radial component of Eq. (3).

When there is more than one valence electron the wave equation may be given as[10]

$$\left(-\frac{1}{2}\nabla^2 - \frac{Z}{r} + U_v^{EP\,\prime} + W'\right)\chi_v' = \epsilon_v \chi_v' , \tag{31}$$

where W' includes all interactions between pseudo-orbitals and between pseudo-orbitals and core orbitals. Since the ultimate goal of this procedure is the reasonable and simple representation of the behavior of valence electrons in the valence region, interactions between the pseudo-orbital and the core orbitals may also be included in EP. Then Eq. (31) becomes

$$\left[-\frac{1}{2}\nabla^2 - \frac{Z}{r} + U_v^{EP} + W(\chi_v', \chi_{v'}')\right]\chi_v' = \epsilon_v \chi_v' , \tag{32}$$

where $W(\chi_v', \chi_{v'}')$ is the sum of Coulomb and exchange interactions of a given psuedo-orbital χ_v' with all other pseudo-orbitals $\chi_{v'}'$.

From Eq. (32) it is seen that U_v^{EP} will not be the same for different pseudo-orbitals. Therefore, the EP may be conveniently expressed as products of angular projection operators and radial functions, as are the EP's derived from HF calculations.[10] Since orbitals with different total angular momenta j but with the same orbital angular momentum l are nondegenerate in DHF results, the EP is expressed as [the lj indices of Eq. (33) can also be expressed as one index k, the relativistic quantum number in Eq. (3)]

$$U^{EP} = \sum_{l=0}^{\infty} \sum_{j=|l-1/2|}^{l+1/2} \sum_{m=-j}^{j} U_{lj}^{EP}(r)|\, ljm\rangle\langle ljm| , \tag{33}$$

with the understanding that $|ljm\rangle$ and $\langle ljm|$ are two-component angular bases that are eigenfunctions in the Pauli approximation of the Dirac Hamiltonian.[17] It may be noted that the direct application of the EP of Eq. (33) in atomic and molecular calculations requires the use of jj-coupling, instead of conventional LS-coupling schemes.

C. Pseudo-orbitals

In the previous section it has been shown that EP's can be obtained if suitable pseudo-orbitals in the form of Eq. (20) are found. Because of orthogonalities of angular basis sets with distinct angular quantum numbers, the radial part of the pseudo-orbitals may be separated and expressed as

$$P_{nk}^{ps}(r) = \sum_{q=1}^{n} a_{qk} P_{qk}(r) , \tag{34}$$

where n is a principal quantum number, k is the relativistic quantum number previously defined, and the P_{qk}'s are the large components of DHF radial wavefunctions.

For our purposes the following properties are desired for $P_{nk}^{ps}(r)$: (i) The $P_{nk}^{ps}(r)$ should be nodeless and (ii) the form of $P_{nk}^{ps}(r)$ should be similar to that of the original valence wavefunction $P_{nk}(r)$ in the valence region. Property (i) is essential to prevent the occurrence of singularities, since the derivation of $U_{lj}^{EP}(r)$ from the radial part of Eq. (32) [Eq. (37) of the following section] involves division by $P_{nk}^{ps}(r)$, whereas (ii) will hopefully prevent the generation of physically unreasonable pseudo-wavefunctions.

TABLE II. a_{qk}'s of 6s (Au) for different choices of t.

t	n_i[a]	a_{1s}	a_{2s}	a_{3s}	a_{4s}	a_{5s}	a_{6s}
1.0	1	0.0031	0.0192	0.0634	0.1730	0.4650	0.8658
1.5	1	0.00060	0.0066	0.0331	0.1246	0.4303	0.8935
2.0	1	0.0001	0.0024	0.0176	0.0901	0.3976	0.9131
2.5							
.							
.		Nodes remain after eight iterations.					
.							
5.0							

[a] n_i is the number of iterations required to eliminate all nodes from $P_{nk}(r)$.

Although many methods developed for the nonrelativistic case[9,10] may be applicable, we pursue the following procedure for finding the appropriate coefficients a_{qk}: Initially, the a_{qk}'s are obtained from a least squares fit of a pseudo-orbital of Eq. (34) to a given Slater-type function $C_\zeta r^t e^{-\zeta r}$. C_ζ and ζ are selected so that the maximum of the function $C_\zeta r^t e^{-\zeta r}$ coincides with that of the original valence orbital $P_{nk}(r)$. This choice of ζ and C_ζ probably satisfies our condition (ii) better than the procedure based on the orbital energy adopted by Goddard and Melius.[9] For this choice of the standard function the least squares fit is accomplished by minimizing the function

$$F = \sum_{i=1}^{N_p} w(r_i)\left[\sum_{q=1}^{n} a_{qk} P_{qk}(r_i) - C_\zeta r_i^t e^{-\zeta r_i}\right]^2 , \tag{35}$$

where N_p is the number of grid points for the numerical radial valence orbital $P_{nk}(r)$[13] and $w(r_i)$ is a weight of a point r_i.

The starting values of $w(r_i)$'s are unities. When nodes are not eliminated with these weights the $w(r_i)$'s are changed to $[1/P_{nk}^{ps}(r_i)]^2$ of the previous run and the minimization of F in Eq. (35) is repeated until all nodes are eliminated.

In DHF calculations $P_{nk}(r)$ near the origin may be expanded in terms of a polynomial

$$P_{nk}(r) = r^{\gamma_k}(p_0 + p_1 r + p_2 r^2 + \cdots) , \tag{36}$$

where $\gamma_k = \sqrt{k^2 - (\alpha Z)^2}$. Although γ_k may be used as t in Eq. (35), several integer and half-integer values are tried and the one that appears to be most satisfactory in terms of condition (ii) is selected. In the case of the 6s orbital of Au (Table II) $t = 2.0$ was chosen since this value of t yields the largest a_{6s}. The other criterion used in the selection of t is the smoothness of the orbitals, as judged by the presence or absence of oscillations. Typically, values of t range from 1.5 to 3.0 with larger values for orbitals with higher angular momenta.[16]

Large and small radial components of the 6s valence orbital and pseudo-orbital of Au are shown in Figs. 4 and 5, respectively. It is interesting that the small component of the pseudo-orbital becomes quite small compared with that of the original valence orbital in the core re-

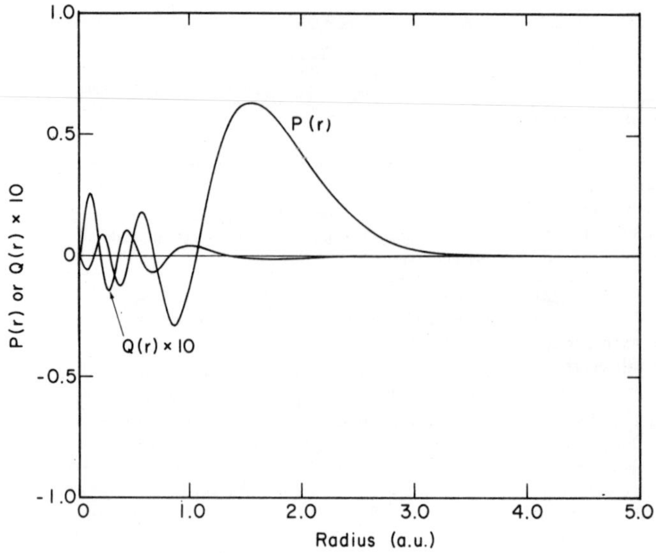

FIG. 4. Radial components of 6s orbitals of Au (DHF calculation).

gion. This seems to be the case for s and p orbitals for other heavy atoms; we found it true also for Bi. Based upon this observation one may expect that the role of the small components is even less important than that anticipated from the analysis of Sec. II, which makes our approximation of eliminating small components even more attractive.

Finally, it should be noted that the above procedure of determining pseudo-orbitals is not unique since the selection of t and ζ is not.

D. Effective potentials

The following radial equation for a closed shell or the average energy of an open shell configuration may be algebraically solved for each term $U_{lj}^{EP}(r)$ of the EP of Eq. (33):

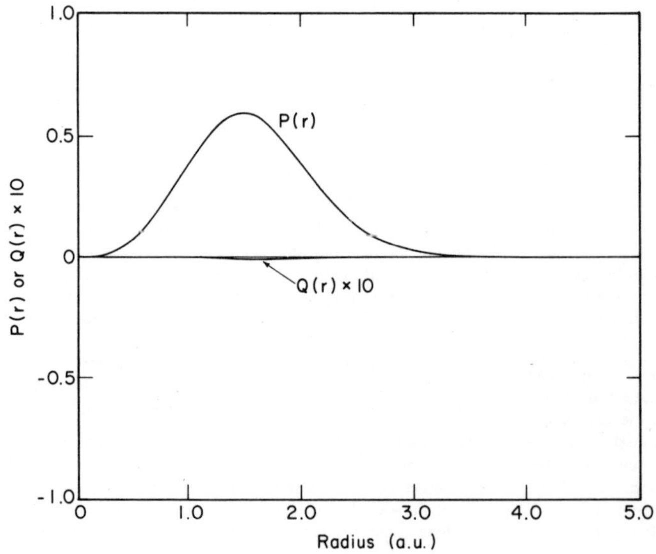

FIG. 5. Radial components of 6s pseudo-orbitals of Au obtained with $t=2.0$ (Table II).

$$\left[-\frac{1}{2}\frac{d^2}{dr^2}+\frac{l_A(l_A+1)}{2r_1^2}+U^{\mathrm{EP}}_{l_A j_A}(r_1)-\frac{z}{r_1}\right]P_A(r_1)+\left[(N_A-1)\int_0^\infty P_A(r_2)^2\frac{1}{r_>}dr_2-\frac{1}{2}\left(\frac{N_A-1}{2j_A}\right)\right.$$
$$\times\sum_{\nu>0}(2j_A+1)\Gamma_{j_A\nu j_A}\int_0^\infty P_A(r_2)^2\frac{r_<^\nu}{r_>^{\nu+1}}dr_2+\left.\sum_{A'\neq A}N_{A'}\int_0^\infty P_{A'}(r_2)^2\frac{1}{r_>}dr_2\right]P_A(r_1)$$
$$-\frac{1}{2}\sum_{A'\neq A}N_{A'}\sum_\nu\Gamma_{j_A\nu j_{A'}}\int_0^\infty P_A(r_2)P_{A'}(r_2)\frac{r_<^\nu}{r_<^{\nu+1}}dr_2P_{A'}(r_1)=\epsilon_A P_A(r_1)+\sum_{A'\neq A}N_{A'}\delta(k_A,k_{A'})P_{A'}(r_1)\epsilon_{AA'}\,, \tag{37}$$

where l_A, j_A, k_A, and N_A denote the orbital angular momentum, the total angular momentum, the relativistic quantum number, and the number of electrons, respectively, of the shells A and A' with the understanding that $P_A(r)$ and $P_{A'}(r)$ refer to pseudo-orbitals and ϵ_A and $\epsilon_{AA'}$ are diagonal and off-diagonal Lagrange multipliers, respectively. The $r_<$ and $r_>$ are the smaller and the larger of r_1 and r_2, respectively, and ν satisfies the triangular condition

$$|j_A-j_{A'}|\leq\nu\leq j_A+j_{A'}\,, \tag{38}$$

with the additional requirement of

$$\begin{aligned} j_A+j_{A'}+\nu &= \text{even}, \quad \text{if } a\neq a'\,,\\ &= \text{odd}\,, \quad \text{if } a=a'\,, \end{aligned} \tag{39}$$

where a and a' are signs of relativistic quantum numbers k_A and $k_{A'}$, respectively. $\Gamma_{j_A\nu j_{A'}}$ is defined in terms of $3j$ symbols by

$$\Gamma_{j_A\nu j_{A'}}=2\begin{pmatrix} j_A & \nu & j_{A'}\\ \frac{1}{2} & 0 & -\frac{1}{2}\end{pmatrix}^2\ . \tag{40}$$

Equation (37) can be obtained from the similar equation of DHF theory.[14]

In theory an infinite number of calculations for highly excited states is required to complete the expansion of the EP given by Eq. (33), since there are only a few occupied valence orbitals in neutral atoms. This difficulty also exists in the nonrelativistic case and is resolved by using the closure property of the projection operator with the assumption that radial parts of EP are the same for all orbitals having higher angular quantum numbers than are present in the core. The same approximation is applicable in our approach if relativistic effects are not too different for electrons in the highly excited orbitals. We expect that this is the case since those electrons spend less time near the nucleus. If this assumption is valid, the EP may be given by

$$U^{\mathrm{EP}}\approx U^{\mathrm{EP}}_{LJ}(r)+\sum_{l=0}^{L-1}\sum_{j=|l-1/2|}^{l+1/2}\sum_{m=-j}^{j}\left[U^{\mathrm{EP}}_{lj}(r)-U^{\mathrm{EP}}_{LJ}(r)\right]\times|ljm\rangle\langle ljm|\,, \tag{41}$$

where L could be found by series of actual calculations. The EP's of Ne (Fig. 6) indicate that the $U_{lj}(r)$ for p, d, and f orbitals are almost identical. EP's of Xe (Fig. 7) and Au (Fig. 8) are nearly indistinguishable for $j=l+\frac{1}{2}$ and $j=l-\frac{1}{2}$ for f orbitals and those with higher angular momenta. EP's of Au shown in Fig. 8 are the results of 11 valence electron calculations. Based upon the above observations the reasonable choice of L appears to be that of one plus, or possibly the same as, the largest l present in the core, as was concluded for nonrelativistic EP's.[9,10]

Although numerical forms of the EP's can be used in applications, it may be more convenient to have them

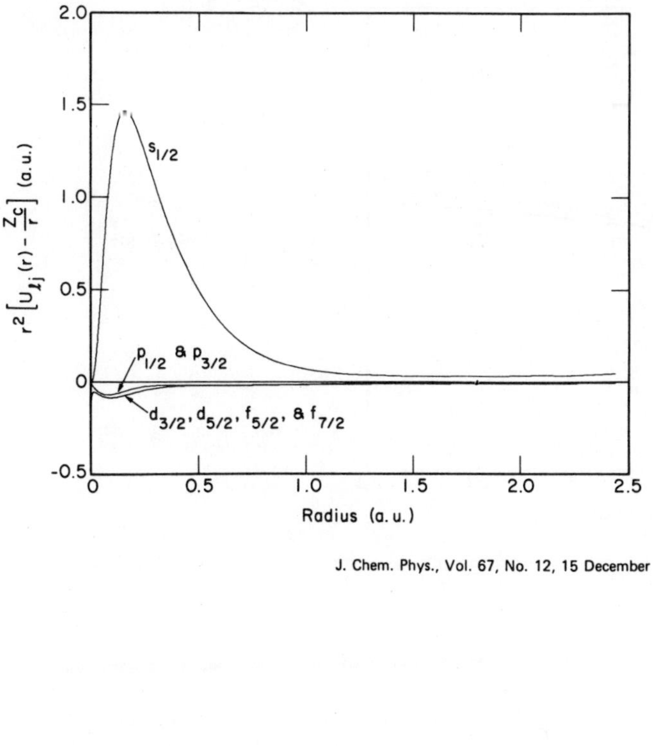

FIG. 6. Effective core potentials of Ne.

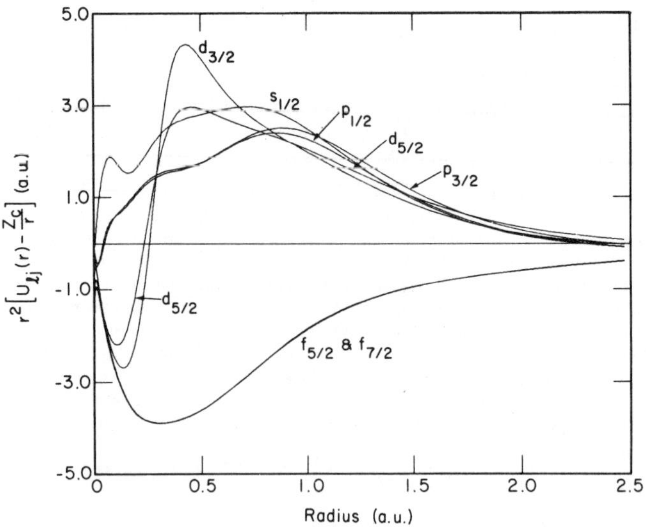

FIG. 7. Effective core potentials of Xe.

expanded in terms of M exponential ($x=1$) or Gaussian ($x=2$) functions

$$[U_\lambda^{\mathrm{EP}}(r) - Z_c/r] = \sum_{i=1}^{M} b_{\lambda i} r^{n_{\lambda i}} e^{-\zeta_{\lambda i} r^x} , \tag{42}$$

where λ represents the quantum numbers j and l of Eq. (41) and Z_c is the number of core electrons. By using the polynomial expansion of $P_{nk}(r)$ [Eq. (36)] the smallest $n_{\lambda i}$ necessary to satisfy Eq. (37) can be obtained. If this is done, one finds $\min(n_{\lambda i}) = -2$ with

$$\sum_{i=1}^{M_{-2}} b_{\lambda i} = \tfrac{1}{2}[\gamma_k(\gamma_k - 1) - k(k+1)] , \tag{43}$$

where M_{-2} is the number of terms with $n_{\lambda i} = -2$ and γ_k and k are previously defined. In practice, we use only one term with $n_{\lambda i} = -2$ and find the corresponding value of $b_{\lambda i}$ by fitting Eq. (42) for the region near the nucleus. There is no guarantee that Eq. (36) is strictly valid after the transformation defined by Eq. (34), but for Xe and Au the $b_{\lambda i}$ as obtained from the numerical EP agree reasonably well with those given by Eq. (43). After the $n_{\lambda i} = -2$ term is determined the numerical EP's are fitted with a number of $n_{\lambda i} = -1$ and $n_{\lambda i} = 0$ terms with the asymptotic condition $U_\lambda^{\mathrm{EP}}(r) \to Z_c/r$ as $r \to \infty$. This asymptotic behavior is ensured by fitting $[U_\lambda^{\mathrm{EP}}(r) - Z_c/r]$ instead of $U_\lambda^{\mathrm{EP}}(r)$.

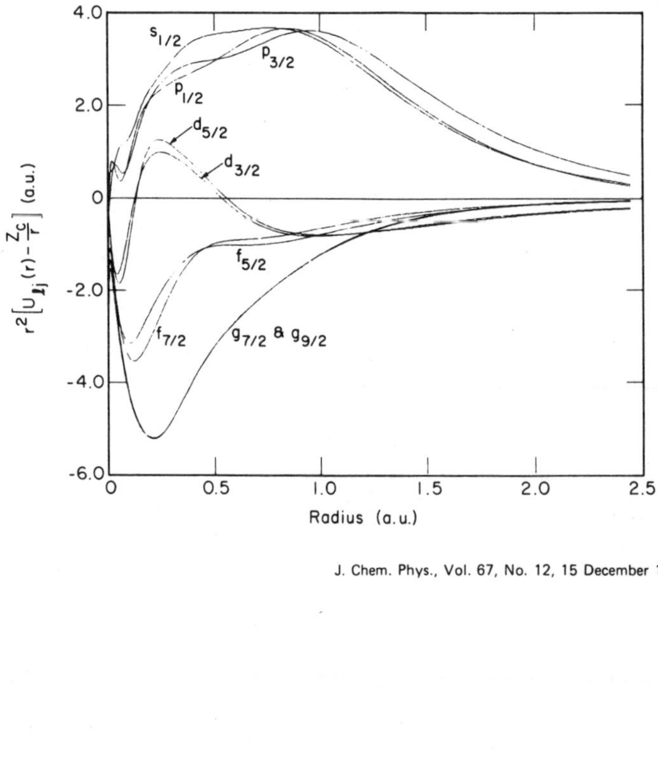

FIG. 8. Effective core potentials of Au.

IV. ATOMIC HARTREE-FOCK EQUATIONS FOR j-j BASIS SETS

We solve the Hartree–Fock self-consistent-field equations for the valence electrons of an atom assuming that the core electrons are represented by an effective potential (EP) derived from numerical Dirac–Hartree–Fock (DHF) calculations for selected states of the atom (Sec. III). It is further assumed that the valence electrons may be described nonrelativistically in terms of only the large component of the atomic orbitals as expressed in terms of expansions in Slater- or Gaussian-type functions.

The Hamiltonian for the N valence electrons is

$$H=\sum_{\mu=1}^{N} h_\mu+\sum_{\mu<\nu}\frac{1}{r_{\mu\nu}} , \tag{44}$$

where

$$h_\mu=-\tfrac{1}{2}\nabla_\mu^2-\frac{Z_v}{r_\mu}+\left[U^{\mathrm{EP}}(\mathbf{r})-\frac{Z_c}{r_\mu}\right] , \tag{45}$$

in which Z_v is the "shielded" nuclear charge (viz. the atomic number minus the number of core electrons) and $U^{\mathrm{EP}}(\mathbf{r})$ is the "relativistic" EP operator of Eq. (41).

The eigenfunctions of H are chosen to be in the form of antisymmetrized products of the large components $(1/r)P_{n\lambda}(r)\chi_{\lambda m}(\theta,\phi)$ of the atomic orbitals $\psi_{n\lambda m}$ of Eq. (3). The radial functions are represented as expansions in B basis functions

$$\frac{1}{r}P_{n\lambda}(r)=\sum_{p=1}^{B} c_{n\lambda p}R_{\lambda p}(r) . \tag{46}$$

The $R_{\lambda p}(r)$ are either Slater-type functions (STO's) or Gaussian-type functions (GTO's).

Atomic self-consistent field equations may be derived by analogy with those of Bagus and Roothaan[19] for the nonrelativistic case and of Kagawa[20] for the Dirac Hamiltonian. In addition to the overlap ($S_{\lambda pq}$), nuclear attraction ($U_{\lambda pq}$), and kinetic energy ($T_{\lambda pq}$) matrix elements those due to the EP are also required:

$$U_{\lambda pq}^{\mathrm{EP}}=\sum_{i=1}^{M} b_{\lambda i}\int_0^{\infty} du\, u^{n_{\lambda i}}e^{-\zeta_{\lambda i}u^x}R_{\lambda p}(u)R_{\lambda q}(u) . \tag{47}$$

In Eq. (47) $x=1$ for STO's and 2 for GTO's and we have assumed that the EP of Eq. (41) has been fitted by a least squares procedure to an expansion of M terms as in Eq. (42). The one-electron matrices enter the SCF iterative procedure as

$$\mathbf{h}=\mathbf{T}-Z'\mathbf{U}+\mathbf{U}^{\mathrm{EP}} . \tag{48}$$

The required Coulomb ($\mathcal{J}^{\nu}_{\lambda pq,\mu rs}$) and exchange ($\mathcal{K}^{\nu}_{\lambda pq,\mu rs}$) matrix elements are collected into supermatrices

$$\mathcal{P}_{\lambda pq,\mu rs}=\mathcal{J}^{0}_{\lambda pq,\mu rs}-\frac{1}{2}\sum_{\nu}A_{\lambda\mu\nu}\mathcal{K}^{\nu}_{\lambda pq,\mu rs} \tag{49a}$$

and

$$\mathcal{Q}_{\lambda pq,\mu rs}=\sum_{\nu}J_{\lambda\mu\nu}\mathcal{J}^{\nu}_{\lambda pq,\mu rs}-\frac{1}{2}\sum_{\nu}K_{\lambda\mu\nu}\mathcal{K}^{\nu}_{\lambda pq,\mu rs} . \tag{49b}$$

[Closed-form formulas for the matrix elements are given by Roothaan and Bagus[19] for STO's and by Huzinaga[21] for GTO's. The EP integrals of Eq. (47) are simple modifications of the $U_{\lambda pq}$ formulas.] The allowed values of ν are given by Eqs. (38) and (39).

The vector coupling coefficients $A_{\lambda\mu\nu}$, $J_{\lambda\mu\nu}$, and $K_{\lambda\mu\nu}$ are chosen to give the proper total symmetry of the atomic electronic state. The closed shell coefficients are given in terms of Clebsch–Gordon coefficients[20]

$$A_{\lambda\mu\nu}=C(j_\lambda j_\mu\nu;\tfrac{1}{2},-\tfrac{1}{2})^2/(2\nu+1) , \tag{50}$$

whereas the open-shell coefficients are given by[22]

$$J_{\lambda\mu\nu}=0 , \tag{51a}$$

$$J_{\lambda\mu\nu}=-f_\nu(\lambda,\mu)/N_\lambda N_\mu \qquad (\nu\neq 0) , \tag{51b}$$

$$K_{\lambda\lambda 0}=A_{\lambda\lambda 0}-2/N_\lambda , \tag{51c}$$

$$K_{\lambda\lambda\nu}=A_{\lambda\lambda\nu}+4f_\nu(\lambda,\lambda)/N_\lambda^2 \qquad (\nu\neq 0) , \tag{51d}$$

$$K_{\lambda\mu\nu}=A_{\lambda\mu\nu}+2g_\nu(\lambda,\mu)/N_\lambda N_\mu \qquad (\lambda\neq 0) \tag{51e}$$

for N_λ and N_μ open shell electrons. The coefficients of Slater integrals needed in Eqs. (51b) and (51c) are derived[23] from the energy expression for the general open shell interaction. Vector coupling coefficients for the cases through $f_{7/2}$ and values of $f_\nu(\lambda,\mu)$ and $g_\nu(\lambda,\mu)$ are given in Ref. 16 for many electronic configurations. The one-electron matrix elements of Eq. (48) and the supermatrices of Eq. (49) are collected into closed and open shell Fock matrices which are used to construct the SCF equations using the coupling operator method exactly as given in Ref. 19.

It should be noted here that the two differences from the nonrelativistic atomic SCF equations as developed in Ref. 19 are that the vector coupling coefficients $A_{\lambda\mu\nu}$ of Eq. (50) and the coefficients of Slater integrals $f_\nu(\lambda,\mu)$ and $g_\nu(\lambda,\mu)$ of Eqs. (51) are chosen to properly account for the jj-coupling scheme as required by the form of our EP's. Thus, the basis functions $R_{\lambda p}(r)$ of Eq. (46) represent the (n, l, j) set of quantum numbers and are designated $1s_{1/2}$, $2s_{1/2}$, $2p_{1/2}$, $2p_{3/2}$, $3p_{1/2}$, $3p_{3/2}$, $3d_{3/2}$, $3d_{5/2}$, etc. for (n, l, j) values of (1, 0, 1/2), (2, 0, 1/2), (2, 1, 1/2), (2, 1, 3/2), (3, 1, 1/2), (3, 1, 3/2), (3, 2, 3/2), (3, 2, 5/2), etc. (The notation $s_{1/2}\equiv s$, $p_{1/2}\equiv\bar{p}$, $p_{3/2}\equiv p$, $d_{3/2}\equiv\bar{d}$, $d_{5/2}\equiv d$, etc. is often used.)

V. RESULTS OF ATOMIC SCF CALCULATIONS USING EP'S

We have chosen the Xe and Au atoms to serve as test cases for the application of the EP and atomic SCF formalisms presented in the previous sections. Pseudo-orbitals and EP's were generated for Xe through $f_{7/2}$ and for Au through $g_{9/2}$ and fits to exponential and Gaussian expansions were computed. These fits were used in atomic SCF calculations for the eight valence electrons of Xe and for choices of one, 11, 17, 19, and 33 valence electrons for Au. In each instance ground, positive ion, and several excited electronic states were investigated. Both average energies of configurations and individual multiplet energies were computed.

A. Xenon

Table III contains the parameters necessary to describe the Xe EP's in terms of Gaussian functions ac-

TABLE III. Gaussian representations of Xe effective core potentials.[a]

n	ζ	$b(s_{1/2})$	$b(p_{1/2})$	$b(p_{3/2})$
0	3101.0580	3300.6124	−3614.3979	1815.4878
0	1656.9400	1803.2145	1415.6830	582.3106
0	400.1800	116.3312	201.2012	−638.3250
0	213.5110	−952.3547	1186.5252	532.7408
0	172.2200	815.8060	−360.3208	−340.7842
0	8.7230	11.6117	19.4800	28.1350
0	5.3170	5.1503	−18.6829	−31.6556
0	1.4510	7.3448	1.5500	−4.2792
−1	1377.9700	−77.5880	6.1604	−70.8533
−1	185.4100	42.3396	−50.3378	4.5462
−1	12.9100	3.8923	0.1701	−3.2248
−1	1.0300	2.1115	5.6172	10.0200
−2	50000.0000	0.0000	−0.9650	0.0000

n	ζ	$b(d_{3/2})$	$b(d_{5/2})$	$b(f_{5/2})$	$b(f_{7/2})$
0	60.1000	−389.2453	−415.5537	−256.2135	−247.5275
0	18.0000	−23.1888	−100.3266	−71.5177	−72.6316
0	5.4000	29.5468	56.3986	−36.6237	−36.1695
0	1.3300	6.2846	5.0261	−5.4657	−5.5721
0	0.3000	0.2034	0.1704	−0.3310	−0.3215
0	0.0446	−0.0084	−0.0128	−0.0032	−0.0040
−1	7257.5000	−312.5458	−93.3057	−380.5876	−7.6836
−1	5425.0000	209.5445	84.5263	234.8154	−9.7199
−1	1000.0000	−34.9143	−34.0789	−26.0590	−5.5409
−1	217.0000	−24.3855	−17.5347	−32.7976	−33.1078
−2	165000.0000	−1.9000	−0.0800	−2.7700	−0.0800

[a]Relative to Eq. (42).

cording to Eq. (42) for all symmetries through $f_{7/2}$ (see also Fig. 7). The EP's for $s_{1/2}$, $p_{1/2}$, and $p_{3/2}$ are given with respect to a set of 13 functions and the EP's for $d_{5/2}$, $d_{7/2}$, $f_{7/2}$, and $f_{9/2}$ with respect to a different set of 11 functions.

Basis sets of GTO's for use in describing the eight valence electrons of Xe are given in Table IV. These were determined by taking three term least squares fits of the $5s_{1/2}$, $5p_{1/2}$, and $5p_{3/2}$ pseudo-orbitals as initial guesses and optimizing the ground state valence energy. To provide adequate flexibility in the basis set to describe the excited (Rydberg) states of Xe two $s_{1/2}$, $p_{1/2}$, and $p_{3/2}$ and three $d_{3/2}$, $d_{5/2}$, $f_{5/2}$, and $f_{7/2}$ GTO's were added to the basis and the exponents were optimized to give the lowest average energies of respective excited electronic configurations. Also shown in Table IV are the atomic orbital expansion coefficients for the ground state as determined by the SCF procedure and those derived by a least squares fit of the same basis set to the numerical pseudo-orbitals.

The valence basis set, as obtained from three term fits to the respective pseudo-orbitals, showed significant differences from the set optimized by means of the SCF procedure for the eight valence electrons with the EP. For example, the fitted ζ's for the $p_{1/2}$ pseudo-orbital were 1.296, 0.417, and 0.134 as compared with the optimized values of 3.191, 0.517, and 0.186 (Table IV). The ground state valence energy resulting from the fitted basis set was 0.034 a.u. higher than that of the optimized set. This implies that merely obtaining the best fit to the pseudo-orbital does not, in general, yield a valence basis set that is optimum with respect to the valence energy. In particular, it appears that additional functions having larger ζ's are required for describing residual corelike characteristics that may not have been totally removed by the EP. However, the use of the energy optimized basis to fit the pseudo-orbital does yield coefficients that are nearly the same as the SCF atomic orbital coefficients (Table IV).

Results of atomic SCF calculations using the basis set of Table IV and the EP parameters of Table III are given for the average energies of configurations in Table V. The excitation energies are compared with weighted averages of the experimentally observed[24] multiplet energies in Table VI. Results of selected numerical DHF calculations are given in Tables V and VI for comparison. In Table VI the excitation energies have been adjusted to give the experimental values for the lowest states arising from the promotion of a $5p_{1/2}$ or a $5p_{3/2}$ electron. This adjustment was done to provide some estimate of the ability of the SCF calculations to predict term values, since the absolute excitation energies are expected to contain large errors due to the neglect of electron correlation.

The SCF multiplet energies for all of the possible states of Xe involving the excitation of one electron from the $5p_{1/2}$ or $5p_{3/2}$ shell to all higher shells through $3d_{3/2}$ have also been computed.[16]

The results in Tables V and VI indicate that the EP's of Table III and the valence basis set of Table IV are capable of reproducing the DHF calculations to reasonable accuracy and yield excitation energies that are of comparable quality to all election SCF calculations. When adjusted to approximately account for the neglect of electron correlation the excitation energies are gen-

TABLE IV. Xenon GTO valence basis set.

Type	ζ	Ground state AO coefficients	Fit of psuedo-orbital
$s_{1/2}$	33.091	−0.00956	−0.00978
	2.221	−0.11806	−0.11680
	0.272	0.96220	0.95847
	0.055	0.15334	0.16190
	0.021	−0.03893	−0.00797
$2p_{1/2}$	3.191	0.04373	0.04505
	0.517	0.44008	0.43131
	0.186	0.58198	0.58852
	0.036	0.10023	0.10543
	0.013	−0.03031	−0.03703
$2p_{3/2}$	3.080	0.04097	0.04230
	0.481	0.42319	0.41569
	0.174	0.58843	0.59424
	0.036	0.11816	0.12197
	0.013	−0.03315	−0.03940
$3d_{3/2}$, $3d_{5/2}$	0.310		
	0.058		
	0.014		
$4f_{5/2}$, $4f_{7/2}$	0.100		
	0.010		
	0.003		

TABLE V. Xe and Xe^+ average energies of configurations (a.u.).

Configuration	Valence energy (+16)	$-\epsilon_{5s}$		$-\epsilon_{5\bar{p}}$		$-\epsilon_{5p}$		$-\epsilon_{Ry}$	
	EP	EP	DHF	EP	DHF	EP	DHF	EP	DHF
$5s^2 5\bar{p}^2 5p^4$	−0.4953	0.997	1.010	0.481	0.493	0.430	0.440		
$5p^3 6s$	−0.2363	1.149		0.643		0.596		0.133	
$5\bar{p}6s$	−0.1914	1.155		0.673		0.593		0.134	
$5p^3 6\bar{p}$	−0.1905	1.195	1.198	0.691	0.688	0.643	0.637	0.087	0.084
$5\bar{p}6p$	−0.1434	1.204		0.724		0.645		0.085	
$5p^3 5\bar{d}$	−0.1643	1.201	1.179	0.697	0.670	0.649	0.619	0.061	0.065
$5p^3 5d$	−0.1649	1.198	1.187	0.694	0.675	0.648	0.628	0.061	0.063
$5\bar{p}5\bar{d}$	−0.1175	1.212		0.733		0.652		0.059	
$5\bar{p}5d$	−0.1209	1.198		0.718		0.639		0.063	
$5p^3 4\bar{f}$	−0.1344	1.279		0.773		0.727		0.031	
$5p^3 4f$	−0.1344	1.279	1.271	0.773	0.759	0.727	0.709	0.031	0.031
$5\bar{p}4\bar{f}$	−0.0889	1.285		0.805		0.725		0.031	
$5\bar{p}4f$	−0.0889	1.285		0.805		0.725		0.031	
$5p^3$	−0.1036	1.339	1.330	0.833	0.820	0.787	0.772		
$5\bar{p}$	−0.0582	1.345		0.865		0.785			

erally within 1%–2% of the observed values. Since the excited states of Xe are all Rydberg states, an alternative procedure of using EP's derived from Xe^+ DHF calculations may be desirable. However, since the EP's are ultimately intended for use in molecular calculations, those due to neutral atom DHF results are generally more appropriate.

B. Gold

As the nuclear charge increases it becomes more difficult to intuitively specify the size of the core. In order to gain some insight into this problem Au has been studied as an atom comprised of one, 11, 17, 19, and 33 valence electrons having cores of $[Xe]4f^{14}5d^{10}$, $[Xe]4f^{14}$, $[Kr]4d^{10}4f^{14}5s^2$, $[Kr]4d^{10}4f^{14}$, and $[Kr]4d^{10}$, respectively. Numerical EP's obtained for each case have been fitted with exponential and Gaussian functions.[16] The EP's for the 11-electron case are plotted in Fig. 8 and the expansions in terms of exponential functions are given in Table VII.

Several atomic states have been calculated using the modified atomic SCF program described in Sec. IV and a valence basis set of five STO's for each symmetry (Table VIII). Results of the ground state calculations are summarized in Table IX. The original DHF results are reasonably reproduced for most orbital energies except for 6s of 19- and 33-valence electron cases. Since the 6s orbital for those cases is calculated as the excited orbital from the EP of 5s orbitals, the magnitudes of valence interactions are expected to be slightly different between those contained in Eq. (37) and those recovered from the atomic SCF calculations. The differences between 17- and 19-valence electron cases are likely to be the result of these valence-interaction differences. Several excitation energies of Au and Au^+ with ionization potentials of Au are shown in Table X.

TABLE VI. Xe and Xe^+ excitation energies for averages of configurations (cm^{-1}).

Configuration	ΔE_{EP}	ΔE_{DHF} [a]	ΔE_{exp} [b]	ΔE_{EP} (adj)	% Error	ΔE_{DHF} (adj)	% Error
$5p^3 6s$	60 060		67 433	(67 433)[c]			
$5\bar{p}6s$	66 705		76 938	(76 938)[d]			
$5p^3 6\bar{p}$	66 902	72 253	78 484	77 484	1.3	(77 484)[e]	
$5\bar{p}6p$	77 251		88 869	87484	1.6		
$5p^3 5\bar{d}$	72 653	76 474	81 206	83 235	−2.5	81 705	−0.6
$5p^3 5d$	72 523	76 864	81 206	83 105	−2.3	82 095	−1.1
$5\bar{p}5\bar{d}$	82 922		92 078	93 155	−1.5		
$5\bar{p}5d$	82 169		91 622	92 751	−1.2		
$5p^3 4\bar{f}$	79 213	83 675	90 893	89 795	1.2	88 906	2.2
$5p^3 4f$	79 213		90 893	89 795	1.2		
$5\bar{p}4\bar{f}$	89 199			99 432			
$5\bar{p}4f$	89 197			99 430			
$5p^3$	85 966	94 741	97 834	95 618	−2.3	99 972	−2.2
$5\bar{p}$	95 950		108 371	106 183	−2.0		

[a]DHF calculations correspond to freezing the core orbitals in their ground state form and allowing the valence orbitals to readjust self-consistently.
[b]Reference 24.
[c]Configurations with a $\bar{p}^2 p^3$ core are adjusted to give ΔE_{exp} of $5p^3 6s$.
[d]Configurations with a $\bar{p}p^4$ core are adjusted to give ΔE_{exp} of $5\bar{p}6s$.
[e]DHF results adjusted to give ΔE_{exp} of $5p^3 6\bar{p}$.

TABLE VII. Exponential representations of Au effective core potentials (11 valence electrons).[a]

n	ζ	$b(s_{1/2})$	$b(p_{1/2})$	$b(p_{3/2})$	
0	288.000	−55132.8875	161942.9160	99102.9091	
0	139.000	−1126007.6516	−488214.6251	346906.3741	
0	67.900	350739.7434	254486.0761	−121125.7849	
0	43.600	101375.4578	64545.6590	−29266.4214	
0	34.000	159878.3808	104073.4087	−50335.4471	
0	20.000	86375.6631	60132.1583	−37509.2820	
0	10.700	3692.5438	4137.6930	−9741.2788	
0	7.800	508.8848	331.9913	4181.2495	
0	2.500	52.5487	55.8331	66.9236	
0	1.000	−0.0607	−0.1412	0.1069	
−1	445.000	−207.4701	806.8294	461.5918	
−1	162.000	−24196.2727	−12894.4984	6969.6166	
−1	107.000	34397.7420	18718.3059	−10767.4468	
−1	17.600	−8191.0833	−5058.0944	851.3200	
−1	11.000	−1662.5439	−1491.5899	2784.4263	
−1	4.500	−80.6511	−95.9327	−318.6402	
−2	150.000	−0.0700	−1.0900	−0.1300	
		$b(d_{3/2})$	$b(d_{5/2})$	$b(f_{5/2})$	$b(f_{7/2})$
0	288.000	−29791.2213	−41660.1347	21134.9048	11416.2125
0	139.000	−536727.6606	−779167.3713	100114.0225	123399.0555
0	67.900	186104.6251	263398.0757	−41332.1685	−34933.8001
0	43.600	43457.4982	66226.5964	−14197.2483	−14388.6127
0	34.000	84112.0336	119667.5528	−13102.4835	−11431.5069
0	20.000	42249.5755	59924.4156	−14581.1430	−13073.7665
0	10.700	1721.6716	2804.7098	−4013.3200	−3907.9384
0	7.800	−288.6855	−547.8473	972.6197	957.2748
0	2.500	−13.4315	−13.8406	−8.1344	−6.0217
0	1.000	−0.2032	−0.2342	−0.0583	−0.0784
−1	445.000	135.9162	−102.8095	446.3665	51.1775
−1	162.000	−11339.3845	−16724.1787	2380.0388	2575.8297
−1	107.000	16508.3572	24037.0173	−3467.8867	−3651.3703
−1	17.600	−4346.8703	−6193.1419	20.9689	21.7522
−1	11.000	−795.2709	−1148.9599	1070.2421	976.4872
−1	4.500	45.6128	61.8165	−52.1616	−53.5363
−2	150.000	−2.0300	−0.1480	−2.9900	−0.1540
n		$b(g_{7/2})$	$b(g_{9/2})$		
0	288.000	52869.6404	21708.3758		
0	139.000	300211.3275	332789.1202		
0	67.900	−115144.1812	−115730.4061		
0	43.600	−30106.8123	−32391.7507		
0	34.000	−46200.0618	−46726.6570		
0	20.000	−32443.0002	−33404.2206		
0	10.700	−6368.6435	−6367.4487		
0	7.800	1645.9473	1595.9459		
0	2.500	−8.6614	−9.0948		
0	1.000	−0.0077	−0.0093		
−1	445.000	727.6671	67.8364		
−1	162.000	6645.2738	7137.2683		
−1	107.000	−9762.0198	−10373.3260		
−1	17.600	1268.5763	1336.9776		
−1	11.000	1863.9734	1881.9489		
−1	4.500	−141.9801	−137.1714		
−2	150.000	−3.9500	−0.1580		

[a]Relative to Eq. (42).

Since there is rather close agreement between 19- and 33-electron calculations, it appears reasonable that the $4f_{5/2}$ and $4f_{7/2}$ electrons may be included in the core without introducing serious errors in Au valence electron calculations. The discrepancy between 19- and 17-valence cases is caused by the lowering of the ground state of the 19-valence-electron case (and the 33-valence-electron case) due to the irreproducibility of the valence interactions as mentioned above.

If these effects (about 0.05 a.u.) are taken into account, 17- and 19-electron results are almost identical

TABLE VIII. Au STO valence basis set.

Symmetry	ζ
$2s_{1/2}$	9.2
	4.7
	2.96
	0.898
	0.25
$2p_{1/2}$	5.5478
	2.78
	1.0048
	0.552
	0.2675
$2p_{3/2}$	5.2857
	2.43
	1.0
	0.491
	0.245
$3d_{3/2}$	3.4873
	2.5973
	1.1575
	0.3196
	0.0639
$3d_{5/2}$	3.2186
	2.493
	1.1207
	0.3164
	0.0633
$4f_{5/2}$	18.6706
	10.3406
	5.7646
	1.2524
	0.2505
$4f_{7/2}$	18.0529
	10.1093
	5.6835
	1.2524
	0.2505

as expected. Although there are some differences between 11- and 17-electron cases (especially for $6p_{1/2}$ states), an 11-valence-electron model appears to be the most reasonable one for Au. A one-electron representation of Au is the most simple and may be useful in some applications, but meaningful descriptions of chemical bonds involving Au are not expected when the participation of $5d$ electrons is completely neglected.

Table XI gives the orbital energies and mean values for the radius from a numerical DHF calculation for Au. It is apparent that the orbital energies for the $5d$ levels are not enough greater than that for $6s$ to justify the omission of the $5d$ orbitals from the valence shell. With a 2 a.u. difference for $5p$ from $5d$, however, it seems clear that the $5p$ electrons, and those more tightly bound will be little affected by valence shell changes. Arguments based on radii are less clear cut but support the same conclusion. Since the radius of the outer part of the distribution is of primary interest, values of $\langle r^4 \rangle^{1/4}$ are more meaningful than those of $\langle r \rangle$. From Fig. 4 one notes that the radius of the outermost node in the $6s$ orbital is at about 1.1 a.u. Hence, one should include in the valence shell other orbitals with effective radii substantially greater than this value. Clearly, the $5d$ orbitals should be included. While the case for the $5p$ orbitals may be marginal, it seems reasonable to omit them from the valence shell on this basis as well as on the basis of orbital energies. Thus, we believe that the 11-electron valence shell will be appropriate for gold but more definite conclusions will be possible after molecular calculations have been carried out.

VI. APPLICATIONS TO MOLECULAR CALCULATIONS

The EP's described in Sec. III were developed primarily for use in calculations on molecules containing one or more heavy atoms. The direct use of the EP's of Eq. (41) in molecular calculations requires that the basis set used for describing the valence electrons be in terms of functions appropriate for the jj-coupling scheme. The molecular orbitals for the valence electrons would then correspond to two-component basis functions analogous to the large component P_{nk} multiplied by the proper angular factor and the Pauli spinors [see Eq. (3)]. The molecular integrals necessary for wavefunctions of this form have been discussed by Malli and Oreg[25] in the context of an extension of the generalization of DHF theory to molecules. They have shown that the required two-electron integrals can be written in terms of nonrelativistic integrals.

The use of jj coupling in the atomic basis sets has the advantage that the molecular wavefunction obeys Hund's case c coupling and yields a self-consistent result. Thus, the spin orbit coupling in the MO's would not have to be imposed as a perturbation since the effects would have been carried throughout the SCF procedure as part of the EP.

In many applications it may be reasonable to assume

TABLE IX. Ground state calculations for Au([Xe] $5d^4_{5/2}\ 5d^6_{7/2}\ 6s_{1/2}$).[a]

		Orbital energies ($-\epsilon$)							
	$-$T.E.	$4f_{7/2}$	$4f_{9/2}$	$5s_{1/2}$	$5p_{1/2}$	$5p_{3/2}$	$5d_{5/2}$	$5d_{7/2}$	$6s_{1/2}$
DHF	19018.1992	3.8675	3.7202	4.6873	3.1893	2.5588	0.4934	0.4286	0.2919
One-electron	0.2912								0.2912
11-electron	35.3827						0.4932	0.4254	0.2936
17-electron	109.4704				3.2294	2.5473	0.5106	0.4368	0.2985
19-electron	145.2199			4.6462	3.2661	2.5843	0.4856	0.4176	0.3395
33-electron	610.9816	3.9555	3.8123	4.6494	3.2376	2.5760	0.4972	0.4254	0.3438

[a]Due to the EP's expanded in STF's and the valence basis set of Table VIII. All energies are in a.u.

TABLE X. Au and Au^+ excitation energies with ionization potentials of Au.

		Au excitation energies[a]					
		one-electron	11-electron	17-electron	19-electron	33-electron	Exp.
[Xe] $5d^4_{3/2}\,5d^6_{5/2}$	$6s$	0.0	0.0	0.0	0.0	0.0	0.0
[Xe] $5d^4_{3/2}\,5d^6_{5/2}$	$6p_{1/2}$	0.1500	0.1442	0.1309	0.1799	0.1854	0.1702
	$6p_{3/2}$	0.1646	0.1553	0.1499	0.1986	0.2026	0.1876
	$6d_{3/2}$	0.2367	0.2227	0.2268	0.2755	0.2799	0.2823
	$6d_{5/2}$	0.2370	0.2230	0.2273	0.2757	0.2801	0.2826
	$5f_{5/2}$	0.2599	0.2463				0.3075
	$5f_{7/2}$	0.2599	0.2463				0.3075
[Xe] $5d^4_{3/2}\,5d^6_{5/2}$		0.2919	0.2776	0.2820	0.3301	0.3346	0.3390
		Au^+ excitation energies					
		zero-electron	10-electron	16-electron	18-electron	32-electron	Exp.
$5d^4_{3/2}\,5d^6_{5/2}$		0.0	0.0	0.0	0.0	0.0	0.0
$5d^4_{3/2}\,5d^5_{5/2}\,6s$	(Av.)		0.0689	0.0767	0.0126	0.0171	0.0735
	$(J=3)$		0.0519	0.0596	0.0053	0.0097	0.0685
	$(J=2)$		0.0920	0.1000	0.0228	0.0273	0.0804
$5d^3_{3/2}\,5d^6_{5/2}\,6s$	(Av.)		0.1254	0.1397	0.0694	0.0776	0.1318
	$(J=1)$		0.1087	0.1229	0.0628	0.0710	0.1265
	$(J=2)$		0.1352	0.1496	0.0734	0.0815	0.1350

[a]See footnote a of Table IX.

TABLE XI. Au orbital energies and radial expectation values from a DHF calculation (a.u.).[a]

	$4d_{5/2}$	$4f_{5/2}$	$4f_{7/2}$	$5s$	$5p_{1/2}$	$5p_{3/2}$	$5d_{3/2}$	$5d_{5/2}$	$6s$
O.E.	13.17	3.87	3.72	4.69	3.19	2.56	0.493	0.429	0.292
$\langle r\rangle$	0.462	0.492	0.499	0.943	1.02	1.11	1.54	1.62	3.06
$\langle r^4\rangle^{1/4}$	0.557	0.646	0.655	1.100	1.21	1.32	1.99	2.12	3.77

[a]Reference 1.

that LS coupling is dominant in the valence shell of the molecule even though it may contain one or more heavy atoms. In such cases the most important relativistic effects may be due to the mass velocity and Darwin terms in the Hamiltonian [see Eqs. (7)–(9)]. Spin-orbit splitting may then be introduced as a correction. Approaches based on this premise are being investigated by Kahn, Hay, and Cowan[26] and Das and Wahl.[27] The EP's discussed in this work may also be applied at this level of approximation by taking the weighted averages of the two EP's having the same value of the quantum number $l(l>0)$. These averaged relativistic EP's may then be applied in the same way as the nonrelativistic EP's[10] in conjunction with standard nonrelativistic molecular programs. Calculations on the ground and excited states of Xe_2 and Xe_2^+ using this approach yield quite reasonable results.[28]

An abstract has also appeared concerning relativistic HF calculations by Datta, Ewig, and Van Wazer.[29]

ACKNOWLEDGMENT

This research was sponsored by the Energy Research and Development Administration.

[1]J. P. Desclaux, At. Data Nucl. Data Tables 12, 311 (1973).

[2]C. C. Lu, T. A. Carlson, F. B. Malik, T. C. Tucker, and C. W. Nestor, J. At. Data 3, 1 (1971).

[3]P. I. Pavlik and S. M. Blinder, J. Chem. Phys. 46, 2749 (1967).

[4]J. P. Desclaux and P. Pyykko, Chem. Phys. Lett. 29, 534 (1974); 39, 300 (1976).

[5]W. C. Mackrodt, Mol. Phys. 18, 697 (1970).

[6]R. L. Matcha, J. Am. Chem. Soc. 95, 7505 (1973); R. L. Matcha, J. Chem. Phys. 65, 1962 (1976).

[7]J. S. Cohen and B. J. Schneider, J. Chem. Phys. 61, 3230 (1974); P. J. Hay, T. H. Dunning, Jr., and R. C. Raffenetti, *ibid.* 65, 2679 (1976); P. J. Hay and T. H. Dunning, Jr., *ibid.* 66, 1306 (1976).

[8]D. E. Ellis, Int. J. Quantum Chem. Symp. 9, 351 (1975); D. E. Ellis and A. Rosén, Chem. Phys. Lett. 27, 595 (1974); D. E. Ellis and A. Rosén, J. Chem. Phys. 62, 3039 (1975); D. E. Ellis, D. D. Koelling, and R. J. Bartlett, *ibid.* 65, 3331 (1976).

[9]W. A. Goddard III and C. F. Melius, Phys. Rev. A 10, 1528 (1974).

[10]L. R. Kahn, P. Baybutt, and D. G. Truhlar, J. Chem. Phys. 65, 3826 (1976). (See references therein for other effective potential methods.)

[11]J. C. Phillips and L. Kleinman, Phys. Rev. 116, 287 (1959); see also, J. D. Weeks, A. Hazi, and S. A. Rice, Adv. Chem. Phys. 16, 283 (1969).

[12](a) J. P. Desclaux, Int. J. Quantum Chem. 6, 25 (1972); (b) J. P. Desclaux and Y. K. Kim, J. Phys. B 8, 1177 (1975).

[13]J. P. Desclaux, Comput. Phys. Commun. 9, 31 (1975).

[14]I. P. Grant, Adv. Phys. **19**, 747 (1970) (see references therein for the detailed theory of atomic DHF calculations); also, P. Pyykkö, Adv. Quantum Chem. (to be published).

[15]J. B. Mann, Los Alamos Laboratory Reports LA-3690 and LA-3691 (1967).

[16]Y. S. Lee, Ph.D. thesis, University of California, Berkeley, 1977.

[17]H. A. Bethe and E. E. Salpeter, *Quantum Mechanics of One- and Two-electron Atoms* (Academic, New York, 1957); see also A. Fröman, Rev. Mod. Phys. **32**, 317 (1961).

[18]F. Herman and S. Skillman, *Atomic Structure Calculations* (Prentice-Hall, Englewood Cliffs, 1963).

[19]C. C. J. Roothaan and P. S. Bagus, Meth. Comp. Phys. **2**, 47 (1963).

[20]T. Kagawa, Phys. Rev. A **12**, 2245 (1975).

[21]S. Huzinaga, J. Chem. Phys. **42**, 1293 (1965).

[22]G. Malli and J. P. Olive, J. Chem. Phys. **43**, 861 (1965).

[23]I. P. Grant, Comput. Phys. Commun. **5**, 263 (1973).

[24]C. E. Moore, Natl. Bur. Stand. Circ. **467**, Vol. I (1949), Vol. II (1952), Vol. III (1958).

[25]G. Malli and J. Oreg, J. Chem. Phys. **63**, 830 (1975).

[26]L. R. Kahn, P. J. Hay, and R. D. Cowan, "Relativistic *Ab Initio* Effective Core Potentials," Abstracts of Thirty-Second Symposium on Molecular Spectroscopy, The Ohio State University (1977).

[27]G. Das and A. C. Wahl, J. Chem. Phys. **64**, 4672 (1976).

[28]N. W. Winter, Y. S. Lee, W. C. Ermler, and K. S. Pitzer (to be published).

[29]S. N. Datta, C. S. Ewig, and J. R. Van Wazer, "Application of Pseudopotentials to Molecular Relativistic Hartree–Fock Calculations," Abstracts of Tenth Midwest Theoretical Chemistry Conference (1977).

Ab initio effective core potentials including relativistic effects. II. Potential energy curves for Xe_2, Xe^+_2, and Xe^*_2

Walter C. Ermler, Yoon S. Lee, and Kenneth S. Pitzer

Department of Chemistry and Lawrence Berkeley Laboratory, University of California, Berkeley, California 94720

Nicholas W. Winter

Theoretical Atomic and Molecular Physics Group, Lawrence Livermore Laboratory, Livermore, California 94550

(Received 10 April 1978)

Potential energy curves for the ground $^1\Sigma_g^+$ state of Xe_2, the first four states of the Xe_2^+ ions, and the eight Xe^*_2 excimer states corresponding to the addition of a $6s\sigma_g$ Rydberg electron to these ion cores have been computed using averaged relativistic effective core potentials (AREP) and the self-consistent field approximation for the valence electrons. The calculations were carried out using the LS-coupling scheme with the effects of spin–orbit coupling included in the resulting potential energy curves using an empirical procedure. A comparison of nonrelativistic and averaged relativistic EP's and subsequent molecular calculations indicates that relativistic effects arising from the mass–velocity and Darwin terms are not important for these properties of Xe_2 molecules. Spectroscopic constants for Xe_2^+ are in good agreement with all electron CI calculations suggesting that the computed values for Xe^*_2 excimers should be reliable. The lifetime for the O_u^+ state of the Xe_2^* is computed to be 5.6 nsec which is in the range of the experimentally determined values.

I. INTRODUCTION

Laser oscillation has been observed for each of the rare gases except Ne, and for rare gas mixtures.[1] These lasers make use of the bound–free transition between the excited rare gas dimer (excimer) and the repulsive ground state. The excimer and dimer ion states also play an important role in the kinetics of rare-gas-halide lasers. The kinetic and radiative models designed to describe the fluorescence and coherent emission of these laser systems requires knowledge of radiative lifetimes and cross sections for photoionization, optical absorption, and stimulated emission.

The Xe_2 excimer has received considerable attention due to its importance as an intense source of vacuum ultraviolet radiation with applications in laser photochemistry and as a pump for fusion lasers. Extensive theoretical modeling and experimental investigation have given insight into the difficulties of the Xe_2 laser, including scaling with pressure, photoionization, and accessing the excimer states with long radiative lifetimes.[2] In order to further refine the modeling of these processes more precise knowledge of the states of Xe_2 and Xe_2^+ is required.

Previous theoretical studies on the excited states of Xe_2 and Xe_2^+ are limited to the empirical potential energy curves derived by Mulliken[3,4] and the all electron configuration interaction calculations of Wadt[5] on the ground state of Xe_2 and four states of Xe_2^+. Concurrent with this work Wadt *et al.*[6] have carried out further calculations on Xe_2^+ using effective core potentials to replace the inner shell electrons. Our results extend quantitative calculations to the excited states of neutral Xe_2^* and provide interesting comparisons of the different methods for the states of the ion Xe_2^+.

The calculation of the potential energy curves for the heavier rare gas dimers presents two obvious difficulties. The first is the large number of electrons to be treated (most of which are not directly involved in determining the properties of interest). In addition, relativistic effects such as spin–orbit interactions increase significantly for higher Z. While relativistic effects become larger for still heavier elements, they are large enough for xenon to make their consideration worthwhile. Recently an approach has been developed[7] to simplify the calculations on heavy atoms by replacing the core electrons with an effective potential that also includes relativistic effects. These potentials are used in the present SCF calculations for Xe_2, Xe_2^+, and Xe_2^*.

These calculations involve several approximations, the most obvious being the SCF approximation. However for other rare gas dimer ions it has been shown that SCF calculations benefit from a fortuitous cancellation of errors and yield good results.[8,9] While the excimer states do not formally dissociate correctly at the Hartree–Fock level, the potential curves should be about as accurate as the ion curves for values of the internuclear distance near the potential minima. Another approximation discussed in more detail in the next section is the use of LS coupling in the calculations and the inclusion of spin–orbit interaction as a later correction. The next section presents the details of the calculations and the results. The final section is devoted to a discussion of the SCF potential energy curves and radiative lifetimes.

II. CALCULATIONS

All-valence-electron self-consistent-field (SCF) calculations on Xe_2, Xe_2^+, and Xe_2^* were carried out using Gaussian expansions of effective core potentials (EP)

0021-9606/78/6903-0976$01.00

derived[7] using pseudo-orbitals extracted from numerical Dirac–Hartree–Fock atomic wavefunctions.[10] To make use of existing computer programs based on the LS-coupling scheme for constituent atoms the EP's (U_{lj}^{EP}) based on the formalism of Ref. 7 are averaged to yield "averaged relativistic effective core potentials (AREP)." Thus each radial component of the total EP is given by

$$U_l^{AREP}(r) = \frac{1}{2l+1}[lU_{ll^-}^{REP}(r) + (l+1)U_{ll^+}^{REP}];$$

$$l > 0,\ l^- = l - \tfrac{1}{2},\ l^+ = l + \tfrac{1}{2}. \quad (1)$$

The total EP now is of the form

$$U^{AREP} = U_L^{AREP}(r) + \sum_{l=0}^{L-1}\sum_{m=-l}^{l}[U_l^{AREP}(r) - U_L(r)]|lm\rangle\langle lm|, \quad (2)$$

where U_L is termed the "residual potential" and L is (ideally) one greater than the highest angular momentum occupied in the core. The angular momentum projection operators act on basis functions defined in the LS-coupling reference frame. The EP of Eq. (2) may be used in molecular calculations in the same manner as described by Kahn, Baybutt, and Truhlar[11] when expanded in M Gaussian functions as

$$[U_l^{EP}(r) - Z_c/r] = \sum_{i=1}^{M} b_{li} r^{n_{li}} \exp(-\zeta_{li} r^2), \quad (3)$$

where Z_c is the number of core electrons.

The EP of Eq. (2) includes the effects due to the mass-velocity and Darwin terms in the Dirac Hamiltonian, whereas the spin–orbit effects have been averaged out by virtue of Eq. (1). To gauge the magnitude of these effects EP's were derived from *nonrelativistic* numerical Hartree–Fock wavefunctions[12] following the procedures analogous to those described in Ref. 7 for defining the appropriate pseudo-orbitals.

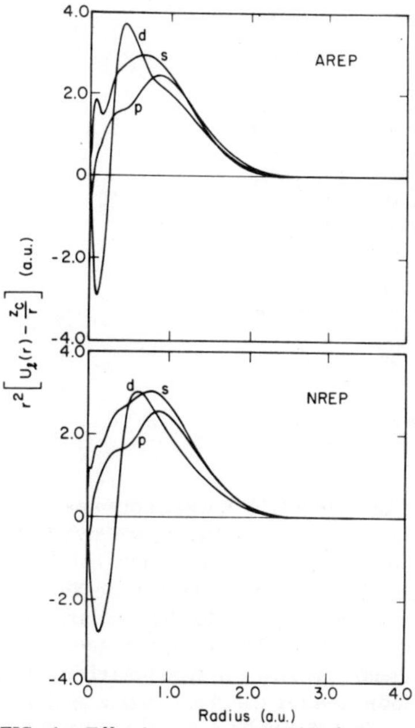

FIG. 1. Effective core potentials of Xe.

The averaged relativistic EP's (AREP) and the nonrelativistic EP's (NREP) are shown in Fig. 1 for $l = 0, 1, 2$. Expansions [Eq. (3)] for the NREP's and AREP's are given in Table I.

A valence basis set of Gaussian-type functions (GTF) was derived by optimization of the valence energy via SCF calculations for the $5s^2 5p^6$ configuration of the Xe atom in the field of the AREP. The orbital exponents of four s-type and three p-type GTF's in a basis set

TABLE I. Gaussian representations of Xe effective core potentials.[a]

		NREP			AREP[b]		
n	ζ	b_s	b_p	b_d	b_s	b_p	b_d
2	3101.0580	10 001.2468103	−2029.8635555	−1261.8724877	3364.8358389	6400.8867608	9715.9126229
2	1656.9400	12 967.2532014	−2499.3599334	−1648.5722662	4714.1097300	7181.5586116	10 450.5903027
2	400.1800	118.6876949	13.6186006	−308.8579700	408.4806715	−76.8423330	−816.3229420
2	213.5110	−1434.1926222	600.2008336	2117.2249998	−1247.2965220	76.1691758	−3178.2001049
2	172.2200	1139.0541750	−75.0987624	−1683.9676399	1181.7588767	106.4682087	2043.2887265
2	60.1000	253.8248434	17.1748795	−1082.1976210	357.2926530	19.3297540	1843.9936166
2	18.0000	191.0803909	23.5399847	−878.3333781	279.7260211	45.6185250	1431.1254621
2	8.7230	30.9549072	32.9978847	−70.6569080	24.0259285	22.7569696	215.1953603
2	5.3170	−17.6599173	−41.4064247	21.4792185	1.6859524	−31.1191342	−30.3926954
2	1.4510	−2.6418222	−9.3991444	−5.9585241	4.1317641	−6.6611822	−4.5418327
2	0.3000	−0.0320120	−0.0480999	−0.0354254	−0.0318452	−0.0426607	−0.0230183
2	0.0446	0.0003275	0.0004805	0.0003504	0.0006014	0.0006215	0.0004836
1	7257.5000	−35.3180702	−43.4291158	−37.0120256	−6.3065139	66.8451734	101.4665571
1	5425.0000	296.9633229	−14.5635957	−12.4908299	67.9439075	65.0477623	107.4413732
1	1377.9700	−335.1700806	64.4242171	21.2430405	−122.9570159	−246.0919993	−341.0505213
1	185.4100	59.0731434	−35.1381251	−159.4208658	71.5587958	−4.9626120	304.1740320
1	12.9100	41.4527070	−10.2430487	152.4005053	−57.3043753	−12.1157615	−358.3400709
1	1.0300	9.7418941	13.7243283	9.3478387	4.2814496	11.4916197	8.9996454
0	16 5000.0000	0.0000000	−0.0800000	0.0000000	0.0000000	−0.3216667	−0.8080000

[a]Relative to Eq. (3). [b]See Eq. (1).

TABLE II. Xe(6s5p1d)/[4s4p1d] GTF basis set.

Symmetry	Exponent	Coefficient
s	28.738697	−0.015324
s	1.960972	−0.192789
s	0.318521	1.099335
s	0.123316	1.0
s	0.055000	1.0
s	0.021000	1.0
p	2.821521	0.084105
p	0.435800	0.964806
p	0.137219	1.0
p	0.036000	1.0
p	0.013000	1.0
d	0.220000	1.0

comprised of six *s* and five *p* GTF's were optimized while the two functions of each type having the smallest exponents were held fixed at values previously optimized[7] for the 6*s* and 6*p* Rydberg orbitals of the Xe atom. The three *s* and two *p* GTF's having the largest exponents were contracted together using the atomic SCF orbital coefficients. A single *d*-type polarization function that was optimized for the ground state of Xe_2^+ was added to yield a final (6*s*5*p*1*d*)/[4*s*4*p*1*d*] contracted GTF valence basis set. The ζ values and contraction coefficients are given in Table II. The analogous procedure for the NREP resulted in a basis set that was sufficiently like that due to the AREP that same GTF's were used with both EP's.

The molecular SCF calculations were carried out using the POLYATOM[13] molecular integrals program, the GTF effective potential integrals program of Kahn,[14] and an open-shell SCF program developed at Caltech.[15]

Although EP's for Xe through $l=3$ were computed,[7] the present limitation of the EP integral program[14] to symmetries $l \leq 2$ required that the residual potential be $U_d^{AREP}(r)$. Earlier investigations have indicated, however, that the EP's for higher *l* have appreciable effects only for cases where basis functions of that symmetry are centered on the same nucleus as the EP.[11] When this is not the case the interaction with EP's of higher *l* comes about only through two- and three-center interactions. Further studies of a proper definition of the residual potential are still warranted because of large differences seen in the shapes of, for example, U_d^{EP} and U_f^{EP} of Xe and U_f^{EP} and U_g^{EP} of Au.[7]

Representative atomic SCF results are given in Table III. Also shown are orbital energies from all-electron AHF[12] and DHF[10] calculations, where the $5p_{1/2}$ and $5p_{3/2}$ orbital energies have been averaged as in Eq. (1). Results[6] based on the Cowan and Griffin (CG) approximation,[16] whereby the mass-velocity and Darwin terms are included in the atomic Hamiltonian and LS coupling is retained by excluding the spin-orbit term, are also shown. The excellent agreement between the averaged DHF and CG results indicates that for Xe the shift of the 5*s* orbital energy is nearly independent of the presence of spin-orbit coupling operators. It may also be concluded that the 5*p* orbital is not radically changed by the presence of the mass-velocity and Darwin terms. This is also reflected in the small difference in the ionization potentials as computed with the AREP and NREP.

SCF calculations using the basis set of Table II and the NREP and AREP of Table I were carried out for the ground $^1\Sigma_g^+$ state of Xe_2 and the ground $^2\Sigma_u^+$ state of Xe_2^+. The results are plotted in Figs. 2 and 3 together with curves due to all-electron calculations using a contracted Gaussian basis set by Wadt.[5] Also shown in Fig. 2 is a curve based on the electron-gas model of Gordon and Kim and Rae[17] that was reported by Wadt.[5]

Based on the atomic results of Table III and the near indistinguishability of the AREP and NREP results shown in Figs. 2 and 3 we conclude that for diatomic Xe molecules the relativistic effects due to mass-velocity and Darwin type terms are unimportant. The AREP curves are only slightly displaced to the left of the NREP curves. This is traceable mainly to the contraction of the 5*s* orbital in the AREP calculation.

Table IV lists the valence energies for the ground state of Xe_2 $(1\sigma_g^2 1\sigma_u^2 2\sigma_g^2 \pi_u^4 \pi_g^4 2\sigma_u^2)$, the four states of Xe_2^+

TABLE III. Xenon atomic energies.

State	Approximation	Orbital energies (a.u.) 5s	5p	6s	ΔE(eV)[a]
Xe $5s^2 5p^6 (^1S)$	DHF(Av)	−1.010	−0.457		
	AHF	−0.944	−0.457		
	NREP	−0.938	−0.451		
	AREP	−1.004	−0.451		
	CG[b]	−1.010	−0.457		
Xe^+ $5s^2 5p^5 (^2P)$	DHF(Av)	−1.330	−0.788		
	NREP	−1.247	−0.803		11.203
	AREP	−1.345	−0.806		11.187
Xe* $5s^2 5p^5 6s (^3P)$	AREP	−1.156	−0.616	−0.134	7.559
(^1P)	AREP	−1.162	−0.621	−0.126	7.776

[a]Relative to the ground state valence energies $E(\text{NREP}) = -16.20431$ a.u. and $E(\text{AREP}) = -16.49885$ a.u.

[b]Numerical atomic results using the method of Cowan and Griffin (Ref. 9).

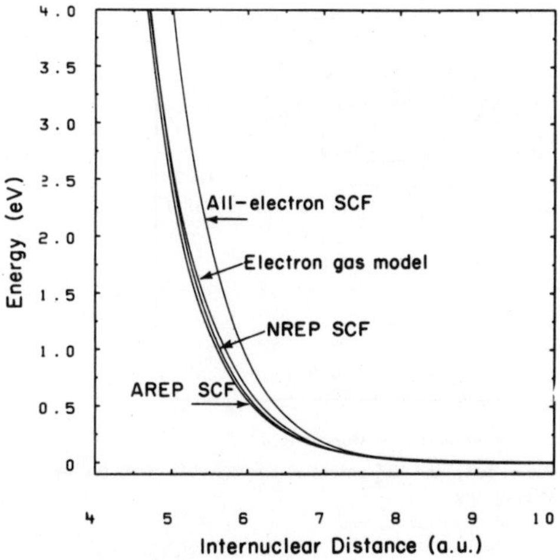

FIG. 2. Xe_2 $X\,^1\Sigma_g^+$ potential energy curves. All-electron SCF and electron gas results are taken from Ref. 5.

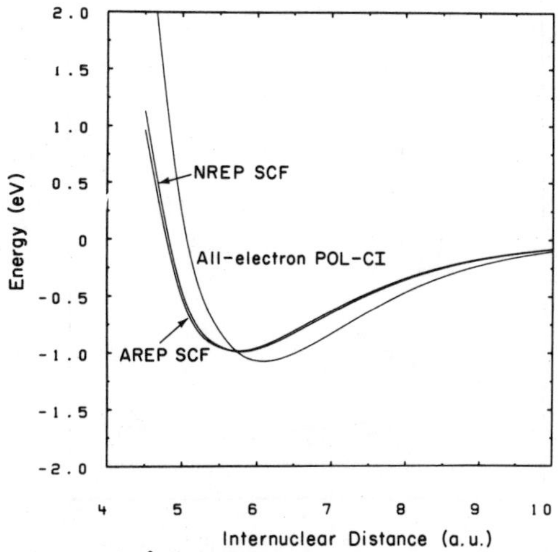

FIG. 3. Xe_2^+ $X\,^2\Sigma_u^+$ potential energy curves. All-electron POL–CI results are taken from Ref. 5.

that arise when an electron is removed from each of the outer molecular orbitals (MO), and the eight states of Xe_2^* due to the addition of an electron into the $\sigma_g 6s$ MO relative to each of the four ion cores. Valence energies at 13 internuclear distances for each of the 13 states were computed in separate SCF calculations using the AREP of Table I and the valence GTF basis set of Table II. The Xe_2^+ potential energy curves based on these data are shown in Fig. 4 and those for Xe_2^* in Fig. 5.

Spin–orbit coupling effects were approximated for the Xe_2^+ and Xe_2^* curves using the empirical model described by Cohen and Schneider.[18] In this procedure the experimental splittings of the Xe^+ and Xe^* are used to determine the matrix elements of the spin–orbit Hamiltonian. This matrix is added to the diagonal matrix of energies of the states that interact to yield a given total angular momentum Ω; viz $^2\Sigma^+$ and $^2\Pi$ for $\Omega = 1/2$, $^3\Pi$ and $^1\Sigma^+$ for $\Omega = 0^+$, $^3\Pi$ and $^3\Sigma^+$ for 0^-, and $^3\Pi$, $^3\Sigma^+$, and $^1\Pi$ for $\Omega = 1$.

Spin–orbit parameters ζ for Xe^+ $(5p^5)$ and Xe^* $(5p^5 6s)$, derived from the atomic tables of Moore,[19] are $2/3[J(1/2) - J(3/2)] = 7024$ cm^{-1} and $2/3[J(0) - J(2)] = 6086$ cm^{-1}, respectively. These parameters were used to calculate the energy for the ω–ω coupling case[3] at each internuclear distance for the data in Table IV. The 23 resulting states are plotted and labeled in Fig. 6, where we have shifted the Xe_2^+ curves vertically such that the experimental (rather than the SCF) ionization potentials of the Xe atom are reproduced. No adjustments were made in the Xe_2^* curves. Because the resolution of many of the curves in Fig. 6 is rather poor, the states of Xe_2^+ are plotted separately in Fig. 7, and the states of Xe_2^* in Fig. 8.

Values of selected spectroscopic constants for the (strongly) bound Xe_2^+ and Xe_2^* states are given in Table V for the curves both with and without spin–orbit coupling. Results based on the all-electron Xe_2^+ calculation

TABLE IV. Valence SCF energies of electronic states of diatomic Xe.[a]

	Xe_2	Xe_2^+				Xe_2^*							
R	$^1\Sigma_g^+$	$^2\Sigma_u^+$	$^2\Pi_g$	$^2\Pi_u$	$^2\Sigma_g^+$	$^3\Sigma_u^+$	$^1\Sigma_u^+$	$^3\Pi_g$	$^1\Pi_g$	$^3\Pi_u$	$^1\Pi_u$	$^3\Sigma_g^+$	$^1\Sigma_g^+$
4.50	0.81444	0.53110	0.42507	0.34045	0.27521	0.64589	0.64179	0.54394	0.53932	0.46150	0.45212	0.32973	0.38823
5.00	0.90935	0.58558	0.50808	0.45424	0.39509	0.69706	0.69303	0.62280	0.61849	0.57027	0.56233	0.50881	0.50497
5.50	0.95534	0.60159	0.54494	0.51093	0.46317	0.71099	0.70680	0.65688	0.65272	0.62362	0.61664	0.57446	0.57078
5.75	0.96839	0.60283	0.55440	0.52742	0.48561	0.71151	0.70718	0.66530	0.66118	0.63888	0.63223	0.59607	0.59227
6.00	0.97740	0.60176	0.56040	0.53902	0.50286	0.70983	0.70540	0.67044	0.66634	0.64948	0.64309	0.61269	0.60867
6.25	0.98361	0.59936	0.56409	0.54717	0.51621	0.70712	0.70244	0.67343	0.66934	0.65681	0.65064	0.62560	0.62125
6.50	0.98789	0.59630	0.56629	0.55290	0.52662	0.70381	0.69893	0.67504	0.67096	0.66189	0.65588	0.63570	0.63098
7.00	0.99288	0.58972	0.56814	0.55979	0.54124	0.69699	0.69173	0.67598	0.67192	0.66773	0.66206	0.64996	0.64447
7.50	0.99527	0.58377	0.56844	0.56324	0.55043	0.69099	0.68542	0.67557	0.67155	0.67040	0.66503	0.65894	0.65290
8.00	0.99643	0.57898	0.56821	0.56499	0.55629	0.68618	0.68040	0.67471	0.67077	0.67148	0.66641	0.66456	0.65835
9.00	0.99730	0.57271	0.56755	0.56633	0.56252	0.67945	0.67369	0.67264	0.66897	0.67137	0.66694	0.66977	0.66426
10.00	0.99755	0.56952	0.56711	0.56666	0.56515	0.67486	0.66954	0.67032	0.66701	0.66981	0.66600	0.67056	0.66618
20.00	0.99703	0.56655	0.56653	0.56653	0.56655	0.64309	0.64067	0.64242	0.64024	0.64241	0.64020	0.64304	0.64098

[a]All quantities in a.u. Energies are negative and are relative to -32.00000.

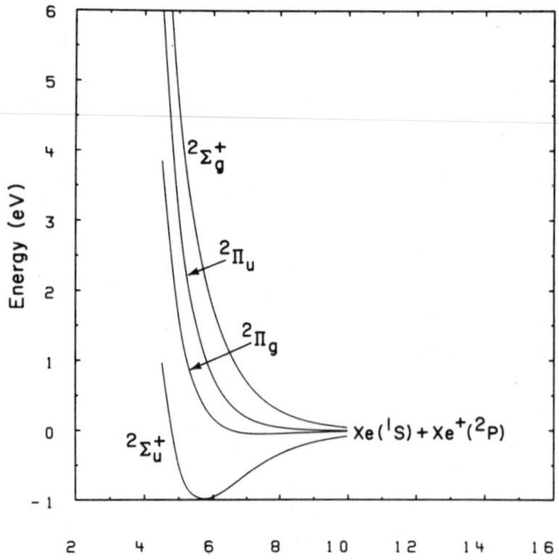

FIG. 4. Xe_2^+ potential energy curves without spin–orbit coupling.

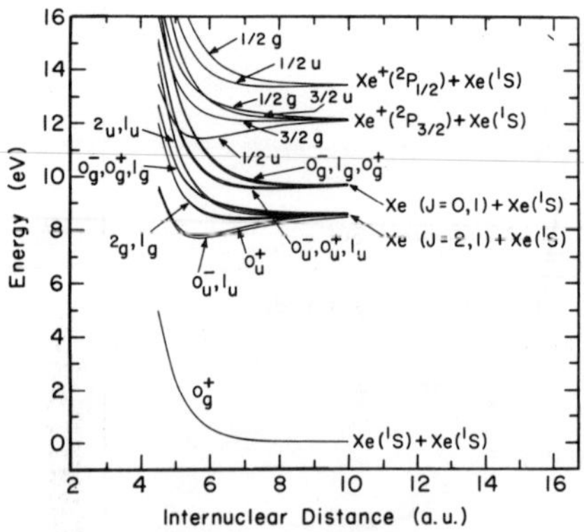

FIG. 6. Xe_2, Xe_2^*, and Xe_2^+ potential energy curves including spin–orbit coupling.

by Wadt[5] and those due to the use of a relativistic EP[6] for Xe derived from[20] the Cowan and Griffin[16] atomic formalism are also given in Table V for comparison. Some vertical transition energies for Xe_2^+ and Xe_2^* are listed in Table VI.

III. DISCUSSION

Although it has been shown that the relativistic effects other than spin–orbit coupling are not important for the properties calculated here for Xe_2, some discussion of the differences between the all-electron calculations of Wadt and the results using the EP's is necessary. It is seen in Figs. 2 and 3 that the EP curves are displaced to the left of the all-electron curves. Presumably the curves from all-electron calculations are more nearly correct, although the limited basis set used for the core electrons in these particular calculations may yield curves that are too repulsive at short distances. Other calculations with effective potentials have yielded curves which also appear to be not repulsive enough.[6,21] Further studies are needed to determine the extent to which this is a difficulty inherent in the effective core potential method as presently employed or to which it may arise from numerical approximations in the use of basis functions to represent EP's or in other steps in the calculations. The choice of the residual potential U_L (see above) may also contribute to discrepancies. We hope to be able to offer more definite conclusions in the near future.

Finally it is noted that the results using the electron gas approximation[17] are in close agreement with the EP

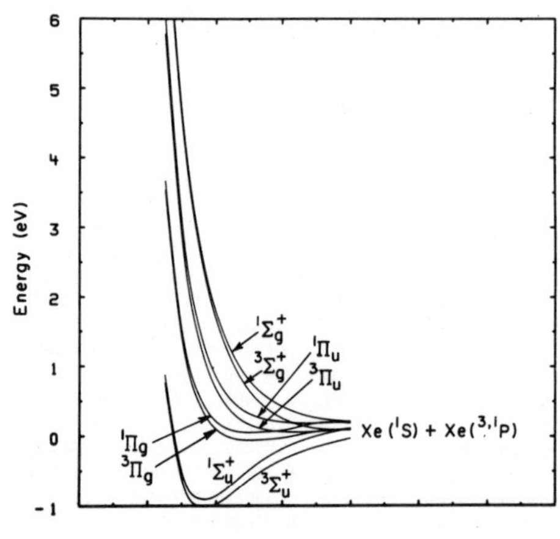

FIG. 5. Xe_2^* potential energy curves without spin–orbit coupling.

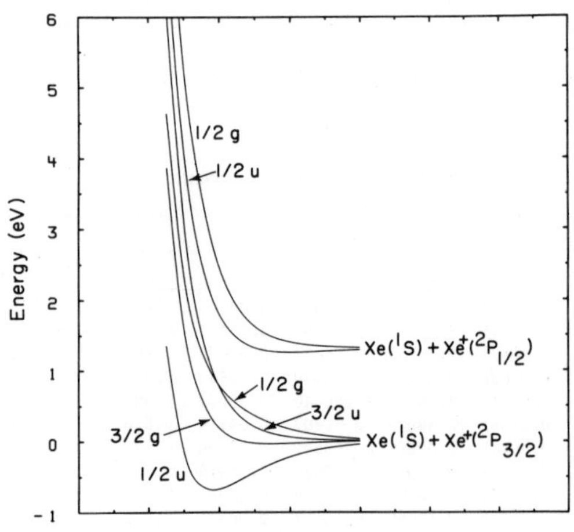

FIG. 7. Xe_2^+ potential energy curves including spin–orbit coupling.

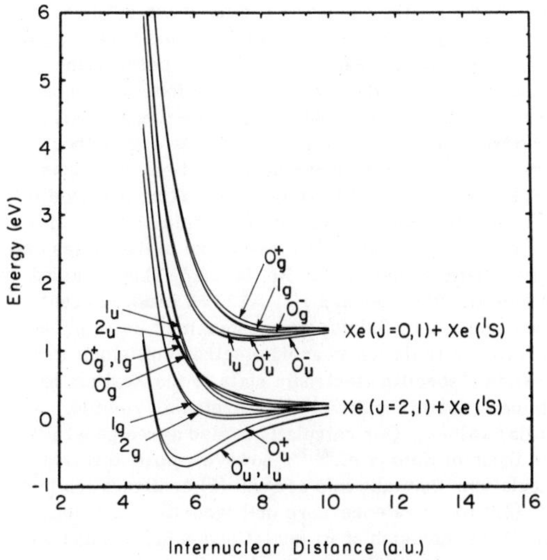

FIG. 8. Xe_2^* potential energy curves including spin–orbit coupling.

results as shown in Fig. 2. This may be related to the inherent frozen core approximation in the electron gas methods.

The results for potential energy curves of rare gas dimer ions have been shown to be consistently well represented by the SCF approximation due to the cancellation of errors arising from the following effects.[8,9] The SCF wavefunctions for homopolar ions formally dissociate to two atoms having a +1/2 charge. This is essentially an orbital relaxation effect and reflects the fact that the orbitals describe averages of a neutral atom and a positive ion. Hence, the total energy at large internuclear distance is higher than the sum of

TABLE VI. Selected Xe_2^+ and Xe_2^* vertical transition energies (eV).

		This work	CI[a]	CI–EP[b]
Xe_2^+	$(1/2)_u \rightarrow (3/2)_g$	1.03	0.99	0.96
	$\rightarrow I(1/2)_g$	1.67	1.60	1.57
	$\rightarrow II(1/2)_g$	3.56	3.31	3.32
Xe_2^* [c]	$0_u^+ \rightarrow X0_g^+$	6.99		
	$1_u \rightarrow X0_g^+$	6.88		

[a]Reference 5.
[b]Reference 6.
[c]A bound–free emission is observed at 7.3 eV [Ref. 1(b)].

the separate atom and ion SCF energies by an amount designated by Gilbert and Wahl[8] as the "left–right" correlation energy. The additional configurations required to acquire the proper atom plus ion limiting value have been discussed in detail elsewhere.[8,5] This orbital relaxation error is balanced by an error of nearly equal magnitude due to the neglect of correlation energy in the bonding region in the use of the SCF approximation. The fact that the approximation is more serious for Xe_2^* is seen in the results in Table IV which show a large increase in energy from $R = 10$ to $R = 20$.

The use of SCF wavefunctions for the Rydberg Xe_2^* states should be reasonable since they correspond to the addition of a single $6s\sigma_g$ electron to the respective Xe_2^+ cores. However, the dissociation products are not as reasonable as in the case of Xe_2^+ since the average of ground and excited Xe orbitals are not as similar as are those for Xe and Xe^+. It appears from our calculations that the results for the Xe_2^* states are valid for internuclear distances as large as about 9–10 a.u. as shown in Table IV. Spectroscopic constants for the Xe_2^* states given in Table V are, as expected, close to those for Xe_2^+, suggesting that the Xe_2^* states are reasonably described by the SCF calculations.

TABLE V. Spectroscopic constants for Xe_2^+ and Xe_2^*.[a]

		$R_e(a_0)$	D_e(eV)	ω_e(cm^{-1})	$\omega_e X_e$(cm^{-1})	B_e(cm^{-1})	α_e(cm^{-1})
$Xe_2^+\,{}^2\Sigma_u^+$	This work	5.74	0.99	122.5	0.45	0.02786	0.00011
	CI[b]	6.08	1.08	124			
	CI–EP[c]	5.84	1.04	125			
$(1/2)_u^d$	This work	5.82	0.70	110.4	0.53	0.02704	0.00013
	CI[a]	6.18	0.79	112			
	CI–EP[b]	5.91	0.76	112			
$Xe_2^*\,{}^3\Sigma_u^+$	This work	5.67	1.00	128.3	0.62	0.02856	0.00012
${}^1\Sigma_u^+$	This work	5.65	1.03	129.2	0.45	0.02868	0.00011
0_u^+	This work	5.72	0.77	118.5	0.53	0.02800	0.00013
0_u^-	This work	5.73	0.78	117.3	0.59	0.02793	0.00014
1_u	This work	5.73	0.79	118.0	0.59	0.02796	0.00014

[a]A spin–orbit correction as described in the text was incorporated for the $(1/2)_u$, 0_u^+, 0_u^-, and 1_u states.
[b]Reference 5.
[c]Reference 6.
[d]Empirical estimate of $D_0 = 1.03$ eV, $R_e = 6.14\ a_0$ (Ref. 4).

TABLE VII. Variation of the magnitude of the transition moment with internuclear distance (a.u.).

State	$^1\Sigma_u^+$	$^1\Pi_u$	0_u^+	1_u
$R(a_0)$				
5.5	0.691	0.953	0.678	0.115
5.75	0.723	0.944	0.705	0.130
6.25	0.785	0.922	0.752	0.159
7.5	0.901	0.866	0.804	0.207
8.0	0.926	0.841	0.803	0.209

It is important to note that while the $\sigma_g 6s$ Rydberg orbital is the lowest in energy, there will be bound excimer states due to excitation to the $6p$ and $5d$ (σ, π, and δ) Rydberg orbitals.[3,4] These will result in avoided curve crossings traceable to the change in character from $\sigma_g 6s$ to the other Rydberg orbitals. In such cases the use of the empirical spin–orbit coupling procedure is no longer valid since the nature of the excited state changes from $5p^5 6s$ to $5p^5 6p$ or $5p^5 5d$.[18] The present SCF calculations do not, of course, allow for a change of orbital character and may be considered as describing the $\sigma_g 6s$ states unperturbed by the presence of nearby states of the same symmetry. Estimates of the curves for these other Xe_2^* states are given in Refs. 3 and 4.

The inclusion of spin–orbit coupling into the curves shown in Figs. 4 and 5 yields curves (Figs. 7–8) that show trends similar to those reported by Cohen and Schneider[18] for Ne_2^+ and Ne_2^* and those of Wadt[5] for Xe_2^+. The magnitude of the spin–orbit interaction is considerably greater for Xe than for Ne as noted by the shift from a ^1P term and a ^3P manifold to the pattern of j–j coupling with the pairs of terms which $J=1,2$ (for $5p_{1/2}^2 5p_{3/2}^3 6s$) and $J=0,1$ (for $5p_{1/2} 5p_{3/2}^4 6s$).

Dipole transition moments for the $^1\Sigma_u^+ \to {}^1\Sigma_g^+$ and $^1\Pi_u \to {}^1\Sigma_g^+$ emissions calculated using the respective SCF wavefunctions are reported in Table VII for several internuclear distances. The transition moments of the states including spin–orbit coupling are related to these through the eigenvectors of the spin–orbit matrices defining the 0_u^+ and 1_u states[18(b)]; viz,

$$\mu(0_u^+) = a_1\mu(^1\Sigma_u^+) + a_2\mu(^3\Pi_u),$$

$$\mu(1_u) = b_1(^1\Pi_u) + b_2\mu(^3\Sigma_u^+) + b_3\mu(^3\Pi_u).$$

Since the transition moments for the triplet states are identically zero the only coefficients required are a_1 and b_1 ($a_1 = 0.9753$ and $b_1 = -0.1375$ for $R = 5.75\ a_0$). The other low-lying spin-orbit state, 0_u^-, is strictly forbidden to radiate to the ground state. The resultant values for the transition moments including spin–orbit coupling for $R = 5.75\ a_0$ are $\mu(0_u^+) = 0.7049$ a.u. and $\mu(1_u) = 0.1297$ a.u. The radiative lifetime τ for a bound-continuum emission can, to a good approximation, be expressed in terms of the vertical energy between the two states ΔE(eV) and the transition moment μ(a.u.) near R_e using the expression $A = \tau^{-1} = 1.063 \times 10^6 |\Delta E|^3 |\mu|^2$ sec for the Einstein coefficient of spontaneous emission.[22] Using the SCF values $\Delta E(0_g^+ - 0_u^+) = 6.987$ eV and $\Delta E(0_g^+ - 1_u) = 6.878$ eV the computed lifetimes are $\tau(0_u^+) = 5.55$ nsec and $\tau(1_u) = 172$ nsec. These lifetimes have proven difficult to measure experimentally[23–29] because the fluorescence from the low-lying vibrational levels of these two states overlap strongly producing a single broad band at 1720 Å. The experiments are further complicated by electron and heavy particle collisions mixing the 0_u^+, 0_u^-, and 1_u states and making it very difficult to isolate the population in a single state.[29] Measurements by Keto *et al.*[24,29] are close to the values used in the kinetic model for the Xe_2 laser.[2] They obtained $\tau(0_u^+) = 6.22 \pm 0.8$ nsec and $\tau(1_u) = 100 \pm 2$ nsec. Recent measurements by Shirley and associates[23] using pulsed synchrotron radiation to avoid electron collisions and to excite a specific electronic state of the Xe atom have obtained results tentatively interpreted as yielding similar values. Our calculations also agree generally with those of Keto *et al.*[24,29]; however more accurate calculations that take into account (1) configuration mixing, (2) proper valence-core orthogonality conditions,[30] and (3) the inclusion of Franck–Condon factors and the continuum contributions in the A coefficient need to be carried out.

The variation of the transition moments with respect to changes in internuclear distance (see Table VII) are not large as would be expected for these Rydberg states. Because the SCF wavefunctions do not have the flexibility to allow for proper dissociation, the values at larger separation are expected to be less reliable. The variation of the excitation energy with internuclear distance is expected to have a more significant influence on the lifetime because the emission is to the steep repulsive wall of the ground state potential energy curve.

In order to calibrate the molecular lifetimes, calculations were carried out on the $^3P_1$ and $^1P_1$ states of the Xe atom using exactly the same methods to calculate the wavefunctions and transition moments as were used for the molecular case. The calculated excitation energies are $\Delta E(^1S_0 - {}^3P_1) = 7.316$ eV and $\Delta E(^1S_0 - {}^1P_1) = 8.395$ eV, and the corresponding transition moments are $\mu(^3P_1) = 0.6133$ a.u. and $\mu(^1P_1) = 0.5280$ a.u. These values lead to the lifetimes $\tau(^3P_1) = 6.39$ nsec and $\tau(^1P_1) = 5.70$ nsec. Anderson[31] has determined the lifetimes of these states to be 3.79 ± 0.12 nsec and 3.17 ± 0.19 nsec. while recent measurements of Matthias *et al.*[32] yielded 3.5 ± 0.1 nsec and 3.4 ± 0.1 nsec, respectively.

It would appear that there is considerable discrepancy between theory and experiment. However, the majority of the error can be attributed to the use of the SCF values for the excitation energy. Using the observed[19] values $\Delta E(^1S_0 - {}^3P_1) = 8.436$ eV and $\Delta E(^1S_0 - {}^1P_1) = 9.570$ eV the calculated lifetimes are $\tau(^3P_1) = 4.17$ nsec and $\tau(^1P_1) = 3.85$ nsec, in reasonable agreement with experiment. Since the SCF excitation energies used in the molecular lifetime calculations are close to the observed values (Table VI), the lifetimes are lowered by only ~10% if the latter energies are used. Finally, it is expected that the calculated lifetime for the 1_u state is less reliable than that for the 0_u^+ state because of potentially large effects on the $^1\Pi_u$–$^1\Sigma_g^+$ transition moment due to the presence of a bound Π_u state involving a $d\pi_g$

Rydberg orbital and coming from the atomic Xe($5p^5 6p$) asymptotic state.[4] The $^1\Pi_u$ state employed in the present calculations, restricted to having Xe($5p^5 6s$) parentage due to the use of the SCF approximation, is purely repulsive (Fig. 5).

The results presented here are intended to place the low lying Xe_2^+ and Xe_2^* potential energy curves on a somewhat more quantitative foundation than the estimates of Mulliken.[3,4] They have also served to demonstrate the reliability of using EP's for molecular calculations involving heavy atoms. It appears that for Xe dimers the incorporation of the effects of spin–orbit coupling by means of an empirical procedure[18] using experimental atomic term splittings is reliable in cases where strong configuration mixing or change in orbital character is not expected. Atomic and molecular lifetimes computed using the respective SCF wavefunctions and AREP's yield results that are in reasonable agreement with experiment.

In future work we plan to compute potential energy curves for Xe_2^* states including configuration mixing to insure proper dissociation properties and to allow for the interactions among the $6s$, $6p$, and $5d$ Rydberg orbitals. The transition moments as a function of internuclear distance will be recomputed and the theoretical emission spectra for Xe_2^* excimers reported.

ACKNOWLEDGMENT

We thank Dr. W. R. Wadt, Dr. P. J. Hay, and Dr. L. R. Kahn for helpful discussions and for providing us with their results prior to publication. We are grateful to Dr. M. Yoshimine for his assistance in the transition moment calculations.

This work was supported by the U.S. Department of Energy under Contract No. W-7405-ENG-48. The support for the part at Berkeley was through the Division of Chemical Sciences, Office of Basic Energy Sciences.

[1](a) N. G. Basov, V. A. Danilychev, Yu. M. Popov, Sov. J. Quantum Electron. **1**, 18 (1971), (b) H. A. Koehler, L. J. Ferderber, D. L. Redhead, and P. J. Ebert, Appl. Phys. Lett. **21**, 198 (1972); J. B. Gerardo and A. W. Johnson, IEEE J. Quantum Electron. **9**, 748 (1973), (c) P. W. Hoff, J. C. Swingle, and C. K. Rhodes, Appl. Phys. Lett. **23**, 245 (1973); W. M. Hughes, J. Shannon, and R. Hunter, Appl. Phys. Lett. **24**, 488 (1974), (d) A. W. Johnson and J. B. Gerardo, J. Appl. Phys. **45**, 867 (1974).

[2]C. W. Werner, E. V. George, P. W. Hoff, and C. K. Rhodes, Appl. Phys. Lett. **25**, 235 (1974); C. W. Werner, E. V. George, P. W. Hoff, and C. K. Rhodes, Report UCRL-77412, Lawrence Livermore Laboratory, Livermore, CA, 1975.

[3]R. S. Mulliken, J. Chem. Phys. **52**, 5170 (1970).

[4]R. S. Mulliken, Radiat. Res. **59**, 357 (1974).

[5]W. R. Wadt, J. Chem. Phys. **68**, 402 (1978).

[6]W. R. Wadt, P. J. Hay, and L. R. Kahn, J. Chem. Phys. **68**, 1752 (1978).

[7]Y. S. Lee, W. C. Ermler, and K. S. Pitzer, J. Chem. Phys. **67**, 586 (1977).

[8]T. L. Gilbert and A. C. Wahl, J. Chem. Phys. **55**, 5247 (1971).

[9]W. J. Stevens, M. Gardner, A. Karo, and P. Julienne, J. Chem. Phys. **67**, 2860 (1977).

[10]J. P. Desclaux, Comput. Phys. Commun. **9**, 31 (1975).

[11]L. R. Kahn, P. Baybutt, and D. G. Truhlar, J. Chem. Phys. **65**, 3826 (1976), and references therein.

[12]C. Froese-Fischer, Computer Phys. Commun. **1**, 151 (1969); **4**, 107 (1972); **7**, 236 (1974).

[13]D. B. Neumann, H. Basch, R. L. Kornegay, L. C. Snyder, J. W. Moskowitz, C. Hornback, and S. P. Liebmann, Program 199, Quantum Chemistry Program Exchange, Indiana University, 1972.

[14]L. R. Kahn (private communication).

[15]W. J. Hunt, T. H. Dunning, Jr., and W. A. Goddard III, Chem. Phys. Lett. **3**, 606 (1969); W. A. Goddard III, T. H. Dunning, Jr., and W. J. Hunt, *ibid.* **4**, 231 (1969); W. J. Hunt, W. A. Goddard III, and T. H. Dunning, Jr., *ibid.* **6**, 147 (1970).

[16]R. D. Cowan and D. C. Griffin, J. Opt. Soc. Am. **66**, 1010 (1976).

[17]R. G. Gordon and Y. S. Kim, J. Chem. Phys. **56**, 3123 (1972); A. I. M. Rae, Chem. Phys. Lett. **18**, 574 (1973).

[18](a) J. S. Cohen and B. Schneider, J. Chem. Phys. **61**, 3230 (1974), (b) B. Schneider and J. S. Cohen, *ibid.* 3240 (1974).

[19]C. E. Moore, Circ. Natl. Bur. Stand. No. 467, Vol. III **68**, 2386 (1978).

[20]L. R. Kahn, P. J. Hay, and R. D. Cowan, J. Chem. Phys. **68**, 2386 (1978).

[21]P. J. Hay, W. R. Wadt, and L. R. Kahn, J. Chem. Phys. **68**, 3059 (1978).

[22]G. Herzberg, *Molecular Spectra and Molecular Structure. I Spectra of Diatomic Molecules* (Van Nostrand Reinhold, Princeton, 1950). For bound–continuum transitions see F. H. Mies, Mol. Phys. **26**, 1233 (1973).

[23]D. A. Shirley (private communication).

[24]J. W. Keto, R. E. Gleason, Jr., and G. K. Walters, Phys. Rev. Lett. **33**, 1365 (1974).

[25]A. W. Johnson and J. B. Gerardo, J. Chem. Phys. **59**, 1738 (1974).

[26]S. C. Wallace, R. T. Hodgson, and R. W. Dreyfus, Appl. Phys. Lett. **23**, 672 (1973).

[27]D. J. Bradley, M. H. R. Hutchinson, and H. Koester, Opt. Commun. **7**, 187 (1973).

[28]H. A. Koehler, L. J. Ferderber, D. L. Redhead, and P. J. Ebert, Phys. Rev. A **9**, 768 (1973).

[29]J. W. Keto, R. E. Gleason, Jr., T. D. Bonifield, G. K. Walters, and F. K. Soley, Chem. Phys. Lett. **42**, 125 (1976).

[30]R. N. Euwema and L. R. Kahn, J. Chem. Phys. **66**, 306 (1977).

[31]D. K. Anderson, Phys. Rev. A **137**, 21 (1965).

[32]E. Matthias, R. A. Rosenberg, E. D. Poliakoff, M. G. White, S.-T. Lee, and D. A. Shirley, Chem. Phys. Lett. **52**, 239 (1977).

Ab initio effective core potentials including relativistic effects. III. Ground state Au_2 calculations

Yoon S. Lee, Walter C. Ermler,[a] and Kenneth S. Pitzer

Department of Chemistry and Lawrence Berkeley Laboratory, University of California, Berkeley, California 94720

A. D. McLean

IBM Research Laboratory, San Jose, California 95193

(Received 29 August 1978)

SCF calculations have been carried out for the ground ${}^1\Sigma_g^+(0_g^+)$ state of Au_2 using a variety of *ab initio* effective core potentials (EP). The effective core potentials studied both include a two-component relativistic EP (REP) that includes spin–orbit effects and also averaged relativistic EP (AREP) and a nonrelativistic EP (NREP). All-electron nonrelativistic calculations were also performed. The values of spectroscopic constants obtained from these calculations indicate that relativistic effects account for a decrease in R_e of over 0.3 Å and an increase in the bond energy of the order of 1 eV. Various intercomparisons indicate the general validity of effective potential methods, properly applied, but also show certain limitations. In particular, the NREP results agree well with the all-electron, nonrelativistic calculations. Also, various relativistic effective-potential methods agree for SCF calculations provided both the basis sets and the EP are carried to sufficiently high order in angular quantum number. The bond distance calculated relativistically agrees very well with experiment.

I. INTRODUCTION

In the first paper of this series[1] (hereafter referred to as paper I), the general basis was established for a treatment of molecules containing heavy atoms by relativistic quantum mechanics and effective potentials (EP). This treatment parallels the *ab initio* core potential methods based on nonrelativistic quantum mechanics as developed recently by several authors.[2] Our procedure starts with the fully relativistic Dirac–Hartree–Fock treatment of each atom. Thus, the EP derived from these atomic data contains all relativistic effects on the core electrons as well as the valence electrons except for Breit interactions and are labeled REP. The results include spin–orbit effects and follow the j–j coupling system, i.e., there are separate EP for $p_{1/2}$ and $p_{3/2}$ states, etc. In paper I, the REP were derived for xenon and gold atoms. In this paper, we use the results for gold to treat the ground state of Au_2 in the SCF and MCSCF approximations and compare the results with those from an all-electron nonrelativistic treatment with more approximate relativistic calculations and with experiment. A paper to follow will present calculations for excited states, as well as the ground state, of Au_2 based on extensive configuration interaction.

Rigorous relativistic molecular calculations must be made in the ω–ω coupling system (equivalent to j–j coupling for atoms). Although we are developing programs for diatomic molecules in ω–ω coupling for configurations with open shells, these are not complete and hence not available for the present calculations. A SCF calculation was made, however, for the single configuration closed shell ground state of Au_2 with two-component relativistic spinors and EP's.

A further approximation allows the use of molecular programs in L–S coupling; this requires the deletion of spin–orbit coupling. We have followed this method by taking the weighted average of the two EP's for a given nonzero l value, i.e., for $p_{1/2}$ and $p_{3/2}$ or $d_{3/2}$ and $d_{5/2}$. Such results are labeled AREP for "averaged relativistic effective potentials." There are semiempirical methods for the approximate reintroduction of spin-orbit effects.[3] Paper II[4] presents results by this method for Xe_2 in the ground and excited states and for Xe_2^+. Kahn, Hay, and Cowan[5] have also developed a method for AREP calculations in which they start with partially relativistic atomic calculations which, in the Pauli approximation, include only the mass–velocity and Darwin terms and omit the spin–orbit term. Comparison of results from the two AREP methods shows little difference between them; hence, one concludes that the additional approximations of Kahn *et al.* are not important in this case, at least.

II. CALCULATIONS

The EP's for Au were obtained using the methods described in paper I. Programs to compute the necessary molecular integrals over the original numerical EP with respect to a basis set of Slater-type functions (STF) were developed based on an earlier program that employed a zeta-function approach.[6] In addition, nonrelativistic (NREP) and averaged relativistic (AREP) SCF calculations have been carried out using EP's derived[7] using the methods of Refs. 2 and 5 for Au expressed as fits to Gaussian-type functions together with corresponding[7] valence electron basis sets. Finally, to gauge the EP approximation of the core electrons, all-electron SCF calculations were carried out for Au_2.

A. Calculations with AREP and NREP

It is instructive to describe certain features of the program[6] used to compute the matrix elements over the

[a]Present address: Department of Chemistry and Chemical Engineering, Stevens Institute of Technology, Hoboken, NJ 07030.

 0021-9606/79/010288-05$01.00

EP with respect to an STF basis. An advantage in using numerical integration procedures is that it allows us to input the EP in any convenient form, viz., as numerical functions, or as fits to Slater (STF) or Gaussian (GTF) functions, from which the appropriate quadrature points may be generated internally. The numerical EP's computed by the methods of paper I, being relative to the logarithmic grid dictated by the numerical Dirac–Hartree–Fock procedure, are matched with the Gauss–Legendre or Gauss–Laguerre quadratures used by the EP integral program by interpolation using a cubic spline technique.

In the case of the relativistic EP's the behavior in the region near the nucleus has been accounted for by matching the expected behavior [paper I, Eq. (43)] as $r \to 0$ with the value and its first derivative of the EP at some value $r_{\min}$ where the EP is still well behaved, i.e., numerical inaccuracies have not yet become evident.

The questions to be addressed here involve the problems inherent in the representation of the EP's for a given atom in the various forms that derive from the general equation[2]

$$U^{\mathrm{AREP}} = U_L^{\mathrm{AREP}} + \sum_{l=0}^{L-1} (U_l^{\mathrm{AREP}} - U_L^{\mathrm{AREP}}) \,|lm\rangle\langle lm| \ . \tag{1}$$

In this part, we treat only those EP's that correspond to the use of LS coupling in the molecular calculations. Relativistic EP's generated by the method described in paper I have been transformed to the appropriate form by taking the weighted average of the two components corresponding to a given value of the angular quantum number l. These averaged relativistic effective core potentials (AREP) yield good results for Xe_2 ground, excited, and positive ion states[4] and compare well with results[8] based on the AREP of Kahn *et al.*[5] The value of L corresponding to the largest angular momentum used in deriving U^{EP} defines the "residual potential" and is, ideally, one greater than the highest l of the core.

We made SCF calculations for Au_2 based upon our AREP functions, which are designated AREP(I), and upon the AREP obtained by Hay *et al.*[7] for gold by the method of Kahn *et al.*[5] which omits spin–orbit effects in the atomic calculations; the latter results are designated AREP(II). We also used the nonrelativistic EP of Hay *et al.*[7] for calculations on Au_2 listed as NREP. For AREP(II) and NREP, we used the basis sets chosen by Hay *et al.*[7]

B. Calculations with REP

The application of the j-dependent relativistic effective core potentials (REP) to molecular calculations requires the computation of matrix elements over a basis set of two-component spinors corresponding to four-component Dirac spinors without small components. The Hamiltonian is expressed as

$$H = \sum_{\mu=1}^{n_v} h_\mu + \sum_{\mu>\nu} \frac{1}{r_{\mu\nu}} , \tag{2}$$

where n_v is the total number of valence electrons and $r_{\mu\nu}$ is the distance between electron μ and ν. In Eq. (2), h_μ is given by

$$h_\mu = -\tfrac{1}{2}\nabla_\mu^2 + \sum_{\alpha=1}^{N} \left(-\frac{Z_\alpha}{r_{\alpha\mu}} + U_\alpha^{\mathrm{REP}} \right) , \tag{3}$$

where Z_α is the nuclear charge of nucleus α and N is the number of nuclei. U_α^{REP} can be approximated by [paper I, Eq. (41)]

$$U_\alpha^{\mathrm{REP}} = U_{LJ}^{\mathrm{REP}}(r_{\alpha\mu}) + \sum_{l=0}^{L-1} \sum_{j=|l-(1/2)|}^{l+(1/2)} \sum_{m=-j}^{j} [U_{lj}^{\mathrm{REP}}(r_{\alpha\mu}) - U_{LJ}^{\mathrm{REP}}(r_{\alpha\mu})] \,|ljm\rangle_\alpha \langle ljm|_\alpha \ , \tag{4}$$

where the projection operators are defined by

$$|ljm\rangle\langle ljm| = \left[\sum_{\sigma=\pm(1/2)} C(l\tfrac{1}{2}j;\, m-\sigma) \,|Y_\lambda^{m-\sigma}(\theta,\phi)\phi_{1/2}^{\sigma}\rangle \right] \times \left[\sum_{\sigma'=\pm(1/2)} C(l\tfrac{1}{2}j;\, m-\sigma', \sigma') \langle Y_\lambda^{m-\sigma'}(\theta,\phi)\phi_{1/2}^{\sigma'}| \right] , \tag{5}$$

with the notation the same as in paper I. All one and two electron matrix elements of the above Hamiltonian with two component basis functions can be expressed as linear combinations of appropriate nonrelativistic integrals. The detailed expressions for these matrix elements may be obtained from the work by Malli and Oreg[9] by eliminating parts dependent on the small components of the Dirac spinors.

Matrix elements over angular projection operators [Eq. (5)] are calculated as linear combinations of the nonrelativistic integrals described in the previous section and additional integrals that correspond to the case $\sigma \neq \sigma'$. The formalism for the SCF procedure is essentially parallel to that developed by Malli and Oreg for relativistic theory for closed-shell molecules.[9] Although their formalism is limited to the minimal basis set representation of the Dirac–Hartree–Fock calculations, this restriction disappears in our application because of the absence of small components. The symmetry properties of these molecular orbitals are the same as those of the relativistic molecular orbitals, viz., orbitals are only distinguished by their total angular momentum like Hund's case (c).

At present, the program is limited to closed-shell configurations of diatomic molecules.

C. All-electron nonrelativistic calculations

A double-zeta basis set generated for the $(\ldots 5d^9\, 6s\, 6p)\, {}^4F$ state of the gold atom was used in single configuration SCF calculations on $(\ldots 27\sigma^2 \ldots 16\pi^2 \ldots 8\delta^2 .. 2\phi^2)\, {}^1\Sigma_g^+$ Au_2. The atomic energy with this basis set is -17864.62328 a.u. compared with a value of -17864.62372 a.u.[10] for a thoroughly optimized double-zeta basis. Simply because of the large number of occupied shells, these double-zeta bases for heavy atoms are better able to contain orbital expansions which approach the Hartree–Fock limit than is the case in light atoms, at least for s and p symmetries. Also, because of the large number of occupied shells, the basis set contains, without further addition, functions whose extent and symmetry type are those required for describing polarization effects in the molecular environment. The results of this double-zeta Au_2 calculation, then,

TABLE I. SCF results for Au_2.

Basis set	EP	L_{max}[a]	R_e(Å)	$E_{20}-E_e$(eV)	ω_e(cm^{-1})
STF (12*s* 10*p* 6*d* 2*f*)	All-electron		2.84	3.00	105
GTF (3*s* 3*p* 3*d*)/[2*s* 2*p* 2*d*]	NREP(GTF)	*d*	2.83	2.86	91
GTF (3*s* 3*p* 3*d*)/[2*s* 2*p* 2*d*]	AREP(II)(GTF)	*d*	2.48	4.00	138
STF (2*s* 2*p* 2*d* 1*f*)	AREP(I) (Numerical)	*g*	2.48	4.01	145
STF (2*s* 2*p* 2*d*)	REP (Numerical)	*g*	2.50	3.93	142

[a]See Eq. (1).

will be close to the Hartree–Fock limit for the shape of the interaction potential. The energy difference between $R=20.$ bohr and $R=5.$ bohr should be within ~ 0.2 eV of the Hartree–Fock limit. Absolute energies, of course, will have very much larger errors, but this is inconsequential.

III. RESULTS AND DISCUSSION

Presented in Tables I and II are a series of results for the ground $^1\Sigma_g^+$ state of Au_2. The calculations are single-configuration SCF (Table I) and two-configuration MCSCF (Table II), where the σ_u^2 term necessary for the proper description of two ^2S atomic states at dissociation has been included. The 2*s* 2*p* 2*d* 1*f* basis set used in our EP calculations is given in Table III. Since the SCF wave functions for Au_2 do not allow for the proper asymptotic behavior, dissociation energies are not available. Instead, we report the values of $(E_{20}-E_e)$, the decrease in total SCF energy from 20 a.u. to the minimum at R_e. These values may be used for comparison among the SCF calculations, but it should be noted that the tendency for these wave functions to admixtures of atom pair and ion pair limits may not have proceeded to the same extent at $R=20.0$ a.u. for all of the cases. Hence, the D_e values for the SCF results should only be taken as qualitative comparisons. The values of R_e and ω_e, on the other hand, should be consistently represented and were derived using five points near R_e.

The SCF calculations are intended to show the degree of consistency of the use of EP's derived from different formalisms (Refs. 1, 2, and 5) and to give an indication of the magnitude of the relativistic effects in Au_2. Table I shows the agreement between the all-electron and valence electron nonrelativistic EP (NREP) results to be quite reasonable. The differences (rows 1 and 2) between values of R_e, $(E_{20}-E_e)$, and ω_e are 0.01 Å, 0.14 eV, and 14 cm^{-1}, respectively. Similarly, the relativistic calculations AREP(I) and REP using a $L=4$ residual potential and a numerical form of U_l determined by the methods of paper I yield results (rows 4 and 5) in good agreement with each other and with those from AREP(II) in row 3. This agreement between AREP and REP indicates that the averaging procedure is reasonable for this case.

Thus, it appears from the results of Table I that the relativistic effects in Au_2 yield a decrease in R_e in excess of 0.3 Å and about a 25% increase in ω_e. The relativistic calculations of R_e are in excellent agreement with the experimentally determined[12] value $R_e=2.47$ Å. The relativistic calculations for ω_e are smaller than the experimental value of 191 cm^{-1} but much closer than the nonrelativistic results. The energy values in Table I are not so clearly interpretable, but they indicate a relativistic increase in bond energy of the order of 1 eV.

Orbital energies at $R=4.75$ a.u. are shown in Fig. 1. Comparing $6s\sigma_g$ orbital energies, one may conclude that 6*s* electrons are responsible for most of the relativistic effects in Au_2, a large increase in binding energy from the nonrelativistic to the relativistic values. The 5*d* orbital energies are decreased slightly from the nonrelativistic values to those of AREP(I), but this effect is small as compared to the splitting from either interatomic interaction or the spin–orbit effect (in REP). Orbitals of a given ω value in the last column arise from different λ values in the other columns, i.e., $\omega=1/2$ from either σ or π and $\omega=3/2$ from either π or δ. Consequently, there can be some mixing of these 1/2 or 3/2 orbitals and the differences of individual REP values from AREP(I) values are not pure spin–orbit effects, although that was the basic cause.

Wave functions for Au_2 including more extensive configuration mixing have been computed[13] using the numer-

TABLE II. MCSCF results for Au_2 with AREP(L).

		L_{max}[a]	R_e(Å)	D_e(eV)	ω_e(cm^{-1})
(1) STF (2*s* 2*p* 2*d* 1*f*)	AREP (Numerical)	*g*	2.54	0.95	107
(2) STF (2*s* 2*p* 2*d* 1*f*)	AREP (Numerical)	*f*	2.50	1.12	120
(3) STF (2*s* 2*p* 2*d* 1*f*)	AREP (Numerical)	*d*	2.18	3.18	218
(4) STF (2*s* 2*p* 2*d* 1*f*)	AREP (STF)[b]	*g*	2.56	0.85	111
(5) STF (2*s* 2*p* 2*d* 1*f*)	AREP (GTF)[c]	*g*	2.56	0.76	107
(6) STF (2*s* 2*p* 2*d* 1*f*)	AREP (GTF)	*d*	2.19	1.49	174
(7) STF (2*s* 2*p* 2*d*)	AREP (Numerical)	*f*	2.56	1.02	116
(8) STF (2*s* 2*p* 2*d*)	AREP (Numerical)	*d*	2.24	2.90	195

[a]See Eq. (1). [b]Reference 1. [c]Reference 11.

TABLE III. Slater basis set of Au for the REP[a] and AREP(I) calculations.

l	n	ζ
s	2	1.17
	2	0.75
p	2	0.82
	2	0.41
d	3	2.70
	3	1.26
f	4	2.00

[a]Identical basis functions were used for $j = l - (1/2)$ and $j = l + (1/2)$ orbitals with the same l.

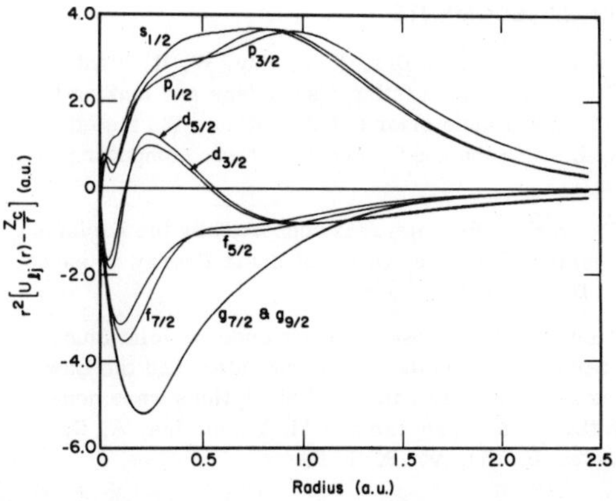

FIG. 2. Relativistic effective core potentials (REP) of the 11-valence electron Au.

ical AREP and the same valence STF basis set for all electronic states arising from the atomic asymptotic limits $^2S + ^2S$ and $^2S + ^2D$. The improved ground state potential energy curve has $R_e = 2.37$ Å, $D_e = 2.27$ eV, and $\omega_e = 165$ cm^{-1}, in good agreement with the experimentally determined values.

Table II presents comparisons of properties of Au_2 derived using the AREP(I) defined by the pseudo-orbital transformation for an 11-valence-electron Au atom as detailed in paper I. There are three principal variances in the calculations of Table II yielding important comparisons. The first is the deletion of the f-type polarization function from the basis set; the second is in the form of the EP as either numerical $U_l(r)$'s or STF or GTF least squares fits of these, and the third is in the choice of the residual potential U_L [Eq. (1)] as $L = 2$, 3, or 4.

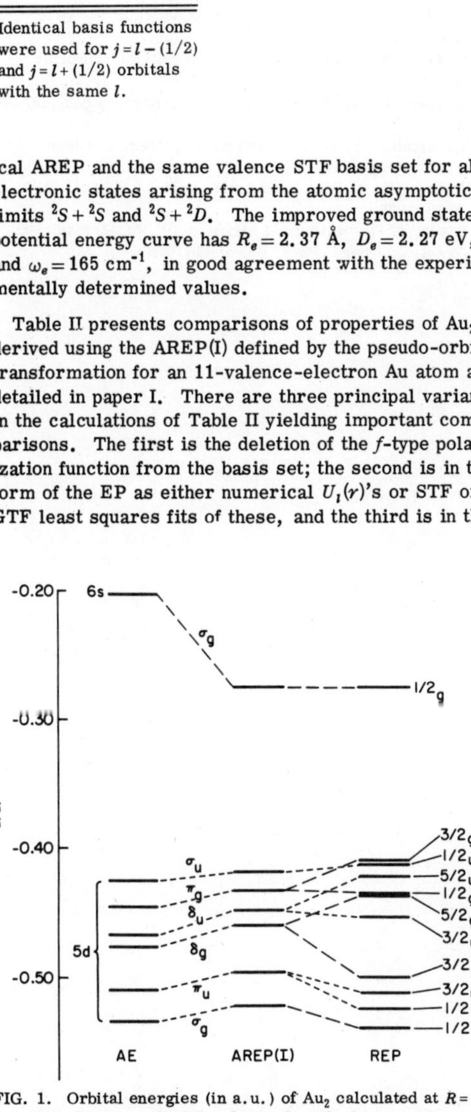

FIG. 1. Orbital energies (in a.u.) of Au_2 calculated at $R = 4.75$ a.u. AE, AREP, and REP refer to the all-electron, the averaged relativistic, and the relativistic EP, respectively. Lines connecting the orbital energies calculated with AREP(I) and REP indicate the probable correlations of the ω–ω coupled orbitals.

The effect of the f-type STF basis function, obtained by comparing rows 2 and 3 with rows 7 and 8, respectively, is about 0.1 eV in D_e and 0.06 Å in R_e. Since this is a rather substantial effect for the addition of polarization-type functions for mostly nonbonding d orbitals, an appreciable part may be attributed to the superposition problem for a double-zeta basis set (i.e., a basis function centered on one nucleus alleviating deficiencies in the atomic basis set at the other nucleus).

The effect of using a least squares fitted AREP as opposed to the original numerical form is seen by comparing rows 1, 4, and 5. The differences of 0.1 and 0.2 eV in D_e of STF and GTF fitted AREP's from the numerical form are a reasonable reflection of accuracies expected from the fitting method.

The most obvious discrepancy in Table II is that among the calculations where the residual potentials differ. Whereas the changes resulting from choosing $L = 3$ instead of $L = 4$ are small, but not negligible, those due to the use of $L = 2$ are unacceptably large (compare the pairs of rows 1–3; 5 and 6; and 7 and 8). The use of U_d as the residual potential for the AREP leads to potential energy curves that are much too attractive with R_e too small by 0.3 Å, and D_e and ω_e too large by about 2 eV and 100 cm^{-1}, respectively. This discrepancy is the consequence of the attractive tail in the residual d potential shown in Fig. 2. This problem has also been encountered in other work and corrected by shortening the region of fitting[4,14,15] or by generating the residual potential from highly positive ions.[14] In our case, the choice of a sufficiently large L value for the residual potentials yielded satisfactory results, but we cannot as yet draw firm general conclusions on this matter.

In conclusion, we note that the calculations reported here show large relativistic effects shortening and strengthening the bond in Au_2 and indicate that the primary source of these changes is in the contraction of the 6s orbitals.

ACKNOWLEDGMENTS

We are grateful to Drs. P. J. Hay, W. R. Wadt, and L. R. Kahn for interesting discussions and making the EP of Au available prior to publication. We also thank Drs. L. R. Kahn and G. Malli for their computer programs.

The work at Berkeley was supported by the Division of Chemical Sciences, Office of Basic Energy Sciences, U.S. Department of Energy.

The work at San Jose was done under a Joint Study Agreement between the IBM Corporation and the Lawrence Berkeley Laboratory. Calculations were done with the ALCHEMY programs of M. Yoshimine, A. D. McLean, B. Liu, and P. S. Bagus.

[1]Y. S. Lee, W. C. Ermler, and K. S. Pitzer, J. Chem. Phys. 67, 5861 (1977).

[2]See, for example, L. R. Kahn, P. Baybutt, and D. G. Truhlar, J. Chem. Phys. 65, 3826 (1976).

[3]See, for example, J. S. Cohen and B. Schneider, J. Chem. Phys. 61, 3230 (1974).

[4]W. C. Ermler, Y. S. Lee, K. S. Pitzer, and N. W. Winter, J. Chem. Phys. 69, 976 (1978).

[5]L. R. Kahn, P. J. Hay, and R. D. Cowan, J. Chem. Phys. 68, 2368 (1978).

[6]Modified version of the program originally developed by L. R. Kahn; L. R. Kahn, Ph.D. Thesis, California Institute of Technology, 1971 and private communication. See also, M. P. Barnett, Methods Comput. Phys. 2, 95 (1963).

[7]P. J. Hay, W. R. Wadt, L. R. Kahn, and F. W. Bobrowicz, J. Chem. Phys. 69, 984 (1978).

[8]W. R. Wadt, P. J. Hay, and L. R. Kahn, J. Chem. Phys. 68, 1752 (1978).

[9]G. Malli and J. Oreg, J. Chem. Phys. 63, 830 (1975).

[10]A. D. McLean, IBM Research Report No. RJ2295, IBM Corporation, San Jose, CA, 1978.

[11]Y. S. Lee, Ph.D. Dissertation, University of California, Berkeley, 1977.

[12]L. L. Ames and R. F. Barrow, Trans. Faraday Soc. 63, 39 (1967).

[13]W. C. Ermler, Y. S. Lee, and K. S. Pitzer, J. Chem. Phys. 70, 288 (1979).

[14]P. J. Hay, W. R. Wadt, and L. R. Kahn, J. Chem. Phys. 68, 3059 (1978).

[15]H. Basch, M. D. Newton, J. Jafri, J. W. Moskowitz, and S. Topiol, J. Chem. Phys. 68, 4005 (1978).

Ab initio effective core potentials including relativistic effects. IV. Potential energy curves for the ground and several excited states of Au_2

Walter C. Ermler,[a] Yoon S. Lee, and Kenneth S. Pitzer

Department of Chemistry and Lawrence Berkeley Laboratory, University of California, Berkeley, California 94720

(Received 29 August 1978)

The ground and excited states of Au_2 are studied using *ab initio* averaged relativistic effective core potentials (AREP) and MCSCF–CI procedures. Spin–orbit effects are included in the excited states derived from $^2S_{1/2}+^2D_{3/2}$ and $^2S_{1/2}+^2D_{5/2}$ atomic states using an empirical procedure. The ground state dissociation energy is calculated to be 2.27 eV as compared to the experimental value of 2.31 eV. The calculated energies for the two spectroscopically allowed 0_g^+ to 0_u^+ transitions and other molecular parameters also agree reasonably well with experiment.

I. INTRODUCTION

There were several reasons for the choice of diatomic gold for an early study in relativistic molecular quantum mechanics. The dissociation energy is anomalously large, larger even than for Cu_2, whereas Ag_2 has a much smaller dissociation energy. Also, Au_2 is isoelectronic with the mercurous ion Hg_2^{++} which is unique in its stability as a dimeric dipositive species. Also, our present theoretical methods appeared to be adequate at this point for the treatment of a molecule where the primary bond was based on *s* orbitals as is explained below.

All-valence-electron calculations on the ground and several excited states of Au_2 are presented here as the fourth paper in a series dealing with the use of the *ab initio* effective core potentials (EP). Paper I[1] describes the formal procedure for deriving the EP's and reports results for the Xe and Au atoms; paper II[2] presents SCF calculations on the ground, positive ion, and $6s\sigma_g$ Rydberg states of Xe_2; and paper III[3] contains the first calculations using these methods on the ground state of Au_2 employing relativistic basis functions in which the results include the effects of spin–orbit coupling directly by the use of an EP defined in terms of two-component spinors. In addition, comparisons with other forms of EP's, both relativistic and nonrelativistic, are discussed in paper III.

The Au_2 molecule was chosen as a test system because, as discussed in paper I, it may be logically treated as a 22-valence electron system with the ground state predominantly being a $6s$–$6s$ σ_g bonding state and the low-lying excited states coming from the $5d^9 6s^2$ configuration leading to 12 states (in standard LS coupling notation) $^1\Sigma_g^+$, $^1\Sigma_u^+$, $^3\Sigma_g^+$, $^3\Sigma_u^+$, $^1\Pi_g$, $^1\Pi_u$, $^3\Pi_g$, $^3\Pi_u$, $^1\Delta_g$, $^1\Delta_u$, and $^3\Delta_u$. The $^2D_{5/2}$ and $^2D_{3/2}$ states of Au are 1.136 and 2.558 eV, respectively, above the ground ^2S state, whereas the next levels ($^2P_{1/2}$ and $^2P_{3/2}$) are at 4.632 and 5.105 eV, resepctively.[4] Hence, the treatment of the ^2D valence states should constitute a reasonably "clean" set in the sense of not being strongly perturbed by other low-lying states of the same symmetry. The availability of experimental results for the ground and some of the excited states allows us to make some comparisons with various spectroscopic constants.

Because the spin–orbit splitting of the $^2D_{5/2}$ and $^2D_{3/2}$ levels of Au is so great (1.522 eV), the description of excited electronic states of Au_2 in terms of ω–ω coupling would appear to be most appropriate. The 12 *LS* symmetries above are combined using an empirical model[6] based on the atomic spin–orbit splitting to yield states of 0_g^+, 0_u^+, 0_g^-, 0_u^-, 1_g, 1_u, 2_g, 2_u, 3_g, and 3_u symmetry in Hund's coupling case c (ω–ω) notation. The observed molecular electronic states may then be compared directly with calculated results.

The following two sections describe the form of the wave functions used in this study and the methods of introducing spin–orbit effects. The resulting potential energy curves are discussed in the last section. Some comments regarding analogous calculations on Ag_2 and Cu_2 are also included.

II. CALCULATIONS

The wave functions for the 12 *LS*-coupling symmetries arising from the interaction of $^2S+{}^2S$ and $^2S+{}^2D$ atomic states were computed using a model of the Au atom whereby the 68 core electrons were replaced by an averaged relativistic effective core potential (AREP), as described in paper III, and the 11 valence electrons ($\ldots 5d^{10}6s$) were described by a valence basis set of two $2s$, two $2p$, two $3d$, and one $4f$ Slater-type functions (STF). The AREP was represented as a numerical function including angular symmetries up to $l=4$ (g-type) as the "residual potential" (see paper III).

The ($2s$ $2p$ $2d$ $1f$) STF set (see paper III) was determined by first fitting the s-, p-, and d-type pseudo-orbitals as two-term expansions, and then fully optimizing the exponents in atomic calculations for various *LS* states in the field of the AREP. The $4f$-type STF was optimized in ground state SCF calculations on Au_2 near R_e and resulted in an energy lowering of about 0.003 a.u.

[a] Present address: Department of Chemistry and Chemical Engineering, Stevens Institute of Technology, Hoboken, NJ 07030.

TABLE I. Ground state Au_2 valence energies (a.u.).

R	E(SCF)	E(MCSCF)	E(CI)
3.0	−70.272820	−70.291183	−70.387333
3.5	0.618155	0.632900	0.719504
4.0	0.738262	0.751744	0.825669
4.25	0.760430	0.773288	0.841041
4.5	0.767895	0.782712	0.844318
4.75	0.769278	0.785526	0.841337
5.0	0.766328	0.784477	0.835026
5.25	0.760811	0.781305	0.827011
5.5	0.754218	0.777429	0.818684
5.75	0.746473	0.773001	0.809870
6.0	0.739423	0.769366	0.802415
7.0	0.712174	0.758443	0.779374
8.0	0.690422	0.753765	0.768405
9.0	0.673762	0.751926	0.764050
10.0	0.661266	0.751150	0.762353
20.0	0.621691	0.750536	0.761078

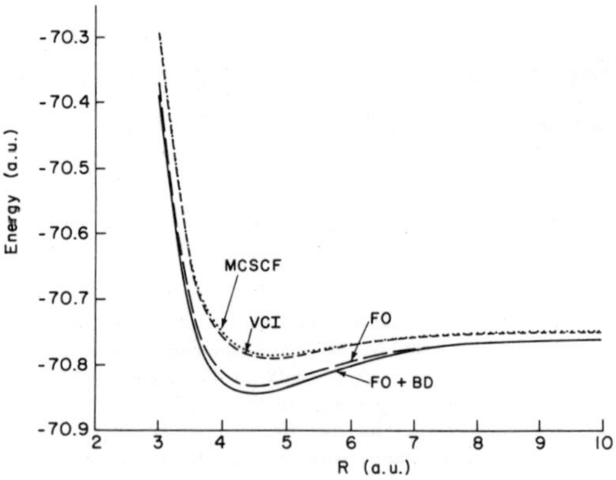

FIG. 1. Potential curves of 0_g^+ of Au_2 calculated with the various levels of configuration interactions (CI).

To insure that the molecular wave function for the ground state of Au_2 was sufficiently flexible to allow for the proper dissociation to two ^2S Au atoms, it was necessary to include the two configurations allowed from the combination of the 6s electrons: $6s\sigma_g^2$ and $6s\sigma_u^2$. MCSCF calculations were carried out incorporating these two configurations at 16 internuclear separations and are reported in Table I together with the analogous SCF (single-configuration) and CI (see below) results.

The molecular integrals for each of the 16 internuclear distances shown in Table I, including the matrix elements involving the AREP (see paper III), were transformed over the MCSCF orbitals for the ground state and used in first-order[7] CI calculations designed as follows: The MCSCF orbitals were placed in three classes, the first corresponding to those derived from the 5d shell of Au, viz., $5d\sigma_g$, $5d\sigma_u$, $5d\pi_u$, $5d\pi_g$, $5d\delta_g$, and $5d\delta_u$. These orbitals were restricted to be fully occupied in the MCSCF calculations but their respective linear coefficients were optimized for each internuclear separation. The second class consists of the $6s\sigma_g$ and $6s\sigma_u$ orbitals that were optimized with respect to configuration mixing at each internuclear separation. The $6s\sigma_g$ orbital is dominant near R_e but the two configurations have equal coefficients as dissociation is approached The third class is made up of the remaining orbitals that can be constructed from the basis set. This class is usually termed the "external" or "virtual" class of orbitals whereas the first two are tremed the "valence" space. The inclusion of only the 5d and 6s shells in the valence space may be rationalized on the grounds that the ^2P states of Au lie about 2 eV higher than ^2D states, as mentioned above,

To construct a list of configurations for the CI calculations, excitations are defined in terms of the distribution of the 22 valence electrons among the orbitals in the three classes. Table II shows how the configurations were defined for the 12 LS symmetries desired and the number of configurations for each electron distribution for $^1\Sigma_g^+$ symmetry. The final electron distribution in Table II corresponds to the double excitations of the two 6s electrons and would serve to comprise a complete CI for the two electron (frozen 5d) system. The importance of including these "bond doubles" as well as the successive improvement of the ground state wave function of Au_2 as more electron distributions are included is shown in Fig. 1. The 6s valence (VCI) 6s first-order (FOCI) binding energies $[E(R=4.75)-E(R=20.0)\cong D_e = 0.91\text{ eV}]$ are the same since only the two MCSCF configurations are allowed. The additional ten single excitations, which would allow for orbital relaxation effects,[7] yield no contribution because the two valence configurations were optimized in the MCSCF procedure. The curve for the complete VCI (distributions 1, 3, and 4) is slightly lower (0.1 eV) than the 6s VCI due to some effects coming from inclusion of 5d (class I) excitations into the 6s orbitals (class II). The addition of the 5d FOCI configurations (distributions 5, 6, and 7) causes a significant increase in binding energy due to the effects of relaxation and polarization of the orbitals arising from the 5d shells of the Au atoms. Finally, it is seen in Fig. 1 that the bond doubles (distribution 8) result in an additional increase in the binding energy of about 0.3 eV to yield a D_e value of 2.27 eV. This change

TABLE II. Electron distributions for CI calculations for 22-valence electron Au_2.

Electron distribution	I	II	III	No. of $^1\Sigma_g^+$	Description
1	20	2	0	2	6s Valence CI(VCI)
2	20	1	1	10	6s First-order (FOCI)
3	19	3	0	2	5d VCI
4	18	4	0	6	5d VCI
5	19	2	1	88	5d FOCI
6	18	3	1	296	5d FOCI
7	17	4	1	402	5d FOCI
8	20	0	2	56	6s "Bond doubles"
Total				862	

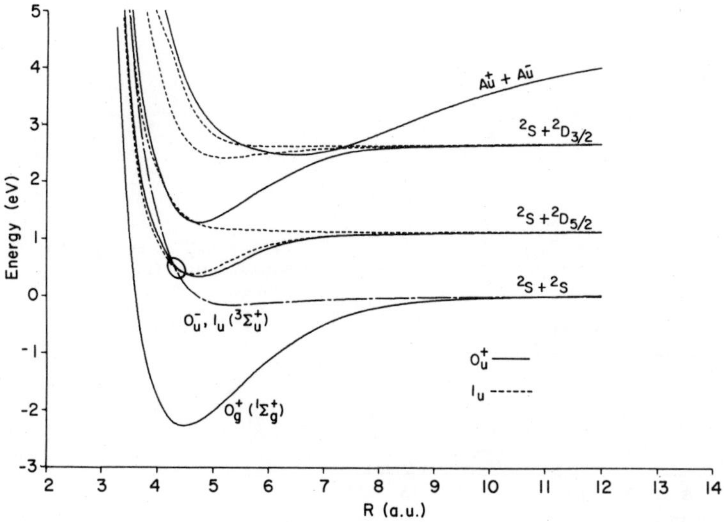

FIG. 2. Molecular states of Au_2 derived from the $^2S + ^2S$ atomic asymptote and the spectroscopically allowed excited states from $^2S + ^2D$ asymptote. Spin–orbit effects and ion-pair mixing are included in the calculation.

shows the importance of including this class of configurations, or conversely, the degree of reliability of the first-order method (distributions 1–7) in describing the ground state of Au_2. It should also be pointed out that the inclusion of configurations that correspond to excitations of one or more electrons from the 5*d* shells with one electron residing in the external space result in the inclusion of 0.3 eV of atomic correlation energy (lower two curves of Fig. 1). This effect is due to atomic excitations and should be nearly constant for all internuclear distances. The upper two curves have dissociation limits that are precisely the sum of two ^2S atomic SCF energies.

III. SPIN-ORBIT INTERACTION

In order to obtain the states corresponding to Hund's coupling case c (ω–ω coupling), the calculated *LS* states from the $^2S + ^2D$ atomic asymptote were corrected with an empirical model[6] based on the experimental atomic spin–orbit splittings $^2D_{3/2}$ and $^2D_{5/2}$. All the ω–ω coupled states from $^2S + ^2D_{3/2}$ and $^2S + ^2D_{5/2}$ atomic asymptotes can be obtained as roots to the spin–orbit matrices given in Appendix A with $\zeta = 0.6087$ eV. Since the contribution from the ion pair ($Au^+ + Au^-$) accounts for all the bonding of the first $^1\Sigma_u^+$ state, it was necessary to modify the spin–orbit matrix for 0_u^+ states that come from the mixing of $^3\Pi_u$ and $^1\Sigma_u^+$ states. Instead of solving the 2×2 matrix for 0_u^+ states, three *LS* states were included in the spin–orbit matrix, as shown in Appendix B. This approximation is based upon two additional assumptions. First, the spin–orbit splitting of ion pair states is zero, which is rigorously true for atomic ions. Second, the two lower-lying $^1\Sigma_u^+$ states are composed of ion pairs and ^2S and ^2D atoms, which is a reasonable approximation, as can be estimated from CI coefficients. Then the contributions from $^2S + ^2D$ atoms can be calculated using CI coefficients as shown in Appendix B. This modification is not used for 0_g^+ states since the ion pair contributes little to the first excited $^1\Sigma_g^+$ state.

In order to obtain the experimental energy separation for the atomic limits, the energies of all the excited states are shifted by 0.305 eV relative to the ground state.

IV. RESULTS AND DISCUSSION

The calculated potential curves for the excited states that have allowed transition moments from the ground state are shown in Fig. 2. The circle in Fig. 2 indicates the site of an avoided crossing of 1_u states if a CI calculation were done in the ω–ω coupled configurations. All other *u* states are shown in Fig. 3. Potential curves of calculated *g* states appear in Fig. 4. Vertical transition energies of these states calculated at the internuclear distance of 4.5 a.u. are given in Table III. They can be of some interest for certain applications and are useful as an additional guide for identifying the curves in Figs. 2–4.

Spectroscopic constants were calculated for states that have D_e larger than 0.3 eV and summarized in Table IV with the experimental results.

The spectroscopic information on Au_2 is given by Ames and Barrow[5] and by Kleman *et al.*,[8] who also cite earlier work. The fact that gold has but a single isotope allowed resolution of rotational structure[5] and yielded the precise interatomic distances which are given in Table IV together with other spectroscopic parameters. The dissociation energy of the ground state is obtained from mass spectrometric experiments, most recently by Kordis *et al.*[9] Their result $D_0 = 52.9 \pm 0.5$ kcal/mole agrees well with earlier work including the value 51.5 ± 2.2 kcal/mole from Ackerman *et al.*[10] With correction for zero point energy, one obtains $D_e = 2.305 \pm 0.02$ eV.

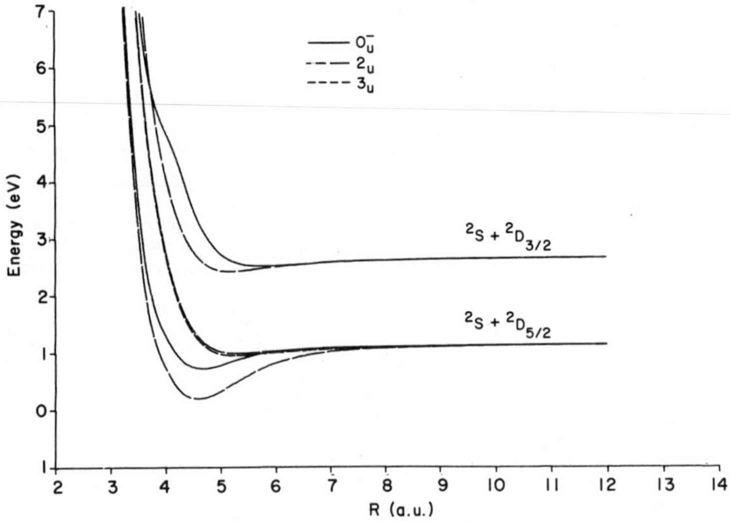

FIG. 3. 0_u^-, 2_u, and 3_u states of Au_2 derived from $^2S + ^2D$ atoms.

The values of 1.00 eV for the $A\ 0_u^+$ state and 1.78 eV for the $B\ 0_u^+$ state are calculated from the experimental T_e values of corresponding states[5] using the dissociation energy of the ground state and the atomic excitation energies to $D_{5/2}$ and $D_{3/2}$ states, respectively. It may be noted that simple extrapolations of vibrational levels yield D_e values of 1.41 and 1.47 eV for the $A\ 0_u^+$ and $B\ 0_u^+$ states, respectively.

One might have expected to observe also transitions to the 1_u state at 2.64 eV but this feature is evidently less intense than the X–A bands at nearly the same energy. Since the large transition dipole arising from the molecular ion-pair state falls only into the 0_u^+ symmetry, the lower intensity for the transition to the 1_u state is not surprising.

The agreement of calculated and experimental dissociation energies for the ground state is fortuitously good; that for R_e and ω_e reasonable. Although there is as much as a 0.1 Å difference in calculated and experimental R_e values, it is interesting to note that the calculations reproduce the correct order with $A\ 0_u^+$ longest and $B\ 0_u^+$ intermediate. The total energies of excitation to the A and B states are also fitted quite well. Indeed, the entire pattern of agreement with experimental values is remarkable and probably somewhat fortuitous. All important aspects of the lower energy states of Au_2 appear to be appropriately taken into account by the relativistic EP methods adopted.

TABLE III. Vertical transition energies[a] (in eV) of selected excited states of Au_2 calculated at $R = 4.50$ a.u.

States	Vertical transition energies (eV)	States	Vertical transition energies (eV)
0_g^+	0.0	0_u^+	2.6529
	3.4550		3.6157
	5.6131		6.0626
0_g^-	2.5616	0_u^-	2.4482
	5.5161		3.0125
1_g	2.5710		5.7679
	3.9543	1_u	2.4482
	5.2359		2.6602
	5.7027		3.6495
2_g	3.2056		5.0216
	3.9443		5.8219
	5.1911	2_u	2.4688
3_g	3.1508		3.6563
			5.0316
		3_u	3.6002

[a]The energy scale shown in the figures can be obtained by adding the energy of the ground 0_g^+ state (−2.265 eV).

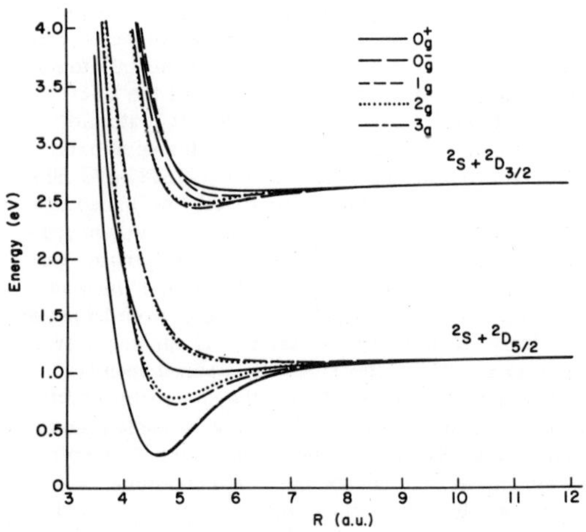

FIG. 4. 0_g^+, 0_g^-, 1_g, 2_g, and 3_g states of Au_2 derived from $^2S + ^2D$ atoms.

TABLE IV. Au_2 spectroscopic constants.

State		R_e(Å)	D_e(eV)	ω_e(cm^{-1})	T_e(eV)
$X\ 0_g^+$	Calc.	2.37	2.27	165	0
	Exptl.[a]	2.47	2.31[b]	191	0
$A\ 0_u^+$	Calc.	2.51	0.79	121	2.61
	Exptl.[a]	2.57	1.00	142	2.44
$B\ 0_u^+$	Calc.	2.50	1.38	146	3.55
	Exptl.[a]	2.51	1.78	180	3.18
1_u	Calc.	2.44	0.76	138	2.64
2_u	Calc.	2.43	0.94	143	2.57
0_g^-	Calc.	2.44	0.85	140	2.66
1_g	Calc.	2.44	0.84	140	2.67
2_g	Calc.	2.61	0.35	110	3.16
3_g	Calc.	2.61	0.41	112	3.10

[a]From Ref. 5. [b]From Ref. 9.

It is also recognized that most states coming from the $^2S+{}^2D_{3/2}$ asymptote and some repulsive states from the $^2S+{}^2D_{5/2}$ could be changed through the configuration mixing in ω–ω coupling case since the energy of the $^2D_{5/2}+{}^2D_{5/2}$ atomic asymptote is lower than that of $^2S+{}^2D_{3/2}$. However, there are no 0_u^+ or 0_g^- states arising from $^2D_{5/2}+{}^2D_{5/2}$; hence, the calculated curves for these symmetries are unaffected. This last point is important for the $B\ 0_u^+$ state.

Calculations were also initiated for Cu_2 and Ag_2 following the same general system. While the results for the ground states give qualitatively correct sequences of R_e values and D_e values, several difficulties arose which made the results unsatisfactory on a quantitative basis. The same problems became even more serious for excited states.

For copper, the most serious difficulty is the failure of SCF atomic calculations to reproduce, even approximately, the relative energies of the ^2S and ^2D states. Similar difficulties are known to arise for nickel. We understand that very extensive CI calculations[11] are required to remedy this situation and such extensive atomic CI would be beyond the scope of our present pattern.

For silver, the problem arises from the fact that the ^2P excited state is somewhat lower than the ^2D state for the atom. Thus, excitations involving both d and p orbitals must be included in an adequate treatment. Futhermore, the methods used here for spin–orbit effects in Au_2 would be inadequate for Ag_2. In view of the more extensive efforts required to properly deal with Cu_2 and Ag_2, it seemed better to present just the results for Au_2 at this time.

ACKNOWLEDGMENTS

Most of the calculations were carried out with the IBM San Jose ALCHEMY programs of M. Yoshimine, A. D. McLean, B. Liu, and P. S. Bagus, extended by the present authors to include averaged relativistic core potentials. We thank A. D. McLean and B. Liu for their help, and the IBM Corporation for their support of this work under a Joint Study Agreement with the Lawrence Berkeley Laboratory.

This research was supported by the Division of Chemical Sciences, Office of Basic Energy Sciences, U.S. Department of Energy.

APPENDIX A. SPIN-ORBIT MATRICES FOR $d^9s^2+d^{10}s$ MOLECULES

$$\Omega=3: \quad [E(^3\Delta)+\zeta],$$

$$\Omega=2: \quad \begin{bmatrix} E(^3\Delta) & & \\ \zeta/\sqrt{2} & E(^3\Pi)+\zeta/2 & \\ \zeta & -\zeta/\sqrt{2} & E(^1\Delta) \end{bmatrix},$$

$$\Omega=1: \quad \begin{bmatrix} E(^3\Delta)-\zeta & & & \\ \zeta/\sqrt{2} & E(^3\Pi) & & \\ 0 & \sqrt{3}\zeta/2 & E(^3\Sigma^+) & \\ \zeta/\sqrt{2} & \zeta/2 & -\sqrt{3}\zeta/2 & E(^1\Pi) \end{bmatrix},$$

$$\Omega=0^-: \quad \begin{bmatrix} E(^3\Pi)-\zeta/2 & \\ \sqrt{6}\,\zeta/2 & E(^3\Sigma^+) \end{bmatrix},$$

$$\Omega=0^+: \quad \begin{bmatrix} E(^3\Pi)-\zeta/2 & \\ \sqrt{6}\,\zeta/2 & E(^1\Sigma^+) \end{bmatrix}.$$

APPENDIX B

In an attempt to account for ion-pair character, let

$$\beta^2=(C_{II}^2+C_{III}^2)/(C_I^2+C_{II}^2+C_{III}^2),$$

where the C's are mixing coefficients corresponding to the respective (dominant) configurations in the first $^1\Sigma_u^+$ wave function

I: $(s\sigma_g)(s\sigma_u)$, ion pair configuration

II: $(d\sigma_u)(s\sigma_g)(s\sigma_u)^2$, III: $(d\sigma_g)(s\sigma_g)^2(s\sigma_u)$. } $^2S+{}^2D$ configurations

The energies of the 0_u^+ states are roots of

$$\begin{bmatrix} E(^3\Pi_u)-\zeta/2 & & \\ \beta(\sqrt{6}/2)\zeta & E_1(^1\Sigma_u^+) & \\ (1-\beta)(\sqrt{6}/2)\zeta & 0 & E_2(^1\Sigma_u^+) \end{bmatrix}.$$

[1]Y. S. Lee, W. C. Ermler, and K. S. Pitzer, J. Chem. Phys. 67, 586 (1977).
[2]W. C. Ermler, Y. S. Lee, K. S. Pitzer, and N. W. Winter, J. Chem. Phys. 69, 976 (1978).
[3]Y. S. Lee, W. C. Ermler, K. S. Pitzer, and A. D. McLean, J. Chem. Phys. 70, 288 (1979).

[4]C. E. Moore, Natl. Bur. Stand. (U.S.) Cir. **467**, Vol. III (1958).
[5]L. L. Ames and R. F. Barrow, Trans. Faraday Soc. **63**, 39 (1967).
[6]See, J. S. Cohen and B. Schneider, J. Chem. Phys. **61**, 3230 (1974).
[7]See H. F. Schaefer III, *The Electronic Structure of Atoms and Molecules: A Survey of Rigorous Quantum Mechanical Results* (Addison-Wesley, Reading, MA, 1972).
[8]B. Kleman, S. Lindqvist, and L. E. Selin, Ark. Fys. **8**, 505 (1954).
[9]J. Kordis, K. A. Gingerich, and R. J. Seyse, J. Chem. Phys. **61**, 5114 (1974).
[10]M. Ackerman, F. E. Stafford, and J. Drowart, J. Chem. Phys. **33**, 1284 (1960).
[11]W. A. Goddard III (private communication).

Improved *ab initio* effective core potentials for molecular calculations

Phillip A. Christiansen, Yoon S. Lee, and Kenneth S. Pitzer

Department of Chemistry and Lawrence Berkeley Laboratory, University of California, Berkeley, California 94720

(Received 29 June 1979; accepted 24 August 1979)

We have investigated the sources of error in bond lengths and dissociation energies computed from *ab initio* effective potentials derived from Phillips–Kleinman type pseudo-orbitals. We propose an alternate pseudo-orbital, effective potential treatment with the primary objective of agreement with all-electron molecular calculations. This new treatment forces the pseudo-orbitals to match precisely the Hartree–Fock orbitals in the valence region and thereby eliminates the major cause of error in the earlier calculations. Effective core potentials derived from these revised pseudo-orbitals were used to compute potential energy curves for the ground states of F_2, Cl_2, and LiCl and the results are compared with previous all-electron and effective potential calculations. Our effective potentials yield dissociation energies and bond lengths which are in excellent agreement with the all-electron values. Furthermore, in contrast to other procedures, our revised effective potentials result in an excellent description of the inner repulsive walls of the dissociation curves.

I. INTRODUCTION

Since the pioneering pseudopotential studies of Phillips and Kleinman[1] and of Weeks and Rice[2] a great deal of effort has gone into the development of *ab initio* effective potential procedures for investigating the valence structure of molecular systems containing large numbers of electrons.[3–12] These procedures are based on the assumptions that, first, the polarization of the core orbitals upon molecular formation is of negligible importance relative to the valence effects (this is the frozen core approximation) and secondly, that the effects that the core electrons do have upon the molecular valence orbitals (including overlap effects) can be well approximated using a potential expressed as a sum of local functions multiplied by projection operators. Although these procedures have tremendously broadened the scope of molecular calculations, inconsistencies relative to all-electron calculations have become apparent in the last few years. The most troublesome of these is the shortening of bond lengths and deepening of potential wells (relative to all-electron results) sometimes seen in potential surface calculations. This problem became discouragingly apparent in the calculations of Kahn *et al.*[12] on F_2 and in the recent calculations of Hay, Wadt, and Kahn[13] on F_2 and Cl_2. The problem has also been observed in calculations on heavier diatomic molecules.[14,15] For Cl_2 Hay *et al.*, using effective potentials developed by Kahn, obtained a bond length that was short by nearly 0.3 bohr and a dissociation energy that was too large by about 70%. They obtained similar results for F_2, though not as exaggerated. SCF calculations for F_2 and Cl_2 in which only the valence orbitals were optimized (the core orbitals having been obtained from atomic calculations) indicated that, for the bond lengths employed in the studies by Kahn *et al.* and Hay *et al.*, the error in the potential curves due to the frozen core approximation was relatively small.[16]

During the course of their study, Hay, Wadt, and Kahn concluded that the error in their Cl_2 and F_2 calculations arose from long-range negative tails in their chlorine $3d$ and fluorine $2p$ effective potentials and suggested *ad hoc* procedures for eliminating them. Using their revised potentials they obtained greatly improved dissociation curves for F_2 and Cl_2, but errors in the neighborhood of 0.1 bohr still remained in their computed bond lengths for Cl_2 and LiCl. This error, along with the somewhat inconsistent nature of their corrections (relative to their effective potential formalism) indicated to us that the problem needed to be investigated in a more fundamental manner.

In the present paper we have investigated the origins of the difficulties discussed by Hay *et al.* in terms of their effective potential formalism. We then suggest an alternative effective potential (EP) procedure which is formally more satisfactory than the corrections of Hay *et al.* and which in practice yields dissociation curves for F_2, Cl_2, and LiCl which are in excellent agreement with all-electron (AE) calculations, even for relatively small internuclear separations.

Although our discussion is based primarily on the Kahn procedure, most of our findings are equally applicable to other formalisms in which the EP's are defined in terms of Phillips–Kleinman type psuedo-orbitals.

II. LONG RANGE EFFECTIVE-POTENTIAL BEHAVIOR

In the Kahn formalism,[12] the first step is to solve[17] numerically the atomic Hartree–Fock (HF) equations for accurate representations of the HF orbitals and orbital energies. The normalized valence pseudoorbitals χ_v are then defined as linear combinations of the valence and core HF orbitals ϕ_v, ϕ_c of a given l quantum number (as in Phillips and Kleinman),

$$\chi_v = C_v\phi_v + \sum_c^{\text{core}} C_c\phi_c \ . \tag{1}$$

The expansion coefficients in (1) are chosen such that χ_v is smooth and nodeless near the origin. Note that, since χ_v must be normalized and the ϕ_v and ϕ_c are all orthonormal orbitals, the coefficient C_v must be smaller than unity unless all of the C_c are zero — a situation which arises only when there are no core obitals of the same symmetry as ϕ_v.

The effective potential U^{EP} is obtained by requiring that it reproduce the pseudoorbital χ_v and the true eigenvalue ϵ_v of the Hartree–Fock orbital,

$$\left(-\frac{1}{2}\nabla^2 - \frac{Z}{r} + U^{EP} + W^{PS}_{val}\right)\chi_v = \epsilon_v\chi_v , \qquad (2)$$

where Z is the atomic number and W^{PS}_{val} is the potential (comprising the usual Coulomb and exchange terms) due to the interaction of an electron in χ_v with the other electrons in the occupied pseudoorbitals. The local potential is then given by

$$U^{EP}(r) = \left[\left(\epsilon_v + \frac{1}{2}\nabla^2 + \frac{Z}{r} - W^{PS}_{val}\right)\chi_v\right]\Big/\chi_v . \qquad (3)$$

If we keep in mind that the core orbitals become negligible in the outer valence region and if C_v is less than unity, then it is apparent from Eq. (1) that the amplitude of the pseudoorbital in the tail region must be smaller than that of the corresponding HF orbital. If r_c is the radius at which the most diffuse core orbital becomes negligibly small, then, for r larger than r_c, the pseudoorbital is given by

$$\chi_v(r) \approx C_v\phi_v(r) . \qquad (4)$$

If we substitute (4) into Eq. (2) we see that U^{EP} becomes

$$U^{EP} \approx W^{HF}_{core} + W^{HF}_{val} - W^{PS}_{val} , \quad r > r_c . \qquad (5)$$

where W^{HF}_{core} and W^{HF}_{val} are the sums of the Coulomb and exchange potentials arising from the HF core and valence orbitals respectively. However, for large values of r, W^{HF}_{core} must have the form

$$W^{HF}_{core} \approx N_c/r , \qquad (6)$$

where N_c is the number of core electrons. Thus, any explanation for the excessive long-range negative tails in the fluorine and chlorine EP's must be in terms of W^{HF}_{val} and W^{PS}_{val}. (Hay *et al.* eliminated the long-range tail in their chlorine d potential by replacing the s and p pseudo-orbitals in W^{PS}_{val} with HF orbitals, in which case $W^{PS}_{val} = W^{HF}_{val}$. However the d potential, though it decays as N_c/r for large r, is then inconsistent with the s and p potentials.)

From Eq. (4) one can see that, in effect, by forming Phillips–Kleinman pseudoorbitals one transfers valence electron density from the valence region into the core. For this reason the Phillips–Kleinman pseudoorbitals alone cannot accurately account for the interaction of electrons in the valence region. This deficiency introduces, through Eq. (3), a long-range tail in the EP. One can see this effect, in part, in the Coulomb potential arising from a Phillips–Kleinman pseudoorbital. From (4) it can be easily shown that the difference of the spherically averaged Coulomb potentials[18] for corresponding HF orbitals and Phillips–Kleinman pseudoorbitals is given by

$$J^{HF}_v(r) - J^{PS}_v(r) \approx \frac{1}{r}(C_v^2 - 1)\int_r^\infty |\phi_v(t)|^2(t^2 - rt)\,dt, \quad r > r_c . \qquad (7)$$

The expression in Eq. (5) $(W^{HF}_{val} - W^{PS}_{val})$ is given by the sum of terms $(J^{HF}_v - J^{PS}_v)$ over all other occupied valence orbitals together with corresponding aspherical Coulomb and exchange terms. Since the integral in Eq. (7) is necessarily positive and the factor $(C_v^2 - 1)$ is negative, all of the $(J^{HF}_v - J^{PS}_v)$ terms are negative. Thus, from Eq. (5), one sees that these terms introduce a negative contribution to the effective potential arising from the difference between the pseudo and HF orbitals in the outer part of the atom. These terms account for a large fraction of the excessive negative potential that Hay *et al.* found in the Kahn EP for chlorine. The fact that the tail is at least partially Coulombic implies that, in general, it cannot be ignored as an artifact of the local potential approximation. Furthermore, since the expansions for the corresponding exchange and aspherical Coulomb potentials will, for most cases, contain powers of r which are different from those that occur in the spherical Coulomb potential, it appears that perfect cancellation with Eq. (7) is unlikely.

From Eqs. (2) and (3) it is clear that the negative tail in the EP is a necessary part of its long-range behavior if the EP is to accurately reproduce the Hartree–Fock atomic orbital energies. However, for orbitals other than those from which it was derived, the potential is only an approximation and may lead to erroneous results. One would expect this to be particularly true in molecular calculations where the tail of the potential extends into the bonding region and into the space of other atoms.

The above discussion suggests that there are fundamental problems associated with the use of Phillips–Kleinman type pseudoorbitals for generating effective potentials for molecular calculations. These problems are due to the deviation of the tail behavior of the EP's relative to the "point charge" potential one would expect from the HF core orbitals. From Eqs. (4) and (7) it is clear that at least part of this deviation is due to the mismatch of the pseudo and HF orbitals in the valence region and can in general be eliminated only if χ_v and ϕ_v are required to be identical in that region.

For these reasons, we abandon the Phillips–Kleinman method of generating pseudoorbitals and adopt the simple objective of close agreement with all-electron molecular calculations for a system of pseudoorbitals and effective potentials. We retain Eq. (3) and subsequent portions of the usual EP treatment, e.g., Kahn *et al.*,[12] but abandon Eq. (1) and require instead that the pseudoorbital match exactly the HF valence orbital outside the core region. Thus we define

$$\chi_v = \phi_v + f_v , \qquad (8)$$

where the function f_v is zero outside the core region and is otherwise chosen to cancel the oscillations of ϕ_v in the core region so that χ_v will be smooth and have no radial nodes. Thus we effectively partition ϕ_v into core and valence terms. For molecular calculations one directly optimizes only the molecular counterpart of the valence term, χ_v.

Although there are no explicit core–valence orthogonality conditions which the orbitals χ_v must satisfy, the partitioning in Eq. (8) suggests that there are implicit overlap requirements. In the EP approximation one assumes that these requirements are enforced by the

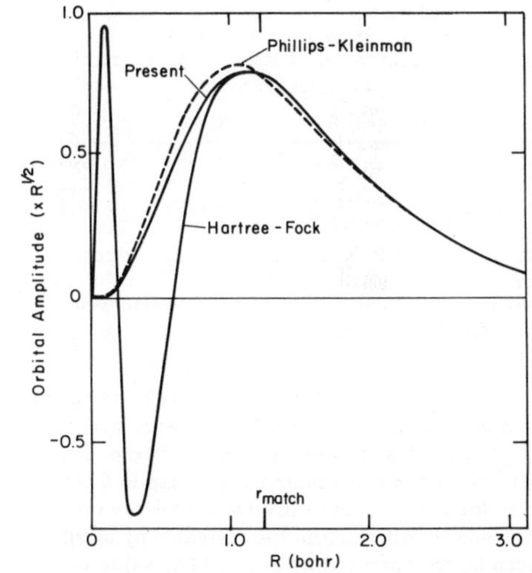

FIG. 1. A comparison of our chlorine 3s pseudoorbital with the Phillips–Kleinman type pseudoorbital of Refs. 12 and 13, and the HF orbital.

shape of the potential in the core region. However, for molecular calculations, if the potential is not sufficiently large, one might expect perturbations from other centers to alter the form of the molecular orbitals in the core region such that a breakdown in the implied overlap conditions might occur. Since the magnitude of the potential depends roughly on the reciprocal of χ_v, one can force the EP to be large in the core region by choosing χ_v to be small near the origin. Furthermore, if χ_v were large near the origin or contained unnecessary oscillations then one would need a significantly larger basis set for molecular calculations. For these reasons, in addition to the requirement that χ_v and ϕ_v be identical in the valence region, we also wish χ_v to be smooth and as small as possible in the core region.

Although pseudoorbitals of the above form have been proposed in the past,[19,20] the advantages for molecular calculations have not been well documented.

III. IMPROVED PSEUDOORBITALS AND POTENTIALS

In this section we will describe a procedure for generating pseudoorbitals which retain exactly the behavior of the HF valence orbitals in the valence region, but are smooth and nodeless, with $\chi_v(r)/r^l$ going to zero at the origin. We will also discuss the long-range behavior of the resulting effective potentials.

We generate the radial factor for our pseudoorbitals in two sections. The outer one, which spans the region from some matching point r_{match} to infinity is just the numerical HF valence orbital in that region. In the inner region (from zero to r_{match}) we define our pseudoorbital by a five-term polynomial expansion in r with a leading power of $l+2$. The expansion coefficients are chosen so as to match the amplitude and first three derivatives of the two sections at r_{match} and also to make the total pseudoorbital normalized. r_{match} is chosen to be the innermost point at which the matching results in a nodeless radial pseudoorbital with no more than two inflexions in its entire range. This procedure gives our pseudoorbitals a unique definition which limits the deviation from the exact HF orbital to as small a region around the origin as possible.

It is clear from this definition of the pseudoorbital that for r values greater than r_{match}, the C_v and r_c in Eqs. (4) and (7) must be replaced by unity and r_{match} respectively. Therefore for r greater than r_{match}, the difference between the spherical parts of the pseudo and HF Coulombic potentials will be zero.

The calculation of the corresponding quantities for the exchange and aspherical Coulomb potentials is not nearly so simple. However for the cases we have examined, these terms are very small for values of r that are much greater than r_{match}.

IV. CALCULATIONS AND RESULTS

We have used the pseudoorbital selection procedure described in the last section along with the EP generating programs of Kahn *et al.*[12] to compute revised EP's for fluorine and chlorine. We used these potentials, expanded in terms of Gaussian functions, along with Guassian basis sets to compute potential curves for F_2, Cl_2, and LiCl and compare our results with other EP calculations.[12,13] We assess the reliability of our potentials by comparison with AE calculations.[13]

Our s and p atomic orbitals (and indirectly pseudoorbitals) for fluorine and chlorine were obtained from numerical HF calculations[17] on the ground states of the neutral atoms. As in Ref. 12, the chlorine $3d$ orbital was obtained from a similar calculation using the ^4P multiplet of the configuration (Ne, $3s\ 3p^5\ 3d$). In solving our pseudoorbital equations for fluorine and chlorine we found the r_{match} values to be identical for the s and p orbitals. The values of r_{match} for fluorine and chlorine were, respectively, 0.64 and 1.26 bohr. To give an idea of the differences between our pseudoorbitals and the Phillips–Kleinman type pseudoorbitals used in Refs. 12 and 13 we have plotted in Fig. 1 the chlorine 3s pseudoorbitals of both types along with the HF orbital.

Our s and p basis sets were obtained by a least-squares fitting of our atomic pseudoorbitals with three or four primitive Gaussian functions. For F and Cl we

TABLE I. Bond lengths (bohr) and dissociation energies (kcal/mole) for the ground states of F_2, Cl_2, and LiCl from 2-configuration MCSCF wave functions.

	F_2		Cl_2		LiCl	
Potential	R_e	D_e	R_e	D_e	R_e	D_e
All-electron[a]	2.84	14.8	4.04	27.0	4.02	86.6
Hay EP[a]	2.82	16.1	3.94	24.6	3.95	87.1
Kahn EP[b]	2.70	21.2	3.76	45.7	4.00	85.7
Present	2.86	14.6	4.04	25.7	4.03	86.9

[a]Values were taken from Hay *et al.*, Ref. 13.
[b]Values were computed by Hay *et al.*, using potentials from Kahn *et al.*, Ref. 12.

TABLE II. Potential energy curves for the ground state of F_2 from 2-configuration MCSCF wave functions.

	Energies (hartree)		
R (bohr)	All-electron[a]	Hay[a]	Present
2.2	0.0570	0.0501	0.0592
2.4	0.0031	−0.0018	0.0052
2.6	−0.0183	−0.0215	−0.0170
2.8	−0.0234	−0.0256	−0.0230
3.0	−0.0215	−0.0228	−0.0213
3.4	−0.0124	−0.0129	−0.0124
20.0	0.0	0.0	0.0

[a]Values taken from Hay *et al.*, Ref. 13.

included a single set of primitive *d* functions with exponents of 0.9 and 0.837 respectively.[13] In addition, for calculations on LiCl we added a set of diffuse *p* functions (the exponent was 0.049[21]) for the chlorine atom. For Li we used the (3s2p) contracted Gaussian basis set of Dunning and Hay.[21] Although our fluorine and chlorine *s* and *p* basis sets differ considerably from those employed by Hay *et al.*[13] the more diffuse functions are similar enough that we believe a comparison of molecular results is valid. This was verified for F_2 by calculation of a few points using the basis set of Ref. 13 but with the primitive *s*-type Gaussian functions recontracted appropriately for our fluorine *s* pseudoorbital. All of our calculations were carried out in terms of two configuration MCSCF wave functions identical in form to those used by Hay[13] and Kahn.[12]

In Table I we list our calculated equilibrium bond lengths and dissociation energies for F_2, Cl_2, and LiCl along with all-electron (AE) and EP results from Refs. 12 and 13. For F_2 our bond length differs from the AE value by about the same amount as did that of Hay *et al.*, 0.02 bohr, but in the opposite direction. Both values are in much better agreement with the AE results than are those based on the EP of Kahn *et al.* (The fluorine EP's of Hay and Kahn differ only in that Hay *et al.* truncated the tail behavior of the potential.) Our value for D_e for F_2 is considerably better than those computed from the EP's of either Hay or Kahn.

TABLE III. Potential energy curves for the ground state of Cl_2 from 2-configuration MCSCF wave functions.

	Energies (hartree)			
R (bohr)	All-electron[a]	Hay[a]	Kahn[b]	Present
3.0	0.1147	0.0813	0.0247	0.1166
3.5	−0.0210	−0.0260	−0.0676	−0.0199
3.756	−0.0387	−0.0378	−0.0729	−0.0369
4.0	−0.0430	−0.0391	−0.0687	−0.0408
4.25	−0.0408	−0.0353	−0.0603	−0.0388
4.5	−0.0353	−0.0292	−0.0503	−0.0336
5.0	−0.0217	−0.0165	−0.0315	−0.0208
6.0	−0.0045	−0.0029	−0.0098	−0.0042
20.0	0.0	0.0	0.0	0.0

[a]Values were taken from Hay *et al.*, Ref. 13.
[b]Values were computed by Hay *et al.*, using potentials from Kahn *et al.*, Ref. 12.

TABLE IV. Potential energy curves for the ground state of LiCl from 2-configuration MCSCF wave functions.

	Energies (hartree)			
R (bohr)	All-electron[a]	Hay[a]	Kahn[b]	Present
3.75	−0.1351	−0.1375	−0.1330	−0.1353
4.00	−0.1380	−0.1387	−0.1365	−0.1385
4.25	−0.1360	−0.1356	−0.1330	−0.1366
20.0	0.0	0.0	0.0	0.0

[a]Values were taken from Hay *et al.*, Ref. 13.
[b]Values were computed by Hay *et al.*, using potentials from Kahn *et al.*, Ref. 12.

For Cl_2 and LiCl we obtained significantly better bond lengths than did Hay *et al.* For both cases our errors, relative to the AE values, were around 0.01 bohr or smaller (Hay's values were in error by close to 0.1 bohr). Also for both Cl_2 and LiCl our values for the dissociation energy differ from the AE value by about half as much as do those of Hay *et al.* (The value for D_e for Cl_2 derived from the Kahn EP is too large by nearly a factor of two.)

The difference between our calculations and the previous EP calculations becomes most apparent at small internuclear separations. In Tables II–IV we list the dissociation potential energy curves (relative to separated atoms) for F_2, Cl_2, and LiCl. Although Hay *et al.* obtained reasonable bond lengths and dissociation energies for F_2, as can be seen in Table II, their curve deviates considerably from the AE curve for bond distances shorter than about 2.6 bohr. At 2.2 bohr the Hay EP underestimates the energy by about 0.007 hartree. The problem is much more serious for Cl_2 where at 3.0 bohr the Hay EP underestimates the energy by more than 0.03 hartree. This is contrary to what one would expect from the original frozen core approximation since core polarization should reduce the energy.

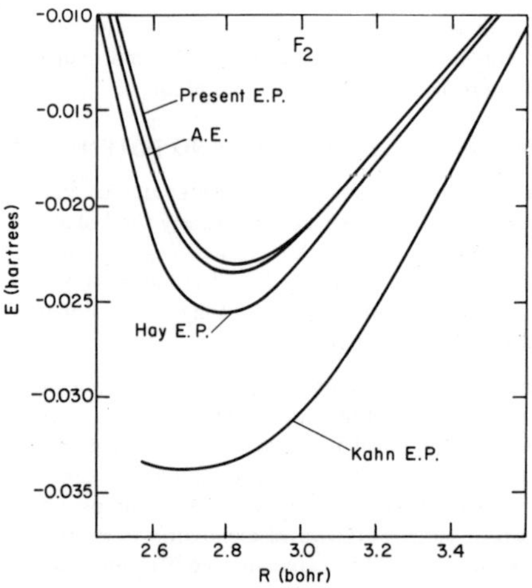

FIG. 2. Potential energy curves for the ground state of F_2 from all-electron (AE) and effective potential (EP) calculations using 2-configuration MCSCF wave functions.

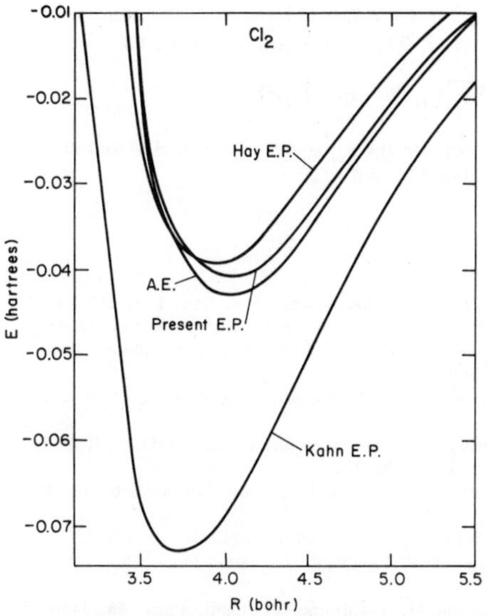

FIG. 3. Potential energy curves for the ground state of Cl_2 from all-electron (AE) and effective potential (EP) calculations using 2-configuration MCSCF wave functions.

By comparison, for both Cl_2 and F_2 our EP overestimates the energy by about 0.002 hartree at the shortest bond lengths tested — a very small error in the expected direction. Not only do we get better behavior in the equilibrium region but we also get a much better representation of the inner repulsive wall of the potential curve.

In Figs. 2–4 we have plotted the EP and AE dissociation curves for F_2, Cl_2, and LiCl. Clearly, in all cases, our dissociation curves match more uniformly the AE curves than do the EP dissociation curves of either Hay or Kahn.

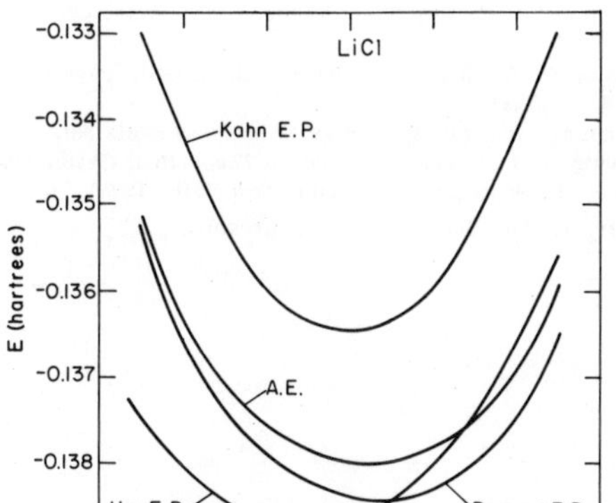

FIG. 4. Potential energy curves for the ground state of LiCl from all-electron (AE) and effective potential (EP) calculations using 2-configuration MCSCF wave functions.

V. SUMMARY

We have shown that effective potential calculations of molecular properties can reliably yield results in close agreement with all-electron calculations only if the atomic pseudoorbitals retain the correct radial distribution of charge in the valence region given by all-electron atomic calculations.

Pseudoorbitals obtained by the Phillips–Kleinman method move some electron density from the valence to the core region (if there are core orbitals of the same symmetry) and this shift of charge leads to excessive long-range negative tails in the resulting effective potentials. In turn, in molecular calculations, the interatomic potential curve is too low at short distances leading to an excessive dissociation energy and too short an equilibrium bond distance.

With the objective that calculations of "valence level" molecular properties based on effective potentials should lead, as nearly as possible, to the same results as all-electron calculations, we adopt a simple, unambiguous procedure for generating pseudoorbitals from HF atomic orbitals in a manner that retains exact equality in the outer portion of the atom. The usual equation is retained for the generation of effective potentials from pseudoorbitals, orbital energies, etc. The use of our pseudoorbitals with Kahn's effective potential formalism in molecular calculations results in potential energy curves which are in excellent agreement with all-electron calculations. The most noticeable improvement relative to recent results of Hay *et al.* and Kahn *et al.* is in the region of short internuclear separation.

ACKNOWLEDGMENTS

In the revision to final form of this manuscript, the authors had the benefit of a preprint[22] from and discussion with Dr. Alex Zunger as well as discussion with Dr. Luis Kahn which informed us of their respective research in this field. This work was supported by the Division of Chemical Sciences, Office of Basic Energy Sciences, U. S. Department of Energy under Contract No. W-7405-ENG-48.

APPENDIX

We have listed here in Tables AI–AIV the contracted Gaussian *s* and *p* basis sets for fluorine and chlorine, as well as the analytical Gaussian expansions for our

TABLE AI. Fluorine (3*s*, 2*p*) contracted Gaussian basis set used in the present calculations.

s functions		*p* functions	
Exponent	Contraction coefficient	Exponent	Contraction coefficient
41.3380	0.1130623	30.42310	0.0496707
8.59848	0.9221688	5.19374	0.3463772
...	...	1.24693	0.7527373
1.14569	1.0	...	...
...	...	0.30613	1.0
0.33689	1.0		

TABLE AII. Chlorine (3*s*, 3*p*) contracted Gaussian basis set used in the present calculations.

s functions		*p* functions	
Exponent	Contraction coefficient	Exponent	Contraction coefficient
24.92230	0.0027723	6.641690	1.0
2.05467	0.9989255	···	···
···	···	0.616684	1.0
0.545234	1.0	···	···
···	···	0.174587	1.0
0.186656	1.0	···	···

TABLE AIII. Gaussian expansions for the fluorine effective potential used in the present calculations.

		Expansion coefficients	
n	ζ	$U_s - U_p$	U_p
2	6.026	47.4266985	50.5141198
2	6.503	−205.2735370	−133.3101020
2	6.967	184.7092921	88.8294726
2	35.980	109.9937919	6.1436341
1	7.888	−0.1849250	−2.0
0	74.090	3.0	

TABLE AIV. Gaussian expansions for the chlorine effective potentials used in the present calculations.

		Expansion coefficients		
n	ζ	$U_s - U_d$	$U_p - U_d$	U_d
2	5.00	−2700.3115080	−149.0917485	−56.3781074
2	5.50	6004.3634238	−5920.8673168	266.4739971
2	6.00	−12937.7374504	−2774.3316746	−171.1705515
2	6.72	8235.8932107	6517.4611787	2.3506943
2	28.89	101.6005388	−35.8188407	2.5515185
2	96.63	199.3749764	286.0375129	5.5724718
1	5.30	6448.5383150	9714.1549630	−119.0912105
1	6.00	−6312.6984959	−9481.0135841	108.1034976
1	33.87	−112.9121148	−211.1595365	0.9877129
0	6.16	13.0730405	26.0712424	
0	67.33	−10.0730405	−21.0712424	

fluorine and chlorine effective potentials. The U_l in Tables AIII and AIV are defined as

$$U_l = U_l^{EP}(r) - (N_c/r) ,$$

where $U^{EP}(r)$ is from Eq. (3) and N_c is the number of core electrons. The U_l are then expanded as

$$U_l = r^{-2} \sum_i C_{il} r^{n_{il}} \exp(-\zeta_{il} r^2) .$$

We have tabulated the n_{il} and ζ_{il} for fluorine and chlorine in Tables AIII and AIV.

[1]J. C. Phillips and L. Kleinman, Phys. Rev. **116**, 287 (1959).
[2]J. D. Weeks, S. A. Rice, J. Chem. Phys. **49**, 2741 (1968); J. D. Weeks, A. Hazi, and S. A. Rice, Adv. Chem. Phys. **16**, 283 (1969).
[3]L. R. Kahn and W. A. Goddard III, Chem. Phys. Lett. **2**, 667 (1968).
[4]C. F. Melius, B. D. Olafson, and W. A. Goddard III, Chem. Phys. Lett. **28**, 457 (1974).
[5]J. C. Barthelat and Ph. Durand, Chem. Phys. Lett. **16**, 63 (1972).
[6]W. H. E. Schwartz, Theor. Chim. Acta **24**, 29 (1972).
[7]V. Bonifacic and S. Huzinaga, J. Chem. Phys. **60**, 2779 (1974).
[8]S. Huzinaga and M. Yoshimine, J. Chem. Phys. **68**, 4486 (1978).
[9]P. Coffey, C. S. Ewig, and J. R. Van Wazer, J. Am. Chem. Soc. **97**, 1656 (1975).
[10]C. S. Ewig and J. R. Van Wazer, J. Chem. Phys. **63**, 4035 (1975).
[11]G. Das and A. C. Wahl, J. Chem. Phys. **64**, 4672 (1976).
[12]L. R. Kahn, P. Baybutt, and D. G. Truhlar, J. Chem. Phys. **65**, 3826 (1976), and references therein.
[13]P. J. Hay, W. R. Wadt, and L. R. Kahn, J. Chem. Phys. **68**, 3059 (1978).
[14]W. C. Ermler, Y. S. Lee, K. S. Pitzer, and N. W. Winter, J. Chem. Phys. **69**, 976 (1978).
[15]W. R. Wadt, P. J. Hay, and L. R. Kahn, J. Chem. Phys. **68**, 1752 (1978).
[16]P. A. Christiansen (unpublished results); see also the results for NaH from E. S. Sachs, J. Hinze, and N. H. Sabelli, J. Chem. Phys. **62**, 3393 (1975).
[17]C. F. Fischer, *The Hartree–Fock Method for Atoms, A Numerical Approach* (Wiley-Interscience, New York, 1977).
[18]J. C. Slater, *Quantum Theory of Atomic Structure* (McGraw-Hill, New York, 1960), Vol. I.
[19]Ph. Durand and J. C. Barthelat, Theor. Chim. Acta, **38**, 283 (1975).
[20]A. Redondo, W. A. Goddard III, and T. C. McGill, Phys. Rev. B **15**, 5038 (1977).
[21]T. H. Dunning, Jr., and P. J. Hay, "Gaussian Basis Sets for Molecular Calculations," in *Modern Theoretical Chemistry*, edited by H. F. Schaefer III (Plenum, New York, 1977), Vol. 3.
[22]A. Zunger, J. Vac. Sci. Technol. (in press).

[Reprinted from ACCOUNTS OF CHEMICAL RESEARCH, 12, 271 (1979).]

Relativistic Effects on Chemical Properties

KENNETH S. PITZER

Department of Chemistry and Lawrence Berkeley Laboratory, University of California, Berkeley, California 94720

Received November 8, 1978

While it is generally recognized that a rigorous theoretical foundation for chemistry would rest on relativistic quantum theory, the great majority of quantum mechanical calculations of quantities of chemical interest are based on the nonrelativistic Schrödinger equation, or its equivalent, together with the qualitative rules related to spin and the Pauli exclusion principle. If fine structure features are of interest, spin–orbit, spin–spin, and other terms are added as perturbations, but for light atoms these terms yield very small energies that may be of spectroscopic interest but have no effect on most chemical properties.

For very heavy atoms, however, this situation changes. Certain terms which were very small compared to thermal energy for light atoms increase very rapidly with atomic number and become comparable to chemical bond energies for elements in the range gold–bismuth.

The most familiar relativistic property is probably the increase in particle mass as the velocity approaches that of light. It is the s electrons with no orbital angular momentum that approach the nucleus most closely and therefore gain the highest velocity. For these heavy atoms the inner 1s and 2s electrons have average velocities close enough to that of light to have a substantial increase in mass. As a result these electrons spend even more time near the nucleus, have contracted mean radii, and have orbital energies which are increased (more negative). With this contraction of the inner s shells, outer s electrons also get closer to the nucleus and are affected in the same way. The relativistic effects on p, d, and f electrons are more complex and will be discussed below. But the essential result is that calculations based on the Schrödinger equation are not adequate for a discussion of the chemical properties of heavy atoms. Also extrapolations downward in the periodic table for the properties of possible superheavy elements are very hazardous since these relativistic effects increase very rapidly with increase in atomic number.

Kenneth S. Pitzer, born in 1914 in Pomona, CA, received his B.S. from the California Institute of Technology and his Ph.D. from the University of California, Berkeley. He is a member of the faculty at Berkeley, having served in that capacity from 1937 to 1961 and again since 1971. He has also held various administrative positions including that of Director of Research for the U.S. Atomic Energy Commission (1949–1951), President of Rice University (1961–1968), and President of Stanford University (1968–1970). His research interests include thermodynamics, statistical mechanics, and quantum mechanics as applied to chemistry, as well as experimental work in spectroscopy and thermochemistry.

Table I
Energy Terms for Fourth Group Elements (Electronvolts)

element X	ionization potential	X_2 bond D_0	spin–orbit $^3P_2$–$^3P_0$
C	11.26	6.1	0.005
Si	8.15	3.2	.03
Ge	7.88	2.8	.17
Sn	7.34	2.0	.42
Pb	7.42	1.0	1.32

My principal purpose in this Account is to examine the special or anomalous chemical effects that have been noted for heavy elements in relation to periodic-table trends and to determine the degree to which these are relativistic effects. To the extent that these anomalies are explained in most inorganic chemistry books, they are usually ascribed to the introduction of the 4f shell. It is found that the effects of the 4f shell and the effects of relativity are of comparable magnitude, with the latter becoming more important for elements substantially heavier than the lanthanides.

The mathematical structure of relativistic quantum mechanics[1] is considerably more complex than that of the nonrelativistic theory; hence, one uses fully relativistic theory based on the Dirac equation only when it is needed. For work with elements of intermediate atomic number, there is a well-developed perturbation theory, due largely to Pauli, which allows one to examine the various relativistic terms on the basis of the Schrödinger wavefunctions and to include those corrections which may be significant in a particular situation. But for very heavy elements the validity, or even the convenience, of this perturbation approach is doubtful, and it is better to start with the Dirac equation.

After submitting the initial version of this Account I learned of the Account of Pyykkö and Desclaux[2] which is being published simultaneously. These authors outline the mathematical formulation of relativistic quantum theory, including the Dirac equation; hence the corresponding material has been deleted from the present paper. Also available is a more detailed review by Pyykkö[3] with an extensive bibliography.

(1) See, for example, H. A. Bethe and E. E. Salpeter, "Quantum Mechanics of One- and Two-Electron Atoms", Academic Press, New York, 1957; or I. P. Grant, *Adv. Phys.*, 19, 747 (1970).
(2) P. Pyykkö and J. P. Desclaux, *Acc. Chem. Res.*, 12, 271 (1979).

In a sense the spin of electrons and other particles and the exclusion principle are themselves relativistic effects in that they are integral parts of relativistic quantum theory whereas they are appended as separate postulates in the conventional theory which starts with the Schrödinger equation. But the effects arising from spin alone or from the exclusion principle are not the subject of this Account; indeed they are part of conventional "nonrelativistic" theory. The spin–orbit energy, which is often mentioned as an effect splitting spectral lines, is a relativistic effect in our present sense and is one of those terms which becomes large for heavy elements. Indeed it is convenient to take this spin–orbit energy as a measure of the need to use relativistic theory for chemical phenomena. In Table I the spin–orbit energies are compared with ionization potentials and bond dissociation energies for the fourth group elements.

A chemical bond energy for relatively large atoms is roughly 2 eV (46 kcal or 190 kJ). From Table I we see that the spin–orbit term is small compared to a chemical bond energy for fourth group elements through tin but becomes larger than the bond energy in lead. Also there is an anomalous drop in the bond energy for lead below that of any reasonable extrapolation of the trend from silicon to germanium to tin. This illustrates the hazard in a simple extrapolation for the bond energy of eka-lead, element 114, which is of interest as a possible synthetic superheavy element.[4]

From data such as that in Table I, one can estimate that the perturbation approach starting with nonrelativistic theory will probably suffice through elements such as tin for chemical bonding calculations as well as for spin–orbit energies, but that for heavier elements one should use fully relativistic theory. Experience to date on molecular calculations is in agreement with that estimate.

Let us turn now to a more comprehensive consideration of the anomalous effects that have been noted for heavy elements. Cotton and Wilkinson,[5] in their third edition of "Advanced Inorganic Chemistry", do not mention relativistic effects explicitly, but they do note: (1) the lanthanide contraction, (2) the inert pair effect, (3) the unique properties of gold (as compared to Ag and Cu), of mercury (as compared to Cd and Zn), and of U^{VI}, Np^{VI}, and Pu^{VI} (as compared to the corresponding lanthanides), and (4) the effect on magnetic properties of the large spin–orbit interaction in compounds of Os^{IV}, etc. Let us consider first the role of relativity as compared to other possible factors for (1) through (3). The spin–orbit effect in (4) is purely relativistic, and we shall subsequently comment briefly on it.

We consider first the conclusions that can be drawn from comparisons of relativistic and nonrelativistic calculations for atoms. There are complete tables of self-consistent-field calculations[6–8] yielding orbital energies, radii of the various orbitals, and other properties. Also the orbital energies are approximately related to experimental ionization potentials.[9] Thus it is convenient to use this information concerning the atoms to a maximum extent.

(3) P. Pyykkö, *Adv. Quantum Chem.*, **11**, 353 (1978).
(4) G. T. Seaborg, W. Loveland, and D. J. Morrissey, *Science*, **203**, 711 (1979).
(5) F. A. Cotton and G. Wilkinson, "Advanced Inorganic Chemistry", 3rd ed., Interscience, New York, 1972.
(6) J. P. Desclaux, *Atom. Data Nucl. Data*, **12**, 311 (1973).
(7) J. B. Mann, Los Alamos Laboratory Reports LA-3690 (1967) and LA-3691 (1968).
(8) C. Froese Fisher, *Atom. Data*, **4**, 301 (1972); *Atom. Data Nucl. Data*, **12**, 87 (1973).

Table II
Lanthanide Contraction: 5p Mean Radii (4p for Yttrium) and Observed Ionic Radii (Å)

	calculated ⟨r⟩				observed ionic radius
	nonrel	av rel	$p_{3/2}$	$p_{1/2}$	
Y	0.778	0.774	0.779	0.763	0.88
La	0.970	0.958	0.971	0.931	1.061
Lu	0.774	0.745	0.764	0.706	0.848

Not very many relativistic calculations have been made, as yet, for the electronic motion in molecules. In a later section a few of these calculations, which yield results of general chemical interest, will be considered, but we make no pretense of giving a complete account of all relativistic calculations on the electronic structure of atoms and molecules.

Conclustions Based on Atomic Properties

In addition to atomic spectral data we have complete tables from Desclaux[6] based on solution of the Dirac equation and yielding orbital energies, radii, and other properties. Comparable nonrelativistic tables are available from several sources.[7,8] Except for s orbitals, where there is no orbital angular momentum, the spin–orbit interaction divides a shell of a given l into subshells with total angular momentum $j = l - {}^1/_2$ and $j = l + {}^1/_2$. Relativistic calculations yield separate energies and radii for each subshell. The difference in energy is just the spin–orbit energy. Where it is appropriate to average the properties of the subshells, they are weighted as $2j + 1$.

Electronic velocities are highest in the region close to the nucleus. Thus it is not surprising that relativistic effects are greatest for s orbitals which have the greatest density near the nucleus and in the direction of decreasing the radius and increasing the ionization potential. The effect on p orbitals is in the same direction, but smaller. These effects on s and p orbitals increase the shielding of the nuclear charge for d and f electrons; hence the net relativistic effect may be reversed, i.e., increased radius and decreased ionization potential.

It is well-known that the radii of the lanthanide ions decrease from La to Lu and that this reduction in radius persists for the following elements. This "lanthanide contraction" is cited as the immediate cause of the near equality of radii (in comparable oxidation states) for Hf and Zr, Ta and Nb, etc., through Au and Ag and possibly further. The underlying cause is commonly stated as the incomplete shielding of the nucleus by the 14 4f electrons; thus a larger effective core charge contracts the 5p, 5d, 6s orbitals. This is the single most important cause of the contraction. But we shall see that the relativistic equation yields smaller radii than the nonrelativistic equation in some cases.

Table II gives both the relativistic and nonrelativistic calculated radii for the 5p shell as well as the experimental ionic radii[5] for lanthanum and lutetium and for comparison also for yttrium (4p shell). One would not

(9) C. E. Moore, "Atomic Energy Levels", NBS Circular No. 467, Vol. I–III, U.S. Government Printing Office, Washington, D.C., revised 1971.

Table III
Radii ($\langle r \rangle$, Å) and Weighted Averages for Relativistic 5d and 6p

	pseudoatom nonrelativistic			real atom nonrelativistic			real atom relativistic		
	5d	6s	6p	5d	6s	6p	5d	6s	6p
Hf	1.325	2.484		1.179	2.153		1.263	1.955	
Re	1.076	2.231		0.952	1.955		0.992	1.734	
Au	0.912	2.235		0.817	1.958		0.839	1.620	
Hg	0.849	1.984		0.758	1.761		0.779	1.515	
Tl		1.762	2.245		1.570	2.078		1.365	2.036
Pb		1.613	1.986		1.442	1.829		1.266	1.783
Bi		1.499	1.805		1.373	1.660		1.187	1.614

Table IV
Calculated Orbital Energies and Experimental Ionization Potentials in eV for Valence Shell Electrons

	$s_{1/2}$	$d_{5/2}$		$d_{3/2}$
Au, exptl	9.22	11.22		12.81
calcd relativ	7.94	11.66		13.43
rel wt av	7.94		12.37	
nonrelativ	6.01		14.17	
pseudoatom	5.18		14.62	
Ag, exptl	7.58	12.51		13.18
calcd relativ	6.45	13.64		14.31
rel wt av	6.45		13.91	
nonrelativ	5.99		14.62	

Table V
The Inert Pair Effect: Orbital Energies for Ge, Sn, and Pb (eV)

element	Ge	Sn	Pb
s, relativ	15.52	13.88	15.41
s, nonrel	15.16	13.04	12.49
p, wt av rel	7.29	6.71	6.48
p, nonrel	7.33	6.76	6.52
$\Delta\epsilon$, relativ	8.23	7.17	8.93
$\Delta\epsilon$, nonrel	7.83	6.28	5.97
$\Delta\epsilon$, pseudoatom	-	-	4.78

expect that exact agreement of the experimental ionic radius with the mean value of r for the outer shell of the ion, but there should be a close relationship of relative values. In fact the ionic radius exceeds the average relativistic radius by 0.10_3 Å in all cases. The ionic radius of lutetium is significantly less than that of yttrium. It is also apparent that there is a relativistic contraction which is trivial for yttrium, small for lanthanum, but significant for lutetium. While the relativistic contribution to the total contraction from La to Lu is only about 10%, it is the relativistic effect that reduces the radius of Lu significantly below that for yttrium.

In order to measure the effect of the 4f shell for elements following the lanthanides, nonrelativistic calculations were made[10] for pseudoatoms in which the nuclear charge was reduced by 14 units and the 4f orbitals were deleted. These results are compared with the nonrelativistic and the relativistic results for the real atoms in Tables III–V. For the radius of the 6s orbital the contraction caused by the 4f shell decreases through the series Hf to Bi from 0.33 to 0.13 Å, while the relativistic effect remains about constant at 0.2 Å. Thus, while both effects have the same general magnitude and the same direction, the relativistic effect becomes the more important for the heavier elements. In the case of the 5d orbitals, however, the relativistic effect is in the opposite direction; the 4f shell contracts the 5d orbital, but relativistic effects expand it. From Tl on, the 6p orbitals begin to be occupied and one finds that the effects on the 6p radii are similar to those on the 6s orbital but of decreased magnitude.

Table IV gives energy values for the valence orbitals in gold and silver. The orbital energies calculated on a self-consistent-field basis are only rough approximations to the ionization potentials because electron correlation and relaxation effects are omitted. But with these factors considered, the relativistic values agree well with experiment. Now comparing gold with silver, the s electron is more tightly bound in gold than in silver by 1.5 eV, whereas the $d_{5/2}$ electrons are more loosely bound in gold by about the same amount. But within about 0.5 eV the nonrelativistic values for gold or pseudogold are the same as those for silver where relativistic effects are not very large. Thus this large shift toward more strongly bound s and more loosely bound d electrons is primarily a relativistic effect. But these shifts explain very well the differences in chemical behavior. With the high ionization potential for the 6s electron even in the large atom, gold is very difficult to oxidize. But in compounds it forms several strong covalent bonds since 5d orbitals can be involved as well as 6s and 6p. Not only is the difference in energy between d and s orbitals smaller in gold than in silver but the differences in radii are smaller also.

This tendency toward strong covalent bonds is also shown in the diatomic molecule Au_2 whose dissociation energy,[11] 2.29 eV, exceeds that of either Ag_2 (1.65 eV) or Cu_2 (1.95 eV). Also note the trend in D_0 down from Cu_2 to Ag_2, which is a normal pattern, and then anomalously up to Au_2. An extensive theoretical study has been made[12] for Au_2 which is discussed below.

The electron affinities of these atoms are[13] Cu 1.226 eV, Ag 1.202 eV, and Au 2.308_6 eV; again the anomaly for gold is striking and is understandable from the relativistic effect in contracting the 6s orbital and making it more strongly bound. The compounds CsAu and RbAu are unusual in being nonmetallic semiconductors[14] with the CsCl structure. They are presumably based on an ionic M^+Au^- model, and the high electron affinity of gold is essential to their nonmetallic character.

In the case of mercury the combined relativistic and 4f-shell effect contracts the 6s orbital and strengthens

(10) P. S. Bagus, Y. S. Lee, and K. S. Pitzer, *Chem. Phys. Lett.*, **33**, 408 (1975).

(11) K. A. Gingerich, *J. Cryst. Growth*, **9**, 31 (1971); J. Kordis, K. A. Gingerich, and R. J. Seyse, *J. Chem. Phys.*, **61**, 5114 (1974).

(12) Y. S. Lee, W. C. Ermler, K. S. Pitzer, and A. D. McLean, *J. Chem. Phys.*, **70**, 288 (1979); W. C. Ermler, Y. S. Lee, and K. S. Pitzer, *ibid.*, **70**, 293 (1979).

(13) H. Hotop, R. A. Bennett, and W. C. Lineberger, *J. Chem. Phys.*, **58**, 2373 (1973); H. Hotop and W. C. Lineberger, *ibid.*, **58**, 2379 (1973).

(14) W. E. Spicer, A. H. Sommer, and J. G. White, *Phys. Rev.*, **115**, 57 (1959).

Table VI
Valence Orbital Energies (eV) and Radii (Å) for Uranium

	relativ wt av	nonrelativ
ϵ, 5f	9.01	17.26
ϵ, 6d	5.09	7.25
ϵ, 7s	5.51	4.54
$\langle r \rangle$, 5f	0.76	0.67
$\langle r \rangle$, 6d	1.71	1.52
$\langle r \rangle$, 7s	2.30	2.67

its bonding capacity, as was the case in gold. Thus mercury is a more noble metal than zinc or cadmium and its compounds show stronger covalent bonding. The unexpected volatility of mercury arises from the increase in 6s to 6p promotion energy as compared to zinc and cadmium. The unusual dimeric, dipositive ion Hg_2^{2+} is isoelectronic with Au_2, and we have already noted the anomaly in the strength of the bond in Au_2 which will be shown later to be primarily of relativistic origin. The 5d orbitals play a lesser and the 6p orbitals a greater role in mercury than in gold; the relativistic and 4f-shell effects on radii are shown in Table III. In summary, the anomalous properties of mercury arise from the sum of relativistic and 4f-shell contributions which are of comparable magnitude and reinforce one another in most respects.

The appearance of compounds, primarily in groups 4 through 7, with oxidation number two less than the group number is ascribed to an "inert pair" of s electrons. This effect is enhanced for the 6s, 6p valence shell and is extended to thallium in group 3. Thus the energy required to remove the 6s pair of electrons is much greater for Tl, Pb, and Bi than for the 5s pair in In, Sn, and Sb. The pertinent atomic orbital energies for Ge, Sn, and Pb are given in Table V. On a nonrelativistic basis, even with the effect of the 4f shell, all of the energies decrease along this sequence and no grossly anomalous trend is indicated. But on a relativistic basis, the s electron in Pb is bound as strongly as in Ge and much more than in Sn, while the normal trend is maintained for the weighted average for the p electrons. Thus the difference in binding for s as compared to p electrons shows a very anomalous trend from Sn to Pb. Although the 4f shell effect is in the same direction, as indicated by the value for pseudolead, the major part of the anomaly arises from relativity in this case. The data for the third or fifth group show the same pattern.

There is also an enhanced inert pair effect for the elements Ge through Br as compared to those of the elements in the rows above and below, but this is not a relativistic effect; rather it is associated with the introduction of the first d shell. Just as in the case of the 4f shell in the lanthanides, the shielding of 4s and 4p electrons by the 3d shell is incomplete and less complete for s than for p orbitals. Hence the orbital energies are increased more for 4s than for 4p and the inert pair effect is enhanced in relation to expectations related in part to elements Si through Cl.

In various discussions[15,16] of the bonding in the uranyl ion, it is suggested that this unique ion has its great stability because 5f, 6d, 7s, and 7p orbitals can all contribute to the remarkably short and strong U–O

(15) R. E. Connick and Z. Z. Hugus, Jr., *J. Am. Chem. Soc.*, 74, 6012 (1952).
(16) C. Y. Yang, K. H. Johnson, and J. A. Horsley, *J. Chem. Phys.*, 68, 1001 (1978), and references cited therein.

Table VII
Dirac Atomic Orbitals[a]

$s_{1/2}$, $m = +^1/_2$

$$\begin{bmatrix} g(r) \\ 0 \\ f(r)\cos\theta \\ f(r)\sin\theta\, e^{i\phi} \end{bmatrix}$$

$s_{1/2}$, $m = -^1/_2$

$$\begin{bmatrix} 0 \\ g(r) \\ f(r)\sin\theta\, e^{-i\phi} \\ -f(r)\cos\theta \end{bmatrix}$$

$p_{1/2}$, $m = +^1/_2$

$$\begin{bmatrix} g(r)\cos\theta \\ g(r)\sin\theta\, e^{i\phi} \\ f(r) \\ 0 \end{bmatrix}$$

$p_{1/2}$, $m = -^1/_2$

$$\begin{bmatrix} g(r)\sin\theta\, e^{-i\phi} \\ -g(r)\cos\theta \\ 0 \\ f(r) \end{bmatrix}$$

$p_{3/2}$, $m = +^3/_2$

$$\begin{bmatrix} g(r)\sin\theta\, e^{-i\phi} \\ 0 \\ f(r)\sin\theta\cos\theta\, e^{i\phi} \\ f(r)\sin^2\theta\, e^{2i\phi} \end{bmatrix}$$

$p_{3/2}$, $m = -^3/_2$

$$\begin{bmatrix} 0 \\ g(r)\sin\theta\, e^{-i\phi} \\ f(r)\sin^2\theta\, e^{-2i\phi} \\ -f(r)\sin\theta\cos\theta\, e^{-i\phi} \end{bmatrix}$$

$p_{3/2}$, $m = +^1/_2$

$$\begin{bmatrix} 2g(r)\cos\theta \\ -g(r)\sin\theta\, e^{i\phi} \\ f(r)(\cos^2\theta - ^1/_3 \\ f(r)\sin\theta\cos\theta\, e^{i\phi} \end{bmatrix}$$

$p_{3/2}$, $m = -^1/_2$

$$\begin{bmatrix} g(r)\sin\theta\, e^{-i\phi} \\ 2g(r)\cos\theta \\ f(r)\sin\theta\cos\theta\, e^{-i\phi} \\ -f(r)(\cos^2\theta - ^1/_3 \end{bmatrix}$$

[a] The angular factors are shown explicitly; other factors are included in $g(r)$ and $f(r)$ in each case.

bonds. Clearly the mean radius of the 5f orbital is much less than that for the others, but in contrast to the 4f orbital, the 5f has a radial node and has significant amplitude at radii much beyond the average. From Table VI it is clear that relativistic effects substantially narrow the ranges of orbital energies and radii from 5f through 7s. Thus it is reasonable to conclude that special properties of oxidation state VI for U, Np, and Pu arise at least in large part because of the relativistic effects in making 7s (and 7p) orbitals more compact and strongly bound while making 5f orbitals larger and much more loosely bound.

We have now shown that most of the anomalous properties of various heavy elements are to be ascribed in substantial part to relativistic effects and that this conclusion can be based on simple arguments from atomic radii and orbital energies (or ionization potentials). In a few situations it is important, however, to recognize differences in the bonding characteristics of relativistic orbitals, especially p orbitals.

Bonding Characteristics of Relativistic Orbitals

While Schrödinger wavefunctions are ordinary scalar functions, Dirac orbitals[1,17] are fourth-order vectors. The electron probability density is given by the sum of the squares (in absolute magnitude) of the four components. Two of these components vanish in the nonrelativistic limit $c \to \infty$; these are called "small components" and are always very small in the outer portion of an atom but may become important near the nucleus. The other two "large components" can be thought to correspond to the Schrödinger wavefunctions for + and − spin, respectively, and it is these functions that must be considered in estimating the bonding characteristics.

Powell[17] has given a simple and clear description of relativistic orbitals for atoms where various properties are discussed. The angular properties of orbitals are especially important for chemical bonding, and these are shown for $s_{1/2}$, $p_{1/2}$, and $p_{3/2}$ orbitals in Table VII.

(17) R. E. Powell, *J. Chem. Educ.*, 45, 558 (1968).

Here various factors are included within the functions $g(r)$ and $f(r)$ which are eigenfunctions of the radial equations for the particular atom and principal quantum number. The large components are listed first with the factor $g(r)$; the small components in third and fourth places with the factor $f(r)$.

As one can see from Table VII, the angular properties of s orbitals are unchanged by relativity. The large components have no angular dependence, i.e., are spherical, and only the first or the second component is non-zero. Thus they can form strong bonds in any direction. The only significant relativistic effect on s orbitals is the quantitative change in $g(r)$ whereby they are contracted and more tightly bound. This increases the strength of bonds based on s orbitals.

In contrast, there are profound differences in angular properties for those relativistic p orbitals with z component of angular momentum $m = \pm^1/_2$. Both large components are non-zero, and one component has the angular dependence of a pσ orbital whereas the other has that of a pπ orbital. If one forms a diatomic molecular orbital from $p_{1/2}$ atomic orbitals for each atom with signs appropriate for σ bonding, then the other components yield π antibonding. More exactly, if the $p_{1/2}$ orbitals are combined with g symmetry one has one-third σ-bond and two-thirds π-antibond character. Correspondingly, the combination with u symmetry yields one-third σ-antibond and two-thirds π-bond character. Thus $p_{1/2}$ orbitals do not form strong bonds. By taking linear combinations of $p_{1/2}$ and $p_{3/2}$ orbitals, the angular and bonding characteristics of Schrödinger p orbitals can be regained, but this requires substantial promotion energy in many cases.

This relativistic effect for p bonding should be most prominent for Tl_2 where the atom has a single $p_{1/2}$ valence electron. Tl_2 is very weakly bound; indeed its bond energy is still uncertain.[18]

In TlH this effect will be less severe since the hydrogen contributes an s orbital to the bond. With negligible π interaction from the hydrogen, the π-antibonding effect is unimportant, and it is only the reduced σ-bonding effect of a $p_{1/2}$ orbital (as compared to a nonrelativistic p orbital) that remains.

In some exploratory calculations[19] I made the very crude assumption of full promotion of $p_{1/2}$ electrons or $p_{3/2}$ vacancies to the nonrelativistic $^1/_3(p_{1/2}) + ^2/_3(p_{3/2})$ combinations for optimum bonding. Of course, in reality the bonding molecular orbitals will involve a compromise with partial promotion and somewhat weaker bonding than would correspond to the $^1/_3(p_{1/2}) + ^2/_3(p_{3/2})$ combination, but as yet there is no simple method of estimation for the intermediate state. More complex calculations[20,21] of the types discussed in the next section indicate that the proportion of $p_{1/2}$ and $p_{3/2}$ participation in TlH is about 60% $p_{1/2}$ and 40% $p_{3/2}$. This is evidently a compromise between the 100% $p_{1/2}$ for minimum energy of the Tl atom and the 33% $p_{1/2}$ for optimum bonding.

On the basis of simplified calculations assuming the full promotion energy to $^1/_3(p_{1/2}) + ^2/_3\,(p_{3/2})$ orbitals, it was estimated[19] that elements 112, 114, and 118 would be relatively stable as atoms and approach "inert gas" (actually inert volatile liquid) characteristics. Also it was suggested[22] that radon fluoride may be an ionic compound (Rn^+F^- or $Rn^{2+}F^-{}_2$) rather than a covalently bonded structure and that element 118 would show an even stronger tendency in that direction. Refined calculations would change the numerical values, but it is not likely that the qualitative indications would be changed.

Finally, it may be noted that similar problems arise for bonding with relativistic d or f orbitals in relation to their angular properties, but it is not yet clear what the impact is on chemical properties. This matter deserves future study.

Table VIII
Properties of Au_2

state	R_e, Å	D_e, eV	T_e, eV
X $0_g{}^+$, calcd	2.37	2.27	0
exptl[a]	2.47	2.31[b]	0
A $0_g{}^+$, calcd	2.51	0.79	2.61
exptl[a]	2.57	1.00	2.44
B $0_u{}^+$, calcd	2.50	1.38	3.55
exptl[a]	2.51	1.78	3.18

[a] From ref 26. [b] From ref 11.

Relativistic Calculations for Molecules

A full exact relativistic treatment of electronic motion in a molecule is very difficult, and none has been completed for a molecule which includes heavy, many-electron atoms where the effects we have been considering are chemically significant. But approximate calculations of useful accuracy have been made on two different bases. One method[12,23–25] assumes that the core electrons remain unchanged in molecule formation and replaces their detailed influence on valence electrons by effective potentials. This method was used to treat diatomic gold[12] with the results shown in Table VIII, which show that the properties[11,26] of the two excited spectroscopic states which have been observed as well as those of the ground state are obtained with useful accuracy from the relativistic calculations. Comparison with nonrelativistic calculations for the ground state indicates that the relativistic effects strengthen the bond by 1.0 eV and shorten it by 0.3_5 Å. The anomalous trend in dissociation energies in the series Cu_2, Ag_2, Au_2 was stated above; if the bond energy in Au_2 is decreased by 1.0 eV, the anomaly disappears. Thus the anomaly is primarily caused by the relativistic effects.

Another approximation for relativistic molecular calculations is the one-center method of Desclaux and Pyykkö.[20,27] Here all electrons are considered, but only in orbitals centered on the heavy atom. The method gives useful results for compounds of a single heavy atom with hydrogen; there are only valence electrons in the vicinity of the protons. These one-center cal-

(18) J. Drowart and R. E. Honig, *J. Phys. Chem.*, **61**, 980 (1957); S. A., Shuchukarev, G. A. Semenov, and I. A. Rat'kovskii, *J. Inorg. Chem. USSR*, **7**, 240 (1962); D. S. Ginter, M. L. Ginter, and K. K. Innes, *J. Phys. Chem.*, **69**, 2480 (1965).
(19) K. S. Pitzer, *J. Chem. Phys.*, **63**, 1032 (1975).
(20) P. Pyykkö and J. P. Desclaux, *Chem. Phys. Lett.*, **42**, 545 (1976).
(21) Y. S. Lee, to be published.
(22) K. S. Pitzer, *J. Chem. Soc., Chem. Commun.*, 760 (1975).
(23) Y. S. Lee, W. C. Ermler, and K. S. Pitzer, *J. Chem. Phys.*, **67**, 5861 (1977); **69**, 976 (1978).
(24) P. J. Hay, W. R. Wadt, L. R. Kahn, and F. W. Bobrowicz, *J. Chem. Phys.*, **69**, 984 (1978).
(25) G. Das and A. C. Wahl, *J. Chem. Phys.*, **64**, 4672 (1976).
(26) L. L. Ames and R. F. Barrow, *Trans. Faraday Soc.*, **63**, 39 (1967).
(27) J. P. Desclaux and P. Pyykkö, *Chem. Phys. Lett.*, **29**, 534 (1974); **39**, 300 (1976); revised values by private communication.

Table IX
Molecular Constants for AuH

R_e, calcd nonrel, Å	1.807,[a] 1.747[b]
calcd relativ, Å	1.522,[a] 1.622[b]
exptl, Å	1.524[c]
D_e, calcd relativ, eV	2.66[a]
exptl, eV	3.37[c]

[a] Effective potential method.[24] [b] One-center method.[27] [c] Experimental.[28]

culations can be made relativistically as well as nonrelativistically, and the difference may give a useful measure of the relativistic effect even where the absolute value of the calculated property is not very accurate. Other approximate calculations including relativistic features are reviewed by Pyykkö.[3]

It is interesting to compare the relativistic effects in AuH with those in Au_2. There are effective potential calculations of Hay et al.[24] and one-center calculations of Desclaux and Pyykkö[27] which are given in Table IX together with experimental values.[28] The effective potential method yields excellent agreement with experiment for the bond distance, with a relativistic shortening of 0.28 Å as compared to 0.35 Å in Au_2.

Direct Consequences of Spin-Orbit Energies

In addition to the effects on bond energies, there are other consequences when spin–orbit energies become large. Cotton and Wilkinson[5] discuss the relationship of spin–orbit coupling to magnetic properties. If the spin–orbit splitting becomes large as compared to thermal energy, paramagnetic effects can be largely suppressed. Another very important consequence is the breakdown of spin selection rules and the enhancement of rates of singlet–triplet interconversion. This is well-known to those working with heavy elements but is sometimes ignored by others. The very strong spectral line of mercury at 2537 Å is a "spin-forbidden" $^3P_1 \rightarrow {}^1S_0$ transition. The intensity of this line is a vivid reminder that spin is not separately quantized in heavy atoms.

The magnetic properties and the related splittings in the spectra of many compounds involving heavy atoms were initially interpreted with appropriate consideration of spin–orbit energies. For example, Moffitt et al.[29] showed that the spectra of the molecules ReF_6 to PtF_6 are closer to j–j coupling than to L–S coupling but that ligand field effects are even more important. Since most of the literature presently gives full consideration to the relativistic (spin–orbit) effects for these properties, there is no need for further comment here.

Summary

While spin–orbit effects have been recognized in interpreting magnetic properties of molecules containing heavy atoms, more elaborate calculations are required to establish the contribution of relativistic terms to bond energies, ionization potentials, and various chemical properties. Relativistic quantum mechanical calculations have now been made for atoms and for a few molecules, and one can show that many of the anomalous departures from periodic table trends for heavy atoms can be attributed to relativistic effects.

The research reviewed in this Account received support from the Energy Research and Development Administration and the Division of Chemical Sciences, Office of Basic Energy Sciences, U.S. Department of Energy, and was carried out in part in cooperation with I.B.M. Laboratories. I gratefully acknowledge the contributions of my collaborators, Y. S. Lee, P. S. Bagus, W. C. Ermler, and A. D. McLean, and valuable discussions with Professor N. Bartlett.

(28) B. Rosen, "Spectroscopic Data Relative to Diatomic Molecules", Pergamon Press, Oxford, 1970.

(29) W. Moffitt, G. L. Goodman, M. Fred, and B. Weinstock, *Mol. Phys.*, **2**, 109 (1959); see also B. Weinstock, *Chem. Eng. News*, **42**, 86 (Sept 21, 1964).

Ab initio effective core potentials including relativistic effects. V. SCF calculations with ω–ω coupling including results for Au_2^+, TlH, PbS, and PbSe

Yoon S. Lee,[a] Walter C. Ermler,[b] and Kenneth S. Pitzer

Department of Chemistry and Lawrence Berkeley Laboratory, University of California, Berkeley, California 94720

(Received 20 November 1979; accepted 17 March 1980)

Ab initio self-consistent field calculations are reported for a series of diatomic molecules using relativistic effective core potentials (REP) and basis sets appropriate for ω–ω coupling. The molecular orbitals are expressed as linear combinations of two-component analogs of Dirac spinors. The unique feature of the present approach is the retention of the spin–orbit operator in the generation of the REP's and the propagation of its effects into the molecular wave functions in a totally consistent fashion. The nature of bonding in the molecules Au_2^+, TlH, PbSe, and PbS is discussed with consideration of the orbital energies, spectroscopic constants, and population analyses. Comparisons with recently obtained photoelectron spectra of PbSe and PbS are made. It is noted that $6p_{1/2}$ and $6p_{3/2}$ orbitals exhibits bonding characteristics that are different from the nonrelativistic $p\sigma$ and $p\pi$ molecular orbitals.

I. INTRODUCTION

In recent years, there has been increasing interest in the inclusion of relativistic effects in electronic structure calculations of molecules containing heavy atoms. The most straightforward method is the use of the Dirac one-electron operator in the calculations. However, with four-component Dirac spinors a very large number of integrals must be evaluated for all electron Dirac–Hartree–Fock calculations; hence, most methods are based on an approximation that reduces the volume of computation.

In the previous papers[1–4] of this series we have reported a method of generating relativistic effective core potentials (REP)[1] from atomic Dirac–Hartree–Fock (DHF) calculations and their applications to the diatomic molecules Xe_2 [2] and Au_2.[3,4]

In our earlier applications,[2–4] we averaged the $p_{1/2}$ and $p_{3/2}$ potentials (also $d_{3/2}$ and $d_{5/2}$, etc.) with the proper weights so that the spin–orbit splittings are removed from the potentials and standard programs based on LS coupling can be used for the molecular calculations. The use of these averaged relativistic effective core potentials (AREP) in molecular calculations is parallel to various relativistic-effective-core-potential methods developed by other workers.[5–8] These procedures differ in the stage at which the average is taken. Apart from our method, one can start from atomic calculations that have been averaged[5] or employ averaged orbitals in calculating the potentials.[6–8] All of these methods yield potentials that include most of the one-electron relativistic effects of atoms except for the spin-orbit splitting.

Since, in molecules containing heavy atoms, the spin-orbit splitting is often larger than other one-electron relativistic effects, it is theoretically more consistent to avoid this averaging and to carry out molecular calculations in ω–ω coupling. In this report we describe a method for self-consistent-field (SCF) molecular orbital calculations using the nonaveraged REP defined in our earlier treatment,[1] thereby retaining the spin–orbit effects. The application of our REP in molecular calculations requires a SCF formalism that is based upon the use of two-component molecular spinors. A brief discussion of the method has been reported in an earlier paper[3] for diatomic molecules with closed-shell configurations. In this paper, we describe this SCF theory more fully and extend it to diatomic molecules having no more than one open shell of a given symmetry. Electronic structure calculations for Au_2^+, TlH, PbS, and PbSe are reported as examples of this procedure together with related calculations on the atoms Tl, Pb Se, and S.

II. THE LCAS–MS SCF THEORY WITH THE REP AND TWO COMPONENT MOLECULAR SPINORS (TCMS)

We use linear combinations of atomic spinors (LCAS) to form molecular spinors (MS). While Dirac spinors have four components, it was shown in earlier papers of this series that the small components were negligible for valence pseudo-orbitals and that molecular calculations could, therefore, be based on two-component molecular spinors formed from the two large components of the atomic spinors.

When the relativistic effective core potentials (REP) derived by the method described in Ref. 1 are used for a diatomic molecular calculation, the Hamiltonian of the system is given by [Eqs. (2) and (3) of Ref. 3]

$$\mathcal{H}=\sum_{\mu=1}^{n_v}\left[-\tfrac{1}{2}\nabla_\mu^2+\sum_{\alpha=1}^{2}\left(-\frac{Z_\alpha}{r_{\alpha\mu}}+U_\alpha^{\text{REP}}\right)\right]+\sum_{\mu>\nu}^{n_v}\frac{1}{r_{\mu\nu}}\,, \tag{1}$$

where n_v is the total number of valence electrons, $r_{\mu\nu}$ is the distance between electrons μ and ν, $r_{\alpha\mu}$ is the distance between the electron μ and the nucleus α, Z_α is the charge of the nucleus α, and U_α^{REP} is the REP due to core electrons of the nucleus α. U_α^{REP} can be ap-

[a] Present address: James Franck Institute, University of Chicago, Chicago, IL 60637.

[b] Present address: Department of Chemistry and Chemical Engineering, Stevens Institute of Technology, Hoboken, NJ 07030.

 0021-9606/80/130360-07$01.00

proximated by [Eqs. (4) and (5) of Ref 3]

$$U_\alpha^{\rm REP} = U_{LJ}^{\rm REP}(r_{\alpha\mu}) + \sum_{l=0}^{L-1} \sum_{j=|l-1/2|}^{l+1/2} \sum_{m=-j}^{j} \times [U_{lj}^{\rm REP}(r_{\alpha\mu}) - U_{LJ}^{\rm REP}(r_{\alpha\mu})] \, |ljm\rangle_\alpha \langle ljm|_\alpha \ , \tag{2}$$

where L and J are, typically, the smallest angular momentum quantum numbers that are not present for the core electrons. The projection operators are defined by

$$|ljm\rangle\langle ljm| = \left[\sum_{\sigma=\pm 1/2} C(l\tfrac{1}{2}j;\ m-\sigma, \sigma)\ |\, Y_l^{m-\sigma}(\theta,\phi)\,\phi_{1/2}^{\sigma}\rangle\right] \times \left[\sum_{\sigma'=\pm 1/2} C(l\tfrac{1}{2}j;\ m-\sigma', \sigma')\langle Y_l^{m-\sigma'}(\theta,\phi)\,\phi_{1/2}^{\sigma'}|\right] \ , \tag{3}$$

with the same notation as in Ref. 1. In Eq. (3) the C's are Clebsch–Gordan coefficients, the Y's are spherical harmonic functions, and the ϕ's are Pauli spinors with

$$\phi_{1/2}^{1/2} = \begin{pmatrix}1\\0\end{pmatrix}, \quad \text{and} \quad \phi_{1/2}^{-1/2} = \begin{pmatrix}0\\1\end{pmatrix} .$$

The projection operators introduce two-by-two matrices into the $U_\alpha^{\rm REP}$ in Eq. (2). The symmetry properties of the operator $U_\alpha^{\rm REP}$ and of the eigenfunctions of this operator can be determined from those of the double group.

The symmetry properties of diatomic molecules in the double group representation may be found in the work by Malli and Oreg.[9] However, the molecular spinors which are eigenfunctions of the REP of Eq. (2) have only two components while the Dirac spinors have four components. In linear molecules, the angular symmetry of a state of a molecule is defined by the total electronic angular momentum Ω as in the Hund case c[10] and that of ith molecular orbital by $m_i = \pm\,\omega_i$. The two orbitals with $m_i = +\,\omega_i$ and $m_i = -\,\omega_i$ are degenerate and constitute a shell. Because spin is incorporated in the orbital each molecular orbital can accommodate only one electron. With these characteristics, one can derive the SCF theory for the two-component molecular spinors (TCMS) for diatomic molecules by following the procedures that have been formulated for conventional nonrelativistic molecular calculations.[11]

For a molecular state that may be represented by a single Slater determinant, the total wave function of the system is approximated by

$$\Phi = \mathcal{A}(\phi^1\phi^2\cdots\phi^n) \ , \tag{4}$$

where $\mathcal{A}$ is the antisymmetry operator and the ϕ's are one-electron molecular spinors. The total energy of the molecular state is expressed by

$$E = \langle\Phi|\mathcal{H}|\Phi\rangle \tag{5}$$

provided Φ is normalized. As in the nonrelativistic case each one-electron molecular spinor, $\phi_{i\lambda}$, is expanded in a basis set of two-component atomic spinors (TCAS), $\chi_{\lambda p}$,

$$\phi_{i\lambda}^{\pm} = \sum_p c_{i\lambda,p}^{\pm}\,\chi_{\lambda p}^{\pm} \ , \tag{6}$$

where λ is the symmetry index ω, i is the index for orbitals of the same symmetry, and $+$ and $-$ refer to the sign of m_i. The TCAS's are defined by

$$\chi_{\lambda p}^{\pm}(n_p, l_p, j_p, \pm\lambda) = \chi_{\lambda p} = R_{\lambda p}(r) \sum_{\sigma=\pm 1/2} C(l\tfrac{1}{2}j;\ \pm\lambda-\sigma, \sigma)\, Y_{lp}^{\pm\lambda-\sigma}\,\phi_{1/2}^{\sigma} \ , \tag{7}$$

where $R(r)$ is a radial function. When Slater-type functions (STF) are used as the basis, $\chi_{\lambda p}$ becomes

$$\chi_{\lambda p}(n_p, l_p, j_p, \pm\lambda) = N_p\, r^{n_p-1} e^{-\zeta_p r} \begin{pmatrix} C(l_p\tfrac{1}{2}j_p,\ \pm\lambda-\tfrac{1}{2}, \tfrac{1}{2})\, Y_{l_p}^{\pm\lambda-1/2} \\ C(l_p\tfrac{1}{2}j_p,\ \pm\lambda+\tfrac{1}{2}, -\tfrac{1}{2})\, Y_{l_p}^{\pm\lambda+1/2} \end{pmatrix} , \tag{8}$$

where N_p is the normalization factor and the angular parts are rewritten to demonstrate the two-component properties of the basis spinors.

Following the nonrelativistic theory for open shells,[11] we express the total energy as

$$E = D_T^\dagger H D_T + \tfrac{1}{2} D_T^\dagger\, \mathcal{P} D_T - \tfrac{1}{2} D_0^\dagger\, \mathcal{Q} D_0 \ , \tag{9}$$

where H is composed of matrix elements of the one-electron operator [Eq. (1)] over TCAS's, D_T and D_0 are the total and the open-shell density matrices, respectively, and $\mathcal{P}$ and $\mathcal{Q}$ are the closed and the open-shell supermatrices. The summations implied in Eq. (9) and the definition of density matrices were given by Roothaan and Bagus.[11] The computational methods for one-electron matrix elements are described in Ref. 3.

Matrix elements of the supermatrices $\mathcal{P}$ and $\mathcal{Q}$ are defined by

$$\mathcal{P}_{\lambda pq,\mu rs} = J_{\lambda pq,\mu rs} - \tfrac{1}{2}[K_{\lambda pq,\mu rs}^{+} + K_{\lambda pq,\mu rs}^{-}] \tag{10}$$

and

$$\mathcal{Q}_{\lambda pq,\mu rs} = A_{\lambda\mu}\, J_{\lambda pq,\mu rs} + B_{\lambda\mu}\, K_{\lambda pq,\mu rs}^{+} + C_{\lambda\mu}\, K_{\lambda pq,\mu rs}^{-} \ , \tag{11}$$

where A, B, and C are the open-shell vector-coupling coefficients and J and K are Coulomb and exchange integrals. In Eqs. (10) and (11),

$$J_{\lambda pq,\mu rs} = \int [\chi_{\lambda p}^{+}(1)]^* [\chi_{\mu r}^{+}(2)]^* \frac{1}{r_{12}}\, \chi_{\lambda q}^{+}(1)\, \chi_{\mu s}^{+}(2)\, d\tau \tag{12}$$

and

$$K_{\lambda pq,\mu rs}^{\pm} = \int [\chi_{\lambda p}^{+}(1)]^* [\chi_{\mu r}^{\pm}(2)]^* \frac{1}{r_{12}}\, \chi_{\mu s}^{\pm}(1)\, \chi_{\lambda p}(2)\, d\tau \ , \tag{13}$$

where the order of signs is preserved. J and K can be calculated from the two-electron integrals over STF's by taking the proper linear combinations.[3,11]

Energy expressions given by Eqs. (9)–(13) were possible because the shell structure was assumed. This equivalence constraint may be expressed as

$$c_{i\lambda p}^{+} \equiv c_{i\lambda p}^{-} \quad \text{for all } p \ . \tag{14}$$

If the shell structure is not conserved, the number of matrix elements to be included in the energy expression will be increased considerably and yield nonequivalent + and − orbitals for open-shell configurations. When the variational procedure is applied to the energy, one obtains equations that are essentially the same as the nonrelativistic equations [Eqs. (12)–(16) of Ref. 11]. Since all the deviations from the conventional SCF equations are included in Eqs. (9)–(14), the SCF equations for the TCMS are omitted.

The open-shell configurations which can be treated

TABLE I. Open-shell vector coupling coefficients for the states represented by a single-Slater determinant.

	$A_{\lambda\mu}$	$B_{\lambda\mu}$	$C_{\lambda\mu}$
$\lambda=\mu$	1	$-1/2$	$-1/2$
$\lambda\neq\mu$			
$\Omega_i=\lambda+\mu$[a]	0	1/2	$-1/2$
$\Omega_i=\lvert\lambda-\mu\rvert$[a]	0	$-1/2$	1/2

[a]If $\Omega_i=\lambda+\mu$ for the interaction of two open shells, the signs of the m's are the same for both orbitals. If $\omega_i=\lvert\lambda-\mu\rvert$, the signs of m's are opposite.

with the present formalism are limited to the cases where $\mathfrak{Q}$ defined by Eq. (11) is adequate for the energy expression; specifically, each open shell must belong to a different symmetry. Table I shows the vector coupling coefficients $A_{\lambda\mu}$, $B_{\lambda\mu}$, and $C_{\lambda\mu}$. Since the only type of relativistic open shell is one with a single electron, Table I may be used for any open-shell case as long as each open shell belongs to a different symmetry.

Once the proper transformation is carried out for one- and two-electron matrix elements, an existing SCF program can be modified for use with TCMS's.

III. CALCULATIONS AND RESULTS

Relativistic effective core potentials (REP) were generated for the atoms, Tl, Pb, S, and Se by the method described in Ref. 1. In order to study the effect of the core size, we generated two sets of REP's for Tl and Pb, one treating $5d$ electrons as valence electrons and the other including those in the core. In addition, nonrelativistic effective core potentials were generated for Tl for comparison. All of the cases are described in Table II. In each case, a double zeta set of STF's is obtained from the energy optimization of the ground state atoms with numerical AREP's. These basis sets are

TABLE II. Separation of the core and the valence electrons for the various atoms.

Atom	Potential[a]	Core electrons	Valence electrons[a]
Tl	REP3	[Xe] $5d^4_{3/2}\,5d^6_{5/2}$	$6s^2 6p_{1/2}$
	REP13	[Xe]	$5d^4_{3/2}\,5d^6_{5/2}\,6s^2 6p_{1/2}$
	NREP3	[Xe] $5d^{10}$	$6s^2 6p$
	NREP13	[Xe]	$5d^{10}\,6s^2\,6p$
Pb	REP4	[Xe] $5d^4_{3/2}\,5d^6_{5/2}$	$6s^2\,6p_{1/2}6p_{3/2}$
	REP14	[Xe]	$5d^4_{3/2}\,5d^6_{5/2}\,6s^2 6p_{1/2}6p_{3/2}$
S	REP6	[Ne]	$3s^2\,3p^1_{1/2}\,3p^3_{3/2}$
Se	REP6	[Ar] $3d^{10}$	$4s^2\,4p^1_{1/2}\,4p^3_{3/2}$

[a]Numerics refer to the total number of valence electrons.

TABLE III. Slater-type functions (STF) and two-component spinors (TCAS) for three-electron Tl.[a]

STF		TCAS	
$n\,l\,m_l$	ζ	$n\,l\,j\,m_l$	ζ
3 0 0	1.7838	$3\,0\,\frac{1}{2}\,\frac{1}{2}$	1.7838
3 0 0	1.7441	$3\,0\,\frac{1}{2}\,\frac{1}{2}$	1.7441
4 1 0	1.6897	$4\,1\,\frac{1}{2}\,\frac{1}{2}$	1.6897
4 1 0	0.9483	$4\,1\,\frac{1}{2}\,\frac{1}{2}$	0.9483
4 1 1	1.6897	$4\,1\,\frac{3}{2}\,\frac{1}{2}$	1.6897
4 1 1	0.9483	$4\,1\,\frac{3}{2}\,\frac{1}{2}$	0.9483
		$4\,1\,\frac{3}{2}\,\frac{3}{2}$	1.6897
		$4\,1\,\frac{3}{2}\,\frac{3}{2}$	0.9487

[a]Tl REP3 of Table II.

used for both atomic and molecular calculations. In our molecular studies, it is possible to use different basis functions for orbitals with the same l but different j quantum numbers. However, we choose to use the same basis set for these orbitals with the condition that it be of double-zeta or better quality to allow for the differences in size between the $l+\frac{1}{2}$ and $\lvert l-\frac{1}{2}\rvert$ components. Also, the selection of different functions for each orbital would greatly increase the size of the basis set with only a slight improvement in accuracy of the results. (See below and Table III.)

After the basis set is selected, the procedure for the SCF calculation is the following:

(1) Calculation of one- and two-electron integrals over STF basis functions;

(2) Transformation of integrals to obtain the matrix elements over TCAS's;

(3) Computation of matrix elements of REP's for TCAS's;

(4) SCF procedure to obtain molecular wave functions, orbital energies, and total energies.

The net effect of the transformation from STF's to TCAS's is to expand all the basis functions with nonzero m values. An example for this expansion is shown in Table III for Tl. In the case chosen, the number of basis functions increases from six STF's (four σ and two π functions) to eight TCAS's (six 1/2 and two 3/2 functions). Therefore, some extra computation is necessary compared with the nonrelativistic calculations for the same basis functions. This extra computational effort corresponds to a modest integral transformation and the necessity to process the additional integrals at the SCF stage. However, the increase in the computation time is justified since the spin–orbit splittings are properly included in the SCF calculations.

The REP's are used in numerical form in all the calculations. We thereby eliminate the possible error introduced in the process of fitting the potentials with analytic functions. This is possible because the REP integral program can treat both analytical and numerical effective core potentials.[3]

A. Atomic calculations

Atomic calculations were carried out for the low-lying states of neutral atoms and the ground state of the ion for Tl and Pb with double zeta basis sets of STF's and the averaged REP's (AREP) obtained from the configurations shown in Table II. The results, summarized in Table IV, indicate that the excitation energies of the atomic states are almost the same for n electron REP's (5d electrons in the core) and $n+10$ electron REP's (5d electrons in the valence shell). For these types of excitations or ionizations, the 5d electrons seem to behave as core electrons. This shows the REP's behave consistently although it is not evident that the 5d electrons are corelike in bonding situations. The excitation energies agree qualitatively with those obtained from experiment.[14] The $np^1_{1/2}np^3_{3/2}$ configuration of S and Se (Table II) used for the generation of REP is close to the average of LS configurations.[15] Excitation energies obtained with these AREP's of S and Se are in good agreement with the nonrelativistic, all-electron calculations of Clementi and Roetti[16] as shown in Table IV.

B. $(1/2)_g$ state of Au_2^+

The molecular energies were computed for Au_2^+ in its $(1/2)_g$ ground state at several internuclear distances to test the open-shell program with the REP and TCMS. The basis set and the REP are the same as those used in Ref. 3 for the O_g^+ state of Au_2. The total energy on a 21 valence electron basis is given for several interatomic distances in Table V as are vertical ionization energies calculated by subtracting the energies of Au_2 from those of Au_2^+. These vertical transition energies are compared with the orbital energies of the least-bound $(1/2)_g$ orbital of Au_2. The agreement between the two sets of values is evidence that the open-shell formalism is correct for a configuration with one open shell and that Koopman's theorem is approximately valid in this case.

TABLE IV. Excitation and ionization energies of Tl and Pb calculated with the double zeta basis sets and AREP's (all in eV).

Atoms	State	Configuration	Excitation energies		
			AREP13[a]	AREP3[a]	Exptl[b]
Tl	^2P	$6s^26p$	0.0	0.0	0.0
Tl^+	^1S	$6s^2$	4.87	4.79	5.47
			AREP14[a]	AREP4[a]	
Pb	^3P	$6s^26p^2$	0.0	0.0	0.0
	^1D	$6s^26p^2$	0.98	0.98	1.60
	^1S	$6s^26p^2$	2.42	2.42	2.61
Pb^+	^2P	$6s^26p$	6.53	6.50	7.81
			AREP	All electron[c]	
S	^3P	$3s^23p^4$	0.0	0.0	
	^1D	$3s^23p^4$	1.33	1.31	
	^1S	$3s^23p^4$	3.43	3.26	
Se	^3P	$4s^24p^4$	0.0	0.0	
	^1D	$4s^24p^4$	1.50	1.43	
	^1S	$4s^24p^4$	3.67	3.54	

[a]Averages are taken from the corresponding REP (Table II).
[b]From Ref. 14.
[c]From Ref. 16.

TABLE V. Total valence energies of Au_2^+, vertical ionization energies of Au_2 (calculated from the total valence energies of Au_2 and Au_2^+) and the orbital energies of the least-bound $(1/2)_g$ electron of Au_2 at various interatomic distances. (Total energies in a.u. and ionization energies in eV. All are based on 11-electron REP's for Au.)

R	$-E$ (a.u.)	$E(Au_2^+)-E(Au_2)$	$-\epsilon\,[(\frac{1}{2})_g]$ of Au_2
4.5	70.5181	7.07	7.64
4.75	70.5266	6.91	7.46
5.0	70.5300	6.77	7.26
5.5	70.5292	6.48	6.94
8.0	70.5022	5.50	5.74

As would be expected, the $(1/2)_g$ state of Au_2^+ has a slightly larger equilibrium bond length (2.74 Å) and a smaller vibrational frequency (94 cm^{-1}) compared with the O_g^+ state of Au_2, which we calculated[3] as 2.50 Å and 142 cm^{-1}. The dissociation energy is expected to be smaller for Au_2^+ than for Au_2. However, an accurate estimate of D_e is not possible due to the improper dissociation behavior of the SCF approximation where the charge is not allowed to localize for ions like Au_2^+.[17] Experimental values of D_e for Au_2^+ are not available.

C. TlH

The ground state of wave function TlH has been calculated with 3 and 13 electron valence shells and with REP, AREP, and NREP for the Tl atom. In each case, a double zeta basis set is derived from the energy optimization of the ground state of Tl. In addition to the two s functions, one p function is added to the basis set of the hydrogen. The spectroscopic constants are summarized in Table VI, including the results of calculations of Pyykko and Desclaux[18] and experimental values.[19]

From Table VI it is apparent that relativistic effects reduce the interatomic distance significantly; this may be ascribed to the relativistic contraction of the 6s orbital on Tl and to a smaller contraction of the $6p_{1/2}$ orbital. Also it appears that the bond is weakened, but when the spin–orbit splitting of the atomic states of Tl is considered, that conclusion is less clear. The average ^2P energy is 0.65 eV higher than the true ground state $^2P_{1/2}$. It is not possible, without ambiguity, to correct the AREP and NREP calculations for this spin–orbit term. The empirical procedure for introducing spin–orbit effects using the atomic splittings that was used in our earlier work[2,4] would be ill defined for molecules like TlH where large charge transfer is observed. These effects may be reasonably approximated if all the CI coefficients of the relevant configurations are known.[4] However, this is not available since the present calculations are based on single determinants. But any plausible correction to the AREP result will reduce the true dissociation energy to the $^2P_{1/2}$ atom of Tl from the value in Table VI. Hence, one can conclude that the inclusion of spin–orbit effects does decrease D_e for TlH. The dif-

TABLE VI. Spectroscopic constants obtained from the SCF calculation of TlH with various EP's.

	Effective core potentials					One center[a]		
	REP13	REP3	AREP3	NREP13	NREP3	*R*	*NR*	Exptl[b]
R_e (Å)	1.84	1.79	1.79	1.94	1.88	1.867	2.008	1.868
D_e (eV)	1.55[c]	1.36[c]	1.77[d]	1.69[d]	1.74[d]			1.96
ω_e (cm^{-1})	1450	1380	1380	1380	1410	1500	1390	1391
$(p_{3/2}/p_{1/2})$[e]	0.37	0.50	2.0	2.0	2.0	0.65	2.0	...
I.P. (eV)[f]	8.11	7.89	7.78	7.24	7.26			

[a]From Reference 18.
[b]From Reference 19.
[c]$E(\text{Tl}\ ^2P_{1/2}) + E(\text{H}) - E(\text{TlH at } R_e)$.
[d]$E(\text{Tl}\ ^2P) + E(\text{H}) - E(\text{TlH at } R_e)$.
[e]From the atomic population analysis.
[f]Ionization potential from Koopman's theorem and orbital energies at near-equilibrium distance.

ference in R_e between the 3 and 13 electron bases for Tl is surprisingly large. Presumably the outer radius of the 5*d* orbitals is important in determining R_e and this should be more reliably established when these 5*d* electrons are included explicitly in the calculation.

The TCMS–MO's have a rather complicated character at the equilibrium internuclear distance. They are much simpler at $R = 6.0$ a.u.; here the lower valence orbital is almost purely Tl-6*s*. In the REP basis the higher energy orbital has a population 1.45 on the hydrogen atom with σ bonding participation by $6p_{1/2}$ and $6p_{3/2}$ orbitals. The ratio of populations $(p_{3/2}/p_{1/2})$ is 0.9 at $R = 6$ a.u., whereas it would be 2.0 in the absence of the spin–orbit term. The decrease in this ratio from 2.0 to 0.9 is caused by spin–orbit interaction, and it introduces a small $p\pi$ population. The more interesting aspect is the large ionic character Tl^+H^- with the transfer of 0.47 electronic charge. Thus the bonding arises primarily from the upper orbital and is very substantially ionic.

The population analyses at the equilibrium bond distance are given in Table VII. At this distance the lower energy orbital has become a bonding orbital with a substantial population on hydrogen. The total charge on hydrogen is reduced, but remains substantial (approximately 0.35 *e*). The Tl-6*s* participation in the higher energy orbital is antibonding and cancels a large portion of the $p\sigma$ bonding from the 6*p* orbitals. Thus, in a simplified description, the covalent bonding comes from the lower orbitals with 6*s*(Tl)–1*s*(H) character, and the remaining contribution to the bonding is primarily ionic. The precise values of the atomic populations should not, of course, be considered accurate measurements of charge transfer due to the approximations inherent in the formalism and to the possible sensitivity of the results to basis set imbalance. (Part of the excess charge on H is undoubtedly due to the presence of the polarization function.) However, the bond appears to be more ionic in TlH than in BH.[20]

The nonbonding character of the 2-1/2 orbital is confirmed by the near constancy of its orbital energy: 0.2892 a.u. at $r = 3.50$ a.u., 0.2894 a.u. at $r = 3.25$ a.u. In contrast, the orbital energy of the 1-1/2 orbital increases from 0.4891 a.u. to 0.5016 a.u. over the same range.

Since both the ionic bonding and the 6*s*(Tl)-1*s*(H) bonding character are not appreciably changed by relativistic effects, it is not surprising that the calculated bond energy values in Table VI are nearly equal. The value of D_e calculated with the REP basis is smaller than the experimental value by an amount that is reasonably attributed to the electron correlation energy not included in the SCF approximation.

D. PbS and PbSe

In addition to the calculations described above on Au_2^+ and TlH, exploratory calculations were made for PbS and PbSe for comparison with photoelectron spectra being measured currently by Shirley and associates.[21]

TABLE VII. Population analyses for TlH ($R = 3.5$ a.u.).

	REP-3			AREP-3			NREP-3		
Orbital	1-1/2	2-1/2	Total	1σ	2σ	Total	1σ	2σ	Total
Tl *s*	1.40	0.43	1.83	3.60	0.30	1.90	1.17	0.70	1.87
$p_{1/2}$	0.00	0.53	0.53 }	0.00	0.74	0.74	0.08	0.73	0.81
$p_{3/2}$	0.02	0.27	0.29 }						
H *s*	0.58	0.77	1.35	0.40	0.96	1.36	0.75	0.56	1.32

TABLE VIII. Orbital energies and ionization energies for PbS (in eV).

		$X\alpha$		
Orbital	REP[a]	A[b]	B[c]	Exptl[d]
4-1/2	8.87	6.18	9.09	9.20 (broad)
1-3/2	9.36	6.61	9.38	
3-1/2	9.77	7.07	9.61	9.80
2-1/2	16.08	13.88	15.40	
1-1/2	24.05	17.20	···	

[a]This research, $R = 2.25$ Å.
[b]Reference 22.
[c]Reference 23, improved $X\alpha$ calculation for transition states with sphere radii of neutral atoms.
[d]Reference 21.

The use of a single relativistic configuration is a rather severe approximation, expecially for orbitals primarily located on the S or Se atom, but it does reflect the large spin–orbit term and the other relativistic effects for Pb. A double zeta basis set was used with other details essentially the same as have been described for the other examples. Fortunately, the ground state configuration for the molecules $(1\text{-}1/2)^2(2\text{-}1/2)^2(3\text{-}1/2)^2(4\text{-}1/2)^2(1\text{-}3/2)^2$ allows dissociation into neutral atoms $6s_{1/2}^2 6p_{1/2}^2$ for Pb and $ns_{1/2}^2 np_{1/2}^2 np_{3/2}^2$ for S or Se, although these states are not the true ground states of the atoms.

Tables VIII and IX show the calculated orbital energies for PbS and PbSe, respectively, at near equilibrium interatomic distances and compare our REP results with those from $X\alpha$ calculations by Yang and Rabii[22] and the experimental results of Shirley *et al.*[21] For both molecules the lower frequency photoelectron peak is broad and presumably includes both the lowest $\Omega = 1/2$ state and the $\Omega = 3/2$ state for the positive ion. The agreement of our calculated energies with experiment is as good as could be expected. Also our results show a rapid variation of orbital energies with interatomic distance for the 4-1/2 orbital and hence predict a broad band which would overlap the 1-3/2 band. The detailed band contour does not agree; however, that is not unexpected in view of our approximations. For the second transition removing an electron from the 3-1/2 orbital, the agreement between calculation and experiment is perfect—doubtless in some degree by accident.

TABLE IX. Orbital energies and ionization energies for PbSe (in eV).

		$X\alpha$		
Orbital	REP[a]	A[b]	B[c]	Exptl[d]
4-1/2	8.43	6.37	8.79	8.90 (broad)
1-3/2	8.63	6.39	8.79	
3-1/2	9.47	7.02	9.20	9.52
2-1/2	15.97	14.37	15.46	
1-1/2	23.21	17.71		

[a]This research, $R = 4.50$ a.u.
[b]Reference 22.
[c]Reference 23, improved $X\alpha$ calculation for transition states with sphere radii of neutral atoms.
[d]Reference 21.

TABLE X. Calculated and experimental quantities for PbS and PbSe.

		PbS	PbSe
R_e (Å)	REP	2.26	2.41
	Exptl[a]	2.287	2.402
ω_e (cm^{-1})	REP	351.	210.
	Exptl[a]	429.4	277.6
D_e (eV)	REP	1.17	0.90
	Exptl[a]	3.48	3.07

[a]From Ref. 19.

The $X\alpha$ calculations of Yang and Rabii[22] yield the correct order of orbital energies but the quantitative values do not agree. More recent and improved $X\alpha$ calculations by Yang and White[23] exhibit agreement that is comparable to our REP calculations.

In Table X are shown our calculated values of R_e, ω_e, and D_e together with experimental data.[19] The agreement with the experimental values of R_e is quite satisfactory. Since electron correlation is not included in our calculation, it is expected that our calculated D_e will be too small, and this effect will also influence the ω_e values in the same direction.

The $\omega = 3/2$ orbital is, of course, a pure π orbital, and the ion state with this vacancy can be described as $^2\Pi_{3/2}$. The atomic p orbitals with $m = \pm 1/2$, however, contain both σ and π components and the molecular orbitals based thereon, 4-1/2 and 3-1/2, likewise include both σ and π components. If one approaches these same states from a nonrelativistic basis, the $^2\Sigma_{1/2}$ and $^2\Pi_{1/2}$ states are mixed and contribute the σ and π components, respectively. Based upon REP calculations, we estimate the following characteristics for molecular orbitals of PbS and PbSe. Molecular orbitals 4-1/2 and 1-3/2 are π bonding, although the population in the 1-3/2 orbital is almost entirely on the Se atom. The 3-1/2 orbital is σ bonding and π antibonding. The 4-1/2 orbital energy changes rapidly with interatomic distance, as expected for a strongly bonding orbital, whereas the orbital energies for the other orbitals show little change in agreement with their weakly bonding (or antibonding) character.

IV. DISCUSSION

It should be realized that a SCF calculation for a single relativistic (ω–ω) configuration may not be a good approximation even for a molecule with heavy atoms.[24] Such molecules (or atoms) usually fall in the range of intermediate coupling (between L–S and j–j for atoms). Electron–repulsion and spin–orbit terms are of comparable magnitude. Unless the molecular orbitals occupied in the selected configuration allow appropriate adjustment for these two effects, a single configuration will not provide a good description of the ground state of the molecule. If the valence shell of the molecule involves only s orbitals, then there is no spin–orbit term, and a single configuration suffices, e.g., Au_2 or Au_2^+. With only a single p orbital involved in TlH, the molecular orbital can include a linear combination of $p_{1/2}$ and

$p_{3/2}$ orbitals from Tl and thus provide the needed adjustment for a good initial approximation. But for Tl_2, for example, the ground state cannot be reasonably approximated without CI including at least the two configurations $(1\text{-}1/2)^2_g(1\text{-}1/2)^2_u(2\text{-}1/2)^2_g$ and $(1\text{-}1/2)^2_g(1\text{-}1/2)^2_u(2\text{-}1/2)^2_u$ which we have found to be of very nearly equal energy from the SCF calculations. In addition, in the SCF approximation many molecules do not dissociate to the proper atomic states. Therefore, configurations accounting for both of these characteristics should be included in most molecular calculations with ω-ω coupling for a reasonable description of even the ground state. Work on the needed program to include multiple configurations is now in progress. However, the ground states of the molecules presented in this paper are reasonably described near the equilibrium bond distance with our present approximation.

The present method does not include the Breit term or other expressions of many-electron relativistic effects Such terms can be important for small splittings[25] for nearly degenerate states and even for larger energy differences[26] between the states of highly ionized heavy atoms. Layzer and Bahcall[27] give useful expressions for the trends with atomic number for various relativistic terms in many-electron atoms. The frozen-core effective-potential approximation precludes accurate calculations of these terms, but we believe that they would be very small for the quantities presented here which relate primarily to the energy dependence on interatomic distance for molecules.

ACKNOWLEDGMENTS

Certain of the calculations were done with the ALCHEMY programs of M. Yoshimine, A. D. McLean, B. Liu, and P. S. Bagus. We appreciate their help regarding the use of the programs.

We are grateful to Dr. M. G. White, Dr. C. Yang, and Dr. D. A. Shirley for interesting discussions and for providing their results for PbS and PbSe prior to publication.

We thank Dr. S. Hagstrom for his help with the two-electron integral program, Professor G. Malli for helpful discussions on open shells, and Dr. P. Christiansen for discussions on various related topics.

This work was supported by the Division of Chemical Sciences, Office of Basic Energy Sciences, U.S. Department of Energy under Contract No. W-7405-ENG-48.

[1]Y. S. Lee, W. C. Ermler, and K. S. Pitzer, J. Chem. Phys. 67, 5861 (1977).
[2]W. C. Ermler, Y. S. Lee, K. S. Pitzer, and N. W. Winter, J. Chem. Phys. 69, 976 (1978).
[3]Y. S. Lee, W. C. Ermler, K. S. Pitzer, and A. D. McLean, J. Chem. Phys. 70, 288 (1979).
[4]W. C. Ermler, Y. S. Lee, and K. S. Pitzer, J. Chem. Phys. 70, 293 (1979).
[5]L. R. Kahn, P. J. Hay, and R. D. Cowan, J. Chem. Phys. 68, 2386 (1978).
[6]G. Das and A. C. Wahl, J. Chem. Phys. 64, 4672 (1976).
[7]S. N. Datta, C. S. Ewig, and J. R. Van Wazer, Chem. Phys. Lett. 57, 83 (1978).
[8]H. Basch and S. Topiol, J. Chem. Phys. 71, 802 (1979).
[9]G. Malli and J. Oreg, J. Chem. Phys. 61, 4349 (1974).
[10]G. Herzberg, *Spectra of Diatomic Molecules* (Van Nostrand, Princeton, NJ, 1950).
[11]C. C. J. Roothaan and P. S. Bagus, Meth. Comp. Phys. 2, 47 (1963).
[12]G. Malli and J. Oreg, J. Chem. Phys. 63, 830 (1975).
[13]See, for example, A. C. Wahl, P. J. Bertonicini, K. Kaiser, and R. H. Land, Argonne National Laboratory Report No. ANL-7271, 1968.
[14]C. E. Moore, Natl. Bur. Stand. Circ. 467, Vol. I (1949), Vol. II (1952), Vol. III (1958).
[15]J. P. Desclaux, At. Data Nucl. Data Tables 12, 311 (1973).
[16]E. Clementi and C. Roetti, At. Data Nucl. Data Tables 14, 177 (1974).
[17]T. L. Gilbert and A. C. Wahl, J. Chem. Phys. 55, 5247 (1971).
[18]P. Pyykkö and J. P. Desclaux, Chem. Phys. Lett. 42, 545 (1976).
[19]K. P. Huber and G. Herzberg, *Molecular Spectra and Molecular Structure. IV. Constants of Diatomic Molecules* (Van Nostrand Reinhold, New York, 1979).
[20]R. S. Mulliken and W. C. Ermler, *Diatomic Molecules; Results of Ab Initio Calculations* (Academic, New York, 1977).
[21]D. A. Shirley, M. G. White, and associates (private communication).
[22]C. Y. Yang and S. Rabii, J. Chem. Phys. 69, 2497 (1978).
[23]M. G. White and C. Y. Yang (private communication).
[24]P. Pyykkö, Adv. Quantum Chem. 11, 353 (1978); K. S. Pitzer, Acc. Chem. Res. 12, 271 (1979).
[25]S. R. Langhoff and C. W. Kern, *Modern Theoretical Chemistry*, edited by H. F. Schaefer III (Plenum, New York, 1977), Vol. 4.
[26]J. P. Desclaux, K. T. Cheng, and Y. K. Kim, J. Phys. B 12, 3819 (1979).
[27]D. Layzer and J. Bahcall, Ann. Phys. 17, 177 (1962).

Electronic structure for the ground state of TlH from relativistic multiconfiguration SCF calculations

Phillip A. Christiansen and Kenneth S. Pitzer

Department of Chemistry and Lawrence Berkeley Laboratory, University of California, Berkeley, California 94720

(Received 18 July 1980; accepted 4 August 1980)

The dissociation curve for the ground state of TlH was computed using a relativistic ω–ω coupling formalism. The relativistic effects represented by the Dirac equation were introduced using effective potentials generated from atomic Dirac–Fock wave functions using a generalization of the improved effective potential formulation of Christiansen, Lee, and Pitzer. The multiconfiguration SCF treatment used is a generalization of the two-component molecular spinor formalism of Lee, Ermler, and Pitzer. Using a five configuration wave function we were able to obtain approximately 85% of the experimental dissociation energy. Our computations indicate that the bond is principally sigma in form, despite the large spin–orbit splitting in atomic thallium. Furthermore the bond appears to be slightly ionic (Tl^+H^-) with about 0.3 extra electron charge on the hydrogen.

INTRODUCTION

In recent years several relativistic effective-potential formalisms have been proposed[1–5] and employed in electronic structure studies of molecules containing heavy elements. In most of these, relativistic contributions, except for spin–orbit, are introduced into the molecular calculations via the effective potentials (EP). The spin–orbit effects are added later as some sort of perturbation. However, for molecules containing some of the sixth row elements such as Tl or Pb the spin–orbit contributions may be substantially larger than correlation effects and perturbation methods may be inadequate. Thus, it seems preferable to include the spin–orbit effects in the SCF step of the calculation. This is the procedure employed in the two-component molecular spinor formulation of Lee, Ermler, and Pitzer.[1] Hafner and Schwarz give an excellent description of a similar method.[2] In these procedures the molecular SCF computations are done directly in ω–ω coupling. Unfortunately, both procedures employ EP formalisms of questionable reliability. Furthermore, little has been done within the ω–ω framework to include molecular correlation effects. Therefore, though quite adequate for qualitative studies, these methods could be misleading if quantitative molecular results are desired.

In the present work we have extended the molecular spinor formalism of Lee *et al.* in two important ways. First, we have generalized the previous ω–ω SCF procedure to include a multiconfiguration SCF treatment (MCSCF). This allows us to include the major effects of electron correlation and to obtain proper dissociation in all cases. In addition, we have replaced the previously used Phillips–Kleinman type EP formalism[6] with the frozen-core EP method reported recently by Christiansen, Lee, and Pitzer[7] which was shown to be much more reliable for the calculation of potential curves or dissociation energies. In that work we were able to show that by retaining in the valence portion of the pseudo-orbital, the true form of the original atomic orbital, *ab initio* effective potentials can be generated which reproduce all-electron results to a high degree of accuracy. Calculations using EP's based on Phillips–Kleinman pseudo-orbitals sometimes give results which differ seriously from those of all-electron calculations.[8–11] Our recent paper identified the cause of this unreliability and demonstrated a procedure which eliminates the problem.

Since most of the important mathematical details of our calculations have been stated in Refs. 1 and 7, they will not be repeated here.

For a first test of this procedure we chose the thallium hydride molecule, for which good spectroscopic data are available.[12,13] In our molecular calculations we treat explicitly only the outermost 13 electrons of thallium (the $5d$, $6s$ electrons and single $6p_{1/2}$ electron). The $p_{1/2}$–$p_{3/2}$ spin–orbit splitting is approximately 1 eV. Thus, TlH should be an excellent test of our relativistic MCSCF formalism. Also, the relativistic contraction of the $6s$ atomic spinor causes a substantial energy gap between the $6s$ and $6p$ atomic spinor energies, resulting in the so-called inert pair effect for the $6s$ electrons. Thus, in terms of molecular bonding with hydrogen, Tl is somewhat similar to the alkali metals with essentially only one valence electron. However, in this case the electron is assigned to a p rather than s type spinor. It is well established for the ground states of the light alkali metal dimers and hydrides[14] that one can obtain a good approximation to the experimental bond energies by correlating only the two valence electrons using a relatively small MCSCF wave function. We expect that this may also be true for the thallium dimer and hydride.

The behavior in chemical bonding of $p_{1/2}$ and $p_{3/2}$ spinors, with a large spin–orbit energy difference, is a matter of great interest. A $p_{1/2}$ orbital is one third sigma and two thirds pi in its character, and in a diatomic molecule such as Tl_2, the molecular orbitals can be either sigma bonding and pi antibonding, or the reverse. For TlH any pi bonding or antibonding effects will be very small, but the behavior of the p orbital in sigma bonding is still of considerable interest. We are currently studying Tl_2, where the pi bonding or antibonding aspects will also become important.

 0021-9606/80/225160-04$01.00

METHOD AND CALCULATIONS

The effective potentials for the present calculations were generated as in Ref. 7; only for Tl is it necessary that we use numerical atomic Dirac–Fock spinors[15] rather than Hartree–Fock spin orbitals to form our pseudo-orbitals (in this case pseudospinors). Since the two small components of the four component Dirac spinors have essentially negligible amplitudes in the valence region, they are neglected in the pseudospinors. Thus, our valence pseudospinors have only two components, greatly simplifying our molecular calculations. The $5d_{3/2}$, $5d_{5/2}$, $6s_{1/2}$, and $6p_{1/2}$ pseudospinors were obtained from a ground state atomic calculation. The $6p_{3/2}$, $5f_{5/2}$, $5f_{7/2}$, $5g_{7/2}$, and $5g_{9/2}$ pseudospinors were obtained from atomic calculations where the single $6p$ electron was transferred from the $p_{1/2}$ spinor into the various excited spinors. In all, we used nine atomic pseudospinors and ultimately nine relativistic EP's in our molecular calculations.

As before [see Eqs. (1)–(3), Ref. 1], the complete effective potential for thallium has the general form

$$U_{\mathrm{Tl}}^{\mathrm{EP}} = U_{LJ}(r) + \sum_{l=0}^{L} \sum_{j=l-(1/2)}^{l+(1/2)} \sum_{m=-j}^{j} [U_{lj}(r) - U_{LJ}(r)] \, | ljm \rangle \langle ljm | \; ,$$

where $L=4$ and $J=9/2$ and the U_{lj} are the EP's derived from the individual pseudospinors with angular quantum numbers l and j. The projection operator on the right insures that the U_{lj} operate only on spinors of the proper angular symmetry. Previous work[1] has indicated that provided the L and J quantum numbers are not present for the core electrons (which precludes any radial "exclusion" effects due to core orbitals of the same symmetry), then U_{LJ} is probably a good approximation to the potentials of higher angular symmetry.

For the determination of the pseudospinors from the spinors, a few minor refinements were added to the procedure described in Ref. 7. As suggested by P. J. Hay,[16] in addition to limiting the maximum number of inflections in the radial pseudo-orbital or pseudospinor to two, we also require that there be no more than three inflexions in the first derivative. This has only a slight effect on the match point (the radial point outside which the spinor and pseudospinor are identical; see Ref. 7), but ultimately results in a smoother potential. Also, we found that for some of the more diffuse excited spinors, one could substantially reduce the value of the match point by adjusting the lead power in the analytical portion of the radial pseudospinor. This adjustment seems to have little effect for the more tightly bound spinors.

The basis set for Tl was double zeta in form (2s, 2p, and 2d) made up of Slater type functions. The exponents were obtained by optimizing the energy of the Tl ground state using the EP's after averaging out the spin–orbit effects. This will, of course, result in somewhat less than double-zeta quality for the individual spinors, but it greatly reduces the number of molecular integrals which need to be generated (to within our program limitations). In these calculations we have assumed that the important bonding spinor on Tl, the $p_{1/2}$, will be adequately polarized by the excess 6s and 5d functions. Unfortunately, at the time of these calculations our molecular integral programs were limited to maximum n quantum numbers of 4. Although this is probably adequate for Phillips–Kleinman[6] type calculations where the pseudo-orbitals contain substantial core character, higher n values might have been more appropriate for the present work. The hydrogen basis set was double-zeta plus polarization as in Lee *et al.*[1] Therefore, we expect that the overall quality of our molecular basis set is somewhat less than double-zeta plus polarization.

Once the exponents have been selected, the Slater type spin orbitals are then transformed into two-component atomic spinors to be used as basis functions in our molecular calculations. The molecular spinor formalism employed here is essentially that of Lee *et al.*[1] However, we wish to correlate the two bonding electrons. For the 0^+ state this requires double excitations from the bonding spinor into the correlating spinors. It can be easily shown that the coupling between configurations is of the form $K_{ij}^{+} + K_{ij}^{-}$, where the + refers to integrals resulting from the exchange of electrons with angular momentum of the same sign, and the − signifies angular momentum of opposite sign. The i and j refer to the two spinors by which the configurations differ. Thus, for the 0^+ state, the correlation coupling is analogous to that of nonrelativistic singlet sigma states as in the ground state alkali metal hydrides. We modified a version of the Bison MCSCF code of Das and Wahl[17] to perform the SCF calculations. To our knowledge this is the first time a molecular MCSCF calculation in ω–ω coupling using accurate relativistic EP's has been attempted.

RESULTS AND DISCUSSION

Using the effective potentials and basis sets described above, we computed total molecular energies for the ground state of TlH using single-determinant (essentially closed-shell Dirac–Fock), two-determinant (to obtain proper dissociation), and five-determinant MCSCF wave functions formed from two-component spinors. In the five-determinant calculation, various plausible configurations were introduced into exploratory calculations and the five most important were used in the final calculations. In addition to the two configurations required for proper dissociation, it was found that two configurations generated by double excitations to $\omega=\frac{1}{2}$ spinors and one configuration with a double excitation to an $\omega=\frac{3}{2}$ spinor made substantial contributions. The internuclear separation was varied from three to 15 atomic units (a.u.). The resulting molecular dissociation curves are listed in Table I. Spectroscopic constants for the single determinant and MCSCF wave functions are listed in Table II along with single determinant values from previous calculations and the experimental values. Plots of our dissociation curves are given in Fig. 1.

First, note the large discrepancy between our single determinant dissociation energy and the value obtained using EP's generated from Phillips–Kleinman type pseudospinors.[1] Our single determinant calculation gives about 47% of the experimental dissociation energy as opposed to about 79% for the Phillips–Kleinman type

TABLE I. Potential curves for TlH (all values are in a.u.).

R	Single configuration[a]	Two configuration[a]	Five-configuration MCSCF[a]
3.0		−0.0275	−0.0377
3.25	−0.0273	−0.0434	−0.0532
3.5	−0.0332	−0.0510	−0.0600
3.75	−0.0336	−0.0532	−0.0614
4.0	−0.0303	−0.0519	−0.0593
4.5	−0.0175	−0.0434	−0.0493
5.0		−0.0321	−0.0366
6.0		−0.0125	−0.0138
8.0		−0.0010	−0.0010
10.0		−0.0001	−0.0001

[a]Energies relative to the value obtained at $R = 15.0$ a.u. for the two- and five-configuration wave functions.

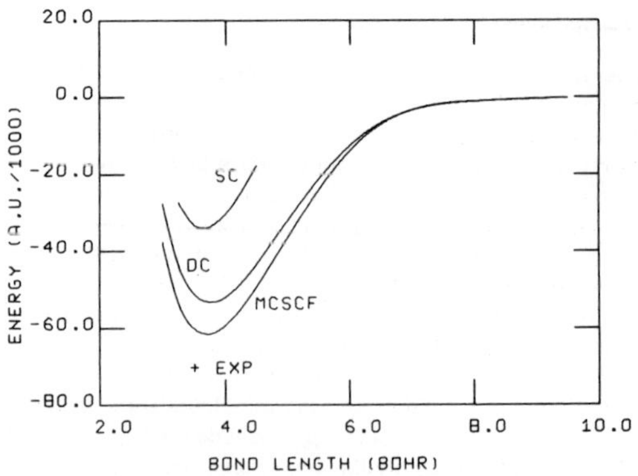

FIG. 1. Dissociation curves for single configuration (SC), double configuration (DC), and five configuration (MCSCF) calculations for the ground state of TlH. The minimum in the experimental curve is also indicated (+).

calculation. Nonrelativistic, all-electron, single-determinant calculations usually give only about one half or less of the dissociation energy.[14] This discrepancy is very similar to that seen in the earlier nonrelativistic EP calculations on F_2 and Cl_2 where Phillips–Kleinman EP's gave substantially larger dissociation energies than the corresponding all-electron calculations.[9] Note also that both our single configuration and MCSCF calculations predict bond lengths which are slightly too long relative to experiment. Though we have no proof, we suspect that this is the result of our slightly inadequate basis set. The fact that the MCSCF bond length is slightly longer than the single determinant value closely parallels previous nonrelativistic calculations.[14]

With our five-determinant MCSCF wave function we were able to obtain more than 85% of the experimental dissociation energy. Although similar calculations[14] for the light alkali metal dimers frequently obtain 90% or more of the bond energy, we believe that the combination of a somewhat poorer basis set and the smaller energy difference between the Tl 6s and 6p spinors (compared to that between the alkali metal valence s and p electrons) could easily explain the differences. It is the large s–p separation which allows us to treat correlation in TlH as though there were only two valence electrons. In this respect Cs_2 would have been an easier test if the calculation of correlation energy in heavy-atom molecules had been our only objective. However, the Cs_2 bond involves primarily 6s spinors with no spin-orbit effect. Since the inclusion of spin–orbit effects was also an important aim, TlH seemed a more appropriate example.

The general picture of the valence molecular spinors for TlH obtained in this research supercedes that reported earlier from less accurate calculations.[1] There are surprisingly large differences in some aspects as well as good agreement in others. Where there are differences, the present results yield a much simpler picture.[1] The present values were taken from our five-configuration MCSCF calculations for $R = 3.75$ a.u. The 6s spinor on thallium remains essentially unchanged in the molecule, the hydrogen and the thallium 6p participation remaining very small. In contrast, this spinor attained a bonding nature with nearly 30% population on hydrogen in the earlier calculations.[1] Since in our present calculations we found little difference between our single-configuration and five-configuration spinors, we expect that the above differences are primarily the result of the improved EP's used in the present work. In the principal bonding spinor (from our five-configuration calculations) the $6p_{1/2}$ to $6p_{3/2}$ population ratio is approximately 0.6_6—compared to 0.65 reported by Pyykkö and Desclaux[18] from one-center calculations and about 0.5 from the earlier EP results.[1] However, in

TABLE II. Spectroscopic constants for TlH.

	Present				
	MCSCF	Single configuration	Single configuration[a]	Single configuration[b]	Experimental[c]
R_e (Å)	1.96	1.93	1.84	1.87	1.87
D_e (eV)	1.66	0.93	1.55	···	1.97
ω_e (cm^{-1})	1330	1450	1450	1500	1391

[a]13 electron calculations employing Phillips–Kleinman type relativistic EP's from Ref. 1.
[b]One-center Dirac–Fock, Ref. 18.
[c]Reference 12.

our present calculations this ratio shows a marked radial dependence relative to the Tl origin, varying from about 0.5 at the Tl nucleus to about 3.5 at infinity. (The ratio for a nonrelativistic sigma orbital is 2.0.) In the region near the H nucleus, the ratio is such that the Tl p contribution to the spinor is more than three quarters sigma Since the contribution from the hydrogen is almost exclusively 1s, the bond appears to be principally sigma in character. Finally, our results indicate some ionic character (Tl^+H^-) with about 0.3 extra electron population on the hydrogen, in close agreement with the earlier calculation. Pyper[19] has shown that in TlH there is a bonding interaction between the pi component of the $6p_{1/2}$ spinor on Tl and the 1s orbital on H analogous to that in the $^3\Pi$ state of nonrelativistic third group hydrides BH, etc. Since this $^3\Pi$ state is only weakly bound in GaH and InH, we do not believe this is a major factor for TlH. Furthermore, Pyper's analysis assumes that the bonding spinor is essentially $p_{1/2}$ in form whereas our calculations indicate that the $p_{1/2}$ and $p_{3/2}$ mixing is substantial, the signs being such that the pi component nearly disappears in the bonding region. Thus, the overwhelming bonding interaction appears to be the sigma bonding as in $^1\Sigma$ BH, etc.

CONCLUSION

We have shown that the combination of our previously developed effective potential procedure[7] with the molecular spinor formalism of Lee *et al.*[1] and its subsequent generalization to include the important electron correlation effects provides a method which is capable of yielding near-quantitative numerical results when used to study bonding in molecules containing heavy elements.

ACKNOWLEDGMENTS

The authors are indebted to Dr. Carl Melius for supplying us with a version of the BISON-MC code, and to Dr. Yoon S. Lee for useful suggestions during the course of this work. This research was supported by the Division of Chemical Sciences, Office of Basic Energy Science, U.S. Department of Energy under Contract No. W-7405-ENG-48.

[1] Y. S. Lee, W. C. Ermler, and K. S. Pitzer, J. Chem. Phys. **73**, 360 (1980); also, earlier papers in that series there cited.

[2] P. Hafner and W. H. E. Schwarz, Chem. Phys. Lett. **65**, 537 (1979).

[3] G. Das and A. C. Wahl, J. Chem. Phys. **69**, 53 (1978).

[4] W. R. Wadt, P. J. Hay, and L. R. Kahn, J. Chem. Phys. **68**, 1752 (1978).

[5] S. N. Datta, C. S. Ewig, and J. R. Van Wazer, Chem. Phys. Lett. **57**, 83 (1978).

[6] J. C. Phillips and L. Kleinman, Phys. Rev. **116**, 287 (1959).

[7] P. A. Christiansen, Y. S. Lee, and K. S. Pitzer, J. Chem. Phys. **71**, 4445 (1979).

[8] L. R. Kahn, P. Baybutt, and D. G. Truhlar, J. Chem. Phys. **65**, 3826 (1976).

[9] P. J. Hay, W. R. Wadt, and L. R. Kahn, J. Chem. Phys. **68**, 3059 (1978).

[10] G. Das, Chem. Phys. Lett. **71**, 202 (1980).

[11] W. C. Ermler, Y. S. Lee, K. S. Pitzer, and N. W. Winter, J. Chem. Phys. **69**, 976 (1978).

[12] These studies are summarized by K. P. Huber and G. Herzberg, *Molecular Spectra and Molecular Structure, IV. Constants of Diatomic Molecules* (Van Nostrand, Reinhold, New York, 1979).

[13] M. L. Ginter and R. Battino, J. Chem. Phys. **42**, 3222 (1965).

[14] H. F. Schaefer III, *The Electronic Structure of Atoms and Molecules* (Addison-Wesley, Reading, Mass., 1972).

[15] J. P. Desclaux, Comput. Phys. Commun. **9**, 31 (1975).

[16] P. J. Hay (personal communication).

[17] G. Das and A. C. Wahl, Argonne National Laboratory Report No. ANL-7955 (1972).

[18] P. Pyykkö and J. P. Desclaux, Chem. Phys. Lett. **42**, 545 (1976).

[19] N. C. Pyper, Chem. Phys. Lett. 73, 385 (1980).

Electronic structure and dissociation curves for the ground states of Tl_2 and Tl_2^+ from relativistic effective potential calculations

Phillip A. Christiansen and Kenneth S. Pitzer

Department of Chemistry and Lawrence Berkeley Laboratory, University of California, Berkeley, California 94720
(Received 2 September 1980; accepted 8 October 1980)

The dissociation curves for the ground states of Tl_2 and Tl_2^+ were computed using a generalization of the molecular relativistic ω–ω coupling formalism of Lee, Ermler, and Pitzer. Relativistic effects, as represented by the Dirac equation, were introduced using effective potentials generated from atomic Dirac–Fock wave functions using a generalization of the improved effective potential formulation of Christiansen, Lee, and Pitzer. Our calculations show that the ground state of Tl_2^+ is $1/2_g$ with computed D_e and R_e values of 0.58 eV and 3.84 Å. For Tl_2 we find that the ground state is 0_u^- but the 0_g^+ and the 1_u states are only slightly higher in energy; the potential curves for these states are repulsive to about 3.5 Å and then essentially flat beyond that radius. While corrections for correlation will increase D_e somewhat, Tl_2 is only weakly bound in any of these states which dissociate to normal atoms. The cause is undoubtedly related to the large spin-orbit splitting between the $6p_{1/2}$ and $6p_{3/2}$ thallium spinors.

INTRODUCTION

Although dissociation energies and other spectroscopic parameters for the ground states of most of the group III homonuclear diatomic molecules have been determined experimentally,[1] to our knowledge this is not the case for Tl_2. Drowart and Honig[2] suggest an upper limit to the bond energy of Tl_2 of 0.9 eV. This is based on the failure to observe any Tl_2 signal in Tl vapor at approximately 900 °K,[3] and therefore implies no lower limit. Considering this upper limit and the bond energies of Hg_2 and Pb_2, they estimated the Tl_2 bond energy to be about 0.6 eV. This value has been reported in some subsequent reviews as if it were an experimental result. The only positive experimental data concerning Tl_2 appear to be the spectra of Ginter *et al.*,[4] which include emission bands in the range 15 300 to 16 000 cm^{-1} and adsorption bands in the range 23 000 to 23 500 cm^{-1} for Tl vapor at about 1600 °K. Unfortunately, a detailed interpretation of the spectra appears unfeasible for the present; indeed, it is not absolutely certain that these bands arise from Tl_2. There is considerable interest in the properties of Tl_2 in relation to recent experiments on parity nonconservation in transitions in atomic thallium.[5] In an effort to explore this situation we have computed dissociation curves for the three states 0_g^+, 0_u^-, and 1_u of Tl_2 which dissociate to neutral atoms in their ground states. In addition, we have computed dissociation curves for the lowest $1/2_g$ and $1/2_u$ states of Tl_2^+. Although this ion has been detected experimentally,[6] its molecular parameters have not been reported.

Since for many purposes it can be considered to have only a single valence electron (due to the relativistic contraction, the 6s shell is substantially lower in energy), and because of the large spin–orbit splitting, almost 1 eV between the $6p_{1/2}$ and $6p_{3/2}$ spinors, Tl is a particularly interesting subject for heavy element bonding studies. Unlike light atom bonding which is usually described in terms of sigma, pi, etc., spin orbitals, bonding between heavy atoms[7] must be described in terms of spinors which are usually combinations of spin orbitals of differing angular symmetry. This mixing of sigma, pi, etc., spin orbitals is the result of the spin–orbit coupling as given by the Dirac equation. The $6p_{1/2}$ spinor is composed essentially of $6p$ sigma and pi type spin orbitals with coefficients of $\sqrt{1/3}$ and $\sqrt{2/3}$, respectively. Therefore, a bonding spinor formed as a sum or difference of two such atomic spinors on different centers, retaining the overall g or u symmetry, must be either sigma bonding and pi antibonding or else sigma antibonding and pi bonding. For this reason one would not expect the $p_{1/2}$ spinors, alone, to form substantial molecular bonds in Tl_2. In our previous TlH calculations,[8] we found that the bonding spinor, though predominantly $p_{1/2}$ near the thallium nucleus, became progressively sigma-like towards the hydrogen, forming a reasonably strong sigma bond with the hydrogen 1s orbital. For Tl_2 one would expect the pi-antibonding component to play a more important role. Furthermore, in contrast to simple molecular orbital theory, this mixture of bonding and antibonding character in our molecular spinor model for $p_{1/2}$ bonds makes a prediction of the relative stabilities of the bonds in the corresponding ions difficult. Thus it is not obvious whether Tl_2^+ should be more or less strongly bonded than Tl_2.

METHOD

All of our present molecular calculations were done using the same molecular spinor formalism as was used for TlH.[8] In this formalism the molecular calculations are carried out in terms of ω–ω coupling, the spin–orbit effects being included explicitly in the SCF optimizations. The important relativistic effects (those implied by the Dirac equation), were introduced using effective potentials (EP) derived from full Dirac–Fock atomic wave functions. The important differences between our present calculations and those of Lee *et al.*[9] include the redefinition of the EP's, using the formulation of Christiansen *et al.*,[10] which greatly improves the

 0021-9606/81/021162-04$01.00

accuracy relative to all-electron calculations, and the inclusion of correlating configurations[8] which allows us to obtain a reasonable representation of molecular dissociation curves. Since the important mathematical details were given in Refs. 8–10, they will not be repeated here.

As in our previous calculations on TlH, we treat explicitly only the outer 13 electrons of atomic thallium (the ten $5d$ electrons, two $6s$ electrons, and the single $6p_{1/2}$ electron). The atomic spinors from which we derived our pseudospinors and EP were obtained from atomic calculations using a version of the numerical Dirac–Fock program written by Desclaux.[11] The double-zeta Tl basis set employed in these calculations is identical to that used in our calculations on TlH.[8] The exponents were optimized for the ground state atom after averaging the spin–orbit effects out of the relativistic EP. The excess d and s basis functions will allow some polarization for the $p_{1/2}$ bonding spinor. Also, as for TlH, the maximum n quantum number allowed by our molecular integral program was 4. Because of this, and because the p and d basis functions were optimized for spin orbitals rather than for spinors, we expect that the overall quality of the basis is somewhat less than double zeta.

CALCULATIONS AND RESULTS

Using the above relativistic EP's and basis set, potential energy curves were computed for the lowest 0_g^+, 0_u^-, and 1_u states of Tl_2 which dissociate to ground state atoms. For the 0_u^- and 1_u states these were single configuration calculations. For the 0_g^+ state this requires two configurations, the difference being a double excitation from a $1/2_g$ to a $1/2_u$ spinor. For this case the coupling is the same as in Ref. 8, i.e., $K_{gu}^+ + K_{gu}^-$, where the K^+ and K^- refer to the exchange of electrons with angular momentum of the same and opposite signs, respectively. Plots of the dissociation curves for these three states are shown in Fig. 1. As can be seen, these curves show very little bonding character, and in fact the 1_u curve shows no minimum in this approximation.

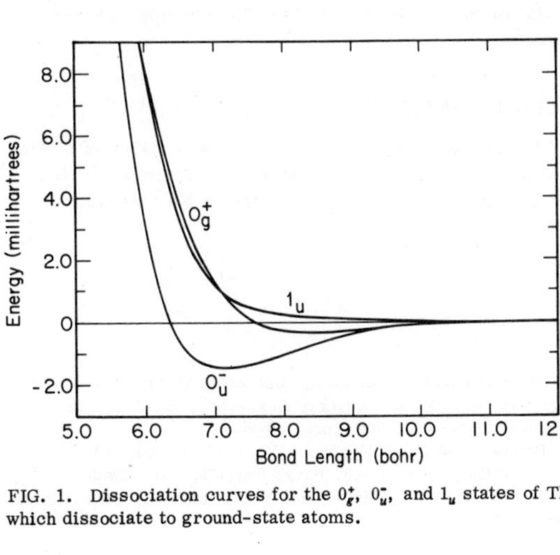

FIG. 1. Dissociation curves for the 0_g^+, 0_u^-, and 1_u states of Tl_2 which dissociate to ground-state atoms.

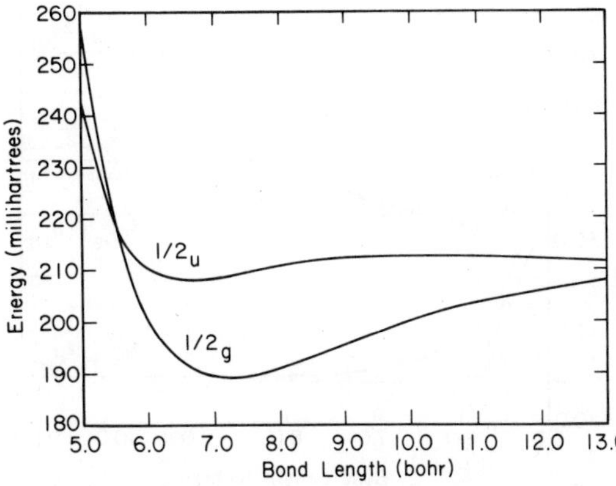

FIG. 2. Dissociation curves for the lowest $1/2_g$ and $1/2_u$ states of Tl_2^+.

Though there is no clear experimental evidence that the Tl_2 ground state is substantially bound, there is evidence that Tl_2^+ is stable.[6] For further comparison we have computed curves for the lowest $1/2_g$ and $1/2_u$ states of Tl_2^+ using single configuration wave functions. It should be noted that, although the single configuration functions dissociate to neutral and positively charged atoms, the core spinors on both are forced to be equivalent. However, calculations on Ar_2^+, where this problem should be more severe,[12] suggest that the resulting error is probably very small. Also, since no electron pairs are formed during bonding, we expect that correlation corrections will be relatively small. Plots of these two curves are given in Fig. 2. The D_e and R_e values are 0.58 eV and 3.84 Å for $1/2_g$ and 0.06 eV and 3.50 Å for $1/2_u$, respectively. In Table I we list the energies at various distances for the 0_g^+, 0_u^-, and 1_u states of Tl_2 as well as for the $1/2_g$ and $1/2_u$ states of Tl_2^+.

DISCUSSION

Further correlation corrections will certainly increase the well depths for the 0_g^+, 0_u^-, and 1_u states of Tl_2. How-

TABLE I. Dissociation curves for the lowest 0_g^+, 0_u^-, and 1_u states of Tl_2 and for the $1/2_g$ and $1/2_u$ states of Tl_2^+.[a]

	Tl_2			Tl_2^+	
R (bohr)	0_g^+	0_u^-	1_u	$1/2_g$	$1/2_u$
5.0	0.0349	0.0405	0.0478	0.2576	0.2420
5.5	0.0162	0.0132	0.0196	0.2186	0.2186
6.0	0.0081	0.0028	0.0079	0.2002	0.2102
6.5	0.0036	0.0007	0.0031	0.1921	0.2080
7.0	0.0012	−0.0015	0.0011	0.1892	0.2083
7.5	0.0001	−0.0014	0.0004	0.1892	0.2095
8.0	−0.0003	−0.0011	0.0002	0.1907	0.2106
8.5	−0.0004	−0.0007	0.0000	0.1928	0.2116
10.0	−0.0001	−0.0001	0.0001	0.2000	0.2126
15.0	(0.0)[a]	0.0000	0.0000	0.2097	0.2109

[a]All energies are in atomic units relative to the 0_g^+ energy at $R = 15.0$ bohr (−99.8202 a.u.).

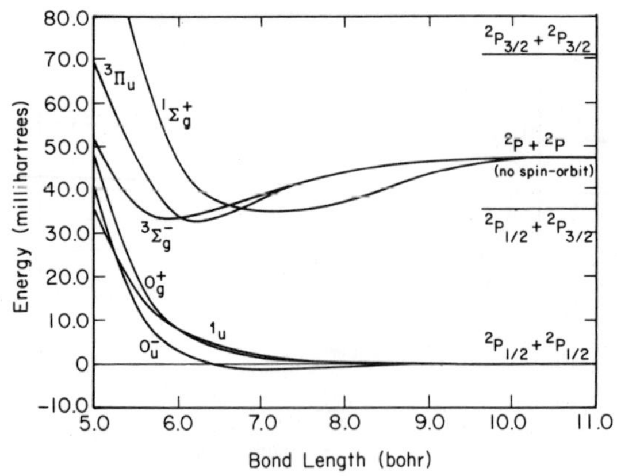

FIG. 3. Potential curves for the 0_g^+, 0_u^-, and 1_u states of Tl_2 with $^3\Sigma_g^-$, $^3\Pi_u$, and $^1\Sigma_g^+$ curves (computed without spin–orbit coupling) for comparison.

ever, because of the relatively large bond lengths, and because the molecular spinors are distorted only slightly from primitive combinations of atomic spinors, we expect the corrections will be fairly small. (Indeed, further calculations on the 0_g^+ state using an eight configuration MCSCF wave function lowered the potential minimum by only about 0.01 eV.) Nevertheless, the shallowness of the wells and the minimal separation between the three curves make detailed conclusions regarding the dissociation energies, the locations of the minima, and even the relative order of these three states impossible for the present. However, we can conclude that the bonding in Tl_2 is very weak.

This conclusion that the bond in Tl_2 is weak comes as no surprise considering our earlier discussion of bonding between $p_{1/2}$ spinors. The underlying cause is illustrated in Fig. 3, where we have plotted the potential curves for 0_g^+, 0_u^-, and 1_u states of Tl_2 along with curves for the corresponding $^1\Sigma_g^+$, $^3\Sigma_g^-$, and $^3\Pi_u$ terms without spin–orbit coupling. These latter curves were computed using the same EP's as before but with the individual $p_{1/2}$ and $p_{3/2}$, etc., components averaged to eliminate the spin–orbit effect; thus, other relativistic effects were retained. The $^1\Sigma_g^+$, $^3\Sigma_g^-$, and $^3\Pi_u$ curves dissociate to spin-averaged ^2P atoms. They have been placed at the proper energies with reference to the true $^2P_{1/2}$ and $^2P_{3/2}$ atoms and our calculated curves for the 0_u^-, 0_g^+, and 1_u molecular states. It is evident that in the absence of the spin–orbit effect, Tl_2 shows significant but rather weak bonding in any of the $^1\Sigma_g^+$, $^3\Sigma_g^-$, or $^3\Pi_u$ states. However, the potential minima for these states are well above the energy for a pair of $^2P_{1/2}$ atoms. Hence, the absence of strong bonding in Tl_2 appears to arise from the spin–orbit term and is a purely relativistic effect. Although the inclusion of the spin–orbit term also lowers the energy of the Tl_2 molecule, this is the result of the transition from molecular spin orbitals with considerable bonding nature to molecular spinors which closely approximate simple combinations of atomic $p_{1/2}$ spinors.

In the $^1\Sigma_g^+$ state the two valence electrons are sigma bonding, in the $^3\Sigma_g^-$ term both are pi bonding, while in the $^3\Pi_u$ term one is sigma and the other pi bonding. Correction for electron correlation will probably lower the singlet state more than the triplets; hence, sigma and pi bonding strengths appear to be about equal for the two-electron system. For Tl_2^+, however, sigma bonding is stronger as indicated by the lower energy of the $1/2_g$ state as compared to $1/2_u$. With the introduction of the spin–orbit term, both the $^3\Sigma_g^-$ and $^1\Sigma_g^+$ terms yield 0_g^+ states. The actual (lowest) 0_g^+ state shows very little departure from the $p_{1/2}$ atomic spinors. The $^3\Sigma_g^-$ term also includes a 1_g state which dissociates to one $^2P_{1/2}$ and one $^2P_{3/2}$ atom. We did not calculate a full potential curve for the 1_g state but a few single-configuration calculations at bonding distances indicated that its energy was well above that for 0_g^+. Similarly, the $^3\Pi_u$ term includes 0_u^+ and 2_u states (dissociating to one $^2P_{1/2}$ and one $^2P_{3/2}$ atom) as well as the 0_u^- and 1_u states. Exploratory calculations indicated that the 0_u^+ and 2_u states also had energies substantially above those for the 0_u^- and 1_u states. In view of the higher energies for their dissociated atoms, the 1_g, 0_u^+, and 2_u states may be more strongly bound with smaller R_e and larger D_e and ω_e values than for the lower energy states. The tentative R_e values are 3.02, 3.26, and 3.30 Å for the 1_g, 0_u^+, and 2_u states, respectively.

It is known that for the alkali metals the M_2^+ ions are more strongly bound than the neutral M_2 molecules. Thus, it is not too surprising that Tl_2^+ is more strongly bonded than Tl_2. This phenomenon has not been fully explained for the alkali metals, to the best of our knowledge, and the situation is more complex for thallium. We are investigating further this and other aspects of these states of Tl_2 and Tl_2^+; also, we hope to treat various excited states of Tl_2.

While we see possible assignments and therefore no conflict with the experimental spectra attributed to Tl_2, it is not worthwhile to suggest definite assignments until our calculations for excited states are completed. At that time we expect to discuss both the spectral interpretation and various aspects of the bonding in these species.

ACKNOWLEDGMENT

This research was supported by the Division of Chemical Sciences, Office of Basic Energy Science, U. S. Department of Energy under Contract No. W-7405-ENG-48.

[1]K. P. Huber and G. Herzberg, *Molecular Spectra and Molecular Structure IV. Constants of Diatomic Molecules* (Van Nostrand Reinhold, New York, 1979).

[2]J. Drowart and R. E. Honig, Phys. Chem. 61, 980 (1957).

[3]R. C. Miller and P. Kush, Phys. Rev. 99, 1314 (1955).

[4]D. S. Ginter, M. L. Ginter, and K. K. Innes, J. Phys. Chem. 69, 2480 (1965).

[5]E. D. Commins, P. Buckshaum, L. Hunter, and P. Drell, Phys. Rev. Lett. 42, 343 (1979); also Phys. Lett. A 60, 96 (1978).
[6]J. Berkowitz and T. A. Walter, J. Chem. Phys. 49, 1184 (1968).
[7]K. S. Pitzer, Acc. Chem. Res. 12, 271 (1979).
[8]P. A. Christiansen and K. S. Pitzer, J. Chem. Phys. 73, 5148 (1980).
[9]Y. S. Lee, W. C. Ermler, and K. S. Pitzer, J. Chem. Phys. 73, 360 (1980).
[10]P. A. Christiansen, Y. S. Lee, and K. S. Pitzer, J. Chem. Phys. 71, 4445 (1979).
[11]J. P. Desclaux, Comput. Phys. Commun. 9, 31 (1975).
[12]W. R. Wadt, J. Chem. Phys. 68, 402 (1978).

Dissociation energies of molecules with very heavy atoms from mass spectrometry

Kenneth S. Pitzer

Department of Chemistry and Lawrence Berkeley Laboratory, University of California, Berkeley, California 94720

(Received 4 November 1980; accepted 17 November 1980)

In a series of excellent mass-spectrometric investigations, Gingerich and associates have measured the dissociation energies of many molecules including heavy atoms. In the "third law" method of treating the data, the partition functions are required. Thus, the electronic partition function of the molecule is needed. For molecules with very heavy atoms, relativistic effects yield large spin–orbit splittings and in some cases this reduces the electronic partition function by a significant factor. This is illustrated below for the two cases Pb_2[1] and BiLi.[2]

In each of these cases lighter analogs Si_2 and SbH have $^3\Sigma^-$ ground states; hence, the assumption made[1,2] of an electronic multiplicity of three was reasonable, *prima facie*. However, for BiH it is known[3] that the $J=1$ component of $^3\Sigma^-$ is 4917 cm^{-1} higher in energy than the $J=0$ component. Thus, the case *c* or relativistic notation 1 and 0^+ is more appropriate and there is a single 0^+ ground state. Even at the temperature of experiment (1100 K), the population of the doubly degenerate 1 state is less than 1%. This spin–orbit splitting is essentially a property of the bismuth atom, and it should not be much different in BiLi. Thus, the electronic partition function should be reduced from three to one and the function $-(G_T-H_0^\circ)/T$ reduced by $R\ln 3$. The effect of this change is shown in Table I, which includes both the original and revised values of ΔH_0° of dissociation from the third law. In addition, the temperature dependence of the equilibrium constant yields a less precise value of ΔH at the temperature of the experiment and this can be converted to 0 K. This is called the second law result and is not affected by this change in molecular parameters. While the second law value for BiLi still differs from the more precise third law value by more than the apparent uncertainty, the deviation is reduced by 10 kJ mol^{-1} by the use of the correct electronic multiplicity.

In the case of Pb_2 there are recent spectroscopic data from Bondybey and English[4] which confirm the ground state as 0_g^+, a single state. While there is no explicit value for the energy of excitation to the 1_g state (the other component of $^3\Sigma_g^-$), it is expected to be comparable to the spin–orbit splitting ($^3P_1-^3P_0$) in the atomic spectrum of 7819 cm^{-1}. Thus, it is expected that the 1_g population at 1000 K will be negligible. For Pb_2 there is also a small revision[4] in the vibration frequency of 110 instead of 119 cm^{-1}. These two changes yield the results shown in Table I. The third law result for Pb_2 now agrees very well with the second law value and gives much greater assurance of the correctness of the new selected value of 83 ± 1 kJ mol^{-1} (0.86 ± 0.01 eV).

Since the vibration frequency and bond distance in BiLi were both estimated, there is further uncertainty in the third law result. I have no reason to change these estimates, but the additional uncertainty is recognized by the (?) in Table I. Similarly, large spin–orbit splittings may be expected for other molecules with very heavy atoms; thus, one should be cautious in adopting the electronic partition function of a lighter analog.

 0021-9606/81/053078-02$01.00

TABLE I. Revision of the dissociation energies of Pb_2 and BiLi.

	Pb_2	BiLi
Approximate temperature (K)	1000	1000
Original ΔH_0° of dissociation (third law) (kJ mol^{-1})	74 ± 1	135.7 ± 1.1
Revised ΔH_0° of dissociation (third law) (kJ mol^{-1})	82.5 ± 1	145.7 ± (?)
ΔH_0° of dissociation (second law) unchanged (kJ mol^{-1})	84 ± 2	160 + 7
Selected value (kJ mol^{-1})	83 ± 1	150 ± 5
Selected value (eV)	0.86 ± 0.01	1.55 ± 0.05

This research was supported by the Division of Chemical Sciences, Office of Basic Energy Science, U. S. Department of Energy under Contract No. W-7405-ENG-48.

[1]K. A. Gingerich, D. L. Cooke, and F. Miller, J. Chem. Phys. 64, 4027 (1976).

[2]A. Neubert, H. R. Ihle, and K. A. Gingerich, J. Chem. Phys. 73, 1406 (1980).

[3]K. P. Huber and G. Herzberg, *Molecular Spectra and Molecular Structure IV. Constants of Diatomic Molecules* (Van Nostrand Reinhold, New York, 1979).

[4]V. E. Bondybey and J. H. English, J. Chem. Phys. 67, 3405 (1977).

Volume 81, number 1 CHEMICAL PHYSICS LETTERS 1 July 1981

AB INITIO EFFECTIVE CORE POTENTIALS INCLUDING RELATIVISTIC EFFECTS. A PROCEDURE FOR THE INCLUSION OF SPIN–ORBIT COUPLING IN MOLECULAR WAVEFUNCTIONS

Walter C. ERMLER
Department of Chemistry and Chemical Engineering, Stevens Institute of Technology, Hoboken, New Jersey 07030, USA

Yoon S. LEE
James Franck Institute, Department of Chemistry, University of Chicago, Chicago, Illinois 60637, USA

and

Phillip A. CHRISTIANSEN and Kenneth S. PITZER
Department of Chemistry and Lawrence Berkeley Laboratory, University of California, Berkeley, California 94720, USA

Received 23 January 1981; in final form 20 March 1981

The first ab initio procedure for the treatment of spin–orbit coupling in molecules based on the use of relativistic effective potentials derived from Dirac–Fock atomic wavefunctions is presented. A rigorous definition for the spin–orbit operator is given and its use in molecular calculations discussed.

1. Introduction

The inclusion of relativistic effects in electronic structure calculations for molecules containing heavy elements has recently received a great deal of attention [1–18]. Inasmuch as relativistic corrections are predominantly core effects, the effective potential (EP) scheme [19] offers a particularly simple approach to the inclusion of relativistic effects with (at least in principle) remarkably little loss in accuracy [20]. Pitzer and co-workers [7–13] have developed EPs which are derived from atomic Dirac–Fock wavefunctions [21] and include explicitly the relativistic corrections as given by the Dirac formalism [22]. This EP procedure greatly reduces the difficulties associated with the large number of core electrons and at the same time eliminates the need to treat explicitly (at least in molecular calculations) the small components of the Dirac wavefunction. The relativistic and core effects (including the non-negligible two-electron contributions) appear in the form of a one-electron operator which is added to the Schrödinger equation for the valence electrons.

Although these methods have been successfully employed in the study of very heavy molecular systems, the explicit inclusion of the spin–orbit effects makes such calculations difficult. Furthermore, for the somewhat lighter elements such as Xe, Kr, etc., the spin–orbit term is probably small enough that it can be adequately treated as a perturbation following an initial SCF or CI calculation.

For the above reasons most workers [5–9, 14, 16–18] have opted to eliminate together the spin–orbit terms from the effective potential. This is done either by deriving the EPs using atomic states from which the spin–orbit splitting has been averaged out [16,18] or averaging the fully relativistic EPs themselves [7–9] to obtain a spin averaged relativistic effective potential (AREP). With AREPs molecular calculations can be carried out using standard non-relativistic formalisms. The spin–orbit correction is then added as a perturbation. In practice this has been done either by semi-empirically estimating the spin–orbit matrix elements which couple the various L–S type molecular states [7–9] or else the matrix elements are evaluated using some convenient spin–orbit operator [4] such as Z^{eff}/r^3 where the adjustable parameter Z^{eff} is chosen to reproduce the atomic splittings.

The above procedures, though possibly inadequate for the heaviest elements [13], provide a particularly convenient technique for treating molecules containing somewhat lighter atoms. Unfortunately, selecting matrix elements semi-empirically becomes difficult if not impossible if very many states are involved. Furthermore, the form of the operator, Z^{eff}/r^3, though justifiable in terms of atomic all-electron calculations, may be seriously inappropriate for calculations involving pseudo-orbitals whose behavior in the heavily weighted core region differs dramatically from that of the original atomic orbitals or spinors from which they were derived [19,20]. In fact, at first inspection it is not obvious how one would go about defining, ab initio, a spin–orbit operator for such applications. Thus, until now there has been no alternative to the semi-empirical procedures discussed above.

In section 2 we propose the first ab initio procedure by which a spin–orbit operator appropriate for use in molecular effective potential calculations may be rigorously determined and we discuss its use in molecular calculations.

2. Formal procedure

The effective potentials of Pitzer and co-workers [7–13] (and also of Hafner and Schwarz [15]) may be written in the general form

$$U^{\mathrm{REP}} = U_{LJ}^{\mathrm{REP}}(r) + \sum_{l=0}^{L} \sum_{j=|l-\frac{1}{2}|}^{l+\frac{1}{2}} [U_{lj}^{\mathrm{REP}}(r) - U_{LJ}^{\mathrm{REP}}(r)] \sum_{m=-j}^{j} |ljm\rangle\langle ljm|, \tag{1}$$

where the $U_{lj}^{\mathrm{REP}}(r)$ are the EPs derived from individual pseudospinors with angular quantum numbers l and j. For the "residual" potential, $U_{LJ}^{\mathrm{REP}}(r)$, L is ideally chosen to be at least one greater than the highest angular quantum number for the core electrons. The projection operators on the right ensure that the U_{lj}^{REP} operate only on spinors of the proper angular symmetry. Previous work has shown that, when the U_{lj} are properly defined in terms of all-electron atomic wavefunctions, the effective potential formalism is capable of reproducing molecular all-electron calculations to a high degree of reliability. Unfortunately, to date, calculations using the full potential of eq. (1) have been limited to single-configuration SCF [11] or relatively simple MC SCF [12,13] calculations. Whenever large-scale configuration interaction has been included, the spin–orbit effects implicit in U^{REP} were first averaged out resulting in an averaged relativistic effective potential, U^{AREP}, of the form

$$U^{\mathrm{AREP}} = U_{L}^{\mathrm{AREP}}(r) + \sum_{l=0}^{L} \sum_{m=-l}^{l} [U_{l}^{\mathrm{AREP}}(r) - U_{L}^{\mathrm{AREP}}(r)]\, |lm\rangle\langle lm|, \tag{2}$$

where the U_{l}^{AREP} are defined as

$$U_{l}^{\mathrm{AREP}}(r) = (2l+1)^{-1}\, [l U_{l,l-1/2}^{\mathrm{REP}}(r) + (l+1)\, U_{l,l+1/2}^{\mathrm{REP}}(r)], \tag{3}$$

and the projector on the right is now defined in terms of orbitals with the usual angular quantum numbers, l and m. The effective potential operator, U^{AREP}, therefore includes all relativistic effects except for spin–orbit. This potential can therefore be added to the usual non-relativistic hamiltonian and molecular calculations, including configuration interaction, carried out in the non-relativistic formalism. The U^{AREP} of Pitzer and co-workers [7–9] (or of Hafner and Schwarz [15]) is then roughly equivalent to the "relativistic" EPs of refs. [16,18] where the spin–orbit effects were averaged out prior to determining the EPs.

From the above discussion it is clear that the spin–orbit operator appropriate for use with molecular pseudo-orbitals can be defined simply as the difference of the U^{REP} and U^{AREP},

$$H^{\mathrm{so}} = U^{\mathrm{REP}} - U^{\mathrm{AREP}} = U_{LJ}^{\mathrm{REP}}(r) - U_{L}^{\mathrm{AREP}}(r) + \sum_{l=0}^{L} \sum_{j=|l-\frac{1}{2}|}^{l+\frac{1}{2}} \sum_{m=-j}^{j} [U_{lj}^{\mathrm{REP}}(r) - U_{l}^{\mathrm{AREP}}(r) - U_{LJ}^{\mathrm{REP}}(r) + U_{L}^{\mathrm{AREP}}(r)]\, |ljm\rangle\langle ljm|. \tag{4}$$

Note that the projection operators of eq. (2) have been expanded in terms of the corresponding operators for two-component spinors in order to combine terms with eq. (1). With increase in l the difference between the effective potentials for $j = l + \frac{1}{2}$ from that for $j = l - \frac{1}{2}$ decreases rapidly. Thus the difference $U_{LJ}^{\mathrm{REP}}(r) - U_{L}^{\mathrm{AREP}}(r)$ in most cases will be very small and can be neglected. If we also introduce the relationship of eq. (3), we obtain as an excellent approximation

$$H^{\mathrm{so}} = \sum_{l=1}^{L-1} \Delta U^{\mathrm{REP}}(r) \left[\frac{l}{2l+1} \sum_{-l-\frac{1}{2}}^{l+\frac{1}{2}} |l, l+\tfrac{1}{2}, m\rangle\langle l, l+\tfrac{1}{2}, m| - \frac{l+1}{2l+1} \sum_{-l+\frac{1}{2}}^{l-\frac{1}{2}} |l, l-\tfrac{1}{2}, m\rangle\langle l, l-\tfrac{1}{2}, m| \right], \tag{5}$$

with

$$\Delta U_{l}^{\mathrm{REP}}(r) = U_{l,l+1/2}^{\mathrm{REP}}(r) - U_{l,l-1/2}^{\mathrm{REP}}(r). \tag{6}$$

The matrix elements of H^{so} with respect to the atomic orbital basis set will have the form,

$$H_{pq}^{\mathrm{so}}(\rho_r \rho_s) = \langle \chi_p \rho_r | H^{\mathrm{so}} | \chi_q \rho_s \rangle, \tag{7}$$

where the χ are spatial basis functions and the Pauli spinors ρ define the α and β spins of the electron such that $\rho_i = \alpha = \binom{1}{0}$ or $\rho_i = \beta = \binom{0}{1}$. The matrix elements of H^{so} between the various L–S type states for a given molecule can then be obtained as an expansion over one-electron integrals employing the various CI, MO, and spin symmetry coefficients. An outline of this procedure follows.

The hamiltonian matrix element corresponding to the L–S states that correlate to the same M atomic asymptotes $n, m = 1, 2, \ldots M$ is defined by

$$H_{nm}^{\mathrm{so}} = \langle \Psi_n | H^{\mathrm{so}} | \Psi_m \rangle, \tag{8}$$

where Ψ_i is the CI wavefunction in L–S coupling. This is then expressed in terms of the sum of Slater determinants, D_i, with the appropriate CI coefficients C_I^n and individual Slater determinant coefficients a_i^I that define the spin eigenfunctions in L–S coupling,

$$H_{nm}^{\mathrm{so}} = \sum_I \sum_J C_I^n C_J^m \sum_i \sum_j a_i^I a_j^J \langle D_i | H^{\mathrm{so}} | D_j \rangle. \tag{9}$$

Substitution for the Slater determinants using the fact that H^{so} is a one-electron operator together with the properties of the antisymmetrizer gives

$$H_{nm}^{\mathrm{so}} = \sum_I \sum_J C_I^n C_J^m \sum_i \sum_j a_i^I a_j^J \sum_k^{2n_{\mathrm{p}}} \sum_l^{2n_{\mathrm{p}}} f_k^i f_l^i \langle \phi_k \rho_k | H^{\mathrm{so}} | \phi_l \rho_l \rangle, \tag{10}$$

where $\phi_k \rho_k$ is a one-electron molecular spin orbital (ρ_k is either α or β depending on the electron configuration), f_k^i is the occupation number of $\phi_k \rho_k$ in determinant D_i, and the sums on the one-electron spin orbitals go to $2n_{\mathrm{p}}$ to include both α and β spins. n_{p} is the number of spatial orbitals. The integrals in eq. (10) can be obtained by a straightforward transformation of the integrals from eq. (7) using the molecular orbital expansion coefficients.

Although (10) is formally identical to standard configuration-interaction property expansions, the limits for the k and l summations are now $2n_{\mathrm{p}}$ rather than n_{p} as is typical in CI programs. Furthermore, the integrals on the

Volume 81, number 1 CHEMICAL PHYSICS LETTERS 1 July 1981

right may be complex. Except for the now rigorous definition of the spin–orbit operator and our inclusion of two-center integrals, the matrix constructed from the above elements is of a form similar to that used by Hay et al. [6]. Diagonalization of $\mathbf{H}^{so}$ will yield the appropriate energies and coupling vectors for the desired manifold of spin–orbit states in the ω–ω coupling framework. It is also noted that this general development in terms of CI wavefunctions can be used to treat the special cases of MC SCF or SCF wavefunctions. Thus correlation and spin–orbit phenomena can be considered separately or simultaneously at varied levels of approximation.

3. Concluding remarks

Understanding the chemistry of molecules comprised of heavy atoms is inexorably bound to the proper inclusion of spin–orbit and other relativistic effects in the molecular wavefunctions. The operator proposed above is to our knowledge the only ab initio spin–orbit operator rigorously applicable to effective potential calculations. At the same time, being in the form of a one-electron operator, it circumvents the necessity for the costly large-scale calculations required when the full microscopic spin–orbit hamiltonian is retained. The present ab initio approach thus emerges as a promising alternative to the existing procedures.

Applications to the homonuclear inert-gas excimer systems are among the initial objectives owing to the widespread interest in our earlier, limited accuracy SCF calculations on the Xe_2^* states [7] in which one of the empirical approaches for the inclusion of spin–orbit coupling was employed. Calculations on molecules where the bonding interactions are significantly stronger are also planned to investigate the dependence of molecular spin–orbit coupling as a function of internuclear distance.

Acknowledgement

This research was partially supported by the Office of Energy Research, Office of Basic Energy Sciences, Chemical Sciences Division of the US Department of Energy through Contract W-7405-ENG-48. In addition, one of us (WCE) acknowledges the Research Corporation and the Donors of the Petroleum Research Fund, administered by the American Chemical Society, for partial support.

References

[1] K.S. Pitzer, Accounts Chem. Res. 12 (1979) 271.
[2] P. Pyykkö, Advan. Quantum Chem. 11 (1978) 353.
[3] G. Malli and J. Oreg, J. Chem. Phys. 63 (1975) 1032;
O. Matsuoka, N. Suzuki, T. Aoyama and G. Malli, J. Chem. Phys. 73 (1980) 1320;
T. Aoyama, H. Yamakawa and O. Matsuoka, J. Chem. Phys. 73 (1980) 1329.
[4] J.S. Cohen, W.R. Wadt and P.J. Hay, J. Chem. Phys. 71 (1979) 2955.
[5] W.R. Wadt and P.J. Hay, J. Am. Chem. Soc. 101 (1979) 5198.
[6] P.J. Hay, W.R. Wadt, L.R. Kahn, R.C. Raffenetti and D.H. Phillips, J. Chem. Phys. 71 (1979) 1767.
[7] W.C. Ermler, Y.S. Lee, K.S. Pitzer and N.W. Winter, J. Chem. Phys. 69 (1978) 976.
[8] Y.S. Lee, W.C. Ermler, K.S. Pitzer and A.D. McLean, J. Chem. Phys. 70 (1979) 288.
[9] W.C. Ermler, Y.S. Lee and K.S. Pitzer, J. Chem. Phys. 70 (1979) 293.
[10] Y.S. Lee, W.C. Ermler and K.S. Pitzer, J. Chem. Phys. 67 (1977) 5861.
[11] Y.S. Lee, W.C. Ermler and K.S. Pitzer, J. Chem. Phys. 73 (1980) 360.
[12] P.A. Christiansen and K.S. Pitzer, J. Chem. Phys. 73 (1980) 5160.
[13] P.A. Christiansen and K.S. Pitzer, J. Chem. Phys., to be published.
[14] G. Das and A.C. Wahl, J. Chem. Phys. 64 (1976) 4672; 69 (1978) 53.
[15] P. Hafner and W.H.E. Schwartz, J. Phys. B11 (1978) 217, 2975; Chem. Phys. Letters 65 (1979) 537.
[16] L.R. Kahn, P.J. Hay and R.D. Cowan, J. Chem. Phys. 68 (1978) 2386.

[17] S.N. Dutta, C.S. Ewig and J.R. van Wazer, Chem. Phys. Letters 57 (1978) 83.
[18] H. Basch and S. Topiol, J. Chem. Phys. 71 (1979) 802.
[19] L.R. Kahn, P. Baybutt and D.G. Truhlar, J. Chem. Phys. 65 (1976) 3826;
R.N. Dixon and I.L. Robertson, in: Theoretical chemistry, eds. R.N. D ixon and C. Thomson (Chem. Soc., London, 1978).
[20] P.A. Christiansen, Y.S. Lee and K.S. Pitzer, J. Chem. Phys. 71 (1979) 4445.
[21] J.P. Desclaux, Computer Phys. Commun. 9 (1975) 31.
[22] H.A. Bethe and E.E. Salpeter, Quantum mechanics of one- and two-electron atoms (Springer, Berlin, 1957).

Relativistic *ab initio* molecular structure calculations including configuration interaction with application to six states of TlH

Phillip A. Christiansen,[a] K. Balasubramanian, and Kenneth S. Pitzer

Department of Chemistry and Lawrence Berkeley Laboratory, University of California, Berkeley, California 94720

(Received 22 December 1981; accepted 4 February 1982)

A system for the inclusion of spin-orbit coupling along with moderate scale CI in calculations for molecules containing very heavy elements is demonstrated. In this effective potential procedure rigorous *ab initio* spin-orbit integrals are computed and added to the conventional integral set after the SCF and integral transformation steps of the calculation. This avoids the use of complex coefficients in the integral transformation and yet includes spin-orbit corrections on an equal footing with electron correlation. The diagonalization of the resulting complex CI plus SO matrix requires only about twice the time of a real CI diagonalization. Our present calculations on the two lowest 0^+ and 1 states and the lowest 0^- and 2 states of TlH indicate that this procedure allows adequate flexibility in the electronic coupling, resulting in bonding curves which are in good agreement with the experimentally established curves. The results also help to understand and to confirm previously conjectural interpretations of other spectral data.

INTRODUCTION

In a series of papers,[1-9] *ab initio* relativistic core potentials (REP) have been defined and applied to the calculation of the electronic structure of molecules which include one or more very heavy atoms. The exact method of defining the effective potentials (EP) and calculating them from atomic Hartree–Fock (HF) or Dirac–Fock (DF) results required considerable investigation,[10] but a set of essential principles was established and a detailed procedure was found which reproduces accurately the results from all-electron calculations[10,11] for molecules such as Cl_2, Kr_2, Kr_2^+, Xe_2, and Xe_2^+. All-electron LCAO relativistic calculations have not been[12] available for molecules containing very heavy atoms, but this analysis of pertinent principles confirmed by these examples gives us confidence in the validity of the method for even heavier atoms.

The spin-orbit (SO) effect is large for most very heavy atoms. Gold is an exception, where the only valence electron is in an s orbital, and Au_2 was treated[3,4] by methods not generally applicable to molecules containing such heavy atoms. Where valence-level p or d electrons are present, the SO energy is of a magnitude comparable to that of valence-level electron repulsion integrals, and the SO matrix elements should be included on an equal basis. One method is to set up the molecular wave function in terms of relativistic spinors, i.e., in $\omega-\omega$ (type c) coupling. Actually two-component spinors suffice; the two small components of four component Dirac spinors are negligible for valence orbitals in an effective potential treatment. Such calculations in $\omega-\omega$ coupling were made, first on a single configuration basis,[5] and then on a multiconfiguration (MCSCF) basis.[7,8] The results[7,8] for TlH, Tl_2^+, and Tl_2 on these bases appear to be reliable and as accurate as can be expected where only a few configurations are included. Thus, the bond distances are quite well determined, but the calculated dissociation energies are at best semiquantitative.

While it would be possible to develop high-order configuration interaction (CI) methods in terms of two-component spinors, an alternate procedure has been devised which is much easier to implement and is probably more desirable in other respects. This is based upon an *ab initio* spin-orbit operator proposed by Ermler *et al.*,[13] which is defined in terms of the difference between the REP for $j=l+1/2$ and $j=l-1/2$. This paper constitutes the first application of the new procedure to a molecule.

Many of the relativistic, molecular calculations which have been reported,[14-19] including the first results from this laboratory,[2-4] were obtained by procedures which omit the SO effect until the final step. The wave function is set up in $\Lambda-S$ (type a) coupling as it would be for a nonrelativistic calculation. The appropriate EP's are the averaged REP's (AREP's), i.e., the weighted average REP's for $j=l+1/2$ and $j=l-1/2$. Alternatively, some workers use atomic calculations in the Pauli approximation, but with the SO terms deleted and thus obtain approximate AREP's directly.[17] Familiar calculational procedures are then available to obtain molecular potential curves, with extensive CI, for the $\Lambda-S$ states without SO. If the SO effect is small, as it is for atoms of intermediate atomic number and may be in some other cases, it suffices to add the SO effect as a perturbation. A semiempirical method is commonly used for the SO contribution; each coefficient is evaluated from atomic spectral data.[16]

While some results obtained by the semiempirical method are probably quite accurate, the method becomes much more doubtful when the orbitals contributing large SO terms are intimately involved in bonding. Furthermore, we wished to maintain fully the *ab initio* basis of our method. With the recognition[13] that an *ab initio* SO operator can be derived from differences in the REP for $j=l+1/2$ and $j=l-1/2$, it became possible to assemble a complete calculational program which begins

[a]Present address: Department of Chemistry, Clarkson College of Technology, Potsdam, N.Y. 13676.

with molecular orbitals in $\Lambda-S$ coupling and which includes electron correlation via a high-order CI step. Indeed the SO terms are introduced at the CI step on an equal basis with the electron repulsion terms.

METHOD

The REP's are obtained from DF orbitals for the atoms by methods already well described.[1,7,10] These REP's are averaged and differenced with respect to spin as follows:

$$U_l^{\mathrm{AREP}} = (2l+1)^{-1}\left[lU_{l,l-1/2}^{\mathrm{REP}} + (l+1)R_{l,l+1/2}^{\mathrm{REP}}\right], \tag{1}$$

$$\Delta U_l^{\mathrm{REP}} = U_{l,l+1/2}^{\mathrm{REP}} - U_{l,l-1/2}^{\mathrm{REP}}. \tag{2}$$

The complete effective potential without SO is

$$U^{\mathrm{AREP}} = U_L^{\mathrm{AREP}} + \sum_{l=0}^{L-1}\sum_{m=-l}^{l}(U_l^{\mathrm{AREP}} - U_L^{\mathrm{AREP}})|lm\rangle\langle lm|, \tag{3}$$

with the final factor the usual projection operator. The potentials cease to depend appreciably on the angular quantum number as soon as it exceeds that represented in the core of the atom; thus L can be chosen on that basis. This EP is introduced into the Hamiltonian and applied to the wave function in a now familiar manner.

The wave function can be set up in Slater or Gaussian basis functions with real or complex angular factors. For linear molecules such as we consider in this paper, there would be considerable advantage in the use of the complex angular factors which are eigenfunctions of angular momentum. Nevertheless, we chose to use real Cartesian Slater-type functions with a view toward future calculations for nonlinear molecules (in which Cartesian Gaussian functions would be used).

The spin-orbit operator is just the difference between the complete, relativistic EP and the AREP. As shown in detail in Ref. 13, this may be simplified to

$$H^{\mathrm{SO}} = \sum_{l=1}^{L-1}\Delta U_l^{\mathrm{REP}}\left[\frac{l}{2l+1}\sum_{-l-1/2}^{l+1/2}|l,l+1/2,m\rangle\langle l,l+1/2,m| - \frac{l+1}{2l+1}\sum_{-l+1/2}^{l-1/2}|l,l-1/2,m\rangle\langle l,l-1/2,m|\right], \tag{4}$$

where the projection operators are now based on two-component relativistic spinors $|l,j,m\rangle$. The $\Delta U_l^{\mathrm{REP}}$ was defined in Eq. (2). The SO matrix elements for an atomic spin-orbital basis set will have the form

$$H_{pq}^{\mathrm{SO}}(\rho_r\rho_s) = \langle\chi_p\rho_r|H^{\mathrm{SO}}|\chi_q\rho_s\rangle, \tag{5}$$

where the χ are spatial basis functions and the Pauli spinors ρ define the α and β spins such that

$$\rho_i = \alpha = \binom{1}{0}, \quad \rho_i = \beta = \binom{0}{1}.$$

In this system, the molecular calculation starts in the usual way with a choice of basis functions and the evaluation of the various integrals including those involving the AREP and the SO. An SCF calculation in $\Lambda-S$ coupling is carried out and the resulting molecular orbitals may then be used to define the occupied and virtual space for a subsequent integral transformation and CI calculation. If, following the integral transformation, the spin-orbit integrals [Eq. (5)] are added into the normal CI matrix, and the matrix diagonalized, one is able to avoid the difficulties involved in a complex two-electron transformation, while treating the spin-orbit corrections on an equivalent level with electron correlation. Within this general plan, the CI matrix may be formulated in either $\Lambda-S$ or $\omega-\omega$ coupling by making the proper spatial and spin combinations. However, for convenience and because the final wave function is probably intermediate in coupling, in the present work we have defined the CI matrix in terms of primitive determinants formed directly from the molecular orbital basis. Within such a crude formalism, particular care must be taken in selecting determinants to ensure that the full range of coupling, from $\Lambda-S$ to $\omega-\omega$ is accounted for.

Given the integrals obtained from the transformation step, one can evaluate the elements of the CI matrix in the usual manner. One should note that, since this integral set does not include the spin-orbit integrals, many of the determinants will not couple to the dominant configuration. At this stage, the matrix could be written out as block diagonal, the different block, contributing to states, which in $\Lambda-S$ coupling, do not mix due to space or spin symmetry differences. The spin-orbit integrals are now added to the matrix. In general, the SO integral matrix may be complex leading to complex elements in the CI matrix. However, in the EP formalism the SO operator is a strictly one-electron operator. Therefore, the number of such integrals will be small. Furthermore, since a one electron operator can only couple determinants which differ by single promotions, the number of complex CI matrix elements will also be relatively small. This fact can be taken advantage of in the diagonalization algorithm.

Using conventional techniques (Givens or Jacobi) to diagonalize a complex Hermitian matrix is a nontrivial task both in terms of developing computer codes and in actual computational time relative to a real symmetric matrix. However, for very large matrices iterative techniques are typically employed. For algorithms such as the Nesbet procedure,[20] or Davidson procedure,[21] or the method of optimal relaxation (MOR),[22] the required modifications in code to allow for complex elements are trivial. Furthermore, due to the small number of such elements, the time required for diagonalization is only about a factor of 2 longer than that required for a comparable real matrix. (If one were to take advantage of the fact that for many cases the matrix elements are not complex but either real or imaginary, this factor could be reduced even further.)

The algorithm employed in the present work is a modification of the Davidson procedure. This scheme has the advantage that it is relatively stable even if the initial guess for the vector is poor. Furthermore, since the CI matrix is defined in terms of primitive determinants, in the initial stages of the diagonalization the Nesbet or MOR procedures may tend to mix symmetries and thereby prolong the iterative process, whereas the

TABLE I. Potential energy curves for several low lying states of TlH.

R	0^+ (I)	0^-	1 (I)	0^+ (II)	2	1 (II)
3.0	−0.0398	+0.0250	+0.0330	+0.0272	+0.0477	+0.0736
3.25	−0.0564	···	···	0.0155	···	···
3.5	−0.0645	0.0093	0.0166	0.0115	0.0327	0.0526
3.75	−0.0667	0.0089	0.0154	0.0116	0.0329	0.0482
4.0	−0.0648	0.0100	0.0153	0.0133	0.0344	0.0457
4.5	−0.0541	···	···	0.0167	···	···
5.0	−0.0398	0.0127	0.0115	0.0183	0.0385	0.0410
5.5	−0.0257	···	···	0.0195	···	0.0401
6.0	−0.137	0.0126	0.0070	0.0214	0.0394	0.0399
7.0	···	···	0.0048	···	0.0402	0.0402
∞	0.0	0.0	0.0	0.0339	0.0339	0.0339

Davidson technique will preserve the symmetry of the starting vector to within round-off error.

CALCULATIONS AND RESULTS

The effective potentials used in the present work are identical to those employed in Refs. 7 and 8. Thus, on thallium we treated explicitly the outer 13 electrons ($10d$, $2s$, and $1p$), the frozen core comprising the remainder. The basis set for hydrogen was composed of two $1s$ Slater type orbitals (STO) and a single set of $2p$ STO for polarization. The exponents were taken from the AlH calculations of Cade and Huo.[23] For thallium, we optimized a $(2s, 2p, 2d)$ STO basis set using the ground state of the isolated atom. The optimization was carried out in $L-S$ coupling using the spin-averaged potentials. The values of the n quantum numbers for the s, p, and d basis functions were respectively 6, 6, and 5. This is a substantially better basis than was used in the previous Tl_2[13] and TlH[12] calculations, where the maximum n value was 4. In subsequent molecular SCF calculations, we found that using a triple zeta basis insignificantly altered the shape of the bonding curve for the ground state.

All calculations were carried out using real STO's. Although this complicates the CI diagonalization by forcing some of the SO integrals to be imaginary, as stated earlier these problems cannot be easily avoided in polyatomic calculations. Also most available SCF and integral transformation programs are not compatible with complex basis functions.

Using the SO averaged EP's and the above basis set, molecular single configuration SCF calculations were carried out for the singlet sigma ground state of TlH. We then generated SO integrals and transformed the atomic orbital (AO) basis set integrals (including SO) into MO form. Since our previous MCSCF calculations suggested that any significant effects on the thallium d shell were already included at the SCF level, that shell was frozen at the integral transformation stage of the calculation. Thus, the CI calculation included explicitly only the outer three electrons of thallium and the $1s$ hydrogen electron.

Our CI wave function for the 0^+ state was generated from seven reference configurations with occupations (ignoring $1\sigma^2$), σ^2, $\sigma\pi_x\alpha\alpha$, $\sigma\pi_x\beta\beta$, $\sigma\pi_y\alpha\alpha$, $\sigma\pi_y\beta\beta$, $\pi_x\pi_y\alpha\beta$, and $\pi_x\pi_y\beta\alpha$. All normal single and double promotions were allowed from the first five references. In view of the small magnitude of the CI coefficients for the sixth and seventh references, single and double promotions were allowed from the π orbitals only. This results in a total of approximately 1700 determinants. These seven references are required to allow the wave function the flexibility of intermediate coupling. The wave function formed in this manner will not give a fully balanced description of the separated atoms relative to the molecule; hence, we have not attempted to compute a bond energy from this wave function alone. However, we have computed the energy for the separated atoms for comparison. For thallium we used a CI wave function generated using all single and double promotions from the three references $6s^2 6p_\sigma\alpha$, $6s^2 6p_x\beta$, and $6s^2 6p_y\beta$. For the $^2P_{1/2}$ state, we obtained the total energy −50.6827 a.u. The $^2P_{3/2}$ state was higher in energy by 0.0339 a.u. or 0.92 eV, which is in reasonable agreement with the experimental splitting of 0.97 eV.[24]

For the 0^- state, the first reference listed above (for the 0^+ state) is eliminated and certain sign relationships between the other terms are reversed. Also a $\sigma\sigma'\alpha\beta$ reference must be added for proper dissociation. Similar methods yield the appropriate references for the 1 and 2 states. From a $\Lambda-S$ coupling basis, the $^1\Sigma^+$ state related to the lowest 0^+ state and the $^3\Pi$ term is split to yield the second 0^+ state and the lowest 0^-, 1, and 2 states. The $^1\Pi$ term yields the second 1 state, while the highly repulsive $^3\Sigma^+$ term yields the second 0^- and third 1 states.

The calculated energies, relative to ground-state atoms, are listed in Table I and plotted in Fig. 1. Included are results for the first excited states of 0^+ and 1 symmetry. These states are related to the $^3\Pi$ and $^1\Pi$ terms and should be reasonably well described by the basis of these calculations.

The experimental evidence for TlH was discussed by Ginter and Battino,[25] whose potential curves for the two 0^+ states are compared in Fig. 2 with our calculations. Other data and references are summarized by Huber and Herzberg.[26] Calculated and experimental spectroscopic constants are given in Table II. As noted by Ginter and Battino there is some uncertainty in the absolute energy scale, i.e., in the relationship of the atomic energies to the molecular curves in the mea-

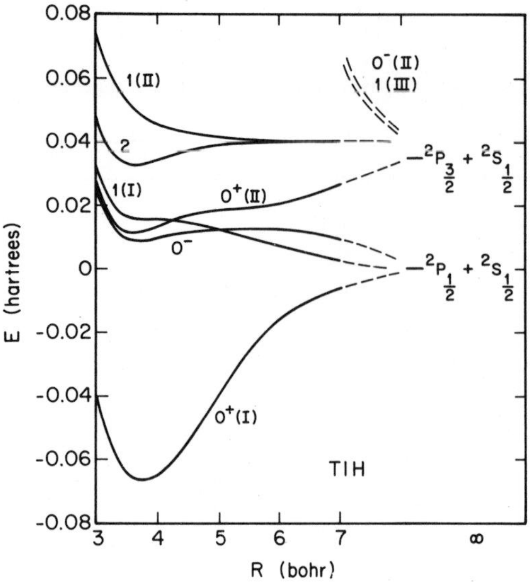

FIG. 1. Computed bonding curves for the two lowest 0^+ and 1 states and the lowest 0^- and 2 states of TlH. The dashed curves at large R are estimates.

sured region. The close agreement of our calculated curve for 0^+(I) with the experimental curve in the range 4.5–6 bohr supports the present interrelationship of atomic and molecular energies.

Our calculated potential curve for the ground state is somewhat too high at short interatomic distances. We believe that the cause is the absence of intershell correlation involving thallium d-shell electrons together with valence-shell electrons. Expansion of the CI to include all d-shell excitations of this type would exceed the capacity of our present program. Also, to properly include these effects, one would have to expand the basis by the addition of f orbitals. The very recent nonrelativistic calculations of McLean[27] for AgH with very extensive CI lend support to this view. He finds about 0.2 bohr shortening of R_e from that for a MCSCF calculation to the values for any of a number of calculations with high order CI including these intershell correlation terms. McLean also reports similar but less extensive results for AuH. It is clear that our

TABLE II. Spectroscopic constants for some low lying bound states of TlH.

State	R_e (Å)	D_e (eV)	ω_e(cm^{-1})	T_e(cm^{-1})
0^+ (I) Theory	1.99	1.81	1300	0
0^+ (I) Experiment	1.87	1.97	1391	0
0^- Theory	1.95	0.24	795	16 600
0^+ (II) Theory	1.91	0.61	~1000	17 100
0^+ (II) Experiment	1.91	0.74	~760	17 723
2 Theory	1.90	0.035	740	21 800
1 (II) Theory	3.1	···	~200	23 400
1 (II) Experiment	2.9	···	~140	24 180

wave function for TlH is somewhat deficient at these short interatomic distances, but further work will be required to remedy this situation. For distances greater than about 4.5 bohr, where d-electron effects on the potential curve should be negligible, the agreement is excellent and confirms our general procedures.

The wave functions for the two 0^+ states, as expected, are dominated by singlet sigma and triplet pi character. As noted in our earlier work,[7] in the bonding region the molecular ground state is essentially singlet sigma. However, at very large distances the triplet pi slightly dominates. The reverse is true for the excited state; at shorter distances the wave function is heavily dominated by triplet pi character, with the singlet sigma slightly dominating at very large distances. This interchange of sigma and pi character is apparently responsible for the peculiar behavior of the excited state around 5 to 7 bohr. Figure 2 shows the striking agreement of the shapes of the calculated and experimental curves for this 0^+(II) state.

With this substantial confirmation of our calculations for the two 0^+ states where the experimental evidence is unambiguous, it is interesting to consider our predictions for the 0^-, 1, and 2 states in relationship to the minimal experimental data for these states for TlH and in comparison with the data for InH where the spin-orbit splitting is much smaller but still significant. First, we note that the inner well at about 3.5 bohr in the 0^+(II), 0^-, 1(I), and 2 states appears to be at least partially the result of an avoided crossing which has

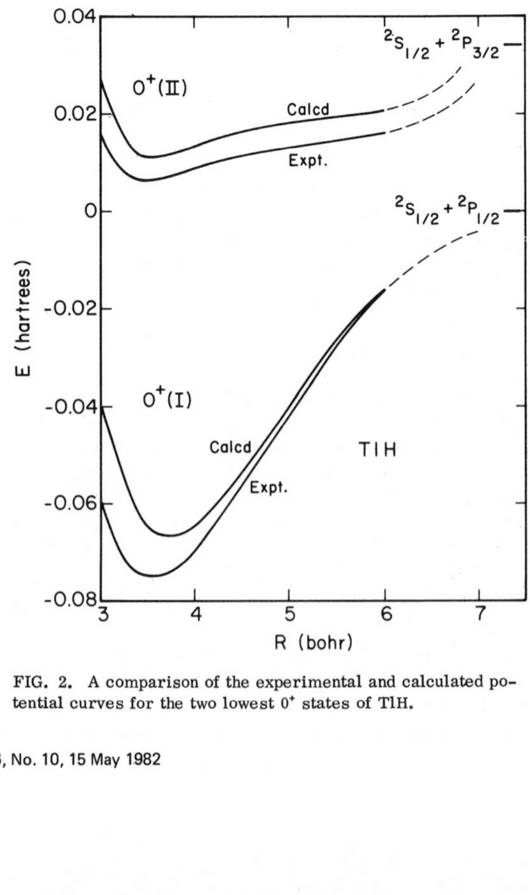

FIG. 2. A comparison of the experimental and calculated potential curves for the two lowest 0^+ states of TlH.

been previously observed for the lowest $^3\Pi$ state of BH.[28] In the region outside the inner well, the wave function is dominated by configurations which correspond roughly to the s^2p isolated thallium atom. However, in the region of the inner well, there is considerable sp^2 character, thereby allowing substantial sigma bonding of H with the s orbital on thallium. As noted above, this unusual shape of the excited 0^+ state agrees very well with the experimentally known potential.[25]

The inner portion of the potential curves for the 0^-, 0^+(II), 1(I), and 2 states are all very similar, hence their relationship to the $^3\Pi$ state in $\Lambda-S$ coupling is pertinent. This is confirmed by an examination of the wave functions which are dominantly $^3\Pi$ in the range 3.0–3.5 bohr. The spin-orbit energies simply shift the absolute energies in this region, and the pattern is similar to that found for InH, where the order is the same and the spacings also increase in the same sequence $(0^+-0^-)<(1-0^+)<(2-1)$. But the very large spin-orbit separation of the atomic energies for Tl has a profound effect at larger R. The curve for the 1(I) state has no significant minimum; this agrees with the failure to observe discrete spectra for this state in TlH (in contrast to InH where it is observed).

Selection rules make direct observation of the 0^- state difficult, and it has not been measured for any of the molecules GaH, InH, or TlH, although there are reasonable estimates[26] for the first two. We predict a low but significant hump in the 0^- potential near 5 bohr. The relative shapes of the 0^- and 1(I) curves in the vicinity of 6 bohr can be understood from the details of the wave functions. At long distance, the Tl atom must approach a $s^2p_{1/2}$ configuration where the $p_{1/2}$ spinor is $1/3\ p_\sigma$ and $2/3\ p_\pi$. For the 1 state, the s orbital on H can immediately have a bonding interaction with the p_σ on Tl, whereas this is not possible for the 0^- state. Thus, the initial interaction of the atoms is more repulsive in the 0^- state than in the 1 state.

There are several spectral lines observed by Larsson and Neuhaus[29] for TlH and TlD, which have been interpreted as arising from transitions from the ground state to the 2 and 1(II) states. They conclude that their "explanations are largely conjectural" and that further experiments are needed. There is little doubt that these lines lie close to the dissociation limit to $^2P_{3/2}$ and $^2S_{1/2}$ atoms; indeed this interpretation is the basis for the interrelationship of atomic and molecular energies presently accepted by various authors.[26] Larsson and Neuhaus find for the 1(II) state the remarkably low and anharmonic sequence of vibrational spacings of 98 and 56 cm^{-1}, with $\omega_e = 140$ cm^{-1} and an R_e value about 2.9 Å or 5.5 bohr. Our calculated curve for the 1(II) state shows a nearly flat region from 5 to 7 bohr; a cubic equation through the four points in this region yields the results given in Table II, $\omega_e \cong 200$ cm^{-1} and $R_e \cong 3.1$ Å. Thus, the agreement is remarkably good for such a sensitive feature in an excited state.

Larsson and Neuhaus also assign a few lines to transitions $2 \leftarrow 0^+$(I) appearing in violation of selection rules because of a perturbation with the 1(II) state. Indeed our calculations yield a crossing of the 2 and 1(II) curves at 7 bohr. However, the curves are so flat that a very small shift in their relative energy would cause a large change in the R value of the crossing.

In their second paper, Larsson and Neuhaus also propose a crossing between the curves for, in their terms, the $^1\Pi$ and $^3\Pi_1$ states. This is unacceptable since both of these states have the same symmetry in type c coupling, i.e., they are the 1(I) and 1(II) states.

Our calculations indicate low potential maxima at large distances for the 2 and 1(II) states. The exact heights of these maxima are rather uncertain, however, because the electron correlation included in our atomic calculations is not exactly equivalent to that in the molecular calculations at large R values. The calculations of Blint and Goddard for BH also show maxima at long distances for both the $^3\Pi$ and $^1\Pi$ states which lends some support for the existence of similar maxima in TlH.

Thus, our calculated results are fully consistent with the experimental measurements, if allowance is made for uncertainties in accuracy, and tend to confirm some previously speculative aspects of interpretation while suggesting revision of other aspects.

CONCLUSION

We have demonstrated a method by which spin-orbit and electron-correlation effects are simultaneously introduced into the wave function for molecules containing very heavy atoms. This is the first calculation of this type that we are aware of. As opposed to most earlier calculations in which the spin-orbit corrections were added semiempirically, in the present work we have employed the rigorously defined (within the effective potential approximation) spin-orbit operator of Ref. 13. Our work demonstrates that the use of imaginary matrix elements (as may be required in future polyatomic calculations) introduces only minor complications into the configuration interaction calculation.

ACKNOWLEDGMENT

This work was supported by the Director, Office of Energy Research, Office of Basic Energy Sciences, Chemical Sciences Division of the U.S. Department of Energy under Contract No. W-7405-ENG-48.

[1]Y. S. Lee, W. C. Ermler, and K. S. Pitzer, J. Chem. Phys. 67, 5861 (1977).
[2]W. C. Ermler, Y. S. Lee, K. S. Pitzer, and N. W. Winter, J. Chem. Phys. 69, 976 (1978).
[3]Y. S. Lee, W. C. Ermler, K. S. Pitzer, and A. D. McLean, J. Chem. Phys. 70, 288 (1979).
[4]W. C. Ermler, Y. S. Lee, and K. S. Pitzer, J. Chem. Phys. 70, 293 (1979).
[5]Y. S. Lee, W. C. Ermler, and K. S. Pitzer, J. Chem. Phys. 73, 360 (1980).
[6]K. S. Pitzer, Acc. Chem. Res. 12, 271 (1979).
[7]P. A. Christiansen and K. S. Pitzer, J. Chem. Phys. 73, 5160 (1980).
[8]P. A. Christiansen and K. S. Pitzer, J. Chem. Phys. 74, 1162 (1981); see also P. Hafner, P. Habitz, Y. Ishikawa, E. Wechsel-Trakowski, and W. H. E. Schwarz, Chem. Phys. Lett. 80, 311 (1981).

[9]K. S. Pitzer and P. A. Christiansen, Chem. Phys. Lett. 77, 589 (1981).
[10]P. A. Christiansen, Y. S. Lee, and K. S. Pitzer, J. Chem. Phys. 71, 4445 (1979).
[11]P. A. Christiansen, K. S. Pitzer, Y. S. Lee, J. H. Yates, W. C. Ermler, and N. W. Winter, J. Chem. Phys. 75, 5410 (1981).
[12]An all-electron relativistic SCF calculation for AuH has just been completed: Y. S. Lee and A. D. McLean, J. Chem. Phys. 76, 735 (1982).
[13]W. C. Ermler, Y. S. Lee, P. A. Christiansen, and K. S. Pitzer, Chem. Phys. Lett. 81, 70 (1981); see also P. Hafner and W. H. E. Schwarz, Chem. Phys. Lett. 65, 537 (1979).
[14]G. Das and A. C. Wahl, J. Chem. Phys. 64, 4672 (1971); 69, 53 (1978).
[15]J. S. Cohen, W. R. Wadt, and P. J. Hay, J. Chem. Phys. 71, 2955 (1979).
[16]P. J. Hay, W. R. Wadt, L. R. Kahn, R. C. Raffenetti, and D. H. Phillips, J. Chem. Phys. 71, 1767 (1979).
[17]L. R. Kahn, P. J. Hay, and R. D. Cowan, J. Chem. Phys. 68, 2386 (1978).
[18]S. N. Datta, C. S. Ewig, and J. R. Van Wazer, Chem. Phys. Lett. 57, 83 (1978).
[19]H. Basch and S. Topiol, J. Chem. Phys. 71, 802 (1979).
[20]R. K. Nesbet, J. Chem. Phys. 43, 311 (1965).
[21]E. R. Davidson, J. Comp. Phys. 17, 87 (1975).
[22]I. Shavitt, C. F. Bender, A. Pipano, and R. P. Hasteny, J. Comp. Phys. 11, 90 (1973).
[23]P. Code and W. Huo, J. Chem. Phys. 47, 649 (1967).
[24]C. E. Moore, *Atomic Energy Levels*, Natl. Bur. Stand. Circ. 467, Vols. I–III.
[25]M. L. Ginter and R. Battino, J. Chem. Phys. 42, 3222 (1965).
[26]K. P. Huber and G. Hertzberg, *Molecular Spectra and Molecular Structure, IV, Constants of Diatomic Molecules* (Van Nostrand, New York, 1979).
[27]A. D. McLean (private communication).
[28]R. J. Blint and W. A. Goddard III, Chem. Phys. 3, 297 (1974).
[29]T. Larsson and H. Neuhaus, Ark. Fys. 23, 461 (1963); 31, 299 (1966).

D. Other Papers on Molecules

APRIL, 1939 JOURNAL OF CHEMICAL PHYSICS VOLUME 7

The Symmetry Number and Thermodynamic Functions for Molecules Having Double Minimum Vibrations

KENNETH S. PITZER
Chemical Laboratory, University of California, Berkeley, California

(Received January 12, 1939)

The energy levels and thermodynamic functions for a double minimum vibrational degree of freedom are presented, and their application illustrated for the case of ammonia. A simple formula is given for the range of validity of the classical expressions for rotational heat capacity and entropy. Also the symmetry number, which is troublesome in examples of double minimum vibration, is discussed briefly.

IN CALCULATING thermodynamic functions for gases through the use of molecular structure data it is usual to classify the various degrees of freedom as translational, rotational or vibrational in nature. Furthermore, in approximate treatments one assumes the rotation to be that of a rigid body and the vibration to be harmonic. In treating ethane it was necessary to consider also a restricted rotational degree of freedom, i.e., a system whose motion at low energies is vibration about one or more positions of minimum potential energy, while at higher energies becomes perturbed rotation. The writer has, in fact, prepared tables for the contribution of such a degree of freedom to the various thermodynamic functions and has made several applications.[1]

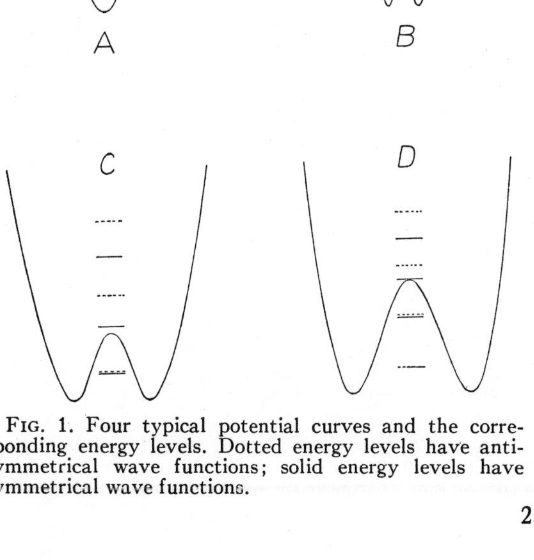

FIG. 1. Four typical potential curves and the corresponding energy levels. Dotted energy levels have antisymmetrical wave functions; solid energy levels have symmetrical wave functions.

Another type of motion which appears in some molecules is a double minimum vibration. The best known example of this motion is found in ammonia, where, although the molecule tends to be pyramidal, the nitrogen can pass through the plane of the three hydrogens without too great an increase in potential energy. In this paper the thermodynamic functions associated with such a motion are discussed. Since the presence of this type of motion is likely to lead to complications in the symmetry number, this aspect is considered briefly. Also a simple expression is set up for a temperature, above which the classical calculations of rotational entropy and heat capacity are satisfactory.

ENERGY LEVELS FOR THE DOUBLE MINIMUM VIBRATOR

The quantum mechanical problem of the symmetrical double minimum vibration has been discussed by a number of authors, who have made various assumptions for the mathematical expression of the potential. The treatment of Manning[2] (who gives references to earlier work) is fundamentally by far the most satisfactory, its only weakness being the fact that the energy levels appear as the roots of a relatively complicated continued fraction. The simple treatment of Wall and Glockler[3] is useful for cases where the central potential barrier is high, while

[1] K. S. Pitzer, J. Chem. Phys. 5, 469, 473 (1937).
[2] M. F. Manning, J. Chem. Phys. 3, 136 (1935).
[3] F. T. Wall and G. Glockler, J. Chem. Phys. 5, 314 (1937).

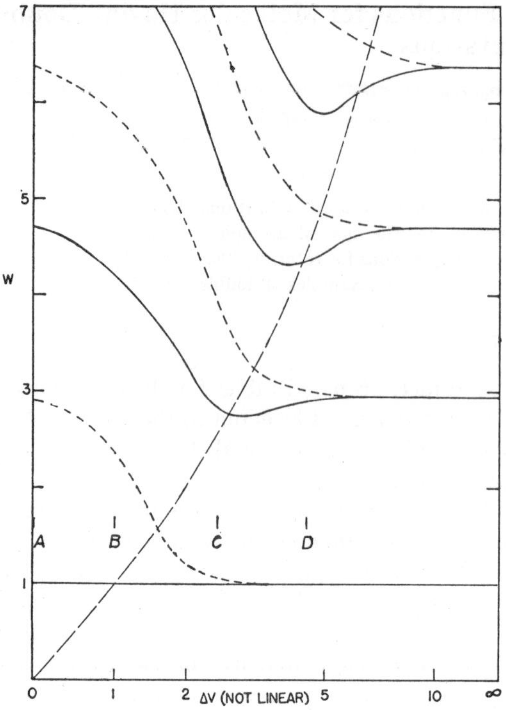

FIG. 2. The energy levels of the double minimum oscillator as a function of the height of intervening potential barrier, DV (dashed line rising to the right). The symmetries of the corresponding wave functions are indicated as in Fig. 1 (dotted=antisym.; solid=sym.). Energy scale is purely for comparison with later figures. A, B, C and D locate the examples in Fig. 1.

for low barriers the writer made a simple calculation in which an error function was added as a perturbation to the usual parabolic potential energy curve of the harmonic oscillator.

Figure 1 presents several potential curves and the resulting energy levels, while Fig. 2 shows how the energy levels behave when the central potential peak is gradually raised. On the right side where the separating potential hill is very large, each of the lower energy levels are doubly degenerate and the corresponding wave functions may be described as symmetrical and antisymmetrical linear combinations of the wave functions for vibration about each minimum. (If, as in optical isomers, the two positions of minimum potential are physically distinguishable, then wave functions representing vibration about only one minimum will have significance.) In the middle of Fig. 2 the pairs of energy levels have separated, the lower one in each case corresponding to the symmetrical wave function. These symmetries are, of course, maintained to the left side where we have the ordinary single minimum oscillator. Because of the arbitrary nature of the assumed potential functions and the lack of immediate applications other than ammonia, these energy levels have been calculated only approximately and are presented only in the graphs.

Thermodynamic Functions for the Double Minimum Vibrator

The heat capacity and entropy were calculated for the four examples of Fig. 1 as a function of temperature and are shown in Fig. 3. It is noteworthy that at the higher temperatures the heat capacities of "A" and "D" are the same, but that the latter has a peak at very low temperatures. This peak gives "D" an entropy $R \ln 2$ greater than "A" at ordinary temperatures, a fact which will be associated below with the different symmetry numbers ordinarily assigned the corre-

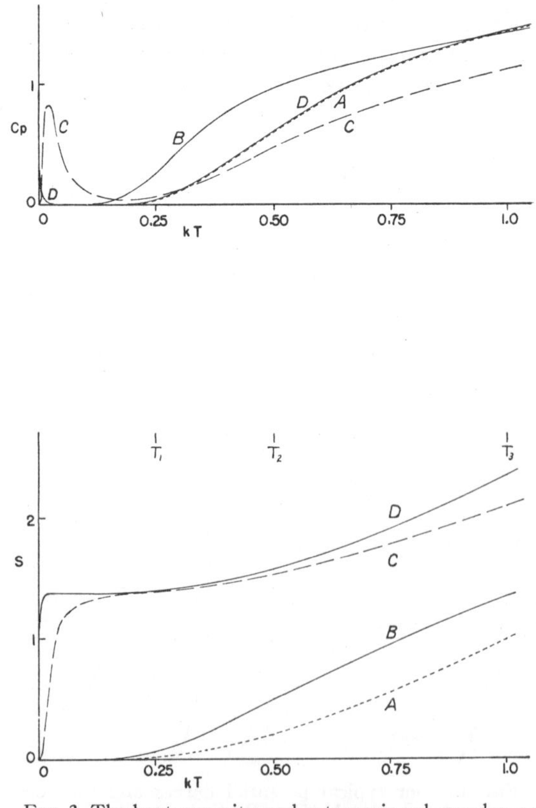

FIG. 3. The heat capacity and entropy in cal. per deg. as a function of temperature for the four systems in Fig. 1. The units of kT are the same as of W in Fig. 2.

sponding molecules. The heat capacity and entropy curves for "B" and "C" differ from those of harmonic oscillators ("A" or "D"), however, the entropies always increase in the order "A," "B," "C," "D."

Figure 4 shows the entropy for three temperatures as a function of the central potential peak (same scale as in Fig. 2). It is apparent that the entropy increases gradually and continuously from left to right, the difference between the two extremes being $R \ln 2$.

Energy Levels and Thermodynamic Functions for the Entire Molecule

Thus far we have considered the isolated double minimum vibrational degree of freedom; now we must fit these results into the picture of a whole molecule. Two general methods will be considered: first, the exact treatment, setting up the complete wave functions of acceptable symmetry; and second, the modified classical treatment of rotation using the symmetry number.[4]

Taking ammonia as an example, one finds that Hund[5] and Wilson[6] have worked out the various acceptable products of the spin, vibration, rotation, translation and electronic wave functions. There are eight spin functions (for the hydrogens) which fall into two groups of four each, designated by Wilson[6] as $4A$ and $2E$ (the symbol E representing a double degeneracy). Fig. 5 shows first the rotational energy levels of the symmetrical top using the moments of inertia for ammonia. The second and third columns show the acceptable energy levels for the $4A$ and $2E$ groups of spin functions. The dotted lines represent states in which the double minimum vibration wave function is antisymmetric, and which have a slightly higher energy as shown. The usual quantum numbers are indicated on the left side of the diagram and the degeneracies on the right side of each column. The calculation of thermodynamic functions is straight forward, although possibly laborious, once these energy level data are available.[7]

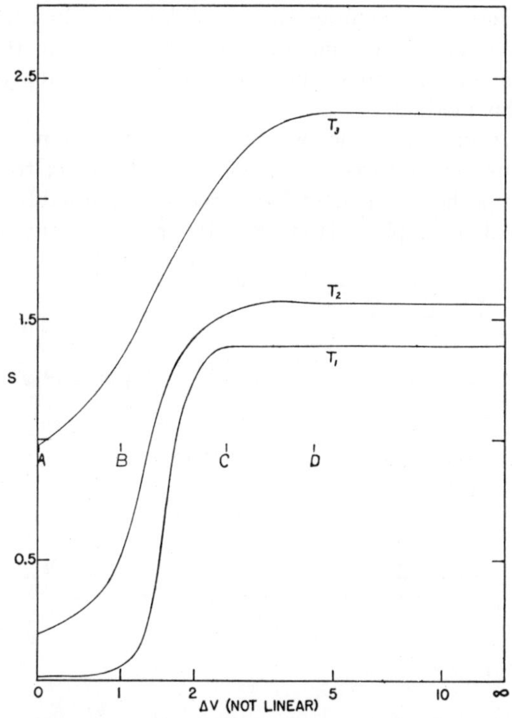

FIG. 4. The entropy of a double minimum oscillator as a function of the intervening potential barrier for the three temperatures indicated on Fig. 3. The abscissa scale is the same as Fig. 2.

At sufficiently high temperatures the effects due to the finite separation of energy levels disappear and the treatment can be correspondingly simplified. The temperature at which this simplification becomes possible varies with the type of motion, and as a result hybrid treatments are usually used. Specifically, translation can, almost without exception, be treated classically (with the usual modification of weight $1/h$ for a unit volume of phase space). Vibration, on the contrary must usually be treated by quantum methods at all ordinary temperatures. It is in the case of rotation that both methods are commonly used. An example of a detailed quantum treatment has been given above, which will now be compared with the classical method.

The formulas for rotational entropy are readily available and need not be repeated here.[8] The rotational heat capacity in the classical treatment is, of course, $\frac{1}{2}R$ per degree of freedom.

[4] It is assumed that the motion in this degree of freedom can be separated from all others. This will not be the case for a triatomic, almost linear molecule, in which one moment of inertia goes to zero as the molecule vibrates through the linear configuration.

[5] F. Hund, Zeits. f. Physik **43**, 788, 805 (1927).

[6] E. B. Wilson, J. Chem. Phys. **3**, 276 (1935).

[7] W. F. Giauque, J. Am. Chem. Soc. **52**, 4808 (1930).

[8] (a) L. S. Kassel, J. Am. Chem. Soc. **55**, 1351 (1933); Chem. Rev. **18**, 277 (1936); (b) J. E. Mayer, S. Brunauer and M. G. Mayer, J. Am. Chem. Soc. **55**, 37 (1933).

The two difficulties that arise have to do with the symmetry number,[8(b)] which appears in the entropy formula, and the range of validity of this method.

Ordinarily one would assign a symmetry number of three to pyramidal NH_3 or six to a hypothetical planar NH_3 (SO_3 is presumably a real example). However, if the pyramid is

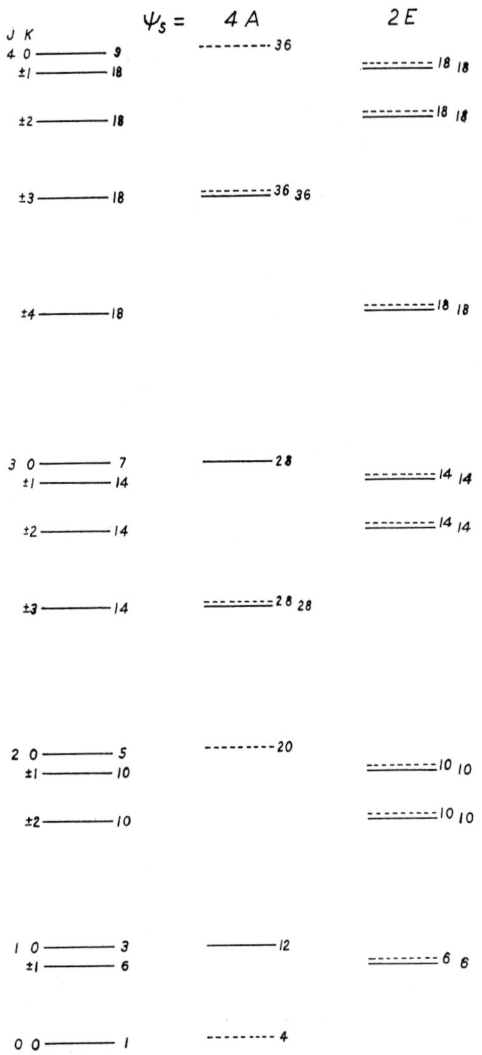

FIG. 5. The rotational energy levels for ammonia. Left column, energy levels for the symmetrical top with moments of inertia for ammonia; middle column, actual energy levels for the four nuclear spin wave functions of symmetry "A;" right column, same for two pairs of symmetry "E." Numbers on the left are the rotation quantum numbers, on the right of each column, the degeneracies. Dotted lines represent states antisymmetrical to the double minimum vibration, and have slightly higher energies as shown.

rapidly reversing itself the effective symmetry should also be six. Although the question as to how rapid reversing is needed to change the symmetry number would seem troublesome, it turns out to be otherwise.

Thus one finds that for any height of central potential peak one may consider the possibility of reversal taking the larger symmetry number and the vibration energy levels and thermodynamic functions indicated in Figs. 1, 2, 3 and 4. As an alternate treatment for very high peaks one may neglect the possibility of reversal and take the smaller (half the larger) symmetry number, giving $R \ln 2$ greater rotational entropy, and at the same time take single instead of doubly degenerate vibration energy levels, giving $R \ln 2$ less vibration entropy, or the identical total entropy obtained by the first treatment. The exact, quantum treatment is of course unique and these two methods are to be regarded simply as two equally satisfactory approximations.

It is interesting, moreover, to see how the symmetry number effects enter in Fig. 5. One observes that each of the right-hand columns have 4/3 as many energy levels as the left column, corresponding to the four nuclear spin functions included in each group and the symmetry number of three. If the molecule were planar, all the "dotted" states would be raised to much higher energies and the right-hand columns would have 4/6 as many levels, i.e., a symmetry number of six.

THE USE OF THERMODYNAMIC DATA TO ANSWER QUESTIONS CONCERNING MOLECULAR STRUCTURE

The entropy and heat capacity calculations can be reversed in principle and the experimental thermodynamic data used in the determination of molecular structure. A recent example has been the use of the entropy and heat capacity of ethane in the calculation of the height of barrier restricting internal rotation in that molecule. Since the calculation is not readily reversed mathematically, the usual method is to calculate the thermodynamic quantity for all plausible molecular structures and then decide by comparison with the experimental result.

The symmetry of a molecule can frequently be

determined by this method. If all internal motions can be assumed to be harmonic oscillations (single minimum) and their frequencies, together with the mass and moments of inertia are independently known, then these quantities can be substituted in the usual formula and the symmetry number calculated.[9] It seems, however, that this situation will be a rare one. The interpretation of the spectroscopic data usually depends on the assumed structure of the molecule, particularly if the observed values are compared with those calculated from bond force constants. Sometimes a few frequencies are not observed and are known only from such a calculation. The moments of inertia for large molecules are usually calculated from interatomic distances and may or may not depend appreciably on the assumed structure. Also, for example, if one is deciding between a plane or slightly pyramidal XY_n molecule or between a plane or slightly puckered ring, then one must consider the double minimum vibration functions discussed above, since the molecule will be able to "reverse the puckering" without surmounting any very high potential barrier. Thus one finds that the entropy increases continuously as the "puckering" appears and increases.

A completely certain decision on this basis concerning the benzene ring is not possible at present; however, Lord and Andrews[10] have shown that the entropy of benzene agrees with a value calculated by assuming a plane ring and very plausible vibration frequencies.

The Range of Validity of the Classical Formulas

The question of the range of validity of the classical formulas for rotational entropy and heat capacity seems worthy of some discussion. The fundamental test is, of course, a comparison of the result of the exact, quantum calculation with that from the classical formula. From a practical point of view, however, this criterion is as useless as it is safe. One wishes an expression involving only the quantities entering the classical formula, i.e., the temperature, moments of inertia and symmetry number.

The asymptotic expansions given by Kassel[11] and others, in which the first term corresponds to the classical formula, served as a point of departure. The second term in the state sum expansion corresponds to a constant correction to the energy content at the absolute zero and so does not enter the heat capacity or entropy formulas.[12] The third term appears in all functions and may be taken as a criterion of the validity of the classical expressions.

Further consideration, including comparison with the exact calculations now available for a number of molecules, leads to the following as the most convenient expression. If

$$T>T_0=1.5\times10^{-38}n/I$$

where I is the moment of inertia in g cm^2 and n the symmetry number for rotation about that axis, then the errors in the classical expressions for S and C are less than 0.05 cal. per deg. (Also the errors in expressions for $(H_0{}^0-F)/T$ and $H^0-H_0{}^0)T$ which include the second term mentioned above are less than 0.05 cal. per deg.) In polyatomic molecules the maximum value of n/I must be used. The error drops rapidly as the temperature increases above T_0 (as $1/T^2$), however, in actual molecules corrections for stretching soon amount to several hundredths of a cal. per degree. The above formula is, of course, unnecessarily severe in some cases, but the differences did not appear sufficient to justify further complications.

To give some specific examples, in methane, I is the same for any axis and is approximately 5.4×10^{-40} while the maximum n for a single axis is 3. One then finds $T_0=83°$K. For ethane $T_0=42°$K, and for ethylene $T_0=53°$K. Further calculations lead to the general conclusion that with the single exception of methane, no organic substance has sufficient vapor pressure for present-day experimental measurements on the gas at temperatures below the limit of validity of the classical formulas for rotational heat capacity and entropy. The closest case (other than methane) is ethylene where the pressure at T_0 is 10^{-9} mm of mercury.

[9] V. Dietz and D. H. Andrews, J. Chem. Phys. **1**, 62 (1933) suggested this procedure.

[10] R. C. Lord and D. H. Andrews, J. Phys. Chem. **41**, 149 (1937).

[11] L. S. Kassel, J. Chem. Phys. **1**, 576 (1933).

[12] From a practical point of view this "second term" correction may often be omitted from the free energy and heat content functions, since the error in $E_0{}^0$ (or $H_0{}^0$) is a constant and does not affect calculations involving data and applications at high temperatures.

Comment: The Symmetry Number and Thermodynamic Functions for Molecules Having Double Minimum Vibrations

The original purpose of this paper is indicated in the last two sections. Other investigators were unwisely trying to determine the structure of a molecule of uncertain symmetry by a calculation using a simplified formula valid only for rigid molecules. As shown in the paper, there is a continuous change in a thermodynamic quantity as a structure gradually loosens or simplifies, e.g. from pyramidal to planar. Ammonia is now so well understood that the detailed information for that example is obsolete, but the general concept and the warning about oversimplification of statistical thermodynamic formulas may still be of interest.

[Reprinted from the Journal of the American Chemical Society, 65, 803 (1943).]

[CONTRIBUTION FROM THE CHEMICAL LABORATORY OF THE UNIVERSITY OF CALIFORNIA]

The Thermodynamics and Molecular Structure of Benzene and Its Methyl Derivatives[1]

BY KENNETH S. PITZER AND DONALD W. SCOTT[2]

Besides the obvious importance of thermodynamic data on benzene and its simpler methyl derivatives, this series of compounds is interesting because of the various possibilities for rotation of the methyl groups with respect to the ring. Since the benzene ring is planar while the methyl group has a three-fold axis of symmetry, the potential energy for internal rotation in toluene must have a six-fold symmetry (six maxima and minima per rotation). In the only other case of this type yet investigated, nitromethane, the potential barrier was found to be very small, leading to essentially free rotation above room temperature.[3,4] In *o*-xylene, however, interactions between the two methyl groups may be expected.

Other interesting points investigated include the freezing behavior of the ternary system, *o*, *m*, and *p*-xylene, the equilibrium composition of the xylenes, and the equilibrium of the reaction: 2 toluene = benzene + xylene. The latter data lead to quite accurate energy relationships between these hydrocarbons.

Part I. Thermodynamic Measurements on the Pure Substances. The Heat Capacities, Heats of Fusion and Vaporization, Vapor Pressures and Entropies of Benzene, Toluene and the Xylenes

In this part thermodynamic measurements are described which include typical low temperature calorimetric studies of the three xylenes. In addition, the vapor pressures up to 60°, and the heat capacities of the vapors have been measured. Insofar as such measurements could not be found in the literature, they were also made for benzene and toluene.

Purification of Materials.—The xylenes used for the measurements on the solid and liquid were purified by fractional crystallization and distillation. The device used for the fractional crystallization is shown schematically in Fig. 1. It was made of glass and was sufficiently compact to fit into a large dewar containing the cooling bath. A mixture of the material to be crystallized and a suitable solvent was introduced into bulb A. A slow stream of dry air entering tube C and bubbling up through the sintered glass plate, E, prevented any liquid from flowing down through the sintered glass plate while crystallization was taking place. After crystallization had reached the desired state, the air stream was stopped and air pressure was applied through tube D, forcing the mother liquor out through the sintered glass plate and over into bulb, B, leaving the crystals in bulb, A. The apparatus was then removed from the cooling bath and disconnected at the ground glass joint, F. As the crystals melted, the liquid drained from the bulb and was collected. After several such crystallizations, the product obtained was refluxed with molten sodium metal and then distilled in order to remove water and solvent which had adhered to the crystals.

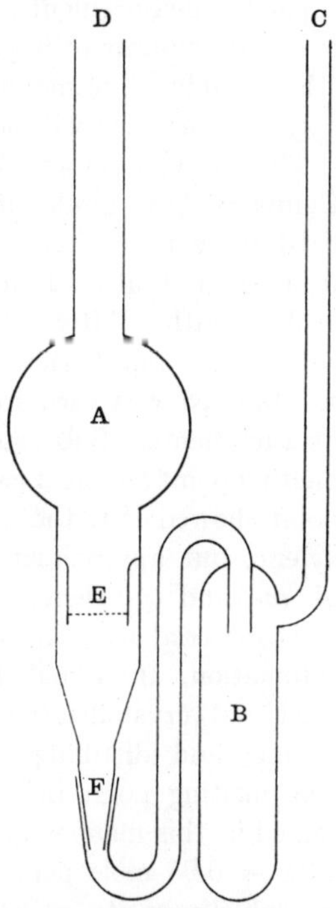

Fig. 1.—The apparatus for purification by fractional crystallization.

The *o*-xylene was Eastman Kodak Company best grade product. Heat capacity measurements on a sample of this original material in the region below the melting point indicated considerable impurity. In addition, there was an isothermal absorption of heat at about 210°K., presumably caused by the melting of a eutectic mixture with *m*-xylene impurity. This material was crystallized three times, using acetone as a sol-

(1) Portions of this paper were presented in the Symposium on Low Temperature Research at the Buffalo meeting of the American Chemical Society, September, 1942.

(2) Du Pont Fellow in Chemistry, 1941–1942. Part I of this paper is essentially an abstract of the Ph.D. Dissertation of Donald W. Scott, University of California, 1942.

(3) K. S. Pitzer and W. D. Gwinn, THIS JOURNAL, 63, 3313 (1941).

(4) T. De Vries and B. T. Collins, *ibid.*, 64, 1224 (1942).

vent and an acetone–dry-ice mixture for the cooling bath. Heat capacity measurements below the melting point on a sample of the material so obtained showed less premelting than previously and the absorption of heat at 210°K. was considerably diminished. This material was further purified by crystallizing it three times from methyl alcohol solvent, again using an acetone–dry-ice cooling bath. This final product, after drying and distilling in a 2-ft. glass bead packed column, was used in all measurements on the solid and liquid. The premelting effect and the change of melting point with per cent. melted indicated 0.12 mole per cent. liquid soluble, solid insoluble impurity. The absorption of heat at 210°K. had virtually disappeared, amounting to only about four calories per mole. This can be accounted for by assuming that 0.10 mole per cent. or 5/6 of the total impurity was *m*-xylene. A portion of the material removed from the calorimeter after the completion of the heat capacity measurements was employed in the vapor pressure experiments. For the gas heat capacity measurements, which are not sensitive to small amounts of impurity, and which require a larger amount of material, Eastman Kodak Co. *o*-xylene was used without further purification.

The *m*-xylene was likewise Eastman Kodak Company best grade product. It was crystallized three times from acetone and three times from methyl alcohol, using an acetone–dry-ice cooling bath. After drying and distilling in a column of about 25 theoretical plates, the product so obtained was used for the gas heat capacity measurements. Subsequent measurements of the melting point and heat capacity below the melting point showed that the material was not pure *m*-xylene, but approximately the eutectic mixture of *m*- and *p*-xylenes (about 13% *p*-xylene). This material was further purified by fractional sulfonation, fractional hydrolysis, followed by fractional crystallization from acetone solvent, drying, and distilling in the 25-plate column. The melting point behavior of the product obtained in this manner indicated that it contained 2.34 ± 0.05 mole per cent. liquid soluble, solid insoluble impurity, of which all except 0.22 mole per cent. was *p*-xylene. Since time did not permit further purification, this material was used for all measurements on the solid and liquid. After completion of the heat capacity measurements, a portion of the material removed from the calorimeter was employed in the vapor pressure experiments. Since the impurity present in the sample was predominantly *p*-xylene, it was possible to make corrections for it, exactly in the case of the heat capacities of the solid, and to a satisfactory approximation in the case of the other thermal quantities. More will be said about these corrections in later paragraphs.

The *p*-xylene was also Eastman Kodak Company best grade product. It was crystallized three times from acetone and three times from methyl alcohol, using ice water as a cooling bath. It was then dried and distilled in a column of about 25 theoretical plates, one portion of the middle fraction being used for the gas heat capacity measurements and another portion for the measurements on the solid and liquid. This latter contained 0.02 mole per cent. of liquid soluble, solid insoluble impurity, as shown by the premelting effect and the change of melting point with per cent. melted. A portion of the material removed from the calorimeter after completion of the heat capacity measurements was used in the vapor pressure experiments.

Toluene was obtained from a commercial source. It was purified by freezing it on a coil of copper tubing through which liquid air was circulated. When about a half or two-thirds of the original material was frozen, the solid was allowed to melt and was again subjected to the same process. After several repetitions of this process, about a fifth of the starting material remained as a final product. This was dried by refluxing with molten sodium and distilled in the 25-plate column. Separate portions of the middle fraction, which distilled within a tenth degree range, were used for the gas heat capacity and the vapor pressure measurements.

Commercial benzene was refluxed with metallic sodium and distilled, the middle fraction being used for the gas heat capacity measurements. The carbon tetrachloride sample was likewise the middle fraction obtained by distilling a commercial product.

Melting Points.—The melting points were observed as a function of the per cent. melted. The data are shown in Table I. The "fraction melted" in Table I is strictly the fraction of the heat of fusion of the sample which has been introduced. The difference is significant only for *m*-xylene with its large impurity. The true melting points were calculated assuming Raoult's law. For purposes of comparison, values taken from the recent literature are also included in the table.

The data for *m*-xylene are shown graphically in Fig. 2 where the temperature is plotted *vs.* the reciprocal of the fraction melted. The latter quantity is proportional to the concentration of the solute since the quantity of solute is constant and its concentration varies inversely with the fraction of the solvent melted. The upper branch of the curve represents the melting of pure *m*-xylene and extrapolates to the true melting point, 225.27°K. Its slope gives the total impurity 2.34%. The lower branch of the curve represents the melting of the *m-p* eutectic which extrapolates to 220.2°K., and its slope gives the impurity other than *p*-xylene, 0.22%. From the point of intersection and the heat of fusion data given below one may calculate the composition of the eutectic to be 12.9% *p*-xylene.

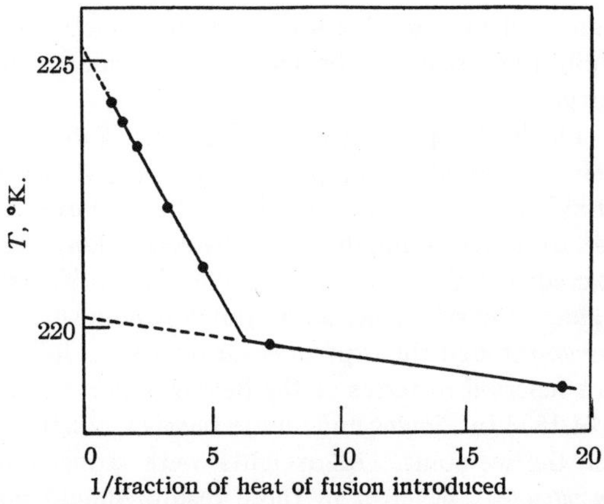

Fig. 2.—The melting behavior of the *m*-xylene sample.

TABLE I

THE MELTING POINTS OF THE XYLENES (ICE POINT = 273.16°K.)

o-Xylene

Per cent. melted	Observed temperature, °K.
31	247.66
55	247.71
78	247.75
True melting point	247.82 ± 0.1°K.
Parks, Huffman and Daniels[5]	247.8°K.
White and Rose[6]	247.86°K.

m-Xylene

Per cent. melted	Observed temperature, °K.
5.3	218.85
13.8	219.67
21.4	221.12
30.0	222.26
46.5	223.40
62.8	223.86
85.4	224.23
Melting point of sample	224.38°K.
Melting point of pure *m*-xylene	225.27 ± 0.15°K.
Parks, Huffman and Daniels[5]	219.6°K.
White and Rose[6]	225.27°K.
Timmermans and Hennaut-Roland[7]	225.8°K.
Nakatsuchi[8]	224.2°K.

p-Xylene

Per cent. melted	Observed temperature, °K.
20	286.36
38	286.37
56	286.38
True melting point	286.39 ± 0.05°K.
Parks, Huffman and Daniels[5]	286.3°K.
White and Rose[6]	286.37°K.

Solid and Liquid Heat Capacities.—The apparatus was a low temperature, vacuum insulated calorimeter which has been used in previous work in this Laboratory and has been described in the literature.[9] It was used without change for the measurements on *o*-xylene. For the measurements on *m*- and *p*-xylene, the thermometer was replaced with a new pure platinum strain-free resistance thermometer This consists of a spiral of 0.1 mm. C. P. platinum wire wound on a mica cross and has an ice-point resistance of 83.318 ohms. It was calibrated at the oxygen, ice, steam and sulfur points according to the international temperature scale. The equation obtained was

$$R = 83.318\ [1 + 0.00397202t - 0.59115 \times 10^{-6}t^2 - 4.384 \times 10^{-12}t^3(t - 100)]$$

Below 90°K. it was calibrated by comparison with the hydrogen triple point, hydrogen vapor pressures, oxygen transitions and triple point, and oxygen vapor pressures. The vapor pressure equations and melting points for hydrogen and oxygen given by Henning and Otto[10] were used. The value, 43.76°K., for the oxygen transition has been reported by both Giauque and Johnson,[11] and Henning and Otto.[10] Using the resistance–temperature tables of Hoge and Brickwedde[12] for pure platinum in this range as a

(5) Parks, Huffman and Daniels, THIS JOURNAL, **52**, 1547 (1930).

(6) White and Rose, *Bur. Standards J. Research*, **9**, 711 (1932).

(7) Timmermans and Hennaut-Roland, *J. chim. phys.*, **27**, 401 (1930).

(8) Nakatsuchi, *J. Soc. Chem. Ind., Japan*, **32**, suppl. bind., 333 (1929).

(9) (a) Latimer and Greensfelder, THIS JOURNAL, **50**, 2202 (1928). (b) Brown, Smith, and Latimer, *ibid.*, **58**, 1758 (1936). (c) Pitzer and Coulter, *ibid.*, **60**, 1310 (1938).

(10) Henning and Otto, *Physik. Z.*, **37**, 633 (1936).

(11) Giauque and Johnston, THIS JOURNAL, **51**, 2300 (1929).

(12) Hoge and Brickwedde, *J. Research, Nat. Bur. Standards*, **22**, 351 (1939).

base, a deviation plot was made on which these calibration points satisfactorily determined the curve.

The heat capacity data are listed in Tables II, III and IV and shown graphically in Fig. 3. The *m*-xylene heat capacities have been corrected for the *p*-xylene impurity. This correction never exceeded 0.05 cal./deg. mole. Table V lists values at even temperatures taken from a smooth curve through the experimental points. All data are reported in terms of the defined calorie equal to 4.1833 Int. joules and on the basis of 273.16°K. for the ice-point. All weights were reduced to *in vacuo*. The error in these results should not exceed 0.2% except at the highest and lowest temperatures and just below the melting points. In the case of *m*- and *p*-xylene the measurements were extended to temperatures above room temperature. In this region the accuracy falls off rapidly because of large radiation losses and the uncertainty in the results may well exceed 1%. It must be remembered that the data in this region are not used in the entropy calculation.

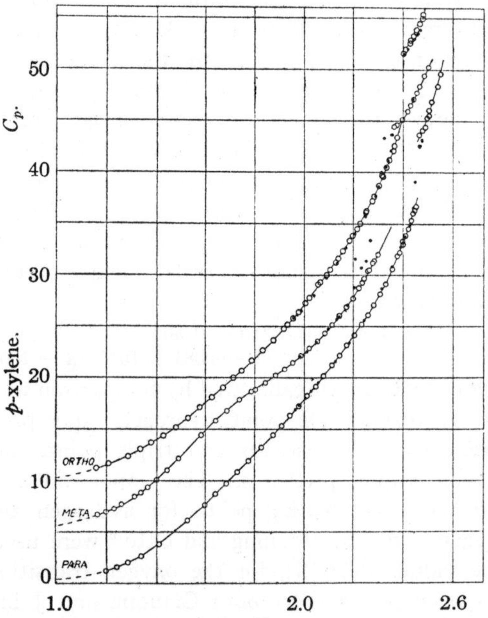

Fig. 3.—The experimental heat capacity data for solid and liquid xylenes. Open circles, this research; solid circles, Parks, Huffman, and Daniels.[5] Vertical scales are offset for clarity.

The heat capacities of these substances above 90°K. have been measured previously by Parks, Huffman and Daniels.[5] Their data are plotted in Fig. 3 for comparison. The melting point which they report for *m*-xylene, 219.6°K., seems to indicate that their sample of this substance was not pure *m*-xylene but approximately the eutectic mixture with *p*-xylene. This probably accounts for the large deviations in their points near the melting point.

Figure 4 shows the smooth curves for all three xylenes for the purpose of comparison. The heat capacities at 20°K. are in the inverse order of the melting points. This would be expected if both effects result from varying degrees of firmness of crystal binding. The relatively marked hump in *m*-xylene near 50°K. and the more gradual one in *p*-xylene near 100°K. may be caused by the beginning of rotation of methyl groups. It will be shown later that the methyl groups rotate freely in these molecules in the gas phase. The data of Kelley[13] show a similar hump for toluene as would be expected if this idea is correct. Internal rotation is well restricted in *o*-xylene so that no hump should appear.

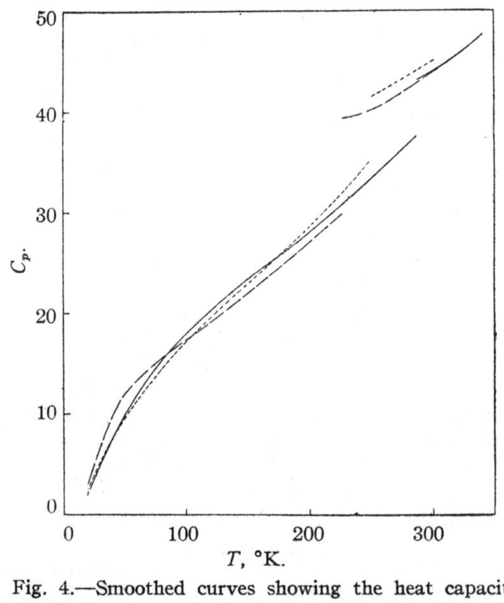

Fig. 4.—Smoothed curves showing the heat capacity of the three xylenes: ortho, dotted curve; meta, dashed; para, solid.

The heat capacities of *m*- and *p*-xylenes are essentially equal in both liquid and gaseous states. The data for the gases are given below. The heat capacity of liquid *o*-xylene is about one calorie per degree higher and a similar difference is shown by the gas. Thus if one removes internal heat capacity effects, all liquid xylenes are

(13) Kelley, THIS JOURNAL, 51, 2733 (1929).

TABLE II

THE HEAT CAPACITY OF *o*-XYLENE

Sample: 52.00 g. (0.4898 mole)

T, °K.	C_p, cal./deg. mole	T, °K.	C_p, cal./deg. mole
14.42	1.290	152.77	23.28
16.15	1.704	157.35	23.78
19.32	2.417	164.13	24.51
21.76	3.113	168.52	25.00
23.88	3.698	177.71	26.08
26.69	4.393	178.76	26.18
30.10	5.219	188.33	27.25
33.78	6.127	191.54	27.67
37.86	7.149	199.91	28.73
42.43	8.126	205.98	29.71[b]
47.00	9.060	207.71	29.64
51.87	9.970	210.38	30.36[b]
56.05	10.72	215.18	30.47[a]
61.48	11.68	220.59	31.22[a]
67.93	12.68	221.07	31.23[a]
75.06	13.72	229.14	32.15[a]
82.41	14.78	230.92	32.65[a]
85.26	15.22	237.06	33.51[a]
90.20	15.85		
94.85	16.44	247.82 melting point	
101.25	17.26		
105.07	17.80	251.65	41.66
114.09	19.07	256.79	42.04
115.54	19.19	268.19	42.65
123.93	20.05	268.81	42.92
127.36	20.43	276.52	43.31
134.52	21.24	284.82	43.96
137.80	21.60	293.52	44.56
144.04	22.32	301.31	45.05
		301.86	45.39

[a] Corrected for premelting. [b] Includes heat to melt the eutectic mixture.

TABLE III

THE HEAT CAPACITY OF *m*-XYLENE

Sample: 66.14 g. (0.6230 mole)

T, °K.	C_p, cal./deg. mole	T, °K.	C_p, cal./deg. mole
14.62	1.686	148.49	21.89
15.76	1.938	157.61	22.72
16.96	2.283	167.15	23.73
18.19	2.712	171.79	24.25
20.54	3.501	175.85	24.66
21.71	3.881	181.56	25.39
23.10	4.389	183.93	25.46
25.03	5.072	186.72	25.66
27.63	6.053	189.84	26.15
30.69	7.019	191.66	26.35
38.40	9.433	198.42	26.97
43.80	10.80		
48.78	11.77		
54.79	12.83	225.27 melting point	
61.01	13.63		
63.45	13.80	231.40	39.45
69.57	14.49	237.18	39.62
76.38	15.21	249.43	40.12
83.11	15.82	262.62	40.92
89.83	16.47	268.83	41.49
97.07	17.11	276.80	42.10
104.89	17.85	276.97	42.02
113.15	18.57	284.83	42.78
121.54	19.32	292.93	43.40
130.26	20.14	305.27	44.36
139.78	21.03	318.16	45.26

TABLE IV

THE HEAT CAPACITY OF *p*-XYLENE

Sample: 68.40 g. (0.6443 mole)

T, °K.	C_p, cal./deg. mole	T, °K.	C_p, cal./deg. mole
15.76	1.210	211.11	29.18
17.42	1.588	224.90	30.64
19.32	2.039	236.25	31.89
21.80	2.674	240.69	32.28
25.68	3.809	247.48	33.13
33.58	6.154	254.85	33.94
39.73	7.692	261.10	34.66
43.86	8.732	262.89	34.73
48.62	9.776	267.69	35.44[a]
53.68	10.88	272.16	36.16[a]
59.56	11.98	275.28	36.40[a]
66.77	13.22	276.58	36.44[a]
74.40	14.46	276.74	36.44[a]
80.30	15.33	279.66	36.84[a]
86.95	16.33		
94.32	17.26	286.39 melting point	
102.27	18.24		
110.72	19.17	292.02	43.62
119.64	20.24	301.10	43.98
129.01	21.25	310.04	44.42
138.65	22.24	314.69	45.23
148.69	23.23	318.47	45.99
159.04	24.22	319.24	45.50
169.68	25.23	327.36	46.85
179.55	26.19	338.96	48.38
189.86	27.16	345.49	48.42
200.15	28.21	354.65	49.62

[a] Corrected for premelting.

the same, which supports the idea that Raoult's law should apply accurately to these systems.

Heat of Fusion.—The heats of fusion were measured with different rates of heating and no trend was observed. Table VI lists the results and, for comparison, the values reported by other investigators. In correcting the heat of fusion of the *m*-xylene sample for the *p*-xylene impurity, the heat of fusion of *p*-xylene at 225.27°K. was calculated from the value obtained at 286.39°K. using values of ΔC_p calculated on the assumption that the apparent molal heat capacity of *p*-xylene in the *m*-xylene is 1.004 times that of liquid *m*-xylene. This correction is not very sensitive to the assumption made in regard to ΔC_p. For example, if we had instead assumed a constant ΔC_p of fusion of 5.4, this would have changed the value for the heat of fusion of pure *m*-xylene by only 2 calories.

TABLE V

THE HEAT CAPACITIES OF THE XYLENES AT EVEN TEMPERATURES

T, °K.	Heat capacity, cal./deg. mole: *o*-Xylene	*m*-Xylene	*p*-Xylene
15	1.30	1.74	1.03
20	2.63	3.28	2.25
25	3.97	5.08	3.72
30	5.19	6.80	5.17
35	6.42	8.43	6.50
40	7.60	9.86	7.76
45	8.65	11.06	8.94
50	9.63	12.04	10.07
60	11.41	13.25	12.05
70	12.98	14.54	13.76
80	14.43	15.54	15.27
90	15.82	16.49	16.68
100	17.13	17.41	17.96
110	18.41	18.28	19.14
120	19.59	19.15	20.26
130	20.72	20.11	21.32
140	21.86	21.03	22.34
150	22.96	21.97	23.33
160	24.03	22.96	24.28
170	25.18	24.04	25.26
180	26.32	25.06	26.22
190	27.47	26.11	27.17
200	28.65	27.15	28.13
210	29.85	28.19	29.13
220	31.13	29.22	30.13
230	32.48	39.40	31.18
240	33.82	39.71	32.28
250	41.52	40.17	33.40
260	42.29	40.78	34.54
270	42.87	41.50	35.68
280	43.58	42.53	36.83
290	44.30	43.13	43.27
300	45.02	43.93	44.07
320		45.6	45.6
340			48.0
360			50.4

It is difficult to reconcile the value reported by Parks, Huffman and Daniels for *o*-xylene with the results of the present research. Such a large discrepancy could scarcely have arisen from a different manner of treating the premelting heat.

Vapor Pressures and Heats of Vaporization.—The vapor pressure apparatus consisted of a sample bulb connected to a mercury manometer, both of which were completely immersed in a water thermostat with a plate glass window. The difference in height of the mercury column was read with a cathetometer of 0.05 mm. sensitivity. Temperatures were determined with a mercury-in-glass thermometer calibrated by the U. S. Bureau of Standards which could be read to 0.01°. To thoroughly dry and degas the samples, the following procedure was employed. The sample was introduced into a bulb connected to the sample bulb through a phosphorus pentoxide tube. It was frozen with liquid air and the system was evacuated to a pressure of 10^{-5} mm. The sample was distilled into the sample bulb, then distilled back into the original bulb, and finally distilled back into the sample bulb, thus having passed through the phosphorus pentoxide tube three times. At the end of each distillation the sample was again frozen with liquid air and the system evacuated to a pressure of 10^{-5} mm. The phosphorus pentoxide tube and the second bulb were then sealed off from the remainder of the system. Finally, about a third of the sample was allowed to evaporate by pumping on it while it was at room temperature. The fact that the measured vapor pressures did not change as further quantities of the sample were pumped off was taken as evidence of ample drying and degassing and the absence of significant amounts of impurity of different volatility.

TABLE VI

THE HEATS OF FUSION

Time of energy input, minutes	Heat of fusion, cal./mole
***o*-Xylene**	
19	3250
24	3245
27	3255
Average	3250 ± 10
Parks, Huffman, and Daniels[5]	3116
Rossini[14]	3330 ± 70
Nakatsuchi[8]	3100
***m*-Xylene (impure sample)**	
27	2780
32	2796
39	2776
Average	2784 ± 15
Heat of fusion of pure *m*-xylene	2765 ± 20
Parks, Huffman, and Daniels[5]	2734
Rossini[14]	2760 ± 50
Nakatsuchi[8]	2800
***p*-Xylene**	
33	4089
51	4085
54	4097
Average	4090 ± 20
Parks, Huffman, and Daniels[5]	4047
Nakatsuchi[8]	4100

The manometer readings were reduced to 0° and standard acceleration of gravity. The large diameter of the manometer arms, 15 mm., made capillary depression corrections negligible. The

(14) Rossini, *Bur. Stand. J. Res.*, **11**, 533 (1933).

data were fitted to three-constant equations. The coefficient of the logarithmic term was obtained by a process of trial and error and the remaining two constants were calculated by the method of least squares. The equations obtained are

(1) Toluene 0–50°

$$\log_{10}p = -2866.53/T - 6.7 \log_{10}T + 27.6470$$

(2) *o*-Xylene 0–60°

$$\log_{10}p = -3327.16/T - 8.0 \log_{10}T + 31.7771$$

(3) *m*-Xylene 0–60°

$$\log_{10}p = -2871.66/T - 5.0 \log_{10}T + 22.9270$$

(4) *p*-Xylene 13.23–60° (liquid)

$$\log_{10}p = -3080.31/T - 6.7 \log_{10}T + 27.8581$$

(5) *p*-Xylene 0–13.23° (solid)

$$\log_{10}p = -3141.33/T + 11.6092$$

Equation 5 was calculated from equation 4 and the measured heat of fusion. Table VII shows a comparison of the observed and calculated values. It can be seen that the equations reproduce the data to within the sensitivity of the cathetometer (0.05 mm.).

TABLE VII

VAPOR PRESSURES

Substance	T, °C.	p (obs.) mm. Hg	p (calcd.) mm. Hg
	0.0	6.75	6.75
	12.5	14.40	14.38
Toluene	25.0	28.40	28.45
	37.5	52.70	52.65
	50.0	92.00	91.98
	0.0	1.30	1.27
	12.5	3.00	3.04
o-Xylene	25.0	6.60	6.64
	37.5	13.50	13.45
	50.0	25.50	25.48
	60.0	40.60	40.63
	0.0	1.75	1.71
	12.5	3.90	3.94
m-Xylene	25.0	8.35	8.39
	37.5	16.65	16.66
	50.0	31.20	31.17
	60.0	49.50	49.49
	0.0	1.35[a]	1.29[a]
	12.5	4.05[a]	4.10[a]
p-Xylene	25.0	8.90	8.87
	37.5	17.50	17.55
	50.0	32.60	32.61
	60.0	51.35	51.32

[a] Solid.

The vapor pressures of toluene given by the "I. C. T." are in good agreement with the results of this research, the former being higher by a few tenths of a per cent. The vapor pressures of the xylenes have been reported by Linder,[15] Rintelen, Saylor and Gross,[16] Kassel,[17] and Stuckey and Saylor.[18] The results reported by the last-named authors are in excellent agreement with those reported here. This is not true for the others

The heats of vaporization at 298.16°K. were calculated by means of the Clapeyron equation. Corrections for gas imperfections were made by means of the Berthelot equation of state, using critical constants from the "I. C. T." for toluene and from the measurements of Altschul[19] for the xylenes. Because of the low vapor pressures of these substances, this correction is rather small, being 26, 10, 12 and 11 calories for toluene, *o*-, *m*- and *p*-xylenes, respectively. The calculated heats of vaporization at 298.16°K. are

Toluene	9,115 ± 50 cal./mole
o-Xylene	10,470 ± 50 cal./mole
m-Xylene	10,160 ± 50 cal./mole
p-Xylene	10,110 ± 50 cal./mole

The correction of the *m*-xylene heat of vaporization for the *p*-xylene impurity in the sample was not significant.

Entropies.—The entropy calculations are summarized in Table VIII. The calculation for benzene uses the value for the liquid at 298.16°K. given by Ahlberg, Blanchard and Lundberg[20] together with vaporization data from the work of Fiock, Ginnings and Holton.[21] The value of the entropy of liquid toluene at 298.16°K. is that reported by Kelley.[13] The double Debye functions (*i. e.*, for six degrees of freedom) used to extrapolate the xylene heat capacities to 0°K. fit the measured heat capacities up to 25 or 30°K. Calculations showed that corrections for gas imperfections were negligible. The entropy of *m*-xylene was calculated in two separate ways. In the first, the entropy was calculated in the conventional manner, using values of the thermal quantities corrected for the *p*-xylene impurity. In the second, the entropy of a mole of substance of the composition of the sample was first calculated and then corrected for the entropy of mixing and the *p*-xylene impurity. The two methods give concordant results. The second method avoids any assumptions which could possibly be subject to question and therefore is to be preferred insofar as there is any difference at all.

(15) Linder, *J. Phys. Chem.*, **35**, 351 (1931).

(16) Rintelen, Saylor and Gross, THIS JOURNAL, **59**, 1129 (1937).

(17) Kassel, *ibid.*, **58**, 670 (1936).

(18) Stuckey and Saylor, *ibid.*, **62**, 2922 (1940).

(19) Altschul, *Z. physik. Chem.*, **11**, 577 (1893).

(20) Ahlberg, Blanchard and Lundberg, *J. Chem. Phys.*, **5**, 539 (1937).

(21) Fiock, Ginnings and Holton, *J. Research Natl. Bur. Standards*, **6**, 881 (1931).

TABLE VIII
ENTROPY CALCULATIONS
Units: cal. per degree mole

	Benzene		
298.16°K.	Liquid (Ahlberg, *et al.*[20])	41.49	
	Vaporization (Fiock, *et al.*[21])	27.05	
	Compression R ln (0.1235)	−4.15	
	Gas, 1 atm., 298.16°K.		64.39 ± 0.3
	Toluene		
298.16°K.	Liquid (Kelley[13])	52.40	
	Vaporization 9115/298.16	30.57	
	Compression $-R$ ln 760/28.45	−6.53	
	Gas, 1 atm., 298.16°K.		76.44 ± 0.3
	***o*-Xylene**		
0.00–14.13°K.	Extrapolation $2D(131.5/T)$	0.40	
14.13–247.82	Solid, graphical integration	37.32	
247.82	Fusion, 3250/247.82	13.12	
247.82–298.16	Liquid, graphical integration	7.96	
	Entropy of liquid at 298.16°K.		58.80
298.16	Vaporization 10,470/298.16	35.12	
	Compression $-R$ ln 760/6.64	−9.42	
	Gas, 1 atm., 298.16°K.		84.50 ± 0.3
	***m*-Xylene (1st method)**		
0.00–14.13°K.	Extrapolation $2D(117/T)$	0.54	
14.13–225.27	Solid, graphical integration	36.13	
225.27	Fusion 2765/225.27	12.26	
225.27–298.16	Liquid, graphical integration	11.49	
	Entropy of liquid at 298.16		60.42
298.16	Vaporization 10,160/298.16	34.08	
	Compression $-R$ ln 760/8.39	−8.95	
	Gas, 1 atm. 298.16°K.		85.55 ± 0.4
	***m*-Xylene (2nd method)**		
0.00–14.13°K.	Extrapolation $2D(118/T)$	0.53	
14.13–224.38	Solid, graphical integration	36.02	
220.20–224.38	Fusion	12.46	
224.38–298.16	Liquid, graphical integration	11.65	
298.16	Vaporization 10,160/298.16	34.08	
	Compression $-R$ ln in 760/8.39	−8.95	
	Entropy of one mole of substance of the composition of the sample	85.79	
	Entropy of mixing	−0.22	
		85.57	
	Entropy of 0.0234 mole *p*-xylene	−1.97	
	Entropy of 0.9766 mole *m*-xylene	83.60	
	Entropy of one mole *m*-xylene, gas, 1 atm. 298.16°K., 83.60/0.9766		85.60 ± 0.3
	***p*-Xylene**		
0.00–15.85°K.	Extrapolation $2D(142/T)$	0.44	
15.85–286.39	Solid, graphical integration	42.73	
286.39	Fusion 4090/286.39	14.28	
286.39–298.16	Liquid, graphical integration	1.75	
	Entropy of liquid at 298.16°K.		59.20
298.16	Vaporization 10,110/298.16	33.91	
	Compression $-R$ ln 760/8.87	−8.84	
	Gas, 1 atm., 298.16°K		84.27 ± 0.3

Gas Heat Capacities.—The gas heat capacities were measured in a flow type calorimeter which has been described previously.[3,22] The high boiling points of toluene and the xylenes made it necessary to carry out the measurements under reduced pressure except at the higher temperatures. Corrections for gas imperfections were made by means of the Berthelot equation of state, using critical constants from the "I. C. T." for benzene, toluene and carbon tetrachloride and those reported by Altschul[19] for the xylenes. The data are given in Table IX and are shown graphically in Fig. 7 of Part V. Within experimental error, meta and para xylenes have the same gaseous heat capacity. This fact makes unnecessary any correction for the *p*-xylene impurity in the *m*-xylene sample used for these measurements. It is difficult to estimate the uncertainty in these gas heat capacity values; they have been somewhat arbitrarily assigned an uncertainty of 1%. The measurement on carbon tetrachloride was made after the completion of the measurements on the other substances. The fact that the value obtained agrees with that calculated from the vibrational assignment of Vold[23] (21.84 cal./deg. mole at 393°K.) serves as a check on the reliability of these measurements. The heat capacity of gaseous benzene has also been measured by Dr. W. D. Gwinn of this Laboratory, using a different calorimeter. His data agree well with the present results and have been included in Table IX. Bennewitz and Rossner's[24] value for the heat capacity of toluene gas (33.6 cal./deg. mole at 410°K.) is significantly lower than the results reported here (about 2%). These results are compared with calculated values in Fig. 7, Part V.

TABLE IX

GAS HEAT CAPACITIES

Substance	T, °K.	P, atm.	C_p, cal./deg. mole	$C_p - C_p^0$	C_p^0, cal./deg. mole
	393	1	26.7	0.3	26.4 ± 0.3
Benzene	393	0.19	26.5	.1	26.4 ± .3
	428	1	28.6	.2	28.4 ± .3
	463	1	30.5	.2	30.3 ± .3
Benzene	388	1	26.3	.3	26.0 ± .3
by Dr. W.	417	1	28.4	.3	28.1 ± .3
D. Gwinn	481	1	31.6	.2	31.4 ± .3
	393	0.53	33.0	.2	32.8 ± .3
Toluene	428	1	36.0	.3	35.7 ± .4
	463	1	38.3	.3	38.0 ± .4
	393	0.22	40.3	.1	40.2 ± .4
o-Xylene	428	0.73	43.7	.2	43.5 ± .4
	463	1	46.3	.3	46.0 ± .5
	393	0.22	39.2	.1	39.1 ± .4
m-Xylene	428	0.69	42.7	.3	42.4 ± .4
	463	1	45.7	.3	45.4 ± .4
	393	0.22	39.1	.1	39.0 ± .4
p-Xylene	428	0.73	42.9	.3	42.6 ± .4
	463	1	45.5	.3	45.2 ± .4
CCl_4	393	1	22.2	.3	21.9 ± .2

Part II Thermodynamic Measurements on Mixed Systems

Equilibrium and Heat Content Relationships between Benzene, Toluene, and Each of the Xylenes. The Freezing Point Diagram of the Ternary System *o*-, *m*-, and *p*-Xylenes

Inspection of the available heat of combustion data for the xylenes led to the conclusion that the differences between isomers could not be determined with any reliability by this method. By comparison, the data of Norris and Vaala[25] indicated that satisfactory values could probably be obtained by direct equilibrium studies using aluminum halide–hydrogen halide catalysts.

The freezing point studies were undertaken primarily as an aid in the analysis of xylene mixtures. However, they offer an interesting measure of the accuracy to which these mixtures follow Raoult's law.

The Freezing Point of Xylene Mixtures.—By use of the accurate melting points and heats of fusion of Part I, together with thermodynamics and Raoult's law, the complete freezing point diagram of the xylenes was calculated. A value of $3R$ (5.96 cal./degree) was assumed for ΔC_p of fusion for each xylene, by comparison of the various heat capacity curves. Its exact value is unimportant. Kravchenko[26] has already published such a calculation but did not have as accurate physical constants available. The results of the calculations of this research are shown in Table X and in Fig. 5.

The data on the impure *m*-xylene sample provide an excellent check on these calculations at

(22) Pitzer, THIS JOURNAL, **63**, 2413 (1941).
(23) Vold, *ibid.*, **57**, 1192 (1935).
(24) Bennewitz and Rossner, *Z. physik. Chem.*, **B39**, 126 (1938).
(25) Norris and Vaala, THIS JOURNAL, **61**, 2131 (1939).
(26) Kravchenko, *Acta physicochim. U. S. S. R.*, **14**, 279 (1941).

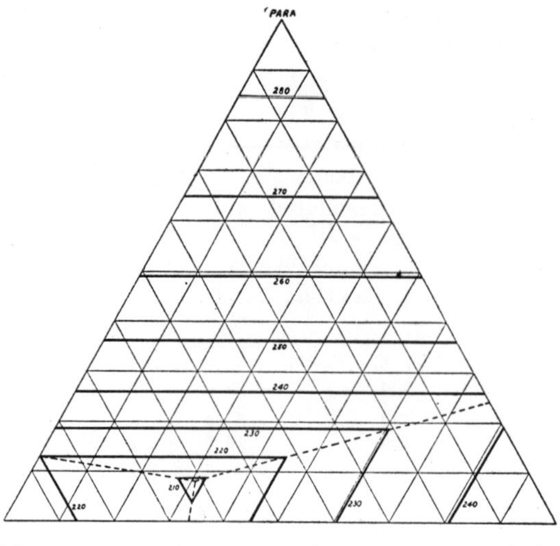

Meta Ortho

Fig. 5.—The freezing point diagram for the ternary system o-. m- and p-xylene.

the meta–para eutectic as shown in Table X.

TABLE X

EUTECTIC POINTS IN THE TERNARY SYSTEM o-, m-, p-XYLENE

Composition calcd.	Temperature, °K. Calcd.	exp. (this research)	exp. (Nakatsuchi[9])
13.0% para, 87.0% meta[a]	220.3	220.2 ± 0.2	216.2 ± 3
24.3% para, 75.7% ortho	237.6	...	237.7 ± 0.4
66.6% ortho, 33.4% meta	211.2	211.4 ± 1	210.2 ± 1
30.5% o-, 61.4% m-, 8.1% p-	208.5	209 ± 1	...

[a] Experimental composition (this research) 12.9% para, 87.1% meta.

Analysis of Xylene Mixtures.—Norris and Vaala analyzed their xylene mixtures for p-xylene by the depression of the freezing point of pure p-xylene on addition of the unknown sample, and for o-xylene by oxidation to phthalic acids and separation of phthalic anhydride. The percentage of meta was then obtained by difference.

It was found possible to simplify this procedure considerably by the substitution of a refractive index measurement for the oxidation to phthalic acids. The refractive indices of the pure xylenes are (at 25°) ortho, 1.5030; meta, 1.4946; para, 1.4931. Thus the refractive index is primarily dependent on the ortho concentration although the para content must be known independently for exact work. Actual analyses were carried out by first determining the para content, then making up knowns with the same para content and with ortho contents near that of the unknown. Since all samples of interest were near the equilibrium composition, only that region was actually investigated experimentally.

A simple Abbe refractometer sufficed for analyses to one or two per cent. The final results were checked with an interferometer. The latter instrument had such sensitivity as to have permitted analyses to a few hundredths of one per cent. if there had been any need of such accuracy (and if the freezing point measurements, for para content, were correspondingly accurate).

If analyses were desired covering the entire composition range, it would be a simple matter to set up a graph showing constant refractive index curves on the ternary composition axes.

With the Abbe instrument it is barely possible to detect the difference between the true refractive index of a mixture and that calculated as the sum of the mole fractions times the refractive indices of the pure substances.

Equilibrium among the Xylenes.—An examination of the data of Norris and Vaala,[25] reproduced in Table XI, shows that these workers definitely obtained equilibrium between meta and para-xylene. In fact their data indicate the reaction to be going past equilibrium which is of course impossible and must be due to experimental error, probably in analysis. With respect to o-xylene, however, their results are less certain. Starting with either meta or para, 14% ortho was obtained after five hours while seventy-five hours gave only 16% ortho. This strongly suggests that equilibrium was attained, but this composition was never reached starting from pure ortho. Only one experiment was done with ortho, of only five-hour duration.

TABLE XI

EQUILIBRIUM COMPOSITION OF THE XYLENES FROM EXPERIMENTS BY NORRIS AND VAALA

(Temperature, 50°C.)

Initial composition	Expts.	Time, hr.	% still xylene	Final composition, % ortho	meta	para
meta	1	5	88	14	74	12
meta	7	18–75	72–85	16	65–69	15–19
para	1	5	83	14	76	10
ortho	1	5	94	40	55	5

It seemed desirable to show definitely that starting with pure o-xylene the concentration of that isomer would approach the value near 14%. In our experiments starting with pure *o*-xylene excessive side reaction occurred before this end was attained. However, by starting with a mixture of *o*-xylene with toluene and benzene, the net side reactions were reduced to a small percentage and the desired result demonstrated. The numerical data are shown in Table XII together with other results which confirm the work of Norris and Vaala. As "best" values of the equilibrium composition of the liquid at 50°, one may take 12 ± 3% ortho, 71 ± 5% meta, 17 ± 2% para. These figures will not change very rapidly with temperature. Considering the vapor pressures of each isomer (given in Part I) the values for the gas phase become 10% ortho, 72% meta and 18% para with the same limits of error as in the liquid. The equilibrium composition at other temperatures will be presented below.

TABLE XII

EQUILIBRIUM COMPOSITION OF THE XYLENES FROM EXPERIMENTS OF THIS RESEARCH

(Catalyst: $AlBr_3$ + HBr anhydrous)

Initial composition	Temp., °C.	Time, days	Side reaction, %	Final composition, % ortho	meta	para
p-Xylene	30	2	20[a]	12	72	16
Benzene + toluene	50	5	15[a]	13	71	16
o-Xylene (excess) + benzene + toluene	50	6	20[a]	12	71	17

[a] In the first experiment "side reactions" include the formation of toluene, in the second and third this is, of course, excluded.

Equilibrium between Benzene, Toluene and Xylene.—So far as we are aware, a measure of the equilibrium of this reaction has not been attained previously. Table XIII shows the results of two final experiments (following many exploratory runs) which limit the equilibrium constant

$$K = N_{\text{Benzene}}N_{\text{Xylene}}/N^2_{\text{Toluene}} \quad (N = \text{mole fraction})$$

to the range 0.15 to 0.22. In this expression the total xylene is considered. The xylenes present were of equilibrium composition at the end of both experiments, as would be expected.

Considering the various vapor pressures involved and assuming Raoult's law, one obtains values for the equilibrium constant in the gas phase which do not differ appreciably from those for the liquid.

TABLE XIII

EQUILIBRIUM MEASUREMENTS ON THE REACTION

2 Toluene = Benzene + Xylene

(Temperature, 50°C.; Catalyst, $AlBr_3$ + HBr anhydrous)

Initial composition, % Benzene	Toluene	Xylene	Time, days	Side reactions, %	Final composition, % Benzene	Toluene	Xylene	K
50	50	0	5	15	54	41	4.7	0.15
55	36	9	6	20	58.5	36.5	5.0	0.22

The analyses for benzene, toluene and total xylene were made by fractional distillation in a column of approximately 25 theoretical plates.

Catalysts and Side Reactions.—Norris and Vaala used anhydrous aluminum chloride with and without hydrogen chloride for catalysts. A few experiments employed ferric chloride. All gave essentially the same results. It was found in this research that anhydrous aluminum chloride gave little or no reaction in the absence of hydrogen chloride. Exploratory experiments also indicated that anhydrous aluminum bromide and hydrogen bromide was a more effective catalyst than the chlorides. Therefore all final experiments were made with the latter substances.

The aluminum bromide was prepared by direct union of the elements; the hydrogen bromide by the action of bromine on phenol. The concentration used was about 5 mole per cent. aluminum bromide (as Al_2Br_6). For each mole of aluminum bromide present, approximately two moles of hydrogen bromide were passed into the mixture. At the end of the desired time the catalyst was destroyed by pouring the mixture on cracked ice and subsequent shaking of the hydrocarbon phase with portions of pure water.

It was found in general that tar-forming side reactions proceeded fastest with the more highly methylated compounds. Thus in the last experiment in Table XIII the ratio of total methyl groups present to benzene rings present decreased during the course of the reaction. Another side reaction which it was desirable to suppress was that forming tri- and tetramethylbenzenes. This would tend to consume xylene and to prevent equilibrium measurements on other reactions until equilibrium amounts of these substances were formed.

By starting with mixtures rich in benzene and

TABLE XIV

THERMODYNAMIC CONSTANTS FOR CERTAIN REACTIONS

All values for gases at 50°C., 323°K.; units, cal. and cal./degree

Reaction	ΔF	ΔH	ΔS
m-Xylene = *p*-xylene	890 ± 100	460 ± 150	−1.33 ± 0.3
m-Xylene = *o*-xylene	1270 ± 150	950 ± 200	−1.00 ± .3
p-Xylene = *o*-xylene	380 ± 250	490 ± 250	+0.33 ± .3
m-Xylene = xylene (equil. mixt.)	−210 ± 40	180 ± 40	1.20 ± .2
2 Toluene = benzene + xylene (equil.)	1100 ± 200	600 ± 300	−1.54 ± .6
2 Toluene = benzene + *m*-xylene	1310 ± 200	420 ± 300	−2.74 ± .6

toluene it was possible to obtain, at equilibrium, measurable amounts of xylene but practically negligible amounts of trimethylbenzenes. This condition also tended to suppress the tar forming side reactions.

Free Energies and Heat Contents.—Using the data presented in the preceding sections together with the entropy values obtained in Part I, one obtains the results shown in Table XIV. Although the values are strictly for the gas phase at 50°, they are only slightly different in the liquid phase. A change to 25° has similarly only a small effect. Further thermodynamic calculations and discussion will be presented after the vibration frequencies and statistical mechanical analysis is completed.

Part III. The Vibration Frequencies of Benzene

Before considering the vibration frequencies of the methyl derivatives, it is desirable to make a complete, although not absolutely certain, vibrational assignment for benzene itself. The most recent researches in this field are those by Ingold and his collaborators[27] and by Langseth and Lord[28] on the spectra of the deuterated benzenes, making use of the Teller–Redlich product rule. These authors give extensive discussions of the theoretical aspects of this problem, hence they will not be repeated here. The present status may be described briefly as follows. Of the twenty fundamental frequencies of benzene, eleven are active in either the Raman or infrared spectra and are definitely established. According to Wilson's[29] numbering scheme, see Fig. 6, these are 1, 2, 6, 7, 8, 9, 10, 11, 18, 19, 20. Vibrations 12 and 13 have been definitely placed by Langseth and Lord through consideration of the deuterated benzene spectra. These authors also established rather definitely the product of frequencies 4 and 5 by the same method and assigned no. 16 to the weak Raman line at 400 cm.$^{-1}$ which appears in violation of the selection rules. This leaves four frequencies for which tentative values were suggested in the cases of no. 3 and 17 and none given for no. 14 and 15.

Although no new spectral data are available, reliable heat capacity values for gaseous benzene are presented in Part I of this paper which can be used to check a tentative complete assignment. Also, it is believed that several of the missing fundamental frequencies can be found among the weak lines in the Raman and infrared spectra of liquid benzene. The line at 400 cm.$^{-1}$ has already been assigned as no. 16. These frequencies are forbidden by the selection rules but do appear, presumably because of the perturbing effect of liquid forces. Particularly to be considered are lines which occur in the infrared absorption of the liquid but are absent for the gas.

Several normal coördinate analyses were made using approximate but reasonable potential functions.[30] These served to indicate approximately where a given line should appear, but did not give exact enough agreement to justify presentation in detail. The normal coördinates shown in Fig. 6 resulted from these calculations. Although these data are only of semi-quantitative significance, they were valuable in connection with the assignments for the methyl derivatives.

Planar Hydrogen Bending Vibrations 3 and 15.—To start the assignments, let us consider first nos. 3 and 15. Normal coördinate calculations place these frequencies in the range 1150–1350 cm.$^{-1}$ with no. 3 probably the higher. In Table XV the lines in the Raman and infrared spectra of benzene are listed. All lines are included except those so weak as to be of doubtful existence. On this basis 1298 is an obvious choice for no. 3. Before adopting it, however, the spectra of the para benzenes d_2 and d_4 must be considered, because no. 3 becomes Raman active and Langseth and Lord suggested the value 1195 on this basis. Table XVI gives these data in the region of interest. No. 3 falls in the symmetry group B_{2g} along with another hydrogen planar bending mode, 9*b*. These two modes become hopelessly mixed in the deuterated compounds. We follow Langseth and Lord in arbitrarily calling 3 the higher frequency in d_2, and 9*b* the higher in the d_4 compound. Following Langseth and Lord for all other frequencies in this symmetry group but using 1298 for no. 3 in benzene, one calculates 1266 for no 3 in para benzene d_2. Considering anharmonicity, this is in satisfactory

(27) Angus, Bailey, Hale, Ingold, Leckie, Raisin, Thompson and Wilson, *J. Chem. Soc.*, 912–987 (1936); *Nature*, **139**, 880 (1937).

(28) Langseth and Lord, *Kgl. Danske Videnskab. Selskab, Math-fys. Medd.*, **16**, No. 6, 1–85 (1938).

(29) E. B. Wilson, *Phys. Rev.*, **45**, 706 (1934).

(30) Use was made of the formulas of Manneback, *Am. Soc. Sci. Bruxelles*, Ser. B, **55**, 129 (1935); Barnard Manneback and Verleysen, *ibid.*, Ser. I, **59**, 376 (1939); **60**, 45 (1940).

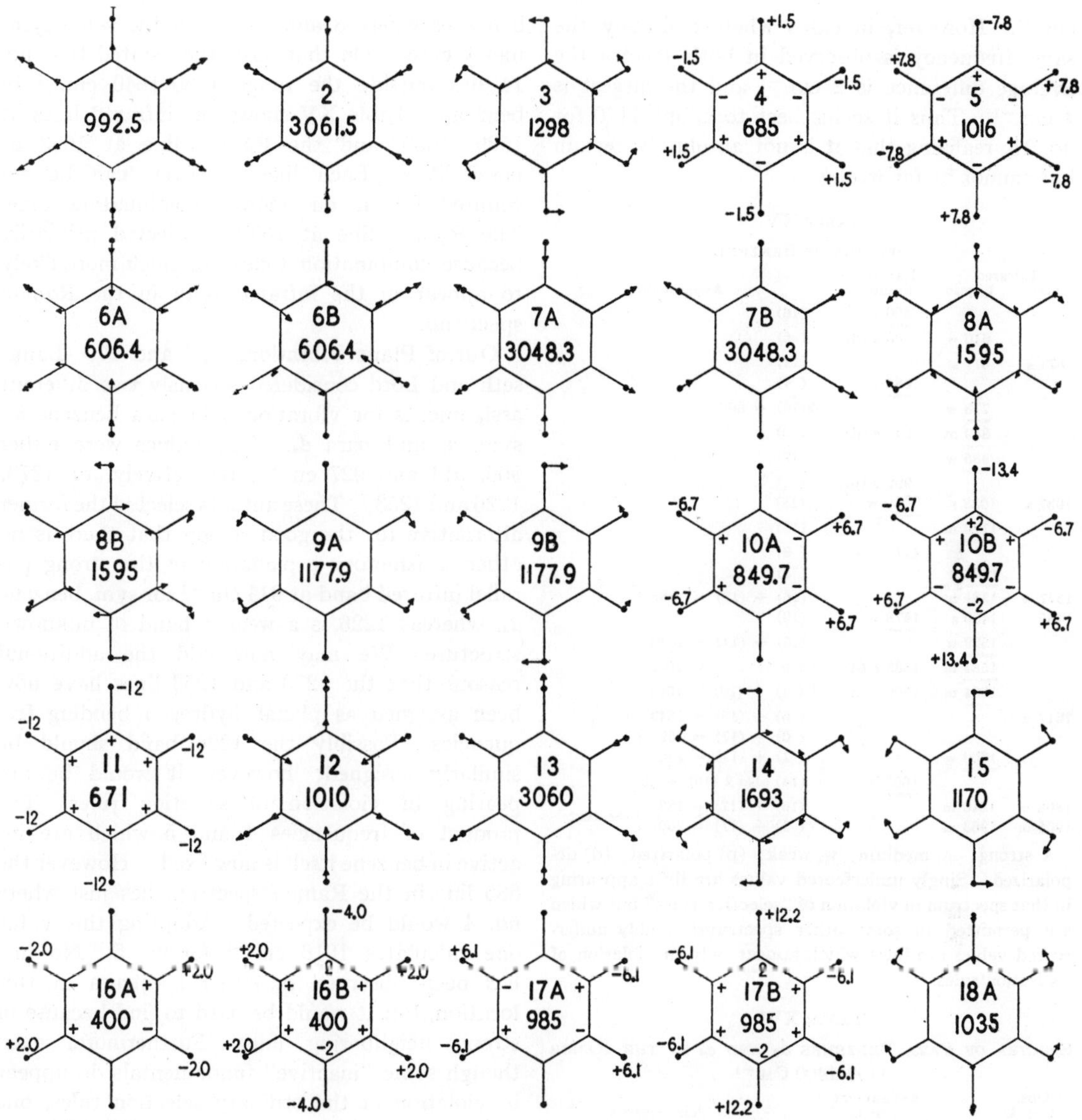

Fig. 6.—The frequencies and normal modes of vibration of benzene; + and − refer to motions perpendicular to the plane of the ring.

agreement with the observed line at 1273 cm.$^{-1}$. For no. 9*b* in para benzene d_4 one calculates 1242 cm.$^{-1}$ as compared to the observed line at 1253 cm.$^{-1}$. Finally, the discarded lines at 1175 and 1137 cm.$^{-1}$ can be satisfactorily accounted for. Thus the value 1298 is quite satisfactory for no. 3.

The only other available line in this range for no. 15 is at 1170 cm.$^{-1}$. This lies almost within experimental error of the Raman line at 1178 cm.$^{-1}$. However, in cases where definitely the same frequency is observed in both spectra the average difference is 2 cm.$^{-1}$ and the largest is 4 cm.$^{-1}$. Thus it seems best to adopt 1170 for no. 15, realizing that it is not absolutely certain but cannot be far wrong.

TABLE XV

SPECTRA OF BENZENE

Infrared Gas	Infrared Liquid	Raman liquid	Assignment
		400 w	(16)
	610 w	606 s (d)	(6)
671 s	671 s		(11)
		685 w	(4)
	773 w		2(16) = 800
	849 m	849 s (d)	(10)
	985 w		(17)
		992 s (p)	(1)
1037 s	1033 s	1030 w	(18)
	1170 m		(15)
		1178 s (d)	(9)
	1298 w		(3)
1377 w	1381 m		(16) + (17) = 1385 ?
1485 s	1480 s	1478 w	(19)
	1529 w		(10) + (11) = 1520
	1584 m	1585 s (d)	{ (8) = 1596
	1604 m	1606 s (d)	{ (1) + (16) = 1598
1617 w			(6) + (18) = 1643, or (6) + (17) = 1591 ?
	1669 w		(1) + (11) = 1663
		1693 w	(14) also 2 (10) = 1698
1808 m	1810 m		(10) + (17) = 1834 } see text
1965 m	1963 m		(5) + (17) = 2001 } see text

s. strong; m, medium; w, weak; (p) polarized; (d) depolarized. Singly underscored values are lines appearing in that spectrum in violation of "selection rules" but which are permitted in some other spectrum; doubly underscored values are lines which appear only in violation of "selection rules."

TABLE XVI

SPECTRA OF PARA BENZENES d_2 AND d_4 IN THE RANGE 1100–1400 CM.$^{-1}$

Obs. L + L	Assignment L + L	This research
	para-$C_6H_4D_2$	
1167 (2)	(3)	resonance (9*a*) with 2(6*a*) = 1186
1175 (4)	(9*a*)	
1273 (0*d*)	2(4) = 1268	(3)
	para-$C_6H_2D_4$	
1137 (1)	(9*b*)	2(6*b*) ≤ 1170; (10b) + (16*a* or *b*) ≅ 1125
1253 (1*d*)	(6) + (10*a*) = 1247	(9*b*)

Carbon Bond Stretching Vibration 14.—This frequency may be expected in the range 1500–1900 cm.$^{-1}$, probably above 1600 cm.$^{-1}$. Langseth and Lord found that no. 8 (1596 in benzene) showed no sign of interaction with a nearby fundamental in those unsymmetrically deuterated benzenes where 8 and 14 fall in the same symmetry class. On that basis they stated that no. 14 lies outside the range 1560–1640 cm.$^{-1}$ in benzene. Table XV shows the infrared lines at 1529, 1669, and the Raman line at 1693 as possibilities. Each line, however, can be accounted for as an allowed combination tone. The Raman line at 1693 is selected primarily because combination tones are much more likely to appear in the infrared than in the Raman spectrum.

Out of Plane Vibrations 4, 5 and 17.—Langseth and Lord considered seriously two different assignments for vibration 5 in para benzene d_2, sym. d_3 and para d_4. The values were either 966, 914 and 927 cm.$^{-1}$, respectively, or 1273, 1226 and 1253. These authors selected the former alternative for the good reason that there is no other satisfactory explanation of the strong parallel infrared band at 914 cm.$^{-1}$ for sym. benzene d_3, whereas 1226 is a weaker band of unknown structure. We may now add the additional reasons that the 1273 and 1253 lines have now been assigned as planar hydrogen bending frequencies. Possibly the 1226 band should be similarly assigned; however, it would be appearing in violation of selection rules. The product of frequencies 4 and 5 which are inactive in benzene itself is now fixed. However the 685 line in the Raman spectrum lies just where no. 4 would be expected. Adopting this value one calculates 1016 cm.$^{-1}$ for no. 5. No line has been observed in either spectrum at this location, but it would be hard to find because of strong neighboring lines. Furthermore, even though some "inactive" fundamentals do appear in violation of the ordinary selection rules, one certainly cannot as yet be sure that all should.

In order to locate no. 17 one considers the spectrum of sym. benzene d_3, where it becomes Raman active, in addition to that of benzene itself. Langseth and Lord suggested 815 cm.$^{-1}$ for no. 17 in the sym. d_3. This value for no. 17 is unsatisfactory because the calculated heat capacity for benzene is then too high and because it leads to an unreasonable potential function. Further evidence is found in the very recent data on the heat capacity of solid benzene d_6 obtained by Ziegler and Andrews[30a] which led them to suggest that the Langseth and Lord assignment (815 cm.$^{-1}$) for no. 17 was too low.

(30a) Ziegler and Andrews, THIS JOURNAL, **64**, 2482 (1942).

TABLE XVII

FUNDAMENTAL FREQUENCIES OF BENZENE, SYM. BENZENE d_3 AND BENZENE d_6

Symmetry classes		No.	C_6H_6			sym. $C_6H_3D_3$			C_6D_6		
A_1'	A_{1g}	1	992	R(p)	—	955	R(p)	—	945	R(p)	(IR)
		2	3062	R(p)	—	3055	R(p)	—	2292	R(p)	—
		12	(1011)	—	—	1002	R(p)	—	977	(R)	(IR)
	B_{1u}	13	(3046)	—	—	2284	R(p)	—	(2268)		
A_2'	A_{2g}	3	1298	—	(IR)	1226	—	(IR)?	1010	(R)?	(IR)?
	B_{2u}	14	1693	(R)	—	—			(1665)	—	—
		15	1170	—	(IR)	—			(841)	—	—
E'	E^+_g	6	606	R(d)	(IR)	592	R	IR	578	R(d)	(IR)
		7	3047	R(d)	—	2272	R	IR	2265	R(d)	—
		8	1596	R(d)	(IR)	1574	R	IR	1556	R(d)	(IR)
		9	1178	R(d)	—	1100	R	IR	868	R(d)	(IR)
	E^-_u	18	1037	(R)	IR	833	R	IR	813	—	IR
		19	1485	(R)	IR	1405	R	IR	1333	(R)	IR
		20	3080	—	IR	3055	R	IR	2294	—	IR
A_2''	B_{2g}	4	685	(R)	—	691	—	IR	—		
		5	(1016)	—	—	914	—	IR	—		
	A_{2u}	11	671	—	IR	533	—	IR	503	—	IR
E''	E^-_g	10	849	R(d)	(IR)	710	R(d)	—	661	R(d)	(IR)
	E^+_u	16	400	(R)		374	R		~350	(R)	
		17	985	—	(IR)	(950)	R?	—	790	(R)	—

Parentheses indicate appearance in spectra of liquids in violation of the simple selection rules for gases. Letters indicate appearance in the indicated spectrum, dashes indicate absence, no mark indicates that spectrum was not taken in that region. (p), polarized, (d), depolarized.

A satisfactory assignment for vibration 17 is the weak infrared band at 985 cm.$^{-1}$ in benzene and the weak Raman line at 790 cm.$^{-1}$ in benzene d_6. This places the line at 950 in sym. benzene d_3 where it should be Raman active but would be masked by the very strong line at 955 cm.$^{-1}$. The 985 infrared band in benzene is a mere shoulder on the strong band at 1037, and might easily be explained as no. 1 (992 cm.$^{-1}$). However, it is very difficult to explain the rather strong infrared bands at 1810 and 1965 cm.$^{-1}$ which appear for both gas and liquid except as the allowed combination tones (10) + (17) and (5) + (17), respectively. Both calculated values are a bit high, but this could be easily corrected by lowering 17 to about 970 cm.$^{-1}$ which would still be consistent with the shoulder on the 1037 infrared band. It might also be desirable to lower 5 a few units, a perfectly permissible change. These last considerations offer considerable support to an otherwise doubtful assignment.

Conclusion.—The complete Raman and infrared spectra (barring lines so extremely weak as to be doubtful) as listed in Table XV are now satisfactorily explained. All of the fundamental frequencies are reasonable in terms of approximate potential functions. Table XVII lists these frequencies. A majority of these values have been given previously.[27,28] The heat capacity calculated using this assignment is in excellent agreement with the experimental values as shown in Fig. 7 of Part V. Even with all this evidence one cannot definitely conclude that this assignment is correct, but rather one must say that it is by far the most probable in view of the available information.

Several authors[27,28,29] have given sets of symmetry coördinates for benzene. So far as the writer is aware, however, Fig. 6 gives the true normal coöordinates of benzene for the first time. The values are of course only semi-quantitative, having arisen from approximate potential functions. They will be very useful, nevertheless, in discussing substituted benzenes.

Part IV. The Vibration Frequencies of Toluene, *o*-, *m*- and *p*-Xylenes and Mesitylene

The Raman spectra, including polarization data, and the infrared absorption spectra, have been measured for one or more of these substances by several workers.[31–40] Some of these authors

(31) Ananthakrishnan, *Proc. Indian Acad. Sci.*, **3A**, 52 (1936) (Raman for toluene).

(32) Kohlrausch, *Wien Ber.*, **142**, 637 (1933) (Raman for toluene).

identified certain lines, but no attempt at complete assignments has been made heretofore. We have available, in addition to the spectra, accurate values of the heat capacity and entropy for all substances except mesitylene and a complete vibrational assignment for benzene including the normal coördinates.

A very important aid has been the use of an approximate product rule *analogous* to the Teller–Redlich product rule for isotopic substitution. With the substitution of methyl groups for hydrogen atoms, not only the mass but also the force constants and bond distances change. In addition the methyl group has internal vibrations which may interact with ring motions. However, these complicating effects should be nearly the same for each substituted methyl group. Therefore, although the product ratio cannot be calculated accurately from theory, its value should remain closely constant for similar groups of vibrations in the various molecules. Actually the values obtained are very close to those calculated by considering only changes in mass and bond distance.

The product ratio may be defined as follows

$$\tau = \frac{\prod_i \nu_{s_i} \prod_j I_{u_j}^{1/2} \prod_k M_{u_k}^{1/2}}{\prod_i \nu_{u_i} \prod_j I_{s_j}^{1/2} \prod_k M_{s_k}^{1/2}}$$

where s indicates the value for the substituted compound and u that for the unsubstituted, ν is a vibration frequency, I a moment of inertia and M a mass. The products are to be taken over all the degrees of freedom of a given symmetry class. This definition differs from that commonly used for the isotopic product rule by including the masses and moments of inertia in the observed product. This is desirable because no exact theoretical formula for τ can be given in this case, hence we wish to put all observed τ's on a directly comparable basis.

(33) (a) Kohlrausch and Pongratz, *Monatsh*, **63**, 427 (1933) (Raman for the xylenes); (b) *Wien Ber.*, **143**, 358 (1934) (Raman for mesitylene).

(34) Cabannes and Rousset, *Ann. Phys.*, **19**, 229 (1933), (Raman for toluene).

(35) Narkiewicz, *Z. Physik*, **96**, 177 (1935) (Raman for xylenes).

(36) Desreux, *Bull. soc. chim.*, [5] **3**, 1659 (1936) (Raman for all subst.).

(37) Venkateswaran, *Phil. Mag.*, **15**, 263 (1933); *Proc. Roy. Soc.* (London), **128A**, 252 (1930) (Raman for mesitylene).

(38) Lambert and Lecompte, *Ann. phys.*, [10], **18**, 329 (1932); [11] **10**, 503 (1938) (infrared for all subst.); Lecompte, *J. Phys. Radium*, **8**, 489 (1937); **9**, 13 (1938).

(39) Coblenz, *Astrophys. J.*, **20**, 207 (1904) (infrared for all but *m*-xylene)

(40) Barchewitz and Perodi, *J. Phys. Radium*, **10**, 142 (1939).

One can calculate approximate theoretical τ's for methyl substitution as follows. For each methyl-ring stretching motion the value is the square root of the mass ratio, $(1/15)^{1/2} = 0.258$. For each bending motion one considers the moments of inertia of the attached group with respect to the point of attachment on the ring obtaining $(1.2/41)^{1/2} = 0.171$.

Throughout this work we have ignored the spectrum in the region around 3000 cm.$^{-1}$ because, although there are always several lines, there are never enough resolved to account for all expected. We have arbitrarily assigned benzene-H stretching modes the value from benzene and C-H stretching within a methyl group the value 2950 cm.$^{-1}$.

Mesitylene.—Let us consider first the substance of highest symmetry and therefore greatest simplicity, *sym*-trimethylbenzene. Table XVIII gives first the symmetry classes and their selection rules, second the benzene and methyl group (M) vibrations falling in these various classes, third the frequencies for benzene, fourth the assigned frequencies for mesitylene, fifth the appearance of these lines in the spectra, and finally the product ratios. The various benzene vibrations are depicted in Fig. 6. The methyl group motions are as follows: M1, sym. stretching; M2, unsym. stretching; M3, sym. bending; M4, unsym. bending within the methyl group; M6, wagging of the whole methyl group. The even numbered motions are degenerate in a methyl halide. In these compounds the degeneracy is removed, the vibration appearing in two different symmetry groups. The letters R and IR indicate appearance in the corresponding spectrum, —indicates absence, no mark indicates that the spectrum was not taken or not examined in this region and parentheses indicate appearance in violation of selection rules. The state of polarization is given for the Raman lines where it is known.

Although the actual modes become so mixed up as to make numbering arbitrary, there can be no doubt about the members of the totally symmetric, A_1, class. For the calculation of the product ratio one must decide between 1300 and 1380 as no. 13 and M3. Since a strong, polarized line always appears at 1380 cm.$^{-1}$ in methyl substituted benzenes, whereas the line below it varies from 1200 to 1300 cm.$^{-1}$, the assignment of 1380 to M3 is definite. One then obtains the

TABLE XVIII

VIBRATIONAL ASSIGNMENT FOR MESITYLENE

(Methyl groups in positions 1, 3, 5)

Symmetry class	No.	Benzene	Mesitylene			Ratio of products
	1	992	998	R(p)	—	0.244
	2	3062	(3062)			
A_1	12	1011	575	R(p)	?	
$R_{(p)}$ —	13	3046	1300	R(p)	—	
	M3	—	1380	R(p)	—	
	M1	—	(2950)			
	6	606	515	R(d)	IR	.0425 ‖ (.171) (.248)
	7	3047	(3047)			
	8	1596	1610	R(d)	IR	
	9	1178	1160	R	IR	
	18	1037	230	R(d)		
E′	19	1485	1385	?	IR	
$R_{(d)}$, IR	20	3080	930	R	IR	
	xy trans.	M = 78	M = 120			
	M6	—	1025	—	IR	
	M4	—	1470	—	IR	
	M3	—	1380	R	—	
	M2	—	(2950)			
	M1	—	(2950)			
	10	849	275	R(d)		.171
E″	16	400	(439)	(obs. in meta xylene)		
$R_{(d)}$ —	17	985	847	R(d)	—	
	xy rot.	I = 145	I = 481			
	M6	—	(1200)	—	—	
	M4	—	1440	R	—	
	M2	—	(2950)			
	4	685	690	—	IR	.179
	5	1016	840	—	IR	
A″	11	671	182	(R)		
— IR	z trans.	M = 78	M = 120			
	M6	—	1025	—	IR	
	M4	—	1480	—	IR	
	M2	—	(2950)			
	3	1298	(1220)	—	—	(.171)
	14	1693	(1650)	—	—	
A_2'	15	1170	(390)	—	—	
— —	z rot.	I = 291	I = 947			
	M6	—	(1200)			
	M4	—	(1450)			
	M2	—	(2950)			

product ratio, 0.244, which is somewhat lower than the theoretical, 0.258, due almost certainly to a slightly lower force constant for C–C than for C–H bond stretching.

In assigning the degenerate modes it is desirable to note that certain of the frequencies should be either exactly or approximately the same as in the *m*-xylene. There the degeneracy is removed, one component changing and the other not. From Fig. 6, Part III, it is apparent nos. 6b, 9a, 10a, 16a, 17a and 18a should fall in this general category, some changing very slightly, 10, 16 and 17 not at all. With this in mind, and the selection rules, the assignment of Table XVIII is quite definite. No line appears in the mesitylene spectra where no. 16 would be expected. A very weak line is observed at 439 cm^{-1} in the Raman for *m*-xylene, which is assigned to 16a there, and can be taken for 16 in mesitylene. The product rule is well satisfied.

In the A_2'' class which is only active in the infrared, one would expect no. 4 to be raised slightly. 5 and 11 will recombine to give an out of plane motion of the hydrogens (arbitrarily called 5) and one of the methyl groups. The former should be near the average of 5 and 11 in benzene, the latter very low. Thus 690 and 840 are very satisfactory for 4 and 5. The infrared spectrum does not continue to long enough wave lengths to find 11. Two doubtful Raman lines have been reported, each by only one investigator, at 145 and 182 cm.$^{-1}$. The higher fits the product rule better and is therefore adopted.

The frequencies in the A_2' group are quite arbitrary since they are inactive in both the Raman and infrared. However, they are of such a nature as to make reasonably accurate estimates possible.

Probably the most important results arising from the mesitylene assignment are the reliable values of the product ratios in the A_1, E', and E'' classes.

***p*-Xylene.**—The assignments for this substance, and those to follow, were made by the same general methods. With the aid of the polarization data and considerations of general reasonableness the A_{1g} class is quite definite. The product ratio seems a trace small but is satisfactory. The only point requiring comment in remaining Raman active classes is the use of 313 for both no. 9b and 10b. One might wish to assign the faint line at 389 cm.$^{-1}$ to one or the other of these vibrations. The product rule prevents this unless utterly unreasonable values are assigned to other frequencies.

In the infrared active classes the assignments are quite satisfactory. There are no data below 500 cm.$^{-1}$ but the frequencies (18b) = 232 and (11) = 170 appear in combination tones: 540 obs. = (9b) + (18b) = 545 calcd. and 635 obs. = (6a) + (11) = 630 calcd. Vibration 16b must lie fairly near 400 cm.$^{-1}$; thus the faint Raman line at 389 is quite acceptable. Nos. (11) and (16b) appear in the Raman in violation of selection rules. The completely inactive class A_{1u} fortunately comprises vibrations essentially unchanged from benzene so that no real doubt arises here.

On the whole the assignment for *p*-xylene may be classed as very satisfactory.

***m*-Xylene.**—As has been already mentioned, a number of the frequencies in the *m*-xylene should be the same as in mesitylene and have been so assigned. The symmetry C_{2v} of *m*-xylene (and of toluene and *o*-xylene) permits all lines in the Raman and all but one class in the infrared. Thus, although all frequencies may be expected

TABLE XIX

VIBRATIONAL ASSIGNMENT FOR *p*-XYLENE

(Methyl groups in positions 1 and 4)

Symmetry class	No.	Benzene	*p*-Xylene			Ratio of products
A_{1g} $R_{(p)}$ –	1	992	826	R(p)	—	0.238
	2	3062	(3062)			
	6a	606	460	R(p)		
	7a	3047	1204	R(p)	—	
	8a	1596	1616	R	—	
	9a	1178	1182	R	—	
	M3	—	1377	R(p)	—	
	M1	—	(2950)			
B_{1g} $R_{(d)}$ –	10a	849	811	R	—	.95
	x rot.	I = 145	I = 156			
	i rot.	—	I = 5.09			
B_{2g} $R_{(d)}$ –	4	685	697	R	—	.179
	5	1016	959	R	—	
	10b	849	~313	R ?	—	
	y rot.	I = 145	I = 578			
	M6	—	1152	R	—	
	M4	—	1451	R(d)	—	
	M2	—	(2950)			
B_{3g} $R_{(d)}$ –	3	1298	~1200	R ?	—	.170
	6b	606	644	R(d)	—	
	7b	3047	(3047)			
	8b	1596	1575	R(d)	(IR?)	
	9b	1178	313	R(d)		
	z rot.	I = 291	I = 724			
	M6	—	1152	R	—	
	M4	—	1451	R ?	—	
	M2	—	(2950)			
A_{1u} — —	16a	400	(400)	—		.99
	17a	985	971	(R)	(IR?)	
	i rot.	—	I = 5.09			
B_{1u} — IR	12	1010	720	—	IR	.245
	13	3046	(3046)			
	18a	1037	1030	—	IR	
	19a	1485	1525	—	IR	
	20a	3080	1200	—	IR	
	x trans.	M = 78	M = 106			
	M3	—	1380	—	IR	
	M1	—	(2950)			
B_{2u} — IR	14	1693	1640	—	IR	.171
	15	1170	1100	—	IR	
	18b	1037	(232)	—	IR ?	
	19b	1485	1450	—	IR	
	20b	3080	(3080)			
	y trans.	M = 78	M = 106			
	M6	—	~1100	—	IR ?	
	M4	—	~1450	—	IR ?	
	M2	—	(2950)			
B_{3u} — IR	11	671	170	(R)		.172
	16b	400	389	(R)		
	17b	985	800	—	IR	
	z trans.	M = 78	M = 106			
	M6	—	~1100	—	IR	
	M4	—	~1450	—	IR ?	
	M2	—	(2950)			

in the spectra, the selection rules are of little help in sorting them out. This must be compensated by reasoning which may be described as interpolation between benzene and mesitylene. Everything considered the assignment is quite satisfactory, although it is certainly not unique.

***o*-Xylene.**—By far this is the worst assignment of the group. Two Raman active frequencies (16a and 18a) must lie in the 300–500 range where none are observed. Nor can the remaining fre-

TABLE XX

VIBRATIONAL ASSIGNMENT FOR *m*-XYLENE

(Methyl groups in positions 3 and 5)

Symmetry class	No.	Benzene	*m*-Xylene			Ratio of products
A_1 $R_{(p)}$ IR	1	992	998	R(p)	—	
	2	3062	(3062)			
	6a	606	538	R	—	
	7a	3047	(3047)			
	8a	1596	1592	R		
	9a	1178	1167	R	—	(0.0422) ‖ (.171)(.247)
	12	1011	724	R	—	
	13	3046	(3046)			
	18a	1037	224	R		
	19a	1485	(1320)	—		
	20a	3080	1247	R(p)	—	
	x trans.	M=78	M=106			
	M6	—	1090	R	IR	
	M4	—	1448	R	IR	
	M3	—	1375	R(p)		
	M2	—	(2950)			
	M1	—	(2950)			
B_1 $R_{(d)}$ IR	3	1298	1262	R	—	
	6b	606	514	R	IR	
	7b	3047	945	R	IR	
	8b	1596	1613	R(d)		
	9b	1178	1152	R	IR	
	14	1693	(1660)	—		
	15	1170	1197			(.0419) ‖ (.171)(.245)
	18b	1037	310	R	IR	
	19b	1485	(1470)	—		
	20b	3080	(3080)			
	y trans.	M=78	M=106			
	z rot.	I=291	I=699			
	M6	—	1031	R	IR	
	M4	—	1448	R	IR	
	M3	—	(1350)	--	—	
	M2		(2950)			
	M1		(2950)			
A_2 $R_{(d)}$ —	10a	849	276	R		
	16a	400	439	R		.169
	17a	985	830	R ?	—	
	x rot.	I=145.3	I=475			
	i rot.	—	I=5.23			
	M6	—	1090	R	—	
	M4	—	(1470)	?	?	
	M2	—	(2950)			
B_2 $R_{(d)}$ IR	4	685	690	—	IR	
	5	1016	890	—	IR	
	10b	849	770	R	IR	
	11	671	200	R		
	16b	400	485	R	IR	.177
	17b	985	880	—	IR	
	y rot.	I=145.3	I=233			
	z trans.	M=78	M=106			
	i rot.	—	I=5.23			
	M6	—	1031	R	IR	
	M4	—	(1470)	?	?	
	M2	—	(2950)			

quencies be reliably enough assigned to permit definite product rule calculations for these two. Many of the individual frequencies are definite, but many others are merely reasonable. In a

TABLE XXI

VIBRATIONAL ASSIGNMENT FOR *o*-XYLENE

(Methyl groups in positions 2 and 3)

Symmetry class	No.	Benzene	*o*-Xylene			Ratio of products
	1	992	733	R(p)	?	
	2	3062	(3062)			
	6a	606	582	R(p)	IR	
	7a	3047	1223	R(p)	?	
	8a	1596	1584	R	IR	
	9a	1178	1117	R	IR	(0.409)
	14	1693	(1680)	—		‖
A_1	15	1170	1156	R	—	(.170)(.240)
$R_{(p)}$, IR	18b	1037	180	R		
	19b	1485	(1550)		—	
	20b	3080	(3080)			
	y trans.	M=78	M=106			
	M6	—	1051	R(p)	?	
	M4	—	1450	R	IR	
	M3	—	1386	R(p)	—	
	M2	—	(2950)			
	M1	—	(2950)			
	3	1298	1240	—	IR	
	6b	606	506	R(d)	IR	
	7b	3047	1275	—	IR	
	8b	1596	1606	R(d)	IR	
	9b	1178	1015	—	IR	(.0420)
	12	1011	740	?	IR	‖
	13	3046	(3046)			(.172)(.244)
B_1	18a	1037	(360)	—		
R, IR	19a	1485	1450	R	IR	
	20a	3080	(3080)			
	z rot.	I=291	I=639			
	x trans.	M=78	M=106			
	M6	—	1156	R	—	
	M4	—	1450	R	IR	
	M3	—	1376	R	—	
	M2	—	(2950)			
	M1	—	(2950)			
	4	685	(685)	—	—	
	5	1016	862	R	—	
	10b	849	256	R		(.173)
A_2	16a	400	(430)	—	—	
$R_{(d)}^{-}$	17a	985	823	R	—	
	y rot.	I=145.3	I=261			
	i rot.	—	I=5.17			
	M6	—	1156	R	—	
	M4	—	1450	R	IR	
	M2	—	(2950)			
	10a	849	770	—	IR	
	11	671	640	—	IR	
	16b	400	148	R		
B_2	17b	985	982	R	IR	.170
$R_{(d)}$, IR	x rot.	I=145.3	I=388			
	z trans.	M=78	M=106			
	i rot.	—	I=5.17			
	M6	—	1156	R	—	
	M4	—	1450	R	IR	
	M2	—	(2950)			

statistical sense it is supported by the heat capacity and entropy calculations.

Toluene.—This assignment is quite definite. The Raman polarization data are more extensive, which is very helpful. On the basis of polarization data and general reasonableness one may consider the A_1 class as definitely correct. Vibration 9a should be unchanged from benzene, where it is depolarized. Thus it is not surprising that it should be reported as depolarized in toluene even though it falls in the totally symmetric-polarized class. The A_2 class, although Raman active by group theory, actually consists of vibrations unchanged from benzene. Hence, one expects to find 10a of moderate intensity, 16a very weak, and 17a not observed at all. This is exactly the actual state of affairs. The remaining classes B_1 and B_2 are not quite as definite but are entirely satisfactory. The very weak 467 cm.$^{-1}$ line assigned to 16b is reported as polarized, whereas it should be depolarized. However, it cannot be a totally symmetric fundamental, nor can it be accounted for as an overtone of that symmetry. Therefore we feel justified in disregarding the polarization data.

Methyl Group Vibrations.—Discussion of the internal vibrations of the methyl groups has been largely omitted above because they are essentially the same from one molecule to another. M1 and M2 are near 2950 cm.$^{-1}$ and in virtually every case the lines overlap one another so that too few are resolved. M3 lies very close to 1380 cm.$^{-1}$ in all cases and is easily identified. Similarly M4 is found very near 1450 cm.$^{-1}$ in every case. The various components are virtually never resolved. In *o*-xylene Desreux reports continuous absorption from 1442 to 1464 cm.$^{-1}$. Only M6 appears to vary appreciably in frequency and its identification is in many cases difficult. The expected range is 1000–1200 cm.$^{-1}$. These values are all entirely reasonable by comparison to the methyl halides and other simple molecules containing methyl groups.

TABLE XXII

VIBRATIONAL ASSIGNMENT FOR TOLUENE

(Methyl group in position 1)

Symmetry class	No.	Benzene	Toluene			Ratio of products
A_1 $R_{(p)}$, IR	1	992	1002	R(p)	IR?	0.246
	2	3062	(3062)			
	6a	606	521	R(p)	IR ?	
	7a	3047	1210	R(p)	—	
	8a	1596	1603	R	IR	
	9a	1178	1175	R(?)	IR	
	12	1011	785	R(p)	IR	
	13	3046	(3046)			
	18a	1037	1030	R(p)	IR	
	19a	1485	1483	R(p)	IR	
	20a	3080	(3080)			
	x trans.	M = 78	M = 92			
	M3	—	1380	R(p)	IR	
	M1	—	(2950)			
B_1 $R_{(d)}$, IR	3	1298	1282	R(d)	—	.177
	6b	606	622	R(d)	IR	
	7b	3047	(3047)			
	8b	1596	1586	R	IR	
	9b	1178	1155	R	IR	
	14	1693	1630	R	—	
	15	1170	1070	—	IR	
	18b	1037	340	R(d)		
	19b	1485	1310	R	IR	
	20b	3080	(3080)			
	y trans.	M = 78	M = 92			
	z rot.	I = 291	I = 494			
	M6	—	1060		IR ?	
	M4	—	1495	R	IR	
	M2	—	(2950)			
A_2 $R_{(d)}$ –	10a	849	842	R	?	1.0
	16a	400	405	R(d)	—	
	17a	985	(985)	—	—	
	x rot.	I = 145.3	I = 150.6			
	i rot.	—	I_r = 5.08			
B_2 $R_{(d)}$, IR	4	685	695	—	IR	.179
	5	1016	943	—	IR	
	10b	849	730	R	IR	
	11	671	216	R		
	16b	400	467	R(?)		
	17b	985	890	R(d)	IR	
	y rot.	I = 145.3	I = 350			
	z trans.	M = 78	M = 92			
	M6	—	1190	R(d)	IR	
	M4	—	1455	R	IR	
	M2	—	(2950)			

Discussion

For the sake of brevity, tables of all observed lines and their assignments have been omitted. This is quite justifiable since the difficulty is not to account in some way for all observed lines, but rather to show which of the many possible assignments is the correct one.

One very striking feature of the Raman spectra of these and other benzene derivatives is the behavior of the strong line near 1000 cm.$^{-1}$ Kohlrausch[41] first discussed certain features of this line, but Langseth and Lord have explained it more completely. We wish to add certain other features. Vibrations 1 and 12 are separated by symmetry in benzene itself (or any completely substituted derivative), but their frequencies are so nearly equal that any small perturbation will cause them to recombine with one another. Specifically, if the perturbation affects one or more of the 1, 3 and 5 positions but does not affect the 2, 4 or 6 positions, then two "trigonal vibrations" arise from the sum and difference of 1 and 12. Here in one case only the 1, 3, 5 carbons move; in the other the 2, 4, 6 atoms move. The 2, 4, 6 trigonal mode then takes almost exactly the average of frequencies 1 and 12 which is 1001.5 cm.$^{-1}$. Thus toluene, *m*-xylene, and

(41) Kohlrausch, *Naturwissenschaften*, **25**, 635 (1937).

mesitylene have lines at 1002, 998, and 998 cm.$^{-1}$, respectively. Similarly, iodo, bromo, chloro, *m*-dichloro and *sym*-trichlorobenzenes[42,43] show, respectively, 1001, 1002, 1004, 999 and 995 cm.$^{-1}$ lines. Intensity and polarization data make this situation even more striking. This line is always one of the most intense lines of the spectrum and is highly polarized.

Similar results arise when three of the positions remain substituted with a given group. Thus the 1, 3, 5 trigonal mode in mesitylene gives a line 575 cm.$^{-1}$ which is very intense and strongly polarized. In 1,2,3,5-tetramethylbenzene[36] (isodurene) we find a line at 573 cm.$^{-1}$. In hexamethylbenzene[36] the motion changes to no. 1 and takes the slightly lower value 550 cm.$^{-1}$. Likewise in the chloro compounds, 1,3,5-trichloro, 1,2,3,5-tetrachloro, pentachloro, and hexachlorobenzenes show lines at 375, 380, 383, and 375 cm.$^{-1}$, all very intense.

Compounds in which both the 1, 3, 5 and 2, 4, 6 sets of positions are partially substituted show no comparable lines in this region. Thus for *o*-xylene they are at 827 and 1204 cm.$^{-1}$. In view of these facts it appears as if this feature could be used safely for determining the positions of substituents under suitable conditions.

Comparisons of calculated and observed heat capacities and entropies are made in Part V. It is sufficient to state here that the agreement is everywhere satisfactory.

Part V. Statistically Calculated Thermodynamic Quantities. Internal Rotation of Methyl Groups Attached to a Benzene Ring. General Discussion

As was stated in the general introduction, one of the primary purposes of this research was the determination of the barriers to internal rotation in this series of substances. We are now ready to calculate these quantities. Certain other phases also deserve discussion.

Statistically Calculated Entropies.—The molecules in this series fall in the class of rigid frames with attached symmetrical tops. Therefore, their moments of inertia for over-all rotation are independent of internal rotation. Furthermore, the tops are very light compared to the frame. Hence they may be treated by the methods of Pitzer and Gwinn.[44] It is convenient to calculate the various quantities assuming free rotation (zero potential barrier), and then to take the corrections for restriction of rotation from the tables given there. Other necessary parameters, masses, moments of inertia, and vibration frequencies are given in the tables of Part IV. The entropy results are summarized in Table XXIII. The complete symmetry number for the whole molecules is readily computed by taking the symmetry number for the corresponding benzene substituted with single atoms (such as chlorine) and multiplying it by three for each methyl group. Thus for benzene $\sigma = 12$, for toluene $\sigma = 6$, for *o*- and *m*-xylene, $\sigma = 18$, for *p*-xylene $\sigma = 36$, and for mesitylene $\sigma = 162$. The factoring of these numbers for the various degrees of freedom is much more complex and will not be discussed since it is of no importance for the desired calculations. It affects the itemization in Table XXIII but not the totals.

Values of the quantity $S + R \ln \sigma$ are also given in Table XXIII since they permit comparisons on a more equitable basis.

One may note from Table XXIII first, that the entropy of benzene checks well within the estimated error. Second, the entropies of toluene, *m*-xylene and *p*-xylene all agree with the values calculated for free internal rotation within the respective experimental errors. However, the calculated entropies are all higher than the experimental values, which suggests a slight restricting barrier. Third, the calculated entropy of *o*-xylene is definitely too large, indicating considerable restriction of rotation. Finally, the values of $S + R \ln \sigma$ increase by increments of about 11 cal. per degree per methyl group, excepting of course *o*-xylene. If one assumes this value, and that the entropy decrease for *o*-xylene (1.7 cal./deg.) is generally applicable whenever two methyl groups are adjacent, one may estimate the entropies of two more highly substituted benzenes as shown in Table XXIV.

When three or more methyl groups are adjacent even higher restrictions to rotation with correspondingly larger entropy decreases may be expected. Quantitative predictions, however, are not yet possible.

(42) Murray and Andrews, *J. Chem. Phys.*, **1**, 406 (1933); Swaine and Murray, *ibid.*, **1**, 513 (1933).
(43) Dadieu, Pongratz and Kohlrausch, *Monatsh*, **61**, 426 (1932).
(44) Pitzer and Gwinn, *J. Chem. Phys.*, **10**, 428 (1942).

TABLE XXIII

STATISTICALLY CALCULATED ENTROPIES

All values for hypothetical ideal gases at 298.16°K. and 1 atm. Units: cal. per degree mole

Entropy terms	Benzene	Toluene	*o*-Xylene	*m*-Xylene	*p*-Xylene	Mesitylene
Translation and over-all rot.	59.73	66.60	66.53	66.72	66.56	65.92
Vibr.	4.70	7.94	12.47	11.97	12.26	15.62
Free internal rotation	...	2.19	7.17	7.19	5.76	10.80
Calcd.-free rot.	64.43	76.73	86.17	85.88	84.58	92.34
Exper.	64.39 ± .3	76.44 ± .3	84.50 ± .3	85.60 ± .3	84.27 ± .3	....
$S_f - S$	...	0.29	1.67	0.28	0.31	...
S calcd. (V_0 = 2000)			84.53			
S calcd. + $R \ln \sigma$	69.37	80.26	91.91	91.61	91.68	102.43
S exp. + $R \ln \sigma$	69.33 ± .3	79.97 ± .3	90.24 ± .3	91.34 ± .3	91.39 ± .3	....

TABLE XXIV

ESTIMATED ENTROPIES FOR SOME METHYL BENZENES

(All values for hypothetical ideal gases at 298.16°K. and 1 atm.)

Subst.	$S + R \ln \sigma$	σ	S (cal./deg.)
1,2,4-Trimethylbenzene (pseudocumene)	100.7	27	94.2
1,2,4,5-Tetramethylbenzene (durene)	110.0	324	98.5

Statistically Calculated Heat Capacities.—The calculated heat capacity curves for the various gases are compared with the experimental points in Fig. 7. Potential barriers of 500 cal./mole were assumed except for *o*-xylene where the value is 2000 cal./mole. Considering all the data it appears likely that the experimental points for each substance show too much curvature because of some systematic error, possibly arising from the thermometer. The values for benzene below 400°K. from some of the other experimenters are likely to be too high because of gas imperfections.

The agreement is evidently satisfactory for benzene and for toluene, *m*-, and *p*-xylene assuming relatively free rotation of the methyl groups. In the case of *o*-xylene it is apparent as it was for the entropy that considerable restriction of rotation must be present.

Potential Barriers Restricting Internal Rotation.—The discussion above shows that for *m*- and *p*-xylene and toluene a very low potential barrier agrees with the experimental data. The gas heat capacity data set the more stringent upper limit, namely, about 1000 cal./mole.

It is of course true that the vibrational assignments are subject to question; however, no simple adjustment there will allow a higher potential barrier and still give agreement with both heat capacity and entropy. Thus increasing the potential barrier to internal rotation lowers the calculated entropy and raises the heat capacity. Lowering one or more vibration frequencies would increase the entropy but would also increase the heat capacity.

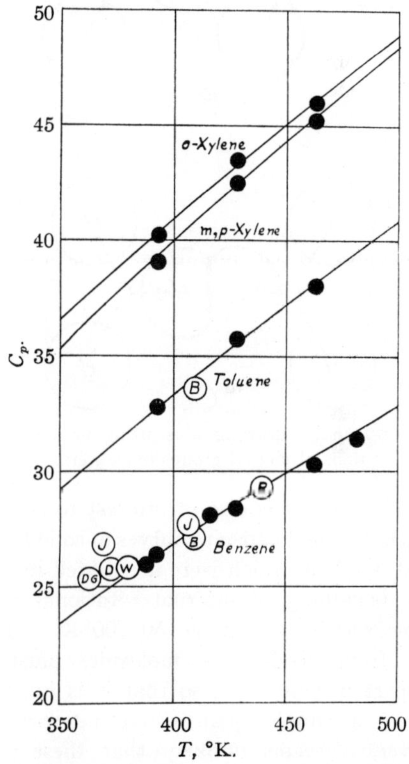

Fig. 7.—The heat capacity in the gas state of benzene, toluene, and the three xylenes: circles are experimental values; solid, this research; open, see footnotes. Curves are calculated: B, Bennewitz and Rossner[24]; D, Déjardin, *Ann. Phys.*, [9] **11**, 253 (1919); DG., Dixon and Greenwood, *Proc. Roy. Soc.* London, **105A**, 199 (1924); J, Jatkar, *J. Ind. Inst. sci.*, **22A**, Pt. II, p. 19 (1939); R, Regnault, *Mém. de l'acad. sci. France*, **26**, 1 (1862); W, Wiedemann, *Ann. chim. phys.*, **2**, 195 (1877). The very recent results of Montgomery and De Vries (THIS JOURNAL, **64**, 2375 (1942)) are not shown on the graph. They are about 1% higher than the curve for benzene and 2–3% for toluene.

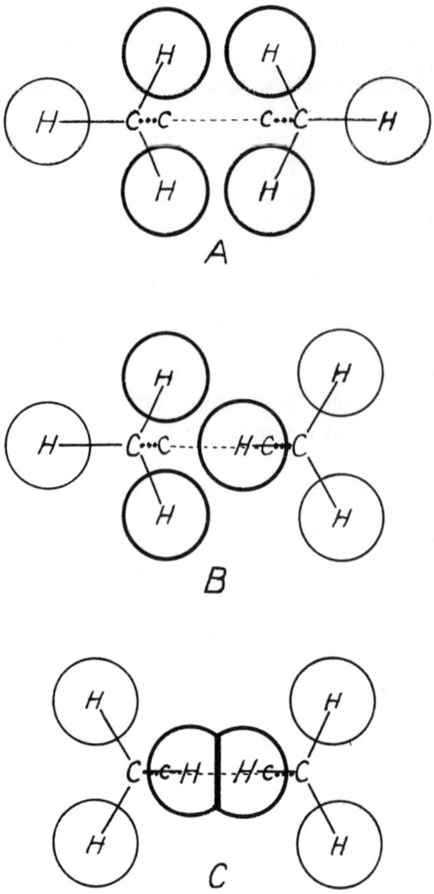

Fig. 8.—Schematic drawing illustrating various orientations of methyl groups in *o*-xylene.

At this point it is also of interest to recall the humps in the heat capacity curves of solid toluene, *m*- and *p*-xylene which were attributed in Part I to the beginning of internal rotational motion of some sort in the range 50–100°K. Forces arising from neighboring molecules must play some part in this effect so that it is impossible to make a simple quantitative interpretation. However, it seems unlikely that these results could be consistent with potential barriers in the free molecules much larger than about 500 cal. per mole.

In toluene or *m*- or *p*-xylene a very small potential barrier is quite reasonable, because the symmetry indicates six maxima and minima per revolution and no particularly large forces would be expected. A low barrier was also found for nitromethane where the symmetry calls for six maxima.

It is difficult to set up a really satisfactory potential function for *o*-xylene. Two extreme approximations may be described as (I)—no interaction term coupling the rotation of one methyl group to the other, just cosine type potential barriers acting on each methyl separately and (II)—only an interaction term such that the two methyls are like gears which can rotate freely in mesh but resist stripping. Examination of a scale model shows that neither approximation is reasonable. As can be seen from Fig. 8, methyl groups are not good gears. The meshed position, B in Fig. 8, involves no serious steric repulsions and the out of mesh position A is also reasonable on the same basis. Only positions near C should be of high energy on this basis. However, if only positions near C are set at high energy then agreement cannot be obtained with the experimental entropy. If one takes approximation (I) with A as the minimum position then B is higher in energy by V_0 while C is higher by $2V_0$. Actually position C should probably be still higher in energy but the effect of this would not be too large. Thus it seems best to use approximation (I) even though it cannot be entirely correct.

Another difficulty in the way of any exact analysis for *o*-xylene is the uncertainty of the vibrational assignment.

Using approximation (I) one finds from the entropy value a potential barrier height, V_0 = 2000 cal./mole. This gives the curve on Fig. 7 for the heat capacity which is in good agreement with experimental points. Therefore, this approximation should be reasonably satisfactory for further thermodynamic calculations.

Calculated Thermodynamic Functions.—The free energy function, $(H^0_0 - F^0_T/T$, the heat content function, $H^0_T - H^0_0$, and the heat capacity have been calculated for gaseous benzene, toluene, all three xylenes, and mesitylene at various temperatures up to 1500°K. and are listed in Tables XXV to XXVII. The vibrational analyses of Parts III and IV are used together with a mean value of 500 cal. per mole

TABLE XXV

THE FREE ENERGY FUNCTION

$(H^0_0 - F^0_T)/T$ IN CAL. PER DEGREE MOLE

T	Benzene	Toluene	Xylene ortho	Xylene meta	Xylene para	Mesitylene
298.16	52.98	62.20	65.74	67.89	66.48	71.07
400	56.75	66.97	71.88	73.75	72.36	77.98
500	60.32	71.44	77.57	79.21	77.82	84.38
600	63.81	75.75	82.93	84.43	83.03	90.46
800	70.42	84.02	93.18	94.42	92.96	102.02
1000	76.66	91.75	102.63	103.72	102.23	112.87
1500	90.60	109.03	123.52	124.44	122.89	136.82

TABLE XXVI

THE HEAT CONTENT FUNCTION $H^0_T - H^0_0$ IN KCAL.

T	Benzene	Toluene	Xylene ortho	Xylene meta	Xylene para	Mesitylene
298.16	3.415	4.314	5.604	5.330	5.364	6.290
400	5.78	7.27	9.31	8.91	8.91	10.46
500	8.79	10.98	13.84	13.36	13.33	15.59
600	12.32	15.35	19.10	18.57	18.51	21.61
800	20.66	25.64	31.42	30.86	30.72	35.80
1000	30.22	37.50	45.57	44.97	44.80	52.16
1500	57.5	71.3	86.1	85.4	85.2	99.2

TABLE XXVII

THE HEAT CAPACITY

C_p IN CAL. PER DEGREE MOLE

T	Benzene	Toluene	Xylene ortho	Xylene meta	Xylene para	Mesitylene
298.16	19.64	24.80	31.93	30.38	30.24	35.65
400	26.84	33.25	41.08	40.01	39.68	46.32
500	32.96	40.58	49.16	48.46	48.10	55.87
600	37.87	46.64	56.01	55.56	55.22	64.08
800	45.14	55.78	66.70	66.51	66.22	76.94
1000	50.25	62.27	74.36	74.29	74.08	86.14
1500	57.77	71.81	85.98	85.95	85.85	100.03

for the potential barriers to internal rotation for toluene, *m* and *p*-xylene, and mesitylene. For the rotation of the two methyl groups in *o*-xylene two independent three-fold potential barriers of 2000 cal. per mole height were assumed as discussed above.

The entropy is readily obtainable from the tabulated functions by the equation

$$S = \frac{H^0_T - H^0_0}{T} + \frac{H^0_0 - F^0_T}{T}$$

We may now compute the change in heat capacity with vaporization and compare the result with the coefficients of the log terms in the vapor pressure equations. The values of ΔC_p from heat capacity data are, for *o*-, *m*-, and *p*-xylene, respectively, −13.0, −13.4 and −13.7, while from the vapor pressure equations the values are −15.9, −13.3 and −9.9 cal. per degree. A consideration of the possible errors shows that the vapor pressure measurements must have been of high accuracy in order to have obtained ΔC_p values as satisfactory as these. The values from heat capacity data are, of course, the more reliable.

It is now possible to use the more reliable ΔC_p values to fix the coefficients of the log terms in the vapor pressure equations and evaluate the remaining constants to best fit the data. Such equations should be more reliable for extrapolation beyond the measured range but would not differ significantly within that range. However, increasing gas imperfection makes extrapolation to higher temperatures impractical while extrapolation to lower temperatures seems unimportant. Consequently it was not deemed worthwhile to revise the vapor pressure equations.

Heat Contents at the Absolute Zero.—Using the equilibrium results of Part II together with the values of $(H^0_0 - F^0_T)/T$ from Table XXV one obtains the results in Table XXVIII. Also included there are values of the heats of formation from the elements derived from heat of combustion data. The values 782.0 ± 1.0 and 935.5 ± 1.5 kcal. per mole were adopted for the heats of combustion of benzene and toluene, respectively, after considering the data of Richards and co-workers[45] and those of Roth and von Auwers.[46] Using the present results for the toluene–benzene–xylene reaction, one calculates for *o*-, *m*- and *p*-xylene, respectively, 1090.4, 1089.4 and 1089.9 kcal., which may be compared with revised values[45d] from Richards of 1090.8, 1090.8 and 1087.0 and with Roth's value for *m*-xylene 1089.5 kcal. It is apparent that the differences between the isomers as found by Richards bear no connection to the true differences. This is not surprising when one considers the very large total heat of combustion being measured.

Calculated Values for Certain Equilibria.—By using the data of Tables XXV and XXVIII

TABLE XXVIII

HEAT CONTENT DIFFERENCES OF GASES AT THE ABSOLUTE ZERO

Reaction	ΔH^0_0, kcal./mole
m-xylene = *p*-xylene	0.43 ± 0.2
m-xylene = *o*-xylene	.62 ± 0.2
p-xylene = *o*-xylene	.19 ± 0.3
2 toluene = benzene + *m*-xylene	.23 ± 0.3
elements = benzene	48.6 ± 1.0
elements = toluene	51.6 ± 1.5
elements = *m*-xylene	54.8 ± 2.0

TABLE XXIX

EQUILIBRIA AMONG THE XYLENES AND BETWEEN TOLUENE, BENZENE, AND XYLENE AT VARIOUS TEMPERATURES IN THE GAS STATE

$$K_m = \frac{P_{\text{benzene}} P_{m\text{-xylene}}}{P^2_{\text{toluene}}}, \quad K_t = \frac{P_{\text{benzene}} P_{\text{total xylene}}}{P^2_{\text{toluene}}}$$

T	Equil., % ortho	Equil., % meta	Equil., % para	K_m	K_t
298.16	9	74	17	0.12	0.16
400	12	68	20	.13	.19
500	15	64	21	.15	.23
600	17	62	21	.16	.26
800	21	58	21	.17	.30
1000	24	55	21	.19	.34
1500	27	52	21	.20	.39

(45) (a) Richards and Jesse, THIS JOURNAL, **32**, 268 (1910); (b) Richards and Barry, *ibid.*, **37**, 993 (1915); (c) Richards and Davis, *ibid.*, **42**, 1599 (1920); (d) Swietoslawski and Bobinska, *ibid.*, **49**, 2476 (1927).

(46) Roth and von Auwers, *Ann. chim.*, **407**, 109 (1914).

the equilibrium composition of the xylenes was computed for various temperatures. The equilibrium constants for the benzene—toluene—xylene reaction were also obtained. These are given in Table XXIX. It is interesting to note that the percentage of para stays almost constant while ortho increases in abundance threefold. The restriction of internal rotation and higher energy content of ortho are the factors causing this.

Discussion.—It is now interesting to see what explanation may be offered for the energy differences among the xylenes. Toluene has a small but definite dipole moment, the methyl group (particularly the hydrogen atoms) being positive with respect to the ring. This might be formulated in terms of hyperconjugation[47] or resonance[48] structures such as

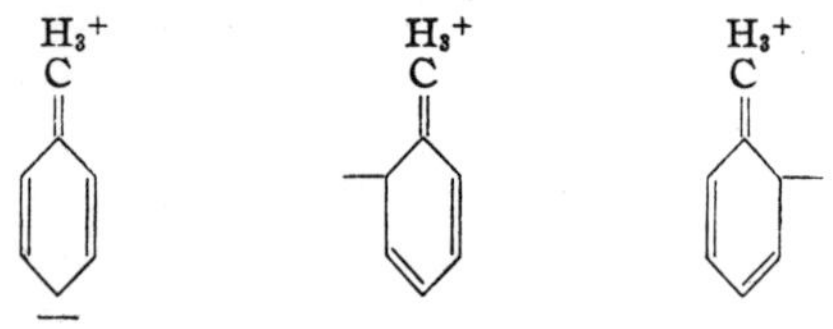

It is apparent that the negative charge appears only in ortho and para positions. Such resonance forms are of course connected to the ortho–para orienting effect of a methyl group with respect to substitution reactions.

Now in *m*-xylene the two methyl groups lead to resonance forms placing the negative charges in the same places on the ring. In *o*- or *p*-xylene on the contrary the two methyl groups give resonance forms which place the negative charges on different sets of atoms in the ring. According to the experimental results the former is the preferred form, *i. e.*, of lower energy. Although it is not difficult to suggest "theories" in accord with this, now that the facts are known, the writers are not aware of any well defined theory that clearly predicts this result. Rather we prefer now to predict from this result that in other situations of this type the meta form will be more stable.

Among the trimethylbenzenes 1,3,5 should have lowest energy being "all meta" with 1,2,4 and 1,2,3 about equal. It is likely, however, that the 1,2,3 isomer is subject to some steric repulsion among the methyl groups so that it would be definitely less stable than 1,2,4.

Similar reasoning would apply to other di- and tri-substituted benzenes so long as either identical or similar groups are present. However, in a case such as nitroaniline when the groups have resonance effects of opposite sign the ortho–para orientation would be preferable since both substituents

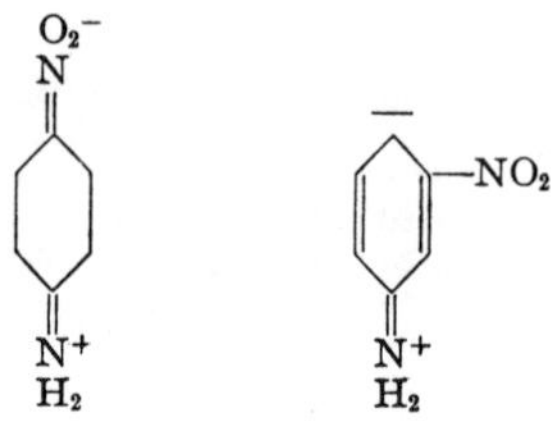

can enter into the same resonance form.

Summary

1. **Thermodynamic Data.**—The following thermodynamic quantities were measured, calculated, or taken from the literature for each of the substances, benzene, toluene, *o*-, *m*- and *p*-xylene: the heat capacity of solid, liquid and gas in appropriate ranges from 15 to 470°K.; the heats of fusion and vaporization; the melting points and vapor pressures; the entropies of liquid and gas; and the free energy function $(H^0_0 - F^0_T)/T$, the heat content, and the heat capacity over the range 298–1500°K.

The freezing point diagram for the ternary system *o*-, *m*- and *p*-xylene was calculated and certain points experimentally verified.

The direct isomerization reaction of the xylenes, catalyzed by aluminum bromide and hydrogen bromide, led to a direct measure of the equilibrium composition. Using the same catalyst, the equilibrium constant for the reaction 2 toluene = benzene + xylene (equil.) was found to be between 0.15 and 0.22 at 323°K. From these data the various thermodynamic constants for the isomerization and the toluene, benzene, xylene reactions were calcuated.

A few of the most important values for the xylenes are listed in the following table.

MISCELLANEOUS THERMODYNAMIC CONSTANTS FOR THE XYLENES

	Ortho	Meta	Para
Melting point, °K.	247.82	225.27	286.39
Heat of fusion, cal./mole	3250	2765	4047
Heat of vapn. (298°K.), cal./mole	10470	10160	10110

(47) Mulliken, Rieke and Brown, THIS JOURNAL, **63**, 41 (1941).

(48) See, for example, Pauling, "The Nature of the Chemical Bond," Cornell University Press, Ithaca, N. Y., 1940, or Branch and Calvin, "The Theory of Organic Chemistry," Prentice-Hall, New York, N. Y., 1941.

	Ortho	Meta	Para
Vapor press. (298°K.), mm.	6.64	8.39	8.87
Entropy of vap. (298°K.) (liq. to hypoth. gas 1 atm.), cal./deg.	25.70	25.13	25.07
Entropy of the gas (298°K.), cal./deg.	84.50	85.60	84.27
Equil. comp. (323°K.), %	10	72	18

Various thermodynamic functions were calculated for mesitylene from molecular structure data, the gas entropy at 298°K. being 92.34 cal. per degree. The entropies of pseudocumene (1,2,4-trimethylbenzene) and durene (1,2,4,5-tetramethylbenzene) were estimated to be 94.2 and 98.5 cal. per degree, respectively, in the gas state at 298°K.

2. Molecular Structure Data.—By consideration of the lines appearing in the infrared and Raman spectra of liquid benzene (but not of gaseous benzene) in violation of the simple selection rules, it was possible to complete the assignment of fundamental vibration frequencies. Although a few values are not completely certain, all are reasonable from the point of view of the force constants of the benzene molecule. The spectra of the simpler deuterated benzenes were also considered.

The fundamental vibration frequencies were selected for toluene, all three xylenes, and mesitylene using Raman and infrared spectra from the literature. A modification of the Teller–Redlich product rule appropriate for methyl substitution was of great assistance in these assignments. The results vary from reasonably certain selections for some substances to somewhat arbitrary ones for others.

Attention is called to the striking behavior of the 1000 cm.$^{-1}$ line in benzene derivatives which are substituted in only 1,3,5-positions. Corresponding effects are observed where the trigonal positions are all similarly substituted. It is suggested that the Raman spectra might be used to determine the position of substituents in favorable cases.

Using these frequencies the vibrational entropy and heat capacity contributions can be calculated and the corresponding quantities for internal rotation of methyl groups obtained. The results indicate relatively free rotation of methyl groups in toluene, *m*- and *p*-xylene but about a 2000 cal. potential barrier for *o*-xylene. These results are shown to be reasonable in terms of steric effects.

The xylene equilibrium data show the meta compound to have lowest energy and suggest that meta orientation of identical or similar groups may be generally the most stable.

BERKELEY, CALIFORNIA RECEIVED NOVEMBER 11, 1942

Comment: The Thermodynamics and Molecular Structure of Benzene and Its Methyl Derivatives

This paper is included as a single paper which includes the various parts of the investigations that were otherwise usually reported separately. Here there are sections on the thermodynamic measurements, including sample purification, the assignment of normal modes from vibrational spectra including estimates for some unobserved or unresolved modes, the comparison of statistically calculated and experimental entropies, and the final conclusions about potentials restricting the rotation of methyl groups.

[Reprinted from the Journal of the American Chemical Society, 73, 2606 (1951).]

[CONTRIBUTION FROM THE DEPARTMENT OF CHEMISTRY AND CHEMICAL ENGINEERING, UNIVERSITY OF CALIFORNIA]

Methyl Alcohol: The Entropy, Heat Capacity and Polymerization Equilibria in the Vapor, and Potential Barrier to Internal Rotation

BY WILLIAM WELTNER, JR., AND KENNETH S. PITZER

The heat capacity of gaseous methanol was measured in a flow calorimeter over the range 345 to 521°K. at pressures of roughly $^1/_3$, $^2/_3$ and 1 atm. The results at 345° K. show a very large and distinctly non-linear trend with pressure. This is interpreted as indicating the presence of methanol tetramers in the gas, somewhat analogous to the polymers of hydrogen fluoride. The heat of dissociation of this tetramer is found to be 24,200 cal./mole which corresponds to a hydrogen bond energy of 6050 cal./mole if the tetramer is a ring. Experimental values from the literature and from this research for the heat capacity and entropy of methanol are compared with values calculated from spectroscopic data, including particularly a potential barrier to internal rotation of 932 cal./mole. Agreement is obtained, but there remain uncertain elements in the spectroscopic data which could affect this result. The apparent variance of this conclusion from that of earlier investigators is due to the unsuspected type of gas imperfection. In addition values are reported for the heat capacity of benzene vapor at several temperatures and for the heat of vaporization of benzene and methyl alcohol.

The potential barrier to internal rotation in methyl alcohol has been the subject of a long series of investigations.[1] The infrared spectrum was first interpreted by Koehler and Dennison[2] to indicate a barrier of about 450 cm.$^{-1}$ (1300 cal./mole). The microwave spectrum, together with further infrared measurements, have been found by Burkhard and Dennison[3] to indicate a barrier of 326 cm.$^{-1}$ (932 cal./mole) which was used by Halford[1] in calculations. However, the experimental value of the entropy has appeared to be inconsistent with as low a barrier as this unless a considerable residual entropy at 0°K. were assumed. Values of the heat capacity of the ideal gas obtained by Eucken and Franck[4] and by Rowlinson[5] were less definitely inconsistent with the spectroscopic barrier but also indicated higher values. Furthermore, the gas heat capacity measurements at 1 atm. of DeVries and Collins[6] show a most unusual increase with decreasing temperature near the saturation curve.

In the present paper measurements are reported of the heat capacity of gaseous methanol at a series of pressures and temperatures. The pressure dependence near the saturation curve indicates a polymerization phenomenon somewhat similar to that observed in hydrogen fluoride. This special type of gas imperfection requires revision of earlier calculations[1] which were based on more conventional behavior.

Experimental

Gaseous Flow Calorimeter.—The calorimeter and its appendages were newly constructed and followed closely the design of Waddington, Todd and Huffman,[7] reference also being made to the apparatus of Pitzer[8] and that of Coleman and DeVries.[9] There were several significant differences which are described in detail elsewhere.[10] These differences will be only briefly mentioned here.

A magnetically actuated valve, constructed from spherical ground glass joints, was employed to divert the vapor stream for a timed interval into a collection vessel. Also it was arranged so that the flow continued as vapor through the heated valve system to the collection vessel which contained the condensing surfaces cooled with ice. The electrical circuit was such that the relay which actuated the magnetic valve also actuated a stop clock for timing the sample take-off time.

A floating heater was employed in the vaporizer. This consisted of a thin glass float and chromel wire heater arranged to slide vertically on tungsten wire guides which served also as electrical lead wires. Also in the tube returning liquid to the vaporizer an appropriate side tube allowed the liquid to flash off superheat or entrained vapor bubbles.

The platinum resistance thermometers were initially calibrated in the usual way at the ice, steam and sulfur points. After assembly into the flow calorimeter unit they were compared with N.B.S. calibrated mercury thermometers. This second calibration indicated small but significant changes and was adopted for our calculations.

Measurement Procedure.—The procedure for measurement of heat capacities was, for the most part, that described by Waddington, Todd and Huffman.[7] However, many of the heat capacity runs were made simultaneously with heat of vaporization determinations so that any error due to incorrect flow rate measurements was eliminated here. Upon applying the corrections of Waddington, *et al.*, to the heats of vaporization, agreement within ±0.1% was obtained between heat capacities measured simultaneously with heats of vaporization and those measured in the usual way.

The heats of vaporization were measured by collecting 20 to 40 ml. of liquid in the trap cooled by ice-water over a period of 1.5 to 8 minutes depending upon the rate of flow which varied from 0.05 to 0.16 mole/min. Readings of the power supplied to the vaporizer and the resistances of the thermometers were made during the collection period; at slow flows, readings were made every minute; at fast flows, as rapidly as possible. A correction for change in the depth of immersion of the vaporizer heater during the measurement was unnecessary since a floating heater was employed. Also, the variation in boiling temperature was small enough to be neglected. However, the corrections for heat dissipated in the tungsten leads of the vaporizer and for increasing vapor volume, due to decreasing liquid volume in the vaporizer, amounted to about 1% of the heat of vaporization and were therefore quite significant.

The estimated limits of error are given along with the results at various points. The average deviation was usually much less and was about 5 cal./mole for heats of vaporization. The precision of the heat capacity measurements was in general quite good except for methanol at 72.4°. Divergence from a straight line amounted to about 0.02 cal./mole deg. when the apparent measured heat capacities were plotted *versus* the reciprocal of the rate of flow. At infinite rate of flow ($1/F = 0$) the heat capacities as determined by thermometers T_2 and T_3 agreed within 0.03 cal./mole deg. except for those measured for methanol at 72.4° and 755 and 500 mm. where the discrepancy was about 0.1 cal./mole deg.

Materials.—C.P. thiophene-free benzene was distilled from freshly cut sodium through an all-glass column, having 30 perforated plates. Only the middle cut was used.

Commercial absolute, acetone-free methanol was dried

(1) J. O. Halford, *J. Chem. Phys.*, **18**, 361, 1051 (1950), reviews the situation and gives earlier references.

(2) J. S. Koehler and D. M. Dennison, *Phys. Rev.*, **57**, 1006 (1940).

(3) Private communication from Professor D. M. Dennison.

(4) A. Eucken and E. U. Franck, *Z. Elektrochem.*, **52**, 195 (1948).

(5) J. S. Rowlinson, *Nature*, **162**, 820 (1948).

(6) T. DeVries and B. T. Collins, THIS JOURNAL, **63**, 1343 (1941).

(7) G. Waddington, S. S. Todd and H. M. Huffman, *ibid.*, **69**, 22 (1947).

(8) K. S. Pitzer, *ibid.*, **63**, 2413 (1941).

(9) C. F. Coleman and T. DeVries, *ibid.*, **71**, 2839 (1949).

(10) Ph.D. Dissertation of William Weltner, Jr., University of California, 1950.

with sodium by the method of Gillo[11] and distilled through the same all-glass column with a continuous flow of dry nitrogen bubbling through the methanol and up the column. It was stored in the dark in a vessel closed by stopcocks greased with Apiezon M. This grease was found to be quite insoluble in methanol.

Results with Benzene.—Measurements were made on benzene to check the operation of the new calorimeter. Table I gives the heat capacities of benzene measured at 1 atm. pressure as compared with those of Scott, Waddington, Smith and Huffman.[12]

TABLE I

HEAT CAPACITY (CAL./MOLE DEG.) OF BENZENE AT 1 ATM. PRESSURE

T, °K.	This research	Scott, *et al.*	Difference
375.0	25.99	25.84	+0.15
403.0	27.60	27.52	+ .08
463.0	30.90	31.24	− .34

The results of Scott, *et al.*, were plotted to obtain the values corresponding to the temperatures of measurement in this research. The differences in Table I seem too large to be accounted for by expected magnitudes of error but are not excessive in comparison with most measurements of gaseous heat capacities.

The heat of vaporization of benzene at 1 atm. pressure was determined to be 7358 ± 15 cal./mole as an average of seven measurements made over the range of flow rates. This compares excellently with the values of Scott, *et al.*,[12] and Fiock, Ginnings and Holton,[13] which are 7349 and 7353 cal./mole, respectively.

Results with Methanol.—The heat capacities were measured at four temperatures from 345 to 521°K. and at 755, 500 and 260 mm. pressure at each temperature. The values obtained are given in Table II together with the estimated accuracy which is the same for the three pressures at a given temperature. The values in Table II for the ideal gas were obtained from those at 260 mm. by use of the gas imperfection equations to be described below. The values in Table II are shown in Fig. 1 along with the values of DeVries and Collins.[6] The curves in Fig. 1 are theoretical and will be discussed later but it is apparent that our results at 755 mm. are quite consistent with those of DeVries and Collins.

TABLE II

HEAT CAPACITY (CAL./MOLE DEG.) OF METHYL ALCOHOL

T, °K.	Pressure (mm.) 755	500	260	Est. uncertainty	Ideal gas
345.6	21.03	14.72	12.18	±0.2	11.46 ± 0.3
403.2	12.96	12.75	12.58	± .05	12.43 ± .1
464.0	13.94	13.84	13.74	± .1	13.67 ± .1
521.2	14.68	14.62	14.48	± .2	14.44 ± .3

The heats of vaporization obtained for methanol to saturated vapor at 755, 500 and 260 mm. pressure are given in Table III along with the values of Fiock, Ginnings and Holton.[13] Except for that at 260 mm., the values given are an average of at least four measured heats of vaporization over the range of flow rates. The measured heats of vaporization at 260 mm. pressure showed a marked decrease with increase in the rate of flow of vapor. The value given in Table III corresponds to the intercept at $1/F = 0$ when the heat of vaporization is plotted against the reciprocal of the rate of flow. The heats of vaporization measured in this research agree with those of Fioch, *et al.*, well within the estimated uncertainties.

TABLE III

HEATS OF VAPORIZATION (CAL./MOLE) OF METHANOL TO SATURATED VAPOR

T, °K.	P (mm.)	This research	Fiock, *et al.*,
337.8	755	8442 ± 15	8429
327.9	500	8579 ± 15	8576
313.1	260	8768 ± 25	8774

(11) L. Gillo, *Ann. chim.*, **12**, 281 (1939).

(12) D. W. Scott, G. Waddington, J. C. Smith and H. M. Huffman, *J. Chem. Phys.*, **15**, 565 (1947).

(13) E. F. Fiock, D. C. Ginnings and W. B. Holton, *J. Research Natl. Bur. Standards*, **6**, 881 (1931).

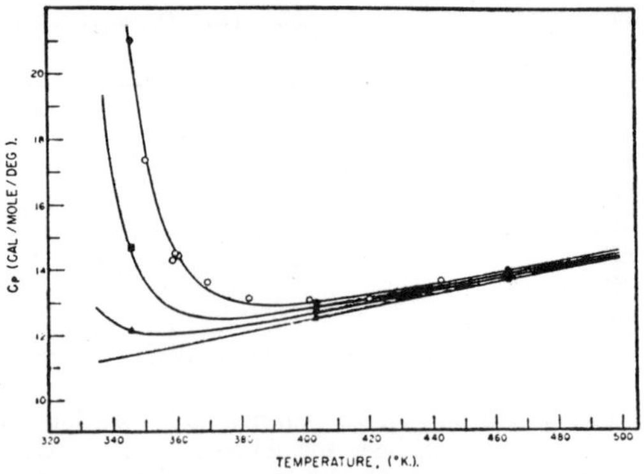

Fig. 1.—Heat capacity of methanol vapor at 755, 500, 260 mm. and in the ideal gas state: O, experimental values of DeVries and Collins at 750 mm.; ●, experimental values of this research at 755 mm.; ■, at 500 mm.; ▲, at 260 mm. Solid lines indicate calculated values at 755, 500, 260 mm. and in the ideal gas state using Stepanov's vibrational assignment and the gas imperfection equations of this research.

Calculations and Discussion

Gas Imperfection.—From Table II or Fig. 1 it is apparent that at 345°K. the trend of heat capacity with pressure is far from linear. Considering that all the data are for pressures below one atmosphere, this is very unusual and is suggestive of the behavior of hydrogen fluoride in forming polymers in the gas. The formation of reasonably strong hydrogen bonds is the point of similarity between hydrogen fluoride and methanol which is undoubtedly the cause of these polymerizations.

However, by 464°K. the curve of C_p *vs.* P is quite clearly linear, showing that the second virial coefficient (which corresponds to dimerization) is here the only significant form of gas imperfection. The data on P–V–T measurements for methanol are relatively few. Eucken and Meyer[14] made measurements from 352 to 374°K. at a single gas concentration. The old work of Ramsay and Young[15] covers a wide range but is not accurate enough to serve as more than a general check in the low pressure, low temperature region.

Since we have few data, it was essential to select a representation of the complex behavior containing a minimum of disposable constants. The system selected contains five: the heat and entropy of formation of the gaseous dimer, the number of monomer units in the higher polymer and the heat and entropy of formation of the higher polymer. A value (the same as Eucken and Meyer's) was assumed for the covolume, which is unimportant at these temperatures and pressures anyway.

Several authors[12,16] have recently used the form of equation for the second virial coefficient which

(14) A. Eucken and L. Meyer, *Z. physik. Chem.*, **5B**, 452 (1929).

(15) W. Ramsay and S. Young, *Roy. Soc. London, Phil. Trans.*, **A178**, 313 (1887).

(16) J. O. Hirschfelder, F. T. McClure and I. F. Weeks, *J. Chem. Phys.*, **10**, 201 (1942).

corresponds to a constant energy of dimerization, *i.e.*, ΔC of dimerization is zero. This is given by a square well potential model. If the walls of the potential well are sloped, or if the rotation of the component monomer units is restricted in the polymer, the relative heat capacity of the polymer will be increased as compared to the square model (and a given monomer). Thus it seemed to us to be a better (though still arbitrary) assumption to make the heat of polymerization (or dimerization) constant and ΔC_p of polymerization zero. The equations yielded by this model in the virial coefficient form (appropriate for small fractions of polymer) are

$$PV = RT + BP + DP^{n-1} \tag{1}$$

$$B = b - RT/K_2 = b - RTe^{-\Delta S_2/R}e^{\Delta H_2/RT} \tag{2}$$

$$D = -(n-1)RT/K_n = -(n-1)RTe^{-\Delta S_n/R}e^{\Delta H_n/RT} \tag{3}$$

where b is the covolume, n is the number of monomer units in the higher polymer, K_2 and K_n are pressure equilibrium constants of dissociation for dimer and higher polymer, respectively. The corresponding expressions for the heat capacity are

$$C_p = C_p^0 + aP + cP^{n-1} \tag{4}$$

$$a = \frac{\Delta H_2^2}{RT^2} e^{-\Delta S_2/R}e^{\Delta H_2/RT} \tag{5}$$

$$c = \frac{\Delta H_n^2}{RT^2} e^{-\Delta S_n/R}e^{\Delta H_n/RT} \tag{6}$$

The heat capacity data at 345.6°K. were clearly fitted better by $n = 4$ than any other integral value, consequently the higher polymer was assumed to be a tetramer. The following successive approximation method was employed to evaluate the other constants. ΔH_2 and ΔS_2 were obtained from the Eucken and Meyer P–V–T data assuming only dimerization and the resulting value of a in equations (4) and (5) calculated. The values of C_p at about one atmosphere from 340 to 390°K. (both DeVries and Collins measurements and those of this research) were then used[17] to evaluate ΔH_4 and ΔS_4. Then ΔH_2 and ΔS_2 were recalculated recognizing the presence of tetramer and the cycle repeated until no significant changes occurred on further repetition. The resulting equations for the second and fourth virial coefficients are

$$B = 80 - 0.0206Te^{3220/RT} \text{ cc./mole} \tag{7}$$

$$D = -4.36 \times 10^{-16}\, Te^{24200/RT} \text{ cc./mole atm.}^2 \tag{8}$$

The thermodynamic constants at standard states corresponding to these virial coefficients are

$\Delta H_2 = 3220$ cal./mole $\quad \Delta S_2 = 16.5$ cal./deg. mole

$\Delta H_4 = 24200$ cal./mole $\quad \Delta S_4 = 81.3$ cal./deg. mole

The agreement of our calculated values of $v - RT/P$ with Eucken and Meyer's measurements is shown in Fig. 2. Our calculated P–V–T behavior has also been checked with the data of Ramsay and Young and yields agreement within experimental error in the range of temperature of interest here, *i. e.*, up to about 400°K.

It is interesting to note that the heat of dissociation of the tetramer, 24,200 cal./mole, corresponds to a hydrogen bond energy of 6,050 cal. if the tetramer is a ring molecule.

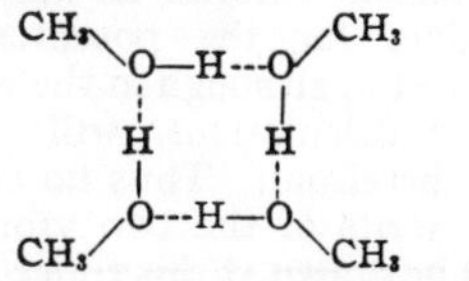

Actually our data do not distinguish between the formation of tetramers and the formation of an appropriate mixture of trimer, tetramer, pentamer, etc. Thus it is difficult to estimate the reliability of this value of the hydrogen bond energy.

Our value of 6.05 kcal./mole for the hydrogen bond energy in the gaseous polymer is very close to the value 6.2 kcal./mole estimated by Pauling[18] from the heat of sublimation of methanol.

For hydrogen fluoride, Long, Hildebrand and Morrell[19] found a heat of dissociation of about 41,000 cal./mole for a hexamer corresponding to a hydrogen bond energy of 6800 cal.

The essential conditions for the formation of polymers of this type would appear to be as follows: (1) The principal intermolecular attraction is hydrogen bonding. (2) There is only one hydrogen available per molecule. (3) The geometry is such that a dimer with two hydrogen bonds is not possible. Failure of condition (3) leads to principally dimeric vapor as in the simpler carboxylic acids while the failure of either (1) or (2) makes the liquid with its larger effective coördination number so stable that the ring type polymers cannot be detected.

Gas Heat Capacity.—Halford[1] has calculated the contribution to various thermodynamic functions from internal rotation based on the 932 cal./mole potential barrier and 1.01×10^{-40} g. cm.[2] moment of inertia for the OH group found by Burkhard and Dennison.[3] Halford's values for the other moments of inertia appear to be sound but the

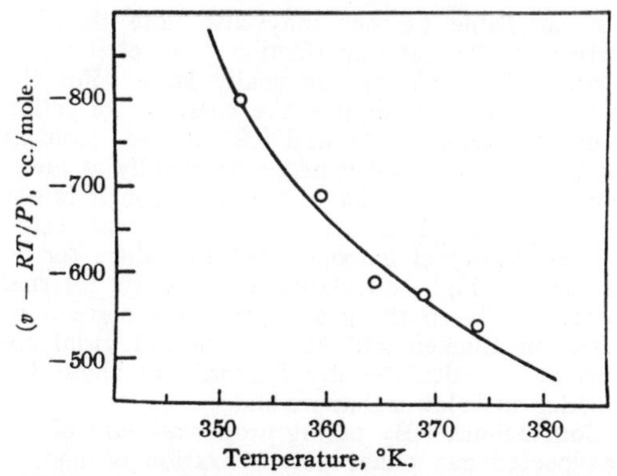

Fig. 2.—($v - RT/P$) of methanol vapor as a function of temperature. Circles are experimental results of Eucken and Meyer. Solid line indicates results calculated from equation of state found in this research.

(17) It was necessary to assume a curve for the ideal gas heat capacity at this point. However, the gas imperfection effect is so large that the result is not sensitive to the exact curve assumed.

(18) L. Pauling, "The Nature of the Chemical Bond," second edition, Cornell University Press, Ithaca, N. Y., 1945, p. 304.

(19) R. W. Long, J. H. Hildebrand and W. E. Morrell, This Journal, 65, 182 (1943).

vibration frequency assignment still has its doubtful aspects. Halford states, "Noether's assignment has been used in the present (Halford's) calculations, merely because it is the later one, although it is not obvious that it is the correct one." After a careful review of the situation we believe Noether[20] and Stepanov[21] are probably correct with respect to the nine frequencies which they assign in essentially the same way. They differ as to the two methyl rocking motions and we do not believe the present spectroscopic data are adequate to justify a firm choice. However, the difference is not large, for thermodynamic calculations, Noether's values being 1250 and 1209 cm.$^{-1}$ while Stepanov's are 1034 and 1104 cm.$^{-1}$. In our calculations we have obtained values for each of these vibrational assignments.

In Fig. 3 are shown the various experimental values for the heat capacity of gaseous methanol reduced to the ideal gas. DeVries and Collins values were reduced by the use of our equations for gas imperfection given above. It is apparent that agreement is obtained within the range of uncertainty in vibrational assignment. However, Professor Dennison[3] informs us that the final spectroscopic values for the potential barrier and moments of inertia, although in the vicinity of those used in these calculations, will probably differ significantly therefrom. Thus no conclusion as to the relative merits of the two vibrational assignments should be drawn at this time.

In Fig. 1 are shown the heat capacity data at various pressures as compared with curves calculated from ideal gas values with Stepanov's frequencies and our gas imperfection equations. It is apparent that the agreement is good, confirming in this case principally the assumptions as to polymerization in the gas.

Entropy.—The necessary low temperature heat capacity measurements have been made by Kelley[22] who calculated the entropy of the liquid at 298.16° K. Combining this result with the heat of vaporization data of Fiock, Ginnings and Holton[13] yields the values of the experimental entropy of methanol listed in Table IV covering the range 298.16 to 380° K. The entropy of the ideal gas was calculated by Halford[1] for the 932 cal./mole potential barrier. We have used Stepanov's vibrational assignment at this point because of the somewhat better agreement shown for the gas heat capacity. Our treatment of the gas imperfection given above was employed to convert from the ideal gas to the real gas at the vapor pressure.

$$S^0 - S_p = R \ln P + \frac{dB}{dT}P + \frac{dD}{dT}\cdot\frac{P^2}{3} \tag{9}$$

where S^0 is standard entropy of ideal gas at 1 atm. and S_p is the entropy of the real gas at pressure P.

The last two columns in Table IV show agreement between calculated and experimental entropies well within the estimated error of about 0.3 cal./deg. mole. Actually, the use of Noether's vibration frequencies would have given even more

TABLE IV

ENTROPY OF METHYL ALCOHOL

Units of entropy; cal./deg. mole; P = vapor pressure

T, °K.	S^0, calcd. ideal gas	$R \ln P$	$\frac{dB}{dT}P$	$\frac{dD}{dT}\cdot\frac{P^2}{3}$	S_p calcd.	S_p exptl.
298.16	57.21	−3.60	0.08	0.31	60.42	60.28
320	57.97	−1.51	.15	.44	58.89	58.78
340	58.63	+0.16	.24	.54	57.69	57.63
360	59.29	1.60	.35	.62	56.72	56.64
380	59.92	2.88	.49	.69	55.86	55.74

perfect agreement. Again the preliminary nature of the spectroscopic potential barrier and moments of inertia makes detailed conclusions premature. Nevertheless, there is no evidence for a significant residual entropy in the crystals at 0°K.

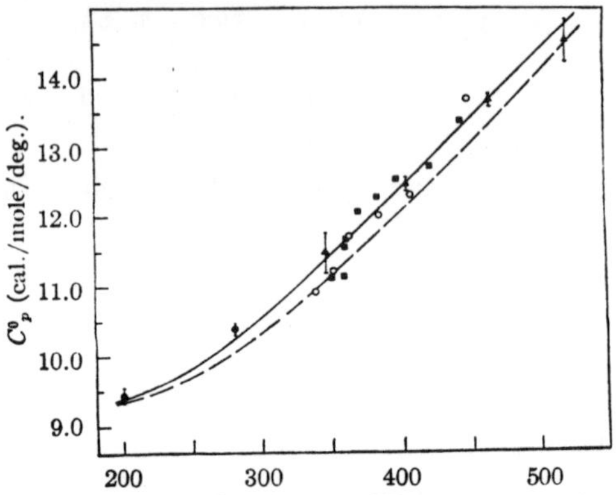

Fig. 3.—Heat capacity of methanol vapor in the ideal gas state: ●, experimental results of Eucken and Franck; ■, experimental results of DeVries and Collins reduced to the ideal gas by the gas imperfection equations of this research; O, experimental results of Rowlinson; ▲, results of this research from Table II. Vertical lines through points indicate estimated limits of error: —, calculated heat capacities using Stepanov's vibrational assignment; — — —, using Noether's vibrational assignment.

From Table IV one may also note that the corrections for gas imperfection (sum of items in columns 4 and 5) are unusually large. For the vapors of normal liquids the entropy correction is usually between 0.1 and 0.2 cal./deg. mole at the boiling point and it decreases rapidly at lower temperatures. In view of this situation it is not surprising that earlier attempts to fit the calculated and experimental entropies led to failure (or to excessively high calculated barriers to internal rotation). Even the gas imperfection corrections based on Eucken and Meyer's second virial coefficient as calculated by Halford are about 0.3 cal./deg. mole lower than we find.

Conclusions.—By taking proper account of the unexpected gas phase polymerization of methyl alcohol vapor, it has been possible to bring the various experimental thermodynamic quantities into agreement with values calculated from the tentative spectroscopic barrier (932 cal./mole) to internal rotation and vibration frequency assignment. Measurements of the heat capacity of a gas at different pressures are a powerful tool in

(20) H. D. Noether, *J. Chem. Phys.*, **10**, 693 (1942).

(21) B. E. Stepanov, *J. Phys. Chem.* (*U. S. S. R.*), **19**, 497 (1945).

(22) K. K. Kelley, THIS JOURNAL, **51**, 181 (1929).

investigating gas imperfection and were the essential element in the solution of this long baffling problem.

While it has been shown that the thermodynamic data are consistent with these spectroscopic constants, other values of the spectroscopic quantities within certain limits are allowed. It seems hardly worth while to investigate this extensively at present. Nevertheless, even though the principal dilemma concerning the potential barrier and entropy of methanol is now removed, the exact 932 cal./mole value for the barrier is only as certain as the spectroscopic data indicate. Considering the uncertainty in the vibration frequency assignment, the thermodynamic data might be consistent with any barrier within a few hundred calories of this 932 cal. figure. Moreover, if the moment of inertia of the OH group (for torsion) is substantially larger than the 1.01×10^{-40} here assumed, then an appreciably higher barrier would be indicated. This moment of inertia requires either an abnormally large C–O–H angle or an abnormally short O–H distance.

Acknowledgments.—We wish to acknowledge the great assistance of Dr. Gordon M. Barrow in the construction of the calorimeter and in making available to us the second set of thermometer calibrations mentioned above. We also wish to thank Professor W. D. Gwinn for his helpful advice and guidance, particularly while one of us (K. S. Pitzer) was away from Berkeley.

BALTIMORE, MD. RECEIVED DECEMBER 7, 1950

Comment: Methyl Alcohol: The Entropy, Heat Capacity, and Polymerization Equilibria in the Vapor and the Potential Barrier to Internal Rotation

This paper yielded a significant value of the potential barrier to rotation about the C–O bond and might have been placed in the first group. But the more novel, interesting, and important discovery was the clear presence of a tetrameric molecule — presumably a hydrogen-bonded, four-molecule ring. Also the heat capacity contribution from the heat of dissociation of this tetramer, which was the key to this discovery, is a phenomenon that we assumed to be well known, but it was omitted in research of others a few years later and then rediscovered as if it had not been known before.

[Reprinted from the Journal of the American Chemical Society, **81**, 4477 (1959).]

[CONTRIBUTION FROM THE LAWRENCE RADIATION LABORATORY AND COLLEGE OF CHEMISTRY, UNIVERSITY OF CALIFORNIA, BERKELEY]

Large Molecules in Carbon Vapor

BY KENNETH S. PITZER AND ENRICO CLEMENTI

RECEIVED APRIL 20, 1959

The molecular orbital theory is used in appropriate semi-empirical forms to predict the properties of carbon vapor. The results indicate that linear polyatomic molecules :C=C=C---C=C: are the important species. Experimental results from the literature for C_3 are combined with the calculated conjugation or resonance energies and with the heats of formation of allene and ethylene to predict heats of formation for all larger carbon molecules. It is found that the odd species have closed shell structures and lower energies than the even species but that the even species should show greater electron affinity. Both of these results are consistent with the mass spectrometric results of Honig and of Chupka and Inghram. Molecular spectroscopic data on C_3O_2 are used to estimate the free energy function increments for the species above C_3. The calculated partial vapor pressures predict C_5 to be the most abundant species in the saturated vapor even at 2000°K. with C_7 becoming comparably abundant in the 2500 to 3000°K. range. At higher temperatures even larger molecules should become important. The results are shown to be generally consistent with all reliable vaporization data provided the evaporation coefficients decrease rapidly for increasing molecular size and vary for different crystal surfaces of graphite. The calculated electronic energy levels for C_2 and C_3 agree satisfactorily with the observed spectra, and trends are predicted for both even and odd larger species. It is proposed that liquid carbon consists of essentially infinite linear chains of this type. Both entropy and energy considerations lead to predicted heats of fusion of about 10 kcal./g. atom at 4000°K.; the agreement between the two values indicates at least the absence of any serious inconsistency.

Mass spectrometric studies[1-3] have clarified the previously confused situation with respect to the heat of sublimation of graphite to monatomic carbon and have established the presence of C_2, C_3 and probably C_5 in the vapor. The effusion measurements of Thorn and Winslow[4] are plausibly interpreted on the basis of C_1, C_2 and C_3 with heats of vaporization of approximately 170, 194 and 184 kcal./mole, respectively. There remains, however, the evidence of the effusion experiments of Doehaerd, Goldfinger and Waelbroeck[5] with very small hole-to-interior surface ratios which indicates some additional vapor species with higher equilibrium pressure but an evaporation coefficient about 10^{-3} in the temperature range 2500–2700°K. In addition, the measurements of Bégué[6] and Basset[7] indicate vapor pressures reaching 1 atmosphere substantially below 4000°K. which likewise require additional vapor species besides C_1, C_2 and C_3.

In seeking an explanation for these latter observations, we were led to a general consideration of the bi-coördinate state of carbon. A long chain of localized double bonds, alternately in the x- and y-planes, is quite unstable as compared to graphite. However, complete resonance or conjugation such that each bond is axially symmetric yields a large stabilization. Carbon dioxide provides the most familiar example of this type of binding. Carbon suboxide O=C=C=C=O provides a better example and use will be made of data on this compound below. The simplest form of molecular orbital theory is first applied to the linear C_n species and qualitative conclusions drawn. Then quantitative calculations are made and the results compared with the experimental data. Ring molecules are also considered and found to be unimportant under most conditions.

Simple Molecular Orbital Theory.—First we consider the general pattern of orbital energies in order to determine which orbitals are occupied in the stable states of the various molecules. Orbital energies were calculated by the methods of Mulliken and his collaborators[8] in which the overlap integral is used as a measure of bond strength. Since we seek only a semi-quantitative estimate of orbital energies, no attempt was made to refine the empirical parameters. For sigma orbitals $\beta = -8.0S$ and for pi orbitals $\beta = -10.5S$, where S is the overlap integral and the units are e.v.; the promotion energy from 2s to 2p atomic orbitals was taken as 8.0 e.v. Overlap integrals are taken for Slater orbitals with $Z = 3.25$. The resulting energy levels for C_2 and C_3 are shown in Figs. 1 and 2, respectively. In Fig. 4 is the band pattern for a very long linear carbon chain. The reduced bond distance $\rho = ZR/2a_0$ has a value near 4 for the low energy states we are considering.

The situation for C_2 has been discussed admirably by Mulliken,[9] but it seems desirable to review it briefly for comparison with the larger molecules. There are 8 valence electrons which fill the lowest σ_g and σ_u levels and leave 4 electrons to be placed in the π_u and the σ_g' orbitals. Since the π_u level is doubly degenerate with respect to orbital angular momentum as well as spin, it can accommodate 4 electrons. At the equilibrium C–C distance of approximately 1.3 Å. or $\rho = 4.0$, the calculated energies of the π_u and σ_g' levels are nearly the same and we expect several low lying electronic states as

$$\text{-- } \pi_u(4)\text{: } {}^1\Sigma_g^+$$
$$\text{-- } \pi_u(3)\,\sigma_g'(1)\text{: } {}^3\Pi_u,\ {}^1\Pi_u$$
$$\text{-- } \pi_u(2)\,\sigma_g'(2)\text{: } {}^3\Sigma_g^-,\ {}^1\Delta_g,\ {}^1\Sigma_g^+$$

It is now quite well established that the $^3\Pi_u$ state is lowest with the $^3\Sigma_g^-$ state observed at 5656 cm.$^{-1}$ (0.70 e.v., or 16.2 kcal./mole) by Ballik and Ramsay[10] and the $^1\Sigma_g^+$ state, based

(1) W. A. Chupka and M. G. Inghram, *J. Chem. Phys.*, **22**, 1472 (1954).

(2) R. E. Honig, *ibid.*, **22**, 126 (1954).

(3) W. A. Chupka and M. G. Inghram, *J. Phys. Chem.*, **59**, 100 (1955).

(4) R. J. Thorn and G. H. Winslow, *J. Chem. Phys.*, **26**, 186 (1957).

(5) T. Doehaerd, P. Goldfinger and F. Waelbroeck, *Bull. soc. chim. Belg.*, **62**, 498 (1953).

(6) (a) J. Bégué, Thesis, Paris, 1945; (b) discussion by P. Goldfinger and F. Waelbroeck, *Bull. soc. chim. Belg.*, **62**, 545 (1953).

(7) J. Basset, *Chemie and industrie*, **45**, No. 3f, 7 (1941); *Brennstoff-Chem.*, **23**, 127 (1942); *J. phys. radium*, **10**, 217 (1939).

(8) R. S. Mulliken, *J. Phys. Chem.*, **56**, 295 (1952); R. S. Mulliken, C. A. Rieke, D. Orloff and H. Orloff, *J. Chem. Phys.*, **17**, 1248 (1949).

(9) R. S. Mulliken, *Phys. Rev.*, **56**, 778 (1939).

(10) E. A. Ballik and D. A. Ramsay, *J. Chem. Phys.*, **29**, 1418 (1958).

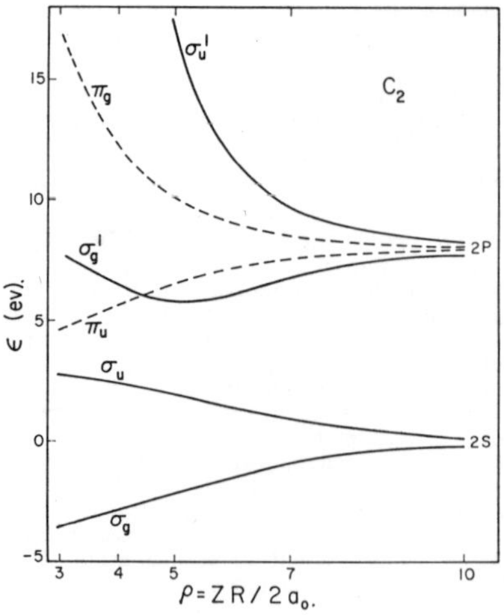

Fig. 1.—Electron orbital energy pattern for C_2.

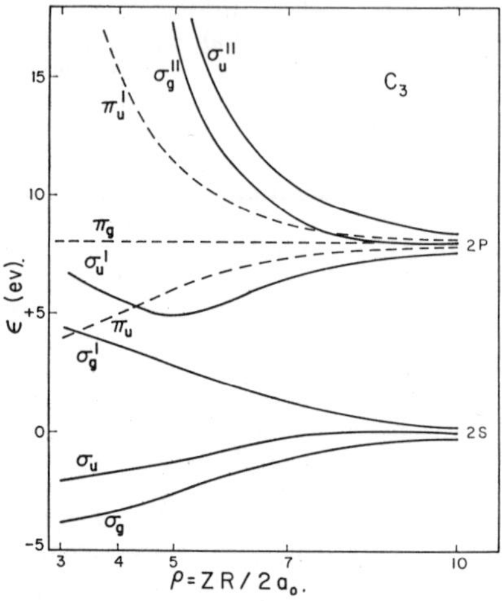

Fig. 2.—Electron orbital energy pattern for C_3.

upon π_u (4), at 8 ± 4 kcal./mole (0.35 e.v.) as observed by Hicks.[11] The other states are at higher energies.

Since the triplet states allow somewhat better electron correlation in the complete wave function, the observed states are consistent with an orbital diagram like Fig. 1 in which the π_u level is a little lower than the σ_g' level at the observed distance. The separation at $\rho = 4.0$ in Fig. 1 is probably a bit too large, however. One may also note that the slope of the π_u level is positive and that of the σ_g' negative near $\rho = 4$. Consequently, one may predict, as Mulliken did, that the C–C distance will be least in the $^1\Sigma_g{}^+$ state, intermediate in the $^3\Pi_u$ state and longest in the $^3\Sigma_g{}^-$ state.

With this confirmation of the method of calculation we turn now to the diagram for C_3, Fig. 2. The 12 valence electrons completely fill the orbitals as follows: σ_g (2), σ_u (2), σ_g' (2), π_u (4), σ_u' (2). Thus one predicts a $^1\Sigma_g{}^+$ ground state. Two configurations should yield excited states at energies in the vicinity of 3 e.v. as

--- π_u (4), σ_u' (1), π_g (1): $^3\Pi_u$, $^1\Pi_u$

--- π_u (3), σ_u' (2), π_g (1): $^3\Delta_u$, $^1\Delta_u$, $^1\Sigma_u{}^+$, $^3\Sigma_u{}^+$, $^1\Sigma_u{}^-$, $^3\Sigma_u{}^-$

The energies of these various states for C_3 were obtained from the orbital energies of Fig. 2 and the electron repulsion terms which were estimated from corresponding terms of C_2 and O_2 but reduced for the greater distances in C_3. The energy level diagram in Fig. 3 shows also the two permitted

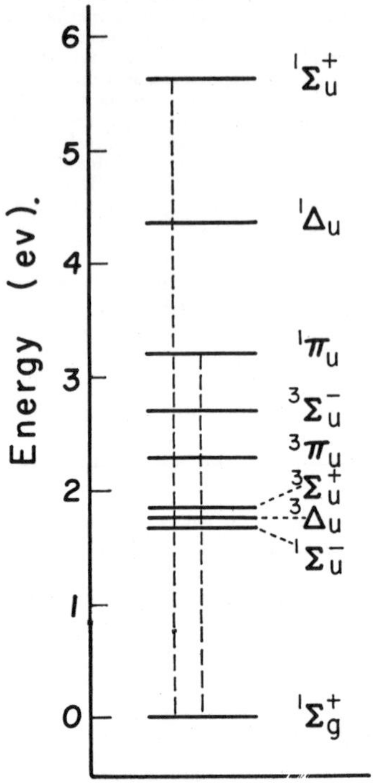

Fig. 3.—Estimated energies of the lower energy states of C_3. The diagram is incomplete about 4 e.v. The dotted lines indicate allowed transitions.

transitions, $^1\Pi_u - {}^1\Sigma_g{}^+$ and $^1\Sigma_u{}^+ - {}^1\Sigma_g{}^+$. The observed absorption near 4050 Å. (3.06 e.v.) is presumably to be ascribed to the $^1\Pi_u \leftarrow {}^1\Sigma_g{}^+$ transition.[12]

The lowest states based upon configurations with two electrons excited were estimated to lie above 4 e.v., consequently the diagram in Fig. 3 may be regarded to be complete up to that level.

As the length of the molecule increases, the levels rapidly fall into the band pattern of the infinitely long chain, Fig. 4. Shockley[13] has shown that, in a case such as this, there are two terminal orbitals (hereafter written t orbitals) which split off, one each from the upper and lower σ-bands, as ρ decreases below the point where the 2s and 2p bands intersect. These t orbitals have most of their amplitude on the terminal atoms and decrease rapidly toward the middle of the chain. They are

(11) W. T. Hicks, Ph.D. Dissertation, University of California, 1957.

(12) N. H. Kiess and H. P. Broida, *Can. J. Phys.*, **34**, 1971 (1956); K. Clusius and A. E. Douglas, *ibid.*, **32**, 319 (1954); A. E. Douglas, *Astrophys. J.*, **114**, 466 (1951); G. Herzberg, *Mém. soc. roy. sci. Liège*, **15**, 291 (1955).

(13) W. Shockley, *Phys. Rev.*, **56**, 317 (1939).

degenerate for an infinite chain. The σ_g' and σ_u' levels of C_3 are already beginning to show the behavior of t orbitals.

For a molecule C_n there are 4 n valence electrons. The lower σ-band has $n-1$ orbitals and accommodates $2n-2$ electrons and the t orbitals accommodate 4 electrons. There remain $2n - 2$ electrons to be accommodated in the π-band. At a distance near $\rho = 4$, the σ'-band is, of course, empty in the ground state or low-lying excited states.

Since the orbitals of the π-band accommodate 4 electrons each, molecules with n odd will just fill the lowest $(n - 1)/2$ of the π-orbitals and will have ${}^1\Sigma_g{}^+$ ground states. Molecules with an even number of carbon atoms will completely fill the lowest $(^1/_2\ n - 1)$ of the π-orbitals and place 2 electrons in the next higher π-orbital. This will yield a ${}^3\Sigma_g{}^-$ ground state with ${}^1\Delta_g$ and ${}^1\Sigma_g{}^+$ states based on the same configuration at slightly higher energies.

Low-lying excited states (for n even or odd) will be obtained by exciting an electron from either the highest filled π-orbital or a t-orbital to the lowest empty or partially filled π-orbital. The required energy for t-electron excitation should be approximately 2 e.v. for n odd or large and even but may be less than 1 e.v. for C_4. The energy for π-electron excitation will decrease from about 2 e.v. for C_5 or C_6 to very small values for long chains which will be one dimensional metals.

The discussion has implicitly assumed approximate uniformity of bond distance. Even n species might be considered in terms of an alternating triple bond single bond structure with single electrons at each end, but our estimates place this structure at substantially higher energy.

Let us restate the description of the long chain C_n molecules in more conventional chemical valence language. The bond structure is

$$:C{=}C{=}C{=}C{=}\ \text{- - - -}\ C{=}C:$$

where all intermediate atoms are contributing 2 electrons to the σ-bonding system to form bonds with each adjacent atom and the terminal atoms use 3 electrons to form one σ-bond and an unshared pair. The σ-bonds comprise the σ-band of Fig. 4 and the unshared pairs the two t-orbitals. Thus there are $2n - 2$ electrons forming π-bonds which are delocalized with respect to the x- and y-planes. Thus the bonding is analogous to O=C=O or O=C=C=C=O rather than to allene.

Quantitative Energy Calculations.—It seems best to use some semi-empirical system of energy calculation which has proven successful for conjugated hydrocarbon molecules, since these are most analogous to the C_n molecules. The method of Mulliken, Rieke and Brown[14] (M.R.B.) was chosen together with the improved modification by Lofthus.[15] In the latter, both Slater and self-consistent field atomic orbital overlap integrals were used and are designated (L-S) and (L-F), respectively. These treatments apply only to the π-electron system, and we use the methods and constants without change to obtain the conjugation energy of the π-system.

(14) R. S. Mulliken, C. A. Rieke and W. G. Brown, THIS JOURNAL, **63**, 41 (1941).

(15) A. Lofthus, *ibid.*, **79**, 24 (1957).

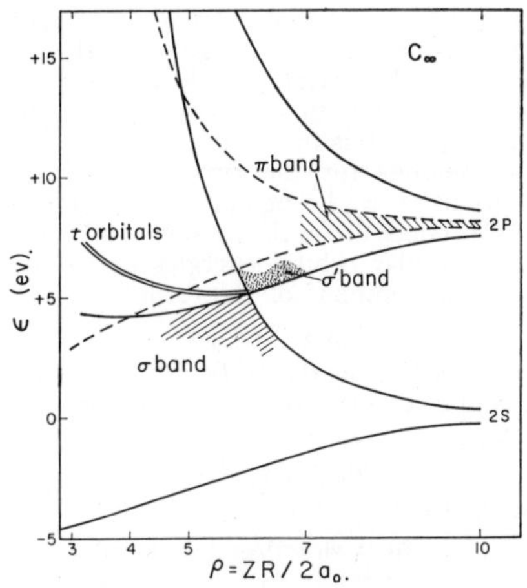

Fig. 4.—Electron orbital energy bands for long linear C_n. The curves marked t-orbitals are for the pair of orbitals associated primarily with the terminal carbon atoms.

First, however, we must obtain the energy of the hypothetical molecule with localized double bonds. This is done in two stages. The energy of interior atoms in a chain is taken from the energy difference allene-ethylene. The heats of formation of these compounds are accurately known[16] and those values, in turn, are corrected for the hyperconjugation present by the (MRB), (L-S) and (L-F) methods. The results are given in Appendix I and each is used self-consistently with the corresponding conjugation energy for the C_n molecule.

There is no satisfactory example of the terminal carbon atom of C_n in an ordinary substance of accurately known heat of formation. Carbon monoxide is the best available example, but it seems unlikely that the large resonance energy of that molecule can be calculated accurately enough at present to make it a useful point of reference. Consequently, the energy of C_3 is taken as a reference point. It is assumed that the mass-spectrometer experiments of Chupka and Inghram[3] with an effusion cell source gave an equilibrium ratio of C and C_3 molecules. Their assumption of a 1 to 3 ratio of ionization cross sections is also retained. Any effect of the small difference in appearance potential between C and C_3 is ignored. The vapor pressure of C is computed from the free energy function discussed later and the value 169.58 kcal./mole for the heat of formation from graphite at 0°K.[17] Use of the free energy function of C_3 then yields a value of $\Delta H_0°$ for the formation of gaseous C_3 from each vapor pressure measurement. The value of $\Delta H_0°$ is practically constant over the temperature range of Chupka and Inghram's measurements. The average is $\Delta H_0^0 =$ 186.5 kcal./mole. The uncertainty from the scatter of results at different temperatures is 0.2 kcal./mole, but other possible errors raise the absolute uncertainty to two or three kcal./mole or possibly even more.

(16) F. D. Rossini, *et al.*, "Selected Values of Physical and Thermodynamic Properties of Hydrocarbons and Related Compounds," Carnegie Press, Pittsburgh, Pa., 1953.

(17) L. Brewer and A. W. Searcy, *Ann. Rev. Phys. Chem.*, **7**, 259 (1956).

The calculation of molecular orbital energies for C_n linear molecules is analogous to that for hydrocarbons with alternate single and double bonds. Assuming the molecular axis is in the z-direction, the two $2p_\pi$-orbitals for each atom point in the x- and y-directions (or the two π-orbitals may be taken to have $\pm$ 1 angular momentum). Thus the double bonds are in the xz- and yz-planes.

The molecular orbital energies are solutions of a secular determinant of the form[15]

$$|\beta_{ij} + S_{ij}\, x_{ij}| = 0 \quad (1)$$

where S_{ij} is the overlap integral. The bond integral β_{ij} is defined in terms of the resonance integral

$$\gamma_{ij} = \int \varphi_i^* H \varphi_j \, d\tau \quad (2)$$

by the equation

$$\beta_{ij} = \gamma_{ij} - {}^1/_2 S_{ij}(\alpha_{ii} + \alpha_{jj}) \quad (3)$$

also

$$x_{ij} = {}^1/_2(\alpha_{ii} + \alpha_{jj}) - \epsilon \quad (4)$$

where the atomic integrals α_{ii} depend in general upon the electronegativities for various atoms and ϵ is the energy of the molecular orbital. For the case of a C_n chain $\alpha_{ii} = \alpha_0$ because all the C atoms have the same electronegativity.

If zero overlap integral is assumed (M.R.B.), then $\gamma_{ij} = \beta_{ij}$. The integral β_{ij} varies with internuclear distance, and its value is fixed for a standard bond length and then varied through a coefficient ρ_{ij}. Lofthus takes ρ_{ij} as the ratio of the overlap integral for the bond in consideration to the standard overlap integral.

Three sets of calculations for the molecular orbital energies were carried out. The first (M.R.B.) assumes zero overlap integral, the second (L-S) uses Slater's atomic orbitals for the overlap integrals, the third (L-F) uses self-consistent field atomic orbitals for the overlap integrals.

With zero overlap integral and considering only nearest neighbor interaction, the secular equation 1 is satisfied by the solutions of the form[18,19,20]

$$\epsilon_i = \beta_0 K_i = \beta_0 \left(2 \cos \frac{i\pi}{2n+1}\right) \; (i = 1,2, \cdots, n) \quad (5)$$

Because of the equivalent of the x- and y-directions for π-electrons, each orbital is doubly degenerate and can accommodate four electrons.

If the overlap integral is evaluated and if only nearest neighbors are considered to interact, the solutions of eq. 1 are of the form

$$\epsilon_i = \frac{K_i}{1 + SK_i}\beta_0 \quad (6)$$

The values for overlap integral at 1.28 Å. are 0.299 for Slater orbitals and 0.395 for self-consistent field atomic orbitals.

For each of the three sets of calculations, the conjugation energy is obtained as the difference between the delocalized total energy for the bond distance 1.28 Å. (observed for C_3) and the localized total energy which is $(n - 1)$ times the standard double bond energy at the standard bond distance. The ρ-values for 1.28 Å. are 1.155, 1. 125 and 1.090 in the (M.R.B.), (L-S) and (L-F) systems, respectively, and the corresponding standard β_0-values (and standard distances) are 44.5 kcal./mole (for 1.33 Å.), 81 kcal./mole (for 1.353 Å.) and 102 kcal./mole (for 1.353 Å.). Figure 5 shows the calculated molecular orbital energies in the (L-F) system for C_1 to C_9. The highest energy levels for $C_4 - C_9$ fall above the top of the figure.

As discussed in the Appendix the contribution to $\Delta H_0{}^0$ of formation per interior carbon atom in a molecule with localized π-bonds is 48.323, 43.112 and 43.408 kcal./mole for the (M.R.B.) (L-S) and (L-F) methods, respectively. Assuming the value of 186.5 kcal./mole as $\Delta H_0{}^0$ of formation for C_3, the heat of formation for C_n is 186.5 plus $(n - 3)$ times the heat of formation for an interior carbon atom less the difference of conjugation energy between C_n and C_3.

For chains with an even number of carbon atoms the calculation yields a value appropriate for the $^1\Delta$ state. The correction to the $^3\Sigma$ ground state is estimated to be $-23/n$ kcal./mole (see discussion on C_2 energy levels). This value decreases very rapidly and reaches a value less than 0.7% of the heat of formation at C_{10}.

While there is a gradually increasing divergence between the three series of calculated values of ΔH_0 as shown in Table I, nevertheless the same general pattern is given in each case with the odd species more stable. The (L-F) series has the best theoretical basis, and it seems likely that these values are correct within 15 kcal./mole through C_7 but may err by increasing amounts for larger n.

TABLE I

CALCULATED VALUES FOR $\Delta H_0{}^0$ OF FORMATION[a] (KCAL./MOLE)

C_n	(M.R.B.)	(L-S)	(L-F)
C_4	212.6	218.3	224.5
C_5	190.2	187.9	193.3
C_6	214.3	220.4	232.9
C_7	198.6	196.9	208.8
C_8	220.0	227.2	246.3
C_9	208.7	209.1	228.0
C_{10}	228.0	236.9	263.3
C_{11}	219.8	222.8	249.2
C_{12}	237.6	248.8	282.4
C_{13}	230.9	237.2	271.7

[a] For C(gas), C_2 and C_3, the observed value of $\Delta H_0{}^0$ of formation are 169.58, 200 and 186.5 kcal./mole, respectively.

In addition to the results for molecules through C_{13} shown in Table I, $\Delta H_0{}^0$ of formation values were calculated by the L-F method for C_{14} (302.9), C_{15} (294.1), C_{16} (324.8), C_{17} (318.1 kcal./mole), and for the infinite chain, 12.7 kcal./g. atom of carbon.

Free Energy, Entropy and Enthalpy Functions.— In order to calculate the equilibrium vapor pressures of the various C_n species, it is necessary to obtain the function $-(F_T{}^0 - H_0{}^0)/T$ for each. The values for atomic carbon and graphite are well established.[16]

The functions for C_2 were calculated by Gordon.[21] To these values a small correction was added for the $^1\Sigma_g{}^+$, $^3\Sigma_g{}^-$ and $^1\Pi_u$ states where energies

(18) E. Hückel, *Z. Physik*, **70**, 204 (1931); **76**, 628 (1932).

(19) J. E. Lennard-Jones and J. Turkevich, *Roy. Soc. Proc.* (*London*), **A158**, 280 (1937).

(20) C. A. Coulsen, *ibid.*, **A169**, 413 (1939).

(21) A. R. Gordon, *J. Chem. Phys.*, **5**, 350 (1937).

have been recently determined. The values of Kelley[22] and of Stull and Sinke[23] err by a factor of two in the multiplicity of the $^3\Pi_u$ state.

Glockler[24] calculated the free energy function for C_3 and Stull and Sinke adopt his values. However, the analysis above indicates a multiplicity of one rather than five for the ground state of C_3. Also, a comparison of the analogous potential constants with those of C_3O_2, ketene and allene suggested somewhat higher vibration frequencies which were adopted: $\nu_1 = 1300$, $\nu_2 = 550$ (degenerate), and $\nu_3 = 2200$ cm.$^{-1}$.

The most important is ν_2 where evidence for a value near 550 cm.$^{-1}$ comes also from the emission spectrum of C_3, observed by Kiess and Broida.[12] This shows a 523 cm.$^{-1}$ frequency difference between band heads 21 times, and these authors suggested interpretation as the bending frequency of C_3.

The rotational constant of C_3 is known accurately from the spectrum to be 0.4280 ± 0.0005 cm.$^{-1}$ which corresponds to a C–C distance of 1.281 Å.

The best basis now available for estimating the increment to a thermodynamic function such as $-(F^0 - H_0^0)/T$ from C_n to C_{n+1} is one-half the difference in that function between C_3O_2 and CO_2. The functions of the latter are well known[16]; those of C_3O_2 were calculated from the data of Long, Murfin and Williams[25] and of Mackle and Sutton.[26] Unpublished infrared measurements of Williams in this Laboratory give an improved value of 215 cm.$^{-1}$ for the lowest bending frequency.

The ground electronic state of C_4 and the larger molecules with n even is predicted to be a $^3\Sigma_g^-$ state whose multiplicity is 3. Other states for even molecules lie within the range of thermal energy, but it would be an unjustified elaboration of our calculations to consider them in detail. It is estimated the contribution to the partition function from these excited states equals that from the ground electronic state, *i.e.*, that the effective multiplicity is 6. Consequently, $R \ln 6 = 3.56$ cal./deg. mole must be added to the $-(F^0 - H_0^0)/T$ and S_0 values for all even species. If the $^3\Pi_u$ state of C_4 lies relatively close to the ground state, the effective multiplicity in that case might approach 9.

The resulting thermodynamic functions are presented in Tables II, III and IV. The error in S or $-(F_T^0 - H_0^0)/T$ for C and C_2 should be small and for C_3 not over 0.5 cal./deg. mole. The first row in each case is for solid graphite[16] which is included for convenience in calculations. The first set of values for the increment per carbon atom above C_3 is based upon the C_3O_2 spectrum as described above. A modified set of values for this increment in $-(F^0 - H_0^0)/T$ and S^0 was obtained by adjustment to fit the observed triple point pressure of graphite, approximately 100 atm. at 4000°K.[7] These modified values correspond to an increase in two of the three bending frequencies of C_3O_2 by about 30%. While this shift seems rather large, the possible errors in the use of C_3O_2 properties to predict the increment per additional carbon atom in a long chain are certainly of this magnitude.

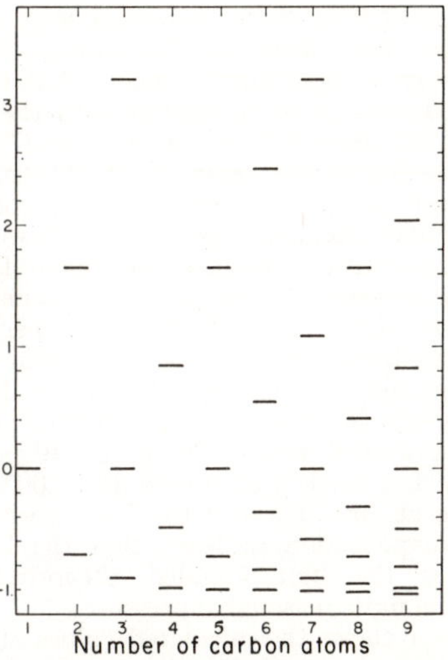

Fig. 5.—Calculated molecular orbital energies for linear molecules C_n.

TABLE II

THE FREE ENERGY FUNCTION[c] $-(F^0 - H_0^0)/T$ IN CAL./DEG. MOLE

Substance	2000	2500	T, °K. 3000	3500	4000
Graphite	5.37	6.39	7.28	8.07	8.78
C_1	42.215	43.334	44.251	45.029	45.707
C_2	56.39	58.27	59.82	61.22	63.35
C_3	61.82	64.59	66.93	68.95	70.73
Increment per C atom *a*	9.57	10.60	11.49	12.24	12.91
Increment per C atom *b*	8.32	9.35	10.24	10.99	11.66

[a] Increment per carbon atom based upon C_3O_2. [b] Increment per carbon atom adjusted to triple point pressure. [c] 3.56 cal./deg. mole is to be added to the values for even n species (except C_2), see text.

TABLE III

THE HEAT CONTENT FUNCTION $(H^0 - H_0^0)/T$ IN CAL./DEG. MOLE

Substance	2000	2500	T, °K. 3000	3500	4000
Graphite	4.39	4.74	5.01	5.22	5.40
C_1	5.014	5.019	5.036	5.062	5.092
C_2	8.14	8.20	8.33	8.35	8.41
C_3	12.21	12.66	12.95	13.23	13.41
Increment per C atom	4.55	4.77	4.90	5.01	5.08

Calculated Vapor Pressures and Vapor Enthalpies.—By combination of the ΔH_0 of formation values with the $-(F_T^0 - H_0^0)/T$ values of Table II, the vapor pressure at equilibrium of each species C_n may be calculated. The results for each of the three conjugation energy methods are given in Table V through C_7.

When the vapor pressure calculations were first completed, it was evident that the predicted

(22) K. K. Kelley, U. S. Bureau of Mines Bulletin 477 (1950).

(23) D. R. Stull and G. C. Sinke, "Thermodynamic Properties of the Elements," American Chemical Society, Washington, D. C., 1956.

(24) G. Glockler, *J. Chem. Phys.*, **22**, 159 (1954).

(25) D. A. Long, F. S. Murfin and R. L. Williams *Proc. Roy. Soc.* (*London*), **A223**, 251 (1954).

(26) H. Mackle and L. E. Sutton, *Trans. Faraday Soc.*, **47**, 937 (1951); R. L. Livingston and C. N. R. Rao, THIS JOURNAL, **81**, 285 (1959).

pressures of long chain molecules were extremely large in the 3500–4000°K. range. The L-F energies gave the lowest pressures, but even these were too high to be consistent with the observed sublimation pressure[6] of 1 atm. near 3700°K. and the triple point pressure[7] of about 100 atm. at 4000°K. While these data are subject to considerable uncertainty, the most likely errors[27] would raise the temperatures assigned and make the inconsistency greater. Consequently, it seemed likely that our $-(F^0 - H_0^0)/T$ increment per C atom, estimated from the C_3O_2 spectrum, was too large and that this error, multiplied by $(n - 3)$, was the primary cause of the disagreement. As stated above, a revised value of the increment per C atom was selected to yield approximately 100 atm. total pressure at 4000°K. with the L-F energies. It is some compensation, at least, that the L-F series, which is the best founded theoretically, also yields the most reasonable pressures.

Figure 6 shows the calculated values of log P for the various species C_n based upon the (L-F) energies and the revised free energy function values.

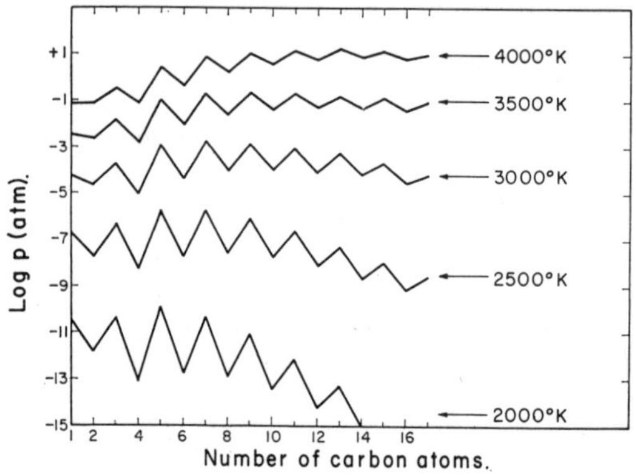

Fig. 6.—Calculated partial pressures of the various vapor species C_n in equilibrium with graphite at several temperatures.

The alternation of the values for the pressure between even and odd carbon species is quite evident. This result is due to the alternation in the pattern of filling of molecular orbitals for the odd and even species as discussed above. The mass spectrographic data[2] show this alternation through C_5, the highest species observed.

TABLE IV

THE ENTROPY[c] IN CAL./DEG. MOLE

Substance		2000	2500	T, °K. 3000	3500	4000
Graphite		9.76	11.13	12.29	13.29	14.18
C_1		47.223	48.353	49.287	50.090	50.798
C_2		64.532	66.470	68.147	69.569	70.764
C_3		74.03	77.25	79.89	82.18	84.14
Increment per C atom	a	14.12	15.37	16.39	17.25	17.99
	b	12.87	14.12	15.14	16.00	16.74

[a] Increment per carbon atom based upon C_3O_2. [b] Increment per carbon atom adjusted to triple point pressure. [c] 3.56 cal./deg. mole is to be added to the values for even n species (except C_2), see text.

(27) See L. Brewer, *J. Chem. Phys.*, **16**, 1165 (1948).

TABLE V

CALCULATED VALUES OF LOG P (ATM.) OF VARIOUS CARBON VAPOR SPECIES IN EQUILIBRIUM WITH GRAPHITE

Substance		2000	2500	T, °K. 3000	3500	4000
C_1		−10.478	−6.75	−4.274	−2.512	−1.195
C_2		−11.87	−7.76	−4.67	−2.64	−1.14
C_3		−10.40	−6.39	−3.74	−1.87	−0.49
C_4	a	−11.81	−7.17	−4.12	−1.97	−0.37
	b	−12.18	−7.47	−3.37	−2.17	−0.55
	c	−13.03	−8.20	−4.97	−2.70	−1.06
	d	−13.12	−8.28	−5.08	−2.83	−1.12
C_5	a	− 9.19	−5.05	−2.32	−0.41	1.00
	b	− 8.73	−4.68	−2.01	−0.14	1.23
	c	− 9.32	−5.14	−2.39	−0.47	0.94
	d	− 9.84	−5.67	−2.93	−1.01	0.42
C_6	a	−10.11	−5.44	−2.37	−0.21	1.37
	b	−10.54	−5.79	−2.66	−0.46	1.16
	c	−11.89	−6.87	−3.56	−1.23	1.08
	d	−12.74	−7.71	−4.39	−2.07	−0.34
C_7	a	− 8.26	−3.94	−1.09	0.89	2.35
	b	− 7.87	−3.62	−0.89	1.24	2.55
	c	− 9.16	−4.66	−1.68	0.38	1.90
	d	−10.89	−5.74	−2.77	−0.70	0.83

[a,b,c] Values for M.R.B., L-S and L-F methods, respectively, with free energy function increment based upon C_3O_2. [d] Values for L-F method with free energy function increment adjusted to triple point pressure.

In addition to the calculations of the partial vapor pressures, the total vapor pressure, the mean number of carbon atoms per molecule $\bar{n}$, and the enthalpy per g. atom and per mole were calculated on the basis of the L-F energies and the adjusted free energy function increments. These values, which are for the saturated vapor in equilibrium with solid graphite, are given in Table VI. More significant figures are retained for $\bar{n}$ in Table VI than the absolute accuracy would justify in order to avoid errors in the use of the molal enthalpy values. The negative apparent heat capacity per gram atom of the saturated vapor is an interesting result of the rapid increase of molecular size with increase in temperature.

TABLE VI

PROPERTIES OF CARBON VAPOR IN EQUILIBRIUM WITH GRAPHITE

(Ref. point for H, graphite at 0° K.)

T, °K.	log P (total) atm.	$\bar{n}$ = mean no. atoms	H_T, kcal./g. atom	H_T, kcal./mole
2000	−9.54	4.777	53.3	255
2500	−5.25	6.231	44.2	276
3000	−2.21	8.137	41.3	336
3500	0.14	10.87	40.3	438
4000	2.01	13.45	39.2	527

It is evident from Table V and Fig. 6 that at 2000°K. and above, our calculations predict C_5 and C_7 to be very important species in equilibrium carbon vapor. At high temperatures even larger molecules become important, although the uncertainties of calculation make this more a qualitative than a quantitative prediction. For comparison with the effusion experiments of Doehaerd, Goldfinger and Waelbroeck[5] it is convenient to restate our results in terms of an apparent heat of sublimation of monoatomic carbon, the only species considered by the above authors. If we denote this quantity by ΔH_{ap}, then for effusion experiments

$$\Delta H_{ap} = 169.5 - RT \ln \frac{\Sigma_n n^{1/2} P_n}{P_1} \quad (7)$$

where the P_n are the partial pressures of the species C_n. The approximate temperature of the experiments is 2600°K. Using the partial pressures given by (L-F) method adjusted to the triplet point pressure, one obtains from eq. 7 ΔH_{ap} = 145.6 as compared with a value of 143.2 given by Doehaerd, Goldfinger and Waelbroeck.[5] Any of the other sets of calculated pressures yields a lower ΔH_{ap}. This agreement is all that could be expected and clearly removes any disagreement between the effusion data and theoretical expectations based on the 169.58 kcal./mole heat of sublimation of monatomic carbon.

Taken at face value this agreement of apparent ΔH values also indicates that the evaporation coefficients of all species through C_7 at least are large enough to have yielded near equilibrium pressures with the aperture to surface ratio of 1 to 23,400. However, Doehaerd, *et al.*, assumed the value of 0.2 for the sticking coefficient of carbon vapor on a relatively hot receiver. This value was obtained by Simpson, Thorn and Winslow[28] under similar but not identical conditions. While Doehaerd, *et al.*, realized that there was uncertainty in transferring the sticking coefficient value to their experiments, that uncertainty seemed moderate if carbon vapor was purely monatomic. We now know that the vapor is a complex mixture of several gaseous species each of which has its own sticking coefficient. Consequently the Doehaerd, *et al.*, value of ΔH_{ap} is subject to as much as 7 kcal./mole uncertainty, and the possibility remains that their conditions yielded less than equilibrium amounts of C_5 and larger species.

The effusion cells used by Thorn and Winslow[4] had aperture to surface ratios of about 1 to 2000 (about ten times that of Doehaerd, *et al.*) but cooled receivers eliminated the sticking coefficient uncertainty. Their ΔH_{ap} = 161.0 kcal./mole at about 2450°K. corresponds quite closely to an equilibrium vapor through C_3 with higher species absent. Thorn and Winslow interpreted their data on this basis and calculated properties of C_3 very similar to those we selected. From this result one may conclude that the evaporation coefficients for C_5 and larger species are distinctly less than 5×10^{-4} under the conditions of the Thorn and Winslow experiments.

Also there is the older work of Brewer, Gilles and Jenkins[29] and of Brewer and Mastick[30] which Thorn and Winslow[4] described in 1957 as "probably the most reliable effusion experiment prior to presently reported one." Brewer, *et al.*, used a cooled receiver and found a sticking coefficient of 0.45. The aperture to evaporating surface ratio is not available because they partially filled their cell with graphite powder of unknown surface area. The values of ΔH_{ap} range from 150.9 to 166.8 kcal./mole at 2600°K. but the low values always were obtained on the first run with a fresh sample of graphite powder. Later runs with the sample gave higher values eventually reaching the range 165 ± 2 kcal./mole which one interprets as indicating an equilibrium amount of C and C_2 with some C_3.

Brewer, Gilles and Jenkins interpreted the drift in their ΔH_{ap} values as due to "volatilization of tars remaining in the material from the binder used in the graphitization process." Since the evaporation coefficients of the polyatomic carbon molecules probably vary greatly for different surfaces of graphite, it now seems more likely that in the early runs the evaporation rates of C_3 and C_5 were substantial but that these rates fell off as the reactive (high evaporation coefficient) surfaces were depleted. This explanation seems reasonable in terms of the structure of graphite and the possible mechanisms for breaking off a C_3 or C_5 molecule.

Finally, there are the measurements of Bégué[6a] which are interpreted to indicate a total vapor pressure of carbon one atmosphere at about 3750°K. The values in Table VI show a total pressure of one atmosphere at a little below 3500°K. These results are not necessarily inconsistent because the evaporation coefficients of the very large carbon molecules may still be small at these temperatures. Since the observed temperatures are for conditions of rather rapid evaporation, they do not necessarily indicate the equilibrium vapor pressure curve. The same objection can be raised with respect to the sublimation point measurements of Basset,[7] but the error should decrease with rise in temperature and pressure.

Rings.—It is possible to use many of these same methods to estimate the abundance of ring molecules in carbon vapor. While a =C=C=C=C= linkage tends to be linear, it is somewhat flexible and a long linear molecule C_n may be bent into a ring whereupon the terminal unshared electron pairs will form another double bond. Thus all atoms have the basic valence state of interior atoms in the linear chain, and the conjugation energy of the π-system is calculated readily for the ring by the same methods as for the chain. If $n = 4k + 2$ with k an integer, then a $^1\Sigma_g$ ground state results. The bending force constants for the central portion of the O=C=C=C=O molecule may be used to estimate the strain energy of the C_n ring to be approximately $(1800/n)$ kcal./mole.

The entropy or free energy function for a ring is substantially smaller than that for a chain. By comparison with values for hydrocarbons we estimate this difference to be at least 20 and 25 cal./mole degree for C_{10} and C_{14}, respectively. In the M.R.B. method the ring molecular orbital energies[20] are ±2, ±1.618, ±0.618, for C_{10} and ±2, ±1.802, ±1.247 and ±0.445 for C_{14} in units of β. The M.O. of energy ±2 accommodates two electrons; the others are doubly degenerate with respect to orbital angular momentum and accommodate

(28) O. C. Simpson, R. J. Thorn and G. Winslow, AECD-2680 (1949).

(29) L. Brewer, P. W. Gilles and F. A. Jenkins, *J. Chem. Phys.*, **16**, 797 (1948).

(30) L. Brewer and D. F. Mastick, University of California Radiation Laboratory report, UCRL-572 (1949); D. F. Mastick, Ph.D. thesis, University of California (1950); see also Ref. 6b.

four electrons. The pattern of ring orbitals is repeated for the $2p_\pi$ orbitals perpendicular to the plane and for those radially oriented in the plane of the ring. Since ring strain was considered separately, we used the same β for both perpendicular and radial orbitals, and therefore the degeneracies stated above are doubled. The resulting $\Delta H_0{}^0$ of formation values are 264 and 297 kcal./mole for C_{10} using (L-S) and (L-F) methods, respectively, and 271 and 322 for C_{14} ring with the corresponding methods.

In the range 2000–4000°K. the calculated pressures of C_{10} and C_{14} rings are smaller than the pressures of the corresponding chains by factors ranging from 10^{-6} to 10^{-8}. Since the even-numbered linear species are already less abundant than the adjacent odd species, it is evident that ring molecules play no significant role in this region. Possibly rings would become important for very large n values but only under conditions where these exceedingly long chains were also abundant.

Negative Ions.—Honig[2] noted the presence of substantial quantities of negative ions C_n^- in carbon vapor and found the even species to be the more abundant. This result follows clearly from the simple molecular orbital theory. The even neutral species C_n have a half-filled bonding π-orbital which can accommodate another electron whereas in the odd species the extra electron must go into the empty non-bonding orbital.

The electron affinity calculations of Honig can be revised and extended by the use of the improved energy of C_3, the calculated energy of C_4, and the electronic multiplicities which we feel to be quite certain. The negative ions have $^2\Pi_u$ ground states, except C^- which is ^4S and C_2^- where the $^2\Sigma_u$ state may be slightly below $^2\Pi_u$. Honig[2] calculated electron affinities, where possible, by two methods: (a) the *difference* in the heats of formation derived from the temperature derivative of the abundances and (b) the *ratio* of the absolute intensities interpreted on the basis of assumed cross sections, etc., and equal electronic multiplicities for neutral and ion in each case. Introduction of our data yields the results of Table VII. The uncertainty in these values is about 1 e.v. in most cases.

TABLE VII

ELECTRON AFFINITY OF CARBON MOLECULES (E.V.)

Species	Difference value	Ratio value
C	..	1.4
C_2	4.0	3.2
C_3	2.4	1.7
C_4	3.6	..

With increase in length of the carbon molecule, the oscillation of electron affinity between even and odd species will diminish because the available orbitals differ less in energy. This too is consistent with Honig's results. The limiting electron affinity for a very long chain should approach that of graphite, 4.4 e.v.

Liquid Carbon.—Since the calculated properties of carbon vapor (after the adjustment in the free energy function) seem to be in reasonable accord with all observed properties, it is interesting to speculate about the nature of liquid carbon. It is difficult to picture a liquid based on the graphite structure, but the long linear polymer molecules certainly could form a liquid somewhat analogous to long n-paraffins or sulfur in the range above 160° where the equilibrium shifts from S_8 ring molecules to long chains. The carbon chains do not have the zig-zag structure of sulfur or a n-paraffin, with the possibility of rotation about single bonds. Consequently, carbon chains would be less flexible in terms of a given thermal energy, but at 4000°K. carbon chains would be sufficiently flexible to form a disordered-liquid-like phase.

Let us assume for liquid carbon the entropy per added carbon atom of long chains in the gas, which from Table IV is 16.82 cal./deg. g. atom at 4000°K. The heat of formation of liquid at 4000°K. is taken as the heat of formation per added carbon atom (given above for 0°K. and corrected to 4000°K.) of 11.4 kcal/g. atom less an estimated value of 1.0 kcal./g. atom for the Van der Waals attractive forces between chains in the liquid state. The latter figure was estimated from a comparison of the heats of vaporization of C_3O_2 and CO_2. The corresponding quantity for long n-paraffins[16] is also about 1.0 kcal. per mole of CH_2 units. The calculations of Coulson and Davies[31] indicate a larger value, possibly 3 to 4 kcal./g. atom, at an interchain distance of 3.4 Å. corresponding interplanar separation in graphite, but we estimate that the inclusion of overlap integrals neglected by Coulson and Davies would reduce this value to the range 1 to 2 kcal./g. atom.

Now we have two estimates of the heat of fusion. From the energy calculations, $\Delta H_f = 10.4$ kcal./g. atom; but from the entropy calculations, $\Delta H_f = T\Delta S_f = 4000(16.82 - 14.18)/1000 = 10.5$ kcal./g. atom. This agreement, which must be regarded as fortuitously good, nevertheless certainly supports the general model of long linear chains (and very large rings) for liquid carbon.

It is also interesting to calculate the heat of sublimation at 4000°K., which is 17.6 kcal./g. atom, and then by deducting the heat of fusion one obtains for the heat of vaporization of the liquid 7 kcal./g. atom. If these values are at all close to the true properties, the vapor pressure curve will show a very large decrease in slope at the melting point because the heat of fusion is larger than the heat of vaporization of the liquid.

Liquid carbon comprising long linear polymer molecules should be relatively viscous as is sulfur when it is in the polymeric state. Also, although the linear polymeric chains may be expected to be one dimensional metals, the macroscopic electrical conductance would be much less than that of graphite with its two dimensional metallic character.

Acknowledgments.—We are greatly indebted to Professor Leo Brewer for many discussions. Dr. W. Weltner, Jr., has given us the benefit of his study of this problem and told us of recent work at the National Carbon Research Laboratories. This work was performed under the auspices of the U. S. Atomic Energy Commission.

(31) C. A. Coulson and P. L. Davies, *Trans. Faraday Soc.*, **48**, 777 (1952).

Appendix

The conjugation energy (c.e.) for the interior atoms in the C_n chain is obtained from the difference in c.e. for allene and ethylene with a correction for compressional energy corresponding to contraction from the standard double bond to the 1.28 Å. bond assumed for the chain.

The c.e. is calculated following the M.R.B. method and the Lofthus modification of that method. M.R.B.[14] assume zero overlap integral, a reference double bond of 1.33 Å. (ethylene double bond) and a standard resonance integral $\beta_0 = -44.5$ kcal./mole. Lofthus[15] uses overlap integrals calculated with Slater or S.C.F. atomic orbitals. The reference double bond distance is 1.353 Å. The β_0 values are -81 and -102 kcal./mole for Slater orbitals or S.C.F. orbitals, respectively.

An additional correction of 0.5 β_0 is introduced by Lofthus to account for the different electronegativity of the terminal hydrogen group as compared to the carbon atom. The secular equation 1 for allene and ethylene is solved for localized and delocalized bonds. The values for the β_0 coefficients and for conjugation energy are given in Table VIII.

Compressional energies are obtained from a Morse potential assuming a dissociation energy $D = 141.5$ kcal./mole and a bond force constant of 8.2×10^5 dyne cm.$^{-1}$.

The contribution to ΔH_0^0 of formation of the hypothetical C_n with localized bonds per interior carbon atom is then the difference in ΔH_0^0 of formation[16] between allene and ethylene (33.178 kcal./mole) corrected for the difference in their conjugation energies and compressional energies. If the compression energy to the 1.28 Å. bond distance of C_3 (and presumably of longer C_n molecules) is added, the final values are 48.323, 43.112 and 43.408 kcal./mole for the (M.R.B.), (L-S) and (L-F) methods. These values are used in the calculations leading to Table I.

TABLE VIII

ORBITAL AND CONJUGATION ENERGIES OF ALLENE AND ETHYLENE

Method		Localized M.O. coefficients		Delocalized M.O. coefficients		Conjugation energy
(M.R.B.)	Ethylene	4.000	...	4.531	3.531	0.124
	Allene	4.000	1.058	4.146	1.023	.444
(L-S)	Ethylene	1.275	...	1.508	1.048	.012
	Allene	1.277	0.845	1.432	0.710	.121
(L-F)	Ethylene	0.945	...	1.177	.718	.008
	Allene	0.946	0.770	1.140	.601	.098

BERKELEY, CALIFORNIA

Comment: Large Molecules in Carbon Vapor

The linear geometry and qualitative characteristics of the electronic structures suggested here have been confirmed unambiguously for several of the odd-numbered species C_3, C_5, C_7, ... But the numerical values for the energies are only rough approximations. Also, for even-numbered species, alternate nonlinear structures have nearly the same energies, and neither experimental nor theoretical methods have clearly resolved all questions. For the competition between linear and ring structures, the bending force constant was estimated from a value from the molecule $0 = C = C = C = 0$. This value was later found to be much too large; hence, the prediction against significance of rings in carbon vapor was incorrect.

Possibly the most important contribution of this paper was the general concept that relatively large polyatomic molecules might be, and probably were, present in carbon vapor at very high temperatures. The discovery of particular very large molecules, and especially of C_{60} and its relatives, has been a fascinating story.

INTER- AND INTRAMOLECULAR FORCES AND MOLECULAR POLARIZABILITY

KENNETH S. PITZER, *Department of Chemistry and Radiation Laboratory, University of California* *

CONTENTS

I. INTRODUCTION

The correlation of electron motion in molecular systems is responsible for many important effects, but its theoretical treatment has proved to be very difficult. Thus many quantum valence calculations use wave functions which are adjusted to optimize kinetic energy effects and the potential energy of interaction of nuclei and electrons but which do not adequately allow for electron correlation and hence yield excessive electron repulsion energy. This problem may be subdivided into cases of overlapping and nonoverlapping electron distributions. Both are very important but we shall concern ourselves here with only the nonoverlapping case.

F. London opened this field with his classic treatment[24] of the

* This paper was prepared under the auspices of the U. S. Atomic Energy Commission.

attractive intermolecular forces which arise because of correlation of electron motion in the two molecules. These are commonly called London forces or dispersion forces. This work explained the previously unaccounted for attraction between nonpolar molecules and rare gas atoms. The importance of this qualitative success obscured for a time discrepancies from quantitative agreements for molecules larger than H_2 and He. Recognition of the need for improved quantitative prediction of London forces[34,17] led to additional work to be reviewed in this paper.

The key element in London's approach is the expansion of the electrical potential energy in multipole series. Since neutral molecules or portions of molecules are involved, the leading term is that for dipole-dipole interaction. While attention has been given to higher-order terms, these are usually small, and the greater need seems to be for improved treatment of the dipole-dipole terms. London used second order perturbation theory in his treatment, but Slater and Kirkwood[38,21] soon followed with a variation method treatment which yielded similar results. Other individual papers will be mentioned later, but the excellent review of Margenau[26] should not be overlooked.

In the first approximation London forces in complex systems may be taken as the sum over all interacting pairs. Higher-order corrections for triple interactions have been given by Axilrod and Teller[2,3] and by Jansen and McGinnies.[18]

We shall consider only forces between systems of moderate size which are in their ground electronic states. Dahler and Hirschfelder[8] have discussed the additional complications when excited states are introduced. Yos, Bade, and Jehle[42] gave special consideration to London forces between macromolecules but found it difficult to determine the magnitude of the special effects they noted.

In polyatomic molecules there are separate regions of essentially nonoverlapping electron clouds whose interaction may be treated by London's methods. This was recognized by many authors, but relatively few calculations have emphasized this factor. Indeed sometimes it may have been erroneously dismissed as of negligible magnitude. Born and Mayer[5] considered London forces in their treatment of ionic crystals and Davydov[9] and Simpson[37] have considered these effects in conjugated π electron systems. Pauling and Simonetta[28a] included London energy terms in calculations on phosphorous. Pitzer[33] suggested that the anomalous sequence of bond energies in the halogens might be explained on this basis and Pitzer and Catalano[32] showed that the isomerization energies of the lighter paraffin hydrocarbons were predicted satisfactorily by the London theory. This work will be reviewed and some further considerations discussed.

II. THEORY FOR MANY-ELECTRON SYSTEMS

It was mentioned in the introduction that theoretical and experimental values of London energies agree reasonably well for He and H_2. But for larger molecules serious diagreement (frequently about a factor of two) arises between various theoretical formulas and between theory and experiment. Let us review the theory for many-electron systems with particular attention to approximations which may become increasingly important as the number of electrons increases. This discussion follows largely that of Donath[11] and, in part, that of Buckingham.[6]

Atomic units are used throughout this section; the most important formulas are rewritten in other terms at the beginning of Section III.

We write for the ground state wave function of the closed shell atom or molecule a Slater determinant for the N electrons

$$\Psi = (N!)^{-\frac{1}{2}} \det \{\psi_1 \psi_2 \ldots \psi_N\} \tag{1}$$

It is presumed that this function is optimized in the Hartree-Fock approximation.

A. Polarizability

Consider first the polarization of the system in an electric field F along the z axis. The perturbing potential is

$$V = F \sum_{p=1}^{N} z_p \tag{2}$$

and the variational wave function is taken to be

$$\Psi' = \Psi + \sum_{\lambda=1}^{N} a_\lambda \Phi_\lambda \tag{3a}$$

$$\Phi_\lambda = (N!)^{-\frac{1}{2}} \det \{\psi_1 \psi_2 \ldots \phi_\lambda \ldots \psi_N\} \tag{3b}$$

$$\phi_\lambda = z_\lambda \psi_\lambda f_\lambda \tag{4}$$

where f_λ is either unity or a function such as $(1+c_1 r+c_2 r^2+\ldots)$ adjusting ϕ slightly from $z\psi$. A variation method adjustment of the parameters in f_λ is possible in principle.

Next we define the three quantities

$$h_\lambda = \langle \Psi^* | \sum z_p | \Phi_\lambda \rangle \tag{5}$$

$$Q_{\lambda\mu} = \langle \Phi_\lambda^* | H | \Phi_\mu \rangle - E_0 g_{\lambda\mu} \tag{6}$$

$$g_{\lambda\mu} = \langle \Phi_\lambda^* \Phi_\mu \rangle \tag{7}$$

Where H is the complete Hamiltonian operator for the unperturbed system and the usual quantum-mechanical integrals over all space are indicated.

The energy change upon imposition of the field is

$$h' = 2F \sum_\lambda a_\lambda h_\lambda + \sum_\lambda \sum_\mu a_\lambda a_\mu Q_{\lambda\mu} \tag{8}$$

which is to be minimized with respect to each a_λ.

$$\frac{\partial h'}{\partial a_\lambda} = 0 = 2Fh_\lambda + 2\sum_\mu a_\mu Q_{\lambda\mu} \tag{9}$$

From determinant theory the solution is

$$a_\lambda = -A_\lambda F \tag{10}$$

$$A_\lambda = \frac{\begin{vmatrix} Q_{11} & \ldots & h_1 & \ldots & Q_{1N} \\ \cdot & & & & \\ \cdot & & & & \\ Q_{\lambda 1} & \ldots & h_\lambda & \ldots & Q_{\lambda N} \\ \cdot & & & & \\ \cdot & & & & \\ Q_{N1} & \ldots & h_N & \ldots & Q_{NN} \end{vmatrix}}{\begin{vmatrix} Q_{11} & \ldots & Q_{1\lambda} & \ldots & Q_{1N} \\ Q_{\lambda 1} & \ldots & Q_{\lambda\lambda} & \ldots & Q_{\lambda N} \\ Q_{N1} & \ldots & Q_{N\lambda} & \ldots & Q_{NN} \end{vmatrix}} \tag{11}$$

and the minimized energy is

$$h' = -F^2\left(2\sum_\lambda A_\lambda h_\lambda - \sum_\lambda \sum_\mu A_\lambda A_\mu Q_{\lambda\mu}\right) \tag{12}$$

After several steps this simplifies to

$$h' = -F^2 \sum_\lambda A_\lambda h_\lambda \tag{13}$$

and the polarizability is

$$\begin{aligned}\alpha &= -2h'/F^2 \\ &= 2 \sum_\lambda A_\lambda h_\lambda \end{aligned}\tag{14}$$

While it is not beyond the realm of possibility to evaluate the g's, h's, and Q's for appropriate wave functions, some further simplification is necessary for our present arguments. We assume that the ϕ's are orthogonal to all ψ's. This is not necessarily true although the f_λ's can be selected to make it true. Even without specific selection of f_λ's it is probably a good approximation for those ϕ's which contribute substantially to the polarizability. Then

$$g_{\lambda\mu} = 0 \qquad \text{if } \lambda \neq \mu \tag{15}$$

$$h_\lambda = \langle \psi_\lambda^* z_\lambda \phi_\lambda \rangle \tag{16}$$

The calculation of the Q's is more complex. If f_λ in equation (4) is unity and Hartree wave functions are used, $Q_{\lambda\lambda} = \frac{1}{2}$. This value was taken by Slater and Kirkwood[38] and by Buckingham,[6] and is probably a reasonable approximation. Buckingham gives values for other simple f_λ's. Donath[11] points out, however, that the off-diagonal terms which have been neglected previously are not without significance. They reduce to

$$\begin{aligned} Q_{\lambda\mu} = \Bigg[&\left\langle \phi_\lambda^*(2)\, \psi_\lambda(2) \left| \frac{1}{r_{12}} \right| \psi_\mu^*(1)\, \phi_\mu(1) \right\rangle \\ &- \left\langle \phi_\lambda^*(2)\, \phi_\mu(2) \left| \frac{1}{r_{12}} \right| \psi_\mu^*(1)\, \psi_\lambda(1) \right\rangle \Bigg], \quad \text{for } \lambda \neq \mu \end{aligned}\tag{17}$$

in the present approximation.

The importance of the off-diagonal $Q_{\lambda\mu}$ terms is most easily demonstrated in the case of a shell of N equivalent orbitals. Then A_λ may be reduced to

$$A_\lambda = h_\lambda / \left(\sum_\mu Q_{\lambda\mu}\right) \tag{18}$$

and

$$\alpha = 2 \sum_{\lambda} h_\lambda^2 / (\sum_{\mu} Q_{\lambda\mu}) = 2Nh^2 / \sum_{\mu} Q_{\lambda\mu} \tag{19}$$

In the noble gas atoms the outermost electron subshell is most important for the polarizability and may be arranged in a fashion to make all orbitals equivalent. Donath[11] calculated approximately the net effect of the off-diagonal terms for the $2p$ shell for Ne with Slater orbitals. Two effective Z values were used,[25] 6.31 and 4.77, and the resulting values of the sum of off-diagonal Q's are 0.325 and 0.430, respectively. These are to be compared with a value of about $\frac{1}{2}$ for the diagonal term. It is evident that, if the approximate results for neon are typical, the use of all Q's will substantially reduce the calculated polarizability of many-electron systems. Earlier calculations[6] yielded too large values; consequently, the improved theory will yield better agreement. More extensive numerical calculations are now being made to test the improved theory and to eliminate certain approximations.

B. London Forces

Since the polarizability is so closely related to the London force, all approximate formulas for the latter are obtained by replacing arrays of unevaluated terms which approximate the polarizability by that quantity itself. We shall examine the several ways in which this has been done with reference to the magnitude of error which may have been introduced.

If we limit ourselves to spherically symmetric systems, the perturbing potential for the interaction of atoms A and B is

$$H' = 6^{\frac{1}{2}} R^{-3} (\sum_{i} z_{A,i}) (\sum_{j} z_{B,j}) \tag{20}$$

where R is the distance between the two atom centers. The wave function generalized for two nonoverlapping atoms is

$$\Psi' = \Psi^A \Psi^B + \sum_{\lambda} \sum_{\mu} b_{\lambda\mu} \Phi_\lambda{}^A \Phi_\mu{}^B \tag{21}$$

where the various quantities were defined in Eqs. 1 and 3b.

It is an assumption to use the same Φ_λ functions, or more specifically, the same f_λ of Eq. 4, for London forces and for polarizability. It is easy to check this point for hydrogen atoms. Let ψ be the regular 1s function and $f = e^{cr}$. If c is adjusted for the polarizability, one finds $c = 0.204$, $\alpha = 4.438$ (as compared to the exact value 4.500), and the London energy $E_L = -6.19R^{-6}$. But if c is adjusted to optimize the London energy directly, $c = 0.134$ and $E_L = -6.45R^{-6}$ (as compared to the exact value $-6.50R^{-6}$). If $c = 0$, $f = 1$, $E_L = -6.00R^{-6}$.

This result indicates that in strictly theoretical calculations, the f_λ functions may almost as well be omitted unless they can be optimized for the London energy itself. For the purpose of semi-empirical calculations, however, the f_λ functions from the polarizability must be retained for the substitution in the London energy. The error for hydrogen atoms is only about 4 per cent, however, and there does not appear to be any reason that it would increase greatly in more complex systems.

We now proceed to write out the London energy with the same symbols and approximations as were made before. In particular we retain the approximation that the ϕ's are orthogonal to the ψ's.

$$E_L = \frac{2\sqrt{6}}{R^3} \sum_{k,p} b_{kp}\, h_k{}^A\, h_p{}^B + \sum_{h,p} \sum_{l,q} b_{kp}\, b_{lq}\, R_{kl,pq} \tag{22}$$

$$R_{kl,pq} = Q_{kl}{}^A\, g_{pq}{}^B + Q_{pq}{}^B\, g_{kl}{}^A \tag{23}$$

The variational treatment follows similar steps to those given above and yields

$$E_L = -\frac{6}{R^6} \sum_{k,p} B_{kp}\, h_k{}^A\, h_p{}^B \tag{24}$$

$$B_{kp} = \det |R_{kl,pq} + (h_k{}^A\, h_p{}^B - R_{kl,pq})\, \delta_{kl}\, \delta_{pq}| \,/\, \det |R_{kl,pq}| \tag{25}$$

Again it is a useful approximation to consider just the set of equivalent orbitals of the outermost subshell. Then

$$B_{kp} = h_k{}^A\, h_p{}^B \,/ \sum_{l,q} R_{kl,pq} = h_k{}^A\, h_p{}^B / \left(g_{pp}{}^B \sum_l Q_{kl}{}^A + g_{kk}{}^A \sum_q Q_{pq}{}^B\right) \tag{26}$$

where the sums are now limited to the orbitals of the outer subshell. Equation 24 may be rearranged to

$$E_L = -\frac{6}{R^6}\left\{\sum_k \frac{(h_k{}^A)^2}{\sum_l Q_{kl}{}^A}\right\}\left\{\sum_p \frac{(h_p{}^B)^2}{\sum_q Q_{pq}{}^B}\right\}\frac{[\sum_l Q_{kl}{}^A/g_{kk}{}^A]\cdot[\sum_q Q_{pq}{}^B/g_{pp}{}^B]}{[\sum_l Q_{kl}{}^A/g_{kk}{}^A]+[\sum_q Q_{pq}{}^B/g_{pp}{}^B]} \tag{27}$$

The quantities in braces are $\frac{1}{2}\alpha_A$, $\frac{1}{2}\alpha_B$, respectively, and, if we define the energies in brackets, for example,

$$U_A = \sum_l Q_{kl}{}^A/g_{kk}{}^A \tag{28}$$

then the London energy takes the familiar form

$$E_L = -\frac{3\alpha_A\alpha_B}{2R^6}\cdot\frac{U_A U_B}{U_A+U_B} \tag{29}$$

This is, of course, just a formula first presented by London but with a new meaning to the characteristic energies U_A, U_B.

If in Eq. 28 only $Q_{kk}{}^A$ were important, then U_A would be approximately the ionization energy, which was London's interpretation. For many-electron systems, however, we have seen that the off-diagonal Q's make a substantial contribution. Consequently, Donath has suggested that the value of U for a typical many-electron atom may be about twice the ionization energy and that this accounts for the fact that observed London energies are about double those calculated by the London formula. Donath also suggests that U be regarded as an empirical parameter for a semi-empirical system for estimation of London forces. We shall return to this system in the next section.

In many electron atoms the maximum contributions to the polarizability and to London forces arise from configurations with more than one electron contributing to the net dipole moment of the atom. But in such configurations the electronic repulsion is especially high. The physical meaning to be attributed to the Q_{kl} terms is just the additional electron repulsive energy which these configurations require.

At this point let us introduce throughout the approximation $f_\lambda = 1$ which we have used in part heretofore. Then

$$g_{\lambda\lambda} = h_\lambda = \langle\psi_\lambda{}^* z_\lambda{}^2 \psi_\lambda\rangle \tag{30}$$

and for just a subshell of equivalent orbitals

$$\begin{aligned}\alpha_A &= 2\sum_\lambda h_\lambda{}^A/U_A \\ &= 2Nh^A/U_A\end{aligned} \tag{31}$$

We shall use these results in various derivations to follow.

There is another effect[31] leading to an increase in the effective energy U_i above even the value of Eq. 28. The treatment for that equation considered only the outer subshell of electrons. Other, more tightly bound electrons must make some contribution to both the polarizability and the London energy. But Eqs. 11, 14, 28, and 30 show that the polarizability depends on terms of the type $(h^2/Q) \cong (h/U)$ while the London energy terms from Eqs. 24 and 25 are of the order $(h^4/gQ) \cong (h^2/U)$. Thus, if other electrons still have large h values but larger Q values than the outer subshell electrons do, they contribute more to the London energy than the product $\alpha_A\alpha_B$ predicts. This has the effect of yielding a higher effective excitation energy or raising U_i even more above the ionization energy.

Let us now derive the Slater-Kirkwood[38] formula in terms of our present quantities. A single subshell of equivalent electrons is assumed. Equation 29 may be rearranged to

$$E_L = -3\alpha_A\alpha_B/2R^6 \cdot (U_A{}^{-1} + U_B{}^{-1})^{-1} \tag{32}$$

where from Eqs. 28 and 30

$$U_A{}^{-1} = h^A/\sum_l Q_{kl}{}^A = (\alpha_A/2N\sum_l Q_{kl}{}^A)^{\frac{1}{2}} \tag{33}$$

and Eq. 19 was used in the last transformation. Then adoption of the value $\frac{1}{2}$ for $\sum Q$ yields the Slater-Kirkwood formula

$$E_L = -3\alpha_A\alpha_B/2R^6\,[(\alpha_A/N_A)^{\frac{1}{2}} + (\alpha_B/N_B)^{\frac{1}{2}}] \tag{34}$$

Our earlier discussion of off-diagonal Q's, however, indicated that $\sum Q$ would substantially exceed $\frac{1}{2}$ in many-electron systems. Consequently, if Eq. 34 is used with N an empirical factor,[31] we may expect the N values to exceed substantially the actual number of electrons of the outer subshell.

Another generally useful equation for the London energy is that of Kirkwood and Müller[27] which uses the diamagnetic susceptibility as well as the polarizability of the individual atoms or molecules. The *susceptibility* per atom or molecule of a *spherically symmetrical* system is given by

$$\chi = -(2mc^2)^{-1} \sum_k \langle z_k^2 \rangle \tag{35}$$

where c is the velocity of light and m the electronic mass. But in view of Eq. 30 and for a subshell of equivalent electrons

$$\chi_A = -Nh^A/2mc^2 \tag{36}$$

and with Eq. 31

$$U_A^{-1} = \alpha_A/2Nh^A = -(4mc^2)^{-1}(\alpha_A/\chi_A) \tag{37}$$

Then the London energy becomes

$$E_L = 6mc^2\alpha_A\alpha_B/R^6\,[(\alpha_A/\chi_A) + (\alpha_B/\chi_B)] \tag{38}$$

Margenau[26] thought that the cancellation of the errors of various approximations would be poor for Eq. 38, but the Kirkwood-Müller formula does have the great advantage that $\sum Q$ has been eliminated and the result does not depend on any assumptions about the off-diagonal Q_{kl} type integrals. On the other hand, as Müller emphasized, the diamagnetic susceptibility of nonspherical systems contains terms in addition to those in Eq. 35, the second-order paramagnetic terms. Consequently, great care must be used in applying Eq. 38 to nonspherical systems.

This completes the discussion of the theory for many-electron systems from which we believe we have elucidated the reason for the failure of earlier approximate formulas for London forces. Many approximations remain and we summarize a few that seem most important.

(*1*) In Eq. 4 the use of the same perturbing function for both polarizability and London energy and in some cases the further assumption $\phi_\lambda = z_\lambda \psi_\lambda$, or $f_\lambda = 1$. In hydrogen the former assumption yields about 4 per cent error and the latter 12 per cent in polarizability and 9 per cent in London energy.

(*2*) The assumption of orthogonality of all ϕ_λ with all ψ_λ.

(*3*) The assumption that only the outermost subshell of electrons contributes to either α or E_L.

(*4*) The use of Hartree-Fock wave functions for the unpolarized atoms (or molecules). The principal deficiency is the lack of correlation terms for electrons of unlike spin in the wave function. The inclusion of such terms will be necessary before purely theoretical calculations yield quantitative agreement with experiment. For example, Buckingham's calculation[6] including exchange of α for Cl^- from Hartree-Fock wave functions yielded 6.9 ($\times 10^{-24}$ cm^3) as compared to the range 3.0 to 3.6 from experimental sources. Donath[11] found that angular correlation for the $3p$ shell approximately halved the polarizability which yields good agreement. Donath also explored the effect of correlation on the semi-empirical formulas for London energy in terms of polarizability, etc. His results indicate that the net effect in various cases is probably of the order of assumptions (*1*), (*2*), and (*3*) preceding and smaller than that of the omission of off-diagonal Q_{kl} terms. Further study of electron correlation is much to be desired.

III. EXPERIMENTAL INTERMOLECULAR ENERGIES

In this section the several theoretical or semi-empirical formulas for the prediction of London energies are compared with experimental values. The formulas to be tested are summarized in conventional units both for unlike atom pairs and in the simplified form for pairs of like atoms as follows:

London, Eq. 29

$$E_L = (3\alpha_A \alpha_B / 2R^6)\, U_A U_B / (U_A + U_B) \qquad (29)$$

$$E_L = -3\alpha^2 U / 4R^6 \quad \text{if } A = B \qquad (29a)$$

London theory: $U = I$ = ionization energy; empirical: evaluate U from E_L and correlate (U/I) with number of electrons in system.

Slater-Kirkwood, Eq. 34

$$E_L = -(3e\hbar/2m^{\frac{1}{2}} R^6)\, \alpha_A \alpha_B / [(\alpha_A/N_A)^{\frac{1}{2}} + (\alpha_B/N_B)^{\frac{1}{2}}] \qquad (39)$$

$$E_L = -3e\hbar\alpha^{\frac{3}{2}} N^{\frac{1}{2}} / 4m^{\frac{1}{2}} R^6 \quad \text{if } A = B \qquad (39a)$$

Slater-Kirkwood theory: N = no. of electrons of outer subshell; empirical: evaluate N from E_L and correlate with actual number of electrons of both outer subshell and of whole system.

Kirkwood-Müller (valid only for *spherical* systems, i.e., atoms)

$$E_L = 6mc^2\alpha_A\alpha_B/N_0R^6\,[(\alpha_A/\chi^A_{\text{mol}})+(\alpha_B/\chi^B_{\text{mol}})] \quad (40)$$

$$E_L = 3mc^2\alpha\chi_{\text{mol}}/N_0R^6 \quad \text{if } A = B \quad (40a)$$

where N_0 is Avogadro's number and χ_{mol} is a molal susceptibility in cm^3. This equation does not adapt itself so well to empirical use although a correction factor could be introduced and correlated with the number of electrons in the system.

Intermolecular potential functions have been fitted to various experimental data, such as second virial coefficients, viscosities, and sublimation energy. The use of data from dense systems involves the additional assumption of the additivity of pair interactions. The viscosity seems to be more sensitive to the shape of the potential than the second virial coefficient; hence data from that source are particularly valuable. These questions are discussed in full by Hirschfelder, Curtiss, and Bird[17] whose recommended potentials based primarily on viscosity data are given in the tables of this section.

Another problem arises from the presence of higher terms in the multipole expansion of the electrostatic interaction. While theoretical formulas exist for these also, they are even more approximate than those for the dipole-dipole term. Also, there is the uncertainty about the exact form of the repulsive interaction. Quite arbitrarily we shall group the higher multipole terms with the true repulsive interaction and assume that the empirical repulsive term accounts for both. The principal merit of this assumption is simplicity; the theoretical and experimental coefficients of the R^{-6} term are compared without adjustment. Since the higher multipole terms are known to be attractive and have been estimated to amount to about 20 per cent of the total attractive potential at the minimum, a rough correction for their possible effect can be made if it is believed that this is a preferable assumption.

The polarizability, ionization potential, and magnetic susceptibility data from the last edition of the Landolt-Börnstein tables are given in Table I for the inert gases and for H_2, N_2, Cl_2, and CH_4. The coefficients of the dipole-dipole or R^{-6} potential term are given in Table II; Table III gives effective U values and U/I ratios; while Table IV gives effective N values for Eq. 39 and a comparison with both the number N_{os} of outer shell electrons and the total number of electrons Z.

TABLE I. Properties of Atoms and Molecules[a]

Subst.	$\alpha \times 10^{24}$ cm^3	$\chi_{mol} \times 10^6$ cm^3	I, ev
He	0.21	1.94	24.56
Ne	0.40	7.2[b]	21.56
Ar	1.63	19.4	15.76
Kr	2.48	28	14.0
Xe	4.01	43	12.13
H_2	0.79	4.0	15.43
N_2	1.76	12.0	15.51
Cl_2	4.61	42	13.2
CH_4	2.60	12.2	13.04

[a] Source, Landolt-Börnstein, 6. Auflage, I, 1 and 3.

[b] This value was based upon the discordant values 7.77 ± .01 and 6.62 ± .18.

TABLE II. Theoretical and Empirical R^{-6} Potential Coefficients (units: erg cm$^6 \times 10^{-60}$; all are negative)

Subst.	Exp.	London Eq. 29	S. and K. Eq. 39	K. and M. Eq. 40
He	1.57[a]	1.30	1.71[b]	1.66
Ne	9.50	4.14	7.8	12.8
Ar	109.2	50.3	64	129
Kr	232	103	120	283
Xe	561	235	217	703
H_2	12.6	11.6	12.5	(12.9)
N_2	125.6	58	72	(86)
Cl_2	957	337	394	(790)
CH_4	258	106	150	(130)

[a] Experimental potentials are from Hirschfelder, Curtiss, and Bird;[17] of the several potentials given there for He the Lennard-Jones potential fitted by DeBoer, Michels, and Lunbeck was selected.

[b] The N values used in Eq. 39 are 2 for He and H_2, 6 for Ne, Ar, Kr, Xe, and N_2, 8 for CH_4, and 10 for Cl_2.

Table II shows clearly the large differences between various theories for many-electron systems. The Kirkwood-Müller equation always yields somewhat too large coefficients for the atoms which are the only spherical systems; but the London equation deviates by a greater amount on the low side. The Slater-Kirkwood equation gives a high value for He but yields coefficients smaller than the empirical ones for all other cases.

TABLE III. Effective U Values for Equation 29

Subst.	U, ev	U/I
He	29.6	1.20
Ne	49.	2.27
Ar	34.2	2.17
Kr	31.4	2.24
Xe	29.0	2.39
H_2	16.8	1.09
N_2	33.7	2.17
Cl_2	37.5	2.84
CH_4	31.8	2.44

The results in Table III give promise of a useful method of prediction and at the same time confirm the importance of the off-diagonal Q_{kl} terms discussed in the previous section. It appears that U/I may be taken as about 1.15 for two-electron systems and about 2.25 for systems with *sp* closed shells. The latter value is also satisfactory for N_2 and CH_4. Treatment of Cl_2 as a spherical system is such a poor approximation that the U/I value should not be given much significance in that case.

The discussion in the previous section indicated that N_{eff} values would be expected to exceed substantially the actual number of outer shell electrons. Table IV amply confirms this conclusion. Since N appears to the $\frac{1}{2}$ power in the Slater-Kirkwood equation, the deviations are exaggerated. Thus, in making a very similar treatment a few years ago in which slightly different empirical potentials were used, the writer[31] found substantially smaller effective N values. For the same reason a relatively crude effective

N will yield a fair estimate of the potential coefficient. Selection of the mean of the total number of electrons and of the number in the outer shell would appear to be a reasonable formula for the rare gas atoms.

TABLE IV. Effective N Values in Equation 39

Subst.	N_{eff}	N_{os}	Z
He	1.7	2	2
Ne	8.8	6	10
Ar	17.3	6	18
Kr	22.2	6	36
Xe	30.6	6	54
H_2	2.0	2	2
N_2	18.1	6	14
Cl_2	59	10	34
CH_4	24	8	12

The empirical potentials for the molecules were obtained on the assumption of single attraction centers. This assumption is probably good for H_2, fair for CH_4 and N_2, and very poor for Cl_2. Even for molecules such as CH_4 which are relatively spherical in shape, the fact that some atoms are near the outer surface rather than the center has an important effect. The closest interatomic distances are emphasized by the R^{-6} dependence of the potential. This point has been considered by several authors who worked out examples showing the net intermolecular potential for several models.

Kihara[20] used a core model in which the Lennard-Jones potential is assumed to hold for the shortest distance between the molecular cores instead of molecular centers. By use of linear, tetrahedral, and other shapes of cores, various molecules can be approximated. Thomaes,[41] Rowlinson,[35] Hamann, McManamey, and Pearse,[14] Atoji and Lipscomb,[1] Pitzer,[30] and Balescu,[4] have used other models of attracting centers and other mathemtical methods, but obtain similar conclusions. The primary effect is to steepen the potential curve so that in terms of inverse powers of the intermolecular distance the effective exponent exceeds six. Depending on the shape of the molecule, angular dependency of the potential may or may not be important.

In view of the complications of the intermolecular potential (as compared to the interatomic potential of the rare gas atoms) the comparisons for molecules in Tables II, III, and IV should be judged with caution. The apparent discrepancies from the theories for single atoms can be misleading. An example is the calculation for CH_4 on the Slater-Kirkwood theory where Table IV shows the absurd value of 24 for the effective number of electrons. Pitzer and Catalano[32] have applied the Slater-Kirkwood equation to the intermolecular potential of CH_4 by addition of all the individual atom interactions and, with $N = 4$ for carbon and 1 for hydrogen, obtained agreement within 5 per cent for the London energy at the potential minimum.

IV. INTRAMOLECULAR APPLICATIONS

Since electrostatic forces propagate through a molecule as through space, the motion of electrons in different parts of the same molecule will interact in at least the same way as electrons in different molecules. There may be additional intramolecular interactions arising from effects propagated through the bonding system of the molecule, such as in conjugated or mesomeric systems. *But we wish particularly to emphasize that the electron correlation energy effects of nonoverlapping portions of molecules can be treated by the same methods as intermolecular energies.*

A. Saturated Hydrocarbons

A relatively unambiguous intramolecular application of London energy theory is that of Pitzer and Catalano[32] to the paraffin hydrocarbons. They assumed that C—H and C—C single-bond energies were constant and that the time average polarity of these bonds was negligible. Since there are no multiple bonds in the paraffins, no conjugation or first-order hyperconjugation effects are possible. The London interaction energies were summed for all pairs of nonbonded atoms. These nonbonded atoms are far enough separated to make the nonoverlapping approximation satisfactory in all cases except a very few where steric repulsion must also be recognized. Then any correction to the London energy can be included with the repulsive energy which is treated empirically.

The Slater-Kirkwood equation (Eq. 39) was selected with $N = 4$ for carbon and $N = 1$ for hydrogen. The success of the equivalent calculation for the intermolecular interaction of CH_4 molecules was mentioned in the previous section. Atoms, rather than bonds, were chosen as the basis for the calculation because the location of the atom centers is unambiguous and the approximation of isotropic polarizability is better for an atom than for a bond. Possible deviations from isotropic polarizability are discussed in Section V. Ketelaar[19] gives for the atomic polarizabilities of hydrogen and carbon: $\alpha = 0.42$ and 0.93×10^{-24} cm^3, respectively. The resulting equation for the London energy is

$$\Delta E_L = -[22.6\Sigma R_{\text{C—C}}^{-6} + 8.68\Sigma R_{\text{C—H}}^{-6} + 3.42\Sigma R_{\text{H—H}}^{-6}] \times 10^{-60} \text{ ergs} \tag{41}$$

Pitzer and Catalano compared calculated heats of formation of the various paraffins through the pentanes with experimental values corrected to 0°K. The known vibrational frequencies were used to calculate the zero-point vibrational energy. The agreement in most cases was well within the experimental error of the heat of formation although the difference did reach 1 kcal/mole in one case and 0.8 in another. It is particularly interesting to consider the energies of isomerization from *n*-butane to isobutane and from *n*-pentane to neopentane because these differences arise exclusively from the London energy and the zero-point vibrational energy. The results are given in Table V.

TABLE V. Isomerization Energies for Butane and Pentane (units: kcal/mole)

Isomerization	ΔE_L	ΔE_{vib}	ΔE_{calc}	ΔE_{exp}
Iso- to *n*-butane	1.28	0.24	1.52	1.63
Neo- to *n*-pentane	3.8	1.0	4.8	4.1

Donath[11] has extended these calculations to cyclohexane and cyclopentane and obtained comparably good agreement. It seems safe to conclude that these long-range electron correlation effects account for the isomerization energies of the paraffins which had heretofore remained unexplained.

B. Unsaturated Hydrocarbons

The application of these methods to unsaturated hydrocarbons involves certain complications. Unsaturated hydrocarbons show an additional polarizability[19] of 0.58×10^{-24} cm^3 per double bond and 0.86×10^{-24} cm^3 per triple bond in the molecule. Similarly the polarizability of a molecule containing a benzene ring exceeds that computed for the atoms present by about 1.28×10^{-24} cm^3. These results are most readily explained on the basis that oscillations of charge from atom to atom are significant when double bonds are present.

The concept of an *extra* polarizability associated with a double bond can be checked from the spectrum of ethylene. The transition at 61,000 cm^{-1} for that substance seems clearly to be associated with the presence of the double bond. Hammond and Price[15] found the intensity of this transition to yield an f value of 0.29. The component of polarizability at low frequency associated with this transition is given by the formula[26, 31]

$$\alpha_k = (e^2/4\pi^2 m)\, f_k/\nu_k^2 \tag{42}$$

This yields 0.51×10^{-24} cm^3 which agrees approximately with the value 0.58×10^{-24} ascribed to the double bond on an empirical basis. Also we may note that this f value corresponds to a transition dipole of 3.16×10^{-18}, if a one-dimensional charge oscillation is assumed, and that this dipole yields an apparent charge of 0.49 times the electronic charge for a C=C distance of 1.34 A.

The usual multipole expansion is inappropriate for the treatment of London energies associated with the interatomic charge oscillation. However, London[23] has given an appropriate method, known as the monopole method. Coulson and Davies[7] and Haugh and Hirschfelder[16] have applied the monopole method to the intermolecular forces between molecules containing large amplitude oscillators.

Simpson[37] discussed the conjugated polyenes in terms of internal London interactions and showed that the results from this theory alone gave a self-consistent explanation of both the conjugation energy of the ground state and the pattern of low lying excited electronic states in terms of a single empirical parameter. Since these phenomena had been previously ascribed to resonance with different bond structures, it became of interest to try to determine the magnitude of each effect.

We assume that the double bonds in 1,3-butadiene would be the same as in ethylene if they did not interact with one another. Introduction of the known geometry of 1,3-butadiene in the *s-trans* conformation and the monopole charge of 0.49 e on each carbon yields an interaction energy $\delta = 0.48$ ev between the two double bonds. Simpson found the empirical value $\delta = 1.91$ ev from his assumption that only a London interaction was present. Hence it appears that only a small part of the interaction between double bonds in 1,3-butadiene is a London type of second-order electrical effect and the larger part is a conjugation or resonance associated with the structure with a double bond in the central position.

The London interaction of saturated portions of a hydrocarbon with a double bond may also be calculated by the use of equations given by Haugh and Hirschfelder.[16] The writer has made exploratory calculations for simple olefins by this method; the results do not fit the observed heat of formation values on the assumption of constant bond energies. The extra stability of alkyl groups adjacent to double bonds, which is commonly attributed to hyperconjugation, is somewhat increased after London energies have been consistently included. There are aspects of this calculation, however, which are on less firm ground than those already described, hence the conclusion should be regarded as somewhat tentative.

C. Halogens

Possible explanations of the peculiarly low heat of dissociation of fluorine as compared to the other halogens have been discussed by Mulliken[28] and the writer.[29,33] The electron correlation or London energy of interaction between the unshared pairs of electrons on the two halogen atoms is a very substantial term. While the assumption of lack of overlap of the electron distributions and the dipole-dipole approximation are both subject to question in this case, nevertheless, the simple formulas should indicate whether the trend down the halogen series is anomalous or not. A simple calculation[33] using the London equation (Eq. 29) indicated that this trend was indeed anomalous with a low energy for F_2. Calculations which are revised to use a recently observed polarizability value for F_2 [13] and which use the (U/I) value of 2.25 recommended above are given in Table VI.

TABLE VI. Intramolecular London Energies for the Halogens

Calculation	F_2	Cl_2	Br_2	I_2
$\alpha_X = \frac{1}{2}\alpha_{X_2}$, 10^{-24} cm^3	0.64	2.30	3.34	5.11
$R(X-X)$, 10^{-8} cm	1.435	1.988	2.284	2.667
$U = 2.25\ I$, ev	39.0	29.1	26.6	23.9
$-E_L$ (dipole-dipole), kcal/mole	32.	43.	36.	30.
ΔE_{dissoc}, kcal/mole	39.	57.1	45.4	35.6

In addition to the calculations in Table VI which are based upon Eq. 29, one may use the Slater-Kirkwood equation (Eq. 39). If one takes effective N values less by one than those given in Table IV for the rare gas atom adjacent to each halogen, the resulting London energies are very nearly the same as those in Table VI.

The departure of the E_L values from a smooth trend is somewhat over half as large as that of the dissociation energy values in the last row of Table VI. Until the London energy calculations are refined to eliminate the dipole-dipole approximation and other uncertainties, it is not possible to say whether that effect accounts for the entire anomaly or not. In any event a substantial portion of the anomaly may be ascribed to the correlation of the motion of the unshared electron pairs in the valence shell.

V. ANISOTROPIC EFFECTS

The rare gas atoms are strictly spherical but other systems which have been considered, such as hydrogen, carbon, or halogen atoms, may be anisotropic. These have been treated in preceding sections as isotropic and it is one purpose of the present section to consider the magnitude of possible error.

A. Polarizability

We consider first the polarizability of a molecule consisting of two or more polarizable parts which may be atoms, bonds, or other units. When the molecule is placed in an electric field the effective field which induces dipole moments in various parts is not just the external field but rather the local field which is influenced by the induced dipoles of the other parts. The classical theory of this interaction of polarizable units was presented by Silberstein[36] and others and is summarized by Stuart in his monograph.[40] The writer has examined the problem in quantum theory and finds that the same results are obtained to the order of approximation being considered.

Let us take the problem of a diatomic molecule, such as Cl_2, with the two atoms as the two polarizable parts. Each atom has a polarizability α_1 parallel to and α_2 perpendicular to the interatomic axis, respectively. The Silberstein theory yields for the effective polarizability of the molecule the values b_1 and b_2, respectively, parallel and perpendicular to the axis.

$$b_1 = 2\alpha_1 \left[\frac{1+(2\alpha_1/R^3)}{1-(2\alpha_1/R^3)^2}\right] \tag{43a}$$

$$b_2 = 2\alpha_2 \left[\frac{1-(\alpha_2/R^3)}{1+(\alpha_2/R^3)^2}\right] \tag{43b}$$

The mean polarizability b is

$$b = \tfrac{1}{3}b_1 + \tfrac{2}{3}b_2 = 2\left(\frac{\alpha_1}{3} + \frac{2\alpha_2}{3}\right) + \frac{2(\alpha_1{}^2 - \alpha_2{}^2)}{3R^3} + \ldots \tag{44}$$

where the expansion is made on the assumption that (α_i/R^3) is less than unity. We note that the first-order term is just the appropriate sum of component polarizabilities and that the second-order term is very small if either $\alpha_1 \simeq \alpha_2$ or if (α_i/R^3) is small. Thus interactions have much less effect on the mean polarizability than on the anisotropy.

The values for Cl_2 are $b_1 = 6.60$, $b_2 = 3.62$, both $\times 10^{-24}$ cm³. When these values are substituted in Eq. 43 together with $R = 1.99 \times 10^{-8}$ cm, one obtains $\alpha_1 = 1.79$ and $\alpha_2 = 2.36$, both $\times 10^{-24}$ cm³. Thus, although the polarizability of the chlorine molecule is almost twice as great parallel than perpendicular to the axis, the atom is nearly isotropic with slightly greater polarizability perpendicular to the axis.

Recently this important interaction of polarizable parts of a molecule seems to have been forgotten. Denbigh[10] and Le Fèvre and Le Fèvre[22] have presented extensive tables of bond polarizabilities which are so selected that simple addition of the polarizability components for the various bonds yields the components of the polarizability of the entire molecule. These schemes have been surprisingly successful in an empirical sense and may have considerable value on that basis. But it is evident from the theory of Silberstein that the anisotropy of these bond polarizabilities cannot be directly interpreted as properties of the bonds. In particular the nearly zero transverse polarizability reported for the C—C single bond must be spurious. It seems most likely that there is some fortuitous cancellation of conflicting factors which allows the Denbigh-Le Fèvre scheme to succeed over a limited range of substances, and the failure of the scheme to fit an additional array of molecules may not indicate any error in the assumed structures of the molecules but only that the range of this accidental cancellation has been exceeded. The failure of Stein[39] to obtain good agreement with the anisotropy of n-paraffin polarizabilities may be an example of this sort, since certain of the structures that he considered fit all theoretically sound experimental criteria.

Unfortunately the complete calculation of the principal polarizabilities of a polyatomic molecule including the Silberstein interactions is very complex. Possibly electronic calculators will make this computation feasible for an array of molecules such as the saturated hydrocarbons. Fortunately, however, these interactions drop out of the first approximation to the mean polarizability of the entire molecule. Consequently simple sum rules for the mean polarizability of a molecule in terms of component atoms or bonds are sound to that approximation.

B. London Forces

London[23] has treated the case of the attractive force between anisotropic molecules on the dipole-dipole interaction basis as well as on the monopole basis mentioned above. The small anisotropy found for the chlorine atom makes the dipole-dipole formulation appropriate. For the symmetrical orientation in the Cl_2 molecule the London formula is

$$-E_L = \frac{2\alpha_1{}^2 U_1 + \alpha_2{}^2 U_2}{4R^6} \tag{45}$$

where U_1 and U_2 are the characteristic energy quantities for polarization parallel to and perpendicular to the axis, respectively. In London's theory these were thought to be approximately the ionization potential but we have found that both theory and experiment indicate much larger values, possibly about 2.25 times I for many electron systems.

There is no reason to believe that the off-diagonal Q integrals which yield this increase in U over I should be equal for longitudinal and transverse polarization but neither is there at present any basis for selecting different values. Hence we assume $U_1 = U_2$ and by use of the α values given above for the chlorine atom, the value of $-E_L$ is calculated to be about 30 per cent smaller than was obtained in Table VI. Since the effect of anisotropy would be expected to be about the same for all of the halogens, the qualitative conclusions drawn from the results in Table VI are not affected. It is clear that anisotropy may be important, however, and must be considered in quantitative work.

In the more complex case of the saturated hydrocarbons, the anisotropic equations have not been solved. Rough estimates from approximate solutions indicated that the anisotropy of hydrogen and of saturated carbon atoms is small but an accurate check on this matter would be desirable.

References

1. Atoji, M., and Lipscomb, W. N., *J. Chem. Phys.* **21**, 1480 (1953).
2. Axilrod, B. M., *J. Chem. Phys.* **19**, 719 (1951).
3. Axilrod, B. M., and Teller, E., *J. Chem. Phys.* **11**, 299 (1943).
4. Balescu, R., *Physica* **22**, 224 (1956).
5. Born, M., and Mayer, J. E., *Z. physik. Chem.* **75**, 1 (1932).
6. Buckingham, R. A., *Proc. Roy. Soc. (London)* **A160**, 94, 113 (1937).
7. Coulson, C. A., and Davies, P. L., *Trans. Faraday Soc.* **48**, 777 (1952).
8. Dahler, J. S., and Hirschfelder, J. O., *J. Chem. Phys.* **25**, 986 (1956).
9. Davydov, A. S., *Zhur. Eksptl, Teoret. Fiz.* **18**, 201 (1948).
10. Denbigh, K. G., *Trans. Faraday Soc.* **36**, 936 (1940).
11. Donath, W. E., Thesis, University of California, 1958, also to be submitted for journal publication.
12. Eisenschitz, R., and London, F., *Z. Physik* **60**, 491 (1930).
13. Franck, E. U., *Naturwiss.* **41**, 37 (1954).
14. Hamann, S. D., Mc Manamey, W. J., and Pearse, J. F., *Trans. Faraday Soc.* **49**, 351 (1953).
15. Hammond, V. J., and Price, W. C., *Trans. Faraday Soc.* **51**, 605 (1955).
16. Haugh, E. F., and Hirschfelder, J. O., *J. Chem. Phys.* **23**, 1778 (1955).
17. Hirschfelder, J. O., Curtiss, C. F., and Bird, R. B., *Molecular Theory of Gases and Liquids*, John Wiley and Sons, New York, 1953, p. 966.
18. Jansen, L., and McGinnies, R. T., *Phys. Rev.* **101**, 1301 (1956); **104**, 961 (1956).
19. Ketelaar, J. A. A., *Chemical Constitution*, Elsevier Pub. Co., Amsterdam and Houston, 1953, p. 90.
20. Kihara, T., *Rev. Mod. Phys.* **25**, 831 (1953).
21. Kirkwood, J. G., *Physik Z.* **33**, 57 (1932).
22. Le Fèvre, C. G., and Le Fèvre, R. J. W., *Rev. Pure Appl. Chem.* **5**, 261 (1955).
23. London, F., *J. Phys. Chem.* **46**, 305 (1942).
24. London, F., *Z. physik. Chem.* **B11**, 222 (1930); *Z. Physik* **63**, 245 (1930).
25. Löwdin, P.-O., *Phys. Rev.* **94**, 1600 (1954).
26. Margenau, H., *Rev. Mod. Phys.* **11**, 1 (1939).
27. Müller, A., *Proc. Roy. Soc. (London)* **A154**, 624 (1936).
28. Mulliken, R. S., *J. Am. Chem. Soc.* **77**, 884 (1955).

28a. Pauling, L., and Simonetta, M., *J. Chem. Phys.* **20**, 29 (1952).

29. Pitzer, K. S., *J. Am. Chem. Soc.* **70**, 2140 (1948).
30. Pitzer, K. S., *J. Am. Chem. Soc.* **77**, 3427 (1955).
31. Pitzer, K. S., *J. Am. Chem. Soc.* **78**, 4565 (1956).
32. Pitzer, K. S., and Catalano, E., *J. Am. Chem. Soc.* **78**, 4844 (1956).
33. Pitzer, K. S., *J. Chem. Phys.* **23**, 1735 (1955).

34. Pitzer, K. S., *Quantum Chemistry*, Prentice-Hall, New York, 1953, p. 339.
35. Rowlinson, J. S., *J. Chem. Phys.* **20**, 337 (1953); *Trans. Faraday Soc.* **50**, 647 (1954).
36. Silberstein, L., *Phil. Mag.* **33**, 92, 215, 521 (1927).
37. Simpson, W. T., *J. Am. Chem. Soc.* **73**, 5363 (1951); **77**, 6164 (1955).
38. Slater, J. C., and Kirkwood, J. G., *Phys. Rev.* **37**, 682 (1931).
39. Stein, R. S., *J. Chem. Phys.* **21**, 1193 (1953).
40. Stuart, H. A., *Molekülstruktur*, Springer, Berlin, 1934, p. 184.
41. Thomaes, G., *J. chim. phys.* **49**, 323 (1952).
42. Yos, J. M., Bade, W. L., and Jehle, H., *Proc. Nat. Acad. Sci.* **43**, 341 (1957); Jehle, H., *Proc. Nat. Acad. Sci.* **43**, 847 (1957).

Molecular structure of XeF_6

Kenneth S. Pitzer and Lawrence S. Bernstein

Department of Chemistry and Inorganic Materials, Research Division, Lawrence Berkeley Laboratory, University of California, Berkeley, California 94720
(Received 23 June 1975)

The full array of structure-related experimental data for XeF_6 is considered simultaneously. Electron diffraction, infrared, Raman and ultraviolet spectra, electric field deflection of molecular beam, and calorimetrically determined entropy data are all found to be consistent with a structural model involving substantial distortion in the T_{1u} bending mode from octahedral symmetry and a pseudorotation in two dimensions of this distortion. Several other vibration frequencies are assigned, and the mean value of three unobserved frequencies is determined from the entropy. No quantitative value could be given for the potential restricting pseudorotation, but various arguments agree in indicating that it is small. There is no reason to question previous proposals that this peculiar structure arises from a pseudo-Jahn–Teller effect associated with the relatively low energy of excitation of a xenon $5s$ electron to the $5p$ level.

The xenon hexafluoride molecule has been a puzzle since its discovery[1] in 1962. While certain properties were consistent with a simple octahedral structure, many other properties were not. Bartell and Gavin[2] made a major advance in the resolution of these problems by showing that their electron diffraction results and certain other properties could be reconciled with a family of structural models implying a large amplitude in the T_{1u} bending mode coupled with the T_{2g} mode. These models were discussed in terms of a pseudo-Jahn–Teller distortion. With the publication of an accurate entropy value[3] and additional spectral data,[4] it seemed desirable to determine first, whether this complete array of experimental structural information was consistent with any plausible model for this molecule, and second, if such consistency was attained, to define that model as closely as is feasible.

An important step in this process was presented[5] separately in an analysis of the Stark effect for pseudorotating molecules. These results indicate the range of molecular characteristics for XeF_6 which are consistent with the absence of focusing in electric-field-deflection experiments.[6]

Wang and Lohr[7] have very recently presented a crystal-field model study for XeF_6 which supports the plausibility of the model we had simultaneously developed. While they give a substantial list of the properties indicating deviation from a simple octahedral structure, they do not attempt a quantitative comparison in most cases.

An alternate proposal of Goodman,[8] that XeF_6 exists in electronic isomers including triplet states, has been shown to be improbable by several workers.[9] We agree in dismissing this proposal, but note that much of the discussion of the postulated triplet states suffers from an implicit assumption of a very small spin–orbit interaction typical of light elements. In contrast, the spin-orbit interaction for xenon is large, much larger than the energy differences Goodman considers. Thus, if this sort of proposal were to be seriously considered, it would require reformulation with a realistic spin-orbit interaction included.

We note also the report of Claassen *et al.*[4] of changes in spectra with temperature and of apparent slowness in reaching equilibrium after temperature change. These effects were cited in support of Goodman's proposal[8] of electronic isomers. The changes in spectra with temperature are small, and can be explained by the expected changes in population of excited vibrational and pseudorotational states. The delay in reaching equilibrium after temperature change, however, is puzzling, but the electronic-isomer explanation is not convincing. With a large spin-orbit energy, singlet and triplet states will be strongly mixed and intersystem conversion should be rapid. Also with relatively small changes in geometry, the potential barriers between isomers could hardly be high enough to yield such a slow conversion. Thus we see no explanation of this latter effect in terms of any plausible model for XeF_6 and are forced to suggest that the delay in obtaining "equilibrium" spectra arose from some source in the experimental apparatus.

STRUCTURAL MODEL

Our model for XeF_6 is formulated in terms of the symmetry coordinates for the octahedral O_h geometry as indicated in Fig. 1. We shall be concerned especially with conformations involving large amplitudes in the T_{1u} bending coordinates S_4, but all other coordinates will be of interest in one or another respect.

The electronic structure of XeF_6 is diagramed in Fig. 2. From the pattern of molecular-orbital energies we expect that states based on exciting one electron to the T_{1u}^* orbital lie not too far above the $^1A_{1g}$ ground state, whereas other excited states lie much higher. With substantial spin–orbit coupling, the excited states are not pure singlet or triplets, but the large interaction with the ground state will relate to the $^1T_{1u}$ component.

This electronic interaction can be related to distortion of the molecule in one or more of its normal coordinates; the results of such a calculation are commonly termed the pseudo-Jahn–Teller effect.[10] The details of such an analysis are given elsewhere.[11] Expansion of the secular equation in powers of the normal coordinates yields expressions indicating the effect of this electronic interaction on each vibrational motion. If the interaction is strong, the potential for the three T_{1u} bending modes (ν_4) ceases to have its minimum at the origin; instead, one obtains an essentially spherical

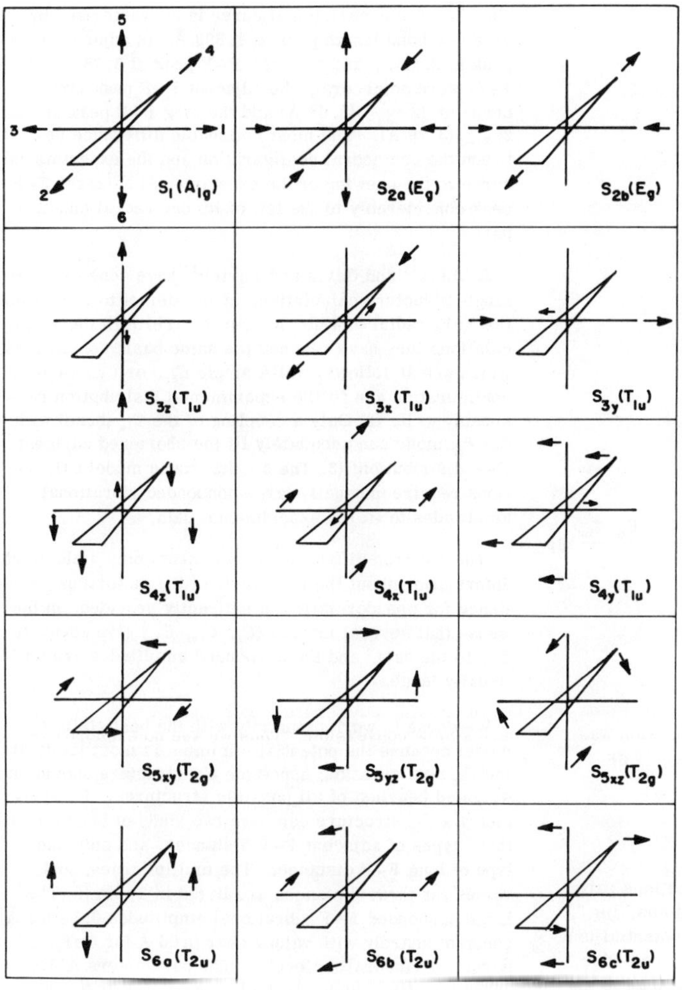

FIG. 1. Symmetry coordinates for an octahedral MX_6 molecule.

potential trough which yields a two-dimensional pseudorotation and a single vibrational mode. Thus, if the three coordinates S_{4x}, S_{4y}, and S_{4z} are transformed to a spherical coordinate system, one obtains

$$V=-aR^2+bR^4-cR^4[\sin^4\theta(\tfrac{1}{4}\sin^2 2\phi-1)+\sin^2\theta], \quad (1)$$

where

$$S_{4x}=R\sin\theta\cos\phi$$

$$S_{4y}=R\sin\theta\sin\phi$$

$$S_{4z}=R\cos\theta.$$

The pseudo-Jahn–Teller analysis gives expressions for each of the coefficients a, b, c in Eq. (1), but we are not now able to evaluate these quantities theoretically with any rigor and accuracy. We can predict that b and c are positive and that a will also be positive if the electronic interaction is strong enough. We did attempt to evaluate empirically from spectra, etc., all of the quantities arising in the pseudo-Jahn–Teller equations. While plausible arrays of values were obtained,[11] no set was fully satisfactory. Also one may question the accuracy of an expansion about the origin of octahedral symmetry when the distortion is large. Consequently, we shall present an interpretation of the experimental properties of XeF_6 using Eq. (1) for the pseudorotational motion and assuming the usual, nearly harmonic, potential for other vibration modes.

Before proceeding to the experimental data, however, we consider further Eq. (1). The first two terms in this potential for the T_{1u} bending motion have spherical symmetry, while the third term has octahedral symmetry. This octahedral field term arises from the coupling of the T_{2g} bending mode with the T_{1u} bending motion.[2,10,11] The result gives a directional preference to the unshared or antibonding pair of valence-shell electrons. When this "lone" pair is centered on a face of the octahedron formed by the fluorine ligands, its repulsive interaction with the ligands is minimized,

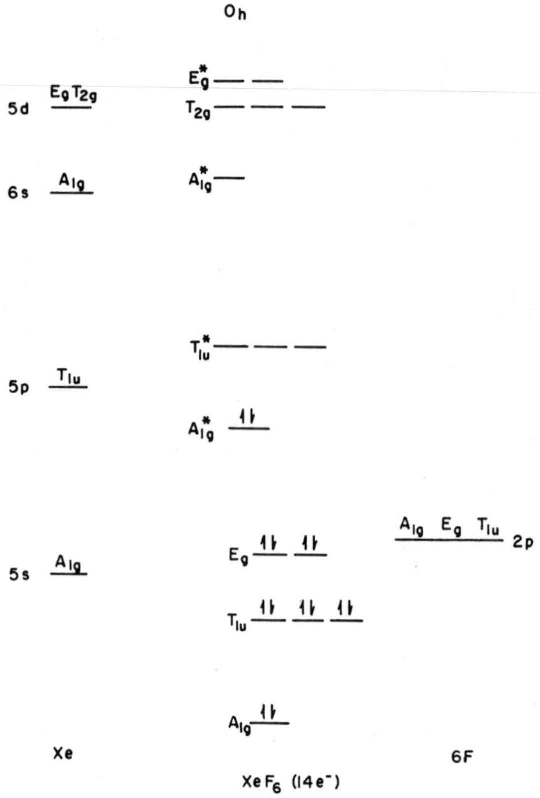

FIG. 2. Molecular orbital diagram XeF_6.

hence the potential is a minimum. When the lone pair is directly in line with one of the fluorine atoms, its repulsive interaction is largest, hence the potential is a maximum. When the lone pair is centered on an edge of the octahedron, it is closer to the ligands than in the face centered situation and farther than in the corner case, which makes the potential a saddle point here.

It is informative to write the one-dimensional potential at these three types of locations on the octahedron:

$$V_{C_{3v}} = -aR^2 + (b - c/3)R^4, \tag{2a}$$

$$V_{C_{2v}} = -aR^2 + (b - c/4)R^4, \tag{2b}$$

$$V_{C_{4v}} = -aR^2 + bR^4. \tag{2c}$$

This demonstrates that the potential is softest in the C_{3v} direction, steepest in the C_{4v} direction, and intermediate in the C_{2v} direction. In the limit that the angular barrier vanishes ($c=0$), the potential becomes spherically symmetric and identical in form to that of a rotating–vibrating diatomic molecule.

ELECTRON DIFFRACTION

A comparison of the experimental radial distribution curve (20 °C)[12,13] with that of a hypothetical octahedral XeF_6 configuration of the same bond length is found in Fig. 3. The experimental curve is characterized by an Xe–F bond length peak at 1.895 Å, an adjacent F–F peak at 2.51 Å, and a "long" F–F peak at 3.76 Å. If XeF_6 were octahedral, the adjacent F–F peak would occur at $\sqrt{2}\, r_{Xe-F}$(2.68 Å) and the long F–F peak at $2r_{Xe-F}$(3.79 Å). The most noticeable difference between the octahedral configuration and the experimental curve is the skewing of the experimental adjacent F–F peak considerably to the left of its octahedral counterpart.

Hedberg[13] and Gavin and Bartell[12] have done extensive single structure calculations in an attempt to understand the XeF_6 radial distribution curve. Through these calculations they have reached the same basic conclusions, which are as follows: (1) A single C_{2v}, or C_{3v}, or C_s configuration can fit the experimental distribution reasonably well, (2) Only a coupling of the T_{1u} (bend) with the T_{2g} mode can adequately fit the shortened adjacent F–F distribution; (3) The single, static model calculations require unusually large nonbonded vibrational amplitudes to fit the experimental data.

The electron diffraction experiments only yield direct information about the magnitude of the distortion. Evidence for pseudorotation is indirectly provided, in the sense that several models (C_s, C_{2v}, C_{3v}) give adequate fits to the data, and the nonbonded amplitudes are unusually large.

We chose to work primarily with the best static C_{3v} model because the potential minimum is most likely at the C_{3v} configuration; hence the C_{3v} structure should be weighted heaviest of all possible structures. By symmetry a C_{3v} structure can have two kinds of bond lengths, three types of adjacent F–F distances, and only one type of long F–F distance. The multiplicities, and values for these distances are listed in Table I. The large nonbonded F–F vibrational amplitudes 0.1–0.3 Å contrast sharply with values near 0.06 Å for TeF_6, a normal octahedral molecule. The parameters ANG4 and ANG5 are angular variables which can be directly related to the S_4 and S_5 symmetry coordinates:

$$S_{4x} = S_{4y} = S_{4z} = \sqrt{8}\,r \text{ ANG4}, \tag{3a}$$

$$S_{5xy} = S_{5xz} = S_{5yz} = 2r \text{ ANG5}. \tag{3b}$$

The radial coordinate becomes

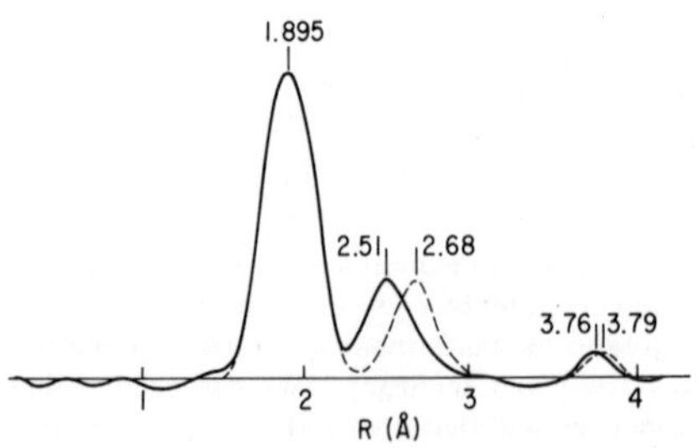

FIG. 3. Comparison of experimental radial distribution curve for XeF_6 (solid line) with a hypothetical octahedral XeF_6 configuration (dashed line).

TABLE I. Structural parameters[13] for the best static C_{3v} configuration of XeF_6.

Parameter	Multi-plicity	R (Å)	Ampli-tude (Å)
r_{Xe-F}	3	1.850	0.036
r_{Xe-F}	3	1.941	0.036
Adjacent F–F	6	2.498	0.108
Adjacent F–F	3	2.535	0.208
Adjacent F–F	3	3.106	0.302
Long F–F	3	3.757	0.070

Also: ANG4 = 5.348°, ANG5 = 7.260°

$$R=(S_{4x}^2+S_{4y}^2+S_{4z}^2)^{1/2}=2\sqrt{6}r \text{ ANG4}. \tag{4}$$

Using the experimental values for the C_{3v} model, we find for each component $S_4=0.502$ Å, $S_5=0.482$ Å. Also the value of R at the potential minimum is $R_0=0.870$ Å. Using the C_{3v} potential, Eq. (2a), the first derivative defines the potential minimum (R_0):

$$R_0^2=a/2(b-c/3). \tag{5}$$

Knowledge of the value of R_0 yields one relationship; hence, if two additional independent relations among a, b, and c can be found by analysis of other experimental data of XeF_6, the potential can be quantitatively defined.

It is also interesting to note the large value of S_5 and the small but significant difference in Xe–F bond lengths. These further indicate a substantial departure from octahedral symmetry.

The parameters calculated so far have been based on the best static C_{3v} model. It would seem desirable to recalculate these parameters using the best static C_{2v}, C_s, or C_{4v} model to see if similar results are obtained. We can eliminate the C_{4v} model from consideration, as it seems incapable of fitting the experimental data nearly as well as the other symmetries. Unfortunately, with the C_{2v} and C_s geometries, their low symmetries permit so many independently adjustable geometric parameters that a unique best static model cannot be defined. Instead there are a variety of C_{2v} and C_s structures which fit the experimental data equally well. The net result is that we are forced to look elsewhere for further information.

It would be even better to remove the assumption of the C_{3v} (or other) conformation and to relate the electron-diffraction data directly to a pseudorotating model. While such calculations are by no means impossible, and initial steps were taken in collaboration with Professor K. Hedberg, it was not possible to carry out these more complex calculations[11] before the remainder of this research was completed.

ULTRAVIOLET SPECTRUM

The ultraviolet spectrum of XeF_6 shows[4] a strong band at 2300 Å and a somewhat weaker band at 3500 Å. This pattern, including the band shapes, is similar to that found for the isoelectronic complex ions[14–16] $TeCl_6^{-2}$, $TeBr_6^{-2}$, as well as for In^+ or Sn^{+2} impurity centers[17] in alkali halide crystals.

From Fig. 2 one can readily determine that the lowest excited state will be based on the excitation of one electron from A_{1g}^* to T_{1u}^*. It is useful to note also that in atomic terms (for xenon) these are $5s$ and $5p$ orbitals. The resulting states (first in atomic terms, then in O_h symmetry) are ($^3P_0$–A_{2u}), ($^3P_1$–T_{1u}), ($^3P_2$–$T_{2u}+E_u$), and ($^1P_1$–T_{1u}). With substantial spin-orbit coupling, the two T_{1u} states are mixed, but the upper will be primarily a spin singlet and the lower primarily a triplet. The selection rules in O_h allow dipole transitions from A_{1g} only to T_{1u}, hence the two observed transitions correlate perfectly: the lower, primarily triplet T_{1u} at 3500 Å and the upper, primarily singlet at 2300 Å.

The accepted interpretation[14,16] of the isoelectronic complex ion spectra is just that stated above. Stufkens[14] extends this explanation to include the theory of Toyozawa and Inoue[18] concerning the band shapes. These authors show that the T_{1u} excited states will have a Jahn–Teller splitting with distortion in the T_{2g} and possibly other modes and that this leads to the observed band shapes. Since XeF_6 is already distorted in the ground state, in contrast to these complex ions which are octahedral in the ground state, we cannot make quantitative use of the equations of Toyozawa and Inoue. The bands for XeF_6 have shapes which are similar but broader and less distinct than those of the isoelectronic complex ions, and we believe that an appropriate extension of the theory of Toyozawa and Inoue would account for these band contours.

INFRARED AND RAMAN SPECTRA OF XeF_6

If XeF_6 were octahedral, its 15 vibrational degrees of freedom would belong to the species $A_{1g}+E_g+2T_{1u}$

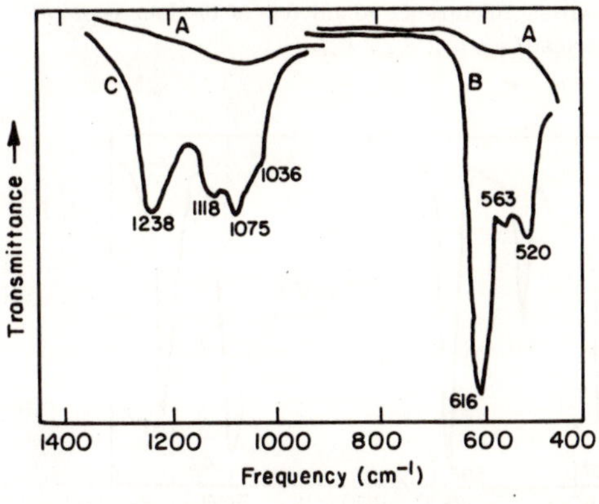

FIG. 4. Infrared spectrum of gaseous XeF_6 at room temperature: (A) background; (B) 10 cm cell, vapor pressure at −10°; (C), 60 cm path cell, vapor pressure at 10°. From the work by Claassen, Goodman, and Kim.[4]

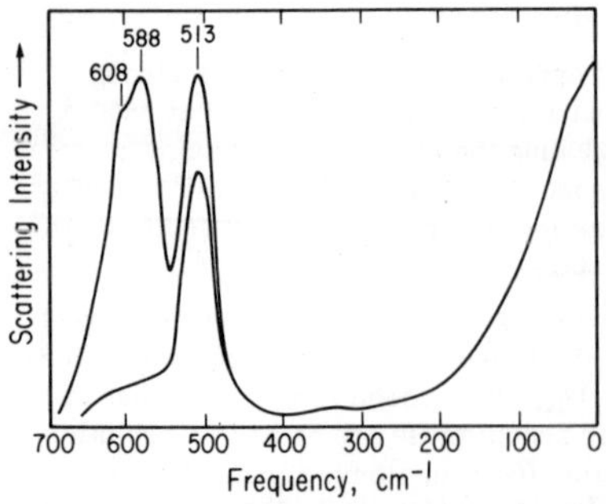

FIG. 5. Raman spectrum of XeF_6 vapor at 1 atm pressure and 85 °C. 6471 Å excitation. (A) incident polarization perpendicular; (B) incident polarization parallel. From the work of Claassen, Goodman, and Kim.[4]

$+T_{2g}+T_{2u}$. The species A_{1g}, E_g, T_{2g} are Raman active, T_{1u} is infrared active, and T_{2u} is inactive. The infrared spectrum would show two bands, a T_{1u} stretch, and a T_{1u} bend, while the Raman would show three bands, a polarized A_{1g} stretch, a depolarized E_g stretch, and a depolarized T_{2g} bend. Some of the experimental XeF_6 Raman and infrared spectra[4,19] are reproduced in Figs. 4–6. The stretching region (500–700 cm^{-1}) in the infrared spectra, Fig. 4, indicates three, possibly four bands. In the Raman, the stretching region, Fig. 5, is composed of three bands. These observations argue strongly against an octahedral structure for XeF_6. It is useful, however, to consider distorting an octahedral MX_6 structure and correlating the fundamental frequencies for various configurations. It is seen from Fig. 7 that the general effect of lowering the symmetry from O_h is to increase the number of fundamentals by splitting the degenerate T and E type bands. In addition, the number of observed infrared and Raman transitions increases, with many absorptions now being coincident; for example, in C_{3v} both A_1 and E modes are active in both infrared and Raman. Because of the breadth of the absorptions in the room temperature Raman and infrared XeF_6 spectra (Figs. 4, 5), it is impossible to give a strong preference to any one of the four possible geometries (C_{3v}, C_{2v}, C_{4v}, C_s).

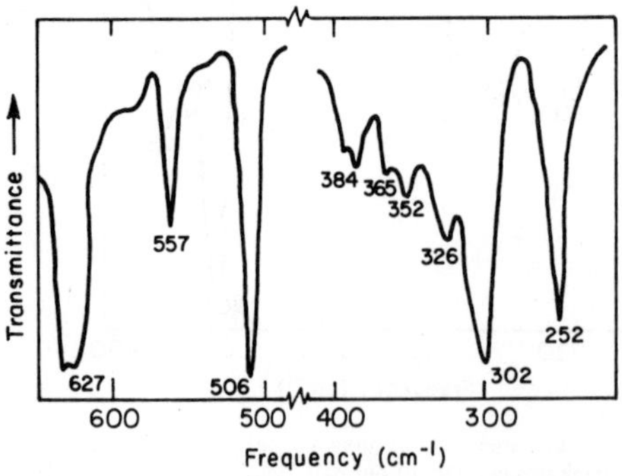

FIG. 6. Matrix-isolation infrared spectrum of XeF_6 in argon at ~7 °K. From the work by Claassen, Goodman, and Kim.[4]

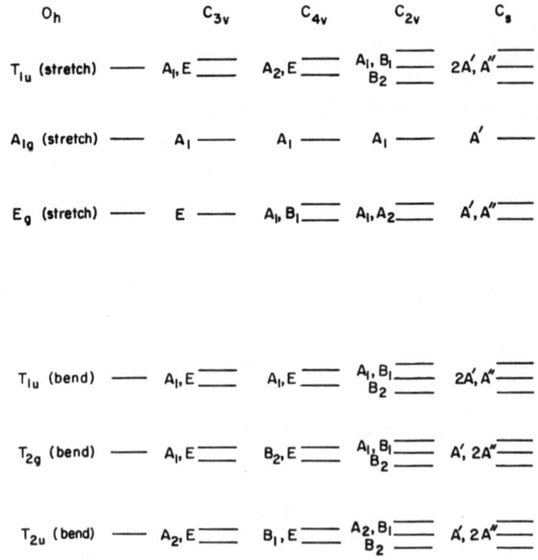

FIG. 7. Correlation diagram for the vibrational modes of a MX_6 molecule of varying symmetry.

Even more striking is the complete absence of distinct bands in the region of bending vibrations. It is reported[19] that there is continuous absorption in the infrared through the 100–300 cm^{-1} region with very broad increases in intensity near 100 and 275 cm^{-1}. Likewise, the Raman spectrum shows gradually increasing absorption below 300 cm^{-1} but no discernible details. This broadening of bands into essentially continuous absorption is more consistent with a pseudorotating model than with any single distorted configuration.

The infrared and Raman spectra have also been measured for XeF_6 isolated in argon matrices at very low temperatures. It is not possible to be certain about the matrix cavity and its influence on the structure of XeF_6. The most likely postulate is a cavity of six argon vacancies in octahedral geometry, and this should tend to moderate rather than accentuate the distortion of XeF_6. Nevertheless, both infrared and Raman spectra of the matrices show a nonoctahedral pattern. The infrared spectrum[4] in Fig. 6 shows three and probably four bands in the stretching region, while the Raman spectrum[4] shows three peaks at 507, 555, and 628 cm^{-1}. There appears to be practically complete coincidence between the Raman and infrared in this region. The infrared spectrum also shows structure in the bending region.

In view of our expectation that the pseudorotational potential minima have C_{3v} symmetry, we shall discuss the vibrational assignment on that basis. Thus a symbol such as $A_1(T_{1u}$ stretch) or $A_1(\nu_3)$ means the mode of symmetry A_1, which is derived from the T_{1u} stretching mode ν_3. Some estimates were made[11] of the magnitudes expected for the splittings of the degenerate octahedral frequencies, but these will not be presented in detail. Since the pseudorotational potential has no

TABLE II. Assignment of low temperature XeF_6 spectrum based on a C_{3v} geometry.

Mode	Frequency (cm^{-1})
$A_1(T_{1u}$ stretch)	630
$A_1(A_{1g})$	624
$E(T_{1u}$ stretch)	557
$E(E_g)$	506
$E(T_{2g})$	252
$A_1(T_{1u}$ bend)	302
$A_1(T_{2g})$	(156)[a]
$A_2(T_{2u})$	(156)[a]
$E(T_{2u})$	(156)[a]

[a]The value given is a mean for these three vibrations determined to fit the entropy.

minima of symmetry C_{2v} or C_s, we do not expect further splitting of the C_{3v} bands, but rather broadening of the bands, especially in those cases where lower symmetry would yield splitting. It is evident from the gas phase spectra that this broadening is substantial for the stretching bands and is so extreme as to yield no discernible band structure in the binding region.

Table II gives an assignment of the low-temperature spectra. In making this assignment, the results for other hexafluoride molecules[20] were kept in mind. The polarization data for the room-temperature Raman spectrum indicate that the A_1 modes are near or above 600 cm^{-1}, while the depolarized E modes are at lower frequency. The braces indicate pairs of modes of the same symmetry in C_{3v} that are probably strongly mixed with respect to their octahedral parentage.

Figure 6 shows two relatively strong bands in the bending frequency range, 302 and 252 cm^{-1}. The former has a complex fine structure on the high frequency side which presumably arises from pseudorotational combinations. Also, we recall that the electron diffraction results indicate a strong mixing of $A_1(\nu_5)$ with $A_1(\nu_4)$. Thus, it seems likely that the 302 cm^{-1} band is the higher frequency member of the two arising from $A_1(\nu_4)$ and $A_1(\nu_5)$, with the lower frequency member unrecognized near or below 200 cm^{-1}. The 252 cm^{-1} band is assigned to $E(\nu_5)$, which is expected in this range and will have much less interaction since the E component of ν_4 has become the low-frequency pseudorotation.

The remaining bending mode ν_6 is inactive in both Raman and infrared in octahedral symmetry and is expected[20] in the range 130–200 cm^{-1}. The A_2 mode remains inactive in C_{3v} symmetry, but the E mode could become observable. There is no convincing spectroscopic evidence for either of these modes. An estimate of the mean frequency of these modes together with the lower of $A_1(\nu_4)$ and $A_1(\nu_5)$ as determined from the entropy is given in Table I. There is no implication that these modes all have the same frequency; however, the 156 cm^{-1} average is a plausible value for all of them.

The gas phase frequencies for the stretching modes show only moderate shifts from the low-temperature values, but the greater bandwidths make detailed discussion of little value. Apparently, the bending-mode bands are so broadened as to become a continuum at low resolution. Presumably, the extensive thermal excitation of pseudorotation is responsible for these effects.

THERMODYNAMICS OF XeF_6

The heat capacity of XeF_6 has been measured by Schreiner and co-workers[21] from 5–350 °K. Their data yielded for the standard entropy of XeF_6 at 335 °K the result $S° = 96.22$ cal/deg. Weinstock and co-workers[22] calculated $S°_{298°K}$ for various static geometries of XeF_6. Their results for a C_1 structure, which because of its low symmetry should have the highest entropy of all static structures, was $S° = 87.72$ cal/deg at 298 °K or about 91 cal/deg at 335 °K. The inability of a static structure to approach the experimental entropy confirms that there is either a very low frequency vibration or a pseudorotational motion occurring. Thermodynamically, the situation in XeF_6 parallels that of cyclopentane,[23] where the experimental entropy exceeded that of any plausible static model, but a free pseudorotational model reproduced the experimental entropy and heat capacity quite well.

The classical approximation will be satisfactory for the calculation of the pseudorotational entropy in view of the close spacing of quantum states, or alternatively, in view of the large mass and low curvature of the potential function. The partition function for the three-dimensional T_{1u} motion may be written

$$Q = \frac{(2\pi m_4 kT)^{3/2}}{h^3} \int_0^{2\pi} \int_0^{\pi} \int_0^{\infty} e^{-[-aR^2+bR^4-cR^4 f(\theta,\phi)]/kT} \times R^2 \sin\theta dR d\theta d\phi . \qquad (6)$$

where $f(\theta, \phi)$ is an abbreviation for the octahedral field term in the potential, Eq. (1), and m_4 is the effective mass for the T_{1u} motion. The entropy $S(T_{1u})$ was calculated by

$$S(T_{1u}) = R\left[T\left(\frac{\partial \ln Q}{\partial T}\right) + \ln Q\right] , \qquad (7)$$

where R is the molar gas constant.

The partition function Q and the related quantity $(\partial \ln Q/\partial T)$ were evaluated by numerical integration for various values of the potential parameters a, b, and c. In addition, the root mean square displacement $\langle R^2\rangle^{1/2}$ was also computed:

$$\langle R^2\rangle = \frac{(2\pi m_4 kT)^{3/2}}{Qh^3} \times \int_0^{2\pi} \int_0^{\pi} \int_0^{\infty} e^{-[-aR^2+bR^4-cR^4 f(\theta,\phi)]/kT} R^4 \sin\theta dR d\theta d\phi . \qquad (8)$$

We must now choose appropriate values for the parameters of the potential function. We know that $\langle R^2\rangle^{1/2}$ or R_0 is approximately 0.87 Å from the electron diffraction results. Somewhat arbitrarily, we chose the 302 cm^{-1} value for the radial frequency $A_1(T_{1u})$ which determines the second radial derivative of the potential. From Eq. (2a) for the potential in C_{3v} orientation, we obtain for the equilibrium value of R

$$a = 2(b - c/3)R_0^2 \quad (9)$$

and for the radial vibration frequency ν of $A_1(T_{1u})$

$$\left(\frac{\partial^2 V}{\partial R^2}\right)_{R_0} = -2a + 12(b - c/3)R_0^2 = 4\pi^2 m_4 \nu^2 . \quad (10)$$

Combination of these equations yields

$$a = 2(b - c/3)R_0^2 = \pi^2 m_4 \nu^2. \quad (11)$$

Substitution of $R_0 = 0.87$ Å, $\nu = 302$ cm^{-1}, and the effective mass $m_4 = 6.56$ amu yields $a = 0.088$ mdyn/Å and $(b - c/3) = 0.058$ mdyn/Å^3.

A series of numerical integrations of Q, $\partial \ln Q/\partial T$, and $\langle R^2\rangle$ were carried out[11] with this value of a and with values of b and c consistent with this relationship. (The preliminary values actually used were $a = 0.0873$ and $b - c/3 = 0.0576$, but the difference is not significant.) Since the use of the 302 cm^{-1} radial frequency was rather arbitrary, the entropy of a classical 302 cm^{-1} oscillator was subtracted from these results, and the remaining entropy for the two-dimensional pseudorotation is shown in Table III. At the higher values of c, the root mean square R value drops significantly below 0.87 Å which was given by electron diffraction for the best C_{3v} static model. While this is a correct reflection of the potential function chosen, Eq. (1), we do not know whether it is consistent with the observed electron diffraction intensities. If the parameters were readjusted to keep $\langle R^2\rangle^{1/2}$ more nearly constant at 0.87 Å, the entropy values would be increased for the larger c values. Thus the decrease in entropy with increase of c would be less than that given in Table III.

Various arguments concerning the limits on the parameter c will be discussed below, where it will be concluded that a small value is more likely to be correct. Consequently, for further calculations we take $c = 0.02$, where the reduction in entropy from free pseudorotation ($c = 0$) is within experimental error. With the molecular dimensions given in Table I and the vibration frequencies in Table II, one finds the entropy contributions listed in Table IV. Since we have considered the full range of solid angle 4π in the pseudorotation, we must assign the symmetry number of 24 for over-all rotation. As noted above, the remaining 11.25 cal/deg for four low-frequency vibrational motions yields a plausible mean frequency of 156 cm^{-1}.

TABLE III. Calculated entropy of the two-dimensional pseudorotational motion at 335 °K.

a	b	c	S(cal/deg)	$\langle R^2\rangle^{1/2}$ Å
0.0873[a]	0.0576[a]	0.00[a]	10.35	0.878
0.0873	0.0610	0.01	10.30	0.869
0.0873	0.0643	0.02	10.23	0.862
0.0873	0.0676	0.03	10.15	0.856
0.0873	0.0710	0.04	10.07	0.850
0.0873	0.0743	0.05	9.99	0.846
0.0873	0.0776	0.06	9.91	0.841
0.0873	0.0810	0.07	9.83	0.838
0.0873	0.0843	0.08	9.77	0.834
0.0873	0.0876	0.09	9.67	0.832
0.0873	0.0910	0.10	9.60	0.829

[a]The units of a are mdyn/Å and of b and c mdyn/Å^3.

TABLE IV. Entropy of XeF_6 at 335 °K from molecular parameters.

Translation and rotation	65.30 cal/deg · mole
Vibration ($\nu_1, \nu_2, \nu_3, \nu_5$)[a]	9.44 cal/deg · mole
Pseudorotation (two-dimensional)	10.23 cal/deg · mole
Vibration [ν_6 and $A_1(\nu_4)$][a]	11.25 cal/deg · mole
Total	96.22 cal/deg · mole

[a]Actually, $A_1(\nu_4)$ and $A_1(\nu_5)$ are mixed; the higher (302 cm^{-1}) is included here as ν_5 and the lower as $A_1(\nu_4)$.

For thermodynamic purposes, it is convenient to have values for the ideal gas at 298 °K. We estimate from our model $C_P^\circ = 29.5$ cal/deg · mole and using this value to correct the experimental entropy from 335 to 298 °K obtain $S^\circ = 92.78$ cal/deg · mole at 298.15 °K.

DISCUSSION

We have shown that a model exists for XeF_6 which is consistent with various experimental properties. This model is more detailed and specific than any previously published, but it is certainly within the range of models suggested by Bartell and Gavin,[2] by Wang and Lohr,[7] and by Bartlett and others.[24,25] While there are many aspects in which one would like to make this model more precise and more detailed, the magnitude of the potential restricting pseudorotation seems most important.

Our earlier paper[5] showed that the electric field deflection results[6] for XeF_6 could be reconciled with a pseudorotating model without assuming an unreasonably small dipole moment for the distorted molecule. The earlier results are summarized in Table V, where the barrier restricting pseudorotation is given both by the parameter c of Eq. (1) in the present paper and by the potential difference between C_{2v} and C_{3v} conformations. Also given is the maximum allowable dipole moment for each case. Distortion of XeF_6 from octahedral symmetry moves negative fluorines in one direction and an unshared electron pair in the opposite direction. Hence the dipole moment may be very small—indeed, practically zero if the cancellation is nearly exact. But it seems more likely that the dipole moment is a few tenths

TABLE V. Restricting potential for pseudorotation in XeF_6.

c	$V(C_{2v}) - V(C_{3v})$	Max. μ
0.00	0.0 cm^{-1}	1.2 D
0.02	48.1 cm^{-1}	0.6 D
0.05	120.1 cm^{-1}	0.3 D
0.08	192.2 cm^{-1}	0.2 D

of a debye, and therefore, that the parameter c is small, possibly in the range 0.0–0.04.

The diffuseness of the Raman and infrared spectral features also support a small c, i.e., nearly free pseudorotation at room temperature. Likewise, the large nonbonded amplitudes of motion reported from the electron diffraction studies support the conclusion that pseudorotation is relatively free. But it is not easy to obtain a quantitative limit on c from either of these sources.

A possibility for more precise determination of this parameter lies in the millimeter-range microwave or the far infrared spectra. Table I of our earlier paper[5] shows that the detailed energy level pattern for pseudorotation is very sensitive to the potential parameter c even though the entropy at 335 °K is not. One effort[19] to find absorptions in the 3–8 cm^{-1} range was unsuccessful. Coupling of over-all rotation with pseudorotation, however, may lead to such an extensive fine structure and so little intensity to each microwave line that detection will be difficult. Far infrared techniques are difficult, but the pseudorotational transitions in that range are also sensitive to the restricting potential.

In conclusion, we have no doubt that our pseudorotational model for XeF_6 is qualitatively and semiquantitatively correct and that the cause of the distortion from octahedral symmetry has been correctly ascribed to the pseudo-Jahn–Teller effect. Further experiments, possibly in the far infrared or millimeter microwave range, are needed to quantitatively specify the potential restricting pseudorotation.

ACKNOWLEDGMENTS

We thank Professor K. Hedberg for providing his electron diffraction results in advance of publication and for helpful discussion. This research was sponsored by the Energy Research and Development Administration.

[1]See, *Noble-Gas Compounds*, edited by H. H. Hyman (University of Chicago Press, Chicago, IL, 1963), and references there cited.
[2]L. S. Bartell and R. M. Gavin, Jr., J. Chem. Phys. **48**, 2466 (1968).
[3]F. Schreiner, D. W. Osborne, J. G. Malm, and G. N. McDonald, J. Chem. Phys. **51**, 4838 (1969).
[4]H. H. Claassen, G. L. Goodman, and H. Kim, J. Chem. Phys. **56**, 5042 (1972).
[5]L. S. Bernstein and K. S. Pitzer, J. Chem. Phys. **62**, 2530 (1975).
[6]W. E. Falconer, A. Büchler, J. Stauffer, and W. Klemperer, J. Chem. Phys. **48**, 312 (1968).
[7]S. Y. Wang and L. L. Lohr, Jr., J. Chem. Phys. **60**, 3901, 3916 (1974); **61**, 4110 (1974); **62**, 2013 (1975).
[8]G. L. Goodman, J. Chem. Phys. **56**, 5038 (1972).
[9]U. Nielsen, R. Haensel, and W. H. E. Schwartz, J. Chem. Phys. **61**, 3581 (1974); also Ref. 7.
[10]L. S. Bartell, J. Chem. Ed. **45**, 754 (1968), gives an elementary presentation of the pseudo-Jahn–Teller effect and numerous references.
[11]L. S. Bernstein, PhD. thesis, University of California, 1974; also Lawrence Berkeley Laboratory Rep. LBL-2701.
[12]R. M. Gavin and L. S. Bartell, J. Chem. Phys. **48**, 2460 (1968).
[13]K. Hedberg, private communication. Gavin and Bartell's results[12] differ slightly; Professor Bartell suggested that his sample may have been contaminated with $XeOF_4$ and that Professor Hedberg's results were to be preferred.
[14]D. J. Stufkens, Rec. Trav. Chim. Pays-Bas **89**, 1185 (1970).
[15]D. J. Stufkens, Rec. Trav. Chim. Pays-Bas **90**, 190 (1971).
[16]D. A. Couch, C. J. Wilkins, G. R. Rossman, and H. B. Gray, J. Am. Chem. Soc. **92**, 307 (1970).
[17]A. Fukuda, K. Inohana, and R. Onaka, J. Phys. Soc. Japan **19**, 1274 (1964).
[18]Y. Toyozawa and M. Inoue, J. Phys. Soc. Japan **21**, 1663 (1966).
[19]H. Kim, H. H. Claassen, and E. Pearson, Inorg. Chem. **7**, 616 (1968).
[20]H. H. Claassen, G. L. Goodman, J. H. Holloway, and H. Selig, J. Chem. Phys. **53**, 341 (1970).
[21]F. Schreiner, D. W. Osborne, J. G. Malm, and G. N. McDonald, J. Chem. Phys. **51**, 4838 (1969).
[22]B. Weinstock, E. Eugene Weaver, and C. Piknop, Inorg. Chem. **5**, 2189 (1966).
[23]J. E. Kilpatrick, K. S. Pitzer, and R. Spitzer, J. Am. Chem. Soc. **69**, 2483 (1947).
[24]R. D. Burbank and N. Bartlett, Chem. Commun. 1968, 645.
[25]N. Bartlett and F. O. Sladky, in *Comprehensive Inorganic Chemistry*, edited by J. C. Bailar (Pergamon, Oxford, 1973), Vol. I, pp. 213–330.

PART II
DENSE PHASES

A. Extended Corresponding States and the Acentric Factor

Introduction

The general background for this research is very well expressed by an invited "introduction" to a symposium on phase equilibria; hence, it is placed out of chronological order at beginning of this segment.

With the advance of computer capacity, it has become convenient to use rather complex equations of state rather than the graphical methods and extensive tables of the original publications. Two recent papers are of interest in this regard. For pure fluids Schreiber and Pitzer [*Fluid Phase Equilibria* **46**, 113–130 (1989)] give a very accurate equation and include results indicating the accuracy to which the acentric factor method is valid for normal fluids. This equation is inconvenient, however, for applications to mixed fluids. For mixtures, a somewhat simpler equation [A. Anderko and K. S. Pitzer, *AIChE J.* **37**, 1379–1391 (1991)] is recommended for reasons there explained.

Reprinted from ACS SYMPOSIUM SERIES, No. 60
PHASE EQUILIBRIA AND FLUID PROPERTIES IN THE CHEMICAL INDUSTRY
Truman S. Storvick and Stanley I. Sandler, Editors

1

Origin of the Acentric Factor

KENNETH S. PITZER

University of California, Berkeley, Calif. 94720

It was a pleasure to accept Dr. Sandler's invitation to open this conference by reviewing the ideas and general point of view which led me to propose the acentric factor in 1955. Although I had followed some of the work in which others have used the acentric factor, the preparation of this paper provided the incentive to review these applications more extensively, and I was most pleased to find that so much has been done. I want to acknowledge at once my debt to John Prausnitz for suggestions in this review of recent work as well as in many discussions through the years.

Beginning in 1937, I had been very much interested in the thermodynamic properties of various hydrocarbon molecules and hence of those substances in the ideal gas state. This arose out of work with Kemp in 1936 on the entropy of ethane (1) which led to the determination of the potential barrier restricting internal rotation. With the concept of restricted internal rotation and some advances in the pertinent statistical mechanics it became possible to calculate rather accurately the entropies of various light hydrocarbons (2). Fred Rossini and I collaborated in bringing together his heat of formation data and my entropy and enthalpy values to provide a complete coverage of the thermodynamics of these hydrocarbons in the ideal gas state (3). As an aside I cite the recent paper of Scott (4) who presents the best current results on this topic.

But real industrial processes often involve liquids or gases at high pressures rather than ideal gases. Hence it was a logical extension of this work on the ideal gases to seek methods of obtaining the differences in properties of real fluids from the respective ideal gases without extensive experimental studies of each substance.

My first step in this direction came in 1939 when I was able to provide a rigorous theory of corresponding states (5) on the basis of intermolecular forces for the restricted group of substances, argon, kryptron, xenon, and in good approximation also methane. This pattern of behavior came to be called that of a simple fluid. It is the reference pattern from which the acentric factor measures the departure. Possibly we should recall the key ideas. The intermolecular potential must be given by a universal function with scale factors of energy and distance for each substance. By then it was well-known that the dominant attractive force followed an inverse sixth-power potential for all of these substances. Also the repulsive forces were known to be very sudden. Thus the inverse sixth power term will dominate the shape of the potential curve at longer distances. Even without detailed theoretical reasons for exact similarity of shorter-range terms, one could expect that a universal function might be a good approximation. In addition one assumed spherical symmetry (approximate for methane), the validity of classical statistical mechanics, and that the total energy was determined entirely by the various intermolecular distances.

I should recall that it was not feasible in 1939 to calculate the actual equation of state from this model. One could only show that it yielded corresponding states, i.e., a universal equation of state in terms of the reduced variables of temperature, volume, and pressure.

One could postulate other models which would yield a corresponding-states behavior but different from that of the simple fluid. However, most such molecular models were special and did not yield a single family of equations. Rowlinson (6) found a somewhat more general case; he showed that for certain types of angularly dependent attractive forces the net effect was a temperature dependent change in the repulsive term. From this a single family of functions arose.

I had observed empirically, however, that the family relationship of equations of state was much broader even than would follow from Rowlinson's model. It included globular and effectively spherical molecules such as tetramethylmethane (neopentane), where no appreciable angular dependence was expected for the intermolecular potential, and for elongated molecules such as carbon dioxide the angular dependence of the repulsive forces seemed likely to be at least as important as that of the attractive forces. Thus the core model of Kihara (7) appealed to me; he assumed that the Lennard-Jones 6-12 potential applied to the shortest distance between cores instead of the distance between molecular centers. He was able to calculate the second virial coefficient for various shapes of core. And I was able to show that one obtained in good approximation a single family of reduced second virial coefficient functions for cores of all reasonable shapes. By a single family I mean that one additional parameter sufficed to define the equation for any particular case. While this did not prove that all of the complete equations of state would fall into a single family, it gave me enough encouragement to go ahead with the numerical work--or more accurately to persuade several students to undertake the numerical work.

Let me emphasize the importance of fitting globular molecules into the system. If these molecules are assumed to be spherical in good approximation, they are easy to treat theoretically. Why aren't they simple fluids? Many theoretical papers ignore this question. In fluid properties neopentane departs from the simple fluid pattern much more than propane and almost as much as n-butane. But propane is much less spherical than neopentane. The explanation lies in the narrower attractive potential well. The inverse-sixth-power attractive potential now operates between each part of the molecule rather than between molecular centers. Thus the attractive term is steeper than inverse sixth power in terms of the distance between molecular centers. This is shown in Figure 1, taken from my paper (8) in 1955. We need not bother with the differences between the models yielding the dotted and dashed curves for the globular molecule. The important feature is the narrowness of the potential well for either of these curves as compared to the solid curve for the molecules of a simple fluid.

It was easy to show that the intermolecular potential curves for spherical molecules would yield a single family of reduced equations of state. If one takes the Kihara model with spherical cores, then the relative core size can be taken as the third parameter in addition to the energy and distance scale factors in the theoretical equation of state.

With an adequate understanding of globular molecule behavior, I then showed as far as was feasible that the properties of other non-polar or weakly polar molecules would fall into the same family. It was practical at that time only to consider the second virial coefficient. The Kihara model was used for nonpolar molecules of all shapes while Rowlinson's work provided the basis for discussion of polar molecules. Figure 2 shows the reduced second virial coefficient for several cases. Curves labeled a/ρ_o refer to spherical-core molecules with a indicating the core size, correspondingly ℓ/ρ_o indicates a linear molecule of core length ℓ, while y refers to a dipolar molecule with $y = \mu^4/\varepsilon_o{}^2 r_o{}^6$ where μ is the dipole moment. The non-polar potential is

$$\varepsilon = \varepsilon_o[(\frac{\rho_o}{\rho})^{12} - 2(\frac{\rho_o}{\rho})^6] \tag{1}$$

where ρ is the shortest distance between cores. For the polar molecules I omitted the core, thus $\rho = r$.

While the curves in Figure 2 appear to fall into a single family, this is investigated more rigorously in Figure 3 where the reduced second virial coefficient at one reduced temperature is compared with the same quantity at another temperature. T_B is the Boyle temperature which is a convenient reference temperature for second virial coefficients. One sees that the non-polar core molecules fall accurately on a single curve (indeed a straight line). While the polar molecules deviate, the difference is only 1% at $y = 0.7$ which I took as a reasonable standard of accuracy at that time. For comparison the y values of chloroform, ethylchloride, and ammonia are 0.04, 0.16, and 4, respectively. Thus the first two fall well below the 0.7 value for agreement of polar with non-polar effects while ammonia is beyond that value.

The next question was the choice of the experimental basis for the third parameter. The vapor pressure is the property most sensitive to this third parameter; also it is one of the properties most

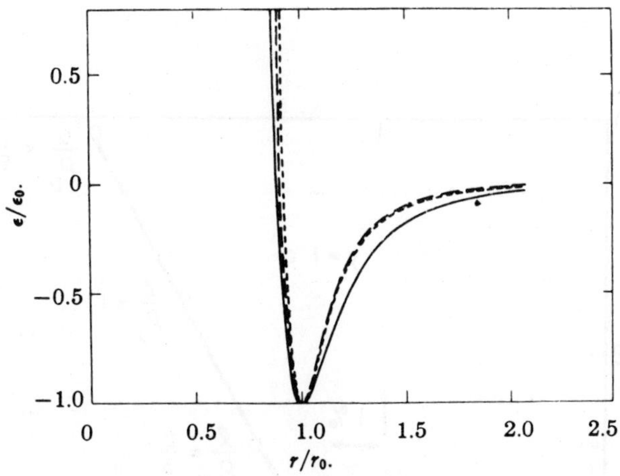

Figure 1. Intermolecular potential for molecules of a simple fluid, solid line; and for globular molecules such as $C(CH_3)_4$, dashed lines

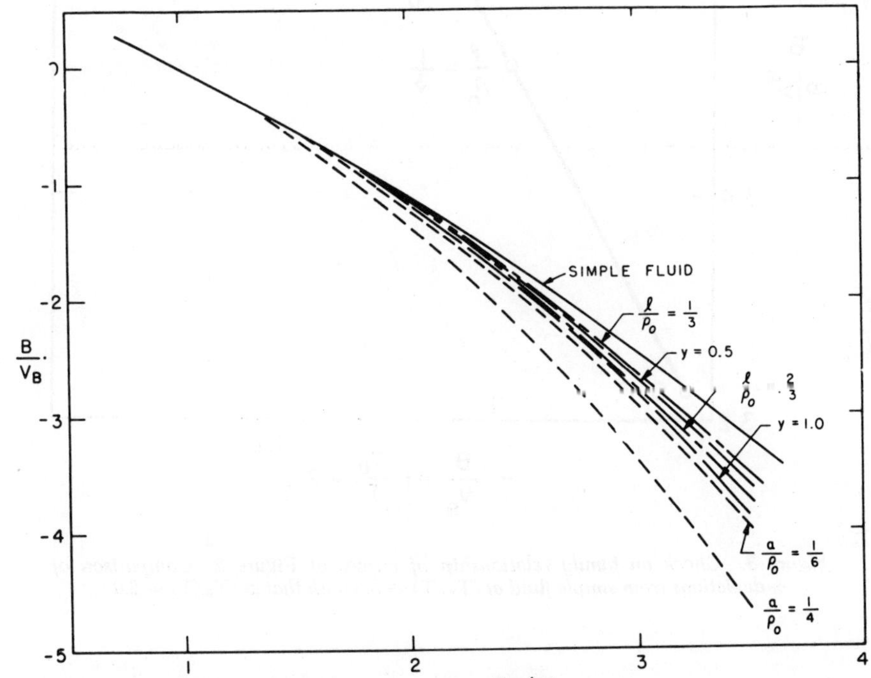

Figure 2. Reduced second virial coefficients for several models: solid curve, simple fluid; curves labeled by a/ρ_0, spherical cores of radius a; curves labeled by l/ρ_0, linear cores of length l; curves labeled by y, molecules with dipoles

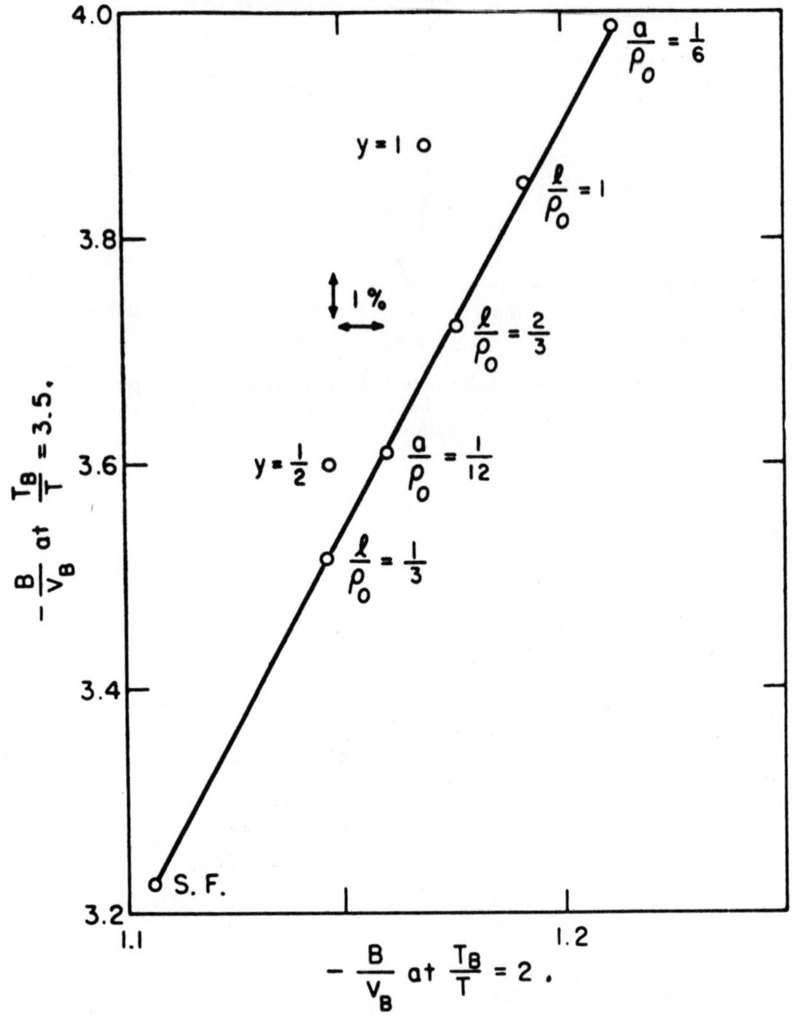

Figure 3. Check on family relationship of curves of Figure 2. Comparison of deviations from simple fluid at $(T_B/T) = 3.5$ with that at $(T_B/T) = 2.0$

widely measured at least near the normal boiling point. Thus both the availability of data and the accuracy of the data for the purpose strongly indicated a vapor pressure criterion. Since the critical data have to be known for a reduced equation of state, the reduced vapor pressure near the normal boiling point was an easy choice for the new parameter. The actual definition

$$\omega = -\ell og\ P_r - 1.000 \tag{2}$$

with P_r the reduced vapor pressure at $T_r = 0.700$ seemed convenient, but the actual determination of ω can be made from any vapor pressure value well-removed from the critical point.

Here I should note the work of Riedel (9) which was substantially simultaneous with mine but whose first paper preceded slightly. His work was purely empirical, but was excellent and fully complementary. He chose for his third parameter also the slope of the vapor pressure curve, but in his case the differential slope at the critical point. That seemed to me to be less reliable and accurate, empirically, although equivalent otherwise. Fortunately Riedel and I chose to emphasize different properties as our respective programs proceeded; hence the full area was covered more quickly with little duplication of effort.

Also I needed a name for this new parameter, and that was difficult. The term "acentric factor" was suggested by some friendly reviewer, possibly by a referee; I had made a less satisfactory choice initially. The conceptual basis is indicated in Figure 4. The intermolecular forces between complex molecules follow a simple expression in terms of the distances between the various portions of the molecule. Since these forces between non-central portions of the molecules must be considered, the term "acentric factor" seemed appropriate.

It is assumed that the compressibility factor and other properties can be expressed in power series in the acentric factor and that a linear expression will usually suffice.

$$\frac{pv}{RT} = z = z^{(0)} + \omega z^{(1)} + \ldots$$

$$z^{(0)} = z^{(0)}(T_r, P_r)$$

$$z^{(1)} = z^{(1)}(T_r, P_r)$$

The preference of P_r over V_r as the second independent variable is purely empirical; the critical pressure is much more accurately measurable than the critical volume.

The empirical effectiveness of this system was first tested with volumetric data as shown on Figure 5. Here pv/RT at a particular

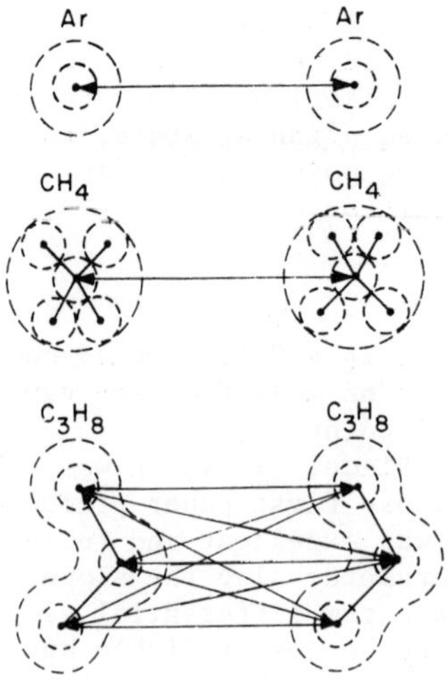

Figure 4. Intermolecular forces operate between the centers of regions of substantial electron density. These centers are the molecular centers for Ar and (approximately) for CH_4, but are best approximated by the separate CH_3 and CH_2 groups in C_3H_8—hence the name acentric factor for the forces arising from points other than molecular centers.

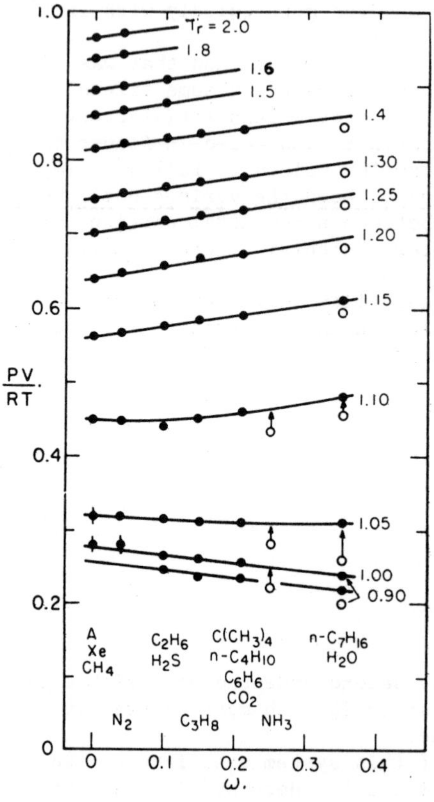

Figure 5. Compressibility factor as a function of acentric factor for reduced pressure 1.6 and reduced temperature shown for each line. Where several substances have approximately the same acentric factor, the individual points are indistinguishable except for n-C_7H_{16} (●) *and* H_2O (○).

reduced temperature and pressure is plotted against ω. The most important result appears only by implication; the results for $C(CH_4)_4$, $n\text{-}C_4H_{10}$, C_6H_6, and CO_2 are so nearly equal that they appear as single points on these plots. Here we have four widely different shapes of molecules which happen to have about the same acentric factor, and they follow corresponding states accurately among themselves.

Also to be noted from Figure 5 is the fact that the highly polar molecules NH_3 and H_2O depart from the system. Furthermore the dependence on ω is linear except for the critical region.

My immediate research group used graphical methods in dealing with the experimental data and reported all of our results in numerical tables (10). At that time the best analytical equation of state was that of Benedict, Webb and Rubin (11) which employed eight parameters and still failed to fit volumetric data within experimental accuracy. Bruce Sage suggested fitting this equation to the data for the normal paraffins both directly for each substance and within the acentric factor system. This work (12) was done primarily by J. B. Opfell at Cal Tech. The results showed that the acentric factor system was a great advance over the simple postulate of corresponding states, but the final agreement was inferior to that obtained by graphical and numerical methods.

Thus we continued with numerical methods for the fugacity, entropy, and enthalpy functions (13), although we did present an empirical equation for the second virial coefficient (14). This work was done by Bob Curl; he did an excellent job but found the almost interminable graphical work very tiresome. Thus I was pleased that the British Institution of Mechanical Engineers included Curl in the award of their Clayton Prize for this work. A fifth paper with Hultgren (15) treated mixtures on a pseudocritical basis, and a sixth with Danon (16) related Kihara core sizes to the acentric factor.

Naturally, I am very pleased to note that others have extended the accuracy and range of our tables and equations with consideration of more recent experimental results. Of particularly broad importance is the 1975 paper by Lee and Kessler (17) which presents both improved tables and analytical equations for all of the major functions including vapor pressures, volumetric properties, enthalpies, entropies, fugacities, and heat capacities. Their equation is an extension of that of Benedict, Webb, and Rubin now containing twelve parameters. They considered more recent experimental data as well as a number of papers which had already extended my earlier work in particular areas. I refer to their bibliography (17) for most of this more detailed work, but I do want to note the improved equation of Tsonopoulos (18) for the second virial coefficient. This equation deals also with effects of electrical polarity.

In addition to references cited by Lee and Kessler there is the work Lyckman, Eckert, and Prausnitz (19) dealing with liquid volumes; they found it necessary to use a quadratic expression in ω. Also Barner and Quinlan (20) treated mixtures at high temperatures and pressures, and Chueh and Prausnitz (21) treated the compressibility of liquids. Reid and Sherwood (22) give an extensive table including acentric factors as well as critical constants for many substances.

On the theoretical side, one great advance has been in the development of perturbation theories of a generalized van der Waals type. Here one assumes that the molecular distribution is determined primarily by repulsive forces which can be approximated by hard cores. Then both the softness of the cores and the atractive forces are treated by perturbation methods. Barker and Henderson (23) have recently reviewed theoretical advances including their own outstanding work. Rigby (24) applied these modern Van der Waals' methods to non-spherical molecules which represent one type of molecules with non-zero acentric factors. In a somewhat similar manner Beret and Prausnitz (25) developed equations applicable even to high polymers and related the initial departures from simple fluids to the acentric factor.

But in my view the most effective approach would concentrate first on globular molecules. These could be modeled by Kihara potentials with spherical cores or by other potentials allowing the well to be narrowed. The great advantage would be the retention of spherical symmetry and its theoretical simplicity. Rogers and Prausnitz (26) made an important beginning in this area with calculations based on Kihara models appropriate for argon, methane, and neopentane with excellent agreement for the properties studied. While they do not discuss these results in terms of the acentric factor, the transformation of spherical core radius to acentric factor is well established (16, 27), Rogers and Prausnitz were also able to treat mixtures very successfully although those calculations were burdensome even with modern computers. I believe further theoretical work using spherical models for globular molecules would be fruitful.

The move to an analytical equation by Lee and Kessler was undoubtedly a wise one in view of the marvelous capacity of modern computers to deal with complex equations. I would expect future work to yield still better equations.

There remains the question of the ultimate accuracy of the acentric factor concept. How accurately do molecules of different shapes but with the same acentric factor really follow corresponding states? Apparently this accuracy is within experimental error for most, if not all, present data. Thus the acentric factor system certainly meets engineering needs, and it is primarily a matter of scientific curiosity whether deviations are presently measurable.

It has been a pleasure to review these aspects of the "acentric factor" with you and I look forward to your discussion of recent advances in these and other areas.

Literature Cited

1. Kemp, J. D. and Pitzer, K. S., J. Chem. Phys., (1936) 4, 749; J. Am. Chem. Soc. (1937) 59, 276.

2. Pitzer, K. S., J. Chem. Phys., (1937) 5, 469, 473, 752; (1940) 8, 711; Chem. Rev. (1940) 27, 39.
3. Rossini, F. D., Pitzer, K. S., Arnett, R. L., Braun, R. M. and Pimentel, G. C., "Selected Values of the Physical and Thermodynamic Properties of Hydrocarbons and Related Compounds," Carnegie Press, Pittsburgh (1953).
4. Scott, D. W., J. Chem. Phys. (1974) 60, 3144.
5. Pitzer, K. S., J. Chem. Phys. (1939) 7, 583.
6. Rowlinson, J. S., Trans. Faraday Soc. (1954) 50, 647; "Liquids and Liquid Mixtures," 2nd ed. Chapter 8, Butterworth, London (1969).
7. Kihara, T., Rev. Mod. Phys. (1953) 25, 831 and papers there cited.
8. Pitzer, K. S., J. Am. Chem. Soc. (1955) 77, 3427.
9. Riedel, L., Chem. Ing. Tech. (1954) 26, 83, 259, 679; (1955) 27, 209, 475; (1956) 28, 557.
10. Pitzer, K. S., Lippman, D. Z., Curl, Jr., R. F., Huggins, C. M. and Petersen, D. E., J. Am. Chem. Soc. (1955) 77, 3433.
11. Benedict, M., Webb, G. B. and Rubin, L. C., J. Chem. Phys. (1940) 8, 334.
12. Opfell, J. B., Sage, B. H. and Pitzer, K. S., Ind. Eng. Chem. (1956) 48, 2069.
13. Curl, Jr., R. F., and Pitzer, K. S., Ind. Eng. Chem. (1958) 50, 265.
14. Pitzer, K. S. and Curl, Jr., R. F., J. Am. Chem. Soc., (1957) 79, 2369.
15. Pitzer, K. S. and Hultgren, G. O., J. Am. Chem. Soc. (1958) 80, 4793.
16. Danon, F. and Pitzer, K. S., J. Chem. Phys. (1962) 36, 425.
17. Lee, B. I. and Kesler, M. G., A.I.Ch.E. Journal (1975) 21, 510.
18. Tsonopoulos, C., A.I.Ch.E. Journal (1974) 20, 263.
19. Lyckman, E. W., Eckert, C. A. and Prausnitz, J. M., Chem. Engr. Sci. (1965) 20, 703.
20. Barner, H. E. and Quinlan, C. W., I. and E.C. Proc. Des. Dev. (1969) 8, 407.
21. Chueh, P. L. and Prausnitz, J. M., A.I.Ch.E. Journal (1969) 15, 471.
22. Reid, R. C. and Sherwood, T. K., "The Properties of Gases and Liquids," 2nd ed., McGraw-Hill Book Co., New York (1966).
23. Barker, J. A. and Henderson, D., Rev. Mod. Phys. (1976) 48, 587.
24. Rigby, M., J. Phys. Chem (1972) 76, 2014.
25. Beret, S. and Prausnitz, J. M., A.I.Ch.E. Journal (1975) 21, 1123.
26. Rogers, B. L. and Prausnitz, J. M., Trans. Faraday Soc. (1971) 67, 3474.
27. Tee, L. S., Gotoh, S., and Stewart, W. E., Ind. Eng. Chem. Fund. (1966) 5, 363.

AUGUST, 1939 JOURNAL OF CHEMICAL PHYSICS VOLUME 7
Printed in U. S. A.

Corresponding States for Perfect Liquids

KENNETH S. PITZER
Chemical Laboratory, University of California, Berkeley, California

(Received April 18, 1939)

On the basis of certain assumptions concerning the properties of the molecules, and by using classical statistics, the van der Waals theory of corresponding states is derived. No mathematical approximations are involved. Argon, krypton and xenon, which conform closely to the assumptions, show the expected correspondence in behavior and are taken as standards. The name *perfect liquid* is suggested for this behavior. In a second part, the deviations from perfection are discussed for a wide variety of substances. Particularly considered are heat capacities and entropies of vaporization which should be compared at points where the vapor to liquid volume ratio has the same value. The observed deviations are explained in a general way.

INTRODUCTION

THE idea of corresponding states arose from the equation of state proposed by van der Waals and was suggested by him in 1873. The theory of corresponding states is of course more general than the equation and holds that in terms of reduced temperatures (T/T_{crit}), volumes, etc., the behavior of all substances should be the same. At that time the available knowledge of the properties of individual molecules was insufficient for a really satisfactory theoretical derivation in these terms or to indicate what substances should follow this theory. Recently such knowledge has become available, particularly due to the work of London and others on the nature of van der Waals forces. The following is a derivation of the theory of corresponding states on the basis of assumptions compatible with the new knowledge of molecular forces.

Recently de Boer and Michels[1] have discussed the theory of corresponding states with respect to gas imperfection, but the writer is not aware of any recent consideration of the liquid from this point of view.

In addition to the theory of corresponding states, in terms of the critical point, there is Trouton's rule and its various modifications, most notably that of Hildebrand.[2] While our derivation will lead to the van der Waals theory of corresponding states, it will be shown that for certain common groups of substances it becomes equivalent to the Hildebrand rule.

On the basis of modern molecular knowledge, Eyring and co-workers,[3] Lennard-Jones and Devonshire,[4] Mayer and co-workers,[5] and Rice[6] have recently presented theoretical treatments of liquids which have been successful in accounting for the properties of that state. The aim of these investigations was somewhat different from that of the present one. While we shall hope only to establish theoretically the proper basis for comparison of liquid properties, these investigators attempted to calculate the properties themselves. On the other hand, we shall avoid some of the approximations which have been made in the more penetrating discussions.

After completing in Part I our derivation of the law of corresponding states and comparing the data for a group of substances fitting the assumptions, a brief discussion will be given in Part II of the various deviation from the law. Professor J. H. Hildebrand[7] has recently considered the deviations of a somewhat different group of substances from this point of view.

PART I

Perfect liquids

The following assumptions must be made for our theoretical derivation. First, classical sta-

[1] J. deBoer and A. Michels, Physica 5, 945 (1938).
[2] J. H. Hildebrand, J. Am. Chem. Soc. 37, 970 (1915); 40, 45 (1918).
[3] H. Eyring and Hirshfelder, J. Phys. Chem. 41, 249 (1937); Newton and H. Eyring, Trans. Faraday Soc. 33, 73, 80 (1936).
[4] J. E. Lennard-Jones and A. F. Devonshire, Proc. Roy. Soc. London A165, 1 (1938).
[5] J. E. Mayer, J. Chem. Phys. 5, 63 (1937); J. E. Mayer and P. G. Ackerman, *ibid.* 5, 74 (1937); J. E. Mayer and S. F. Harrison, *ibid.* 6, 87, 101 (1938).
[6] O. K. Rice, J. Chem. Phys. 5, 353 (1937) and earlier papers.
[7] J. H. Hildebrand, J. Chem. Phys. 7, 233 (1939).

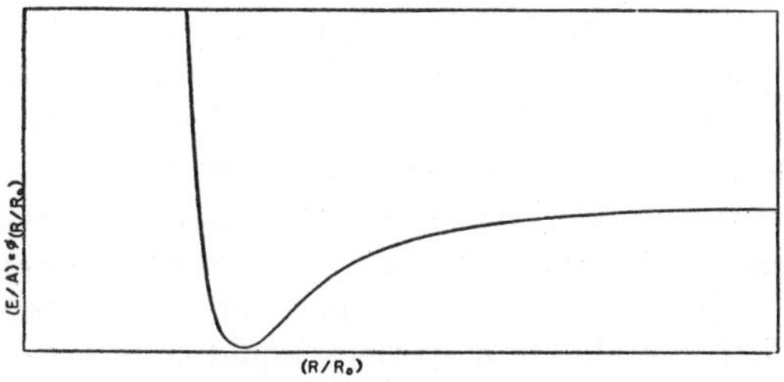

FIG. 1. The approximate curve of the function φ.

tistical mechanics will be used. As an example, this assumption for a harmonically vibrating solid gives the Dulong and Petit value of the heat capacity. Since most solids attain this condition before melting, this assumption should be acceptable for their liquids. However, this will exclude hydrogen and helium from consideration and make doubtful a few other cases including neon. Second, it will be assumed that the molecules are spherically symmetrical, either actually or by virtue of rapid and free rotation. Third, the nature of any intramolecular vibrations will be assumed to be the same whether the molecules are in the liquid or gas states. Fourth, the potential energy of an assemblage of molecules will be taken as a function only of the various intermolecular distances. As Mayer[5] has pointed out, this last consideration excludes any substances tending to form a limited number of valence bonds, but is consistent with the properties of saturated molecules where the only attractive forces are of the van der Waals-London type. Fifth, it is postulated that the potential energy for a pair of molecules can be written: $E=A\varphi(R/R_0)$ where φ is a universal function, R is the intermolecular distance, and A and R_0 are constants to be specified for each substance. Probably no group of molecules fulfills this last assumption precisely but at least the heavier rare gases satisfy it to a high degree of accuracy. Fig. 1 shows qualitatively the form of the function φ. The inverse sixth power attraction is characteristic of most, if not all, molecules forming normal liquids. Furthermore, provided that the repulsive branch is very steep, minor variations in slope should not affect the results greatly. Nothing need be said concerning the exact nature of the function φ; the only assumption is its universality. In their discussion of corresponding states for gas imperfection de Boer and Michels[1] were led to conditions essentially the same as these.

While a mathematical derivation will be given presently it seems worth while to present the following simple argument first. If two systems fulfilling the above conditions and containing equal numbers of molecules are enclosed in containers with linear dimensions proportional to the R_0's (thus volumes proportional to R_0^3), it is apparent that spacial correspondence has been attained. Since energies of motion and temperature are proportional, the corresponding temperatures will be proportional to the A's. Thus if the systems are compared at volumes bearing the same ratio to the values of R_0^3 and at temperatures thus related to the A's, then these systems should present corresponding behavior.

In more mathematical terms one may set up a classical partition function or phase integral.

$$Q=\int_0^\infty\cdots\int_0^\infty\int_0^{X_0}\cdots\int_0^{Z_0} e^{-(K.E.+P.E.)/kT} \times dp_{x_1}\cdots dp_{z_n}dx_1\cdots dz_n.$$

This partition function differs from those commonly set up for gases in that the integration is to be over the coordinates and momenta for all molecules instead of only one. The complex potential energy makes it impossible to factor out a contribution from each molecule. The kinetic energy can of course be written

$$\text{K.E.}=\frac{1}{2m}\sum_{i=1}^{n}(p_{x_i}^2+p_{y_i}^2+p_{x_i}^2).$$

The integration over momenta is then easily carried out giving:

$$Q=(2\pi mkT)^{3n/2}\int_0^{X_0}\cdots\int_0^{Z_0} e^{-P.E./kT}dx_1\cdots dz_n.$$

The potential energy is given by

$$\text{P.E.}=\tfrac{1}{2}A\sum_{i=1}^{n}\sum_{j=1}^{n}{}'\,\varphi(\Delta R_{ij}/R_0),$$

in which $\Delta R_{ij}=[(x_i-x_j)^2+(y_i-y_j)^2+(z_i-z_j)^2]^{\frac{1}{2}}$ is the distance between molecules i and j, the prime indicates omission of terms with i and j equal, and the factor $\frac{1}{2}$ arises because each term is taken twice. The integral for Q is thus extremely complex and no attempt will be made to evaluate it. We may rewrite it as follows however:

$$Q=\left[2\pi mkA\left(\frac{T}{A}\right)\right]^{3n/2}R_0^{3n}\int_0^{(X_0/R_0)}\cdots\int_0^{(Z_0/R_0)}$$

$$\times e^{-(A/T)(1/2k)\Sigma\Sigma\varphi(\Delta R_{ij}/R_0)}d\left(\frac{x_i}{R_0}\right)\cdots d\left(\frac{z_n}{R_0}\right).$$

The integral now depends only on the value of the ratio (A/T) and upon the upper limits of integration, the latter amounting to the ratio of the volume $(V=X_0Y_0Z_0)$ to R_0^3 since the shape of the container should not be a factor. We may then write:

$$Q=(2\pi mkAR_0^2)^{3n/2}F(V/R_0^3,\ A/T),$$

where the (T/A) factor before the integral is now included in F. Since the thermodynamic properties of interest (excepting the absolute, third law entropy) are determined by this partition function and its temperature and volume derivatives, and since constant factors such as those before F are of no consequence, it is thus established that the behavior of all substances conforming to the above assumptions will be the same, provided they are compared at equal values of (T/A) and (V/R_0^3).

Nothing has been said concerning the exact definition of R_0 and A. Since the critical point is certainly a corresponding point, it seems logical to take the critical temperature, T_c, for A and the critical volume, V_c, for R_0^3. Then the reduced temperatures and volumes will be those in common usage. Recent experiments of Maass and co-workers[8] and theoretical considerations of Mayer and Harrison[5] show that the behavior of substances near the critical point is more complex than had been assumed earlier. Nevertheless, this seems the best point to take as a standard. As will be shown later, the behavior of actual liquids can be compared conveniently without reference to critical point data.

The present and principal interest in the above considerations is with respect to the liquid state, and the term "perfect liquid" is proposed for substances following this behavior. Although the vapor of a perfect liquid is not necessarily a perfect gas, nevertheless this usage of terms will be very similar. Thus the term perfect gas, suggests a behavior representable by an equation which all gases follow more or less closely. Similarly the term perfect liquid, refers to a behavior which all normal liquids follow more or less closely.

The heavier rare gases are the substances which may be expected to show most perfect behavior in the above sense, and their properties will now be considered in order both to verify the theory and to prepare tables describing perfect behavior.

First considering vapor pressure, precise data covering the entire liquid range are available for argon and krypton, the latter from recent measurements by Meihuizen and Crommelin.[9] For argon, and in fact wherever other references are not given, the data have been taken from the *International Critical Tables*, or in a few cases from the Landolt Börnstein *Tabellen*. Fig. 2 shows these vapor pressure data, together with less precise data for xenon. Also included are the values for neon which show a deviation attributable to the failure of classical mechanics, and the values for methane which show good agreement due possibly to a cancellation of small errors arising from quantum effects and restriction of rotation. The agreement between the

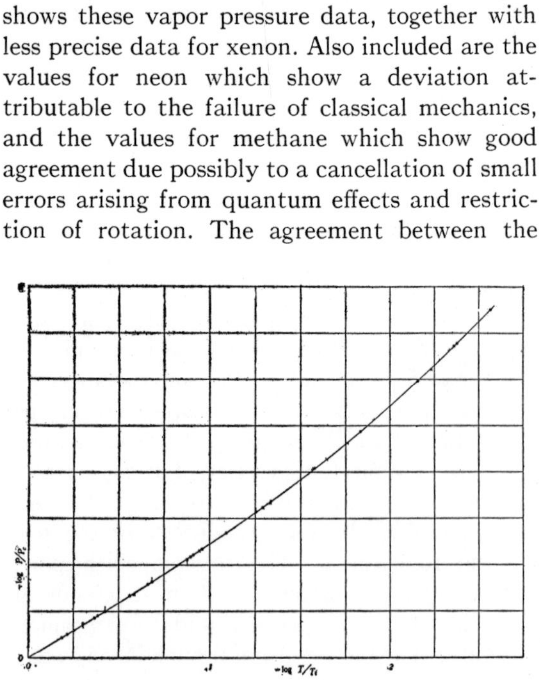

FIG. 2. Reduced vapor pressure data: 0 argon, ● krypton, x xenon, ♀ neon, ძ methane.

[8] Maass and Geddes, Phil. Trans. Roy. Soc. London **A236**, 303 (1937) and earlier work there listed.

[9] Meihuizen and Crommelin, Physica **4**, 1 (1937)

precise data for argon and krypton is remarkable, the largest deviations being considerably less than 0.1 percent if attributed to the temperature. The equation given for krypton by Meihuizen and Crommelin[9] was put into the reduced form and taken as representative.

$$\log_{10}\pi = \frac{4.29810}{\theta} - 12.55400 \log \theta + 0.63158 + 3.666520\theta,$$

where $\pi = p/p_c$ and $\theta = T/T_c$.

While the exact correspondence of the vapor pressure curves requires that the entropies of vaporization also correspond, it is worth while to present independent calorimetric values. The data in Table I are in part from calorimetric heats of vaporization and in part calculated from statistical entropies of the vapor and third law entropies of the liquid. In both cases they are for vaporization to the real gas. For comparison, values are given at the boiling points (Trouton's Rule) and at constant vapor density (Hildebrand's rule). The basis of comparison selected for the present theory is $\theta(=T/T_c)$ equal to 0.56 or the equivalent, a gas to liquid volume ratio of 335. The marked disagreement in the case of neon is to be ascribed to a failure of classical mechanics. The heat capacities of the liquids are also included, the value for neon being again low. The heat capacity of argon also appears to be slightly low.

Figure 3 shows the gas and liquid densities as a function of reduced temperature and Fig. 4 shows the gas to liquid volume ratio. These data are less precise than those on vapor pressure. However, the agreement appears to be rather satisfactory.

TABLE I. *Entropies of vaporization of perfect liquids.**

SUBSTANCE	ΔS_{VAP}			C_p LIQ
	$V_{GAS}/V_{LIQ}=335$	B. P.	$V_{GAS}=10$ l.	
Neon	(17.95)	15.0	(19.3)	8.9
Argon	18.67	18.0	19.0	10.1
Krypton	18.60	18.1	18.5	10.6
Xenon	18.66	18.4	18.0	10.7

* (Ne, A, Kr) K. Clusius, Zeits. f. physik. Chemie **B31**, 459 (1936); (Kr) Clusius, Krius and Konnertz, Ann d. Physik (S)**33**, 642 (1938); (Xe) K. Clusius and Riccoboni, Zeits. f. physik. Chemie **B38**, 81 (1937). The values for Ne at 10 l. gas volume and at $V_{gas}/V_{liq}=335$ are extrapolated.

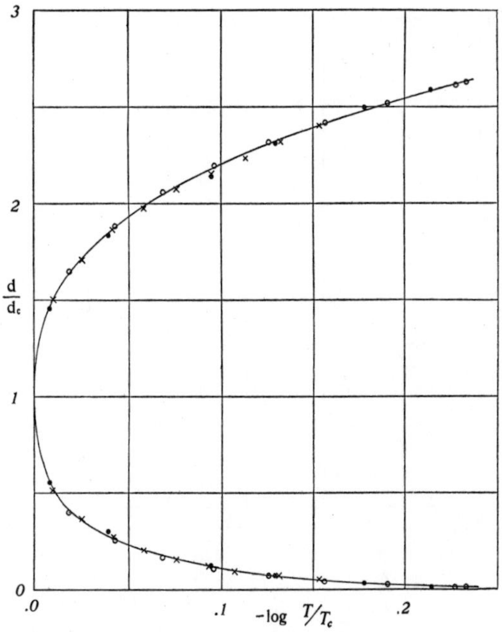

FIG. 3. Reduced density data: ● argon, 0 methane, x xenon.

These various data are combined to give the behavior of the perfect liquid. The vapor pressure equation has been given already. Other data are included in Table II. The accuracy of the various values is roughly indicated by the number of significant figures given, the error being usually confined to the last figure. It should be remembered that the true entropies of vaporization are given; these values are from three to ten percent lower than would be obtained from vapor pressures if the perfect gas law were assumed and the liquid volume neglected.

While Table II is presented in terms of reduced temperature, θ, certain of the other quantities are equally satisfactory for specifying corresponding states. In particular the gas to liquid volume ratio (last column) varies rapidly with temperature and involves no measurements at the critical point. Thus it is to be especially recommended for the comparison of various substances, inasmuch as the critical point data are likely to be unavailable or inaccurate in some cases.

So long as classical mechanics still holds, the behavior of substances fulfilling our conditions should also correspond in the solid state. Actually quantum-mechanical effects enter so soon that

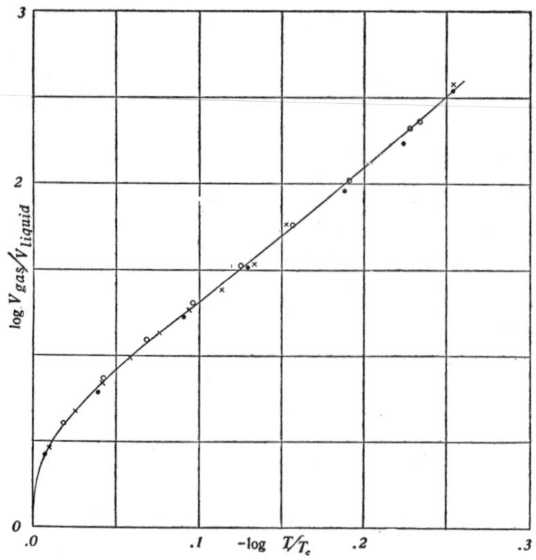

FIG. 4. Gas to liquid volume ratio as a function of reduced temperature: ● argon, 0 methane, x xenon.

this idea will have little practical interest. However, a few valid conclusions can be drawn. First, all perfect liquids will freeze into cubic closest packed (face-centered) crystals. Also, the entropy of fusion, and the triple point temperatures show regularities as exhibited in Table III.

In closing, Part I may be summarized by saying that when certain conditions are met, statistical mechanics requires a rather complete correspondence in the behavior of the liquid and gaseous substances. While it is realized that the available group of substances may not exactly fulfill the required conditions, nevertheless the observed correspondence is sufficiently exact that this idea of a perfect liquid behavior will probably be useful for some time.

PART II

Imperfect liquids

In Part I it was shown that there would be a rather complete correspondence in the behavior of liquids fulfilling certain assumptions, and the name perfect liquids was suggested for these. The deviation in behavior will now be considered for substances which fail to satisfy these conditions completely. These deviations can be best classified in terms of the requirement violated. The five requirements of perfection may be briefly restated as follows: (1) The validity of classical statistical mechanics, (2) spherical or freely rotating molecules, (3) intramolecular vibration same in liquid and gas, (4) potential energy depending only on intermolecular distances, and (5) universal shape of potential curve.

Since we were unable to calculate the perfect behavior theoretically we shall not expect to derive exact expressions for the deviation in the various cases. The discussion of these deviations is further complicated by the fact that whenever a source of imperfection exists, it usually causes all the properties of the liquid to deviate, so that none are left to specify the proper conditions for comparison. The best that can be done is to select some properties that are relatively sensitive to imperfection and insensitive to temperature, and compare these properties at points chosen by the use of properties of the opposite character. Also, it is possible to be consistent, which will assure that the more imperfect substances will deviate more.

Since the entropy of a gas depends particularly on its volume, and because imperfection cannot change the liquid volume very greatly, it seems best to compare entropies of vaporization at an equal vapor to liquid volume ratio. This basis is, moreover, a very convenient one. The change in heat capacity with vaporization is relatively insensitive to the temperature, and will be taken at the same point. Unless it is stated otherwise, the volume ratio 335 will be used.

These two properties (heat capacity and entropy of vaporization) will receive most attention in the following discussions. Since either may be

TABLE II. *The behavior of the perfect liquid.*

$\theta=T/T_c$	$-\text{LOG }\theta$	$-\text{LOG }\pi$ ($\pi=p/p_c$)	ΔS_{VAP}	$C_{S(\text{TR})}$	ϕ LIQ ($\phi=V/V_c$)	$\frac{V_{\text{GAS}}}{V_{\text{LIQ}}}=\frac{d_{\text{LIQ}}}{d_{\text{GAS}}}$
0.56	0.25181	1.8291	18.68	10.7	0.376	335
.57	0.24413	1.7542	18.20	10.7	0.379	290
.58	0.23657	1.6825	17.72	10.7	0.382	250
.59	0.22915	1.6134	17.25	10.7	0.385	215
.60	0.22185	1.5469	16.82	10.7	0.387	187
.62	0.20761	1.4213	15.95		0.392	142
.64	0.19382	1.3044	15.18		0.398	110
.66	0.18046	1.1953	14.42		0.404	84
.68	0.16749	1.0933	13.73		0.410	67
.70	0.15490	0.9974	13.11		0.417	53
.75	0.12494	0.7808			0.437	31
.80	0.09691	0.5912			0.461	19
.85	0.07058	0.4224			0.490	12
.90	0.04576	0.2697			0.531	7.3
.95	0.02228	0.1298			0.596	4.2
1.00	0.00000	0.0000			1.000	1.0

$p_cV_c/T_c=23.9$ cc atmos. per degree Kelvin.

TABLE III. *Data on the freezing of perfect liquids.**

SUBSTANCE	ΔS_{FUSION} CAL./DEG.	$\theta = T/T_c$ T.P.
Argon	3.35	0.557
Krypton	3.37	0.554
Xenon	3.40	0.557

* See footnote to Table I.

taken as a measure of imperfection, a check of one against the other will offer a check on the explanations offered. In Fig. 5 deviations in entropy are plotted against deviations in heat capacity for several substances of various types. An abnormally large entropy of vaporization, and an abnormally large liquid heat capacity are indicated as positive deviations, respectively.

Quantum-mechanical effects

The failure of neon to follow perfect behavior exactly has already been ascribed to quantum-mechanical effects. The deviation of hydrogen and helium are of course even more marked. As pointed out by Byk,[10] the following parameter can be said to measure the extent of quantum deviations: $(mkT)^{\frac{1}{2}}V^{\frac{1}{3}}/h$. An equivalent quantity which is convenient for our purposes is $M^{\frac{1}{2}}T_c^{\frac{1}{2}}V_c^{\frac{1}{3}}$. Table IV gives values of this parameter for a representative group of substances.

Some of the substances in Table IV are, of course, imperfect for other reasons than quantum-mechanical deviations. They are included now so that an estimate may be made of the quantum mechanical contribution to their total imperfection.

While, as stated above, we shall not expect to predict the exact deviations caused by quantum effects, some approximate arguments can be given. In many respects the nature of a liquid and solid are similar. It has been commonly assumed in treating liquids that most of the time a molecule vibrates within a cage of neighboring molecules in about the same manner as in the solid. Thus if the ability of molecules in the liquid to occasionally jump from cage to cage is of no particular significance to the property of interest, the difference between liquid and solid may be slight. With this idea in mind, we shall take the well-known Debye functions for a somewhat idealized solid, and apply them to the liquid. The last two columns of Table IV were obtained in this manner, by taking the -1.8 cal. per degree deviation in the heat capacity of neon as a standard. The changes in C_p and S are in opposite directions. The heat capacity becomes too small and the entropy of the liquid becomes too large. This leaves the entropy of vaporization too small.

The liquid volume is also affected similarly. Quantum-mechanical effects cause the volume to remain too large and the coefficient of expansion to become too small. This, of course, changes the gas to liquid volume ratio which was taken as a standard for comparing entropies. The liquid volume of neon is about three percent too large at its triple point, and this change affects the entropy by less than 0.1 cal. per degree, which is almost negligible. The experimental entropy of vaporization of neon deviates from the perfect by 0.7 cal. per degree as compared to the calculated 1.0, which is probably as good agreement as can be expected.

Hydrogen and helium are so much affected by quantum considerations that it is not worth while to discuss them in detail here.

TABLE IV. *Quantum-mechanical effects.*

SUBSTANCE	$(M^{\frac{1}{2}}T_c^{\frac{1}{2}}V_c^{\frac{1}{3}})$	$-\delta C_{\text{DEBYE}}$	δS_{DEBYE}
Helium	17.6		
Hydrogen	32.6		
Neon	104	(1.8)	1.0
Methane	256	0.4	0.2
Nitrogen	266	0.3	0.2
Oxygen	295	0.3	0.1
Argon	328	0.2	0.1
Ethane	491	0.1	0.1
Lithium	500	0.1	0.1
Krypton	600	0.1	0.0
Xenon	940	0.0	0.0
Mercury	2000	0.0	0.0

Restriction of rotation

Turning now to substances whose principal imperfection arises from a restriction of rotation of the molecules in the liquid, we shall again present only a rough theoretical consideration. Thermodynamic functions for restricted rotational degrees of freedom are available[11] for the case of one-dimensional rotation. The results for

[10] Byk, Ann. d. Physik **66**, 167 (1921); **69**, 161 (1922).

[11] K. S. Pitzer, J. Chem. Phys. **5**, 469 (1937).

the two- and three-dimensional cases are probably similar within the degree of accuracy that a simple assumed potential function fits the case at hand. The curves rising to the right in Fig. 5 were drawn by taking two or three times the deviation for a single degree of freedom, respectively.

The points on Fig. 5 for such substances as the hydrogen halides, nitrogen, carbon monoxide, carbon disulfide, chlorine and bromine fit the curve for two degrees of restricted rotation reasonably well. Similarly, phosphine, hydrogen sulfide, ethane ethylene and tetramethyl methane fit the curve for three degrees of restricted rotation satisfactorily.

It is noteworthy that the substances in which restriction arises from dipolar effects fall in Fig. 5 along with those in which restriction arises from nonspherical shape.

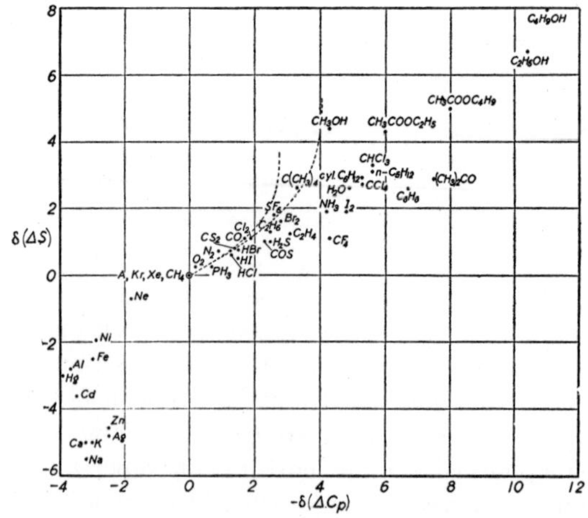

FIG. 5. The behavior of imperfect liquids. The deviations from the perfect entropy of vaporization are plotted against the derivations from perfect change of heat capacity with vaporization. The dotted lines show the expected position of points if the only imperfection is restricted rotation of molecules in the liquid. The left curve is for two-dimensional rotation, the right curve, for three. The points for all perfect substances lie at the origin.

Change of intramolecular vibrations

It is apparent that many of the points on the right side of Fig. 5 fall below and to the right of the curves for restricted rotation. This is explicable in a general way when the numerous intramolecular motions of complex molecules are considered. While intermolecular forces have little effect on the principal valence bond vibrations, these forces may seriously change the torsional motion about bonds and the bending of long chains. On the whole, the frequencies of these vibrations should be lowered in the liquid, and this will lead to a higher liquid heat capacity (larger $-(\Delta C_p)$) and a larger liquid entropy (smaller (ΔS)).

This effect is particularly marked for long chain molecules, also in *n*-pentane as compared to neo-pentane (tetramethyl methane).

At this point a few remarks seem in order concerning the very normal position taken by water in Fig. 5. The gas to liquid volume ratio becomes 335 at slightly above 150°C and at this temperature most of the molecules are probably rotating. The observed normal behavior is then to be expected. It seems that the small molal volume of water as compared to that of most substances boiling near 100°C (18 cc compared to about 100 cc) led to the view that water was very abnormal.

The position O_2 in Fig. 5 suggests almost perfect behavior for this paramagnetic molecule. This view would be confirmed by the volume and coefficient of expansion behavior. However, the melting point of O_2 is much too low. In this respect it deviates from perfect behavior much more than N_2, CO, HI, and other similar substances. Thus the apparent perfection of O_2 is probably to be regarded as an accidental cancellation of deviations in different directions.

Metals

While it may be doubtful which assumptions the metals violate, in any case they form a group and should be considered together. Experimentally, their entropies of vaporization are much less than perfect and vary over a considerable range. The decrease of heat capacity with vaporization is also less than perfect, but is about the same for most metals. Another notable "imperfection" of the metals is their relatively low melting points. Many melt at temperatures less than half that corresponding to the perfect behavior. The coefficient of expansion is considerably smaller for metals than for chemically saturated substances.

Table IV shows that even lithium is essentially unaffected by quantum considerations, and it seems unlikely that lack of spherical symmetry can be an important consideration in many cases. On the other hand, the possibility of thermal excitation of electronic motion may be important. Also, metallic binding forces are quite different from the London-van der Waals forces of the saturated molecules. The assumption that the potential energy can be expressed in terms of intermolecular distances alone is no longer valid.

A recent paper by Pauling[12] presents a most interesting and enlightening view with respect to metallic binding. He considers electron pair bonds of the ordinary type to be resonating among a large number of possible positions. Thus, when a given atom has as many neighbors as it has electrons to share, the full number of bonds can form. Further, neighbors lead to decrease in energy due to resonance, but this effect will probably soon show a saturation. In other words, a given atom has a greater affinity for its first few neighbors than for the eleventh or twelfth. Since arrangements with somewhat less than twelve coordination probably afford more freedom of motion and consequently greater entropy,[13] they will be assumed at the higher temperatures. Also, because of this saturation of binding energy, they will be assumed sooner in the case of the metal than in the perfect substance. Thus the low melting points of metals and the larger liquid entropies (lower ΔS vap.) are to be expected.

The relative entropies within the metals are in general accord with this idea. Thus iron and nickel, to which Pauling attributes about six bonding electrons, are more nearly perfect than the alkali metals with but one. On the other hand, zinc, cadmium and mercury with similar electronic structures show considerable variation.

The entropy attributable to electronic motion through the metal will be removed by vaporization, and should therefore reduce the total entropy of vaporization. The magnitude of this effect should be about one cal. per degree or less[14] and is alone insufficient to explain the observed results. It undoubtedly contributes, however.

While it seems likely that a single function φ will be quite general for van der Waals forces, there is no reason to believe that this same function will be satisfactory for metallic binding. In particular, if the function for metals were more symmetrical near the minimum, the smaller coefficient of expansion and smaller liquid heat capacity would be explained.

On the whole it seems as if the nature of liquid metals will have to be treated as an almost independent problem.

Before concluding, it seems desirable to discuss briefly the Hildebrand rule[2] for entropies of vaporization. The great success of this rule together with the encouragement of its author led to this investigation. In the light of the above results, the selection of a constant molal gas volume rather than a constant pressure is completely justified. On the other hand, it appears that one should actually take a constant gas to liquid volume ratio as the criterion. Many simple liquids boiling somewhat above room temperature, have about the same molal volumes, and in such cases the Hildebrand rule will be essentially the same as the one here proposed. It was such a group of substances that Professor Hildebrand recently discussed.[7]

In conclusion, it is hoped that the ideas of Part I will serve as a foundation for relatively precise discussions of the behavior of actual liquids. The second part of this paper has merely outlined some more obvious principles. Further experimental investigations of the properties of the perfect substances would also seem to be very worth while.

[12] L. Pauling, Phys. Rev. **54**, 899 (1938).

[13] O. K. Rice, J. Chem. Phys. **7**, 136 (1939).

[14] W. M. Latimer, J. Am. Chem. Soc. **44**, 2136 (1922).

[Reprinted from the Journal of the American Chemical Society, **77**, 3427, 3433 (1955).]

[CONTRIBUTION FROM THE DEPARTMENT OF CHEMISTRY AND CHEMICAL ENGINEERING, UNIVERSITY OF CALIFORNIA, BERKELEY]

The Volumetric and Thermodynamic Properties of Fluids. I. Theoretical Basis and Virial Coefficients[1]

BY KENNETH S. PITZER

RECEIVED JANUARY 21, 1955

The basic theory of the volumetric and thermodynamic properties of fluids is examined in relation to the statistical mechanical proof of the theory of corresponding states. It is shown that the general range of non-polar or slightly polar "normal" liquids cannot be expected to conform to the hypothesis of corresponding states but may well conform to a slightly more complex equation involving just one additional parameter. Rigorous theoretical test can be made only for the second virial coefficient. There it is found that linear and globular non-polar molecules fit this theory very exactly and that slightly polar molecules also conform in reasonable approximation.

The quantitative representation of the volumetric behavior of fluids over both gas and liquid regions has proven to be an unusually difficult problem. It is now over eighty years since van der Waals[2] first proposed his famous equation of state. Yet today we are still in serious need of improvements in the prediction and correlation of the volumetric properties of fluids. This situation is in marked contrast to the extensive theoretical advances with respect to the thermodynamic properties of ideal gases.

Our understanding of the forces operating between molecules has developed rapidly since the emergence of a satisfactory quantum theory in 1926. Particularly important is the work of London on the attractive forces between molecules. Also it became clear that certain anomalous properties of hydrogen and helium arise from quantum effects in the translational motion of those molecules. However, our understanding of intermolecular forces also indicated a bewildering complexity of effects which probably tended to discourage attempts at systemization.

The most generally useful method of prediction of the volumetric properties of fluids is the hypothesis of *corresponding states*, which also came originally from van der Waals. Engineers have used this method extensively to obtain estimated properties for design purposes, and many authors have presented charts of both volumetric and related thermodynamic properties on the basis of corresponding states. Hougen and Watson[3] give a particularly extensive discussion of this method, together with estimates of the accuracy to be expected. They indicate that the maximum error in predicted volume is about 15% while errors in other predicted properties range from 5 to 35%. These percentages indicate the degree to which actual substances conform to the hypothesis of corresponding states.

It is the purpose of this series of papers to present a correlative and predictive scheme which is only slightly more complex but which yields results about one order of magnitude more accurate. This scheme involves one parameter in addition to the critical constants for each substance. It is postulated that any group of substances having the same value of this additional parameter will conform within the group to the corresponding states principle.

The possibility of a scheme of this type has been mentioned on several past occasions. Nernst suggested it in the 1907 edition of "Theoretischen Chemie"[4] but dropped it in later editions. Recently Meissner and Seferian[5] discussed such a scheme, but we shall see that their attempt to include highly polar substances such as water necessarily impaired the accuracy of the results. However, Riedel[6] has discussed vapor pressures on a basis very similar to that which we have selected.

(1) This research was a part of the program of Research Project 50 of the American Petroleum Institute.

(2) J. H. van der Waals, Dissertation, Leiden, 1873.

(3) O. A. Hougen and K. M. Watson, "Chemical Process Principles, Part Two, Thermodynamics," John Wiley and Sons, Inc., New York, N. Y., 1947, Chapter VII.

(4) See quotation in ref. 6.

(5) H. P. Meissner and R. Seferian, *Chem. Eng. Progress*, **47**, 579 (1951).

(6) L. Riedel, *Chem. Ing. Tech.*, **26**, 83 (1954).

He excluded highly polar or associated liquids and the quantum liquids (H_2 and He) and showed that the remaining *normal* liquids had reduced vapor pressure curves falling accurately into a single family. We shall return to Riedel's work in a later paper when other properties of the two phase region are discussed. Rowlinson[7] has also presented certain theoretical work which yields a correlation scheme similar to the present one.

In this paper we present the theoretical background in terms of modern molecular data and discuss theoretical second virial coefficients calculated from appropriate models.

Classification of Substances; Intermolecular Forces

Simple Fluids.—Our present theoretical knowledge is of great aid in selecting the substances which may reasonably be expected to conform to a particular equation. The theoretical requirements for conformity to the postulate of corresponding states have been given in several places including a paper by the present author[8] which includes a more detailed presentation of several topics essential to the present development. The simplest class of substances following corresponding states behavior comprises the heavier rare gases (A, Kr, Xe) and methane. The principal characteristics are *spherical shape* (which methane gains by rapid and relatively free rotation) and an *inverse sixth power attractive potential.* London's theory yields this force law and applies to cases where the valence shells of all atoms are filled. We shall adopt the term *simple fluid*[9] for this class of substances.

Since the intermolecular potential curve is of primary importance, it is wise to discuss it a little further. We distinguish between time average or permanent electric moments (the usual dipole moments) and instantaneous moments arising from the momentary location of the electrons. The inverse sixth power attractive term arises from the interaction of instantaneous dipole moments. Higher power terms arise from higher electrical moments, but these are small and their effect may be absorbed in our ignorance of the repulsive potential. We know that molecules repel one another on close approach, that this repulsion arises very suddenly, and that it is a manifestation of the Pauli exclusion principle. However, we do not have a good simple mathematical expression for this repulsive effect. While exponential terms are believed to be somewhat superior expressions of this repulsive potential, inverse power terms are much easier to use and have proven almost as satisfactory. The best compromise of accuracy and simplicity is probably the Lennard-Jones or 6–12 potential given by the expression

$$\epsilon(r) = \epsilon_0 \left[\left(\frac{r_0}{r} \right)^{12} - 2 \left(\frac{r_0}{r} \right)^{6} \right] \quad (1)$$

where ϵ_0 is the depth of the potential minimum, r is the intermolecular distance, and r_0 its value at the minimum. We shall also define σ as the collision diameter which is the point of zero energy on the repulsive branch of the curve. Its value is

$$\sigma = (2^{-1/6})r_0$$

The curve of equation 1 is shown as the solid line in Fig. 1.

Another important characteristic of London forces is their additivity. To a first approximation, equation 1 may be applied to each intermolecular distance in a large assemblage of molecules. The accuracy of this approximation has been studied by Axilrod.[10]

We have assumed a simple fluid to be one having an intermolecular potential essentially like equation 1. Each substance has its own value of ϵ_0 and r_0. These are related to the critical temperature and critical volume, respectively. But the shape of the curve is the same for all simple fluids.

An additional limitation on the simple fluid is the absence of quantum effects for molecular translation. Helium and hydrogen must be excluded on this account. Neon also deviates at low temperatures from the simple fluid behavior for this reason. Since we have already named all of the substances likely to deviate because of quantization of molecular translation, there is little need to extend general correlation schemes to cover this area. The volumetric behavior of these substances is known and does not need to be predicted from a practical point of view. We shall ignore quantum fluids hereafter and turn our attention to other types.

Globular Molecules.—Next let us consider large molecules of nearly spherical shape, globular molecules. Neopentane, $C(CH_3)_4$, is a simple example. The attractive forces are still of the London type but now the effective centers are the peripheral atoms or the methyl groups. The intermolecular potential may be taken as the sum of inverse sixth power terms summed over the various distances between methyl groups and is no longer an inverse sixth power potential between molecular centers. This effect is probably responsible for most of the deviation of neopentane from simple fluid behavior, but there is also probably some deviation from spherical shape.

Figure 1 shows the effect of this globular character on the intermolecular potential curve. Two approximations are given. The dashed line is based on the core model of Kihara.[11] He assumes a core inside each molecule and then takes the Lennard-Jones potential for the shortest distance between molecular cores. The potential is thus

$$\epsilon = \epsilon_0 \left[\left(\frac{\rho_0}{\rho} \right)^{12} - 2 \left(\frac{\rho_0}{\rho} \right)^{6} \right] \quad (2)$$

where ρ is the shortest distance between the cores, ρ_0, this distance at the potential minimum, and a is the radius of the spherical core. In the present case of spherical cores, $\rho = r - 2a$. For the curve in Fig. 1 we take a spherical core of diameter equal to one fourth of the intermolecular distance at the potential minimum, *i.e.*, $a = 0.125r_0 = 0.140\rho_0$.

Since attractive effects between the more distant parts of the molecule are not negligible, al-

(7) J. S. Rowlinson, *Trans. Faraday Soc.*, **50**, 647 (1954).

(8) K. S. Pitzer, *J. Chem. Phys.*, **7**, 583 (1939). The symbols E, A and R_0 of that paper become ϵ, ϵ_0 and r_0 in the present text.

(9) In the earlier paper[8] the term perfect liquid was used. Extension of the treatment to the imperfect gas region makes a change obviously desirable.

(10) B. M. Axilrod, *J. Chem. Phys.*, **19**, 719, 724 (1951).

(11) T. Kihara, *Rev. Mod. Phys.*, **25**, 831 (1953); and earlier papers there cited; see also J. S. Rowlinson, *J. Chem. Phys.*, **20**, 337 (1952).

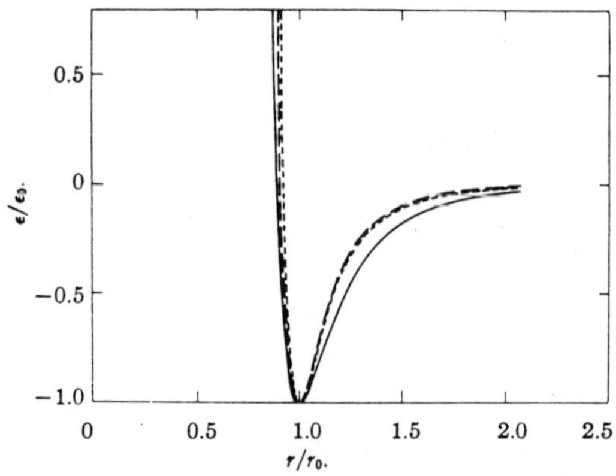

Fig. 1.—The intermolecular potentials for globular molecules; see text for details.

though repulsive effects presumably are, we also investigate additional models. Repulsion will be taken as the inverse twelfth power of the minimum distance between the surfaces of spherical cores, like the Kihara model. The attractive force will be taken as arising from centers distributed equally on the surface or through the volume of the spherical core. By expanding the integral for the attractive potential in a Maclaurin series in the radius of the core, a, the following expressions result. They may be arranged as

$$V(r) = -cr^{-6}\psi(a/r) \quad (3)$$

where the surface model yields

$$\psi_s = 1 + 10\left(\frac{a}{r}\right)^2 + \left(\frac{148}{3}\right)\left(\frac{a}{r}\right)^4 + 480\left(\frac{a}{r}\right)^6 \quad (4)$$

and the volume model yields

$$\psi_v = 1 + 6\left(\frac{a}{r}\right)^2 + \left(\frac{148}{7}\right)\left(\frac{a}{r}\right)^4 + 160\left(\frac{a}{r}\right)^6 \quad (5)$$

The complete potential becomes

$$\epsilon = \frac{c_1}{(r-2a)^{12}} - \frac{c_2}{r^6}\psi(a/r) \quad (6)$$

where c_1 and c_2 are constants which can be evaluated in terms of ϵ_0 and r_0 for any particular value of a/r_0.

In the important region near the potential minimum these two models give similar results provided the core radius of the surface model is taken as about 0.8 of that for the volume model. The dotted curve in Fig. 1 shows the potential for the surface model with $a = 0.2r_0$. The nearly equivalent volume model would have $a = 0.25r_0$. It is apparent from Fig. 1 that the Kihara model with $a = 0.125r_0$ is very similar in its net effect—a little steeper outside the minimum and a little less steep inside. Thus there seems little need to pursue these additional models further except to note that the Kihara core radius should be taken considerably smaller than that of the polarizable electron cloud.

The important effect of the globular character of a molecule on its potential function, as is evident from any of the models, is a narrowing of the potential minimum. This decreases the probability of any given clustering of molecules at a given ratio, ϵ_0/kT, of the intermolecular potential energy to thermal energy. In statistical mechanics probabilities are proportional to volumes in phase space, and in this case the important dimension is the range of intermolecular distance within the potential curve at a height approximately kT above the minimum. Likewise decreased probabilities lead to decreased entropies. Thus we expect the decrease in entropy from the ideal gas state to some comparable state in the imperfect gas or liquid region to be greater for globular molecules than for those of simple fluids.

Since the liquid state represents the essentially complete clustering of the molecules, the vapor–liquid equilibrium should give the most marked manifestation of the effect just mentioned. With a greater decrease in entropy on condensation for globular molecules, the temperature must be lower, relatively, before condensation takes place. Also the change in vapor pressure with temperature will be more rapid for substances with globular molecules than for simple fluids. This is shown in Fig. 2. Eventually, we shall take this increase in slope of the vapor pressure curve over that of a simple fluid as the essential third parameter of our theory.

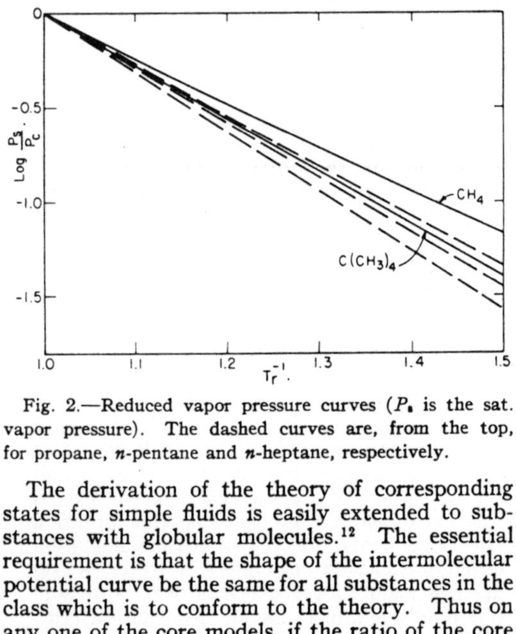

Fig. 2.—Reduced vapor pressure curves (P_s is the sat. vapor pressure). The dashed curves are, from the top, for propane, *n*-pentane and *n*-heptane, respectively.

The derivation of the theory of corresponding states for simple fluids is easily extended to substances with globular molecules.[12] The essential requirement is that the shape of the intermolecular potential curve be the same for all substances in the class which is to conform to the theory. Thus on any one of the core models, if the ratio of the core radius to the intermolecular distance at the potential minimum (a/r_0) is the same for a group of substances, then that group will follow corresponding states behavior. The behavior of this group will differ, however, from that of simple fluids where the core radius is zero.

Alternatively, we may say that the volumetric behavior of a fluid with globular molecules is a function of three variables: (1) the depth of the potential minimum ϵ_0, (2) the intermolecular distance at the minimum r_0, and (3) the relative core size a/r_0. Simple fluids form the special case with $a/r_0 = 0$. In terms of macroscopic properties one

(12) See ref. 8. There is now a different function φ but the proof is exactly the same.

may identify the critical temperature, the critical volume and the surplus slope of the vapor pressure curve (over that of a simple fluid) as yielding the equivalent information.

Our general correlation scheme is essentially that defined in the preceding paragraph. However, to be useful it must apply to a broader class of substances than have been considered so far.

Non-spherical, Non-polar Molecules.—Let us turn now to non-polar molecules of irregular shape. Kihara[11] has extended his core model to such cases by taking cores of non-spherical shape. We shall show later that the first-order effect of the introduction of a core on the second virial coefficient is the same for all shapes of core. This arises because the essential effect is a narrowing of the potential minimum even though it must now be studied in multidimensional space.

The Kihara model assumes the same minimum energy for a pair of molecules at their optimum separation regardless of their orientation. This is probably an oversimplification. Thus it seems likely that two long molecules exert greater attraction when lying side by side at the optimum distance apart than when they are oriented end to end. However, the introduction of this effect will again merely further narrow the potential curve in this generalized sense. In other words, the minimum energy is attainable only for optimum orientation as well as linear separation. Consequently, the volume in multidimensional configuration space just above the minimum (by the order of kT) is relatively smaller than it would be for a spherical molecule.

It is not pretended that the exact effect of non-spherical shape is the same for every shape or the same as for globular character. However, the macroscopic properties are calculated by averaging appropriate functions over all space and are frequently insensitive to details. Consequently it seems likely that the behavior of a substance with non-polar molecules of non-spherical shape will conform closely to that of globular molecules with some value of the (a/r_0) parameter. This postulate will be tested in the comparisons to follow.

Polar Molecules.—The interaction potential of permanent dipoles of fixed orientation varies as the inverse third power of the distance between their centers. Also it is highly dependent on their orientation—indeed it is repulsive for half of the possible solid angle. Consequently, it is apparent that a cluster of non-rotating dipolar molecules will have properties significantly different from those of non-polar molecules.

If the dipolar effects are small compared to thermal energies, then one does obtain an inverse sixth power interaction potential because of a partial excess of favorable orientations. The formula for this "orientation effect" is

$$\epsilon_\mu = -\frac{2}{3}\frac{\mu^4}{kTr^6} + \ldots \tag{7}$$

where μ is the dipole moment.

For many substances the interaction of their permanent dipole moments constitutes only a small perturbation to the London force already present. Examples include H_2S, CH_3Cl and HCl. Even in HCl the calculations indicate that the orientation effect constitutes less than 20% of the total interaction energy at room temperature.[13] In such cases the effect of the permanent dipole on the intermolecular potential function is again that of a narrowing. Just as for non-spherical, non-polar molecules, the lowest energy arises only in the most favorable angular orientation. Thus the relative volume in a general configuration space just above the potential minimum will be reduced in the polar case as compared to that for a spherical non-polar molecule. We will leave it to later comparisons to show the degree of similarity in the effect on volumetric properties of small dipole moments and of non-polar types of deviation (*i.e.*, globular or non-spherical).

Virial Coefficients

While it is impractical at the moment to carry our general theoretical studies much further, the second virial coefficient has been calculated for several of these models. We shall draw chiefly upon the work of Kihara[11] on his core models and that of Hirshfelder, McClure and Weeks[14] and that of Rowlinson[15] on molecules with permanent dipoles in addition to a Lennard-Jones 6–12 potential. The latter potential is

$$\epsilon = \epsilon_0\left[\left(\frac{r_0}{r}\right)^{12} - 2\left(\frac{r_0}{r}\right)^6\right] - \frac{\mu^2}{r^3}(2\cos\theta_1\cos\theta_2 - \sin\theta_1\sin\theta_2\cos\varphi) \tag{8}$$

Here it must be remembered that ϵ_0 and r_0 have lost the precise significance they had in the non-polar case; also θ_1 and θ_2 are the angles the dipoles make with the axis connecting them while φ is the difference in their azimuthal angle.

Kihara's final formula[11] for all of his core models gives for the second virial coefficient

$$B = \frac{2\pi}{3}N_0\rho_0^3F_3(Z) + M_0\rho_0^2F_2(Z) + \left(S_0 + \frac{M_0^2}{4\pi}\right)F_1(Z) + \left(V_0 + \frac{M_0S_0}{4\pi}\right) \tag{9}$$

where $Z = (\epsilon_0/kT)$, ρ_0 is the core separation for minimum energy, see eq. 2, and M_0, S_0 and V_0 are the mean curvature, the surface area and the volume of the core, respectively. He gives expressions for many shapes of cores. We shall use only the two extreme shapes, as follows

thin rod of length l:

$$M_0 = \pi l,\ S_0 = V_0 = 0$$

sphere of radius a:

$$M_0 = 4\pi a,\ S_0 = 4\pi a^2,\ V_0 = 4\pi a^3/3$$

For these two cases the virial coefficient formulas become

Thin rod:

$$B = \frac{2\pi}{3}\rho_0^3\left[F_3(Z) + \frac{3}{2}\left(\frac{l}{\rho_0}\right)F_2(Z)\right] \tag{10a}$$

Sphere:

$$B = \frac{2\pi}{3}\rho_0^3\left[F_3(Z) + 6\left(\frac{a}{\rho_0}\right)F_2(z) + 12\left(\frac{a}{\rho_0}\right)^2F_1(z) + 8\left(\frac{a}{\rho_0}\right)^3\right] \tag{10b}$$

(13) F. London, *Trans. Faraday Soc.*, **33**, 8 (1937).

(14) J. O. Hirshfelder, F. T. McClure and I. F. Weeks, *J. Chem. Phys.*, **10**, 201 (1942). This paper uses r_0 for the collision diameter which is here designated σ.

(15) J. S. Rowlinson, *Trans. Faraday Soc.*, **45**, 974 (1949).

In the absence of a core, ρ_0 becomes r_0, and F_3 gives the virial coefficient for a Lennard-Jones gas which is our "simple fluid." It is evident that, for small values of l or a, the effect in each case is to introduce a small component of F_2. All shapes of core give this same result for the first order deviation

$$B = \frac{2\pi}{3} N_0\rho_0^3 \left[F_3(Z) + xF_2(Z)\right] \quad (11)$$

Turning now to the permanent dipole molecules, the second virial coefficient is[14]

$$B = \frac{2\pi}{3} N_0 r_0^3 \left[F_3(Z) + yH(Z) + y^2J(Z) \ldots\right] \quad (12)$$

where[16] r_0 is the distance at the minimum in the non-polar part of the potential function, see equation 8, and $y = \mu^4/\epsilon_0^2 r_0^6$ is the parameter which measures the relative importance of the dipole. This is a series and additional terms are available but are not needed for our purposes.

For ease in our calculations we fitted third-order equations to the tabulated values of the virial coefficient functions over the range from the lowest useful temperatures to about four times the critical temperature. Our four-constant expressions for the various functions are

$$F_3 = 0.5188 - 1.6223Z - 0.5235Z^2 - 0.1684Z^3 \quad (13a)$$
$$F_2 = 0.6426 - 0.7237Z - 0.3677Z^2 - 0.0936Z^3 \quad (13b)$$
$$F_1 = 0.8007 - 0.2443Z - 0.1993Z^2 - 0.0372Z^3 \quad (13c)$$
$$H = 0.0168 - 0.165Z - 0.119Z^2 - 0.630Z^3 \quad (13d)$$
$$J = 0.0135 - 0.119Z + 0.319Z^2 - 0.302Z^3 \quad (13e)$$

No unusual effort was made to maximize the accuracy of these expressions, since the original functions are available for precision work. It is believed that they are accurate to at least 1% or 0.001 (whichever is larger) in the range $0.2 < Z < 2$ for F_1, F_2 and F_3 and $0.2 < Z < 1$ for H and J. Rowlinson[15] gives tables for dipolar molecules in the range $Z > 1$.

Our interest is to determine the extent to which shape or dipole factors cause deviation from the postulate of corresponding states. Since the critical properties have not been calculated for these models we shall use the Boyle point as the reference point for reduced variables. However, to avoid possible confusion with the usual reduced variables we shall always write the explicit ratios, T/T_B, Z/Z_B, B/v_B, etc. We take the reference volume to be

$$v_B = \left(T\frac{\partial B}{\partial T}\right)_{T=T_B} = -\left(Z\frac{\partial B}{\partial Z}\right)_{Z=Z_B} \quad (14)$$

in accordance with Kihara.

Since the addition of functions F_2 and H in equations 11 and 12 affects the reference points, the final reduced equation had to be corrected for this effect. To the first order, the result is

$$\frac{B}{v_B} = \left[\frac{B}{v_B}\right]_0 + x\left[\frac{\partial}{\partial x}\left(\frac{B}{v_B}\right)\right]_0 + y\left[\frac{\partial}{\partial y}\left(\frac{B}{v_B}\right)\right]_0 \quad (15)$$

$$\left[\frac{B}{v_B}\right]_0 = 0.908 - 0.824\left(\frac{T_B}{T}\right) - 0.0771\left(\frac{T_B}{T}\right)^2 - 0.00720\left(\frac{T_B}{T}\right)^3 \quad (15a)$$

$$\left[\frac{\partial}{\partial x}\left(\frac{B}{v_B}\right)\right]_0 = -0.079 + 0.148\left(\frac{T_B}{T}\right) - 0.0597\left(\frac{T_B}{T}\right)^2 - 0.0096\left(\frac{T_B}{T}\right)^3 \quad (15b)$$

$$\left[\frac{\partial}{\partial y}\left(\frac{B}{v_B}\right)\right]_0 = -0.0444 + 0.065\left(\frac{T_B}{T}\right) + 0.0039\left(\frac{T_B}{T}\right)^2 - 0.0244\left(\frac{T_B}{T}\right)^3 \quad (15c)$$

The effect of the reduction is to make each function zero at $T = T_B$. The slope of the base function is unity at this point while the slope of each derivative function is zero. The range of validity of these equations is for (T_B/T) from 0.7 to 6 for (15a and b) and to 3.5 for (15c).

While the derivative functions 15b and 15c do not look very much alike in equation form, it is apparent from Fig. 3 that they are actually quite similar. It is a coincidence, of course, that x and y have been defined in a manner that these two functions have similar magnitude as well as the same shape.

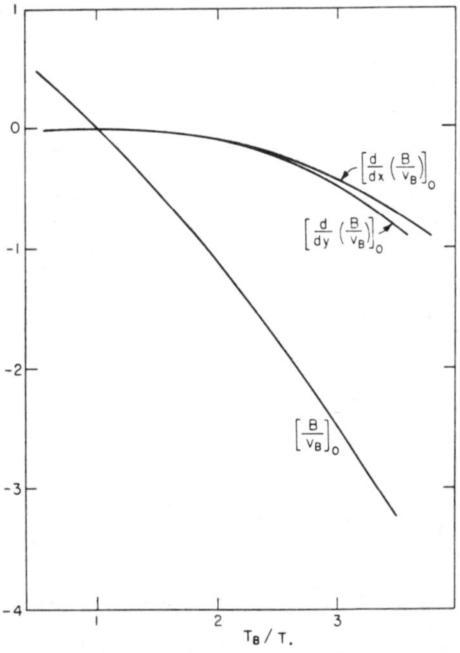

Fig. 3.—Reduced second virial coefficient functions.

Our basic correlation scheme as stated above is to involve just one variable in addition to the energy and distance parameters. For the moment we take this variable to be the x defined by equation 11 for the core model. Now we may determine the extent to which dipole effects may be substituted for core effects without exceeding some limit of error. Let us consider a limit such as 1%, which is a high accuracy for virial coefficients, and take the range of temperature above $T_B/3.5$. We find that the two effects are interchangeable within the limit of validity of the first-order equation 15. We turn, therefore, to a comparison of curves for finite values of the parameters for linear, spherical and dipolar molecules. Such a set of curves is shown in Fig. 4. A more sensitive comparison is given in Fig. 5

(16) Note that our functions are related to those of reference 14 as follows: $y = x/2$, $F_3 = F_{12}/\sqrt{2}$, $H = \sqrt{2}\,H_{11}$, $J = 2\sqrt{2}\,J_{12}$. These changes were made for consistency with Kihara's equations.

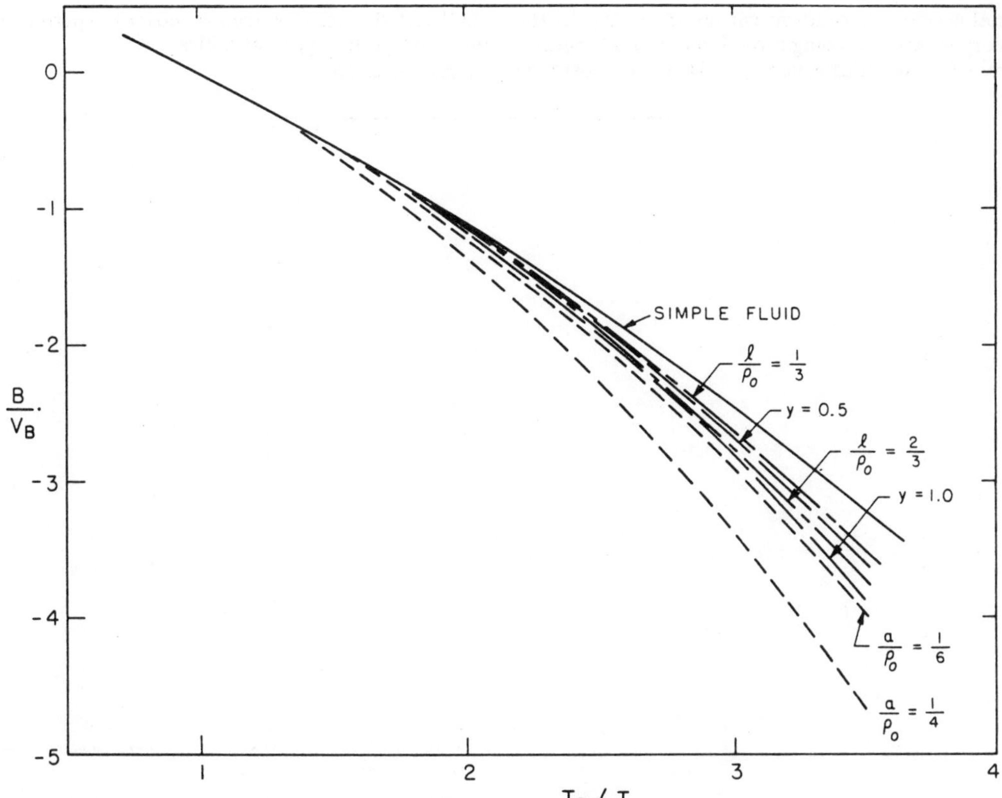

Fig. 4.—Reduced second virial coefficients for several models.

where the reduced virial at $T_B/T = 3.5$ is plotted against the value at $T_B/T = 2$ for various models. It is apparent that the points for linear and spherical core molecules fall on the same curve within very small limits. The points for dipolar molecules are on a definitely different curve. However, y can have values up to about 0.7 before a 1% difference is exceeded at each temperature.

The available experimental second virial coefficients were examined in terms of these effects. While the data generally conformed to these families of theoretical curves, the experimental accuracy is not high enough to allow precise determinations of the parameters related to the core or the dipole moment. Kihara assumed reasonable cores and obtained agreement with the experimental virial coefficients—a result consistent with this conclusion. For dipolar molecules meaningful values of the parameter y can be obtained by use of the dipole moment value in addition to the virial coefficient data provided the latter cover a wide enough temperature range. Rowlinson[15] has made such analyses in a number of cases. For example he finds for ammonia values yielding $y = 4$ which is far above the limit of 0.7 for conformity to our scheme within 1%. By contrast the y value for chloroform is only 0.04 and that for ethyl chloride is 0.16.

In conclusion, we have shown that for non-polar molecules of various shapes and for molecules with modest dipoles ($y \leq 0.7$), the theoretical second

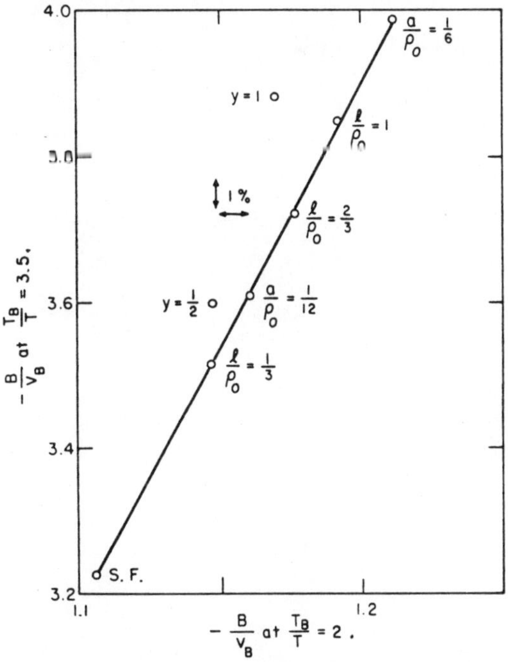

Fig. 5.—Deviation plot of the data of Fig. 4. Deviations from the simple fluid (S.F.) at $(T_B/T) = 3.5$ and 2, respectively, are compared.

virial coefficients conform within about 1% to the theory we are proposing for volumetric and thermodynamic properties generally. In later papers we shall test this scheme against various experimental data and present general tables.

BERKELEY, CALIF.

[CONTRIBUTION FROM THE DEPARTMENT OF CHEMISTRY AND CHEMICAL ENGINEERING, UNIVERSITY OF CALIFORNIA, BERKELEY]

The Volumetric and Thermodynamic Properties of Fluids. II. Compressibility Factor, Vapor Pressure and Entropy of Vaporization[1]

BY KENNETH S. PITZER, DAVID Z. LIPPMANN,[2] R. F. CURL, JR., CHARLES M. HUGGINS AND DONALD E. PETERSEN

RECEIVED JANUARY 21, 1955

The theoretical considerations of Part I suggested that the compressibility factor of a normal liquid in either gas or liquid state should be expressible as a function of just one parameter in addition to the reduced temperature and reduced pressure. The additional parameter is defined in terms of the vapor pressure at $T_r = 0.7$. This third parameter is required because the intermolecular force in complex molecules is a sum of interactions between various parts of the molecules—not just their centers—hence the name *acentric factor* is suggested. The theory requires that any group of substances with equal values of the acentric factor should conform among themselves to the principle of corresponding states. This result is verified with relatively high accuracy. While a completely analytical expression for the compressibility factor was not obtained, power series expressions in the acentric factor proved satisfactory and the coefficients are tabulated for a wide range of reduced temperature and pressure. The reduced vapor pressure and the entropy of vaporization are also treated similarly. Agreement is obtained to 0.5% over most regions with maximum deviations of about 2%.

A general introduction together with discussion of the theoretical basis for the correlation scheme has been presented in Part I of this series.[3] Although exact theory was available only for the second virial coefficient, those results together with general arguments indicated that a three parameter correlation might well yield at least a factor of ten greater accuracy than is obtained from the simple hypothesis of corresponding states. As was indicated in Part I, we take two critical constants for two of our three independent parameters for each substance. The critical temperature is readily selected as the first parameter which characterizes the intermolecular interaction energy.

While the critical volume would be the simple measure related to intermolecular distance, it is unsatisfactory from the empirical viewpoint. The differential compressibility is infinite at the critical point. Consequently the critical volume is not directly measureable with any accuracy. The values commonly given are extrapolated. The critical pressure is a much more accurately determinable quantity and it suffices just as well for correlation purposes. Hence we choose the critical pressure as our second parameter.

The third parameter is to measure the deviation of the intermolecular potential from that of a simple fluid. An important deviation arises from the fact that the sum of the inverse sixth power terms applying to the various portions of a pair of complex molecules cannot be replaced by a single inverse sixth power term in the distance between molecular centers. Since these forces between noncentral portions of the molecules must be considered, the term *acentric factor* is suggested.

The most convenient empirical quantity is the reduced vapor pressure at a point well removed from the critical point. This in effect gives the slope of the vapor pressure curve, see Fig. 2, Part I. For a *simple fluid*[3] (*e.g.*, A, Kr, Xe, CH_4) the reduced vapor pressure is almost precisely 0.1 at a reduced temperature of 0.7. This point is well removed from the critical yet above the melting point for almost all substances. Consequently it is convenient to take 0.7 as our standard value of reduced temperature for the determination of the *acentric factor* which we define as

$$\omega = -\log P_r - 1.000$$

with P_r the reduced vapor pressure (P/P_c) at $T_r = 0.7$.

The slope of the vapor pressure curve is, of course, closely related to the entropy of vaporization. Thus we may regard our acentric factor as a measure of the increase in the entropy of vaporization over that of a simple fluid. It was also shown in Part I that the acentric factor would depend upon the core radius of a globular molecule, the length of an elongated molecule, or the dipole moment of a slightly polar molecule.

Table I lists the essential parameters for the various substances which were given substantial consideration in our correlations. On the basis of the arguments in Part I it was not expected that the highly polar molecules, ammonia and water, would conform to our scheme. Points for these substances are included on some graphs to illustrate the magnitude of the deviations; however, these points were given no weight in preparing the final tables. The references to Table I include the sources of data for the respective substances for the other tables of this paper.[4]

Compressibility Factor.—The compressibility factor was interpolated graphically to even values

(1) This research was a part of the program of Research Project 50 of the American Petroleum Institute.

(2) A portion of this paper is abstracted from the Ph.D. Dissertation of David Z. Lippmann, University of California, 1953.

(3) K. S. Pitzer, THIS JOURNAL, **77**, 3427 (1955).

(4) In addition to the substances listed in Table I, supplementary use was made of data for *n*-hexane in certain areas. The sources are S. Young, *Scient. Proc. Roy. Dub. Soc.*, New Series **12**, 374 (1909–1910); E. A. Kelso with W. A. Felsing, *Ind. Eng. Chem.*, **34**, 161 (1942); E. A. Kelso with W. A. Felsing, THIS JOURNAL, **62**, 3132 (1940).

TABLE I

THE ACENTRIC FACTOR AND CRITICAL DATA

(P_r is reduced vapor pressure at $T_r = 0.7$)

Substance	$-\log P_r$	ω	T_c (°K.)	P_c (atm.)
Argon[a]	0.998	−0.002	150.72	48.00
Krypton[b]	0.998	− .002	209.39	54.182
Simple fluid	(1.000)	(.000)		
Xenon[c]	1.002	+ .002	289.75	57.636
Methane[d]	1.013	.013	190.66	45.795
Nitrogen[e]	1.040	.040	126.26	33.540
Hydrogen sulfide[f]	1.100	.100	373.55	88.868
Ethane[g]	1.105	.105	305.75	48.864
Propane[g]	1.152	.152	369.99	42.011
Neopentane[h]	1.195	.195	433.76	31.57
n-Butane[g]	1.201	.201	425.17	37.470
Benzene[i]	1.215	.215	562.66	48.664
Carbon dioxide[j]	1.225	.225	304.16	72.800
n-Pentane[g]	1.252	.252	470.60	33.628
Ammonia[k]	1.250	.250	406.00	112.31
Water[l]	1.348	.348	647.66	219.515
n-Heptane[m]	1.352	.352	539.94	26.882

[a] A. M. Clark, F. Din, J. Robb, A. Michels, T. Wassenaar and Th. Zwietering, *Physica*, **17**, 876 (1951); A. Michels, Hub. Wijker and Hk. Wijker, *ibid.*, **15**, 627 (1949); L. Holborn and J. Otto, *Z. Physik*, **33**, 1 (1925). [b] E. Mathias, C. A. Crommelin and J. J. Meihuizen, *Physica*, **4**, 1200 (1937); J. J. Meihuizen and C. A. Crommelin, *ibid.*, **4**, 1 (1937). [c] A. Michels and T. Wassenaar, *ibid.*, **16**, 253 (1950); J. A. Beattie, R. J. Barriault and J. S. Brierly, *J. Chem. Phys.*, **19**, 1219 (1951); H. W. Habgood and W. G. Schneider, *Can. J. Chem.* **32**, 98 (1954). [d] W. H. Corcoran, R. R. Bowles, B. H. Sage and W. N. Lacey, *Ind. Eng. Chem.*, **37**, 825 (1945); A. Stock, F. Henning and E. Kuss, *Ber.*, **54**, 1119 (1921); A. Michels and G. W. Nederbragt, *Physica*, **2**, 1000 (1935); A. Michels and G. W. Nederbragt, *ibid.*, **3**, 569 (1936); R. H. Olds, H. H. Reamer, B. H. Sage and W. N. Lacey, *Ind. Eng. Chem.*, **35**, 922 (1943). [e] A. S. Friedman and D. White, THIS JOURNAL, **72**, 3931 (1950); O. T. Bloomer and K. N. Rao, "Thermodynamic Properties of Nitrogen," Institute of Gas Technology, Technology Center, Chicago (1952). [f] H. H. Reamer, B. H. Sage and W. N. Lacey, *Ind. Eng. Chem.*, **42**, 140 (1950). [g] B. H. Sage and W. N. Lacey, "Thermodynamic Properties of the Higher Paraffin Hydrocarbons and Nitrogen," American Petroleum Institute, New York, N. Y., 1950; B. H. Sage, D. C. Webster and W. N. Lacey, *Ind. Eng. Chem.*, **29**, 658 (1937). [h] J. A. Beattie, D. R. Douslin and S. W. Levine, *J. Chem. Phys.*, **19**, 948 (1951). [i] E. J. Gornowski, E. H. Amick, Jr., and A. N. Nixon, *Ind. Eng. Chem.*, **39**, 1348 (1947). [j] A. Michels, T. Wassenaar, Th. Zweitering and P. Smits, *Physica*, **16**, 501 (1950); A. Michels, B. Blaisse and C. Michels, *Proc. Roy. Soc. (London)* **A160**, 358 (1937); A. Michels and C. Michels, *ibid.*, **A153**, 201 (1936); A. Michels, C. Michels and H. Wouters, *ibid.*, **A153**, 214 (1936); B. J. Kendall and B. H. Sage, *Petroleum (London)*, **14**, 184 (1951). [k] F. G. Keyes and R. B. Brownlee, THIS JOURNAL, **40**, 25 (1918); F. G. Keyes, *ibid.*, **53**, 965 (1931); J. A. Beattie and C. K. Lawrence, *ibid.*, **52**, 6 (1930). [l] J. H. Keenan and F. G. Keyes, "Thermodynamic Properties of Steam," John Wiley and Sons, Inc., New York, N. Y., 1936. [m] L. B. Smith, J. A. Beattie and W. C. Kay, THIS JOURNAL, **59**, 1587 (1937); S. Young, *Sci. Proc. Roy. Dub. Soc.*, new series **12**, 374 (1909–1910).

of reduced temperature and pressure for the substances listed in Table I. The region T_r from 0.8 to 4.0 and P_r from 0 to 9 was selected for detailed study. This region includes the interesting behavior near the critical point and covers the area of greatest practical interest. Outside of this region the available data become too sparse to give a good test of our theory. Indeed the data are relatively sparse above $P_r = 3$ and $T_r = 2$. For each point in a closely spaced matrix of values of P_r and T_r, the compressibility factor was plotted as a function of the acentric factor. Most of the points fell on smooth curves (and usually on straight lines) within a few tenths of 1%. The largest deviation of experimentally well established points is about 2%. The highly polar H_2O and NH_3 are excepted from this agreement, of course. Two typical sets of these curves are shown in Figs. 1 and 2 which are for $P_r = 1.6$ and 3.0, respectively, and the various values of T_r indicated. Several sets of substances have nearly equal values of the confining factor. In order to avoid undue confusion in Figs. 1 and 2, single points are shown for each set as indicated at the bottom of the figure. The agreement of the individual substances within each set is excellent. For the data in Figs. 1 and 2, the average deviation is between 0.1 and 0.2%.

Figure 1 shows very clearly one advantage of the present system over that of Meissner and Seferian.[5] Any attempt to include such highly polar substances as water and ammonia in a three parameter system will necessarily lead to deviations such as are shown by the open circles in Fig. 1. If such highly polar substances are to be included, a fourth parameter will be necessary for an accurate correlation.

The compressibility factor is expressed generally as a function of three variables.

$$(PV/RT) = z(T_r, P_r, \omega) \tag{2}$$

The functional dependence on temperature and pressure is very complicated. Indeed this complexity has retarded advances in this field very considerably. No simple analytical equation is adequate. The eight constant Benedict[6] equation is the simplest one which even approaches the desired accuracy. While we expect to examine completely analytical representations further, for the present we will expand the compressibility factor function as a power series in the acentric factor

$$z = z^{(0)} + \omega z^{(1)} + \dots \tag{3}$$

where $z^{(0)}$, $z^{(1)}$, etc., are each functions of T_r and P_r.

In almost all regions the first two terms in equation 3 are sufficient. This result corresponds to the straight lines in Figs. 1 and 2. An attempt was made to evaluate the quadratic term $z^{(2)}$ for the small region in which it appeared to be significant. While there was no difficulty at any particular value of T_r and P_r (such as $T_r = 1.10$ in Fig. 1), the resulting values of $z^{(2)}$ showed such irregular behavior as functions of T_r and P_r that there arose considerable doubt as to their validity. Consequently, no values of $z^{(2)}$ are reported at this time, and the data in regions of apparent curvature are fitted with the best straight lines.

The values of $z^{(1)}$ were plotted as functions of T_r and P_r and were adjusted within the limits of experimental error to yield a reasonably smooth function. The values of $z^{(0)}$ were initially much more accurate than those of $z^{(1)}$ and were found to be smooth functions of temperature and pressure without further adjustment. However, large graphs

(5) H. P. Meissner and R. Seferian, *Chem. Eng. Progress*, **47**, 579 (1951).

(6) M. Benedict, G. B. Webb and L. C. Rubin, *J. Chem. Phys.*, **8**, 334 (1940).

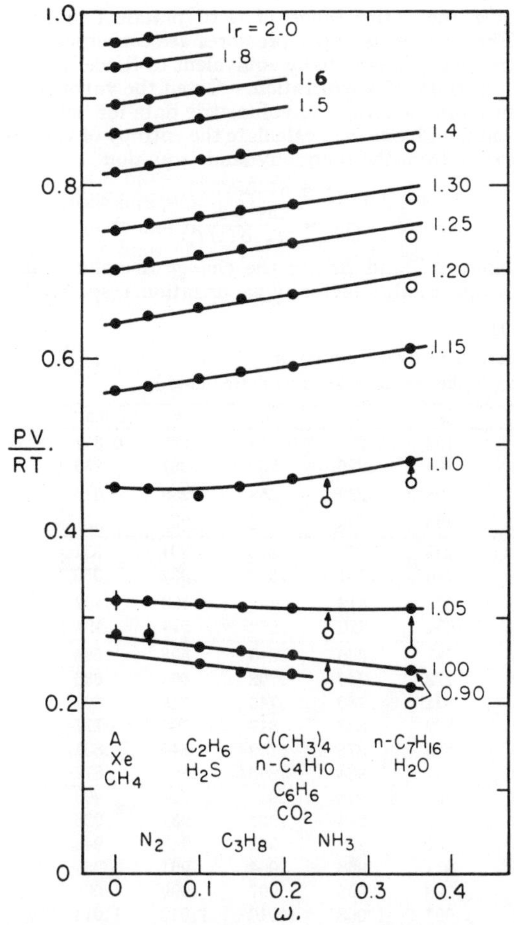

Fig. 1.—The compressibility factor as a function of the acentric factor, ω, at P_r = 1.6 and the values of T_r indicated.

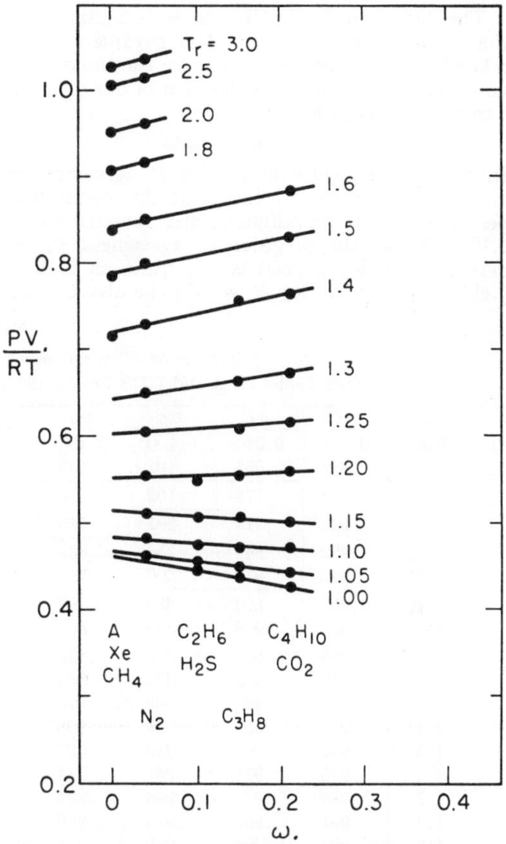

Fig. 2.—The compressibility factor as a function of the acentric factor at P_r = 3.0 and the indicated values of T_r.

were used to interpolate values of $z^{(0)}$ at intermediate points. The final values of the functions in equation 3 are given in Tables II–V. In view of the omission of the quadratic term one must regard the results in the region T_r = 1.05–1.10; P_r = 1.4–2.0 as less precise.

Outside of the critical region the spacing of points in Tables II and IV are close enough to allow linear interpolation without significant error. Tables III and V provide more closely spaced values in regions of large curvature. Even this close spacing is not sufficient to allow linear interpolation at all points, but simple graphs should suffice.

In certain regions either the data are poor or more commonly the data deviate from the linear correlation in ω. In such regions values are given to only the second decimal place and correspondingly lower accuracy must be expected in the calculated results.

Coexisting Phases.—Separate studies were made of the vapor pressure and of the compressibility factor for both liquid and gas along the saturation curve. The vapor pressure data are generally quite accurate and log P_r *vs.* ω plots give good straight lines even at temperatures well removed from 0.7 where ω is defined. The values are summarized in Table VI where log P_r for ω = 0 and $(\partial \log P_r/\partial\omega)_T$ are tabulated. Second-order terms are not needed for the vapor pressure. The volumetric data are less accurate and the correlation plots show random deviations as high as 3% but averaging less than 1%. The greatest deviations occur in the range where the values are rounded to the hundredths place.

The volumetric data for the two-phase region are repeated for convenience in Table VII as a function of reduced pressure. It should be noted that the functions $z_T^{(1)}$ and $z_P^{(1)}$ are partial derivatives where the subscript indicates the variable held constant, thus

$$\begin{aligned} z_T^{(1)} &= (\partial z/\partial\omega)_T \\ z_P^{(1)} &= (\partial z/\partial\omega)_P \\ &= z_T^{(1)} + (\partial z/\partial T)_\omega(\partial T/\partial\omega)_P \end{aligned} \quad (4)$$

The last equation gives a relationship between $z_P^{(1)}$ and $z_T^{(1)}$; however, we found it more convenient to evaluate each of the z functions directly from volumetric data.

We expect to treat the volumetric data for liquids at low temperatures in greater detail later. The present approximate values in this region suffice, however, for calculations of Δz of vaporization.

The values in Table VI can be compared with the tabulations of Riedel.[7] His parameter α_k is related to our ω since both measure the slope of the vapor pressure curve. Application of our equation 1 to his formula yields

$$\alpha_k = 5.808 + 4.93\omega \tag{5}$$

Riedel's tables yield values of log P_r agreeing with our results within 0.001 except at the lowest temperatures where the difference still remains within 0.1%. While the precision of agreement in the derivative (∂ log $P_r/\partial\omega$) is not quite as high, it yields agreement in log P_r within the above limits over the entire range of ω of practical interest. Thus, so far as vapor pressures are concerned, our treatment is essentially equivalent to Riedel's.

Entropy of Vaporization.—Given the vapor pressure function and the volumetric data for both gas and liquid, one may calculate the entropy of vaporization from the thermodynamic equation

$$\Delta S = \Delta V(\partial P/\partial T) = \frac{R\Delta z}{T_r}\frac{\partial \ln P}{\partial (1/T_r)} \tag{6}$$

where ΔV and Δz are the change in volume and compressibility factor on vaporization, respectively.

Table II

Values of $z^{(0)}$ for Compressibility Factor Calculation

(See Tables III-A and III-B for additional data in the region enclosed by dotted lines.)

T_r	P_r 0.2	0.4	0.6	0.8	1.0	1.2	1.4	1.6	1.8	2.0
0.80	0.851	0.066	0.100	0.133	0.164	0.192	0.225	0.258	0.287	0.318
.85	.882	.067	.101	.134	.165	.194	.226	.258	.287	.316
.90	.904	.778	.102	.135	.167	.198	.229	.258	.288	.316
.95	.920	.819	.697	.145	.176	.205	.235	262	.292	.321
1.00	.932	.849	.756	.638	.291	.231	.250	.278	.304	.331
1.05	.942	.874	.800	.714	.609	.470	.341	.320	.332	.350
1.10	.950	.893	.833	.767	.691	.607	.512	.442	.408	.402
1.15	.958	.908	.858	.805	.746	.684	.620	.562	.514	.484
1.20	.963	.921	.879	.835	.788	.737	.690	.640	.598	.568
1.25	.968	.930	.896	.858	.820	.778	.740	.702	.664	.636
1.30	.971	.940	.909	.878	.846	.811	.780	.749	.718	.691
1.4	.977	.952	.929	.908	.883	.859	.838	.817	.795	.777
1.5	.982	.963	.945	.927	.909	.892	.875	.859	.844	.831
1.6	.985	.971	.957	.944	.930	.917	.904	.893	.882	.872
1.7	.988	.977	.966	.956	.946	.936	.926	.919	.911	.903
1.8	.991	.982	.974	.966	.958	.950	.944	.937	.931	.926
1.9	.993	.986	.980	.974	.968	.962	.958	.952	.948	.944
2.0	.995	.989	.984	.979	.975	.971	.968	.964	.961	.959
2.5	1.000	.999	.999	.998	.998	.998	.998	.997	.999	1.000
3.0	1.001	1.002	1.003	1.004	1.005	1.007	1.008	1.010	1.012	1.014
3.5	1.002	1.004	1.006	1.008	1.011	1.013	1.015	1.018	1.020	1.022
4.0	1.003	1.005	1.008	1.010	1.013	1.015	1.017	1.020	1.022	1.024

T_r	P_r 2.0	2.2	2.4	2.6	2.8	3.0	3.2	3.4
0.8	0.318	0.347	0.376	0.405	0.433	0.461	0.490	0.519
.85	.316	.345	.374	.403	.431	.459	.487	.515
.9	.316	.345	.373	.402	.430	.458	.485	.512
.95	.320	.347	.375	.403	.430	.457	.484	.510
1.0	.331	.356	.381	.407	.433	.458	.484	.509
1.05	.351	.372	.393	.417	.441	.466	.489	.512
1.10	.408	.410	.421	.440	.462	.484	.504	.525
1.15	.486	.477	.478	.485	.498	.513	.529	.546
1.2	.569	.553	.545	.544	.548	.554	.563	.574
1.25	.636	.618	.606	.599	.597	.598	.602	.609
1.3	.691	.671	.657	.649	.644	.642	.642	.645
1.4	.777	.759	.745	.734	.725	.720	.718	.718
1.5	.831	.819	.808	.800	.794	.790	.785	.784
1.6	.872	.863	.855	.848	.843	.840	.836	.834
1.7	.903	.896	.889	.883	.879	.875	.873	.872
1.8	.926	.921	.916	.913	.910	.908	.907	.906
1.9	.944	.940	.936	.933	.931	.930	.929	.929
2.0	.959	.956	.954	.953	.953	.952	.952	.953
2.5	1.000	1.001	1.001	1.002	1.004	1.006	1.008	1.009
3.0	1.014	1.016	1.019	1.022	1.025	1.028	1.030	1.033
3.5	1.022	1.024	1.027	1.030	1.033	1.036	1.039	1.042
4.0	1.024	1.026	1.029	1.032	1.035	1.038	1.041	1.044

It is well known that the logarithm of the vapor

(7) L. Riedel, *Chem. Ing. Tech.*, **26**, 83 (1954).

TABLE II (*Continued*)

T_r	P_r 3.6	3.8	4.0	4.5	5.0	6.0	7.0	8.0	9.0
0.8	0.547	0.576	0.605	0.675	0.746	0.883	1.017	1.15	
.85	.542	.569	.597	.663	.730	.861	0.990	1.115	
.9	.538	.565	.591	.655	.718	.842	.966	1.089	1.21
.95	.536	.561	.587	.647	.709	.828	.947	1.066	1.185
1.0	.534	.557	.582	.642	.702	.819	.932	1.048	1.166
1.05	.535	.557	.580	.639	.700	.814	.923	1.032	1.147
1.10	.547	.567	.589	.643	.699	.810	.916	1.019	1.129
1.15	.563	.581	.600	.651	.705	.809	.911	1.008	1.113
1.2	.587	.601	.618	.664	.714	.810	.907	1.000	1.100
1.25	.618	.629	.643	.682	.726	.816	.907	0.994	1.088
1.3	.651	.659	.668	.701	.740	.824	.910	.992	1.078
1.4	.722	.727	.734	.754	.781	.844	.921	.994	1.071
1.5	.784	.786	.790	.805	.826	.877	.934	1.000	1.070
1.6	.833	.834	.835	.844	.860	.904	.953	1.010	1.075
1.7	.872	.873	.874	.882	.895	.930	.972	1.023	1.082
1.8	.906	.907	.908	.914	.925	.955	.993	1.039	1.091
1.9	.930	.932	.934	.941	.950	.976	1.010	1.051	1.097
2.0	.954	.954	.956	.962	.972	.996	1.027	1.064	1.106
2.5	1.012	1.014	1.018	1.026	1.035	1.055	1.079	1.105	1.136
3.0	1.036	1.038	1.041	1.049	1.058	1.077	1.10	1.124	1.150
3.5	1.045	1.048	1.051	1.058	1.067	1.086	1.105	1.126	1.148
4.0	1.047	1.050	1.053	1.060	1.068	1.086	1.104	1.124	1.143

TABLE IIIA

VALUES OF $z^{(0)}$ NEAR THE TWO-PHASE REGION

T_r	P_r 0.4	0.5	0.6	0.7	0.8	0.9	1.0
0.90	0.778	0.701	0.102	0.118	0.135	0.151	0.167
.91	.787	.715	.104	.120	.136	.152	.168
.92	.796	.728	.650	.122	.138	.153	.169
.93	.805	.740	.666	.124	.140	.155	.170
.94	.812	.751	.681	.125	.142	.157	.173
.95	.819	.762	.697	.612	.145	.160	.176
.96	.826	.772	.711	.632	.149	.164	.180
.97	.832	.782	.724	.652	.56	.170	.186
.98	.838	.791	.735	.669	.591	.177	.193
.99	.844	.800	.746	.685	.616	.514	.205
1.00	.849	.807	.757	.699	.638	.554	.291
1.01	.854	.813	.767	.713	.654	.583	.476
1.02	.860	.820	.776	.726	.672	.608	.525
1.03	.865	.826	.784	.737	.687	.630	.558
1.04	.870	.833	.793	.748	.701	.648	.586
1.05	.874	.838	.800	.758	.714	.665	.000

The substitution of our functional relationships for z and P yields a quadratic formula for the entropy.

$$\Delta S = \Delta S^{(0)} + \omega \Delta S^{(1)} + \omega^2 \Delta S^{(2)} \quad (7)$$

Each of the coefficients is readily derived and, upon appropriate numerical calculations, yields the values given in the last three columns of Table VI. The unit is cal./degree mole.

Since the available data for the volume of the saturated vapor at very low reduced temperatures are scanty, use was made of calorimetric values of the entropy of vaporization in this region. However, once $z^{(0)}$ and $z^{(1)}$ are fixed at the lowest temperature to agree with entropy data, the remainder of their values follow from volumetric data and a requirement of reasonable smoothness.

TABLE IIIB

VALUES OF $z^{(0)}$ IN THE CRITICAL REGION

T_r	P_r 1.0	1.1	1.2	1.3	1.4	1.5	1.6	1.7	1.8	1.9	2.0
0.98	0.193	0.204	0.217	0.230	0.244	0.257	0.270	0.284	0.299	0.313	0.326
0.99	.205	.210	.223	.235	.247	.260	.273	.287	.301	.315	.328
1.00	.291	.220	.231	.241	.250	.265	.278	.290	.304	.317	.331
1.01	.476	.283	.243	.248	.259	.271	.283	.294	.307	.319	.331
1.02	.525	.402	.273	.260	.270	.278	.291	.300	.311	.323	.334
1.03	.558	.466	.34	.29	.283	.288	.297	.306	.316	.328	.339
1.04	.586	.509	.41	.33	.307	.302	.307	.314	.324	.334	.343
1.05	.609	.543	.470	.375	.341	.324	.320	.323	.332	.341	.350
1.06	.628	.572	.505	.423	.370	.349	.336	.333	.343	.348	.358
1.07	.645	.597	.534	.468	.408	.379	.358	.349	.356	.358	.367
1.08	.663	.618	.562	.504	.445	.412	.385	.373	.370	.369	.375
1.09	.677	.636	.587	.535	.480	.443	.412	.396	.387	.383	.387
1.10	.691	.652	.607	.561	.512	.473	.442	.422	.408	.400	.402
1.11	.703	.667	.625	.584	.538	.502	.469	.448	.428	.418	.417
1.13	.726	.693	.658	.621	.584	.549	.520	.494	.472	.456	.450
1.15	.746	.715	.684	.652	.620	.589	.562	.536	.514	.495	.484

pressure is very nearly a linear function of the reciprocal of the temperature, hence the second form of the equation is most convenient.

Discussion.—The advantage of this scheme over simple corresponding states correlations is readily seen from Figs. 1 and 2. On the simpler basis

Table IV

Values of $z^{(1)}$ for Compressibility Factor Calculation

(See Tables VA and VB for additional data in the region enclosed by dotted lines.)

T_r	P_r 0.2	0.4	0.6	0.8	1.0	1.2	1.4	1.6	1.8	2.0
0.80	−0.095	−0.028	−0.044	−0.058	−0.07	−0.08	−0.10	−0.11	−0.12	−0.13
.85	− .067	− .031	− .049	− .064	− .08	− .09	− .11	− .12	− .13	− .14
.90	− .042	− .09	− .053	− .068	− .085	− .10	− .11	− .12	− .13	− .14
.95	− .025	− .050	− .10	− .072	− .091	− .10	− .11	− .12	− .12	− .13
1.00	− .012	− .016	− .020	− .05	− .080	− .090	− .099	− .108	− .115	− .123
1.05	.000	+ .001	+ .005	+ .015	+ .02	+ .01	− .01	− .04	− .06	− .07
1.10	+ .002	.008	.016	.030	.055	.082	+ .11	+ .082	+ .035	.000
1.15	.004	.012	.021	.040	.064	.093	.12	.140	.136	+ .100
1.20	.006	.014	.025	.043	.07	.10	.13	.16	.17	.17
1.25	.007	.016	.028	.046	.07	.10	.13	.16	.18	.19
1.30	.008	.018	.031	.05	.07	.10	.13	.16	.18	.20
1.4	.012	.025	.039	.05	.07	.10	.13	.16	.18	.19
1.5	.016	.032	.051	.06	.08	.10	.13	.15	.17	.18
1.6	.018	.035	.051	.07	.08	.10	.12	.14	.16	.17
1.7	.018	.035	.051	.07	.08	.10	.11	.13	.15	.16
1.8	.018	.035	.051	.07	.08	.10	.11	.13	.14	.15
1.9	.018	.035	.051	.07	.08	.10	.11	.13	.14	.15
2.0	.018	.035	.051	.07	.08	.10	.11	.13	.14	.15
2.5	.018	.035	.051	.07	.08	.10	.11	.13	.14	.15
3.0	.018	.035	.051	.07	.08	.10	.11	.13	.14	.15
3.5	.018	.035	.051	.07	.08	.10	.11	.13	.14	.15
4.0	.018	.035	.051	.07	.08	.10	.11	.13	.14	.15

T_r	P_r 2.2	2.4	2.6	2.8	3.0	4.0	5.0	6.0	7.0	8.0	9.0
0.80	−0.14	−0.15	−0.16	−0.17	−0.18	−0.23	−0.26	−0.29	−0.32	−0.35	−0.37
.85	− .15	− .16	− .17	− .18	− .18	− .22	− .25	− .28	− .31	− .34	− .36
.90	− .15	− .16	− .17	− .17	− .18	− .21	− .24	− .27	− .30	− .32	− .35
.95	− .14	− .15	− .15	− .16	− .17	− .20	− .22	− .25	− .28	− .31	− .34
1.00	− .13	− .13	− .14	− .14	− .15	− .17	− .20	− .23	− .26	− .30	− .33
1.05	− .08	− .09	− .10	− .10	− .11	− .14	− .17	− .20	− .24	− .28	− .31
1.10	− .02	− .03	− .05	− .06	− .07	− .10	− .13	− .16	− .21	− .25	− .28
1.15	+ .07	+ .04	+ .02	.00	− .01	− .04	− .08	− .12	− .16	− .20	− .24
1.20	.16	.14	.12	+ .09	+ .07	.00	− .04	− .08	− .12	− .16	− .19
1.25	.19	.18	.16	.14	.12	+ .05	.00	− .03	− .07	− .11	− .13
1.30	.20	.20	.20	.19	.18	.10	+ .04	.00	− .04	− .07	− .09
1.4	.20	.21	.21	.21	.20	.15	.11	+ .07	+ .04	+ .01	− .01
1.5	.20	.20	.21	.21	.21	.20	.17	.14	.11	.09	+ .07
1.6	.18	.19	.20	.20	.21	.22	.21	.19	.17	.15	.14
1.7	.17	.18	.19	.20	.21	.24	.25	.26	.25	.24	.22
1.8	.17	.18	.19	.20	.21	.26	.29	.31	.32	.32	.30
1.9	.17	.18	.19	.20	.21	.26	.30	.35	.38	.40	.40
2.0	.17	.18	.19	.20	.21	.26	.30	.35	.40	.43	.45
2.5	.17	.18	.19	.20	.21	.26	.30	.35	.40	.45	.50
3.0	.17	.18	.19	.20	.21	.26	.30	.35	.40	.45	.50
3.5	.17	.18	.19	.20	.21	.26	.30	.35	.40	.45	.50
4.0	.17	.18	.19	.20	.21	.26	.30	.35	.40	.45	.50

horizontal lines would have to be drawn through each set of points at a given reduced temperature, since the acentric factor is not considered. At the lowest temperatures, the range of values of the compressibility factor from methane to heptane is about 15% even when water and ammonia are excluded. Thus a corresponding states correlation could not possibly yield a maximum error less than 7% at this point. We see that our formulation in terms of the acentric factor, which is determined independently from the vapor pressure data, yields agreement in this region within about 1%. At higher temperatures the agreement on both bases is better, but we have still reduced the error by about a factor of ten.

It should be emphasized that this improvement is attained at very little expense in terms of complexity. Vapor pressure data near the normal boiling point are even more generally available than are reliable critical data. Thus there is no difficulty in obtaining the acentric factor. As is illustrated later, the vapor pressure datum need not be at the point of definition, $T_r = 0.7$. Two coefficients must be interpolated instead of a single quantity, but this is a simple procedure. The principal increase in effort required by the present system

TABLE VA

VALUES OF $z^{(1)}$ NEAR THE TWO PHASE REGION

T_r	P_r 0.4	0.6	0.8	1.0
0.90	−0.09	−0.053	−0.068	−0.085
.91	− .08	− .053	− .069	− .087
.92	− .072	− .18	− .070	− .089
.93	− .066	− .15	− .071	− .090
.94	− .058	− .12	− .072	− .091
.95	− .050	− .10	− .072	− .091
.96	− .042	− .08	− .072	− .091
.97	− .035	− .065	− .14	− .091
.98	− .027	− .050	− .11	− .090
.99	− .021	− .033	− .08	− .087
1.00	− .016	− .020	− .05	− .080
1.01	− .012	− .012	− .02	− .02
1.02	− .008	− .006	.00	− .01
1.03	− .005	− .001	+ .005	.00
1.04	− .002	+ .002	+ .010	+ .01
1.05	+ .001	+ .005	+ .015	+ .02

TABLE VB

VALUES OF $z^{(1)}$ IN THE CRITICAL REGION

T_r	P_r 1.0	1.2	1.4	1.6	1.8	2.0
0.98	−0.090	−0.099	−0.109	−0.118	−0.125	−0.130
0.99	− .087	− .095	− .104	− .114	− .121	− .127
1.00	− .080	− .090	− .099	− .108	− .115	− .123
1.01	− .02	− .080	− .091	− .102	− .10	− .100
1.02	− .01	− .065	− .082	− .095	− .09	− .09
1.03	.00	− .047	− .068	− .085	− .08	− .09
1.04	+ .01	− .025	− .050	− .073	− .07	− .08
1.05	+ .02	+ .01	− .01	− .04	− .06	− .07
1.06	+ .03	+ .06	+ .07	− .02	− .05	− .073
1.07	+ .04	+ .08	+ .09	.000	− .038	− .059
1.08	+ .047	+ .08	+ .10	+ .030	− .015	− .041
1.09	+ .050	+ .08	+ .11	+ .056	+ .012	− .022
1.10	+ .055	+ .082	+ .11	+ .082	+ .035	.000
1.11	+ .057	+ .085	+ .12	+ .099	+ .062	+ .020
1.13	+ .062	+ .089	+ .12	+ .123	+ .105	+ .060
1.15	+ .064	+ .093	+ .122	+ .140	+ .136	+ .100

coefficients of that equation in terms of our three fundamental variables, the critical temperature, the critical pressure and the acentric factor. Atten-

TABLE VI

DATA FOR COEXISTING PHASES

$(PV/RT)_T = z^{(0)} + \omega z_T^{(1)}$ $\log P_r = (\log P_r)^{(0)} + \omega(\partial \log P_r/\partial\omega)_T$ $\Delta S = \Delta S^{(0)} + \omega\Delta S^{(1)} + \omega^2\Delta S^{(2)}$

T_r	$-(\log P_r)^{(0)}$	$-\left(\frac{\partial \log P_r}{\partial\omega}\right)_T$	Liquid $z^{(0)}$	Liquid $z_T^{(1)}$	Gas $z^{(0)}$	Gas $z_T^{(1)}$	$\Delta S^{(0)}$	$\Delta S^{(1)}$	$\Delta S^{(2)}$
1.00	0.000	0.000	0.291	−0.080	0.291	−0.080	0.00	0.00	0.0
0.99	.025	.021	.202	− .090	.43	− .030	2.57	2.83	.6
.98	.050	.042	.179	− .093	.47	.000	3.38	3.91	.9
.97	.076	.064	.162	− .095	.51	+ .020	4.00	4.72	1.1
.96	.102	.086	.148	− .095	.54	.035	4.52	5.39	1.3
.95	.129	.109	.136	− .095	.565	.045	5.00	5.96	1.4
.94	.156	.133	.125	− .094	.59	.055	5.44	6.51	1.5
.92	.212	.180	.108	− .092	.63	.075	6.23	7.54	1.8
.90	.270	.230	.0925	− .087	.67	.095	6.95	8.53	2.0
.88	.330	.285	.0790	− .080	.70	.110	7.58	9.39	2.2
.86	.391	.345	.0680	− .075	.73	.125	8.19	10.3	2.4
.84	.455	.405	.0585	− .068	.756	.135	8.79	11.2	2.5
.82	.522	.475	.0498	− .062	.781	.140	9.37	12.1	2.6
.80	.592	.545	.0422	− .057	.804	.144	9.97	13.0	2.7
.78	.665	.620	.0360	− .053	.826	.144	10.57	13.9	2.8
.76	.742	.705	.0300	− .048	.846	[illegible]	[illegible]	14.9	2.9
.74	[illegible]	.800	.0250	− .043	.864	.137	11.84	16.0	2.9
.72	.909	.895	.0210	− .037	.881	.131	12.49	17.0	2.8
.70	1.000	1.00	.0172	− .032	.897	.122	13.19	18.1	2.8
.68	1.096	1.12	.0138	− .027	.911	.113	13.89	19.3	2.7
.66	1.198	1.25	.0111	− .022	.922	.104	14.62	20.5	2.6
.64	1.308	1.39	.0088	− .018	.932	.097	15.36	21.8	2.5
.62	1.426	1.54	.0068	− .015	.940	.090	16.12	23.2	2.4
.60	1.552	1.70	.0052	− .012	.947	.083	16.92	24.6	2.3
.58	1.688	1.88	.0039	− .009	.953	.077	17.74	26.2	2.2
.56	1.834	2.08	.0028	− .007	.959	.070	18.64	27.8	2.1

arises directly from the greater accuracy expected. Modest sized graphs no longer suffice. Also the interpolations must be made with greater care to avoid loss in accuracy.

The principles of thermodynamics allow the calculation of a number of functions from the compressibility factor and its derivatives. It is this point which emphasizes the desirability of analytical representation. A program is currently under way in coöperation with Professor B. H. Sage of the California Institute of Technology which involves the application of the Benedict equation. Expressions will be obtained for each of the eight tion will be given to such derived functions as the fugacity, entropy, and heat content as soon as this analytical function has been obtained.

The class of substances for which this correlation succeeds is just that which has been designated *normal liquids*[8] and it would seem desirable to use that term in the present connection. We know that highly polar or hydrogen bonding molecules such as H_2O and NH_3 do not form normal

(8) See J. H. Hildebrand, "Solubility," First edition, The Chemical Catalog Co., New York, 1924, pp. 90–95, or F. Daniels, "Outlines of Physical Chemistry," John Wiley and Sons, Inc., New York, N. Y., 1948, pp. 182–186.

Table VII

Data for Coexisting Phases

$(PV/RT)_P = z^{(0)} + \omega z_P^{(1)}$

P_r	Liquid $z^{(0)}$	Liquid $z_P^{(1)}$	Gas $z^{(0)}$	Gas $z_P^{(1)}$
1.00	0.291	−0.080	0.291	−0.080
0.99	.244	− .074	.35	− .083
.98	.228	− .071	.38	− .085
.97	.218	− .069	.40	− .087
.96	.210	− .067	.41	− .088
.95	.203	− .065	.42	− .089
.94	.197	− .063	.43	− .089
.92	.188	− .060	.45	− .090
.90	.180	− .058	.47	− .091
.85	.164	− .055	.50	− .090
.80	.150	− .053	.53	− .087
.75	.137	− .050	.56	− .081
.70	.125	− .047	.59	− .075
.65	.114	− .044	.615	− .069
.60	.104	− .041	.64	− .063
.55	.0945	− .038	.665	− .056
.50	.0850	− .036	.688	− .049
.45	.0758	− .033	.711	− .041
.40	.0670	− .030	.734	− .033
.35	.0584	− .027	.758	− .025
.30	.0500	− .023	.783	− .018
.25	.0416	− .020	.809	− .012
.20	.0334	− .017	.835	− .008
.15	.0253	− .013	.864	− .005
.10	.0175	− .010	.896	− .002
.05	.0093	− .006	.935	.000
.00	.0000	.000	1.000	.000

liquids. In Part I theoretical methods were applied to the questions of the maximum polarity allowable for conformity to our system. The resulting criterion involved molecular parameters not generally available. If we substitute appropriate powers of the critical temperature and pressure for these molecular quantities, we obtain the expression $\mu^4 P_c^2/T_c^4$ as a measure of the ratio of polar to non-polar intermolecular forces. Here μ is the dipole moment. Also, if we take our units as 10^{-18} e.s.u., atm., and degrees K., respectively, we obtain for this ratio the values 0.3, 2.3 and 3.2, all $\times\ 10^{-6}$ for H_2S, NH_3, and H_2O in that order. We know that H_2S conforms well with our correlation whereas NH_3 and H_2O do not. Hence the maximum value of $\mu^4 P_c^2/T_c^4$ for a normal liquid will be in the vicinity of 0.5×10^{-6}.

Example.—Finally, it seems desirable to present an example both to illustrate the method of use of these tables and to illustrate the accuracy obtained. We consider isobutane[9] which was not used in the construction of the tables. To determine ω we select a single vapor pressure value well removed from the critical point: for example the datum at 0°F. or 459.7°R. Since T_c is 734.65°R, we find $T_r = 0.6257$; and similarly $P_c = 529.1$, $P = 11.53$ lb./sq. in. and $-\log P_r = 1.6617$. Then interpolating to this T_r in Table VI we find values for the equation

$$-\log P_r = -\log P_r^{(0)} + \omega\left(\frac{-\partial \log P_r}{\partial \omega}\right)_T$$

$$1.6617 = 1.392 + 1.50\omega$$

$$\omega = 0.180$$

A similar calculation with the vapor pressure datum[9b] at 100°F. yields $\omega = 0.184$, which is reasonable agreement.

For the calculation of the compressibility factor of isobutane we take an example at 340°F. and 800 lb./sq. in. since this is in the sensitive region a little above critical temperature and pressure. The reduced variables are $P_r = 1.5533$, $T_r = 1.0919$. Interpolation in Tables IIB and VB yields $z^{(0)} = 0.433$ and $z^{(1)} = 0.074$, and substitution into equation 3 yields a calculated value $z = 0.447$ which may be compared with an experimental value[9b] of 0.4517.

Note Added in Proof:—In additional papers Riedel (*Chem. Ing. Tech.*, 26, 259, 679 (1954)) has treated the volumetric behavior of liquid and vapor along the saturation curve and the entropy of vaporization. While the agreement with our results in these cases is not as precise as for vapor pressures,[7] it is satisfactory.

Berkeley, Calif.

(9) (a) F. D. Rossini, *et al.*, "Selected Values of the Physical and Thermodynamic Properties of Hydrocarbons and Related Compounds," Carnegie Press, Pittsburgh, Pa., 1953; (b) Ref. *g* to Table I.

[Reprinted from the Journal of the American Chemical Society, **79**, 2369 (1957).]

[CONTRIBUTION FROM THE DEPARTMENT OF CHEMISTRY AND CHEMICAL ENGINEERING, UNIVERSITY OF CALIFORNIA, BERKELEY]

The Volumetric and Thermodynamic Properties of Fluids. III. Empirical Equation for the Second Virial Coefficient[1]

BY KENNETH S. PITZER AND R. F. CURL, JR.

RECEIVED DECEMBER 22, 1956

An equation has been developed to represent the second virial coefficient of a normal fluid: $BP_c/RT_c = (0.1445 + 0.073\omega) - (0.330 - 0.46\omega)T_r^{-1} - (0.1385 + 0.50\omega)T_r^{-2} - (0.0121 + 0.097\omega)T_r^{-3} - 0.0073\omega T_r^{-8}$ where ω is the acentric factor (defined by $\omega = -\log (P/P_c)_{sat} - 1.000$ at $T_r = 0.7$) which was discussed in an earlier paper. This equation not only fits the volumetric data with considerable accuracy but its second derivative also yields agreement with measured values of the pressure derivative of the gas heat capacity.

In 1907 Berthelot[2] proposed a simple reduced equation for the second virial coefficient which has been of great value in correcting the properties of real gases at low pressures to the ideal gas state. We know, however, that real gases do not follow accurately the principle of corresponding states, hence any such equation is of limited accuracy. In the earlier papers of this series[3] we have shown that the introduction of the acentric factor as an additional parameter increases the accuracy of possible correlation by about one order of magnitude.

In this paper we present an empirical equation for the second virial coefficient in reduced form which includes the acentric factor dependence.

Basic Data.—In addition to the second virial coefficient values based on volumetric measurements,[4] we calculated the volume of the saturated vapor from vapor pressure, heat of vaporization and liquid volume data in several cases.[5]

We also used the information on the change with pressure of the gas heat capacity.[6] If the second virial coefficient is B, thermodynamics yields the equation

$$\lim_{P \to 0} \left(\frac{\partial C_p}{\partial P}\right)_T = -T\frac{\partial^2 B}{\partial T^2} \quad (1)$$

Consequently, the pressure dependence of the heat capacity offers a very sensitive test of any equation for B.

(1) This research was part of the program of Research Project 50 of the American Petroleum Institute.

(2) D. Berthelot, *Trav. et Mém. Bur. Int. Poid et Mes.*, 13 (1907).

(3) K. S. Pitzer, D. Z. Lippmann, R. F. Curl, Jr., C. M. Huggins and D. E. Petersen, THIS JOURNAL, **77**, 3427, 3433 (1955).

(4) A, Kr, Xe: E. Whalley and W. G. Schneider, *J. Chem. Phys.*, **23**, 1644 (1955); N_2: values of Holborn and Otto given by T. Kihara, *Rev. Mod. Phys.*, **25**, 831 (1953); CO_2: K. E. MacCormack and W. G. Schneider, *J. Chem. Phys.*, **18**, 1269 (1950); *n*-heptane: J. O. Hirschfelder, *et al.*, *ibid.*, **10**, 201 (1942); benzene: P. G. Francis, M. L. McGlashan, S. D. Hamann and W. J. McManamey, *ibid.*, **20**, 1341 (1952).

(5) Kr: A. Michels, T. Wassenaar and Th. N. Zwietering, *Physica*, **18**, 63 (1952), K. Clusius, A. Kruis and F. Konneitz, *Ann. Physik*, **33**, 642 (1938); Xe: A. Michels and T. Wassenaar, *Physica*, **16**, 253 (1950); K. Clusius and L. Riccoboni, *Z. physik. Chem.*, **B38**, 81 (1937); hydrocarbons: F. D. Rossini, *et al.*, "Selected Values of Physical and Thermodynamic Properties of Hydrocarbons and Related Compounds," Carnegie Press, 1953; *n*-heptane: G. Thodos, *Ind. Eng. Chem.*, **42**, 1514 (1950); G. Waddington, S. Todd and H. Huffman, THIS JOURNAL, **69**, 22 (1947); benzene: D. W. Scott, G. Waddington, J. C. Smith and H. M. Huffman, *J. Chem. Phys.*, **15**, 565 (1947).

(6) *n*-Heptane: G. Waddington, S. Todd and H. Huffman, THIS JOURNAL, **69**, 22 (1947); C_6H_6: D. W. Scott, G. Waddington, J. C. Smith and H. M. Huffman, *J. Chem. Phys.*, **15**, 565 (1947); 2,2,3-trimethylbutane: G. Waddington, S. Todd and H. Huffman, THIS JOURNAL, **69**, 22 (1947); ethanethiol: J. P. McCullough *et al.*, *ibid.*, **74**, 2801 (1952).

Selection of the Equation.—Our earlier studies[3] indicated that an equation linear in the acentric factor was usually adequate and we find this to be the case for the second virial coefficient. We have then, in reduced dimensionless form

$$\frac{BP_c}{RT_c} = \frac{B^{(0)}P_c}{RT_c} + \omega\frac{B^{(1)}P_c}{RT_c}$$

where $B^{(0)}$ and $B^{(1)}$ are functions of the reduced temperature, and ω is the acentric factor. Although ω was defined as

$$\omega = -\log P_r - 1.000 \quad (2)$$

where P_r is the reduced vapor pressure (P/P_c) at $T_r = 0.7$, it may be evaluated readily from any vapor pressure value well removed from the critical point by the use of Table VI of Paper II of this series.[3]

The simple fluids of zero acentric factor are the heavier inert gases A, Kr, Xe. We know their properties are quite well represented by a Lennard-Jones intermolecular potential. Consequently, we took the equation of Stockmayer and Beattie[7] as a point of departure. This was converted from the molecular potential constants to the macroscopic critical constants and then modified slightly to improve the fit at the lowest temperatures by the addition of a small T^{-3} term and other small adjustments. The resulting equation is

$$\frac{B^{(0)}P_c}{RT_c} = 0.1445 - \frac{0.330}{T_r} - \frac{0.1385}{T_r^2} - \frac{0.0121}{T_r^3} \quad (3)$$

Figure 1 shows the reduced virial coefficient data for a number of substances together with calculated curves. The agreement of the curve for $\omega = 0$ with the points for A, Kr and Xe demonstrates the adequacy of equation 3.

It is impractical to show all of the data considered in the development of the equation for $B^{(1)}$. The resulting fit with the data for benzene and for *n*-heptane is shown on Fig. 1 for B and on Fig. 2 for the pressure derivative of the heat capacity. Similar agreement is obtained for other substances. The resulting equations are

$$B = B^{(0)} + \omega B^{(1)} \quad (4)$$

$$\frac{B^{(1)}P_c}{RT_c} = 0.073 + \frac{0.46}{T_r} - \frac{0.50}{T_r^2} - \frac{0.097}{T_r^3} - \frac{0.0073}{T_r^8} \quad (5)$$

The selection of a term in T_r^{-8} is, of course, somewhat arbitrary, but a very high power is required to fit the data at low temperatures.

(7) W. H. Stockmayer and J. A. Beattie, *J. Chem. Phys.*, **10**, 476 (1942).

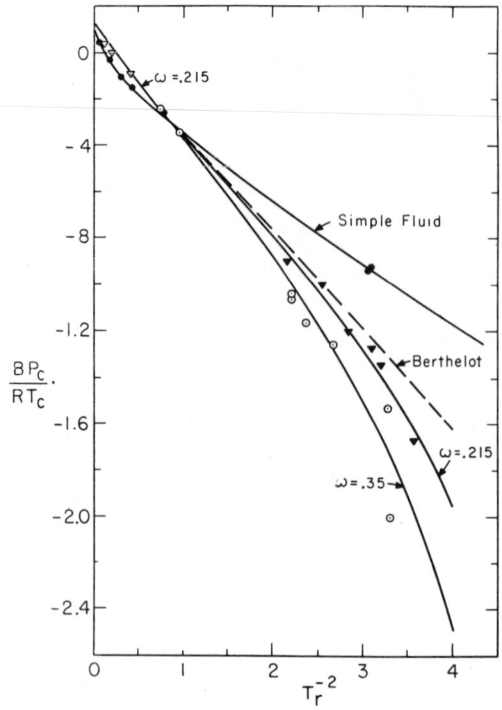

Fig. 1.—A comparison of calculated and experimental values of the second virial coefficient. Solid circles are A, Kr or Xe points and the line through these points is that of equation 3. The open triangles are CO_2 and solid triangles are benzene points. These are to be compared with the line for $\omega = 0.215$. The open circles are points for n-heptane ($\omega = 0.35$). It is to be noted that the Berthelot equation yields excellent agreement for ω in the range 0.15 to 0.20.

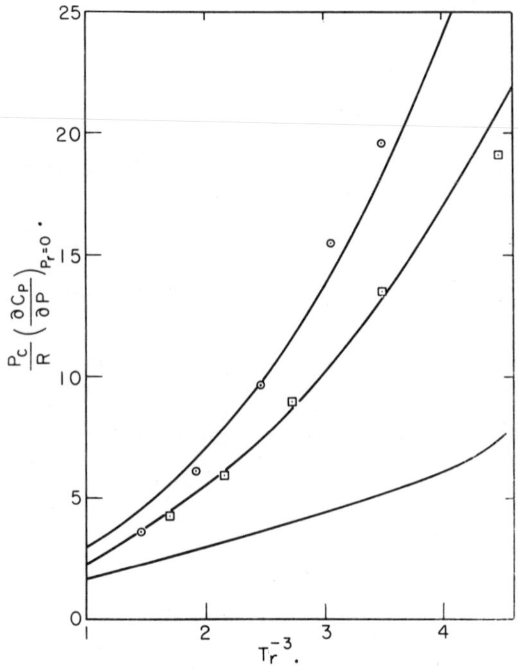

Fig. 2.—A comparison of calculated and experimental values of the pressure derivative of the heat capacity. The circles and upper curve are for n-heptane ($\omega = 0.35$); the squares and intermediate curve are for benzene ($\omega = 0.215$). The bottom curve is for $\omega = 0$, but no experimental data are available for comparison.

The work leading to the selection of equation 5 indicated certain small changes in the previously published values of $z^{(1)}$, the term in the compressibility factor expression which is multiplied by ω. The revised values are given in Table I (the upper right-hand portion of each section is unchanged but included for convenience).

TABLE I

REVISED VALUES OF Z^1 FOR COMPRESSIBILITY FACTOR CALCULATION (SEE TABLE IV, PAPER II)

T_r	P_r 0.2	0.4	0.6	0.8	1.0
1.20	0.009	0.018	0.028	0.044	0.069
1.25	.011	.023	.036	.050	.069
1.3	.013	.027	.041	.055	.072
1.4	.016	.032	.049	.065	.082
1.5	.017	.035	.052	.070	.088
1.6	.018	.036	.054	.072	.090

T_r	P_r 1	2	3	4	5	6	7	8	9
1.7	0.09	0.16	0.21	0.24	0.25	0.26	0.25	0.24	0.22
1.8	.09	.16	.21	.26	.29	.31	.32	.32	.30
1.9	.09	.16	.21	.26	.30	.35	.38	.40	.40
2.0	.08	.15	.21	.26	.30	.35	.40	.43	.45
2.5	.07	.13	.20	.25	.30	.35	.40	.45	.50
3.0	.06	.11	.17	.23	.28	.34	.38	.45	.50
3.5	.05	.09	.14	.19	.24	.28	.33	.38	.42
4.0	.04	.08	.12	.16	.20	.23	.27	.31	.35

Discussion.—This equation should be valuable for the estimation of gas imperfection corrections to various data on gases at low pressures. The substances which can be expected to conform are those forming normal liquids. Metals and hydrogen bonding substances must be excluded. The limitations are discussed in greater detail in Papers I and II.[3] The final equations for various properties are readily derived from thermodynamics. In these equations H^*, S^*, C_p^* refer to the properties of the hypothetical ideal gas at the same pressure as the real gas; f is the fugacity.

$$\frac{BP_c}{RT_c} = (0.1445 + 0.073\omega) - (0.330 - 0.46\omega)T_r^{-1} - (0.1385 + 0.50\omega)T_r^{-2} - (0.0121 + 0.097\omega)T_r^{-3} - 0.0073\omega T_r^{-8} \quad (6)$$

$$\log\left(\frac{f}{P}\right) = \left(\frac{P_r}{2.303}\right)[(0.1445 + 0.073\omega)T_r^{-1} - (0.330 - 0.46\omega)T_r^{-2} - (0.1385 + 0.50\omega)T_r^{-3} - (0.0121 + 0.097\omega)T_r^{-4} - 0.0073\omega T_r^{-9}] \quad (7)$$

$$\frac{H - H^*}{RT} = P_r\,[(0.1445 + 0.073\omega)T_r^{-1} - (0.660 - 0.92\omega)T_r^{-2} - (0.4155 + 1.50\omega)T_r^{-3} - (0.0484 + 0.388\omega)T_r^{-4} - 0.0657\omega T_r^{-9}] \quad (8)$$

$$\frac{S - S^*}{R} = P_r[-(0.330 - 0.46\omega)T_r^{-2} - (0.2770 + 1.00\omega)T_r^{-3} - (0.0363 + 0.29\omega)T_r^{-4} - 0.0584\omega T_r^{-9}] \quad (9)$$

$$\frac{C_p - C_p^*}{R} = P_r[(0.660 - 0.92\omega)T_r^{-2} + (0.831 + 3.0\omega)T_r^{-3} + (0.145 + 1.16\omega)T_r^{-4} + 0.526\omega T_r^{-9}] \quad (10)$$

BERKELEY, CALIF.

Reprinted from INDUSTRIAL AND ENGINEERING CHEMISTRY, Vol. 50, February 1958, Page 265

R. F. CURL, Jr., and KENNETH S. PITZER

Department of Chemistry and Chemical Engineering, University of California, Berkeley, Calif.

Volumetric and Thermodynamic Properties of Fluids—Enthalpy, Free Energy, and Entropy

THE PREDICTION of the volumetric and thermodynamic properties of pure fluids has been the subject of many studies over the last 100 years. Although the underlying principles in terms of intermolecular forces are understood, the calculation or even the quantitative empirical representation of the resulting macroscopic properties has proved to be unusually difficult. The failure to find a satisfactory analytical equation of state arises in the authors' opinion, not from any lack of skill in selecting the appropriate combination of common mathematical functions, but rather in the nonexistence of any suitable function. Consequently, the authors have, in effect, let nature generate mathematical functions and have developed an equation in which the minimum number of such functions are combined with other precisely defined parameters to yield the desired volumetric and thermodynamic properties for a broad class of pure substances.

These introductory ideas have been presented in preceding articles in this series. The first article (*11*) gives the theoretical basis with particular reference to the modern picture of intermolecular forces. A group of substances will conform to the principle of corresponding states only if their intermolecular potentials are identical except for distance and energy scale factors characteristic of each substance (*12*). Also their intermolecular motion must be classical—i.e., quantum effects must be negligible. The only group of substances which may be expected to conform to these criteria are the heavier rare gases argon, krypton, and xenon which do conform accurately to corresponding states behavior. These are called simple fluids. Various types of

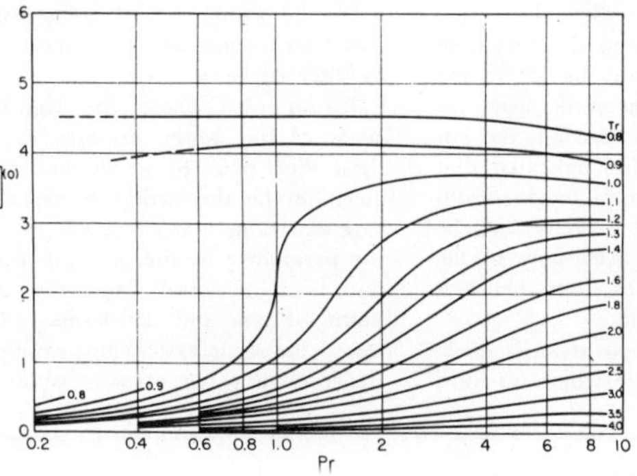

Figure 1. Enthalpy decrease from the ideal gas for a simple fluid

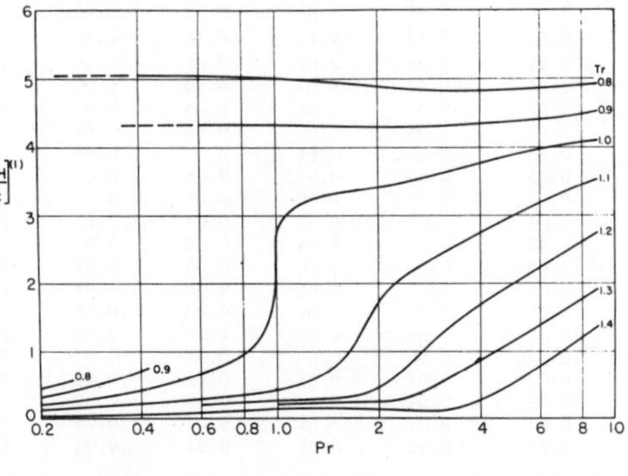

Figure 2. Change with acentric factor of the enthalpy decrease from the ideal gas

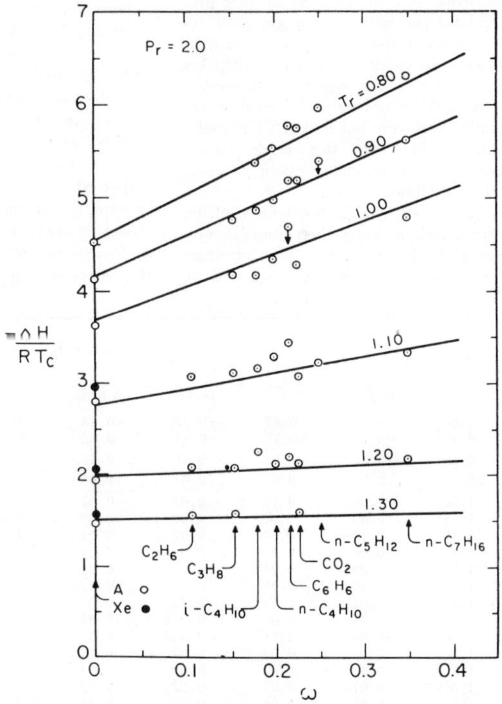

Figure 3. A comparison of calculated and observed values of the enthalpy difference from the ideal gas at the same pressure and temperature. All values are for reduced pressure of 2.0 and at the reduced temperatures indicated. The acentric factor is a constant for a particular substance, consequently the points lie in vertical columns as labeled

Table I. Values of $[\log (f/P)]^0$

T_r	P_r 0.2	0.4	0.6	0.8	1.0	1.2	1.4	1.6	1.8	2.0	2.2	2.4	2.6
0.80	−0.060	−0.262	−0.425	−0.535	−0.618	−0.683	−0.736	−0.780	−0.817	−0.849	−0.877	−0.901	−0.922
0.85	−0.046	−0.120	−0.281	−0.392	−0.474	−0.539	−0.592	−0.636	−0.673	−0.705	−0.733	−0.757	−0.779
0.90	−0.0425	−0.0875	−0.163	−0.273	−0.356	−0.421	−0.474	−0.517	−0.554	−0.587	−0.614	−0.639	−0.660
0.95	−0.033	−0.070	−0.112	−0.173	−0.255	−0.319	−0.372	−0.415	−0.452	−0.483	−0.511	−0.535	−0.557
1.00	−0.028	−0.059	−0.094	−0.131	−0.175	−0.237	−0.287	−0.330	−0.367	−0.398	−0.425	−0.449	−0.470
1.05	−0.024	−0.051	−0.079	−0.109	−0.142	−0.178	−0.218	−0.257	−0.292	−0.322	−0.349	−0.372	−0.393
1.10	−0.021	−0.044	−0.067	−0.093	−0.120	−0.147	−0.177	−0.207	−0.237	−0.264	−0.289	−0.311	−0.331
1.15	−0.018	−0.037	−0.058	−0.079	−0.101	−0.123	−0.146	−0.170	−0.194	−0.217	−0.238	−0.258	−0.276
1.20	−0.016	−0.032	−0.050	−0.067	−0.086	−0.104	−0.124	−0.143	−0.163	−0.182	−0.200	−0.217	−0.233
1.25	−0.014	−0.029	−0.044	−0.059	−0.075	−0.091	−0.107	−0.123	−0.139	−0.155	−0.171	−0.186	−0.199
1.3	−0.012	−0.025	−0.038	−0.051	−0.065	−0.078	−0.092	−0.106	−0.119	−0.133	−0.146	−0.159	−0.171
1.4	−0.010	−0.021	−0.031	−0.041	−0.052	−0.062	−0.072	−0.082	−0.092	−0.102	−0.111	−0.120	−0.130
1.5	−0.008	−0.016	−0.024	−0.032	−0.040	−0.047	−0.055	−0.063	−0.070	−0.078	−0.085	−0.092	−0.099
1.6	−0.007	−0.013	−0.019	−0.026	−0.032	−0.038	−0.044	−0.050	−0.056	−0.062	−0.067	−0.072	−0.077
1.7	−0.005	−0.010	−0.015	−0.020	−0.025	−0.030	−0.034	−0.039	−0.043	−0.047	−0.051	−0.056	−0.059
1.8	−0.004	−0.008	−0.012	−0.015	−0.019	−0.022	−0.026	−0.030	−0.033	−0.036	−0.039	−0.042	−0.045
1.9	−0.003	−0.006	−0.009	−0.012	−0.015	−0.018	−0.020	−0.023	−0.025	−0.028	−0.030	−0.033	−0.035
2.0	−0.002	−0.004	−0.007	−0.009	−0.011	−0.013	−0.015	−0.017	−0.019	−0.021	−0.023	−0.025	−0.026
2.5	0.000	0.000	0.000	0.000	−0.001	−0.001	−0.001	−0.001	−0.001	−0.001	−0.001	−0.001	−0.001
3.0	0.000	+0.001	+0.001	+0.002	+0.002	+0.003	+0.003	+0.004	+0.004	+0.005	+0.005	+0.006	+0.007
3.5	+0.001	0.002	0.003	0.003	0.004	0.005	0.006	0.007	0.008	0.009	0.010	0.011	0.012
4.0	0.001	0.002	0.003	0.005	0.006	0.007	0.008	0.009	0.010	0.011	0.012	0.013	0.014

molecular shapes and molecular dipole moments might be expected to cause different deviations of macroscopic properties from that of simple fluids. It was shown, however, in that article (*11*) that the reduced theoretical second virial coefficients for a wide variety of these molecular types fall into a single family of curves which may be characterized by a single parameter. The only exception was molecules with large dipole moments, although there are unusual types of abnormality that were not tested. The molecules falling into this single family are those commonly called normal liquids or fluids.

This theory led to an extension of the corresponding states correlation involving a third parameter. The introduction of a third parameter suggested by Meissner and Seferian (*8*) has been considered subsequently by several authors. Lydersen, Greenkorn, and Hougen (*7*) adopted the use of the compressibility factor at the critical point for the third parameter in a purely empirical analysis. There are two weaknesses. First, the critical volume is a peculiarly inaccurate datum and consequently its use in the third parameter reduces the accuracy of subsequent results. Second, the authors' theoretical studies indicated that highly polar or hydrogen bonding substances such as water, could not be expected to conform accurately to the system for normal substances, hence it is better to exclude them.

Rowlinson (*25*) proposed an interesting theoretical model in which his third parameter has the effect of scale changes of temperature, volume, and other variables. This approach has been successful in the area of gases of moderate density, but it fails at higher densities and low temperatures. Specifically, it predicts the same critical compressibility factor for all substances in contradiction to observed facts.

Riedel (*16*) chose the logarithmic slope of the vapor pressure curve at the critical point α_k, as his third parameter, but he then adopted vapor pressure equations and effectively evaluated the parameter by use of vapor pressure data well removed from the critical point. Pitzer and coworkers (*14*) followed the same system but avoided reliance on a vapor pressure equation to

Table II. Values of $[\log (f/P)]^{(1)}$

T_r	P_r 0.2	0.4	0.6	0.8	1.0	1.2	1.4	1.6	1.8	2.0	2.2
0.80	−0.04	−0.47	−0.48	−0.48	−0.48	−0.49	−0.50	−0.50	−0.51	−0.51	−0.52
0.85	−0.03	−0.31	−0.31	−0.32	−0.33	−0.33	−0.34	−0.35	−0.35	−0.36	−0.37
0.90	−0.02	−0.04	−0.18	−0.20	−0.20	−0.21	−0.21	−0.22	−0.23	−0.23	−0.24
0.95	−0.01	−0.02	−0.03	−0.09	−0.10	−0.11	−0.12	−0.12	−0.13	−0.13	−0.14
1.00	−0.01	−0.01	−0.01	−0.02	−0.03	−0.03	−0.04	−0.05	−0.05	−0.06	−0.06
1.05	0.00	0.00	0.00	0.00	+0.01	+0.01	+0.01	+0.01	0.00	0.00	0.00
1.10	0.00	0.00	0.00	0.01	0.01	0.02	0.02	0.03	0.03	0.03	0.03
1.15	0.00	0.00	0.00	0.01	0.02	0.02	0.03	0.04	0.04	0.05	0.05
1.20	0.00	0.01	0.01	0.01	0.02	0.03	0.04	0.05	0.05	0.06	0.07
1.25	0.00	0.01	0.01	0.02	0.03	0.03	0.04	0.05	0.06	0.07	0.07
1.3	0.01	0.01	0.02	0.02	0.03	0.04	0.04	0.05	0.06	0.07	0.08
1.4	0.01	0.01	0.02	0.03	0.04	0.04	0.05	0.06	0.07	0.08	0.08
1.5	0.01	0.02	0.02	0.03	0.04	0.05	0.05	0.06	0.06	0.07	0.08
1.6	0.01	0.02	0.02	0.03	0.04	0.05	0.05	0.06	0.06	0.07	0.08
1.7	0.01	0.02	0.02	0.03	0.04	0.05	0.05	0.06	0.06	0.07	0.08
1.8	0.01	0.02	0.02	0.03	0.04	0.05	0.05	0.06	0.06	0.07	0.08
1.9	0.01	0.02	0.02	0.03	0.04	0.05	0.05	0.06	0.06	0.07	0.08
2.0	0.01	0.01	0.02	0.03	0.04	0.05	0.05	0.06	0.07	0.07	0.08
2.5	0.01	0.01	0.02	0.02	0.03	0.04	0.04	0.05	0.06	0.06	0.07
3.0	0.00	0.01	0.01	0.02	0.02	0.03	0.04	0.04	0.05	0.05	0.05
3.5	0.00	0.01	0.01	0.02	0.02	0.02	0.03	0.03	0.04	0.04	0.04
4.0	0.00	0.01	0.01	0.02	0.02	0.02	0.02	0.03	0.03	0.03	0.04

Table I. Values of $[\log (f/P)]^0$ (*Continued*)

T_r	P_r 2.8	3.0	3.2	3.4	3.6	3.8	4.0	4.5	5.0	6.0	7.0	8.0	9.0
0.80	−0.941	−0.957	−0.972	−0.985	−0.997	−1.007	−1.016	−1.035	−1.048	−1.064	−1.067	−1.063	−1.052
0.85	−0.797	−0.814	−0.829	−0.842	−0.854	−0.864	−0.874	−0.893	−0.907	−0.924	−0.929	−0.926	−0.917
0.90	−0.679	−0.696	−0.710	−0.724	−0.736	−0.746	−0.756	−0.775	−0.789	−0.807	−0.814	−0.813	−0.805
0.95	−0.575	−0.592	−0.607	−0.621	−0.632	−0.643	−0.652	−0.672	−0.687	−0.706	−0.713	−0.713	−0.707
1.00	−0.489	−0.505	−0.520	−0.534	−0.545	−0.556	−0.566	−0.586	−0.601	−0.620	−0.629	−0.630	−0.624
1.05	−0.411	−0.428	−0.442	−0.455	−0.467	−0.478	−0.488	−0.508	−0.523	−0.543	−0.552	−0.553	−0.549
1.10	−0.348	−0.364	−0.378	−0.391	−0.403	−0.413	−0.422	−0.442	−0.457	−0.477	−0.487	−0.489	−0.486
1.15	−0.293	−0.307	−0.321	−0.333	−0.344	−0.354	−0.363	−0.383	−0.397	−0.417	−0.427	−0.429	−0.426
1.20	−0.247	−0.261	−0.273	−0.285	−0.295	−0.305	−0.314	−0.332	−0.346	−0.366	−0.375	−0.378	−0.376
1.25	−0.212	−0.224	−0.236	−0.246	−0.256	−0.264	−0.273	−0.290	−0.304	−0.322	−0.331	−0.334	−0.332
1.3	−0.182	−0.193	−0.203	−0.212	−0.221	−0.229	−0.237	−0.253	−0.266	−0.283	−0.292	−0.295	−0.294
1.4	−0.138	−0.146	−0.154	−0.162	−0.169	−0.175	−0.181	−0.194	−0.205	−0.220	−0.228	−0.231	−0.229
1.5	−0.104	−0.112	−0.117	−0.124	−0.129	−0.134	−0.139	−0.149	−0.158	−0.170	−0.176	−0.178	−0.176
1.6	−0.082	−0.087	−0.092	−0.096	−0.100	−0.104	−0.108	−0.116	−0.123	−0.132	−0.137	−0.138	−0.136
1.7	−0.063	−0.067	−0.071	−0.074	−0.077	−0.080	−0.083	−0.089	−0.094	−0.101	−0.105	−0.105	−0.102
1.8	−0.048	−0.051	−0.053	−0.056	−0.058	−0.060	−0.063	−0.067	−0.071	−0.076	−0.078	−0.077	−0.074
1.9	−0.037	−0.039	−0.041	−0.043	−0.045	−0.046	−0.048	−0.051	−0.054	−0.057	−0.057	−0.055	−0.051
2.0	−0.028	−0.029	−0.031	−0.032	−0.033	−0.034	−0.035	−0.037	−0.039	−0.040	−0.039	−0.038	−0.034
2.5	−0.001	−0.001	−0.001	−0.001	0.000	0.000	0.000	+0.001	+0.003	+0.006	+0.011	+0.016	+0.022
3.0	+0.007	+0.008	+0.009	+0.010	+0.011	+0.012	+0.012	0.015	0.017	0.023	0.028	0.035	0.042
3.5	0.013	0.014	0.015	0.016	0.017	0.018	0.020	0.022	0.025	0.031	0.038	0.044	0.051
4.0	0.015	0.016	0.017	0.019	0.020	0.021	0.022	0.025	0.028	0.034	0.040	0.047	0.054

give the slope at the critical point. Consequently, the reduced vapor pressure was taken at a point well removed from the critical point as our fundamental definition of the third parameter, ω, specifically defined as

$$\omega = -\log (P/P_c)_{sat.} - 1.000 \quad \text{at } T_r = 0.700$$

The name "acentric factor" was adopted to emphasize that the factor measures the deviation of the intermolecular potential function from that of the simple spherical molecules of the simple fluids, argon, krypton, and xenon. This parameter is empirically advantageous because the vapor pressure is easily measured with high accuracy and is a very sensitive function of the deviation from simple fluid behavior.

The second article (*14*) presented a general test of this correlation method with volumetric data over the range of 0 to 9 for reduced pressure and 0.8 to 4.0 for reduced temperature. It was found that an equation linear in the acentric factor was adequate. Hence the compressibility factor is given by the equation

$$z = z^{(0)}(P_r, T_r) + \omega z^{(1)}(P_r, T_r) \qquad (1)$$

where $z^{(0)}$ and $z^{(1)}$ are universal functions of the reduced temperature and pressure. These functions were developed and tabulated (*14*). The maximum deviations between experiment and prediction were reduced by about a factor of ten as compared to simple corresponding states correlations.

Opfell, Sage, and Pitzer (*10*) presented the results of an application of the Benedict equation (*1*) to this system. The functions $z^{(0)}$ and $z^{(1)}$ were represented as separate functions of the Benedict type. Several values were assumed for the reduced coefficient in the exponent, γ, and the remaining 14 coefficients, 7 for $z^{(0)}$ and 7 for $z^{(1)}$, were adjusted by least squares to represent the *n*-paraffins, methane through decane. The accuracy of prediction was found to be nearly as good as that when the Benedict equation is fitted separately to the data for each substance. Nevertheless, the error in some regions of pressure and temperature is still substantial, consequently the authors prefer to use the directly evaluated functions of $z^{(0)}$ and $z^{(1)}$ (*14*).

A summary is presented for the properties already treated by the acentric factor method or the equivalent method of Riedel.

Table II. Values of $[\log (f/P)]^{(1)}$ (*Continued*)

T_r	P_r 2.4	2.6	2.8	3.0	4.0	5.0	6.0	7.0	8.0	9.0
0.80	−0.52	−0.53	−0.53	−0.54	−0.56	−0.59	−0.61	−0.63	−0.65	−0.67
0.85	−0.37	−0.38	−0.38	−0.39	−0.41	−0.44	−0.46	−0.48	−0.50	−0.51
0.90	−0.24	−0.25	−0.26	−0.26	−0.29	−0.31	−0.33	−0.35	−0.36	−0.38
0.95	−0.15	−0.15	−0.16	−0.16	−0.18	−0.20	−0.22	−0.24	−0.26	−0.27
1.00	−0.07	−0.07	−0.08	−0.08	−0.10	−0.12	−0.13	−0.15	−0.17	−0.18
1.05	0.00	−0.01	−0.01	−0.01	−0.03	−0.05	−0.06	−0.07	−0.09	−0.11
1.10	0.03	+0.03	+0.03	+0.03	+0.02	0.00	−0.01	−0.02	−0.03	−0.05
1.15	0.05	0.06	0.06	0.06	0.05	0.05	+0.04	+0.02	+0.01	0.00
1.20	0.07	0.08	0.08	0.08	0.09	0.09	0.08	0.07	0.07	0.06
1.25	0.08	0.09	0.09	0.10	0.11	0.11	0.11	0.10	0.10	0.09
1.3	0.08	0.09	0.10	0.10	0.12	0.13	0.13	0.13	0.12	0.12
1.4	0.09	0.10	0.11	0.11	0.13	0.15	0.15	0.16	0.16	0.16
1.5	0.08	0.09	0.10	0.11	0.13	0.15	0.16	0.17	0.17	0.18
1.6	0.08	0.09	0.10	0.11	0.14	0.16	0.18	0.19	0.20	0.21
1.7	0.08	0.09	0.10	0.11	0.14	0.16	0.18	0.20	0.21	0.23
1.8	0.08	0.09	0.10	0.11	0.14	0.16	0.19	0.21	0.23	0.24
1.9	0.08	0.09	0.10	0.11	0.14	0.16	0.19	0.21	0.23	0.25
2.0	0.08	0.09	0.09	0.10	0.13	0.16	0.19	0.21	0.23	0.26
2.5	0.07	0.08	0.08	0.09	0.12	0.14	0.17	0.19	0.22	0.24
3.0	0.06	0.06	0.07	0.07	0.10	0.12	0.15	0.17	0.20	0.22
3.5	0.05	0.05	0.06	0.06	0.08	0.10	0.13	0.15	0.17	0.19
4.0	0.04	0.04	0.05	0.05	0.07	0.09	0.10	0.12	0.14	0.15

Table III. Values of $\left[\frac{H^* - H}{RT_c}\right]^{(0)}$

T_r	P_r 0.2	0.4	0.6	0.8	1.0	1.2	1.4	1.6	1.8	2.0	2.2	2.4	2.6
0.80	0.37	4.52	4.52	4.52	4.52	4.52	4.52	4.52	4.52	4.52	4.52	4.52	4.52
0.85	0.32	4.35	4.35	4.35	4.34	4.34	4.33	4.32	4.32	4.32	4.32	4.32	4.32
0.90	0.27	0.60	4.06	4.10	4.14	4.14	4.15	4.15	4.14	4.13	4.13	4.12	4.12
0.95	0.23	0.52	0.86	3.69	3.80	3.85	3.87	3.88	3.89	3.90	3.90	3.90	3.90
1.00	0.21	0.45	0.76	1.15	2.35	3.09	3.32	3.44	3.52	3.57	3.60	3.63	3.65
1.05	0.19	0.40	0.64	0.95	1.35	1.94	2.54	2.86	3.07	3.21	3.30	3.36	3.39
1.10	0.17	0.36	0.57	0.82	1.10	1.44	1.83	2.25	2.55	2.75	2.89	3.00	3.08
1.15	0.14	0.30	0.49	0.70	0.93	1.19	1.48	1.78	2.07	2.33	2.52	2.67	2.78
1.20	0.13	0.27	0.44	0.63	0.83	1.03	1.25	1.49	1.73	1.95	2.13	2.30	2.44
1.25	0.12	0.25	0.39	0.56	0.73	0.91	1.09	1.29	1.50	1.70	1.87	2.03	2.17
1.3	0.11	0.23	0.36	0.50	0.66	0.81	0.97	1.14	1.32	1.49	1.64	1.79	1.93
1.4	0.09	0.19	0.31	0.42	0.54	0.67	0.80	0.94	1.08	1.23	1.36	1.47	1.59
1.5	0.09	0.18	0.29	0.39	0.49	0.59	0.70	0.80	0.93	1.04	1.15	1.26	1.36
1.6	0.09	0.18	0.27	0.36	0.45	0.54	0.62	0.71	0.81	0.91	1.00	1.09	1.18
1.7	0.08	0.16	0.25	0.33	0.41	0.48	0.56	0.64	0.71	0.80	0.87	0.95	1.02
1.8	0.07	0.15	0.23	0.30	0.37	0.44	0.51	0.58	0.64	0.71	0.78	0.84	0.90
1.9	0.06	0.13	0.19	0.26	0.33	0.40	0.46	0.51	0.57	0.63	0.68	0.73	0.78
2.0	0.06	0.12	0.18	0.24	0.30	0.36	0.42	0.46	0.51	0.55	0.59	0.64	0.69
2.5	0.04	0.08	0.12	0.16	0.19	0.22	0.25	0.28	0.31	0.34	0.37	0.40	0.43
3.0	0.03	0.05	0.07	0.09	0.11	0.14	0.16	0.18	0.20	0.22	0.24	0.26	0.28
3.5	0.02	0.04	0.05	0.06	0.07	0.09	0.10	0.11	0.12	0.13	0.15	0.16	0.17
4.0	0.01	0.02	0.03	0.04	0.04	0.05	0.06	0.07	0.08	0.08	0.09	0.10	0.10

Table IV. Values of $\left[\frac{H^* - H}{RT_c}\right]^{(0)}$ in the Critical Region

T_r	P_r 0.8	0.9	1.0	1.1	1.2	1.3	1.4	1.5	1.6	1.7	1.8	1.9	2.0
0.98	1.25	3.2	3.35	3.43	3.50	3.55	3.59	3.63	3.65	3.67	3.68	3.70	3.71
0.99	1.19	1.6	3.1	3.20	3.31	3.39	3.46	3.51	3.55	3.58	3.61	3.63	3.64
1.00	1.15	1.5	2.35	2.9	3.09	3.23	3.32	3.39	3.44	3.48	3.52	3.54	3.57
1.01	1.11	1.38	1.8	2.5	2.86	3.05	3.18	3.27	3.34	3.39	3.44	3.47	3.50
1.02	1.07	1.31	1.6	2.1	2.59	2.85	3.02	3.14	3.23	3.30	3.36	3.40	3.44
1.03	1.03	1.24	1.52	1.88	2.33	2.63	2.87	3.02	3.12	3.21	3.27	3.32	3.36
1.04	0.99	1.18	1.43	1.74	2.12	2.47	2.71	2.87	2.99	3.09	3.17	3.23	3.28
1.05	0.95	1.13	1.35	1.62	1.94	2.27	2.54	2.73	2.86	2.98	3.07	3.14	3.21
1.06	0.92	1.08	1.29	1.53	1.80	2.09	2.37	2.58	2.74	2.88	2.98	3.06	3.13
1.07	0.89	1.04	1.23	1.45	1.69	1.96	2.22	2.43	2.61	2.76	2.88	2.97	3.04
1.08	0.87	1.02	1.18	1.37	1.59	1.83	2.07	2.29	2.49	2.64	2.77	2.86	2.95
1.09	0.84	0.99	1.14	1.33	1.51	1.72	1.94	2.16	2.37	2.54	2.66	2.77	2.86
1.10	0.82	0.96	1.10	1.26	1.44	1.64	1.83	2.05	2.25	2.42	2.55	2.66	2.75
1.11	0.80	0.93	1.06	1.21	1.38	1.55	1.74	1.94	2.15	2.31	2.45	2.57	2.66
1.13	0.74	0.87	1.00	1.14	1.28	1.44	1.59	1.77	1.95	2.11	2.25	2.38	2.50
1.15	0.70	0.82	0.93	1.06	1.19	1.33	1.48	1.62	1.78	1.92	2.07	2.20	2,33

Table V. Values of $\left[\frac{H^* - H}{RT_c}\right]^{(1)}$

T_r	P_r 0.2	0.4	0.6	0.8	1.0	1.2	1.4	1.6	1.8	2.0	2.2
0.80	0.44	5.05	5.04	5.02	4.98	4.97	4.94	4.93	4.90	4.88	4.87
0.85	0.37	4.74	4.71	4.70	4.67	4.65	4.63	4.63	4.63	4.60	4.57
0.90	0.31	0.71	4.33	4.32	4.31	4.31	4.30	4.30	4.30	4.30	4.29
0.95	0.25	0.55	1.01	3.83	3.80	3.81	3.84	3.84	3.87	3.87	3.88
1.00	0.20	0.41	0.68	0.95	2.66	3.17	3.27	3.33	3.38	3.41	3.45
1.05	0.14	0.27	0.40	0.54	0.68	1.22	1.77	2.19	2.45	2.59	2.71
1.10	0.12	0.22	0.29	0.36	0.42	0.52	0.69	0.92	1.32	1.71	1.97
1.15	0.09	0.17	0.23	0.28	0.32	0.36	0.39	0.45	0.58	0.81	1.08
1.20	0.08	0.14	0.20	0.24	0.28	0.29	0.29	0.32	0.37	0.47	0.59
1.25	0.06	0.13	0.18	0.21	0.25	0.25	0.26	0.28	0.28	0.30	0.36
1.3	0.03	0.06	0.10	0.13	0.17	0.19	0.24	0.23	0.24	0.25	0.27
1.4	0.02	0.04	0.06	0.08	0.10	0.11	0.11	0.12	0.12	0.12	0.14
1.5	0.01	0.01	0.02	0.03	0.04	0.4	0.04	0.04	0.04	0.04	0.05
1.6	0	0	0	0	0	0	0	0	−0.01	−0.01	−0.01
1.7	−0.01	−0.01	−0.02	−0.03	−0.04	−0.04	−0.05	−0.05	−0.06	−0.06	−0.07
1.8	−0.01	−0.02	−0.03	−0.04	−0.06	−0.07	−0.08	−0.09	−0.10	−0.10	−0.11
1.9	−0.02	−0.03	−0.05	−0.07	−0.09	−0.11	−0.13	−0.15	−0.17	−0.19	−0.20
2.0	−0.02	−0.04	−0.06	−0.08	−0.10	−0.13	−0.16	−0.19	−0.22	−0.25	−0.25
2.5	−0.03	−0.06	−0.10	−0.13	−0.17	−0.20	−0.23	−0.26	−0.29	−0.32	−0.35
3.0	−0.04	−0.08	−0.13	−0.17	−0.21	−0.25	−0.29	−0.33	−0.37	−0.41	−0.45
3.5	−0.04	−0.08	−0.13	−0.17	−0.21	−0.25	−0.29	−0.34	−0.37	−0.42	−0.47
4.0	−0.04	−0.08	−0.13	−0.17	−0.21	−0.25	−0.29	−0.34	−0.37	−0.42	−0.46

Table III. Values of $\left[\frac{H^* - H}{RT_c}\right]^{(0)}$ (Continued)

T_r	P_r 2.8	3.0	3.2	3.4	3.6	3.8	4.0	4.5	5.0	6.0	7.0	8.0	9.0
0.80	4.52	4.52	4.51	4.51	4.50	4.50	4.49	4.46	4.43	4.38	4.34	4.29	4.24
0.85	4.32	4.32	4.32	4.31	4.31	4.31	4.30	4.28	4.24	4.20	4.16	4.10	4.04
0.90	4.12	4.11	4.10	4.10	4.10	4.10	4.10	4.09	4.05	4.02	3.99	3.92	3.85
0.95	3.90	3.90	3.90	3.90	3.89	3.89	3.89	3.85	3.85	3.84	3.81	3.76	3.71
1.00	3.68	3.70	3.71	3.71	3.70	3.70	3.70	3.69	3.68	3.67	3.64	3.60	3.57
1.05	3.42	3.44	3.46	3.47	3.49	3.50	3.51	3.51	3.50	3.49	3.48	3.46	3.45
1.10	3.15	3.20	3.24	3.27	3.29	3.30	3.32	3.34	3.34	3.34	3.33	3.32	3.32
1.15	2.86	2.93	2.98	3.02	3.05	3.09	3.12	3.17	3.18	3.18	3.19	3.19	3.20
1.20	2.56	2.66	2.73	2.78	2.82	2.87	2.91	2.99	3.02	3.05	3.07	3.07	3.08
1.25	2.29	2.39	2.48	2.55	2.61	2.67	2.72	2.82	2.87	2.92	2.93	2.95	2.98
1.3	2.05	2.16	2.24	2.32	2.39	2.45	2.52	2.63	2.72	2.79	2.81	2.84	2.88
1.4	1.70	1.79	1.88	1.96	2.04	2.11	2.18	2.32	2.43	2.53	2.58	2.62	2.65
1.5	1.45	1.53	1.61	1.68	1.75	1.82	1.88	2.01	2.12	2.25	2.33	2.39	2.41
1.6	1.26	1.33	1.39	1.44	1.51	1.57	1.64	1.78	1.87	2.01	2.12	2.19	2.21
1.7	1.08	1.15	1.21	1.27	1.33	1.39	1.45	1.57	1.66	1.79	1.91	2.00	2.03
1.8	0.95	1.01	1.06	1.12	1.18	1.24	1.31	1.42	1.50	1.62	1.74	1.83	1.86
1.9	0.82	0.87	0.92	0.97	1.02	1.08	1.13	1.23	1.32	1.44	1.54	1.63	1.67
2.0	0.74	0.78	0.82	0.85	0.89	0.93	0.97	1.07	1.14	1.25	1.33	1.43	1.46
2.5	0.45	0.47	0.50	0.52	0.54	0.56	0.58	0.63	0.67	0.75	0.81	0.87	0.91
3.0	0.30	0.31	0.33	0.34	0.36	0.37	0.38	0.41	0.44	0.50	0.53	0.56	0.58
3.5	0.18	0.19	0.20	0.21	0.22	0.23	0.23	0.25	0.26	0.28	0.30	0.30	0.29
4.0	0.11	0.11	0.11	0.12	0.12	0.12	0.13	0.13	0.14	0.14	0.14	0.13	0.12

Property	Source
Volume (unsaturated and saturated gas and liquid)	Pitzer, others (14)
Volume (saturated liquid)	Riedel (17)
Volume (saturated vapor)	Riedel (18)
Vapor pressure, heat and entropy of vaporization	Pitzer, others (14)
	Riedel (16, 18)
Second virial coefficient and thermodynamic functions derived therefrom	Pitzer, Curl (13)
Surface tension, liquid compressibility	Riedel (19)

Calculation of Thermodynamic Functions

In this article the enthalpy functions $\left(\frac{H^* - H}{RT_c}\right)^{(0)}$ and $\left(\frac{H^* - H}{RT_c}\right)^{(1)}$, the entropy functions $\left(\frac{S^* - S}{R}\right)^{(0)}$ and $\left(\frac{S^* - S}{R}\right)^{(1)}$, and the fugacity functions $[\log (f/P)]^{(0)}$ and $[\log (f/P)]^{(1)}$ are calculated from values of $z^{(0)}$ and $z^{(1)}$ and the entropy of vaporization (14). The reference state (*) is the hypothetical ideal gas at the same temperature and pressure. An equation for the second virial coefficient in terms of reduced variables and ω has been developed (13) and was used to obtain zero pressure intercepts. The complete thermodynamic functions are obtained from the equations

$$\log (f/P) = [\log (f/P)]^{(0)} + \omega\, [\log (f/P)]^{(1)} \tag{2a}$$

$$\frac{S^* - S}{R} = \left(\frac{S^* - S}{R}\right)^{(0)} + \omega \left(\frac{S^* - S}{R}\right)^{(1)} \tag{2b}$$

$$\frac{H^* - H}{RT_c} = \left(\frac{H^* - H}{RT_c}\right)^{(0)} + \omega \left(\frac{H^* - H}{RT_c}\right)^{(1)} \tag{2c}$$

The thermodynamic equations used in the calculation of the thermodynamic functions were

$$\ln (f/P) = \int_0^{P_r} \frac{(z - 1)}{P_r}\, dP_r \tag{3}$$

Table V. Values of $\left[\frac{H^* - H}{RT_c}\right]^{(1)}$ (Continued)

T_r	P_r 2.4	2.6	2.8	3.0	4.0	5.0	6.0	7.0	8.0	9.0
0.80	4.86	4.84	4.83	4.82	4.83	4.85	4.86	4.88	4.90	4.91
0.85	4.57	4.56	4.56	4.56	4.57	4.61	4.63	4.65	4.68	4.71
0.90	4.28	4.29	4.30	4.30	4.36	4.39	4.43	4.43	4.50	4.53
0.95	3.90	3.90	3.93	3.93	4.04	4.12	4.19	4.23	4.28	4.32
1.00	3.49	3.53	3.58	3.61	3.76	3.88	3.97	4.02	4.07	4.10
1.05	2.79	2.88	2.94	2.99	3.26	3.48	3.57	3.66	3.75	3.81
1.10	2.11	2.24	2.30	2.39	2.74	3.01	3.17	3.30	3.42	3.53
1.15	1.32	1.52	1.67	1.78	2.24	2.52	2.73	2.93	3.09	3.23
1.20	0.74	0.91	1.08	1.21	1.70	1.98	2.21	2.41	2.59	2.79
1.25	0.43	0.51	0.63	0.76	1.26	1.56	1.79	1.98	2.15	2.41
1.3	0.32	0.39	0.46	0.53	0.85	1.11	1.43	1.60	1.83	2.10
1.4	0.16	0.19	0.22	0.25	0.45	0.73	0.98	1.15	1.40	1.66
1.5	0.06	0.07	0.08	0.09	0.20	0.47	0.68	0.86	1.10	1.36
1.6	−0.01	−0.02	−0.02	−0.02	0.05	0.28	0.45	0.64	0.86	1.10
1.7	−0.08	−0.09	−0.10	−0.10	−0.07	0.11	0.25	0.43	0.63	0.85
1.8	−0.12	−0.13	−0.14	−0.16	−0.16	−0.03	0.08	0.23	0.42	0.61
1.9	−0.20	−0.21	−0.21	−0.21	−0.21	−0.15	−0.07	0.06	0.22	0.39
2.0	−0.25	−0.26	−0.26	−0.26	−0.28	−0.26	−0.19	−0.10	0.03	0.18
2.5	−0.39	−0.43	−0.45	−0.48	−0.53	−0.55	−0.57	−0.59	−0.63	−0.61
3.0	−0.49	−0.53	−0.57	−0.60	−0.74	−0.85	−0.95	−1.05	−1.14	−1.21
3.5	−0.51	−0.55	−0.59	−0.63	−0.83	−1.02	−1.19	−1.37	−1.50	−1.67
4.0	−0.50	−0.54	−0.58	−0.62	−0.84	−1.03	−1.25	−1.47	−1.66	−1.86

Table VI. Values of $\left[\frac{H^* - H}{RT_c}\right]^{(1)}$ in the Critical Region

T_r	P_r 0.8	1.0	1.2	1.4	1.6	1.8	2.0
0.98	1.15	3.28	3.39	3.48	3.55	3.57	3.59
0.99	1.0	3.05	3.28	3.39	3.44	3.47	3.51
1.00	0.95	2.66	3.17	3.27	3.33	3.38	3.41
1.01	0.84	1.47	2.97	3.11	3.18	3.24	3.29
1.02	0.75	1.19	2.61	2.92	3.00	3.08	3.14
1.03	0.68	0.99	2.10	2.67	2.82	2.91	3.00
1.04	0.60	0.82	1.60	2.30	2.55	2.66	2.80
1.05	0.54	0.68	1.22	1.77	2.19	2.45	2.59
1.06	0.49	0.61	0.96	1.40	1.89	2.18	2.42
1.07	0.44	0.57	0.80	1.12	1.58	1.95	2.25
1.08	0.42	0.51	0.69	0.95	1.35	1.71	2.07
1.09	0.38	0.46	0.59	0.83	1.10	1.52	1.91
1.10	0.36	0.42	0.52	0.69	0.92	1.32	1.71
1.11	0.33	0.40	0.47	0.61	0.80	1.08	1.47
1.13	0.29	0.34	0.40	0.47	0.59	0.78	1.04
1.15	0.28	0.32	0.36	0.39	0.45	0.58	0.81

$$\left(\frac{H^* - H}{RT_c}\right) = T_r^2 \int_0^{P_r} \left(\frac{1}{P_r}\right)\left(\frac{\partial z}{\partial T_r}\right) dP_r \quad (4)$$

$$\left(\frac{S^* - S}{R}\right) = \int_0^{P_r} \left(z - 1 + T\frac{\partial z}{\partial T}\right)\left(\frac{1}{P_r}\right) dP_r \quad (5)$$

and also note that

$$\left(\frac{S^* - S}{R}\right) = \left(\frac{H^* - H}{RT_c}\right)\left(\frac{1}{T_r}\right) + \ln (f/P) \quad (6)$$

$z^{(0)}$ and $z^{(1)}$ were differentiated graphically with respect to the T_r. Then, the integrations were made graphically which was satisfactory except in the neighborhood of the critical point. In the critical region it is still possible to obtain accurate integrals for the fugacity function but the derivatives are very uncertain. Consequently the enthalpy and entropy functions are of low accuracy near the critical point.

In calculating $\left(\frac{H^*-H}{RT_c}\right)^{(1)}$, $\left(\frac{S^*-S}{R}\right)^{(1)}$, and $[\ln(f/P)]^{(1)}$ it was necessary to remember that, as ω increases, the condensation occurs at lower pressure for the same temperature. This results in a special contribution to these quantities for the liquid region. For example, this special contribution to $[\ln(f/P)]^{(1)}$ is given by Equation 7,

$$\Delta[\ln (f/P)]^{(1)} = \left[\left(\frac{\partial[\ln (f/P)]^{(0)}}{\partial P_r}\right)^{\text{gas}} - \left(\frac{\partial[\ln (f/P)]^{(0)}}{\partial P_r}\right)_{\text{liquid}}\right]\left(\frac{\partial P_r}{\partial \omega}\right)_{T,\ \text{sat.}} \quad (7)$$

Analogous equations apply to the other functions.

If one is near the two-phase region a little caution is required in using these tables for fluids with a high ω. One should first determine whether a fluid is a liquid or a gas by using the $-(\log P_r)^{(0)}$ and $-(\log P_r)^{(1)}$ functions [Table VI (*14*)]. As a fluid with a high ω condenses at a lower P_r than a simple fluid ($\omega = 0$), the liquid side of the tables must be extrapolated. This extrapolation should be done with respect to pressure at constant temperature.

The resulting fugacity, enthalpy, and entropy functions are presented in Tables I through VIII. The enthalpy functions in Figures 1 and 2 show the qualitative similarity of the two functions.

More closely spaced tables of $\left(\frac{H^* - H}{RT_c}\right)^{(0)}$ and $\left(\frac{H^* - H}{RT_c}\right)^{(1)}$ are presented for the critical region. Simple graphs are desirable for accurate interpolation in this region. The $[\log(f/P)]^{(0)}$ and $[\log(f/P)]^{(1)}$ can be interpolated easily in the critical region, and $\left(\frac{S^*-S}{R}\right)^{(0)}$ and $\left(\frac{S^* - S}{R}\right)^{(1)}$ can be found from Equation 6.

Enthalpy values from these tables were compared with reported enthalpies of benzene (*3*), xenon (*9*), and some hydrocarbons (*4, 26*) by a series of graphs similar to Figure 3. Also, from the results of Joule-Thomson experiments reported by Roebuck on argon (*23*), nitrogen (*24*), and carbon dioxide (*22*) graphs of $\left(\frac{H^* - H}{RT_c}\right)$ were constructed by use of the ideal gas heat capacities and ΔS of vaporization from tables (*14*). Figure 3 shows that the need for the third parameter, ω, increases markedly as the liquid region is approached.

Lydersen, Greenkorn, and Hougen (*7*) have presented tables of thermodynamic properties of fluids based upon a modification of the corresponding

Table VII. Values of $\left[\frac{S^* - S}{R}\right]^{(0)}$

T_r	P_r 0.2	0.4	0.6	0.8	1.0	1.2	1.4	1.6	1.8	2.0	2.2	2.4	2.6
0.80	0.33	5.04	4.66	4.41	4.23	4.08	3.94	3.85	3.76	3.69	3.63	3.56	3.53
0.85	0.27	4.85	4.47	4.21	4.01	3.86	3.73	3.62	3.53	3.46	3.39	3.34	3.29
0.90	0.20	0.47	4.13	3.92	3.78	3.63	3.52	3.42	3.32	3.23	3.18	3.11	3.06
0.95	0.17	0.39	0.64	3.48	3.41	3.32	3.22	3.13	3.05	2.99	2.93	2.87	2.82
1.00	0.14	0.31	0.54	0.85	1.95	2.55	2.66	2.68	2.68	2.65	2.62	2.60	2.57
1.05	0.12	0.27	0.43	0.66	0.96	1.44	1.91	2.13	2.25	2.31	2.34	2.34	2.32
1.10	0.11	0.23	0.36	0.53	0.73	0.97	1.25	1.57	1.77	1.89	1.96	2.01	2.04
1.15	0.08	0.17	0.30	0.43	0.57	0.75	0.95	1.16	1.36	1.53	1.64	1.73	1.18
1.20	0.07	0.15	0.25	0.37	0.49	0.62	0.76	0.91	1.07	1.21	1.32	1.42	1.50
1.25	0.07	0.14	0.21	0.31	0.41	0.52	0.62	0.75	0.88	1.00	1.10	1.20	1.28
1.30	0.06	0.12	0.19	0.27	0.36	0.44	0.53	0.63	0.74	0.84	0.92	1.01	1.09
1.4	0.04	0.09	0.15	0.21	0.26	0.34	0.41	0.49	0.56	0.64	0.71	0.77	0.84
1.5	0.04	0.09	0.14	0.19	0.23	0.28	0.34	0.39	0.45	0.51	0.57	0.63	0.68
1.6	0.04	0.08	0.12	0.17	0.21	0.25	0.29	0.33	0.38	0.43	0.47	0.51	0.56
1.7	0.03	0.07	0.11	0.15	0.18	0.21	0.25	0.29	0.32	0.36	0.39	0.43	0.46
1.8	0.03	0.06	0.09	0.13	0.16	0.19	0.22	0.26	0.28	0.31	0.34	0.37	0.39
1.9	0.03	0.05	0.08	0.11	0.14	0.17	0.19	0.22	0.24	0.27	0.29	0.31	0.33
2.0	0.03	0.05	0.08	0.10	0.13	0.15	0.18	0.19	0.21	0.23	0.24	0.27	0.29
2.5	0.02	0.03	0.05	0.06	0.07	0.09	0.10	0.12	0.13	0.14	0.15	0.16	0.17
3.0	0.01	0.02	0.03	0.04	0.04	0.05	0.06	0.07	0.08	0.08	0.09	0.10	0.11
3.5	0.01	0.02	0.02	0.03	0.03	0.04	0.04	0.05	0.05	0.06	0.07	0.08	0.08
4.0	0.005	0.01	0.01	0.02	0.02	0.03	0.03	0.04	0.04	0.05	0.05	0.05	0.06

states hypothesis somewhat similar to the present one. They adopted the procedure of Meissner and Seferian (8) that the compressibility factor at the critical point be taken as the third parameter.

Table IX compares the authors' predictions of the enthalpy difference with those of Lydersen, Greenkorn, and Hougen and with experimental values. The list is an abridgment of the points chosen by Lydersen and others. The standard deviations for the predictions are: Lydersen and others: 0.165; this paper 0.075 in dimensionless units of (H/RT_c).

Surface Tension as a Criterion of a Normal Liquid

Up to this point the classification of fluids as normal or abnormal has been assumed. While the general class of normal liquids or fluids has been recognized for several decades (2, 5) it is desirable at this point to provide a definite criterion of normality. Such a criterion was proposed (14) to be the function $\delta = \mu^4 P_c^2/T_c^4$, where μ is the dipole moment of the molecule in Debye, P_c, the critical pressure in atmosphere, and T_c the critical temperature in ° K. If $\delta < 0.5 \times 10^{-6}$, then the substance was supposed to be a normal fluid, excluding hydrogen, helium, metals, and salts. Riedel (27) pointed out that δ is an unsatisfactory criterion of the normal fluid: as ethyl alcohol having δ of 0.5×10^{-6} is not a normal fluid; and sulfur dioxide having a δ of 1.2×10^{-6} is a normal fluid. Chemically this is understandable, since ethyl alcohol is known to hydrogen bond whereas sulfur dioxide cannot.

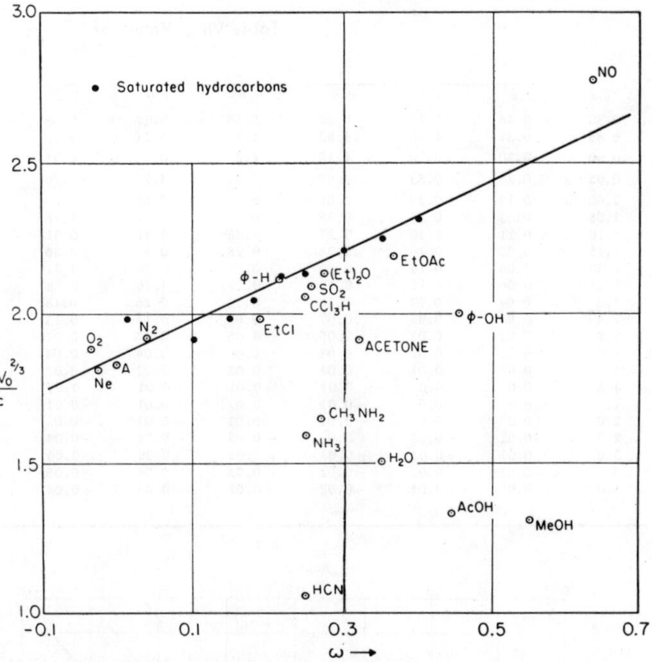

Figure 4. A criterion for a normal fluid. The reduced surface tension parameter $(\sigma_o V_o^{2/3}/T_c)$ as a function of the acentric factor ω. Points for normal fluids fall within 5% of the line of Equation 10

Table VII. Values of $\left[\frac{S^* - S}{R}\right]^{(0)}$ (Continued)

T_r	P_r 2.8	3.0	3.2	3.4	3.6	3.8	4.0	4.5	5.0	6.0	7.0	8.0	9.0
0.80	3.48	3.44	3.39	3.36	3.33	3.30	3.26	3.19	3.12	3.01	2.96	2.91	2.88
0.85	3.25	3.21	3.18	3.13	3.11	3.08	3.05	2.98	2.90	2.82	2.76	2.70	2.64
0.90	3.01	2.97	2.92	2.89	2.86	2.83	2.81	2.76	2.68	2.61	2.56	2.48	2.42
0.95	2.78	2.74	2.70	2.67	2.64	2.61	2.58	2.57	2.48	2.43	2.37	2.31	2.29
1.00	2.55	2.54	2.51	2.48	2.44	2.42	2.40	2.34	2.30	2.24	2.19	2.14	2.13
1.05	2.31	2.30	2.28	2.26	2.25	2.23	2.22	2.17	2.13	2.08	2.05	2.02	2.02
1.10	2.06	2.07	2.07	2.07	2.06	2.05	2.05	2.02	1.98	1.94	1.91	1.89	1.90
1.15	1.82	1.84	1.85	1.86	1.86	1.87	1.88	1.88	1.85	1.81	1.79	1.78	1.80
1.20	1.57	1.62	1.64	1.66	1.67	1.69	1.70	1.72	1.72	1.70	1.69	1.68	1.70
1.25	1.34	1.39	1.44	1.47	1.50	1.53	1.55	1.59	1.60	1.59	1.58	1.59	1.62
1.30	1.15	1.22	1.25	1.29	1.33	1.35	1.39	1.44	1.48	1.49	1.48	1.51	1.54
1.4	0.89	0.94	0.99	1.03	1.07	1.11	1.14	1.21	1.26	1.30	1.32	1.34	1.37
1.5	0.72	0.76	0.81	0.83	0.87	0.91	0.93	0.99	1.05	1.11	1.15	1.18	1.20
1.6	0.60	0.63	0.66	0.68	0.71	0.74	0.78	0.84	0.89	0.95	1.01	1.05	1.07
1.7	0.49	0.52	0.55	0.58	0.61	0.64	0.66	0.72	0.76	0.82	0.88	0.94	0.96
1.8	0.42	0.44	0.47	0.49	0.52	0.55	0.58	0.63	0.67	0.73	0.79	0.84	0.86
1.9	0.35	0.37	0.39	0.41	0.43	0.46	0.48	0.53	0.57	0.63	0.68	0.73	0.76
2.0	0.31	0.33	0.34	0.35	0.37	0.39	0.41	0.45	0.48	0.54	0.58	0.63	0.65
2.5	0.18	0.19	0.20	0.21	0.22	0.23	0.23	0.25	0.27	0.31	0.35	0.37	0.41
3.0	0.12	0.12	0.13	0.14	0.15	0.16	0.16	0.18	0.19	0.22	0.24	0.27	0.29
3.5	0.09	0.09	0.10	0.10	0.11	0.11	0.11	0.12	0.13	0.15	0.17	0.19	0.20
4.0	0.06	0.06	0.07	0.07	0.07	0.08	0.08	0.09	0.10	0.11	0.13	0.14	0.15

Table VIII. Values of $\left[\frac{S^* - S}{R}\right]^{(1)}$

T_r	P_r 0.2	0.4	0.6	0.8	1.0	1.2	1.4	1.6	1.8	2.0	2.2
0.80	0.46	5.23	5.20	5.16	5.12	5.08	5.03	5.00	4.96	4.92	4.89
0.85	0.37	4.86	4.82	4.79	4.74	4.70	4.66	4.63	4.61	4.58	4.54
0.90	0.30	0.70	4.38	4.36	4.33	4.31	4.29	4.27	4.26	4.24	4.22
0.95	0.24	0.53	0.98	3.81	3.77	3.76	3.76	3.76	3.76	3.76	3.76
1.00	0.19	0.39	0.65	0.90	2.60	3.09	3.18	3.23	3.26	3.28	3.31
1.05	0.13	0.26	0.38	0.51	0.67	1.18	1.71	2.11	2.34	2.47	2.58
1.10	0.11	0.20	0.27	0.35	0.41	0.51	0.68	0.91	1.27	1.63	1.86
1.15	0.09	0.16	0.22	0.28	0.32	0.36	0.41	0.48	0.60	0.81	1.06
1.20	0.08	0.14	0.20	0.24	0.28	0.31	0.33	0.38	0.44	0.53	0.65
1.25	0.06	0.12	0.17	0.22	0.26	0.28	0.30	0.33	0.35	0.39	0.46
1.3	0.04	0.07	0.10	0.13	0.16	0.18	0.20	0.22	0.24	0.26	0.31
1.4	0.03	0.05	0.07	0.09	0.11	0.13	0.14	0.15	0.16	0.17	0.20
1.5	0.02	0.03	0.05	0.06	0.07	0.07	0.08	0.09	0.09	0.10	0.11
1.6	0.01	0.02	0.03	0.04	0.04	0.04	0.05	0.05	0.06	0.06	0.07
1.7	0.00	0.01	0.01	0.02	0.02	0.02	0.02	0.03	0.03	0.03	0.04
1.8	0.0	0.0	0.01	0.01	0.01	0.01	0.01	0.01	0.01	0.01	0.01
1.9	0.0	0.0	−0.01	−0.01	−0.01	−0.01	−0.02	−0.02	−0.03	−0.03	−0.02
2.0	0.0	0.0	−0.01	−0.01	−0.01	−0.02	−0.03	−0.04	−0.05	−0.06	−0.05
2.5	−0.01	−0.02	−0.02	−0.03	−0.04	−0.04	−0.05	−0.06	−0.06	−0.07	−0.08
3.0	−0.01	−0.02	−0.03	−0.04	−0.05	−0.06	−0.07	−0.08	−0.09	−0.09	−0.10
3.5	−0.01	−0.02	−0.02	−0.03	−0.04	−0.05	−0.06	−0.07	−0.08	−0.08	−0.09
4.0	−0.01	−0.01	−0.02	−0.02	−0.03	−0.04	−0.05	−0.06	−0.07	−0.08	−0.08

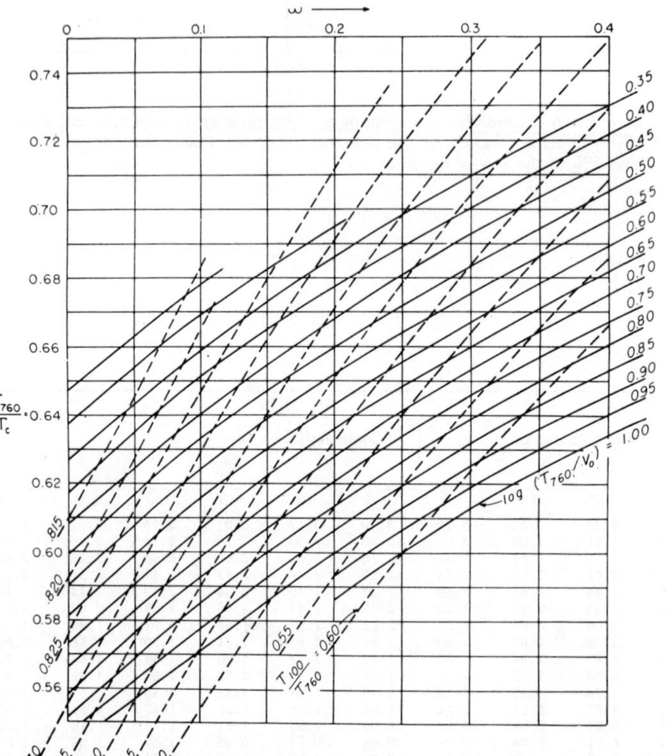

Figure 5. T_{760}/T_c and ω as functions of T_{100}/T_{760} and $\log (T_{760}/V_o)$. The lines running steeply upward from left to right are constant values of T_{100}/T_{760}. The lines running upward less steeply from left to right are constant values of $\log (T_{760}/V_o)$ with T_{760} in ° K. and V_o in cc./mole. By finding the point given by the values of T_{100}/T_{760} and $\log (T_{760}/V_o)$ the values of T_{760}/T_c and ω may be read from the ordinate and abscissa, respectively

In a hydrogen bonding substances the hydrogen bond raises the critical temperature a great deal, but it does not raise the surface energy very much, because good hydrogen bonds can be formed in the surface by orientation of the molecules. Therefore a dimensionless quantity involving the ratio of surface tension to critical temperature might be expected to indicate the presence of hydrogen bonding or other specific bonds of limited number which yield abnormal behavior (nitrogen dioxide, for example).

Riedel (*79*) has defined a hypothetical surface tension of the liquid at absolute zero, σ_0, and developed equations for the dependence of σ/σ_0 on temperature and $(\sigma_0 V_0^{2/3}/T_c)$ on α_k (or ω). Here, σ_0 is the surface tension and V_0 the hypothetical volume of the liquid at absolute zero. Riedel (*77*) presented a correlation of liquid densities in terms of a hypothetical density of the liquid at 0° K. Table X is an abridgment of Riedel's tables for the region of principal interest. Now V_0 is calculated from a liquid density using Table X and the equation

$$V_0 = (M/d)(d/d_0) \quad (8)$$

σ/σ_0 is given by Riedel for normal liquids as

$$\sigma/\sigma_0 = (1 - T_r)^{11/9} \quad (9)$$

The calculated values of $(\sigma_0 V_0^{2/3}/T_c)$ for an array of substances are shown in Figure 4 as a function of the acentric factor. The equation of the line characteristic of the normal substance is

$$\sigma_0 V_0^{2/3}/T_c = 1.86 + 1.18\,\omega, \text{ (dynes/cm.)(cc./mole)}^{2/3}(\text{° K})^{-1} \quad (10)$$

This is the same equation as Riedel

Table VIII. Values of $\left[\frac{S^* - S}{R}\right]^{(1)}$ *(Continued)*

T_r	P_r 2.4	2.6	2.8	3.0	4.0	5.0	6.0	7.0	8.0	9.0
0.80	4.86	4.83	4.81	4.79	4.74	4.71	4.68	4.65	4.62	4.60
0.85	4.52	4.50	4.48	4.46	4.43	4.41	4.39	4.37	4.36	4.35
0.90	4.20	4.19	4.19	4.18	4.18	4.17	4.17	4.16	4.16	4.15
0.95	3.76	3.76	3.77	3.77	3.83	3.87	3.90	3.90	3.91	3.92
1.00	3.34	3.37	3.41	3.42	3.53	3.61	3.66	3.67	3.68	3.68
1.05	2.65	2.72	2.77	2.82	3.03	3.20	3.26	3.32	3.36	3.39
1.10	1.99	2.11	2.16	2.23	2.53	2.75	2.86	2.96	3.04	3.10
1.15	1.28	1.45	1.58	1.68	2.07	2.30	2.46	2.60	2.71	2.81
1.20	0.79	0.94	1.09	1.21	1.62	1.85	2.03	2.18	2.31	2.38
1.25	0.53	0.61	0.71	0.83	1.25	1.50	1.68	1.82	1.95	2.02
1.3	0.36	0.41	0.46	0.51	0.77	0.98	1.23	1.36	1.53	1.74
1.4	0.22	0.24	0.26	0.29	0.45	0.67	0.85	0.98	1.16	1.34
1.5	0.12	0.13	0.15	0.17	0.26	0.46	0.61	0.74	0.90	1.09
1.6	0.08	0.08	0.09	0.10	0.17	0.33	0.46	0.59	0.74	0.90
1.7	0.04	0.04	0.05	0.05	0.10	0.22	0.33	0.45	0.59	0.73
1.8	0.01	0.01	0.02	0.02	0.05	0.14	0.23	0.34	0.46	0.58
1.9	−0.02	−0.01	−0.01	0	0.03	0.08	0.15	0.24	0.35	0.46
2.0	−0.05	−0.04	−0.04	−0.03	−0.01	0.03	0.09	0.16	0.25	0.35
2.5	−0.09	−0.09	−0.10	−0.10	−0.09	−0.08	−0.06	−0.05	−0.03	0
3.0	−0.11	−0.11	−0.12	−0.13	−0.15	−0.16	−0.17	−0.18	−0.18	−0.18
3.5	−0.09	−0.10	−0.11	−0.12	−0.16	−0.19	−0.21	−0.24	−0.26	−0.29
4.0	−0.09	−0.10	−0.10	−0.11	−0.14	−0.17	−0.21	−0.25	−0.28	−0.32

obtained (in terms of α_k) after consideration of a somewhat different group of substances.

If preferred, one may convert to the dimensionless parameter

$$\frac{\sigma_0 V_0^{2/3}}{RT_c} = 2.24 \times 10^{-8} + 1.42 \times 10^{-8}\,\omega \qquad (11)$$

The values of $\sigma_0 V_0^{2/3}/T_c$ for hydrogen bonding liquids are as low as 50% of the predicted value for a normal fluid. Also hydrogen, helium, and highly

Table IX. Comparison of Predicted and Experimental Enthalpy Values $(H^* - H)/RT_c$

n-Pentane

P_r	$T_r = 0.804$ C., P.[a]	Exptl.	L., G., H.[b]	$T_r = 0.875$ C., P.	Exptl.	L., G., H.	$T_r = 0.946$ C., P.	Exptl.	L., G., H.	$T_r = 1.017$ C., P.	Exptl.	L., G., H.	$T_r = 1.088$ C., P.	Exptl.	L., G., H.
0.41	5.77	5.79	5.72	5.32	5.29	5.22	0.68	0.65		0.54	0.56		0.44	0.51	
2.04	5.71	5.76	5.71	5.33	5.35	5.32	4.90	4.91	4.87	4.26	4.40	4.20	3.46	3.50	3.13
3.06	5.64	5.72	5.67	5.32	5.34	5.32	4.91	4.96	4.90	4.46	4.57	4.35	3.76	3.95	3.71
4.08	5.63	5.68	5.62	5.33	5.32	5.28	4.91	4.98	4.90	4.55	4.65	4.45	4.01	4.12	3.90
6.12	5.57	5.58	5.51	5.24	5.26	5.21	4.91	4.95	4.90	4.71	4.68	4.52	4.20	4.26	4.05

Carbon Dioxide

P_r	$T_r = 1.1$			$T_r = 1.2$			$T_r = 1.3$			$T_r = 1.4$		
0.50	0.53	0.48	0.47	0.40	0.39	0.36	0.33	0.34	0.29	0.27	0.29	0.24
1.00	1.19	1.17	1.17	0.89	0.89	0.87	0.71	0.71	0.69	0.57	0.60	0.56
1.50	2.24	2.215	2.16	1.44	1.45	1.46	1.10	1.12	1.12	0.90	0.91	0.90
2.00	3.14	3.09	2.88	2.06	2.07	2.02	1.55	1.54	1.54	1.26	1.24	1.20
3.00	3.74	3.69	3.60	2.93	2.90	2.82	2.28	2.27	2.25	1.81	1.81	1.80

iso-Butane

P_r	$T_r = 0.844$			$T_r = 0.926$			$T_r = 1.007$			$T_r = 1.089$		
0.378	5.28	5.14		0.65	0.52	0.53	0.49	0.40	0.47	0.39	0.35	0.31
0.756	5.23	5.16	4.95	4.62	4.57	4.48	1.15	1.15	1.15	0.81	0.80	0.77
1.134	5.25	5.16	5.04	4.76	4.62	4.48	3.15	3.27	2.77	1.49	1.45	1.41
1.512	5.24	5.16	5.46	4.78	4.65	4.92	3.91	4.05	3.98	2.35	2.43	2.31
1.890	5.22	5.16	5.49	4.73	4.68	5.00	4.13	4.26	4.23	3.13	3.24	2.97
2.835	5.20	5.15	5.48	4.74	4.71	5.03	4.32	4.41	4.45	3.69	3.69	3.60
3.780	5.20	5.12	5.45	4.78	4.72	5.02				3.87	3.86	3.85
5.670	5.08	4.88	5.19	4.71	4.56	4.88				3.94	3.95	4.03

[a] Present authors.
[b] (7).

Table X. Values of (d/d_0)

T_r	0	1	2	3	4	5	6	7	8	9	10
0.3	0.875	0.870	0.866	0.862	0.858	0.853	0.849	0.844	0.840	0.835	0.830
0.4	0.830	0.826	0.821	0.816	0.812	0.807	0.802	0.797	0.792	0.787	0.782
0.5	0.782	0.777	0.772	0.767	0.762	0.757	0.752	0.747	0.742	0.737	0.731
0.6	0.731	0.726	0.720	0.715	0.709	0.703	0.698	0.692	0.686	0.680	0.674
0.7	0.674	0.668	0.662	0.656	0.649	0.642	0.636	0.629	0.622	0.615	0.608

polar substances have low values of $(\sigma_0 V_0^{2/3}/T_c)$. On the other hand nitric oxide has a distinctly high surface tension. Thus all substances whose properties deviate from those of a normal fluid seem to show this deviation markedly in an alteration of the surface tension ratio $(\sigma_0 V_0^{2/3}/T_c)$. A deviation of over 5% from Equation 10 indicates significant abnormality.

Estimation of Critical Constants

Many systems for estimating unknown critical constants have been proposed (*6*, *15*, *18*). One method proposed by Riedel (*20*) is directly applicable to the estimation of the acentric factor also. In this article he gives a graphical scheme for estimating T_c, P_c, and his parameter α_k, which is closely related to the acentric factor $\omega(\alpha_k = 5.808 + 4.93\ \omega)$, from various sets of fairly accessible data (boiling point, liquid density, parachor, enthalpy of vaporization, temperature at 100-mm. vapor pressure). The authors have adapted and condensed the most useful of his schemes to the ω system.

To estimate T_c, P_c, and ω for a fluid, three pieces of information are required—its liquid density at some temperature near or below the boiling point; the normal boiling point, T_{760}; and the temperature T_{100} at which the vapor pressure is 100 mm. T_{100} may be calculated from a vapor pressure equation for the substance.

The density is used to calculate a characteristic molal volume for the substance. This is the same V_0 discussed in the preceding section.

The calculation of a set of critical constants is a rapidly converging sequence of approximations. One starts with the assumption that $T_c = T_{760}/0.65$, and calculates an approximate value of T_r at which the liquid density is known. One then obtains from Table X the value of (d/d_0) at this reduced temperature and the volume $V_0 = M(d/d_0)/d$ where M is the molecular weight. Next the values of $\log(T_{760}/V_0)$ and (T_{100}/T_{760}) are calculated and their intersection found on Figure 5. In the first expression T_{760} is in ° K. and V_0 in cc./mole. This gives a new estimate of (T_{760}/T_c) and a value of ω.

This estimate of (T_{760}/T_c) is used to calculate a new value of T_r for the temperature at which the density is measured. This T_r is used to look up d/d_0. Then $\log(T_{760}/V_0)$ is recalculated and a new estimate of (T_{760}/T_c) and ω is found. This process is continued until no further change is found.

T_c is given by $T_c = \left(\frac{T_{760}}{T_c}\right)^{-1} T_{760}$. Then P_c is found by using Table XI, the calculated values of T_c and ω, and the equation $\log P_c = (\log P_c)^{(0)} + \omega(\log P_c)^{(1)}$. Riedel calculated the critical parameters of 26 substances. The root-mean-square percentage error for these substances for T_c = 0.8%, for P_c = 3.3%.

Table XI. Dependence of P_c(atm.) on T_{760}/T_c and ω

$\log P_c = (\log P_c)^{(0)} + \omega(\log P_c)^{(1)}$

T_{760}/T_c	$(\log P_c)^{(0)}$	$(\log P_c)^{(1)}$
0.76	0.742	0.705
0.74	0.823	0.800
0.72	0.909	0.895
0.70	1.000	1.00
0.68	1.096	1.12
0.66	1.198	1.25
0.64	1.308	1.39
0.62	1.426	1.54
0.60	1.552	1.70
0.58	1.688	1.88
0.56	1.834	2.08

The calculation of the critical constants of toluene is given as an example of this procedure.

$$T_{760} = 383.8° \text{ K.} \quad T_{100} = 325.1° \text{ K.}$$

$$d = 0.866 \text{ g./cc. at } 293.2° \text{ K.}$$

$$\frac{T_{100}}{T_{760}} = 0.847$$

Assume $T_{760} = 0.65\ T_c$

$$T_r = 0.65\left(\frac{293.2}{383.8}\right) = 0.496 \text{ for the density value}$$

from Table X

$$\log\left(\frac{T_{760}}{V_0}\right) = \log\left(\frac{383.8 \times 0.866}{0.784 \times 93}\right) = 0.659$$

From Figure 2

$$\frac{T_{760}}{T_c} = 0.643 \qquad \omega = 0.233$$

$$T_r = 0.643\left(\frac{293.2}{383.8}\right) = 0.491 \text{ for the density value}$$

From Table X

$$\log\left(\frac{T_{760}}{V_0}\right) = \log\left(\frac{383.8 \times 0.866}{93 \times 0.786}\right) = 0.658$$

There is no change in this approximation, therefore

$$\omega = 0.233$$

$$T_c = \frac{383.8}{0.643} = 597° \text{ K.}$$

From Table XI

$$\text{at } T_r = 0.643$$

$$\log P_c = 1.291 + \omega(1.37)$$

$$= 1.610$$

$$P_c = 40.8 \text{ atm.}$$

In addition, let us calculate V_c.

$$z_c = z_c^{(0)} + \omega z_c^{(1)}$$

$$= 0.291 + \omega(-0.080)$$

$$= 0.272$$

$$V_c = \frac{z_c R T_c}{P_c} = \frac{(0.272)(0.082)(597)}{(40.8)}$$

$$= 0.326 \text{ liter}$$

	T_c, ° K.	P_c, Atm.	V_c, Liter
Estimate	597	40.8	0.326
Experimental	594.0	40.	0.32

Acknowledgment

This research was a part of the program of Research Project 50 of the American Petroleum Institute. The aid of a National Science Foundation Fellowship to R. F. Curl is gratefully acknowledged.

Literature Cited

(1) Benedict, M., Webb, G. B., Rubin, L. C., *J. Chem. Phys.* **8**, 334 (1940).
(2) Daniels, F., "Outlines of Physical Chemistry," pp. 182–6, Wiley, New York, 1948.
(3) Gilliland, E. R., Lukes, R. V., IND. ENG. CHEM. **32**, 957 (1940).
(4) Gilliland, E. R., Parekh, M. P., *Ibid.*, **34**, 360 (1942).
(5) Hildebrand, J. H., "Solubility," 1st ed., pp. 90–5, The Chemical Catalog Co., New York, 1924.
(6) Lydersen, A. L., Rept. No. 3, Engineering Experiment Station, University of Wisconsin, Madison, Wis., 1955.
(7) Lydersen, A. L., Greenkorn, R. A., Hougen, O. A., "Generalized Thermodynamic Properties of Pure Fluids," *Ibid.*, 1955.
(8) Meissner, H. P., Seferian, Ralph, *Chem. Eng. Progr.* **47**, 579 (1951).
(9) Michels, A., Wassenaar, T., Wolkers, G. J., Dawson, J., *Physica* **22**, 17 (1956).
(10) Opfell, J. B., Sage, B. H., Pitzer, K. S., IND. ENG. CHEM. **48**, 2069 (1956).
(11) Pitzer, K. S., *J. Am. Chem. Soc.* **77**, 3427 (1955).
(12) Pitzer, K. S., *J. Chem. Phys.* **7**, 583 (1939).
(13) Pitzer, K. S., Curl, R. F., *J. Am. Chem. Soc.* **79**, 2369 (1957).
(14) Pitzer, K. S., Lippmann, D. Z., Curl, R. F., Huggins, C. M., Petersen, D. E., *Ibid.*, **77**, 3433 (1955).
(15) Riedel, L., *Chem.-Ing.-Tech.* **24**, 353 (1952).
(16) *Ibid.*, **26**, 83 (1954).
(17) *Ibid.*, p. 259.
(18) *Ibid.*, p. 679.
(19) *Ibid.*, **27**, 209 (1955).
(20) *Ibid.*, p. 475.
(21) *Ibid.*, **28**, 557 (1956).
(22) Roebuck, J. R., Murrell, T. A., Miller, E. E., *J. Am. Chem. Soc.* **64**, 400–11 (1942).
(23) Roebuck, J. R., Osterberg, H., *Phys. Rev.* **46**, 785 (1934).
(24) *Ibid.*, **48**, 450 (1935).
(25) Rowlinson, J. S., *Trans. Faraday Soc.* **50**, 647 (1954); **51**, 1317 (1955).
(26) Sage, B. H., Lacey, W. N., "Thermodynamic Properties of Higher Paraffin Hydrocarbons and Nitrogen," American Petroleum Institute, New York, 1950.

RECEIVED for review February 11, 1957
ACCEPTED August 10, 1957

Division of Petroleum Chemistry, 131st Meeting, ACS, Miami, Fla., April 1957.

Reprinted from the JOURNAL OF CHEMICAL PHYSICS, Vol. 36, No. 2, 425–430, January 15, 1962
Printed in U. S. A.

Volumetric and Thermodynamic Properties of Fluids. VI. Relationship of Molecular Properties to the Acentric Factor

F. DANON AND KENNETH S. PITZER*

Department of Chemistry and Lawrence Radiation Laboratory, University of California, Berkeley, California

(Received July 5, 1961)

A method is developed for the derivation of improved intermolecular potentials from empirical second virial coefficient data together with the acentric factors of the fluids. The Kihara core model is considered first, and relationships are derived between the acentric factor and the core size for several shapes. The resulting cores for CH_4, CF_4, $C(CH_3)_4$, C_6H_6, and N_2 are reasonable in view of semiquantitative theoretical expectations, but the core for CO_2 is somewhat larger than expected. An approximate treatment is given which combines the Kihara core model with an electric quadrupole interaction. The core calculated for CO_2 on this model including the quadrupole moment is of a reasonable size.

INTRODUCTION

THE empirical success of the numerical equation of state based upon critical properties and the acentric factor led us to explore further the relationship of the intermolecular potential to these macroscopic properties. The previous work is presented in a series of five papers.[1] In the present work we will deal almost exclusively with the second virial coefficient and the corresponding interaction between pairs of molecules. The empirical second virial coefficient is presented in paper III while paper I gives a general discussion of the nature of intermolecular forces between the molecules of normal fluids.

Several mathematical models[2] have been proposed which may be expected to yield good approximations to the intermolecular forces in particular cases of normal fluids and for which the second virial coefficient has been calculated.

Most of these proposals are limited in their application to a particular class of normal fluids, such as those with spherical molecules, but the core model of

* Present address: Rice University, Houston, Texas.

[1] Papers in the series will be cited by the roman numerals as follows: I. K. S. Pitzer, J. Am. Chem. Soc. **77**, 3427 (1955); II. K. S. Pitzer, D. Z. Lippmann, R. F. Curl, Jr., C. M. Huggins, and D. E. Petersen, *ibid.* **77**, 3433 (1955); III. K. S. Pitzer and R. F. Curl, Jr., *ibid.* **79**, 2369 (1957); IV. R. F. Curl, Jr., and K. S. Pitzer, Ind. Eng. Chem. **50**, 265 (1958); V. K. S. Pitzer and G. O. Hultgren, J. Am. Chem. Soc. **80**, 4793 (1958).

[2] J. O. Hirschfelder, F. T. McClure, and I. F. Weeks, J. Chem. Phys. **10**, 201 (1942); J. S. Rowlinson, Trans. Faraday Soc. **45**, 974 (1949).

TABLE I. Comparison of the equations based on the acentric factor and the core size parameter.

$\frac{T_B}{T}$	$-\frac{B^{(1)}}{V_B}$	$-7.0\left[\frac{\partial}{\partial x}\left(\frac{B}{V_B}\right)\right]_{x=0}$
0.8	0.027	0.026
1.0	0.001	0.002
1.5	0.168	0.166
2.0	0.690	0.690
2.5	1.611	1.625
3.0	3.010	3.020

Kihara[3] is quite general in nature. Kihara assumes a core of any shape within each molecule and takes the intermolecular potential to be the Lennard-Jones potential for the shortest distance between the outer surfaces of molecular cores. Thus the potential is

$$U=U_0[(\rho_0/\rho)^{12}-2(\rho_0/\rho)^6], \tag{1}$$

where ρ is the shortest distance between cores and ρ_0 is this distance for the potential minimum.

In paper I Kihara's functions for this model were reduced on the Boyle point basis to equations for the second virial coefficient. Consequently, a comparison of these equations based on the core model with the empirical second virial coefficients of paper III provides a convenient way of relating the acentric factor to this molecular model.

Our results indicate in most cases smaller cores than those assumed by Kihara from molecular structure information and this conclusion finds support in recent independent reports on benzene[4] and methane[5] as well as the general considerations of the intermolecular potential in paper I.

VIRIAL COEFFICIENT EQUATIONS

The equation for the second virial coefficient given in III in dimensionless form is

$$BP_c/RT_c=(B^{(0)}P_c/RT_c)+\omega(B^{(1)}P_c/RT_c), \tag{2}$$

where

$$\frac{B^{(0)}P_c}{RT_c}=0.1445-\frac{0.330}{T_r}-\frac{0.1385}{T_r^2}-\frac{0.0121}{T_r^3}, \tag{2a}$$

$$\frac{B^{(1)}P_c}{RT_c}=0.073+\frac{0.46}{T_r}-\frac{0.50}{T_r^2}-\frac{0.097}{T_r^3}-\frac{0.0073}{T_r^8}, \tag{2b}$$

and the acentric factor

$$\omega=-\log P_r-1.000, \tag{2c}$$

where T_r is the reduced temperature T/T_c and P_r in Eq. (2c) is the reduced vapor pressure P/P_c at $T_r=0.7$. It is convenient to transform Eq. (2) from the critical point basis to the Boyle point basis, whereupon the result may be compared with Eqs. (15), (15a), and (15b) of paper I. At the Boyle point, $B=0$, $T=T_B$, and we define the Boyle volume

$$V_B=[T(dB/dT)]_{T=T_B}. \tag{3}$$

In the special case of zero acentric factor Eq. (2a) readily yields the result $T_r=2.656$ at the Boyle point and $V_B=0.1651RT_c/P_c$ (with $\omega=0$). Substitution of these results gives

$$B^{(0)}/V_B=0.8736-0.7512(T_B/T)-0.1185(T_B/T)^2-0.0039(T_B/T)^3. \tag{4}$$

The values for T_B and V_B for $\omega=0$ may be substituted into Eq. (2b), but we find then that the general equation for nonzero values of the acentric factor indicates that T_B and V_B depend on ω as well as the critical constants. Hence the criterion for the Boyle point must be reapplied to the general Eq. (2) and yields

$$T_B=2.656T_c/(1+1.028\omega)$$

and

$$V_B=(RT_c/P_c)(0.1651+0.1981\omega).$$

Now the second virial coefficient is

$$B=B^{(0)}+\omega B^{(1)}+\cdots, \tag{5}$$

with

$$B^{(1)}/V_B=-0.6087+1.11760(T_B/T)-0.5299(T_B/T)^2-0.0386(T_B/T)^3-1.79\times10^{-5}(T_B/T)^8. \tag{6}$$

The last term in Eq. (6) is negligible except at low temperature and can usually be omitted. In these transformations only linear terms in ω were retained.

We now wish to compare Eqs. (4), (5), and (6) with the corresponding equations for the core model. Kihara's general expression for the second virial coefficient is

$$B=(2\pi/3)\rho_0^3F_3(Z)+M_0\rho_0^2F_2(Z)+[S_0+(M_0^2/4\pi)]\rho_0F_1(Z)+[V_0+(M_0S_0/4\pi)], \tag{7}$$

TABLE II. Comparison of constants for the core model with those derived from the acentric factor theory for tetrahedral molecules.

Molecule		Kihara a	Kihara b	Present work
CH_4	ρ_0 A	1.92	3.15	3.75
	U_0/k °K	378	226	175
	d A	1.09	0.596	0.28
CF_4	ρ_0 A	2.48		3.08
	U_0/k °K	372		350
	d A	1.32		1.01

[a] Reference 3.
[b] T. Kihara and S. Koba, J. Phys. Soc. Japan **14**, 247 (1959).

[3] T. Kihara, Revs. Modern Phys. **25**, 831 (1953) and references given therein.
[4] J. F. Connolly and G. A. Kandalic, Phys. Fluids **3**, 463 (1960).
[5] G. Thomaes and R. Van Steenwinkel, Nature **187**, 229 (1960).

where $Z=U_0/kT$, M_0 is the surface integral of the mean curvature of the core, S_0 and V_0 are the surface area and the volume of the core, respectively, F_1, F_2, F_3 are functions given in Kihara's paper[3] that have recently been extended by Connolly and Kandalic.[4] Considering only first-order deviation, we obtain for all shapes of cores

$$B=(2\pi/3)\rho_0^3[F_3(Z)+xF_2(Z)] \quad (8)$$

with

$$x=3M_0/2\pi\rho_0. \quad (9)$$

Thus x is proportional to a linear dimension of the core.

In paper I, Eq. (8) was converted to the Boyle point basis and expressed as

$$B/V_B=[B/V_B]_0+x[\partial/\partial x(B/V_B)]_{x=0}. \quad (10)$$

Equations (15a) and (15b) of I give the expressions for the two terms as functions of (T_B/T). The first term is just the usual Lennard-Jones function since $x=0$ corresponds to zero core size. The empirical equation for zero acentric factor, i.e., for Ar, Kr, and Xe, is very similar to the Lennard-Jones function but a distinctly better fit is obtained with Eq. (10) if $x=0.24$. The agreement is then within 0.5% over the range of validity of the empirical equation.

If the core model is in satisfactory agreement with the experimental data, the second term in Eq. (10) must equal $B^{(1)}/V_B$ given by Eq. (6) for some value of x. Table I gives this comparison for $x=7.0\omega$ and shows good agreement over the range of interest. Thus the comparison of the empirical equations with those from Kihara's theory yields the relationship

$$x=3M_0/2\pi\rho_0=7.0\omega+0.24. \quad (11)$$

This result, together with the fact that V_0 and S_0 are expressable in terms of M_0 for a particular shape of core, enables us to write Eq. (7) depending parametrically on ω. By comparison between this function (7) and the experimental values of $B=B(T)$ a new set of parameters for the Lennard-Jones type potential function can be obtained.

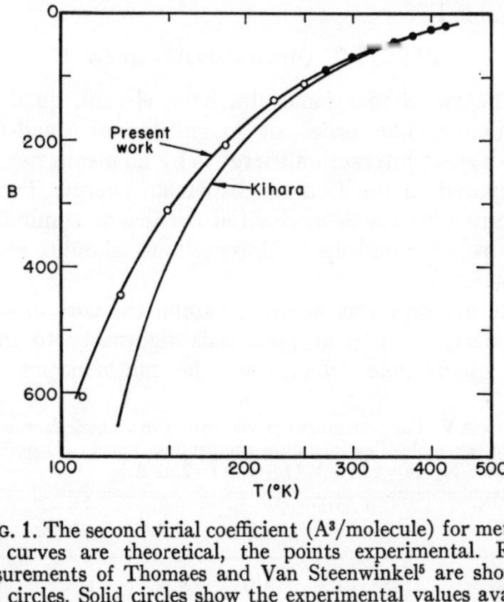

FIG. 1. The second virial coefficient (A^3/molecule) for methane. The curves are theoretical, the points experimental. Recent measurements of Thomaes and Van Steenwinkel[5] are shown as open circles. Solid circles show the experimental values available to Kihara. [A. Michels and G. W. Nederbragt, Physica **3**, 569 (1936).]

TABLE III. Comparison of constants for the core model with those derived from the acentric factor theory for benzene.

Molecule		Kihara	Present work
C_6H_6	ρ_0 A	3.4[a]	3.0[b]
	U_0/k °K	830	990
	l A	1.93	1.16

[a] From the lattice constant of graphite.
[b] From second virial coefficient data.

For each particular substance with fixed ω, Eq. (7) depends on U_0/k and ρ_0 only. By plotting this theoretical expression for $B(T)$ vs log (kT/U_0) on transparent paper for different given values of ρ_0 a family of curves is obtained which can be superimposed over a graph of the experimental curve $B(T)$ vs log T. This gives ρ_0 and U_0/k. Thus for any given shape of core the intermolecular potential parameters are determined by the macroscopic properties.

The various transformations which led up to Eq. (11) were carried out only through the first order in ω. The adequacy of Eq. (11) for large ω values was tested by direct comparison of the calculated second virial corfficient curves for particular cores corresponding to large x values with the curves given by Eq. (2) with the corresponding ω. While this test was not extremely precise, it showed that the linear development was adequate in the range of ω values here considered and for the present accuracy of experimental data.

EXAMPLES

We consider first molecules with zero or negligible dipole or quadrupole electric moments.

(i) Tetrahedral cores:

l: Length of one edge of the tetrahedron;
$M_0=5.733l$;

$S_0=1.732l^2$; $\quad l=2.737^{-1}\rho_0 x$; $\quad x=7.0\omega+0.24$;
$V_0=0.1178l^3$;

$$B=\rho_0^3[(2\pi/3)F_3(Z)+(2\pi/3)xF_2(Z)+0.5802x^2F_1(Z)+0.3911x^3].$$

The distance d from center to vertex of the tetrahedron is given by $d=l/2\sin(\frac{1}{2}\alpha)$; α is H—C—H bond angle in methane. Results of the calculations together with Kihara's values are given in Table II. For comparison the C—H distance in methane is 1.09A and the C—F distance in CF_4 is 1.32 A. Figure 1 shows the curves for methane corresponding to our parameters of Table II and to those of Kihara. Although both curves fit the data above 273°K, the curve based upon the

TABLE IV. Comparison of constants for the core model with those derived from the acentric factor theory for linear molecules.

Molecule		Kihara a	Kihara b	Present work Thin rod core	Present work Cylindrical core
N_2	ρ_0 A	3.47		3.40	3.30
	U_0/k °K	124		132	125
	l A	1.09		1.19	0.73
	a A	0.00		0.00	0.13
CO_2	ρ_0 A	3.36	3.7	2.72	2.6
	U_0/k °K	279	2.79	400	405
	l A	2.20	2.30	3.29	2.82
	a A	0.00	0.00	0.00	0.10

[a] Reference 3.
[b] T. Kihara, J. Phys. Soc. Japan 6, 289 (1951).

acentric factor agrees much better with the recent experimental data of Thomaes and Van Steenwinkel[5] at lower temperatures.

(ii) Hexagonal cores:

l: length of one side of the hexagon;

$$M_0=3\pi l;$$

$$S_0=5.2l^2;$$

$$l=\tfrac{2}{9}x\rho_0;$$

$$B=\rho_0^3[(2\pi/3)F_3(Z)+\tfrac{2}{3}\pi xF_2(Z)+0.6057x^2F_1(Z)+0.0428x^3].$$

At the time Kihara's work was published (1953) there were not sufficient observed values of the second virial coefficient for benzene to determine both U_0 and ρ_0. His value of ρ_0 is taken from the lattice constant of graphite. Recently, however, Andon *et al.*[6] reported measurements on the second virial coefficient for benzene between 67.1° and 164.6°C. By use of their results together with the older values of Francis and McGlashan[7] we have calculated both ρ_0 and U_0 from second virial coefficient data only.

Results are given in Table III. For comparison, the C—C distance in benzene is 1.397A.

(iii) Spherical core:

If we assign a spherical core for CH_4, which is a permissible approximation considering its rapid and relatively free rotation, we obtain for the radius of the spherical core $a=0.21$ A.

We also consider the globular-type molecule, neopentane, and assume a spherical core of radius a.

$$M_0=4\pi a, \qquad a=\tfrac{1}{6}\rho_0 x$$

$$B=\tfrac{2}{3}\pi\rho_0^3[F_3(Z)+xF_2(Z)+\tfrac{1}{3}x^2F_1(Z)+\tfrac{1}{27}x^3].$$

[6] R. J. L. Andon, J. D. Cox, E. F. Herington, and J. F. Martin, Trans. Faraday Soc. 53, 1074 (1957); J. D. Cox and R. L. Andon, *ibid.* 54, 1622 (1958).

[7] P. G. Francis, M. L. McGlashan, S. D. Hamann, and W. J. McManamey, J. Chem. Phys. 20, 1341 (1952).

There results

$$\rho_0=3.96;$$

$$U_0/k=590°K;$$

$$a=1.06 \text{ A}; \qquad \text{C—C bond length}=1.536 \text{ A}.$$

(iv) Thin rod-shaped cores:

We consider in this case N_2 and CO_2.

$$M_0=\pi l,\ l=\tfrac{2}{3}\rho_0(7.0\omega+0.24),$$

where l is the length of the rod.

$$B=(2\pi/3)\rho_0^3[F_3(Z)+xF_2(Z)+\tfrac{1}{6}x^2F_1(Z)]. \quad (12)$$

The results obtained are listed in Table IV. For comparison the N—N distance is 1.095 A in N_2 and the O—O distance in CO_2 is 2.32 A. The cores for N_2 and CO_2 seem peculiarly large, particularly so for CO_2. The fact that Eq. (11) gives nonzero core constants for $\omega=0$ indicates that there is some small core effect even in the simple fluids (Ar, Xe, etc.). It is then reasonable to consider cylindrical cores in N_2 and CO_2 with the radius of the cylinder a obtained from the relationship between a and ρ_0 for simple fluids.

From Eq. (11) for $\omega=0$ one gets a spherical core radius $a=0.04\rho_0$. This relationship for a enters the expressions for a cylindrical core which are

$$M_0=\pi(\pi a+l)=(2\pi/3)x\rho_0, \qquad S_0=2\pi a(a+l),$$

$$V_0=\pi a^2 l, \qquad l=\tfrac{2}{3}x\rho_0-\pi a.$$

One then obtains

$$B=(2\pi/3)\rho_0^3\{F_3(Z)-0.010F_1(Z)-0.0003+[F_2(Z)+0.080F_1(Z)]x+[\tfrac{1}{6}F_1(Z)+0.0133]x^2\}. \quad (13)$$

The results considering cylindrical core are also given in Table IV.

ELECTRIC QUADRUPOLE EFFECTS

The two linear molecules have electric quadrupole moments. The order of magnitude of quadrupole-quadrupole interaction energy is by no means negligible compared to the London dispersion energy. The contribution to the second virial coefficient coming from quadrupole-quadrupole interactions should also be considered.

An attempt was made to extend the core model for angularly dependent potentials rigorously to include the quadrupole effect, but the mathematical com-

TABLE V. The constants derived from the acentric factor theory for linear molecules including quadrupole effect. (Experimental distance N—N=1.095 A, O—C—O=2.32 A.)

	θ (esu)	S	ρ_0 A	U_0/k °K	l A
N_2	$1.488\cdot10^{-26}$	0.05	3.60	115	0.86
CO_2	$3.12\cdot10^{-26}$	0.33	3.42	270	1.73

plexity of the problem involving nonspherical cores makes an exact and rigorous treatment unduly complex.

An approximate treatment including quadrupole as well as core effects may be obtained by combining the first-order correction term for each effect with the equation for a spherical Lennard-Jones potential.

The results obtained by Pople[8] allow one to write the virial coefficient for molecules interacting by a Lennard-Jones potential together with permanent quadrupole moments as

$$B(T)=\tfrac{2}{3}\pi N r_0^3[F(y)-\tfrac{7}{320}(\Theta^2/U_0 r_0^5)^2 H_{10}(y)], \quad (14)$$

where r_0 is the intermolecular distance at which the nonangular (Lennard-Jones) part of the potential energy function vanishes, and $F(y)$ is the usual Lennard-Jones function. U_0 is the most negative value of the same part of the potential, and y is defined as

$$y=2(U_0/kT)^{\frac{1}{2}}.$$

Θ is the quadrupole moment[9] and $H_{10}(y)$ is the function expressing the quadrupole interaction which has been given as a series and is known in tabular form.[10]

We fitted a third degree polynomial to the tabulated values of $H_{10}(y)$ and by transformation to our set of variables obtained

$$H(Z)=0.07089-0.78915Z-0.05201Z^2-2.40478Z^3. \quad (15)$$

If we define the dimensionless parameter

$$s=(\Theta^2/U_0\rho_0^5)$$

depending on the quadrupole moment, we can write for the second virial coefficient including first-order core and quadrupole effects

$$B=(2\pi/3)\rho_0^3[F_3(Z)+xF_2(Z)+sH(Z)] \quad (16a)$$

or in reduced form

$$\frac{B}{V_B}=\left[\frac{B}{V_B}\right]_0+x\left[\frac{\partial}{\partial x}\left(\frac{B}{V_B}\right)\right]_0+s\left[\frac{\partial}{\partial s}\left(\frac{B}{V_B}\right)\right]_0. \quad (16b)$$

As we want to compare Eq. (16b) with the empirical Eqs. (4) and (6), we reduced the function $H(Z)$ to Boyle point basis using the definition of Boyle volume given by (3). The result is

$$[(\partial/\partial s)(B/V_B)]_0=-0.1020+0.1105(T_B/T)$$
$$+0.0842(T_B/T)^2-0.0932(T_B/T)^3. \quad (17)$$

Comparing Eq. (17) with the corresponding equation for x, Eq. (15b) in paper I, one finds that in the low-temperature range, where the quadrupole effect is most marked, the relationship

$$[(\partial/\partial s)(B/V_B)]/[(\partial/\partial x)(B/V_B)]=3.2 \quad (18)$$

is a fair approximation.

Substitution of Eq. (18) in (16b) and use of Eq. (11) yields

$$x+3.2s=7.0\omega+0.24. \quad (19)$$

This relationship between core size, quadrupole moment and acentric factor enables us to write Eq. (16a) in terms of ω. As the parameter s depends on ρ_0 and U_0 by definition, a successive approximation method is necessary for the evaluation of ρ_0 and U_0. The results obtained[11] for CO_2 and N_2 are listed in Table V. These results show that, when the quadrupole effect is considered, the sizes of the cores are again smaller than the ones assumed by Kihara from molecular structure information. A few years after Kihara's work was published a measurement of the second virial coefficient for CO_2 at low temperatures was reported.[12] These values, together with the older results reported by Schäfer[13] and MacCormack and Scheneider[14] are shown in Fig. 2. A distinctly better fit to this experimental data is obtained from the parameters of Table V of this paper than from the expression given by Kihara.

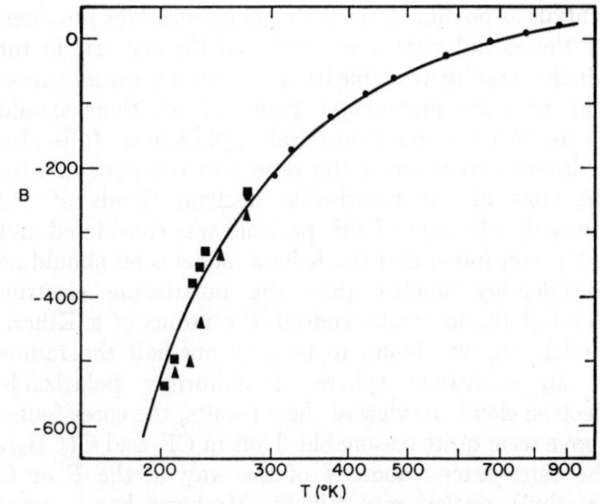

FIG. 2. The second virial coefficient (A^3/molecule) for CO_2. The curve is calculated using the results of this paper. Solid circles correspond to MacCormack and Schneider measurements,[14] triangles to Cook,[12] and squares to Schafer.[13]

DISCUSSION

Our aim has been to develop a system for the calculation of as accurate and realistic intermolecular po-

[8] J. A. Pople, Proc. Roy. Soc. (London) **A221**, 498 (1954); **A221**, 508 (1954).

[9] The quadrupole moment definition used in this paper is $\Theta=\Sigma_i e_i(z_i^2-x_i^2)$, where z_i is measured along the molecular axis.

[10] Tables for $H_{10}(y)$ are given by A. D. Buckingham and J. A. Pople, Trans. Faraday Soc. **51**, 1173 (1955).

[11] The numerical value for the quadrupole moment of N_2 is taken from W. Gordy, W. V. Smith, and R. Trambarulo, *Microwave Spectroscopy* (John Wiley & Sons, Inc., New York, 1953), p. 345; that for CO_2 from W. V. Smith and R. Howard, Phys. Rev. **79**, 132 (1950).

[12] D. Cook, Can. J. Chem. **35**, 369 (1957).

[13] K. Schäfer, Z. Phys. Chem. **B36**, 85 (1937).

[14] K. E. MacCormack and W. G. Schneider, J. Chem. Phys. **18**, 1269 (1950).

tentials as possible from the macroscopic data provided by the second virial coefficient and the acentric factor which is readily available from the vapor pressure curve. The resulting microscopic potential functions should be useful for other theoretical applications. It is also of interest to consider the results in comparison with the sizes of the polarizable electron clouds of the molecules. In paper I this problem was considered and it was concluded that the Kihara model cores should be considerably smaller than the polarizable electron cloud of the molecule. Indeed, the radius of a Kihara model core was found to be only one-half the radius of an equivalent sphere of uniformly polarizable electron cloud. In view of these results, the cores found above seem quite reasonable. Thus in CF_4 and $C(CH_3)_4$ the cores extend about $\frac{2}{3}$ of the way to the F or C (methyl) nuclei, respectively. Methane has a very small core since all of the polarizable electrons are partially centered on the single carbon atom. The hexagonal core for benzene extends almost to the ring carbon atoms. The cores chosen by Kihara in 1953 seem to have been selected rather arbitrarily; for example his core for methane extended to the hydrogen nuclei whereas in benzene the core extended only to the midpoints of the C—H bonds. Consequently it is not surprising that our results differ rather substantially. In the case of methane the results in Fig. 1 show clearly that the core selected from the acentric factor is the correct one to fit the wide range of data now available.

The tetrahedral molecules have zero electric quadrupole moment by symmetry but the linear molecules N_2 and CO_2 have significant moments. While our treatment including the quadrupole effect is only an approximate one for the model, it does show that the linear cores are thereby reduced to reasonable sizes.

Also of considerable interest is the result that a Kihara core of radius $a=0.04\rho_0$ yields considerably better results for the rare gases, Ar, Kr, Xe, than the simple Lennard-Jones potential. This is in agreement with the conclusions of Brown and Rowlinson[15] based upon totally different arguments.

With these results as a guide one should be able to estimate reasonable core sizes for other molecules if the acentric factor or equivalent data are unavailable. Conversely, the acentric factor and other data may be used, as has been illustrated in this paper, to derive an intermolecular potential for use in other calculations.

ACKNOWLEDGMENTS

This work has been performed under the auspices of U.S. Atomic Energy Commission. The aid of a Fulbright Travel grant to one of us (F.D.) is gratefully acknowledged.

[15] W. B. Brown and J. S. Rowlinson, Mol. Phys. **3**, 35 (1960)

B. Spin Species Conversion in Methane

Introduction

In 1966 the very interesting properties of ortho and para hydrogen had been well known for about three decades. The underlying characteristic which allowed the physical separation of the two species was the slowness of interconversion of nuclear spin alignment and the significant difference in internal energy at low temperature. Among others, I had speculated about another molecule that might show this spin-species separability; required were symmetrically placed hydrogens, at most only one atom heavier than hydrogen, and volatility at very low temperature. Methane, CH_4, seemed the most likely candidate. It proved to be impossible to prepare separated gaseous species but very interesting properties, including slow interconversion, were observed in solid CH_4. And the associated theory exposed some other interesting properties.

Reprinted from THE JOURNAL OF CHEMICAL PHYSICS, Vol. 44, No. 12, 4636–4637, 15 June 1966
Printed in U. S. A.

Spin Statistics Isomerization in Methane

R. F. CURL, JR., JEROME V. V. KASPER,* KENNETH S. PITZER,
AND KRISHNAN SATHIANANDAN†

Department of Chemistry, Rice University, Houston, Texas

(Received 21 January 1966)

WHILE many substances exist in different species with respect to nuclear spin orientation, only ortho- and parahydrogen and deuterium have been obtained in separated form. In this Communication we report an unsuccessful attempt to prepare methane enriched in one of its spin species and, more important, the theoretical pattern of interspecies conversion rates for methane, obtained subsequently, which is strikingly different from that for hydrogen.

The important difference in conversion rate concerns the gas phase and arises from the existence of rotational states of different symmetry but identical energy for methane but not for hydrogen, or for any molecule other than a spherical top.

In H_2 (or any symmetrical linear molecule) there are no energy degeneracies between ortho and para quantum states. One species has all even values of the rotational quantum number, and the other species all odd values. Consequently, ortho and para states are separated in energy by at least one rotational quantum unit. In CH_4 the $J=0$ state has symmetry A (meta species) and all $J=1$ states have the symmetry F (ortho species). For the $J=2$ states, however, a different situation arises in that some of the individual states are of symmetry F and others are of symmetry E (para species). Likewise, for all higher values of J there are states of at least two symmetry species (for $J\geq 6$, of all three symmetry species) for each J.

In the case of methane both the spin–spin and spin–rotation interactions are not symmetric in the spin operators and lead to mixing of the different I wavefunctions. The spin–spin interaction can lead to mixing of $I=0$ and $I=2$ (or ortho and meta species) wavefunctions. The spin–rotation interaction can lead to mixing of $I=1$ and $I=0$ (or ortho and para species) wavefunctions and $I=1$ and $I=2$ (or ortho and meta species) wavefunctions. When the wavefunctions being mixed are nearly degenerate, mixing is large.

The gas-phase equilibration is extremely slow in H_2 because large differences in rotational energy prevent appreciable mixing of wavefunctions of the ortho and para species. The mixing described above for methane allows the nuclear spin to be changed in all states with $J\geq 2$. The calculation of the rate is rather complicated and will be presented in detail later, but our present results are that the conversion time for CH_4 at 1 atm and room temperature is of the order of 1 min or less.

The degeneracy of rotational energies for $J\geq 2$ which allows rapid conversion in gaseous methane is broken by the forces to be expected in the solid (and presumably also for the liquid). The symmetry of the methane molecule itself assures that the dominant term in the potential restricting rotation in the solid will be one of tetrahedral symmetry. Nagamiya[1] has shown that such a potential splits the rotational states with $J=2$ into several subgroups and that states of different nuclear spin species are no longer degenerate. Consequently, the same sort of qualitative difference between hydrogen and methane as is described here for the gas phase is not expected for the solid phase.

While it is improbable that we would have attempted the experimental preparation of enriched metamethane if we had already reached these theoretical conclusions, it may be worthwhile to describe the experiments very briefly. The possibility that interspecies conversion might be taking place in methane was suggested to us by the thermal anomaly in methane below 10°K reported by Colwell, Gill and Morrison.[2] Subsequently, Wolf and Whitney[3] reported nuclear magnetic resonance experiments and presented this interpretation.

In our experiments we rapidly evaporated solid methane which had been held at approximately 5°K for several days. The analysis was carried out by high-resolution infrared spectroscopy. The three nuclear spin species contribute in very different amounts to the intensities of the low-order P and R lines in the principal bands; hence there should be no difficulty in detecting any substantial change from the high-temperature equilibrium mixture of ortho, meta, and para species. We observed no change from the normal mixture. In view of our theoretical results, there would have been a rapid reconversion in the gaseous state and our experiments yield no conclusions at all concerning possible conversion in the solid below 10°K.

The details of this theory together with its application to other substances will be presented in another paper.

* Welch Foundation Postdoctoral Fellow.

† U.S. Atomic Energy Commission Research Associate.

[1] T. Nagamiya, Progr. Theoret. Phys. **6,** 702 (1951).

[2] J. H. Colwell, E. K. Gill, and J. H. Morrison, J. Chem. Phys. **36,** 2223 (1962); **39,** 635 (1963).

[3] R. P. Wolf and W. M. Whitney, Proc. Intern. Conf. Low Temp. Phys. 9th Columbus, Ohio, 1964, 1118 (1965).

Reprinted from THE JOURNAL OF CHEMICAL PHYSICS, Vol. 46, No. 8, 3220–3228, 15 April 1967
Printed in U. S. A.

Nuclear Spin State Equilibration through Nonmagnetic Collisions*

R. F. CURL, JR., JEROME V. V. KASPER,† AND KENNETH S. PITZER
Department of Chemistry, Rice University, Houston, Texas
(Received 1 August 1966)

For nonlinear molecules with equivalent identical nuclei an important pathway leading to equilibration of nuclear spin statistics isomers is provided by wavefunction mixing induced by the spin–rotation interaction and in some cases the spin–spin interaction. For CH_4 this mixing is very large and provides for very rapid equilibration of the three spin statistics isomers. For asymmetric rotor molecules such as H_2O and CH_2O the rapidity of the equilibration is sensitive to the exact rotational energy level pattern. The spin–rotation mixing may be very important if there is an accidental near degeneracy of the right sort. Then most of the isomerization "funnels" through the near-degenerate levels.

INTRODUCTION

IN 1927 Dennison[1] suggested that the even and odd rotational levels of H_2 might not equilibrate rapidly with each other. The following year, Giauque and Johnston[2] observed that the vapor pressure of hydrogen which had been held at 85°K for 6 months and subsequently liquefied was different from that of ordinary hydrogen.

Subsequent work has convincingly demonstrated the existence of ortho- and parahydrogen, and the theory of the phenomenon is well understood. A review of the early results is given in the book by Farkas.[3]

Theories of ortho–para conversion in the presence and absence of catalysts in gas and solid have been proposed[4] and appear to fairly satisfactorily explain the results.

Nonequilibrium distributions of spin (statistics) isomers have been obtained for only H_2, D_2, and T_2. For most other molecules, rotational motion will cease at temperatures above those required to give a distribution of isomers significantly different from that at very high temperature. However, a slow transition in methane at about 4°K might be a spin-isomerization transition.[5] Unsuccessful efforts[6] have been made to observe the low-temperature isomer distribution in the room temperature gas obtained by rapid heating of solid held below the transition temperature. This work shows that such an experiment must fail. Rapid conversion from one isomer to the other must occur since the spin wavefunctions are mixed by spin–rotation interaction. The mixing is large because of the K degeneracy of the rotational states of the spherical top.

In addition to spherical tops there are other molecules in which the spin–rotation interaction might lead to equilibration of nuclear spin isomers. Two such are H_2O and CH_2O. Of course, H_2O would certainly equilibrate rapidly by exchange. Here the spin–rotation matrix elements connect different rotational energy levels. The rate is much slower than that in CH_4 and depends sensitively upon whether two levels which can interact are accidentally near degenerate.

Observation of isomerization rates would require the preparation of a nonequilibrium isomeric mixture which would be virtually impossible to achieve by cooling of pure H_2CO or H_2O. However, a nonequilibrium mixture

* This work was supported by Grants C-092 and C-071 of the Robert A. Welch Foundation.
† Welch Foundation Postdoctoral Fellow.
[1] D. M. Dennison, Proc. Roy. Soc. (London) **A115**, 483 (1927).
[2] W. F. Giaque and H. L. Johnston, J. Am. Chem. Soc. **50**, 3221 (1928).
[3] A. Farkas, *Orthohydrogen, Parahydrogen, and Heavy Hydrogen* (Cambridge University Press, Cambridge, England, 1935).
[4] See K. Motizuki and T. Nagamiya, J. Phys. Soc. Japan **11**, 93 (1956).
[5] J. H. Colwell, E. K. Gill, and J. H. Morrison, J. Chem. Phys. **36**, 2223 (1962); **39**, 635 (1963).
[6] R. F. Curl, Jr., J. V. V. Kasper, K. S. Pitzer, and K. Sathianandan, J. Chem. Phys. **44**, 4636 (1966).

might be prepared by the matrix isolation technique. Water is found to rotate in dilute matrices at very low temperatures[7] with the low-temperature isomer distribution. In addition it might be possible though difficult to prepare such a mixture by photolysis of one state of an isomer with monochromatic light.

CALCULATION OF THE MIXING FOR METHANE

The spin statistics for CH_4 have been calculated using group theory by Wilson.[8] The appropriate group for a spherical top is T which has three irreducible representations A, E (a complex conjugate set), and F. The four $I=\frac{1}{2}$ proton spins couple to give $I=0$ (of representation E, two states total), $I=1$ (of representation F, nine states total), and $I=2$ (of representation A, five states total). The total number of states is $2+9+5=16=2^4$.

In order to satisfy the Pauli exclusion principle for protons, the total wavefunction must transform as A since all operations of T interchange even numbers of protons. If only the ground electronic and vibrational wavefunctions are considered this means that the product $\psi_{\text{rot}}\psi_{\text{nuc}}$ must belong to A. The rotational wavefunctions have quantum numbers JKM with the energy depending only on J. For $J>1$ there will be linear combinations of the K wavefunctions corresponding to more than one irreducible representation. The irreducible representations occurring for low values of J are given in Table I. The coupling of I and J to give the total momentum F is also shown.

The coupling between I and J is caused by the spin-rotation and the spin–spin (magnetic dipole–magnetic dipole) interactions. These interactions also cause mixing of the spin wavefunctions of different isomers.

The spin–rotation Hamiltonian for CH_4 is

$$H_{1a}=(A'+\tfrac{1}{3}A)(\mathbf{I}\cdot\mathbf{J}), \tag{1a}$$

$$\begin{aligned}H_{1b}=(\tfrac{1}{3}A)\{&J_x[(I_{1y}+I_{2y}-I_{3y}-I_{4y})+(I_{1z}-I_{2z}-I_{3z}+I_{4z})]\\&+J_y[(I_{1x}+I_{2x}-I_{3x}-I_{4x})+(I_{1z}-I_{2z}+I_{3z}-I_{4z})]+J_z[(I_{1x}-I_{2x}-I_{3x}+I_{4x})+(I_{1y}-I_{2y}+I_{3y}-I_{4y})]\},\end{aligned} \tag{1b}$$

where A' and A are constants (if the spin–rotation interaction is caused by point magnetic dipole at the carbon nucleus, $A=-3A'$). The spin–spin Hamiltonian for CH_4 is

$$\begin{aligned}H_{2a}=\tfrac{1}{2}B\{&[(I_{1z}I_{2z}+\epsilon I_{1x}I_{2x}+\epsilon^*I_{1y}I_{2y})+(I_{1z}I_{2z}+\epsilon^*I_{1x}I_{2x}+\epsilon I_{1y}I_{2y})]\\&+[\epsilon^*(I_{1z}I_{3z}+\epsilon I_{1x}I_{3x}+\epsilon^*I_{1y}I_{3y})+\epsilon(I_{1z}I_{3z}+\epsilon^*I_{1x}I_{3x}+\epsilon I_{1y}I_{3y})]\\&+[\epsilon(I_{1z}I_{4z}+\epsilon I_{1x}I_{4x}+\epsilon^*I_{1y}I_{4y})+\epsilon^*(I_{1z}I_{4z}+\epsilon^*I_{1x}I_{4x}+\epsilon I_{1y}I_{4y})]\\&+[\epsilon(I_{2z}I_{3z}+\epsilon I_{2x}I_{3x}+\epsilon^*I_{2y}I_{3y})+\epsilon^*(I_{2z}I_{3z}+\epsilon^*I_{2x}I_{3x}+\epsilon I_{2y}I_{3y})]\\&+[\epsilon^*(I_{2z}I_{4z}+\epsilon I_{2x}I_{4x}+\epsilon^*I_{2y}I_{4y})+\epsilon(I_{2z}I_{4z}+\epsilon^*I_{2x}I_{4x}+\epsilon I_{2y}I_{4y})]\\&+[(I_{3z}I_{4z}+\epsilon I_{3x}I_{4x}+\epsilon^*I_{3y}I_{4y})+(I_{3z}I_{4z}+\epsilon^*I_{3x}I_{3x}+\epsilon I_{3y}I_{4y})]\},\end{aligned} \tag{2a}$$

$$H_{2b}=-\tfrac{3}{2}B[(I_{1x}I_{2y}+I_{1y}I_{2x}-I_{3x}I_{4y}-I_{3y}I_{4x})+(I_{1y}I_{3z}+I_{1z}I_{3y}-I_{2y}I_{4z}-I_{2z}I_{4y})+(I_{1x}I_{4z}+I_{1z}I_{4x}-I_{2x}I_{3z}-I_{2z}I_{3x})], \tag{2b}$$

where $B=(g_I\mu_N)^2r^{-3}$, where r is the distance between protons and $\epsilon=\exp(2\pi i/3)$. The coordinate system and location of atoms is shown in Fig. 1.

The matrix elements of the Hamiltonian in the coupled scheme of Table I are required. The most powerful technique for evaluation of these matrix elements which can be brought to bear involves combining the method of Racah vector recoupling coefficients on the group $O(3)$[9] with the finite-group, irreducible-tensor method[10] on the internal symmetry.

The over-all symmetry group is that of the direct product of $O(3)$ (reflecting the invariance of the Hamiltonian with respect to rotation of axes fixed in space) with T (reflecting the invariance of the Hamiltonian with respect to the operations of the tetrahedral group of axes fixed in the molecule). Unfortunately, the group T has the undesirable property that some direct products of irreducible representations contain some irreducible representations more than once ($E\times E=2A+E$, $F\times F=2F+E+A$). Coupling coefficients for such groups have not been tabulated. It is possible to bring the method of irreducible tensorial sets to bear by using the group T_d instead of T. The group T_d has representations A_1, A_2, E, F_1, F_2. All that is necessary to get the correct answer is to recognize that A_1 and A_2 of T_d correlate with A of T and F_1 and F_2 correlate with F and to ignore the subscripts. Rather than follow this somewhat confusing procedure the internal symmetry will be treated by more elementary methods.

Since the methane molecule has symmetry T_d rather than T, it appears that T_d should be the natural choice

[7] R. L. Redington and D. E. Milligan, J. Chem. Phys. **37**, 2162 (1962).

[8] E. B. Wilson, J. Chem. Phys. **3**, 276 (1935).

[9] A. R. Edmonds, *Angular Momentum in Quantum Mechanics* (Princeton University Press, Princeton, N.J., 1957), see especially Chap. 7.

[10] J. S. Griffith, *The Irreducible Tensor Method for Molecular Symmetry Groups* (Prentice-Hall, Inc., Englewood Cliffs, N.J., 1962).

TABLE I. Symmetry of wavefunctions for CH_4.

Nuclear spin functions

Total I	T Symmetry	Number of states
0	E	2
1	F	9
2	A	5

Rotational wavefunctions

J	T Symmetry	Number of states
0	A	1
1	F	9
2	E F	10 15
3	A $2F$	7 42
4	A E $2F$	9 18 54
5	E $3F$	22 99

Over-all wavefunctions (all of symmetry A)

J	I	F	Number of states
$0A$	$2A$	2	5
$1F$	$1F$	2 1 0	5 3 1
$2E$	$0E$	2 (twofold degenerate)	2×5
$2F$	$1F$	3 2 1	7 5 3

of group. However, the use of T_d introduces the complication of destroying the natural factorization of the group operations into a subgroup (T) operating on the molecule-fixed axis system and a subgroup [$O(3)$] operating on the space-fixed axis system. The improper operations of T_d must operate on both molecule and space fixed systems simultaneously in order that these coordinate systems be superimposable after the operation.

The rotational wavefunctions form a basis for irreducible representations of the direct product group $T\times O(3)$. They can be labeled by $a\alpha JM_J$ where a gives the irreducible representation of T and α the component of that irreducible representation. The $a\alpha$ part of the wavefunction is formed by taking the appropriate linear combination of the $2J+1$ wavefunctions of different K. Similarly the nuclear spin wavefunctions form a basis for irreducible representation of the direct product group $T\times O(3)$. The labeling scheme is essentially the same, $a\alpha IM_I$ with the possibilities $A\alpha 2M_I$, $E\alpha 00$, $F\alpha 1M_I$. As noted previously the Pauli exclusion principle requires that the over-all wavefunction be of symmetry A under the operations of the group T.

The various components of the irreducible representations of T are shown in Table II. The nuclear spin wavefunctions and rotational wavefunctions for $J=2$ which are basis functions for irreducible representations of T are also shown in Table II. (The choice of phase is that of Edmonds.[9]) The over-all wavefunctions of symmetry A for $J=2$ are combinations e_1 (nuclear spin) e_2(rotation), e_2(nuclear spin) e_1(rotation), and $(1/\sqrt{3})(xx+yy+zz)$.

All the Hamiltonian terms are of the form,

$$[\mathbf{T}_a(k)\cdot\mathbf{U}_a(k)]_A,$$

where $\mathbf{T}_a(k)$ operates only on the rotational functions and is of symmetry a under T and is a spherical tensor of rank k. $\mathbf{U}_a(k)$ operates only on I. The form and nonzero matrix elements of the Hamiltonian terms are shown in Table III.

The case of $J=2$ is now considered in some detail. This is the lowest value of J for which mixing can occur. The rotational wavefunctions of symmetry $F(J=2)$ combine with the spin wavefunctions of symmetry $F(I=1)$ to give basis functions of total angular momentum $F=3$, 2, and 1. In a similar manner $2F=2$ basis functions are obtained from the E wavefunctions. The Hamiltonian matrix consists of seven identical 1×1 blocks for $F=3$ corresponding to the seven values of M_F, five identical 3×3 blocks for $F=2$, and three identical 1×1 blocks for $F=1$. The mixing of $I=1$ and $I=0$ functions occurs in the diagonalization of the 3×3 blocks for $F=2$. By using the linear combinations $(1/\sqrt{2})(e_1+e_2)$ and $(1/\sqrt{2}i)(e_1-e_2)$ as the basis of the E functions, the 3×3 block factors into a 2×2 and a 1×1.

As shown in Table III, the Hamiltonian terms H_{1a}, H_{1b}, H_{2a}, H_{2b} have nonzero diagonal elements for $I=1$. All diagonal elements for $I=0$ are zero. The term H_{1b} gives rise to the matrix element connecting $I=0$ to $I=1$.

The evaluation of element H_{1a} may easily be carried out by elementary techniques and is not discussed here. The procedure for the evaluation of the remaining elements is as follows. These elements are of the form

$$\langle Ja'I'a'FAM_F\,|\,[\mathbf{T}_c(k)\cdot\mathbf{U}_c(k)]_A\,|\,JaIaFAM_F\rangle,$$

where a', a, c, and A all specify irreducible representations of T.

These matrix elements can be expressed in terms of elements uncoupled on T as follows:

$$\langle Ja'I'a'FAM_F\,|\,[\mathbf{T}_c(k)\cdot\mathbf{U}_c(k)]_A\,|\,JaIaFAM_F\rangle$$
$$=\sum_{\alpha,\beta,\alpha',\beta',\gamma,\delta} C_{\alpha\beta}{}^{A}C_{\alpha'\beta'}{}^{A}C_{\gamma\delta}{}^{A}\langle Ja'\alpha'I'a'\beta'FM_F\,|\,[\mathbf{T}_{c\gamma}(k)\cdot\mathbf{U}_{c\delta}(k)]\,|\,Ja\alpha Ia\beta FM_F\rangle, \tag{3}$$

where α, β, α', β', γ, δ refer to particular components of a, a', and c.

The usual vector recoupling methods on $O(3)$ may now be used to evaluate the terms in the sum {see Edmonds [Ref. 9, p. 111, Eq. 7.1.6]}.

$$\langle Ja'\alpha'I'a'\beta'FM_F \mid \mathbf{T}_{c\gamma}(k)\cdot\mathbf{U}_{c\delta}(k)\mid Ja\alpha Ia\beta FM_F\rangle$$

$$=(-1)^{J+I'+F}\begin{Bmatrix} F & I' & J \\ k & J & I\end{Bmatrix}\langle Ja'\alpha' \| \mathbf{T}_{c\gamma}(k)\| Ja\alpha\rangle\langle I'a'\beta' \| U_{c\delta}(k)\| Ia\beta\rangle. \quad (4)$$

The reduced matrix elements on $O(3)$ $\langle Ja'\alpha' \| \mathbf{T}_{c\gamma}(k)\| Ja\alpha\rangle$ and $\langle I'a'\beta' \| \mathbf{U}_{c\gamma}(k)\| Ia\beta\rangle$ need now be evaluated. The tensor operator $\mathbf{T}_{c\gamma}(k)$ is the product of two operators for the Hamiltonian $H_{1b}(k=1)$:

$$\mathbf{T}(1)=-(\tfrac{5}{3})^{1/2}[\mathbf{V}(2)\times\mathbf{W}(1)]_1. \quad (5)$$

The operator $\mathbf{W}$ is simply $\mathbf{J}$.

An expression for such a product is given by Edmonds[9] (p. 109, Eq. 7.1.1):

$$\langle Ja'\alpha' \| \mathbf{T}_{c\gamma}(k)\| Ja\alpha\rangle=(5)^{1/2}\begin{Bmatrix} 2 & 1 & 1 \\ J & J & J\end{Bmatrix}\langle Ja'\alpha' \| \mathbf{V}_{c\gamma}(2)\| Ja\alpha\rangle\langle J \| \mathbf{J}\| J\rangle, \quad (6)$$

where $\langle J \| \mathbf{J} \| J\rangle=[J(J+1)(2J+1)]^{1/2}$.

Pushing the process of calculating matrix elements onward we next indicate how $\langle Ja\alpha' \| \mathbf{V}_{c\gamma}(2)\| Ja\alpha\rangle$ is evaluated. The function $\mid Ja\alpha\rangle$ may be expressed as linear combinations of $\mid JK\rangle$ functions. This was done for $J=2$ in Table II. For the operator $\mathbf{V}_{c\gamma}(2)$, $c=F$, and $\gamma=x, y, z$. It is the off-diagonal part of a symmetric second-rank Cartesian tensor with V_{Fx} corresponding to I_{yz}, V_{Fy} to I_{xz}, and V_{Fz} to I_{xy}, where I_{yz}, I_{xz}, and I_{xy} are numerically equal to $A/3$.

The methods of Edmonds[9] for evaluating symmetric-top matrix elements are now used (pp. 116–117):

$$\langle JK' \| \mathbf{V}_{c\gamma}(2)\| JK\rangle$$

$$=\sum_{q'}V(2,q')\begin{pmatrix} J & 2 & J \\ -K' & q' & K\end{pmatrix}(2J+1)(-1)^{J'+K'}. \quad (7)$$

Evaluation of Eq. (7) gives

$$\langle x \| \mathbf{V} \| e_1\rangle=\langle y \| \mathbf{V} \| e_2\rangle=(\tfrac{10}{7})^{1/2}\epsilon^* A/3,$$
$$\langle y \| \mathbf{V} \| e_1\rangle=\langle x \| \mathbf{V} \| e_2\rangle=(\tfrac{10}{7})^{1/2}\epsilon A/3,$$
$$\langle z \| \mathbf{V} \| e_1\rangle=\langle z \| \mathbf{V} \| e_2\rangle=(\tfrac{10}{7})^{1/2}A/3,$$
$$\langle x \| \mathbf{V} \| y\rangle=\langle y \| \mathbf{V} \| z\rangle=\langle z \| \mathbf{V} \| x\rangle=-(\tfrac{15}{7})^{1/2}A/3. \quad (8)$$

The reduced matrix elements of the nuclear spin operator U are obtained by explicit evaluation of a nonreduced element, e.g., $\langle 1x-1 \mid U_z(0) \mid 1y-1\rangle$, which is related to the reduced element by the Wigner–Eckart theorem (Edmonds 5.4.1)

$$\langle 1x-1 \mid U_z(1,0) \mid 1y-1\rangle$$

$$=(-1)^0\begin{pmatrix} 1 & 1 & 1 \\ 1 & 0 & -1\end{pmatrix}\langle 1x \| \mathbf{U}_z(1)\| 1y\rangle \quad (9)$$

yielding

$$\langle x \| \mathbf{U}_x \| e_1\rangle=\langle y \| \mathbf{U}_y \| e_2\rangle=2\epsilon^*,$$
$$\langle x \| \mathbf{U}_x \| e_2\rangle=\langle y \| \mathbf{U}_y \| e_1\rangle=2\epsilon,$$
$$\langle z \| \mathbf{U}_z \| e_1\rangle=\langle z \| \mathbf{U}_z \| e_2\rangle=2,$$
$$\langle x \| \mathbf{U}_z \| y\rangle=\langle x \| \mathbf{U}_y \| z\rangle=\langle y \| \mathbf{U}_x \| z\rangle=-(6)^{1/2}. \quad (10)$$

The reduced matrix elements of $\mathbf{V}$, Eq. (8), upon substitution into Eq. (6) yield the reduced matrix elements of $\mathbf{T}$. These elements and the reduced elements of $\mathbf{U}$, Eq. (10), are inserted in Eq. (4). The resulting elements when placed in Eq. (3) give the matrix elements of H_{1b} in the desired basis.

TABLE II. Basis functions under the tetrahedral group.

	Nuclear spin ($M_I=0$)	$J=2$ K combinations
A	$[1/(6)^{1/2}][\alpha_1\alpha_2\beta_3\beta_4+\beta_1\beta_2\alpha_3\alpha_4+\alpha_1\beta_2\beta_3\alpha_4+\beta_1\alpha_2\alpha_3\beta_4+\alpha_1\beta_2\alpha_3\beta_4+\beta_1\alpha_2\beta_3\alpha_4]$	None
E	$e_1[1/(6)^{1/2}][\alpha_1\alpha_2\beta_3\beta_4+\beta_1\beta_2\alpha_3\alpha_4+\epsilon(\alpha_1\beta_2\beta_3\alpha_4+\beta_1\alpha_2\alpha_3\beta_4)+\epsilon^*(\alpha_1\beta_2\alpha_3\beta_4+\beta_1\alpha_2\beta_3\alpha_4)]$	$(1/\sqrt{2})[\mid 0\rangle-(i/\sqrt{2})\mid +2\rangle-(i/\sqrt{2})\mid -2\rangle]$
	$e_2=e_1^*$	$=e_1^*$
F	$x(1/\sqrt{2})[\alpha_1\beta_2\alpha_3\beta_4-\beta_1\alpha_2\beta_3\alpha_4]$	$-(i/\sqrt{2})[\mid +1\rangle+\mid -1\rangle]$
	$y(1/\sqrt{2})[\alpha_1\beta_2\beta_3\alpha_4-\beta_1\alpha_2\alpha_3\beta_4]$	$-(1/\sqrt{2})[\mid +1\rangle-\mid -1\rangle]$
	$z(1/\sqrt{2})[\alpha_1\alpha_2\beta_3\beta_4-\beta_1\beta_2\alpha_3\alpha_4]$	$(i/\sqrt{2})[\mid +2\rangle-\mid -2\rangle]$

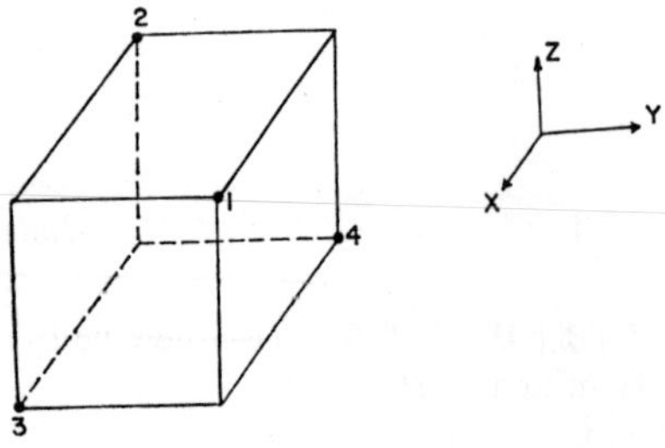

FIG. 1. Choice of proton numbering and coordinate system for methane.

The procedure for the evaluation of the elements of H_{2a}, H_{2b} is exactly analogous. The final 3×3 Hamiltonian matrix for $F=2$ is

$$\begin{pmatrix} \frac{5}{4}B-(A'+A/3)-A/3 & \frac{2}{3}A & \frac{2}{3}A \\ \frac{2}{3}A & 0 & 0 \\ \frac{2}{3}A & 0 & 0 \end{pmatrix}.$$

In order to obtain a numerical answer for the mixing of $I=1$ and $I=0$ wavefunctions the values of A, A', and B are required. The value of B depends only on the geometry and is ~21 kc/sec for CH_4. Unfortunately the values of A and A' are not known. By analogy to the constants found for NH_3[11] they should be of the order of 10 kc/sec. If $A'+A/3$ is assumed to be zero (point dipole at carbon) and A is assumed to be 10 kc/sec, then the two wavefunctions are

$$\psi_a = 0.94\psi_1 + 0.34\psi_0,$$

$$\psi_b = -0.34\psi_1 + 0.94\psi_0,$$

where ψ_1 is pure spin 1 and ψ_0 is pure spin zero. The third level is pure $I=0$ with $E=0$.

ASYMMETRIC ROTORS

The spin–rotation interaction can lead to isomerization in the case of asymmetric rotors such as H_2O or H_2CO. The requirement is that the spin-rotation interaction tensors of the two nuclei not be oriented in the same way. Thus in either H_2O or H_2CO, the plane of the molecule is perpendicular to the c axis. The nonzero spin–rotation interaction tensor components are T_{aa}, T_{bb}, T_{cc}, and T_{ab}. The values of T_{aa}, T_{bb}, and T_{cc} are the same for both nuclei and give rise to no isomerization. However, $T_{ab}^{(1)}$ for Nucleus 1 equals $-T_{ab}^{(2)}$ for Nucleus 2.

TABLE III. Contributions of various Hamiltonian terms to matrix elements.

Form of terms			
H_{1a}	$T_A(1)\cdot U_A(1)$		
H_{1b}	$[T_F(1)\cdot U_F(1)]_A$		
H_{2a}	$[T_E(2)\cdot U_E(2)]_A$		
H_{2b}	$[T_F(2)\cdot U_F(2)]_A$		
I[a]	0	1	2
0	0	H_{1b}	H_{2a}
1	H_{1b}	$H_{1a}+H_{1b}+H_{2b}+H_{2a}$	$H_{1b}+H_{2b}$
2	H_{2a}	$H_{1b}+H_{2b}$	H_{1a}

[a] A nonzero contribution of a term to a matrix element is indicated by the term name appearing in the appropriate position.

The Hamiltonian of interest is

$$H = H_{\text{rot}} + T_{ab}[(I_{1a}-I_{2a})J_b + (I_{1b}-I_{2b})J_a]. \quad (11)$$

The term in T_{ab} connects rotational states of different symmetry with respect to the C_{2b} axis (C_{2a} axis for H_2CO) and it connects states of different nuclear spin. The symmetry of T_{ab} on the D_4 group of rotational symmetry is B_c. It therefore connects rotational states A to B_c and B_a to B_b. The selection rules for the term in T_{ab} are rigorously $\Delta F=0$, $\Delta I=\pm1$, $\Delta J=0, \pm1$, $A\leftrightarrow B_c$, $B_b\leftrightarrow B_a$. In the prolate top limit there is the additional rule $\Delta K_{-1}=\pm1$. In the oblate top limit there is the additional rule $\Delta K_{+1}=\pm2$.

The matrix elements for the interaction in the prolate symmetric-top Wang wavefunction limit for nuclear spin of $\frac{1}{2}$ is

$$\langle J'_{K'_-K'_+}I'FM_F \mid H \mid J_{K_-K_+}IFM_F\rangle = (-1)^{J+J'+K'_-+I'+F}\{15[J(J+1)(2J'+1)]\}^{1/2}(2J+1) \times \begin{Bmatrix} F & I' & J' \\ 1 & J & I \end{Bmatrix} \times \begin{Bmatrix} J' & J & 2 \\ 1 & 1 & J \end{Bmatrix} \times \begin{pmatrix} J' & 2 & J \\ -K'_- & 1 & K_- \end{pmatrix} \times T_{ab}, \quad (12)$$

where $I'=1$, $I=0$, $F=J$, and K'_+ has the same parity as K_+. K'_- has been assumed to be K_-+1. A similar formula holds when $K'_-=K_--1$.

The matrix elements for an asymmetric rotor of finite asymmetry may be obtained by finding the transformation which diagonalizes the asymmetric rotor Hamiltonian and applying it to the matrix of interactions.

Thus if one knew the three rotational constants A, B, C of the molecule and the interaction parameter T_{ab} it would be possible to calculate the mixing of $I=0$ and $I=1$ wavefunctions for any pair of levels. Unfortunately T_{ab} is not known for any molecule. It should be approximately the same order of magnitude as T_{aa}

[11] G. R. Gunther-Mohr, C. H. Townes, and J. H. Van Vleck, Phys. Rev. **94**, 1191 (1954).

which is usually ~10 kc/sec for fairly light molecules. Rotational spacings are ~10^5 Mc/sec for H_2O or H_2CO. Thus the fractional mixing would be about ~10^{-7} for most rotational levels. However, two interacting rotational levels may be accidentally nearly degenerate. The mixing would then be much larger. The possibility thus exists that spin statistics isomerization "funnels" through a single pair of strongly interacting levels.

The rotational energy levels of H_2O and H_2CO were calculated using the observed rotational constants.[12,13] No corrections for centrifugal distortion could be made. The corrections would be of importance only if one wished to fit experimental isomerization rates for these molecules. The present purpose is to investigate the isomerization phenomena for a sufficiently realistic case. Of course, H_2O would probably equilibrate by chemical exchange.

In looking at the predicted rotational transitions it is immediately apparent that a systematic near degeneracy occurs. The levels J_{0J} and J_{1J} rapidly become nearly degenerate as J increases. For H_2O the $12_{0,12}$ and $12_{1,12}$ differ by only 75 Mc/sec and the $18_{0,18}$ and $18_{1,18}$ differ by only 0.18 Mc/sec. These near degeneracies are the K type doublets of the oblate near symmetric top. The near degeneracy occurs for higher values of J for H_2CO because it is more prolate. Thus for H_2CO $18_{0,18}$ and $18_{1,18}$ are different in energy by 41 000 Mc/sec. The $40_{0,40}$ and $40_{1,40}$ levels differ by 318 Mc/sec.

The interaction which leads to isomerization becomes smaller as J increases. This is clear because the oblate selection rule for the interaction should eventually apply and $\Delta K_{+1}=0$ is not allowed. These systematic degeneracies lead to an interesting limit problem of the 0/0 type. The wavefunction mixing is given by $\langle J_{0J} \mid H_{ab} \mid J_{1J}\rangle/[E(J_{0J})-E(J_{1J})]$.

Numerical calculations of the ratio indicate that it remains approximately constant as J increases. For H_2O the ratio decreases by nearly a factor of 2 going from $J=4$ to $J=14$. The energy denominator changes from 95 540 Mc/sec for $J=4$ to 10 Mc/sec for $J=14$.

These degeneracies therefore do not contribute significantly to the isomerization by nonmagnetic collisions. It is not clear however whether the rate of isomerization by magnetic collisions would be greatly increased by the presence of nearly degenerate levels of different I, but it is possible. This point deserves further consideration which is not given here.

Some accidental near degeneracies of the right type to give mixing of the nuclear spin wavefunctions do occur and are listed in Table IV for H_2O and H_2CO.

KINETICS OF SPIN ISOMERIZATION

In order for a collision to change I, magnetic forces which interact with the magnetic moments of the protons must be involved. Since such forces are weak the probability of a change in I due to a collision is small. Indeed the conversion of H_2 gas at pressures of the order of 1 atm and liquid-air temperature has a half-life of greater than a year.[3] Most of the conversion that does occur probably involves a heterogeneous reaction.[3]

However, we have shown that states of different I are mixed by the spin–rotation interaction for molecules such as CH_2O and CH_4. In the case of methane the spin–spin interaction also causes mixing. No mixing from either source is possible for H_2 by symmetry. This mixing must lead to equilibration of spin isomers even in the absence of magnetic collisions.

Let us now turn our attention to the kinetics of this isomerization process. Our goal is to arrive at a qualitative understanding of the isomerization process, estimate the order of magnitude of the rate and its dependence on pressure.

Suppose all molecules are initially in a particular pure I state (e.g., $I=0$, $J=0$ for CH_2O or $I=2$, $J=0$ for CH_4). Regardless of the initial distribution over rotational states, thermal equilibrium among rotational states of this particular value of I will be rapidly established (~$10t_c$, where t_c is the mean time between collisions).

At the time of a collision the Hamiltonian for a single molecule will have the form

$$H(t)=H_R+H_I+H_{IR}+H_{CR}(t), \qquad (13)$$

with H_R symbolizing the rotational Hamiltonian, H_I the nuclear spin Hamiltonian, H_{IR} the interaction between nuclear spin and rotation (which includes the portion of the spin–spin interaction which is orientation dependent as well as the direct spin–rotation interaction), and $H_{CR}(t)$ the collision Hamiltonian which has the subscript R to indicate that it is assumed to act on the rotation only. The effect of the collision can be described by a scattering operator

$$S=\exp\left[-i\int_0^{\tau_c}\frac{H(t)dt}{\hbar}\right], \qquad (14)$$

where τ_c is the duration of a collision.

The wavefunction immediately after collision, ψ', is given by

$$\psi'=S\psi, \qquad (15)$$

where ψ is the wavefunction just before collision.

A reasonable assumption for gas-phase collision is that $H_{CR}(t)\gg H_I$, H_{IR} during the collision so that S may be written to a very good approximation as

$$S=\exp\left(-i\int_0^{\tau_c}\frac{[H_R+H_{CR}(t)]dt}{\hbar}\right). \qquad (16)$$

Thus the effect of the collision is to change the rotational state without changing I. This implies that if the molecule is in a state of pure I before collision it will be in a state of pure I after collision, even if the latter is a nonstationary state.

[12] H_2O, G. Herzberg, *Infra-red and Raman Spectra* (D. Van Nostrand Co., Inc., Princeton, N.J. 1945).

[13] H_2CO, T. Oka, J. Phys. Soc. Japan **15**, 2274 (1960).

In CH_2O, for example, two rotational levels may interact via the spin–rotation interaction giving the following wavefunctions for the energy levels a and b:

$$\psi_{a,R}=\cos\theta\psi_{0,R}+\sin\theta\psi_{1,R'},$$

$$\psi_{b,R'}=-\sin\theta\psi_{0,R}+\cos\theta\psi_{1,R'},$$

$$\tan 2\theta=2\langle 0R \mid H_{IR} \mid 1R'\rangle/(E_{0R}-E_{1R'}), \quad (17)$$

where $\psi_{0,R}$ is the $I=0$, rotational-state R wavefunction in the absence of interaction and $\psi_{1,R'}$ has $I=1$, rotational state R'. The angle θ will usually be small. If CH_2O is in the $I=0$ state, a collision scattering into state R will give the nonstationary state $\psi_{0,R}$ immediately after collision rather than $\psi_{a,R}$. The state $\psi_{0,R}$ then evolves with time:

$$\psi_{0,R}(t)=[C_0(t)\psi_{0,R}+C_1(t)\psi_{1,R'}]\exp(-iEt/\hbar),$$

$$C_0(t)=\cos(\omega t/2)-i\cos 2\theta\sin(\omega t/2),$$

$$C_1(t)=+i\sin 2\theta\sin(\omega t/2), \quad (18)$$

where

$$E=\tfrac{1}{2}(E_a+E_b),$$

$$\omega=(E_a-E_b).$$

This time evolution is terminated by a second collision at time $t=t_0$. Immediately after the second collision the coefficients of $I=1$ in the various rotational states $C'_{1,R''}(t_0)$ will be proportional to $C_1(t_0)$. However, the relative phases of $C'_{1,R''}(t_0)$ and $C'_{0,R''}(t_0)$ will be random. If states $I=0$, R''' and $I=1$, R'' interact the time evolution of this portion of ψ'

$$\psi'_{0,R'''1,R''}=C'_{0,R'''}\psi_{0,R'''}+C_{1,R''}\psi_{1,R''}$$

may lead to a larger or smaller coefficient of $I=1$.

The time evolution of the spin isomerization might be followed experimentally by observing the intensity of a rotation component of an infrared band which originates from an $I=1$ level grow with time. If the state being observed is called $\psi_{1,R(\mathrm{obs})}$ then this intensity is proportional to $|C_{1,R(\mathrm{obs})}|^2$ summed over all molecules. There will be thermal equilibrium among all the $I=1$ levels so that

$$\sum_{\text{molecules}} |C_{1,R(\mathrm{obs})}|^2$$

will be proportional to

$$\sum_{\text{molecules}}\sum_R |C_{1,R}|^2.$$

The proportionality constant is $Z_1^{-1}\exp(-E_{R(\mathrm{obs})}/kT)$, where Z_1 is the rotational partition function for spin 1. Since time evolution after a collision may lead to a larger or smaller value of C_{1R}, the quantity $\sum_R |C_{1,R}|^2$ will grow in a random-walk manner. Each collision and the subsequent time evolution is one step in the walk which takes place in the complex plane. The steps will be of varying lengths. The length of steps will vary both because they start in different rotational states (with different interactions with states of different I) and because the times between collisions vary. It is necessary to average the length squared over both the rotational states available (weighted by the Boltzmann distribution) and the distribution of collision times:

$$\Big(\sum_{\text{molecules}}\sum_R |C_{1R}|^2\Big)N^{-1}=\left(\frac{t}{t_c}\right)\frac{\sum_R \exp(-E_R/kT)}{Z}\times\int_0^\infty |C_{1R}(t')|^2 f(t')dt', \quad (19)$$

where $f(t)$ is the distribution of collision times.

$$f(t)=(1/t_c)\exp(-t/t_c). \quad (20)$$

The number of steps per molecule is (t/t_c). The rate of change of this system to State 1 is

$$k=\Big(\sum_{\text{molecules}}\sum_R |C_{1R}|^2\Big)(Nt)^{-1}. \quad (21)$$

The integrations may be easily carried out yielding

$$t_c k_R=\int_0^\infty |C_{1R}(t)|^2 f(t)dt$$

$$=\sum_{R'}\left(\frac{2t_c^2}{1+\omega^2_{RR'}t_c^2}\right)\frac{|\langle 0R' \mid H_{IR} \mid 1R\rangle|^2}{\hbar^2}, \quad (22)$$

where

$$\omega_{RR'}=(E_R-E_{R'})/h,$$

$$k=Z^{-1}\sum_R k_R\exp(-E_R/kT). \quad (23)$$

The cross terms are omitted in the sum over R' in Eq. (22) because the sign of their contribution will depend on the exact nature of the subsequent collision with zero average contribution.

Now the two cases of interest, CH_2O and CH_4, may be considered. For pressures of about 10 mm Hg ($t_c=10^{-8}$ sec), $t_c\omega_{RR'}\gg 1$ for all states of interest in CH_2O and $t_c\omega_{RR'}\ll 1$ for the K degenerate states of CH_4. If $t_c\omega_{RR'}\gg 1$ the rate is inversely proportional to t_c and thus proportional to pressure. If $t_c\omega_{RR'}\ll 1$ the rate is proportional to t_c and thus inversely proportional to pressure.

For methane then $k_R=2t_c\hbar^{-2}|\langle I \mid H \mid I'\rangle_R|^2$; since for most levels in methane which interact

$$|\langle I \mid H \mid I'\rangle_R|/\hbar \simeq 10^4\ \mathrm{sec}^{-1},$$

an approximation to the rate is

$$k=Z^{-1}(10^{+4})^2 10^{-8}\sum_R \exp(-E_R/kT)$$

$$=ZZ^{-1}=1\ \mathrm{sec}^{-1}. \quad (24)$$

In CH_2O only levels with near degeneracies are of importance at 10 mm Hg. Then,

$$k_R=(2/t_c)|C_{1R}|^2 Z^{-1}\exp(-E_R/kT),$$

where

$$|C_{1R}|^2=\sum_{R'}|\langle 0R \mid H \mid 1R'\rangle|^2/\hbar^2\omega^2_{RR'}. \quad (25)$$

TABLE IV. Example calculations of isomerization rates of H_2CO and H_2O.

H_2CO

Rotational constants used

$A = 282\ 029.22$ Mc/sec

$B = 38\ 835.28$

$C = 34\ 003.16$

Rotational partition function at 300°K

$Q_r = 721$

Rotational near degeneracies

	ΔE[a] (Mc/sec)	$\vert\langle H\rangle/T_{ab}\vert$	$\vert C_1\vert$[b]$\times 10^5$	P_0[c]$\times 10^3$	Rate[d]$\times 10^5$ (sec^{-1})
$3_{03}\rightarrow 2_{11}$	34 699	1.51	0.043	1.7	0.005
$8_{26}\rightarrow 9_{18}$	8 830	3.64	0.41	3.3	1.1
$13_{2,12}\rightarrow 14_{1,14}$	16 309	5.46	0.34	2.6	5.9
$17_{3,15}\rightarrow 18_{2,17}$	3 901	6.92	1.77	1.4	9.0
$28_{5,23}\rightarrow 29_{4,25}$	4 021	8.98	2.24	0.06	0.66
$36_{6,32}\rightarrow 35_{6,30}$	4 853	11.4	2.35	0.004	0.04

Total rate 1.7×10^{-4} sec^{-1}

H_2O

Rotational constants used

$A = 8.332\times 10^5$ Mc/sec

$B = 4.347\times 10^5$

$C = 2.985\times 10^5$

Rotational partition function at 300°K

$Q_r = 42.4$

Near degeneracies	ΔE[a] (Mc/sec)	$\vert\langle H\rangle/T_{ab}\vert$	$\vert C_1\vert$[b]	P_0[c]	Rate[d] (sec^{-1})
$8_{63}\rightarrow 7_{71}$	7 982	1.36	1.7×10^{-6}	9.7×10^{-5}	5.6×10^{-8}
$8_{62}\rightarrow 7_{70}$	7 572	1.27	1.7×10^{-6}	9.7×10^{-5}	5.6×10^{-8}
$12_{11,1}\rightarrow 13_{10,3}$	41 573	1.33	3.2×10^{-7}	1.5×10^{-9}	3.1×10^{-14}
$12_{11,2}\rightarrow 13_{10,4}$	41 610	1.28	3.1×10^{-7}	1.5×10^{-9}	3.1×10^{-14}

Total rate $= 1.2\times 10^{-7}$ sec^{-1}

[a] ΔE is the energy of the level on the right minus the energy of the level on the left.
[b] Assuming $T_{ab} = 10$ kc/sec.
[c] P_0 is the probability of being in the $I=0$ state specified.
[d] Assuming the time between collision is 10^{-8} sec.

The interactions between near degenerate levels are given in Table IV. The over-all isomerization rate calculated assuming $T_{ab} = 10$ kc/sec is 1.7×10^{-4} sec^{-1}.

It is interesting to note that if higher pressures are considered where $\omega_{RR'}t_c \ll 1$ for both CH_4 and CH_2O, the rate for both molecules would be about the same. The presence of a near degeneracy is no longer of importance.

This kinetic treatment can also be applied to the longitudinal relaxation of nuclear magnetic polarization in a gas. The theory of this has already been worked out for H_2 by Bloembergen[14] and exactly the same result is obtained by this method in the limit $t_c\omega_0 \ll 1$.

If separated isomers were available for CH_2O this mechanism could be tested in several ways. First the maximum in the rate as the pressure is raised should be readily observable. Secondly, the isomerization rates of $CH_2{}^{16}O$ and $CH_2{}^{18}O$ could be compared. The substitution of ^{18}O would shift the asymmetric rotor energy levels and destroy some near degeneracies while possibly creating others. In the low-pressure region the rate would be sensitive to these changes. Only a careful centrifugal distortion analysis of the spectrum could allow the direction of the shift in the rate to be predicted. However, it seems likely that the shift would be large.

[14] N. Bloembergen, *Nuclear Magnetic Relaxation* (W. A. Benjamin, Inc., New York, 1961), p. 104.

ACKNOWLEDGMENTS

The authors are grateful to Professor Z. W. Salsburg and Professor George Trammell for discussions clarifying the basic elements of the problem.

APPENDIX

Derivation of the Hamiltonian for CH_4

H_1: Spin–Rotation Interaction

Consider Nucleus 1. Its interaction with the rotation is a symmetric second-rank tensor form

$$H_1(1)=\mathbf{J}\cdot\mathsf{T}^{(1)}\cdot\mathbf{I}_1. \quad (A1)$$

The tensor T should be symmetric about the CH_1 line [(1, 1, 1) vector]. That is, the tensor should be invariant to the C_3 operation which causes $x\rightarrow y$, $y\rightarrow z$, $z\rightarrow x$ or

$$\begin{pmatrix} T_{xx}^{(1)} & T_{xy}^{(1)} & T_{xz}^{(1)} \\ T_{xy}^{(1)} & T_{yy}^{(1)} & T_{yz}^{(1)} \\ T_{xz}^{(1)} & T_{yz}^{(1)} & T_{zz}^{(1)} \end{pmatrix} = \begin{pmatrix} T_{zz}^{(1)} & T_{zx}^{(1)} & T_{zy}^{(1)} \\ T_{zx}^{(1)} & T_{xx}^{(1)} & T_{xy}^{(1)} \\ T_{zy}^{(1)} & T_{xy}^{(1)} & T_{yy}^{(1)} \end{pmatrix}. \quad (A2)$$

Therefore

$$T_{xx}=T_{zz}=T_{yy}=(A'+A/3),$$
$$T_{xy}=T_{zx}=T_{zy}=A/3. \quad (A3)$$

If the interaction arises from a point dipole, T is traceless hence $3(A'+A/3)=0$. Now consider the spin–rotation interaction of Nucleus 2. The Hamiltonian of Nucleus 2 may be obtained from that of Nucleus 1 by applying the operation that takes Nucleus 1 into Nucleus 2: for C_{2z}.

$$1\rightarrow 2, \quad x\rightarrow -x, \quad y\rightarrow -y, \quad z\rightarrow z, \quad (A4)$$

$$H_1(2)=\mathbf{J}\cdot\mathsf{T}^{(2)}\cdot\mathbf{I}_2, \quad (A5)$$

$\mathsf{T}^{(2)}=C_{2z}\mathsf{T}^{(1)}$,

$$\mathsf{T}^{(2)}=\begin{pmatrix} (A'+A/3) & A/3 & -A/3 \\ A/3 & (A'+A/3) & -A/3 \\ -A/3 & -A/3 & (A'+A/3) \end{pmatrix}. \quad (A6)$$

By applying C_{2x} and C_{2y} to $H_1(1)$, the Hamiltonians for Nuclei 3 and 4 may be obtained:

$$H_1=H_1(1)+H_1(2)+H_1(3)+H_1(4). \quad (A7)$$

The result is given in Eq. 1.

H_2: Spin–Spin Interaction

The magnetic-dipole–magnetic-dipole interaction between the nuclei is expected to dominate over the indirect coupling. Consider the interaction of Nucleus 1 with Nucleus 2:

$$H_2^{(12)}=[(g_I\mu_N)^2r_{12}^{-3}][(\mathbf{I}_1\cdot\mathbf{I}_2)-3(\mathbf{I}_1\cdot\hat{\mathbf{r}}_{12})(\mathbf{I}_2\cdot\hat{\mathbf{r}}_{12})]. \quad (A8)$$

The vector $\hat{\mathbf{r}}_{12}$ has components $(-1/\sqrt{2}, -1/\sqrt{2}, 0)$. Let $B=[(g_I\mu_N)^2r_{12}^{-3}]$:

$$H_2^{(12)}=B(\mathbf{I}_1\cdot\mathsf{T}_2^{(12)}\cdot\mathbf{I}_2),$$

$$\mathsf{T}_2^{(12)}=\begin{pmatrix} -\frac{1}{2} & -\frac{3}{2} & 0 \\ -\frac{3}{2} & -\frac{1}{2} & 0 \\ 0 & 0 & 1 \end{pmatrix}=\frac{1}{2}\begin{pmatrix} \epsilon+\epsilon^* & -3 & 0 \\ -3 & \epsilon+\epsilon^* & 0 \\ 0 & 0 & 2 \end{pmatrix},$$

$$H_2=H_2^{(12)}+H_2^{(13)}+H_2^{(14)}+H_2^{(23)}+H_2^{(24)}+H_2^{(34)}. \quad (A9)$$

The term $H_2^{(13)}$ may be calculated by performing the operation which interchanges Nuclei 2 and 3: C_3 about CH_1

$$2\rightarrow 3, \quad x\rightarrow y, \quad y\rightarrow z, \quad z\rightarrow x,$$

$$\mathsf{T}^{(13)}=\frac{1}{2}\begin{pmatrix} 2 & 0 & 0 \\ 0 & \epsilon+\epsilon^* & -3 \\ 0 & -3 & \epsilon+\epsilon^* \end{pmatrix}. \quad (A10)$$

Similar permutations give the other terms. Collection of the terms on the diagonal gives H_{2a}. Collection of terms off the diagonal gives H_{2b}.

Reprinted from THE JOURNAL OF CHEMICAL PHYSICS, Vol. 47, No. 2, 864–866, 15 July 1967
Printed in U. S. A.

Oxygen Catalysis of Nuclear Spin Species Conversion in Solid Methane*

HARRY P. HOPKINS, JR., PAUL L. DONOHO, AND KENNETH S. PITZER
Departments of Chemistry and Physics, Rice University, Houston, Texas

(Received 13 January 1967)

We wish to report here some nuclear magnetic resonance studies as evidence for the oxygen catalysis of nuclear spin species conversion in solid methane. These data confirm the preliminary observations of R. P. Wolf and W. M. Whitney[1] and suggest a re-evaluation of the explanation of the anomalous thermal behavior found in solid methane below 15°K by Colwell, Gill, and Morrison.[2]

The proton magnetic resonance intensity of solid methane was observed at 4.2°K as a function of time. During the periods between the measurements, the sample was maintained at various temperatures between 4.2° and 10°K by an electrical heater circuit controlled by a germanium thermometer supplied by the Texas Instruments Corporation. Samples of several oxygen concentrations were sealed in glass cells and placed in a coil constructed on a Teflon base. To avoid being misled by slow drifts in the electronic circuits, the ratio of the proton signal to the fluorine-19 signal of the Teflon coil base was followed. The derivative was recorded and the total area of both portions of the derivative curve was measured and multiplied by the linewidth to calculate the ratio of proton to fluorine signals on a basis proportional to the intensity.

Figure 1 shows the measured ratio as a function of time for a sample kept at 4.2°K and containing 0.01 mole % oxygen. The data for higher concentrations of oxygen show the initial portion of the change occurring more rapidly and show less change after the first few hours. A sample of pure methane (supplied by Matheson and Company with a stated purity of 2 ppm) did not exhibit any appreciable change over a period of 12 h. Measurements on samples both with and without oxygen kept at higher temperatures were approximately the same as those at 4.2°K. On a sample containing 0.98% oxygen, sufficient time elapsed to allow the initial large increases to occur. Then the temperature was raised to 10°K for several hours. Upon lowering the temperature to 4.2°K, the ratio was seen to have decreased, but quickly grew back into its earlier value

Motizuki and Nagamiya[3] present a theory for the oxygen catalyzed conversion of orthohydrogen in the solid state and find the rate of conversion to be proportional to the inverse twelfth power of the distance from the impurity center. Although the situation is far more complicated in solid methane, the conversion between any two of three spin species ($I=2$, $I=1$, and $I=0$) should follow the same pattern. This proportionality predicts a curve very similar to that of Fig. 1. Methane molecules immediately around an oxygen molecule will be converted almost instantaneously while successive layers of methane at greater distances will convert less and less rapidly.

The theoretical treatment of the oxygen catalysis in solid hydrogen was made under the assumption of complete oxygen isolation and negligible diffusion of the hydrogen molecules. The nuclear magnetic resonance measurements for solid methane[1] at higher temperatures indicate that molecular diffusion becomes very slow below 60°K. At higher concentration of oxygen, the action spheres of the oxygen molecules will overlap and practically all of the methane will be converted quickly.

These results are pertinent to the interpretation of the anomalously high heat capacity values reported for methane below 15°K by Colwell, Gill, and Morrison.[2] These authors note that all of the deuterated methanes have two lambda-type transitions above 15°K, whereas methane itself has only one, and suggest that the anomaly below 15°K in methane corresponds to the lower transition in the deuterated species. There are several objections to this explanation.

Thermal equilibrium is rapid in the lower transitions in all of the deuterated species whereas Colwell, Gill, and Morrison[2] report very long times for equilibrium in methane. Substitution of a lighter isotope should increase rather than decrease rates of change. Also Sperandio[4] measured the heat capacity of methane in this range and did not observe any anomaly in methane below 15°K.

Bartholome, Driko, and Eucken[5] investigated the

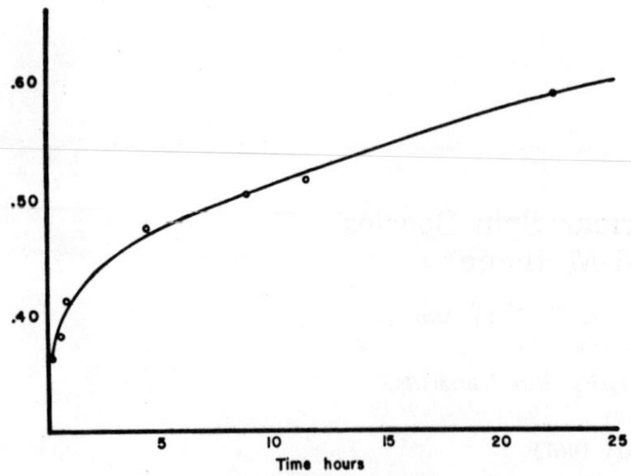

FIG. 1. Proton magnetic resonance intensity as a function of time for methane containing 0.01 mole % oxygen at 4.2°K. See text for explanation of the ordinate scale.

transitions in various solid solutions of CH_4 and CD_4; they found that the total energy change of the lower transition decreased rapidly as the fraction CD_4 decreased and that the lower transition had disappeared when the fraction of CD_4 was reduced to 15%. The effect of pressure[5,6] on CH_4 has an effect very similar to that of the substitution of D for H; at pressures above 300 atm CH_4 shows two transitions, but the lower transition disappears as the pressure is reduced and is not detectable at 150 atm.

Comparison of the statistical-spectroscopic entropy of methane with the third-law value based upon a normal heat capacity curve below 15°K yields agreement within experimental error. Similar agreement is obtained for CD_4 with both transitions included in the calculations. Thus the only remaining entropy in CH_4 which could be appearing in the anomaly below 15°K is that associated with nuclear spin, rather than entropy associated with the same molecular rotational effects that cause the lower transition in deuterated methanes

We believe the sluggish anomaly reported by Colwell, Gill, and Morrison[2] may be explained as nuclear spin species conversion since they report an oxygen concentration of 0.005% in the specimen for which the anomalously high heat capacities were measured below 15°K. Our results indicate that this is enough oxygen to catalyze conversion of a significant portion of the sample.

* Financially assisted by the Robert A. Welch Foundation and the U.S. Atomic Energy Commission.

1 R. P. Wolf and W. M. Whitney, *Low Temperature Physics*, J. G. Daunt *et al.*, Eds. (Plenum Press, Inc., New York, 1965), p. 1118.

2 J. H. Colwell, E. K. Gill, and J. A. Morrison, J. Chem. Phys. **39**, 635 (1963).

3 K. Motizuki and T. Nagamiya, J. Phys. Soc. Japan **11**, 93 (1956).

4 A. Sperandio, thesis, University of Zurich, 1961.

5 E. Bartholome, G. Driko, and A. Eucken, Z. Physik. Chem. Leipzig **B39**, 371 (1938).

6 J. S. Rosenshein and W. M. Whitney, *Low Temperature Physics*, J. G. Daunt *et al.*, Eds. (Plenum Press, Inc., New York, 1965), p. 1114.

Reprinted from THE JOURNAL OF CHEMICAL PHYSICS, Vol. 48, No. 7, 2959–2965, 1 April 1968
Printed in U. S. A.

Infrared Matrix-Isolation Studies of Nuclear-Spin-Species Conversion*

H. P. HOPKINS, JR.,† R. F. CURL, JR., AND KENNETH S. PITZER
Chemistry Department, Rice University, Houston, Texas

(Received 10 October 1967)

Studies of the time evolution of intensities in the fine structure of infrared bands of CH_3D, H_2O, and NH_3 isolated in rare-gas matrices in the temperature region 6.5°–20°K have been made. The molecule CH_3D exhibits a fine structure at 6.5°K with time-dependent intensities. The data are interpreted in terms of nuclear-spin-species conversion. Because of rapid reconversion at 20°K it is concluded that CH_3D cannot be obtained at room temperature with a nonequilibrium distribution of spin isomers by low-temperature equilibration. The time dependence of the fine structure of the H_2O bending mode at 6.5°K has been reinvestigated. The essence of the previous work is confirmed. The rate of nuclear-spin conversion is found to be rapid at 30°K. The fine structure of the NH_3 umbrella motion exhibits a time dependence which indicates that most of the fine structure observed is not due to rotation.

INTRODUCTION

Since the discovery of ortho- and parahydrogen by Giauque and Johnston,[1] the phenomenon of nuclear-spin-species conversion has been of considerable interest. It has been possible to prepare separated isomers in the gas phase at room temperature only for H_2, D_2, and T_2. There is a considerable body of theory and experimental results for these molecules.[2] However, it is of interest to consider other molecules.

Ideally spin-species conversion should be studied in the gas phase because the nature of the energy levels is well understood and it is easiest to develop a theory[3] of the conversion process for the gas phase. However, it has not been possible to prepare a nonequilibrium distribution except for the hydrogen isotopes mentioned above.

The infrared bands of several hydride molecules isolated in rare-gas matrices at low temperature have been observed to exhibit a fine structure. This fine structure is often interpreted as arising from some form of hindered rotation of the molecule in the matrix cage. The nuclear-spin-species conversion in the matrix may be studied by following the intensity changes of the fine-structure components. The first attempt to do this was by Glasel[4] who studied H_2O at 20°K, but he was not able to conclude that he had actually observed conversion. Redington and Milligan[5] studied the time dependence of the fine structure of the infrared bands of H_2O at 5°K. They observed intensity changes occurring over about 3 h. The process was catalyzed by O_2. Redington and Milligan[5] were able to assign rotational quantum numbers for many of the observed lines. The results clearly indicated that nuclear-spin-species conversion was observed. Recently CH_4 has been carefully

* This work was supported by the Robert A. Welch Foundation Grants C-071 and C-092, by the Atomic Energy Commission, and by National Science Foundation Grant GP 6305X.

† Present address: Department of Chemistry, Georgia State College, Atlanta, Ga.

[1] W. F. Giauque and H. L. Johnston, J. Am. Chem. Soc. **50**, 3221 (1928).

[2] A. Farkas, *Orthohydrogen, Parahydrogen, and Heavy Hydrogen* (Cambridge University Press, Cambridge, England, 1935).

[3] R. F. Curl, J. V. V. Kasper, and K. S. Pitzer, J. Chem. Phys. **46**, 3220 (1967).

[4] J. A. Glasel, J. Chem. Phys. **33**, 252 (1960).

[5] R. L. Redington and D. E. Milligan, J. Chem. Phys. **39**, 1276 (1963).

studied at temperatures down to 2.2°K by Frayer and Ewing.[6a,6b] They were able to assign completely the fine structure observed with rotational quantum numbers and to follow accurately the kinetics of the spin isomerization process in several experiments with different concentrations of O_2 as a conversion catalyst. Frayer and Ewing's results for CH_4 are so clearcut as to leave little possibility for doubt that spin-species conversion was observed.

It has been shown[3,7] that CH_4 will come to equilibrium in a few seconds in the gas phase at room temperature. The investigation being reported here began with an attempt to prepare CH_3D in the gas phase with a nonequilibrium spin distribution. General considerations[3] suggested that CH_3D should not equilibrate for at least a few minutes at about 10 torr pressure and room temperature. In this experiment, CH_3D was held at 4.2°K for several weeks to allow conversion from the $I_H=\frac{1}{2}$ to the $I_H=\frac{3}{2}$ states to take place. On examining the gas-phase infrared bands immediately after rapid warmup no lack of spin equilibrium was observed. The question then arose: did the conversion not take place at 4.2°K or did reconversion take place during warmup? An investigation of the infrared spectrum of CH_3D in argon, krypton, and xenon matrices near liquid-He temperature was begun to resolve this question. After positive results were obtained, the spectra of H_2O and NH_3 were studied as well.

EXPERIMENTAL

The cell used had a cold CsI plate on which the matrix was deposited by spray on. The CsI plate was sandwiched in a copper block to which good thermal contact was ensured by using indium gaskets between the window and block. The block was surrounded, except for the light path through the CsI window, by a liquid-N_2-cooled shield. The block could be cooled either by allowing liquid He to flow into it or by forcing cold He gas through it. The temperature of the block could be fairly easily controlled between 6.5° and 77°K by adjusting the He gas flow rate. The temperature was measured by a germanium resistance thermometer mounted on the block at the bottom of the window where it was exposed to the spectrometer source radiation. In addition the temperature could be measured by a copper–constantan thermocouple mounted in the block. At the lowest window temperature the He boiloff rate was about $\frac{1}{3}$ liter/h.

The CH_3D was obtained from Volk Radiochemical Corporation and used without analysis or purification. The H_2O was ordinary deionized water which was thoroughly degassed by alternate freezing and thawing under vacuum. The NH_3, CH_3, argon, and krypton were manufactured by Matheson Company. The xenon was manufactured by Air Reduction Company.

The impurity of most concern was O_2. Because the matrices were dilute, the purity of the rare gases was most critical. The O_2 reported by the manufacturer was 1 ppm for krypton and less than 5 ppm for xenon. The argon was Matheson high-purity grade (less than 0.005% O_2).

The matrices were prepared in a glass vacuum system attached directly to the cryostat which was pumped down to about 5×10^{-5} torr. The low-concentration species (CH_3D, H_2O, or NH_3) plus O_2 in some experiments were introduced into the storage bulb first, then the matrix material was added. The experimental arrangement was such that any departure from perfect mixing would give higher M/A values than those calculated. The matrices had M/A in the range 100 to 350 for all experiments.

Frayer and Ewing[6] have pointed out the disadvantages of the matrix spray-on experiment as contrasted with the equilibrium matrix technique which they used. Spray-on matrix spectra often have additional absorption lines arising presumably from different matrix cavities (such as the 3033-cm^{-1} line of CH_4 in Ar) and these features are not reproducible from experiment to experiment. Particularly frustrating is the difficulty in controlling spray-on conditions so as to avoid highly scattering matrices. The only generalization the present authors reached is that matrices with M/A lower than 100 are apt to be highly scattering.

On the other hand, two of the species studied here, H_2O and NH_3, could not be studied by the equilibrium matrix technique because of polymerization through hydrogen bonding.

The spectra of CH_4 in argon and krypton were observed during the course of this investigation. In general, there was complete agreement with the results of Frayer and Ewing[6] except for the additional band at 3033 cm^{-1} in argon which was originally observed in spray-on experiments by Cabana, Savitsky, and Hornig.[8] The graph of optical density of $R(0)$ vs time obtained by us for CH_4 in krypton (the time dependence in argon was not investigated) had a slope of almost 50% more than Frayer and Ewing's[6] result for oxygen-free CH_4 in argon. There appeared to be about the same mole fraction $J=0$ molecules immediately after cool down. These incidental observations for CH_4 are consistent with the conclusions drawn by Frayer and Ewing[6] for that system.

RESULTS

CH_3D

The CH_3D has two nuclear-spin species corresponding to proton spin I_H of $\frac{1}{2}$ and $\frac{3}{2}$. The rotational levels for

[6] (a) F. H. Frayer and G. E. Ewing, J. Chem. Phys. **46**, 1994 (1967); (b) **48**, 781 (1968).

[7] I. Ozier and P. Yi, J. Chem. Phys. **47**, 5458 (1967).

[8] A. Cabana, G. B. Savitsky, and D. F. Hornig, J. Chem. Phys. **39**, 2942 (1963).

which K is divisible by 3 have $I_H=\frac{3}{2}$ while those levels for which K is not divisible by 3 have $I_H=\frac{1}{2}$. If CH_3D rotates freely in the matrix three levels will have large populations at 7°K, $J=0$, $K=0$, $J=1$, $K=0$; and $J=1$, $K=\pm 1$. The first two belong to $I_H=\frac{3}{2}$, the last to $I_H=\frac{1}{2}$. As in CH_4, four lines will be present in a parallel band such as the 1300-cm^{-1} band. These are $P(1)$, $Q(1)$, $R(0)$, and $R(1)$. In CH_4, $R(0)$ has $I=2$ and $P(1)$, $Q(1)$, and $R(1)$ have $I=1$. In contrast, for CH_3D, $R(0)$ and $P(1)$ have $I_H=\frac{3}{2}$, $Q(1)$ has $I_H=\frac{1}{2}$, and $R(1)$ has both $I_H=\frac{1}{2}$ and $I_H=\frac{3}{2}$. As a result of these differences, the intensity changes expected for CH_3D are much less than those of CH_4.

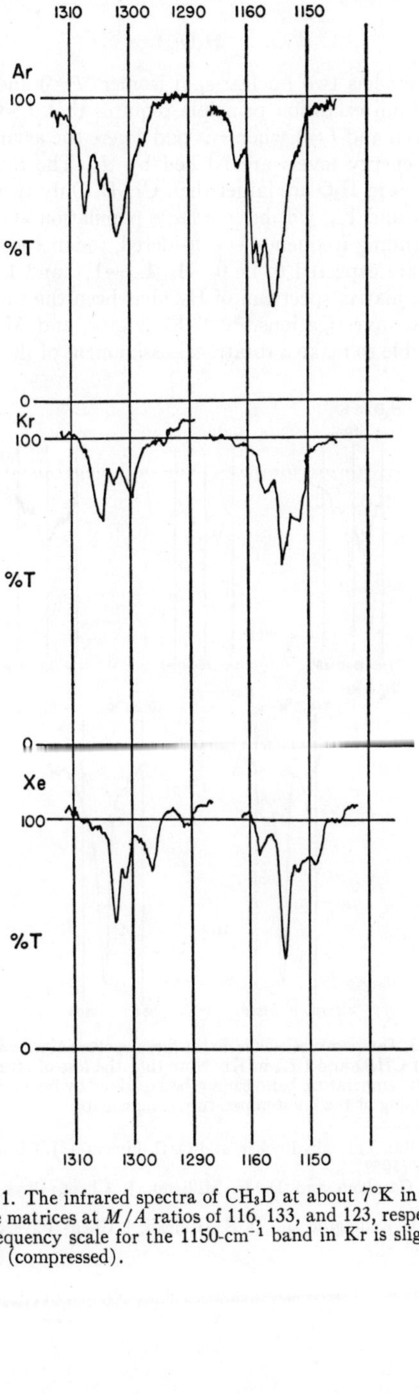

FIG. 1. The infrared spectra of CH_3D at about 7°K in Ar, Kr, and Xe matrices at M/A ratios of 116, 133, and 123, respectively. The frequency scale for the 1150-cm^{-1} band in Kr is slightly incorrect (compressed).

The absorption bands in CH_3D at 1155, 1300, and 3010 cm^{-1} were observed in argon, krypton, and xenon matrices. All bands exhibited some fine structure. The fine structure of the 3010-cm^{-1} band was very poorly resolved. The spectra of the 1155- and 1300-cm^{-1} bands observed in the various matrices are shown in Fig. 1. There are significant differences between the fine structure observed in different matrices, but the fine structure appears to change uniformly going from Ar to Kr to Xe. The positions of the various bands are given in Table I. Because the fine structure of the 1300-cm^{-1} band was best resolved, attention was focused on it in the time-dependence studies.

In the time-dependence studies the sample was warmed to about 25°K by reducing the He flow and held there for several minutes. Then the temperature was reduced in about 30 sec to the ~7°K temperature which was the lowest obtainable. A succession of scans of the 1300-cm^{-1} band were made. A small but real change in the relative intensities of the two prominent bands (1300 cm^{-1} and 1306 cm^{-1} in Kr) was observed to take place over a time span of about 20 min. The time dependence for CH_3D was most apparent for the Kr matrix as shown in Fig. 2.

The time scale was short enough to cause some concern that the changes might be the result of thermal lag. In order to rule out this possibility the cool down was made with the spectrometer source shuttered. The temperature then overshot rapidly by about $\frac{1}{2}$°K. Upon removing the shutter the same time dependence was observed.

The observed fine-structure pattern for the 1300-cm^{-1} band of CH_3D appears to be incompatible with free rotation. Frayor and Ewing[6] have tentatively established a barrier to rotation for CH_4 of about 1.1 kcal/mole. In principle it is possible to calculate the energy levels of CH_3D from this barrier and predict the expected spectrum, but this has not been attempted. The time-dependent intensity changes suggest that the 1306-cm^{-1} (Kr) band is of species $I_H=\frac{3}{2}$, and the 1300 cm^{-1} (Kr) band is $I_H=\frac{1}{2}$.

$R(0)$ and $P(1)$ are expected to increase with time; $Q(1)$ is expected to decrease with time; $R(1)$ will decrease slightly with time. Therefore, it appears likely that $Q(1)$ is the 1300-cm^{-1} line. Frequency and intensity considerations strongly suggest that the 1306-cm^{-1} line is $R(0)$ rather than $P(1)$. The line at 1295 cm^{-1} is weak enough to make its time dependence somewhat uncertain although its intensity appears to be time independent. It could be assigned to $P(1)$.

The line at 1303.5 cm^{-1} is time independent and is probably a matrix band. Therefore, no assignment is proposed for $R(1)$.

TABLE I. Frequencies (cm^{-1}) of the distinguished lines in the 1150 and 1300 bands of CH_3D.

Ar	1156.5	1158	1159.5	1163	1298	1302.5	1305	1308.5
Kr	1150	1154	1157	1160	1295	1300.0	1303.5	1306.0
Xe	1145	1149	1154.5	1158.2	1291.5	1297.3	1302	1303.5

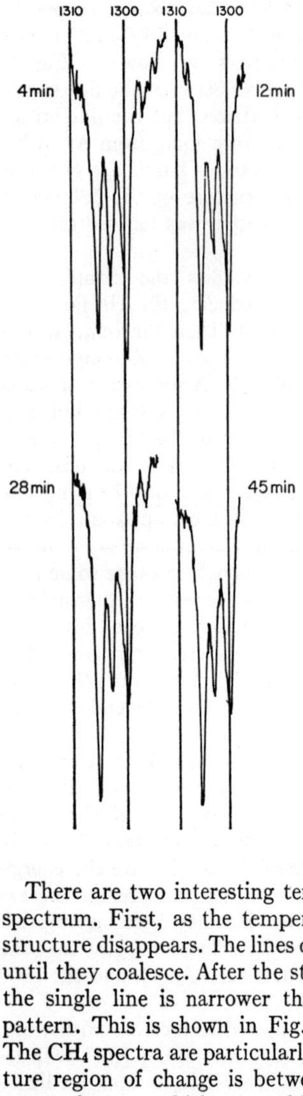

FIG. 2. The time dependence of the fine structure of the infrared band at 1300 cm^{-1} of CH_3D in Kr. The vertical scale is expanded in order to emphasize the effect.

There are two interesting temperature effects in the spectrum. First, as the temperature is raised the fine structure disappears. The lines do not appear to broaden until they coalesce. After the structure has disappeared the single line is narrower than the original over-all pattern. This is shown in Fig. 3 for CH_3D and CH_4. The CH_4 spectra are particularly striking. The temperature region of change is between 10° and 15°K. The exact changes which are taking place deserve more study.

Second, the rate of spin equilibration appears to be fairly rapid near 20°K. If a matrix equilibrated at 7°K is warmed to 20°K and quenched, the sample has the original nonequilibrium intensity pattern. The experimental arrangement requires a minimum time of about 3 min for warming and quenching. Our efforts to prepare a nonequilibrium distribution of spin isomers by cooling CH_3D to 4°K, allowing equilibration to take place, and warming rapidly presumably failed because of rapid re-equilibration in the solid on warming.

H_2O

Water has two nuclear-spin isomers $I=0$ and $I=1$. The Pauli exclusion principle requires that $I=0$ when τ is even and $I=1$ when τ is odd where the asymmetric rotor energy levels are labeled by J_τ. The rotational spacings in H_2O are larger than CH_3D. Only two levels, the 0_0 and 1_{-1}, will have a large population at 7°K. If the bending frequency is considered, the intense transitions are expected to be $0_0\to1_0$, $1_{-1}\to1_{+1}$, and $1_{-1}\to2_{-1}$.

The matrix spectrum of H_2O has been the subject of several investigations.[4,5,9–11] Redington and Milligan[11] were able to make a rotational assignment of the promi-

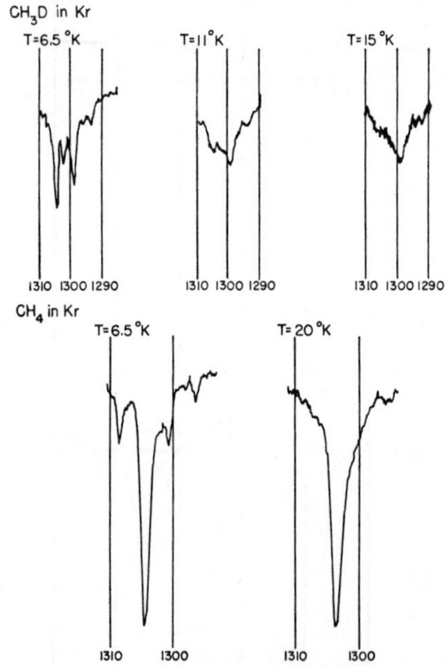

FIG. 3. Temperature effects on the fine structure of the 1300-cm^{-1} band of CH_3D and CH_4 in Kr. Note that the loss of structure in the high-temperature band cannot be explained by broadening and overlapping of the low-temperature components.

[9] M. Van Thiel, E. Becker, and G. C. Pimentel, J. Chem. Phys. **27**, 486 (1957).

[10] E. Catalano and D. E. Milligan, J. Chem. Phys. **30**, 45 (1959).

[11] R. L. Redington and D. E. Milligan, J. Chem. Phys. **37**, 2162 (1962).

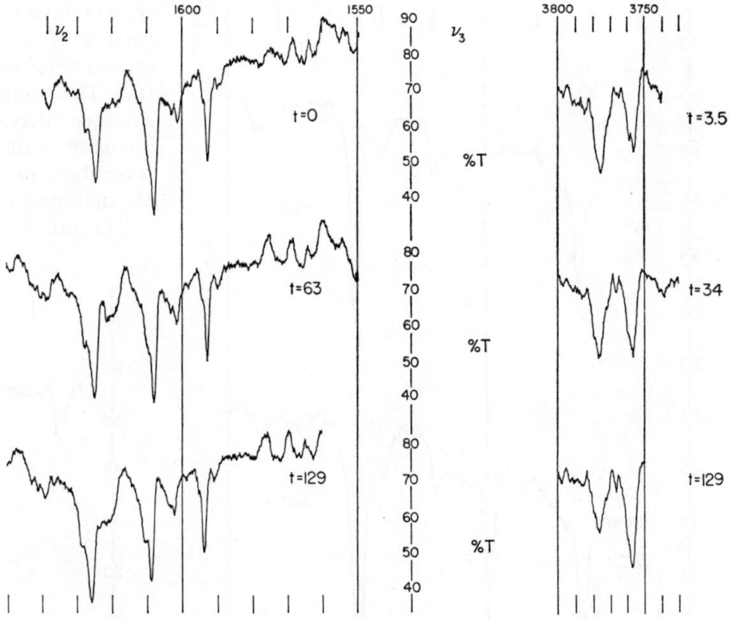

FIG. 4. Time changes in the fine structure of the v_2 and v_3 fundamentals of H_2O in an Ar matrix at 7°K, with M/A 176.

nent features in the ν_2(bend) and ν_3(asymmetric stretch) bands.

They then studied[5] the time dependence of these features at 5°K. They found intensity changes taking place over a period of about 3 h which were consistent with nuclear-spin-species conversion and their original assignment. They found the process catalyzed by O_2 as would be expected for the nuclear-spin-species conversion process and proposed a theory of the oxygen-catalyzed kinetics.

In the course of the present investigation a study of the time dependence of the ν_2 and ν_3 bands of H_2O in Ar was undertaken. In these experiments the temperature of the sample was raised to about 30°K for several minutes and then quenched rapidly to 7°K. This treatment gave the 30°K distribution of spin isomers. The spectra were then followed as a function of time. Both H_2O in pure Ar and in O_2-doped Ar matrices were studied. The results for the undoped matrices are shown in Fig. 4. It is clear that the prominent peaks at 1625 and 3758 cm^{-1} are growing with time while the prominent peaks at 1608 and 3776 cm^{-1} are decreasing with time. The intensity of the peak at 1593 cm^{-1} and the broad absorption at 3710 cm^{-1} are time independent. Except for the 3710-cm^{-1} feature, these observations are in complete agreement with the assignment of Redington and Milligan,[11] who assigned 1625 cm^{-1} to $0_0\rightarrow1_0$, 1608 cm^{-1} to $1_{-1}\rightarrow1_{+1}$, 1593 cm^{-1} to a nonrotating H_2O band, 3758 cm^{-1} to $0_0\rightarrow1_{-1}$, and 3776 cm^{-1} to $1_{-1}\rightarrow2_{-2}$. They assigned the 3710-cm^{-1} band to $1_{-1}\rightarrow0_0$.

The spectra obtained in this investigation do not agree in other details with those of Redington and Milligan. This may result from one or more of three causes: the general lack of reproducibility of spray-on matrix spectra, the lower M/A values used here, and the low light level in these regions in the double-beam spectrometer which occurs because of atmospheric H_2O absorption and which makes the balance adjustment critical.

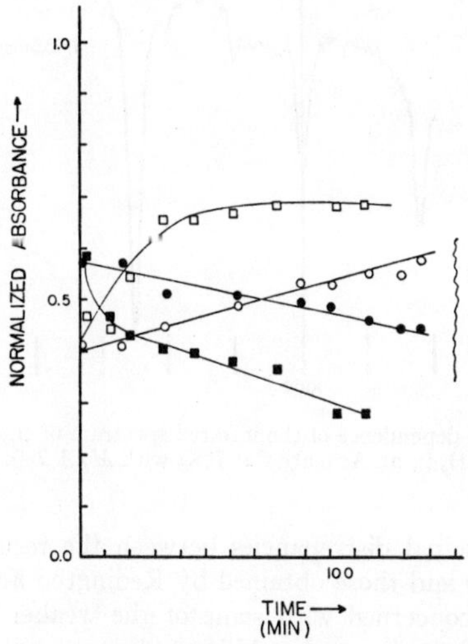

FIG. 5. Time dependence of normalized peak optical densities of fine-structure components for H_2O in Ar at 7°K with and without added O_2. The sum of the peak optical densities of the 1625- and 1609-cm^{-1} bands is approximately constant. The ratio of the optical density of one component to this is plotted. The circles show the the intensities without O_2, the squares the corresponding values with about 1% added O_2; the open symbols are for the 1625-cm^{-1} line, the closed symbols are for the 1609-cm^{-1} line. The points marked with an asterisk correspond to intensities immediately after warmup to 30°K and cool down.

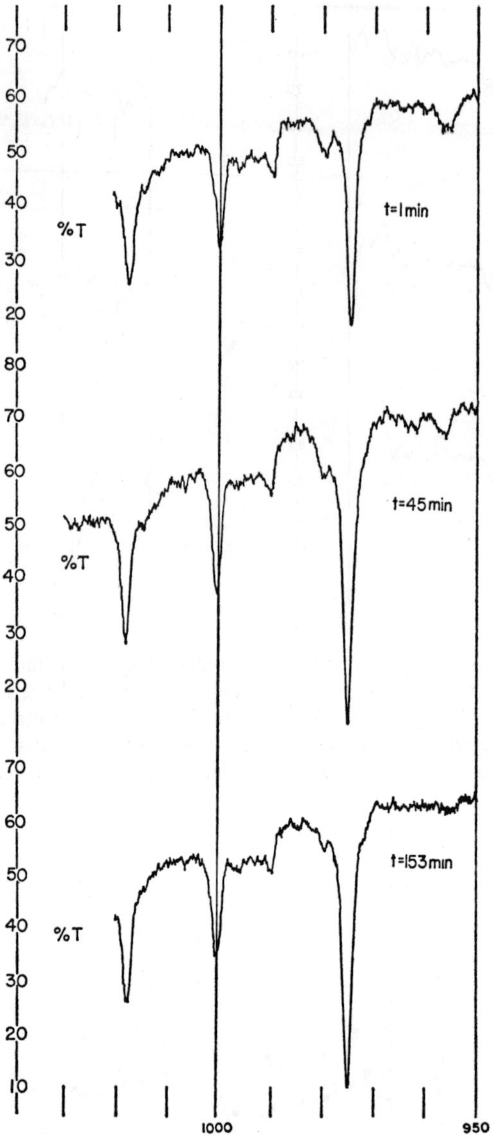

FIG. 6. Time dependence of the infrared spectrum of the ν_2 band of NH_3 in an Ar matrix at 7°K, with M/A 200.

The principal discrepancies between the results obtained here and those obtained by Redington and Milligan[5] are concerned with some of the weaker bands. The band they observed at 1574 cm^{-1} was not observed here. The bands at 1558, 1638, 1663, and 3710 cm^{-1} did not appear to be time dependent in this study. Because of the difficulties cited above it is difficult to assess the origin of these discrepancies.

Redington and Milligan[5] could not vary and control the temperature or conveniently do warmup and quenching experiments. Warmup studies made in the present investigation indicate that the rate of reconversion of a 7°K partially equilibrated sample increases more and more rapidly with increasing temperature. The reconversion becomes quite rapid above 25°K. Another interesting feature of the warmup is that the intensity of the rotating bands decreased sharply (but reversibly) on warming to 30°K but the intensity of the nonrotating band at 1593 cm^{-1} was unchanged.

The rate of evolution is strongly catalyzed by O_2

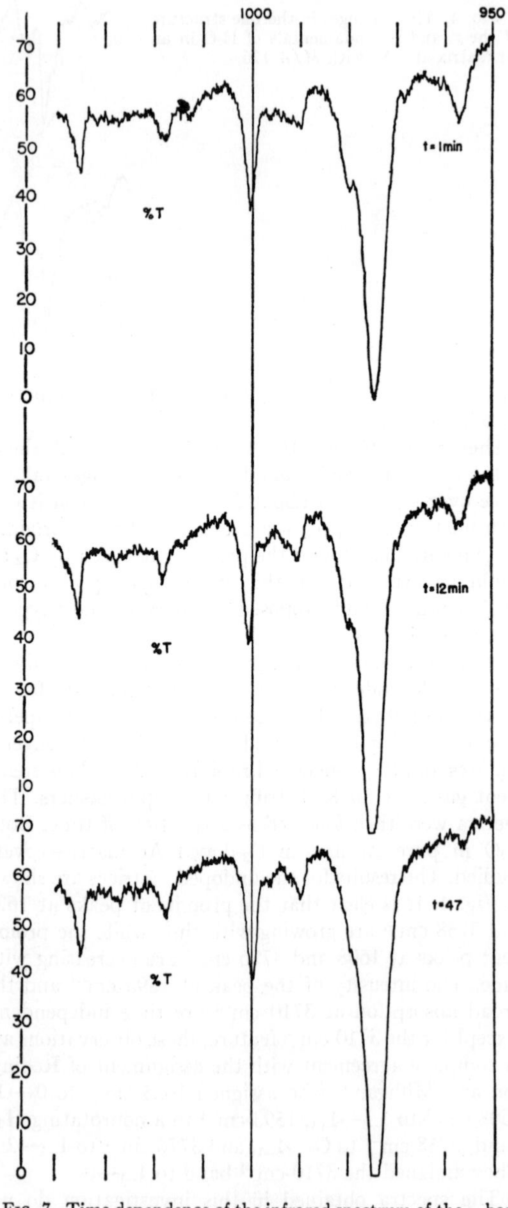

FIG. 7. Time dependence of the infrared spectrum of the ν_2 band of NH_3 in an Ar matrix at 7°K with 0.03% added O_2. The M/A ratio was 200.

added to the matrix indicating that the process taking place is indeed a spin conversion. The peak optical density of the two changing bands of υ_2 with and without added O_2 are plotted versus time in Fig. 5. Note that the original relative intensity of the bands is restored by warming to 30°K and quenching. The oxygen-free curve appears to be following first-order kinetics. The oxygen-doped curves are similar to the CH_4 oxygen-doped curves of Frayer and Ewing[6] and to the theoretical curves of Redington and Milligan.[5] Similar graphs of the experimental time dependence found by Redington and Milligan[5] are complicated because the spray on was taking place during the first hour in which they were trying to follow the time dependence.

NH_3

The correlation of nuclear-spin states with rotation for NH_3 is almost exactly analogous to CH_3D. The only difference is that the existence of inversion causes some states to disappear. The infrared spectra of NH_3 in argon and nitrogen matrices were observed by Milligan, Hexter, and Dressler.[12] The υ_2 (umbrella motion) band exhibited a sharp well-resolved fine structure which was interpreted in terms of rotation and inversion structure. A nearly complete assignment of the structure was made. This interpretation was contradicted for N_2 matrices by the work of Pimentel, Bulanin, and Van Thiel[13] who argued that the separate features arise from dimers and trimers.

In the present investigation a study of the time evolution of the fine structure in Ar was undertaken. Again the matrix was allowed to warm to approximately 30°K for several minutes and then was rapidly quenched to 7°K. The spectrum was then repeatedly scanned as a function of time. The result of such a study is shown in Fig. 6.

Only two lines show a clear time dependence. These are the strongest line at 974 cm^{-1} which grows in intensity and a weak line at 956 cm^{-1} which decreases in intensity. An experiment with 0.03% added oxygen demonstrated that the process was catalyzed by O_2. These experiments are shown in Fig. 7.

[12] D. E. Milligan, R. M. Hexter, and K. Dressler, J. Chem. Phys. **34**, 1009 (1951).
[13] G. C. Pimentel, M. O. Bulanin, and M. Van Thiel, J. Chem. Phys. **36**, 500 (1962).

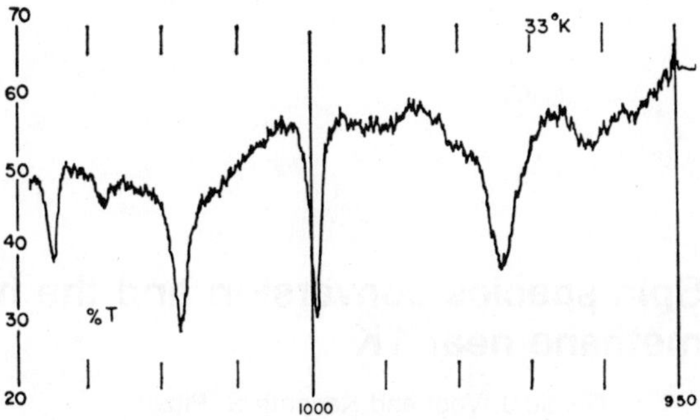

FIG. 8. The ν_2 band of NH_3 at 33°K for the same matrix as Fig. 6 showing the temperature dependence.

Because the lines are not time dependent it appears likely that most of the fine structure in NH_3 does not arise from rotation. This conclusion is supported by the lack of temperature dependence of the lines in question which is shown in Fig. 8. In this respect the lines behave similarly to the 1593-cm^{-1} band of H_2O which was assigned to nonrotating H_2O. Further support for this conclusion is offered by the consideration that the population of the $J=2$ level at 4.2°K would be very small regardless of whether spin conversion has taken place and the line assigned to $R(2)$ should have a much smaller intensity than was observed.

Presumably the strong line at 974 cm^{-1} which grows with time is $R(0)$. The weak line at 956 cm^{-1} is probably $Q(1)$. The line $R(1)$ which should be more intense than $Q(1)$ is missing. Another point difficult to interpret is that the line at 974 cm^{-1} increases so much more in intensity than the line at 956 cm^{-1} decreases.

DISCUSSION

The study of the time dependence of fine-structure features of matrix-isolation infrared bands has proved to be a powerful aid in discriminating rotational structure from structure arising from different matrix sites or agglomeration. The oxygen catalysis of the spin-species conversion process causing this time dependence provides a valuable specific check of an interpretation of time dependence in terms of rotational structure.

The directions of the intensity changes often give a clue to the assignment of rotational quantum numbers.

Spin species conversion and the heat capacity of solid methane near 1°K

Gerald J. Vogt and Kenneth S. Pitzer

Department of Chemistry and Inorganic Materials Research Division of the Lawrence Berkeley Laboratory, University of California, Berkeley, California 94720

(Received 1 July 1975)

In order to further explore[1] the energy levels and interconversion properties of the various nuclear-spin species of methane, we have measured the heat capacity of samples containing controlled amounts of a spin conversion catalyst, oxygen. With approximately 0.8% O_2, the spin-species conversion is fast enough to allow measurements of the equilibrium heat capacity. Our results, which start at 0.4 °K, are shown in Fig. 1. An anomalous peak near 1 °K is apparent. The time to reach thermal equilibrium after each introduction of heat ranged from 20 to 40 min and was longest near 0.8 °K. These long equilibration times reduce the accuracy, of course, but there is no doubt about the essential character of the results.

It was impractical to measure the heat capacity of pure methane below about 2.2 °K. The time to reach spin-species equilibrium was impossibly long, but the heat release from conversion caused a rapid temperature drift. Above 2.2 °K we were able to measure the nonequilibrium heat capacity for methane with no change in spin-species composition; these results are generally similar to those of earlier investigators.[2]

Our results for the equilibrium heat capacity above 5.5 °K agree satisfactorily with those of Colwell, Gill, and Morrison.[2c]

Kataoka, Okada, and Yamamoto[3] have recently proposed a model for CH_4 (below 20 °K) that appears to be consistent with our results. This model assumes the eight-sublattice antiferrorotational structure with three-quarters of the molecules in sites of symmetry D_{2d} and one-fourth in sites of O_h symmetry. The lowest energy levels of spin species A, T, and E in O_h sites are so widely separated that conversion to the A species is practically complete below 2 °K. Kataoka *et al.* suggest that the energy separations for the D_{2d} sites for T and E molecules are 1.9 and 2.8 °K, respectively, above that for A molecules. The respective degeneracies of these lowest levels are 5, 9, and 2 for A, T, and E species.

Our equilibrium heat capacities through 1.5 °K are fitted quite well by this pattern with three-fourths of the molecules in a set of levels with separations of 1.951 and 2.89 °K and the degeneracies stated above and with a small correction for the heat capacity of oxygen. If all of the molecules had access to such a set of energy states the heat capacity would be four-thirds as large—a definite conflict with our data. This provides confirmation of that aspect of the structure placing three-quarters of the molecules in a set of sites with smaller energy separations.

Since the oxygen molecules are presumably in cavities of approximately cubic or octahedral symmetry, their rotation and spin energy-level pattern is estimated to be approximately the same as in the gas. The heat capacity of oxygen calculated on this basis was subtracted from the experimental values to obtain equilibrium heat capacities of methane.

Recent experiments of other types have been interpreted on the basis of energy separations of 1.69 and 2.05 °K for the level-crossing method, Glättli *et al.*,[4] and 1.66 and 2.51 °K for neutron scattering, Press and Kollmar.[5] Our results are definitely in disagreement

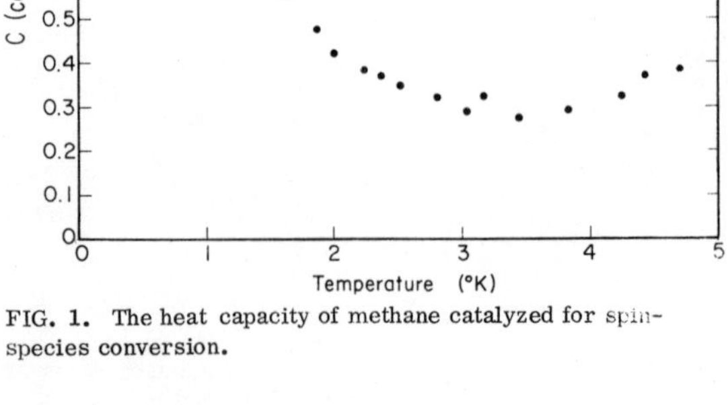

FIG. 1. The heat capacity of methane catalyzed for spin-species conversion.

with an energy separation as small as 1.7 °K for all nine T states (above the A level). Among possibilities for resolution of this discrepancy are more complex models involving the splitting of the group of nine T states.

Our measurements were made with calorimetric techniques similar to those described by Finegold and Phillips.[6] Helium-3 was used for cooling below 1 °K. A full report of our measurements will be given later together with a discussion of various calculations and interpretation.

We thank Professor Norman Phillips for the use of calorimetric equipment and for advice on procedures. This research was sponsored by the Energy Research and Development Administration.

[1]For earlier work involving one of the present authors, see R. F. Curl, Jr., J. V. V. Kasper, and K. S. Pitzer, J. Chem. Phys. **46**, 3220 (1967); and H. P. Hopkins, P. L. Donoho, and K. S. Pitzer, *ibid.* **47**, 864 (1967).

[2](a) K. Clusius, Z. Phys. Chem. Abt. B 3, 41 (1929); (b) A. Sperandio, thesis, University of Zurich, 1961; (c) J. H. Colwell, E. K. Gill, and J. A. Morrison, J. Chem. Phys. **36**, 2223 (1962); **37**, 635 (1963).

[3]Y. Kataoka, K. Okada, and Y. Yamamoto, Chem. Phys. Lett. **19**, 365 (1973).

[4]H. Glättli, A. Sentz, and M. Eisenkremer, Phys. Rev. Lett. **28**, 871 (1972).

[5]W. Press and A. Kollmar, Bull. Am. Phys. Soc. 20, 462 (1975).

[6]L. Finegold and N. E. Phillips, Phys. Rev. **177**, 1383 (1969).

J. Chem. Thermodynamics **1976,** 8, 1011–1031

Entropy and heat capacity of methane; spin-species conversion

GERALD J. VOGT AND KENNETH S. PITZER

Department of Chemistry and Materials and Molecular Research Division of the Lawrence Berkeley Laboratory, University of California, Berkeley, California 94720, U.S.A.

(*Received 25 February 1976*)

The heat capacity of methane from 0.4 to 28 K was measured for both pure methane and samples doped with about 0.8 per cent of oxygen as a spin-species conversion catalyst. With the conversion catalyst it was possible to obtain equilibrium with the spin system within about 30 min and to measure heat capacities down to 0.4 K. For pure methane the spin-species conversion becomes very slow below 10 K; however, one can measure non-equilibrium heat capacities on the basis of no change in spin-species composition. Entropy values are calculated from the heat capacities on the basis of both fixed and equilibrium spin-species composition and compared with the statistical calculation for gaseous methane. The results agree within possible experimental error with the expected zero-point entropies for the two spin situations.

1. Introduction

The low-temperature properties of methane and its deuterated derivatives have long been of interest for their λ-transitions[1–4] associated, presumably, with the beginning of rotational disorder. With the extension of measurements into the liquid helium range, additional complications appeared[5] in the properties of methane which were later established as related to interconversion of nuclear-spin species.[6–9] With four protons and tetrahedral structure there are three spin-species of the following symmetry and total spin: A(I = 2), T(I = 1), E(I = 0). The high-temperature distribution[10] is 5 A, 9 T, 2 E. Since the lowest rotational state, $J = 0$, is allowed only for the A species, the equilibrium composition at $T = 0$ will be 100 per cent A for the gas and presumably also for the solid since some torsional motion will be allowed by quantum tunneling. In contrast to the behavior of ortho and para hydrogen, interconversion of spin-species in methane is rapid in the gas[11] because of degeneracies for higher states of different species. Interconversion rates decrease with decreasing temperature in the solid, and Colwell *et al.*[5] were unable to measure equilibrium heat capacities below 5 K. At that point the thermal relaxation time had become several hours.

This research was designed particularly to explore the thermal properties of methane below 5 K as well as to refine knowledge in related areas. It was hoped that by addition of a relatively high concentration of a paramagnetic species, the methane interconversion could be catalyzed and equilibrium measurements extended to much lower temperatures. About 0.8 per cent of oxygen proved to be a satisfactory catalyst.

Equilibrium heat-capacity measurements revealed a striking peak near 1 K which has been reported briefly.[12]

In this paper we present a full report on our experimental measurements together with their thermodynamic interpretation. Further discussion in terms of detailed models for the crystalline state will be presented elsewhere together with the details of our calculations related to the properties of oxygen in a methane lattice and the mechanism of the spin conversion catalysis.[13–15] Even for our present purposes, however, a brief summary of the present picture of the structure of crystalline methane is needed.

James and Keenan[16] proposed several structures for methane and analyzed them on a classical basis more appropriate for CD_4 than for CH_4. Yamamoto, Kataoka, and others[17,18] have published an extensive series of papers on the quantum statistical mechanics of methane in these structures. They conclude that methane below the λ-transition near 20 K has the "eight sublattice antiferrorotational structure", which places three-fourths of the molecules in sites of D_{2d} symmetry while the remaining one-fourth are in sites of O_h symmetry.

The molecules of O_h sites will remain essentially disordered orientationally and have rotational energy levels similar to those for the gas; the lowest level has symmetry A; the next set of nine levels ($J = 1$) at estimated energies 12 k K to 15 k K are of symmetry T; while the 25 levels with $J = 2$ include 10 of E and 15 of T symmetry and are at energy 30 k K to 45 k K. With these relatively large energy separations, conversion preferentially to the A spin-species on cooling will begin above 10 K and will be nearly complete below 2 K. It is primarily this process that yields the rather high heat capacity near 6 K observed by Colwell *et al.*[5]

The molecules on the D_{2d} sites interlock with one another, as cog wheels, and their torsional energy-level pattern will be collapsed and distorted from that of the gas. Kataoka *et al.*[17] have developed an intermolecular potential for torsional motion which appears to fit our results very well. It yields at lowest energy a state of symmetry A and degeneracy 5 from spin. Next are a set of T states at about 1.9 k K with total degeneracy 9 followed by a doubly degenerate E level at about 2.8 k K. Then there is a large energy interval between this array of "ground" torsional states and the librationally excited states above about 70 k K.

Excitation of the molecules on D_{2d} sites from the lowest A state to the T and E states at 1.9 k K and 2.8 k K, respectively, yields the heat capacity peak near 1 K. In our earlier note we showed that our heat capacities were well fitted on this model and obtained the more precise values of 1.951 k K and 2.89 k K for the A–T and A–E separations.

The statistical theory of Yamamoto, Kataoka, and others, considers the cooperative nature of these torsional phenomena, and this is important at higher temperatures; but below 2 K it is probably a good approximation to assume fixed energy separations as was indicated above.

There have also been extensive investigations of nuclear magnetic resonance[6–9] for solid methane which confirmed the interconversion of spin-species below 10 K. We shall relate these results to our work in the paper to follow. Bloom and Morrison[19] have reviewed many topics related to solid methane.

2. Experimental methods and apparatus

The measurements were made in a calorimeter designed for use down through the helium-3 range to about 0.3 K. It is generally similar to that described by Finegold and Phillips.[20] The helium-3-cooled cryostat was inserted in a dewar system, consisting of two concentric glass dewars, the outer for liquid nitrogen and the inner with a pumpable jacket for liquid helium-4. The pumping speed was such that the temperature of the helium-4 bath could be reduced to 1.1 K. In addition, the pumping system for the helium-4 bath was equipped with a pressure regulator so that the temperature of the bath could be controlled between 4.2 and 1.1 K.

The calorimeter and helium-3 chamber were enclosed in a vacuum can. Thermal contact could be established between the helium-4 bath and the helium-3 chamber and between the helium-3 chamber and the calorimeter by air-actuated mechanical clamps (heat switches). The helium-3 could be cooled from 1.1 to 0.3 K by reducing the vapor pressure.

The calorimeter was mechanically supported by monofilament nylon threads; also, the calorimeter was connected to the laboratory gas-handling system by a thin-wall stainless-steel capillary. This filling capillary and the electrical leads were thermally anchored to the helium-3 chamber before contacting the calorimeter.

The calorimeter, shown in figure 1, was constructed primarily of copper. The cylindrical section has a length of 37 mm, an internal diameter of 44 mm with a wall thickness of 1.0 mm, and with an additional thickness of 0.6 mm at each reinforcing ring. The cylindrical section and the bottom dished end were one integral piece of metal. The dished ends had a wall thickness of 1.0 mm and a depth of 12 mm. Inside the cell a spiral coil of 0.3 mm thick copper foil, uniformly spread over the volume of the cell, was thermally anchored to the outer wall to improve the thermal

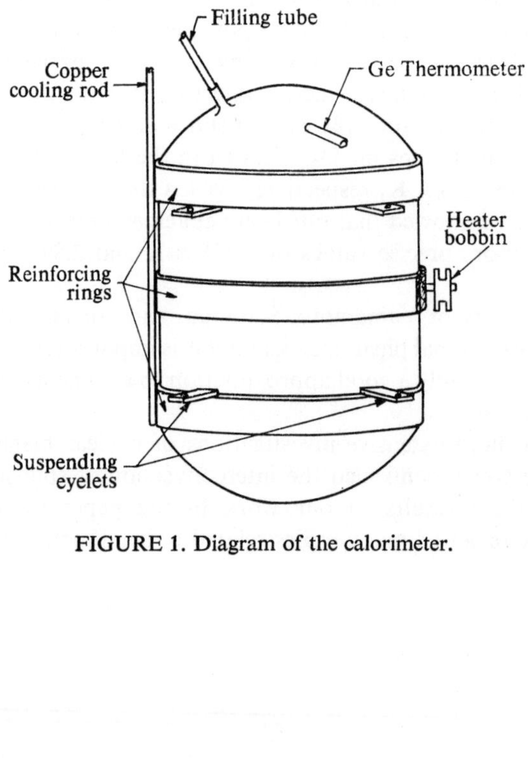

FIGURE 1. Diagram of the calorimeter.

contact with a sample. Midway between the dished ends a copper screw was attached to the outer wall of the cell so that a bobbin wound with a heater coil could be readily connected to the calorimeter. In order to provide a means of attaching the thermometer to the cell, a piece of foil was soldered to the top dished end. A copper rod, 1.6 mm in diameter, was soldered to the outside of the calorimeter and was extended above the calorimeter to lie in between the jaws of the mechanical heat switch. All the pieces, including the top dished end, were silver-soldered together; the interior and exterior surfaces were gold-plated. The calorimeter had an internal volume of 79 cm^3 and a mass consisting of 125 g of copper and 6 g of other materials.

An encapsulated four-lead doped Ge resistance thermometer[(21)] was attached to the calorimeter by rolling the thermometer in a foil soldered to the cell and using General Electric 7031 varnish for thermal contact and mechanical stability. To ensure good thermal contact with the cell, each lead from the thermometer was connected to a length of fine copper wire (0.3 mm diameter), wound and varnished in place about the copper cooling rod. To the other end of the fine copper wires a free-standing coil of (Pt+9 mass per cent of W) wire (0.06 mm diameter) was connected for thermal isolation. These coils were then connected to leads from the measuring circuit.

The Ge thermometer (Ge 782) was calibrated for the region 0.3 to 30 K on a temperature scale, designated T_η, in the laboratory of Professor Norman E. Phillips. The calibration was derived from a combination of Pt-resistance thermometry (T_{55}) between 15 and 30 K, gas thermometry between 4.2 and 20 K, the vapor-pressure of ^{4}He (T_{58}) between 4.2 and 1.1 K, and magnetic thermometry below 3 K with a single crystal of $Ce_2Mg_3(NO_3)_{12}\cdot 12H_2O$ (CMN).

The heater attached to the calorimeter was constructed by non-inductively winding and thermally anchoring with Stycast 2850GT a given length of (Pt+9 mass per cent of W) wire of 0.022 mm diameter upon the heater bobbin. The resistance of the heater, made from approximately 4 m of wire, was 5470 Ω at 4.2 K.

The heat capacity of the empty calorimeter was measured several times over the temperature range of interest, 0.3 to 30 K. The contribution of the empty calorimeter to the total heat capacity was never over 50 per cent and was less than 1 per cent in the interesting region below 2.5 K for the samples containing oxygen.

The mass of calorimetric samples was determined by weighing in a small pressure vessel.[(23)] Since methane has a rather high freezing temperature, a special refrigerant was needed for the transfer of a sample to the calorimeter without obstruction of the filling tube by a frozen plug. Natural gas, which is impure methane, served very well as this refrigerant; it was in turn liquefied by use of liquid nitrogen.

Our pure methane was obtained from the Pacific Oxygen Company with a reported purity of 99.999 moles per cent. Analysis in a mass spectrometer showed mole fractions less than 5×10^{-6} of O_2 or other impurity. Our samples of (CH_4+O_2) were intended to have about 1 mol per cent of O_2. Some fractionation occurred in transfer to the calorimeter; hence each sample was analyzed by gas chromatography after calorimetric measurement. Table 1 gives the size and composition of the samples containing oxygen.

Some additional details concerning the apparatus are given elsewhere.[(15)]

TABLE 1. Samples of $CH_4 + O_2$: *m*, sample mass; *x*, mole fraction; *n*, amount of substance

Sample no.	m/g	$10^2x(O_2)$	$n(CH_4)$/mol [a]
100	6.380	0.68	0.3923_4
101	7.287	0.86	0.4465_6
102	6.561	0.85	0.4021_4
103	3.358	0.80	0.2060_2
104	2.671	0.87	0.1636_4

[a] Molar mass of CH_4: 16.043 g mol^{-1}.

3. Experimental measurements

The heat-capacity measurements were made by the usual heat-pulse method. The effective surroundings of the calorimeter to which heat exchange took place was the helium-3 chamber and associated structure which will be called the "block". In the

TABLE 2. Heat capacity of oxygen-doped methane (cal_{th} = 4.184 J)

T/K	C/cal_{th} K^{-1} mol^{-1}
Sample 101	
1.5760	0.5980[a]
1.6557	0.5605[a]
1.8710	0.4763[a]
2.0088	0.4245
2.2451	0.3795[a]
2.5230	0.3556[a]
2.8135	0.3397[a]
3.1661	0.3410[a]
3.4410	0.3446[a]
3.8261	0.3540[a]
6.9561	0.756
7.5233	0.795[a]
7.7342	0.824
8.3367	0.904[a]
8.9974	1.013
Sample 102	
0.70885	1.126[a]
0.76893	1.142[a]
0.84133	1.142[a]
0.90087	1.106[a]
0.93992	1.109[a]
0.99886	1.071[a]
1.0884	1.002[a]
1.2001	0.904[a]
1.2969	0.814[a]
11.440	1.515[a]
12.639	1.879
13.331	2.047[a]
14.139	2.368
16.510	3.502

T/K	C/cal_{th} K^{-1} mol^{-1}
17.291	4.069
18.025	4.847
18.721	6.071
21.105	4.636
21.787	4.685
22.527	4.797
23.401	4.908
24.388	5.027
25.472	5.234
26.654	5.395
Sample 103	
0.45619	0.6857[a]
0.49313	0.7955[a]
0.50345	0.8324
0.51258	0.8556
0.52364	0.8911
0.53647	0.9249
0.55029	0.9585
0.56482	0.9984
0.58028	1.0379
0.59569	1.0750
1.4194	0.7237[a]
1.4691	0.6871
15.446	2.9297
15.784	3.0926
16.237	3.3558
16.802	3.7275
17.429	4.2325
18.136	5.0637
18.791	6.3291
19.302	8.9500

T/K	C/cal_{th} K^{-1} mol^{-1}
19.668	13.12
19.918	22.22
20.117	24.60
20.644	5.1927
21.586	4.7151
22.661	4.8224
23.811	4.9660
25.079	5.1890
26.406	5.4125
27.767	5.4809
Sample 104	
0.43673	0.603[a]
0.43744	0.604[a]
0.44461	0.633
0.44796	0.657
0.45327	0.665
0.46001	0.691
0.46319	0.699
0.47276	0.734
0.47427	0.740
0.48641	0.779
0.48765	0.780
0.49947	0.819
0.50281	0.838
0.51123	0.856
0.52497	0.895
0.53920	0.942
0.56771	1.008
0.58067	1.039
0.59560	1.062
0.62389	1.101

T/K	C/cal_{th} K^{-1} mol^{-1}
0.64427	1.131
0.66558	1.169
0.68774	1.195
0.70877	1.200
0.72915	1.214
0.75262	1.217
0.77991	1.219
0.81133	1.205
0.84733	1.188
0.87870	1.169
0.90186	1.163
0.92699	1.142
0.95467	1.120
0.98463	1.093
1.0186	1.066
1.0559	1.035
1.0957	0.999
1.1389	0.954
1.1860	0.912
1.2396	0.857
1.3019	0.806
1.3725	0.749
1.4474	0.696
1.5274	0.645
6.3880	0.715
6.8760	0.744
7.1087	0.761
7.5268	0.837
7.8374	0.859
8.2514	0.925
8.6723	0.991
9.1150	1.066
9.6040	1.143

[a] These values may be slightly too small; reason given in text.

principal series of measurements reported in tables 2 and 3 the temperature of the block was controlled by one or another automatic temperature regulator; either a regulator of the vapor pressure of helium-3 or helium-4 or a regulated electrical heater yielding constant temperatures above the boiling temperature of helium-4. Prior to the use of these regulators, however, some measurements were made in which the block was allowed to drift thermally—presumably at a nearly constant temperature. Since there was no accurate measurement of block temperature in these preliminary measurements, their interpretation must be based entirely on calorimeter drifts and is somewhat uncertain. We report here in table 4 only those preliminary values that constitute an important supplement to the regular series with a regulated block temperature.

Interpretation of heat-exchange corrections in this research was complicated by the slowness of thermal equilibration between the nuclear-spin-species conversion and the lattice-vibration energy. While we shall later discuss this situation for various temperature ranges for both pure and oxygen-doped methane, we consider first a general case and two limiting simplifications. The calorimeter is assumed to reach rapid thermal equilibrium with the lattice energy of the methane. There remains, then, slow heat transfer with rate constants k_B and k_s respectively with the block at temperature T_B and the nuclear spin system at temperature T_s. In general the calori-

TABLE 3. Heat capacity of pure methane (cal_{th} = 4.184 J)

$\frac{T}{K}$	$\frac{C}{cal_{th}\ K^{-1}\ mol^{-1}}$	$\frac{T}{K}$	$\frac{C}{cal_{th}\ K^{-1}\ mol^{-1}}$	$\frac{T}{K}$	$\frac{C}{cal_{th}\ K^{-1}\ mol^{-1}}$	$\frac{T}{K}$	$\frac{C}{cal_{th}\ K^{-1}\ mol^{-1}}$
		Non-equilibrium results				Equilibrium results	
Series 9		3.2046	0.01412	6.2856	0.1804	Series 7	
5.9835	0.178	3.3008	0.01587	6.4343	0.1997	10.8742	1.305
6.9203	0.284	3.3997	0.01834	6.6152	0.2226	11.9787	1.5541
8.0141	0.449	3.4425	0.01848	6.7438	0.2431		
9.9042	0.793	3.5442	0.02067	6.8989	0.2651	Series 13	
		3.7191	0.02479	7.1134	0.2955	7.9301	0.855
Series 14A		3.7880	0.02698	7.2737	0.3213	9.8299	1.143
2.1945	0.00381	3.9253	0.03036	7.5524	0.3607	11.3526	1.407
2.2338	0.00404			7.9043	0.4143		
2.3238	0.00439	Series 14B		8.2415	0.4717	Series 14C	
2.4174	0.00496	4.3029	0.0382	8.6328	0.5367	7.7669	0.801
2.4855	0.00539	4.5015	0.0463	9.1080	0.6273	9.7402	1.114
2.5321	0.00555	4.7078	0.0559	9.6137	0.7269		
2.5378	0.00587	4.8996	0.0652			Series 14D	
2.6051	0.00640	5.0945	0.0761	Series 14C		21.4026	4.599
2.6750	0.00704	5.2832	0.0879	4.4985	0.0575	22.7433	4.738
2.7503	0.00780	5.4318	0.0988	4.5898	0.0612	24.5678	4.976
2.9013	0.00960	5.6000	0.1128	5.3604	0.1072	26.5792	5.295
2.9314	0.01007	5.7705	0.1271	5.6860	0.1367	28.7184	5.581
3.0055	0.01097	5.9444	0.1435	7.9392	0.4257		
3.1107	0.01259	6.1294	0.1620	9.8450	0.7723		

TABLE 4. Additional heat-capacity measurements for methane
($cal_{th} = 4.184$ J)

T/K	C/cal_{th} K^{-1} mol^{-1}	T/K	C/cal_{th} K^{-1} mol^{-1}	T/K	C/cal_{th} K^{-1} mol^{-1}	T/K	C/cal_{th} K^{-1} mol^{-1}
Sample 100, oxygen-doped		8.8695	1.009±0.01	16.976	3.363	16.333	2.946
		9.5878	1.146±0.01	17.372	3.611	16.332	2.958
2.3706	0.366±0.03	10.216	1.230±0.02	17.730	3.922	16.931	3.305
3.0316	0.411±0.07	10.995	1.399±0.01	18.058	4.067	17.487	3.654
4.2448	0.371±0.05	12.109	1.676±0.01	18.396	4.619	17.921	4.032
4.4105	0.458±0.09	13.282	2.002±0.02	18.751	4.999	18.334	4.459
4.6907	0.463±0.08	14.067	2.238±0.03	18.753	4.904	18.754	4.976
5.0129	0.497±0.08	15.512	2.871±0.01	19.225	5.877	19.086	5.472
5.3223	0.555±0.08	Series 4, pure methane		19.712	7.164	20.989	4.522
5.7910	0.596±0.08			20.221	10.511	21.571	4.510
6.2778	0.633±0.06	15.251	2.486	20.481	52.044	22.275	4.592
6.8211	0.716±0.05	15.614	2.661	Series 5, pure methane		23.168	4.702
7.1847	0.736±0.01	16.034	2.847				
7.7809	0.910±0.06	16.322	2.930	15.131	2.365		
8.2777	0.928±0.02	16.656	3.154	15.798	2.662		

meter temperature drift after a heating interval will be governed by the equation:

$$C\,\mathrm{d}T/\mathrm{d}t = \dot{q} = k_B(T_B - T) + k_s(T_s - T), \tag{1}$$

where t is the time, C is the heat capacity of the calorimeter plus methane lattice, and $\dot{q}$ is the rate of heat flow into that system.

If the heat transfer from the spin system is either negligibly slow or so rapid that it has already attained equilibrium, the last term in equation (1) can be omitted and the remainder integrated, with T_B constant, to yield:

$$T = T_B + (T_0 - T_B)\exp(-k_B t/C). \tag{2}$$

Here T_0 is the temperature at zero time. If the spin system is in equilibrium, C includes the spin heat capacity. If the exponent is small, this approximates to a linear equation which was found to be adequate in most but not all cases.

Another simplified case arises if the heat exchange with the block can be neglected. In this case we cannot assume T_s constant; rather it will vary in accordance with heat transfer and the spin-related heat capacity C_s which can be taken to be constant for a given measurement. After considerable manipulation one obtains

$$T = (CT_{s,0} + C_s T_0)/(C + C_s) + C\{(T_0 - T_{s,0})/(C + C_s)\}\exp\{-k_s(C + C_s)t/C^2\}, \tag{3}$$

where $T_{s,0}$ is the spin-system temperature at zero time. This equation is seen to have the same general form as equation (2). In either case the logarithm of the drift rate, $\log(\mathrm{d}T/\mathrm{d}t)$, decreases linearly with time and one can extrapolate to find T_0. But if equation (3) is assumed, the extrapolation also yields an unchanged spin temperature at zero time.

While we have a few examples where both terms on the right side of equation (1) are comparably important at some time during the after drifts, it did not seem worthwhile to attempt to interpret these results in terms of a general integration of that equation. In these cases the heat transfer to the spin system dominated the early drifts while the late drifts were well-fitted on the basis of a small heat transfer with the block after equilibration with the spin system.

Any errors arising from energy and temperature measurements should not lead to errors in heat capacities exceeding 1 per cent, which is less, in most regions, than the uncertainty arising from interpretation of the after drifts.

Let us now turn to a discussion of the actual results in various temperature regions. For oxygen-doped methane we always obtain equilibrium with the spin system but in some cases only after about 30 min. Colwell *et al.*[5] reported equilibration times of several hours near 5 K for a sample containing mole fraction 50×10^{-6} of O_2. In all cases the temperature history included about 10 h at 4.2 K after initial cooling with helium-4; also the calorimeter was sometimes held overnight at that temperature between sets of measurements. For work below 1.5 K the calorimeter was cooled by use of helium-3 and measurements followed immediately after equilibration.

For the first heat-capacity measurement after rapid cooling there is always some additional uncertainty about thermal equilibration before the heat input. This is especially likely to be a problem in the region below 6 K where equilibration is very slow with the spin system. The effect would be to yield too low a heat capacity since the initial spin energy would be higher than the equilibrium value for the initial temperature. All measurements of this type below 6 K, *i.e.* immediately preceded by cooling, are marked [a] in table 2. Those in series 102 below 1 K are definitely too low and are omitted from figure 2. It is not clear that others are in error, but the entire series from 1.5 to 4.0 K are subject to this error and may indicate too low a curve in figure 3.

The most interesting result is the peak in the heat capacity near 0.8 K shown in figure 2 and concerning which a preliminary report[12] has been published. Although this is a region of slow equilibrium the measurements are apparently reproducible within a single series to within ± 1 per cent. However, as just discussed, lack of perfect thermal equilibration can lead to uncertainties somewhat larger than 1 per cent.

In the 2 to 10 K region shown in figure 3 the heat capacity continues to show a spin-related contribution, but the lattice contribution increases rapidly and dominates near 10 K. The heat capacity in the 2 to 5 K region is rather small, equilibrium is still slow, and the results are less accurate than below 2 K. Between 4 and 6 K we have measurements only from the less accurate preliminary series without block regulation. The estimated uncertainty, which is given for each value in table 4 for this series, is based on the range of reasonable interpretations of the after drifts. Several of these early points overlap the region of more accurate measurement above 6 K. In this range the agreement is good enough to support the validity of the results from 4 to 6 K. The results of Colwell *et al.*[5] begin near 5 K and agree quite well with our results up to 10 K.

We turn now to the results for pure methane below 10 K. The behavior of pure methane was explored by cooling for 9 h at 1.1 K and then breaking the thermal contact (by the heat switch). The calorimeter temperature rose rapidly to 2.0 K in 10 min and reached 2.3 K in 1 h. After another 6 h of cooling at 1.1 K and again

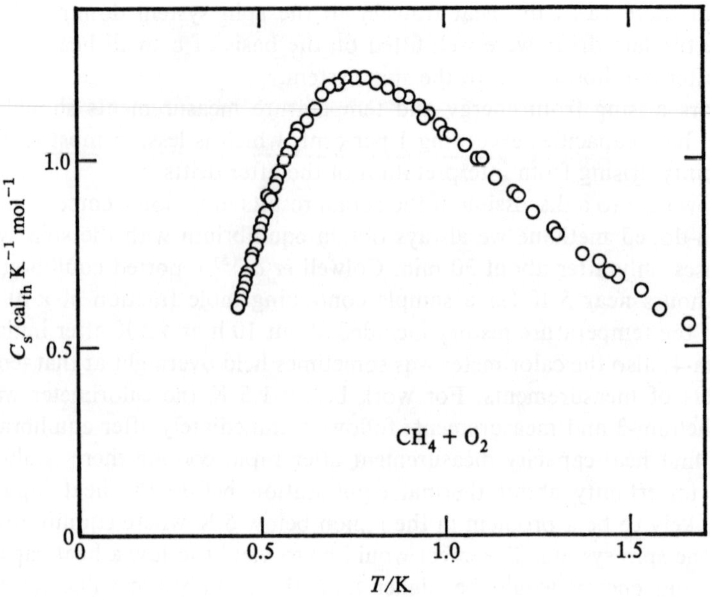

FIGURE 2. Equilibrium heat capacity of methane (doped with oxygen) below 1.7 K.

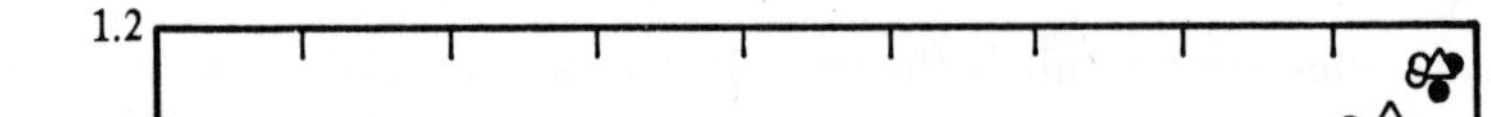

FIGURE 3. Equilibrium heat capacity of methane from 1 to 10 K. ○, Methane doped with oxygen; ●, pure methane, from this research; △, from reference 5.

breaking thermal contact, the calorimeter temperature again rose rapidly to 2.0 K and approached 2.2 K after 2 h. It was impractical to make heat-capacity measurements in this temperature range because the lattice heat capacity is so small that the heat flow from the spin system dominates the situation.

Beginning at about 2.2 K one can measure the "non-equilibrium" heat capacity of the lattice system alone by extrapolation of the after drifts with equation (3) which considers the heat transfer from the spin system. Although this extrapolation removes the heat capacity arising from change in spin-species composition, it does not determine the spin-species composition or the equivalent spin temperature. Since there is at least some rotational or librational contribution to the "lattice" heat capacity which will be different for different spin-species, there is some residual uncertainty associated with these non-equilibrium heat capacities.

Figures 4 and 5 show the non-equilibrium results for the 2 to 4 and 4 to 10 K regions, respectively, together with the results of Colwell *et al.*[5] The results of the

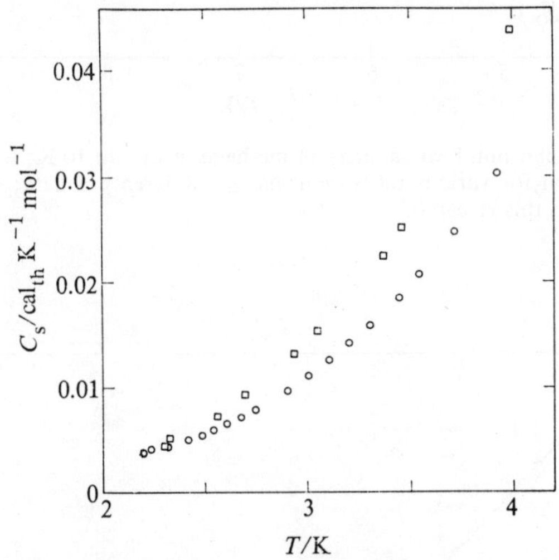

FIGURE 4. Non-equilibrium heat capacity of methane (with frozen spin-species composition) from 2 to 4 K. □, Reference 5 data; ○, series 14A data, this research.

two investigations agree reasonably well near 2.3 K and again near 6.5 K, but the data of Colwell *et al.* are considerably higher than ours from 3.5 to 6 K while they are significantly lower from 7.5 to 10 K. On the basis of the model of Kataoka *et al.*,[17] we have made rough calculations of the librational heat capacity for various spin-species distributions. This effect is very small at 2 K, but rises rapidly for the T (and E) spin-species.[24] For the A spin-species the librational heat capacity becomes significant only about 6 K.

Before the series 14B measurements the sample had been held below 4.2 K for a very long time while the series 14A measurements were made and between series 14A and 14B. On the other hand, the sample had been heated above 18 K prior to cooling again to 4.2 K for 11 h prior to series 14C. The heat-capacity values in series 14C

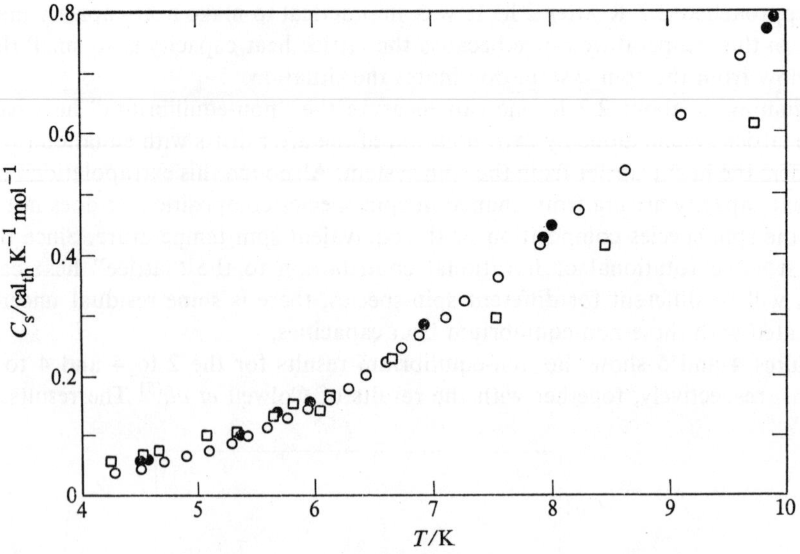

FIGURE 5. Non-equilibrium heat capacity of methane from 4 to 10 K; see text for discussion of spin-species composition for various measurements. □, Reference 5 data; ○, Series 14B and ●, Series 14C and 13, from this research.

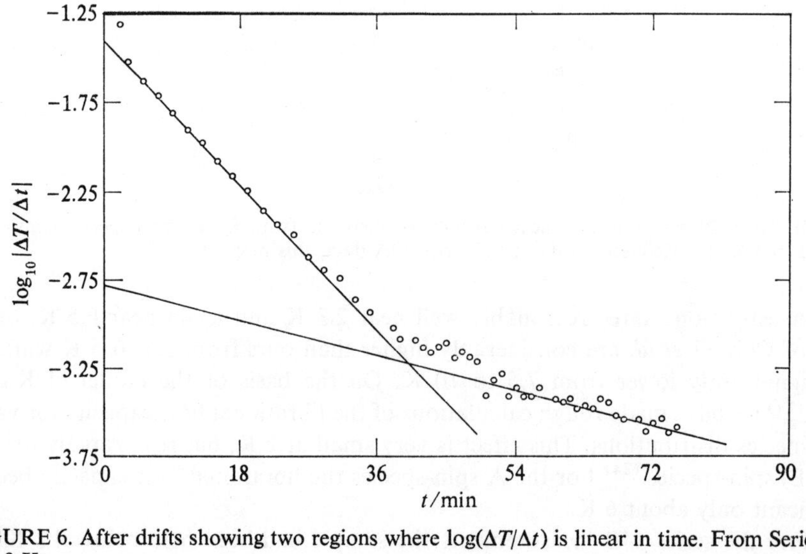

FIGURE 6. After drifts showing two regions where $\log(\Delta T/\Delta t)$ is linear in time. From Series 14C near 7.9 K.

are a little larger than those in series 14B which would agree with a higher spin temperature (more species T and E) in series 14C. Presumably, if the sample had been held an even shorter time at 4.2 K, even higher heat capacities would have been obtained. Hence, if Colwell *et al.*[5] did not hold their sample as long below 4.2 K as we did, they would have had more T (and E) species remaining and their higher heat capacities from 3.5 to 6 K are understandable.

In the vicinity of 6 K the rate of spin-species conversion becomes large enough that the spin temperature will not differ much from the lattice temperature at the beginning of a heat-capacity measurement. Correspondingly, the extrapolation of after drifts becomes more uncertain. At slightly higher temperatures one obtains after drifts such as are shown on figure 6. There are two regions in which the $\log(dT/dt)$ is linear in t. In the earlier period presumably the spin relaxation dominates and one may apply equation (3) to yield the non-equilibrium heat capacity while in the later period one assumes spin equilibrium and that equation (2) may be applied for heat transfer to the block to yield an equilibrium heat capacity. There are five examples between 7.5 and 10 K for which both equilibrium and non-equilibrium heat capacities were derived and listed in separate sections of table 3.

Above 10 K measurements on both pure methane and the oxygen-doped samples yield equilibrium heat capacities shown on figures 7 and 8. Our only values for pure methane are from preliminary series without temperature regulation of the block. We report these in table 4 only above 15 K where complete thermal equilibrium after heating is reasonably rapid. As can be seen on figure 8 our results for pure methane agree with those of Colwell *et al.* above 18 K, but below that temperature our results are slightly lower (and our unreported results below 15 K depart even

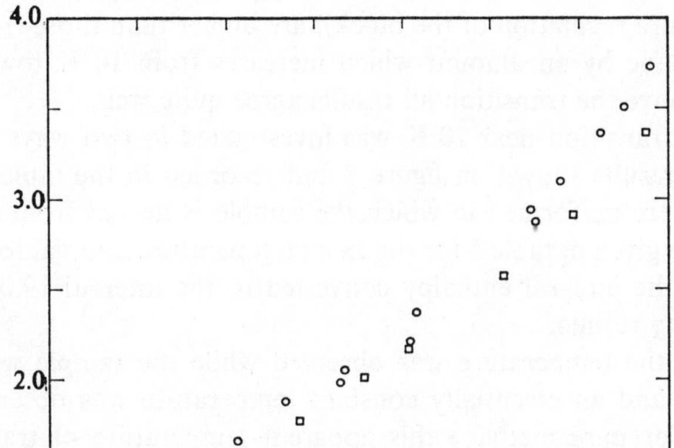

FIGURE 7. Equilibrium heat capacity of methane from 10 to 17 K. □, Reference 5 data; ○, Methane with O_2; ●, Pure methane, from this research.

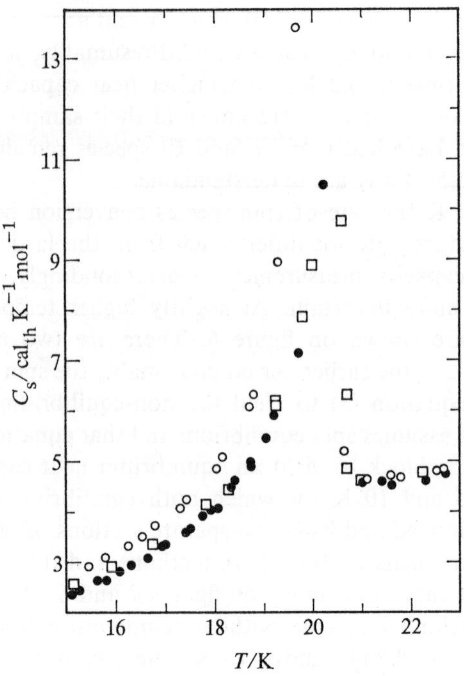

FIGURE 8. Heat capacity of methane in the region of the transition near 20 K. □, Reference 5 data. ○, Methane with O_2; ●, Pure methane, from this research.

further on the low side). Our results for oxygen-doped methane (which were made with temperature regulation of the block), are higher than those from either research for pure methane by an amount which increases from 10 K toward the transition near 20 K. Above the transition all results agree quite well.

The λ-type transition near 20 K was investigated in two ways in addition to the heat-capacity results shown on figure 8 and recorded in the tables. Integral heating experiments were performed in which the sample is heated from about 19 to 21 K. The results are given in table 5 for the exact temperature interval for each experiment. Also given is the integral enthalpy converted to the interval 19.0 to 21.0 K by use of heat-capacity results.

In addition the temperature was observed while the sample was heated through the transition and an essentially constant temperature was observed for an appreciable time. For pure methane this apparent temperature of transition was found to be 20.53 K. Colwell *et al.*[5] measured this temperature by heating into the transition and then measuring the temperature at zero heat input. They found 20.49 K; the difference of 0.04 K could easily be caused by thermal gradients in our experiment. We shall use Colwell *et al.*'s value for our calculations. Sperandio[25] reported 20.55 K for this transition temperature in reasonable agreement with the other values.

For oxygen-doped methane we obtained 20.30 K for the transition temperature which is 0.23 K lower than that for pure methane. Also note that the enthalpy of

TABLE 5. Enthalpy change for the λ-transition (cal_{th} = 4.184 J)

Temperature range		$H(T_2) - H(T_1)$	$H(21.0\ K) - H(19.5\ K)$
T_1/K	T_2/K	$cal_{th}\ mol^{-1}$	$cal_{th}\ mol^{-1}$
	Pure methane		
19.251	20.654	21.949	22.27
18.427	20.929	27.847	22.36
18.819	21.300	27.753	22.45
		average:	22.36 ± 0.1
	Oxygen-doped methane		
19.557	21.265	19.167	18.55
19.417	20.778	17.627	17.84
18.681	20.793	23.451	18.08
18.746	20.834	23.369	18.18
		average:	18.17 ± 0.3

transition $\{H(21.0\ K) - H(19.5\ K)\}$ is 4.2 $cal_{th}\ mol^{-1}$ less, whereas the heat capacity below 19 K is considerably larger, for the oxygen-doped sample than for pure methane.†

In this connection it is interesting to note the results of Eucken and Veith[26] for krypton mixtures with methane. For a sample with 3.70 moles per cent Kr the λ-transition has shifted down in temperature by 1 K and the heat capacity of the mixture below the transition has increased considerably as for our oxygen mixtures. The height of the transition has, however, decreased significantly from some 80 $cal_{th}\ K^{-1}\ mol^{-1}$ to under 20 $cal_{th}\ K^{-1}\ mol^{-1}$. For further increases of the mole fraction of krypton the transition shifted to lower temperatures and generally broadened until the transition was no longer detected at 32 moles per cent Kr.

Thus, it seems plausible that the introduction of the oxygen impurities (0.8 to 0.9 mole per cent) has so affected the intermolecular forces that the λ-transition has been shifted down in temperature and broadened.

Since the heat capacity for pure methane and the oxygen-doped samples agree to within experimental uncertainty at 9 and 28 K, with considerable disparity between these temperatures, the changes in enthalpy and entropy between 9 and 28 K were computed and compared. The following results were obtained:

	$H(28\ K) - H(9\ K)$	$S(28\ K) - S(9\ K)$
	$cal_{th}\ mol^{-1}$	$cal_{th}\ K^{-1}\ mol^{-1}$
$CH_4 + O_2$	84.80 ± 0.30	4.368 ± 0.014
Pure CH_4	84.85 ± 0.21	4.341 ± 0.005

The enthalpy increments are indeed nearly equal, but the value for the oxygen-doped samples should have some contribution from the oxygen. As determined

† Throughout this paper cal_{th} = 4.184 J.

from the free-molecule model for an additive oxygen heat capacity, the dissolved oxygen could possibly contribute as much as 0.38 cal_{th} mol^{-1}. Taking this estimate into account, the two values of ΔH still agree to within 0.5 per cent.

The entropy increments likewise tally with each other quite closely. In this case, the maximum additive contribution by oxygen is 0.021 cal_{th} K^{-1} mol^{-1} for the free-molecule model. With this correction $\{S(28\ K) - S(9\ K)\}$ for the $CH_4 + O_2$ sample remains slightly higher by 0.14 per cent. Thus, despite the considerable effect of the dissolved oxygen on the λ-transition, the overall change in enthalpy and entropy for the affected region remains the same within experimental uncertainty.

4. Entropy comparisons; discussion

As indicated in the introduction, we shall present here the calculations of experimental entropy differences based upon our experimental results and make comparisons with gas-phase statistical calculations of the entropy and zero-point entropy for appropriate models. We leave to another place the discussion of our results in relation to more detailed statistical calculations for the solid state.

For comparison we recalculated the statistical entropy of methane in the ideal-gas state using the usual equations[(27)] with recent constants and molecular parameters.[(28, 29)] The result for 298.15 K and 101.325 kPa is 44.51 cal_{th} K^{-1} mol^{-1} on the practical basis excluding nuclear spin. This does not differ significantly from literature values, for example a 1944 calculation[(30)] reported 44.50 cal_{th} K^{-1} mol^{-1}. This statistically calculated entropy for the gas remains the best value for use in thermodynamic calculations. If nuclear spin entropy is included this is increased by $R \ln 16 = 5.51$ cal_{th} K^{-1} mol^{-1} to yield a total of 50.02 cal_{th} K^{-1} mol^{-1}. For calculations related to the experimental measurements on liquid methane a convenient set of conditions is 99.54 K and 32.81 kPa. The statistical entropy without nuclear spin for these conditions is 37.93 cal_{th} K^{-1} mol^{-1}.

We shall use as a basis of comparison the entropy of the solid at 21.0 K. The experimental data on solid methane above this temperature and on the liquid are uncomplicated by the peculiarities at lower temperatures and are quite accurately known.[(1, 5, 25)] Literature values for the enthalpies of fusion and vaporization are given in table 6. There is good agreement for the enthalpy of fusion. The calorimetric enthalpy of vaporization of Clusius and Frank [(4b)] agrees well with that calculated from the vapor-pressure equation by the exact Clapeyron equation. The liquid[(31)] and vapor[(32)] volumes are accurately known; hence this calculation has high reliability. The equation of state of the vapor also yields an accurate correction from the ideal to the real gas. The equation of Pitzer and Curl[(33)] with parameters for methane from Pope *et al.*[(32b)] yields -0.05_7 cal_{th} K^{-1} mol^{-1} while the equation of McGlashan and Potter[(32a)] yields -0.05_3 cal_{th} K^{-1} mol^{-1}. Other equations[(34, 35)] which should yield a good approximation for methane confirm a value near -0.06 cal_{th} K^{-1} mol^{-1} (which differs from the value -0.14 cal_{th} K^{-1} mol^{-1} adopted without explanation by Colwell *et al.*[(5)]). The result for the entropy of solid methane at 21 K is given in table 7.

For most substances the nuclear-spin species distribution remains unchanged

TABLE 6. Enthalpies of fusion and vaporization for methane (cal_{th} = 4.184 J)

Enthalpy of fusion: T_{TP} = 90.67 K

Source	T_1/K	T_2/K	ΔH_f/cal_{th} mol^{-1}
Sperandio[25]	87.62	92.32	225.63
	87.20	90.91	224.72
	88.70	93.44	225.15
Clusius[1]	89.20	90.92	225.7
	90.15	91.92	222.3
	88.33	95.27	224.5
	88.71	93.49	223.3
Average:			224.4_7

Enthalpy of vaporization:	T = 99.54 K, p = 32.81 kPa
Source	ΔH_{vap}/cal_{th} mol^{-1}
Clusius and Frank[4b]	2036 ± 2
Calculation from vapor-pressure data	2037.5

TABLE 7. Calculated entropy of solid methane at 21 K from data at higher temperatures (cal_{th} = 4.184 J)

Process	S/cal_{th} K^{-1} mol^{-1}
$S°$ (ideal gas at 99.54 K, 32.81 kPa)	37.93
Correction to real gas, 99.54 K	−0.06
Condensation (2037 cal_{th} mol^{-1}/99.54 K)	−20.46
Liquid heat capacity (99.54 to 90.67 K)	−1.21
Solidification (224.47 cal_{th} mol^{-1}/90.67 K)	−2.48
Solid heat capacity (90.67 to 21.0 K)	−10.66
$S°$ (solid at 21.9 K, without nuclear spin)	3.06

down to the lowest temperature of heat-capacity measurements; then the usual equations and third law of thermodynamics yield an entropy value excluding nuclear spin. Use of the non-equilibrium heat capacities for methane (or extrapolation from above 12 K) should yield an approximation to this case for methane, but there are complicating factors to be discussed. Figure 9 shows a selection of pertinent data below 20 K presented as C/T values whose integral yields the entropy. We show the values of Colwell *et al.*[5] from 12 to 20 K since we believe their data are probably better than ours in this range. Below 10 K we show our values from series 13 and 14C together with a few points from Colwell *et al.* The entropy values corresponding to several curves through this region are given in table 8.

Parenthetically we first note that if one had data only above 12 K, as is typical when liquid hydrogen is the refrigerant, one would find an excellent fit with the Debye

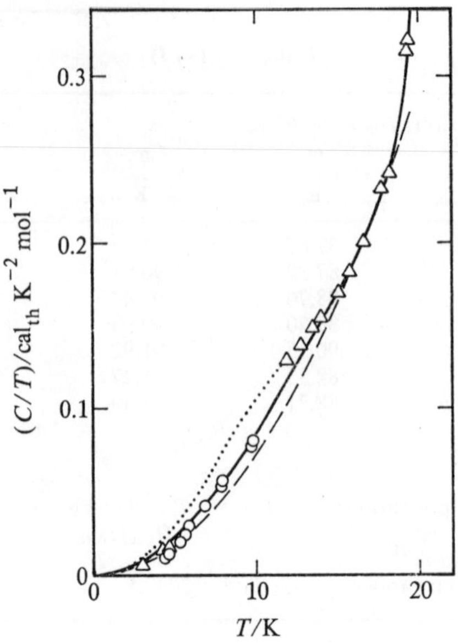

FIGURE 9. Entropy for non-equilibrium methane from C/T against T plot; see text for explanation of various curves. △, Data of reference 5; ○, data of this research.

function in the region 12 to 14 K with $\Theta_D = 72.8$ K. Then if one adds the Debye entropy at 12 K to that for the experimental curve above 12 K, the result at 19.5 K is 2.06 $\mathrm{cal_{th}\,K^{-1}\,mol^{-1}}$ which slightly exceeds the value of 1.97 $\mathrm{cal_{th}\,K^{-1}\,mol^{-1}}$ calculated from the statistical entropy of the gas and experimental data above 19.5 K. This calculation yields agreement to better than 0.1 $\mathrm{cal_{th}\,K^{-1}\,mol^{-1}}$, which one would interpret as quite satisfactory since either the extrapolation below 12 K or the value based on data at higher temperatures could be in error by 0.05 and possibly by 0.1 $\mathrm{cal_{th}\,K^{-1}\,mol^{-1}}$. While the data then available were slightly less accurate in some regions, the situation remained substantially as stated above until the measurements were extended into the liquid-helium range.

TABLE 8. Entropy of methane at 19.5 K on the basis of frozen spin-species composition ($\mathrm{cal_{th}} = 4.184$ J)

$\frac{T}{\mathrm{K}}$	Source	$\frac{S}{\mathrm{cal_{th}\,K^{-1}\,mol^{-1}}}$
21.0	From table 7	3.06
	S(trans) = 22.36 $\mathrm{cal_{th}\,mol^{-1}}$/20.49 K	−1.09
19.5	From data at higher temperatures	1.97
19.5	From "best curve" below 19.5 K	1.93
19.5	From Debye function ($\Theta_D = 72.8$ K) plus experimental curve above 12 K	2.06
19.5	From pure methane, results of this research	1.81

Next we note that a straightforward calculation based upon our non-equilibrium heat capacities below 10 K and our pure-methane results above 10 K yields $S(19.5\ \mathrm{K}) = 1.81\ \mathrm{cal_{th}\ K^{-1}\ mol^{-1}}$ (which is too small by $0.16\ \mathrm{cal_{th}\ K^{-1}\ mol^{-1}}$). Although these data yield a reasonably smooth curve (not shown on figure 9), we believe this is accidental. The higher values of Colwell *et al.* are to be preferred above 10 K because in this region our results are not based on a controlled block temperature. Below 10 K our experimental quality is excellent, but there is a variation of result with thermal history which indicates that the spin-species composition is varying and affects the heat capacity.

We show as a solid line on figure 9 our estimate of the true heat capacity for frozen spin-species composition. This is based on the non-equilibrium values near 2 K but rises above the non-equilibrium values from 3 to 10 K by an amount calculated from estimates about the spin-species composition and its effect on the liberational heat capacity calculated from the model of Kataoka *et al.*[17] Near 12 K the heat capacity is reduced from the measured equilibrium value by an amount similarly calculated for change in spin-species composition. This last correction decreases rapidly with rising temperature. The entropy based on this "best curve" is $1.93\ \mathrm{cal_{th}\ K^{-1}\ mol^{-1}}$ at 19.5 K in good agreement with the value $1.97\ \mathrm{cal_{th}\ K^{-1}\ mol^{-1}}$ derived from data at higher temperatures.

There is an interesting theoretical possibility that the remaining entropy at $T = 0$ on the frozen-spin-species basis may not be exactly $R \ln 16$. This situation is discussed in an appendix.

It is now interesting to re-examine figure 8. The solid curve is our best estimate of the true heat capacity with the spin-species composition frozen at the high-temperature pattern. The Debye extrapolation (dotted curve) is too high below 15 K and the fit in the 12 to 15 K region is accidental since the experimental results include a small contribution from spin-species conversion.

Also shown for comparison on figure 9 is a dashed curve for $C = (\mathrm{const})T^3$ fitted in the 16 to 18 K range. Actually, the fit to the T^3 curve in the 16 to 18 K range is accidental since the λ-transition is beginning to have some effect here.

We turn now to consideration of the equilibrium heat-capacity curve below 10 K including our results for oxygen-doped methane. For our comparison we add the nuclear spin entropy $R \ln 16 = 5.509$ to our earlier value for the entropy at 21.0 K obtaining $8.57\ \mathrm{cal_{th}\ K^{-1}\ mol^{-1}}$. Also the expected zero-point entropy on this basis is $R \ln 5 = 3.198\ \mathrm{cal_{th}\ K^{-1}\ mol^{-1}}$ since all of the methane will be converted to the A spin-species with spin $I = 2$ and $g_0 = 2I+1$.

The various increments in this calculation are shown in table 9. The extrapolation below 0.44 K is calculated from the energy-level pattern given above which fits the heat-capacity peak near 1 K. The entropy introduced in spin-species redistribution for the molecules at D_{2d} sites is $(3/4)R \ln(16/5) = 1.733\ \mathrm{cal_{th}\ K^{-1}\ mol^{-1}}$. It may be noted that this is approximately the entropy at 3 K. Since the heat capacity remains above $0.3\ \mathrm{cal_{th}\ K^{-1}\ mol^{-1}}$ at its minimum near 3 K, there is considerable overlap between the peak near 1 K associated with the spin-species redistribution for the methane on D_{2d} sites and the high heat capacity in the 4 to 6 K region associated with the redistribution for that on O_h sites. Lattice heat capacity also begins to be significant above 2 K; hence there is no distinct peak associated with the spin for the O_h sites.

The comparison with the entropy integrated from our heat capacities for oxygen-doped methane is shown in table 9.

Furthermore, the oxygen impurity must contribute some entropy which should be subtracted from the 8.44 cal_{th} K^{-1} mol^{-1} value in table 9. We have estimated this correction for the oxygen[15] to be (0.06 ± 0.02) cal_{th} K^{-1} mol^{-1} which increases the difference shown in table 9 to 0.19 cal_{th} K^{-1} mol^{-1}. This discrepancy is at the upper limit of experimental error for all of the increments included in table 7 as well as table 9 and seems improbably but not impossibly large. However, it is also possible that the oxygen was not uniformly distributed in our doped samples when crystallized

TABLE 9. Entropy including nuclear spin for solid methane ($cal_{th} = 4.184$ J)

T/K	Process	S or ΔS / cal_{th} K^{-1} mol^{-1}
0 to 0.44	Extrapolation	0.167
0.44 to 3.0	Heat capacity	1.52 ± .05
3.0 to 19.5	Heat capacity	2.66 ± .05
19.5 to 21.0	Transition (18.17 cal_{th} mol^{-1}/20.30 K)	0.89_5
	Zero point entropy (R ln 5)	3.198
21.0	S(sample containing 0.8 per cent O_2)	8.44
	Estimated entropy of O_2	−0.06
21.0	S, for pure methane	8.38
21.0	S, from higher temperature data	8.57

and that a small portion of the methane failed to attain spin-species equilibrium at very low temperatures. If this last-stated situation arose, it could explain part or all of the entropy difference between the two values for S(21.0 K) in table 9, and in that case our heat capacities in the spin-species conversion region below about 6 K should be increased slightly.

In conclusion, it appears that the various peculiarities in the low-temperature thermal properties of methane are now understood in terms of spin-species composition which becomes frozen below about 10 K unless catalyzed with paramagnetic molecules such as oxygen. With the catalyst present the methane converts to species A with a zero-point entropy of R ln 5. Without the catalyst the methane retains most of its total nuclear spin entropy of R ln 16. The detailed shape of the spin-equilibrium heat capacity curve, including the peak near 1 K, is generally consistent with the model presented by Kataoka, Okoda, and Yamamoto,[17] but we await with interest calculated heat capacities based on full use of their theory including cooperative effects.

This research was sponsored by the Energy Research and Development Administration.

5. Appendix

When methane is cooled without a spin-species conversion catalyst some species T and E molecules will remain at $T = 0$. On the assumption of the eight-sublattice antiferrorotational structure as discussed by James and Keenan[16] and by Kataoka *et al.*,[17] the zero-point degeneracy is 5 for the A species and 9 for the T species for either the D_{2d} or O_h sites without ambiguity. Also for the E species the degeneracy is clearly 2 at the D_{2d} sites, but at the O_h sites the E molecules are expected to have a 6-fold degeneracy in their lowest energy level. This is shown most clearly by Nishiyama and Yamamoto[18] for a methane molecule in a rare-gas lattice with sites also of O_h symmetry. Kataoka *et al.*[17] indicate a similar pattern for the E species at the O_h sites in solid methane. Our extrapolation below 2 K implied by the "best curve" of figure 9 and table 8 relates to a state including some E molecules on O_h sites. If we take the high-temperature fraction of such molecules, $(1/4) \times (2/16) = 1/32$, the extra zero-point entropy (above the $R \ln 16$ for proton spin) would be $(1/32)R \ln(6/2) = 0.07\ \mathrm{cal_{th}\ K^{-1}\ mol^{-1}}$ on the basis of the 6-fold degeneracy for the E species on O_h sites. This value is of the same sign and slightly larger than the difference (1.97 to 1.93) shown in table 8, but the uncertainty in these experimental values is too large to make this comparison really significant.

REFERENCES

1. Clusius, K. *Z. Phys. Chem.* **1929,** B3, 41.
2. Clusius, K.; Perlick, A. *Z. Phys. Chem.* **1934,** B24, 313.
3. Clusius, K.; Popp, L.; Frank, A. *Physica* **1937,** 4, 1105.
4. Frank, A.; Clusius, K. *Z. Phys. Chem.* **1937,** B36, 291; **1939,** B42, 395.
5. Colwell, J. H.; Gill, E. K.; Morrison, J. A. *J. Chem. Phys.* **1962,** 36, 2223; **1963,** 39, 635.
6. Wolf, R. P.; Whitney, W. M. *Proceedings of the Ninth Conference on Low Temperature Physics.* (Plenum Press: New York **1965,**), p. 1118.
7. Hopkins, H. P.; Donoho, P. L.; Pitzer, K. S. *J. Chem. Phys.* **1967,** 47, 864.
8. Wong, K. P.; Noble, J. D.; Bloom, M.; Alexander, S. *J. Magn. Reson* **1969,** 1, 55.
9. van Hecke, P.; Grobet, P.; van Gerven, L. (a) *Phys. Lett.* **1970,** 33A, 379; (b) *J. Magn. Reson* **1972,** 7, 117; (c) *Physica* **1973,** 68, 359.
10. Wilson, E. B. *J. Chem. Phys.* **1935,** 3, 276.
11. Curl, Jr., R. F.; Kasper, J. V. V.; Pitzer, K. S. *J. Chem. Phys.* **1967,** 46, 3220.
12. Vogt, G. J., Pitzer, K. S. *J. Chem. Phys.* **1975,** 63, 3667.
13. A preliminary presentation of this material may be found in the theses of Janice J. Kim and Gerald J. Vogt (references 14 and 15, respectively).
14. Kim, J. J. Ph.D. Thesis, University of California, **1975,** also Lawrence Berkeley Laboratory Report LBL 3955.
15. Vogt, G. J. Ph.D. Thesis, University of California, **1976,** also Lawrence Berkeley Laboratory Report LBL 4149.
16. James, H. M.; Keenan, T. A. *J. Chem. Phys.* **1959,** 31, 12.
17. Kataoka, Y.; Okada, K.; Yamamoto, T. *Chem. Phys. Lett.* **1973,** 19, 365.
18. Nishiyama, K.; Yamamoto, T. *J. Chem. Phys.* **1973,** 58, 1001 and earlier papers there cited.
19. Bloom, M.; Morrison, J. A. Orientational Order and disorder in the solid isotopic methanes. In *Surface and Defect Properties of Solids* (Specialist Periodical Reports of the Chemical Society, London) **1973,** 2, 140.
20. Finegold, L.; Phillips, N. E. *Phys. Rev.* **1969,** 177, 1383.
21. Cryocal, Inc., Riviera Beach, Florida.
22. Ho, J.; Phillips, N. E. *Rev. Sci. Instr.* **1965,** 36, 1382.
23. Finegold, L. J. *Sci. Instr.* **1967,** 44, 789.
24. See reference 15, pp. 149–153.

25. Sperandio, A. Thesis, University of Zurich, **1961**,.
26. Eucken, A.; Veith, H. *Z. Phys. Chem.* **1936,** B34. 275.
27. Pitzer, K. S.; Brewer, L. revision of *Thermodynamics* by G. N. Lewis and M. Randall, Chapter 27. McGraw Hill Book Co. New York, **1961.**
28. Herzberg, G. *Molecular Spectra and Molecular Structure, II. Infrared and Raman Spectra of Polyatomic Molecules* Van Nostrand Co. New York. **1945,** p. 307.
29. Herranz, J.; Stoicheff, B. P. *J. Mol. Spect.* **1963,** 10, 448.
30. Pitzer, K. S. *Ind. Eng. Chem.* **1944,** 36, 829.
31. Terry, M. J.; Lynch. J. T.; Bunclark, M.; Mansell, K. R.; Staveley, L. A. *J. Chem. Thermodynamics* **1969,** 1, 413.
32. (a) McGlashan, M. L.; Potter, D. J. B. *Proc. R. Soc. London* **1962,** A267, 478. (b) Pope, G. A.; Chappelear, P. S.; Kobayashi, R. *J. Chem. Phys.* **1973,** 59, 423.
33. Pitzer, K. S.; Curl, Jr., R. F. *J. Am. Chem. Soc.* **1957,** 79, 2369.
34. see Dymond, J. H.; Smith, E. B. *The Virial Coefficients of Gases: A Critical Compilation.* Clarendon Press: Oxford **1969,** 26–31.
35. Douslin, D. R.; Harrison, R. H.; Moore, R. T.; McCullough, J. P. *J. Chem. Eng. Data* **1964,** 9 358.

Crystal field effects on oxygen in solid methane and the catalysis of spin-species conversion of methane

Janice J. Kim and Kenneth S. Pitzer

Department of Chemistry and Materials and Molecular Research Division of the Lawrence Berkeley Laboratory, University of California, Berkeley, California 94720

(Received 14 September 1976)

The energy states are calculated for an oxygen molecule in a crystal field of cubic symmetry that would be appropriate for a substitutional site in solid methane. The resulting heat capacity is also calculated and discussed in relation to data for oxygen-doped methane. The catalytic effect of oxygen on the spin-species conversion of methane is treated by consideration of the dipole–dipole interaction of the electron spin of oxygen with the proton spin of methane. The matrix elements are calculated for the low-energy states of methane and of oxygen. It is found that oxygen should be an effective catalyst for spin-species conversion above about 0.3°K but that its effectiveness may decrease rapidly below that temperature.

I. INTRODUCTION

The addition of a small amount of paramagnetic substance such as oxygen has been found to facilitate the conversion of nuclear-spin species. Early studies on the kinetics of the ortho–para hydrogen conversion process noted that oxygen and other paramagnetic molecules greatly enhanced the conversion rate.[1]

The phenomenon of nuclear-spin-species conversion in solid methane has been observed in the low-temperature phase (1–20 °K) using a variety of techniques including nuclear magnetic resonance, level-crossing experiments, and heat capacity measurements.[2] The Pauli principle allows the existence of only certain nuclear spin states with a given rotational state to preserve the overall symmetry of the total wavefunction. In solid methane above the transition near 20.4 °K all of the molecules undergo somewhat restricted rotation.[3] Below this transition,[4] however, three-fourths of the molecules are rather strongly oriented in a cog-wheel manner in sites of D_{2d} symmetry, while the remaining one-fourth of the molecules rotate rather freely in sites of O_h symmetry. The molecular centers lie on a face-centered cubic lattice both above and below 20.4 °K.

While various methane–methane interactions, e.g., spin–spin and spin–rotation, induce interspecies conversion, these processes are very slow at very low temperatures. In an effort to reduce relaxation times in low-temperature studies of methane, small quantities of oxygen were often introduced into the lattice to obtain more rapid equilibrium.

Vogt and Pitzer[2(f),(g)] have recently reported low temperature heat capacity studies of methane containing approximately 0.8% oxygen. An anomaly observed in the region about 1 °K is attributed to spin-species conversion. The present theoretical work[5] was undertaken to further interpret these and other experimental studies of oxygen-doped methane.

In Sec. II calculations are presented concerning the properties of oxygen in a cubic lattice such as methane. Included are calculations of the heat capacity for various cubic fields as well as for freely rotating molecules. These results indicate that corrections to the heat capacity of typical oxygen-doped methane samples will be small and that values from the freely rotating model will be a good approximation for cubic fields of moderate strength.

Section III treats the problem of spin-species conversion of methane by interaction with oxygen. The results are discussed in Sec. IV with particular reference to conversion rates below 1 °K.

II. PROPERTIES OF O_2 IN A CUBIC LATTICE

As oxygen–methane samples are solidified, it is likely that oxygen will occupy a methane site in the lattice structure. The carbon atoms of the methane lattice form a face-centered cubic structure. Since oxygen is smaller than a methane molecule, it may be rather freely rotating in the methane lattice, and we choose as our model that of a rotating oxygen molecule perturbed by a crystal field of cubic symmetry.

The spectrum of the ground electronic state of gaseous oxygen ($^3\Sigma_g^-$) shows a fine structure splitting of the rotational levels.[6] Kramers[7] has shown that the dipole–dipole interaction between the two unpaired electron spins, when averaged over the rotation, removes the degeneracy of the rotational levels in the observed manner. There is also a weaker Hund's case (b)[6] coupling of the spin with the total angular momentum apart from spin, but this term is negligible for our purposes.

Our total Hamiltonian for oxygen in methane includes the contributions from rotation, crystal field, and spin–spin interactions.

$$\mathcal{H}_{\rm tot} = \mathcal{H}_{\rm rot} + \mathcal{H}_{\rm cryst\ field} + \mathcal{H}_{\rm ss} \, . \tag{1}$$

The spin–spin coupling arising from the interaction of the two magnetic-dipole moments is of the form

$$\mathcal{H}_{\rm ss} = \left(\frac{\mu_1 \cdot \mu_2}{|\mathbf{r}|^3} - \frac{3(\mu_1\cdot\mathbf{r})(\mu_2\cdot\mathbf{r})}{|\mathbf{r}|^5}\right) \, , \tag{2}$$

where $|\mathbf{r}|$ is the relative distance between the two dipoles μ_1 and μ_2. This spin–spin interaction may be expressed in the form[8]

$$\begin{aligned}\mathcal{H}_{\rm ss} = A\{&[S_z^2 - \tfrac{1}{4}(S^+S^- + S^-S^+)](3\cos^2\theta - 1) \\ &+ \tfrac{3}{2}\sin\theta\cos\theta[(S_zS^+ + S^+S_z)e^{-i\phi} + (S_zS^- + S^-S_z)e^{i\phi}] \\ &+ \tfrac{3}{4}\sin^2\theta[(S^+)^2e^{-i2\phi} + (S^-)^2e^{i2\phi}]\} \, ,\end{aligned} \tag{3}$$

TABLE I. Energy levels of oxygen in a cubic field for the rotational levels $K=1,3$.

	Energy (cm^{-1})					
XO_h(cm^{-1}) =	10.0	0.0	−10.0	−20.0	−100.0	−500.0
$K=1$	0.00(1)[a]	0.00(1)	0.00(1)	0.00(1)	0.00(1)	0.00(1)
	1.97(2)	2.11(5)	2.02(3)	1.94(3)	1.52(3)	0.75(3)
	2.20(3)		2.24(2)	2.35(2)	2.94(2)	3.50(2)
	3.97(3)	3.97(3)	3.97(3)	3.96(3)	3.88(3)	3.73(3)
$K=3$	16.09(1)	16.24(5)	15.94(3)	15.43(3)	11.95(3)	5.50(3)
	16.09(2)		16.19(2)	16.11(2)	16.65(2)	29.16(2)
	16.22(3)	16.43(9)	16.27(3)	16.25(3)	16.97(3)	29.61(3)
	16.26(3)		16.63(3)	16.82(3)	18.21(3)	31.27(3)
	16.54(2)		16.66(2)	17.04(2)	18.91(1)	31.71(1)
	16.76(3)		16.85(1)	17.34(1)	21.77(2)	41.00(1)
	18.34(3)	18.35(7)	18.25(1)	18.20(1)	22.47(1)	41.20(3)
	18.49(1)		18.46(3)	18.69(3)	23.02(3)	41.24(2)
	18.55(3)		18.48(3)	18.83(3)	23.65(3)	41.46(3)

[a]The degeneracy of each level is denoted in parentheses. The constants $B_0=1.43\,777$ cm^{-1}, $\lambda=1.984$ cm^{-1} were used in the calculation.

where A is a constant measuring the strength of the interaction. Here the spin–spin Hamiltonian is expressed in terms of the total spin $\mathbf{S}=\mathbf{S}_1+\mathbf{S}_2$ with $S^{\pm}=S_x \pm iS_y$ and the polar coordinates refer to the orientation of the vector from spin 1 to spin 2.

The basis $|Km_K m_S\rangle$ proves to be convenient, where K is the rotational quantum number and m_K and m_S are the components of K and S on the z axis, respectively. The kinetic energy yields only the familiar diagonal contributions $B_0 K(K+1)$. The spin term $\mathcal{H}_{ss}$ yields nonzero matrix elements only between states where $m_J=m_K+m_S=$ constant. These have been evaluated by others[7,9] and were verified.[5]

If the cubic-crystal-field potential is expressed as an expansion in spherical harmonics, the matrix elements $\langle K'm_K'm_s'|V|Km_K m_s\rangle$ can easily be calculated. The first nonconstant set of spherical harmonics (Y_{lm}) which yield linear combinations totally symmetric (A_1) in an octahedral field are the $l=4$ states. The appropriate linear combination[10] is given by

$$V(\theta,\phi)=XO_h\left(\frac{-2\sqrt{\pi}}{15\sqrt{7}}\right)\times\left(\sqrt{7}\,Y_{40}(\theta,\phi)+\frac{\sqrt{5}}{\sqrt{2}}\left[Y_{44}(\theta,\phi)+Y_{4-4}(\theta,\phi)\right]\right), \tag{4}$$

where XO_h is a measure of the strength of the field.

Devonshire,[11] the first to consider the rotation of diatomic molecules in fields of octahedral symmetry, constructed a potential equivalent to that in Eq. (4). For a positive potential constant there are six equivalent C_{4v} minima and eight equivalent C_{3v} maxima. For a face-centered cubic lattice one would expect less repulsive interaction for an oxygen pointing along the diagonal of the cube. This implies a negative value for the constant XO_h and yields eight equivalent C_{3v} potential minima for an oxygen occupying a methane site.

The crystal field contributions were readily evaluated using the relationship of spherical harmonics

$$\langle l'm'|Y_{LM}|lm\rangle=(-1)^{-m'}\{[(2l'+1)(2l+1)(2L+1)/4\pi]\}^{1/2}\times\begin{pmatrix} l' & L & l\\ -m' & M & m\end{pmatrix}\begin{pmatrix} l' & L & l\\ 0 & 0 & 0\end{pmatrix}, \tag{5}$$

where the 3-j symbols

$$\begin{pmatrix} j_1 & j_2 & j_3\\ m_1 & m_2 & m_3\end{pmatrix},$$

have been tabulated by Rotenberg *et al.*[12] and are also available in computer subroutines. One notes that the crystal field does not affect the spin states. Hence, $\langle K'm_K'm_s'|V|Km_K m_s\rangle$ vanishes unless $m_s'=m_s$.

With the methods described, the Hamiltonian matrix was constructed and diagonalized to obtain the energy eigenvalues and eigenfunctions. The values[6] $B_0=1.437\,777$ cm^{-1} and $\lambda=3/2\,A=1.984$ cm^{-1} were used in calculating the energy levels for various values of the potential parameter XO_h. Only odd values of K are allowed for $^{16}O_2$, the dominant isotopic species.

Table I summarizes the resulting energies relative to the ground state. Figure 1 shows the fine structure splitting calculated for oxygen in the gas phase ($XO_h=0$), and the additional splitting that arises in a weak octahedral field.

The calculations were made using a basis set with a maximum rotational value of $K=7$. The $3(2K+1)$ states for each K value yielded a total of 108 basis functions. This was a sufficiently large basis set to give an accurate description of the behavior of the levels which are populated at low temperatures.

The heat capacity for rotational motion of oxygen in a cubic field was calculated for various values of XO_h using the usual statistical mechanical equations, and is shown in Fig. 2. In interpreting these results it is useful to note that the maximum difference in potential for molecular rotation is $1/3\,|XO_h|$. For negative values of this constant the lowest potential barrier between minima is $|XO_h|/12$. Thus an XO_h value of -100 cm^{-1} implies a barrier between minima of 8.3 cm^{-1}.

Also included in Fig. 2 is the heat capacity curve for an infinite potential barrier that represents the case where the molecular rotation is reduced to torsional vi-

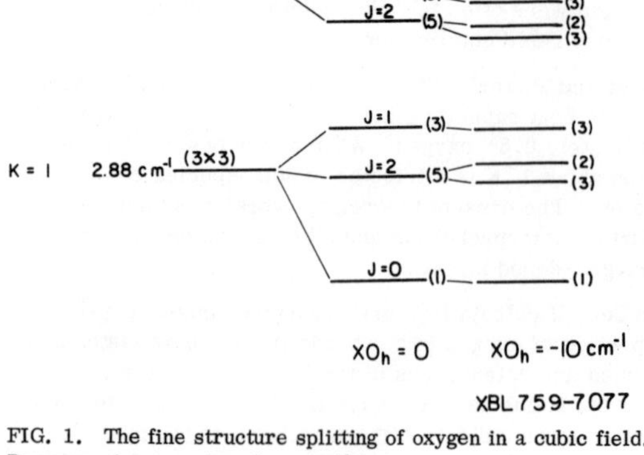

FIG. 1. The fine structure splitting of oxygen in a cubic field. Degeneracies are given in parentheses.

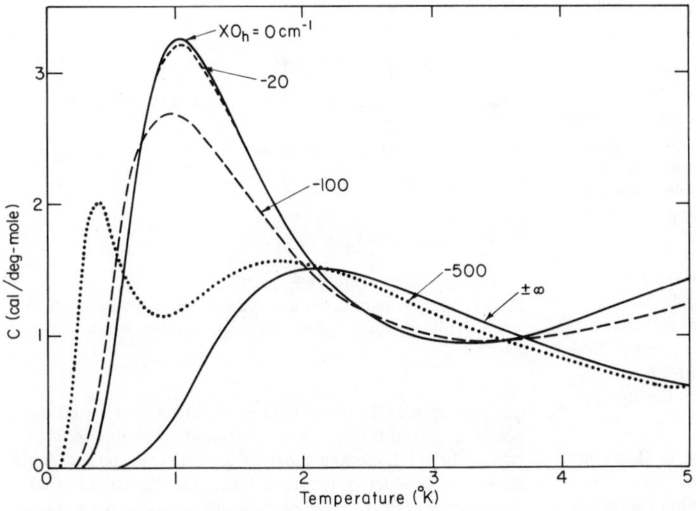

FIG. 2. The heat capacity of oxygen for rotational motion in a cubic field.

bration of small amplitude. This is essentially the situation found for O_2 in lattices of diatomic molecules such as N_2 and CO. Heat capacity measurements by Burford and Graham[13] for these systems show a broad maximum in the oxygen contribution near 2 °K with a shape very similar to that shown for $\pm\infty$ in Fig. 2. In the diatomic-molecule lattices there is undoubtedly an elongated cavity, which yields a single lowest level and a doubly degenerate level at $2\lambda = 3.57$ cm^{-1}. With the infinite cubic field, one obtains the same pattern but with degeneracies 4 and 8 for large negative XO_h or 3 and 6 for large positive XO_h corresponding to the number of potential minima for a symmetrical linear molecule. The approach to this pattern can be seen in the column for $XO_h = -500$ cm^{-1} in Table I.

In EPR studies of O_2 in an N_2 matrix at 4.2 °K, Simoneau, Harvey, and Graham[14] confirmed an axial substitutional site for O_2. In preliminary experiments with O_2 in a CO lattice, they found results almost identical to those in a N_2 lattice. Samples were also prepared with argon and CD_4 matrices; however, these data "showed no distinctive features and were not amenable to analysis." Hence, Simoneau *et al.*, concluded there was no evidence here of an axial site about the O_2 molecule. This result is consistent with our choice of a cubic field for O_2 in methane, but unfortunately it gives only a vague upper limit to the constant for the cubic field.

While it is not possible to set a quantitative value for the field strength, it is apparent from Fig. 1 that the heat capacity contribution from a small percentage of oxygen will not be large and that the value for freely rotating molecules will set an upper limit near 1 °K. Since the peak observed by Vogt and Pitzer[2(f),(g)] for methane containing 0.8% O_2 occurs essentially at 1 °K, as shown in Fig. 3, their use of the free-rotation heat capacity for the small contribution from oxygen seems justified.

In view of the particular interest in this anomaly in methane near 1 °K, we have calculated the population of oxygen in each of several low-energy states. These results are given in Table II.

TABLE II. Relative population of the first six levels at $T = 1$ °K for various field strengths.

K	J	Degen.	XO_h(cm^{-1}) 0	−20	−100	−500
1	0	(1)	0.800	0.793	0.726	0.488
1	2	(3) }	0.192	0.145	0.245	0.499
1	2	(2) }		0.054	0.021	0.006
1	1	(3)	0.008	0.008	0.008	0.007
3	2	(3) }	3×10^{-10}	5×10^{-9}	7×10^{-8}	5×10^{-4}
3	2	(2) }		1×10^{-10}	6×10^{-11}	6×10^{-19}

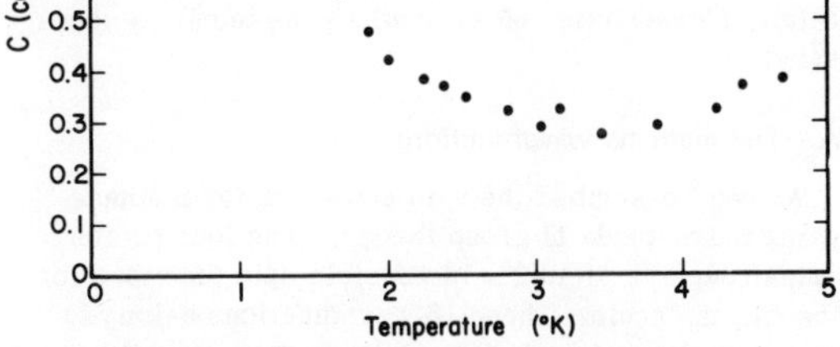

FIG. 3. The heat capacity of methane catalyzed for spin species conversion. From Vogt and Pitzer.[2(f)]

III. MECHANISM FOR OXYGEN-CATALYZED SPIN CONVERSION IN METHANE

Bloembergen[15] first considered the theory of nuclear spin relaxation via paramagnetic ion impurities. It was postulated that nuclear spins within a certain radius R_0 of the impurity relaxed rapidly via a nuclear-spin-dipole interacting with an impurity electronic spin dipole. Then distant spins are relaxed by spin diffusion. This picture has been refined[16] but not changed basically.

Motizuki and Nagamiya[17] have described the theory of oxygen-induced spin conversion in solid ortho-para hydrogen. The description of oxygen-catalyzed spin conversion in methane is formally the same. However, the interaction with oxygen involves four protons in methane and must be averaged over the more complex rotational motion of methane.

In general the spin change for protons in methane may be induced by an oscillating magnetic field. The unpaired electron spins of oxygen yield an inhomogeneous magnetic field. Oscillations in this field at any given proton can arise from rotation of either the methane or oxygen or oscillation in the distance between them. Since we seek only a qualitative picture of the conversion process we have not investigated all of these aspects. Rather we follow Motizuki and Nagamiya[17] and consider only the rotation of the methane together with the energy requirements including phonons. Presumably the detailed consideration of oxygen rotation and lattice vibration would enhance the rate.

Also the rotation of both oxygen and methane molecules is presumably restricted somewhat in the lattice. A restricted rotational wavefunction can be expressed in a series, where the leading term is the function for free rotation of the appropriate symmetry, etc. While a quantitative evaluation of the rate of conversion would require use of the functions for restricted rotation, the qualitative pattern should be correctly given by use of the leading terms alone, and we use this approximation throughout this section.

While our treatment will apply to all states at moderate energy, we have worked out the details only for the low-energy states that are populated near 1 °K and are important in spin-species conversion of methane on D_{2d} sites. We will be particularly interested in the catalytic effectiveness of oxygen in the ground state with $J=0$ since this is a nondegenerate state of zero total angular momentum.

Before considering the nature of the Hamiltonian in detail, a description of the methane molecule is necessary.

A. The methane wavefunctions

Wilson[18] described the spin statistics for methane using the methods of group theory. The four proton spins couple to yield $2^4=16$ possible spin functions for the CH_4 molecule. These 16 wavefunctions belong to one of the three irreducible representations of the T group. There are five A states that correspond to

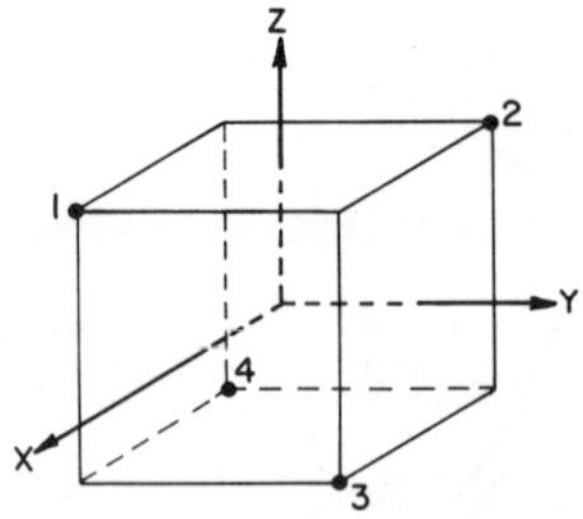

XBL 759-7090

FIG. 4. The orientation of a methane molecule reactive to the molecule-fixed frame. The carbon atom at the origin is not shown.

nuclear spin $I=2$, three sets of T states ($I=1$), and one set of E states ($I=0$). The rotational wavefunctions $|JKM\rangle$ of methane are those of a spherical top molecule. Since the product of rotational and spin functions must have symmetry A, only certain spin-species are allowed for a given rotational level.

More detailed definitions of the rotational and spin wavefunctions for methane have been given by several authors.[19–21] Since we are interested in behavior at very low temperature, we give primary attention to the states with $J=0$ and 1 of spin symmetry A and T, respectively.

The rotational function for $J=0$ is just $1/2\pi$ and the spin functions are the totally symmetric functions[20] with $I=2$ and $m_I=2$, 1, 0, -1, and -2, which we write as $|Am_I\rangle$. The product for $\psi(J, m_J, m_I)$ is then

$$\psi(00m_I)=(1/2\pi)\,|Am_I\rangle\ . \tag{6}$$

For the nine states with $J=1$ and spin symmetry T the situation is more complex. It is convenient to choose both the spin and rotation functions in forms that transform as the Cartesian coordinates in the molecule-fixed frame shown in Fig. 4. The spin functions $|T_x, m_I\rangle$, etc., are given explicitly in Table III.[20]

The rotational wavefunctions of a spherical-top molecule may be expressed in terms of the normalized D functions[21,22]

$$|JKM\rangle=\psi^{(J)}[\{\alpha\beta\gamma\}]_{K,M}=\left[\frac{(2J+1)}{8\pi^2}\right]^{1/2}D^{(J)}[\{\alpha\beta\gamma\}]_{K,M} \tag{7}$$

where the Euler angles $\{\alpha\beta\gamma\}$ describe the molecular orientation with respect to a space-fixed reference axis. We adopt the convention of Edmonds[23] for defining the D function in Eq. (7).

Yamamoto and Katakoa[21] have shown that the appropriate linear combinations of $|JKM\rangle$ which transform as x, y and z for $J=1$ are

$$\psi^{(1)}_{x,M}=\psi^{(1)-}_{1,M}\ , \tag{8a}$$

$$\psi^{(1)}_{y,M}=-i\psi^{(1)+}_{1,M}\ , \tag{8b}$$

$$\psi^{(1)}_{z,M}=\psi^{(1)}_{0,M}\ , \tag{8c}$$

where

TABLE III. The CH_4 spin functions of symmetry T.

$$|T_x 1\rangle = \frac{1}{2}[(\alpha\alpha\alpha\beta+\alpha\beta\alpha\alpha)-(\alpha\alpha\beta\alpha+\beta\alpha\alpha\alpha)]$$

$$|T_x 0\rangle = \frac{1}{\sqrt{2}}(\alpha\beta\alpha\beta-\beta\alpha\beta\alpha)$$

$$|T_x -1\rangle = \frac{1}{2}[(\alpha\beta\beta\beta+\beta\beta\alpha\beta)-(\beta\alpha\beta\beta+\beta\beta\beta\alpha)]$$

$$|T_y 1\rangle = \frac{1}{2}[(\alpha\alpha\alpha\beta+\beta\alpha\alpha\alpha)-(\alpha\alpha\beta\alpha+\alpha\beta\alpha\alpha)],$$

$$|T_y 0\rangle = \frac{1}{\sqrt{2}}(\beta\alpha\alpha\beta-\alpha\beta\beta\alpha)$$

$$|T_y -1\rangle = \frac{1}{2}[(\beta\alpha\beta\beta+\beta\beta\alpha\beta)-(\alpha\beta\beta\beta+\beta\beta\beta\alpha)]$$

$$|T_z 1\rangle = \frac{1}{2}[(\alpha\alpha\alpha\beta+\alpha\alpha\beta\alpha)-(\alpha\beta\alpha\alpha+\beta\alpha\alpha\alpha)]$$

$$|T_z 0\rangle = \frac{1}{\sqrt{2}}(\alpha\alpha\beta\beta-\beta\beta\alpha\alpha)$$

$$|T_z -1\rangle = \frac{1}{2}[(\alpha\beta\beta\beta+\beta\alpha\beta\beta)-(\beta\beta\alpha\beta+\beta\beta\beta\alpha)]$$

$$\psi^{(J)\pm}_{\mu,M} \equiv \frac{1}{\sqrt{2}}\left[\frac{(2J+1)}{8\pi^2}\right]^{1/2}[D^{(J)}_{-\mu,M} \pm D^{(J)}_{\mu,M}], \quad (9a)$$

$$\psi^{(J)}_{0,M} \equiv \left[\frac{(2J+1)}{8\pi^2}\right]^{1/2} D^{(J)}_{0,M}. \quad (9b)$$

M can take on the values 1, 0, −1, for $J=1$, yielding nine rotational wavefunctions, three sets which transform as either x, y, or z.

The product wavefunctions, $\psi_{\rm rot}\psi_{\rm spin}$ of $J=1$ which are of overall A symmetry would transform as $x^2+y^2+z^2$ or for $\psi(J, m_J, m_I)$

$$\psi(1m_J m_I) = 3^{-1/2}(\psi^{(1)}_{x,M}|T_x m_I\rangle + \psi^{(1)}_{y,M}|T_y m_I\rangle + \psi^{(1)}_{z,M}|T_z m_I\rangle). \quad (10)$$

Since $m_J = M$ and independently m_I may take on values −1, 0, or 1, there are 9 states in all.

B. The methane-oxygen dipole-dipole interaction

The interaction between two dipoles has been described in Eq. (2). The magnetic moment μ_1 is that of the oxygen molecule, which interacts with the proton spin magnetic moment μ_2 of a CH_4 molecule. The proton magnetic moment μ_2 is proportional to the nuclear spin moment I. While the O_2 magnetic moment has contributions from both the unpaired electron-spin angular momentum and the rotational angular momentum, the dominant term is the electron-spin magnetic moment which is proportional to the electron spin angular momentum S.

The interaction between an O_2 and CH_4 molecule is the sum of the interactions between the oxygen electron-spin and each of the four protons as labeled in Fig. 4:

$$\mathcal{H}(O_2-CH_4) = \sum_{n=1}^{4} \mathcal{H}(S, I_n). \quad (11)$$

If we recast Eq. (3) in Cartesian coordinates,[8,17] we can write the oxygen interaction with the nth proton as

$$\mathcal{H}(S, I_n) = -g_e\beta g_p\beta_N\left\{[S_z I_{z,n} - \tfrac{1}{4}(S^+I_n^- + S^-I_n^+)]\left(\frac{R_n^2-3Z_n^2}{R_n^5}\right) - \frac{3}{2}\left[(S^+I_{z,n}+S_zI_n^+)\frac{(X_n-iY_n)Z_n}{R_n^5} + (S^-I_{z,n}+S_zI_n^-)\frac{(X_n+iY_n)Z_n}{R_n^5}\right] - \frac{3}{4}\left[S^+I_n^+\frac{(X_n-iY_n)^2}{R_n^5} + S^-I_n^-\frac{(X_n+iY_n)^2}{R_n^5}\right]\right\}, \quad (12)$$

where $g_e\beta S$ and $g_p\beta_N I$ are the electron and proton magnetic moments and the coordinates are for the relative distance from the n-th proton to the electron spin centered on the oxygen.

As we proceed to evaluate this expression for the oxygen–methane interaction, let us consider the oxygen transitions which can occur in the process.

C. Spin-allowed transition for oxygen

The allowed oxygen-induced methane transitions are those nonvanishing matrix elements

$$\langle\psi_i|\mathcal{H}(O_2-CH_4)|\psi_j\rangle \neq 0. \quad (13)$$

The wavefunctions for the system are a product of oxygen and methane wavefunctions

$$\psi_{\rm system} = \psi(O_2)\psi_{\rm rot}\psi_{\rm spin}(CH_4). \quad (14)$$

The appropriate methane wavefunctions have been described above. The oxygen wavefunctions are those obtained from diagonalizing the Hamiltonian for oxygen as discussed in Sec. II. We now consider only free rotation and give in Table IV the results for $K=1$, $J=0$, 1. Note that the complete function is designated by $|Jm_J\rangle$ where the terms on the right are $|m_K m_S\rangle$.

Only the spin part of the interaction Hamiltonian (Eq. 12) will affect the oxygen wavefunction since we have assumed oxygen is fixed relative to the rotating methane. We seek the oxygen "spin-allowed" transitions which are permitted by the $S(O_2)$ operator, where S operates only on the spin projection m_S. The S_z, S^+, and S^- operators yield

$$S_z|m_K m_S\rangle = m_S|m_K m_S\rangle, \quad (15a)$$

TABLE IV. Oxygen wavefunctions for the $K=1$ rotational level with $J=0$ and 1.

$\|J\, m_J\rangle$	$\|m_K m_S\rangle$
$\|00\rangle = 3^{-1/2}(-\|1\,-1\rangle + \|00\rangle - \|-11\rangle)$	
$\|11\rangle = 2^{-1/2}(\|10\rangle - \|01\rangle)$	
$\|10\rangle = 2^{-1/2}(\|1\,-1\rangle - \|-11\rangle)$	
$\|1\,-1\rangle = 2^{-1/2}(\|0\,-1\rangle - \|-10\rangle)$	

$$S^+|m_K m_S\rangle = [S(S+1) - m_S(m_S+1)]^{1/2}|m_K m_S+1\rangle \;, \tag{15b}$$

$$S^-|m_K m_S+1\rangle = [S(S+1) - m_S(m_S+1)]^{1/2}|m_K m_S\rangle \;. \tag{15c}$$

Thus we find nonvanishing matrix elements for $S^\pm$ if $\Delta m_S = \pm 1$ and for S_z if $\Delta m_S = 0$ but $m_S \neq 0$. On substitution of the wavefuctions given in Table III, we find that the following transitions are allowed between the $|00\rangle$ and $|Jm_J\rangle$ states:

$$\langle 10|S_z|00\rangle = (2/3)^{1/2}\;, \quad \Delta m_S = 0\;, \tag{16a}$$

$$\langle 11|S^+|00\rangle = -2/3^{1/2}\;, \quad \Delta m_S = 1\;, \tag{16b}$$

$$\langle 1-1|S^-|00\rangle = 2/3^{1/2}\;, \quad \Delta m_S = -1\;. \tag{16c}$$

Results for some states with higher J are given elsewhere.[5] In general the selection rule for the states of O_2 is $\Delta J = \pm 1$, 0 but with $J = 0$ to $J = 0$ forbidden.

For a state with $J = 0$ the total angular momentum, spin + rotation, add to zero and would seem "spherical." However, the oxygen molecule's coupling with a methane proton is essentially via its electron-spin magnetic dipole which is much stronger than its rotation-induced dipole. The individual components of the $J = 0$ state, spin and rotation, are not zero, and hence this "weighted" interaction provides a mechanism for catalyzing conversion. Thus the $J = 0$ ground state may participate in a dipole–dipole interaction inducing methane spin conversions but requires a simultaneous transition to a $J = 1$ level.

D. Spin-rotation transitions for methane

Our basic assumption that the oxygen molecule was stationary for the time of an oxygen–methane transition allowed us to consider in detail just the spin factor for O_2. This is not possible for methane; rather we must consider in detail both the spin and rotation functions. An appropriate coordinate system is adopted, consistent with Fig. 4, and the corresponding transformations are made for the coordinates of each of the four protons. The details of this lengthy procedure will not be given here but are outlined in the Appendix and described in greater detail elsewhere.[5] The results are presented in Table V for the allowed transitions between $J = 0$ and $J = 1$ levels of methane.

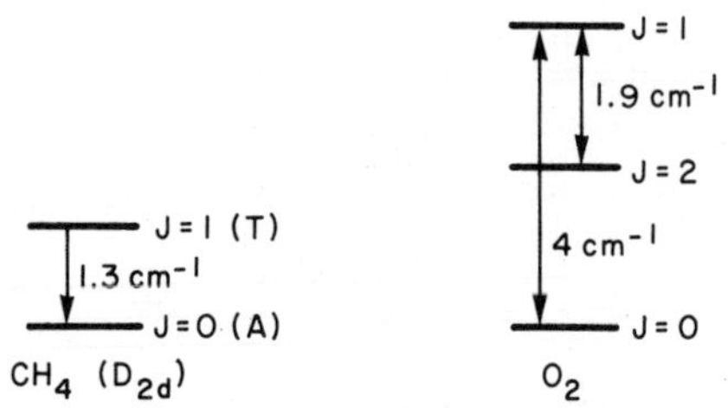

FIG. 5. Energy levels for molecules in solid methane with allowed spin conversion processes.

TABLE V. Oxygen-induced transitions for methane between the $J = 0$ and $J = 1$ levels.

Initial state $J=0$	Final state $J=1$		Transition matrix element
m_I	m_J	m_I	
± 2	0	± 1	$\pm A\langle S^\pm$
± 2	± 1	± 1	$2^{1/2}A\langle S_z\rangle$
± 1	0	± 1	$2A\langle S_z\rangle$
± 1	0	0	$\pm 2^{-1/2}A\langle S^\pm\rangle$
± 1	± 1	± 1	$\mp 2^{-1/2}A\langle S^\mp\rangle$
± 1	± 1	0	$A\langle S_z\rangle$
± 1	∓ 1	± 1	$\pm 2^{-1/2}A\langle S^\pm\rangle$
0	0	0	$(4/3^{1/2})A\langle S_z\rangle$
0	0	± 1	$\mp 6^{-1/2}A\langle S^\mp\rangle$
0	± 1	0	$\mp(2/3)^{1/2}A\langle S^\mp\rangle$
0	± 1	∓ 1	$3^{-1/2}A\langle S_z\rangle$

In Table V the factor arising from the oxygen is given as $\langle S_z\rangle$, $\langle S^+\rangle$, or $\langle S^-\rangle$ which are abbreviations for the expressions in Eq. (16a, b, c). The factor A arises from the methane functions and is

$$A = 6^{1/2} g_e \beta g_p \beta_N r/R^4 \;, \tag{17}$$

where r is the C–H distance in methane, 1.1 Å, and R is the distance from the methane to the oxygen molecule.

Our treatment is incomplete in that we have ignored the phonon absorption or emission and the oscillations in R and possibly other coordinates which would be involved. The phonon energies required for these low-level transitions are well within the elastic or acoustic band for solid methane[24]; hence, the transitions will remain allowed, but the quantitative transition probabilities will be affected. Also it is necessary to consider the phonon population at a given temperature when phonons are absorbed.

IV. CONCLUSIONS AND DISCUSSION

We are now in a position to consider the actual spin-species transformation in oxygen-doped solid methane at low temperature. The structure of methane below 20.4 °K was described in the Introduction. We consider first the methane molecules on D_{2d} sites as the temperature is lowered near 1 °K. The important features for CH_4 and O_2 are shown in Fig. 5; a phonon must also be absorbed or emitted to conserve energy. From Table II it is seen that there are appreciable populations of all of the oxygen states arising from $K = 1$ at 1 °K. Thus the methane transition from T to A spin-species (J from 1 to 0) can arise from several different oxygen transitions with appropriate phonon absorption or emission.

As the temperature drops even lower, at about 0.3 °K, however, this process would become much slower. Now the oxygen molecules are almost entirely in the $J = 0$ state with only 0.02% in the $J = 2$ states at 2.1 cm^{-1} and only one part in 10^8 in the $J = 1$ states at 4 cm^{-1}.

Phonons of these energies are comparably rare. Thus processes starting with oxygen in the $J=1$ state will become very slow. Processes starting with oxygen in the $J=2$ state will be slowed less, but the population in this state is only 0.02%. Processes starting with oxygen in the $J=0$ state require absorption of a phonon of energy $4-1.3=2.7$ cm^{-1} and the population of such phonons will be small.

While our calculations allow only qualitative conclusions, it is apparent that all of the conversion processes considered above will become markedly slower as the temperature decreases below 0.3 °K. Thus it seems likely that on a laboratory time scale the spin-species population of methane would be frozen at a level in the range 0.1 to 0.3 °K even if cooled further in the presence of oxygen. If conversion continued in this temperature range it would have to be by a mechanism different than the one here considered. Above about 0.3 °K, however, a reasonable conversion rate is expected as explained above.

For the molecules on O_h sites the diagram of Fig. 5 is modified by increasing the energy change for methane to about 9 cm^{-1} instead of 1.3. The thermal excitation of these molecules in O_h sites occurs at higher temperatures where the necessary phonon energy will be available. Hence, oxygen catalysis should be effective for the conversion of these molecules.

Our results should be useful in considering other properties of oxygen-doped methane at low temperatures but we will not carry the discussion further.

ACKNOWLEDGMENT

This research was sponsored by the Energy Research and Development Administration.

APPENDIX

Spin-rotation transitions in methane

Since the relative distances change as methane rotates and must be taken into account in the evaluation of the transition matrix elements, it is more convenient to express the relative coordinates (X_n, Y_n, Z_n) in terms of the center of mass coordinates $(0, 0, R)$ and the internal coordinates (x_n, y_n, z_n) which describe CH_4 rotation. For simplicity we have assumed the center of the CH_4 molecule remains a fixed distance R apart along the z axis above the oxygen molecule.

If a proton initially positioned at (x_{n0}, y_{n0}, z_{n0}) at Euler angle (0, 0, 0) [Fig. 4] is rotated through the Euler angles (α, β, γ), the final position of the n-th proton (x_n, y_n, z_n) can be related to its initial coordinates via the rotational transformation matrices which describe the rotation through the angles (α, β, γ). Using Eqs. (7) and (8) we find that the relative coordinates $(X_n \pm iY_n)$ and Z_n may be expressed in terms of the initial proton coordinates and the methane rotational wavefunctions

$$(X_n \pm iY_n) = \mp \frac{4\pi}{\sqrt{3}} (x_{n0}\psi_{x,\pm1} + y_{n0}\psi_{y,\pm1} + z_{n0}\psi_{z,\pm1}) , \quad \text{(A1)}$$

$$Z_n = R + \frac{2\pi\sqrt{2}}{\sqrt{3}} (x_{n0}\psi_{x,0} + y_{n0}\psi_{y,0} + z_{n0}\psi_{z,0}) . \quad \text{(A2)}$$

In addition we assume as a first approximation $R_n = R$ where it appears in the denominator in Eq. (12).

To evaluate the allowed $A \leftrightarrow T$ spin transition matrix elements, let us look at the allowed transition from a $J=0$ $|A2\rangle$ spin state of CH_4 due to the Hamiltonian term in Eqs. (11), (12)

$$-g_e\beta g_p\beta_N \sum_{n=1}^{4} \left(\frac{R_n^2 - 3Z_n^2}{R_n^5}\right) [S_z I_{z_n} - 1/4(S^- I_n^+ + S^+ I_n^-)] . \quad \text{(A3)}$$

Only the I_n^- term allows $|A2\rangle \to T$ transitions. Using the above expression for Z_n (Eq. A2)

$$Z_n^2 = R^2 + \frac{4\pi\sqrt{2}}{\sqrt{3}} (x_{n0}\psi_{x,0} + y_{n0}\psi_{y,0} + z_{n0}\psi_{z,0}) + \frac{2\pi\sqrt{2}^2}{\sqrt{3}} (x_{n0}\psi_{x,0} + y_{n0}\psi_{y,0} + z_{n0}\psi_{z,0})^2 . \quad \text{(A4)}$$

When this expression is substituted in Eq. (A3) the R^2 term will allow only $A \leftrightarrow A$ transitions. In fact, only the second term will allow a $J=0 \to J=1$ transition. This may be seen as follows:

Recall that the $J=1$ product functions are of the form given in Eq. (10). In evaluating the matrix element

$$\langle\psi_{rot}(J=1)\psi_{spin}(Tm_I)|\sum_n (x_{n0}\psi_{x,0} + y_{n0}\psi_{y,0} + z_{n0}\psi_{z,0}) \times I_n^- |\psi_{rot}(J=0)\psi_{spin}(A2)\rangle ,$$

as we perform the sum, the spin term $\langle Tm_I|I_n^-|A2\rangle$ requires that the resultant $m_I = 1$. The orthogonality of the rotational wavefunctions requires $m_J = 0$. The value of the transition matrix element is $\langle\psi'_{O_2}|S^+|\psi_{O_2}\rangle(A)$, where A was given in Eq. (17).

The third term in the expansion of Z_n^2 will not contribute to a $J=1$ transition. If one expands the third term in Eq. (A4) we find no $A \leftrightarrow T$ contributions are allowed for the $\sum_n x_{n0}^2 I_n^-$ terms and similarly for the y_{n0}^2 and z_{n0}^2 terms. Using the definitions in Eqs. (7), (9) the cross terms $(x_{n0}y_{n0}, x_{n0}z_{n0}, y_{n0}z_{n0})$ yield integrals of the product of three D functions. Using various properties of the 3-j symbols[23] the cross terms are also seen to vanish.

The other allowed $A \to T$ transitions (Table V) are evaluated in similar fashion. In fact all of the spin allowed transitions $\langle\psi_{spin}(T)|I_n|\psi_{spin}(A)\rangle$ follow a similar pattern, and any of the allowed methane transitions from a $J=0$ to a $J=1$ rotational level can be readily described in terms of one of three different categories.

[1]A. Farkas, "Orthohydrogen, Parahydrogen, and Heavy Hydrogen" (Cambridge Univeristy, London, 1935), especially Chapter IV.

[2]See for example (a) to (g) below and references cited therein: (a) H. P. Hopkins, Jr., P. L. Donoho, and K. S. Ptizer, J. Chem. Phys. 47, 864 (1967). (b) K. P. Wong, J. D. Noble, M. Bloom, and S. Alexander, J. Magn. Reson. 1, 55 (1969). (c) G. A. de Wit and M. Bloom, Can. J. Phys. 47, 1195 (1969). (d) P. van Hecke, P. Grobet, and L. van Gerven, J. Magn. Reson. 7, 117 (1972). (e) H. Glättli, A. Sentz, and

M. Eisenkremer, Phys. Rev. Lett. **28**, 871 (1972). (f) G. J. Vogt and K. S. Pitzer, J. Chem. Phys. **63**, 3667 (1975). (g) G. J. Vogt and K. S. Pitzer, J. Chem. Thermodyn. (in press). (h) R. F. Code and J. Higinbotham, Can. J. Phys. **54**, 1248 (1976).

[3]K. Tomita, Phys. Rev. **89**, 429 (1953).

[4](a) H. M. James and T. A. Keenan, J. Chem. Phys. **31**, 12 (1959). (b) Y. Kataoka, K. Okada, and T. Yamamoto, Chem. Phys. Lett. **19**, 365 (1973) and references cited therein.

[5]J. J. Kim, Ph.D. thesis (University of California, 1975); also Lawrence Berkeley Laboratory Report LBL 3955.

[6]G. Herzberg, "Molecular Spectra and Molecular Structure. I. Spectra of Diatomic Molecules," 2nd ed. (Van Nostrand Reinhold, New York, 1950).

[7]H. A. Kramers, Z. Phys. **53**, 422 (1929).

[8]A. Carrington and A. D. McLachlan, "Introduction to Magnetic Resonance," (Harper & Row, New York, 1967), Chaps. 3 and 8.

[9]R. Schlapp, Phys. Rev. **51**, 342 (1937).

[10]J. S. Griffith, "The Theory of Transition Metal Ions" (Cambridge University, London, 1961). Especially Appendix 2.

[11]A. F. Devonshire, Proc. Roy. Soc. Ser. A **153**, 601 (1936).

[12]M. Rotenberg, R. Bivins, N. Metropulis, J. K. Wooten, Jr., "The 3-j and 6-j Symbols," (The Technology Press, Massachusetts Institute of Technology, Cambridge MASS, 1959).

[13]J. C. Burford and G. M. Graham, J. Chem. Phys. **49**, 763 (1968).

[14]R. Simoneau, J. S. M. Harvey, and G. M. Graham, J. Chem. Phys. **54**, 4819 (1971).

[15]N. Bloembergen, Physica (Utrecht) 15, 386 (1949).

[16]See, for example, (a) A review by G. R. Khutsuhvili, in "Progress in Low-Temperature Physics," edited by C. J. Gorter (North-Holland, Amsterdam, 1970), Vol. 6, p. 375. (b) J. P. Wolfe, Phys. Rev. Lett. **31**, 907 (1973), and references cited therein.

[17]K. Motizuki and T. Nagamiya, J. Phys. Soc. Jpn, **11**, 93 (1956).

[18]E. B. Wilson, J. Chem. Phys. **3**, 276 (1935).

[19]R. F. Curl, Jr., J. V. V. Kasper, and K. S. Pitzer, J. Chem. Phys. **46**, 3220 (1967).

[20]P. Yi, I. Ozier, and C. H. Anderson, Phys. Rev. **165**, 92 (1968).

[21]T. Yamamoto and Y. Kataoka, J. Chem. Phys. **48**, 3199 (1968).

[22]E. P. Wigner, "Group Theory and its Application to the Quantum Mechanics of Atomic Spectra," Translated by J. J. Griffin (Academic, New York, 1959), Especially Chapters 15 and 19.

[23]A. R. Edmonds, "Angular Momentum in Quantum Mechanics," (Princeton University, Princeton, N. J., 1957). Note that Wigner and Edmonds definitions of D are complex conjugates of one another:

$$D^{(J)}_{KM}\,(\text{Wigner}) = \langle JM \mid D \mid JK \rangle$$

$$D^{(J)}_{KM}\,(\text{Edmonds}) = \langle JK \mid D \mid JM \rangle .$$

[24](a) B. Dorner and H. Stiller, Phys. Status. Solidi. **5**, 511 (1964). (b) H. Stiller and S. Hautecler, "Inelastic Scattering of Neutrons in Solids and Liquids," Vol. 2, Vienna 1963 (p. 281).

C. Semiempirical Equations for Aqueous Electrolytes

Introduction

Although the behavior of electrolytes in the very dilute range has been known since the work of Debye and Hückel in 1923, equations for higher concentration had been either purely empirical or not accurate enough to represent measured data within experimental accuracy. Sufficient information had been obtained by 1970, primarily from Monte Carlo simulations of a simple ionic model, to indicate the character of terms needed in addition to that of Debye and Hückel. This made it possible to design an equation combining the theoretical Debye-Hückel term with a smaller number of higher-order terms of theoretically guided form but containing empirical parameters. The equation is valid for mixed as well as pure electrolytes.

The first seven papers give the original development of the equation and an initial array of applications. Others have adopted this equation for many application. The final paper in this sequence is a geologically oriented application to mineral solubility at high temperatures. This paper also contains a useful generalization of the contribution of each H_2O in hydrated solid salts to the heat capacity and the entropy.

The current status of work in this area is summarized in several chapters of a recent book which I edited (Activity Coefficients in Electrolyte Solutions, 2nd Edition, CRC Press, Boca Raton, Florida, 1991). Chapter 6 by S. L. Clegg and M. Whitfield gives a comprehensive report on various applications to natural waters.

[Reprinted from the Journal of Physical Chemistry, 77, 268 (1973).]

Thermodynamics of Electrolytes. I. Theoretical Basis and General Equations

Kenneth S. Pitzer

Inorganic Materials Research Division of the Lawrence Berkeley Laboratory and Department of Chemistry, University of California, Berkeley, California 94720 *(Received May 12, 1972)*

Publication costs assisted by The U.S. Atomic Energy Commission

A system of equations for the thermodynamic properties of electrolytes is developed on the basis of theoretical insights from improved analysis of the Debye–Hückel model as well as recently published numerical calculations for more realistic models. The most important result is the recognition of an ionic strength dependence of the effect of short-range forces in binary interactions. By modifying the usual second virial coefficients to include this feature, one obtains a system of equations which are only slightly more complex than those of Guggenheim but yield agreement within experimental error to concentrations of several molal instead of 0.1 M. If one compares instead with the recently proposed equations of Scatchard, Rush, and Johnson, the present equations are very much simpler for mixed electrolytes (and somewhat simpler for single electrolytes) yet appear to yield comparable agreement with experimental results for both single electrolytes and mixtures.

The thermodynamic properties of aqueous electrolytes have been extensively investigated both experimentally and theoretically. The monographs of Harned and Owen[1] and of Robinson and Stokes[2] provide excellent summaries. While the detailed nature of these solutions is so complex that an *ab initio* quantum-statistical theory is not feasible, the data appear to relate to few enough independent parameters to make relatively exact semiempirical representation possible. It is the present objective to develop equations which reproduce the measured properties substantially within experimental accuracy, which are compact and convenient in that only a very few parameters need be tabulated for each substance and the mathematical calculations are simple, which have appropriate form for mixed electrolytes as well as for solutions of a single solute, and whose parameters have physical meaning as far as possible.

In 1960 Brewer and the writer[3] selected as the best available system one proposed and applied to dilute solutions by Guggenheim[4] with modifications suggested by Scatchard[5] for concentrated solutions. While this system was useful in providing a simple and compact summary of experimental data, it did not fully satisfy the other desired qualities. Recent theoretical advances of Friedman and collaborators[6,7] and of Card and Valleau[8] provide important insights and support for the greatly improved semiempirical treatment proposed below. Indeed it is interesting that a key idea can be obtained by introducing the Debye–Hückel model and distribution function into modern equations relating such functions to the osmotic pressure. But first we review the Guggenheim equations.

(1) H. S. Harned and B. B. Owen, "The Physical Chemistry of Electrolytic Solutions," Reinhold, New York, N. Y., 1950.
(2) R. A. Robinson and R. H. Stokes, "Electrolytic Solutions," Butterworths, London, 1959.
(3) K. S. Pitzer and L. Brewer, revised edition of "Thermodynamics" by G. N. Lewis and M. Randall, McGraw-Hill, New York, N. Y., 1961.
(4) (a) E. A. Guggenheim, *Phil. Mag.*, [7] **19**, 588 (1935); (b) E. A. Guggenheim and J. C. Turgeon, *Trans. Faraday Soc.*, **51**, 747 (1955); (c) E. A. Guggenheim, "Applications of Statistical Mechanics," Clarendon Press, Oxford, 1966, pp 166–170; here Guggenheim surprisingly abandons the limitation to 0.1 M for validity of his equations for the difference in φ from a reference salt. As is obvious from Figure 1 or the figures and tables of ref 4, this is a crude approximation.
(5) G. Scatchard, *Chem. Rev.*, **19**, 309 (1939); also in W. J. Hamer, Ed., "The Structure of Electrolytic Solutions," Wiley, New York, N. Y., 1959, p 9.
(6) J. C. Rasaiah and H. L. Friedman, *J. Chem. Phys.*, **48**, 7242 (1968); **50**, 3965 (1969).
(7) P. S. Ramanathan and H. L. Friedman, *J. Chem. Phys.*, **54**, 1086 (1971).
(8) (a) D. N. Card and J. P. Valleau, *J. Chem. Phys.*, **52**, 6232 (1970); (b) J. C. Rasaiah, D. N. Card, and J. P. Valleau, *J. Chem. Phys.*, **56**, 248 (1972).

Guggenheim–Scatchard Equations

Since the Guggenheim[4] equations have had considerable success, it is desirable to recall them at this point and to discuss the aspects that need improvement. The equations for osmotic and activity coefficients are

$$\ln \gamma_{M'X'} = -\frac{A_\gamma |z_M z_X| I^{1/2}}{1 + I^{1/2}} + \frac{2\nu_+}{\nu_+ + \nu_-}\sum_X \beta_{M'X} m_X + \frac{2\nu_-}{\nu_+ + \nu_-}\sum_M \beta_{MX'} m_M \quad (1)$$

$$\varphi - 1 = -\frac{2A_\gamma I^{3/2}\sigma(I^{1/2})}{3\left(\sum_M m_M + \sum_X m_X\right)} + \frac{\sum_M\sum_X \beta_{MX} m_M m_X}{\sum_M m_M + \sum_X m_X} \quad (2)$$

$$\sigma(x) = (3/x^3)\{1 + x - [1/(1+x)] - 2\ln(1+x)\} \quad (3)$$

$$A_\gamma = (2\pi N_0 d_w/1000)^{1/2} l^{3/2} \quad (4)$$

$$l = e^2/DkT \quad (5)$$

Here the sums over M and X cover all positive and negative ions, respectively; A_γ is the usual Debye–Hückel coefficient with N_0 Avogadro's number and d_w the density of the solvent; and l is a useful parameter which expresses the distance at which the electrostatic energy for singly charged ions in the dielectric just equals thermal energy.

The quantities β_{MX} are constants (at given T and D) analogous to second virial coefficients which represent the net effect of various short-range forces between the M and X ions. It should be noted that Guggenheim followed Brønsted's[9] principle of specific interaction and excluded terms related to short-range forces between ions of like sign.

We shall not distinguish between direct interactions of solute species at short distances and changes in solvation with concentration since both phenomena influence the effective interionic potential of average force and thereby the second and higher virial coefficients. Likewise the distinction between molality and concentration will be ignored since the statistical calculations based upon molecular models are used only to suggest appropriate forms for empirical use and to provide qualitative understanding. The advantage of the molality in practical calculations, especially if the temperature varies, is overwhelming.

It is also useful to consider the total excess Gibbs energy (for a solution containing n_w kg of solvent) which is obtained by appropriate integration of the activity or osmotic coefficient.

$$\frac{G^{ex}}{n_w RT} = -\frac{4}{3}A_\gamma I^{3/2}\tau(I^{1/2}) + 2\sum_M\sum_X \beta_{MX} m_M m_X \quad (6)$$

$$\tau(x) = (3/x^3)[\ln(1+x) - x + (x^2/2)] \quad (7)$$

For a single 1–1 electrolyte eq 1 and 2 reduce to

$$\ln \gamma = -[A_\gamma m^{1/2}/(1 + m^{1/2})] + 2\beta_{MX} m \quad (8)$$

$$\varphi - 1 = -(A_\gamma m^{1/2}/3)\,\sigma(m^{1/2}) + \beta_{MX} m \quad (9)$$

Guggenheim and Turgeon[4b] showed that eq 8 and 9 fitted, essentially within experimental error, the data for 1–1 electrolytes in water at 0° and at room temperature at concentrations up to 0.1 M. They acknowledge, however, and others have shown that substantial discrepancies arise at higher concentrations. In seeking equations that can be used at higher concentration one may examine separately

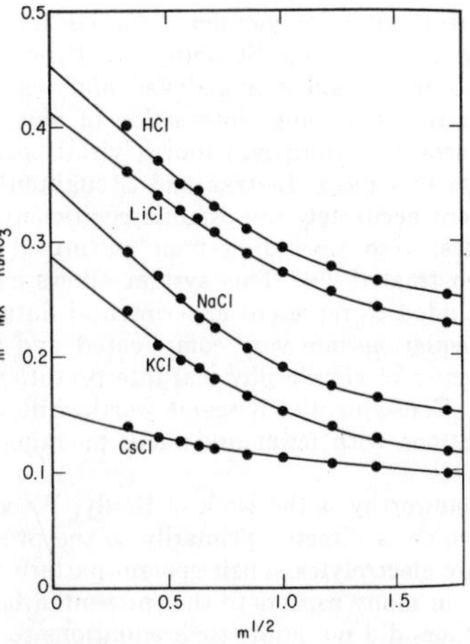

Figure 1. Difference in osmotic coefficient of several electrolytes from that of $RbNO_3$.

the first and second terms on the right side. The first term in eq 6 must remain a general function of ionic strength if these equations are to maintain their simplicity and utility for mixed electrolyte solutions. We shall return later to the question of an improved mathematical form of this first term. Next we note that, by taking differences between the properties of different electrolytes of the same type and at the same concentration, the first term cancels and we have (for 1–1 electrolytes)

$$\ln \gamma_{M''X''} - \ln \gamma_{M'X'} = 2m(\beta_{M''X''} - \beta_{M'X'}) \quad (10)$$

$$\varphi_{M''X''} - \varphi_{M'X'} = m(\beta_{M''X''} - \beta_{M'X'}) \quad (11)$$

These equations (10 and 11) still do not fit the experimental properties above 0.1 M with constant β's.

Following a suggestion of Scatchard,[5] Brewer and the writer[3] considered β to vary slowly with concentration and produced thereby a very compact tabulation of the experimental data for pure electrolyte solutions. If the β's are no longer constants, however, the relationships between eq 1, 2, and 6 and between 8 and 9 are no longer valid. We shall return to the derivation of correct relationships corresponding to the variable β's in a later section but here point out in Figure 1 the qualitative nature of the dependence of β on concentration. The notable (and initially surprising) feature is the substantial change in β at low concentration in contrast to its relative constancy at higher concentration. While the change in β below 0.1 M is not large, there is no reason to doubt that the curves maintain their slope in that region. Thus an improved theory which adequately accounts for these effects at higher concentration will presumably also represent an improvement below 0.1 M.

Recently Scatchard[10] and coworkers[11] have extended and elaborated the Guggenheim equations in several ways. First the Debye–Hückel term in eq 6 is subdivided into a series of terms with different coefficients of $I^{1/2}$ in τ corresponding to different distances of closest approach for the solute components. Appropriate derivatives then

(9) J. N. Brønsted, *Kgl. Dan. Vidensk. Selsk., Mat. Fys. Medd.*, **4**, (4) (1921); *J. Amer. Chem. Soc.*, **44**, 877 (1922); **45**, 2898 (1923).
(10) G. Scatchard, *J. Amer. Chem. Soc.*, **83**, 2636 (1961).
(11) G. Scatchard, R. M. Rush, and J. S. Johnson, *J. Phys. Chem.*, **74**, 3786 (1970).

yield correct but very complex formulas for the osmotic and activity coefficients. Secondly, the Brønsted principle of specific interaction is abandoned and terms are introduced for the short-range interaction of ions of like sign. Finally, arrays of third and fourth virial coefficients are added. On this basis, Lietzke and Stoughton[12] were able to represent accurately the osmotic coefficients of 20 pure electrolytes; also several systems of mixed electrolytes have been treated.[10,12] This system allows enough terms to be included to represent experimental data accurately; but the equations are very complicated and there seems little promise of simple physical interpretation of the parameters. Consequently, it seems worthwhile to seek simpler equations with fewer and more meaningful parameters.

Also noteworthy is the work of Reilly, Wood, and Robinson[13] which is directed primarily to the process of mixing of pure electrolytes. Their general pattern of equations is similar in many aspects to that presented hereafter, but these authors did not apply their equations to the numerical expression of the properties of single electrolytes and thus did not deal with one of the principal subjects of this paper.

Hard-Core Effects in Debye–Hückel Theory

Traditional Debye–Hückel theory of electrolyte solutions[1,2] recognizes the distance of closest approach a in the calculation of the electrostatic energy of the distribution of ions but ignores the kinetic effect of the hard core on the osmotic pressure or other properties. As Kirkwood,[14] among others, noted, this hard core effect cannot be treated rigorously by the traditional charging process methods of calculating free energy. Van Rysselberghe and Eisenberg[15] did add approximate hard-core terms to the traditional Debye–Hückel formulas, but the feature of particular interest was lost in that approximation.

Recently it has been shown[6,16] that there are several equations which relate the intermolecular potential and the radial distribution function (also called the pair correlation function) to thermodynamic properties. Each equation will give the same set of thermodynamic results if the distribution function is exact, but different results arise from approximate distribution functions. The most convenient equation for a hard-core potential is the so-called "pressure" equation which yields the pressure of a pure fluid or the osmotic pressure of a solution.

$$\Pi - ckT = -1/6\sum_i\sum_j c_i c_j \int_0^\infty (\partial u_{ij}/\partial r) g_{ij} 4\pi r^3 \mathrm{d}r \quad (12)$$

Where Π is the osmotic pressure, c_i and c_j, are concentrations of species i and j, c is the total solute concentration Σc_i, u_{ij} is the intermolecular potential, and g_{ij} is the radial distribution function. The sums cover all solute species.

We introduce the potential

$$u_{ij} = \infty,\ r < a \quad (13a)$$

$$u_{ij} = z_i z_j e^2/Dr,\ r \geq a \quad (13b)$$

and obtain

$$\Pi - ckT = (e^2/6D)\sum_i\sum_j c_i c_j z_i z_j \int_a^\infty g_{ij}(r) 4\pi r \mathrm{d}r + 2/3(\pi a^3 kT)\sum_i\sum_j c_i c_j g_{ij}(a) \quad (14)$$

Where the second term, involving $g(a)$, the radial distribution function at the outer surface of contact, constitutes the kinetic effect of the hard core. Note that we have assumed the same core size a for all species of ions but different charges z_i and z_j in accordance with Debye–Hückel theory. The usual Debye–Hückel distance $1/\kappa$ is defined

$$\kappa^2 = (4\pi e^2/DkT)\,\Sigma_i z_i^2 c_i \quad (15)$$

and the function

$$q_{ij}(r) = [z_i z_j e^2/DkT(1 + \kappa a)](e^{-\kappa(r-a)}/r) \quad (16)$$

In the Debye–Hückel derivation the radial distribution of ions around a given ion is obtained from Boltzmann's law with the interaction energy given by the product of the charge on one ion and the average electrical potential surrounding the other ion. Summation over the various ions yields a charge density which, after approximation by its linear term, is substituted into the Poisson equation. The solution of the Poisson equation yields an explicit result for the electrical potential. The quantity $q_{ij}(r)$ defined in eq 16 is the resulting exponent in the Boltzmann factor for the radial distribution function, which is $g_{ij}(r) = \exp[-q_{ij}(r)]$.

The charge density around a given ion i now becomes

$$\rho_i = \Sigma c_j z_j e \exp[-q_{ij}(r)] = \Sigma_j c_j z_j e - \Sigma_j c_j z_j e q_{ij}(r) + \tfrac{1}{2}\Sigma_j c_j z_j e q_{ij}^2(r) + \ldots$$

In the expanded expression the first term vanishes because the solution is electrically neutral. Since q_{ij} contains the factor z_j, the second or linear term is a sum over $c_j z_j^2$ and does not vanish; indeed, it yields the charge distribution in Debye–Hückel theory. The third term involves a sum over $c_j z_j^3$ which vanishes for symmetrical electrolytes and is relatively small in other cases. Thus for symmetrical electrolytes there is no inconsistency with the linearized Poisson–Boltzmann solution in retaining the third term, but it has no effect on the charge density and therefore no effect on the conventional Debye–Hückel results for thermodynamic properties.

Now, however, we wish to include the kinetic effect of the hard core by the use of eq 14 which involves the sum over $c_i c_j g_{ij}(a)$. In discussion of this sum we again consider the expanded expression for the radial distribution function

$$g_{ij}(r) = 1 - q_{ij}(r) + \tfrac{1}{2}q_{ij}^2(r) \ldots \quad (17)$$

for $r \geq a$; $g = 0$ for $r < a$, of course. In this case the sum for the first term is finite while that for the second term is zero. But for the hard-core effect the third term again yields a finite contribution. Thus for symmetrical electrolytes this third term can be included without any inconsistency with the Debye–Hückel derivation, and its inclusion should yield an improved approximation for the hard-core effect. Even for unsymmetrical electrolytes the effect of the third term on the charge density is small; hence the "inconsistency" of its retention is small and it may be desirable to include its contribution to the hard core-effect.

(12) M. H. Lietzke and R. W. Stoughton, *J. Phys. Chem.*, **66**, 508 (1962).
(13) (a) P. J. Reilly and R. H. Wood, *J. Phys. Chem.*, **73**, 4292 (1969); (b) P. J. Reilly, R. H. Wood, and R. A. Robinson, *ibid.*, **75**, 1305 (1971).
(14) J. C. Kirkwood, *Chem. Rev.*, **19**, 275 (1936); J. C. Kirkwood and J. C. Poirier, *J. Phys. Chem.*, **58**, 591 (1954).
(15) P. Van Rysselberghe and S. Eisenberg, *J. Amer. Chem. Soc.*, **61**, 3030 (1939); **62**, 451 (1940).
(16) T. L. Hill, "Statistical Mechanics," McGraw-Hill, New York, N. Y., 1956.

Since the theory is only approximate, however, there is no reason why the exponential expressions for g_{ij} should not be used, if desired, and Card and Valleau[8a] have found good agreement between such distribution functions and their Monte Carlo calculations for 1–1 electrolytes up to about 1 M.

For our purposes it is preferable to use the three-term expansion of eq 17 since this allows simpler mathematical expression of the results while still providing a good approximation whenever the general Debye–Hückel solution is valid. The osmotic coefficient is found, on this basis, to be

$$\varphi - 1 = \frac{\Pi}{ckT} - 1 = \frac{-\kappa^3}{24\pi c(1 + \kappa a)} + c\left[\frac{2\pi a^3}{3} + \left(\frac{1}{48\pi}\right)\left(\frac{\kappa^4 a}{c^2(1 + \kappa a)^2}\right)\right] \quad (18)$$

If all ions have charges $\pm z$ this reduces to

$$\varphi - 1 = \frac{-z^2 l \kappa}{6(1 + \kappa a)} + c\left[\frac{2\pi a^3}{3} + \frac{\pi a z^4 l^2}{3(1 + \kappa a)^2}\right] \quad (19)$$

Now $\kappa = (4\pi l)^{1/2} z c^{1/2}$ and $l = e^2/DkT$ as defined earlier. In eq 18 and 19 the first term on the right arises from the electrostatic energy and the second term from the hard core; each has interesting implications.

This result for the electrostatic effects differs from the traditional Debye–Hückel formula although it reduces to the same limiting law when κa can be neglected as compared to unity. Such a difference is not unexpected since the distribution function is approximate except in the limit of small κ. In Figure 2 we compare the alternate thermodynamic expressions based on the Debye–Hückel distribution function with the theoretically exact but numerically cumbersome Monte Carlo statistical treatment[8] of the same molecular model (a = 4.25 Å, aqueous solution at 25°). The new expression, eq 19, agrees, within computational uncertainty, with the Monte Carlo results for concentrations up to I = 0.5 M. Substitution of the conventional term for the electrostatic effects causes a serious discrepancy. It is also interesting that this theoretical curve of eq 19 (or the Monte Carlo results) fits the experimental data for HBr very well.

In view of the approximations of the molecular model, however, we shall regard both expressions for the thermodynamic effect of the charge distribution as candidates for adoption on the basis of success in fitting experimental data.

Of greater interest is the hard-core term in eq 18 and 19. The first term within the brackets, $2\pi a^3/3$, is independent of concentration and was considered by Van Rysselberghe and Eisenberg.[15] The second term within the brackets, however, does depend, through κ, on the ionic strength of the solution. Furthermore, this expression in brackets varies with concentration in just the same manner as was shown in Figure 1 with an initially rapid decrease from the value at zero concentration to a smaller and relatively constant value at high concentration.

This decrease in the second virial coefficient with ionic strength arises from a similar decrease in the average radial distribution function at contact. Thus it is important to note that the more rigorous statistical treatment[8] by the Monte Carlo method yields a similar behavior of the radial distribution function at contact. Furthermore, the calculations of Ramanathan and Friedman[7] on a model with a soft repulsive potential show a similar decrease

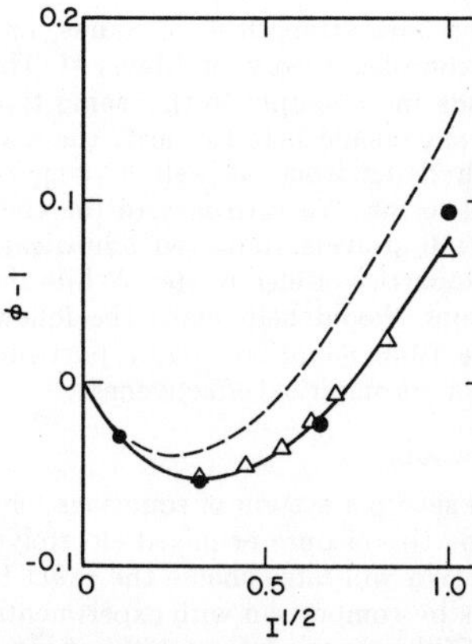

Figure 2. Osmotic coefficient for the Debye–Hückel model (a = 4.25 Å, aqueous solution at 25°): the solid circles are calculated values by the Monte Carlo method,[8a] the solid curve is from the present theory with eq 19, the dashed curve differs from the solid curve by substituting the conventional D–H expression for the electrostatic energy for the corresponding term in eq 19, and the triangles are experimental values for HBr.

with increase in ionic strength in the average radial distribution function at a comparable radius.

The Debye–Hückel distribution function was also inserted in the so-called "compressibility" equation of statistical mechanics.[6] The result is much more complex, mathematically, although similar in qualitative behavior. Hence it does not add anything useful for our present purposes.

While the Debye–Hückel treatment does not allow the consideration of different ionic radii within the same solution, we can compare the properties of different solutions, each of a single solute with a particular radius sum. We consider the difference in osmotic coefficient for symmetrical electrolytes with radius sums a_1 and a_2 and ion charge z.

$$\frac{\varphi_2 - \varphi_1}{c} = \frac{2\pi}{3}\left\{\frac{z^2 l^2 (a_2 - a_1)}{(1 + \kappa a_2)(1 + \kappa a_1)} + (a_2^3 - a_1^3) + \frac{z^4 l^2}{2}\left[\frac{(a_2 - a_1)(1 - \kappa^2 a_2 a_1)}{(1 + \kappa a_2)^2(1 + \kappa a_1)^2}\right]\right\} \quad (20)$$

Here the first term inside the braces arises from the electrostatic effects and the second and third terms from the hard-core effects (*i.e.*, the same order as in eq 19). The quantity $(\varphi_2 - \varphi_1)/c$ corresponds to the quantity $(\varphi_2 - \varphi_1)/m$ which was shown in Figure 1 and would be a constant if the Guggenheim equations were valid. It is apparent from eq 20 that it will yield the same qualitative behavior as was shown in Figure 1; specifically, the second term is constant, but both the first and third terms decrease from a finite value at zero concentration to very small values as κ increases.

The principal consequence of this section is to suggest that the properties of electrolyte solutions can be expressed by an "electrostatic" term plus a virial coefficient series in which the virial coefficients may be functions of the ionic strength of the solution. More particularly it is suggested that the second virial coefficient will vary with ionic strength in the pattern indicated by Figure 1 or eq 18 and 19. An alternate and probably improved form is also proposed for the "electrostatic" term.

The suggestion of a second virial coefficient which is a

function of the ionic strength is, of course, implicit in the much more complex theory of Mayer.[17] The preceding argument leads more simply to this same type of expression and thereby should help to clarify the reasons for this ionic strength dependence as well as suggesting simple mathematical forms. We turn now to the choice of equations which will provide improved empirical representation of the properties of electrolytes. While the discussion up to this point should help make the following choices plausible, the final adoption of any particular equation will be based on its empirical effectiveness.

General Equations

Let us now set up a system of equations for the thermodynamic properties of pure or mixed electrolytes in rather general form. We will later choose the exact form for certain functions by comparison with experimental data. The total excess Gibbs energy for a solution containing n_w kg of solvent and $n_i, n_j, \ldots$ moles of solute species $i, j, \ldots$ is taken to be

$$\frac{G^{ex}}{RT} = n_w f(I) + \frac{1}{n_w}\sum_{ij}\lambda_{ij}(I)n_in_j + \frac{1}{n_w^2}\sum_{i,j,k}\mu_{ijk}n_in_jn_k \tag{21}$$

Here $f(I)$ is a function of ionic strength (also temperature and solvent properties) expressing the effect of the long-range electrostatic forces; $f(I)$ may have the form of the first term of eq 6 or the similar form which can be obtained by integration of eq 18; $\lambda_{ij}(I)$ is a function of ionic strength with the qualitative behavior indicated by Figure 1 on the second term on the right in eq 18 or 19. The effect of short-range forces between species i and j is, of course, the basis for $\lambda_{ij}(I)$. We also include a term for triple ion interaction which may be significant at high concentration but ignore any dependence of μ_{ijk} on ionic strength. We assume that the λ and μ matrices are symmetric, *i.e.*, $\lambda_{ij} = \lambda_{ji}$, etc.

The equations for the activity and osmotic coefficients follow from the appropriate derivatives of G^{ex}

$$\varphi - 1 = -\frac{\partial G^{ex}/\partial n_w}{RT\sum_i m_i} = \frac{(If' - f) + \sum_{i,j}(\lambda_{ij} + I\lambda_{ij}')m_im_j + 2\sum_{i,j,k}\mu_{ijk}m_im_jm_k}{\sum_i m_i} \tag{22}$$

$$\ln\gamma_i = \frac{1}{RT}\frac{\partial G^{ex}}{\partial n_i} = \frac{z_i^2}{2}f' + 2\sum_j\lambda_{ij}m_j + \frac{z_i^2}{2}\sum_{j,k}\lambda_{jk}'m_jm_k + 3\sum_{j,k}\mu_{ijk}m_jm_k \tag{23}$$

where $f' = df/dI$, $\lambda_{ij}' = d\lambda_{ij}/dI$, and $m_i = n_i/n_w$, etc. Combination of terms for the neutral electrolyte $M(\upsilon_M)X(\upsilon_X)$ yields for its activity coefficient

$$\ln\gamma_{MX} = \frac{|z_Mz_X|}{2}f' + \frac{2\upsilon_M}{\upsilon_M + \upsilon_X}\sum_j\lambda_{Mj}m_j + \frac{2\upsilon_X}{\upsilon_M + \upsilon_X}\sum_j\lambda_{Xj}m_j + \frac{|z_Mz_X|}{2}\sum_{j,k}\lambda_{jk}'m_jm_k + \frac{3\upsilon_M}{\upsilon_M + \upsilon_X}\sum_{j,k}\mu_{Mjk}m_jm_k + \frac{3\upsilon_X}{\upsilon_M + \upsilon_X}\sum_{j,k}\mu_{Xjk}m_jm_k \tag{24}$$

In a solution of the single electrolyte MX these equations reduce to

$$\varphi - 1 = \frac{|z_Mz_X|}{2}\left(f' - \frac{f}{I}\right) + \frac{m}{\upsilon}\{2\upsilon_M\upsilon_X(\lambda_{MX} + I\lambda_{MX}') + \upsilon_M^2(\lambda_{MM} + I\lambda_{MN}') + \upsilon_X^2(\lambda_{XX} + I\lambda_{XX}')\} + \frac{6\upsilon_M\upsilon_Xm^2}{\upsilon}(\upsilon_M\mu_{MMX} + \upsilon_X\mu_{MXX}) \tag{25}$$

$$\ln\gamma = \frac{|z_Mz_X|}{2}f' + \frac{m}{\upsilon}\{2\upsilon_M\upsilon_X(2\lambda_{MX} + I\lambda_{MX}') + \upsilon_M^2(2\lambda_{MM} + I\lambda_{MM}') + \upsilon_X^2(2\lambda_{XX} + I\lambda_{XX}')\} + \frac{9\upsilon_M\upsilon_Xm^2}{\upsilon}(\upsilon_M\mu_{MMX} + \upsilon_X\mu_{MXX}) \tag{26}$$

where $\upsilon = \upsilon_M + \upsilon_X$. The terms for triple interactions with all ions of the same sign (μ_{MMM} and μ_{XXX}) have been dropped in eq 25 and 26 and hereafter since they seem certain to be exceedingly small.

From eq 25 and 26 it is apparent that the properties of single electrolytes are determined by the combinations $(2\upsilon_M\upsilon_X\lambda_{MX} + \upsilon_M^2\lambda_{MM} + \upsilon_X^2\lambda_{XX})$ and $(\upsilon_M\mu_{MMX} + \upsilon_X\mu_{MXX})$. A similar but much more cumbersome examination of the equations for mixed electrolytes indicates additional observable combinations of the λ and μ functions. The following definitions of the more directly observable quantities appear to be most convenient. Quantities appearing only for mixed electrolytes are distinguished by the use of greek instead of latin letters.

$$B_{MX}^{\varphi}(I) = \lambda_{MX} + I\lambda_{MX}' + (\upsilon_M/2\upsilon_X)(\lambda_{MM} + I\lambda_{MM}') + (\upsilon_X/2\upsilon_M)(\lambda_{XX} + I\lambda_{XX}') \tag{27}$$

$$B_{MX}^{\gamma}(I) = 2\lambda_{MX} + I\lambda_{MX}' + (\upsilon_M/2\upsilon_X)(2\lambda_{MM} + I\lambda_{MM}') + (\upsilon_X/2\upsilon_M)(2\lambda_{XX} + I\lambda_{XX}') \tag{28}$$

$$B_{MX}^{\gamma}(I) = B_{MX}^{\varphi}(I) + (1/I)\int_0^I B_{MX}^{\varphi}(x)dx \tag{29}$$

$$C_{MX}^{\varphi} = [3/(\upsilon_M\upsilon_X)^{1/2}][\upsilon_M\mu_{MMX} + \upsilon_X\mu_{MXX}] \tag{30}$$

$$C_{MX}^{\gamma} = \tfrac{3}{2}C_{MX}^{\varphi} \tag{31}$$

$$\theta_{MN} = \lambda_{MN} - (z_N/2z_M)\lambda_{MM} - (z_M/2z_N)\lambda_{NN} \tag{32}$$

$$\theta_{MN}' = d\theta_{MN}/dI = \lambda_{MN}' - (z_N/2z_M)\lambda_{MM}' - (z_M/2z_N)\lambda_{NN}' \tag{33}$$

$$\psi_{MNX} = 6\mu_{MNX} - (3z_N/z_M)\mu_{MMX} - (3z_M/z_N)\mu_{NNX} \tag{34}$$

Also we define

$$f^{\varphi} = \tfrac{1}{2}[f' - (f/I)] \tag{35}$$

$$f^{\gamma} = \tfrac{1}{2}f' \tag{36}$$

In these terms the properties of the single electrolyte take the simple form

$$\varphi - 1 = |z_Mz_X|f^{\varphi} + m(2\upsilon_M\upsilon_X/\upsilon)B_{MX}^{\varphi} + m^2[2(\upsilon_M\upsilon_X)^{3/2}/\upsilon]C_{MX}^{\varphi} \tag{37}$$

$$\ln\gamma = |z_Mz_X|f^{\gamma} + m(2\upsilon_M\upsilon_X/\upsilon)B_{MX}^{\gamma} + m^2[2(\upsilon_M\upsilon_X)^{3/2}/\upsilon]C_{MX}^{\gamma} \tag{38}$$

In the case of symmetrical electrolytes the coefficients of mB and m^2C reduce to unity; m is the usual stoichiometric molality.

(17) J. E. Mayer, *J. Chem. Phys.*, **18**, 1426 (1959), and subsequent work including E. Meeron, *J. Chem. Phys.*, **26**, 804 (1957).

The expressions for mixed electrolytes become somewhat cumbersome and we will restrict ourselves at this time to a few examples. Consider first a mixture of two ions of one sign, M and N, with a common ion X of opposite sign but with unrestricted magnitudes of charge z_M, z_N, and z_X. Electrical neutrality requires that

$$m_X = |z_M/z_X|m_M + |z_N/z_X|m_N$$

The osmotic coefficient for this solution is

$$\varphi - 1 = [2/(m_M + m_N + m_X)]\{If^\varphi + m_M m_X B_{MX}^\varphi + m_N m_X B_{NX}^\varphi + m_M n_N(\theta_{MN} + I\theta_{MN}') + |z_X/z_M|^{1/2} m_M m_X^2 C_{MX}^\varphi + |z_X/z_N|^{1/2} m_N m_X^2 C_{NX}^\varphi + m_M m_N m_X \psi_{MNX}\} \quad (39)$$

and the ionic strength in this case is

$$I = \tfrac{1}{2}[|z_M z_X|(1 + |z_M/z_X|)m_M + |z_N z_X|(1 + |z_N/z_X|)m_N]$$

If the two electrolytes are symmetric this expression can be readily expressed in terms of the molality m of the common ion and the solute mole fraction y of NX. Thus for MX–NX mixtures one obtains for the two activity coefficients as well as the osmotic coefficient

$$\varphi - 1 = z^2 f^\varphi + m\{(1 - y)B_{MX}^\varphi + yB_{NX}^\varphi + y(1 - y)(\theta_{MN} + I\theta_{MN}')\} + m^2\{(1 - y)C_{MX}^\varphi + yC_{NX}^\varphi + y(1 - y)\psi_{MNX}\} \quad (40)$$

$$\ln \gamma_{MX} = z^2 f^\gamma + m\{B_{MX}^\gamma + y(B_{NX}^\varphi - B_{MX}^\varphi + \theta_{MN}) + y(1 - y)I\theta_{MN}'\} + m^2\{C_{MX}^\gamma + y(C_{NX}^\varphi - C_{MX}^\varphi + \tfrac{1}{2}\psi_{MNX}) + \tfrac{1}{2}y(1 - y)\psi_{MNX}\} \quad (41)$$

$$\ln \gamma_{NX} = z^2 f^\gamma + m\{B_{NX}^\gamma + (1 - y)(B_{MX}^\varphi - B_{NX}^\varphi + \theta_{MN}) + y(1 - y)I\theta_{MN}'\} + m^2\{C_{NX}^\gamma + (1 - y)(C_{MX}^\varphi - C_{NX}^\varphi + \tfrac{1}{2}\psi_{MNX}) + \tfrac{1}{2}y(1 - y)\psi_{MNX}\} \quad (42)$$

This formulation shows clearly the quantities θ and ψ which arise for mixtures in contrast to the other terms which can be determined by the properties of the pure MX and NX solutions. It is interesting that θ and θ' show different composition dependence in eq 41 and 42 but the same dependence in eq 40.

The mixture MX–NY where there is no common ion yields more complex equations.

$$\varphi - 1 = z^2 f^\varphi + m\{y(1 - y)(B_{MY}^\varphi + B_{NX}^\varphi + \theta_{MN} + I\theta_{MN}' + \theta_{XY} + I\theta_{NY}') + (1 - y)^2 B_{MX}^\varphi + y^2 B_{NY}^\varphi\} + m^2\{(1 - y)^2 C_{MX}^\varphi + y^2 C_{NY}^\varphi + y(1 - y)(C_{MY}^\varphi + C_{NX}^\varphi) + y(1 - y)^2(\psi_{MNX} + \psi_{MXY}) + y^2(1 - y)(\psi_{MNY} + \psi_{NXY})\} \quad (43)$$

$$\ln \gamma_{MX} = z^2 f^\gamma + m\{(1 - y^2)B_{MX}^\gamma + y(\theta_{MN} + \theta_{XY}) + y(1 - y)(B_{MY}^\varphi + B_{NX}^\varphi - 2B_{MX}^\varphi + I\theta_{MN}' + I\theta_{XY}') + y^2(B_{MY}^\gamma - B_{MY}^\varphi + B_{NX}^\gamma - B_{NX}^\varphi + 2B_{NY}^\varphi - B_{NY}^\gamma)\} + m^2\{(1 - y)^2 C_{MX}^\gamma + y(1 - y)(\psi_{MNX} + \psi_{MXY} + C_{MX}^\varphi + C_{NX}^\varphi + C_{MY}^\varphi) + \tfrac{1}{2}y^2(\psi_{MNY} + \psi_{NXY} + C_{MY}^\varphi + C_{NX}^\varphi + C_{NY}^\varphi)\} \quad (44)$$

The expression for $\ln \gamma_{NY}$ may be obtained from that for $\ln \gamma_{MX}$ by replacement of M by N, X by Y, and y by $(1 - y)$ throughout eq 44. Although the formulas for this case are more complex, no new quantities are introduced as compared to those for the set of solutions with common ions MX–NX, MY–NY, MX–MY, and NX–NY. Hence the measurement of the set of solutions with common ions provides all the parameters needed to calculate the properties of the mixture without a common ion. Reilly and Wood[13a] discuss in greater detail the process of mixing electrolytes with equations whose terms can be related to those presented here. We turn now to the use of these equations for the numerical representation of experimental data.

Single Electrolytes

A comprehensive program is now underway for the selection of optimum functions for f and B_{ij} and the evaluation of constants in B_{ij} as well as C_{ij} for various electrolytes. At this time a few initial results will be reported showing the accuracy of agreement which has already been obtained. The comprehensive tables of osmotic coefficients presented by Robinson and Stokes[2] provide the experimental data for this initial phase of evaluation.

Three functions were considered for f , the electrostatic term

$$f_{DHO}^\varphi = -A_\varphi[I^{1/2}/(1 + bI^{1/2})] \quad (45)$$

$$f_{DHC}^\varphi = -A_\varphi I^{1/2}\sigma(bI^{1/2}) \quad (46)$$

$$f_G^\varphi = -A_\varphi(I^{1/2}/4)\{1 + [3/(1 + bI^{1/2})^2]\} \text{ for } bI^{1/2} < 2.0 \quad (47a)$$

$$= 1/3\, A_\varphi(2I/b)^{1/3} \text{ for } bI^{1/2} > 2.0 \quad (47b)$$

$$A_\varphi = \tfrac{1}{3}A_\gamma = \tfrac{1}{3}(2\pi N_0 d_w/1000)^{1/2} l^{3/2} \quad (48)$$

A_φ is the usual Debye–Hückel constant for the osmotic coefficient and all three forms reduce to the usual limiting law at small I. The first form, DHO for "Debye–Hückel-osmotic," is that derived above, eq 18 and 19, by use of the Debye–Hückel radial distribution function in the "pressure" equation of statistical mechanics which yields the osmotic pressure directly. The second form, DHC for "Debye–Hückel-charging," is the familiar result obtained by initially calculating the activity coefficient from a charging process and subsequent conversion by the Gibbs–Duhem equation to the osmotic coefficient. The third form is one recommended by Glueckauf[18] which goes smoothly from the limiting law into a cube-root dependence on I at high concentration. The parameter b is to be optimized separately for each form.

Two forms were tested for the second virial coefficient

$$B_I^\varphi = \beta^{(0)} + [\beta^{(1)}/(1 + \alpha I^{1/2})^2] \quad (49)$$

$$B_{II}^\varphi = \beta^{(0)} + \beta^{(1)} \exp(-\alpha I^{1/2}) \quad (50)$$

The first form is suggested directly by eq 18 and 19 while the second is considered because it is an especially simple form having the same general properties, namely; (a) finite value at zero ionic strength, (b) rapid change linear in $I^{1/2}$ at low ionic strength, and (c) smooth approach to a constant value at high ionic strength. Although in principle all three parameters α, $\beta^{(0)}$, and $\beta^{(1)}$ may be adjusted for each solute, it was hoped that α might remain the same for broad classes of electrolytes and that there would

(18) E. Glueckauf, *Proc. Roy. Soc., Ser. A*, **310**, 449 (1969).

TABLE Parameters for Thermodynamic Functions for Dilute Aqueous Electrolytes at 25°[a]

Electrolyte	$\beta^{(0)}$	$\beta^{(1)}$	$\sigma \times 10^3$
HCl	0.1802	0.2753	0.6
LiCl	0.1575	0.2811	0.8
NaCl	0.0781	0.2659	0.7
KCl	0.0460	0.2186	0.4
CsCl	0.0320	0.0273	1.5
$NaNO_3$	0.0059	0.1714	0.7
NH_4NO_3	−0.0143	0.1045	1.5
$RbNO_3$	−0.0663	−0.0623	1.9
$Ca(ClO_4)_2$	0.5789	2.5883	1.1
$MgCl_2$	0.4869	2.1062	3.3
$CaCl_2$	0.4162	2.2324	1.6
Na_2CrO_4	0.1186	1.8765	2.1
Na_2SO_4	0.0428	1.3491	1.6

[a] $b = 1.20$, $\alpha = 2.00$, $C^{\varphi} = 0$, range 0–2 *M*.

TABLE II: Parameters for Thermodynamic Functions for Aqueous Electrolytes at 25° from 0 to 6 *M*[a]

Electrolyte	$\beta^{(0)}$	$\beta^{(1)}$	C^{φ}	$\sigma \times 10^3$
HCl	0.18352	0.25503	−0.00059	1.7
NaCl	0.07670	0.26495	+0.00122	0.7
KCl	0.04827	0.20887	−0.00082	0.4
CsCl	0.03449	0.01336	−0.00049	1.5
$NaNO_3$	0.00661	0.17964	−0.00067	1.4
KNO_3	−0.08155	0.04939	+0.00660	1.2
$RbNO_3$	−0.07885	−0.01736	+0.00528	1.0

[a] $b = 1.20$ and $\alpha = 2.00$ throughout.

be a systematic relationship between $\beta^{(0)}$ and $\beta^{(1)}$. In special cases, however, the form of B^{φ} can be modified or terms added without affecting the simple form for most solutes. For this initial test the third virial coefficient C^{φ} was omitted.

The six possible combinations of f^{φ} and B^{φ} were tested systematically with the osmotic coefficient values up to 2 *M* for the 13 electrolytes listed in Table I, by varying b and α independently but holding each the same for all substances. The best general agreement was obtained for f_{DHO} and B_{II}^{φ} with $b = 1.2$ and $\alpha = 2.0$; in this case the over-all standard deviation was 0.0015. The other combinations of f^{φ} and B^{φ} gave standard deviations about twice as large and not significantly differing one from another.

Table I shows the values of $\beta^{(0)}$ and $\beta^{(1)}$ for each substance with f_{DHO}^{φ} and B_{II}^{φ} equation and the standard deviations for each solute. Since osmotic coefficients are ordinarily reported to only 0.001, it is apparent that the agreement is well within experimental uncertainty in most, if not all, cases.

With this choice of form of f^{φ} and B^{φ} and these values of b and α the third virial coefficient C^{φ} was restored to eq 37 and data for several electrolytes were fitted up to 6 *M*. The experimental data are from Robinson and Stokes[2] with recent revisions for $NaNO_3$ and KNO_3 from Bezboruah, Covington, and Robinson.[19] The results, given in Table II, show excellent agreement within experimental error. It is also significant that $\beta^{(0)}$ and $\beta^{(1)}$ have not changed much with the addition of the third virial coefficient and the extension to more concentrated solutions. Thus the second virial coefficient expression appears to be

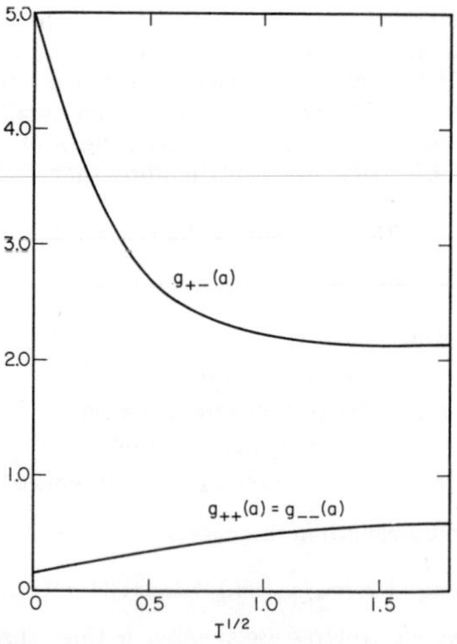

Figure 3. Radial distribution function at hard-core contact as a function of ionic strength from Monte Carlo calculations.[8a]

well established even though its physical interpretation in terms of molecular quantities is not simple.

Lietzke and Stoughton[12] fitted the osmotic coefficient data for several electrolytes to the four constant equation comprising the conventional Debye–Hückel term f_{DHC}^{φ} with adjustable distance of closest approach and second, third, and fourth virial coefficients, each taken as a constant. While the respective calculations of Lietzke and Stoughton and the present research are not precisely comparable, the new results of Table I with two adjustable parameters for each substance fit the data up to 2 *M* as well or better than the earlier work with four adjustable constants fitted to data ranging up to concentrations from 3 to 6 *M*. Likewise the results in Table II with three adjustable constants compare favorably with those of the earlier equation using four adjustable constants.

More definitive results for 1–1 and 1–2 electrolytes as well as calculations for higher valence types will be presented in a subsequent paper.

Comparison of the $\beta^{(0)}$ and $\beta^{(1)}$ values in Table I shows a generally parallel trend but substantial individual variation. In order to understand the difference between $\beta^{(0)}$ and $\beta^{(1)}$ it is desirable to consider the relative values of the radial distribution functions at contact for hard cores. Figure 3 shows $g_{+-}(a)$ and $g_{++}(a) = g_{--}(a)$ as a function of concentration from the Monte Carlo calculations of Card and Valleau[8a] which were for a 1–1 aqueous electrolyte with a = 4.25 Å. The results of Ramanathan and Friedman[7] for soft-core models show the same general trend but are not as directly interpretable. The curves on Figure 3 show clearly that the short-range interaction of unlike charged ions becomes relatively much more important at low ionic strength. At high ionic strength the short-range interactions are weighted approximately 3.5 to 1 for unlike and like charged ions, respectively. This ratio increases to more than 10 to 1 at low ionic strength. Since it is the sum of short-range interaction effects in eq 27 and 28 which constitute the B^{φ} and B^{γ} functions, it is clear that the principal contribution to $\beta^{(1)}$ will come from the short-range interaction of unlike charged ions

(19) C. P. Bezboruah, A. K. Covington, and R. A. Robinson, *J. Chem. Thermodyn.*, **2**, 431 (1970).

while the interactions of like and unlike charged ions will both contribute to $\beta^{(0)}$ in a proportion given by the magnitudes of $g_{+-}(a)$, $g_{++}(a)$, and $g_{--}(a)$ at high concentration. With multiply charged ions the short-range interaction of like-charged ions will be less important, of course, and may be negligible even for $\beta^{(0)}$.

It is apparent from Table I that the ratio of $\beta^{(1)}$ to $\beta^{(0)}$ for 2–1 electrolytes is much greater than for 1–1 solutes. This is in general accord with eq 18 where the two terms in the brackets correspond to the two terms in B^{φ}. In the limit of zero concentration the ratio of the second term to the first is $(l^2/2a^2)$ for 1–1 electrolytes and $(2l^2/a^2)$ for 2–1 solutes. Since $l = 7.1$ Å and a is frequently near 4 Å, ratios near 1.5 and 6 would be plausible for $\beta^{(1)}/\beta^{(0)}$ for 1–1 and 2–1 solutes, respectively. While the values in Table I show too much variability to give much meaning to this comparison, nevertheless, it seems likely that this increase in $\beta^{(1)}$ as compared to $\beta^{(0)}$ arises from the increased ionic charge as it affects the last term in eq 18.

Since the electrostatic interaction function f was not changed for individual solutes, the B^{φ} function also includes a correction to f for the effective distance of closest approach. Equation 20 and the accompanying discussion describe this correction which will contribute to both $\beta^{(0)}$ and $\beta^{(1)}$. If the effective distance of closest approach for a particular solute is less than that implied in f, this contribution to $\beta^{(0)}$ and $\beta^{(1)}$ will be negative. Hence the small negative values in Table I are understandable.

In conclusion of this section the chosen forms of f and B are summarized for excess Gibbs energy and activity coefficient as well as osmotic coefficient. The subscripts are omitted since the alternate forms will not be used further.

$$f^{\varphi} = -A_{\varphi}\,[I^{1/2}/(1 + bI^{1/2})] \quad (51)$$

$$f^{\gamma} = -A_{\varphi}\,[I^{1/2}/(1 + bI^{1/2}) + (2/b)\ln(1 + bI^{1/2})] \quad (52)$$

$$f^{Gx} = -A_{\varphi}(4I/b)\ln(1 + bI^{1/2}) \quad (53)$$

$$B^{\varphi} = \beta^{(0)} + \beta^{(1)}e^{-\alpha I^{1/2}} \quad (54)$$

$$B^{\gamma} = 2\beta^{(0)} + (2\beta^{(1)}/\alpha^2 I) \times [1 - \exp(-\alpha I^{1/2})(1 + \alpha I^{1/2} - \tfrac{1}{2}\alpha^2 I)] \quad (55)$$

$$B^{Gx} = \beta^{(0)} + (2\beta^{(1)}/\alpha^2 I)[1 - \exp(-\alpha I^{1/2})(1 + \alpha I^{1/2})] \quad (56)$$

The superscript Gx indicates the form for the excess Gibbs energy; the value $b = 1.2$ has been selected for use with all substances at 25° aqueous solution; also the value $\alpha = 2.0$ was found to be satisfactory for the 1–1 and 2–1 electrolytes listed in Table I but α may be adjusted for each substance if desired.

Mixed Electrolytes

The thermodynamic treatment of isopiestic data for mixed electrolytes has been simplified considerably by the use of equations which clearly distinguish the effects of mixing from those of the pure components. This is further facilitated by analytical equations representing accurately the properties of the pure components. Recent papers of Rush, Scatchard, and others[11,20] offer excellent illustrations. For mixing solutions of equal ionic strength with a common ion, eq 40–42 show clearly that any nonlinear effect arises only from interactions of ions of the same sign either as pairs in θ_{MN} or in combination with one ion of the opposite sign in ψ_{MNX}. The linear terms for the activity coefficient (but not the osmotic coefficient) also include θ_{MN} and ψ_{MNX} as well as functions for the pure components.

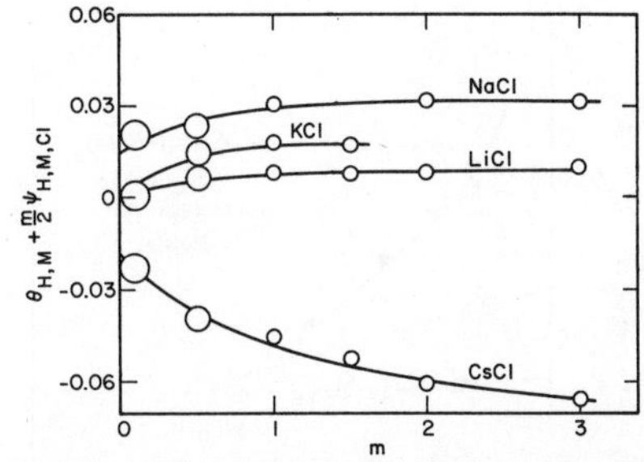

Figure 4. Parameter for the difference in short-range interactions of like charged ions as a function of molality. Data from Harned and Owen[1] for mixtures of HCl with the salt indicated; the sizes of circles show their estimates of experimental error.

It is desirable to estimate theoretically the expected dependence of θ_{MN} on ionic strength. Equation 32 shows that this term is a difference in second virial coefficients for ions of the same sign. Figure 3 shows that g_{++} rises from a very small value at zero concentration to a value somewhat larger but still much less than unity at high ionic strength. Thus we expect θ_{MN} to increase from a smaller value at zero concentration to a relatively constant value at high concentration.

The best test of this picture is the extensive array of galvanic cell measurements which yield the activity coefficient of HCl in mixtures with LiCl, NaCl, KCl, and CsCl. Harned and Owen[1] summarize this work and express the results in terms of coefficients α_{12} and α_{21} which are given in our terminology by the equations (where y is the solute fraction MCl)

$$\ln \gamma_{HCl} = \ln \gamma_{HCl}^0 - 2.303\alpha_{12}my \quad (57a)$$

$$\ln \gamma_{MCl} = \ln \gamma_{MCl}^0 - 2.303\alpha_{21}m(1 - y) \quad (57b)$$

Here γ_{HCl}^0 and γ_{MCl}^0 are the activity coefficients of the pure electrolytes at the molality m.

Comparison of eq 57 with 38, 41, and 42 indicates consistency provided the terms $my(1 - y)I\theta_{MN}'$ and $\tfrac{1}{2}m^2y(1 - y)\psi_{MNX}$ may be neglected. On that basis one obtains

$$\theta_{H,M} + (m/2)\psi_{H,M,Cl} = -(2.303/2)(\alpha_{12} + \alpha_{21}) \quad (58)$$

$$B_{HCl}^{\varphi} - B_{MCl}^{\varphi} + m(C_{HCl}^{\varphi} - C_{MCl}^{\varphi}) - (2.303/2)(\alpha_{12} - \alpha_{21}) \quad (59)$$

Figure 4 shows the values of $\theta_{H,M} + (m/2)\psi_{H,M,Cl}$ obtained from Harned and Owen with the size of the circles indicating their estimates of experimental error. Since the second term will be negligible in dilute solutions, we may interpret the curves below 1 M with respect to $\theta_{H,M}$. It is clear that the general nature of these curves corresponds to the argument given above; the decrease in θ at low concentration corresponds to the decrease in radial distribution function at hard-core contact for pairs of ions of like sign.

In the more concentrated solutions the slope of the curve for CsCl–HCl indicates a significant value of $\psi_{H,Cs,Cl}$. For the other curves the corresponding term is very small.

One can also compare, in accordance with eq 59, our B^{φ} values with the differences $(\alpha_{12} - \alpha_{21})$ from Harned and Owen. The agreement is satisfactory for the constants

(20) Y. C. Wu, R. M. Rush, and G. Scatchard, *J. Phys. Chem.*, **72**, 4048 (1968); **73**, 2047, 4433, 4434 (1969).

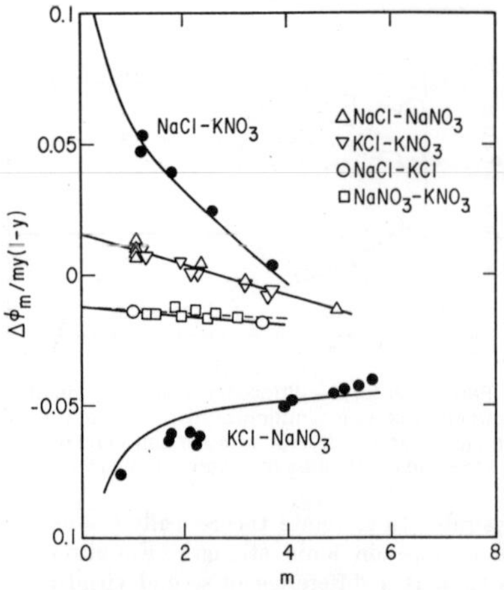

Figure 5. Difference in osmotic coefficient of mixed electrolytes from pure components at the same ionic strength. The curves for $NaCl-KNO_3$ and $KCl-NaNO_3$ are calculated by eq 61 with parameters from pure electrolytes and mixtures with a common ion.

of Table I (through 2 *M*) or those of Table II.

Since we find $\theta_{H,M}$ to vary with ionic strength, $\theta_{H,M}'$ cannot be zero. The resulting nonlinear term in eq 41 is, however, very small and within experimental error at 0.1 *M*. This effect might have been detected at higher concentrations, especially in HCl-CsCl solutions, but the appropriate measurements were not made. Hence there is no conflict between our interpretation and available data.

We consider next the change in osmotic coefficient for mixtures MX-NX with a common ion. From eq 37 and 40 one obtains

$$\Delta\varphi_m = y(1-y)m(\theta_{MN} + I\theta_{MN}' + m\psi_{MNX}) \qquad (60)$$

It is of interest to consider a series of systems with the same pair of ions of like sign, M and N, but with different common ions X, X′, etc. In this situation the θ_{MN} term must be the same throughout but ψ_{MNX} will differ with X. We considered the series NaCl-KCl, NaBr-KBr, and $NaNO_3$-KNO_3 for which osmotic coefficients have been measured by several investigators.[19,21,22] It is found, however, that the bromides differ so slightly from the chlorides that little is gained by their inclusion. The data for the chlorides and nitrates are shown in the middle of Figure 5. Actually there are many points for the system NaCl-KCl; the two shown define the straight line which represents all the points quite well.

It is apparent from Figure 5 that the data for systems with a common ion can be represented by straight lines. But there is so little information at concentration below 1 *M* that one cannot determine whether there is curvature in that region. In view of the theory and the data on the HCl-MCl systems given above one would expect curvature toward zero in $\Delta\varphi_m$ in the dilute region. Since the values of $[\Delta\varphi_m/my(1-y)]$ for mixtures with a common ion are small, this uncertainty from possible curvature in the dilute region is also small and we can omit the terms in ϑ' from osmotic coefficient treatments until such time as improved experimental accuracy justifies their inclusion. On this basis we find $\theta_{Na,K} = -0.012$, $\theta(Cl,NO_3) = +0.016$, $\psi_{Na,K,Cl} = -0.0018$, $\psi(Na,K,NO_3) = -0.012$, $\psi(Na,Cl,NO_3) = \psi(K,Cl,NO_3) = -0.0060$.

These values determined from the mixtures with a common ion suffice, when combined with values from Table II for the pure electrolytes, to calculate $\Delta\varphi_m$ for the mixtures without common ion. The equation for mixing MX with NY, which follows from (43) and (37) and with the omission of θ', is

$$[\Delta\varphi_m/my(1-y)] = [\theta_{MN} + \theta_{XY} + (m/2)(\psi_{MNX} + \psi_{MNY} + \psi_{MXY} + \psi_{NXY})] + [B_{MY}^{\varphi} + B_{NX}^{\varphi} - B_{MX}^{\varphi} - B_{NY}^{\varphi} + m(C_{MY}^{\varphi} + C_{NX}^{\varphi} - C_{MX}^{\varphi} - C_{NY}^{\varphi})] \qquad (61)$$

It is notable that all terms in the first brackets remain unchanged if M and N (or X and Y) are interchanged whereas all terms in the second brackets change sign. Thus the sum of $[\Delta\varphi_m/my(1-y)]$ for mixing MX-NY and for NX-MY is just twice the first brackets and is exactly the sum of the four corresponding quantities for the mixing processes with a common ion. This last equality is just the cross-square rule of Young, Wu, and Krawetz,[23] although it is more commonly applied to heats or Gibbs energies of mixing.

Figure 5 shows the experimental values of $\Delta\varphi_m$ for mixing $NaCl-KNO_3$ and $KCl-NaNO_3$ together with the curves calculated from eq 61. The fit for the $NaCl-KNO_3$ system is almost perfect. For the $KCl-NaNO_3$ mixtures the agreement is less perfect but appears to be within experimental uncertainty. If the data for the mixtures without a common ion had been included in an adjustment of all parameters, a better fit would doubtless be possible, but it seems preferable in most cases, and much easier, to evaluate parameters from the simplest systems to which they are related.

The agreement shown on Figure 5 suggests that the cross-square rule is satisfied for this system whereas Bezboruah, Covington, and Robinson[19] concluded that the difference probably exceeded experimental error. These authors did not compare the $\Delta\varphi$ of mixing quantities directly but rather calculated ΔG of mixing instead. This calculation of ΔG involves the integral of $(\Delta\varphi_m/m)dm$ from zero to the molality of interest. It is evident from Figure 5 the curves we predict in the very dilute region are quite different from the lines one might draw through the points without reference to other information. Application of the cross-square condition directly to $\Delta\varphi_m$ avoids this difficulty and appears to me to yield agreement within the uncertainty of experiments. It should be emphasized that our present approach and eq 61 predict much more than the cross-square rule; they predict the two individual curves for the systems without a common ion instead of predicting only their sum.

Although the point[24] has been made previously that an ion pair interaction quantity such as θ_{MN} must be the same regardless of the ion of opposite sign, most experimental data have been interpreted without considering this limitation. An exception is the very recent work of Scatchard, Rush, and Johnson[11] where it is shown that this limitation can be imposed without difficulty for the various mixed electrolytes formed from NaCl, Na_2SO_4, $MgCl_2$, and $MgSO_4$. This point was also recognized by

(21) A. K. Covington, T. H. Lilley, and R. A. Robinson, *J. Phys. Chem.*, **72**, 2759 (1968).
(22) R. A. Robinson, *J. Phys. Chem.*, **65**, 662 (1961); R. M. Rush and R. A. Robinson, *J. Tenn. Acad. Sci.*, **42**, 22 (1968).
(23) T. F. Young, Y. C. Wu, and A. A. Krawetz, *Discuss. Faraday Soc.*, **24**, 37, 77 (1957).
(24) H. L. Friedman, *J. Chem. Phys.*, **32**, 1134 (1960); "Ionic Solution Theory," Interscience, New York, N. Y., 1962.

Wood and Smith[25] in making their very convincing case that the heat of mixing of 1–1 electrolytes with a common ion arises primarily from differences in the interaction of pairs of ions of the same sign and consequently disproves Brønsted's principle of specific ion interaction.

Summary

It is shown from an improved statistical treatment based upon the Debye–Hückel ion distribution function that the effect of short-range forces between pairs of ions, *i.e.*, the second virial coefficient, should depend also on the ionic strength. Numerical calculations on more realistic models confirm this general effect in which second virial coefficients for pure electrolytes decrease rapidly with increase in ionic strength in the very dilute region yet become nearly constant at higher concentrations. This pattern of behavior was known experimentally and seemed puzzling heretofore.

By including the appropriate ionic strength dependence in the second virial coefficients a relatiely simple system of equations for the thermodynamic properties of electrolytes is developed which appears to represent experimental behavior to much higher concentration than any previous system of comparable complexity. Numerical parameters and comparisons with experimental data are presented for the osmotic coefficients of several pure electrolytes, for the activity coefficients of a series of mixtures HCl–MCl, and for the osmotic coefficient for all possible mixtures of sodium and potassium chlorides and nitrates. A much more extensive treatment of many particular systems is in progress, and it is expected that a very compact summary can be given of the extensive array of experimental information on aqueous electrolytes.

Acknowledgment. I wish to thank Mr. Guillermo Mayorga for carrying out some of the numerical calculations. This research was sponsored by the U. S. Atomic Energy Commission.

(25) R. H. Wood and R. W. Smith, *J. Phys. Chem.*, **69**, 2974 (1965).

[Reprinted from the Journal of Physical Chemistry, 77, 2300 (1973).]

Thermodynamics of Electrolytes. II. Activity and Osmotic Coefficients for Strong Electrolytes with One or Both Ions Univalent

Kenneth S. Pitzer* and Guillermo Mayorga

Inorganic Materials Research Division of the Lawrence Berkeley Laboratory and Department of Chemistry, University of California, Berkeley, California 94720 *(Received February 5, 1973)*

Publication costs assisted by the U. S. Atomic Energy Commission

The system of equations developed in the first paper of this series is successfully applied to the available free energy data at room temperature for 227 pure aqueous electrolytes with one or both ions univalent. The experimental data are represented substantially within experimental error from dilute solutions up to an ionic strength varying from case to case but typically 6 *M*. Where the data extend to high concentration, three parameters are evaluated for each solute, but one of these has negligible effect and is omitted if there are data only for the dilute range. This yields a very compact set of tables from which these important and useful properties can be reproduced. These parameters will also be of importance in treating mixed electrolytes. A simplified graphical presentation is given for activity coefficients of 1-1 electrolytes. In most cases the new equations are fitted to the osmotic coefficient data as recommended by Robinson and Stokes but for hydroxides, zinc halides, hydrogen halides, and a few other cases we have based our evaluation on the original data from several sources. The implications of our parameters are also discussed in terms of solvent structure and interionic forces.

It is the twofold purpose of this research to represent as compactly and as accurately as possible the very extensive array of experimental data on the thermodynamic properties of single aqueous electrolytes at room temperature and to interpret the resulting parameters in terms of interionic forces. In the first paper[1] of this series (cited hereafter as I) the theoretical aspects were considered in order to guide the selection of equations which reproduce the measured properties substantially within experimental accuracy, which are compact and convenient in that only a very few parameters need be tabulated for each substance and the mathematical calculations are simple, which have appropriate form for mixed electrolytes as well as for a single solute, and whose parameters have physical meaning as far as possible. Choices were made between several possible forms of equations on the basis of accuracy of representation of the experimental data for several solutes.

In this paper we present the results for single electrolytes which show no association. Also included are several solutes which are said, on the basis of some properties, usually conductance, to show some association, but whose thermodynamic properties are fitted satisfactorily by the same equations used for nonassociating electrolytes. In later papers we expect to treat single electrolytes showing moderate association as well as mixed electrolytes.

Equations and General Parameters

The equations chosen in I differ from those generally used heretofore in several respects, but most important is the ionic strength dependence of the second virial coefficient. This was shown to be justified theoretically by a simple derivation in which the Debye–Hückel radial distribution is substituted into the osmotic pressure equation of statistical mechanics, although it was also noted that the more complex theory of Mayer[2] previously had indicated such a dependence. This same simple derivation also led to a different form for the long-range electrostatic term which was found to be empirically superior to the conventional Debye-Hückel form derived by a charging process (instead of the osmotic pressure equation).

The pertinent equations for single electrolytes for the excess Gibbs energy G^{ex}, the osmotic coefficient φ, and the activity coefficient γ are as follows

$$(G^{ex}/n_wRT) = f^{Gx} + m^2(2\nu_M\nu_X)B_{MX}{}^{Gx} + m^3[2(\nu_M\nu_X)^{3/2}]C_{MX}{}^{Gx} \quad (1)$$

$$\varphi - 1 = |z_Mz_X|f^\varphi + m\left(\frac{2\nu_M\nu_X}{\nu}\right)B_{MX}{}^\varphi + m^2\,\frac{2(\nu_M\nu_X)^{3/2}}{\nu}\,C_{MX}{}^\varphi \quad (2)$$

$$\ln\gamma = |z_Mz_X|f^\gamma + m\left(\frac{2\nu_M\nu_X}{\nu}\right)B_{MX}{}^\gamma + m^2\,\frac{2(\nu_M\nu_X)^{3/2}}{\nu}\,C_{MX}{}^\gamma \quad (3)$$

where ν_M and ν_X are the numbers of M and X ions in the formula and z_M and z_X give their respective charges in electronic units; also $\nu = \nu_M + \nu_X$, while n_w is the number of kg of solvent and m is the conventional molality. The other quantities have the form

$$f^{Gx} = -A_\varphi(4I/b)\ln(1 + bI^{1/2}) \quad (4)$$

$$f_\varphi = -A_\varphi\frac{I^{1/2}}{1+bI^{1/2}} \quad (5)$$

$$f^\gamma = -A_\varphi\left[\frac{I^{1/2}}{1+bI^{1/2}} + \frac{2}{b}\ln(1 + bI^{1/2})\right] \quad (6)$$

$$B_{MX}{}^{Gx} = \beta_{MX}{}^{(0)} + \frac{2\beta_{MX}{}^{(1)}}{\alpha^2I}[1 - e^{-\alpha I^{1/2}}(1 + \alpha I^{1/2})] \quad (7)$$

$$B_{MX}{}^\varphi = \beta_{MX}{}^{(0)} + \beta_{MX}{}^{(1)}e^{-\alpha I^{1/2}} \quad (8)$$

$$B_{MX}{}^\gamma = 2\beta_{MX}{}^{(0)} + \frac{2\beta_{MX}{}^{(1)}}{\alpha^2I}[1 - e^{-\alpha I^{1/2}}(1 + \alpha I^{1/2} - (1/2)\alpha^2I)] \quad (9)$$

$$C_{MX}^{Gx} = (1/2)C_{MX}^{\varphi} \tag{10}$$

$$C_{MX}^{\gamma} = (3/2)C_{MX}^{\varphi} \tag{11}$$

Note that the superscripts Gx, φ, and γ are labels (not exponents). Also I is the ionic strength, $\frac{1}{2}\Sigma m_i z_i^2$, and A^{φ} is the Debye–Hückel coefficient for the osmotic function, $[\frac{1}{3}(2\pi N_0 d_w/1000)^{1/2}(e^2/DkT)^{3/2}]$, which has the value 0.392 at 25° for water. In order to maintain simple equations for mixed electrolytes, b must remain the same for all solutes and in I the value 1.2 was selected; also the value α = 2.0 was found to be satisfactory for all solutes considered in this paper. For each substance the two parameters $\beta_{MX}^{(0)}$ and $\beta_{MX}^{(1)}$ define the second virial coefficient and C_{MX}^{φ} defines the third virial coefficient which is usually very small and sometimes completely negligible. If necessary to fit unusual behavior, one could add to the second virial coefficient further terms of the same form but with different values of α and β, but this was not required for the present work.

The numerical factors, such as $(2\nu_M\nu_X)$, multiplying the second and third virial coefficients in eq 1–3 are needed in order to retain the simple meaning that the virial coefficients represent the short-range interaction of pairs and triplets of ions, respectively. In eq 2 and 3 the quantities $(2\nu_M\nu_X/\nu)$ and $2(\nu_M\nu_X)^{3/2}/\nu$ are both unity for a symmetrical MX electrolyte. For convenience in working with single electrolytes, we shall tabulate $\frac{4}{3}\beta^{(0)}$, $\frac{4}{3}\beta^{(1)}$, and $(2^{5/2}/3)C_{\varphi}$ for 2–1 electrolytes and the corresponding quantities for other types, but these numerical factors must be removed before the virial coefficients are used in work with mixed electrolytes.

While eq 1–11 are quite simple and should be used by anyone desiring accurate values of the thermodynamic properties, we recognize that for many purposes a rough estimate of the activity coefficient will suffice. Since we shall see that there is a close relationship between $\beta^{(1)}$ and $\beta^{(0)}$ and that C^{φ} is usually very small, one can make a reasonably good estimate of the activity coefficient from $\beta^{(0)}$ alone. For this purpose we present a large graph with curves of γ for various values of $\beta^{(0)}$ from which one can visually interpolate values of the activity coefficient directly for 1–1 electrolytes.

Evaluation of Parameters

Robinson and Stokes critically reviewed the available data in their classic papers and in the various editions of their book. In most cases we have accepted the tables of osmotic coefficients in the second (revised) edition of their book[3] and evaluated by least squares the best values of $\beta^{(0)}$, $\beta^{(1)}$, and C^{φ} which are listed in Tables I-IX. Also listed are the maximum molality for which agreement is attained to 0.01 in φ or for which data are available, and the standard deviation of fit. In Tables IV, V, VII, VIII, and IX the data seemed less accurate, and the departures from ideal behavior are greater; hence the allowable error in φ was increased to 0.02 (and to 0.03 for the 5–1 type salt $K_5P_3O_{10}$).

The last column in the tables gives the reference, three to Robinson and Stokes book, or to a more recent literature source.[4-60] The symbol "t" indicates that several references are involved and these are identified either in the text, in Table X, or in formulas in parentheses after the journal citations. In the preceding column the letters a, b, and c indicate high, intermediate, and low accuracy, re-

Table I: Inorganic Compounds of 1-1 Type.

	$\beta^{(0)}$	$\beta^{(1)}$	C^{ϕ}	Max. m	σ	ref.
HCl	0.1775	0.2945	0.00080	6	a	t
HBr	.1960	.3564	.00827	3	a	t
HI	.2362	.392	.0011	3	b	t
$HClO_4$	.1747	.2931	.00819	5.5	.002	3
HNO_3	.1119	.3206	.0010	3	.001	3
LiCl	.1494	.3074	.00359	6	.001	3
LiBr	.1748	.2547	.0053	2.5	.002	3
LiI	.2104	.373	---	1.4	.006	3
LiOH	.015	.14	---	4	c	t
$LiClO_4$	.1973	.3996	.0008	3.5	.002	3
$LiNO_2$	.1336	.325	-.0053	6	.003	4
$LiNO_3$	.1420	.2780	-.00551	6	.001	3
NaF	.0215	.2107	---	1	.001	3
NaCl	.0765	.2664	.00127	6	.001	3
NaBr	.0973	.2791	.00116	4	.001	3
NaI	.1195	.3439	.0018	3.5	.001	3
NaOH	.0864	.253	.0044	6	b	t
$NaClO_3$	.0249	.2455	.0004	3.5	.001	3
$NaClO_4$	.0554	.2755	-.00118	6	.001	3
$NaBrO_3$	-.0205	.1910	.0059	2.5	.001	3
NaCNS	.1005	.3582	-.00303	4	.001	3
$NaNO_2$	.0641	.1015	-.0049	5	.005	4
$NaNO_3$	.0068	.1783	-.00072	6	.001	3
NaH_2PO_4	-.0533	.0396	.00795	6	.003	3
NaH_2AsO_4	-.0442	.2895	---	1.2	.001	3
$NaBO_2$	-.0526	.1104	.0154	4.5	.004	5
$NaBF_4$	-.0252	.1824	.0021	6	.006	5
KF	.08089	.2021	.00093	2	.001	3
KCl	.04835	.2122	-.00084	4.8	.0005	t
KBr	.0569	.2212	-.00180	5.5	.001	3
KI	.0746	.2517	-.00414	4.5	.001	3
KOH	.1298	.320	.0041	5.5	b	6
$KClO_3$	-.0960	.2481	---	0.7	.001	3
$KBrO_3$	-.1290	.2565	---	0.5	.001	3
KCNS	.0416	.2302	-.00252	5	.001	3
KNO_2	.0151	.015	.0007	5	.003	7
KNO_3	-.0816	.0494	.00660	3.8	.001	3
KH_2PO_4	-.0678	-.1042	---	1.8	.003	3
KH_2AsO_4	-.0584	.0626	---	1.2	.003	3
$KPtF_6$	-.163	-.282	---	0.5	.001	8
RbF	.1141	.2842	-.0105	3.5	.002	9
RbCl	.0441	.1483	-.00101	5	.001	3
RbBr	.0396	.1530	-.00144	5	.001	3
RbI	.0397	.1330	-.00108	5	.001	3
$RbNO_2$	.0269	-.1553	-.00366	5	.002	4
$RbNO_3$	-.0789	-.0172	.00529	4.5	.001	3
CsF	.1306	.2570	-.0043	3.2	.002	9
CsCl	.0300	.0558	.00038	5	.002	t
CsBr	.0279	.0139	.00004	5	.002	3
CsI	.0244	.0262	-.00365	3	.001	3
CsOH	.150	.30	---	---	b	t
$CsNO_3$	-.0758	-.0669	---	1.4	.002	3
$CsNO_2$	.0427	.060	-.0051	6	.004	4
$AgNO_3$	-.0856	.0025	.00591	6	.001	3
$TlClO_4$	-.087	-.023	---	0.5	.001	3
$TlNO_3$	-.105	-.378	---	0.4	.001	3
NH_4Cl	.0522	.1918	-.00301	6	.001	3
NH_4Br	.0624	.1947	-.00436	2.5	.001	10
NH_4ClO_4	-.0103	-.0194	---	2	.004	11
NH_4NO_3	-.0154	.1120	-.00003	6	.001	3

Table II: Salts of Carboxylic Acids (1-1 Type).

	$\beta^{(0)}$	$\beta^{(1)}$	C^{ϕ}	Max. m	σ	ref.
Li Acetate	0.1124	0.2483	-.00525	4	.001	3
Na Formate	.0820	.2872	-.00523	3.5	.001	3
Na Acetate	.1426	.3237	-.00629	3.5	.001	3
Na Propionate	.1875	.2789	-.01277	3	.001	3
NaH Malonate	.0229	.1600	-.00106	5	.002	3
NaH Succinate	.0354	.1606	.00040	5	.001	3
NaH Adipate	.0472	.3168	---	0.7	.001	3
K Acetate	.1587	.3251	-.00660	3.5	.001	3
KH Malonate	-.0095	.1423	.00167	5	.004	3
KH Succinate	.0111	.1564	.00274	4.5	.002	3
KH Adipate	.0419	.2523	---	1	.001	3
Rb Acetate	.1622	.3353	-.00551	3.5	.001	3
Cs Acetate	.1628	.3605	-.00555	3.5	.001	3
Tl Acetate	.0082	.0131	-.00127	6	.001	3

Table III: Tetraalkylammonium Halides

	$\beta^{(0)}$	$\beta^{(1)}$	C^{ϕ}	Max. m	σ	ref.
Me_4NF	0.2677	0.2265	0.0013	3	.002	37
Et_4NF	.3113	.6155	.0349	2	.002	37
Pr_4NF	.4463	.4090	.0537	2	.002	37
Bu_4NF	.6092	.402	-.0281	1.7	.005	37
Me_4NCl	.0149	-.083	.0057	3.4	.005	38
Et_4NCl	.0336	-.153	.0084	3	.002	38
Pr_4NCl	.1065	-.354	.0098	2.5	.002	38
Bu_4NCl	.2058	-.464	-.0588	2.5	.001	38
Me_4NBr	-.0363	-.201	.0084	3.5	.004	38
Et_4NBr	-.0457	-.448	.0135	4	.001	38
Pr_4NBr	.0108	-.826	.0078	.3.5	.003	38
Bu_4NBr	-.0558	-.579	-.0010	4.5	.007	38
Me_4NI	.0345	-.585	---	0.3	.003	38
Et_4NI	-.1930	-.599	.0401	2	.007	38,39
Pr_4NI	-.2839	-.863	---	0.5	.005	38

See text for meaning of a, b, c, t.

Table IV: Sulfonic Acids and Salts (1-1 Type)
(SA = sulfonic acid; S = sulfonate)

	$\beta^{(0)}$	$\beta^{(1)}$	C^{ϕ}	Max. m	σ	ref.
Methane SA	0.1298	0.629	0.0052	4	---	40
Li methane S	.1320	.271	-.0030	4	---	40
Na methane S	.0787	.274	-.0024	4	---	40
K methane S	.0581	.165	-.0046	4	---	40
NH_4 methane S	.0661	.191	-.0041	4	---	40
Me_4N methane S	.1458	.168	-.0043	4	---	40
Et_4N methane S	.1548	.090	-.0034	4	---	40
Bu_4N methane S	.2145	.235	-.0392	4	---	40
Ethane SA	.1536	.341	-.0056	4	---	40
Li ethane S	.1799	.319	-.0118	4	---	40
Na ethane S	.1316	.374	-.0082	4	---	40
K ethane S	.0965	.250	-.0074	4	---	40
NH_4 ethane S	.1142	.179	-.0114	4	---	40
Me_4N ethane S	.1796	.083	-.0116	4	---	40
Et_4N ethane S	.1805	.075	-.0040	4	---	40
Bu_4N ethane S	.1827	.445	-.0374	4	---	40
Benzene SA	.0526	.445	.0036	5	.002	41
Li benzene S	.1134	.466	-.0075	4.5	.002	41
Na benzene S	.0842	.351	-.0181	2.5	.001	41
p-toluene SA	-.0366	.281	.0137	5	.002	3
Li p-toluene S	.0189	.399	.0046	4.5	.004	3
Na p-toluene S	-.0344	.396	.0043	4	.003	3
K p-toluene S	-.0985	.453	.0122	3.5	.002	3
2,5 Me_2benzene SA	-.0965	.141	.0210	4.5	.01	3
Li 2,5 Me_2benzene S	-.0098	.361	.0039	3.5	.002	41
Na 2,5 Me_2benzene S	-.0277	.228	---	1	.005	41
p-Et benzene SA	-.1736	.435	.0383	2	.007	3
Li p-Et benzene S	-.1438	.804	.0317	5	.01	41,42
Na p-Et benzene S	-.2240	.895	.0355	2.5	.01	41,42
Mesitylene SA	-.2209	.248	.0432	2	.01	41
Li Mesitylene S	-.1998	.871	.0456	2	.004	41
Na Mesitylene S	-.2018	.767	---	1	.003	41

Table V: Additional 1-1 Type Organic Salts

	$\beta^{(0)}$	$\beta^{(1)}$	C^{ϕ}	Max. m	σ	ref.
Choline Cl	.0457	-.196	.0008	6	.004	43
Choline Br	-.0066	-.227	.0036	6	.004	43
Me_3BzNCl	-.0821	-.178	.0162	3.5	.01	43
Me_3BzNBr	-.1517	-.545	.0187	3	.01	43
$Me_2OEtBzNCl$	-.0879	-.343	.0134	4	.01	43
$Me_2OEtBzNBr$	-.1518	-.778	.0177	3	.01	43
$(HOC_2H_4)_4NF$	.0938	.128	-.0030	4	.001	44
$(HOC_2H_4)_4NBr$	-.0474	-.259	.0106	3	.002	44
Me_3SCl	.0314	-.184	.0023	6	.005	45
Me_3SBr	-.0228	-.245	.0044	6	.004	45
Me_3SI	-.0601	-.604	.0006	3	.01	45
Bu_3SCl	.0726	-.245	-.0099	6	.01	45
Bu_3SBr	-.0803	-.616	.0053	6	.01	45

Table VI: Inorganic Compounds of 2-1 Type.

	$\frac{4}{3}\beta^{(0)}$	$\frac{4}{3}\beta^{(1)}$	$\frac{2^{5/2}}{3}C^{\phi}$	Max. m	σ	ref.
$MgCl_2$	0.4698	2.242	0.00979	4.5	.003	3
$MgBr_2$	.5769	2.337	.00589	5	.004	3
MgI_2	.6536	2.4055	.01496	5	.003	3
$Mg(ClO_4)_2$	.6615	2.678	.01806	2	.002	3
$Mg(NO_3)_2$	.4895	2.113	-.03889	2	.003	3
$CaCl_2$	.4212	2.152	-.00064	2.5	.003	3
$CaBr_2$	.5088	2.151	-.00485	2	.002	3
CaI_2	.5839	2.409	-.00158	2	.001	3
$Ca(ClO_4)_2$	.6015	2.342	-.00943	2	.005	3
$Ca(NO_3)_2$	.2811	1.879	-.03798	2	.002	3
$SrCl_2$	.3810	2.223	-.00246	4	.003	3
$SrBr_2$	.4415	2.282	.00231	2	.001	3
SrI_2	.5350	2.480	.00501	2	.001	3
$Sr(ClO_4)_2$	.5692	2.089	-.02472	2.5	.003	3
$Sr(NO_3)_2$	.1795	1.840	-.03757	2	.002	3
$BaCl_2$	.3504	1.995	-.03654	1.8	.001	3
$BaBr_2$	.4194	2.093	-.03009	2	.001	3
BaI_2	.5625	2.249	-.03286	1.8	.003	3
$Ba(OH)_2$	.229	1.60	---	0.1	---	36
$Ba(ClO_4)_2$	.4819	2.101	-.05894	2	.003	3
$Ba(NO_3)_2$	-.043	1.07	---	0.4	.001	3
$MnCl_2$	.4363	2.067	-.03865	2.5	.003	3
$FeCl_2$	.4479	2.043	-.01623	2	.002	3
$CoCl_2$	.4857	1.967	-.02869	3	.004	3
$CoBr_2$	.5693	2.213	-.00127	2	.002	3
CoI_2	.695	2.23	-.0088	2	.01	3
$Co(NO_3)_2$	.4159	2.254	-.01436	5.5	.003	3
$NiCl_2$	.4639	2.108	-.00702	2.5	.002	3
$CuCl_2$	.4107	1.835	-.07624	2	.003	3
$Cu(NO_3)_2$	.4224	1.907	-.04136	2	.002	3
$ZnCl_2$	.3469	2.190	-.1659	1.2	.006	t
$ZnBr_2$	.6213	2.179	-.2035	1.6	.007	t
ZnI_2	.6428	2.594	-.0269	0.8	.002	t
$Zn(ClO_4)_2$	.6747	2.396	.02134	2	.003	3
$Zn(NO_3)_2$	.4641	2.255	-.02955	2	.001	3
$Cd(NO_3)_2$	.3820	2.224	-.04836	2.5	.002	3
$Pb(ClO_4)_2$	.4443	2.296	-.01667	6	.004	3
$Pb(NO_3)_2$	-.0482	0.380	.01005	2	.002	3

Table VI: Inorganic Compounds of 2-1 Type (Continued)

	$\frac{4}{3}\beta^{(0)}$	$\frac{4}{3}\beta^{(1)}$	$\frac{2^{5/2}}{3}C^{\phi}$	Max. m	σ	ref.
UO_2Cl_2	0.5698	2.192	-.06951	2	.001	3
$UO_2(ClO_4)_2$	.8151	2.859	.04089	2.5	.003	3
$UO_2(NO_3)_2$	.6143	2.151	-.05948	2	.002	3
Li_2SO_4	.1817	1.694	-.00753	3	.002	3
Na_2SO_4	.0261	1.484	.00938	4	.003	3
$Na_2S_2O_3$	.0882	1.701	.00705	3.5	.002	3
Na_2CrO_4	.1250	1.826	-.00407	2	.002	3
Na_2CO_3	.2530	1.128	-.09057	1.5	.001	3
Na_2HPO_4	-.0777	1.954	.0554	1	.002	3
Na_2HAsO_4	.0407	2.173	.0034	1	.001	3
K_2SO_4	.0666	1.039	---	0.7	.002	3
K_2CrO_4	.1011	1.652	-.00147	3.5	.003	3
$K_2Pt(CN)_4$	.0881	3.164	.0247	1	.005	53
K_2HPO_4	.0330	1.699	.0309	1	.002	3
K_2HAsO_4	.1728	2.198	-.0336	1	.001	3
Rb_2SO_4	.0772	1.481	-.00019	1.8	.001	3
Cs_2SO_4	.1184	1.481	-.01131	1.8	.001	3
$(NH_4)_2SO_4$	.0545	0.878	-.00219	5.5	.004	3
cis$[Co(en)_2NH_3NO_2](NO_3)_2$	-.0928	0.271	---	0.6	.002	54
tr.$[Co(en)_2NH_3NO_2](NO_3)_2$	-.0901	.249	---	.8	.002	54
cis$[Co(en)_2NH_3NO_2]Cl_2$	-.0327	.684	.0121	2.8	.005	54
tr.$[Co(en)_2NH_3NO_2]Cl_2$	.0050	.695	.0066	2.4	.005	54
cis$[Co(en)_2NH_3NO_2]Br_2$	-.1152	.128	.0158	1	.004	54
tr.$[Co(en)_2NH_3NO_2]Br_2$	-.0912	.424	.0223	2.4	.005	54
cis$[Co(en)_2NH_3NO_2]I_2$	-.1820	.594	---	0.6	.004	54
tr.$[Co(en)_2NH_3NO_2]I_2$	-.1970	1.003	---	.3	.003	54

Table VII: Organic Electrolytes of 2-1 Type
(SA = Sulfonic acid; S = Sulfonate)

m-Benzenedi SA	0.5611	2.637	-.0463	1.6	.004	55
Li_2 m-Benzenedi S	.5464	2.564	-.0622	2.5	.004	55
Na_2 m-Benzenedi S	.3411	2.698	-.0419	3	.004	55
4,4'bibenzyldi SA	.1136	2.432	.0705	2	.01	3,t
Li_2 4,4'bibenzyldi S	.1810	1.755	.0462	1.2	.007	55
Na_2 4,4'bibenzyldi S	.0251	1.969	---	0.4	.01	55
Na_2 fumarate	.3082	1.203	-.0378	2	.003	3
Na_2 maleate	.1860	0.575	-.0170	3	.004	3

Table VIII: 3-1 Electrolytes

	$\frac{3}{2}\beta^{(0)}$	$\frac{3}{2}\beta^{(1)}$	$\frac{3^{3/2}}{2}C^{\phi}$	Max. m	σ	ref.
$AlCl_3$	1.0490	8.767	.0071	1.6	.005	3
$SrCl_3$	1.0500	7.978	-.0840	1.8	.005	3
YCl_3	0.9599	8.166	-.0587	1.8	.007	3
$LaCl_3$	.9158	8.231	-.0831	1.8	.007	3
$CeCl_3$	.9187	8.227	-.0809	1.8	.01	3
$PrCl_3$	.9030	8.181	-.0727	2	.006	3
$NdCl_3$	.9175	8.104	-.0737	1.8	.007	3
$SmCl_3$	.9330	8.273	-.0728	1.8	.01	3
$EuCl_3$	.9370	8.385	-.0687	1.8	.007	3
$CrCl_3$	1.1046	7.883	-.1172	1.2	.005	3
$Cr(NO_3)_3$	1.0560	7.777	-.1533	1.4	.004	3
$Ga(ClO_4)_3$	1.2381	9.794	.0904	2	.008	3
$InCl_3$	-1.68	-3.85	---	0.01	---	t
Na_3PO_4	.2672	5.777	-.1339	0.7	.003	3
Na_3AsO_4	.3582	5.895	-.1240	0.7	.001	3
K_3PO_4	.5594	5.958	-.2255	0.7	.001	3
$K_3P_3O_9$	.4867	8.349	-.0886	0.8	.004	60
K_3AsO_4	.7491	6.511	-.3376	0.7	.001	3
$K_3Fe(CN)_6$	.5035	7.121	-.1176	1.4	.003	3
$K_3Co(CN)_6$	.5603	5.815	-.1603	1.4	.008	3
$Co(en)_3Cl_3$	.2603	3.563	-.0916	1	.003	t
$Co(en)_3(NO_3)_3$	.1882	3.935	---	0.3	.01	58
$Co(en)_3(ClO_4)_3$	.1619	5.395	---	0.6	.007	59
$Co(pn)_3(ClO_4)_3$	.2022	3.976	---	0.3	.003	58

Table IX: 4-1 and 5-1 Electrolytes

4-1 Solute	$\frac{8}{5}\beta^{(0)}$	$\frac{8}{5}\beta^{(1)}$	$\frac{16}{5}C^{\phi}$	Max. m	σ	ref.
$ThCl_4$	1.622	21.33	-.3309	1	.006	3
$Th(NO_3)_4$	1.546	18.22	-.5906	1	.01	3
$Na_4P_2O_7$	0.699	17.16	---	0.2	.01	60
$K_4P_2O_7$	.977	17.88	-.2418	0.5	.01	60
$K_4Fe(CN)_6$	1.021	16.23	-.5579	0.9	.008	3
$K_4Mo(CN)_8$	0.854	18.53	-.3499	0.8	.01	3
$K_4W(CN)_8$	1.032	18.49	-.4937	1	.005	53
$Me_4NMo(CN)_8$	.938	15.91	-.3330	1.4	.01	53
5-1 Solute	$\frac{5}{3}\beta^{(0)}$	$\frac{5}{3}\beta^{(1)}$	$\frac{5^{3/2}}{3}C^{\phi}$			
$Na_5P_3O_{10}$	1.869	36.10	-1.630	0.4	.01	60
$K_5P_3O_{10}$	1.939	39.64	-1.055	0.5	.015	60

See text for meaning of a, b, c, t.

TABLE X: Standard Potentials of Cells

Cell	$E°$	Ref
H_2|HCl(m)|AgCl, Ag	0.2224	12
	0.2232	13
H_2|HCl(m)|Hg_2Cl_2, Hg	0.2681$_5$	14
H_2|HBr(m)|Hg_2Br_2, Hg	0.1392	15
H_2|HBr(m)|AgBr, Ag	0.0710	16, 17
	0.0709	18
H_2|HI(m)|AgI, Ag	−0.1522	20
Zn, Hg|$ZnCl_2$(m)|AgCl, Ag	0.9845	46
Zn, Hg|$ZnBr_2$(m)|AgBr, Ag	0.8333	49
Zn, Hg|ZnI_2(m)|AgI, Ag	0.6097	51
In|$InCl_3$(m)|AgCl, Ag	0.5600	57

spectively, in cases where some or all of the experimental data are other than isopiestic measurements.

In evaluating parameters, various weighting schemes were used in order to fit the experimental points within their estimated uncertainty to as high a concentration as possible.

In a number of cases the basic data are not isopiestic measurements but rather cell potentials. Ordinarily we then used these experimental potential values to evaluate our parameters in a manner that also allowed readjustment of the standard cell potential. The resulting standard cell potentials are given in Table X.

Occasionally there are data of several types available for a given solute, and in several cases special procedures were used giving appropriate weights to various data in obtaining best values of the parameters.

The third virial coefficient is omitted in cases where the experimental data do not extend to high enough concentration (usually 2 M) to require that term. Additional details and comments on these evaluations are given elsewhere.[61]

The present values supercede those given in paper I, although there is no significant difference (after an inadvertant error is corrected[1]).

Hydroxides

In contrast to most other cases there are special difficulties in obtaining reliable values of activity coefficients for hydroxides. Data from two types of galvanic cell are available.

A $\quad Pt, H_2|MOH(m')|M_xHg|MOH(m'')|H_2, Pt$

B $\quad Pt, H_2|MOH(m_a)MX(m_b)|AgX, Ag$

where M_xHg indicates an amalgam electrode in cell A and X indicates either Cl or Br in cell B.

Amalgam electrode cells were employed by both Harned and Akerlof and their respective collaborators. Although the amalgam electrodes are troublesome, these investigators have used them successfully in other types of cells where the results can be checked from independent sources. The reaction for cell A involves transfer of solvent as well as solute, hence the equation for the potential includes both the activity and osmotic coefficients

$$E = \frac{RT}{F}\left[2\ln\frac{\gamma''m''}{\gamma'm'} + 0.036(m''\varphi'' - m'\varphi')\right] \quad (12)$$

Since both activity and osmotic coefficients can be expressed in terms of $\beta^{(0)}$, $\beta^{(1)}$, etc., we have evaluated these quantities directly by least squares from the original cell potentials. The values for Li, Na, K, and Cs hydrox-

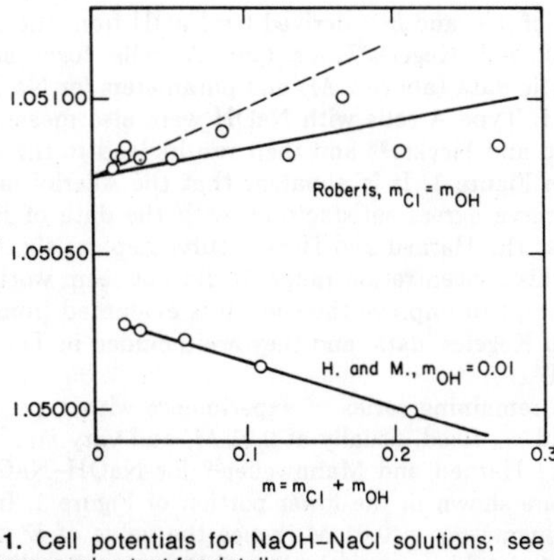

Figure 1. Cell potentials for NaOH–NaCl solutions; see eq 15 and accompanying text for details.

ides presented in Table I come from measurements of amalgam cells but with weighting of data influenced by the measurements of type B cells as discussed below. In the case of sodium hydroxide, the isopiestic measurements of Stokes[24] were used along with cell potentials.

In 1955 Guggenheim and Turgeon[25] showed that, with certain assumptions, the activity coefficients of these hydroxides could also be evaluated from the existing measurements of cells of type B. Since the solutions in these cells are mixed electrolytes, the analysis in general involves additional terms for the interaction of ions of like charge, but Guggenheim and Turgeon accepted Brønsted's principle of specific interaction and ignored these terms. We now have good evidence that such terms are not negligible; hence, we must develop equations for these cell potentials on an adequate basis. First we express the potential for a cell of type B

$$E = E° - \frac{RT}{F}\left[\ln K_w - 0.036(\varphi m) - \ln\frac{m_b}{m_a} - \ln\frac{\gamma_{X^-}}{\gamma_{OH^-}}\right] \quad (13)$$

where $m = m_a + m_b$ and K_w is the ionization constant for the solvent. Since we shall confine our use of data from this type of cell to solutions below 0.3 M, we shall ignore terms related to the third virial coefficient. Also the osmotic coefficient may be estimated with more than sufficient accuracy for the small term $0.036(\varphi m)$. The activity coefficient term may be expanded by substitution of either eq 41 and 42 or of 23, 27, 28, and 32 of paper I to yield

$$\ln(\gamma_{X^-}/\gamma_{OH^-}) = 2m\{(B_{MX}{}^{\gamma} - B_{MX}{}^{\varphi}) - (B_{MOH}{}^{\gamma} - B_{MOH}{}^{\varphi}) + (2y-1)\theta_{X,OH}\} \quad (14)$$

where $y = m_a/m$ and terms in C and ψ have been dropped as indicated above. Also we note that $B^{\gamma} - B^{\varphi} = B^{Gx}$ as defined in eq 56 of I or eq 7 of this paper.

For the particular case $y = 1/2$ which requires $m_a = m_b$, the last term in eq 14 disappears; then we can write

$$E' = E + \frac{RT}{F}\ln\frac{m_b}{m_a} = E° - \frac{RT}{F}\left[\ln K_w - 0.036(\varphi m) + 2m(B_{MOH}{}^{Gx} - B_{MX}{}^{Gx})\right] \quad (15)$$

Roberts[26] measured cells of type B with NaCl–NaOH solutions and $m_a = m_b$. His results are shown in the upper portion of Figure 1. The solid line is calculated from the values of $\beta^{(0)}$ and $\beta^{(1)}$ derived for NaOH from the data of Akerlof and Kegeles[27] for type A cells together with isopiestic data (above 2 M) and parameters for NaCl from Table I. Type A cells with NaOH were also measured by Harned and Hecker[28] and their results lead to the dashed line on Figure 1. It is apparent that the Akerlof and Kegeles curve agrees satisfactorily with the data of Roberts whereas the Harned and Hecker curve departs significantly in this concentration range. It did not seem worthwhile to attempt to improve the constants evaluated from Akerlof and Kegeles' data, and they are included in Table I for NaOH.

The remaining series of experiments with type B cells all hold m_a fixed, usually at 0.01 M, and vary m_b. The results of Harned and Mannweiler[29] for NaOH-NaCl solutions are shown in the lower portion of Figure 1. In these experiments m_a = 0.01 M; hence the value of E' at m = 0.02 M should agree with that of Roberts. The difference of 0.5 mV seems large, and standard cell calibration errors appear to be the only plausible explanation. Our primary interest, however, is in the slopes of the curves, and the difference in slopes gives the value of $\theta_{\mathrm{Cl,OH}}$. This may be seen by substituting eq 14 in 13. The result from Figure 1 is $\theta_{\mathrm{Cl,OH}} = -0.047$.

For the other alkali metals it seems best to consider the difference in potential for a given cell of type B from that of the corresponding cell with NaOH-NaCl or NaOH-NaBr solution of the same m_a and m_b. This has the effect of cancelling the term in $\theta_{\mathrm{Cl,OH}}$ or $\theta_{\mathrm{Br,OH}}$. In most cases there are experimental potentials from the same laboratory for each cell of the pair but there is no difficulty in interpolating the values for the sodium cells when necessary. The difference in potential for such a pair of cells is

$$E_{\mathrm{M}^+} - E_{\mathrm{Na}^+} = \frac{RT}{F}\left[\ln\left(\frac{\gamma_{\mathrm{OH}^-}}{\gamma_{\mathrm{X}^-}}\right)_{\mathrm{M}^+} - \ln\left(\frac{\gamma_{\mathrm{OH}^-}}{\gamma_{\mathrm{X}^-}}\right)_{\mathrm{Na}^+}\right] \quad (16)$$

if we substitute the appropriate expressions cited above for the activity coefficients, we obtain

$$mB_{\mathrm{MOH}}^{\mathrm{Gx}} + \delta = \left(\frac{F}{2RT}\right)(E_{\mathrm{M}^+} - E_{\mathrm{Na}^+}) + m(B_{\mathrm{MCl}}^{\mathrm{Gx}} + B_{\mathrm{NaOH}}^{\mathrm{Gx}} - B_{\mathrm{NaCl}}^{\mathrm{Gx}}) \quad (17)$$

Here we have added a term δ to allow for the possibility of small differences in standard cell calibration between series of experiments carried out of different times, such as apparently arose in the experiments with NaOH-NaCl, or for differences between the AgX,Ag electrodes used in different series. Presumably δ is a very small constant for any series. The parameters for the alkali halides are taken from Table I. The results, which are discussed in detail elsewhere,[62] yield good agreement with data from type A cells for KOH and CsOH. There are small but troublesome conflicts among data for LiOH and further experiments are needed to resolve them.

The data[36] for type B cells with $Ba(OH)_2$-$BaCl_2$ solutions were also evaluated by equations modified as required for the double charge on the cation.

Behavior of Second Virial Coefficients

The values of $\beta^{(0)}$ and $\beta^{(1)}$ are plotted in Figures 2-5. Considering first Figure 2, we note that the values for most inorganic salts, acids, and bases of 1-1 type fall in a rather narrow band. This is not unexpected since $\beta^{(0)}$ and $\beta^{(1)}$ depend on the same properties of the ions and the solvent although we shall see that these properties are

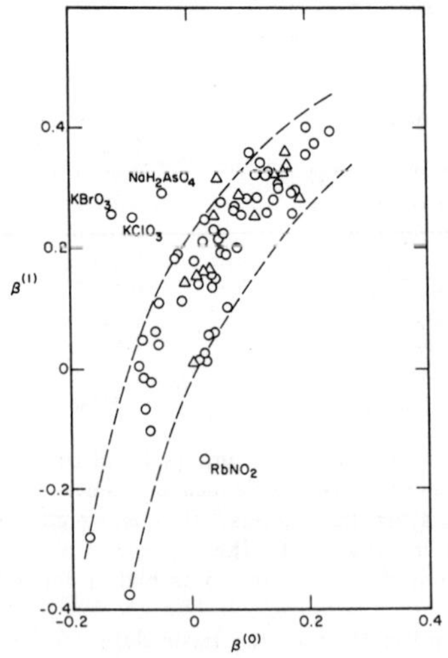

Figure 2. The relationship of $\beta^{(1)}$ to $\beta^{(0)}$ for 1-1 electrolytes. Inorganic solutes appear as circles and carboxylate salts as triangles.

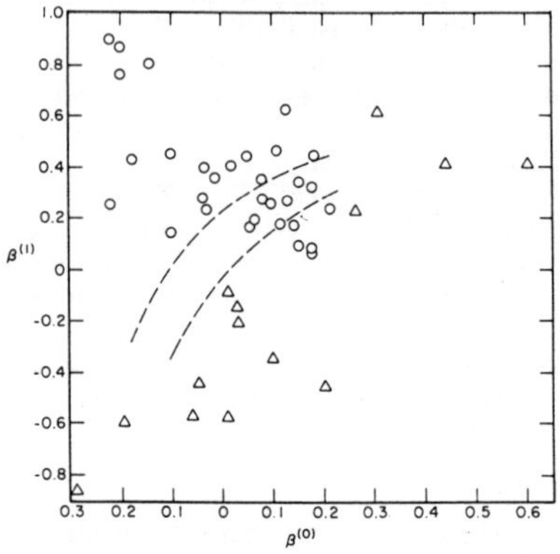

Figure 3. The relationship of $\beta^{(1)}$ and $\beta^{(0)}$ for organic electrolytes of 1-1 type. Sulfonates and sulfonic acids appear as circles and tetraalkylammonium halides as triangles. The dashed curves here show the region similarly designated on Figure 2.

weighted somewhat differently for the two parameters. Careful consideration of several solutes where measurements are precise and numerous indicates that the small but finite range of $\beta^{(1)}$ values for a given $\beta^{(0)}$ is real.

The organic electrolytes in Figure 3 and the higher valence types in Figures 4 and 5 show greater scatter but a clear trend of $\beta^{(1)}$ with $\beta^{(0)}$ is still apparent. These relationships are considered further in the Discussion.

A Convenient Approximation

For many purposes only a rough estimate of an activity coefficient is needed. Since $\beta^{(1)}$ has been shown to be determined within rather narrow limits if $\beta^{(0)}$ is known and the effect of the third virial coefficient is small, a convenient approximation becomes possible. Figure 6 shows the activity coefficient as a function of molality for 1-1 electrolytes for a series of values of $\beta^{(0)}$. In each case $\beta^{(1)}$ was given a value in the middle of the band in Figure 1, and C^{φ} was neglected.

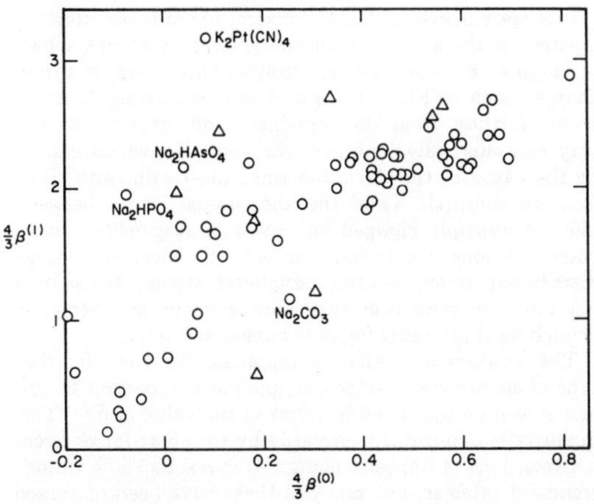

Figure 4. The relationship of $\beta^{(1)}$ to $\beta^{(0)}$ for 2–1 electrolytes. Inorganic solutes appear as circles while those with organic groups as triangles.

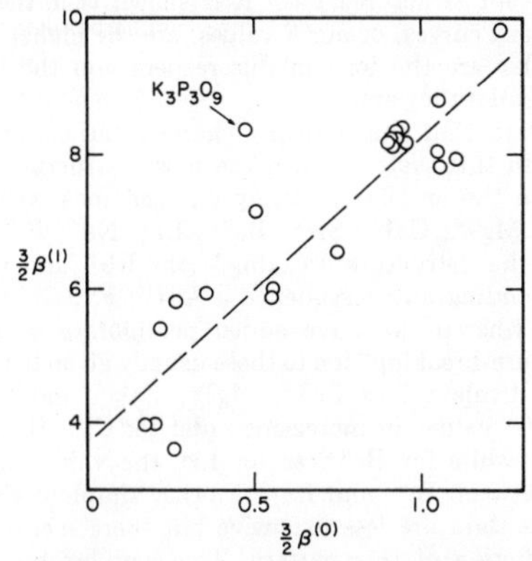

Figure 5. The relationship of $\beta^{(1)}$ to $\beta^{(0)}$ for 3–1 electrolytes.

By taking the value of $\beta^{(0)}$ from Tables I–V and interpolating between the appropriate curves on Figure 6, one can read directly an approximate value of the activity coefficient for any 1–1 electrolyte.

Similar graphs could, of course, be prepared for higher valence types, but the more extreme departures from unity of these activity coefficients reduces the utility of this procedure. Hence such graphs are omitted.

Discussion

First it is interesting to note that the equations, originally selected on the basis of their effectiveness in representing the properties of 1–1 and 2–1 electrolytes, proved to be equally effective for 3–1, 4–1, and even 5–1 type solutes. While the maximum molality values are lower for the high valence types, the maximum ionic strengths are about the same as for 1–1 and 1–2 solutes.

Let us now turn from our emphasis on convenient and accurate representation of data for thermodynamic purposes to the interpretation of the results in terms of interionic forces. In I it was shown that B^{φ} is a second virial coefficient arising from short-range forces between pairs of ions although its interpretation is complicated by the

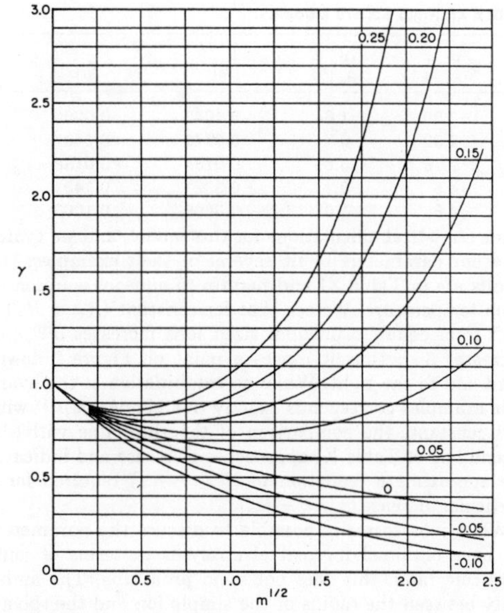

Figure 6. Activity coefficients for 1–1 electrolytes by the approximate method. The value of $\beta^{(0)}$ for each curve is indicated.

presence of the term for long-range electrostatic forces and the ionic strength dependence of the radial distribution functions. Thus much of the discussion will be qualitative in nature.

The second virial coefficient may be either positive or negative depending on the net predominance of repulsive or attractive short-range forces, respectively. Also the second virial coefficient for an electrolyte is a weighted mean of the interactions of pairs of ions with signs ++, +−, and −−, and it was shown in I that the relative weighting of +− interactions is greatest at low ionic strength. At higher ionic strength, the ++ and −− terms become more important but never equal to the weight of the +− interaction. Since multiply charged ions of the same sign will rarely approach one another closely, we can expect that the ++ or −− terms for such ions will be even smaller than for singly charged ions of the same sign and same size.

In view of this difference in weighting of +− compared to ++ and −− interactions, it is possible to determine two parameters for the short-range binary interactions of a single electrolyte (but only two). In our formulation these are $\beta^{(0)}$ and $\beta^{(1)}$. Ramanathan and Friedman[63] included more parameters in their model but, in accordance with these ideas, found that only two could be evaluated from osmotic or activity data and that any others had to be set to zero or to some other arbitrary value unless they were known from different sorts of measurements.

The simplest molecular model presenting effectively two parameters is one of rigid spherical ions of varying size. The thermodynamic properties depend primarily on the sum of radii for unlike ions ($R_+ + R_-$) but there is a secondary dependence on the absolute magnitude of the difference $|R_+ - R_-|$. The excluded volumes depend on the cubes of the distances of closest approach; hence, we expect a larger repulsive effect for like-charged ions when there is a difference in size. Rasaiah and Friedman[64]

TABLE XI: Rigid Sphere Models

$R_+ + R_-$, Å	$\|R_+ - R_-\|$, Å	$\beta^{(0)}$	$\beta^{(1)}$
2.90	1.64	0.1043	0.0200
2.90	0	0.0773	0.0497
2.76	0.86	0.0734	0.0188
4.6	0	0.3221	0.1431
4.6	1.0	0.3506	0.1063

made statistical calculations for this model, and we evaluated our parameters to fit several of their examples; the results are in Table XI and pertain to aqueous solution at room temperature. We see that for constant $(R_+ + R_-)$ a shift from equal to unequal sized ions increases $\beta^{(0)}$ and decreases $\beta^{(1)}$; thus it moves a point on Figure 2 downward and to the right. While no combination of the available examples corresponds exactly to a change of $\beta^{(1)}$ with $\beta^{(0)}$ constant, the comparison of the third line with the second line of Table XI approaches this case and indicates the approximate reduction in $(R_+ + R_-)$ required for a particular increase in $|R_+ - R_-|$.

While it is formally possible to discuss the parameters we have obtained for real electrolytes in terms of ionic hard-core radii, this does not seem promising. The ambiguity between the radius of the simple ion and the solvated ion is well known. Also there seems to be no correlation of the parameters in Table I with the expectations based upon ionic radii determined from crystals.

In their model for the intermolecular potential Ramanathan and Friedman[63] assumed electrostatic and repulsive terms based upon independent knowledge of charges and of radii in crystals and then evaluated empirically coefficients of a term, first introduced by Gurney,[65] which can be considered to represent the net effect of solvation, dispersion forces, and any other effects of similar range. Our results support this general concept that the second virial coefficients for electrolytes are best discussed in terms of a combination of repulsive radii determined from crystals or other sources and slightly longer range forces which arise from solvation effects, from dispersion forces, and in some cases also from such other effects as would arise from permanent dipoles or multipoles (for example, with OH^- or NO_3^-). The contribution of each of these various types of short-range forces to ++ and −− interactions will not follow the same relationship as its contribution to +− interactions; hence $\beta^{(1)}$ will not be exactly determined by $\beta^{(0)}$. But the relationship is complex, and we do not expect any simple pattern related to ionic size.

In Figures 2–5 we have noted that in each case the points for most solutes fall in a band with positive slope; this indicates that these short-range forces follow similar patterns for ++ and −− as compared to +− interionic interactions. In some cases the points outside these bands are individually labeled. Thus in Figure 2 the point for $RbNO_2$ falls below the band while the points for other nitrites are within but near the lower boundary of the band. Since these data for nitrites do not appear to be of very high accuracy, it is probable that $RbNO_2$ is not really anomalous. The results for all nitrites, however, indicate more repulsive or less attractive forces between pairs of nitrite ions than is typical for most anions. Corresponding statements can be made about the labeled points above the band with the conclusion that like ion interactions are less repulsive or more attractive than normal for BrO_3^-, ClO_3^-, and $H_2AsO_4^-$.

It is not surprising that the magnitude of these effects is greater for the salts involving large organic groups shown in Figure 3. For the electrolytes involving multiply charged ions in Figures 4 and 5 it is surprising that the points farthest from the principal band involve only the very common univalent ions Na^+ and K^+ which cannot be the cause of the deviation since most salts with these ions are normal. Thus the short range forces between pairs of multiply charged ions must be responsible. In all cases the ions are polyatomic with the negative charge distributed among several peripheral atoms; hence it is not too surprising that they approach one another close enough for short-range forces to have some effect.

The location of a given point along the band for that type of electrolyte is, of course, the more important factor, and it will be discussed in terms of the value of $\beta^{(0)}$. This quantity is determined primarily by the short-range forces between ions of opposite charge. Our results show various trends of interest, but many of these have been discussed previously by Gurney[65] and Frank[66] among others. These authors emphasize a classification of both cations and anions with respect to a tendency to enhance or to disrupt the solvent structure. Then it is shown that the activity coefficient curves, or our β values, are the higher the more dissimilar are the ions in this respect and the lower the more similar they are.

We find that the multiply charged cations are readily added to this system. Then the most "structure-making" ions are the smaller multiply charged ions with the sequence Mg^{2+}, Ca^{2+}, Sr^{2+}, Ba^{2+}, Li^+, Na^+, K^+ continuing to the "structure-breaking" ions Rb^+ and Cs^+. The corresponding anion sequence is OH^-, F^-, Cl^-, Br^-, I^-, ClO_4^- wherein we have added perchlorate as the most "structure-breaking" ion to those usually given heretofore.

The divalent ions Zn^{2+}, Mg^{2+}, Ca^{2+}, and Sr^{2+} each have $\beta^{(0)}$ values in increasing order for Cl^-, Br^-, I^-, and ClO_4^-; while for Ba^{2+} as for Li^+ the values for ClO_4^- drop between Br^- and I^-. For other simple divalent cations the data are less extensive but there seems to be no contradiction of this pattern. The complex but compact ion UO_2^{2+} also follows this trend, but large cobalt complex ions show the opposite sequence and may be classified as "structure breakers."

When one ion is of intermediate character, for example, Na^+ or K^+, then the sequence may be irregular. For potassium there is a smooth trend of $\beta^{(0)}$ from OH^- down through F^- to Cl^- and then back upward for Br^- and I^-. But for sodium, the hydroxide and perchlorate both take intermediate values. Hydrogen ion in strong acids is a "structure maker" with properties close to those of Li^+. The sulfate ion appears to fall between F^- and Cl^-; its $\beta^{(0)}$ values drop from Li^+ to Na^+ and then rise through K^+, Rb^+, and Cs^+.

The $\beta^{(0)}$ values for nitrates are relatively low in all cases regardless of the structure making or breaking character of the cation. This suggests some tendency toward ion pair formation as is also indicated by spectroscopic data in several cases. Lemley and Plane[67] report a particularly complete spectral study for zinc nitrate while Peleg[68] reports a similar study of magnesium nitrate; these authors refer to work on other nitrates.

The electrolytes involving ClO_3^-, BrO_3^-, $H_2PO_4^-$, $H_2AsO_4^-$, BO_2^-, BF_4^-, and PF_6^- all have low $\beta^{(0)}$ values, but data are available for only one or two cations in most cases. Presumably these large anions are structure breakers and in addition tend toward ion pair formation as do nitrates. Nevertheless, their behavior is well represented by or equations, and it is not necessary to assume incomplete dissociation.

It has been noted previously that acetate is a weak "structure maker," and our results agree. The values for nitrite follow those for chlorides rather closely; hence nitrite is a weak "structure breaker." The electrolytes involving large organic ions have been measured recently, and the papers presenting these results are accompanied in most cases by discussions making use of recent theory; hence we shall not comment further on these cases.

Bromley[69] has independently developed a system of representation and estimation for activity coefficients of strong electrolytes, and we have enjoyed discussions with him. His system, although developed simultaneously, may be described as a simplification of the present system. The third virial coefficient is omitted, and the second virial coefficient is modified into a form, still dependent on ionic strength, but with a single parameter B. This is, in effect, a relationship between our $\beta^{(1)}$ and $\beta^{(0)}$. He further shows that his B values can be approximated from two parameters for each ion B_M and δ_M or B_X and δ_X by the equation

$$B = B_M + B_X + \delta_M\delta_X$$

Structure-making cations have positive δ_M while structure breakers have negative δ_M. For anions the signs are reversed with structure makers having negative δ_X, and the struture breakers having positive δ_X. Thus the last term gives the effect just discussed in which $\delta_M\delta_X$ is the more positive the more dissimilar are pairs of ions with respect to this quality and is negative for pairs in which both ions have the same characteristic (either structure making or structure breaking).

Although less accurate than our equations, Bromley's system is quite effective and will be useful for many purposes. He also compares his results with a somewhat similar system of Meissner and Tester.[62]

Electrolytes of the 2–2 type and others involving ion pairing will be treated in a subsequent paper. Also we are proceeding to calculations for mixed electrolytes based upon the constants here obtained; indeed the convenience of our equations as applied to mixed electrolytes was a primary incentive for this work.

Acknowledgment. This research was sponsored by the U. S. Atomic Energy Commission.

Supplementary Material Available. Additional details concerning the evaluation of data for individual solutes will appear following these pages in the microfilm edition of this volume of the journal. Photocopies of the supplementary material from this paper only or microfiche (105 × 148 mm, 20× reduction, negatives) containing all of the supplementary material for the papers in this issue may be obtained from the Journals Department, American Chemical Society, 1155 16th St., N.W., Washington, D. C. 20036. Remit check or money order for $4.00 for photocopy or $2.00 for microfiche, referring to code number JPC-73-2300.

References and Notes

(1) K. S. Pitzer, *J. Phys. Chem.*, **77**, 268 (1973). In Table I the entries for 2–1 and 1–2 electrolytes should be reduced by the factor ¾ for $\beta^{(0)}$ and $\beta^{(1)}$. Also eq 50 should have a plus rather than a minus sign before $\beta^{(1)}$, and on p 276 $\psi(Na,K,NO_3) = -0.0012$.
(2) J. E. Mayer, *J. Chem. Phys.*, **18**, 1426 (1959).
(3) R. A. Robinson and R. M. Stokes, "Electrolyte Solutions," 2nd ed, revised, Butterworths, London, 1965.
(4) N. P. Chekhunova, P. J. Protsenko, and L. N. Venerovskaya, *Russ. J. Phys. Chem.*, **43**, 1158 (1969).
(5) R. F. Platford, *Can. J. Chem.*, **47**, 2271 (1969).
(6) G. Akerlof and P. Bender, *J. Amer. Chem. Soc.*, **70**, 2366 (1948).
(7) N. P. Chekhunova and P. I. Protsenko, *Russ. J. Phys. Chem.*, **41**, 2266 (1967).
(8) R. A. Robinson, J. M. Stokes, and R. M. Stokes, *J. Phys. Chem.*, **65**, 542 (1961).
(9) H. Ti Tien, *J. Phys. Chem.*, **67**, 532 (1963).
(10) A. K. Covington and D. E. Irish, *J. Chem. Thermodyn.*, **17**, 175 (1972).
(11) O. E. Esval and S. Y. Tyree, *J. Phys. Chem.*, **66**, 940 (1962).
(12) H. S. Harned and R. W. Ehlers, *J. Amer. Chem. Soc.*, **55**, 2179 (1933).
(13) G. Akerlöf and J. W. Teare, *J. Amer. Chem. Soc.*, **59**, 1855 (1937).
(14) S. R. Gupta, G. J. Hills, and D. J. Ives, *Trans. Faraday Soc.*, **53**, 1874 (1963).
(15) S. R. Gupta, G. J. Hills, and D. J. Ives, *Trans. Faraday Soc.*, **53**, 1886 (1963).
(16) H. S. Harned, A. S. Keston, and J. G. Donelson, *J. Amer. Chem. Soc.*, **58**, 989 (1936).
(17) W. J. Bierman and R. S. Yamasaki, *J. Amer. Chem. Soc.*, **77**, 241 (1955).
(18) H. B. Hetzer, R. A. Robinson, and R. G. Bates, *J. Phys. Chem.*, **66**, 1423 (1962).
(19) H. S. Harned and R. A. Robinson, *Trans. Faraday Soc.*, **37**, 302 (1941): (HI).
(20) H. B. Hetzer, R. A. Robinson, and R. G. Bates, *J. Phys. Chem.*, **68**, 1929 (1964).
(21) H. S. Harned and M. A. Cook, *J. Amer. Chem. Soc.*, **59**, 1290 (1937): (KCl).
(22) T. Shedlovsky and D. A. MacInnes, *J. Amer. Chem. Soc.*, **59**, 503 (1937): (KCl).
(23) (a) T. Mussini, P. Longlin, and G. Riva, *J. Chem. Thermodyn.*, **4**, 591 (1972); (b) R. A. Robinson and D. A. Sinclair, *J. Amer. Chem. Soc.*, **56**, 1830 (1934): (CsCl).
(24) R. H. Stokes, *J. Amer. Chem. Soc.*, **67**, 1689 (1945).
(25) E. A. Guggenheim and J. C. Turgeon, *Trans. Faraday Soc.*, **51**, 747 (1955).
(26) E. J. Roberts, *J. Amer. Chem. Soc.*, **52**, 3877 (1930).
(27) G. Akerlof and G. Kegeles, *J. Amer. Chem. Soc.*, **62**, 620 (1940).
(28) H. S. Harned and J. C. Hecker, *J. Amer. Chem. Soc.*, **55**, 4838 (1933).
(29) H. S. Harned and G. E. Mannweiler, *J. Amer. Chem. Soc.*, **57**, 1873 (1935).
(30) H. S. Harned and W. J. Hamer, *J. Amer. Chem. Soc.*, **55**, 2194, 4496 (1933): (KOH).
(31) H. S. Harned and M. A. Cook, *J. Amer. Chem. Soc.*, **59**, 496 (1937): (KOH).
(32) H. S. Harned and O. E. Schupp, Jr., *J. Amer. Chem. Soc.*, **52**, 3886, 3892 (1930): (CsOH).
(33) H. S. Harned and F. E. Swindells, *J. Amer. Chem. Soc.*, **48**, 126 (1926): (LiOH).
(34) H. S. Harned and H. R. Copson, *J. Amer. Chem. Soc.*, **55**, 2206 (1933): (LiOH).
(35) H. S. Harned and J. G. Donelson, *J. Amer. Chem. Soc.*, **59**, 1280 (1937): (LiOH).
(36) H. S. Harned and C. G. Geary, *J. Amer. Chem. Soc.*, **59**, 2032 (1937).
(37) W. Y. Wen, S. Saito, and C. Lee, *J. Phys. Chem.*, **70**, 1244 (1966).
(38) S. Lindenbaum and G. E. Boyd, *J. Phys. Chem.*, **68**, 911 (1964).
(39) V. E. Bower and R. A. Robinson, *Trans. Faraday Soc.*, **59**, 1717 (1963).
(40) H. P. Gregor, M. Rothenberg, and N. Fine, *J. Phys. Chem.*, **67**, 1110 (1963).
(41) O. D. Bonner and O. C. Rogers, *J. Phys. Chem.*, **64**, 1499 (1960).
(42) S. Lindenbaum and G. E. Boyd, *J. Phys. Chem.*, **71**, 581 (1967).
(43) G. E. Boyd, A. Schwarz, and S. Lindenbaum, *J. Phys. Chem.*, **70**, 821 (1966).
(44) W. Y. Wen and S. Saito, *J. Phys. Chem.*, **69**, 3569 (1965).
(45) S. Lindenbaum, *J. Phys. Chem.*, **72**, 212 (1968).
(46) G. Scatchard and R. F. Tefft, *J. Amer. Chem. Soc.*, **52**, 2272 (1930).
(47) R. A. Robinson and R. H. Stokes, *Trans. Faraday Soc.*, **36**, 740 (1940): ($ZnCl_2$).
(48) Calculated from the Gibbs energy values of D. D. Wagman, *et al.*, *Nat. Bur. Stand. (U. S.), Tech. Note*, **No. 270-3** (1968): ($ZnCl_2$).
(49) R. H. Stokes and J. M. Stokes, *Trans. Faraday Soc.*, **41**, 688 (1945).
(50) R. H. Stokes, J. M. Stokes, and R. A. Robinson, *Trans. Faraday Soc.*, **40**, 533 (1944): ($ZnBr_2$).
(51) R. G. Bates, *J. Amer. Chem. Soc.*, **60**, 2983 (1938).
(52) R. M. Stokes, *Trans. Faraday Soc.*, **41**, 12 (1945): (ZnI_2).
(53) K. O. Groves, J. L. Dye, and C. H. Brubaker, *J. Amer. Chem. Soc.*, **82**, 4445 (1960).
(54) W. L. Masterton, T. L. Munnelly, and L. H. Berka, *J. Phys. Chem.*, **71**, 942 (1967).
(55) O. D. Bonner and O. C. Rogers, *J. Phys. Chem.*, **65**, 981 (1961).
(56) D. J. Karl and J. L. Dye, *J. Phys. Chem.*, **66**, 550 (1962): ($Co(en)_3Cl_3$).

(57) A. K. Covington, M. A. Hakeem, and W. F. K. Wynne-Jones, *J. Chem. Soc.*, **4394** (1963).
(58) R. A. Wynveen, J. L. Dye, and C. H. Brubaker, *J. Amer. Chem. Soc.*, **82**, 4441 (1960).
(59) C. H. Brubaker and T. E. Haas, *J. Phys. Chem.*, **65**, 866 (1961).
(60) G. Miller and A. S. Porter, *Trans. Faraday Soc.*, **63**, 335 (1967).
(61) Additional details concerning the evaluation of data for individual solutes are given in Lawrence Berkeley Laboratory Report No. LBL-1434 and will also appear as supplementary material. See paragraph at end of paper regarding supplementary material.
(62) H. P. Meissner and J. W. Tester, *Ind. Eng. Chem., Process Des. Develop.*, **11**, 128 (1972).
(63) P. S. Ramanathan and H. L. Friedman, *J. Chem. Phys.*, **54**, 1086 (1971).
(64) J. C. Rasaiah and H. L. Friedman, *J. Chem. Phys.*, **48**, 2742 (1968); **50**, 3965 (1969).
(65) R. W. Gurney, "Ionic Processes in Solution," McGraw-Hill, New York, N. Y., 1953, Chapter 16.
(66) H. S. Frank, *Z. Phys. Chem. (Leipzig)*, **228**, 364 (1965).
(67) A. T. G. Lemley and R. A. Plane, *J. Chem. Phys.*, **57**, 1648 (1972).
(68) M. Peleg, *J. Phys. Chem.*, **76**, 1019 (1972).
(69) L. A. Bromley, *AIChE J.*, **19**, 313 (1973).

Kenneth S. Pitzer and Guillermo Mayorga: Thermodynamics of Electrolytes. II. Activity and Osmotic Coefficients for Strong Electrolytes with One or Both Ions Univalent.

Page 2301. In Table III the parameters for the tetraalkylammonium chlorides and bromides (but not the fluorides or iodides) should be increased as follows: for $\beta^{(0)}$, +0.0281$_5$; for $\beta^{(1)}$, +0.0542; for C^{ϕ}, +0.00211. For $(Et)_4$-NI corrected constants are $\beta^{(0)} = -0.179$, $\beta^{(1)} = -0.571$, and $C^{\phi} = 0.0412$.

Page 2303. Sign errors should be corrected by reversing the sign of E in eq 12, the signs of the last two terms in eq 13, and the sign preceding $2m$ in eq 15.—Kenneth S. Pitzer

Journal of Solution Chemistry, Vol. 3, No. 7, 1974

Thermodynamics of Electrolytes. III. Activity and Osmotic Coefficients for 2-2 Electrolytes

Kenneth S. Pitzer[1] and Guillermo Mayorga[1]

Received October 29, 1973; revised December 20, 1973

The peculiar behavior of 2–2 and higher valence type electrolytes is discussed in terms of various theories some of which assume, while others do not, an equilibrium between separated ions and ion pairs as distinct chemical species. It is recognized that in some cases a distinct species of inner-shell ion pairs is indicated by spectroscopic or ultrasonic data. Nevertheless, there are many advantages in representing, if possible, the properties of these electrolytes by appropriate virial coefficients and without chemical association equilibria. It is shown that this is possible and is conveniently accomplished by the addition of one term to the equations of Parts I and II of this series. The coefficients of these equations are given for nine solutes. It is also noted that these equations have been successfully applied to mixed electrolytes involving one component of the 2–2-type.

KEY WORDS: Activity coefficient; osmotic coefficient; thermodynamics; electrolytes; aqueous solutions; sulfate solutions.

1. INTRODUCTION

The consideration of 2–2 and higher valence types of aqueous electrolytes was omitted from Parts I(1) and II(2) of the present series (cited hereafter as I and II) because of certain well-known peculiarities of such solutes.

Soon after the publication of the Debye–Hückel theory, Bjerrum(3) showed that an ion association effect could be expected on the basis of purely electrostatic forces for 2–2 electrolytes in water or for lower valence types in solvents of lower dielectric constant. Various authors have dealt with the experimental data for the bivalent metal sulfates and other 2–2 solutes on this basis; recent papers are by Gardner and Glueckauf(4) and by Pitzer.(5)

[1] Inorganic Materials Research Division of the Lawrence Berkeley Laboratory and Department of Chemistry, University of California, Berkeley, California 94720.

Although these ion-association treatments fitted the data well, there are serious disadvantages which lead one to seek a better method.

It has also been shown by Gronwall, LaMer, and Sandved[6] and later with steadily improving rigor and accuracy by Guggenheim,[7] by Gardner and Glueckauf,[8] and by Rasaiah[9] that these 2–2 electrolytes could be treated by statistical theory without the assumption of an association equilibrium provided one avoids the Debye–Hückel approximation of linearization of the exponential in the Boltzmann distribution. The equations now become so complex, however, that one has only a few numerical results for particular ionic models.

Thus we seek equations which are simpler than those now available but which still reflect adequately the complexities of actual thermodynamic properties of the solutions.

2. DISCUSSION—ASSOCIATION OR NONASSOCIATION

While we have noted that the thermodynamic properties of 2–2 electrolytes do not require the assumption of a discrete chemical species of ion pairs, we recognize that spectroscopic[10,11] and ultrasonic[12,13] measurements do

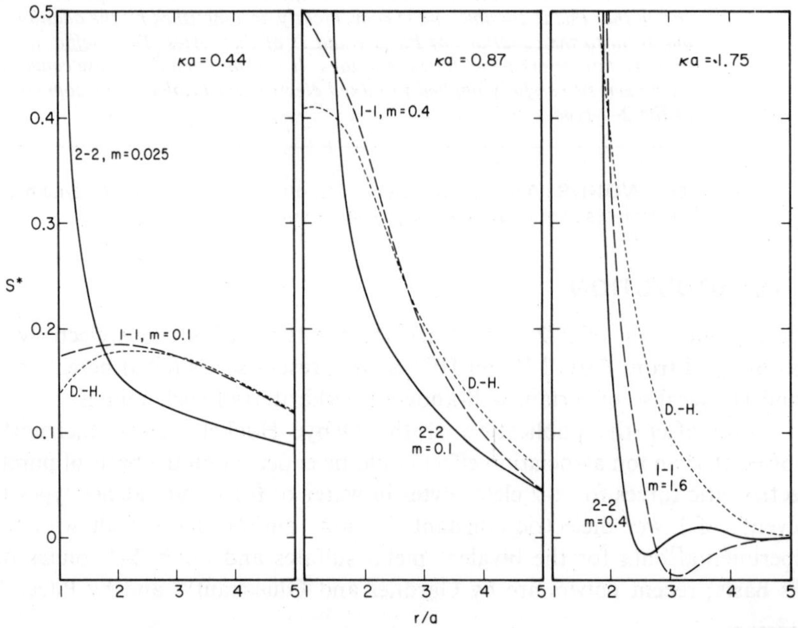

Fig. 1. Radial charge distributions for 1–1 and 2–2 electrolytes at three different ionic strengths.

indicate a discrete species of inner-shell ion pairs in some cases. Thus, for some purposes it is clear that ion association must be recognized. In many cases, however, the proportion of inner-shell ion pairs is never large, and it may be a satisfactory approximation for thermodynamic purposes to ignore the distinction between inner-shell and solvent-separated ion pairs. In that case one has returned to the broader question whether ion pairing need be explicitly recognized as the formation of a distinct species or merely as the inner portion of a continuous radical distribution.

The recent theoretical paper of Rasaiah[9] presents not only his calculations by the HNC method for 2–2 electrolytes but also comparisons with other methods. Especially interesting are his radial charge density curves which are reproduced in Fig. 1. These are for the "primitive" model with a hard-core diameter $a = 4.2$ Å, the macroscopic dielectric constant of water, and a temperature of 25°C. The curves are presented in terms of a relative radius r/a snd a reduced radial charge density defined as

$$s^* = 4\pi r^2 ac(g_{+-} - g_{++}) \tag{1}$$

where c is the concentration (shown as m on the figure) and g_{+-} and g_{++} are the radial distribution functions for unlike and like pairs of ions, respectively. The figure shows the curves calculated by the HNC method for 1–1 as well as 2–2 electrolytes at equal ionic strength and for the Debye–Hückel theory. The latter theory yields the simple equation

$$s^*_{\text{DH}} = \frac{\kappa^2 ar}{1 + \kappa a} e^{-\kappa(r-a)} \tag{2}$$

where κ is the usual Debye–Hückel inverse distance.

The striking aspect of the results shown in Fig. 1 is that the HNC charge distributions for 1–1 and 2–2 electrolytes differ most at a very low, but finite, concentration and then become more nearly alike at higher concentration. All three curves are, of course, exactly alike in the limit of zero concentration. At $\kappa a = 0.44$ the ion association effect for 2–2 solutes is clearly shown by the sudden increase in charge density at r/a just above unity, which is absent in the 1–1 case. But, as the concentration increases further, this region of anomalously large charge density expands while the "normal" diffuse charge cloud is contracting. Thus at $\kappa a = 1.75$ the 1–1 and 2–2 curves are similar in placing almost all the charge in the range 1 to 2 in r/a with only a small oscillating density beyond.

The osmotic coefficient for $CuSO_4$ shown in Fig. 2 displays this same peculiarity in another way. The solid curve fitting the data at low as well as higher concentrations approaches the theoretical Debye–Hückel slope from the lower side at very low concentration. The dashed curve fits the data above 0.1 M equally well but has a simpler form typical of 1–1 or 2–1

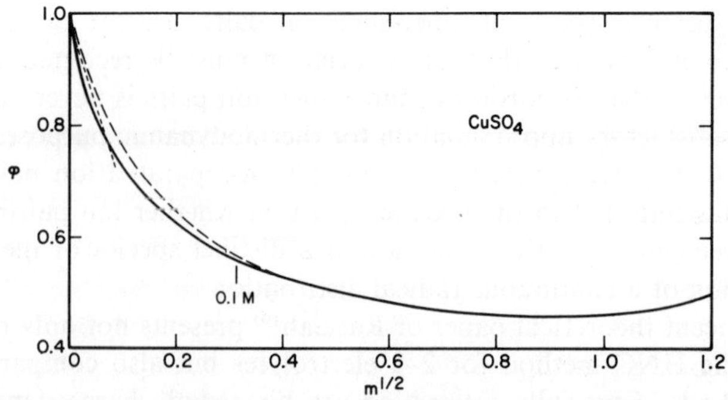

Fig. 2. The osmotic coefficient of $CuSO_4$. The solid line follows all experimental data; the straight dotted line is the Debye–Hückel limiting slope; the dashed line is an approximate curve typical of lower valence electrolytes which fits in this case only above 0.2 *M* and in the limit at infinite dilution.

electrolytes and approaches the theoretical slope from the upper side. Thus, the difference or anomaly in the osmotic coefficient is confined to an intermediate concentration range approximately 0.0001 to 0.1 *M*. Various authors[14] have had no difficulty fitting the data for 2–2 electrolytes above 0.1 *M* to the normal equations used for other types of strong electrolytes.

Further insight into this situation can be obtained from a simple calculation assuming ion association. The equations are familiar and are written for a symmetrical *z–z* electrolyte

$$M^{+z} + X^{-z} = MX$$

$$K_{eq} = [\alpha/(1-\alpha)^2][\gamma_{MX}/m\gamma_M\gamma_X] \tag{3}$$

Here m is the molality, αm is the molality of the ion pair, and the activity coefficients have their usual meaning. For 1–1 electrolytes the second bracketed factor unambiguously decreases with increase in m since changes in the γ's are relatively small. Thus the association steadily increases with molality (even though it may be very small at all molalities of interest). For 2–2 electrolytes it becomes important to specify the distance of closest approach used in an expression for activity coefficients. If this is several times the hard-core diameter, as suggested by Bjerrum and adopted by others, then a similar steady increase in association with concentration is observed.

But it seems also possible to regard the MX species as only the excess of closely paired ions over a "normal" distribution and to use a typical hard-core diameter of 3 or 4 Å in the activity coefficient expression. Then it is found for 2–2 and higher types (in water at room temperature) that the activity coefficients γ_M and γ_X decrease so rapidly that the entire second bracketed factor in Eq. (3) actually passes through a minimum at some low concentration and then increases with further increase in m. Consequently, one would have a maximum degree of association at this intermediate concentration.

Davies[15] has discussed this type of calculation in detail, both on the basis given above and on the basis of conductance data, and shows that maxima of association are found for typical 2–2 electrolytes in the range 0.03 to 0.1 M and for 3–2, 4–2, or 3–3 electrolytes at even lower concentrations.

It is certainly possible to deal with these electrolytes on the basis of association equilibria, but there are at least two objections. From the viewpoint of theoretical elegance, the need to assume an arbitrary maximum distance of separation for ion pairs is distasteful. Association equilibrium constants are meaningless unless this distance is specified as well as the formulas for the individual ion activity coefficients. Probably more serious is the ambiguity in the ionic strength, which also depends on these rather arbitrary assumptions, yet is a quantity one wishes to use in various parts of the calculation. In mixed electrolytes this becomes particularly troublesome since the ionic strength will affect the properties of other components as well. In addition, there is the practical disadvantage of an ionic strength which depends on an association equilibrium; this introduces the complexity of iterative solution of simultaneous equations of complicated types.

We conclude that it is desirable to represent the properties of these 2–2 and higher valence types of solutes without a dissociation equilibirum if possible. The test of success with a particular form of equation will be the accuracy of representation of observed properties of mixed as well as pure electrolytes.

3. EQUATIONS FOR PURE 2–2 ELECTROLYTES

Since the peculiarities under discussion relate to interactions between pairs of ions, we must seek to account for them by the second virial coefficient. Let us consider an equation of the form adopted in I and II but with one additional term

$$B^{\phi} = \beta^{(0)} + \beta^{(1)} e^{-\alpha_1 I^{1/2}} + \beta^{(2)} e^{-\alpha_2 I^{1/2}} \tag{4}$$

If $\beta^{(0)}$ and $\beta^{(1)}$ represent the same effects of short-range forces for 2–2 electrolytes as for other types of solutes, we can seek values for $\beta^{(2)}$ and α_2 which reproduce the anomalous behavior of 2–2 electrolytes in the range below 0.1 M. It is found that this is accomplished with a large negative value of $\beta^{(2)}$ and a large positive α_2. Also one can show theoretically by an appropriate expansion that in the limit of very dilute solutions the ion-pair association constant K is $-2\beta^{(2)}$ and α_2 is $32A_{\phi}$, where A_{ϕ} is the Debye–Hückel coefficient for the osmotic coefficient. For water at room temperature, $32A_{\phi} =$

12.5. Our purpose, however, is to fit the osmotic coefficient over the entire concentration range; hence we regard the results of the expansion for very dilute solutions merely as a confirmation of the plausibility of the mathematical form of Eq. (4).

The available data for 2–2 metal sulfates were discussed in some detail in a recent publication[5] where the correction of osmotic coefficients from 0 to 25°C was also considered. We have also included the data on uranyl sulfate and a recent series of isopiestic measurements[16] on $MgSO_4$. In the present calculations we used osomotic coefficients directly from Robinson and Stokes[17] or those corrected to 25°C from the values of Brown and Prue.[18] In the case of $CdSO_4$ one must depend on electrochemical cell measurements, and we worked with the osmotic coefficients previously reported[5] to be consistent with the cell potentials.

The complete equation, including the Debye–Hückel term and a simple third virial coefficient, is, for a 2–2 electrolyte, as follows

$$\phi - 1 = 4f^{\phi} + mB^{\phi}_{MX} + m^2C^{\phi}_{MX} \tag{5}$$

$$f^{\phi} = -A_{\phi}[I^{1/2}/(1 + bI^{1/2})] \tag{6}$$

$$B^{\phi}_{MX} = \beta^{(0)}_{MX} + \beta^{(1)}_{MX}e^{-\alpha_1 I^{1/2}} + \beta^{(2)}_{MX}e^{-\alpha_2 I^{1/2}} \tag{7}$$

For the activity coefficient the corresponding equation is

$$\ln \lambda = 4f^{\gamma} + mB^{\gamma}_{MX} + m^2C^{\gamma}_{MX} \tag{8}$$

$$f^{\gamma} = -A_{\phi}[I^{1/2}/(1 + bI^{1/2}) + (2/b)\ln(1 + bI^{1/2})] \tag{9}$$

$$\begin{aligned} B^{\gamma}_{MX} = {} & 2\beta^{(0)}_{MX} + (2\beta^{(1)}_{MX}/\alpha_1^2 I)[1 - (1 + \alpha_1 I^{1/2} - \tfrac{1}{2}\alpha_1^2 I)e^{-\alpha_1 I^{1/2}}] \\ & + (2\beta^{(2)}_{MX}/\alpha_2^2 I)[1 - (1 + \alpha_2 I^{1/2} - \tfrac{1}{2}\alpha_2^2 I)e^{-\alpha_2 I^{1/2}}] \end{aligned} \tag{10}$$

$$C^{\gamma}_{MX} = \tfrac{3}{2}C^{\phi}_{MX} \tag{11}$$

Here the terminology is the same as in I and II, and the value $b = 1.20$ is retained. The differences are the addition of the last term in the second virial coefficient and the substitution of numerical values for a 2–2 electrolyte. Note that the superscripts ϕ, γ, (0), etc. are labels rather than exponents. All parameters that are adjusted for each substance are given the subscript MX.

The full array of data for the sulfates of Mg, Ni, Cu, Zn, and Cd were fitted by least-squares adjustment of the four parameters $\beta^{(0)}$, $\beta^{(1)}$, $\beta^{(2)}$, and C^{ϕ} for each substance. A series of values of α_1 and α_2 were used. While the optimum values of α_1 and α_2 varied a little from case to case, the values $\alpha_1 = 1.4$ and $\alpha_2 = 12.0$ fitted all cases very well and were adopted for all 2–2 electrolytes. This value of α_1 is somewhat smaller than the value 2.0 used previously in I and II. The value 12 for α_2 is close to that suggested by the theory mentioned above. The resulting values, the concentration range of

Table I. Parameters for Thermodynamic Functions for 2–2 Electrolytes in Water at 25°C ($b = 1.2$, $\alpha_1 = 1.4$, $\alpha_2 = 12.0$ throughout)

Electrolyte	$\beta^{(0)}$	$\beta^{(1)}$	$\beta^{(2)}$	C^{ϕ}	Range	σ
$MgSO_4$	0.2210	3.343	−37.23	0.0250	0.006–3.0	0.004
$NiSO_4$	0.1702	2.907	−40.06	0.0366	0.005–2.5	0.005
$CuSO_4$	0.2358	2.485	−47.35	−0.0012	0.005–1.4	0.003
$ZnSO_4$	0.1949	2.883	−32.81	0.0290	0.005–3.5	0.004
$CdSO_4$	0.2053	2.617	−48.07	0.0114	0.005–3.5	0.002
$CoSO_4$	0.20	2.70	−30.7	—	0.006–0.10	0.003
$CaSO_4$	0.20	2.65	−55.7	—	0.004–0.011	0.003
$BeSO_4$	0.317	2.914	?	0.0062	0.1–4.0	0.004
$MnSO_4$	0.201	2.980	?	0.0182	0.1–4.0	0.003
UO_2SO_4	0.322	1.827	?	−0.0176	0.1–5.0	0.003

data used, and the accuracy of fit are given in Table I. It is apparent that agreement is obtained substantially within experimental error.

For the first five solutes in Table I the data extend over a wide concentration range, and the resulting parameters should be reliable. Since data for $CoSO_4$ and $CaSO_4$ are limited to dilute solutions, because of solubility in the case of the calcium salt, the third virial coefficient was omitted and $\beta^{(0)}$ was set at 0.20 on the basis of the results for other solutes. The data sufficed to evaluate the two remaining parameters.[2]

There are no data below 0.1 M for the last three solutes, $BeSO_4$, $MnSO_4$, and UO_2SO_4; hence no meaningful value can be given for $\beta^{(2)}$. If an estimate is desired, the rounded value −40 might be chosen as typical of this type of solute.

For the osmotic coefficient the large value of α_2 makes the term in $\beta^{(2)}$ negligible above 0.1 M. We see from Eq. (10), however, that the corresponding term in the logarithm of the activity coefficient approaches the constant value $\beta^{(2)}/288$ at high concentration. Since this term is constant, it does not affect the relative values of the activity coefficient (or other functions) at various concentrations above 0.1 M, but it does affect the relationship to the solute standard state and the absolute values of the activity coefficients. This indicates the category of calculations which are valid even though $\beta^{(2)}$ is not known and the different category where $\beta^{(2)}$ is required.

Even when the data extend to low enough concentration to determine

[2] The freezing point data for $CoSO_4$ appear to depart slightly from the family of curves for other bivalent sulfates. The parameters in Table I suppress this anomaly and yield the best fit within the normal pattern. If all three β parameters are adjusted freely, a set such as 0.70, 1.60, and −25.0 yields better agreement, $\sigma = 0.002$, but these parameters are so inconsistent with the general pattern that we hesitate to recommend their use.

$\beta^{(2)}$ rather accurately, there is some uncertainty in the implied extrapolation to the solute standard state. In this connection it is interesting to compare the present results with those of the association treatment[5] published earlier. For Mg, Ni, Cu, Zn, and Cd sulfates the average difference in activity coefficients at 0.1 M is below 2% with a maximum of 3%. This pattern may be expected to hold in other cases where the experimental data extend down to approximately 0.005 M.

This requirement for experimental data at very low concentration becomes even more stringent for still higher charged electrolytes, e.g., 2–3, 2–4, and 3–3 types, where the $\beta^{(2)}$ term will be important at lower concentrations than for 2–2 solutes. There are, to our knowledge, no cases where this effect can now be adequately evaluated for these higher charge types—hence their omission from this paper.

Mixed electrolytes are currently being investigated with these equations. Solutions involving $MgSO_4$ as one component have been successfully fitted on the basis of these parameters and an ionic strength calculated without assuming any association. These results, which will be reported in detail in part IV of this series, offer further support to this method of representation of the properties of 2–2 electrolytes.

ACKNOWLEDGMENT

This research was sponsored by the U.S. Atomic Energy Commission.

REFERENCES

1. K. S. Pitzer, *J. Phys. Chem.* **77**, 268 (1973).
2. K. S. Pitzer and G. Mayorga, *J. Phys. Chem.* **77**, 2300 (1973).
3. N. Bjerrum, *Kgl. Danske Videnskab, Selskab, Mats.-fys. Medd.* **7**, No. 9 (1926).
4. A. W. Gardner and E. Glueckauf, *Proc. Roy. Soc.* **A313**, 131 (1969).
5. K. S. Pitzer, *J. Chem. Soc., Faraday Trans.* II **68**, 101 (1972).
6. T. H. Gronwall, V. K. LaMer, and K. Sandved, *Phys. Z.* **29**, 358 (1928).
7. E. A. Guggenheim, *Trans. Faraday Soc.* **55**, 1714 (1959); **56**, 1152 (1960).
8. A. W. Gardner and E. Glueckauf, *Proc. Roy. Soc.* **A321**, 515 (1971).
9. J. C. Rasaiah, *J. Chem. Phys*, **56**, 3071 (1972).
10. A. R. Davis and B. G. Oliver, *J. Phys. Chem.* **77**, 1315 (1973).
11. R. Larsson, *Acta Chem. Scand.* **18**, 1923 (1964).
12. G. Atkinson and S. K. Kor, *J. Phys. Chem.* **69**, 128 (1965); **71**, 673 (1967).
13. G. Atkinson and S. Petrucci, *J. Phys. Chem.* **70**, 3122 (1966).
14. M. H. Lietzke and R. W. Stoughton, *J. Phys. Chem.* **66**, 508 (1962).
15. C. W. Davies, *Ion Association* (Butterworths, London, 1962), Chap. 10.
16. Y. C. Wu, R. M. Rush, and G. Scatchard, *J. Phys. Chem.* **72**, 4048 (1968); **73** 2047 (1969).
17. R. A. Robinson and R. M. Stokes, *Electrolyte Solutions*, 2nd rev. ed. (Butterworths, London, 1965).
18. P. G. M. Brown and J. E. Prue, *Proc. Roy. Soc.* **A232**, 320 (1955).

Thermodynamics of Electrolytes. IV. Activity and Osmotic Coefficients for Mixed Electrolytes

Kenneth S. Pitzer* and Janice J. Kim

Contribution from the Inorganic Materials Research Division of the Lawrence Berkeley Laboratory and the Department of Chemistry, University of California, Berkeley, California 94720. *Received April 16, 1974*

Abstract: An equation has been developed with the guidance of recent statistical theories of electrolytes which is designed for convenient and accurate representation and prediction of the thermodynamic properties of aqueous electrolytes including mixtures with any number of components. The three previous papers have given the theoretical background and the evaluation of parameters for pure electrolytes of various charge types. The equation is here applied to a wide variety of mixed aqueous electrolytes at room temperature and at ionic strengths up to 6 *M* in many cases and occasionally even higher. The first objective is the prediction of properties of mixed electrolytes using only the parameters for pure electrolytes; on this basis standard deviations in ln γ or ϕ for 69 sets of mixtures are less than 0.01 in 36 cases and above 0.05 in only seven cases all involving Cs^+ or OH^-. A second objective is the determination of parameters giving the differences in short-range interaction of ions of the same sign where these differ significantly from zero. As expected, these difference terms, while always small, are relatively most important for singly charged ions (and especially for OH^- and Cs^+) and less important for ions of higher charge. The equations, including difference terms where known from binary mixtures with a common ion, were finally tested on 17 sets of mixtures involving four or more ions without any further adjustment of parameters. The standard deviation is less than 0.01 in all cases and is 0.003 or less in 11 cases. Thus these equations appear to yield accurate predictions of properties of mixed aqueous electrolytes.

Many systems of practical, biological or geological as well as chemical, interest involve mixed aqueous electrolytes. One of the primary objectives of the present series of papers is the prediction of the thermodynamic properties of such mixed electrolytes at concentrations of practical interest by equations no more complex than necessary. The three preceding papers in this series, cited hereafter as I,[1] II,[2] and III,[3] respectively, have prepared the basis for the treatment of mixed electrolytes in this paper. The first paper gave the theoretical and empirical bases for the choice of form of equations and some preliminary applications to mixed electrolytes. The evaluation of parameters for the activity and osmotic coefficients of pure electrolytes at room temperature is given in II for 1–1, 2–1, 3–1, and 4–1 types[2] and in III for 2–2 electrolytes[3] where a slightly different but compatible form of equation was required. In II and III the measured osmotic or activity coefficients were fitted substantially within experimental error up to ionic strength about 6 *M* in most cases.

Since the activity or osmotic coefficients of most pure electrolytes of interest have been measured at room temperature, the equations provide primarily greater convenience of interpolation for pure electrolytes. But for mixed electrolytes there are experimental data for only a very limited number of cases in contrast to the enormous range of compositions of potential interest. Hence reasonably accurate and reliable equations covering this enormous range should be of great value in making relatively accurate predictions of these properties.

The form of equation used throughout the series is defined initially for the excess Gibbs energy

$$\frac{G^{ex}}{RT} = n_w f + \frac{1}{n_w}\sum_{ij}\lambda_{ij}n_in_j + \frac{1}{n_w^{\,2}}\sum_{ijk}\mu_{ijk}n_in_jn_k \quad (1)$$

Here n_w is the number of kilograms of solvent and n_i, n_j, etc., are the numbers of moles of the ionic species *i*, *j*, etc. The function *f* depends only on the ionic strength, *I*, and represents in essentially the Debye–Hückel manner the long-range effects of Coulomb forces. In I it was shown that an alternate mathematical form arose from a different but equally sound statistical derivation from the Debye–Hückel distribution and that this form was slightly preferable empirically to the conventional one. Our form yields exactly the same limiting law as the familiar Debye–Hückel form and a similar but somewhat smaller effect of ionic size. Since we want to use a single function *f* regardless of ionic size, we must accept an approximate expression in any case.

The λ_{ij} and μ_{ijk} are, in effect, second and third virial coefficients which represent, respectively, the effects of short-range forces between ions considered two and three at a time. The second virial coefficients, λ_{ij}, depend somewhat on ionic strength; this dependence is implicit in the work of Mayer[4] and is shown simply in the derivation in I. We assume that the third virial

(1) K. S. Pitzer, *J. Phys. Chem.*, 77, 268 (1973). Minor typographical errors are corrected in ref 2.

(2) K. S. Pitzer and G. Mayorga, *J. Phys. Chem.*, 77, 2300 (1973). Sign errors should be corrected by reversing the sign of *E* in eq 12, the signs of the last two terms in eq 13, and the sign preceding 2*m* in eq 15.

(3) K. S. Pitzer and G. Mayorga, *J. Solution Chem.*, in press.

(4) J. E. Mayer, *J. Chem. Phys.*, **18**, 1426 (1959).

coefficients may be taken to be independent of ionic strength and may be neglected if all three ions have the same sign.

The principal features in this system are (1) the use of ions rather than neutral electrolytes as components and (2) the ionic strength dependence of λ_{ij} which makes it feasible to adopt a universal function f and to obtain rapid convergence in the virial series. The 1970 equations of Scatchard, Rush, and Johnson[5] also use ions as components, but in order to obtain comparable agreement with experimental data, they use, instead of a single function f, a series of terms involving parameters determined pairwise for the ions, and they require fourth virial coefficients. These differences enormously complicate their equations.

The virial coefficients λ and μ are not measurable individually but only in certain combinations. Appropriate sums are determined by the properties of pure electrolytes and these are reported in papers II and III. The properties of mixtures involve, in addition to these sums, also certain differences between the interactions of different ions of the same sign from the interactions of like ions of the same sign. If Brønsted's principle[6] of specific interaction held fully, all of these difference terms would be zero. While we shall find that these difference terms are often measurably greater than zero, they are small and can be neglected without serious error. In this paper we evaluate these difference terms when the necessary data are available but also indicate the accuracy attainable without including them. Thus one can estimate with some confidence the accuracy of predictions in other cases where difference terms have not been determined.

Equations for Mixed Electrolytes

Starting from eq 1 as defined and discussed above one may obtain working expressions for the osmotic coefficient of a mixed electrolyte and for the activity coefficient of each neutral electrolyte which can be said to be present. We write $f' = \partial f/\partial I$ and $\lambda_{ij}' = \partial\lambda_{ij}/\partial I$ and for the various ion molalities $m_i = n_i/n_w$. In these terms the equations for the osmotic and activity coefficients are

$$\phi - 1 = -\frac{\partial G^{ex}/\partial n_w}{RT\sum_i m_i} = (\sum_i m_i)^{-1}\{(If' - f) + \sum_{ij} m_i m_j(\lambda_{ij} + I\lambda_{ij}') + 2\sum_{ijk} m_i m_j m_k \mu_{ijk}\} \quad (2)$$

$$\ln \gamma_i = \frac{1}{RT}\frac{\partial G^{ex}}{\partial n_i} = \frac{z_i^2}{2}f' + 2\sum_j m_j\lambda_{ij} + \sum_{jk} m_j m_k\left(\frac{z_i^2}{2}\lambda_{jk}' + 3\mu_{ijk}\right) \quad (3)$$

For the activity coefficient of a neutral electrolyte this yields

$$\ln \gamma_{MX} = \nu^{-1}(\nu_M \ln \gamma_M + \nu_X \ln \gamma_X) = \frac{|z_M z_X|}{2}f' + \frac{2}{\nu}\sum_j m_j(\nu_M\lambda_{Mj} + \nu_X\lambda_{jX}) + \sum_{jk} m_j m_k\left[\frac{|z_M z_X|}{2}\lambda_{jk}' + \frac{3}{\nu}(\nu_M\mu_{Mjk} + \nu_X\mu_{jkX})\right] \quad (4)$$

where ν_M and ν_X are the numbers of ions M and X in the neutral salt and z_M and z_X are the charges on the ions in electronic units. Also $\nu = \nu_M + \nu_X$.

As mentioned above only certain combinations of the λ's and μ's are observable. For the osmotic coefficient the definitions previously used are very convenient.

$$f^\phi = {}^1/_2[f' - (f/I)] = -A_\phi[I^{1/2}/(1 + 1.2I^{1/2})] \quad (5)$$

$$B_{MX}^\phi = \lambda_{MX} + I\lambda_{MX}' + \left|\frac{z_X}{2z_M}\right|(\lambda_{MM} + I\lambda_{MM}') + \left|\frac{z_M}{2z_X}\right|(\lambda_{XX} + I\lambda_{XX}') = \beta_{MX}^{(0)} + \beta_{MX}^{(1)}\exp(-\alpha I^{1/2}) \quad (6)$$

$$C_{MX}^\phi = 3\left[\left|\frac{z_X}{z_M}\right|^{1/2}\mu_{MMX} + \left|\frac{z_M}{z_X}\right|^{1/2}\mu_{MXX}\right] \quad (7)$$

$$\theta_{MN} = \lambda_{MN} - \left(\frac{z_N}{2z_M}\right)\lambda_{MM} - \left(\frac{z_M}{2z_N}\right)\lambda_{NN} \quad (8)$$

$$\psi_{MNX} = 6\mu_{MNX} - \left(\frac{3z_N}{z_M}\right)\mu_{MMX} - \left(\frac{3z_M}{z_N}\right)\mu_{NNX} \quad (9)$$

Here A_ϕ is the Debye–Hückel constant for the osmotic coefficient $[{}^1/_3(2\pi N_0 d_w/1000)^{1/2}(e^2/DkT)^{3/2}]$ and has the value 0.392 for water at 25°. The value 1.2 for the parameter in f^ϕ was chosen empirically in I. Normally, α has the value 2.0. Also in the special case of 2–2 or higher valence types it was necessary to add a term to β^ϕ as follows.

$$B_{MX}^\phi = \beta_{MX}^{(0)} + \beta_{MX}^{(1)}\exp(-\alpha_1 I^{1/2}) + \beta_{MX}^{(2)}\exp(-\alpha_2 I^{1/2}) \quad (10)$$

For 2–2 electrolytes $\alpha_1 = 1.4$ and $\alpha_2 = 12.0$. Also one should remember that the superscript ϕ is a label and not an exponent. In these terms the osmotic coefficient for a mixed electrolyte is

$$\phi - 1 = (\sum_i m_i)^{-1}\Big\{2If^\phi + 2\sum_c\sum_a m_c m_a\left[B_{ca}^\phi + \frac{(\sum mz)}{(z_c z_a)^{1/2}}C_{ca}^\phi\right] + \sum_c\sum_{c'} m_c m_{c'}[\theta_{cc'} + I\theta_{cc'}' + \sum_a m_a\psi_{cc'a}] + \sum_a\sum_{a'} m_a m_{a'}[\theta_{aa'} + I\theta_{aa'}' + \sum_c m_c\psi_{caa'}]\Big\} \quad (11)$$

Here $(\Sigma mz) = \Sigma_c m_c z_c = \Sigma_a m_a|z_a|$; also c and c' are indices covering all cations while a and a' cover all anions. We note that the first term within the braces is the general "Debye–Hückel" term for long-range forces and that the second term comprises a double sum over molalities and the second and third virial coefficients for pure electrolytes. These terms can be evaluated from information on pure electrolytes and these quantities are presented in papers II and III for a large number of solutes. The final two terms include the differences between the second and third virial coefficients for unlike ions of the same sign from the appropriate averages for like ions. As indicated above, these quantities are expected to be small.

Next we convert eq 4 for the activity coefficient into a form based upon these same observable quantities. In this case it is convenient to define the following quantities, some of which were defined in I.

(5) G. Scatchard, R. M. Rush, and J. S. Johnson, *J. Phys. Chem.*, **74**, 3786 (1970).
(6) J. N. Brønsted, *Kgl. Dan. Vidensk. Selsk., Mat.-Fys. Medd.*, **4**, (4) (1929); *J. Amer. Chem. Soc.*, **44**, 877 (1922); **45**, 2898 (1923).

$$f^{\gamma} = {}^{1}\!/_{2}f' = -A_{\phi}\left[\frac{I^{1/2}}{1 + 1.2I^{1/2}} + \left(\frac{2}{1.2}\right)\ln(1 + 1.2I^{1/2})\right] \quad (12)$$

$$B_{MX} = \lambda_{MX} + \left|\frac{z_X}{2z_M}\right|\lambda_{MM} + \left|\frac{z_M}{2z_M}\right|\lambda_{XX} = \beta_{MX}^{(0)} + (2\beta_{MX}^{(1)}/\alpha^2 I)[1 - (1 + \alpha I^{1/2})\exp(-\alpha I)^{1/2}] \quad (13)$$

$$B_{MX}' = \lambda_{MX}' + \left|\frac{z_X}{2z_M}\right|\lambda_{MM}' + \left|\frac{z_M}{2z_X}\right|\lambda_{XX}' = (2\beta_{MX}^{(1)}/\alpha^2 I^2)[-1 + (1 + \alpha I^{1/2} + {}^{1}\!/_{2}\alpha^2 I)\exp(-\alpha I^{1/2})] \quad (14)$$

$$C_{MX} = C_{MX}^{\phi}/2|z_M z_X|^{1/2}$$

Again for 2–2 and higher valence types of electrolytes another term must be added to B_{MX} and B_{MX}' with parameters $\beta_{MX}^{(2)}$ and α_2 but with the same form as the term in $\beta_{MX}^{(1)}$. Then

$$\ln\gamma_{MX} = |z_M z_X|f^{\gamma} + (2\nu_M/\nu)\sum_a m_a[B_{Ma} + (\textstyle\sum mz)C_{Ma} + (\nu_X/\nu_M)\theta_{Xa}] + (2\nu_X/\nu)\sum_c m_c[B_{cX} + (\textstyle\sum mz)C_{cX} + (\nu_M/\nu_X)\theta_{Mc}] + \sum_c\sum_a m_c m_a\{|z_M z_X|B_{ca}' + \nu^{-1}[2\nu_M z_M C_{ca} + \nu_M\psi_{Mca} + \nu_X\psi_{caX}]\} + {}^{1}\!/_{2}\sum_c\sum_{c'} m_c m_{c'}[(\nu_X/\nu)\psi_{cc'X} + |z_M z_X|\theta_{cc'}'] + {}^{1}\!/_{2}\sum_a\sum_{a'} m_a m_{a'}[(\nu_M/\nu)\psi_{Maa'} + |z_M z_X|\theta_{aa'}'] \quad (15)$$

There are many somewhat simplified forms of eq 15 for cases where all solutes are of the same valence type and further where this is a simple type or where there is a common cation or a common anion or where there are only two solutes. Since these transformations are quite straightforward and in many cases do not shorten the expression very much, we will not burden this paper with them. For a mixture of just two symmetrical electrolytes of charge z and with a common anion the expression is considerably simplified. We write y for the solute fraction of the component NX. The activity coefficient of MX is

$$\ln\gamma_{MX} = z^2 f^{\gamma} + m\{(2 - y)B_{MX} + (1 - y)IB_{MX}' + y(B_{NX} + IB_{NX}') + m[({}^{3}\!/_{2} - y)C_{MX}^{\phi} + yC_{NX}^{\phi}] + y(\theta_{MN} + {}^{1}\!/_{2}m\psi_{MNX}) + y(1 - y)[(m/2)\psi_{MNX} + I\theta_{MN}']\} \quad (16)$$

In this case the ionic strength is, of course, $I = mz^2$; also we have substituted C^{ϕ} which in this case is $2zC$. For comparison of this equation with the equivalent eq 41 in paper I, we note that $B = B^{\gamma} - B^{\phi}$ and $B' = (2B^{\phi} - B^{\gamma})/I$.

Another observable combination of activity coefficients is that for the exchange of an amount of one ion by an equal electrical charge of a different ion of the same sign. This occurs, for example, with exchange between two liquid phases when positive ions are complexed to form neutral molecular species in the nonaqueous phase and in certain electrical cells. The pertinent combination of activity coefficients may be written, for M^{+z_M} and N^{+z_N}

$$z_N\ln\gamma_M - z_M\ln\gamma_N = z_N z_M(z_M - z_N)f^{\gamma} + 2\sum_a m_a[z_N B_{Ma} - z_M B_{Na} + (\textstyle\sum mz)(z_N C_{Ma} - z_M C_{Na})] + 2\sum_c m_c(z_N\theta_{Mc} - z_M\theta_{Nc}) + {}^{1}\!/_{2}z_N z_M(z_M - z_N)\sum_c\sum_{c'} m_c m_{c'}\theta_{cc'}' + \sum_c\sum_a m_c m_a[z_N z_M(z_M - z_N)B_{ca}' + z_N\psi_{Mca} - z_M\psi_{Nca}] + {}^{1}\!/_{2}\sum_a\sum_{a'} m_a m_{a'}[z_N\psi_{Maa'} - z_M\psi_{Naa'} + z_N z_M(z_M - z_N)\theta_{aa'}'] \quad (17)$$

The corresponding equation for the difference in activity coefficients of anions is readily obtained by transposing symbols in eq 17.

As indicated above all of these equations include the Debye–Hückel limiting law. While we shall use the extended forms in our applications, it seems worthwhile to express the limiting law forms to which these reduce at very low concentration in mixed electrolytes without limitation as to valence types.

$$\frac{G^{ex}}{n_w RT} = -4A_{\phi}I^{3/2} \quad (18)$$

$$\phi - 1 = -2A_{\phi}I^{3/2}/\sum_i m_i \quad (19)$$

$$\ln\gamma_{MX} = -3A_{\phi}|z_M z_X|I^{1/2} \quad (20)$$

$$z_N\ln\gamma_M - z_M\ln\gamma_N = -3A_{\phi}z_N z_M(z_M - z_N)I^{1/2} \quad (21)$$

There are two "higher order" limiting laws which should be noted. One is the limiting law for mixing discovered by Friedman[7] and discussed also by Robinson, Wood, and Reilly.[8] These authors found no inconsistency between existing activity or osmotic coefficient data and this limiting law for mixing. On the other hand, the use of this law has no significant effect on results for mixed electrolytes at the present level of experimental accuracy. Recent heat of mixing measurements by Falcone, Levine, and Wood[9] have confirmed the corresponding law for heat of mixing. We will show in paper V of this series how the ionic strength dependence of θ may be defined to satisfy this limiting law. In order, however, to avoid complications of no practical significance for most existing data, we shall neglect this ionic strength dependence of θ for the present paper. In other words we take all θ_{MN}' to be zero.

There is also a higher order limiting law[7,10] which pertains only to unsymmetrical electrolytes. This likewise has a negligible effect in most practical applications and was ignored in paper II. Its effect is clearly very small for systems with only univalent and bivalent ions and we confine ourselves to such cases in this paper. For 3–1 or 4–1 electrolytes this effect may

(7) H. L. Friedman, "Ionic Solution Theory," Interscience, New York, N. Y., 1962.

(8) R. A. Robinson, R. H. Wood, and P. J. Reilly, *J. Chem. Thermodyn.*, 3, 461 (1971).

(9) J. S. Falcone, A. S. Levine, and R. H. Wood, *J. Phys. Chem.*, 77, 2137 (1973).

(10) V. K. LaMer, T. H. Gronwall, and L. J. Greiff, *J. Phys. Chem.*, 35, 2245 (1931).

Table I. Binary Mixtures with a Common Ion

System	Exptl	Max I	σ $\theta = \psi = 0$	θ	ψ	σ with θ and ψ	Ref
$HCl–LiCl$	ln γ	5	0.023	0.015	0.000	0.007	*a*
$HBr–LiBr$	ln γ	2.5	0.027	0.015	0.000	0.011	*b, c*
$HClO_4–LiClO_4$	ϕ	4.5	0.006	0.015	−0.0017	0.001	*d*
$HCl–NaCl$	ln γ	3	0.040	0.036	−0.004	0.002	*c*
$HBr–NaBr$	ln γ	3	0.028	0.036	−0.012	0.002	*c*
$HClO_4–NaClO_4$	ϕ	5	0.025	0.036	−0.016	0.002	*d*
$HCl–KCl$	ln γ	3.5	0.014	0.005	−0.007	0.010	*e*
$HBr–KBr$	ln γ	3	0.030	0.005	−0.021	0.008	*e*
$HCl–CsCl$	ln γ	3	0.082	−0.044	−0.019	0.005	*f*
$HCl–NH_4Cl$	ln γ	2		−0.016	0.000		*g*
$HCl–Me_4NCl$	ln γ	0.1	0.003	−0.0		0.003	*h*
$HCl–Et_4NCl$	ln γ	0.1	0.003	−0.0		0.003	*h*
$LiCl–NaCl$	ϕ	6	0.002	0.012	−0.003	0.001	8
$LiNO_3–NaNO_3$	ϕ	6	0.014	0.012	−0.0072	0.002	8
$LiClO_4–NaClO_4$	ϕ	2.6	0.003	0.012	−0.0080	0.001	*d*
$LiOAc–NaOAc$	ϕ	3.5	0.004	0.012	−0.0043	0.002	8
$LiCl–KCl$	ϕ	4.8	0.045	−0.022	−0.010	0.003	*i*
$LiCl–CsCl$	ϕ	5	0.100	−0.095	−0.0094	0.004	*j*
$NaCl–KCl$	ϕ	4.8	0.014	−0.012	−0.0018	0.001	*k*
$NaBr–KBr$	ϕ	4	0.009	−0.012	−0.0022	0.003	*l*
$NaNO_3–KNO_3$	ϕ	3.3	0.008	−0.012	−0.0012	0.001	*m*
$Na_2SO_4–K_2SO_4$	ϕ	3.6	0.011	−0.012	−0.010	0.004	*n*
$NaCl–CsCl$	ϕ	5	0.027	−0.033	−0.003	0.001	*o*
$KCl–CsCl$	ϕ	5	0.003	0.000	−0.0013	0.001	*j*
$HCl–SrCl_2$	ln γ	8	0.034	−0.020	0.018	0.010	*p, q*
$HCl–BaCl_2$	ln γ	3	0.009	−0.036	0.024	0.005	*c*
$HCl–MnCl_2$	ln γ	3	0.008	0.000	0.000	0.008	*r*
$LiCl–BaCl_2$	ϕ	4.3	0.006	−0.070	0.019	0.002	*s*
$NaCl–MgCl_2$	ϕ	5.9	0.002	0.000	0.000	0.002	*t*
$Na_2SO_4–MgSO_4$	ϕ	9	0.005	0.000	0.000	0.005	*t*
$NaCl–CaCl_2$	ϕ	8	0.004	0.000	0.000	0.004	*u*
$NaCl–BaCl_2$	ϕ	5	0.001	−0.003	0.000	0.001	*v*
$NaCl–MnCl_2$	ϕ	5.5	0.004	0.000	−0.003	0.003	*w*
$NaBr–ZnBr_2$	ϕ	0.4	0.007	0.0		0.007	*x*
$KCl–CaCl_2$	ϕ	5	0.025	−0.040	−0.015	0.003	*y*
$KCl–BaCl_2$	ϕ	5	0.018	−0.072	0.000	0.001	*z*
$CsCl–BaCl_2$	ϕ	4	0.024	−0.150	0.000	0.003	*s*
$MgCl_2–CaCl_2$	ϕ	7.7	0.003	0.010	0.000	0.002	*aa*
$NaCl–NaBr$	ϕ	4.4	0.001	0.000	0.000	0.001	*l*
$KCl–KBr$	ϕ	4.4	0.002	0.000	0.000	0.002	*l*
$NaCl–NaOH$	ln (γ/γ')	3	0.155	−0.050	−0.006	0.002	*bb*
$KCl–KOH$	ln (γ/γ')	3.5	0.196	−0.050	−0.008	0.008	*cc*
$NaBr–NaOH$	ln (γ/γ')	3	0.225	−0.065	−0.018	0.009	*cc*
$KBr–KOH$	ln (γ/γ')	3	0.212	−0.065	−0.014	0.012	*cc*
$NaCl–Na_2SO_4$	ϕ	9	0.004	−0.035	0.007	0.002	*t*
$KCl–K_2SO_4$	ϕ	2.3	0.005	−0.035	0.000	0.002	*n*
$MgCl_2–MgSO_4$	ϕ	7	0.011	−0.035	0.000	0.002	*t*
$LiCl–LiNO_3$	ϕ	6	0.008	0.016	−0.003	0.004	*i*
$NaCl–NaNO_3$	ϕ	5	0.007	0.016	−0.006	0.001	*m*
$KCl–KNO_3$	ϕ	4	0.003	0.016	−0.006	0.001	*m*
$MgCl_2–Mg(NO_3)_2$	ϕ	4	0.008	0.016	0.000	0.002	*dd*
$CaCl_2–Ca(NO_3)_2$	ϕ	6	0.014	0.016	−0.017	0.003	*dd*

[a] H. S. Harned and H. R. Copson, *J. Amer. Chem. Soc.*, **55**, 2206 (1933). [b] J. E. Vance, *ibid.*, **55**, 4518 (1933). [c] H. S. Harned and B. B. Owen, "The Physical Chemistry of Electrolyte Solutions," 3rd ed, Reinhold, New York, N. Y., 1958. [d] R. M. Rush and J. S. Johnson, *J. Phys. Chem.*, **72**, 797 (1968). [e] See text for references. [f] H. S. Harned and O. E. Schupp, Jr., *J. Amer. Chem. Soc.*, **52**, 3892 (1930). [g] R. A. Robinson, R. N. Roy, and R. G. Bates, *J. Solution Chem.*, in press. [h] C. J. Downes, *J. Phys. Chem.*, **74**, 2153 (1970). [i] R. A. Robinson and C. K. Lim, *Trans. Faraday Soc.*, **49**, 1144 (1953). [j] R. A. Robinson and C. K. Lim, *ibid.*, **49**, 1147 (1953). [k] R. A. Robinson, *J. Phys. Chem.*, **65**, 662 (1961). [l] A. K. Covington, T. H. Lilley, and R. A. Robinson, *ibid.*, **72**, 2759 (1968). [m] C. P. Bezboruah, A. K. Covington, and R. A. Robinson, *J. Chem. Thermodyn.*, **2**, 431 (1970). [n] R. A. Robinson, R. F. Platford, and C. W. Childs, *J. Solution Chem.*, **1**, 167 (1972). [o] R. A. Robinson, *J. Amer. Chem. Soc.*, **74**, 6035 (1952). [p] J. E. Vance, *ibid.*, **55**, 2729 (1933). [q] H. S. Harned and T. R. Paxton, *J. Phys. Chem.*, **57**, 531 (1953). [r] C. J. Downes, *J. Chem. Soc., Faraday Trans. 1*, **68**, 1964 (1972). [s] S. Lindenbaum, R. M. Rush, and R. A. Robinson, *J. Chem. Thermodyn.*, **4**, 381 (1972). [t] Y. C. Wu, R. M. Rush, and G. Scatchard, *J. Phys. Chem.*, **72**, 4048 (1968); **73**, 2047, 4433, 4434 (1969). [u] R. A. Robinson and V. E. Bower, *J. Res. Nat. Bur. Stand., Sect. A*, **70**, 313 (1966). [v] R. A. Robinson and V. E. Bower, *ibid.*, **69**, 19 (1965). [w] C. J. Downes, *J. Chem. Eng. Data*, **18**, 412 (1973). [x] G. E. Boyd, S. Lindenbaum, and R. A. Robinson, *J. Phys. Chem.*, **75**, 3153 (1971). [y] R. A. Robinson and A. K. Covington, *J. Res. Nat. Bur. Stand., Sect. A*, **72**, 239 (1968). [z] R. A. Robinson and V. E. Bower, *ibid.*, **69**, 439 (1965). [aa] R. A. Robinson and V. E. Bower, *ibid.*, **70**, 305 (1966). [bb] H. S. Harned and G. E. Mannweiler, *J. Amer. Chem. Soc.*, **57**, 1873 (1935). [cc] H. S. Harned and W. J. Hamer, *ibid.*, **55**, 2194, 4496 (1933). [dd] R. F. Platford, *J. Chem. Thermodyn.*, **3**, 319 (1971).

be more significant and we shall consider such examples in paper V along with the appropriate theory.

Comparison with Experiment

In our analysis of experimental data in terms of these equations we seek two principal results. First, we want to learn how accurate our prediction would be on the basis of parameters from pure electrolytes only. Second, we wish to obtain values for the difference parameters θ and ψ, if they are of significant magnitude,

5705

or to determine that they are negligible and can be taken to be zero. It should be remembered that the principal effects on mixing electrolytes arise from differences in the pure electrolyte parameters $\beta^{(0)}$, $\beta^{(1)}$, and C^{ϕ} and that the parameters θ and ψ have only a small effect, if any.

The results are summarized in Table I for binary mixtures with a common ion. The values of θ and ψ were obtained by calculating the difference between the experimental value of ϕ or $\ln \gamma$ and the value calculated with the appropriate values for all pure-electrolyte terms but with zero values for θ and ψ in eq 11 or 15. This difference, when multiplied by a function of composition, is found to be equal to θ plus ψ times another function of composition. For the osmotic coefficient of a MX–NX mixture one obtains

$$\Delta\phi[\Sigma m_i/2m_M m_N] = \theta_{MN} + m_X \psi_{MNX} \quad (22)$$

or the equivalent for mixing anions. For the activity coefficient of MX in this mixture one has

$$(\Delta \ln \gamma_{MX})[\nu/2\nu_M m_N] = \theta_{MN} + {}^1/_2(m_X + m_M|z_M/z_X|)\psi_{MNX} \quad (23)$$

with equivalent expressions for other cases.

We plotted the quantity on the left against the coefficient of ψ on the right. One should obtain a linear plot with intercept θ and slope ψ. This presentation of the results, to which estimated errors could be attached, allowed us to judge whether values of θ or ψ were significantly different from zero and whether the data were consistent with these equations within reasonable limits of error.

Table I gives, in the fourth column, the root mean square average of $\Delta\phi$ or $\Delta \ln \gamma$ when θ and ψ are taken as zero. Then the selected values of θ and ψ are given for the mixing indicated and finally the standard deviation when these values of θ and ψ are included. Thus, for the first entry the system HCl–LiCl relates to $\theta_{H,Li}$ and $\psi_{H,Li,Cl}$; the quantity measured is $\ln \gamma_{HCl}$; the data range up to $I = 5$; and the root mean square $\Delta \ln \gamma$ is 0.023 with $\theta = \psi = 0$. The value $\theta_{H,Li} = 0.015$ is selected from consideration of the first three systems all of which involve mixing of H^+ with Li^+. For the chloride ψ was found to be zero and with these θ and ψ values the standard deviation is reduced to 0.007.

The values of σ with $\theta = \psi = 0$ give an estimate of the accuracy to be obtained without difference parameters. However, these deviations are usually proportional to molality. Hence, a better estimate is obtained by noting the effect of $\theta_{cc'}$ or $\theta_{aa'}$ in eq 11 or 15. Thus, in the former one finds the result $(m_c m_{c'}/\Sigma m_i)\cdot\theta_{cc'}$. And, if one has an equimolal mixture of $HClO_4$ and $LiClO_4$, for example, $(m_c m_{c'}/\Sigma m_i)$ is $m/8$ where m is the total molality of ClO_4^-. With $\theta_{H,Li} = 0.015$ one calculates the effect on the osmotic coefficient to be $0.0019\ m$. This is negligible for most purposes unless m is large (considerably above 1 M). The example chosen has a typical θ; in some cases the effect of θ and ψ is even smaller while in a few cases it is somewhat larger.

In a few cases data are available only for rather dilute solutions and these results can be fitted quite well without the difference terms involving θ and ψ. In these cases we report 0.0 for θ in Table I to indicate that the correct value of θ must be small but is not determined accurately. In the very recent work on $HCl-NH_4Cl$, the data were analyzed using the equations of this series of papers; hence, we did not repeat the calculations and have no σ values. The fit is very good up to 2 M with the θ value in Table I.

Table II contains the results for binary mixtures without a common ion. In this case we show only the standard deviation without the difference terms θ and ψ and that with these terms. It is apparent that in most of these examples quite good results are obtained with only the pure electrolyte terms and in all cases there is good agreement when the θ and ψ values are included. The values of θ and ψ were determined from the mixtures with common ion; hence, there were no adjustable parameters in this treatment of the mixtures without common ion.

Table II. Binary Mixtures without Common Ion

System	Exptl	Max I	σ $\theta = \psi = 0$	σ with θ and ψ	Ref
NaCl–KBr	ϕ	4	0.012	0.002	*b*
KCl–NaBr	ϕ	4	0.012	0.001	*b*
$NaCl-KNO_3$	ϕ	4	0.007	0.001	*c*
$NaNO_3-KCl$	ϕ	4	0.007	0.002	*c*
$LiCl-Na_2SO_4$	ϕ	5	0.008	(0.006)[a]	*d*
$NaCl-K_2SO_4$	ϕ	3.6	0.012	0.003	*e*
$KCl-Na_2SO_4$	ϕ	3.6	0.015	0.004	*e*
$CsCl-Na_2SO_4$	ϕ	5	0.024	(0.007)[a]	*d*
$NaCl-MgSO_4$	ϕ	9	0.008	0.002	*f*
$Na_2SO_4-MgCl_2$	ϕ	7	0.005	0.005	*f*
$NaCl-CuSO_4$	ϕ	2.8	0.003	(0.003)[a]	*g*

[a] Some ψ values were not available from other mixtures and were set at zero; also $\theta_{Na,Cu} = 0$ for $NaCl-CuSO_4$. [b] A. K. Covington, T. H. Lilley, and R. A. Robinson, *J. Phys. Chem.*, **72**, 2759 (1968). [c] C. P. Bezboruah, A. K. Covington, and R. A. Robinson, *J. Chem. Thermodyn.*, **2**, 431 (1970). [d] R. A. Robinson, *J. Solution Chem.*, **1**, 71 (1972). [e] R. A. Robinson, R. F. Platford, and C. W. Childs, *ibid.*, **1**, 167 (1972). [f] Y. C. Wu, R. M. Rush, and G. Scatchard, *J. Phys. Chem.*, **72**, 4048 (1968); **73**, 2047, 4433, 4434 (1969). [g] C. J. Downes, private communication.

In Tables I and II, as well as others to come, the maximum I is usually that of the most concentrated mixed electrolyte measured, but in a few cases the limit of validity of the equations for pure electrolytes determined the maximum I for valid comparison. In cases involving K_2SO_4, where solubility limits pure electrolyte data to $I = 2.1\ M$, good fits were obtained for mixtures to considerably higher concentration as shown in the tables. This suggests that the equation for pure K_2SO_4 is valid somewhat above $I = 2.1\ M$.

A set of systems of particularly great interest is that involving the ions Na^+, Mg^{2+}, Cl^-, and SO_4^{2-} which was investigated thoroughly by Wu, Rush, and Scatchard.[11] They measured the osmotic coefficient over a wide range of compositions. As shown in Tables I and II, our equations fit these data quite well. Indeed there is no need for difference parameters θ and ψ for the Na^+-Mg^{2+} interaction, and the corresponding terms for $Cl^--SO_4^{2-}$ are small. Also this good agreement holds to high ionic strength, typically 6 to 9 M. Since these data for pure Na_2SO_4 differ slightly from the values of Stokes and Robinson[12] which were used in

(11) Y. C. Wu, R. M. Rush and G. Scatchard, *J. Phys. Chem.*, **72**, 4048 (1968); **73**, 2047, 4433, 4434 (1969).

(12) R. A. Robinson and R. M. Stokes, "Electrolyte Solutions," 2nd ed, revised, Butterworth, London, 1965.

paper II, we modified one of the pure electrolyte parameters. Thus, for for Na_2SO_4 we took $C^\phi = 0.0057$.

Scatchard, Rush, and Johnson[5] treated these data for the Na^+, Mg^{2+}, Cl^-, SO_4^{2-} systems using a set of equations considerably more complex than ours. Their results (SRJ) are compared with ours (PK) in Table III.

Table III. Standard Deviations of Fits for the System Na^+, Mg^{2+}, Cl^-, SO_4^{2-}

System	Pure electrolyte terms PK[a]	Pure electrolyte terms SRJ[b]	All terms PK	All terms SRJ
NaCl–$MgCl_2$	0.0016	0.0056	0.0016	0.0009
NaCl–Na_2SO_4	0.0035	0.0054	0.0023	0.0009
$MgCl_2$–$MgSO_4$	0.0108	0.0066	0.0025	0.0012
Na_2SO_4–$MgSO_4$	0.0052	0.0121	0.0052	0.0013
$MgCl_2$–Na_2SO_4	0.0047	0.0074	0.0055	0.0042
NaCl–$MgSO_4$	0.0079	0.0037	0.0021	0.0016

[a] This research. [b] Scatchard, Rush, and Johnson, ref 5.

If only the pure electrolyte terms are used, our results are slightly superior, since our standard deviation is smaller in four out of six cases and has a smaller average and a smaller maximum. When the difference terms are included their fits are all better than ours, but they introduced 18 new parameters as compared to 6 in our case. They adopted nonzero values for 11 of these 18 parameters in their difference terms while we used only two nonzero difference parameters. Since the experimental error is probably as large as their standard deviations and possibly as large as ours, our fits are really quite satisfactory. And the fact that we can obtain agreement using very few parameters suggests that our equations relate more closely to the real physical relationships.

These mixed electrolytes involving both Mg^{2+} and SO_4^{2-} are also of special interest because they include a 2–2 electrolyte. The fact that good agreement was obtained for these mixtures offers further support for the treatment of 2–2 electrolytes in paper III which avoids an explicit association equilibrium and handles the peculiarity of this type with an extra term in the second virial coefficient. As a further check on the effectiveness of this treatment we calculated the activity coefficient of NaCl in mixtures with $MgSO_4$ for comparison with the cell measurements of Wu, Rush, and Scatchard.[11] Excellent agreement is obtained, even for solutions with a large fraction of $MgSO_4$ and at ionic strengths up to 6.16 *M;* the standard deviation for all points is less than 0.01 in $\ln \gamma$ or less than 1% in the activity coefficient.

Although all data for the system HCl–KCl are in reasonable agreement with the parameters in Table I, interpretations of certain of these data have been made which indicated anomalous behavior. These matters are discussed in the Appendix.

The results involving hydroxide ion are based upon measurements of cells of the type Pt, $H_2|MOH(m_a)$, $MX(m_b)|AgX$, Ag which were discussed in II. These data yield $\ln (\gamma_{X^-}/\gamma_{OH^-})$, to which eq 17 applies, and $E_0' = E°(AgX,Ag) - (RT/F) \ln K_w$, where K_w is the dissociation constant for water. The results for the activity coefficient are given in Table I and the values of E_0' and $\log K_w$ are given in Table IV. The measured quantity, $\ln (\gamma_{X^-}/\gamma_{OH^-})$, and the compositions chosen in these examples are very sensitive to the difference

Table IV. Dissociation Constant for Water

System	E_0'	$E°$ (AgX,Ag),[a] V	$\log K_w$	Ref
NaOH–NaCl	1.0503	0.2224	13.995	*b*
KOH–KCl	1.0501	0.2224	13.991	*c*
NaOH–NaBr	0.8996	0.0710	14.007	*c*
KOH–KBr	0.8996	0.0710	14.007	*c*

[a] From Table X of paper II. [b] H. S. Harned and G. E. Mannweiler, *J. Amer. Chem. Soc.*, **57**, 1873 (1935). [c] H. S. Harned and W. J. Hamer, *ibid.*, **55**, 2194, 4496 (1933).

terms; hence, one has unusually large σ values for θ and ψ values which are substantial but not extremely large.

Table V summarizes our results for mixed electrolytes with three or more solutes. The fits without difference terms are good except in the case involving Cs^+. With the difference terms there is excellent agreement in all cases. The results for sea water are especially interesting. These data concern not only sea water at normal concentration but also at much greater concentration as would arise after evaporation of much of the water. Under these conditions $CaSO_4$ will precipitate. Hence, for the measurements at high concentration Ca^{2+} was replaced by Mg^{2+}. The agreement for this complex mixture involving both singly and doubly charged ions of each sign is very good without difference terms and practically perfect when these terms are included. While values for θ and ψ are not available for a few of the interactions, these involve the less abundant species and their omission cannot be significant.

Table V. Mixed Electrolytes with Three or More Solutes

System	Exptl	Max *I*	σ $\theta = \psi = 0$	σ with θ, ψ	Ref
LiCl–NaCl–KCl	ϕ	3.1	0.023	0.004	*a*
LiCl–NaCl–CsCl	ϕ	5.2	0.085	0.004	*a*
LiCl–NaCl–$BaCl_2$	ϕ	3.3	0.002	0.001	*a*
NaCl–KCl–$BaCl_2$	ϕ	4.5	0.015	0.003	*a*
Sea water (no Ca)	ϕ	6	0.004	0.001	*b*
Sea water (with Ca)	ϕ	2	0.002	0.000	*b*

[a] R. J. Reilly, R. H. Wood, and R. A. Robinson, *J. Phys. Chem.*, **75**, 1305 (1971). [b] R. M. Rush and J. S. Johnson, *J. Chem. Eng. Data*, **11**, 590 (1966).

There seems little need to comment on most of the other systems listed in Tables I through V. Among the 69 systems, the standard deviation of fit with pure electrolyte terms only is less than 0.01 in ϕ or $\ln \gamma$ (or 1% in γ) in 36 cases and more than 0.05 in only 7 cases, all of which involve hydroxide ion or cesium ion. Thus one has a good chance of obtaining quite accurate results up to moderate concentration by the use of our equations with only the pure electrolyte terms which are available for almost all electrolytes.

For maximum accuracy, all terms should be included, of course, and values of θ and ψ are given in Table I for many combinations of ions. Since the effect of any interaction is proportional to the concentration of the ion involved, little error will be introduced by omitting terms in θ and ψ for ions present at small concentration even if the total ionic strength is large.

Discussion

The parameters in our equations were selected to be

related to interionic forces, although this relationship is, at times, complex or only approximate. Consideration of such relationships for the difference parameters θ and ψ must begin with eq 8 and 9 which show that these quantities are differences between the virial coefficients for interaction of unlike ions of the same sign from the average of interactions of like ions. Since ions of the same sign repel one another and are seldom close together, one would expect any differences in their interactions to be small and to be especially small for ions with double charge.

This general view is confirmed in that all θ's and ψ's are relatively small. Also the only systems where the σ value for $\theta = \psi = 0$ exceeds 0.05 involve either Cs^+ with H^+ or Li^+ or OH^- with Cl^- or Br^-. These systems involve mixing of only singly charged ions and in particular of ions which differ greatly in their interaction with the solvent. Thus, Cs^+ is a strong "structure breaker" as are Cl^- and Br^- whereas H^+, Li^+, and OH^- are strong "structure makers." Thus it is not surprising to find a relatively large effect in these cases, but the details of solvent structure near these ions are so poorly known that it is not feasible to pursue the matter in greater detail.

Rubidium ion would, presumably, be similar to cesium and F^- similar to OH^-, but no data are available for mixtures involving Rb^+ or F^-. The mixtures of most other ions with potassium ion show effects in the same direction but much smaller than with cesium ion as one would expect. All of the doubly charged positive ions are "structure makers" and their interaction with cesium ion yields significant, negative difference parameters similar to those for H^+ or Li^+ with Cs^+.

As one moves, however, to cases where difference parameters are expected to be smaller, one can conclude only that they are smaller; there seems to be little system or order to their exact magnitude. One may note that, where the ions mixed are singly charged, ψ is either negative or zero, but if singly and doubly charged ions are being mixed, then, with two exceptions, ψ is positive or zero.

When the ions mixed have the same charge, one can show that long range Coulombic forces cannot have any direct contribution to θ or ψ. But when singly and doubly charged ions are mixed, the direct effects of Coulombic forces cancel only in the first order, *i.e.*, in the Debye–Hückel approximation, and there may be higher order effects which make small contributions to θ and ψ. We are investigating this last point and hope to report on it in paper V.

Finally, how well can we expect to predict mixed electrolyte properties on the basis of existing knowledge which includes parameters for almost all pure electrolytes and difference terms for only a limited number of mixtures, mostly among singly charged ions? For ions of the general type considered here one can expect good results for multicomponent mixtures as well as for those of just two solutes. The only difference terms of substantial magnitude are those involving OH^- and Cs^+ (also possibly F^- and Rb^+) and these are known or can be estimated reasonably well. For optimum accuracy, difference terms for other ions should be included, where known, and they are known for many of the pairs of ions likely to be present at high concentration in systems of practical interest. Thus the equations and parameters given in this series of papers should allow effective prediction of the osmotic or activity coefficients of most aqueous electrolytes at room temperature.

We are extending this work to thermal properties and to activity and osmotic coefficients at other temperatures.

Acknowledgment. This research was sponsored by the U. S. Atomic Energy Commission.

Appendix

The Systems HCl–KCl and HBr–KBr. The behavior of the HCl–KCl system as reported by Harned and Gancy[13] seems peculiar; also these results conflict somewhat with those of Harned and Hamer[14] for the HBr–KBr system in the limit of zero concentration where the effect of the anion should disappear. In view of the importance of the ions involved, we investigated these systems with special care.

From eq 16 one readily sees that ln γ_{MX} is linear in the composition fraction y except for the last term which depends on the product $y(1 - y)$. For the corresponding equation for ln γ_{NX} the last term will be the same. Harned defines quantities α and β to express the linear and quadratic dependences of ln γ, respectively. For HCl–KCl, β is found[13] to be zero for ln γ_{HCl} but nonzero for ln γ_{KCl}. This result is inconsistent with our equations and, if correct, requires a very peculiar combination of terms involving the interactions of three positive ions. Since there was no direct measurement of ln γ_{KCl}, however, it seems more likely that errors accumulated in the long calculation in a manner to yield this peculiar result.

The possibility that Harned and Gancy's results may be subject to larger error than they assumed receives support from four other sources. First, the less extensive but more recent work of Lietzke and O'Brien[15] yields zero β for the activity coefficient of KCl as well as for HCl. Second, the exceptionally careful work of Guntelberg[16] at 0.1 M total concentration disagrees with the trend of Harned and Gancy's data at slightly higher concentration; third, the data of Harned and Hamer[14] on HBr–KBr yield an intercept at $m = 0$ for H^+–K^+ mixing more consistent with Guntelberg's data than with Harned and Gancy's for HCl–KCl; and fourth, Bates, *et al.*,[17] have shown that the Ag,AgCl electrode, upon which the data depend, is more erratic than had been thought.

It should be emphasized that these various discrepancies are not large. Indeed, they are smaller than the 0.2 mV suggested[17] as the range of variability of the Ag,AgCl electrode. If one allows these still moderate deviations, all of the data are in agreement with the parameters $\theta_{H,K} = 0.005$, $\psi_{H,K,Cl} = -0.007$, and $\psi_{H,K,Br} = -0.021$ which are included in Table I.

(13) H. S. Harned and A. B. Gancy *J. Phys. Chem.*, **62**, 627 (1958).

(14) H. S. Harned and W. J. Hamer, *J. Amer. Chem. Soc.*, **55**, 2194, 4496 (1933).

(15) M. H. Lietzke and H. A. O'Brien, Jr. *J. Phys. Chem.*, **72**, 4408 (1968).

(16) E. Guntelberg, *Z. Phys. Chem.*, **123**, 199 (1926).

(17) R. G. Bates, *et al.*, *J. Chem. Phys.*, **25**, 361 (1956); see also R. G. Bates and V. E. Bower *J. Res. Nat. Bur. Stand., Sect. A*, **53**, 283 (1954).

Journal of Solution Chemistry, Vol. 4, No. 3, 1975

Thermodynamics of Electrolytes. V. Effects of Higher-Order Electrostatic Terms

Kenneth S. Pitzer[1]

Received July 26, 1974; revised October 14, 1974

The contribution of higher-order electrostatic terms (beyond the Debye–Hückel approximation) to the thermodynamic properties of mixed and pure electrolytes is investigated. It is found that these effects are important for cases of unsymmetrical mixing, especially when one ion has a charge of three units or more. The appropriate correction can be made by a purely electrostatic function since the mutual repulsion of ions of the same sign keeps them far enough apart that short-range forces have little effect. This function is evaluated, and several convenient approximations are also given. Application is made to systems mixing ions of the type 1–2 and 1–3. Higher-order limiting laws exist for symmetrical mixtures and for pure, unsymmetrical solutes, but these effects were not found to be significant in relationship to existing activity or osmotic-coefficient data.

KEY WORDS: Activity coefficient; osmotic coefficient; thermodynamics; electrolytes; solution theory.

1. INTRODUCTION

At finite concentration the properties of an electrolyte solution depend in a very complex way on short-range forces between solute ions, both direct and as modified by the solvent, in addition to the long-range electrostatic forces. However, knowledge of the limiting or idealized behavior arising purely from electrostatic forces, with the solvent as a dielectric, is extremely valuable in the form of the limiting law of Debye and Hückel.

The Debye–Hückel limiting law arises from a mathematical approximation, and various authors have pointed out that, in at least three cases, higher terms in the mathematical expansion yield "higher-order" limiting laws of the same sort, i.e., arising purely from electrostatic effects. In this paper these higher-order electrostatic effects will be explored in connection with the

[1] Inorganic Materials Research Division of the Lawrence Berkeley Laboratory and Department of Chemistry, University of California, Berkeley, California 94720.

general equations presented in Part I[1] of this series and applied extensively to most pure and mixed electrolytes in Parts II,[2] III,[3] and IV[4] (hereafter cited as I, II, etc.). Insofar as possible, however, we shall consider the entire electrostatic effect (in excess of the Debye–Hückel limiting law), rather than merely the next term in the series expansion.

One case arises with unsymmetrical pure electrolytes, e.g., $MgCl_2$ or $AlCl_3$, where the existence of a higher-order term was noted by Gronwall, LaMer, and Sandved[5] in 1928. This effect has never been found to be distinguishable from the effects of short-range forces, although there is no reason to doubt its theoretical existence. Recent freezing-point-lowering data for a 6–1 electrolyte[6] allows us to make a more searching test than heretofore, but the result is similarly negative.

The other cases relate to mixed electrolytes. If ions of different charges of the same sign are mixed, e.g., HCl with $AlCl_3$, there is a higher-order electrostatic function[7] which we shall find to be significant and useful. Most of the paper will be concerned with this effect, which is small for 2–1 mixing and can be omitted without large error, but which must be included to deal with 3–1 or more unsymmetrical types of mixing.

Even for the mixing of ions of the same charge but with different short-range interactions, Friedman[7] has shown that there is a higher-order limiting law; but in this case, as in the first, its effect is too small to be of practical importance.[8]

2. THEORY FOR UNSYMMETRICAL MIXING

The mixing of ions of different charge of the same sign is best discussed theoretically by the cluster-integral method introduced for electrolytes by Mayer[9] and applied with generality by Friedman.[7] Since the theoretical infrastructure is extensive, it will not be repeated here. Including only interactions between pairs of solute ions, Friedman obtains the following equation for the excess free energy.

$$\frac{G^{ex}}{VkT} = -\frac{\kappa^3}{12\pi} + \sum_i \sum_j c_i c_j \left[\frac{2\pi z_i z_j l}{\kappa^2} J_{ij}(\kappa, z_i, z_j, \ldots)\right] \qquad (1)$$

$$l = e^2/DkT \qquad (2)$$

Here V is the volume, κ is the usual Debye–Hückel parameter, c_i is the concentration and z_i the number of charges (sign included) on the ith ion, D is the dielectric constant, and e, k, and T have familiar meanings. The function J depends on the interionic potential and κ. If the potential is

$$v_{ij} = u_{ij} + z_i z_j l/r \qquad (3)$$

where the short-range potential u_{ij} is a function of interionic distance r, then J is

$$J_{ij} = \frac{\kappa^2}{z_i z_j l}\int_0^\infty [\exp(q_{ij} - u_{ij}/kT) - 1 - q_{ij} - \tfrac{1}{2}q_{ij}^2]r^2\,dr \tag{4}$$

with

$$q_{ij} = -(z_i z_j l/r)e^{-\kappa r} \tag{5}$$

In general, the integral and J cannot be evaluated without explicit knowledge of the short-range potential u_{ij}, but for our particular purposes a simplification is possible, as will be shown below.

Since we wish to work in macroscopic units, in particular molalities, these equations are restated below. We omit the secondary terms which arise from differences between concentration and molality, etc., since these can be accommodated within the virial coefficients we add for the effects of all short-range forces.

$$G^{ex}/RT = -4A_\phi I^{3/2} + \sum_i \sum_j m_i m_j (z_i z_j/4I)J_{ij} \tag{6}$$

$$A_\phi = \tfrac{1}{3}(2\pi N_0 d_w/1000)^{1/2} l^{3/2} \tag{7}$$

Here A_ϕ is the Debye–Hückel coefficient for the osmotic coefficient, G^{ex} is for a solution containing 1 kg of solvent of density d_w, and m_i is the molality of the ith species. The ionic strength I is, of course, $\frac{1}{2}\sum m_i z_i^2$.

The corresponding general equation adopted in I is

$$G^{ex}/RT = f(I) + \sum_i \sum_j m_i m_j \lambda_{ij}(I) \tag{8}$$

where

$$f(I) = -A_\phi(4I/b)\ln(1 + bI^{1/2}) \tag{9}$$

Here A_ϕ is the same Debye–Hückel constant defined above, and b was chosen to be 1.2 for all solutions. The second virial coefficients λ_{ij} are functions of ionic strength as indicated.

Clearly, there is a close analogy between the first term on the right in (6) and (8), each of which contains the Debye–Hückel limiting law. Also, each of the second virial coefficients λ_{ij} in (8) is analogous to the corresponding terms involving J_{ij} in (6).

It is possible to divide $f(I)$ into two terms: one is identical with the corresponding term in (6), while the remainder can be further divided into terms which can be combined with the λ's. Indeed, this was done in exploratory calculations, but it was found to offer no practical advantage. It seems preferable to assume that these terms based on the difference of our $f(I)$ from the first term in (6) are combined with effects of short-range forces and of the use of molalities instead of concentrations in the virial coefficients.

If we proceed now to the experimentally observable combinations of λ's as defined in I, the properties of a pure electrolyte MX are related to

$$B_{MX} = \lambda_{MX} + |z_X/2z_M|\,\lambda_{MM} + |z_M/2z_X|\,\lambda_{XX} \tag{10}$$

In mixtures the properties depend also on quantities of the type

$$\theta_{MN} = \lambda_{MN} - (z_N/2z_M)\lambda_{MM} - (z_M/2z_N)\lambda_{NN} \tag{11}$$

These same combinations of the J's will yield the corresponding quantities for the cluster-integral theory.

In order to proceed further, in general, one must have the expression for u_{ij}, the potential for short-range interionic interaction. But for pairs of ions of the same sign, if the electrostatic repulsion is strong enough, the ions will seldom approach one another closely; hence the short-range potential should have little or no effect. This can be seen mathematically in Eq. (4). If q_{ij} is large and negative for the range of r for which u_{ij} differs from zero, then the value of $\exp(q_{ij})$ is extremely small throughout this range. Thus, provided u_{ij} is positive (or if negative, is small), the effect of u_{ij} will be negligible.

In view of this situation, we can evaluate the effect of electrostatic forces on the difference terms θ_{MN} or θ_{XY} without making any detailed assumption about short-range forces. Let us write

$$\theta_{MN} = {}^{S}\theta_{MN} + {}^{E}\theta_{MN} \tag{12}$$

where the first term on the right arises from the combined effects of short-range forces acting directly or through the solvent, of the use of molalities instead of concentration, and of the difference in the Debye–Hückel term in Eq. (8) from that in (6). The second term ${}^{E}\theta_{MN}$ will be calculated from the corresponding terms of the cluster-integral theory with the omission of short-range forces. From the definition of θ_{MN}, we have

$${}^{E}\theta_{MN} = {}^{E}\lambda_{MN} - (z_N/2z_M){}^{E}\lambda_{MM} - (z_M/2z_N){}^{E}\lambda_{NN} \tag{13}$$

$${}^{E}\lambda_{ij} = (z_i z_j/4I)J_{ij} \qquad \text{with} \qquad u_{ij} = 0 \tag{14}$$

$$J_{ij} = \frac{\kappa^2}{z_i z_j l}\int_0^\infty (1 + q_{ij} + \tfrac{1}{2}q_{ij}^2 - e^{q_{ij}})r^2\,dr \tag{15}$$

With the substitutions

$$y = \kappa r \tag{16}$$

$$x = z_i z_j l \tag{17}$$

$$q = -(x/y)e^{-y} \tag{18}$$

$$J(x) = x^{-1}\int_0^\infty (1 + q + \tfrac{1}{2}q^2 - e^q)y^2\,dy \tag{19}$$

The integrals of the second and third terms within the parentheses are straightforward and yield

$$J(x) = \tfrac{1}{4}x - 1 + J_2(x) \tag{20}$$

$$J_2(x) = x^{-1} \int_0^\infty (1 - e^q) y^2 \, dy \tag{21}$$

There is no simple integral for J_2, but this function is readily evaluated by direct numerical integration on modern computers.

For very large x, one finds that J_2 is small, and the first two terms of Eq. (20) give a limiting approximation. For very small x, an alternate expansion is possible, yielding a series whose leading term is

$$J(x) = -\tfrac{1}{6}x^2(\ln x + 0.419711) \tag{22}$$

From Eq. (22) one notes the limiting form (omitting the 0.419711) for very small x which has been stated[7] previously as a "limiting law." We find, however, that this limiting form is not a good approximation in the range of practical application, and use the full expression for J in all of our work.

The derivative of J is also readily computed by numerical integration. The resulting curves for J and J' are shown in Fig. 1. For practical calculations, it is convenient to use simpler approximate expressions for J and J'; several examples are given in the Appendix together with further discussion of the evaluation of these functions and a table of numerical values. In our working units

$$x_{ij} = 6 z_i z_j A_\phi I^{1/2} \tag{23}$$

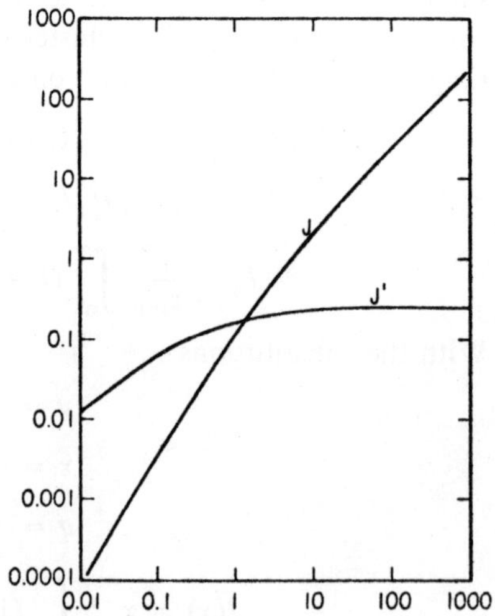

Fig. 1. The functions J and J'.

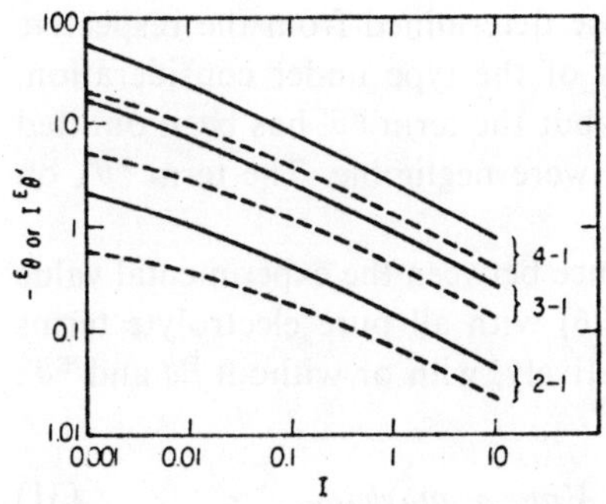

Fig. 2. The functions $-{}^E\theta$ (solid curves) and $I{}^E\theta'$ (dashed curves) for mixing ions of charge types 2–1, 3–1, and 4–1.

and

$$ {}^E\theta_{MN} = (z_M z_N/4I)[J(x_{MN}) - \tfrac{1}{2}J(x_{MM}) - \tfrac{1}{2}J(x_{NN})] \qquad (24)$$

$$ {}^E\theta'_{MN} = -({}^E\theta_{MN}/I) + (z_M z_N/8I^2)[x_{MN}J'(x_{MN}) - \tfrac{1}{2}x_{MM}J'(x_{MM}) - \tfrac{1}{2}x_{NN}J'(x_{NN})] \qquad (25)$$

It should be emphasized that these results apply over the full range of ionic strength and therefore have much greater applicability than an expression which applies only in the limit at low concentration.

Figure 2 shows the values of ${}^E\theta$ for mixing ions of charge types 2–1, 3–1, and 4–1. This quantity may be compared with θ or ${}^S\theta$ values to judge the magnitude of the effect we are discussing. Since typical θ values range up to about 0.1, it is clear that the effect is significant for 2–1 mixing and large for 3–1 or 4–1 systems. Also note the marked increase of ${}^E\theta$ and $I{}^E\theta'$ as ionic strength decreases.

3. DATA FOR UNSYMMETRICAL MIXING

The most interesting systems for application of these equations are those of the types HCl–$AlCl_3$ and HCl–$BaCl_2$ where the activity coefficient of HCl is measured. For the type HX–NX_z the activity coefficient of HX is given by Eq. (15) of IV which reduces to

$$\begin{aligned} \ln \gamma_{HX} = {} & f^\gamma + (m_H + m_X)(B_{HX} + m_X C_{HX}) \\ & + m_N(B_{NX} + m_X C_{NX} + {}^S\theta_{HN} + {}^E\theta_{HN}) \\ & + m_H m_X(B'_{HX} + C_{HX}) + m_N m_X(B'_{NX} + C_{NX} + \tfrac{1}{2}\psi_{HNX}) \\ & + m_H m_N({}^E\theta'_{HN} + \tfrac{1}{2}\psi_{HHX}) \end{aligned} \qquad (26)$$

$$f^\gamma = -A_\phi[I^{1/2}/(1 + 1.2I^{1/2}) + (2/1.2)\ln(1 + 1.2I^{1/2})] \qquad (27)$$

$$B_{ij} = \beta_{ij}^{(0)} + (2\beta_{ij}^{(1)}/\alpha^2 I)[1 - (1 + \alpha I^{1/2})\exp(-\alpha I^{1/2})] \qquad (28)$$

$$B'_{ij} = (2\beta_{ij}^{(1)}/\alpha^2 I^2)[-1 + (1 + \alpha I^{1/2} + \tfrac{1}{2}\alpha^2 I)\exp(-\alpha I^{1/2})] \qquad (29)$$

$$C_{HX} = C_{HX}^\phi/2 \qquad C_{MX} = C_{MX}^\phi/2z^{1/2} \qquad (30)$$

Here the $\beta^{(0)}$, $\beta^{(1)}$, and C^{ϕ} values are those determined from the respective pure electrolytes, and $\alpha = 2$ for all cases of the type under consideration. Also, Eq. (12) for θ has been introduced; but the term ${}^{S}\theta'$ has been omitted since in IV it was found that such terms were negligible. The term ${}^{E}\theta'$, on the contrary, is quite important.

Next we define $\Delta \ln \gamma_{HX}$ as the difference between the experimental value of $\ln \gamma_{HX}$ and that calculated from Eq. (26) with all pure electrolyte terms included, with ${}^{S}\theta$ and ψ zero, and, alternatively, with or without ${}^{E}\theta$ and ${}^{E}\theta'$. The relationship

$$(\Delta \ln \gamma_{HX})/m_N = {}^{S}\theta_{HN} + \tfrac{1}{2}(m_H + m_X)\psi_{HNX} \tag{31}$$

shows that a plot of the left side of this equation against $\frac{1}{2}(m_H + m_X)$ should give a straight line with intercept ${}^{S}\theta$ and slope ψ.

The results for the systems HCl–$SrCl_2$[10,11] and HCl–$AlCl_3$[12,13] are shown in Fig. 3, where the open circles and triangles indicate results including

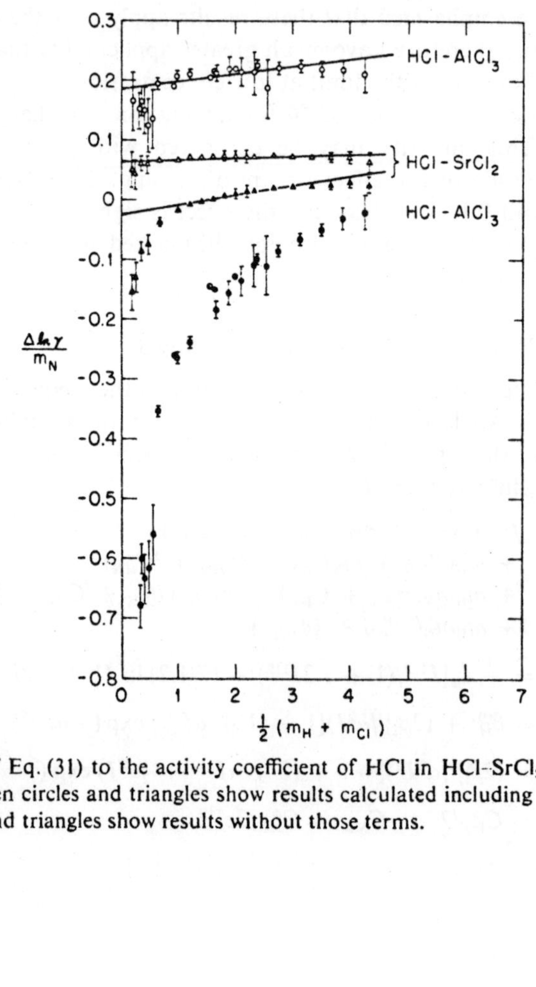

Fig. 3. Fits of Eq. (31) to the activity coefficient of HCl in HCl–$SrCl_2$ and HCl–$AlCl_3$ solutions. Open circles and triangles show results calculated including ${}^{E}\theta$ and ${}^{E}\theta'$, while solid circles and triangles show results without those terms.

$^E\theta$ and $^E\theta'$, and the solid circles and triangles indicate values excluding those terms. The estimated effect of experimental uncertainty is shown by error bars where it is significant.

It is apparent from Fig. 3 that the inclusion of the additional electrostatic terms $^E\theta$ and $^E\theta'$ is essential in order to treat HCl–$AlCl_3$ satisfactorily. For HCl–$SrCl_2$, the discrepancy without those terms is somewhat greater than the experimental uncertainty, whereas by including those terms the agreement is excellent. Similar results are obtained for the system HCl–$BaCl_2$.(14,15) For other systems of these types, the data are much less extensive, but in all the cases examined there is no difficulty fitting the results when $^E\theta$ and $^E\theta'$ are included.

The results shown in Fig. 3 clearly confirm the correctness of the higher-order electrostatic contribution to the interaction of H^+ and N^{+z} ions in these systems. Also, they show that this effect is of a significant magnitude and should be included in practical calculations, especially when z is 3 or greater.

Mixtures of singly and doubly charged ions were treated in IV with considerable success without the inclusion of $^E\theta$ and $^E\theta'$. Most of these calculations were repeated with $^E\theta$ and $^E\theta'$ included, and the results for mixtures with a common ion are shown in Table I. It is apparent that the inclusion of these terms improves the fit significantly for several systems, most notably the first three, while it makes little difference in other cases. The values of

Table I. Systems Mixing Doubly and Singly Charged Ions

	With $^E\theta$ and $^E\theta'$			Without $^E\theta$ and $^E\theta'$			
System	θ	ψ	σ	θ	ψ	σ	Ref.
HCl–$SrCl_2$	0.065	0.003	0.002	−0.020	0.018	0.010	10, 11
HCl–$BaCl_2$	0.072	0.000	0.003	−0.036	0.024	0.005	14, 15
HCl–$MnCl_2$	0.073	−0.007	0.002	0.000	0.000	0.008	16
LiCl–$BaCl_2$	0.000	0.009	0.002	−0.070	0.019	0.002	17
NaCl–$MgCl_2$	0.070	−0.010	0.001	0.000	0.000	0.002	18
Na_2SO_4–$MgSO_4$	0.070	−0.023	0.003	0.000	0.000	0.005	18
NaCl–$CaCl_2$	0.070	−0.007	0.003	0.000	0.000	0.004	19
NaCl–$BaCl_2$	0.067	−0.012	0.001	−0.003	0.000	0.001	20
NaCl–$MnCl_2$	0.082	-0.017_4	0.002	0.000	−0.003	0.003	21
KCl–$CaCl_2$	0.032	−0.025	0.004	−0.040	−0.015	0.003	22
KCl–$BaCl_2$	0.010	−0.017	0.002	−0.072	0.000	0.001	23
CsCl–$BaCl_2$	−0.070	−0.015	0.002_5	−0.150	0.000	0.003	17
NaCl–Na_2SO_4	0.020	0.0014	0.002	−0.035	0.007	0.002	18
KCl–K_2SO_4	0.020	0.000	0.002	−0.035	0.000	0.002	24
$MgCl_2$–$MgSO_4$	0.020	−0.014	0.002	−0.035	0.000	0.002	18

${}^{S}\theta$ are increased over the values of θ (without ${}^{E}\theta$ and ${}^{E}\theta'$) by about 0.07, while the new values of ψ are decreased by about 0.01 from those reported before.

Two systems of mixing without a common ion were also recalculated with ${}^{E}\theta$ and ${}^{E}\theta'$ and with the other parameters determined from the data for mixing with a common ion. The resulting fits for NaCl–$MgSO_4$ and for Na_2SO_4–$MgCl_2$ yielded the same σ values as were obtained without ${}^{E}\theta$ and ${}^{E}\theta'$.

One may conclude that the ${}^{E}\theta$ and ${}^{E}\theta'$ terms should be included for systems mixing singly and doubly charged if maximum accuracy is desired. But in most practical cases of this type, these may be omitted provided the appropriate θ and ψ values are used. It is essential, however, to maintain consistency, i.e., if ${}^{E}\theta$ and ${}^{E}\theta'$ are included, the values of ${}^{S}\theta$ and ψ must be those derived on that basis, whereas if ${}^{E}\theta$ and ${}^{E}\theta'$ are omitted, the corresponding θ and ψ must be used.

For the system HCl–$AlCl_3$ the parameters are ${}^{S}\theta = 0.185$, $\psi = 0.013$, which yield a good fit, $\sigma = 0.006$. There are no other systems involving 3–1 mixing for which comparably accurate and extensive data are available. The results for HCl–$LaCl_3$[25–27] and for HCl–$CeCl_3$[28] show behavior similar to that of HCl–$AlCl_3$ in the dilute range, which fully confirms the need for the additional electrostatic terms ${}^{E}\theta$ and ${}^{E}\theta'$. In the more concentrated range, however, the data are sufficiently scattered that I hesitate to recommend specific values of ${}^{S}\theta$ and ψ. One can use either zero values for these quanities or the values for HCl–$AlCl_3$ if estimates are desired.

The possibility was explored that the discrepancy for the HCl–$AlCl_3$ system (without ${}^{E}\theta$ and ${}^{E}\theta'$) might be explained by ion association to species such as $AlCl^{++}$, $AlCl_2^{+}$, etc. Such equilibria will not explain the discrepancy displayed in Fig. 3; the primary effect of these association reactions would be at high concentration, whereas the anomaly is found at low concentration.

4. SYMMETRICAL MIXING

The higher-order electrostatic terms considered above for unsymmetrical mixtures vanish if the ions mixed have the same charge. Friedman[7] has shown, however, that there is a different and more subtle limiting law applicable to symmetrical mixtures. In terms of our parameters this is

$$\lim_{I\to 0} \frac{\partial \ln |\theta_{MN}|}{\partial I^{1/2}} = 6z^2 A_\phi \tag{32}$$

where z is the number of charges on the ions mixed and A_ϕ is the Debye–Hückel coefficient for the osmotic coefficient. Robinson, Wood, and Reilly[8] gave an alternate derivation of this limiting law in simple terms but were unable to find experimental confirmation of it with isopiestic data. This was not surprising since Friedman's sample calculations indicated an approach to Eq. (32) only below an ionic strength of 0.01.

The physical basis of this law may be related to the change with ionic strength of the radial distribution function at repulsive contact for ions of the same sign of charge. This is shown in Fig. 3 of Part I. It is this change with ionic strength of the radial distribution functions that leads to the ionic strength dependence of the second virial coefficients for pure electrolytes. Thus, it seems reasonable to assume the same mathematical form for θ_{MN} as for B_{MX}. For the combination appearing in the osmotic coefficient, this yields

$$\theta + I\theta' = \theta^{(0)} + \theta^{(1)}e^{-\alpha I^{1/2}}$$

Then one obtains

$$\theta = \theta^{(0)}\{1 + (2\theta^{(1)}/\alpha^2 I\theta^{(0)})[1 - (1 + \alpha I^{1/2})e^{-\alpha I^{1/2}}]\} \tag{33}$$

$$\lim_{I\to 0} \frac{\partial \ln|\theta|}{\partial I^{1/2}} = -\frac{\frac{2}{3}\alpha(\theta^{(1)}/\theta^{(0)})}{1 + (\theta^{(1)}/\theta^{(0)})} = 6z^2 A_\phi$$

$$\theta^{(1)}/\theta^{(0)} = -9z^2 A_\phi/(\alpha + 9z^2 A_\phi) \tag{34}$$

One may now make the reasonable assumption that α retains the same value 2.0 for these expressions as for the second virial coefficients of most pure electrolytes; then the numerical relationship of $\theta^{(1)}$ to $\theta^{(0)}$ is determined.

Equations (33) and (34) thus yield a plausible function for θ at all concentrations which is consistent with the limiting law. These equations were applied to the systems HCl–KCl[29] and HCl–CsCl[30] for which the activity-coefficient measurements appeared most likely to yield confirmation of this effect. While no contradiction was observed, the use of Eqs. (33) and (34) gave so little improvement over the use of constant θ's that the complexity is unjustified.

There is a corresponding limiting law for heats of mixing, and Falcone, Levine, and Wood[31] have presented experimental data on mixing for the systems LiCl–CsCl and n-Bu_4NCl–KCl which tend to confirm that law. While the concepts behind Eqs. (33) and (34) could be applied to the enthalpy, this approach will not be carried further at present.

5. UNSYMMETRICAL PURE ELECTROLYTES

The theory for a higher-order limiting law for unsymmetrical pure electrolytes has been well established by several authors.[5,7] It can be expressed for the osmotic coefficient by the equation

$$\lim_{m\to 0}\left(\frac{\phi - 1 - \text{DHLL}}{I}\right) = \frac{3}{2}\frac{n_3^2}{n_2} A_\phi^2 \ln I \tag{35}$$

with similar expressions for the activity coefficient and the excess Gibbs energy. Here DHLL refers to the appropriate expression for the Debye–Hückel limiting law, A_ϕ is the coefficient in that law for the osmotic coefficient, and n_3 and n_2 are, respectively, the third and second moments of concentration of charge types. In contrast to the case discussed above for unsymmetrical mixing, it is not possible to extend this law over a finite range of concentrations without making some assumption concerning short-range forces.

Friedman[7] showed that the experimental data available in 1962 did not extend to low enough concentration with high enough precision to confirm this law, although there was no conflict with the law. Recently, Stock and Plewinsky[6] presented data for an even more unsymmetrical electrolyte of the 6-1 type, $Na_6(H_2W_{12}O_{30})$. Their freezing-point measurements were shown to be consistent with the Debye–Hückel limiting law for that charge type, but the higher-order law was not considered.

First the experimental data are fitted to the equation employed throughout this series of papers. For a $z - 1$ charge type of pure electrolyte MX_z of molality m it becomes

$$\phi - 1 = zf^\phi + m[2z/(z+1)]B^\phi_{MX} + m^2[2z^{3/2}/(z+1)]C^\phi_{MX} \tag{36}$$

$$B^\phi_{MX} = \beta^{(0)}_{MX} + \beta^{(1)}_{MX} \exp(-\alpha I^{1/2}) \tag{37}$$

$$I = mz(z+1)/2 \tag{38}$$

The experimental data range is from 8×10^{-5} to 1.2×10^{-2} M. Below 0.001 M, however, the freezing-point depression becomes so small that the data do not contribute to the determination of these constants nor to the detection of the higher-order limiting law. Since the highest concentration is relatively low, the third virial coefficient C^ϕ was omitted. A good fit is obtained with the following values

$$\tfrac{12}{7}\beta^{(0)} = -11$$
$$\tfrac{12}{7}\beta^{(1)} = 112$$
$$\sigma = 0.005$$

While the above treatment showed no discrepancies suggestive of the higher-order limiting law, the original data are also presented in Fig. 4 in a form suggested by Friedman. The higher-order limiting law requires the slope to approach that of the dashed line eventually as log I becomes more negative. Again it is apparent that the data above 0.001 M, log I = −1.7, show no confirmation of the higher-order limiting law. Below this concentration, the effect of experimental error becomes so great that nothing can be concluded. Thus, the higher-order law, although undoubtedly valid as $I \to 0$, is not useful at experimentally accessible concentrations, and its effect on measurable properties is negligible by present standards.

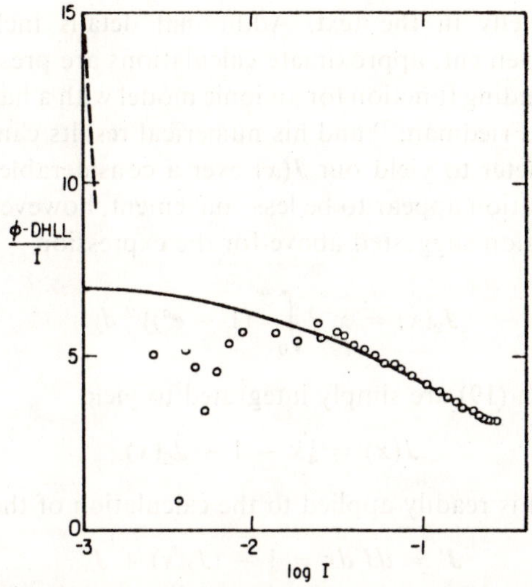

Fig. 4. Freezing-point data for $Na_6(H_2W_{12}O_{40})$ presented as deviations from the Debye–Hückel limiting law. The dashed line shows the slope of the high-order limiting law which must be approached at large negative values of log *I*. The solid curve is defined in the text.

6. DISCUSSION

While there seems no doubt about the theoretical validity of the higher-order limiting laws, and there is good experimental confirmation in the case of unsymmetrical mixing, it is difficult to offer any simple physical picture. Mathematically, these laws arise from additional terms in the expansion in series of the Boltzmann factor after the separation of the first terms which yield the Debye–Hückel limiting law (or are zero by electrical neutrality). The physical concept of a statistical distribution governed by the Boltzmann expression is clear enough, but it is not easy to picture the meaning of individual terms in the series expansion of the exponential. One can say that these effects arise from modifications of the radial distribution function from its linearized (Debye–Hückel) approximation. Possibly, future statistical calculations of distribution functions for appropriate examples will clarify this matter.

APPENDIX

The evaluation of the function $J(x)$

$$J(x) = x^{-1}\int_0^\infty (1 + q + \tfrac{1}{2}q^2 - e^q)y^2\,dy \tag{19}$$

$$q = -(x/y)e^{-y} \tag{18}$$

was discussed briefly in the text. Additional details including empirical equations for convenient, approximate calculations are presented here.

The corresponding function for an ionic model with a hard-core potential was evaluated by Friedman,[7] and his numerical results can be extrapolated to zero core diameter to yield our $J(x)$ over a considerable range of x. His methods of calculation appear to be less convenient, however, than the direct numerical integration suggested above for the expression

$$J_2(x) = x^{-1}\int_0^\infty (1 - e^q)y^2\,dy \tag{21}$$

The other terms in (19) are simply integrated to yield

$$J(x) = \tfrac{1}{4}x - 1 + J_2(x) \tag{20}$$

The same method is readily applied to the calculation of the derivative of J.

$$J' = dJ/dx = \tfrac{1}{4} - (J_2/x) + J_3 \tag{39}$$

$$J_3 = x^{-1}\int_0^\infty e^{-y}e^q y\,dy \tag{40}$$

Some numerical values of J and J' calculated by this method are given in Table II.

It is also possible to obtain a series expression for J valid for small x. One may break the integration into two ranges and use different expansions of Eq. (19) for each range. It is convenient to divide the ranges at $y = x/L$, where L is a constant and $L \gg x$. Then one has

$$J = J_4 + J_5 + J_6 \tag{41}$$

$$\begin{aligned} J_4 &= x^{-1}\int_0^{x/L} (1 + q + \tfrac{1}{2}q^2)y^2\,dy \\ &= x^2/3L^3 - 1 + e^{-x/L}(1 + x/L) + (x/4)(1 - e^{-2x/L}) \end{aligned} \tag{42}$$

$$\begin{aligned} J_5 &= -x^{-1}\int_0^{x/L} e^q y^2\,dy \\ &= -x^{-1}\int_0^{x/L} y^2\,dy(e^{-x/y})(e^x)(e^{-xy/2})(e^{xy^2/6})\cdots \\ &= -x^2e^x[L^{-3}E_4(L) - \tfrac{1}{2}x^2L^{-4}E_5(L) + (\tfrac{1}{6}x^3 + \tfrac{1}{8}x^4)L^{-5}E_6(L)\cdots] \end{aligned} \tag{43}$$

$$\begin{aligned} J_6 &= x^{-1}\int_{x/L}^\infty (1 + q + \tfrac{1}{2}q^2 - e^q)y^2\,dy \\ &= x^2\sum_{p=3}^\infty (L^{p-3}/p!)E_{p-2}(px/L) \end{aligned} \tag{44}$$

Table II. Values of the Functions J and J'

x	J	J'
0.01	0.0000706	0.0127
0.02	0.0002387	0.0207
0.03	0.0004806	0.0275
0.04	0.0007850	0.0333
0.05	0.0011443	0.0385
0.06	0.0015529	0.0432
0.07	0.0020063	0.0475
0.08	0.0025010	0.0514
0.09	0.0030340	0.0551
0.10	0.0036028	0.0586
0.12	0.0048393	0.0649
0.14	0.0061961	0.0706
0.16	0.0076615	0.0758
0.18	0.0092260	0.0806
0.20	0.010882	0.0850
0.24	0.014441	0.0928
0.28	0.018295	0.0997
0.32	0.022409	0.1059
0.36	0.026755	0.1114
0.40	0.031313	0.1164
0.44	0.036061	0.1210
0.48	0.040985	0.1252
0.52	0.046070	0.1291
0.56	0.051306	0.1327
0.60	0.056680	0.1360
0.80	0.085346	0.1499
1.00	0.11644	0.1605
1.20	0.14941	0.1689
1.40	0.18390	0.1758
1.60	0.21965	0.1815
1.80	0.25645	0.1864
2.00	0.29416	0.1906
3.00	0.49283	0.2053
4.00	0.70293	0.2142
5.00	0.92035	0.2202
6.00	1.14288	0.2246
7.00	1.36918	0.2279
8.00	1.59839	0.2304
9.00	1.82990	0.2325
10.00	2.06328	0.2342
12.00	2.53446	0 2368
16.00	3.48916	0.2402
20.00	4.45453	0.2423
24.00	5.57865	0.2374
28.00	6.40378	0.2447

Table II. Continued

x	J	J'
32.00	7.38429	0.2455
36.00	8.36745	.02461
40.00	9.35270	0.2465
50.00	11.82248	0.2474
60.00	14.29890	0.2479
70.00	16.77979	0.2483
80.00	19.26387	0.2485
90.00	21.75033	0.2487
100.00	24.23861	0.2489
200.00	49.17099	0.2496
400.00	99.11907	0.2498
600.00	149.09520	0.2499
800.00	199.08083	0.2499
1000.00	249.07101	0.2500
2000.00	499.04682	0.2500
4000.00	999.03028	0.2500
6000.000	1499.02328	0.2500
8000.00	1999.01925	0.2500
10000.00	2499.01659	0.2500

One may now insert the series expressions for the exponential integrals and collect all terms by powers of x. Terms involving L cancel, and there results:

$$J = -\sum_{p=3}^{\infty} [x^{p-1}p^{p-3}/p!(p-3)!](\ln x + K_p) \tag{45}$$

The quantities K_p are constants given as sums involving Euler's constant, $\ln p$, and integral fractions; the first few values are $K_3 = 0.419710$, $K_4 = -0.292608$, $K_5 = -0.369464$, and $K_6 = -3.00381$.

Equation (45) shows the limiting behavior of J for very small x, but it is convenient for numerical calculation only up to about $x = 0.10$.

Since only approximate values of J are needed for use with experimental data, simpler expressions were developed for convenience in practical calculations. The two most useful expressions are

$$J = -\tfrac{1}{6}x^2(\ln x)e^{-10x^2} + \left(\sum_{k=1}^{6} C_k x^{-k}\right)^{-1} \tag{46}$$

$$J = x[4 + C_1 x^{-C_2} \exp(-C_3 x^{C_4})]^{-1} \tag{47}$$

with the parameters given in Table III. Either Eq. (46) or (47) is fully adequate for practical calculations with experimental data of presently available

Table III. Parameters for Equations (46) and (47)

Parameters	Eq. (46)	Eq. (47)
C_1	4.118	4.581
C_2	7.247	0.7237
C_3	−4.408	0.0120
C_4	1.837	0.528
C_5	−0.251	—
C_6	0.0164	—

accuracy. Equation (46) includes the correct limiting law for low x and is accurate to 2% or better over the important range $0.1 \leq x \leq 80$. It does deviate by as much as 10% near $x = 0.02$. Equation (47) includes the correct limiting form for large x and is accurate to 2% or better for x greater than 0.03, and its maximum deviation for smaller x is 6×10^{-6}. Since this function and its derivative are used only for relatively small terms in the expressions for thermodynamic functions, this accuracy is adequate.

ACKNOWLEDGMENTS

This research was sponsored by the U.S. Atomic Energy Commission. A portion of the work was done while the author was visiting the Oak Ridge National Laboratory, and many interesting discussions with members of the Oak Ridge staff are gratefully acknowledged.

REFERENCES

1. K. S. Pitzer, *J. Phys. Chem.* **77**, 268 (1973).[2]
2. K. S. Pitzer and Guillermo Mayorga, *J. Phys. Chem.* **77**, 2300 (1973).
3. K. S. Pitzer and Guillermo Mayorga, *J. Solution Chem.* **3**, 539 (1974).
4. K. S. Pitzer and Janice J. Kim, *J. Am. Chem. Soc.* **96**, 5701 (1974).
5. T. H. Gronwall, V. K. LaMer, and K. Sandved, *Physik. Z.* **29**, 558 (1928).
6. H. P. Stock and B. Plewinsky, *Ber. Bunsenges. Physik. Chem.* **76**, 694 (1972).[3]
7. H. L. Friedman, *Ionic Solution Theory* (Interscience Publishers, New York, 1962).
8. R. A. Robinson, R. H. Wood, and P. J. Reilly, *J. Chem. Thermodyn.* **3**, 461 (1971).
9. J. E. Mayer, *J. Chem. Phys.* **18**, 1426 (1959).
10. H. S. Harned and R. Gary, *J. Am. Chem. Soc.* **77**, 1994 (1955).
11. H. S. Harned and T. R. Paxton, *J. Phys. Chem.* **57**, 531 (1953).

[2] Minor typographical errors in refs. 1-4 are corrected in subsequent papers of the series.
[3] Supplementary details have also been kindly provided by the authors through correspondence.

12. H. S. Harned and R. Gary, *J. Am. Chem. Soc.* **77**, 4695 (1955).
13. H. S. Harned and C. M. Mason, *J. Am. Chem. Soc.* **53**, 3377 (1931).
14. H. S. Harned and R. Gary, *J. Am. Chem. Soc.* **76**, 5924 (1954).
15. H. S. Harned and C. G. Geary, *J. Am. Chem. Soc.* **59**, 2032 (1937).
16. C. J. Downes, *J. Chem. Soc., Faraday Trans. I* **68**, 1964 (1972).
17. S. Lindenbaum, R. M. Rush, and R. A. Robinson, *J. Chem. Thermodyn.* **4**, 381 (1972).
18. Y. C. Wu, R. M. Rush, and G. Scatchard, *J. Phys. Chem.* **72**, 4048 (1968); **73**, 2047, 4433, 4434 (1969).
19. R. A. Robinson and V. E. Bower, *J. Res, Nat. Bur. Std.* **70A**, 313 (1966).
20. R. A. Robinson and V. E. Bower, *J. Res. Nat. Bur. Std.* **69A**, 19 (1965).
21. C. J. Downes, *J. Chem. Eng. Data* **18**, 412 (1973).
22. R. A. Robinson and A. K. Covington, *J. Res. Nat. Bur. Std.* **72A**, 239 (1968).
23. R. A. Robinson and V. E. Bower, *J. Res. Nat. Bur. Std.* **69A**, 439 (1965).
24. R. A. Robinson, R. F. Platford, and C. W. Childs, *J. Solution Chem.* **1**, 167 (1972).
25. M. Randall and G. F. Breckenridge, *J. Am. Chem. Soc.* **49**, 1435 (1927).
26. M. H. Lietzke and R. W. Stoughton, *J. Phys. Chem.* **71**, 662 (1967).
27. M. H. Lietzke and M. D. Danford, *J. Chem. Eng. Data* **17**, 459 (1972).
28. C. M. Mason and D. B. Kellam, *J. Phys. Chem.* **38**, 689 (1934).
29. H. S. Harned and A. B. Gancy, *J. Am. Chem. Soc.* **62**, 627 (1958).[a]
30. H. S. Harned and O. E. Schupp, Jr., *J. Am. Chem. Soc.* **52**, 3892 (1930).
31. J. S. Falcone, A. S. Levine, and R. H. Wood, *J. Phys. Chem.* **77**, 2137 (1973).

[a] See also appendix of ref. 4.

Journal of Solution Chemistry, Vol. 5, No. 4, 1976

Thermodynamics of Electrolytes. VI. Weak Electrolytes Including H_3PO_4

Kenneth S. Pitzer[1] and Leonard F. Silvester[1]

Received November 6, 1975

Equations previously developed and widely applied to the thermodynamic properties of strong electrolytes are extended to solutions involving a dissociation equilibrium. Excellent agreement is obtained with the data for pure phosphoric acid to 6 M and for phosphate buffer solutions. The parameters of the strong electrolyte components of the buffer solutions are taken from other work, and the remaining parameters for H^+, $H_2PO_4^-$, and H_3PO_4 are evaluated, including a pK of 2.146. The present method avoids ambiguities which formerly arose in treating weak acids with as small pK as this.

KEY WORDS: Activity coefficient; osmotic coefficient; thermodynamics; solution theory; dissociation constants; phosphoric acid.

1. INTRODUCTION

In much of the literature on weak electrolytes, the entire focus is on the determination of the dissociation constant. While data for solutions of finite concentration are employed, the interpretation is directed entirely toward the limiting behavior at zero concentration which determines the dissociation constant. In contrast, we seek to represent the thermodynamic properties of such solutions over a broad range from zero to rather high concentrations Appropriate sets of equations include, of course, the dissociation constant of a weak electrolyte, and also terms for interaction of various solute species, both neutral and charged.

In the earlier papers of this series[1–4] equations were developed which represent effectively the properties of mixed as well as pure strong electrolytes. In the case of 2–2 salts, where some association is indicated by other measurements, it was feasible to treat thermodynamic properties without explicitly considering association.[3] We now consider cases where association is

[1] Department of Chemistry and the Lawrence Berkeley Laboratory, Berkeley, California 94720.

substantial and is explicitly recognized. The general equation adopted earlier is readily extended to include terms for the associated species.

In earlier work[5-8] it was found that dissociation constants smaller than about 10^{-4} were determined without difficulty but that, for stronger acids, pK 2 to 4, an ambiguity arises and becomes greater the larger the dissociation constant. This problem will be considered in some detail.

Phosphoric acid was chosen as an example in view of (1) the availability of excellent experimental data, (2) the fact that its pK is in the range where ambiguities have arisen, and (3) its general importance.

2. GENERAL EQUATIONS

In paper I of this series[1] the statistical-mechanical basis was given for the form of equation adopted. In brief, the general framework is that of the McMillan–Mayer theory of solutions and the equation relating intermolecular forces and distributions to the osmotic pressure.[9] Since we do not expect to make absolute calculations based upon detailed short-range intermolecular forces, we have ignored the differences between concentration and molality and between the McMillan–Mayer reference states and the standard states of chemical thermodynamics. The effects of these differences at finite concentration are absorbed in the virial coefficients for the interaction of solute species.

The basic equation is

$$\frac{G^{\text{ex}}}{RT} = n_{\text{w}}f(I) + \frac{1}{n_{\text{w}}}\sum_{ij}\lambda_{ij}(I)n_i n_j + \frac{1}{n_{\text{w}}^2}\sum_{ijk}\mu_{ijk}n_i n_j n_k \tag{1}$$

where G^{ex} is the excess Gibbs energy for a solution containing n_{w} kg of solvent and n_i, n_j, etc., moles of solute species i, j, etc. Also $f(I)$ is a function of ionic strength (and temperature and solvent properties) expressing the effect of long-range electrostatie forces between ions; $\lambda_{ij}(I)$ is a second virial coefficient giving the effect of short-range forces between species i and j (including indirect forces arising from the solvent); μ_{ijk} is a corresponding third virial coefficient for the interaction of three solute species i, j, and k. The second virial coefficients for ion–ion interactions are expected to depend on ionic strength for theoretical reasons given in paper I[1]; this dependence has been amply justified empirically. While these effects which cause a first-order variation in λ with ionic strength for ion–ion interactions might cause indirect or second-order variations in λ's for ion–neutral or neutral–neutral interactions, these latter variations should be very small and will be ignored. Thus, λ_{ij} is a function of ionic strength if both i and j are charged; otherwise it is a constant (for a particular solvent and temperature). The third virial coefficients are taken as constants even when all species are charged, and are ignored if all species have the same sign of charge. It is assumed that the λ and μ matrices are symmetric, i.e., $\lambda_{ij} = \lambda_{ji}$, etc.

The appropriate derivatives yield the activity and osmotic coefficients

$$\ln \gamma_i = \frac{1}{RT}\frac{\partial G^{ex}}{\partial n_i} = z_i^2 f' + 2\sum_j \lambda_{ij} m_j + \sum_{jk} m_j m_k \left(\frac{z_i^2}{2}\lambda'_{jk} + 3\mu_{ijk}\right) \tag{2}$$

$$\phi - 1 = -\frac{\partial G^{ex}/\partial n_w}{RT\sum_i m_i} = \left(\sum_i m_i\right)^{-1}\left[(If' - f) + \sum_{ij} m_i m_j(\lambda_{ij} + I\lambda'_{ij}) + 2\sum_{ijk} m_i m_j m_k \mu_{ijk}\right] \tag{3}$$

where $f' = \partial f/\partial I$, $\lambda'_{ij} = \partial\lambda_{ij}/\partial I$, and the various molalities $m_i = n_i/n_w$. Unless both species are ions, $\lambda'_{ij} = 0$. The general equations for the measurable combinations of ion activities such as $\gamma_\pm$ are given in papers I[(1)] and IV.[(4)]

In order to simplify these equations as far as possible without ignoring any important types of terms, we now assume a solution comprising a weak acid HA and its ions H^+ and A^- and another positive ion M^+ and negative ion X^-. The composition will be stated in terms of m_{HA}, m_H, m_A, m_M, and m_X. Neutrality requires that $m_H + m_M = m_A + m_X$. The resulting equations can readily be generalized for more complex solutions when desired.

The functions for long-range electrostatic forces were determined in paper I and contain, of course, the Debye–Hückel limiting law. They are

$$f^\phi = \tfrac{1}{2}[f' - (f/I)] = -A_\phi[I^{1/2}/(1 + 1.2I^{1/2})] \tag{4}$$

$$f^\gamma = \tfrac{1}{2}f' = -A_\phi[I^{1/2}/(1 + 1.2I^{1/2}) + (2/1.2)\ln(1 + 1.2I^{1/2})] \tag{5}$$

where A_ϕ is the Debye–Hückel constant for the osmotic coefficient and 1.2 is an empirical value discussed in paper I.

For the virial coefficients for ions, it proved convenient to define the sums which appear in equations for the pure electrolyte MX. For a 1–1 electrolyte these are

$$B_{MX} = \lambda_{MX} + \tfrac{1}{2}\lambda_{MM} + \tfrac{1}{2}\lambda_{XX} \tag{6}$$

$$B'_{MX} = \lambda'_{MX} + \tfrac{1}{2}\lambda'_{MM} + \tfrac{1}{2}\lambda'_{XX} \tag{7}$$

$$C_{MX} = \tfrac{3}{2}(\mu_{MMX} + \mu_{MXX}) \tag{8}$$

The adopted empirical forms which include, the ionic strength dependence of B and B' are

$$B_{MX} = \beta^{(0)}_{MX} + (2\beta^{(1)}_{MX}/\alpha^2 I)[1 - (1 + \alpha I^{1/2})\exp(-\alpha I^{1/2})] \tag{9}$$

$$B'_{MX} = (2\beta^{(1)}_{MX}/\alpha^2 I^2)[-1 + (1 + \alpha I^{1/2} + \tfrac{1}{2}\alpha^2 I)\exp(-\alpha I^{1/2})] \tag{10}$$

$$C^\phi_{MX} = 2C_{MX} = \tfrac{2}{3}C^\gamma_{MX} \tag{11}$$

where the general empirical constant α is 2.0 for all 1–1 and most other electrolytes and the specific constants $\beta^{(0)}_{MX}$, $\beta^{(1)}_{MX}$, and C^{ϕ}_{MX} have been determined in paper II[(2)] for most aqueous electrolytes of interest.

For mixed electrolytes, additional terms appear involving differences in virial coefficients. These have been defined (for mixing singly charged ions A and X, or similarly for H and M) as

$$\theta_{AX} = \lambda_{AX} - \tfrac{1}{2}\lambda_{AA} - \tfrac{1}{2}\lambda_{XX} \tag{12}$$

$$\psi_{MAX} = 6\mu_{MAX} - 3\mu_{MAA} - 3\mu_{MXX} \tag{13}$$

Although the λ's are presumed to vary with ionic strength, it was found to be a satisfactory approximation to take the θ's to be constants.

We turn now to the new relationships arising from association and from the presence of neutral solute species. First is the dissociation equilibrium itself where standard chemical thermodynamics yields

$$\mathrm{HA} = \mathrm{H}^+ + \mathrm{A}^-$$

$$-\Delta G^\circ/RT = \ln K = \ln(m_H m_A \gamma_H \gamma_A / m_{HA} \gamma_{HA}) \tag{14}$$

where K is the usual dissociation constant.

Once the HA association has been assumed, this reaction supercedes the second virial coefficient for H^+–A^- interaction. In practice one cannot determine the $\lambda_{H,H}$ and $\lambda_{A,A}$ terms separately, and hence one assumes $B_{H,A}$ to be zero. This point is discussed further below. For neutral–neutral interactions the second and third virial coefficients $\lambda_{HA,HA}$ and $\mu_{HA,HA,HA}$ can be used directly. There are four second virial coefficients for ion–neutral interactions: $\lambda_{H,HA}$, $\lambda_{M,HA}$, $\lambda_{A,HA}$, and $\lambda_{X,HA}$. Only those sums or differences of these coefficients that correspond to electrical neutrality can be measured, i.e., $(\lambda_{H,HA} + \lambda_{A,HA})$ or $(\lambda_{H,HA} - \lambda_{M,HA})$, etc. There are six such sums or differences, but only three are independent. Hence, only three of these λ's may be adjusted independently. We found it more convenient to set one λ at zero than to work with the various sums and differences and chose to let $\lambda_{X,HA} = 0$. Any desired sum or difference can, of course, be calculated from our results.

There are a large number of ion–neutral third virial coefficients, far more than could be determined from data now available or likely to be obtained. Such terms are included only when they are clearly required; none were required in the present example.

3. PHOSPHORIC ACID

There are experimental data for certain buffer solutions as well as for pure phosphoric acid. For the pure acid the osmotic-coefficient data based on solvent vapor pressure[(10,11)] were used. Among various measurements on buffer solutions, the data of Bates[5] on HCl–KH_2PO_4 mixtures are particularly extensive and of the highest accuracy. For the buffer-solution calculations, the parameters are needed for the pure electrolytes HCl, KCl, and KH_2PO_4; these were obtained in paper II.[2] More recent data[12] on pure KH_2PO_4 were also considered, but they did not justify any change in the previously published parameters. Values are also needed for the parameters $\theta_{H,K}$ and $\theta_{Cl,A}$ (A^- is $H_2PO_4^-$) from appropriate mixtures. The former is given in paper IV[4]; the latter was determined to be $\theta_{Cl,A} = 0.10$ from the isopiestic measurements of Childs, Downes, and Platford[13] which also yield $\psi_{K,Cl,A} = -0.010_5$.

For pure H_3PO_4 the pertinent equations are given below, where ϕ is an osmotic coefficient calculated from the water activity on the basis of dissociation into two ions, $\nu = 2$ (note that Platford reports his data on the basis of $\nu = 4$). If the actual dissociation of acid of molality m yields molality m_H of H^+ and A^-, the osmotic coefficient on that basis ϕ' is obtained from Eq. (3).

$$(\phi' - 1)(m + m_H) = 2m_H f^\phi + 2m_H(m - m_H)(\lambda_{H,HA} + \lambda_{A,HA}) + (m - m_H)^2\lambda_{HA,HA} + 2(m - m_H)^3\mu_{HA,HA,HA} \quad (15)$$

$$\phi'(m + m_H) = 2m\phi \quad (16)$$

In addition, we have the dissociation equilibrium determined from Eq. (14) and the expressions for $\frac{1}{2}\ln(\gamma_H\gamma_A) = \ln\gamma_\pm$ and $\ln\gamma_{HA}$ which may be obtained from Eq. (2) and several that follow it.

$$\tfrac{1}{2}\ln(\gamma_H\gamma_A) = f^\gamma + (m - m_H)(\lambda_{H,HA} + \lambda_{A,HA}) \quad (17)$$

$$\ln\gamma_{HA} = 2m_H(\lambda_{H,HA} + \lambda_{A,HA}) + 2(m - m_H)\lambda_{HA,HA} + 3(m - m_H)^2\mu_{HA,HA,HA} \quad (18)$$

In view of the high concentrations of pure phosphoric acid to be considered, the third virial coefficient $\mu_{HA,HA,HA}$ was included in Eq. (18). Third virial coefficients for ion–neutral interaction were tentatively omitted, and later results justify that decision.

For the buffer solutions with $HCl(m_1)$ and $KH_2PO_4(m_2)$ the measured quantity is the potential E of a cell with hydrogen and silver–silver chloride electrodes. If the hydrogen ion molality is m_H and the standard potential is E°, we obtain

$$E - E^\circ = -(RT/F)\ln(m_H m_1 \gamma_H \gamma_{Cl}) \quad (19)$$

Other symbols have their usual meaning, and we take $E^\circ = 0.2224$ V. Since the buffer-solution measurements extend only to a maximum concentration of 0.4 M, third virial coefficients are omitted throughout.

We obtain, for the ionic strength, $I = m_2 + m_H$, and also $m_{HA} = m_1 - m_H$, $m_{Cl} = m_1$, $m_K = m_2$, and $m_A = m_2 - m_1 + m_H$. For these buffer solutions Eq. (2) and subsequent equations yield

$$\begin{aligned}\ln \gamma_{HCl} &= \tfrac{1}{2}\ln(\gamma_H\gamma_{Cl})\\ &= f^{\gamma} + (m_1 + m_H)B_{HCl} + m_2(B_{KCl} + \theta_{HK})\\ &\quad + m_H m_1 B'_{HCl} + m_1 m_2 B'_{KCl} + m_2(m_2 - m_1 + m_H)B'_{KA}\\ &\quad + (m_2 - m_1 + m_H)\theta_{Cl,A} + (m_1 - m_H)(\lambda_{H,HA} + \lambda_{Cl,HA})\end{aligned} \tag{20}$$

$$\begin{aligned}\tfrac{1}{2}\ln(\gamma_H\gamma_A) &= f^{\gamma} + m_1(B_{HCl} + \theta_{Cl,A}) + m_2 B_{KA}\\ &\quad + m_H m_1 B'_{HCl} + m_1 m_2 B'_{KCl} + m_2(m_2 - m_1 + m_H)B'_{KA}\\ &\quad + m_2\theta_{H,K} + (m_1 - m_H)(\lambda_{H,HA} + \lambda_{A,HA})\end{aligned} \tag{21}$$

$$\begin{aligned}\ln \gamma_{HA} &= 2[m_H\lambda_{H,HA} + m_1\lambda_{Cl,HA} + m_2\lambda_{K,HA}\\ &\quad + (m_2 - m_1 + m_H)\lambda_{A,HA} + (m_1 - m_H)\lambda_{HA,HA}]\end{aligned} \tag{22}$$

For either the pure acid or the buffer solutions, it was necessary to eliminate the hydrogen ion molality m_H by an iterative solution of the appropriate set of equations.

In Eqs. (20) and (21) all the B and θ parameters are independently known. Thus, in the full array of equations (14) to (21) there are six parameters to be evaluated. While all might be simultaneously evaluated from the full array of data for both pure acid and buffer solutions, that calculation would be very complex and the assignment of appropriate weights would be difficult. Instead we chose to evaluate three parameters from each set of data. From the pure-acid data the sum $(\lambda_{H,HA} + \lambda_{A,HA})$ and the second and third virial coefficients for neutral–neutral interaction were determined by appropriate least-squares methods, while K and the individual second virial coefficients for ion–neutral interaction were determined from data on buffer solutions. This process was iterated until all results were consistent. The resulting parameters are given in Table I. The standard deviation of fit for the buffer solutions was 0.0002 V in the cell potential while for the pure acid

Table I. Parameters for Phosphoric Acid Solutions at 25°C (A^- is $H_2PO_4^-$)

$K = 7.142_5 \times 10^{-3}$	$(pK = 2.1461_5)$
$\lambda_{HA,HA} = 0.0503_1$	$\mu_{HA,HA,HA} = 0.0109_5$
$\lambda_{A,HA} = -0.400$	$\lambda_{H,HA} = 0.290$
$\lambda_{K,HA} = -0.070$	$(\lambda_{Cl,HA} = 0)$
Also determined from KCl–KH_2PO_4 mixtures:	
$\theta_{Cl,A} = 0.10$	$\psi_{K,Cl,A} = -0.010_5$

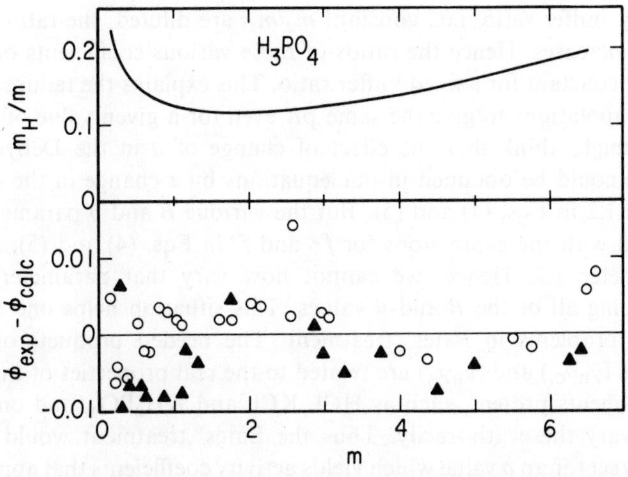

Fig. 1. Fraction dissociation of H_3PO_4 in upper section; deviation of calculated from experimental osmotic coefficients in lower section (circles, Elmore *et al.*[10]; triangles, Platford[11]).

to 6 M it was 0.005 in the osmotic coefficient. In each case, the fit is essentially within the experimental uncertainty. Figure 1 shows in the lower section the deviations in osmotic coefficient as a function of total molality for each set of experimental measurements. In the upper portion of Fig. 1 the fraction dissociated, m_H/m, is shown. The increase in dissociation at very low concentration is expected; the minimum near 2 M and the increase at higher concentration are surprising and are caused primarily by the large, negative value of the sum of ion–neutral second virial coefficients ($\lambda_{H,HA} + \lambda_{A,HA}$).

4. DISCUSSION

First we recall the ambiguities noted by Bates[5] and others[6–8] in determining the dissociation constant of phosphoric acid. Bates used the conventional Debye–Hückel expression for an activity coefficient, including the term involving a distance of closest approach a, and found that the value of pK obtained by linear extrapolation to zero ionic strength varied not only with the value of a but also with the buffer ratio. Also the slopes of the lines for different buffer ratios differed greatly.

The primary effect of buffer ratio can be seen in Eqs. (20) and (21) where the second virial terms in B's and θ's have different coefficients involving m_1 or m_2. Since these second virial coefficients can be measured independently, it is desirable to include them explicitly. Also it is important to note the appearance of m_H in some of the coefficients of B and θ terms. As solutions of constant buffer ratio, i.e., constant m_2/m_1, are diluted, the ratio of m_H to m_2 or m_1 increases. Hence the ratios of these various coefficients of B's and θ's are not constant for a fixed buffer ratio. This explains the failure of Bates' linear extrapolations to give the same pK even for a given value of a.

One might think that the effect of change of a in the Debye–Hückel expression could be obtained in our equations by a change in the empirical parameter 1.2 in Eqs. (4) and (5). But the various B and θ parameters were determined with the expressions for f^{ϕ} and f^{γ} in Eqs. (4) and (5), including the parameter 1.2. Hence, we cannot now vary that parameter without redetermining all of the B and θ values. This situation helps one to understand the problems in Bates' treatment. The needed products of activity coefficients ($\gamma_H\gamma_{Cl}$) and ($\gamma_H\gamma_A$) are related to the real properties of the electrolyte components present, such as HCl, KCl, and KH_2PO_4, and one cannot properly vary these arbitrarily. Thus the Bates' treatment would be most nearly correct for an a value which yields activity coefficients that approximate those of the various component electrolytes. Bates' final choice of pK as an average of extrapolated values for a = 4 Å and a = 6 Å is appropriate on this basis, and his result pK = 2.148 differs only slightly from our value of 2.146.

In our treatment we omitted the second virial coefficient $B_{H,A}$ for H^+ interaction with A^- on the basis that it was redundant once the association equilibrium was introduced. It can be shown that the introduction of a small value of $B_{H,A}$ should change the dissociation constant by an amount $\Delta(1/K) = 2B_{H,A}$. This was verified in a calculation for the buffer solutions with $B_{H,A} = -0.05$ in which the value of K changed by the expected amount and the agreement with the experimental data was unchanged. Not only was the standard deviation unchanged, but the individual differences between calculated and experimental cell potentials were the same to 0.00001 V. Thus, the redundancy of this papameter was confirmed, and its omission justified.

With respect to intermolecular phenomena, the most interesting result is the large negative value ($\lambda_{A,HA} = -0.4$) for the interaction of $H_2PO_4^-$ with H_3PO_4. This indicates a tendency toward association to form $H_5P_2O_8^-$. Presumably this species would be hydrogen bonded in a manner

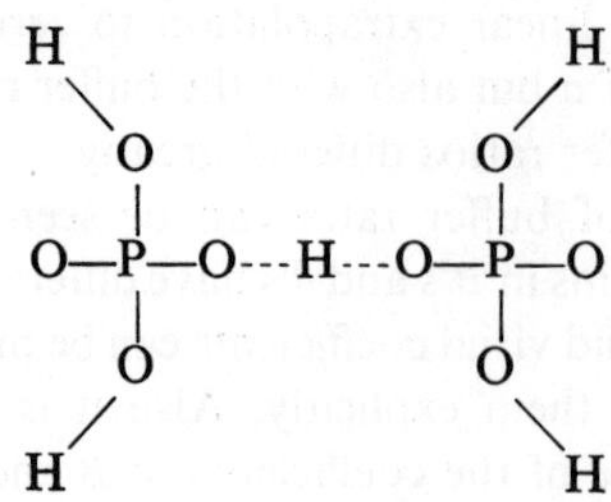

Evidence for this species has been presented by Elmore *et al.*[14] from thermodynamic data, by Selvaratnam and Spiro[15] from conductance and transference data, and also by Childs[16] and by Ivakin and Voronova.[17] It has also been suggested by some of these authors and by Wood and Platford[18] that the ion $H_2PO_4^-$ has a tendency to dimerize to $H_4P_2O_8^{-2}$. This is consistent with the abnormally low $\beta^{(0)}$ and $\beta^{(1)}$ for sodium and potassium salts of $H_2PO_4^-$ and the relatively large, positive $\theta_{Cl,A} = 0.1$ for the Cl^-–$H_2PO_4^-$ mixtures. Since $\theta_{Cl,A} = \lambda_{Cl,A} - \frac{1}{2}\lambda_{Cl,Cl} - \frac{1}{2}\lambda_{A,A}$, if the other λ's are small, $\lambda_{AA} \cong -0.2$. Thus, the tendency of $H_2PO_4^-$ to dimerize is about half as great as the tendency of H_3PO_4 and $H_2PO_4^-$ to associate. Each of these associations is mild enough that the virial-coefficient treatment is satisfactory, as verified by the excellent agreement we obtained.

Elmore *et al.*[14] calculated values for the molality of H^+ in phosphoric acid over a wide range of concentrations using their model which included the $H_5P_2O_8^-$ species. Their values for m_H are somewhat larger than ours, but the general trend is similar. Since m_H is not directly measured, it is quite possible for treatments to differ with respect to this quantity while agreeing on measured quantities such as the solvent vapor pressure.

Finally, the excellent fit with experimental data demonstrates that our equations treating most solute interactions by virial coefficients can be combined with an association equilibrium. Also it seems clear that presently well-established dissociation constants (below the range of ambiguity discussed above) can be used in such treatments if desired.

ACKNOWLEDGMENTS

We thank Mr. Jeffery A. Brown for assistance in programming and calculations. This research was sponsored by the Energy Research and Development Administration.

REFERENCES

1. K. S. Pitzer, *J. Phys. Chem.* **77**, 268 (1973).
2. K. S. Pitzer and G. Mayorga, *J. Phys. Chem.* **77**, 2300 (1973).
3. K. S. Pitzer and G. Mayorga, *J. Solution Chem.* **3**, 539 (1974).
4. K. S. Pitzer and J. J. Kim, *J. Am. Chem. Soc.* **96**, 5701 (1974).
5. R. G. Bates, *J. Res. Nat. Bur. Stand.* **47**, 127 (1951).
6. W. J. Hamer, *The Structure of Electrolyte Solutions* (John Wiley & Sons, New York, 1959), Chap. 15.
7. K. S. Pitzer and L. Brewer, in *Thermodynamics*, G. N. Lewis and M. Randall, eds. (McGraw-Hill, New York, 1961), pp. 587–590.
8. T. H. Lilley and C. C. Briggs, *Electrochim. Acta* **20**, 253 (1975).
9. T. L. Hill, *Introduction to Statistical Mechanics* (Addison-Wesley, Reading, Mass., 1960), Chap. 19.

10. K. L. Elmore, C. M. Mason, and J. H. Christensen, *J. Am. Chem. Soc.* **68**, 2528 (1946).
11. R. F. Platford, *J. Solution Chem.* **4**, 591 (1975).
12. C. W. Childs, C. J. Downes, and R. F. Platford, *Aust. J. Chem.* **26**, 863 (1973).
13. C. W. Childs, C. J. Downes, and R. F. Platford, *J. Solution Chem.* **3**, 139 (1974).
14. K. L. Elmore, J. D. Hatfield, R. L. Dunn, and A. D. Jones, *J. Phys. Chem.* **69**, 3520 (1965).
15. M. Selvaratnam and M. Spiro, *Trans. Faraday Soc.* **61**, 360 (1965).
16. C. W. Childs, *J. Phys. Chem.* **73**, 2956 (1969); *Inorg. Chem.* **9**, 2465 (1970).
17. A. A. Ivakin and E. M. Voronova, *Russ. J. Inorg. Chem.* **18**, 465 (1973).
18. R. H. Wood and R. F. Platford, *J. Solution Chem.* **4**, 977 (1975).

[Reprinted from the Journal of the American Chemical Society, 99, 4930 (1977).]

Thermodynamics of Electrolytes. 7. Sulfuric Acid

Kenneth S. Pitzer,* Rabindra N. Roy, and Leonard F. Silvester

Contribution from the Department of Chemistry and the Lawrence Berkeley Laboratory, University of California, Berkeley, California 94720. *Received March 9, 1977*

Abstract: Although the thermodynamic properties of sulfuric acid above 0.1 M and near 25 °C are well established numerically, they have not been represented accurately by equations which are based upon the ionic species present, H^+, HSO_4^-, and SO_4^{2-}. We have developed and fitted such equations over the range from 0 to 6 M in a system compatible with those for fully dissociated, strong electrolytes. The enthalpy is treated as well as the activity and osmotic coefficients. These equations also establish the solute standard state and the relationship between the properties of sulfuric acid in that state with those for the pure acid. Among the results obtained (for 25 °C) are the dissociation constant 0.0105 and the heat of dissociation −5.61 kcal mol^{-1} for HSO_4^- and the entropy of SO_4^{2-}, 4.2 ± 0.2, and of HSO_4^-, 32.1 ± 0.3 cal K^{-1} mol^{-1}. Also for the reaction $H_2SO_4(l) = 2H^+(aq) + SO_4^{2-}(aq)$, $\Delta H° = -22844$, $\Delta G° = -12871$ cal mol^{-1}.

In view of the great practical importance of sulfuric acid, it is desirable to have the most accurate and convenient expression of its thermodynamic properties. Above 0.1 M these properties are now well established,[1-6] but there is still considerable uncertainty about the properties of very dilute solutions and the related solute standard state. In this research we have the dual purposes first to establish as accurately as possible the thermodynamic properties of dilute sulfuric acid and second to provide a convenient yet accurate analytical representation of the properties of this acid in a form compatible with that used for other electrolytes[7-9] and over as wide a range of concentration as is feasible.

The thermodynamic treatment of sulfuric acid has been unusually difficult because the dissociation constant of HSO_4^- lies in the most troublesome region where methods fail that were successful for weaker acids.[10,11] The preceding paper of this series[12] considered this general problem with phosphoric acid as an example. Sulfuric acid is even more troublesome in view of the higher charge on the sulfate ion and the correspondingly larger changes in its activity coefficient.

General Equations

The statistical mechanical basis for the form of equation for a complex electrolyte was given in the first paper of this series.[7] The general framework is that of the McMillan–Mayer theory of solutions and the equation relating intermolecular forces and distributions to the osmotic pressure.[13] Our basic equation is

$$\frac{G^{ex}}{RT} = n_w f(I) + \frac{1}{n_w}\sum_{ij} \lambda_{ij}(I) n_i n_j + \frac{1}{n_w^2}\sum_{ijk} \mu_{ijk} n_i n_j n_k \quad (1)$$

where G^{ex} is the excess Gibbs energy for a solution containing n_w kg of solvent and n_i, n_j, etc., moles of solute species i, j, etc. Here $f(I)$ is a function of ionic strength (and temperature and solvent properties) expressing the effect of long-range electrostatic forces between ions and including, of course, the Debye–Hückel limiting law. Short-range interactions of solute species lead to the terms $\lambda_{ij}(I)$ for binary interactions and μ_{ijk} for ternary interactions; the theoretical basis for expecting an ionic-strength dependence for λ_{ij} was given earlier[7] and this has been fully confirmed empirically. The λ and μ matrices are symmetric, i.e., $\lambda_{ij} = \lambda_{ji}$, etc.

Equations for Sulfuric Acid

The intermediate thermodynamic derivations of activity and osmotic coefficients and the definitions of experimentally measurable parameters and convenient empirical forms have been given previously.[9] For example, for univalent ions a measurable combination of λ's is $B_{MX} = \lambda_{MX} + \frac{1}{2}\lambda_{MM} + \frac{1}{2}\lambda_{XX}$. We shall move directly to the results appropriate for a solution of sulfuric acid with stoichiometric molality m and molality m_H of H^+, $m_1 = 2m - m_H$ of HSO_4^- and $m_2 = m - m_1$ of SO_4^{2-}. With two different anions this is a mixed electrolyte for our equations even though the second anion arises from a dissociation equilibrium of the first. For the osmotic coefficient ϕ' on a mixed electrolyte basis one finds

$$\phi' - 1 = (\Sigma m_i)^{-1}\{2If^\phi + 2m_H[m_1(B_{H1}{}^\phi + m_H C_{H1}{}^\phi) + m_2(B_{H2}{}^\phi + m_H C_{H2}{}^\phi/2^{1/2})] + m_1 m_2(\theta_{12} + m_H\psi_{H12})\} \quad (2)$$

$$f^\phi = -A_\phi[I^{1/2}/(1 + 1.2I^{1/2})] \quad (3)$$

$$B_{MX}{}^\phi = \beta_{MX}{}^{(0)} + \beta_{MX}{}^{(1)} \exp(-\alpha I^{1/2}) \quad (4)$$

where the ionic strength $I = m + 2m_2$, A_ϕ is the Debye-Hückel limiting law parameter for the osmotic coefficient (see Appendix for numerical values), α is a general empirical parameter equal to 2 throughout this work, while $\beta_{MX}{}^{(0)}$ and $\beta_{MX}{}^{(1)}$ are specific parameters for the appropriate sum of λ_{ij} for binary interactions M–X, M–M, and X–X where the cation M is always H and the anion X is either $HSO_4^- = 1$ or $SO_4^{2-} = 2$. Also $C_{MX}{}^\phi$ is the corresponding third virial coefficient for triple interactions μ_{ijk} for MMX and MXX; θ_{12} is the difference in binary interaction λ_{ij} of HSO_4^- with SO_4^{2-} from the appropriate average of HSO_4^-–HSO_4^- and SO_4^{2-}–SO_4^{2-} interactions; while ψ_{H12} is the corresponding difference in triple interactions μ_{ijk} involving a hydrogen ion together with two anions.

In addition we have the dissociation equilibrium

$$HSO_4^- = H^+ + SO_4^{2-} \quad (5)$$

$$K_2 = \left(\frac{m_H m_2}{m_1}\right)\left(\frac{\gamma_H\gamma_2}{\gamma_1}\right) \quad (6)$$

with the familiar dissociation constant K_2 and the activity coefficients of the various ions. These activity coefficients can be expressed in terms of the same parameters as appear in eq 1 to 4.

These equations contain two different expressions for the short-range interaction of H^+ with SO_4^{2-}: one is the association to HSO_4^- represented by $1/K_2$, the other the second virial coefficient B_{H2} containing $\beta_{H2}{}^{(0)}$ and $\beta_{H2}{}^{(1)}$. If this series of equations is expanded in powers of $m^{1/2}$, the coefficient of the term in m involves a sum with terms in $1/K_2$ and $(\beta_{H2}{}^{(0)} + \beta_{H2}{}^{(1)})$ indicating a redundancy at this order. Higher order terms, however, do not maintain this redundancy, and it is possible to include $\beta_{H2}{}^{(0)}$ or $\beta_{H2}{}^{(1)}$ or both, if desired.

In sulfuric acid above 0.1 M the more abundant anion is HSO_4^- and the osmotic coefficient has been observed to behave very much like that of HCl or a similar acid. Thus we may expect that the parameters most important for this concentration range will be $\beta_{H1}{}^{(0)}$ and $\beta_{H1}{}^{(1)}$ which relate to the short-range interaction of H^+ with HSO_4^-. Thus we expect K_2 to be most important for the very dilute range with $\beta_{H1}{}^{(0)}$ and $\beta_{H1}{}^{(1)}$ becoming important at higher concentrations.

There remain in eq 2 through 4 six additional parameters which might have significant effect at least at very high concentration. In exploratory calculations it was soon discovered that $\beta_{H2}{}^{(1)}$ was so nearly redundant to $1/K_2$ and $\beta_{H2}{}^{(0)}$ that it was best omitted. Also it was found that θ_{12} and ψ_{H12} were not needed. Good results were obtained with either the combination $\beta_{H2}{}^{(0)}$ and $C_{H2}{}^\phi$ or with $C_{H1}{}^\phi$ and $C_{H2}{}^\phi$ but the former proved to be slightly superior and was adopted. The equations then reduce to

$$\phi' - 1 = (\Sigma m_i)^{-1}\{2If^\phi + 2m_H[m_1 B_{H1}{}^\phi + m_2(\beta_{H2}{}^{(0)} + m_H C_{H2}{}^\phi/2^{1/2})]\} \quad (7)$$

with $B_{H1}{}^\phi$ still represented by the two-term expression

$$B_{H1} = \beta_{H1}{}^{(0)} + \beta_{H1}{}^{(1)} \exp(-2I^{1/2}) \quad (4a)$$

The corresponding equations for the two combinations of activity coefficients of interest are

$$\ln(\gamma_H{}^2\gamma_{SO_4}) = 6f^\gamma + 4m_1 B_{H1} + (4m_2 + 2m_H)\beta_{H2}{}^{(0)} + (8m_2 + 2m_H)m_H C_{H2} + 6m_H m_1 B_{H1}' \quad (8)$$

$$\ln(\gamma_H\gamma_{SO_4}/\gamma_{HSO_4}) = 4f^\gamma + 2(m_1 - m_H)B_{H1} + 2(m_2 + m_H)\beta_{H2}{}^{(0)} + 2m_H(2m_2 + m_H)C_{H2} + 4m_1 m_H B_{H1}' \quad (9)$$

$$f^\gamma = -A_\phi[I^{1/2}/(1 + 1.2I^{1/2}) + (2/1.2)\ln(1 + 1.2I^{1/2})] \quad (10)$$

$$B_{H1} = \beta_{H1}{}^{(0)} + (\beta_{H1}{}^{(1)}/2I)[1 - (1 + 2I^{1/2})\exp(-2I^{1/2})] \quad (11a)$$

$$B_{H1}' = (\beta_{H1}{}^{(1)}/2I^2)[-1 + (1 + 2I^{1/2} + 2I)\exp(-2I^{1/2})] \quad (11b)$$

$$C_{H2} = C_{H2}{}^\phi/2^{3/2} \quad (12)$$

The stoichiometric activity and osmotic coefficients for sulfuric acid (on the basis of complete dissociation) are

$$\gamma_\pm{}^3 = (\gamma_H{}^2\gamma_{SO_4})(m_H{}^2 m_2/4m^3) \quad (13)$$

$$\phi = \phi'(\Sigma m_i)/3m \quad (14)$$

The total excess Gibbs energy per mole of solute is given by the thermodynamic relationship

$$G^{ex}/n_2 = 3RT(\ln\gamma_\pm + 1 - \phi) \quad (15)$$

Equations for the various molal enthalpy functions can be derived from the temperature derivatives of the Gibbs energy functions. For example

$$\bar{L}_1 = (M_w m/1000)(3RT^2)\,\partial\phi/\partial T \quad (16)$$

$$\bar{L}_2 = -3RT^2\,\partial\ln\gamma_\pm/\partial T \quad (17)$$

$${}^\phi L = 3RT^2(\partial\phi/\partial T - \partial\ln\gamma_\pm/\partial T) \quad (18)$$

where M_w is the molecular weight of the solvent.

In taking the temperature derivative of eq 7, 8, 13, and 14 one must recognize that I, m_H, m_1, and m_2 are temperature dependent through the change in dissociation of HSO_4^-. Any one can be chosen to express this effect and the others related thereto; we used $\partial m_1/\partial T$ but indicate derivatives at constant composition with the subscript I. One then obtains

$$\bar{L}_1 = \left(\frac{M_w RT^2}{1000}\right)\left\{2i\left(\frac{\partial f^\phi}{\partial T}\right)_I + 2m_H\left[m_1\left(\frac{\partial B_{H1}{}^\phi}{\partial T}\right)_I + m_2\left(\frac{\partial\beta_{H2}{}^{(0)}}{\partial T}\right)_I + \frac{m_2 m_H}{2^{1/2}}\left(\frac{\partial C_{H2}{}^\phi}{\partial T}\right)_I\right] - \left(\frac{\partial m_1}{\partial T}\right)\left[1 + 4f^\phi + 4I\left(\frac{\partial f^\phi}{\partial I}\right)_T - 4m_2 B_{H1}{}^\phi - 4m_H m_1\left(\frac{\partial B_{H1}{}^\phi}{\partial I}\right)_T + 2I\beta_{H2}{}^{(0)} + 2^{1/2}m_H(m_H + 2m_2)C_{H2}{}^\phi\right]\right\} \quad (19)$$

$$\bar{L}_2 = -RT^2\left\{6\left(\frac{\partial f^\gamma}{\partial T}\right)_I + 4m_1\left(\frac{\partial B_{H1}}{\partial T}\right)_I + (4m_2 + 2m_H)\left(\frac{\partial\beta_{H2}{}^{(0)}}{\partial T}\right)_I + m_H(8m_2 + 2m_H)\left(\frac{\partial C_{H2}}{\partial T}\right)_I\right.$$

$$+ 6m_Hm_1\left(\frac{\partial B_{HI}'}{\partial T}\right)_I - \left(\frac{\partial m_1}{\partial T}\right)\left[\frac{2}{m_H} + \frac{1}{m_2} + 12\left(\frac{\partial f^\gamma}{\partial I}\right)_T - 4B_{HI} - (8m + 4m_2)B_{HI}' + 6\beta_{H2}^{(0)} + (8m_2 + 12m_H)C_{H2} - 6m_Hm_1 B_{HI}''\right]\Bigg\} \quad (20)$$

with

$$B_{HI}'' = (\beta_{HI}^{(1)}/I^3)[1 - (1 + 2I^{1/2} + 2I + I^{3/2})e^{-2I^{1/2}}] \quad (21)$$

The change of dissociation with temperature is obtained from eq 6 and 9. After several steps one finds

$$\frac{\partial m_1}{\partial T} = \left[\frac{\partial \ln K_2}{\partial T} - 4\left(\frac{\partial f^\gamma}{\partial T}\right)_I - 2(m_1 - m_H)\left(\frac{\partial B_{HI}}{\partial T}\right)_I + 4m_1m_H\left(\frac{\partial B_{HI}'}{\partial T}\right)_I + 2(m_H + m_2)\frac{\partial \beta_{H2}^{(0)}}{\partial T} + 2m_H(2m_2 + m_H)\frac{\partial C_{H2}}{\partial T}\right] \times\left[-\frac{1}{m_H} - \frac{1}{m_2} - \frac{1}{m_1} - 8\left(\frac{\partial f^\gamma}{\partial I}\right)_T + 4B_{HI} + 16m_2B_{HI}' + 4m_1m_HB_{HI}'' - 4\beta_{H2}^{(0)} - (8m_H + 4m_2)C_{H2}\right]^{-1} \quad (22)$$

The heat of dissociation is contained in the above, since

$$\partial \ln K_2/\partial T = \Delta H_2/RT^2 \quad (23)$$

Even with all parameters known or assumed it is not feasible to solve these equations directly; an iterative solution, however, converges quickly and easily. We found it convenient to use m_1 as the variable and to use the quadratic solution of eq 6. After each cycle the improved m_1 is used to recalculate the ionic strength, which appears throughout the equations, for the next cycle. Thus the solution for known parameters is relatively easy.

The problem of finding the best parameters to fit an array of various types of experimental data is much more difficult, since this is not only a nonlinear problem, but there is also the nonlinear auxiliary condition of eq 6. The nonlinear least-squares equations were used with the derivatives with respect to the various parameters including the indirect dependence through eq 6. While simultaneous optimization of 19 parameters was possible, we found it preferable to use a series of fixed values for K_2 at 25 °C with least-squares optimization of the remaining 18 parameters.

Experimental Data

There are three cells each of which yields the thermodynamic properties of sulfuric acid:

$$\text{Pt, } H_2|H_2SO_4(m)|PbSO_4\text{, Pb, Hg} \quad (A)$$

$$\text{Pt, } H_2|H_2SO_4(m)|Hg_2SO_4\text{, Hg} \quad (B)$$

$$\text{Pt, } H_2|H_2SO_4(m)|PbO_2\text{, }PbSO_4\text{, Pt} \quad (C)$$

For cells A and B the equation for the emf is

$$E = E° - (RT/2F)\ln(4m^3\gamma_\pm^3) \quad (24)$$

while for cell C the corresponding equation is

$$E = E° + (RT/2F)\ln(4m^3\gamma_\pm^3/a_w^2) \quad (25)$$

where a_w is the activity of the water which is given by

$$\ln a_w = -3m\phi/55.5 \quad (26)$$

For cells B and C there are extensive investigations of Wynne-Jones and collaborators[3,4] over a series of temperatures and from 0.1 to over 7 M. In addition at 25 °C there are measurements of Covington, Dobson, and Wynne-Jones[14] for dilute solutions from 0.007 to 0.1 M. These data have been shown to be consistent to high accuracy as between cells B and C and, in the range above 0.1 M, also with the osmotic data of Shankman and Gordon[1] and of Stokes.[2] Furthermore, Giauque and associates[5,6] have measured the entropies of various sulfuric acid hydrates by third law methods and combined these results with other entropy and enthalpy data to check accurately the temperature coefficient of cell C.

Cell A was measured over a range of temperatures and from 0.001 to 0.02 M by Shrawder and Cowperthwaite[15] in 1934. This work has been ignored by many more recent workers but was recently shown by Lilley and Briggs[16] to be consistent with the results of cells B and C in the region of overlap at 25 °C and has been discussed recently by one of us.[17] The early work of Hamer[18] and Harned and Hamer[19] on cells C and B, respectively, has been shown by various workers[1-4,6,14] to be less accurate than the more recent data here considered.

At the very lowest concentrations the solubility of $PbSO_4$[15] or Hg_2SO_4[20] becomes significant and corrections were made for the primary effect of the dissolved solid.

Information about the enthalpy of sulfuric acid comes primarily from heat of dilution measurements. These include the measurements of Lange, Monheim, and Robinson[21] in the very dilute range, of Kunzler and Giauque[22] for concentrated solutions, and of Groenier[23] and Wu[24] (both students of T. F. Young) for differential dilutions at intermediate concentrations. The heat of ionization of HSO_4^- has been measured calorimetrically[25,26] with the results −5.2 ± 0.5 and −5.74 ± 0.2 kcal/mol. These approximate values were not used directly but provided a check on values of ΔH_2 obtained from analysis of heat of dilution data.

In addition to the cell measurements on pure sulfuric acid, we considered the results of Nair and Nancollas[27] on the cell

$$\text{Pt, } H_2|H_2SO_4(m)\ HCl(m_3)|AgCl\text{, Ag} \quad (D)$$

which yields directly the activity of HCl. Dunsmore and Nancollas[28] made further calculations on these data. Application of our equations for mixed electrolytes to these solutions yields equations with additional terms which are given in the Appendix. It suffices to note here that the additional terms which are important are those for the interaction of H^+ with Cl^- which are known from the properties of pure HCl. Interaction terms for Cl^- with SO_4^{2-} are known to be small, and one can estimate that the corresponding terms for Cl^- interaction with HSO_4^- may also be neglected in good approximation at the low concentrations measured for cell D.

In contrast to the situation just described, most of the other thermodynamic data which have been used in calculating the dissociation constant for bisulfate ion involve complex mixed solutions where unknown interactions would now be estimated to be significant and where elimination of these terms by extrapolation is uncertain. The most favorable of these cases, that for mixed solutions of Na_2SO_4 and $NaHSO_4$ in cell B,[29] is discussed in the Appendix. Here the serious uncertainty is the interaction of Na^+ with HSO_4^- which is not known accurately from another source.

Evaluation of Parameters

Our first calculations dealt only with the data on cells A, B, and C for pure sulfuric acid and the osmotic coefficients at 25 °C. It soon became apparent that the osmotic data were less precise than the cell data; consequently, only the smoothed values of Robinson and Stokes[30] and of Rard et al.[31] were retained for ϕ. Excellent fits were obtained for all of these cell potentials, but it was noted that K_2 was being given a value near 0.0120 which is much larger than that found in most recent work. This implied a real conflict with the data on cell D which were interpreted by Nair and Nancollas[28] to yield K_2 = 0.0110 and by Dunsmore and Nancollas[29] to yield K_2 = 0.0105. Further investigation indicated that all data for cells B and C and those for cell A at 0.005 M and above could be fitted with smaller values of K_2, including the value K_2 = 0.0105 which yielded the best agreement with the data on cell D when our more detailed equations were used with specific parameters for HCl (see Appendix).

Table I. Parameters for Equations for $H_2SO_4(aq)$

	Values for 25 °C			Values for 25 °C	
	Adopted set	Alternate set		Adopted set	Alternate set
K_2	0.0105	0.0120	C_{H2}^{ϕ}	0.0416	0.0384
$\beta_{H1}^{(0)}$	0.2103	0.2110	E_0(A)	-0.3528_0	-0.3517_5
$\beta_{H1}^{(1)}$	0.4711	0.4096	E_0(B)	0.6124_2	0.6135_4
$\beta_{H2}^{(0)}$	0.0027	0.0153	E_0(C)	1.6903_8	1.6892_7

Temperature Dependent Equations (Adopted Set)

$$\ln K_2 = -14.0321 + 2825.2/T$$
$$\beta_{H1}^{(0)} = 0.05584 + 46.040/T$$
$$\beta_{H1}^{(1)} = -0.65758 + 336.514/T$$
$$\beta_{H2}^{(0)} = -0.32806 + 98.607/T$$
$$C_{H2}^{\phi} = 0.25333 - 63.124/T$$
$$E_0(A) = 0.00589733T - 0.361098 - 0.00103017T \ln T$$
$$E_0(B) = 0.00487142T + 0.608163 - 0.00085249T \ln T$$
$$E_0(C) = -0.00691441T + 1.910511 + 0.00108398T \ln T$$

The results at this point yielded two sets of parameters fitting most data equally well. Further calculations indicated the heat of dilution data were better fitted by the set of parameters including $K_2 = 0.0105$ together with appropriate temperature derivative parameters. Thus we favor these parameters ($K_2 = 0.0105$) as more probably representing the true properties of sulfuric acid below 0.005 M, but we cannot assert that this choice is absolutely certain.

In further comment on the choice of the last paragraph we note that other work in the same laboratory as that for cell A with cells also involving the $PbSO_4$, Pb electrode has been found[32] to be reliable down to 0.005 M but in significant error below that concentration. Also a very simple interpretation[17] of the data for cell A cast some doubt on the results for the most dilute solutions at the higher temperatures. In contrast, however, the concentration of sulfate ion as measured by Raman spectra by Young, Maranville, and Smith[33] is better fitted by our calculations with the higher $K_2 = 0.0120$. These concentration measurements are not very precise and the difference is not very large; hence we do not believe this is a serious objection to the smaller K_2. Conductance data on sulfuric acid have been interpreted by Kerker[34] to yield $K_2 = 0.0102$ and by Dunsmore and Nancollas[28] to yield $K_2 = 0.0105$ at 25 °C and therefore favor the smaller value of K_2.

Table I gives the adopted parameters both in the form for 25 °C and that for temperature-dependent functions; also given are the alternate parameters for 25 °C as discussed above.

Of the ion-interaction parameters in Table I only $\beta_{H1}^{(0)}$ for the H^+-HSO_4^- interaction is sufficiently free from ambiguity to make comparison meaningful. The value for H^+-HSO_4^-, 0.2103, is similar to the $\beta^{(0)}$ values for other strong acids with large anions, for example, 0.236 for HI and 0.175 for $HClO_4$. For H^+-HSO_4^-, $\beta_{H1}^{(1)}$ (0.47) is higher than that found for the simple acids but not very different. Since we have arbitrarily suppressed some parameters and have the redundancy of $\beta_{H2}^{(0)}$, $\beta_{H2}^{(1)}$, and $1/K_2$, great caution should be used in interpreting these other quantities. None are so large as to be unreasonable, however.

The alternate set of parameters shifts each of the E_0 values by 1.1 mV which corresponds to a 3% increase in activity coefficients.

Table II shows the agreement with experimental cell potentials for values calculated with the adopted set in the range 0.005 to 0.1 M. Table III gives a similar comparison for the range 0.1 to 6 M but in terms of the activity coefficient; also included is the osmotic coefficient.

Table II. Comparison of Calculated and Experimental Properties below 0.1 M

m	a	$\gamma_\pm$	ΔE, mV (cell)
0.005	0.67	0.615	−0.18 (A)
0.0728	0.61	0.560	0.06 (B), 0.22 (C)
0.01	0.56	0.515	−0.04 (A)
0.0103	0.55	0.510	−0.16 (B), 0.01 (C)
0.0171			−0.15 (B)
0.02	0.45	0.419	0.11 (A)
0.0215			0.12 (B)
0.0401			0.36 (B)
0.0571	0.32	0.296	0.32 (B), −0.01 (C)
0.0840			0.31 (B)
0.095	0.27	0.247_7	0.19 (B), 0.15 (C)
0.096			0.27 (B)
0.1	0.27	0.243_8	−0.05 (B), 0.03 (C)

[a] Fraction of HSO_4^- dissociated.

Tables II and III also give the fraction of sulfate dissociated from HSO_4^- to SO_4^{2-}. This quantity is sensitive to small changes in parameters which do not seriously affect the thermodynamic properties and in particular to the choice between C_{H1} or $\beta_{H2}^{(0)}$ as a parameter to be included. Hence only two significant figures are given and these results should be used with caution.

It is clear from Tables II and III that the agreement for cell potentials (or activity coefficients) at 25 °C is within or near to experimental accuracy. Similarly, the calculated osmotic coefficients are probably within experimental uncertainty up to 5 M. At other temperatures the agreement is almost as good with standard deviations of 0.13, 0.28, and 0.16 mV for the potential values for cell A (at and above 0.005 M) and for cells B and C (up to 6.1 M), respectively.

All of the heat of dilution data were fitted directly except for the most dilute point of Lange et al. with a final concentration 0.00005 M, which is necessarily least accurate, and the large ratio dilutions of Groenier for which his calorimeter was much less accurate than Lange's. The standard deviations were: for 24 differential dilutions of Wu,[24] 3.1 cal mol^{-1}; for the earlier but similar work of Groenier,[23] with 42 points, 6.8 cal mol^{-1}; for 10 large ratio dilutions of dilute solutions by Lange et al.,[21] 35 cal mol^{-1}; and for 6 dilutions of concentrated solutions by Kunzler and Giauque,[22] 8.0 cal mol^{-1}. Although σ is largest for Lange's values, it is clearly within experimental error for these dilutions yielding final concentrations in the range 0.00009 to 0.002 M. The accurate, differential dilution data of Wu are well fitted with the adopted parameters, but for the alternate set with $K_2 = 0.0120$ there is a serious discrepancy in the range 0.1 to 0.6 M where all deviations are of the same sign for 11 measurements with $\sigma = 8.4$ cal mol^{-1}.

Thermodynamic Properties of Sulfuric Acid

The thermodynamic properties for cells A, B, and C at 25 °C are compared in Table IV with literature values. The standard potentials for cells B and C are the same within 0.1 and 0.3 mV, respectively, as those chosen by Covington et al.[14] For cell A our present value is very close to the value −0.3526 favored by Lilley and Briggs[16] but differs by 1.5 mV from the value one of us[17] obtained from the data for cell A alone, without reference to the other data considered here. This difference of 1.5 mV is essentially that of the choice between $K_2 = 0.0105$ or 0.0120 where the latter value fits cell A results down to 0.001 M but disagrees with some other results.

A more independent check is possible for the temperature coefficient of the cell potential which yields ΔS for the cell reaction. Here one has entropy values from statistical calculations or third-law methods for all of the substances involved,

Table III. Comparison of Calculated and Experimental Properties above 0.1 M

		$\gamma_\pm$		ϕ		
m	a	PRS	CDW[b]	PRS	RS[30]	RHS[31]
0.1	0.27	0.243_6	0.244	0.676	0.680	0.680
0.2	0.23	0.191_8	0.192	0.666	0.668	0.665
0.3	0.21	0.167_6	0.168	0.667	0.668	0.665
0.4	0.21	0.153_2	0.153	0.672		0.669
0.5	0.21	0.143_4	0.143	0.678	0.676	0.675
0.7	0.21	0.131_1	0.130	0.694	0.689	0.691
1.0	0.22	0.121_5	0.121	0.723	0.721	0.719
1.5	0.24	0.116_0	0.116	0.778	0.780	0.777
2.0	0.25	0.117_1	0.118	0.843	0.846	0.843
2.5	0.25	0.122_4	0.124	0.915	0.916	0.916
3.0	0.24	0.131_1	0.132	0.992	0.991	0.992
3.5	0.23	0.142_5	0.143	1.071	1.071	1.070
4.0	0.20	0.156_3	0.157	1.149	1.150	1.148
4.5	0.18	0.172_4	0.172	1.225	1.226	1.225
5.0	0.15	0.190_7	0.190	1.299	1.303	1.301
5.5	0.13	0.211_3		1.369	1.376	1.374
6.0	0.11	0.234_3	0.234	1.437	1.445	1.444

[a] Fraction of HSO_4^- dissociated. [b] Reference 14 gives three sets of values differing only slightly; this is the set for cell B with E_0 = 612.5 mV.

Table IV. Thermodynamic Properties for Cell Reactions at 25 °C

	E_0, V		ΔS, cal K^{-1} mol^{-1}		
Cell	PRS	CDW	PRS	Lit.[a]	$S°(SO_4^{2-})$
A	−0.3528		-46.2_3	-45.7_7	4.0_4
B	0.6124_2	0.6125	-38.6_6	-38.3_3	4.1_7
C	1.6903_8	1.6901	15.9_4	16.0_7	4.3_7

[a] From literature values, see text, including $S°(SO_4^{2-})$ = 4.50.

in particular 31.208 for H_2(g), 16.71 for H_2O(l), 35.51 for $PbSO_4$(s), 18.17 for Hg(l), and 47.96 for Hg_2SO_4(s), all in cal K^{-1} mol^{-1} from the NBS Tables.[35] Also Duisman and Giauque[6] give 17.156 for PbO_2(s), Gallagher et al.[36] give 35.509 for $PbSO_4$(s), Meads et al.[37] give 15.514 for Pb(s), and CODATA-Part IV[38] recommends 4.50 for SO_4^{2-}(aq) from a variety of sulfate salts. Since cell A involves an amalgamated lead electrode, the value[39] ΔS = 0.93 cal K^{-1} mol^{-1} must be included for conversion of pure solid lead to lead in mercury. The entropy of H^+ ion is zero by definition. Combination of these values yields the results in the next-to-last column of Table IV. The agreement is very good. But it is equally justifiable to use each set of data to determine a value for the entropy of sulfate ion; the last column gives these values.

Our preferred parameters yield for the heat of dissociation of bisulfate ion $\Delta H_2 = -5.61$ kcal mol^{-1} in excellent agreement with the values −5.2 ± 0.5 and −5.74 ± 0.2 cited above.[25,26] Also one obtains $\Delta S_2° = -27.87$ cal K^{-1} mol^{-1}.

Recently Powell[42] and Cabani and Gianni[43] have discussed the interpretation of data from various sources to yield values of the heat of dissociation of bisulfate ion. These authors showed the sensitivity of these calculations to assumptions about the activity coefficients in the mixed electrolytes involved in these experiments, including those cited[25,26] above. While it should now be possible to remove much of the ambiguity in these assumptions about activity coefficients in mixed electrolytes, it seemed best to base our results on the experiments with pure sulfuric acid solutions.

These data do not have sufficient accuracy over their limited temperature range to yield reliable second temperature derivatives and thereby heat capacities. While three-term expressions were used for the cell potentials, the resulting second derivatives do not yield meaningful ΔC_p values because only two-term expressions were used for K_2 and the other parameters for the aqueous phase. One could introduce independent knowledge of the heat capacity of the solution into the treatment, but the increased complexity did not seem justified.

In view of the interrelationship between parameters it is not useful to set limits of error for most of the individual parameters in Table I. The uncertainties in E_0 values are about 0.3 mV provided the "adopted set" is essentially correct. Alternate E_0 values for K_2 = 0.0120 are also given. Even excluding this alternate choice, the value of K_2 is uncertain by a few percent since other parameters can be adjusted to retain the fit at finite concentration.

Table V presents a survey of various properties of aqueous sulfuric acid as calculated from our equations and parameters. For accurate values at other concentrations the equations should be used.

By comparison of our values in Table V at finite concentrations from 1 to 4 M with those of Giauque et al.,[5] one obtains the difference between the properties of sulfuric acid in the solute standard state and those for the pure liquid. Thus for the reaction

$$H_2SO_4(l) = 2H^+(aq) + SO_4^{2-}(aq)$$

we find

$$\Delta G° = -12871 \text{ cal mol}^{-1}$$

$$\Delta H° = -22844 \text{ cal mol}^{-1}$$

$$\Delta S° = -33.45 \text{ cal mol}^{-1} \text{ K}^{-1}$$

If one adds the entropy of pure liquid H_2SO_4, 37.501, to this value for $\Delta S°$, one obtains for the entropy of aqueous SO_4^{2-} the value 4.05 (since the entropy of H^+ is zero by definition). This result agrees reasonably well with those obtained earlier by a different method and we conclude that $S°(SO_4^{2-})$ = 4.2 ± 0.2 cal mol^{-1} K^{-1}. Combination of this value with $\Delta S_2°$ yields for the entropy of HSO_4^- the value 32.1 ± 0.3 cal K^{-1} mol^{-1}.

The consistency of data in the 1-4 M range is excellent with variations of only a few calories in the $\Delta H°$ and $\Delta G°$ values. However, our absolute knowledge for the dilute range is uncertain to the extent of about 15 cal for $\Delta G°$ and 40 cal for $\Delta H°$ if our adopted parameters are essentially correct. If the

Table V. Properties of H_2SO_4 at 25 °C and at Rounded Molalities

m	$-G^{ex}$	γ	ϕ	a_w	$^{\phi}L$	$-\bar{L}_1$	$\bar{L}_2$
0.0001	58	0.9500	0.981	0.99999	119	0.0002	220
0.0002	87	0.9253	0.971	0.99999	214	0.0006	393
0.0005	150	0.8736	0.949	0.99997	451	0.0032	803
0.001	228	0.8151	0.924	0.99995	757	0.0097	1295
0.002	342	0.7382	0.889	0.99990	1200	0.0265	1936
0.005	568	0.6146	0.833	0.99978	1970	0.083	2891
0.01	802	0.5146	0.787	0.99957	2623	0.171	3571
0.02	1092	0.4190	0.744	0.99920	3261	0.317	4140
0.05	1547	0.3098	0.699	0.99811	3991	0.638	4699
0.1	1934	0.2436	0.676	0.99635	4435	1.034	5008
0.2	2341	0.1918	0.666	0.99283	4793	1.679	5259
0.3	2582	0.1676	0.667	0.98925	4972	2.255	5389
0.5	2880	0.1434	0.678	0.9818	5172	3.318	5540
0.7	3069	0.1311	0.694	0.9741	5292	4.343	5636
1.0	3254	0.1215	0.723	0.9617	5412	6.029	5747
1.5	3436	0.1160	0.778	0.9388	5553	10.23	5932
2.0	3534	0.1171	0.843	0.9129	5677	18.03	6177
2.5	3582	0.1224	0.915	0.8837	5809	31.56	6510
3.0	3598	0.1311	0.992	0.8514	5959	52.12	6924
4.0	3564	0.1563	1.149	0.7801	6318	111.9	7871
5.0	3476	0.1907	1.299	0.7041	6720	184.1	8764
6.0	3357	0.2343	1.437	0.6274	7124	257.1	9503

alternate set were correct these differences become about 50 and 200 cal, respectively.

Acknowledgment. This research was sponsored by the Energy Research and Development Administration. Also L.F.S. held an N.S.F. Energy-Related Fellowship for 1975–1976.

Appendix

Values for the Debye–Hückel parameter A_ϕ are well-known;[11,30] the precise values used here were obtained from a careful evaluation of the data on the density and dielectric constant over a very wide temperature range. The resulting equations, valid to 300 °C, are presented elsewhere.[41] A very much simpler equation is adequate for the temperature range 0–55 °C.

$$A_\phi = 0.3770 + 4.684 \times 10^{-4}(T - 273.15) + 3.74 \times 10^{-6}(T - 273.15)^2 \quad \text{(A1)}$$

The value of the parameter for the enthalpy at 25 °C is A_H = 697.5 cal mol^{-1}.

For the mixed electrolyte solutions of cell D, $H_2SO_4(m)$, $HCl(m_3)$, additional terms arise for interactions of Cl^-. Since all of the solutions are very dilute, third virial coefficients may be omitted. Also we neglect terms for interactions of ions of the same sign since these are known[9] to be small for Cl^--SO_4^{2-} and expected to be small also for Cl^--HSO_4^-. One then obtains

$$\tfrac{1}{2}\ln(\gamma_H\gamma_{Cl}) = f^\gamma + (m_3 + m_H)B_{HCl} + m_1 B_{H1} + m_2\beta_{H2}^{(0)} + m_H m_3 B_{HCl}' + m_H m_1 B_{H1}' \quad \text{(A2)}$$

$$\ln(\gamma_H\gamma_{SO_4}/\gamma_{HSO_4}) = 4f^\gamma + 2[m_3 B_{HCl} + (m_1 - m_H)B_{H1} + (m_H + m_2)\beta_{H2}^{(0)}] + 4m_H[m_3 B_{HCl}' + m_1 B_{H1}'] \quad \text{(A3)}$$

Other equations continue to apply but m_H and I must include, of course, the contribution m_3 from the HCl. The potential for cell D is then

$$E - E_0 = -(RT/F)[\ln(m_H m_3) + \ln(\gamma_H\gamma_{Cl})] \quad \text{(A4)}$$

While the parameters for HCl at 25 °C are known,[8] they were reevaluated as a function of temperature by least-squares analysis of the results of Gupta, Hills, and Ives.[40] The results can be expressed as

$$\beta_{HCl}^{(0)} = 0.1754 - 0.0004286(T - 298.15) \quad \text{(A5)}$$

$$\beta_{HCl}^{(1)} = 0.3004 - 0.00406(T - 298.15) + 7.89 \times 10^{-6}(T^2 - 298.15^2) \quad \text{(A6)}$$

These equations (valid 0–50 °C) give slightly different values for the parameters at 25 °C from those given in the earlier treatment[8] which included additional data for HCl; the difference is negligible.

The potential for cell D was calculated with the above parameters for HCl and those of Table I for H_2SO_4. For the set with K_2 = 0.0105 the standard deviation at 25 °C was 0.05 mV while at both 0 and 45 °C the standard deviation was 0.06 mV. In contrast to this excellent agreement, calculations with K_2 = 0.0120 yield systematic disagreement rising from 0.2 to 0.4 mV with increase in concentration.

For the buffer solutions $Na_2SO_4(m_4)$ and $NaHSO_4(m_5)$ in cell B[29] the equations include terms for the interaction of Na^+ with SO_4^{2-}, which are known,[8] and for Na^+ with HSO_4^-, which are not known. For the latter one may take the parameters[8] for $NaClO_4$ as a rough estimate. Now the molality of HSO_4^- is $m_1 = m_5 - m_H$ and for SO_4^{2-} $m_2 = m_4 + m_H$ while for Na^+ $m_{Na} = 2m_4 + m_5$. With very dilute solutions we again omit third virial coefficients and terms for interactions of ions of the same sign and obtain

$$\ln(\gamma_H\gamma_{SO_4}) = 6f^\gamma + 4m_1 B_{H1} + 2[m_{Na}B_{Na2} + (m_H + 2m_2)\beta_{H2}^{(0)}] \quad \text{(A7)}$$

$$\ln(\gamma_H\gamma_{SO_4}/\gamma_{HSO_4}) = 4f^\gamma + 2[(m_1 - m_H)B_{H1} + (m_H + m_2)\beta_{H2}^{(0)} + m_{Na}(B_{Na2} - B_{Na1})] \quad \text{(A8)}$$

The calculated cell potentials for the buffer solutions with $NaClO_4$ parameters assumed for $NaHSO_4$ interactions differ by about 1 mV from those observed. This is essentially the same difference as was noted, but not emphasized, in the original paper of Covington, Dobson, and Wynne-Jones.[29] This disagreement is decreased but not removed by the assumption of larger but still reasonable values for $\beta^{(0)}$ and $\beta^{(1)}$ for $NaHSO_4$; it is not appreciably affected by shift to the alternate set of parameters in Table I for H_2SO_4. This discrepancy is small, fortunately, but it is not removed by any plausible change in ionic-interaction parameters provided the consistent E_0 values are used.

References and Notes

(1) S. Shankman and A. R. Gordon, *J. Am. Chem. Soc.*, **61**, 2370 (1939).
(2) R. H. Stokes, *J. Am. Chem. Soc.*, **69**, 1291 (1947); *Trans. Faraday Soc.*, **44**, 295 (1948).
(3) W. H. Beck, K. P. Singh, and W. F. K. Wynne-Jones, *Trans. Faraday Soc.*, **55**, 331 (1959).
(4) W. H. Beck, J. V. Dobson, and W. F. K. Wynne-Jones, *Trans. Faraday Soc.*, **56**, 1172 (1960).
(5) W. F. Giauque, E. W. Hornung, J. E. Kunzler, and T. R. Rubin, *J. Am. Chem. Soc.*, **82**, 62 (1960).
(6) J. A. Duisman and W. F. Giauque, *J. Phys. Chem.*, **72**, 562 (1968).
(7) K. S. Pitzer, *J. Phys. Chem.*, **77**, 268 (1973).
(8) K. S. Pitzer and G. Mayorga, *J. Phys. Chem.*, **77**, 2300 (1973); **78**, 2698 (1974).
(9) K. S. Pitzer and J. J. Kim, *J. Am. Chem. Soc.*, **96**, 5701 (1974).
(10) W. J. Hamer, "The Structure of Electrolytic Solutions", Wiley, New York, N.Y., 1959, Chapter 15.
(11) K. S. Pitzer and L. Brewer, "Thermodynamics", Revised edition, G. N. Lewis and M. Randall, Ed., McGraw-Hill, New York, N.Y., 1961, pp 587–590.
(12) K. S. Pitzer and L. F. Silvester, *J. Solution Chem.*, **5**, 269 (1976).
(13) T. L. Hill, "Introduction to Statistical Mechanics", Addison-Wesley, Reading, Mass., 1960, Chapter 19.
(14) A. K. Covington, J. V. Dobson, and W. F. K. Wynne-Jones, *Trans. Faraday Soc.*, **61**, 2050 (1965).
(15) J. Shrawder, Jr., and I. A. Cowperthwaite, *J. Am. Chem. Soc.*, **56**, 2340 (1934).
(16) T. H. Lilley and C. C. Briggs, *Electrochim. Acta*, **20**, 257 (1975).
(17) K. S. Pitzer, *J. Phys. Chem.*, **80**, 2863 (1976).
(18) W. J. Hamer, *J. Am. Chem. Soc.*, **57**, 9 (1935).
(19) H. S. Harned and W. J. Hamer, *J. Am. Chem. Soc.*, **57**, 27 (1935).
(20) S. A. Brown and J. E. Land, *J. Am. Chem. Soc.*, **79**, 3015 (1957).
(21) E. Lange, J. Monheim, and A. L. Robinson, *J. Am. Chem. Soc.*, **55**, 4733 (1933).
(22) J. E. Kunzler and W. F. Giauque, *J. Am. Chem. Soc.*, **74**, 3472 (1952).
(23) W. L. Groenier, Thesis, University of Chicago, 1936.
(24) Y. C. Wu, Thesis, University of Chicago, 1957.
(25) K. S. Pitzer, *J. Am. Chem. Soc.*, **59**, 2365 (1937).
(26) J. M. Austin and A. D. Mair, *J. Phys. Chem.*, **66**, 519 (1962).
(27) V. S. K. Nair and G. H. Nancollas, *J. Chem. Soc.*, 4144 (1958); see also C. W. Davies, H. W. H. Jones, and C. B. Monk, *Trans. Faraday Soc.*, **48**, 921 (1952).
(28) H. S. Dunsmore and G. H. Nancollas, *J. Phys. Chem.*, **68**, 1579 (1964).
(29) A. K. Covington, J. V. Dobson, and W. F. K. Wynne-Jones, *Trans. Faraday Soc.*, **61**, 2057 (1965).
(30) R. A. Robinson and R. H. Stokes, "Electrolytic Solutions", Butterworths, London, 1959.
(31) J. A. Rard, A. Habenschuss, and F. H. Spedding, *J. Chem. Eng. Data*, **21**, 374 (1976).
(32) K. S. Pitzer, *J. Chem. Soc., Faraday Trans. 2*, **68**, 101 (1972).
(33) T. F. Young, L. F. Maranville, and H. M. Smith, "The Structure of Electrolytic Solutions", W. J. Hamer, Ed., Wiley, New York, N.Y., 1959, Chapter 4.
(34) M. Kerker, *J. Am. Chem. Soc.*, **79**, 3664 (1957).
(35) D. D. Wagman et al., *Natl. Bur. Stand. (U.S.), Tech. Note*, 270-3 (1968).
(36) K. Gallagher, G. E. Brodale, and T. E. Hopkins, *J. Phys. Chem.*, **64**, 687 (1960).
(37) P. F. Meads, W. R. Forsythe, and W. F. Giauque, *J. Am. Chem. Soc.*, **63**, 1902 (1941).
(38) CODATA, Tentative set of key values for thermodynamics, Part IV (1974).
(39) R. G. Bates, M. Edelstein, and S. F. Acree, *J. Res. Natl. Bur. Stand.*, **36**, 159 (1946).
(40) S. R. Gupta, G. J. Hills, and D. J. G. Ives, *Trans. Faraday Soc.*, **59**, 1874 (1963).
(41) L. F. Silvester and K. S. Pitzer, Lawrence Berkeley Laboratory Report No. 4456 (1976), to be published.
(42) H. K. J. Powell, *J. Chem. Soc., Dalton Trans.*, 1947 (1973).
(43) S. Cabani and P. Gianni, *Anal. Chem.*, **44**, 253 (1972).

[Reprinted from ACCOUNTS OF CHEMICAL RESEARCH, 10, 371 (1977).]

Electrolyte Theory—Improvements since Debye and Hückel

Kenneth S. Pitzer

Department of Chemistry and Lawrence Berkeley Laboratory, University of California, Berkeley, California 94720

Received April 13, 1977

In 1923 Debye and Hückel[1] provided, in a simple theory, the correct equation for the behavior of electrolyte solutions in the limit of very low concentration. This eliminated an anomaly that had troubled physical chemists and allowed a great advance in the semiempirical, semitheoretical treatment of dilute electrolytes at finite concentrations. While the severe approximations of Debye and Hückel were found not to affect the limiting law, great uncertainty[2] remained concerning any higher order terms.

Kenneth S. Pitzer, born in 1914 in Pomona, Calif., received his B.S. from the California Institute of Technology and his Ph.D. from the University of California, Berkeley. He is a member of the faculty at Berkeley, having served in that capacity from 1937 to 1961 and again since 1971. He has also held various administrative positions including that of Director of Research for the U.S. Atomic Energy Commission (1949–1951), President of Rice University (1961–1968), and President of Stanford University (1968–1970). His research interests include thermodynamics, statistical mechanics, and quantum mechanics as applied to chemistry as well as experimental work in spectroscopy and thermochemistry.

There were many intermediate advances[3] through the years, especially in the work of Mayer, Kirkwood, Poirier, and Friedman. However, a really sound and fruitful theory for electrolytes at substantial concentration has only emerged in the last decade. While these contributions have been summarized in reviews[4,5] addressed to specialists in statistical mechanics and electrolyte theory, it seemed desirable to call this advance to the attention of physical chemists and solution chemists more generally. This more recent theory may be too complex for use or for presentation in some cases, but even then it should be realized that a rigorous

(1) P. Debye and E. Hückel, *Phys. Z.*, **24**, 185, 334 (1923); **25**, 97 (1924).
(2) H. S. Frank and P. T. Thompson, *J. Chem. Phys.*, **31**, 1086 (1959).
(3) This work is reviewed by H. L. Friedman, "Ionic Solution Theory", Wiley–Interscience, New York, N.Y., 1962.
(4) H. C. Andersen, *Mod. Aspects Electrochem.*, **No. 11**, 1 (1975).
(5) J. C. Rasaiah, *J. Solution Chem.*, **2**, 301 (1973).

theory now exists. Thus one should evaluate the accuracy of the traditional Debye–Hückel theory by comparison with the rigorous theory for the same model.

It is also possible to extend the Debye–Hückel theory in a simple manner to yield a relatively simple equation which is an excellent approximation.[6] The recent theory suggests new forms of semiempirical equations which have proven to be remarkably successful in representing experimental results and in thermodynamic calculations.

In reviewing and summarizing the recent development it is desirable to divide the complete theory into three parts. This division is equally applicable to Debye–Hückel theory; hence, we can discuss the improvements in each part. The three parts are: (1) the selection of the molecular model, i.e., the interionic potentials of mean force, (2) the calculation of the interionic radial distribution functions from this model, and (3) the calculation of thermodynamic functions from the potentials and the radial distribution functions.

Interionic Potentials

In principle one would like to have an adequate theoretical treatment for water and to use it together with the properties of the ions to derive[7] the interionic potentials of mean (or average) force in water. But there is not yet an adequate model for water. Hence one must start with a model for the interionic potentials of mean force.

The potential of mean force is defined in a manner such that its derivative gives the force on the selected particle averaged over the motion of other particles in the system.[8] Thus the familiar equation for the electrostatic potential in the presence of a dielectric is a potential of mean force; its derivative gives the force on a charged particle averaged over the motion of the particles constituting the dielectric.

Any plausible model can be described by the interionic potential equation

$$u_{ij}(r) = u_{ij}{}^*(r) + z_i z_j e^2/Dr \qquad (1)$$

where z_ie is the charge on the ith ion, r is the distance between the i and j ions, D is the dielectric constant of the solvent, and $u^*(r)$ is a short-range function giving the difference between the true interionic potential of mean force and the electrostatic term.

There is no question concerning the effect of the dielectric constant of the solvent in yielding the last term in eq 1 for sufficiently large interionic separation. All effects yielding departures from this electrostatic term are combined in $u_{ij}{}^*(r)$. These include at intermediate distances a variety of effects related to the molecular nature of the solvent (solvation, dielectric saturation, etc.) as well as dispersion forces between the ions and at shorter distances the direct repulsion of ions as their electron shells begin to overlap.

The simplest assumption for u^*, used by Debye and Hückel and frequently called the primitive model, is

(6) K. S. Pitzer, *J. Phys. Chem.*, **77**, 268 (1973).

(7) An interesting approach toward the derivation of interionic potentials of mean force is given by G. N. Patey and J. P. Valleau, *J. Chem. Phys.*, **63**, 2334 (1975).

(8) T. L. Hill, "An Introduction to Statistical Thermodynamics", Addison-Wesley, Reading, Mass., 1960, p 313.

$$u_{ij}{}^* = +\infty,\ r < a$$
$$u_{ij}{}^* = 0,\ r > a \qquad (2)$$

where the hard-core diameter a is the same for all ions. A first improvement is to make the hard core distance in (2) the sum of radii which may differ for the positive and negative ions in a salt. Further improvements are continuous functions reflecting the softness of the direct interionic repulsion and the short-range effects of the solvent. Since there has been a major advance in the correct solution of the primitive model, we will proceed at once in describing that theory. But we will describe equations valid for other forms of u^* and will eventually discuss some results for another model.

While improvements in part 2, the calculation of the radial distribution function, were important, the most critical advance was for part 3; hence we discuss part 3 next. Debye and Hückel were unable to include the direct effect of the short-range potential u^* on thermodynamic properties; they included only the indirect effect of u^* on the electrostatic term (and even that only partially).

Equations Yielding Thermodynamic Functions

An adequate theory relating intermolecular potentials and radial distribution functions to thermodynamic functions was first developed for simple fluids such as argon. For the energy content the result is rather easily derived:

$$\frac{E}{N} = \frac{3}{2}kT + \frac{1}{2}\frac{N}{V}\int_0^\infty u(r)g(r)4\pi r^2 \mathrm{d}r \qquad (3)$$

where $g(r)$ is the radial distribution function and $u(r)$ is the intermolecular potential. It is considerably more difficult to obtain a useful relationship for free energy, but somewhat indirect approaches were successful. The most direct and useful gives the pressure

$$\frac{PV}{NkT} = 1 - \frac{N}{6VkT}\int_0^\infty r\frac{\partial u}{\partial r}\, g(r)(4\pi r^2)\mathrm{d}r \qquad (4)$$

Given the pressure as a function of volume, other functions can be calculated from thermodynamic relationships. There is also an equation for the compressibility which provides a useful check, but the pressure equation is ordinarily the most convenient.

This theory for simple fluids can be found in a number of books on statistical mechanics; for example, Fowler[9] and Henderson and Davison[10] give derivations based on the virial theorem. A different and relatively simple derivation based on the volume derivative of the partition function is given by Davidson[11] and by Hill.[12] Still another derivation has been given by Hill.[13]

The generalization of these equations to multicomponent fluids is relatively straightforward. But we can hardly deal in detail with all of the solvent molecules as well as the solute; hence some method of averaging over solvent effects is essential. This was provided by McMillan and Mayer[14] who considered a solution in

(9) R. H. Fowler, "Statistical Mechanics", Cambridge University Press, Cambridge, England, 1936, p 286.

(10) D. Henderson and S. G. Davison, in "Physical Chemistry, an Advanced Treatise", Vol. II, H. Eyring, Ed., Academic Press, New York, N.Y., 1967, p 359.

(11) N. Davidson, "Statistical Mechanics", McGraw-Hill, New York, N.Y., 1962, p 473.

(12) T. L. Hill, "Statistical Mechanics", McGraw-Hill, New York, N.Y., 1956, p 190.

(13) Reference 8, p 304.

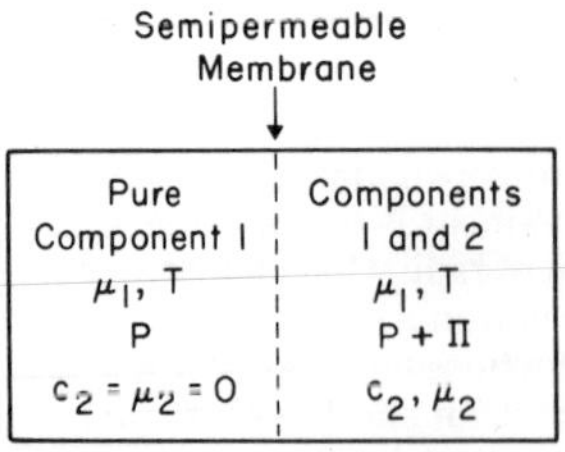

Figure 1. Osmotic equilibrium for the McMillan–Mayer theory. The solute is component 2 and the total pressure on the solution in the right side is $P + \pi$.

osmotic equilibrium with pure solvent through a solvent-permeable, solute-impermeable membrane; see Figure 1. They showed that the entire array of equations for molecules in imperfect gases (fluids) applied to the solute in this system provided one used potentials of mean force at infinite dilution in the solvent and interpreted the pressure as the osmotic pressure. There are some questions of convergence for long-range electrical forces, but these have been answered.

Hill[15] gives both a simplified and a more rigorous account of the McMillan–Mayer theory.

Rasaiah and Friedman[16] were the first to take full advantage of this theory for application to electrolytes. The pressure equation now yields the osmotic coefficient for the solution

$$\phi - 1 = (\Pi/ckT) - 1$$
$$= -(6ckT)^{-1} \sum_i \sum_j c_i c_j \int_0^\infty r \frac{\partial u_{ij}}{\partial r} g_{ij}(r)(4\pi r^2)\mathrm{d}r \quad (5)$$

where c_i is the concentration of the ith ion and c is the total concentration of solute species, $c = \sum c_i$. Once ϕ is known as a function of concentration, the activity coefficient can be calculated from the Gibbs–Duhem equation.

It should be noted (see Figure 1) that these functions apply to the solution under a pressure equal to its own osmotic pressure plus the pressure on the pure solvent rather than to the usual standard pressure of 1 atm. Friedman[17] has discussed the conversion between these states and provides the needed equations. The differences are negligible for our purposes.

Rasaiah and Friedman[16] also derived the analogue of the compressibility equation for the electrolyte system and used it to verify certain calculations.

Although the complete theory underlying eq 5 is complex and lengthy, it is rigorous and provides a valid means to obtain thermodynamic properties provided both the potentials of mean force and the radial distribution functions are available.

Radial Distribution Functions

We turn now to the problem of obtaining the radial distribution functions from the potentials. The basic equation of statistical mechanics[18] for the radial distribution function (or pair correlation function) yields for a pair of like ions in a binary electrolyte

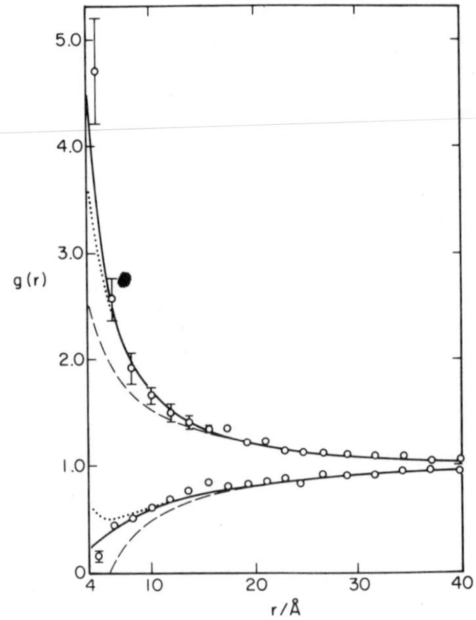

Figure 2. The radial distribution functions (g_{+-} above, $g_{++} = g_{--}$ below) for 0.00911 M aqueous solution (1–1 type, a = 4.25 Å). The points are Monte Carlo calculations, the solid curve is the exponential D–H expression, the dotted and dashed curves are, respectively, the three-term and two-term D–H expressions.

$$g_{ii}(r) = V^2 \frac{\int\ldots\int \exp(-\Phi/kT)\mathrm{d}r_{i3}\ldots\mathrm{d}r_{iN}\mathrm{d}r_{jl}\ldots\mathrm{d}r_{jN}}{\int\ldots\int \exp(-\Phi/kT)\mathrm{d}r_{il}\ldots\mathrm{d}r_{iN}\mathrm{d}r_{jl}\ldots\mathrm{d}r_{jN}} \quad (6)$$

where Φ is the total potential energy for all pairs of ions interacting in accordance with the potentials $u_{ij}(r)$ of eq 1. Note that the numerator differs only in that the first two coordinates for the ith type of ions are not integrated.

For the radial distribution function for a pair of unlike ions the analogous equation is

$$g_{ij}(r) = V^2 \frac{\int\ldots\int \exp(-\Phi/kT)\mathrm{d}r_{i2}\ldots\mathrm{d}r_{iN}\mathrm{d}r_{j2}\ldots\mathrm{d}r_{jN}}{\int\ldots\int \exp(-\Phi/kT)\mathrm{d}r_{il}\ldots\mathrm{d}r_{iN}\mathrm{d}r_{j1}\ldots\mathrm{d}r_{jN}} \quad (7)$$

where now the integral in the numerator omits the integration over the coordinates of one ion of each type.

These basic equations may be evaluated directly by Monte Carlo methods.[19] Such calculations involve numerical approximations and various technical features, but they may be made as accurate as desired with the expenditure of sufficient care and computer time. Vorontsov-Veliaminov et al.[20,21] and Card and Valleau[22] have made significant Monte Carlo calculations for electrolytes. We shall present the results of Card and

(14) W. G. McMillan and J. E. Mayer, *J. Chem. Phys.*, **13**, 276 (1945).

(15) T. L. Hill, ref 8, p 341, is the simplified account of McMillan–Mayer theory. A more rigorous account is in ref 12, p 262. See also Andersen, ref 4, p 7.

(16) J. C. Rasaiah and H. L. Friedman, *J. Chem. Phys.*, **48**, 2742 (1968); **50**, 3965 (1969).

(17) H. L. Friedman, *J. Chem. Phys.*, **32**, 1351 (1960); see also the appendix in ref 29 and ref 3, p 207.

(18) See Davidson, ref 11, Chapter 2, or Hill, ref 8, Section 17-4.

(19) J. A. Barker and D. Henderson, *Rev. Mod. Phys.*, **48**, 587 (1976).

(20) P. N. Vorontsov-Veliaminov, A. M. Eliashevich, and A. K. Kron, *Elektrokhimaya*, **2**, 708 (1966).

(21) P. N. Vorontsov-Veliaminov and A. M. Eliashevich, *Elektrokhimaya*, **4**, 1430 (1968); *Strukt. Rol. Vody, Zhivom. Org.*, **2**, 99 (1968).

(22) D. N. Card and J. P. Valleau, *J. Chem. Phys.*, **52**, 6232 (1970).

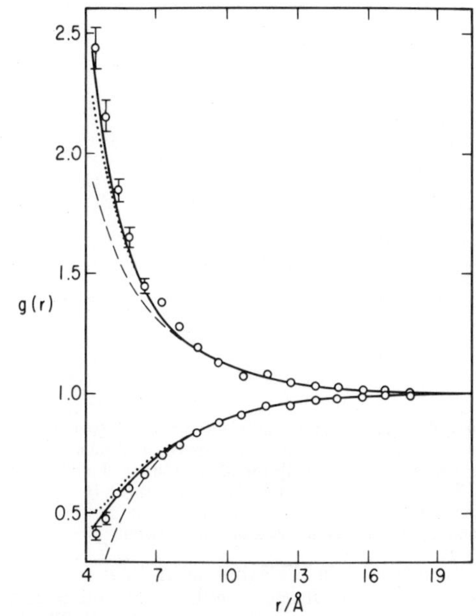

Figure 3. The radial distribution functions for 0.425 M solution; other aspects the same as in Figure 2.

Valleau which are based on the primitive model appropriate for aqueous solutions at room temperature and for a uni-univalent solute with a = 4.25 Å. The results are shown as points on Figures 2 and 3.

In addition to the combination of Monte Carlo calculations with eq 5 to obtain thermodynamic quantities, the method of "molecular dynamics" would also provide a rigorous numerical solution for an electrolyte within the McMillan–Mayer solution theory framework. In this method[19] the motions of a number of particles are followed by integration of the equations of motion subject to the potentials assumed. Then the desired quantities are determined from averages over the motion. It has been successfully applied to nonelectrolyte fluids with results comparable to those of the Monte Carlo method, and an exploratory calculation has just been reported for an electrolyte.[23]

While we shall presently mention some other recent theories, let us first compare results which can be derived from the Debye–Hückel treatment with the Monte Carlo results.

In the conventional Debye–Hückel treatment[24] the concentration of ionic species i near an ion of species j is given by the Boltzmann expression

$$c_i' = c_i \exp(-z_i e\psi_j/kT) \tag{8}$$

where c_i is the average concentration and $\psi_j(r)$ is an electrostatic potential associated with the ion j. This implies a radial distribution function

$$g_{ij}(r) = \exp(-z_i e\psi_j(r)/kT) \tag{9}$$

The Debye–Hückel treatment then combines the concentrations of all charged species to yield the charge density

$$\rho_j(r) = \sum_i z_i e c_i' = \sum_i z_i e c_i \exp(-z_i e\psi_j/kT) \tag{10}$$

The exponentials are expanded, yielding

$$\rho_j(r) = e\sum_i z_i c_i - (e^2\psi_j/kT)\sum_i z_i^2 c_i + (e^3\psi_j^2/2k^2T^2)\sum_i z_i^3 c_i \ldots \tag{11}$$

From electrical neutrality the first term on the right is zero. The second term is clearly nonzero and is the only term used for ρ in the usual treatment, where the Poisson equation now takes a linear form

$$\nabla^2\psi = -(4\pi/D)\rho = \kappa^2\psi \tag{12}$$

Appropriate boundary conditions then yield

$$\psi_j(r) = \frac{z_j e \exp(\kappa a)}{D(1+\kappa a)}\frac{\exp(-\kappa r)}{r} \tag{13}$$

where

$$\kappa^2 = (4\pi e^2/DkT)\sum_i z_i^2 c_i \tag{14}$$

and a is the diameter of the hard core in eq 2.

Let us return to the distribution function which now becomes

$$g_{ij} = \exp(-q_{ij}) \tag{15}$$

$$q_{ij} = \frac{z_i z_j e^2 \exp(\kappa a)}{DkT(1+\kappa a)}\frac{\exp(-\kappa r)}{r} \tag{16}$$

Clearly the required symmetry to exchange of i and j is present. However, for the Poisson equation the exponentials in ρ were expanded[25] and only the second term was used. Thus it can be argued that it is inconsistent to retain the exponential form for g_{ij}. Since this entire treatment is approximate, this objection has limited significance. Alternatively one may note that the third term in the expanded form for the charge density is zero for a symmetrical electrolyte. Thus, without any inconsistency for symmetrical electrolytes, one may retain the third term in the expansion and the corresponding expression for the radial distribution function is

$$g_{ij}(r) = 1 - q_{ij} + q_{ij}^2/2 \tag{17}$$

The original Debye–Hückel approximation can be regarded as the first two terms

$$g_{ij}(r) = 1 - q_{ij} \tag{18}$$

In Figures 2 and 3 we compare the exponential form (15), the three-term form (17), and the two-term form (18) for g_{ij} with the "exact" results of the Monte Carlo calculations for the case a = 4.25 Å and solvent properties for water at 25 °C. For the dilute solution, 0.00911 M, in Figure 2 one notes that the exponential form is, indeed, an excellent approximation which agrees with the Monte Carlo results essentially within the computational uncertainty of the latter. The three-term "extended D–H" approximation agrees very well except for r values just above a. Here g_{+-} is a little too small and g_{++} shows a false increase as r decreases below 6 Å.

(23) P. Turq, F. Lantelme, and H. L. Friedman, *J. Chem. Phys.*, **66**, 3039 (1977).

(24) R. A. Robinson and R. H. Stokes, "Electrolyte Solutions", Butterworths, London, 1955; revised 1959, Chapter 4.

(25) There is an extended literature on solution of the Poisson equation without the linear approximation; see D. M. Burley, V. C. L. Hutson, and C. W. Outhwaite, *Mol. Phys.*, **23**, 867 (1972), and references cited therein.

The two-term approximation is satisfactory for large r but deviates considerably at small r. The function for like ions even goes negative below about 6 Å; this is quite unacceptable. Since the three-term formula involves no inconsistency with the original derivation (for symmetrical electrolytes), there would appear to be no reason for further use of the much poorer two-term approximation.

Figure 3 gives the same functions for a moderately concentrated solution, 0.425 M. The same comments apply as for the dilute solution, although the differences in the various D–H approximations are somewhat reduced.

We can now bypass the many objections[2] to the formulation of D–H theory by noting that it yields good agreement with the correct radial distribution function, particularly in the exponential form.

In applying any of these radial distribution functions with hard-core potentials to eq 5, one faces an anomaly in the integral at $r = a$ where $\partial u/\partial r$ is infinite. This has been resolved mathematically and yields[26] for the osmotic coefficient

$$\phi - 1 = (\Pi/ckT) - 1$$
$$= (6ckT)^{-1} \sum_i\sum_j c_i c_j \int_a^\infty r \frac{\partial u_{ij}}{\partial r} g_{ij}(4\pi r^2)\mathrm{d}r + \frac{2\pi a^3}{3c}\sum_i\sum_j c_i c_j g_{ij}(a) \quad (19)$$

where $g_{ij}(a)$ is the value of $g_{ij}(r)$ for r infinitesimally greater than a.

For the Monte Carlo calculations $g_{ij}(r)$ must be extrapolated to $r = a$; this adds some numerical uncertainty. The various D–H expressions are unambiguous for $g_{ij}(a)$; hence, there is no difficulty in evaluating the expressions in eq 19. Also there is some cancellation of the errors of the $g(a)$ values for the three-term approximation. For the exponential form, the integral can be evaluated only numerically, but for the two- or three-term approximations, eq 18 and 17, the integration yields simple functions. The latter result is

$$\phi - 1 = \frac{-\kappa^3}{24\pi c(1 + \kappa a)} + c\left[\frac{2\pi a^3}{3} + \frac{\kappa^4 a}{48\pi c^2(1 + \kappa a)^2}\right] \quad (20)$$

It is convenient to define

$$w = \sum_i z_i^2(c_i/c)$$

which is an average value of z_i^2 and becomes z^2 for a symmetrical electrolyte. Then

$$\phi - 1 = \frac{-wl\kappa}{6(1 + \kappa a)} + c\left[\frac{2\pi a^3}{3} + \frac{\pi a w^2 l^2}{3(1 + \kappa a)^2}\right] \quad (22)$$

where $l = e^2/DkT$ and $\kappa^2 = 4\pi lwc$. In both eq 20 and 22 the first term, which contains the D–H limiting law, arises from the second term in the expansion of g_{ij}. This is the term that leads to nonzero charge density; it can be thought of as the primary electrostatic term. The first term, unity, in the expressions for g_{ij} yields the term $c(2\pi a^3/3)$ which would be the kinetic effect of the hard core in the absence of electrical effects. The last term in (20) and (22) arises from the third term in the expressions for g_{ij} and would be absent if only the D–H two-term expression were used. This term gives the increased hard-core repulsive effect caused by the electrical interactions.

(26) See Barker and Henderson, ref 19, p 591.

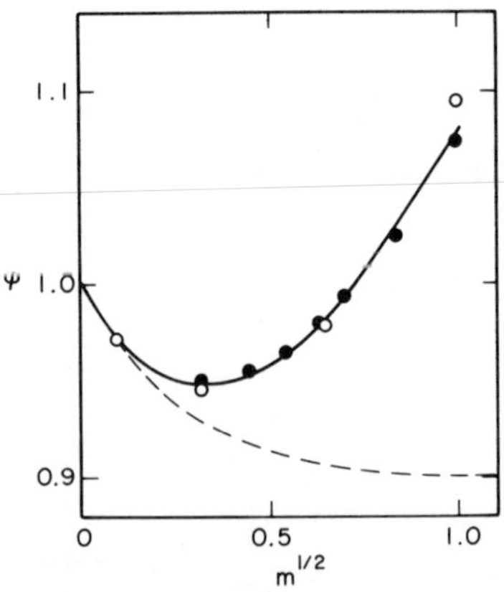

Figure 4. The osmotic coefficient for a 1–1 type aqueous electrolyte at 25 °C. Solid circles are experimental values for HBr. Open circles are calculated by the Monte Carlo method (a = 4.25 Å). The solid curve is based on the three-term D–H expression and the "pressure" equation while the dashed curve is the traditional D–H expression based on the "charging process".

Figure 4 shows, for the case a = 4.25 Å and a 1–1 aqueous electrolyte, the osmotic coefficient from the Monte Carlo calculations, that from the three-term D–H approximation of eq 22, and that from the original D–H formula which was derived by a charging process

$$\phi - 1 = -(\kappa^3/24\pi c)\sigma(\kappa a) \quad (23)$$

where

$$\sigma(x) = (3/x^3)[1 + x - (1 + x)^{-1} - 2\ln(1 + x)] \quad (24)$$

This model fits the observed data for HBr very well, and the experimental osmotic coefficients for HBr are also shown on Figure 4. The three-term D–H expression is an excellent approximation. The original D–H equation is a useful approximation, however, only below 0.01 M. The three-term D–H expression for g_{ij} had been recognized earlier[24] as a good approximation, but so long as the charging process was used to calculate thermodynamic properties it did not yield any improvement over the two-term expression.

Thermodynamic equations may be used to transform eq 22 to the forms for the activity coefficient and the excess Gibbs energy. With the assumption that concentration is proportional to molality, the results are

$$\ln\gamma_{\pm} = -\left(\frac{wl}{6}\right)\left[\frac{2\kappa}{1 + \kappa a} + \frac{1}{a}\ln(1 + \kappa a)\right] + c\left[\frac{4\pi a^3}{3} + \frac{\pi a w^2 l^2}{3(1 + \kappa a)^2}\right] \quad (25)$$

$$G^{\mathrm{ex}}/ckT = -\frac{\mathrm{lw}}{6}\left[\frac{\kappa}{1 + \kappa a} + \frac{1}{a}\ln(1 + \kappa a)\right] + c\left(\frac{2\pi a^3}{3}\right) \quad (26)$$

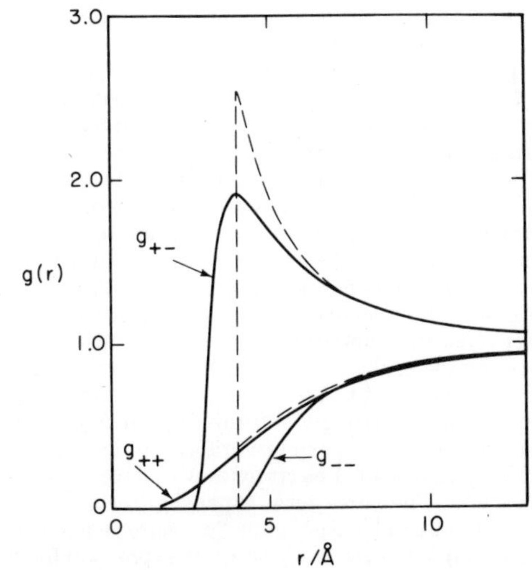

Figure 5. Radial distribution functions for LiBr at 0.4 M. The solid curves are based on a refined model, see text; the dashed curves on the primitive model.

In addition to the Monte Carlo method, which is exact in principle, a number of approximate equations or methods[27] have been proposed and applied recently to the calculation of the radial distribution functions for electrolytes. While these yield equations which usually require less computation than the Monte Carlo method, their derivation is more complex and the evaluation of possible errors from the approximations is even more complex. Indeed, now that Monte Carlo solutions are available, these other methods are tested[28] by comparison with the Monte Carlo results for identical potential models.

The principal advantage of these other methods arises from the possibility of exploring a variety of interionic potentials with less computational cost than the Monte Carlo method would require. This has been done by Friedman and Rasaiah and collaborators.[29,30] They used the "hypernetted chain equation" and tested its accuracy in various ways, including comparison with Monte Carlo results.[28] We shall not discuss these results in detail but show in Figure 5 the radial distribution function for LiBr obtained by Ramanathan and Friedman.[29,31] They assumed the bare-ion repulsive potentials obtained in studies of ionic crystals and adjusted the potentials at intermediate distances to obtain agreement with experiment. The final potentials are smooth functions of radius. It was found that the selection of the three potentials (Li^+–Li^+, Li^+–Br^-, and Br^-–Br^-) at intermediate distances was not unique; many alternatives fitted the osmotic coefficients equally well.

For comparison, the D–H exponential radial distribution functions are also shown in Figure 5 for a value of a = 4.0 Å which fits the experimental data for LiBr reasonably well. There are Monte Carlo results[21] for a = 4.0 Å but not at the molality 0.4 of the other curves in Figure 5. The Monte Carlo results at other concentrations, however, show agreement with the D–H exponential functions.

(27) See ref 16; also several cited in ref 4 and 5.

(28) J. C. Rasaiah, D. N. Card, and J. P. Valleau, *J. Chem. Phys.*, **56**, 248 (1972).

(29) P. S. Ramanathan and H. L. Friedman, *J. Chem. Phys.*, **54**, 1086 (1971).

(30) J. C. Rasaiah, *J. Chem. Phys.*, **52**, 704 (1970).

(31) Numerical values of the distribution functions supplied by Professor H. L. Friedman.

Equations for Use in Chemical Thermodynamics

For chemical thermodynamic calculations for electrolytes it is usually preferable to use semiempirical equations rather than purely theoretical equations since the latter are either very complex or contain significant approximations. Also at present any theoretical equation is based on a model whose accuracy is subject to argument. The semiempirical equations include the D–H limiting law and such other features taken from theory as seem useful.

The traditional D–H result, eq 23 for ϕ or the more familiar

$$\ln \gamma_{\pm} = -wl\kappa/2(1 + \kappa a) \tag{27}$$

is seen from Figure 4 to become inadequate even at very low concentration. Equation 27 is commonly extended with a term analogous to the second term in eq 20 and 22. After rearrangement and for a symmetrical electrolyte, this becomes

$$\ln \gamma_{\pm} = -\frac{z^2 A_{\gamma} I^{1/2}}{(1 + \rho I^{1/2})} + 2\beta m \tag{28}$$

where A_{γ} is the traditional D–H parameter, m is the molality, and I is the ionic strength. Also $\rho = \kappa a/I^{1/2}$ and β are parameters generally related to the diameter of the ions. The corresponding form for the osmotic coefficient is

$$\phi - 1 = -z^2 A_{\phi} I^{1/2} \sigma(\rho I^{1/2}) + \beta m \tag{29}$$

Guggenheim[32] proposed taking $\rho = 1$ for all solutes and letting the specific properties of each solute appear in β. This system has been widely used[33] with considerable success. But with β a constant, this equation still does not fit the better quality data within experimental accuracy over a substantial range in m.

The needed improvement[6] is suggested by the last term in eq 20, 22, and 25. One notes that this term which decreases with increasing ionic strength (increasing κ) is added to the constant term, $2\pi a^3/3$, in (22) to yield the quantity equivalent to β in eq 28 and 29. A directly analogous form would be

$$\beta = \beta^{(0)} + [\beta^{(1)}/(1 + \alpha I^{1/2})^2] \tag{30}$$

while a somewhat simpler form with similar properties is

$$\beta = \beta^{(0)} + \beta^{(1)} \exp(-\alpha I^{1/2}) \tag{31}$$

These forms were tested[6] with a wide range of accurate osmotic coefficient data, and the second (eq 31) was found to be slightly better. In each case $\beta^{(0)}$ and $\beta^{(1)}$ were freely adjusted for each electrolyte while α was optimized but to a single value for all solutes. Similarly the two forms for the electrostatic term, in eq 22 and 23, were compared[6] and the more recent (eq 22) was slightly superior. The resulting practical equation for

(32) E. A. Guggenheim, *Phil. Mag.*, [7] **19**, 588 (1935); E. A. Guggenheim and J. C. Turgeon, *Trans. Faraday Soc.*, **51**, 747 (1955).

(33) K. S. Pitzer and L. Brewer, revised edition of "Thermodynamics" by G. N. Lewis and M. Randall, McGraw-Hill, New York, N.Y., 1961.

the osmotic coefficient of a single symmetrical electrolyte is

$$\phi - 1 = -z^2 A_\phi I^{1/2}/(1 + bI^{1/2}) + m[\beta^{(0)}{}_{MX} + \beta^{(1)}{}_{MX} \exp(-\alpha I^{1/2})] + m^2 C^\phi{}_{MX} \quad (32)$$

where b is kept the same for all solutes with a best value of 1.2, α is 2.0 for all 1–1 and many other electrolytes, and an additional term in m^2 has been added to extend the equation to very high concentrations. Equation 32 fits the experimental osmotic coefficients for over 200 pure aqueous electrolytes essentially within experimental accuracy.[34]

The corresponding equation for the activity coefficient is

$$\ln \gamma_\pm = -z^2 A_\phi [I^{1/2}/(1 + bI^{1/2}) + (2/b) \ln (1 + bI^{1/2})] + m\{2\beta^{(0)}{}_{MX} + (2\beta^{(1)}{}_{MX}/\alpha^2 I)[1 - (1 + \alpha I^{1/2} - \alpha^2 I/2) \exp(-\alpha I^{1/2})]\} + m^2(3C^\phi{}_{MX}/2) \quad (33)$$

These equations become only a little more complex for unsymmetrical solutes. Even for mixed electrolytes[35] the corresponding equations are relatively simple, and all of the more important parameters are determined from the data on single solutes. The result is an extremely general and compact representation of the known properties of aqueous electrolytes which also has the capacity of prediction of properties of mixtures to useful accuracy.

Summary

First we recall that there is not yet an adequate theory for water (or for other ionizing solvents); hence any theory must start with assumptions concerning the interionic potentials of mean force in the solvent. From that point, however, we now have an exact theory which can be applied numerically to obtain as accurately as desired the interionic radial distribution functions and thermodynamic properties.

There is no longer any need to speculate about the net inaccuracies in the Debye Hückel theory, or extensions thereof, or of other approximate theories. These questions can be answered by comparison with calculations based on exact theories. The original D–H expression which contained no contribution from the direct effects of short-range forces is seriously inadequate outside of the range of the limiting law. But the D–H calculation of interionic distribution functions is found to be surprisingly accurate in either the exponential or three-term expansion form (at least for the model for aqueous 1–1 electrolytes with ionic diameters near 4 Å). The three-term expansion form of these distributions can be used in the "pressure" equation to yield a simple but relatively accurate expression for the osmotic coefficient up to about 1 M for this model.

It is now feasible also to investigate more realistic interionic potentials. This should be very valuable in the future, although the requirement of fitting osmotic and activity coefficient data is not likely to yield a unique determination of the short-range potential.

The recent theory suggested important improvements in the form of semiempirical equations. Reasonably simple equations are now available which fit within experimental accuracy a wide variety of data on pure and mixed electrolytes. Also, since the more important parameters are determined from the single solutes, the equations allow prediction of the properties of mixtures.

Financial support of the U.S. Atomic Energy Commission and of the Energy Research and Development Administration is gratefully acknowledged.

(34) K. S. Pitzer and G. Mayorga, *J. Phys. Chem.*, **77**, 2300 (1973); *J. Soln. Chem.*, **3**, 539 (1974).

(35) K. S. Pitzer and J. J. Kim, *J. Am. Chem. Soc.*, **96**, 5701 (1974).

Geochimica et Cosmochimica Acta Vol. 51, pp. 2429–2443
0016-7037/87/$3.00 + .00

Thermodynamics of concentrated electrolyte mixtures and the prediction of mineral solubilities to high temperatures for mixtures in the system Na-K-Mg-Cl-SO_4-OH-H_2O

ROBERTO T. PABALAN and KENNETH S. PITZER

Lawrence Berkeley Laboratory and Department of Chemistry, University of California, Berkeley, CA 94720, U.S.A.

(*Received January* 23, 1987; *accepted in revised form June* 8, 1987)

Abstract—Mineral solubilities in binary and ternary electrolyte mixtures in the system Na-K-Mg-Cl-SO_4-OH-H_2O are calculated to high temperatures using available thermodynamic data for solids and for aqueous electrolyte solutions. Activity and osmotic coefficients are derived from the ion-interaction model of PITZER (1973, 1979) and co-workers, the parameters of which are evaluated from experimentally determined solution properties or from solubility data in binary and ternary mixtures. Excellent to good agreement with experimental solubilities for binary and ternary mixtures indicate that the model can be successfully used to predict mineral-solution equilibria to high temperatures. Although there are currently no theoretical forms for the temperature dependencies of the various model parameters, the solubility data in ternary mixtures can be adequately represented by constant values of the mixing term θ_{ij} and values of ψ_{ijk} which are either constant or have a simple temperature dependence. Since no additional parameters are needed to describe the thermodynamic properties of more complex electrolyte mixtures, the calculations can be extended to equilibrium studies relevant to natural systems. Examples of predicted solubilities are given for the quaternary system NaCl-KCl-$MgCl_2$-H_2O.

INTRODUCTION

EQUILIBRIUM CONSIDERATIONS based on experimental data and thermodynamic calculations provide an important basis for understanding rock-fluid interactions in natural systems. The problems associated with experimental work, particularly in complex and concentrated electrolyte mixtures, underscore the need for a predictive model for calculating mineral-solution equilibria in brine systems. Applications for such a model include studies of seawater and evaporite systems, understanding the origin and evolution of brines in sedimentary basins, rift zones and geothermal systems, and in desalination studies. In addition, predictions of brine-mineral equilibria are important in assessing the possible far-field impact of a repository breach where salt formations are used as storage sites for high-level radioactive waste.

The objective of this study is to investigate the applicability of the ion-interaction or virial-coefficient model developed by PITZER (1973, 1979) to solubility equilibria over a wide range of temperature. The ion-interaction approach models the thermodynamic properties of electrolyte solutions based on a set of theoretically and empirically derived equations that account for the interactions between the particular aqueous ions present and for indirect forces arising from the solvent. This model has been used successfully by HARVIE and WEARE (1980) to predict mineral solubilities at 25°C in the Na-K-Ca-Mg-Cl-SO_4-H_2O system to high concentrations. Several papers applying the model to the study of sea water, deep brines and other saline waters have since been published (*cf.* HARVIE *et al.*, 1982, 1984; GUEDDARI *et al.*, 1983; MEIJER and VAN ROSMALEN, 1984; MONNIN and SCHOTT, 1984; LANGMUIR and MELCHOIR, 1985), but most of these calculations applied only to a temperature of 25°C. However, the occurrence of brine solutions in a variety of geologic environments points to the need to extend the treatment to a wider range of temperature and pressure conditions. DE LIMA and PITZER (1983a,b) have made high-temperature calculations comparing activity coefficients in aqueous mixtures derived from the model with those from solubility data, but these are not true solubility calculations. WOOD (1976) has done solubility calculations to 200°C in the NaCl-KCl-H_2O system using a Scatchard-Harned model, but has not extended his high-temperature calculations to other systems.

In this study, solubilities in binary and ternary electrolyte mixtures in the system Na-K-Mg-Cl-SO_4-OH-H_2O were calculated to high temperatures using available thermodynamic properties of the solids and of aqueous electrolyte solutions, and were compared with experimentally determined values. Although a few parameters were adjusted to fit ternary mixtures, no additional parameters are needed by the model to describe electrolyte properties in quaternary and more complex systems; consequently, the solubility calculations can be extended to compositions relevant to natural fluids. Examples of predicted solubilities are given for the quaternary system NaCl-KCl-$MgCl_2$-H_2O.

THEORETICAL CONSIDERATIONS

Mineral solubilities in concentrated electrolyte solutions can be calculated from thermodynamic considerations provided that equilibrium constants are known and activity coefficients can be obtained. For a hydrated solid, $M_{\nu_M}X_{\nu_X}\cdot\nu_0H_2O$, having ν_M positive ions, M, of charge z_M, and ν_X negative ions, X, of charge z_X, as well as ν_0 molecules of water, the equilibrium constant at a fixed temperature and pressure for the dissolution reaction

$$M_{\nu_M}X_{\nu_X}\cdot\nu_0H_2O=\nu_M M^{+z_M}+\nu_X X^{-z_X}+\nu_0H_2O \quad (1)$$

is given by

$$\ln K=-(\nu_M\mu_M^0+\nu_X\mu_X^0+\nu_0\mu_{H_2O}^0)/RT+\mu_s^0/RT \quad (2)$$

where the μ_i^0's represent the chemical potential of the solid, of water, or of the aqueous ion in a defined standard state. The standard state for the aqueous ions and electrolytes was taken to be the hypothetical one molal solution referenced to infinite dilution at any pressure and temperature. The solid and solvent standard state was taken to be the pure phase at the pressure and temperature of interest.

The equilibrium constant can be calculated provided the chemical potential values are available at the temperature and pressure of interest. The relationships between the dependencies of the chemical potentials on pressure and temperature, and on volumes, entropies (or enthalpies), and heat capacities are well-known. At constant pressure, the temperature dependence of the standard state chemical potential of component i is given by

$$(\partial\mu_i^0/\partial T)_P=-S_i^0 \quad (3)$$

or

$$\mu_{i,T_f}^0-\mu_{i,T_r}^0=-S_{i,T_r}^0(T_f-T_r)+\int_{T_r}^{T_f}C_{p_i}^0dT-T\int_{T_r}^{T_f}(C_{p_i}^0/T)dT \quad (4)$$

where S_i^0 represents the standard entropy of component i, and $C_{p_i}^0$ is the standard state heat capacity (at constant pressure) which itself is expressed as a function of temperature (discussed below). T_r and T_f represent the reference temperature and temperature of interest, respectively. An analogous expression for the temperature dependence of the chemical potential could be written in terms of the enthalpy and heat capacity.

The effect of pressure on the chemical potential is given by the molar or partial molar volume and can be determined from density data. At the saturation pressures considered in this study these effects are usually small, particularly for the solids, but can be significant at higher pressures (ROGERS and PITZER, 1982; PITZER, 1986). In a few cases like NaCl (ROGERS and PITZER, 1982; PITZER *et al.*, 1984) the pressure dependency of the solution properties has been studied. We have assumed that the pressure effects on the calculated solubilities are within the combined uncertainties of all the parameters.

In terms of activities or molalities, the equilibrium constant for the solubility reaction is given by

$$\ln K=\nu_M\ln(m_M\gamma_M)+\nu_X\ln(m_X\gamma_X)+\nu_0\ln(a_{H_2O}) \quad (5)$$

where m_i and γ_i represent the molality and activity coefficient of the aqueous ions, respectively. Of course, absolute values of γ_i's for ions cannot be determined. One procedure is to define a mean activity coefficient $\gamma_\pm$, which is unambiguous,

$$\ln\gamma_\pm=(\nu_M\ln\gamma_M+\nu_X\ln\gamma_X)/\nu \quad (6)$$

with $\nu=\nu_M+\nu_X$. However, it can be shown (PITZER, 1979) that one can write equations for γ_M and γ_X where the only ambiguity is a term which cancels for the $\gamma_\pm$ product for any charge type (see footnote below). It is optional whether to use equations for $\gamma_\pm$ or those for γ_M and γ_X. We have used the latter because in a complex mixture the number of possible $\gamma_\pm$ is much larger than the number of γ_M and γ_X.

The activity of water is related to the osmotic coefficient, ϕ, by the equation

$$\ln a_1=-\phi(M_W/1000)\sum_i m_i \quad (7)$$

where M_W is the molecular mass of water and the sum covers all solute species. The Gibbs-Duhem equation relates the composition dependence of ϕ or a_1 to the various γ_i or $\gamma_\pm$. These activity and osmotic coefficients are calculated from the ion-interaction model of PITZER (1973, 1979) described below.

THE ION-INTERACTION MODEL

The basic theoretical rationale of the ion-interaction model has already been discussed in several publications (*cf.* PITZER, 1973, 1979; PITZER and KIM, 1974; HARVIE and WEARE, 1980). Briefly, the model starts with a virial expansion of the excess free energy, *i.e.*, the actual free energy of the solution minus the free energy of an ideal solution of the same composition: G^{ex}/RT. The expressions for osmotic and activity coefficients follow directly from the equation for G^{ex}/RT through the appropriate derivatives with respect to n_w and m_i, respectively. Equation (8) shows the expression for the osmotic coefficient, while Eqns. (9a) and (9b) give the activity coefficients of cation M and anion X, respectively.[1]

$$(\phi-1)=-\frac{\partial G^{ex}/\partial n_w}{RT\sum_i m_i}=(2/\sum_i m_i)[-A_\phi I^{3/2}/(1+bI^{1/2})$$
$$+\sum_c\sum_a m_cm_a(B_{ca}^\phi+ZC_{ca})+\sum\sum_{c<c'} m_cm_{c'}(\Phi_{cc'}^\phi+\sum_a m_a\psi_{cc'a})$$
$$+\sum\sum_{a<a'} m_am_{a'}(\Phi_{aa'}^\phi+\sum_c m_c\psi_{caa'}) \quad (8)$$

$$\ln\gamma_M=\frac{1}{RT}\frac{\partial G^{ex}}{\partial m_M}=z_M^2F+\sum_a m_a(2B_{Ma}+ZC_{Ma})$$
$$+\sum_c m_c(2\Phi_{Mc}+\sum_a m_a\psi_{Mca})+\sum\sum_{a<a'} m_am_{a'}\psi_{Maa'}$$
$$+|z_M|\sum_c\sum_a m_cm_aC_{ca} \quad (9a)$$

$$\ln\gamma_X=z_X^2F+\sum_c m_c(2B_{cX}+ZC_{cX})+\sum_a m_a(2\Phi_{Xa}+\sum_c m_c\psi_{cXa})$$
$$+\sum\sum_{c<c'} m_cm_{c'}\psi_{cc'X}+|z_X|\sum_c\sum_a m_cm_aC_{ca} \quad (9b)$$

where c and c' are cations and a and a' are anions. The ion charges are indicated by z_M and z_X. The B and C terms can be evaluated empirically from data on binary systems, while the Φ and ψ terms arise only for mixed solutions and can best

---

[1] While Eqns. (9a) and (9b) above give the practical activity coefficients for single ions, the complete expressions are (PITZER, 1979):

$$\ln\gamma_M=[\text{Right side of Eqn. (9a)}]+|z_M|Q$$
$$\ln\gamma_X=[\text{Right side of Eqn. (9b)}]-|z_X|Q$$
$$Q=[\sum_c m_c\lambda_{cc}/z_c-\sum_a m_a\lambda_{aa}/|z_a|$$
$$+(3/2)\sum_c\sum_a m_cm_a(\mu_{cca}/z_c-\mu_{caa}/|z_a|)].$$

Here λ_{ij} and μ_{ijk} are the second and third virial coefficients, respectively, for individual ion interactions; see Eqn. (21) of PITZER (1973) or Eqn. (24) of PITZER (1979). These individual ion coefficients cannot be determined from thermodynamic measurements because the solutions are always electrically neutral. Likewise, the quantity Q drops out from all mean activity coefficients or other measurable combinations of activity coefficients because these combinations maintain electrical neutrality.

be determined from common ion mixtures. The quantity F includes the Debye-Hückel term and other terms as follows:

$$F = -A_\phi[I^{1/2}/(1+bI^{1/2}) + (2/b)\ln(1+bI^{1/2})] + \sum_c\sum_a m_c m_a B'_{ca} + \sum_{c<c'}\sum m_c m_{c'}\Phi'_{cc'} + \sum_{a<a'}\sum m_a m_{a'}\Phi'_{aa'}. \quad (10)$$

Also,

$$Z = \sum_i m_i|z_i| \quad (11)$$

$$B^\phi_{ca} = B_{ca} + IB'_{ca} \quad (12)$$

$$\Phi^\phi_{cc'} = \Phi_{cc'} + I\Phi'_{cc'}. \quad (13)$$

B' and Φ' are the ionic strength derivatives of B and Φ. The sums over i include all solute species; uncharged species do not contribute to I or Z. The double summation indices, $c < c'$ and $a < a'$, denote the sum over all distinguishable pairs of dissimilar cations or anions. The third virial coefficients C_{ij} and ψ_{ijk} in Eqns. (8) and (9) are assumed to be independent of ionic strength. The parameter, C_{MX}, is related to the commonly tabulated parameter (*cf.*, PITZER, 1973; PITZER and MAYORGA, 1973), C^ϕ_{MX}, by the equation

$$C_{MX} = C^\phi_{MX}/2(|z_M z_X|)^{1/2}. \quad (14)$$

The ionic strength dependence of the B terms is shown by the following equations:

$$B^\phi_{MX} = \beta^{(0)}_{MX} + \beta^{(1)}_{MX}\exp(-\alpha_1 I^{1/2}) + \beta^{(2)}_{MX}\exp(-\alpha_2 I^{1/2}) \quad (15a)$$

$$B_{MX} = \beta^{(0)}_{MX} + \beta^{(1)}_{MX} g(\alpha_1 I^{1/2}) + \beta^{(2)}_{MX} g(\alpha_2 I^{1/2}) \quad (15b)$$

$$B'_{MX} = [\beta^{(1)}_{MX} g'(\alpha_1 I^{1/2}) + \beta^{(2)}_{MX} g'(\alpha_2 I^{1/2})]/I \quad (15c)$$

where the functions g and g' are given by

$$g(x) = 2[1-(1+x)\exp(-x)]/x^2 \quad (16a)$$

$$g'(x) = -2[1-(1+x+x^2/2)\exp(-x)]/x^2. \quad (16b)$$

$\beta^{(0)}$, $\beta^{(1)}$, and $\beta^{(2)}$ are solute-specific parameters fit to isothermal-isobaric experimental data on pure-electrolytes at varying concentrations. $\beta^{(2)}$ is important only for 2-2 or higher valence electrolytes that show a tendency towards electrostatic ion-pairing. For electrolytes in which at least one of the ions is univalent, α_1 is taken to be 2.0 kg$^{1/2}$ mol$^{-1/2}$ (PITZER, 1973; PITZER and MAYORGA, 1973). However, other values of α_1 can be used without undue complication (HOLMES and MESMER, 1986). For 2-2 electrolytes at 25°C the optimized values of α_1 and α_2 are 1.4 and 12 kg$^{1/2}$ mol$^{-1/2}$, respectively. These values of α_1 and α_2 can be taken to be independent of T and P for many applications as we have done in this study, but there are theoretical reasons for taking α_2 to be proportional to the Debye-Hückel parameter, A_ϕ, as discussed by PHUTELA and PITZER (1986b).

In the case of the Φ terms there is a large ionic-strength dependence for unsymmetrical mixing such as Na^+ with Mg^{2+}, or Cl^- with SO_4^{2-}, arising from long-range forces and is given by theory (PITZER, 1975, 1983). The complete expressions for Φ_{ij} are:

$$\Phi_{ij} = \theta_{ij} + {}^E\theta_{ij}(I) \quad (17a)$$

$$\Phi'_{ij} = {}^E\theta'_{ij}(I) \quad (17b)$$

$$\Phi^\phi_{ij} = \theta_{ij} + {}^E\theta_{ij}(I) + I\,{}^E\theta'_{ij}(I) \quad (17c)$$

where ${}^E\theta(I)$ and ${}^E\theta'(I)$ account for electrostatic unsymmetrical mixing effects and depend only on the charges of the ions i and j, the total ionic strength, and on the density and dielectric constant of the solvent (hence, on the temperature and pressure). Equations for calculating these terms are derived by PITZER (1975) and a convenient set of Chebyshev approximations to these equations is given by HARVIE (1981). The remaining term θ_{ij} arising from short-range forces is taken as a constant for any particular c, c' or a, a' at a given T and P. Its ionic strength dependence, which can be expressed in the same form as Eqn. (15), is very small and is usually neglected as we have done here, but this effect may be significant in some cases as shown by recent calculations of heats of mixing by PHUTELA and PITZER (1986a).

Thus starting from theoretical considerations, the ion-interaction model gives an expression for the activity and osmotic coefficients of electrolyte mixtures in terms of six types of empirical parameters, namely, $\beta^{(0)}_{MX}$, $\beta^{(1)}_{MX}$, $\beta^{(2)}_{MX}$, C^ϕ_{MX}, θ_{ij}, and ψ_{ijk}. Provided that their temperature and pressure dependencies are known, these parameters permit the calculation of solubilities in binary, ternary and more complex mixtures at different temperatures and pressures.

It is important to recognize that the same equation for total excess Gibbs energy yields other thermodynamic functions such as excess enthalpies and heat capacities by appropriate differentiation. These other functions are directly obtainable from experiment and their evaluation yields accurate data on the temperature-dependencies of the ion-interaction coefficients.

There is, therefore, a wide array of experimental data from which the ion-interaction parameters and their temperature functions can be determined. This includes measurements on

1) freezing point depression
2) boiling point elevation
*3) vapor pressure
*4) isopiestic concentrations
5) E.M.F.
*6) enthalpy of dilution
7) enthalpy of mixing
8) enthalpy of solution
*9) heat capacity
10) solubility.

Those marked with an asterisk are those that so far have been most useful at temperatures greater than 100°C. Enthalpies of dilution measured at 25°C are also important for their relation to the temperature dependency of the Gibbs energy.

Table 1 shows the binary 1:1, 1:2, or 2:1 and 2:2 electrolyte systems for which high temperature data are available. Because there is currently little theoretical guidance for the temperature and pressure dependencies of the ion-interaction parameters, arbitrary analytical forms are being used to describe them. For example, for NaCl(aq) which has the most precisely known thermodynamic data base at high temperature and pressure, the full T and P dependencies of the activity and osmotic properties are described by 37 parameters (Eqns. 43–45, PITZER *et al.*, 1984). For other systems that have a more limited amount of available data, the equations will contain a smaller number of terms. However, there is still not one general form of equation that is being used. Perhaps as more precise measurements become available a common form will be found to be useful for all electrolytes.

Nevertheless, with these parameters for binary systems available at high temperatures, plus the equilibrium constants calculated from the chemical potentials

Table 1. Binary electrolyte solutions with available high temperature data

	Temperature range (°C)	Ref.
1:1 Electrolytes		
HCl	0 - 375	a,b
LiCl	0 - 250	c
NaCl	0 - 300	d
NaI	25 - 100	a
NaOH	0 - 350	e
KCl	0 - 250	c
CsF	25 - 100	a
CsCl	0 - 250	c
CsI	25 - 100	a
1:2 or 2:1 Electrolytes		
Li_2SO_4	0 - 225	f
Na_2SO_4	0 - 225	f
K_2SO_4	0 - 225	f
Cs_2SO_4	0 - 225	f
$MgCl_2$	25 - 200	g,h
$CaCl_2$	25 - 200	g,h
$SrCl_2$	25 - 200	g
2:2 Electrolytes		
$MgSO_4$	25 - 200	i

[a]Saluja et al. (1986).
[b]Holmes et al. (1987).
[c]Holmes and Mesmer (1983a).
[d]Pitzer et al. (1984).
[e]Pabalan and Pitzer (1987).
[f]Holmes and Mesmer (1986).
[g]Phutela et al. (1987).
[h]Holmes et al. (1978).
[i]Phutela and Pitzer (1986b).

of the solid, water, and aqueous species, the solubilities of some salts in binary systems can be calculated and compared with experimentally measured values. For ternary mixtures only the NaCl-KCl-H_2O system has data at high temperatures from which the mixing parameters, $\theta_{Na,K}$ and $\psi_{Na,K,Cl}$, could be determined (HOLMES *et al.*, 1978). Thus for the ternary systems, we have relied mainly on the solubility data in evaluating the mixing parameters.

The approach taken to calculate the equilibrium assemblage is based on solution of a set of non-linear mass-action equations involving equilibrium constants for all the relevant equilibria and an auxiliary set of linear mass-balance and charge-balance equations (*cf.*, CRERAR, 1975). The equations were solved iteratively by matrix inversion (*cf.*, FRANTZ *et al.*, 1981). This is a standard technique used also by other computer codes such as WATEQ (TRUESDELL and JONES, 1974), SOLMNEQ (KHARAKA and BARNES, 1973), and EQ3NR (WOLERY, 1983). Alternatively, the multiphase equilibria can be solved by means of a Gibbs free energy minimization algorithm as used by HARVIE and WEARE (1980).

CHOICE OF THERMODYNAMIC DATA IN THE Na-K-Mg-Cl-SO_4-OH-H_2O SYSTEM

If chemical potentials and entropies (or enthalpies) at the reference temperature and heat capacity functions are known, the chemical potentials at the temperature of interest can be derived using Eqn. (4). In the case of water, the chemical potential at T and P can be conveniently calculated from the equations of HAAR *et al.* (1984). For the solids, although there are some inaccuracies or lack of data at high temperatures, their data base is well-organized and given in the compilations of KELLEY (1960), STULL and PROPHET (1971), ROBIE *et al.* (1978), and WAGMAN *et al.* (1982).

The reference values at 25°C of the chemical potentials, enthalpies and entropies are listed in Table 2 for water, the aqueous ions and the solids considered in this study. The values of μ_f^0/RT were taken mostly from those tabulated by HARVIE and WEARE (1980). Those for the solids $MgCl_2 \cdot nH_2O$ were taken from WAGMAN *et al.* (1982). The reference chemical

Table 2. The chemical potentials, enthalpies of formation, and entropies at 298.15 K of the species and minerals of the Na-K-Mg-Cl-SO_4-OH-H_2O system and the temperature functions of the heat capacity of the solids.

					$C_p/R = a + bT + cT^{-2}$			
Substance	Formula	$-\mu_f^\circ/RT$	$-\Delta_f H^\circ/RT$	S°/R	a	$10^3 b$	$10^{-5} c$	T(K)range
Water	$H_2O(\ell)$	95.6635	115.304	8.409				
Hydroxide ion	$OH^-(aq)$	63.452	92.780	-1.293				
Chloride ion	$Cl^-(aq)$	52.955	67.432	6.778				
Sulfate ion	$SO_4^{2-}(aq)$	300.386	366.800	2.42				
Magnesium ion	$Mg^{2+}(aq)$	183.468	188.329	-16.64				
Sodium ion	$Na^+(aq)$	105.651	96.865	7.096				
Potassium ion	$K^+(aq)$	113.957	101.81	12.33				
Arcanite	$K_2SO_4(c)$	532.39	580.01	21.12	14.48	11.98	-2.14_4	298-856
Bischofite	$MgCl_2 \cdot 6H_2O(c)$	853.1	1008.11	44.03	29.08	29.56	-	298-385
Carnallite	$KCl \cdot MgCl_2 \cdot 6H_2O$	1020.3	1184.85	55.53	96.11	-180.4	-	273-423
Epsomite	$MgSO_4 \cdot 7H_2O(c)$	1157.74	(1366.27)	44.79	11.8	118	-	~273-~473
Halite	$NaCl(c)$	154.99	165.88	8.676	5.525	1.96_3	-	298-1073
Hexahydrite	$MgSO_4 \cdot 6H_2O(c)$	1061.37	(1244.79)	41.87	10.9	104	-	~273-~473
Kieserite	$MgSO_4 \cdot H_2O(c)$	579.18_4	649.34	(14.99)	6.89	31.05	-	~273-~473
Leonhardtite	$MgSO_4 \cdot 4H_2O(c)$	868.55	1007.13	(30.64)	9.39	74.8	-	~273-~473
-	$MgCl_2(c)$	238.74	258.71	10.78	9.511	0.714_6	-1.03_7	298-987
-	$MgCl_2 \cdot H_2O(c)$	347.66	389.94	16.505	10.95	9.788	-	298-650
-	$MgCl_2 \cdot 2H_2O(c)$	451.06	516.24	21.64	15.05	13.74	-	298-500
-	$MgCl_2 \cdot 4H_2O(c)$	654.93	766.06	31.75	22.56	21.65	-	298-450
-	$MgSO_4(c)$	472.26	518.33	11.02	6.71	16.30	-	298-700
Mirabilite	$Na_2SO_4 \cdot 10H_2O(c)$	1471.15	1475.75	71.21	10.65	196.0	-	~200-~450
Pentahydrite	$MgSO_4 \cdot 5H_2O(c)$	965.13	-	(35.7)	10.2	89.3	-	~273-~473
Sylvite	$KCl(c)$	164.84	176.034	9.934	5.575	2.011	-	~298-~700
Thenardite	$Na_2SO_4(c)$	512.39	559.55	17.99	13.16	13.70	-1.666	270-700

potential of the solid $MgSO_4 \cdot 7H_2O$ was adjusted slightly to fit the 25°C solubility in the binary system, while for $MgSO_4 \cdot 6H_2O$ and $MgSO_4 \cdot H_2O$, solids which are stable at higher temperatures, the reference chemical potentials were slightly adjusted to their lowest temperature solubility values in binary mixtures. In these cases the heat capacity and either the entropy and enthalpy, whichever is more accurately known (ROBIE *et al.*, 1978; KO and DAUT, 1979; WAGMAN *et al.*, 1982), were determined from measurements on the solid and were independent of the solubility. Thus their temperature dependence was predicted in each case.

Heat capacity (C_p) data from calorimetric measurements also exist for the majority of the solids considered here. These heat capacities are frequently expressed in the form of the Maier-Kelley equation

$$C_p = a + bT + cT^{-2}. \quad (18)$$

Values of a, b, and c are also given in Table 2, together with their temperature ranges of validity. These heat capacity parameters were taken mostly from KELLEY (1960). The coefficients given by KELLEY (1960) for KCl(s) are incorrect, so we used the equation of HOLMES and MESMER (1983a), which was fit to KCl(s) heat capacities tabulated by STULL and PROPHET (1971). The values for $Na_2SO_4 \cdot 10H_2O$(s) and Na_2SO_4(s) were fitted to data from BRODALE and GIAUQUE (1958, 1972), while those for $MgSO_4$(s) were derived from JANAF data (STULL and PROPHET, 1971). The C_p coefficients for $MgSO_4 \cdot H_2O$(s) and $MgSO_4 \cdot 6H_2O$(s) were derived from linear fits to the low temperature data of FROST *et al.* (1957) and COX *et al.* (1955), respectively.

Heat capacity data for $MgSO_4 \cdot 4H_2O$(s), $MgSO_4 \cdot 5H_2O$(s), and $MgSO_4 \cdot 7H_2O$(s) are not available. However, the contribution of each water molecule to the entropy or heat capacity of a hydrated solid is expected to be about the same. In fact, isothermal plots of the measured entropies and heat capacities of $MgCl_2 \cdot nH_2O$(s) and $MgSO_4 \cdot nH_2O$(s) *versus* the number of hydration waters show a linear trend (Fig. 1). Thus the C_p values for $MgSO_4 \cdot 4H_2O$(s), $MgSO_4 \cdot 5H_2O$(s), and $MgSO_4 \cdot 7H_2O$(s) were estimated from this linear dependence of the solid heat capacity with the number of waters of hydration.

The S^0/R values in Table 2 are mostly from WAGMAN *et al.* (1982). Entropy values enclosed in parentheses were calculated from the corresponding μ_f^0/RT and $\Delta H_f^0/RT$. For pentahydrite, its entropy is not known independently and was estimated from the entropy values of the other $MgSO_4$ hydrous salts. With the exception of $MgSO_4 \cdot H_2O$(s), values of $\Delta H_f^0/RT$ given in the table were calculated from μ_f^0/RT and S^0/R or were taken from WAGMAN *et al.* (1982) when the value of S^0/R is in parentheses. For $MgSO_4 \cdot H_2O$(s), the enthalpy of formation at 25°C was taken from KO and DAUT (1979).

For the double-salt carnallite ($KCl \cdot MgCl_2 \cdot 6H_2O$), we have also accepted the reference value of μ_f^0/RT determined by HARVIE and WEARE (1980) from solubility data. The heat capacity parameters and the reference entropy of carnallite were evaluated from solubility data in the ternary mixture KCl-MgCl$_2$-H$_2$O from 0–150°C tabulated by LINKE and SEIDELL (1965). The enthalpy of formation of carnallite given in Table 2 was calculated from the reference entropy determined above and the reference chemical potential.

In the case of the aqueous ions, additivity rules allow us to use standard state heat capacities of neutral electrolytes in calculating equilibrium constants for dissolution reactions. There is now an extensive array of data for the most important aqueous solutes extending upward in temperature, as shown in Table 1. PITZER *et al.* (1984) evaluated calorimetric data on sodium chloride solutions and gave parameters necessary to calculate standard state heat capacities of NaCl(aq) to 300°C and 1 kb. Similar evaluations have been done on KCl by HOLMES and MESMER (1983a) to 250°C, and on Na_2SO_4(aq) to 200°C by ROGERS and PITZER (1981) and to 225°C by HOLMES and MESMER (1986). We have used the latter values

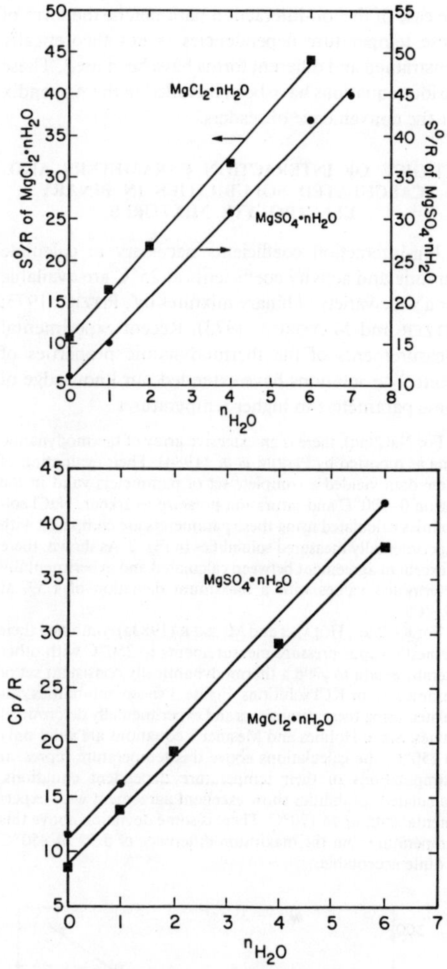

FIG. 1. (a) Entropies (S^0/R) and (b) heat capacities (C_p/R) at 25°C for the hydrous salts $MgCl_2 \cdot nH_2O$ and $MgSO_4 \cdot nH_2O$. The symbols represent data taken from sources given in the text, while the lines are best fits to these data points.

in order to be consistent with the usage of HOLMES and MESMER's (1986) activity coefficient parameters given in the same paper.

$MgSO_4$(aq) heat capacities have been measured by PHUTELA and PITZER (1986b) from 75°C to 200°C. These authors have combined their data with literature values at 25°C and reported the temperature dependence of the $MgSO_4$(aq) standard state heat capacities. A similar evaluation has been done by PHUTELA *et al.* (1987) for $MgCl_2$(aq) based on the data of SALUJA *et al.* (1987), LIKKE and BROMLEY (1973) and PERRON *et al.* (1974, 1981). In the case of K_2SO_4(aq), heat capacity data are not available at high temperatures. Standard state heat capacities for K_2SO_4(aq) were therefore calculated from the known values for NaCl(aq), KCl(aq), and Na_2SO_4(aq) as shown by

$$C_p^0(K_2SO_4) = 2C_p^0(KCl) + C_p^0(Na_2SO_4) - 2C_p^0(NaCl). \quad (19)$$

The above standard state heat capacities are usually reported as functions of temperature. However, as in the case of the ion-interaction parameters, the form of these temperature dependencies is not theoretically constrained and different forms have been used. These various equations have been included in the Appendix for the convenience of readers.

CHOICE OF INTERACTION PARAMETERS AND CALCULATED SOLUBILITIES IN BINARY ELECTROLYTE MIXTURES

Ion-interaction coefficients necessary to calculate osmotic and activity coefficients at 25°C are available for a wide variety of binary mixtures (*cf.*, PITZER, 1973; PITZER and MAYORGA, 1973). Recent experimental measurements of the thermodynamic properties of electrolyte solutions have extended our knowledge of these parameters to higher temperatures.

For NaCl(aq), there is an extensive array of thermodynamic data as reported by PITZER *et al.* (1984). Their evaluation of these data yielded a complete set of parameters valid in the region 0–300°C and saturation pressure to 1 kbar. NaCl solubilities calculated using these parameters are compared with experimentally measured solubilities in Fig. 2. As shown, there is excellent agreement between calculated and experimentally determined values with a maximum deviation of 1.5% at 275°C.

For KCl(aq), HOLMES and MESMER (1983a) combined their isopiestic vapor pressure measurements to 250°C with other literature data to yield a thermodynamically consistent set of parameters for KCl solutions. Figure 3 shows solubilities calculated using these parameters and experimentally determined values. Since Holmes and Mesmer's equations are valid only to 250°C, the calculations above this temperature represent extrapolations of their temperature dependent equations. Calculated solubilities show excellent agreement with experimental data up to 170°C. There is some deviation above this temperature, but the maximum difference of 6.5% at 250°C is quite acceptable.

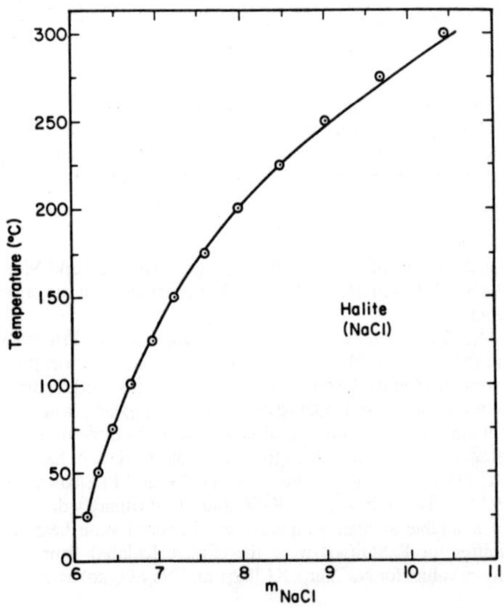

FIG. 2. Calculated halite (NaCl) solubilities in the binary NaCl-H_2O system compared to experiment. The solubility data from 75–300°C are from LIU and LINDSAY (1972), and those below 75°C are from LINKE and SEIDELL (1965).

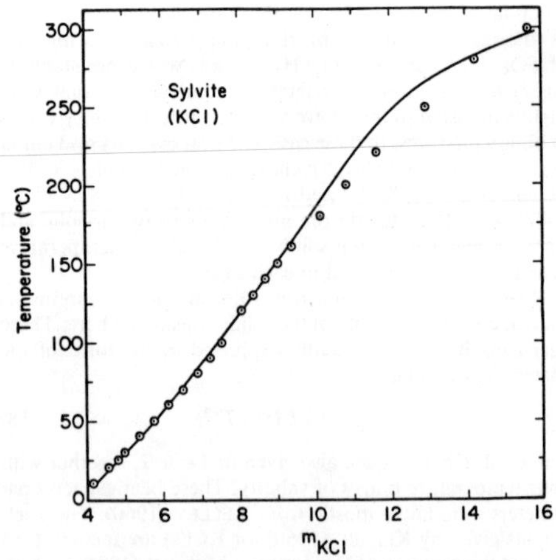

FIG. 3. Calculated sylvite (KCl) solubilities in the KCl-H_2O system compared to experimental data from LINKE and SEIDELL (1965).

For $MgCl_2$(aq), we used the temperature functions for the ion-interaction parameters of DE LIMA and PITZER (1983b), which were fit to the data of HOLMES *et al.* (1978) at 382, 414, 445, and 474 K, and to the 298 K data of RARD and MILLER (1981). Initial calculations using those equations yielded good agreement with experimental values up to 100°C, but above this temperature calculations indicated increasing deviation with increasing temperature, probably because the isopiestic data of HOLMES *et al.* (1978) went up only to 3.5 m $MgCl_2$, while the solubility of $MgCl_2 \cdot nH_2O$(s) is already 7.70 m $MgCl_2$ at 100°C. Because of the m^2 dependence of either osmotic or activity coefficients on C^ϕ, that parameter is much more sensitive to the measured values at higher molality. We therefore adjusted the C^ϕ_{MX} values at the higher temperatures to fit the activity coefficients at saturation concentrations, but still retaining the temperature equations of DE LIMA and PITZER (1983b) for $\beta^{(0)}_{MX}$ and $\beta^{(1)}_{MX}$. The change in C^ϕ_{MX} to fit the solubility also affects the calculated osmotic coefficient of $MgCl_2$ solutions. At 445 and 474 K, the standard deviations of calculated osmotic coefficients versus the data of HOLMES *et al.* (1978) are 0.0091 and 0.0049, respectively, compared with the stated experimental standard deviations of 0.0028 and 0.0029 at 445 and 474 K, respectively. With this adjustment the calculated solubilities show good agreement with experimental data up to 200°C, with a maximum deviation of 4% at 150°C, as shown in Fig. 4.

It should be noted that PHUTELA *et al.* (1987) also gave parameters for activity and osmotic coefficients of $MgCl_2$(aq) based on the integration of their derived heat capacity parameters. This probably yields a more accurate temperature derivative of the ion-interaction parameters at low molalities than simply fitting a temperature function to the osmotic data of HOLMES *et al.* (1978) and RARD and MILLER (1981). PHUTELA *et al.*'s parameters, however, severely underestimate the osmotic and activity coefficients when extrapolated to saturation concentrations, most likely because the heat capacity data used did not extend beyond 1 molal (SALUJA and LEBLANC, 1987; PHUTELA *et al.*, 1987).

In the case of Na_2SO_4(aq) we have used the equations of HOLMES and MESMER (1986) which were fit to data up to 225°C. Figure 5 shows that there is good agreement up to about 185°C, but there is a sudden decrease in calculated solubilities above this temperature, anticipating the measured solubility maximum at about 240°C (LINKE and SEIDELL, 1965). The temperature 185°C corresponds to a change in the stable Na_2SO_4(s) going from form (V) to form (IV) (BRO-

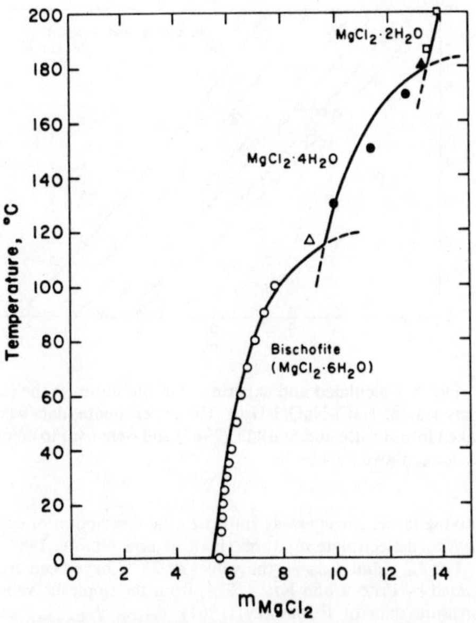

FIG. 4. Calculated solubilities in the $MgCl_2$-H_2O binary system compared to experimental data from LINKE and SEIDELL (1965). The triangles represent experimental data at the triple points. The C^{ϕ}_{MX} values for $MgCl_2$ above 100°C were adjusted on the basis of the solubility data.

DALE and GIAUQUE, 1972). We cannot presently explain this sudden dip in the solubility even if we consider the possibility of other metastable solids. It is possible we need more accurate solution data from which to evaluate the temperature dependency of the ion-interaction parameters at these higher temperatures.

For K_2SO_4(aq), ion-interaction coefficients are given by HOLMES and MESMER (1986) up to 225°C. Our calculated results show relatively good agreement with solubility data to

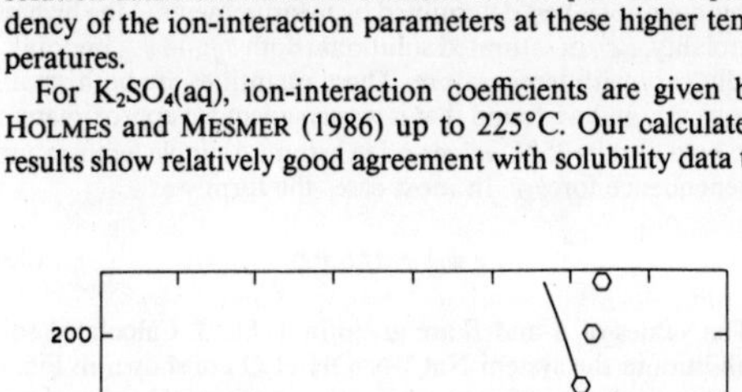

FIG. 5. Calculated solubilities in the Na_2SO_4-H_2O binary system compared to experiment. Solubility data for mirabilite ($Na_2SO_4 \cdot 10H_2O$) are from LINKE and SEIDELL (1965), while those for thenardite (Na_2SO_4) are from POTTER and CLYNNE (1978) below 100°C and SCHROEDER *et al.* (1935) above 100°C.

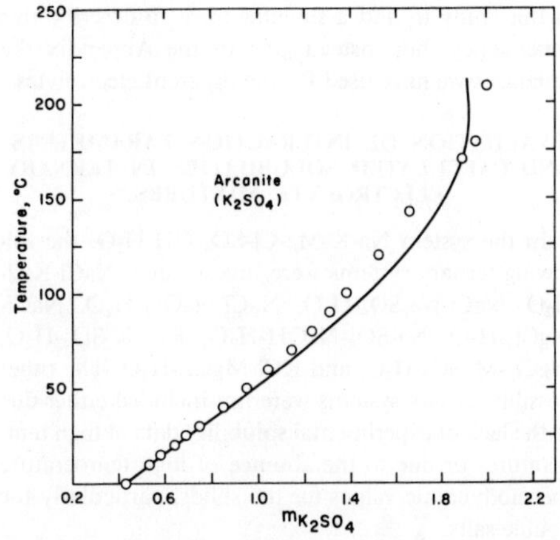

FIG. 6. Calculated and experimental solubilities of arcanite (K_2SO_4) in the binary system K_2SO_4-H_2O. The experimental data are from LINKE and SEIDELL (1965).

208°C, with a maximum deviation of 8% at 143°C, as indicated in Fig. 6. In the case of $MgSO_4$(aq) solutions, a comprehensive regression of heat capacity, enthalpy, and osmotic coefficient data by PHUTELA and PITZER (1986b) yielded parameters that are valid from 25–200°C. Figure 7 shows that calculated solubilities are in very good agreement with experimental data to 200°C.

As mentioned previously, various forms have been used to describe the temperature dependency of the ion-interaction coefficients. We have made no attempt

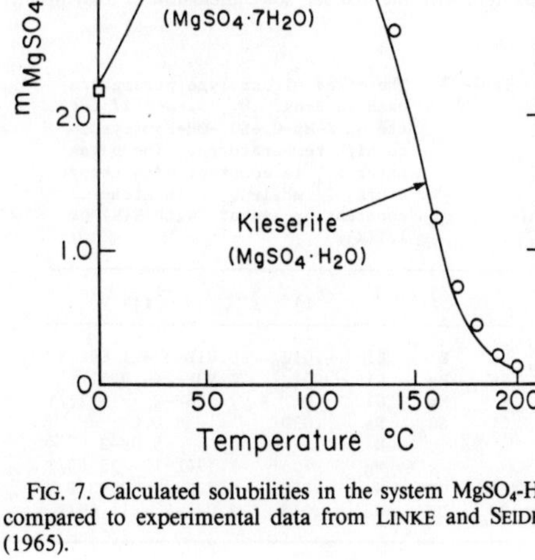

FIG. 7. Calculated solubilities in the system $MgSO_4$-H_2O compared to experimental data from LINKE and SEIDELL (1965).

in this study to find a singular form to describe this dependency, but instead give in the Appendix the equations we have used for the different electrolytes.

EVALUATION OF INTERACTION PARAMETERS AND CALCULATED SOLUBILITIES IN TERNARY ELECTROLYTE MIXTURES

In the system Na-K-Mg-Cl-SO_4-OH-H_2O, the following ternary systems were investigated: NaCl-KCl-H_2O, NaCl-Na_2SO_4-H_2O, NaCl-NaOH-H_2O, NaCl-$MgCl_2$-H_2O, Na_2SO_4-NaOH-H_2O, KCl-K_2SO_4-H_2O, $MgCl_2$-$MgSO_4$-H_2O, and KCl-$MgCl_2$-H_2O. The other possible ternary systems were not included either due to the lack of experimental solubility data at high temperatures or due to the absence of high temperature thermodynamic values for the solids, particularly the double salts.

For these ternary and for more complex systems, the interaction coefficients θ_{ij} and ψ_{ijk} complete the parameters necessary to describe the thermodynamic properties of electrolyte mixtures. These parameters are best evaluated from activity or osmotic coefficient data for common-ion mixtures. For example, the values of $\theta_{Na,K}$ and $\psi_{Na,K,Cl}$ may be determined from osmotic coefficient data for NaCl-KCl solutions. The θ_{ij} parameters are independent of the oppositely charged ions. Thus, all the data for Na^+-K^+ mixing are considered simultaneously (NaCl-KCl, Na_2SO_4-K_2SO_4, $NaNO_3$-KNO_3, etc.) to determine a single value of $\theta_{Na,K}$ and values of $\psi_{Na,K,Cl}$, ψ_{Na,K,SO_4}, etc. The solubility of a solid salt in a common-ion mixture depends on a ψ_{ijk} value, and the best values of ψ_{ijk} for many salts come from solubility data (HARVIE and WEARE, 1980; HARVIE *et al.*, 1984).

The values of these mixing terms at 25°C are available for a large number of systems and have been reported by various investigators (*cf.*, PITZER and KIM, 1974; PITZER, 1975; DOWNES and PITZER, 1976; HARVIE and WEARE, 1980; HARVIE *et al.*, 1984). It should be noted that two sets of mixing parameters are given in the literature for cases of unsymmetrical mixing. One set considers the effects of higher order electrostatic terms as discussed by PITZER (1975), while the other set neglects these effects. It is preferable to use values consistent with the first set, since inclusion of unsymmetric

Table 3. The mixed electrolyte parameters used in Eqns. 8, 9, and 17 for the Na-K-Mg-Cl-SO_4-OH-H_2O system to high temperatures. The parameter θ_{ij} is constant with temperature, while ψ_{ijk} is either constant or linear with T(K) or 1/T(K).

i	j	k	θ_{ij}	ψ_{ijk}
Na	K	Cl	-.012	-6.81E-3 + 1.68E-5T
Na	Mg	Cl	.07	1.99E-2 - 9.51/T
K	Mg	Cl	0	2.586E-2 - 14.27/T
Cl	SO_4	Na	.030	0.0
		K		-5.0E-3
		Mg		-1.174E-1 + 32.63/T
Cl	OH	Na	-.050	2.73E-2 - 9.93/T
SO_4	OH	Na	-.013	3.02E-2 - 11.69/T

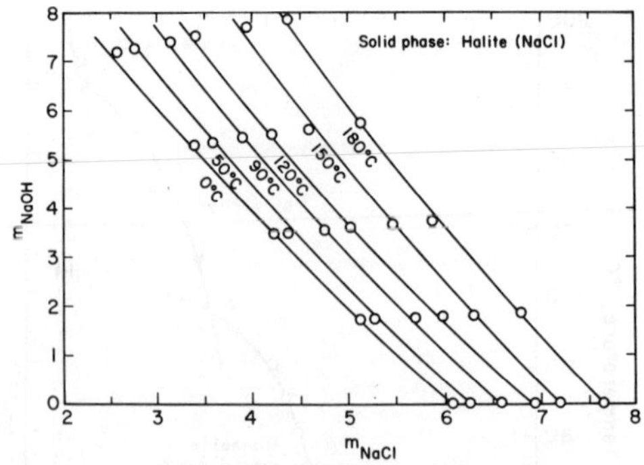

FIG. 8. Calculated and experimental solubilities in the ternary system NaCl-NaOH-H_2O. The experimental data were taken from LINKE and SEIDELL (1965) and were used to adjust $\psi_{Cl,OH,Na}$ above 25°C.

mixing terms considerably improves the description of concentrated electrolyte mixtures (HARVIE and WEARE, 1980).

For $\theta_{Na,K}$ and $\psi_{Na,K,Cl}$, the values at 25°C have been estimated by PITZER and KIM (1974) from the isopiestic vapor pressure data of ROBINSON (1961). θ_{Cl,SO_4}, $\psi_{Cl,SO_4,Na}$, and $\psi_{Cl,SO_4,K}$, which were also reported by PITZER and KIM (1974), were revised by DOWNES and PITZER (1976) based on their new osmotic coefficient data in the common ion-mixtures. For $\psi_{Cl,SO_4,Mg}$ we have used a value which gives good agreement with the solubility data at 25°C. We used the value of $\theta_{Na,Mg}$ reported by PITZER (1975), but we chose to accept the value of −0.012 for $\psi_{Na,Mg,Cl}$ evaluated by HARVIE and WEARE (1980) from solubility data instead of Pitzer's value of −0.010. For $\theta_{K,Mg}$, $\theta_{Cl,OH}$, $\theta_{SO_4,OH}$, $\psi_{K,Mg,Cl}$, $\psi_{Cl,OH,Na}$, and $\psi_{SO_4,OH,Na}$, the 25°C values have been evaluated by HARVIE *et al.* (1984) solely from solubility data.

Since the terms in ψ_{ijk} involve the second power of molality, this quantity is best determined by measurements at the highest molality, *i.e.*, in saturated solutions. Both θ_{ij} and ψ_{ijk} undoubtedly vary with temperature. These quantities are both small, however, and we found that it was an adequate approximation to hold θ_{ij} at its 25°C value and to assume a simple temperature dependence for ψ_{ijk}. In most cases the form was

$$\psi_{ijk} = A/T + B. \tag{20}$$

The values of A and B are given in Table 3. Calculated solubilities in the system NaCl-NaOH-H_2O are shown in Fig. 8 and indicate good agreement with experimental values from 0–180°C. Agreement is good to 200°C for Na_2SO_4-NaOH-H_2O as shown in Fig. 9, as well as for NaCl-$MgCl_2$-H_2O shown in Fig. 10. Figure 11 shows that calculated values are likewise very close to experimental solubilities in the system $MgSO_4$-$MgCl_2$-H_2O where solubility data extend to 100°C. For the KCl-$MgCl_2$-H_2O system, agreement is relatively good to 200°C, at least up to 8 molal $MgCl_2$ (corresponding to $I \geq 24$) as shown in Fig. 12.

In the two cases NaCl-Na_2SO_4-H_2O and KCl-K_2SO_4-H_2O, constant values of $\psi_{Cl,SO_4,Na}$ and $\psi_{Cl,SO_4,K}$ sufficed. Figures 13 and 14 show the results which extend to 200° and 100°C, respectively. At 200°C the small deviation shown for the binary Na_2SO_4-H_2O system appears also for the ternary NaCl-Na_2SO_4-H_2O and NaOH-Na_2SO_4-H_2O.

There are osmotic-coefficient data for the ternary NaCl-KCl-H_2O from HOLMES *et al.* (1979). They reported that $\psi_{Na,K,Cl}$ could be neglected and represented $\theta_{Na,K}$ by Eqn. (20) with A = −6.726 and B = 0.0039. This model fitted the solubility values from LINKE and SEIDELL (1965) up to 150°C but gave large deviations at higher temperature and overflow conditions at 200°C. This problem could have been solved

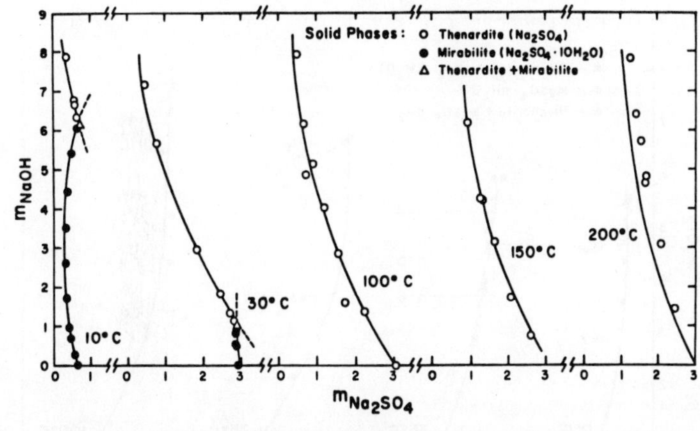

FIG. 9. Calculated solubilities in the Na$_2$SO$_4$-NaOH-H$_2$O system compared with experimental data from LINKE and SEIDELL (1965). The solubility data were used to adjust the values of $\psi_{SO_4,OH,Na}$ above 25°C.

by a small, positive ψ above 150°C, while retaining the equation for $\theta_{Na,K}$. We chose, however, to follow the pattern which was satisfactory for the other ternary systems with θ constant at its 25°C value −0.012 from PITZER and KIM (1974), and the following expression for ψ:

$$\psi_{Na,K,Cl} = -0.00681 + 1.68E-5T \tag{21}$$

at 25°C this equation gives the value −0.0018 determined by PITZER and KIM (1974). Figure 15 shows that there is excellent agreement through 150°C and that the small differences in ternary at 200°C are now no larger than those of the KCl-H$_2$O binary.

The choice of new parameters for the NaCl-KCl-H$_2$O system adjusted to saturation molalities reduces the accuracy of the equations in the lower concentration range. However, a comparison of calculated osmotic coefficients and values taken from HOLMES *et al.* (1978) at the highest experimental ionic strength and at the temperatures 474.15 and 445.25 K indicates only a small effect, with a deviation of 0.0067 and 0.0068 *versus* the stated experimental standard deviations of 0.0044 and 0.0040 at 474.15 and 445.25 K, respectively.

DISCUSSION

Most measurements of the thermodynamic properties of binary electrolyte systems were made on solutions well below the saturation molalities. Thus solubility calculations mostly represent extrapolations of the molality dependence of the Pitzer equations to concentrations outside the fitted range. The results presented above indicate that, in most cases, such extrapolations give a very good prediction of the activity and osmotic coefficients at the saturation concentration, particularly where the triple-ion interaction term,

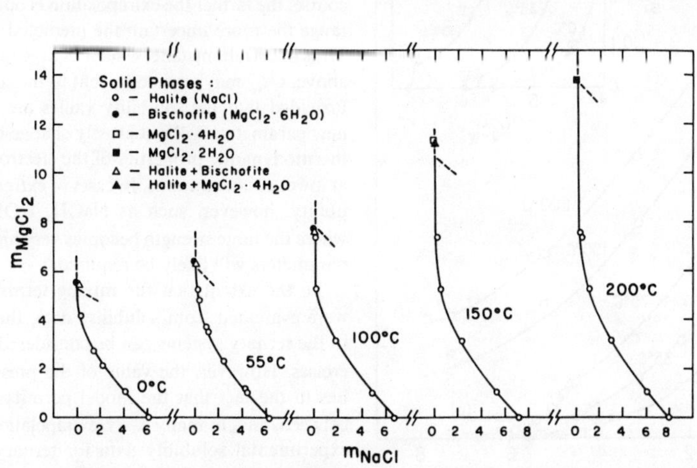

FIG. 10. Calculated and experimental solubilities in the NaCl-MgCl$_2$-H$_2$O system. Experimental data below 100°C are from LINKE and SEIDELL (1965), while those above 100°C are from AKHUMOV and VASIL'EV (1932). The dashed curves are extrapolations of the solubilities of either MgCl$_2 \cdot n$H$_2$O or NaCl into the supersaturated solution concentration of the other solid. The intersection of the curves denotes the calculated triple point. Values of $\psi_{Na,Mg,Cl}$ above 25°C were fit to the solubility data.

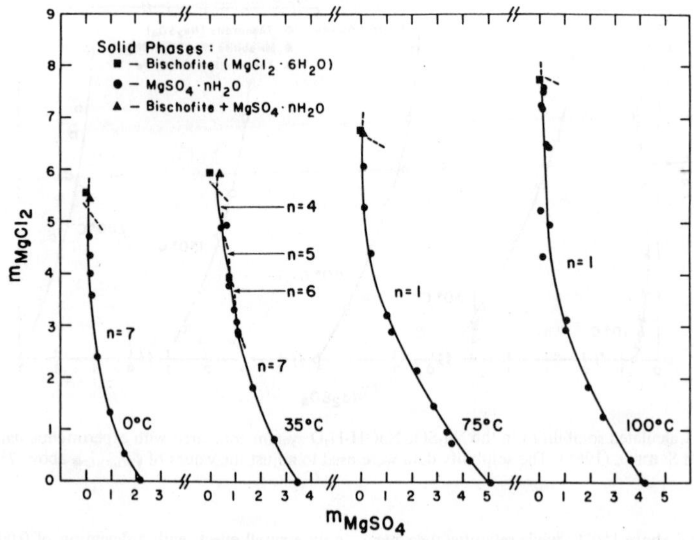

FIG. 11. Calculated solubilities in the system $MgCl_2$-$MgSO_4$-H_2O compared with experimental data taken from LINKE and SEIDELL (1965). The dashed curves are extrapolations of the solubilities of either $MgCl_2 \cdot nH_2O$ or $MgSO_4 \cdot nH_2O$ into the supersaturated solution concentration of the other solid. The intersection of the curves denotes the calculated triple point. The experimental data were used to adjust the values of $\psi_{SO_4,Cl,Mg}$.

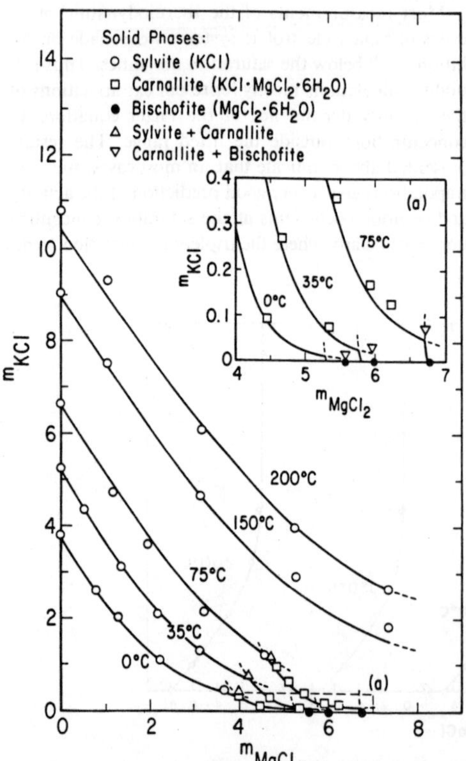

FIG. 12. Calculated solubilities in the system KCl-$MgCl_2$-H_2O compared with experimental data from LINKE and SEIDELL (1965). The inset shows an expanded view of the area delineated by box (a). The values of $\psi_{K,Mg,Cl}$ above 25°C were adjusted on the basis of the experimental data.

C^{ϕ}_{MX}, which essentially determines the molality dependence of the equations at high concentrations, is well-defined by data in moderate concentration solutions. The results for binary and ternary systems involving the 2-2 electrolyte $MgSO_4$ are particularly encouraging because they indicate that the ion-interaction model can successfully predict activity and osmotic coefficients in concentrated electrolyte mixtures to high temperatures without the complication of an explicit accounting for ambiguous populations of ion-pairs. Of course, the farther the extrapolation is outside the fitted range the more uncertain the predicted value will be. Thus in certain instances, such as $MgCl_2$(aq) discussed above, C^{ϕ}_{MX} may have to be refit to the solubility data. Provided that the solubility values are accurate, the new parameters will not grossly change the calculated thermodynamic properties of the electrolyte mixtures at lower concentrations. In cases of extremely high solubility, however, such as NaOH, KOH, or $CaCl_2$, where the ionic strength becomes very high, additional parameters will likely be required.

To the extent that the mixing terms, θ_{ij} and ψ_{ijk}, were evaluated from solubility data, the calculations in the ternary systems can be considered as fitting exercises. However, the value of the present approach lies in the fact that the model permits interpolation between, and, in many cases, extrapolation beyond the experimental solubility data for ternary systems. In addition, since the solubility data are adequately represented by equations using constant θ_{ij} values and ψ_{ijk} terms which are either constant or have a simple temperature dependence, the model offers an elegant approach to solubility calculations at high temperatures. More importantly, since no additional parameters need

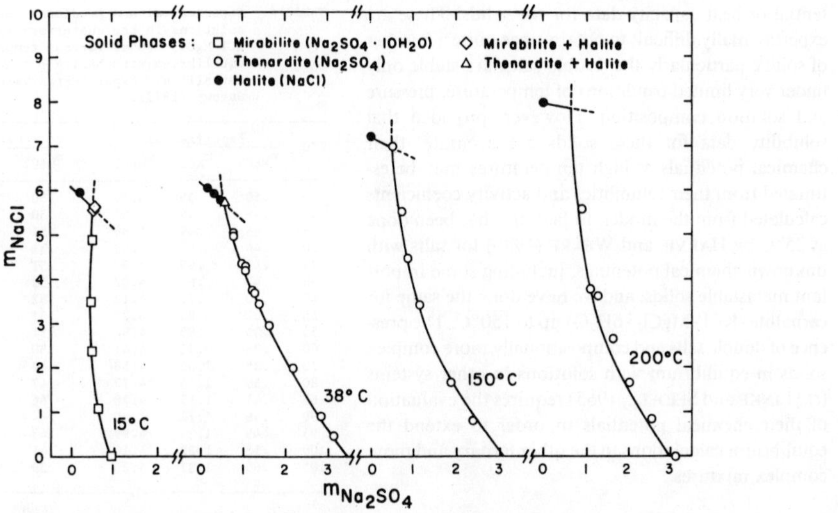

FIG. 13. Calculated solubilities in the NaCl-Na_2SO_4-H_2O system compared with experimental data from LINKE and SEIDELL (1965) to 100°C and from SCHROEDER *et al.* (1935) at 150°C and above. The intersections of isothermal curves represent calculated triple points.

to be determined for quaternary and more complex systems, these calculations can be extended to compositions relevant to natural fluids. As an initial test of the model and of our derived parameters, we calculated the solubility equilibria in the quaternary system NaCl-KCl-$MgCl_2$-H_2O. Figure 16 compares the predicted solubilities of halite and/or sylvite at 20, 55, and 90°C and at various molalities of $MgCl_2$ with experimental data given by KAYSER (1923), while Table 4 compares the predicted solution compositions up to 105°C at the quaternary invariant point halite + sylvite + carnallite + saturated solution with the experimental values of SEROWY (1923). The agreement between predicted and experimental values is quite acceptable considering the uncertainties in the $MgCl_2$ ion-interaction parameters and in the thermodynamic properties of carnallite.

These solubility calculations, of course, also depend on the availability of high-temperature chemical po

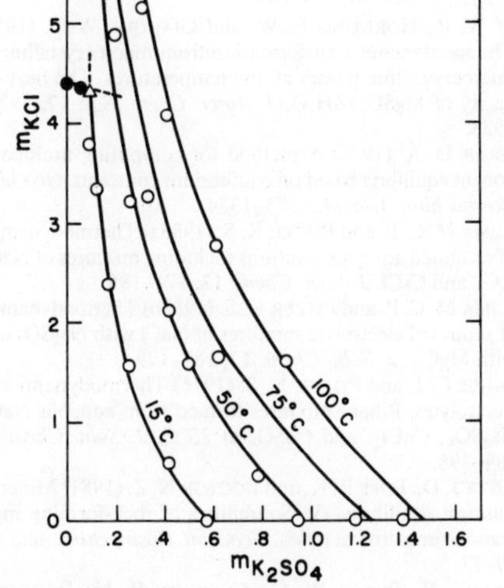

FIG. 14. Calculated and experimental solubilities in the KCl-K_2SO_4-H_2O system. The solubility values are from LINKE and SEIDELL (1965).

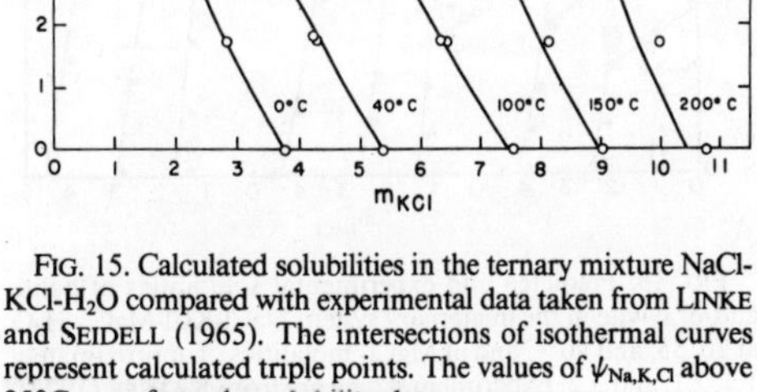

FIG. 15. Calculated solubilities in the ternary mixture NaCl-KCl-H_2O compared with experimental data taken from LINKE and SEIDELL (1965). The intersections of isothermal curves represent calculated triple points. The values of $\psi_{Na,K,Cl}$ above 25°C were fit to the solubility data.

tential or heat capacity data for the solids. These are experimentally difficult to obtain since quite a number of solids, particularly the double salts, are stable only under very limited conditions of temperature, pressure and solution composition. However, provided that solubility data for these solids are accurate, their chemical potentials at high temperatures may be estimated from their solubilities and activity coefficients calculated from the model. In fact, this has been done at 25°C by HARVIE and WEARE (1980) for salts with unknown chemical potentials, including some important metastable solids, and we have done the same for carnallite ($KCl \cdot MgCl_2 \cdot 6H_2O$) up to 150°C. The presence of double salts and compositionally more complex solids in equilibrium with solutions in other systems (*cf.*, LINKE and SEIDELL, 1965) requires the evaluation of their chemical potentials in order to extend the equilibrium calculations to the other ternary and more complex mixtures.

Table 4. Experimental and predicted compositions of solutions in the quaternary system NaCl-KCl-$MgCl_2$-H_2O at various temperatures in equilibrium with halite + sylvite + carnallite. Experimental values are from Serowy (1923).

T°C	Experimental			Calculated		
	m_{NaCl}	m_{KCl}	m_{MgCl_2}	m_{NaCl}	m_{KCl}	m_{MgCl_2}
0	.50	.35	3.77	.50	.31	3.83
10	.46	.45	3.84	.50	.40	3.89
15	.45	.49	3.88	.51	.45	3.92
20	.44	.54	3.92	.51	.50	3.95
25	.43	.59	3.97	.51	.56	3.98
30	.42	.64	4.02	.52	.61	4.02
40	.40	.73	4.13	.52	.73	4.09
50	.39	.83	4.26	.51	.85	4.18
55	.39	.88	4.34	.51	.91	4.23
60	.38	.93	4.41	.50	.98	4.29
70	.39	1.03	4.58	.49	1.10	4.42
80	.39	1.13	4.72	.47	1.21	4.57
83	.41	1.15	4.78	.46	1.24	4.63
90	.46	1.22	4.90	.44	1.31	4.76
95	.49	1.27	4.99	.43	1.35	4.87
100	.55	1.32	5.10	.41	1.39	4.99
105	.60	1.37	5.20	.39	1.43	5.12

CONCLUSION

Comparison of calculated *versus* experimentally determined solubilities of highly soluble minerals in binary and ternary electrolyte solutions show that the ion-interaction model of PITZER (1973, 1979) can be used to calculate brine-mineral equilibria at high temperatures. These calculations make use of virial coefficients derived from experimentally determined solution properties or from solubility data in binary and ternary mixtures. While there are currently no theoretical forms for the temperature dependencies of these coefficients, this study suggests that the solubilities in ternary systems are adequately represented by constant θ_{ij} values and ψ_{ijk} terms which are either constant with temperature or have a simple dependence with temperature. Since no additional parameters need to be determined for quaternary and more complex systems, these equilibrium calculations can be applied to compositions of geological significance.

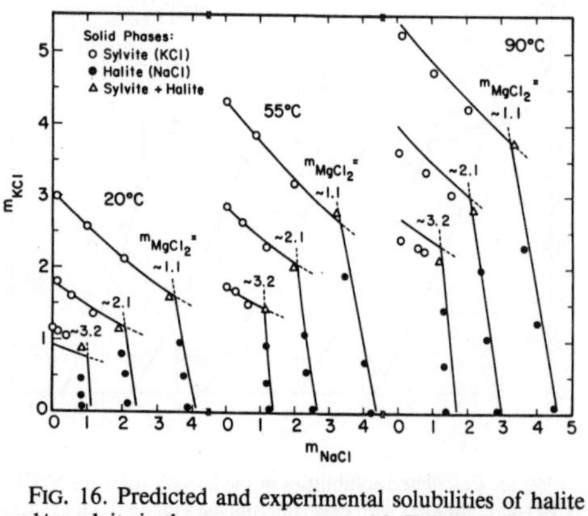

FIG. 16. Predicted and experimental solubilities of halite and/or sylvite in the quaternary system NaCl-KCl-$MgCl_2$-H_2O at 20, 55, and 90°C and at $MgCl_2$ molalities of approximately 1.1, 2.1, and 3.2. Experimental data are from KAYSER (1923).

Acknowledgements—We thank Drs. Robert E. Mesmer and Loren G. Hepler, who served as reviewers, for their valuable suggestions. This work was supported by the Director, Office of Energy Research, Office of Basic Energy Sciences, Division of Engineering and Geosciences of the U.S. Department of Energy under Contract No. DE-AC03-76SF00098.

Editorial handling: J. H. Weare

REFERENCES

AKHUMOV E. I. and VASIL'EV B. B. (1932) A study of aqueous solutions at elevated temperatures. II. *Zh. Obs. Khimii* **2,** 282–289 (in Russian).

BRODALE G. E. and GIAUQUE W. F. (1958) The heat of hydration of sodium sulfate. Low temperature heat capacity and entropy of sodium sulfate decahydrate. *J. Amer. Chem. Soc.* **80,** 2042–2044.

BRODALE G. E. and GIAUQUE W. F. (1972) The relationship of crystalline forms I, III, IV, and V of anhydrous sodium sulfate as determined by the third law of thermodynamics. *J. Phys. Chem.* **76,** 737–743.

COX W. P., HORNUNG E. W. and GIAUQUE W. F. (1955) The spontaneous transformation from macrocrystalline to microcrystalline phases at low temperatures. The heat capacity of $MgSO_4 \cdot 6H_2O$. *J. Amer. Chem. Soc.* **77,** 3935–3938.

CRERAR D. A. (1975) A method for computing multicomponent equilibria based on equilibrium constants. *Geochim. Cosmochim. Acta* **39,** 1375–1384.

DE LIMA M. C. P. and PITZER K. S. (1983a) Thermodynamics of saturated aqueous solutions including mixtures of NaCl, KCl, and CsCl. *J. Soln. Chem.* **12,** 171–185.

DE LIMA M. C. P. and PITZER K. S. (1983b) Thermodynamics of saturated electrolyte mixtures of NaCl with Na_2SO_4 and with $MgCl_2$. *J. Soln. Chem.* **12,** 187–199.

DOWNES C. J. and PITZER K. S. (1976) Thermodynamics of electrolytes. Binary mixtures formed from aqueous NaCl, Na_2SO_4, $CuCl_2$, and $CuSO_4$ at 25°C. *J. Soln. Chem.* **5,** 389–398.

FRANTZ J. D., POPP R. K. and BOCTOR N. Z. (1981) Mineral-solution equilibria—V. Solubilities of rock-forming minerals in supercritical fluids. *Geochim. Cosmochim. Acta* **45,** 69–77.

FROST G. B., BRECK W. G., CLAYTON R. N., REDDOCH A. H. and MILLER C. G. (1957) The heat capacities of the crystalline and vacuum dehydrated form of magnesium sulphate monohydrate. *Can. J. Chem.* **35,** 1446–1453.

GUEDDARI M., MONNIN C., PERRET D., FRITZ B. and TARDY Y. (1983) Geochemistry of the brines of Chott El Jerid in Southern Tunisia: application of Pitzer's equations. *Chem. Geol.* **39,** 165–178.

HAAR L., GALLAGHER J. S. and KELL G. S. (1984) *NBS/NRC Steam Tables:* Thermodynamic and Transport Properties and Computer Programs for Vapor and Liquid States of Water in SI Units. Hemisphere Publishing, 318p.

HARVIE C. E. (1981) Theoretical investigations in geochemistry and atom surface scattering. Ph.D. thesis, Univ. California, San Diego, 266p.

HARVIE C. E. and WEARE J. H. (1980) The prediction of mineral solubilities in natural waters: the Na-K-Mg-Ca-Cl-SO_4-H_2O system from zero to high concentration at 25°C. *Geochim. Cosmochim. Acta* **44,** 981–997.

HARVIE C. E., EUGSTER H. P. and WEARE J. H. (1982) Mineral equilibria in the six-component seawater system, Na-K-Mg-Ca-SO_4-Cl-H_2O at 25°C. II. Compositions of the saturated solutions. *Geochim. Cosmochim. Acta* **46,** 1603–1618.

HARVIE C. E., MØLLER N. and WEARE J. H. (1984) The prediction of mineral solubilities in natural waters: The Na-K-Mg-Ca-H-Cl-SO_4-OH-HCO_3-CO_3-CO_2-H_2O system to high ionic strengths at 25°C. *Geochim. Cosmochim. Acta* **48,** 723–751.

HOLMES H. F. and MESMER R. E. (1983a) Thermodynamic properties of aqueous solutions of the alkali metal chlorides to 250°C. *J. Phys. Chem.* **87,** 1242–1255.

HOLMES H. F. and MESMER R. E. (1983b) Isopiestic studies of aqueous solutions at elevated temperatures VII. $MgSO_4$ and $NiSO_4$. *J. Chem. Thermodyn.* **15,** 709–719.

HOLMES H. F. and MESMER R. E. (1986) Thermodynamics of aqueous solutions of the alkali metal sulfates. *J. Soln. Chem.* **15,** 495–518.

HOLMES H. F., BAES C. F. and MESMER R. E. (1978) Isopiestic studies of aqueous solutions at elevated temperatures: 1. KCl, $CaCl_2$ and $MgCl_2$. *J. Chem. Thermodyn.* **10,** 983–996.

HOLMES H. F., BAES C. F. and MESMER R. E. (1979) Isopiestic studies of aqueous solutions at elevated temperatures: 2. NaCl + KCl mixtures. *J. Chem. Thermodyn.* **11,** 1035–1050.

HOLMES H. F., BUSEY R. H., SIMONSON J. M., MESMER R. E., ARCHER D. G. and WOOD R. H. (1987) The enthalpy of dilution of HCl(aq) to 648 K and 40 MPa. Thermodynamic properties. *J. Chem. Thermodyn.* **19** (in press).

KAYSER E. (1923) Die ersatzzahlen inkonstanter Lösungen über Kaliumchlorid und Natriumchlorid. *Kali* **17,** 1–9.

KELLEY K. K. (1960) Contributions to the data on theoretical metallurgy. XIII. High-temperature heat-content, heat-capacity, and entropy data for the elements and inorganic compounds. *U.S. Bur. Mines Bull. 584,* 232p.

KHARAKA Y. K. and BARNES I. (1973) SOLMNEQ: Solution-Mineral Equilibrium Computations. U.S. Geol. Survey Computer Contributions Publication, 215–899.

KO H. C. and DAUT G. E. (1979) Enthalpies of formation of α- and β-magnesium sulfate and magnesium sulfate monohydrate. *U.S. Bureau of Mines Report of Investigations 8409,* 8p.

LANGMUIR D. and MELCHIOR D. (1985) The geochemistry of Ca, Sr, Ba and Ra sulfates in some deep brines from the Palo Duro Basin, Texas. *Geochim. Cosmochim. Acta* **49,** 2423–2432.

LIKKE S. and BROMLEY L. A. (1973) Heat capacities of aqueous NaCl, KCl, $MgCl_2$, $MgSO_4$, and Na_2SO_4 solutions between 80° and 200°C. *J. Chem. Eng. Data* **18,** 189–195.

LINKE W. F. and SEIDELL A. (1965) *Solubilities of Inorganic and Metal Organic Compounds,* 4th ed. Amer. Chem. Soc., vols. 1 and 2.

LIU C. and LINDSAY W. T. JR. (1972) Thermodynamics of sodium chloride solutions at high temperatures. *J. Soln. Chem.* **1,** 45–69.

MEIJER J. A. M. and VAN ROSMALEN G. M. (1984) Solubilities and supersaturation of calcium sulfate and its hydrates in seawater. *Desalination* **51,** 255–305.

MONNIN C. and SCHOTT J. (1984) Determination of the solubility products of sodium carbonate minerals and an application to trona deposition in Lake Magadi (Kenya). *Geochim. Cosmochim. Acta* **48,** 571–581.

PABALAN R. T. and PITZER K. S. (1987) Thermodynamics of NaOH(aq) in hydrothermal solutions. *Geochim. Cosmochim. Acta* **51,** 829–837.

PERRON G. and DESNOYERS J. E. (1974) Apparent molal volumes and heat capacities of alkaline earth chlorides in water at 25°C. *Can. J. Chem.* **52,** 3738–3741.

PERRON G., ROUX A. and DESNOYERS J. E. (1981) Heat capacities and volumes of NaCl, $MgCl_2$, $CaCl_2$, and $NiCl_2$ up to 6 molal in water. *Can. J. Chem.* **59,** 3049–3054.

PHUTELA R. C. and PITZER K. S. (1986a) Thermodynamics of electrolyte mixtures. Enthalpy and the effect of temperature on the activity coefficient. *J. Soln. Chem.* **15,** 649–662.

PHUTELA R. C. and PITZER K. S. (1986b) Heat capacity and other thermodynamic properties of aqueous magnesium sulfate to 473 K. *J. Phys. Chem.* **90,** 895–901.

PHUTELA R. C., PITZER K. S. and SALUJA P. P. S. (1987) Thermodynamics of aqueous magnesium chloride, calcium chloride and strontium chloride at elevated temperatures. *J. Chem. Eng. Data* **32,** 76–80.

PITZER K. S. (1973) Thermodynamics of electrolytes. I. Theoretical basis and general equations. *J. Phys. Chem.* **77,** 268–277.

PITZER K. S. (1975) Thermodynamics of electrolytes. V. Effects of higher-order electrostatic terms. *J. Soln. Chem.* **4,** 249–265.

PITZER K. S. (1979) Theory: ion interaction approach. In *Activity Coefficients in Electrolyte Solutions* (ed. R. PYTKOWICZ), pp. 157–208, CRC.

PITZER K. S. (1983) Thermodynamics of unsymmetrical electrolyte mixtures: Enthalpy and heat capacity. *J. Phys. Chem.* **87,** 2360–2364.

PITZER K. S. (1986) Theoretical considerations of solubility with emphasis on mixed electrolytes. *Pure Appl. Chem.* **58,** 1599–1610.

PITZER K. S. and MAYORGA G. (1973) Thermodynamics of electrolytes: 2. Activity and osmotic coefficients for strong electrolytes with one or both ions univalent. *J. Phys. Chem.* **77,** 2300–2308.

PITZER K. S. and KIM J. J. (1974) Thermodynamics of electrolytes. IV. Activity and osmotic coefficients for mixed electrolytes. *J. Amer. Chem. Soc.* **96,** 5701–5707.

PITZER K. S., PEIPER J. C. and BUSEY R. H. (1984) Thermodynamic properties of aqueous sodium chloride solutions. *J. Phys. Chem. Ref. Data* **13,** 1–102.

POTTER R. W. II and CLYNNE M. A. (1978) Solubility of highly soluble salts in aqueous media—Part 1. NaCl, KCl, $CaCl_2$, Na_2SO_4, and K_2SO_4 solubilities to 100°C. *J. Res. U.S. Geol. Survey* **6,** 701–705.

RARD J. A. and MILLER D. G. (1981) Isopiestic determination of the osmotic and activity coefficients of aqueous $MgCl_2$ solutions at 25°C. *J. Chem. Eng. Data* **26,** 38–43.

ROBIE R. A., HEMINGWAY B. S. and FISCHER J. R. (1978) Thermodynamic properties of minerals and related substances at 298.15K and 1 bar pressure and at higher temperatures. *U.S. Geol. Survey Bull. 1452,* 456p.

RÓBINSON R. A. (1961) Activity coefficients of sodium chloride and potassium chloride in mixed aqueous solutions at 25°C. *J. Phys. Chem.* **65,** 662–667.

ROGERS P. S. Z. and PITZER K. S. (1981) High-temperature thermodynamic properties of aqueous sodium sulfate solutions. *J. Phys. Chem.* **85,** 2886–2895.

ROGERS P. S. Z. and PITZER K. S. (1982) Volumetric properties of aqueous sodium chloride solutions. *J. Phys. Chem. Ref. Data* **11,** 15–81.

SALUJA P. P. S. and LEBLANC J. C. (1987) Apparent molar heat capacities and volumes of aqueous solutions of $MgCl_2$, $CaCl_2$ and $SrCl_2$ at elevated temperatures. *J. Chem. Eng. Data* **32,** 72–76.

SALUJA P. P. S., PITZER K. S. and PHUTELA R. C. (1986)

High-temperature thermodynamic properties of several 1:1 electrolytes. *Can. J. Chem.* **64,** 1328–1335.
SCHROEDER W. C., GABRIEL A. and PARTRIDGE E. P. (1935) Solubility equilibria of sodium sulfate at temperatures 150 to 350°C. I. Effect of sodium hydroxide and sodium chloride. *J. Amer. Chem. Soc.* **57,** 1539–1546.
SEROWY (1923) Die Polythermen der Viersalzpunkte des Chlorkaliumfeldes im quinären System ozeanischer Salzablagerungen; ihre teilweise Nachprüfung und Vervollständigung bis zu Temperaturen über 100°. *Kali* **17,** 345–350.
STULL D. R. and PROPHET H. (1971) *JANAF Thermochemical Tables,* 2nd ed. Natl. Stand. Ref. Data Ser., U.S. Natl. Bur. Stand. **37,** 1141p.
TRUESDELL A. H. and JONES B. F. (1974) WATEQ, a computer program for calculating chemical equilibria of natural waters. *J. Res. U.S. Geol. Survey* **2,** 233–248.
WAGMAN D. D., EVANS W. H., PARKER V. B., SCHUMM R. H., HALOW I., BAILEY S. M., CHURNEY K. L. and NUTTALL R. L. (1982) The NBS tables of chemical thermodynamic properties. Selected values for inorganic and C_1 and C_2 organic substances in SI units. *J. Phys. Chem. Ref. Data* **11,** Suppl. 2, 392p.
WOLERY T. J. (1983) EQ3NR, A computer program for geochemical aqueous speciation-solubility calculations: User's guide and documentation. Lawrence Livermore Natl. Lab., CA, UCRL-53414, 191p.
WOOD J. R. (1976) Thermodynamics of brine-salt equilibria—II. The system NaCl-KCl-H_2O from 0 to 200°C. *Geochim. Cosmochim. Acta* **40,** 1211–1220.

APPENDIX

The following are the temperature functions for the parameters of the ion-interaction model of PITZER (1973, 1979) and for the standard state heat capacities of aqueous electrolytes reported by various investigators. The pressure dependencies of the ion-interaction coefficients are currently neglected except for NaCl(aq) and NaOH(aq), which have the requisite *PVT* data available in sufficient quantity to allow their evaluation. In the case of standard state heat capacities, only NaCl(aq) have both *P*- and *T*-dependent functions. The standard state heat capacities of NaOH(aq) are not represented since its solubilities are beyond the valid concentration range of the present model. In all the equations, T_R and T refer to the reference temperature of 298.15 K and the temperature of interest, respectively. Standard state heat capacities are in units of J K^{-1} mol^{-1}.

NaCl(aq)

Ion-interaction parameters—PITZER *et al.* (1984), 0–300°C and saturation pressure to 1 kbar. *P* refers to pressure in bars.

$$f(T) = Q1/T + Q2 + Q3P + Q4P^2 + Q5P^3 + Q6\ln(T) + (Q7 + Q8P + Q9P^2 + Q10P^3)T + (Q11 + Q12P + Q13P^2)T^2 + (Q14 + Q15P + Q16P^2 + Q17P^3)/(T - 227) + (Q18 + Q19P + Q20P^2 + Q21P^3)/(680 - T)$$

	$\beta^{(0)}$	$\beta^{(1)}$	C^{ϕ}_{MX}
Q1	−656.81518	119.31966	−6.1084589
Q2	24.86913	−4.8309327E − 1	4.0217793E − 1
Q3	5.3812753E − 5	0	2.2902837E − 5
Q4	−5.5887470E − 8	0	0
Q5	6.5893263E − 12	0	0
Q6	−4.4640952	0	−7.5354649E − 2
Q7	1.1109914E − 2	1.4068095E − 3	1.5317673E − 4
Q8	−2.6573399E − 7	0	−9.0550901E − 8
Q9	1.7460070E − 10	0	0
Q10	1.0462619E − 14	0	0
Q11	−5.3070129E − 6	0	−1.5386008E − 8
Q12	8.6340233E − 10	0	8.6926600E − 11
Q13	−4.1785962E − 13	0	0
Q14	−1.579366	−4.2345814	3.5310414E − 1
Q15	2.2022821E − 3	0	−4.3314252E − 4
Q16	−1.3105503E − 7	0	0
Q17	−6.3813683E − 11	0	0
Q18	9.7065780	0	−9.1871455E − 2
Q19	−2.6860396E − 2	0	5.1904777E − 4
Q20	1.5344744E − 5	0	0
Q21	−3.2153983E − 9	0	0

Standard state heat capacity—the following equation was fit to values from 0–300°C and at 1 bar or saturation pressure tabulated by PITZER *et al.* (1984, Table A-4). For values at other pressures, the reader is referred to the tables and equations given by PITZER *et al.* (1984) which are valid in the range 0–300°C and to 1 kbar pressure.

$$C_p^0 = -1.848175E6 + 4.411878E7/T + 3.390654E5\ln(T) - 8.893249E2T + 4.005770E - 1T^2 - 7.244279E4/(T - 227) - 4.098218E5/(647 - T)$$

KCl(aq)

Ion-interaction parameters—HOLMES and MESMER (1983a), 0–250°C.

$$f(T) = Q1 + Q2(1/T - 1/T_R) + Q3\ln(T/T_R) + Q4(T - T_R) + Q5(T^2 - T_R^2) + Q6\ln(T - 260)$$

	$\beta^{(0)}$	$\beta^{(1)}$	C^{ϕ}_{MX}
Q1	0.04808	0.0476	−7.88E − 4
Q2	−758.48	303.9	91.270
Q3	−4.7062	1.066	0.58643
Q4	0.010072	0	−0.0012980
Q5	−3.7599E − 6	0	4.9567E − 7
Q6	0	0.0470	0

Standard state heat capacity—HOLMES and MESMER (1983a), 0–250°C.

$$C_p^0 = -991.51 + 5.56452T - 0.00852996T^2 - 686/(T - 270)$$

$MgCl_2$(aq)

Ion-interaction parameters—DE LIMA and PITZER (1983b), with equation for C^{ϕ}_{MX} modified to fit the solubility data, 25–200°C.

$$f(T) = Q1T^2 + Q2T + Q3$$

	$\beta^{(0)}$	$\beta^{(1)}$	C^{ϕ}_{MX}
Q1	5.93915E − 7	2.60169E − 5	2.41831E − 7
Q2	−9.31654E − 4	−1.09438E − 2	−2.49949E − 4
Q3	0.576066	2.60135	5.95320E − 2

Standard state heat capacity—PHUTELA *et al.* (1987), 25–180°C.

$$C_p^0 = -7.39872E6/T + 7.96487E4 - 3.25868E2T + 5.98722E - 1T^2 - 4.21187E - 4T^3$$

Na_2SO_4(aq)

Ion-interaction parameters—HOLMES and MESMER (1986), 0–225°C. Note: these parameters are consistent with a value of $\alpha_1 = 1.4$, instead of the usual value of 2.0.

$$f(T) = Q1 + Q2(T_R - T_R^2/T) + Q3(T^2 + 2T_R^3/T - 3T_R^2) + Q4(T + T_R^2/T - 2T_R) + Q5[\ln(T/T_R) + T_R/T - 1] + Q6\{1/(T - 263) + (263T - T_R^2)/[T(T_R - 263)^2]\} + Q7\{1/(680 - T) + (T_R^2 - 680T)/[T(680 - T_R)^2]\}$$

D. Other Condensed-State Papers

[Reprint from the Journal of the American Chemical Society, 59, 2365 (1937).]

[CONTRIBUTION FROM THE CHEMICAL LABORATORY OF THE UNIVERSITY OF CALIFORNIA]

The Heats of Ionization of Water, Ammonium Hydroxide, Carbonic, Phosphoric, and Sulfuric Acids. The Variation of Ionization Constants with Temperature and the Entropy Change with Ionization

BY KENNETH S. PITZER[1]

In addition to their value as thermochemical data, these heats of ionization make possible the calculation of the changes of entropy with ionization, and the entropies of some of the ions involved. The latter calculations will be presented in another paper. Also it is found that certain approximate generalizations can be made concerning the change in entropy and heat capacity with ionization, and that these results lead to equations for the variation of ionization constants with temperature.

Throughout this paper a defined calorie equal to 4.1833 international joules will be used, and extrapolations to infinite dilution will be made consistent with the Debye–Hückel theory. The uncertainties given for the various results are from two to three times the probable errors and should, therefore, outline a region within which the true value almost certainly lies. An attempt has been made to include all possible sources of error, including questionable assumptions and estimates made as part of the calculations.

Apparatus.—The chief aim in the design of this calorimeter was the attainment of high sensitivity in temperature determination. This makes possible heat of reaction measurements at sufficiently low concentration so that heats of dilution may be estimated without too great uncertainty. Another feature is the economy in electrical equipment.

(1) Shell Research Fellow, Academic Year, 1936–1937.

The calorimeter itself consists of a Pyrex vacuum flask with a ground glass joint in the neck. Into this fits a stopper which is also evacuated, and which carries the electrical leads and stirrer through a central tube of small diameter. Thus, when the flask is so filled that the water surface is just up in the small central tube, there is little opportunity for evaporation. The flask is 25 cm. long and has a volume of 885 ml. The entire unit is completely submerged in a water thermostat controlled to within 0.003° during a single run. For heat of solution measurements the sample is placed in a very thin glass bulb sealed with paraffin on to a glass rod which in turn slides through the hollow stirrer shaft. An additional propeller is placed just above the sample where it will be most effective after the bulb has been broken against the bottom of the flask. When two solutions are to be mixed one is used at relatively high concentration and placed in a glass tube with silver disks sealed into its ends by means of paraffin. This is supported just below the stirrer through which a rod may be thrust, breaking the seals.

A resistance thermometer consisting of 96 ohms of no. 38 silk insulated copper wire and a heater composed of 100 ohms of no. 30 manganin are wound on a silver tube 5.2 cm. in diameter and 16 cm. long. The leads are of no. 24 manganin and no. 24 copper, respectively, and single in each case. The windings are afforded protection by a cover of thin silver welded in place. This thermometer serves also as a stirrer tube which with its large surface gives excellent heat exchange.

The electrical circuits are shown in Fig. 1. The three constant arms of the thermometer bridge are of manganin, wound in a manner similar to the thermometer itself and submerged in a three-liter oil-bath. The heat exchange is such that a current of 0.01 amp. may be used if desired. The connections to the thermometer have so low a resistance that moderate temperature changes introduce a negligible error. The only switch contacts in this circuit are in the parallel, adjustable resistance, R, which is a high grade Leeds and Northrup resistance box with 10,000, 1000, 100, and 10-ohm diais. With this set at 20,000 ohms or more, contact resistances are estimated to introduce errors of less than 0.00003° in the temperatures.

Although the use of a measuring current of 0.01 amp. makes the resistance thermometer sensitive to 0.00001°, and the temperature drift before and after heating is consistent to about this amount, the results of calibration runs with a temperature rise of 0.1° spread over about 0.0003°, which suggests that the temperature readings after comparatively large changes are reproducible only to that accuracy. For this reason a smaller measuring current, say 0.003 amp., may be used without loss of accuracy and with a reduction in heating.

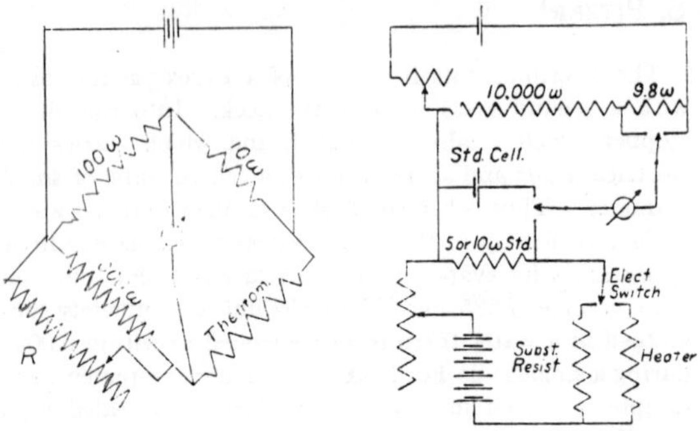

Fig. 1.—Resistance thermometer and heater circuits for the calorimeter.

The heater circuit is very simple, its chief requirements being that the resistances of the heater and the comparing standard be known accurately. These were calibrated by comparison with standard resistances of nearly equal resistance, a White potentiometer being used for this purpose. The electrically actuated timing switch is so connected to a timing circuit from the University Observatory that it may be thrown on any given second with an error of not more than one or two hundredths of a second.

Materials.—The chemicals used were all of "C. P." grade and were purified further if necessary. The hydrochloric acid was redistilled in an all-glass still, diluted with freshly distilled water, and stored in a bottle which previously had contained hydrochloric acid for a period of at least one year. Analysis by weighing precipitated silver chloride gave as the average of three determinations 1.1414 ± 0.0001 moles of hydrochloric acid per kg. of soln. The series of experiments on the heat of ionization of water was completed within three weeks after making up a fresh solution, thus avoiding any possible errors due to reaction of the acid with the glass. The sodium hydroxide solution was carefully freed from carbonate by the addition of a very slight excess of barium hydroxide. A paraffin-lined bottle was used to hold the sodium hydroxide solution, which was analyzed by comparison with the hydrochloric acid. Weight burets were used except for experiments where the precision needed was only about 1%. Unless the presence of carbon dioxide was of no significance, the distilled water was boiled and was protected from the air thereafter. The preparation of the other solutions which were used was carried out in a similar fashion, except that in several cases less precise analyses sufficed.

The Heat of Ionization of Water.—The best available value for the heat of ionization of water is that obtained by Rossini,[2] after a review of the available data, principally that of Richards and co-workers. The value thus obtained for 18° (ΔH° = 13,709 ± 16 cal.) seems quite satisfactory; however, the temperature coefficient is somewhat uncertain. Rossini[3] has also reviewed the available data on the heat capacities of aqueous solutions, but the extrapolations to infinite dilution were carried out in a manner not entirely consistent with the Debye–Hückel theory. The recent work of Gulbransen and Robinson[4] and Young and Machin[5] on the relative heat capacity of sodium chloride solutions establishes the validity of the limiting form of the Debye–Hückel theory for heat capacities in the very dilute region, and, moreover, gives probably the best value for the limiting slope. Consequently, it seemed desirable to recalculate the heat capacity of the pertinent aqueous solutions making use of this more recent information. The data for lithium, sodium, and potassium chlorides and hydroxides, and hydrochloric acid appeared most reliable in the dilute region and were therefore employed. The experimental values of the apparent molal heat capacity (Φ_c) were plotted against the square root of the molality as in Fig. 2. Three requirements were imposed in drawing the curves: first, that all experimental data be fitted within their reasonable uncertainties; second, that the limiting slope $d\Phi_c/d\sqrt{m}$ equal 8.0 (this is an average value obtained from both heat of dilution and dielectric constant data); and, third, that the heat capacities at infinite dilution be the sums of those of the individual ions. These criteria fortunately leave but little uncertainity in the course of the heat

Table I

Molal Heat Capacities of Aqueous Ions at Infinite Dilution and 25°

(Potassium and Chloride Assumed Equal)

Ion	Φ_c^0, cal./deg.	Ion	Φ_c^0, cal./deg.
K^+	−14.1	H^+	−15.9
Na^+	− 7.5	OH^-	−16.1
Li^+	− 2.4	Cl^-	−14.1

(2) Rossini, *Bur. Standards J. Research*, **6**, 847 (1931).

(3) Rossini, *ibid.*, **7**, 47 (1931).

(4) Gulbransen and Robinson, This Journal, **56**, 2637 (1934).

(5) Young and Machin, *ibid.*, **58**, 2254 (1936).

capacity curves. The values for infinite dilution are summarized in Table I.

Combining the values in Table I for hydrogen and hydroxide ions with the heat capacity of water, one obtains for the reaction $H_2O = H^+ + OH^-$, $\Delta C_p^0 = -50.0$ cal./deg. at 25° and, using Rossini's results, $d(\Delta C_p^0)/dT = 0.3$ and $\Delta H^0 = 13{,}709$ cal./mole at 18°, one finds $\Delta H^0 = 13{,}352 - 50.0(t - 25) + 0.15(t - 25)^2$ cal./mole.

While the error in the above value could hardly be very large, nevertheless an accurate experimental value for 25° seemed desirable. Consequently, a single series of experiments was carried out with the greatest care. The reactions whose heats were measured are the following ·

$$HCl{\cdot}46.5H_2O + NaOH{\cdot}1065.5H_2O = NaCl{\cdot}1113H_2O \quad (1)$$

$$HCl{\cdot}46.5H_2O + 1165.5H_2O = HCl{\cdot}1112H_2O. \quad (2)$$

The results for these reactions may then be combined with the heats of dilution of $HCl{\cdot}1112H_2O$,[6] $NaOH{\cdot}1065.5H_2O$,[7] and $NaCl{\cdot}1113H_2O$[4] to give the heat of neutralization at infinite dilution, or the heat of ionization of water.

In the heat of neutralization experiments (Equation 1) the temperature intervals of the reaction and calibration runs were made to coincide as nearly as possible. The calculations were made in the usual manner, using the rate of temperature drift before and after heating periods to correct for heat conduction to the thermostat and other small losses or gains of heat. The average of three experiments gave 13,828 ± 12 cal. for the heat of the reaction in Equation (1) at 25°. The heat effect for Equation (2), being small, was easily measured, the result being 352 ± 8 cal. at 25°. Since this result is subtracted from that for Equation (1), any constant errors will be removed from the final value. Thus the value here obtained for the heat of the dilution of hydrochloric acid indicated in Equation (2) is to be preferred for the present calculations even though equally accurate values have been given previously.[7] The final calculation of the heat of ionization of water is summarized in Table II.

TABLE II

THE HEAT OF IONIZATION OF WATER

$$HCl{\cdot}46.5H_2O + NaOH{\cdot}1065.5H_2O = NaCl{\cdot}1113H_2O + 13{,}828 \pm 12 \text{ cal.} \quad (1)$$

$$HCl{\cdot}46.5H_2O + 1065.5H_2O = HCl{\cdot}1112H_2O + 352 \pm 8 \text{ cal.} \quad (2)$$

$$HCl{\cdot}1112H_2O + \infty H_2O = HCl{\cdot}(\infty + 1112)H_2O + 100 \text{ cal.} \quad (3)$$

$$NaOH{\cdot}1065.5H_2O + \infty H_2O = NaOH{\cdot}(\infty + 1065.5)H_2O + 86 \text{ cal.} \quad (4)$$

$$NaCl{\cdot}1113H_2O + \infty H_2O = NaCl{\cdot}(\infty + 1113)H_2O + 73 \text{ cal.} \quad (5)$$

(1) + (5) − (2) − (3) − (4) gives

$$HCl{\cdot}(\infty + 1112)H_2O + NaOH{\cdot}(\infty + 1065.5)H_2O = NaCl{\cdot}(\infty + 1113)H_2O + (\infty + 1065.5)H_2O + 13{,}363 \pm 16 \text{ cal.} \quad (6)$$

or $H_2O(l.) = H^+(aq.) + OH^-(aq.) \quad \Delta H = 13{,}363 \pm 16$ cal.

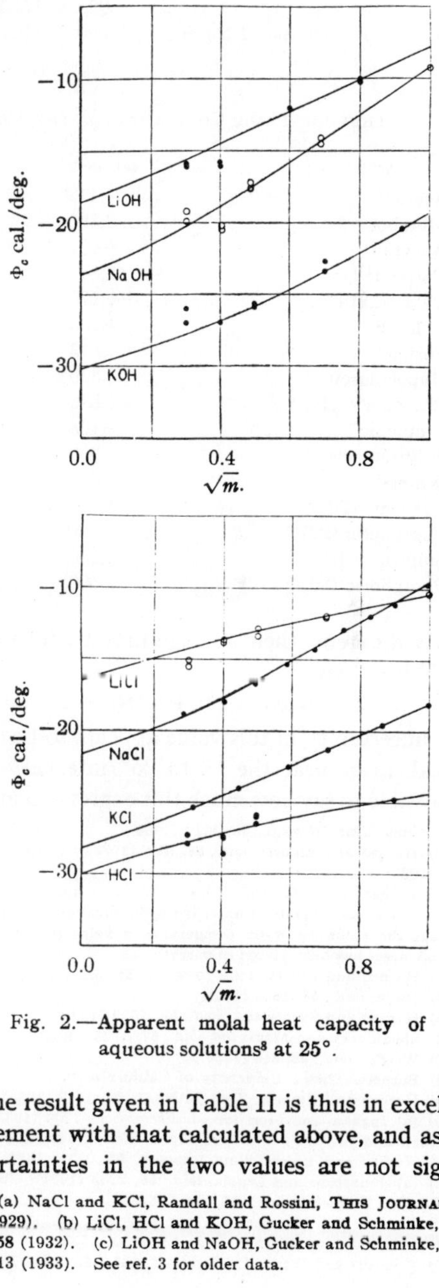

Fig. 2.—Apparent molal heat capacity of aqueous solutions[8] at 25°.

The result given in Table II is thus in excellent agreement with that calculated above, and as the uncertainties in the two values are not signifi-

(6) Landolt-Börnstein, "Tabellen."

(7) Rossini, *Bur. Standards J. Research*, **6**, 791 (1931), converted to 25°.

(8) (a) NaCl and KCl, Randall and Rossini, THIS JOURNAL, **51**, 323 (1929). (b) LiCl, HCl and KOH, Gucker and Schminke, *ibid.*, **54**, 1358 (1932). (c) LiOH and NaOH, Gucker and Schminke, *ibid.*, **55**, 1013 (1933). See ref. 3 for older data.

TABLE III

HEATS OF IONIZATION

Subst.	Vol. (1), ml.	Soln. (1)	Soln. (2) vol. (2) = 885 − vol. (1)	Heat evolved, cal.	Molal ΔH^0, cal. This research	Other research	"Best value"
NH_4OH	35	0.956 M NaOH	0.05 M NH_4Cl	29.4 ± 0.5	865 ± 30	....	865 ± 30
H_2CO_3	10	1.162 M HCl	.015 M $NaHCO_3$	26.6 ± .6	1843 ± 60	....	1843 ± 60
HCO_3^-	10	1.162 M HCl	.015 M $NaHCO_3$ / .015 M Na_2CO_3	46.7 ± .5	3500 ± 100	....	3500 ± 100
H_3PO_4	15	1.000 M HCl	.020 M NaH_2PO_4	−9.0 ± .5	−1950 ± 80	−1750[9]	−1880 ± 80
$H_2PO_4^-$	5	1.000 M HCl	.009 M Na_2HPO_4 / .001 M NaH_2PO_4	6.8 ± .3	800 ± 80	740[10]	800 ± 80
HPO_4^{--}	15	1.000 M HCl	.080 M Na_2HPO_4 / .020 M Na_3PO_4	76.4 ± .5	3500 ± 500	....	3500 ± 500
HSO_4^-	10	0.790 M Na_2SO_4	1.000 M HCl	−45.1 ± 1.0	−5200 ± 500	−2229[11]	−5200 ± 500

TABLE IV

THERMODYNAMIC CONSTANTS FOR THE IONIZATION OF WEAK ACIDS IN AQUEOUS SOLUTION AT 25°

Acid	ΔH^0, cal./mole	ΔF^0, cal./mole	$-\Delta S^0$, cal./deg. mole	$-\Delta C_p^0$, cal./deg. mole
Acetic[13]	−112	6,488	22.1	34
Metaboric[14]	3,360	12,599	31.0	43
Butyric[15]	−691	6,574	24.4	46
Carbonic (1st)[16]	1,843	8,683	22.9	
Chloroacetic[17]	−1,170	3,900	17.0	35
Chlorous[18]	−4,100	2,670	22.7	
Formic[19]	−13	5,120	17.2	41
Hypochlorous[20]	3,320	10,100	22.8	
Phosphoric (1st)[9]	−1,880	2,898	16.0	
Propionic[21]	−168	6,650	22.9	38
Sulfurous (1st)[22]	4,000	2,725	22.6	
Water[23]	13,358	19,093	19.24 (27)[12]	50
Carbonic (2d)[24]	3,500	13,983	35.2	
Phosphoric (2d)[10]	800	9,830	30.3	
Sulfuric (2d)[11]	−5,200	2,620	26.3	
Phosphoric (3d)[25]	3,500	16,300	43	

cantly different, their average may be taken as a final "best" value

$$\Delta H = 13{,}358 - 50.0(t - 25) + 0.15(t - 25)^2 \text{ cal./mole}$$

The uncertainty in this value appears to be about 12 cal. in or near the 18 to 25° interval. It is impossible to say how much this expression may be in error at temperatures considerably above or below this region.

The Heats of Ionization of Ammonium Hydroxide, Carbonic, Phosphoric, and Sulfuric Acids.—Because of the lack of accurate heat of dilution data it was desirable to measure most of these heat effects in very dilute solution, with resulting loss of precision. Otherwise the estimation of heats of dilution would have led to even greater uncertainty. In each case the heat of ionization itself was measured rather than the heat of neutralization of the weak acid or base. The latter method is less satisfactory because it gives the desired heat of ionization, itself relatively small, as the difference between two large quantities. As an example of the former method, the heat of ionization of ammonium hydroxide was obtained by measuring the heat of reaction of sodium hydroxide with ammonium chloride in dilute aqueous solution. The final solution thus contains both sodium chloride (ionized) and am-

(9) Nims, THIS JOURNAL, **56**, 1110 (1934).

(10) Harned and Embree, *ibid.*, **56**, 1050 (1934); Nims, *ibid.*, **55**, 1946 (1933).

(11) Hamer, *ibid.*, **56**, 860 (1934). See also discussion below.

(12) The standard state of water is the 55 molal liquid. In order to make the value for water comparable, a value of the entropy R ln 55 larger has been placed in parentheses.

(13) Harned and Ehlers, THIS JOURNAL, **55**, 652 (1933).

(14) Owen, *ibid.*, **56**, 1695 (1934).

(15) Harned and Sutherland, *ibid.*, **56**, 2039 (1934).

(16) Shedlovsky and MacInnes, *ibid.*, **57**, 1705 (1935).

(17) Wright, *ibid.*, **56**, 314 (1934).

(18) Barnett, Thesis, University of California, 1935.

(19) Harned and Embree, THIS JOURNAL, **56**, 1042 (1934).

(20) (a) Ingham and Morrison, *J. Chem. Soc.*, 1200 (1933); (b) Neumann and Müeller, *Z. anorg. allgem. Chem.*, **185**, 428 (1930).

(21) Harned and Ehlers, THIS JOURNAL, **55**, 2379 (1933).

(22) (a) Johnstone and Leppla, *ibid.*, **56**, 2233 (1934); (b) Roth and Zeumer, *Z. Elektrochem.*, **38**, 164 (1932).

(23) Harned and Mannweiler, THIS JOURNAL, **57**, 1873 (1935).

(24) MacInnes and Belcher, *ibid.*, **55**, 2630 (1933).

(25) Prideaux and Ward, *J. Chem. Soc.*, **125**, 423 (1924).

monium hydroxide (principally un-ionized); however, the latter should not significantly affect the heat of dilution, except by change of the fraction ionized. This is apparent when one considers that only long range, *i. e.*, coulombic, forces should give a heat of dilution in the range of low concentration.

The experimental results are given in Table III together with values of the molal heats of ionization at infinite dilution calculated therefrom. In these calculations, heats of dilution were taken from sources given above, or were estimated using the Debye–Hückel theory. Where the reaction was not complete, the amount of reaction was calculated from the ionization constant, using estimated activity coefficients if necessary.

No comparisons will be made with the early work of Thomsen, Berthelot, and others because of the difficulty of reducing their results to 25° and infinite dilution, and because improved calorimetric methods have rendered their work comparatively inaccurate.

The Entropy of Ionization of Weak Acids

The changes in heat, free energy, entropy, and heat capacity with ionization are tabulated in Table IV for a number of weak acids. While an exhaustive search of the literature has not been made, it is believed that the values listed are in most cases the best available. On inspection of the data in Table IV it is at once apparent that although ΔH and ΔF have values varying over a wide range, ΔS and ΔC_p are always negative, and for a first ionization have roughly the values -22 and -40 cal./deg. mole, respectively. It is of interest to note that rough theoretical considerations also lead to this result.

The entropy change in the reaction $HA = H^+ + A^-$ may be taken as the absolute entropy of the hydrogen ion plus the entropy difference between A^- and HA. The absolute entropy of the hydrogen ion is known accurately enough for the present purpose from the value of Eastman and Young[26] for the absolute entropy of chloride ion, and the sum of the entropies of hydrogen and chloride ions given by Latimer, Schutz, and Hicks.[27] The value thus obtained is -4.6 cal./deg. mole. The difference between the entropies of the negative acid ion and neutral molecule in solution may be considered as arising from the difference in entropy of the gas ion and molecule, and the difference in their entropies of solution.

The entropy of the gas molecule (HA) will differ from that of the ion (A^-) because of several factors. The increased mass of HA will contribute, at most, about 0.1 cal./deg. mole. If A^- is a non-linear polyatomic molecule, then no rotational degrees of freedom will be lost, although the moments of inertia will be slightly changed. This will have only a small effect on the entropy. In water, hypochlorous acid, and analogous molecules the ion is linear and has then only two degrees of rotational freedom. The loss of the other rotation may mean a loss of the order of five cal./deg. mole in the entropy at 25°. In all of the molecules here considered, the removal of the hydrogen will mean the loss of two vibrations, both of such frequency that they contribute but slightly to the entropy, and, where one rotation of the entire molecule was not lost, an additional vibration of torsional character will be lost. This, of course, assumes that the two bonds of the oxygen atom adjacent to the hydrogen form an angle less than 180°, and that rotation of the OH group with respect to the remainder of the molecule is appreciably hindered.[28] This torsional vibration will contribute in the neighborhood of three cal./deg. mole to the entropy. The symmetry number may be changed by the ionization, thus contributing a difference of the order of one cal./deg. mole. It is thus apparent that the entropy of the gas ion will be less than that of the neutral molecule, and in almost every case by an amount between three and six cal./deg. mole.

The difference in the entropies of solution is less easily discussed. In all the acids in Table IV and, in fact, in almost all weak acids the ionizing proton is attached to an oxygen atom. Since the acid is weak, one presumes tha most of the negative charge lies on this oxygen. Then, unless some radical rearrangement of charge occurs on ionization, we may consider the resulting negative ion as an essentially neutral molecule attached to a charged oxygen atom. Applying now the ideas developed by Latimer and co-workers[29] concerning the entropy of solution of simple ions, one finds that the difference in entropies of solution will arise from the compression, due to the electrostatic field, of the water near the charged oxygen atom. Since the effective charge is probably less than one unit, and because the rest of the molecule

(26) Young, M. B., Thesis, University of California, 1935.

(27) Latimer, Schutz and Hicks, *J. Chem. Phys.*, **2**, 82 (1934).

(28) See Kemp and Pitzer, THIS JOURNAL, **59**, 276 (1937); Pitzer, *J. Chem. Phys.*, **5**, 473 (1937).

(29) (a) Latimer and Kasper, *ibid.*, **51**, 2293 (1929); (b) Latimer, *Chem. Rev.*, **18**, 349 (1936).

excludes water from some of the region around the oxygen, this difference in entropies will probably be considerably less than, possibly one-half of, the entropy of solution of OH^- ion which is -26 cal./deg. mole.[30] In cases where the ion has several equivalent oxygens among which the charge is divided, the above description must be suitably amended, although the expected entropy of solution should be very similar. The entropy of ionization is then the sum of three terms: the absolute entropy of the hydrogen ion, -4.6, the difference in the entropies of the gas ion (A^-) and the molecule (HA), about -4, and the difference in the entropies of solution of the gas ion and molecule which will be considerably smaller than -26. If the last term is taken as $-26/2 = -13$, we find the entropy of ionization to be -22, or roughly the average of the values for the first ionizations listed in Table IV. The weaker acids show larger entropies of ionization, as is to be expected due to the greater concentration of negative charge on the oxygen atom. The entropies of second or third ionizations are larger, because the entropy of solution term depends on the square of the charge.

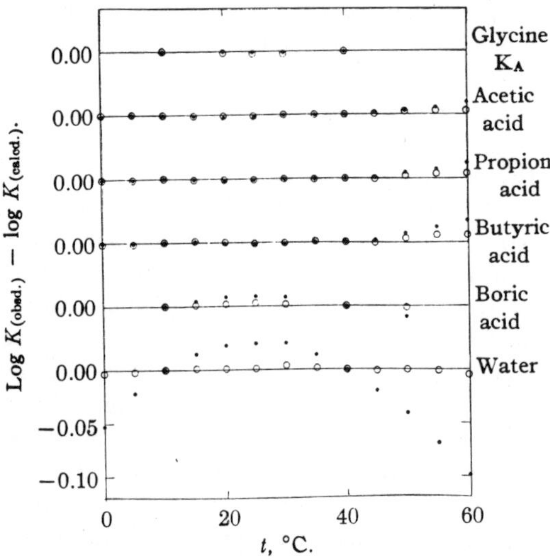

Fig. 3.—Test of equations (10) and (11) with glycine, acetic, propionic, butyric, boric acids and water: open circles, eq. (10); solid circles, eq. (11).

This discussion gives, of course, only a very approximate value for the entropy of ionization, but it does seem to show why this quantity is approximately the same for various weak acids and why it is of the order of magnitude found experimentally. Since the derivation was valid for any ordinary temperature, we may take the derivative of the entropy of ionization with respect to the logarithm of the temperature and obtain the change in heat capacity with ionization. The latter quantity should then have about the same magnitude for different acids, as is found experimentally.

We may now test further the values of the heat of ionization of sulfuric acid given in Table III. It is apparent from Table IV that the value of the entropy of ionization given there, -26.3 cal./deg. mole, is about as small as one would expect. The fact that the heat of ionization found by Hamer[11] gives a still smaller value of the entropy, -16.3, argues strongly against that result.

Variation of Ionization Constants with Temperature.—Using the results of the preceding section, we may now set up approximate expressions which give the ionization constant as a function of temperature, provided a single value of the constant is available. Assuming a constant ΔC_p, thermodynamics gives

$$\Delta H = \Delta H_0 + \Delta C_p T$$
$$\Delta S = \Delta S_0 + \Delta C_p \ln T$$
$$\Delta F = RT \ln K = \Delta H - T\Delta S = \Delta H_0 - T\Delta S_0 + \Delta C_p (T - T \ln T)$$

If K and thereby ΔF is known for some temperature say T_1

$$\Delta F = \Delta F_{T_1} + \Delta S_0 (T_1 - T) + \Delta C_p (T - T_1 + T_1 \ln T_1 - T \ln T) \quad (7)$$

If $T_1 = 298.1$, $\Delta C_p = -40$, $\Delta S_{298.1} = -22$

$$\Delta F = \Delta F_{298.1} + 5425 - 246T + 40T \ln T \quad (8)$$

or

$$\log K = (\text{const.}/T) + 53.78 - 20 \log T \quad (9)$$

This expression should be useful for a great many approximate calculations.

If the data on an ionization constant cover a sufficient range of temperature to determine accurately the first derivative (heat of ionization), nevertheless, the knowledge that ΔC_p is probably near -40 cal./deg. should be useful. This gives an equation for the form

$$\log K = A + B/T - 20 \log T \quad (10)$$

Harned and Embree[10] also have discussed the variation of ionization constants with temperature, and have proposed the equation

$$\log K = \log K_m - 5.0 \times 10^{-5}(t - \theta)^2 \quad (11)$$

in which the constants K_m and θ are the maximum value of the ionization constant and the temperature at which the maximum is attained.

(30) The rotational contribution was excluded from the entropy of the OH^- gas ion since rotation is certainly not present in solution and only the "electrostatic" entropy of solution was desired.

The range of validity of the equation is given as $t - \theta = \pm 75$. The authors make no attempt to interpret this equation theoretically.

It may be noted that both equations (10) and (11) have two undetermined constants, and, since it is known that both can be fitted satisfactorily to the experimental data for many acids, it seems likely that the fundamental assumptions in equation (11) are essentially the same as those of (10), *i. e.*, ΔC_p is a constant and has a value of about -40 cal./deg. Application of the usual thermodynamic equations to (11) yields

$$\Delta C_p = -2.3 \times 10^{-4} RT[T + 2(t - \theta)] \quad (12)$$

or at $T = 298.1$ °K.

$$\Delta C_p = -40.6 - 0.273(25 - \theta)$$

Thus for the case in which both θ and t are near 25° it becomes apparent that equations (10) and (11) are essentially the same, and that the theoretical considerations given above are applicable to equation (11) also.

It remains then to consider which of the two equations is more satisfactory when either t or θ is considerably different from 25°. Figure 3 presents graphically the deviations of the experimental from the calculated values of K, for several acids, when the constants of each equation are evaluated from the data at 10 and 40°. The data for other acids which to save space were omitted from Fig. 3, give results similar to those shown, the greater number being similar to glycine. It is thus apparent that equation (10) gives better agreement in virtually every instance where there is an appreciable difference between the two equations.

The writer wishes to acknowledge with thanks the many valuable suggestions made by Professor W. M. Latimer.

Summary

A calorimeter for the precise measurements of heats of reaction in aqueous solution is described. The available data on the heat of ionization of water are considered, and a revised value for the ΔC_p of the reaction obtained. Accurate measurements of the heat of neutralization of hydrochloric acid with sodium hydroxide at 25° are described, which give a value of the heat of ionization of water that is in good agreement with the revised value calculated from previous measurements. The selected "best" value of ΔH^0 for the ionization of water is: $13{,}358 - 50.0(t - 25) + 0.15(t - 25)^2$ cal./mole. The heat of ionization was measured for several weak acids and ammonium hydroxide with the following results

Substance	ΔH^0 of ionization at 25°, 30 cal./mole
NH_4OH	865 ± 30
H_2CO_3	1843 ± 60
HCO_3^-	3500 ± 100
H_3PO_4	−1880 ± 80
$H_2PO_4^-$	800 ± 80
HPO_4^{--}	3500 ± 500
HSO_4^-	−5200 ± 500

The changes in heat, free energy, entropy, and heat capacity with ionization are tabulated for sixteen weak acids. It is discovered that for all "first" ionizations, ΔS° is near -22 cal./mole and ΔC_p near -40 cal./mole, and an explanation of this fact is offered. Use is also made of these generalizations to develop the following formula which gives approximately the ionization constant over a range of temperature, provided a single experimental value is known: $\log K = \text{const.}/T + 53.78 - 20 \log T$. When a series of values of an ionization constant are available, an equation of the form: $\log K = A + B/T + 20 \log T$ is suggested. It is also pointed out that within its limited range of applicability the equation recently suggested by Harned and Embree is essentially equivalent to the second one given above.

BERKELEY, CALIF. RECEIVED JULY 12, 1937

Comment: The Heats of Ionization of Water, Ammonium Hydroxide, Carbonic, Phosphoric, and Sulfuric Acids. The Variation of Ionization Constants with Temperature and the Entropy Change with Ionization

This modest and primarily experimental paper continues to receive considerable attention, both for the calorimetric values reported for important reactions and for the generalizations about the change in heat capacity and entropy with ionization.

FEBRUARY, 1939 JOURNAL OF CHEMICAL PHYSICS VOLUME 7
Printed in U. S. A.

The Free Energy of Hydration of Gaseous Ions, and the Absolute Potential of the Normal Calomel Electrode

WENDELL M. LATIMER, KENNETH S. PITZER AND CYRIL M. SLANSKY
Department of Chemistry, University of California, Berkeley, California

(Received December 7, 1938)

The free energies of hydration of the alkali and halide ions are found to agree reasonably well with the simple expression of Born ($-\Delta F=(1-1/D)Ne^2/2r_e$) for solution of charged spheres in a dielectric medium, provided the crystal radii are suitably modified so as to correspond to the radii of the cavities in the dielectric medium. The results show that the dielectric constant of water remains large even in the intense field next to the ion. The entropies of hydration are also found to be consistent with these radii. Because of the simplicity of this calculation, the resulting free energies of solution of individual ions are considered to be *a priori* the most probable and are used to calculate a value of -0.50 volt for the absolute potential of the calomel half-cell.

SEVERAL discussions of the energy of hydration of gaseous ions have been published recently.[1-3] In these the simple Born expression ($-\Delta F=Ne^2/2r_e[1-1/D]$) is considered, but declared to be unsatisfactory in some important respects. In most cases modifications were introduced which complicated the treatment and increased the opportunities for adjustment, after which satisfactory agreement with experiment was obtained. The purpose of the present communication is to point out that the simple Born equation alone gives satisfactory results, indicating that refinements are for the present unnecessary. Furthermore, since such a simple treatment suffices, it is likely that the energies of hydration for single ions thus obtained may have some significance, and so they are employed to calculate the absolute potential of the calomel half-cell.

The heats of hydration of pairs of gas ions were obtained from the crystal energies of Huggins[4] and corrected to 298.1°K, and the heats of solution were from Bichowsky and Rossini.[5]

$$M^+(g)+X^-(g)=MX(s) \qquad \Delta H=-U$$
$$MX(s)=M^+(aq)+X^-(aq) \qquad \Delta H=L$$

$$M^+(g)+X^-(g)=M^+(aq)+X^-(aq); \quad \Delta H=L-U.$$

The entropies of gas ions were calculated from the Sackur equation and corrected to one mole per liter concentrations while the values for aqueous ions were taken from the recent summary of Latimer, Pitzer and Smith.[6]

These data were then combined to give the free energies of hydration for the twenty pairs of ions comprising the alkali halide series. Table I contains these values together with the differences between iodide and each of the other halide ions, computed from the data for each metal; also the differences for the metals computed for each halide. These results show the data to be consistent to within a few kcal., which is also the magnitude estimated for their absolute accuracy.

In fitting the Born expression to these data three arbitrary constants must be introduced. First comes the ever present problem of dividing the free energy of hydration, known only for ion pairs, between the two ions of the pair. Once a single absolute value is specified all others are determined. The other two constants arise from the fact that the well-known crystal radii of Pauling[7] cannot be expected to be directly applicable in the Born expression. Here r_e is the radius of the cavity in the dielectric medium and is closely related to the distance from the center of the ion to the center of the nearest water dipoles. Since the negative ions will have the hydrogens of the surrounding water directed in, their effective radii will be but slightly larger

[1] T. J. Webb, J. Am. Chem. Soc. **48**, 2589 (1926).
[2] J. D. Bernal and R. H. Fowler, J. Chem. Phys. **1**, 515 (1933).
[3] Andr. Voet, Trans. Faraday Soc. **32**, 1301 (1936).
[4] M. Huggins, J. Chem. Phys. **5**, 143 (1937).
[5] F. R. Bichowsky and F. D. Rossini, *Thermochemistry of Chemical Substances* (Reinhold Publishing Corporation, New York, 1936).
[6] W. M. Latimer, K. S. Pitzer and W. V. Smith, J. Am. Chem. Soc. **60**, 1829 (1938).
[7] L. Pauling, J. Am. Chem. Soc. **49**, 765 (1927).

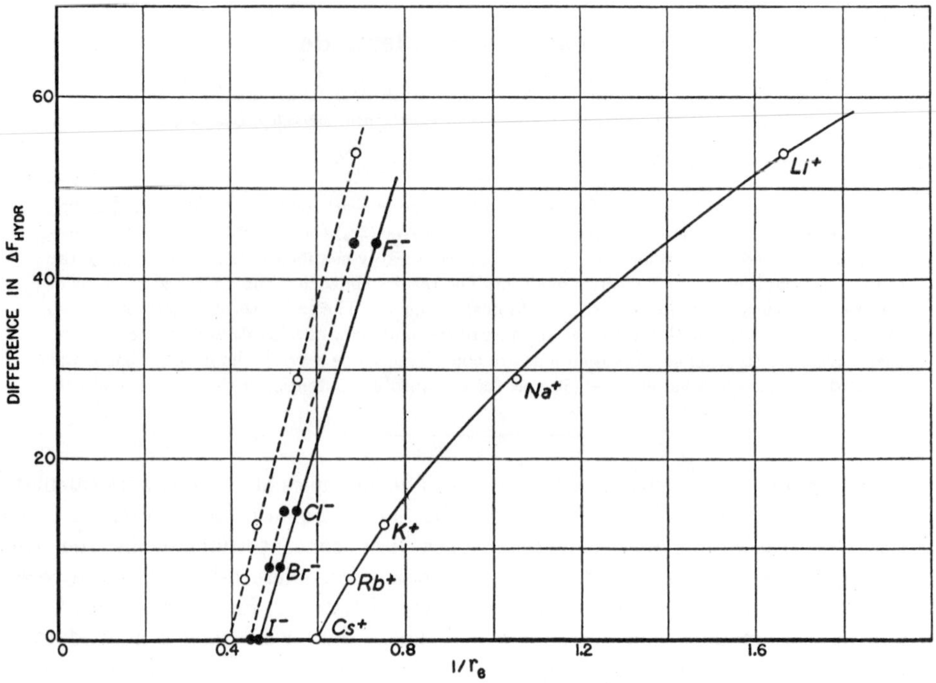

FIG. 1. Differences in free energy of hydration in gaseous ions. Solid curves for crystal radii, dotted for corrected radii.

than the ordinary crystal radii. On the other hand, the positive ions will have the water hydrogens directed out and as a result their effective radii may be considerably larger.

Figure 1 shows the effect of these radii corrections. The differences in free energy of hydration of the negative ions from Table I are first plotted against the crystal radii (solid curve) and a good straight line is obtained. However, it is found that the best fit to the Born expression can be obtained by adding 0.1A to these radii (dotted curve). The similar plot for the positive ions with the crystal radii shows considerable curvature, but upon adding 0.85A to all the radii, a straight line is obtained which, moreover, shows the same slope as the line for the negative ions. Then dividing up the sum of hydration energies for say, cesium and iodide ions so that both positive and negative ions fall on the same curve we obtain Fig. 2, where the line is now the theoretical curve given by the Born equation with the ordinary dielectric constant of water.

TABLE I. *Negative free energies of hydration of pairs of alkali halide gas ions.*

	F^-	$\Delta F_{I^-} - \Delta F_{F^-}$	Cl^-	$\Delta F_{I^-} - \Delta F_{Cl^-}$	Br^-	$\Delta F_{I^-} - \Delta F_{Br^-}$	I^-	MEAN DIFF.
Li^+	229.2	46.8	198.7	16.3	191.5	9.1	182.4	0
$\Delta F_{Cs^+} - \Delta F_{Li^+}$	54.5		55.9		53.3		51.7	53.8±1.4
Na^+	202.6	43.6	173.4	14.4	167.1	8.1	159.0	0
$\Delta F_{Cs^+} - \Delta F_{Na^+}$	27.9		30.6		28.9		28.3	28.9±0.9
K^+	186.3	43.4	157.1	14.2	150.7	7.8	142.9	0
$\Delta F_{Cs^+} - \Delta F_{K^+}$	11.6		14.3		12.5		12.2	12.7±0.9
Rb^+	179.1	41.5	151.5	13.9	145.0	7.4	137.6	0
$\Delta F_{Cs^+} - \Delta F_{Rb^+}$	4.4		8.7		6.8		6.9	6.7±1.1
Cs^+	174.7	44.0	142.8	12.1	138.2	7.5	130.7	0
	0		0		0		0	0
Mean Diff.		43.9 ±1.2		14.2 ±0.9		8.0 ±0.4		0

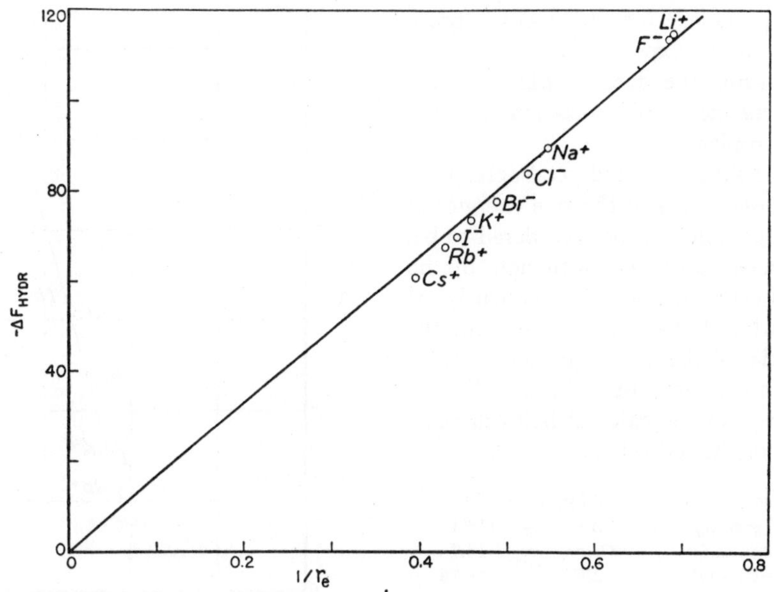

FIG. 2. Free energy of hydration of ions. Straight line represents the Born equation.

Considering that an admittedly over simplified theoretical expression has been used, and that only three arbitrary constants were evaluated while nine experimental data were used (five cations and four anions), the fit in Fig. 2 is to be regarded as quite satisfactory. It should be emphasized that the form of the Born equation is such that the exact value of the dielectric constant is of little significance so long as it remains large compared to unity. The present results show that even in the intense fields next to the ion, the effective dielectric constant of water is fairly high as has been suggested previously.[8, 8a] On the other hand, it is not pretended that the dielectric constant retains its macroscopic value under these conditions.

The foregoing simple assumptions give surprisingly satisfactory results with polyvalent ions; but because of the greater uncertainties in these cases, they will be omitted from the present discussion.

A further test on the above treatment can be made by considering the temperature coefficient of the free energy of hydration, i.e., the entropy of hydration of gas ions. The theoretical expression becomes

$$\Delta S = -\frac{d\Delta F}{dT} = \frac{Ne^2}{2r_e}\frac{1}{D^2}\frac{dD}{dT}.$$

Since this equation is sensitive to small changes in the dielectric constant and its temperature derivative, we cannot expect quantitative agreement; however, we may expect that the entropies of hydration for both positive and negative ions should fall on a single smooth curve.

Using the same effective radii as in the free energy calculations, one still has to divide up the entropy of hydration for a pair of ions between the positive and negative ions. When this is done Fig. 3 is obtained which shows that the positive and negative ions do fall on approximately the same curve when our effective radii are used. The calculation is, of course, essentially that given previously by Latimer.[9]

Since the absolute entropies of the gas ions are known, the absolute entropies of the aqueous ions can now be calculated. The value obtained for chloride ion is 15 cal. per deg. mole which may be compared with the results of Eastman and Young[10] who obtain as a provisional value

[8] W. M. Latimer and C. Kasper, J. Am. Chem. Soc. **51**, 2293 (1929).

[8a] W. J. Dunning and W. J. Shutt, Trans. Faraday Soc. **34**, 467 (1938).

[9] W. M. Latimer, Chem. Rev. **18**, 349 (1936).

[10] M. B. Young, Thesis, University of California, 1935.

18.1 cal. per deg. mole. This agreement is probably as good as can be expected.

Table II contains the data of Fig. 2 and 3 together with the heats of hydration and the absolute ion entropies.

Because of the simplicity and good agreement found above, these values of the free energies of hydration of single ions can be considered to be, *a priori*, the most probable, although by no means conclusively proved. Consequently it seemed worth while to calculate a value for the absolute potential of the normal calomel cell for comparison with experimental values.

The free energy of the calomel half-cell reaction is the sum of the following:

(1)	$Na^+(g)=Na^+(aq)$	$\Delta F_{Hydr}=-\ 89.7$
(2)	$Na(g)=Na^+(g)+E^-(g)$	$\Delta F\ =\ 118.1$
(3)	$Na(s)=Na(g)$	$\Delta F\ =\ 18.7$
(4)	$Hg+Na^+(aq)+Cl^-(aq)$ $=HgCl+Na(s)$	$\Delta F\ =\ 68.76$
(5)	$E^-(g)=E^-(\text{in Hg})$	$\Delta F\ =-104.5$
	$Hg+Cl^-(aq)=HgCl+E^-(\text{in Hg})$;	$\Delta F\ =\ 11.4$ kcal.
		$E_{abs}\ =-0.495$ volt

Equation (2) is from the table of ionization potentials by Herzberg,[11] Eq. (3) gives the free energy of sublimation and is from the data of Kelley,[12] Eq. (5) is the photoelectric effect in mercury and was determined by Kazda.[13] The value of -0.495 volt for the absolute potential of the normal calomel electrode is in reasonable agreement with the old value of -0.56 obtained from the dropping mercury electrode. The more recent determinations by Billitzer[14] and Garrison[15] differ from ours by some 8 kcal.

We may conclude, as did Baur[16] that the old value of -0.56 volt is our best available experimental value of the absolute potential of the calomel half-cell.

TABLE II. *Thermodynamic properties of single alkali and halide ions.*

	r_e	$-\Delta F°$ HYDR. (kcal.)	$-\Delta S°$ HYDR. (cal./deg.)	$-\Delta H°$ HYDR. (kcal.)	$S°$ ABSOLUTE (cal./deg.)
Li^+	1.45	114.6	22	121.2	3
Na^+	1.80	89.7	17	94.6	12
K^+	2.18	73.5	8	75.8	23
Rb^+	2.33	67.5	6	69.2	27
Cs^+	2.54	60.8	4	62.0	30
F^-	1.46	113.9	29	122.6	−1
Cl^-	1.91	84.2	15	88.7	+15
Br^-	2.05	78.0	12	81.4	21
I^-	2.26	70.0	7	72.1	27

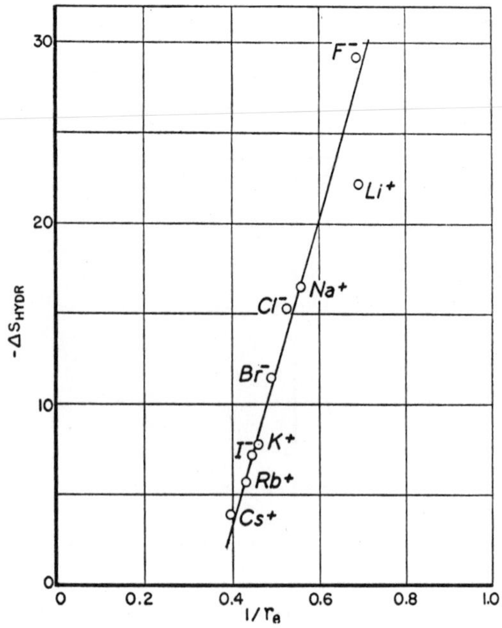

FIG. 3. Entropy of hydration of alkali and halide ions.

[11] G. Herzberg, *Atomic Structure and Atomic Spectra* (Prentice-Hall, 1937).

[12] K. K. Kelley, "Contribution to the Data on Theoretical Metallurgy," Bull. 383, Bureau of Mines (1935).

[13] C. B. Kazda, Phys. Rev. **26**, 643 (1925).

[14] J. Billitzer, Zeits. f. physik. Chemie **48**, 513 (1904).

[15] A. Garrison, J. Am. Chem. Soc. **45**, 37 (1923).

[16] E. Baur, Zeits. Electrochem. **38**, 665 (1932).

Comment: The Free Energy of Hydration of Gaseous Ions and the Absolute Potential of the Normal Calomel Electrode

The Born equation for the Gibbs free energy of solvation was shown to be successful for simple ions in water provided all positive ionic radii are increased by one constant and all negative ion radii are increased by a different constant. This remarkably simple treatment has been elaborated by many others with little empirical improvement for the Gibbs Energy. One must recognize, however, that the temperature derivative does not give a good result for the entropy of hydration. Thus, the agreement for the Gibbs energy must be to some extent accidental.

[Reprinted from the Journal of Physical Chemistry, **60**, 1140 (1956).]

THE ORDER-DISORDER PROBLEM FOR ICE

By Kenneth S. Pitzer and Jan Polissar

Department of Chemistry and Chemical Engineering, University of California, Berkeley, California

Received February 10, 1956

In 1935 Pauling[1] proposed a structure for ice which provided for hydrogen bonds throughout the crystal but involved a disordered pattern of proton locations. The proton was assumed to be unsymmetrically located in each hydrogen bond and about each oxygen atom two protons were assumed to be nearby, comprising a water molecule, and two farther away. This disorder involves an entropy of $R \ln 3/2 = 0.81$ cal./degree mole, which is in excellent agreement with the residual entropy at 0°K., found by Giauque and Stout[2] to be 0.82 ± 0.05 for H_2O and by Long and Kemp[3] to be 0.77 ± 0.1 for D_2O.

Although the agreement in entropy of disorder is excellent and the neutron diffraction results of Wollan, Davidson and Shull[4] are consistent with the Pauling structure, it is also desirable to estimate the interaction energies which would tend to remove the disorder. Bjerrum[5] initiated this sort of study and obtained approximate results, which indicated that ice would not have a disordered state even at its melting point. Several authors[6] have discussed other possible structures which, however, are not acceptable for reasons to be mentioned below.

The dielectric constant measurements on ice[7] yield high values below the freezing point which are roughly an extension of the values for the liquid. At lower temperatures the time required for polarization increases and the apparent dielectric constant for high frequencies drops to a value of 3.1. If ice had an ordered array of dipolar water molecules it would necessarily show one or the other of two effects.

If the crystal pattern was such as to cancel all dipole moments, then the dielectric constant would be small. The other possible alternative for an ordered structure is that with net parallel dipole moment along some axis; this situation yields a ferroelectric such as Rochelle salt or KH_2PO_4. The dielectric properties of ice are clearly of neither of these types, but are those characteristic of a disordered structure with molecules free to reorient themselves in response to the field.

Let us now review Bjerrum's model and the approximations in his calculations. The water molecule in ice is assumed to be a tetrahedron with center to vertex distance of 0.99 Å. and a charge of $0.171e$ at each vertex. Two of these charges are

(1) L. Pauling, *J. Am. Chem. Soc.*, **57**, 2680 (1935).

(2) W. F. Giauque and J. W. Stout, *ibid.*, **58**, 1144 (1936).

(3) E. A. Long and J. D. Kemp, *ibid.*, **58**, 1829 (1936).

(4) E. A. Wollan, W. L. Davidson and C. G. Shull, *Phys. Rev.*, **75**, 1348 (1949).

(5) N. Bjerrum, *Kgl. Danske Videnskab. Selskab, Math-fys. Medd.*, **27**, 1 (1951); *Science*, **115**, 385 (1952).

(6) R. E. Rundle, *J. Chem. Phys.*, **21**, 1311 (1953); **22**, 344 (1954); This Journal, **59**, 680 (1955); W. N. Lipscomb, *J. Chem. Phys.*, **22**, 344 (1954); P. G. Owston, *J. chim. phys.*, **50**, C13 (1953).

(7) C. P. Smyth, "Dielectric Behavior and Structure," Chap. V, McGraw-Hill Book Co., Inc., New York, N. Y., 1955, and references there cited.

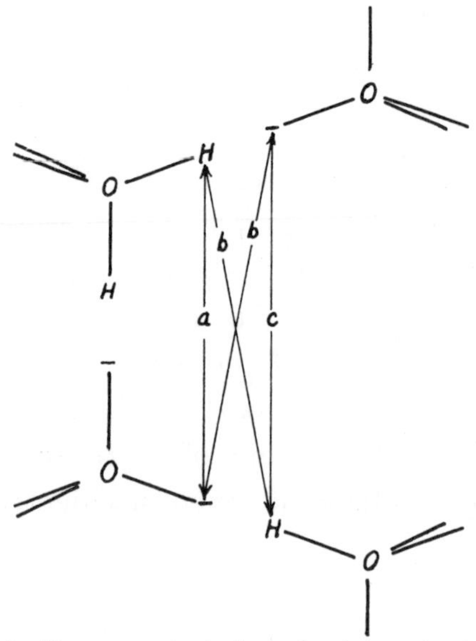

Fig. 1.—Three successive hydrogen bonds in an ice crystal.

positive, corresponding to the hydrogen atoms, and the other two negative, corresponding to the unshared electron pairs. This magnitude of charge yields the correct dipole moment for the water molecule. The ice structure involves tetrahedral orientations of oxygen atoms about a given central oxygen with an O–O distance of 2.76 Å. Thus the hydrogen bond in this model has opposite charges of $0.171e$ spaced 0.78 Å. apart.

The ice crystal has hexagonal symmetry. Bjerrum showed very clearly the possible orientation of pairs of water molecules. One fourth of the hydrogen bonds are parallel to the hexagonal axis, and for these the two H_2O tetrahedra are mirror symmetric. But when signs of charge are considered the orientation may be inverse mirror (*i.e.*, all + charges reflect to − charges) or oblique mirror (which differs by ±120° rotation). All other hydrogen bonds have their pairs of tetrahedra rotated by 60° so that they have a center of symmetry and again when charge signs are considered the orientation may be either inverse center or oblique center symmetric. The Pauling structure requires random distributions between the inverse and oblique orientation for each type of bond.

In his first approximation Bjerrum included only the electrostatic interactions between the charges of a pair of water molecules. In a second approximation he included all interactions between two water molecules hydrogen bonded to an intermediate molecule. Bjerrum remarked that it would be an extremely difficult task to include all interactions between pairs of water molecules separated by two other water molecules. However, reference to Fig. 1 will show that one such interaction is very important. In Fig. 1 we show just a sequence of three hydrogen bonds. The interaction marked *a* is included in Bjerrum's first approximation while his second approximation includes the two terms marked *b*. However, the interaction marked *c* is of nearly the same magnitude but is omitted in Bjerrum's calculation. Since this term will be opposite in sign to the b terms it will tend to reduce the net energy markedly.

By making the calculation in terms of hydrogen bonds instead of water molecules we believe we have obtained a reasonably good approximation without excessive effort. We followed Bjerrum's model exactly, and included all hydrogen bonds adjacent to a given bond. Thus there are three hydrogen bonds adjacent to each end of the central bond, and we included both the positive and the negative charge associated with each of the seven hydrogen bonds. All interactions between this array of 14 charges were considered.

In Table I we present our results together with those of Bjerrum's first approximation. His second approximation yields intermediate values but these are complicated by the orientations about additional hydrogen bonds. The dielectric constant is unity in these calculations; its proper value will be discussed later.

TABLE I

ENERGY DIFFERENCES IN ICE FOR $\epsilon = 1$

	Bjerrum (1st approx.)	This research
Oblique mirror–inverse mirror	7.48×10^{-14}	1.67×10^{-14} erg
Inverse center–oblique center	6.75×10^{-14}	0.95×10^{-14}

We see that the inclusion of the interactions of the c type in Fig. 1 has a profound effect and greatly reduces the energy differences.

We must now consider the effective dielectric constant for these electrostatic interactions. Since all molecular orientations are specified in our model, the contribution of molecular orientation to ϵ must be excluded. Our model assumes an electron cloud distribution for a single molecule but makes no allowance for its readjustment under the influence of neighboring molecules. Consequently these electron cloud readjustments should be introduced through a dielectric constant. We have noted above that the high-frequency value for ice is 3.1. We shall adopt the value 3, not with any pretense that it is precisely correct for the calculation at hand, but merely that the value should be near 3 rather than 1 or 80.

The transition from disordered ice to ordered ice, if it occurred at a detectable rate, would undoubtedly be a coöperative phenomenon. There would be a lambda or Curie type of transition with high heat capacity just below the Curie temperature and almost complete disorder above that temperature. We may estimate the approximate Curie temperature by dividing the energy change from complete order to disorder by the corresponding entropy change. The ordered state would have all mirror symmetric bonds in the inverse orientation and all center symmetric bonds oblique. There are $^1/_3$ inverse and $^2/_3$ oblique of each type for disorder. Per water molecule there is $^1/_2$ mirror symmetric bond and $^3/_2$ center symmetric bond. Consequently

$$\Delta E = \frac{1}{\epsilon}\left(\frac{1.67}{3} + \frac{0.95}{2}\right) \times 10^{-14} \text{ erg}$$

$$= \frac{149}{\epsilon} \text{ cal./mole}$$

$$T_c \cong \frac{149}{0.81\epsilon} = \frac{184°\text{K}}{\epsilon}$$

If ϵ is taken as 3, we have $T_c \cong 60°$K., which is a reasonable value. Giauque and Stout observed that ice reaches thermal equilibrium sluggishly in the range 85–100°K. If this phenomenon is associated with a slight shift toward an ordered structure, the Curie temperature cannot be greatly lower. Nevertheless, it must be appreciably lower because Giauque and Stout were unable to obtain actual transition to the ordered state even on very slow cooling. In other words, for the sluggish equilibrium at 85–100°K. the relaxation time is of the order of an hour but at the Curie point it must exceed a day.

Bjerrum was able to show that, with dielectric constant unity, his model gave a total binding energy in reasonable agreement with the observed heat of sublimation of ice. The introduction of a dielectric constant of three would, of course, spoil this agreement. However, the binding energy arises chiefly from the adjacent charge pairs which constitute the hydrogen bonds themselves. The important phenomenon is then distortion of the electron cloud comprising the unshared pair on the water molecule right next to the bonding hydrogen. For this problem the use of a dielectric constant would be a poor approximation. Indeed this distortion could easily be of a type to increase rather than decrease the bond energy. The interactions which are important for the order-disorder problem are the more distant ones shown in Fig. 1. For these the use of a dielectric constant seems an appropriate method.

For complete disorder one calculates that a cubic ice, analogous to the diamond structure, would have the same energy as actual hexagonal ice. A metastable cubic ice has been reported by König[8] to form from the vapor at −80°. Even the slightest amount of ordering about the mirror symmetric bonds in the hexagonal crystal makes the hexagonal structure more stable. Hence there is no difficulty in understanding the structure which ice actually takes. Near the melting point one might expect another type of disorder, however, in which a large group of normally mirror symmetric bonds in a plane perpendicular to the hexagonal axis are of the center symmetric type instead. This sort of irregularity is observed in silicon carbide.[9] Possibly the disorder X-ray scattering[10] which occurs near the melting point but disappears at low temperatures arises from irregularities of this type.

In conclusion we may remark that all of our evidence is consistent with the Pauling structure and the thermal properties observed by Giauque and Stout.

(8) H. König, *Z. Krist.*, **105**, 279 (1944); *Nature*, **157**, 449 (1946).

(9) A. W. Hull, *Phys. Rev.*, **15**, 545 (1920); see also *Z. Krist, Strukturbericht*, **1**, 144 (1931).

(10) P. G. Owston and K. Lonsdale, *J. Glaciology*, **1**, 118 (1951).

Comment: The Order-Disorder Problem for Ice

In 1984 Y. Tajima, T. Matsuo, and H. Suga (*J. Phys. Chem. Solids* **45**, 1135–1144) measured the temperature of this order-disorder transition in ice doped with KOH. They found the transition at 72 K. The estimate of 60 K in this simple treatment was remarkably close.

Reprinted from the JOURNAL OF CHEMICAL PHYSICS, Vol. 32, No. 5, 1279–1288, May, 1960
Printed in U. S. A.

Interactions between Molecules Adsorbed on a Surface

OKTAY SINANOĞLU* AND KENNETH S. PITZER

Lawrence Radiation Laboratory and College of Chemistry, University of California, Berkeley 4, California

(Received August 26, 1959)

The intermolecular potential energy between two inert gas molecules is considerably altered when these molecules are next to a solid surface as in physical adsorption. The change in the interaction is evidenced by the additional long-range repulsion that is often observed between the molecules of a monolayer and also by the additional attractions that must play a role in multilayer formation.

In this article, the two-molecule-surface potential is derived from quantum mechanical third-order perturbation theory. It is shown that this potential consists of two parts just as the energy giving the van der Waals attraction of a single molecule to a surface does. The first part exists only when the surface has a net electrostatic field and this is equivalent to the classical polarization effect. The second part arises from the fluctuations of the surface fields and is of the same origin as the dispersion forces. The third-order energy, i.e., the new intermolecular interaction caused by the surface, is directly related to the zero-coverage heat of adsorption and except for this experimental quantity, the results do not require specific assumptions about the surface. Thus, the theory is applicable to either metal or insulator surfaces. When both the two-molecule-surface and the one-molecule-surface interactions are available experimentally (for example, from the application of virial coefficients treatment in physical adsorption) the electrostatic field of the surface can be estimated.

The fluctuation or dispersion part of the third-order energy is shown to yield a repulsion between two molecules in a monolayer that amounts to 20–40% of the gas phase Lennard-Jones potential minimum ϵ_0. The same energy yields an additional attraction of about 10–20% of ϵ_0 when the two molecules are on top of one another as in multilayer formation. The theory is applicable also when more than two molecules at a time need be considered on the surface.

A. INTRODUCTION

PHYSICAL adsorption is characterized by relatively weak forces as compared to chemisorption and it might be considered a process in which both the adsorbed molecules and the adsorbent surface remain quite unchanged. Thus the van der Waals forces that cause physical adsorption can be understood for the most part through the use of quantum mechanical perturbation theory. The theory of these forces between molecules has been reviwed by Margenau[1] and some recent considerations by Pitzer.[2] Extensive reviews on the forces that cause physical adsorption also exist.[3,4]

Some interesting phenomena such as the formation of monolayers and multilayers on an adsorbent involve the forces between adsorbed molecules themselves as well as with the surface. Such forces can be investigated through the application of statistical mechanics to experimental measurements especially in physical adsorption, and although this is difficult when the adsorbate is dense, it becomes particularly suitable in the dilute case, i.e., relatively high temperature adsorption making use of a theory similar to the virial coefficients treatment of imperfect gases.[5] By such a method it is possible to investigate the adsorption of essentially one molecule, two molecules, etc., at a time on a surface and the potential energies of interaction involved in each case.[5]

Such experimental evidence has long shown[3,5b] that two inert gas atoms adsorbed into a monolayer display a rather different potential energy between themselves as compared to their energy when they are in the gas phase, i.e., the presence of the surface apparently causes a new long-range repulsion to appear between the two molecules of the monolayer and this repulsion tends to counterbalance their usual London attraction. On the other hand, in multilayer formation additional

[1] H. Margenau, Revs. Modern Phys. **11**, 1 (1939).
[2] K. S. Pitzer, *Advances in Chemical Physics* (Interscience Publishers, Inc., New York, 1959), Vol. II; or see K. S. Pitzer, UCRL–8395 (September, 1958).
[3] J. H. de Boer, Advances in Catalysis **8**, 29 (1956).
[4] S. Brunauer, *The Adsorption of Gases and Vapors* (Princeton University Press, Princeton, New Jersey, 1945), p. 180.
[5] (a) W. A. Steele and G. D. Halsey, Jr., J. Chem. Phys. **22**, 979 (1954); see also (b) M. P. Freeman, J. Phys. Chem. **62**, 729 (1958).

attractions must exist. These observations are not explainable in terms of second-order perturbation theory of long-range forces. This theory predicts that the total interaction between a system of adsorbed molecules and the surface must be pairwise additive; hence the interaction between molecules would be unaltered.[1] The repulsion in the monolayer has usually been ascribed[3] to the dipoles permanently induced in the nonpolar molecules by an electrostatic field that is known to be present on metal[6] and charcoal[3] surfaces. The magnitude of these induced dipoles can be inferred[3] from the nonideal behavior of the adsorbed two-dimensional gas if the entire additional repulsion in the presence of the surface is attributed to this classical electrostatic effect. However the values so obtained usually seem to be too high in comparison to the estimates from measured contact potentials[6] and work function lowering.[7] It is apparent that part of the many-body repulsion must be caused by a mechanism other than ordinary electrostatic polarization.

Many-body forces between spherical, inert molecules have been investigated quite extensively. When overlap effects are negligible these forces arise for the most part from the triple-dipole interaction in third-order perturbation theory. These three-body forces have been reviewed extensively, along with their application to the third virial coefficient of an imperfect gas, by Kihara.[8] A recent discussion with reference to colloidal particles has been given by Sparnaay.[9] Although the three-body force between three molecules is quite small, it is to be expected that forces of similar origin should become appreciable, as compared to the interaction between two molecules, when the third body is a large surface.

If the two molecules overlap, then even in second-order perturbation appreciable many-body forces should arise as shown by Jansen and McGinnies[10] for three atoms. However, we shall assume that this effect is part of the change that must take place in the short-range repulsive region of the intermolecular potential and concern ourselves with nonoverlap, long-range interactions.

Higher-order perturbations have been studied by Bade[11] using the simple Drude or harmonic oscillator model. It is found that for a system of a few molecules the perturbation series converges quite rapidly and more will be said about convergence in Sec. C.

In the following we shall give a general derivation of some forces that arise between spherical (hence nonpolar) molecules in the presence of a semi-infinite solid surface (metal or insulator) based on a third-order perturbation treatment. Since in this order the energy is triplewise additive[11] (see end of Sec. C) it will be necessary to consider only the system of two molecules and the surface. We shall show that, just as in the case of van der Waals energy between one molecule and the surface,[3] the energy of two molecules and the surface is composed of an "electrostatic" part and a quantum mechanical fluctuation or dispersion part. Moreover, it will be found that each part of the three-body energy can be related directly to the corresponding part of the one molecule adsorption energy. Thus it is often possible to obtain the potential energy of interaction between two molecules, physically adsorbed on a surface, in terms of the experimental zero-coverage heat of adadsorption U_0. On the other hand, the present treatment enables one to estimate the "electrostatic" fraction of U_0, when both U_0 and the surface intermolecular potential are available experimentally. The intimate connection between these two energies makes it necessary to discuss first the van der Waals energy of a single molecule on a surface with comments on the application of some of the existing theories before going into the derivation of the "two-molecule-surface" potential.

B. SECOND-ORDER ENERGY

Let us consider the most general derivation of the van der Waals attraction energy between a spherical, nonpolar molecule such as a rare gas atom and any solid which extends from minus to plus infinity in the x and y directions, defining the surface plane, and is semi-infinite in the perpendicular z direction. We shall use straightforward perturbation theory. If the adsorbed molecule is not too large, the perturbation can be written as

$$J=-\boldsymbol{\mu}_1\cdot\mathbf{F}_1, \tag{1}$$

where $\boldsymbol{\mu}_1$ is the instantaneous dipole moment of the molecule and $\mathbf{F}_1$ is the instantaneous electric field produced by all the electrons and nuclei of the solid at the center of molecule 1. The unperturbed Hamiltonian is simply

$$H^0=H_1+H_s, \tag{2}$$

i.e., the sum of the parts describing the molecule and the surface separately. Where overlap is negligible, the system will be represented by

$$\psi_0=\psi_1^0\psi_s^0, \tag{3}$$

i.e., the product of unperturbed wave functions for the molecule and the surface. When (1) is applied as the perturbing potential on (2) and (3), the first-order energy vanishes. But the second-order energy becomes

$$E_{1s}^{(2)}=\sum_{(\rho,\tau)\neq 0}\frac{|\langle 00\,|\,\boldsymbol{\mu}_1\cdot\mathbf{F}_1\,|\,\rho\tau\rangle|^2}{\delta_{0\rho}+\delta_{0\tau}}, \tag{4}$$

[6] J. C. P. Mignolet, Rec. trav. chim. **74**, 701 (1955).

[7] G. Ehrlich and F. G. Hudda, J. Chem. Phys. **30**, 493 (1959).

[8] T. Kihara, *Advances in Chemical Physics* (Interscience Publishers, Inc., New York, 1958), Vol. I, p. 267; Triple-dipole energy was first discussed by B. M. Axilrod and E. Teller, J. Chem. Phys. **11**, 299 (1943); Axilrod, *ibid.* **19**, 719 (1951).

[9] M. J. Sparnaay, Physica **25**, 217 (1959).

[10] L. Jansen and R. T. McGinnies, Phys. Rev. **104**, 961 (1956).

[11] W. L. Bade, J. Chem. Phys. **27**, 1280, 1284 (1957); **28**, 282 (1958).

where ρ, τ refer to the various unperturbed energy states of the molecule and the "surface" (meaning the semi-infinite solid), respectively, and δ's denote the excitation energies to these various states. Now, if we assume that the δ's can be replaced by some "mean excitation energies" and use the relations

$$\sum_{\rho=0}^{\infty} |\langle 0 | \boldsymbol{\mu}_1 | \rho \rangle|^2 = \langle 0 || \boldsymbol{\mu}_1 |^2 | 0 \rangle \equiv \langle | \boldsymbol{\mu}_1 |^2 \rangle_{00}, \quad (5)$$

and

$$\alpha_1 = \alpha_1^{xx} \cong 2(\langle \mu_{1x}^2 \rangle_{00} / \delta_1),$$

for a spherical molecule, Eq. (4) becomes

$$E_{1s}^{(2)} \cong -\tfrac{1}{2}\alpha_1 |\langle 0 | \mathbf{F}_1 | 0 \rangle|^2 - \tfrac{1}{2}\alpha_1 [\delta_1/(\delta_1+\delta_s)] \sum_{\tau\neq 0}^{\infty} |\langle 0 | \mathbf{F}_1 | \tau \rangle|^2. \quad (7)$$

The first term is simply the electrostatic polarization of the molecule due to a net field $\langle \mathbf{F}_1 \rangle_{00}$ over the surface. This field would be zero if the solid were uniform up to its surface, as is roughly the case for an insulator. But on a metal it arises from an effective dipole layer[12] on the surface which may be caused by electron leakage through the surface barrier.[13(a)] (See, however, Mignolet.[13(b)]) The second term corresponds to the usual "dispersion" type of force. The physical meaning of this force becomes clearer if we write

$$\langle | \mathbf{F}_1 |^2 \rangle \equiv \sum_{\tau\neq 0}^{\infty} |\langle 0 | \mathbf{F}_1 | \tau \rangle|^2 = \langle | \mathbf{F}_1 |^2 \rangle_{00} - |\langle \mathbf{F}_1 \rangle_{00}|^2 = \langle | \mathbf{F}_1 - \langle \mathbf{F}_1 \rangle_{00} |^2 \rangle_{00}. \quad (8)$$

Thus the second term of Eq. (7) may be looked upon as arising from the additional polarization of the molecule due to the nonvanishing of the mean square of the fluctuations of the "surface" electric field about its average value. And of course, when the thermal motion of the nuclei in the solid is negligible, these fluctuations are due primarily to the quantum motion of the electrons. In fact, Eq. (8) can be written[14] in terms of first- and second-order density matrices for the electrons after making the Born-Oppenheimer approximation, in a form most useful when the solid is a metal. However we must be cautious with this interpretation of the second term of Eq. (7) due to the presence of the factor $[\delta_1/(\delta_1+\delta_s)]$. If for a metal the free-electron model is adopted and δ_s assumed negligible compared to δ_1, as was done by Prosen and Sachs,[15] this factor is unity making the above interpretation correct, i.e., only the molecule is polarized. However, the presence of a quite large δ_s, as in insulators and real metals, in effect accounts for the polarization of the surface in turn by the molecule.

Now, all the formulas[3,4] that are used for either insulators or metals can be shown to follow directly or indirectly from the general expression Eq. (7). When the surface is an insulator, Eq. (8) can be evaluated by summing (or integrating) the contributions from each molecule or section that makes up the solid considering these as fluctuating dipoles. Then the over-all van der Waals attraction can be obtained directly from one of the formulas[1–4] which give the intermolecular dispersion energy (such as London's and Kirkwood-Müller's) using empirical quantities such as polarizabilities and diamagnetic susceptibilities. Although a metal should be treated in terms of metallic electrons and it seems at first thought that it could not be divided into sections, the consideration of electron correlation[14] has been shown to lead to essentially the same result as though the metal behaved as an insulator made up of sections of oscillating dipoles.[16] With these considerations in mind, an estimate of the total potential between one molecule and the surface (and the heat of adsorption) can be obtained by assuming the usual Lennard-Jones [12–6] potential, frequently used in imperfect gas theory,[17] to hold between a portion of the solid and the molecule, and integrating.[18] The result which we shall need later is

$$U(R) = [3/2(5)^{\frac{1}{2}}]U_0[(r_0/R)^3 - \tfrac{1}{15}(r_0/R)^9],$$
$$U_0 = -[2(5)^{\frac{1}{2}}/9]\pi n \epsilon_0 r_0^3, \quad (9)$$

with R the distance from the molecule to the surface and ϵ_0, r_0 the parameters of the Lennard-Jones [12–6] potential.[17] Notice that at the minimum, R_0, the attractive portion of energy is simply 1.5 times U_0, the heat of adsorption at zero coverage.

The magnitude of the van der Waals attraction between a molecule and a surface can be estimated by the use of the "nonmetallic" or metallic type formulas some of which we have cited above. (These formulas will be mentioned here only by name; the reader is referred to the general references given for details.) Such estimates are usually found to yield values for the heats of adsorption,[4,19] however, which are too low. A similar situation has existed in the comparison of observed and calculated van der Waals attractions between two molecules.[20] Recently, however, in connection with London forces it has been shown[2,21] that the trouble often lies in the δ's, the "mean excitation

[12] W. G. Pollard, Phys. Rev. **61**, 740(A) (1942).

[13] (a) K. Huang and G. Wyllie, Proc. Phys. Soc. (London) **62**, 180 (1949); (b) J. C. P. Mignolet, in *Chemisorption* (Butterworths Scientific Publications, Ltd., London, 1957), p. 181.

[14] J. Bardeen, Phys. Rev. **58**, 727 (1940).

[15] E. J. R. Prosen and R. G. Sachs, Phys. Rev. **61**, 70 (1942).

[16] H. Margenau and W. G. Pollard, Phys. Rev. **60**, 128 (1941).

[17] See for example, Hirschfelder, Curtiss, and Bird, *Molecular Theory of Gases and Liquids* (John Wiley & Sons, Inc., New York, 1954).

[18] T. L. Hill, J. Phys. Chem. **63**, 456 (1959).

[19] R. A. Pierotti and G. D. Halsey, Jr., J. Phys. Chem. **63**, 680 (1959).

[20] K. S. Pitzer, *Quantum Chemistry* (Prentice-Hall, Inc., New York, 1954), p. 339.

[21] W. E. Donath, University of California, Ph.D. thesis (1958) and UCRL-8385 (July 21, 1958).

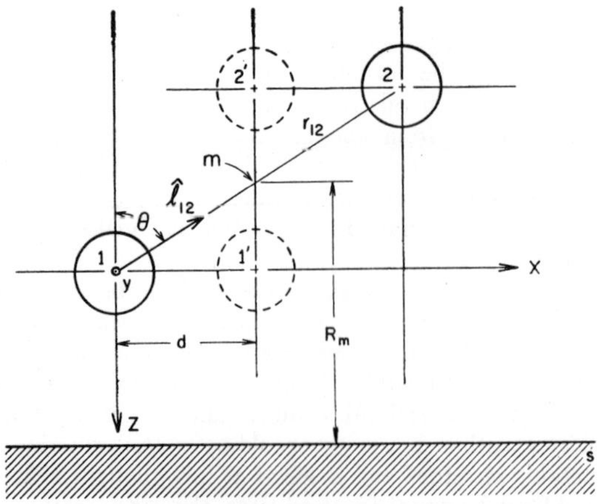

FIG. 1. Coordinates used in the general derivation of $E^{(3)}$.

energies." These also occur in Eq. (7) and in most of the specific formulas mentioned in the foregoing. In adsorption as well as in intermolecular dispersion energies the δ's are usually assumed to be approximately equal to the ionization potentials I. Recent considerations show[2,20,21] that in more than two-electron atoms electron correlation raises the δ by a factor of about 2.1 higher than I. It must be pointed out here that the use of this proper δ (for the molecules and possibly for the solid) will bring the insulator-type adsorption energy formulas such as London's into quite good agreement with experiment. Slater-Kirkwood formula[2,19] involves N, the number of electrons in the outer shell of molecules, instead of δ; here N should be regarded as an empirical parameter and "effective" values used.[2] Kirkwood-Müller formula[2,4,19] does not contain any of these "parameters" and although it is strictly valid for spherical systems,[2] it is still quite successful in adsorption.[19] In the rest of this article, we shall use the appropriate molecular δ's $(2.1I)$ instead of I wherever they occur.

The formulas derived from Eq. (7) and Eq. (8) strictly for metals[14,16] also predict heats which are too small.[19] Here the increase in δ_1 has a smaller effect on $E_{1s}^{(2)}$. However, it should be recalled that the δ's appearing as the main feature in both Bardeen's[14] and Margenau's and Pollard's[16] formulas were derived[14] for monovalent metals neglecting the effect of the periodic potential of the lattice and the effect of the polarizability of the ion cores. Thus, such formulas should not be applied to, for instance, transition metals that have incomplete d shells and might show very large "inner core" effects.

The low values predicted for the heats of adsorption by the theories above, in the way that they are often applied, have induced discussion[6] of another type of force, i.e., the charge-transfer[22] force as contributing to van der Waals adsorption on metals. However, in view of the lack of a sufficiently quantitative treatment of these forces in adsorption and considering the preceding discussion, we shall assume that van der Waals attraction of nonpolar molecules to metals as well as to insulators is given primarily by $E_{1s}^{(2)}$ of Eq. (7).

[22] R. S. Mulliken, J. Am. Chem. Soc. **74**, 811 (1952).

C. THIRD-ORDER ENERGY

Now, if instead of a single molecule on the surface several molecules are considered, no new effects arise in the nonoverlap second-order perturbation. The potential energy of the system is simply the sum of London attractions[1] between molecules and their individual van der Waals attractions to the surface as given by Eq. (7). To see the effect of the surface on the long-range interaction between molecules we must turn to the third-order term of the perturbation series. If, as we had done for the second-order term in Eq. (1), we consider the adsorbed molecules to be quite small, and after expanding the instantaneous potential of the molecules in multipole series in the usual fashion,[1] take only the leading dipole terms, the interaction operator between two molecules and the surface becomes

$$J=-\boldsymbol{\mu}_1\cdot\mathbf{F}_1-\boldsymbol{\mu}_2\cdot\mathbf{F}_2+\boldsymbol{\mu}_1\cdot\mathbf{T}_{12}\cdot\boldsymbol{\mu}_2, \tag{10}$$

where again $\mathbf{F}_1$ and $\mathbf{F}_2$ are the instantaneous electric fields produced at the centers of the molecules 1 and 2, respectively, by a given distribution of all the electrons and nuclei that make up the solid. The $\boldsymbol{\mu}$'s are the instantaneous dipole moments of the molecules (here we always mean inert gas atoms or other small quasi-spherical molecules) given by $e\sum_i \bar{r}_i$, i.e., summing the radius vectors of all the electrons of a given atom. The last term of Eq. (10) is the dipole-dipole operator for the interaction between two molecules with the tensor $\mathbf{T}_{12}$ given by

$$\mathbf{T}_{12}=(\mathbf{I}-\mathbf{l}_{12}\mathbf{l}_{12})/r_{12}^3. \tag{11}$$

Here $\mathbf{l}_{12}$ is the unit vector along r_{12}, the distance between the two molecules as shown in Fig. 1.

With the J of Eqs. (10) or (1) the third-order energy becomes[23]

$$E^{(3)}=\sum_{n=0}\sum_{m\neq n,0}\frac{\langle 0\mid J\mid n\rangle\langle n\mid J\mid m\rangle\langle m\mid J\mid 0\rangle}{(E_0-E_n)(E_0-E_m)}. \tag{12}$$

This energy vanishes for one molecule and the surface, i.e., for J given by Eq. (1). If on the other hand, Eq. (10) is substituted into Eq. (12), each matrix element there can be written as a sum of three terms and since the eigenstates of the system are taken in the product form as in Eq. (3) (with each factor forming an ortho-

[23] In general $E^{(3)}$ will have an additional term which is different in different perturbation theories (see: Morse and Feshbach, *Methods of Theoretical Physics* (McGraw-Hill Book Company, Inc., New York, 1953), Vol. II; however, this term vanishes in our case because of the dipole selection rules.

normal set) a typical matrix element is of the form

$$\langle 0 \mid J \mid n\rangle = -\langle 00 \mid \boldsymbol{\mu}_1\cdot \mathbf{F}_1 \mid \rho\tau\rangle\delta_{0\sigma} - \langle 00 \mid \boldsymbol{\mu}_2\cdot \mathbf{F}_2 \mid \sigma\tau\rangle\delta_{0\rho} + (\langle 0 \mid \boldsymbol{\mu}_1 \mid \rho\rangle\cdot \mathbf{T}_{12}\cdot \langle 0 \mid \boldsymbol{\mu}_2 \mid \sigma\rangle)\delta_{0\tau}, \quad (13)$$

where again σ, ρ, τ label the various eigenstates of molecules 1 and 2 and the entire solid (including all the nuclei and electrons that make it up), respectively. [The $\delta_{0\sigma}$, etc. are Kronecker deltas only here and should not be confused with the excitation energies in Eq. (4), Eq. (5), and in the forthcoming equations.] Then, multiplying out such matrix elements in Eq. (12) and using the dipole selection rules for spherical molecules,[24] it can be seen that only certain combinations of transitions giving rise to such terms as

$$\sum_{\tau\neq 0}\sum_{\sigma'\neq 0}\frac{\langle \boldsymbol{\mu}_1\rangle_{0\rho}\cdot\langle \mathbf{F}_1\rangle_{0\tau}\langle \boldsymbol{\mu}_2\rangle_{0\sigma'}\cdot\langle \mathbf{F}_2\rangle_{\tau 0}\langle \boldsymbol{\mu}_1\rangle_{\rho 0}\cdot\mathbf{T}_{12}\cdot\langle \boldsymbol{\mu}_2\rangle_{\sigma' 0}}{[\delta_1(0\rho)+\delta_s(0\tau)][\delta_1(0\rho)+\delta_2(0\sigma')]}, \quad (14)$$

with $\delta_1(0\rho)=E_1^{\rho}-E_1^{0}$ and similarly δ_s for the surface, contribute to the sum [Eq. (12)]. The derivation follows along quite similar lines to that given by Kihara[8] for three molecules. There are two types of terms designated by transitions such as

$$000\to\rho 00\to\rho\sigma' 0\to 000$$

and

$$\underbrace{000}_{0}\to\underbrace{0\sigma 0}_{n}\to\underbrace{\rho'\sigma 0}_{m}\to\underbrace{000}_{0},$$

in which only the ground state of the solid is involved, and six types of terms corresponding to different double transitions in each matrix element, such as

$$000\to\rho\sigma 0\to\rho 0\tau'\to 000.$$

The numerators of these six are all equivalent [Eq. (14)] but four of the denominators [of the form $(\delta_1+\delta_2)(\delta_1+\delta_s)$] are different from the other two [of the form $(\delta_1+\delta_s)(\delta_2+\delta_s)$]. After using mean excitation energies and polarizabilities for the molecules [Eqs. (5) and (6)] as had been done for $E_{1s}^{(2)}$ previously, the result is

$$E_{12s}^{(3)} = (\alpha_1^2/r_{12}^3)[\langle \mathbf{F}_1\rangle_{00}\cdot\langle \mathbf{F}_2\rangle_{00} - 3\langle \mathbf{F}_1\rangle_{00}\cdot \mathbf{l}_{12}\mathbf{l}_{12}\cdot\langle \mathbf{F}_2\rangle_{00}] + (\alpha_1^2/2r_{12}^3)[(2\delta_1+\delta_s)\delta_1/(\delta_1+\delta_s)^2]\sum_{\tau\neq 0}(\langle \mathbf{F}_1\rangle_{0\tau}\cdot\langle \mathbf{F}_2\rangle_{\tau 0} - 3\langle \mathbf{F}_1\cdot \mathbf{l}_{12}\rangle_{0\tau}\langle \mathbf{l}_{12}\cdot \mathbf{F}_2\rangle_{\tau 0}) \quad (15a)$$

or

$$E_{12s}^{(3)} \equiv E_{es}^{(3)} + E_{fl}^{(3)} \quad (15b)$$

where es means electrostatic and fl fluctuation and where for convenience we have assumed molecules 1 and 2 to be identical. In arriving at this result use has been made of relations of the type:

$$\sum_{\rho\neq 0}\langle \boldsymbol{\mu}_1\rangle_{0\rho}\cdot \mathbf{T}'\cdot\langle \boldsymbol{\mu}_1\rangle_{\rho 0} = \sum_{\rho\neq 0}\langle \boldsymbol{\mu}_1\rangle_{0\rho}\langle \boldsymbol{\mu}_1\rangle_{\rho 0}:\mathbf{T}' = \langle \mu_{1x}\rangle_{00}^2 \mathbf{I}:\mathbf{T}',$$

$$\mathbf{T}' = \langle \mathbf{F}_1\rangle\langle \mathbf{F}_2\rangle\cdot \mathbf{T}_{12}, \quad (17)$$

$$\mathbf{I}:\mathbf{T}' = \mathrm{Trace}\,\mathbf{T}' = (\langle \mathbf{F}_1\rangle\cdot\langle \mathbf{F}_2\rangle - 3\langle \mathbf{F}_2\cdot \mathbf{l}_{12}\rangle\langle \mathbf{F}_1\cdot \mathbf{l}_{12}\rangle)/r_{12}^3,$$

and τ in Eq. (15) still refers to an eigenstate of the semi-infinite solid as a whole.

The meaning of Eq. (15) is similar to that of Eq. (7) discussed previously. Again the first part, $E_{es}^{(3)}$, exists only when the surface possesses a net average electrostatic field $\langle \mathbf{F}\rangle_{00}$ and within the approximation of Eq. (6), i.e., the substitution of atomic polarizability into Eq. (16), it is the same as the classical interaction between two permanently induced dipoles and already has been considered by de Boer.[3] The second part $E_{fl}^{(3)}$, involving the sum

$$\langle \mathbf{F}_1\cdot \mathbf{F}_2 - 3\mathbf{F}_1\cdot \mathbf{l}_{12}\mathbf{l}_{12}\cdot \mathbf{F}_2\rangle_{\tau} \equiv \sum_{\tau\neq 0}(\langle \mathbf{F}_1\rangle_{0\tau}\cdot\langle \mathbf{F}_2\rangle_{\tau 0} - 3\langle \mathbf{F}_1\cdot \mathbf{l}_{12}\rangle_{0\tau}\langle \mathbf{l}_{12}\cdot \mathbf{F}_2\rangle_{\tau 0}) = (\langle \mathbf{F}_1\cdot \mathbf{F}_2\rangle_{00} - 3\langle \mathbf{F}_1\cdot \mathbf{l}_{12}\mathbf{l}_{12}\cdot \mathbf{F}_2\rangle_{00}) - (\langle \mathbf{F}_1\rangle_{00}\cdot\langle \mathbf{F}_2\rangle_{00} - 3\langle \mathbf{F}_1\rangle_{00}\cdot \mathbf{l}_{12}\mathbf{l}_{12}\cdot\langle \mathbf{F}_2\rangle_{00}) \quad (18)$$

is again of the dispersion or fluctuation type. When δ_s is small, it represents the interaction resulting from the additional simultaneous polarization of the molecules by the in phase fluctuations of the surface fields at the points 1 and 2. In fact, Eq. (18) can be simply regarded as the "correlation coefficient" of $\mathbf{F}_1$ and $\mathbf{F}_2$ in the sense of elementary statistics.

The electrostatic or average fields in Eq. (15) must be in the z direction (see Fig. 1) whenever they exist, on account of the symmetry in the xy plane, so that the electrostatic energy in Eq. (15a) is given by

$$E_{es}^{(3)} = (\alpha_1^2/r_{12}^3)\langle F_1^z\rangle_{00}\langle F_2^z\rangle_{00}(1-3\cos^2\theta), \quad (19)$$

and similarly we have from Eq. (7)

$$E_{es}^{(2)} = -\tfrac{1}{2}\alpha_1\langle F_1^z\rangle_{00}^2. \quad (20)$$

When one of these energies is known the other is obtainable by the appropriate elimination of $\langle F^z\rangle$'s.[3]

On insulators Eq. (19) and (20) are usually unimportant[3] and fluctuation energies constitute the primary effects. On metal and charcoal surfaces both es and fl energies must be taken into account. The discussion of $E^{(3)}$ and $E^{(2)}$ in relation to one another when es terms are not negligible will be deferred until the end of this section, i.e., until after a detailed derivation of what we have called the fluctuation energy between adsorbed molecules, to which we now turn.

[24] Landau and Liftschitz, *Non-Relativistic Quantum Mechanics* (Addison-Wesley Publishing Company, Inc., Reading, Massachusetts, 1958).

The fluctuation parts of Eqs. (7) and (15a) are given by

$$E_{f1}^{(2)}=-\tfrac{1}{2}\alpha_1[\delta_1/(\delta_1+\delta_s)]\langle\,|\,\mathbf{F}_1\,|^2\rangle_\tau, \qquad (21a)$$

$$\langle\,|\,\mathbf{F}_1\,|^2\rangle_\tau=\langle(F_1^z)^2\rangle_\tau+2\langle(F_1^x)^2\rangle_\tau, \qquad (21b)$$

and

$$E_{f1}^{(3)}=(\alpha_1^2/r_{12}^3)[\delta_1(2\delta_1+\delta_s)/2(\delta_1+\delta_s)^2](\langle\mathbf{F}_1\cdot\mathbf{F}_2\rangle_\tau -3\langle\mathbf{F}_1\cdot\mathbf{l}_{12}\mathbf{l}_{12}\cdot\mathbf{F}_2\rangle_\tau), \qquad (22)$$

where the summations over the eigenstates of the solid not including the $\tau=0$ ground state have been indicated by $\langle\;\;\rangle_\tau$. Eq. (21b) follows from the equivalence of the x and y coordinates over the surface.

In terms of the coordinates shown in Fig. 1, Eq. (22) becomes

$$E_{f1}^{(3)}=(\alpha_1^2/r_{12}^3)[\delta_1(2\delta_1+\delta_s)/2(\delta_1+\delta_s)^2][\langle F_1^xF_2^x\rangle_\tau \cdot(1-3\sin^2\theta)+\langle F_1^yF_2^y\rangle_\tau+\langle F_1^zF_2^z\rangle_\tau(1-3\cos^2\theta) -3\sin\theta\cos\theta\langle F_1^xF_2^z+F_1^zF_2^x\rangle_\tau], \qquad (23)$$

θ denoting the angle between $\mathbf{l}_{12}$ and the normal to the surface. The coordinate axes have been so chosen that z is along the normal and the xz plane contains $\mathbf{l}_{12}$ (Fig. 1). Notice also that, e.g., $\langle F_1^xF_2^z\rangle_\tau$ is identical with $\langle F_1^xF_2^z\rangle_{00}$ i.e., the expectation value over the ground state of the solid, because $\langle F_1^x\rangle_{00}$ is zero [see Eqs. (8) and (18)]. Equation (23) can be simplified considerably by noting that $\langle F_1^xF_2^z\rangle_{00}$ and $\langle F_1^zF_2^x\rangle_{00}$ vanish because F^x is an odd function in all the x coordinates involved (i.e., all electrons and nuclei of the solid). In addition the matrix elements, $\langle F_1^xF_2^x\rangle_\tau$, $\cdots$, can be replaced to a good approximation by the corresponding quantities $\langle F_{1'}^xF_{2'}^x\rangle_\tau$, $\cdots$ (Fig. 1); i.e., the points 1 and 2 can be brought on top of one another along the normal to the surface passing through the midpoint m of r_{12}. This approximation can be partly justified in the following manner.

Since the states of the solid, $\psi_0, \psi_\tau\cdots$, form an orthonormal set, we have, as in Eq. (18),

$$\langle F_1^zF_2^z\rangle_\tau=\sum_{\tau\neq0}^{\infty}\langle F_1^z\rangle_{0\tau}\langle F_2^z\rangle_{\tau0}, \qquad (24)$$

and similarly for x and y. Now because of translational symmetry the over-all eigenstates describing the solid must be of generalized Bloch function form; i.e.,

$$\psi_\tau=[\exp(i\mathbf{k}_\tau\cdot\mathbf{R})]u_\tau(\mathbf{R}), \qquad (25)$$

where $\mathbf{R}$ denotes the coordinates of *all* the electrons and nuclei, $\mathbf{k}_\tau$ is a generalized wave vector and u_τ a periodic function of $\mathbf{R}$ depending on the symmetry of the solid (In the xy plane) Thus the matrix elements such as $\langle F_1^z\rangle_{0\tau}$ will be of the form

$$\langle F_1^z\rangle_{0\tau}=\int\exp(i\mathbf{k}_\tau'\cdot\mathbf{R})F_1^zu_0^*u_\tau d\mathbf{R}. \qquad (26)$$

Here F_1^z is the sum of the instantaneous electric fields of all the electrons and nuclei in the solid, and these electrons and nuclei may be assumed to be quite uniformly distributed at *all* instances. This assumption allows us to consider $\mathbf{F}_1$ as a nearly periodic operator in the sense that if all the electrons and nuclei were to be translated by a lattice constant in the x or y directions, it would remain quite unchanged. In this case, according to group theory, only those matrix elements [Eq. (26)] for which the $\mathbf{k}_\tau'(x, y)$ equals zero would be nonvanishing.[25,26] Even though this selection rule is not rigorously applicable here because $\mathbf{F}_1$ is not really a "periodic" operator, nevertheless it suggests that the greatest contribution to Eq. (24) will come from excitations for which the generalized wave vector $\mathbf{k}(x, y)$ in the xy plane is approximately unchanged. It is interesting to note that even when the crude free-electron approximation is made when the solid is a metal, a similar conclusion can be shown[15] to follow; in this case where the assumption of the periodicity of $\mathbf{F}_1$ cannot be made at all, Eq. (26) reduces to a single electron matrix element involving ordinary one-electron Bloch functions. Nevertheless it is found[15] that even these matrix elements have appreciable values only when the changes in the wavelength (in the xy plane) of an electron are small compared with the reciprocal of a distance of the order of a lattice constant. [See Eq. (14) of reference 15].

If then, it is further assumed that d (Fig. 1) is either much smaller than or approximately equal to a lattice constant of the "surface," the origin of the coordinates for the wave functions in Eq. (26) can be shifted from 1 and 2 to 1′ or 2′, respectively, and in view of the translational properties just discussed, one gets

$$\langle F_1^xF_2^x\rangle_\tau=\sum_{\tau\neq0}^{\infty}\langle F_1^x\rangle_{0\tau}\langle F_2^x\rangle_{\tau0}\cong\sum_{\tau\neq0}^{\infty}\langle F_{1'}^x\rangle_{0\tau}\langle F_{2'}^x\rangle_{\tau0} =\langle F_{1'}^xF_{2'}^x\rangle_\tau \qquad (27)$$

and similarly for y and z. Eq. (27) makes the x and y matrix elements in Eq. (23) equivalent and hence

$$E_{f1}^{(3)}\approx(\alpha_1^2/r_{12}^3)[\delta_1(2\delta_1+\delta_s)/2(\delta_1+\delta_s)^2][\langle F_{1'}^zF_{2'}^z\rangle_\tau -\langle F_{1'}^xF_{2'}^x\rangle_\tau](1-3\cos^2\theta). \qquad (28)$$

When the two molecules are the same distances from the surface ($\theta=90°$), the points 1′ and 2′ coincide with m the midpoint of r_{12} and we already have $\langle F_{1'}^zF_{2'}^z\rangle_\tau$ and $\langle F_{1'}^xF_{2'}^x\rangle_\tau$ going over into $\langle(F_m^z)^2\rangle_\tau$ and $\langle(F_m^x)^2\rangle_\tau$, which are the same matrix elements appearing in $E_{f1}^{(2)}$. [Eq. (21)]. On the other hand when θ differs from 90° so that 1′ and 2′ are above and below m, by writing

$$\langle\mathbf{F}_{1'}\cdot\mathbf{F}_{2'}\rangle_\tau\sim\langle\,|\,\mathbf{F}_m\,|^2\rangle_\tau \qquad (29)$$

[25] N. F. Mott and H. Jones, *The Theory of the Properties of Metals and Alloys* (reprinted by Dover Publishers, Inc., New York, 1936), p. 59.

[26] M. Tinkham, *Advanced Quantum Mechanics of Atoms, Molecules, and Solids* (A. S. U. C. Store, University of California, Berkeley, 1958), p. 94.

we might be expected to provide a reasonable approximation. Actually somewhat later we shall give another quite different derivation of $E_{\mathrm{fl}}^{(3)}$ which will show among other assumptions made in the foregoing, the validity of Eq. (29).

It is, by this time, apparent that Eqs. (28) and (29) would permit the replacement of the matrix elements in $E_{\mathrm{fl}}^{(3)}$ in terms of $E_{\mathrm{fl}}^{(2)}$ (which might eventually be viewed as an empirical quantity) provided a simple relation can be found among the various components of $\langle|\mathbf{F}|^2\rangle_\tau$. Fortunately such a relation does indeed exist. The same proportionality constant between $\langle(F^x)^2\rangle_\tau$ and $\langle(F^z)^2\rangle_\tau$ is obtained upon an examination of two very different models of a solid and the result is perhaps perfectly general. First, it has been shown[15] that in the nondegenerate or degenerate free-electron gas model of a metal,

$$\langle(F^x)^2\rangle_\tau=\langle(F^y)^2\rangle_\tau=\tfrac{1}{2}\langle(F^z)^2\rangle_\tau. \qquad (30)$$

Secondly, if we consider an insulator or a metal (Margenau-Pollard[16] model) as made up of fluctuating-dipole sections (in the sense of dispersion forces) then $\langle(\mathbf{F}^x)^2\rangle_{00}$ etc. can be calculated from

$$\langle|\mathbf{F}_m|^2\rangle_{00}\cong\int\langle|\mathbf{F}_{m\omega}|^2\rangle_{00}d\omega, \qquad (31)$$

where $d\omega$ refers to an infinitesimal volume element in the solid and $\langle|\mathbf{F}_{m\omega}|^2\rangle_{00}$ is the expectation value of the square of the field produced at m by the section $d\omega$. By using

$$\mathbf{F}_{m\omega}=-\mathbf{T}_{m\omega}\cdot\boldsymbol{\mu}_\omega, \qquad (32)$$

where $\mathbf{T}_{m\omega}$ and $r_{m\omega}$ are analogous to $\mathbf{T}_{12}$ and r_{12} of Eq. (11) we find for the components

$$\langle(F_{m\omega}^x)^2\rangle_{00}=[\langle(\mu_\omega^x)^2\rangle_{00}/r_{m\omega}^6][1+3(l_{\omega m}^x)^2], \qquad (33)$$

and similarly for y and z, $l_{\omega m}^x$ being the x component of the unit vector in the direction from ω to m. Equation (31) can be integrated separately for each component; e.g.,

$$\langle(F_m^x)^2\rangle_{00}\cong\langle(\mu_x)^2\rangle_{00}\int_{R_m}^{\infty}\int_{-\infty}^{\infty}\int_{-\infty}^{\infty}[(1+3x^2/r^2)/r^6]dxdydz; \qquad (34)$$

and $\langle(\mu_x)^2\rangle_{00}$ is related to the polarizability of the solid per unit volume [Eqs. (5), (6)]. Equations (33) and (34) again lead to exactly the same result as Eq. (30) after performing the elementary integrations. Thus we take Eq. (30) as a sufficiently general result.

Finally, combining Eqs. (21), (22), and (28)–(30), we have an approximate relation between $E_{\mathrm{fl}}^{(3)}$ and $E_{\mathrm{fl}}^{(2)}$ analogous to the relation between the es energies in Eqs. (19) and (20):

$$E_{\mathrm{fl}}^{(3)}\cong(\alpha_1/r_{12}^3)\Delta_0\,|\,E_{\mathrm{fl}}^{(2)}(R_m)\,|\,\tfrac{1}{4}(1-3\cos^2\theta), \qquad (35)$$

where

$$\Delta_0=(2\delta_1+\delta_s)/(\delta_1+\delta_s),$$

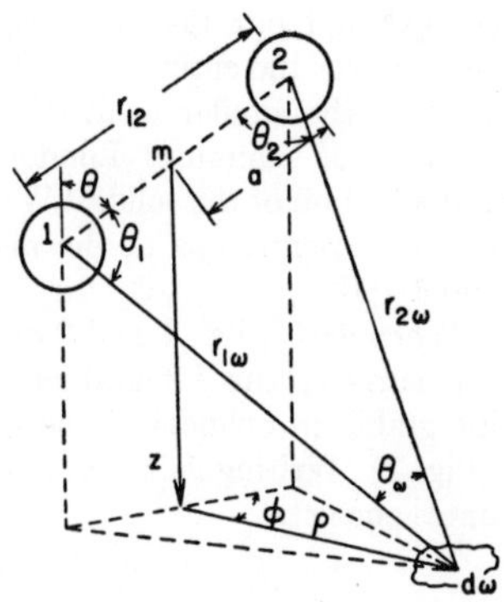

FIG. 2. Derivation of $E^{(3)}$ by integration.

and $E_{\mathrm{fl}}^{(2)}(R_m)$ usually[3,4,14,16] is of the form

$$E_{\mathrm{fl}}^{(2)}=-K_{\mathrm{f}}/R_m^3, \qquad (36)$$

with K_{f} a constant (to be determined empirically when $E_{\mathrm{es}}^{(2)}\sim0$, or from the theories discussed in Sec. B) and R_m the distance from m to the surface.

The derivation presented has the advantage of being general and independent of the nature of the solid. Everything that would specifically involve the solid has been taken care of by $E_{\mathrm{fl}}^{(2)}$ which of course may be viewed if necessary as an experimental quantity. However a crucial assumption has been made in Eq. (27) depending on the uniformity of electron and nuclear distribution at *all* times, an assumption which seems reasonable intuitively but cannot be fully justified. We shall give another derivation of $E_{\mathrm{fl}}^{(3)}$, however, which is quite independent of the first one, yet leads essentially to the same results as Eq. (35).

The matrix elements of the electric fields appearing in Eq. (22) can be evaluated directly if a specific model is adopted for the solid. This evaluation would be possible, for instance, for the free-electron model of a metal[15]; however this model is not suitable because van der Waals interactions are very strongly influenced by electron correlation.[2,14,21] The oscillating-dipoles model which we have already discussed is a valid one for insulators and even for metals.[16] To evaluate $E_{\mathrm{fl}}^{(3)}$ for this model we again consider the solid to consist of small elementary portions $d\omega$ and write as in Eq. (31)

$$\langle\mathbf{F}_1\cdot\mathbf{F}_2\rangle_\tau-3\langle\mathbf{F}_1\cdot\mathbf{l}_{12}\mathbf{l}_{12}\cdot\mathbf{F}_2\rangle_\tau$$
$$\cong\int[\langle\mathbf{F}_{1\omega}\cdot\mathbf{F}_{2\omega}\rangle_{00}-3\langle\mathbf{F}_{1\omega}\cdot\mathbf{l}_{12}\mathbf{l}_{12}\cdot\mathbf{F}_{2\omega}\rangle_{00}]d\omega, \qquad (37)$$

where $\mathbf{F}_{1\omega}$ and $\mathbf{F}_{2\omega}$ are the fields due to the section $d\omega$ at the points 1 and 2. After substituting Eq. (32) and some manipulation the integral in Eq. (37) becomes

$$3\langle(\mu_\omega^x)^2\rangle_{00}\int\frac{(3\cos\theta_1\cos\theta_2\cos\theta_\omega+1)d\omega}{r_{1\omega}^3r_{2\omega}^3} \qquad (38)$$

in terms of the angles and distances defining the triangle 12ω as shown in Fig. 2. Of course the integrand is just the expression that appears in the triple-dipole energy[8] and our task now is the integration of this energy over the entire solid for given positions of 1 and 2, as in the similar evaluation of $E^{(2)}$ by integrating (instead of summing) London forces between 1, say, and a section of the solid.[3,4] We shall integrate Eq. (38) for two orientations of the molecules only, i.e., $\theta=0°$ and $\theta=90°$.

When $\theta=0°$, i.e., 1 and 2 are on the same normal to the surface, the integral A in Eq. (38) is exactly integrable in cylindrical coordinates with origin at m (Fig. 2). Writing θ_1, θ_2, θ_ω, and $r_{1\omega}$, $r_{2\omega}$ in these coordinates one gets

$$A=A_1+3A_c \tag{39}$$

where

$$A_1=\int\frac{dV}{r_{1\omega}^3 r_{2\omega}^3}=2\pi\int_{R_m}^{\infty}\int_0^{\infty}\frac{\rho d\rho dz}{[b^4-4a^2z^2]^{3/2}},$$

$$A_c=\int\frac{\cos\theta_1\cos\theta_2\cos\theta_\omega dV}{r_{1\omega}^3 r_{2\omega}^3}$$

$$=-2\pi\int_{R_m}^{\infty}\int_0^{\infty}\frac{(z^2-a^2)(z^2-a^2+\rho^2)\rho d\rho dz}{[b^4-4a^2z^2]^{5/2}},$$

and $b^2\equiv z^2+\rho^2+a^2$.

The result is

$$A(\theta=0°)=-\pi/6R_m^3. \tag{40}$$

On the other hand, when $\theta=90°$, i.e., when 1 and 2 are in the same monolayer, the integration is somewhat more involved owing to the appearance of $\cos\phi$ in the denominators of

$$A_1=\int\frac{dV}{b^6(1-k^2\cos^2\phi)^{3/2}},$$

and

$$A_c=\int\frac{a^2(1-2a^2/b^2)(1-\rho^2\cos^2\phi/a^2)dV}{b^8(1-k^2\cos^2\phi)^{5/2}},$$

$$(k^2\equiv 4a^2\rho^2/b^4). \tag{41}$$

Nevertheless, by expanding these denominators in binomial series and keeping 3–4 terms, A is obtained to about one percent accuracy:

$$A(\theta=90°)\cong(0.34\pi/R_m^3)f(a,R_m),$$

$$f(a,R_m)=\{[(2R_m^2-a^2)/3(R_m^2+a^2)]$$
$$+0.320[a^2R_m^4/(R_m^2+a^2)^3]-0.422[R_m^4/(R_m^2+a^2)^2]$$
$$+(a^2/5R_m^2)-(a^4/7R_m^4)+(a^6/9R_m^6)-\cdots\}, \tag{42}$$

where the arctan resulting from the final z integration has also been expanded in a power series of (a/R_m).

The connection between these and the result of the former derivation, Eq. (35), becomes apparent when we substitute the troublesome quantity $\langle(\mu_\omega^z)^2\rangle_{00}$, related to the polarizability of the metal, in Eq. (38) from $E^{(2)}$ based on the same model[16] [see Eq. (34)]. One obtains

$$E_{f1}^{(3)}(\theta=0°)=-(\alpha_1/2r_{12}^3)\Delta_0(K_f/R_m^3) \tag{43a}$$

and

$$E_{f1}^{(3)}(\theta=90°)\cong(\alpha_1/r_{12}^3)\Delta_0(K_f/R_m^3)f(a,R_m). \tag{43b}$$

Equation (43a) is exactly the same as Eq. (35) provided a R_m^{-3} law is assumed for $E_{f1}^{(2)}(R_m)$, and this result lends additional support to the assumption of Eq. (29) that had been made in the previous derivation. On the other hand, Eq. (43b) would also be the same as Eq. (35) if $f(a,R_m)$ of the former were to equal the factor of $\frac{1}{4}$ in the latter. An examination of the Eq. (42) reveals that $f(a,R_m)$ is a very slowly varying function of both a and R_m, and its value remains within about $\pm 15\%$ of 0.25 over the relevant ranges of these variables.

The previous derivation of Eq. (35) had been quite independent of the nature of the solid, and, for instance, the **F**'s are not restricted to dipole fields as in the second derivation. Yet, when the proper (e.g., empirical when $\langle\mathbf{F}\rangle_{00}\sim 0$) $E^{(2)}$'s are used in both, the two derivations lead essentially to the same result (at least for the interesting cases, $\theta=0°$ and $\theta=90°$, examined). This is not surprising, however, since any differences that would have arisen between the results of the two derivations as a consequence of the specific model used in the second, are largely cancelled in the elimination of $\langle(\mu_\omega^z)^2\rangle_{00}$ which is common to both $E_{f1}^{(2)}$ and $E_{f1}^{(3)}$.

$E_{es}^{(3)}$ and $E_{f1}^{(3)}$ in Eqs. (19) and (35) have the same dependence on θ and r_{12} and can be combined together giving the total surface intermolecular potential $E_{12s}^{(3)}$,

$$E_{12s}^{(3)}\cong(\alpha_1/r_{12}^3)(1-3\cos^2\theta)\{\alpha_1\langle F_m^z\rangle_{00}^2$$
$$+\tfrac{1}{4}\Delta_0\,|\,E_{f1}^{(2)}(R_m)\,|\,\}. \tag{44}$$

The electrostatic or average field $\langle F^z\rangle_{00}$ is independent of the x, y coordinates on the surface and we have also made the approximation

$$\langle F_1^z\rangle_{00}\langle F_2^z\rangle_{00}\approx\langle F_m^z\rangle_{00}^2 \tag{45}$$

and used the average field at m. The total $E_{12s}^{(3)}$ cannot be related to the total $E_{1s}^{(2)}$, because the relative importance of the es terms are quite different in each case. Nevertheless Eq. (44) can be put into a very useful form after substituting for $\langle F_m^z\rangle_{00}$ from Eq. (20),

$$E_{12s}^{(3)}\cong(\alpha_1/r_{12}^3)(1-3\cos^2\theta)\,|\,E_{1s}^{(2)}(R_m)\,|$$
$$\times[(\Delta_0+\gamma_m\Delta_1)/4], \tag{46}$$

where

$$\Delta_1\equiv[(6\delta_1+7\delta_s)/(\delta_1+\delta_s)],$$

and we have defined the parameter

$$\gamma_m\equiv E_{es}^{(2)}(R_m)/E_{1s}^{(2)}(R_m)$$
$$=E_{es}^{(2)}/(E_{es}^{(2)}+E_{f1}^{(2)}), \tag{47}$$

i.e., the electrostatic fraction of the *total* van der Waals attraction of a molecule at m to the surface. Here γ_m is not a constant but depends on R_m, because $E_{fl}^{(2)}$ varies in general[3,4] as R_m^{-3}, whereas $E_{es}^{(2)}$ falls off exponentially.[13] However if we are interested in the behavior of molecules strictly in a monolayer, then we find a simple proportionality between $E^{(3)}$ and $E^{(2)}$. In general the range of R_m that is important for $E^{(3)}$ is not very large so that we might replace γ_m by an average $\bar{\gamma}$ either over the whole range or at least up to a certain R_m^0 and then assume $\gamma_m=0$ for $R_m > R_m^0$.

If a Lennard-Jones (12–6) potential had been assumed between two molecules (e.g., in inert gases), when the molecules are adsorbed on a surface the new intermolecular potential is modified and now becomes

$$\epsilon'(r_{12}) = \epsilon_0[(r_0/r_{12})^{12} - 2(r_0/r_{12})^6] + [S(R_m)/r_{12}^3](1-3\cos^2\theta), \quad (48)$$

provided the original r_{12}^{-12} repulsion has not been altered appreciably by the surface. Here ϵ_0 and r_0 are unchanged and are the gas phase parameters[17]; $S(R_m)$ includes both the es and fl contributions [Eq. (46)]. When more than two molecules are considered on the surface, the total potential energy of interaction *between* the molecules themselves will be obtained by summing ϵ_{12}' over all pairs. In other words, the *intermolecular* energy is still "pairwise additive" to the same extent as the gas phase Lennard-Jones potential. Bade[11] has found that for a system of molecules the third-order energy is triplewise additive. With a system of molecules and the surface the only additional triplets that we haven't included in $E^{(3)}$ is the triple-dipole energy[8] of three adsorbed molecules and this of course is the same as in the gas phase. Then, pairwise additivity of $\epsilon'(r_{12})$, Eq. (48), to the same extent as the Lennard-Jones (12–6)$\epsilon(r_{12})$ follows. An estimate of the contribution of higher order perturbations to $\epsilon'(r_{12})$ can also be obtained from the work of Bade[11] with reference to the model we have used in the second derivation of $E^{(3)}$. Replacing summation over the solid by integration as in Eq. (37) the effect of higher order perturbations on ϵ_{12}' can be shown to be negligible. This is to be expected since the higher terms involve higher inverse powers of the distances.

The full use of Eqs. (46)–(48) in conjunction with experimental data and relative magnitudes of $E_{fl}^{(3)}$ and $E_{es}^{(3)}$ will be discussed in the next section.

D. APPLICATIONS AND MAGNITUDES

$E^{(3)}$ as given by Eq. (46) can be obtained most directly when the electrostatic effects are quite unimportant as on most insulators.[3] One only has to substitute for $E_{1s}^{(2)}$. Especially in a monolayer the repulsion between two atoms can be estimated simply by using $1.5U_0$ for $E_{1s}^{(2)}$, [Eq. (9)], if we assume the atoms to be restricted to their minimum energy distances (R_0) from the surface and behave like a two-dimensional gas; then one does not need an estimate of this distance $R_m=R_0$ but only U_0, the energy of adsorption at zero coverage.

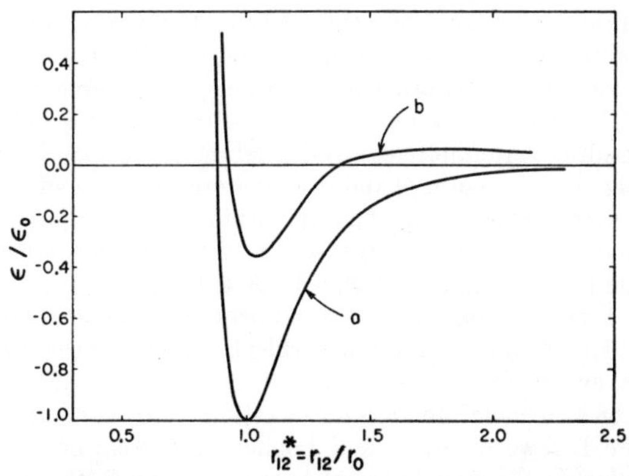

FIG. 3. Potential energy of interaction between two Xe atoms; curve a, Lennard-Jones (12–6) potential in the gas phase; curve b, Lennard-Jones (12–6) plus the third-order repulsion between two Xe atoms in a monolayer on tungsten. Both atoms are the same distance R_0, from the surface.

The most straightforward application of Eq. (46) is in connection with the virial coefficients treatment of physical adsorption.[5] There, $E_{1s}^{(2)}$ can be obtained without the complications of nonzero coverage from B_{1s}, the "second-surface virial coefficient" involving a single molecule and the surface.[5] When $\bar{\gamma}$ is not negligible, the total $E^{(2)}$ might be approximated by a general R_m^{-n} law and its parameters obtained from a best fit to B_{1s} data. Similarly $E_{12s}^{(3)}$ is related to C_{12s}, the two-molecule-surface virial coefficient.[5b] A combination of B_{1s} and C_{12s} would enable one to get an estimate of an average $\bar{\gamma}$. Notice that although in general it is difficult to estimate a reliable value for δ_s, the surface mean excitation energy, its appearance in Δ_0 and Δ_1 of Eq. (46) is not critical. Because Δ_0 varies only from unity to two, and Δ_1 from six to seven as (δ_s/δ_1) ranges from infinity to zero. In general Δ_0 will lie close to 1.5 and Δ_1 to 6.5.

A few applications with actual data will serve to demonstrate the use of some of these ideas and at the same time show the relative magnitudes of $E_{fl}^{(3)}$ and $E_{es}^{(3)}$.

Ehrlich and Hudda[7] have estimated the heat of desorption of Xe on tungsten to be about 8 kcal/mole and $E_{es}^{(2)} \sim 1.6$ kcal/mole. If we neglect any effects of nonzero coverage and assume that $(8-1.6)=6.4$ kcal/mole represents the energy minimum U_0 of a Lennard-Jones (9–3) potential, then $E_{fl}^{(2)}(R_0) \sim 9.6$ kcal/mole.

The same value is obtained from Kirkwood-Müller formula[2,4] at $R_m=3.1$ A, so that this estimate is reasonable. Taking δ_s very roughly as twice the ionization potential I of tungsten atom (see Sec. B about δ_1's), $\Delta_0 \sim 1.59$. With these values one gets $E_{fl}^{(3)} \sim 0.163$ kcal/mole and $E_{es}^{(3)} \sim 0.133$ kcal/mole, at $r_{12}=r_0$ and when $\theta=90°$ and the two molecules are at their equilibrium distance from the surface. Comparing these figures with the gas Lennard-Jones (12–6) potential minimum ϵ_0 between two Xe atoms, we observe that $E_{fl}^{(3)}$ amounts to $\sim 37\%$ and $E_{es}^{(3)} \sim 30\%$ of ϵ_0. The resulting intermolecular potential is the curve b in Fig. 3. On the other hand if we consider the case of the two molecules on top of another as in multilayer formation we get $E_{fl}^{(3)} \sim -0.18\epsilon_0$, i.e., an attraction when the first molecule is at $R_1 \sim 3.5$ A and $r_{12}=r_0$.

Freeman[27] has studied the adsorption of single atoms of Kr on alumina and A on carbon black. Neglecting the es effects for this system and using $E_{1s}^{(2)}=1.5U_0=$ 4.58 kcal/mole[27] and $\delta_s \sim 8.2$ ev estimated by comparing the Kirkwood-Müller and London formulas,[2] one obtains $E_{12s}^{(3)}(R_0) \sim 23.1\%$ of ϵ_0 at $r_{12}=r_0$ and $\theta=90°$ and about -10% ϵ_0 attraction when $\theta=0°$.

Of course very often it is also possible to estimate $E^{(3)}$ using a calculated value of $E_{fl}^{(2)}$, e.g., from Kirkwood-Müller's formula. For instance, the attraction of Kr to tungsten, $E_{fl}^{(2)}$, is found to be ~ 6.9 kcal/mole at a distance of $R_0 \sim 3$ A by this formula. This leads to 0.32 ϵ_0 repulsion in the monolayer at $r_{12}=r_0$. (The Lennard-Jones parameters ϵ_0, r_0 in the above calculations as well as I's have been taken from reference 20, p. 339).

Finally the estimation of $\bar{\gamma}$ from data can be demonstrated on the system: A and carbon black. At the distance of minimum energy R_0 one has

$$E_{fl}^{(2)} \sim 1.5U_0(1-\gamma)/[1+(\gamma/2)] \quad (49)$$

if it is assumed that a Lennard-Jones (9–3) potential has been fitted only to $E_{fl}^{(2)}$ excluding $E_{es}^{(2)}$. We take $\delta_s \sim 2I=22.4$ ev and from Freeman[27] $1.5U_0=5.8$ kcal/mole. Then for a *total* repulsion, $E_{12s}^{(3)}$ between two A atoms in a monolayer that would amount to about 70% of ϵ_0 at $r_{12}=r_0$, as in the Xe–W case, one would get $\bar{\gamma} \sim 38\%$. This number compares favorably with $\bar{\gamma} \sim 30\%$ that one gets from Eq. (49) after calculating $E_{fl}^{(2)}$ from Kirkwood-Müller's formula at $R_0=$ 2.94 A.[27]

The results presented above might be summarized by stating that the third-order perturbation introduces, aside from the "electrostatic" effect, a repulsion between the molecules in a monolayer which amounts to about 20–40% of their gas phase potential minimum, ϵ_0(at $r_{12}=r_0$). The same third-order fluctuation energy gives rise to an attraction, if two molecules lie on top of one another, which might be about 10–20% of ϵ_0. This attraction is among the factors contributing to the formation of multilayers (multimolecular adsorption).

It is well known that with a relatively dense adsorbate resembling a liquid only a small amount of additional attraction above and beyond the heat of liquefaction of the adsorbate is needed to provide for the formation of multilayers.[3] Although the greater portion of this energy is provided by the ordinary van der Waals field of the surface,[28] i.e., by $E_{fl}^{(2)}$, (es effects decreasing exponentially from the surface, might be negligible in the additional layers) $E_{fl}^{(3)}$ as discussed above is also a contributing factor; in the Xe–W example with only two molecules on the surface $E_{12s}^{(3)}$ amounts to roughly ten percent of the $E_{fl}^{(2)}$ acting on the top molecule at $R_2 \sim 8$ A.

ACKNOWLEDGMENT

We are indebted to Dr. Mark P. Freeman who attracted our attention to this problem.

This research was performed under the auspices of the U. S. Atomic Energy Commission.

[27] M. P. Freeman, J. Phys. Chem. **62**, 723 (1958).

[28] T. L. Hill, Advances in Catalysis **4**, 236 (1952); G. D. Halsey, *op. cit.*, p. 265.

[Reprinted from the Journal of the American Chemical Society, **84**, 2025 (1962).]

[CONTRIBUTION FROM THE RICE UNIVERSITY, HOUSTON, TEXAS]

Solubility and the Nature of Bonding in Fused Alkali Halide–Metal Systems[1]

BY KENNETH S. PITZER

RECEIVED NOVEMBER 9, 1961

A solution of an alkali metal in a fused halide of the same metal is discussed in terms of the mixing of the two negative species, halide ions and F-center-like electrons in cavities, within the irregular lattice of positive ions. The positive energy of mixing is related to the excess energy of a hypothetical metal comprising an ionic lattice of F-centers and positive ions over that of the true metal. The theory predicts substantial positive excess entropies of mixing. The solubility data of Bredig and collaborators are shown to be consistent with this model and certain parameters are evaluated for each system.

Systems which show a continuous transition from metallic to non-metallic character are of special interest; examples are solutions of the alkali metals in liquid ammonia and in fused alkali halides. Certain conclusions about the nature of the ammonia solutions were presented recently.[2,3] In the present paper the solutions of metals in fused alkali halides are considered from two aspects. First, a qualitative theory is developed for these solutions and, second, the theory is tested against solubility data and certain parameters are evaluated empirically. These calculations are based upon the extensive experimental results of Bredig and collaborators.[4]

These fused salt–metal systems display thermodynamic properties which, except for the high temperatures, are similar to those expected for mixtures of normal non-polar molecular liquids. These latter systems are commonly called regular solutions. It seems surprising that such qualitatively dissimilar substances as fused salts and liquid metals should dissolve in one another in such a simple way. We shall show first that such behavior is, indeed, to be expected, and second, that these systems differ in behavior from regular solutions of molecular liquids in small but significant details.

Theory.—A homogeneous mixture, either solid or liquid, of an alkali metal with a halide of the same metal may be regarded as a mixture of the two negative species: the halide ions and the valence electrons of the metal. The positive ions are all the same and will comprise a lattice, regular for the solid or irregular for the liquid, with negative species in the cavities between these positive ions. In the solid state there is difficulty in mixing negative species of substantially different size. Liquid solutions generally show no difficulty in accommodating different sized species and the data on solutions of fused alkali halides[5] conform closely to Raoult's law.

In the range where the salt is the abundant species the electrons from the metal must be confined to isolated cavities. These are the well known F-centers in the solid state. In the liquid state the cavities need no longer conform to the size of the halide ions, but otherwise they should be like F-centers. The formation of such a solution may be considered in two steps: first, the conversion of the metal to a hypothetical ionic lattice of positive ions and F-centers, and second, the mixing with the metal halide. The energy change in the second step would be expected to be small, consequently the observed energy of mixing can be related to the excess energy of the hypothetical ionic lattice metal.

(1) This research was initiated at the University of California and was assisted by support from the U. S. Atomic Energy Commission. It was reported in part at the International Congress of Pure and Applied Chemistry, Montreal, Canada, August, 1961.

(2) K. S. Pitzer, *J. Am. Chem. Soc.*, **80**, 5046 (1958).

(3) T. A. Beckman and K. S. Pitzer, *J. Phys. Chem.*, **65**, 1527 (1961).

(4) (a) M. A. Bredig, H. R. Bronstein and W. T. Smith, Jr., *J. Am. Chem. Soc.*, **77**, 1454 (1955); (b) J. W. Johnson and M. A. Bredig, *J. Phys. Chem.*, **62**, 604 (1958); (c) M. A. Bredig and H. R. Bronstein, *ibid.*, **64**, 64 (1960); (d) M. A. Bredig and J. W. Johnson, *ibid.*, **64**, 1899 (1960).

(5) E. Aukrust, *et al.*, *Ann. N. Y. Acad. Sci.*, **79**, 830 (1960); R. P. Tastogi and K. T. Varma, *Trans. Faraday Soc.*, **53**, 1165 (1957).

A model of this sort for metals was proposed by Thomson[6] in 1922 and doubtless by others. In our calculations on this model the treatments of F-center wave functions and other properties by Tibbs[7] and by Gourary and Adrian[8] were used to guide the choice of reasonable parameters. The sodium chloride structure was adopted which is equivalent to a cubic closest packed metal. The positive ion radius (Pauling) together with the molal volume of the metal yield the radius of the spherical cavity. It is assumed that the electron spends 70% of the time within the cavity volume which is taken as that of the sphere touching the nearest positive ions increased to include half of the additional empty space associated with the given lattice site. Some overlap of electron orbitals is permitted in view of the probability that at least half of the neighboring electrons will have antiparallel spin. The exact choice of parameters is not essential; these values are reasonable in view of Tibbs' and Gourary and Adrian's results. The usual Madelung formula is used for the electrostatic lattice energy, and the quantum mechanical formula for the energy of a particle in a spherical box yields the kinetic energy of the electron in the cavity. The experimental lattice energies of the real metals are greater than those calculated on this model by 19 kcal./mole for Li, 18 for Na, 10 for K and 9 for Rb and Cs.

These calculations show that the dissolving of an alkali metal in a fused halide of the same metal with a positive heat of solution of several kcal./mole is reasonable. Let us now consider the entropy change to be expected. There is first the entropy of nearly random mixing of unlike species—the usual term for ideal solutions. In addition the electron spins will be uncoupled in the process yielding a positive term $R \ln 2$. Furthermore, the F-centers probably are less rigid in both size and shape than are halide ions. Consequently the vibration frequencies of the surrounding positive and negative ions will be reduced on formation of the solution and this will yield an additional positive entropy contribution. Consequently we may expect substantial positive excess entropies of mixing.

Let us now turn to the other side of the phase diagram where the salt is the dilute component. In this case the metal electrons remain in a metallic state and the halide ions represent foreign particles. Since these foreign particles carry a negative electrical charge, the metal electrons will tend to avoid them by electrostatic forces. Nevertheless the size and rigidity of the halide ions will interfere with the metallic electron system. The halide ion itself will be surrounded by positive ions much as in the pure alkali halide.

It is not easy to calculate the magnitude of this interference of the halide ion with the metallic electron system. Calculation for a neutral sphere from which the neutral particles of a Fermi gas were excluded yielded much too large an effect. But this is to be expected since the correlation of electron motion by the charge of the halide ion would have reduced the free electron density at the ion below the average value. The calculated electron densities for metallic electrons around defects or fixed charges by March and Murray[9] and by Cowan and Kirkwood[10] are informative in this connection. For example, the curves given by Cowan and Kirkwood show for parameters appropriate for sodium a very low density of other electrons out to a radius of 1.3 Å. from a given negative charge. The approximations made by Cowan and Kirkwood probably yield too complete a correlation of electron motion at short distances, but they show that an ion such at F^- with radius 1.4 Å. should cause only a moderate interference with the metallic electrons.

It is found experimentally that fused alkali fluorides are remarkably soluble in the metals. At the respective melting points of the fluoride salts the solubilities are: Na, 17%; K, 40%; Rb and Cs, miscible in all proportions.

The larger halide ions will interfere more with the metallic electrons, particularly for the lighter metals. While the qualitative picture seems clear, it did not prove feasible to make any numerical estimates of the energy of this interference.

Again the entropy effects must be considered. In this case the electron spins remain mostly paired. But the restraining forces for vibration of the alkali metal ions in the metal are much less than in the halide. For example, the vibration frequency of Na in the metal is about 100 cm.$^{-1}$ as inferred from the heat capacity data whereas the Na^+ *vs.* Cl^- vibration frequency in NaCl is approximately twice as great or 200 cm.$^{-1}$. The metal ions surrounding an isolated halide ion in solution in the metal may be slightly more restrained than in the pure metal but this effect should be small. Hence, we expect an excess entropy of solution of halide in metal corresponding to the decreased restraint to vibration of the metal ion and presumably also of the halide ion. In view of the three directions of vibration and the frequencies mentioned above, the excess entropy may be expected to fall in the range from $3R \ln 2$ to $6R \ln 2$ or 4 to 8 cal./deg. mole for NaCl in Na.

Solubility Equilibria.—Let us now consider the solubility data of Bredig and collaborators[4] in terms of these expectations. Since these very difficult experiments at high temperatures are necessarily of limited precision, it is best to emphasize the solubility itself and to be cautious about derivatives. We take ideal solution behavior (Raoult's law) as a point of departure and deal with excess functions. In particular the excess partial molal Gibbs free energy of the i-th component or the excess chemical potential is

$$\Delta \bar{F}_i^E = \Delta \mu_i^E = \Delta \mu_i - RT \ln x_i = RT \ln \gamma_i$$

where the reference state is the pure liquid component.

In the case of solid salt in equilibrium with solution the usual freezing point equations apply with heats of fusion given by Dworkin and Bredig.[11] The effect on salt activity of any metal dissolved in the solid salt was neglected.

(6) J. J. Thomson, *Phil. Mag.*, **43**, 721; **44**, 657 (1922).

(7) S. R. Tibbs, *Trans. Faraday Soc.*, **35**, 1471 (1939).

(8) B. S. Gourary and F. J. Adrian, *Phys. Rev.*, **105**, 1180 (1957).

(9) N. H. March and A. M. Murray, *Proc. Roy. Soc. (London)*, **A256**, 400 (1960).

(10) R. D. Cowan and J. G. Kirkwood, *Phys. Rev.*, **111**, 1460 (1958).

(11) A. S. Dworkin and M. A. Bredig, *J. Phys. Chem.*, **64**, 269 (1960).

The case of two liquid phases is more complex in that the activity of the abundant component (solvent) in each phase is decreased somewhat from unity. Where the solubility is small, however, it is easy to estimate the activity coefficient γ of the solvent with sufficient accuracy and then to calculate γ for the same component as the solute in the other phase. As the solubilities increase at higher temperatures, it is sometimes necessary to use successive approximations with the assumption at this stage that the regular solution equation

$$RT \ln \gamma_i = \Delta\mu_i^0(1 - x_i)^2 \quad (1)$$

applies accurately enough for the small corrections to solvent activity. $\Delta\mu_i^0$ is the excess chemical potential at zero concentration which is also the difference in chemical potential between the solute and solvent standard states. Where the two components have substantially different volumes the corresponding equation

$$RT \ln \gamma_i = \Delta\mu_i^0(1 - \phi_i)^2 \quad (2)$$

was used where ϕ_i is the volume fraction. The solute activity coefficient is just $\gamma_i = a_i/x_i$ where x_i is the solute mole fraction and a_i has been estimated from the composition of the other phase in which i is the solvent.

The data on sodium systems cover a particularly extensive temperature range and will be treated first. Most of the values come from liquid–liquid equilibria which were corrected by equation 1 except for NaF where equation 2 was used because of the large volume ratio. The results are summarized by the linear expressions for $\Delta\mu_i$ as a function of T in Table I.

TABLE I

EXCESS CHEMICAL POTENTIALS FOR SODIUM HALIDE–SODIUM SOLUTIONS

Salt	Temp. range, °K.	$\Delta\mu^0_{salt}$, kcal./mole	$\Delta\mu^0_{metal}$, kcal./mole	T_c, °K.	$(\Delta\mu^0/RT)$ at T_c Salt	$(\Delta\mu^0/RT)$ at T_c Metal
NaF	1120–1400	10.7–0.004 T	....	1453	1.7	..
NaCl	1070–1230	15.3– .007 T	18.1–0.009 T	1353	2.2	2.2
NaBr	1020–1230	13.7– .006 T	18.1– .010 T	1299	2.3	2.0
NaI	930–1230	14.2– .006 T	16.5– .009 T	[illegible]	[illegible]	[illegible]

There are several interesting features of the results of Table I. First one notes that the $\Delta\mu^0$ values vary quite appreciably over the temperature range of measurement. The negative of the coefficient of T in each expression is, of course, the excess partial molal entropy. These values must be multiplied by 1000 to convert from kcal. to cal., and have the sign and order of magnitude suggested by the discussion above. Also the $\Delta\mu^0_{metal}$ values are in the range indicated for the excess energy of the hypothetical metal comprised of F-centers and positive ions. This excess energy for the hypothetical sodium metal is, of course, the same for each system in Table I, and we find only a modest variation of the μ^0_{metal} values for the various halide systems.

The substantial excess entropies found in Table I constitute a departure from the behavior of regular solutions which were postulated by Hildebrand to have ideal entropies of mixing. The effect of volume differences between components has long been discussed for molecular solutions and considerable evidence has accumulated in favor of equation 2 which shows the activity coefficient dependence on volume fraction. If equation 2 holds for each component, then the ratio of the coefficients is expected to follow the equation

$$\frac{\Delta\mu_1^0}{\Delta\mu_2^0} = \frac{V_1}{V_2} \quad (3)$$

where V_1 and V_2 are the molal volumes of the respective components. Bredig and Bronstein[4c] gave the following molal volumes: Na, 33; NaF, 22; NaCl, 38; NaBr, 45; NaI, 55 cc. While the ratios of $\Delta\mu_i^0$ values do not conform accurately to equation 3, there is approximate agreement. Indeed, the temperature coefficients of $\Delta\mu^0_{salt}$ and $\Delta\mu^0_{metal}$ differ, and it would be surprising if the temperature dependence of the volume ratio was such to fit equation 3 exactly.

It is also interesting to note the values of $(\Delta\mu_i^0/RT)$ at the critical temperature for liquid–liquid mixing which are given in the last two columns of Table I. The regular solution equation 1 is applicable for nearly equal molal volumes and for $\Delta\mu^0_{salt} = \Delta\mu^0_{metal}$. This equation yields the value $(\Delta\mu^0/RT) = 2.0$ at the critical temperature. The value for Na–NaCl in Table I is 2.2 which indicates a small but significant deviation from equation 1.

The data on the heavier alkali metal systems cover narrower temperature ranges and sometimes show no liquid–liquid phase separation. It is difficult or impossible in most of these cases to unambiguously determine the temperature coefficient of $\Delta\mu$, hence the data are summarized in Table II as $\Delta\mu^0$ values assuming no temperature dependence. As many as three values are given; the basis for each is explained in the table. The values are given as $A_{12} = (\Delta\mu_i^0/V_i)$ which is constant in the volume fraction system of regular solution equations. The value from the critical mixing temperature obviously applies to that temperature. The value from γ_{metal} applies at the melting point of the salt while the value from γ_{salt} arises from data at temperatures somewhat below the salt melting point.

TABLE II

EXCESS CHEMICAL POTENTIALS FOR ALKALI HALIDE ALKALI METAL SOLUTIONS

$A_{12} = \frac{\Delta\mu_i^0}{V_i}$ {top value from γ_{salt} at low salt concn.; middle value from T_c of liquid mixing; bottom value from γ_{metal} at low metal concn.}

	Units, cal./cc. F	Cl	Br	I
K	110	90	80	70
	105	78	69	65
	115	94	..	80
Rb	..	70	62	53
	82	64	..	50
	..	..	..	..
Cs	45	55	..	42
	..	..	..	..
	..	..	..	..

Volumetric data for the systems in Table II are sparse at the temperature of interest for these calculations. Literature[12] values were used, but the change to the temperature of interest had to be estimated in many cases. The values adopted were: K, 60; KF, 31; KCl, 49; KBr, 56; KI, 69; Rb,

(12) See J. W. Johnson, P. A. Agron and M. A. Bredig, *J. Am. Chem. Soc.*, **77**, 2734 (1955), and references there cited.

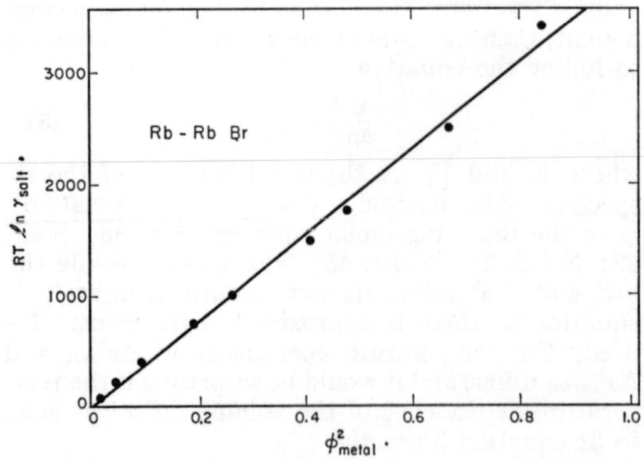

Fig. 1.

70; RbF, 36; RbCl, 54; RbBr, 62; RbI, 76; Cs 85; CsF, 45; CsCl, 60; CsBr, 68; CsI, 81 cc./mole.

The values of A_{12} for the sodium systems from T_c are, respectively, 215, 155, 136, 120 cal./cc., which are considerably larger than those in Table II for the heavier metals. We note that the value of A_{12} from T_c is the lowest for each potassium system and corresponds, of course, to the highest temperature. Presumably this is the result of a large excess partial molal entropy, as was found for the the sodium systems, and which yields a negative temperature coefficient for $\Delta\mu^0$. Where values based upon both γ_{metal} and γ_{salt} are available in Table II, they are nearly equal, which indicates conformity to the volume fraction equations. This is further indicated in Fig. 1 which shows the data for RbBr in a manner to test equation 2.

Discussion

The preceding calculations and arguments have shown that the valence electrons from the alkali metal atoms in a metal–fused salt solution may be considered to be a negative ion species much like the halide ions themselves. The general similarity of the solubility properties of these systems to the corresponding properties of mixtures of non-polar molecules has been noted as well as the significant differences. It is clear that the mixing of the two negative species must be nearly random. However, if the metal concentration is high, we know that the valence electrons assume metallic character and that it is no longer a useful approximation to regard a particular electron localized in a particular cavity. One must now regard the array of cavities as an irregular lattice of sites of low potential energy for electrons and solve the many-electron problem for electronic motion in and among these sites. The work of Lax and Phillips[13] shows that the irregular geometry itself does not qualitatively change the electronic energy level pattern. The decisive matters with respect to metallic *vs.* non-metallic character are those discussed by Mott.[14] The overlap of the wave functions between the F-center cavities is a critical factor. As Mott predicts, intermediate states are unstable at low temperatures, and one finds the continuous transition from metallic to non-metallic character only above a critical mixing temperature.

The metal to non-metal transition composition may be taken to be that of the critical point for phase separation. It is interesting to note that this composition is of the order of 50 mole % metal for the fused salt solutions whereas it is only 4.2% metal in solutions of sodium in liquid ammonia. Apparently the electronic wave functions are much more localized in the fused salts than in ammonia.

Additional properties such as electrical conductance[15] can be shown to be consistent with our general model. We may also note that the system[16] Li–LiH appears to be similar to the halide systems.

(13) M. Lax and J. C. Phillips, *Phys. Rev.*, **110**, 41 (1958).

(14) N. F. Mott, *Phil. Mag.*, [8] **6**, 287 (1961); N. F. Mott and W. D. Twose, *Adv. in Phys.*, **10**, 107 (1961); and earlier papers there cited.

(15) H. R. Bronstein, A. S. Dworkin and M. A. Bredig, *J. Chem. Phys.*, **34**, 1843 (1961); also *J. Am. Chem. Soc.*, **80**, 2077 (1958).

(16) C. E. Meiser, *et al.*, *J. Phys. Chem.*, **62**, 220 (1958).

[Reprinted from the Journal of the American Chemical Society, **84**, 2264 (1962).]

A REVISED MODEL FOR AMMONIA SOLUTIONS OF ALKALI METALS

Sir:

It is well established that, in extremely dilute solutions of alkali metals in liquid ammonia, the metal is completely dissociated into ammoniated metal ions and ammoniated electrons. Each ammoniated electron is believed to exist in a large spherical cavity in the solvent and to be stabilized by the orientation of the ammonia dipoles on the periphery of the cavity. The decrease in the equivalent electrical conductance of these solutions as the concentration is increased to about 0.05 *M* indicates that ion pairing or assembly into larger aggregates takes place. The decrease of the molar paramagnetic susceptibility of the solutions with increasing concentration indicates that the ammoniated electrons associate to form species containing electron pairs. Both of these effects may be reproduced by assuming appropriate constants for these equilibria

$$M^+ + e^- = M$$
$$2M^+ + 2e^- = M_2$$

At $-33°$ these equilibria shift in the vicinity of 0.01 *M* and the associated species predominate at higher concentrations.

It has been proposed by Becker, Lindquist, and Alder[1] that the monomer, M, consists of an ammoniated M^+ ion with the electron located in an expanded orbital on the protons of the coördinated ammonia molecules. Likewise, it has been proposed that the dimer, M_2, consists of two ammoniated M^+ ions held together by a pair of electrons in a bonding molecular orbital located principally between the two ions. These descriptions of the monomer and dimer are inconsistent with certain properties, in part newly measured, of alkali metal–ammonia solutions which are summarized below. We propose a revised description of the monomer and dimer species which more adequately accounts for all the properties of the solutions.

(1) The absorption spectra of sodium–ammonia solutions which are less concentrated than 0.03 *M* follow Beer's law within an experimental uncertainty of ± 2–5% at all wave lengths between 4000 and 25,000 Å. In fact, Beer's law is obeyed with similar precision for wave lengths between 4000 Å. and the absorption maximum (15,000 Å.) for concentrations at least as high as 0.1 *M*.[2] There is no indication of a separate absorption band near 6700 Å. Such a band has been observed in amines by Fowles, McGregor and Symons[3] and was attributed to a diamagnetic species. We have been unable to substantiate the claimed[4] appearance of shoulders at 6700 and 8000 Å. in the absorption spectra of ammonia solutions of sodium and sodium iodide. Thus the three species e^-, Na, and Na_2 have almost identical absorption spectra.

(2) The absorption spectra of solutions of lithium, potassium[5] and cesium are identical with those for sodium solutions of similar concentration within the limits of experimental uncertainty, which for the former solutions was ± 10%. Hence the spectra of the M and M_2 species (in the visible and infrared) do not depend upon the particular metal.

(3) Gunn and Green[7] have found that the apparent molar volume of lithium in ammonia 0° is practically constant at 50 ml./mole from 0.02 to 0.99 *M*. For sodium, the apparent molar volume changes only from 56.3 to 58.6 ml./mole from 0.0093 to 0.34 *M*. The latter data are inconsistent with some preliminary data of Evers and Filbert[8] which indicate a marked minimum in the volume–concentration curve for sodium at 0.03 *M*. However, until the latter findings are confirmed, we shall favor the former data and therefore conclude that the partial molal volumes of $M^+ + e^-$, of M, and of $^1/_2\ M_2$ are nearly the same.

(4) Nuclear magnetic resonance data[9] are available only above 0.05 *M*. In the range 0.05 to 0.2 *M* the unpaired electron spin density on the sodium nuclei is only about 0.1% of that expected for isolated sodium atoms. The n.m.r. shift for nitrogen is rather large and indicates that an appreciable fraction of the unpaired electron spin density extends into the ammonia solvent regions. Since in this concentration range the concentration ratio of M to e^- is not small, the electron spin density on the metal atom nucleus must be small for the species M.

This evidence, taken as a whole,[10] is quite inconsistent with the Becker, *et al.*, model where the volumetric and spectral properties of $(e^- + M^+)$, M, and $^1/_2\ M_2$ would be expected to differ widely. Douthit and Dye[6] and Evers[11] have pointed out that if the monomer were simply an ion-pair consisting of an ammoniated metal ion and an ammoniated electron, the constancy of the spectra in the dilute range could be explained. They make no such proposal for the dimer. However, our spectral data require a new picture for M_2 as well. We picture the M_2 species as a quadrupolar ionic assembly of $2e^- + 2M^+$ in which there is little distortion of either the ammoniated electrons or ammoniated metal ions. Presumably the electrons and ions are held in a square or rhombic configuration. The probability density for the electron in the solvated e^- species extends with decreasing intensity through several solvent layers.[12] Thus the wave functions for the two e^- in M_2 will overlap significantly and it is reasonable that the singlet

(1) E. Becker, R. H. Lindquist and B. J. Alder, *J. Chem. Phys.*, **25**, 971 (1956).

(2) M. Gold and W. L. Jolly, unpublished data.

(3) G. W. A. Fowles, W. R. McGregor and M. C. R. Symons, *J. Chem. Soc.*, 3329 (1957).

(4) H. C. Clark, A. Horsfield and M. C. R. Symons, *ibid.*, 2478 (1959).

(5) We find no evidence for a negative deviation from Beer's law for potassium as reported by Douthit and Dye.[6]

(6) R. C. Douthit and J. L. Dye, *J. Am. Chem. Soc.*, **82**, 4472 (1960).

(7) S. R. Gunn and L. G. Green, *J. Chem. Phys.*, **36**, 363 (1962).

(8) E. C. Evers and A. M. Filbert, *J. Am. Chem. Soc.*, **83**, 3337 (1961).

(9) H. M. McConnell and C. H. Holm, *J. Chem. Phys.*, **26**, 1517 (1957); J. Acrivos and K. S. Pitzer, unpublished data.

(10) W. E. Blumberg and T. P. Das (*J. Chem. Phys.*, **30**, 251 (1959)) treat the n.m.r. data on the basis of the Becker, *et al.*, model with reasonable agreement. The fit is strained, however, and would be easier on our model.

(11) E. C. Evers, *J. Chem. Ed.*, **38**, 590 (1961).

(12) J. Jortner, *J. Chem. Phys.*, **30**, 839 (1959).

state should be lower in energy than the triplet by more than kT. Also a small electron density at the sodium nucleus in the ion pair (M) species is to be expected. Both the volume and the 1s–2p spectral frequency depend primarily on the cavity size for the solvated electron. Thus, if this cavity retains its size through ion pair and quadruplet formation, the results cited in (1), (2), and (3) above become understandable.

Presumably the solvated electron retains its structure in further polymeric species (*e.g.*, M_4) which probably form in more concentrated solutions. However, in highly concentrated solutions, one should expect that the nature of the ammoniated electrons would change if for no other reason than that there are insufficient ammonia molecules to properly coördinate both the metal ions and the electrons. We believe the deviations from Beer's law which occur in sodium solutions more concentrated than 0.03 M indicate the formation of high polymers with incipient metallic bonding.

DEPARTMENT OF CHEMISTRY AND
LAWRENCE RADIATION LABORATORY
UNIVERSITY OF CALIFORNIA
BERKELEY 4, CALIFORNIA
MARVIN GOLD
WILLIAM L. JOLLY
DEPARTMENT OF CHEMISTRY
RICE UNIVERSITY
HOUSTON 1, TEXAS
KENNETH S. PITZER

RECEIVED FEBRUARY 12, 1962

Comment: A Revised Model for Ammonia Solutions of Alkali Metals

Most of the features of this model are still considered to be valid: in particular, the absence of metal atoms or diatomic molecules but instead the presence of solvated alkali metal positive ions, either as separate species or electrostatically associated with negative species which are solvated electrons. The least certain aspects in 1962 was the character of the double electron species which was required by its diamagnetic property. Papers are still being written currently about the nature of this species.

Proc. Natl. Acad. Sci. USA
Vol. 77, No. 6, pp. 3103–3104, June 1980
Chemistry

Phase equilibria for highly unsymmetrical plasmas and electrolytes

(Debye–Hückel theory/critical phenomena)

KENNETH S. PITZER

Department of Chemistry and Lawrence Berkeley Laboratory, University of California, Berkeley, California 94720

Contributed by K. S. Pitzer, March 10, 1980

ABSTRACT **The conclusion of classical Debye–Hückel theory that a phase separation may occur in highly unsymmetrical plasmas or electrolytes is shown to be false and to arise from a serious error in the treatment of the interaction of pairs of the most highly charged ions. After an approximate correction for this error, no phase separation is predicted. Specific application to iron in the solar plasmas is discussed.**

The classical Debye–Hückel equation indicates separation into two phases for a plasma or electrolyte with a highly unsymmetrical charge pattern, such as the solar plasma, including highly ionized iron. Pollock and Alder (1) showed that the Debye–Hückel equation gave this result for solar conditions and discussed its implications for the solar neutrino dilemma. They also noted that Debye–Hückel theory was not valid for the iron-rich phase; thus, a better theory is needed for reliable prediction of the phase behavior of solar plasma.

Although the extension of Debye–Hückel theory to be presented below is approximate, it identifies the major correction needed and shows that the indication of phase separation was false. The present theory is based on classical (rather than quantum) mechanics for the translational motion of the ions (including dissociated electrons); in that domain no phase separation is expected.

In a two-component system, the possibility of phase separation is determined by the sign of the second derivative of the Gibbs energy with respect to the fraction of one component. One may use any of several alternate measures of composition on a self-consistent basis; we shall use the fraction of total particles whereby the z^+ ion and z electrons comprise the $z + 1$ particles assigned to that component. This is equivalent to the volume fraction on an ideal gas basis. Thus, we wish to obtain for a mixture with a fixed total of N particles

$$\frac{\partial^2 G}{\partial x^2} = \frac{\partial^2 G^{\text{id}}}{\partial x^2} + \frac{\partial^2 G^{\text{ex}}}{\partial x^2}, \qquad [1]$$

where G^{id} is the Gibbs energy on an ideal gas (or ideal solution) basis and G^{ex} is the excess Gibbs energy. The ideal term is always positive, and if it dominates, no phase separation is possible. But if the second derivative of the excess Gibbs energy becomes sufficiently negative, it may exceed the positive value of the ideal term and separation into two phases may occur.

It proves easier to discuss initially the excess internal energy E^{ex}. One then calculates the excess Helmholz energy by integrating the thermodynamic relationship $[\partial(A/T)/\partial(1/T)]_V = E$ from zero in $1/T$ to the final value and thereafter one adds the term $VP = -V(\partial A/\partial V)_T$ to obtain G.

The radial distribution or binary correlation functions provide the basis for understanding the present problem. This function, $g_{ij}(r_{ij})$, gives the probability of finding ions i and j at distance r_{ij} as a ratio to the purely random probability. Then the excess internal energy of the system is given by the equation

$$E^{\text{ex}} = 2\pi V \sum_i \sum_j c_i c_j \int_0^\infty u_{ij} g_{ij} r^2 \mathrm{d}r, \qquad [2]$$

where c_i and c_j are the concentrations of the particular species and V is the total volume for N particles. Also, u_{ij} is the interparticle potential, in this case the coulombic potential $z_i z_j e^2/r_{ij}$. Although Debye and Hückel developed their theory in other terms, it may be restated in this form with the results

$$g_{ij} = 1 - (z_i z_j e^2/kT)(e^{-\kappa r}/r) \qquad [3]$$

$$\kappa^2 = (4\pi e^2/kT) \sum z_i^2 c_i \qquad [4]$$

$$E^{\text{ex}} = G^{\text{ex}} = kTV\kappa^3/8\pi. \qquad [5]$$

This expression for g_{ij} is a linear approximation of a Boltzmann exponential in which the exponent is the last term in Eq. 3. Thus, the simple Debye–Hückel treatment fails if this exponent becomes larger than one for any appreciable range in r.

For the solar plasma it is reasonable to assume that the iron is +24 (i.e., retaining only the pair of 1s electrons undissociated), but the exact degree of ionization is not important. We also assume $T = 1.7 \times 10^7$ K (1.5 keV) and $P = 10^5$ Mbar (1 bar = 1.00×10^5 pascals). Then the radial distribution functions on a Debye–Hückel basis are shown on Fig. 1. The H^+-e^-, H^+-H^+, and e^--e^- functions depart only slightly from unity and are

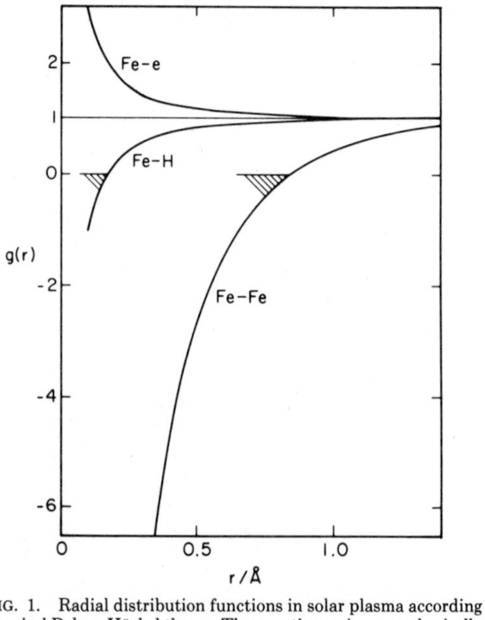

FIG. 1. Radial distribution functions in solar plasma according to classical Debye–Hückel theory. The negative regions are physically impossible and are removed in the corrected calculation.

The publication costs of this article were defrayed in part by page charge payment. This article must therefore be hereby marked "*advertisement*" in accordance with 18 U. S. C. §1734 solely to indicate this fact.

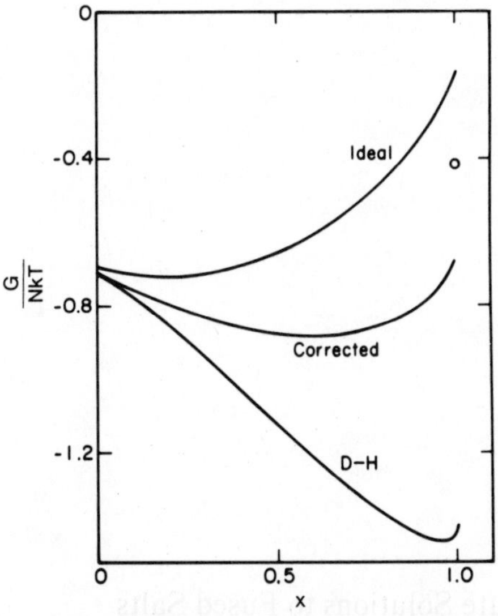

FIG. 2. The Gibbs energy for iron-hydrogen plasma under solar conditions. The curves are explained in the text.

omitted. The Fe^{+24}-e^- and Fe^{+24}-H^+ functions exceed 2 and drop below zero, respectively, only at small r; hence, the error in these cases is moderate. But the Fe^{+24}-Fe^{+24} function becomes large and negative over a major range in r, and this leads to a large and spuriously negative value of the integral in Eq. 2 for this term. A negative value of g_{ij} is, of course, physically impossible, and it arises only from the linear approximation in Eq. 3.

At very low concentration of iron, the factors c_{Fe} in Eq. 4 and c_{Fe}^2 in Eq. 2 are so small that little error is contributed by the Fe-Fe terms to the reciprocal length κ or the total excess energy in Eq. 5. But the second derivative with respect to the fraction iron essentially removes the c_{Fe}^2 factor, and that term assumes full importance in comparison with the other terms in Eq. 2. A greatly improved approximation is obtained by a simple substitution of zero for the false negative value of g. Thus, if g_{ij} is zero at r'_{ij}, the corrected excess internal energy becomes

$$E^{ex} = -\frac{kTV\kappa^3}{8\pi} - 2\pi V \sum_i \sum_j c_i c_j \int_0^{r'} g_{ij} u_{ij} r^2 dr, \qquad [6]$$

where the integral covers only that region where $g_{ij} < 0$.

In view of the approximate nature of this basic procedure, approximations were made in the expression for r' and in the conversion from E^{ex} to G^{ex}, but it was verified that these approximations would not significantly affect the final results. Also it was found that the small region of negative g for the Fe-H term had only a small effect, hardly detectable on Fig. 2, which displays the results for H-Fe plasma under solar conditions. For Fig. 2, the Gibbs energy includes the terms for mixing electrons with positive ions.

Although the curve D–H on Fig. 2 for the simple Debye–Hückel equation shows a negative curvature from small x up to about 0.4 or 0.5, the "corrected" curve shows a substantial positive curvature across the entire composition range. This was verified by calculation of the second derivative; the negative contribution from the corrected excess Gibbs energy is never as much as one-half of the positive contribution from the ideal term. The "ideal" curve is also shown on Fig. 2.

The linearized Poisson–Boltzmann equation, which yields g_{ij} in Eq. 3, becomes a poorer approximation as the fraction of iron increases; hence the "corrected" curve in Fig. 2 is only a rough estimate in the region of substantial fraction of iron. In order to obtain an independent estimate for this region, a different method was used for the case of pure iron. Following in a sense the method of Bjerrum (2), the space was partitioned into an outer region for a Debye–Hückel treatment and inner spheres of radius R around each iron nucleus containing only electrons. A Boltzmann distribution of electrons was assumed, except for the 1s pair, but the coulombic potential for $+24e$ was truncated at about half the thermal de Broglie wavelength. The Debye–Hückel equation for a distance of closest approach of $2R$ was used, and the excess energy was taken as the total for the electrons within the spheres and for those outside as well as the iron ions. The radius was chosen as small as possible subject to the condition that the coulombic energy of two spheres at contact not appreciably exceed thermal energy. Reasonable values are $R = 0.25$ Å and a net charge of 20 protonic units. After integration with respect to $1/T$ (with appropriate variation of R) and addition of PV, the resulting Gibbs energy is shown by the open circle at $x = 1$ on Fig. 2. Because this point is well above the "corrected" curve, one estimates that the true curve toward the right side of Fig. 2 is above the "corrected" curve, and the conclusion of positive curvature (and no phase separation) is reinforced.

This last calculation for pure iron plasma was repeated with the use of a truncated coulomb potential (3) instead of a hard-core in the Debye–Hückel treatment. The result is substantially the same.

Although no mathematical proof is offered for the general absence of phase separation in classical unsymmetrical electrolytes, calculations for various examples are convincing to the writer. If one decreases the temperature from the solar case considered above, the degree of ionization of the iron will decrease. Also, for a given degree of ionization, the region of negative g implied by the Debye–Hückel distribution increases. Thus, although the Debye–Hückel equation for fixed z would yield an increasingly negative value of $\partial^2 G^{ex}/\partial x^2$, the "corrected" value for an appropriately decreased z remains smaller than the positive contribution from the ideal term. Eventually, further decrease in temperature at high density leads to a boundary beyond which all electron motion is in the quantum domain and the present methods cease to apply.

Exploratory calculations for highly unsymmetrical charge mixtures in electrolyte solutions also indicate no phase separation; this conclusion is in agreement with experiment. As for the plasma containing highly charged ions, there is an association of counter ions with a highly charged ion in solution, but this does not lead to phase separation as a function of solute fraction. There are examples in one-component electrolytes at low effective temperature where there is a separation into concentrated and dilute phases. This phenomenon has been discussed by Friedman and Larsen (4). In two-phase equilibrium the dilute phase comprises largely ion pairs whereas the concentrated phase is either fused-salt-like or metallic (e.g., metal-ammonia systems). This last type of phase separation might occur in an unsymmetrical mixed electrolyte, but it would arise in a very different range of temperature and concentration from that of the hypothetical Debye–Hückel separation discussed above and found not to occur.

I appreciated valuable discussions with Dr. B. J. Alder. The research received support from the Division of Chemical Sciences, Office of Basic Energy Sciences, U.S. Department of Energy under Contract W-7405-ENG-48.

1. Pollock, E. L. & Alder, B. J. (1978) *Nature (London)* **275**, 41–43.
2. Bjerrum, N. (1926) *K. Dan. Vidensk. Selsk. Mat. Fys. Medd.* **7** (9), 1–48.
3. Pitzer, K. S. (1979) *J. Chem. Phys.* **70**, 393–399.
4. Friedman, H. L. & Larsen, B. (1979) *J. Chem. Phys.* **70**, 92–100.

[Reprinted from the Journal of the American Chemical Society, 102, 2902 (1980).]

Electrolytes. From Dilute Solutions to Fused Salts

Kenneth S. Pitzer

Contribution from the Department of Chemistry and Lawrence Berkeley Laboratory, University of California, Berkeley, California 94720. *Received September 10, 1979*

Abstract: Solutions with composition extending continuously from molecular liquids such as water to fused salts are relatively unusual but of considerable interest. Conductance and thermodynamic properties are considered for several examples. New equations for the activities of the respective components represent the data more accurately than previous treatments and delineate the similarities and differences between such systems and nonelectrolyte solutions.

Electrolyte systems extending in the liquid phase from a dilute solution in a polar molecular solvent (such as water or an alcohol) to a pure fused salt constitute an interesting but infrequently studied type. In a 1954 article with the same title as this paper, Kraus[1] summarized the information then available concerning such systems. Kraus emphasized measurements of conductance but also considered the few thermodynamic data then available for such systems. In the following 25 years, the vapor pressure and thereby the activity of water have been measured for the systems $(Li,K)NO_3-H_2O$ and $(Ag,Tl)NO_3-H_2O$ over the entire range from pure water to fused salt. In each case there is a fixed, nearly equal ion

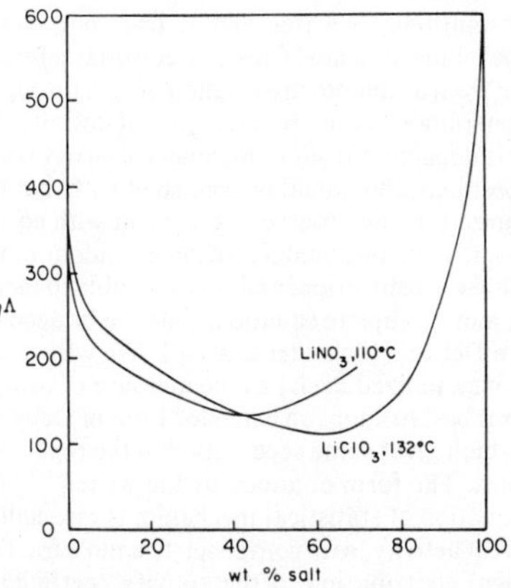

Figure 1. The conductance-viscosity product for water-salt solutions over the full range of composition.

fraction of cations so that either can be regarded as a two-component MNO_3–H_2O system. The mixing of the salts reduces the melting point and allows the experiments to be carried out at low pressure near 100 °C.

With the availability of these thermodynamic data it seemed worthwhile to examine again the nature of these novel systems. A new interpretation will be given to the thermodynamic data which has many advantages over that previously presented, although it uses a form of equation which has been commonly used for nonelectrolytes.

Conductance

Before dealing in detail with the thermodynamic data, it is desirable to review the situation with respect to conductance, both as Kraus described it and as it has changed since. The appropriate function giving a simple picture of the character of the solution is the product of the equivalent conductance and the viscosity, $\eta\Lambda$. Figure 1 shows the results for two systems from the measurements of Campbell and Paterson[2] and Campbell, Debus, and Kartzmark.[3] In the dilute aqueous solution the well-known electrostatic effects cause a reduction in conductance from its limiting value as explained by Debye and Hückel[4] and Onsager.[5] These data follow the theoretical equations for a fully ionized solute as far as they are valid, and it now appears to me, as it did to Kraus, that there is no reason to assume other than complete ionization at higher concentration. In the middle range the $\eta\Lambda$ product passes through a shallow minimum but, in the absence of an exact theory, it is not possible to draw any precise conclusions. The near constancy of the $\eta\Lambda$ product, however, suggests continued full ionization. The increase in $\eta\Lambda$ in approaching the pure fused $LiClO_3$ is striking. For the addition of a little water to the fused salt, the viscosity decreases faster than the conductance increases. The writer is not aware of any theoretical analysis of this region; such a study would be interesting.

There are also organic systems which have been investigated over the full range from dilute electrolyte solution to fused salt. An example is tetra-*n*-butylammonium picrate in *n*-butyl alcohol, measured by Seward[6] and discussed by Kraus.[1] In this case there is ion pairing in the dilute soluion range with a sharper drop in the $\eta\Lambda$ product. The minimum in $\eta\Lambda$ occurs at about 5% salt after which that product rises to about the same value as it had at infinite dilution in the alcohol solvent. Thus there appears to be a redissociation in the intermediate composition range. This redissociation puzzled Kraus but it is explained by Davies.[7]

With this brief review of the situation with respect to conductance, we turn to the thermodynamic data.

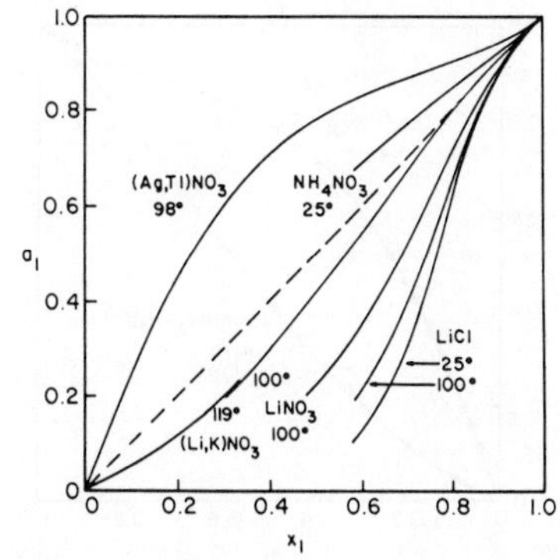

Figure 2. The activity of water for water-salt solutions over the full range of composition.

Activity. Preliminary Survey

The activity of water is given by its vapor pressure above the solution; this has been measured for $(Li,K)NO_3$–H_2O by Tripp and Braunstein,[8] for that system, for $LiNO_3$–H_2O, and for several others by Braunstein and Braunstein,[9] for $LiCl$–H_2O by Gibbard and Scatchard,[10] and for $(Ag,Tl)NO_3$–H_2O by Trudelle, Abraham, and Sangster.[11] Robinson and Stokes[12] summarize earlier results for other systems including NH_4NO_3–H_2O. The activity of water for many of these systems is shown in Figure 2. The composition variable is the mole fraction on an ionized basis, i.e, $x_1 = n_1/(n_1 + \nu n_2)$ where n_1 and n_2 are moles of water and salt, respectively, and ν is the number of ions in the salt. On this basis Raoult's law applies in the very dilute range, with the Debye–Hückel correction applicable as the concentration increases.

The similarity of the curves on Figure 2 to those for nonelectrolyte solutions is striking. The dashed line representing $a_1 = x_1$ can be called "ideal-solution behavior" for these systems, as it is for nonelectrolytes, but it is realized that a statistical model yielding that result would be more complex for the ionic case. Also the Debye–Hückel effect is a departure from this ideal behavior. Nevertheless, it seems worthwhile to explore the use for these systems of the simple equations for nonelectrolytes. One of the simplest and most successful had its origin in the work of van Laar[13] and has been widely used since. Prausnitz[14] discusses this and related equations as well as the contributions of Margules, Hildebrand, Scatchard, Guggenheim, and others to this topic. For the activity of either component, referenced to the pure liquid, one has

$$\ln a_1 = \ln x_1 + w_1 z_2^2 \tag{1a}$$

$$\ln a_2 = \ln x_2 + w_2 z_1^2 \tag{1b}$$

$$z_1 = n_1/[n_1 + \nu n_2(b_2/b_1)] \tag{1c}$$

$$z_2 = \nu n_2/[n_1(b_1/b_2) + \nu n_2] = 1 - z_1 \tag{1d}$$

$$w_2 = (b_2/b_1)w_1 \tag{1e}$$

Note first that, if (b_1/b_2) is unity, z_1 and z_2 reduce to the mole fractions x_1 and x_2. Then one has the even simpler equations

$$\ln a_1 = \ln x_1 + w x_2^2 \tag{2a}$$

$$\ln a_2 = \ln x_2 + w x_1^2 \tag{2b}$$

In either (1) or (2) the nonideality parameter w (sometimes written w/RT) arises from the difference between the intermolecular attraction of unlike species as compared to the mean of the intermolecular attraction for pairs of like species. The second parameter in eq 1, (b_1/b_2), is sometimes ascribed to the

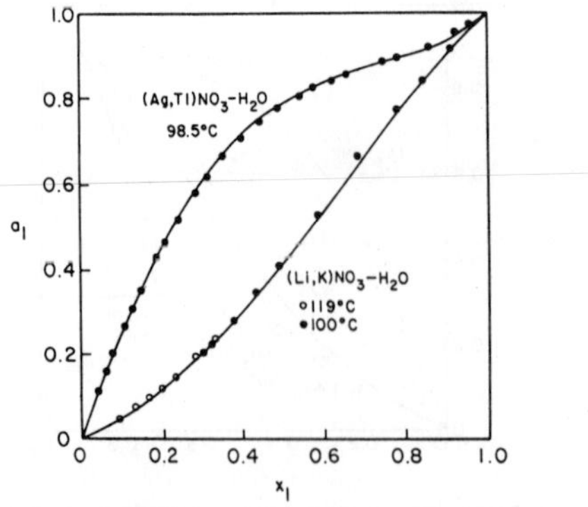

Figure 3. Comparison of the experimental activity of water with that calculated from eq 1.

ratio of the volumes of the molecules or to the ratio of molal volumes in the liquid. (Originally van Laar related b_1 and b_2 to the b quantity in the van der Waals equation, which is in turn related to the volume of the molecules, but this relationship to an approximate equation for the imperfect gas is less useful.) In some systems, especially metallic solutions, eq 1 is still quite satisfactory but (b_1/b_2) departs greatly from the ratio of molal or atomic volumes. For fused salt–water mixtures it seems best to regard (b_1/b_2) as a freely adjustable parameter and subsequently to compare the values with ratios of molal volumes.

Equation 1 was fitted to the two systems remaining liquid over the full range of composition with the results $w_1 = 1.02$, $(b_1/b_2) = 0.50$ for (Ag,Tl)NO_3–H_2O and $w_1 = -0.89$, $(b_1/b_2) = 1.2$ for (Li,K)NO_3–H_2O. Water is component 1 and the salt component 2. For the latter system the simpler equation (2) serves almost as well with $w = -0.80$ (this implies $b_1/b_2 = 1.0$). The calculated curves based on eq 1 are compared with the experimental data in Figure 3, where it is apparent that the agreement is excellent.

This treatment of very concentrated electrolytes in a manner analogous to that of nonelectrolytes seems to me to be the simplest and most useful initial approach, but it has not been used to the writer's knowledge. A favorite method has been the use of the Brunauer–Emmett–Teller adsorption isotherm as proposed by Stokes and Robinson.[15] Here the rationale is that the fused salt can attract water molecules, presumably to the surfaces of the cations, much as water is attracted to the surface of a crystal. As more water is added, the binding energy per molecule decreases gradually to that of pure water in a dilute solution or with a multilayer film. Braunstein and Braunstein[9] and Trudelle, Abraham, and Sangster[11] summarize the results of fitting several sets of recent data to the BET equation which is

$$a_w(1 - x_w)/x_w(1 - a_w) = 1/cr + (c - 1)a_w/cr \qquad (3)$$

Here $a_w = a_1$ is the activity of water, x_w is its mole fraction on an un-ionized basis, and both c and r are empirical parameters. While this equation fits reasonably well up to $x_w = 0.65$, the fit is not so good as that for eq 1. Above $x_w = 0.65$ the BET equation is much less satisfactory. Also eq 1 and 2 have the great convenience of a simple form expressing the activity as an explicit function of the mole fraction and an equally simple expression for the activity of the fused salt component. With this preliminary survey, we now consider a more complete equation including a Debye–Hückel term.

Activity. Further Consideration

Although eq 1 gives a remarkably accurate representation over most of the composition range of the activity of water in a water–fused salt system, it becomes somewhat inaccurate for dilute solutions in water and it does not include the Debye–Hückel limiting law. Clearly electrostatic forces cause a departure from a random distribution of ions throughout the composition range except in the limit of infinite dilution. Possibly this departure from randomness is nearly constant at most compositions; this would be consistent with, but probably is not required for, the observed agreement with eq 1. In any case, the ionic distribution does become random in the limit of zero solute concentration, and it is desirable to include this aspect in a more complete equation. This can be accomplished by adding a Debye–Hückel term to eq 1. We wish to use mole fraction (on an ionized basis) as the measure of composition. Also it seems best to adopt an extended form of Debye–Hückel equation which gives some recognition to the repulsive forces between ions. The form obtained by the writer[16,17] from the pressure equation of statistical mechanics is especially simple for the solvent activity; with conversion to a mole fraction basis, the electrical contribution to the activity coefficient of the solvent becomes

$$\ln \gamma_1^{el} = 2(1000/M_1)^{1/2}A_\phi I_x^{3/2}/(1 + \rho I_x^{1/2}) \qquad (4)$$

where A_ϕ is the usual Debye–Hückel parameter (for molality and the osmotic coefficient)

$$A_\phi = (1/3)(2\pi N_0 d_w/1000)^{1/2}(e^2/DkT)^{3/2} \qquad (5)$$

I_x is the ionic strength on a mole fraction basis

$$I_x = \frac{1}{2}\sum z_i^2 x_i \qquad (6)$$

where x_i is the mole fraction of ion i. Also ρ is a parameter related to the closest approach of ions but increased by the factor $(1000/M_1)^{1/2}$ from the parameter b used on the molality basis.

For our case of a 1:1 electrolyte these equations reduce to $I_x = \frac{1}{2}x_2$ and

$$\ln \gamma_1^{el} = (500/M_1)^{1/2}A_\phi x_2^{3/2}/[1 + (\rho/2^{1/2})x_2^{1/2}] \qquad (7)$$

In our earlier work[16] we found that a value of 1.2 was optimum for the "closest approach" parameter for a variety of simple electrolytes and with the use of particular forms for the terms expressing the effects of short-range forces. Conversion of the 1.2 by the factor $(1000/M_1)^{1/2}$ yields $\rho = 8.94$ on the new basis. But the expression for the effects of short-range forces is now different, eq 1; hence it seems reasonable to allow ρ to be adjusted provided that the result is of the same general magnitude. Also to maintain simplicity in equations for more complex mixed electrolytes, it is desirable that ρ should have the same value for a wide variety of salts. For the two examples considered here a somewhat larger value, 14.9 (corresponding to 2.0 on the old basis), seems quite satisfactory. A somewhat different choice may be desirable eventually, when a larger body of data can be considered.

If this expression for the electrostatic contribution is combined with eq 1, we have (for a 1:1 electrolyte)

$$\ln \gamma_1 = w_1 z_2^2 + (500/M_1)^{1/2}A_\phi x_2^{3/2}/(1 + \rho 2^{-1/2}x_2^{1/2}) \qquad (8a)$$

$$a_1 = x_1\gamma_1 \qquad (8b)$$

This expression fits the experimental data for the two systems (Ag,Tl)NO_3–H_2O and (Li,K)NO_3–H_2O very well with standard deviations in a_1 of 0.003 and 0.002, respectively. The parameters are $w_1 = 0.835$, $(b_1/b_2) = 0.56$ for (Ag,Tl)NO_3–H_2O at 98.5 °C. For (Li,K)NO_3–H_2O the data are at 119 °C near the fused salt and at 100 °C for the less concentrated solutions. Since w represents an energy divided by kT, the w values were adjusted correspondingly, with the result $w_1 = -1.124$ at 100 °C, -1.070 at 119 °C, $(b_1/b_2) = 1.0$. The

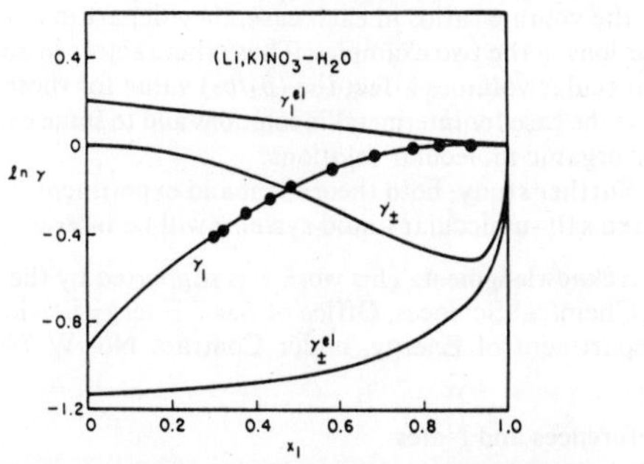

Figure 4. Activity coefficients for both water and salt for the system $(Li,K)NO_3-H_2O$ at 100 °C. It is a numerical accident that the curves for $\gamma_\pm$ based on the pure liquid or the infinitely dilute standard states are indistinguishable on the scale of the graph.

Debye–Hückel parameter, $A_\phi = 0.461$ at 100 °C, was taken from Bradley and Pitzer.[18]

A few words are needed at this point concerning the measure of composition of salt components and the definition of their activities and activity coefficients. One cannot use the familiar molality basis because the molality becomes infinite for the pure salt. Also we wish the sum of mole fractions to be unity; hence we include the number of ions in the salt in the definition of its mole fraction in contrast to the practice for the molality of a salt. We use the mole fraction for each ion as well as that for the complete salt and for the molecular component. Then, if n_i is the number of moles of the ith ion with n_1 the number of moles of molecular component

$$x_1 = n_1/(n_1 + \sum n_j),\ x_i = n_i/(n_1 + \sum n_j) \qquad (9)$$

with the sums including all ionic species. Also the mole fraction of the salt is the sum over its constituent ions

$$x_2 = \sum x_i \qquad (10)$$

One then defines activities and activity coefficients as

$$a_1 = x_1\gamma_1 \qquad (11a)$$

$$a_i = x_i\gamma_i \qquad (11b)$$

However, one does not ordinarily measure the activity of an ion but rather that of a neutral salt; hence one must define for a simple neutral salt MX

$$a_{MX} = a_M a_X = x_M x_X \gamma_{MX}^2 \qquad (12)$$

If there is but a single solute, $\gamma_\pm = \gamma_{MX}$ and $x_M = x_X = \frac{1}{2}x_2$; consequently

$$a_2 = a_{MX} = x_2^2\gamma_\pm^2/4 \qquad (13)$$

Note that the factor of 4 arises because x_2 is the sum of mole fractions of both ions, whereas in the usual definition m is the molality of either ion.

For more complex types of salts such as $CaCl_2$, $AlCl_3$, etc., the formulas become more complex but in a manner analogous to those[19] for the conventional molality basis. In general for three-component systems MX–NX–1, the activities and activity coefficients can be measured separately for MX and NX if the proportion of these components is varied. But the data we are considering is for fixed (and unit) ratio of x_M to x_N; hence only the average chemical potential or the geometric mean of activity can be determined. Thus for the apparent single salt

$$a_2 = (a_M a_N a_X^2)^{1/2} = (x_M x_N x_X^2)^{1/2}(\gamma_{MX}\gamma_{NX}) = x_2^2\gamma_\pm^2/8 \qquad (14)$$

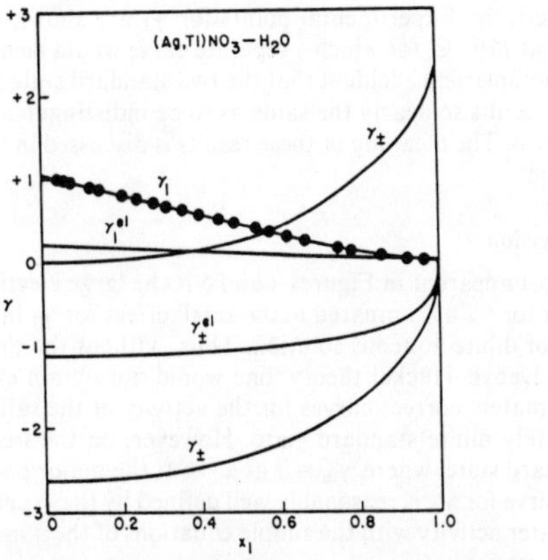

Figure 5. Activity coefficients for both water and salt for the system $(Ag,Tl)NO_3-H_2O$ at 98 °C. Separate curves give $\gamma_\pm$ on the pure liquid standard state (above) and the infinitely dilute standard state (below).

with $x_2 = 2x_X = 4x_M = 4x_N$ and $\gamma_\pm$ the geometric mean of the activity coefficients of MX and NX.

The expression for the electrostatic contribution to the activity coefficient of the water, eq 4, is consistent with the following equation for the electrostatic excess Gibbs energy:

$$G^{ex,el}/RT = -(\Sigma n_k)(1000/M_1)^{1/2}(4A_\phi I_x/\rho)\ln(1 + \rho I_x^{1/2}) \qquad (15)$$

where the sum includes all species, neutral as well as ions.

The derivative yields for the activity coefficient of any species

$$\ln\gamma_i = -(1000/M_1)^{1/2}A_\phi\{(2z_i^2/\rho)\ln(1 + \rho I_x^{1/2}) + (z_i^2 I_x^{1/2} - 2I_x^{3/2})/(1 + \rho I_x^{1/2})\} \qquad (16)$$

which reduces to eq 4 for a neutral component with $z_i = 0$. For various valence types of solutes, the result given in eq 16 can be combined for the appropriate charges of the ions to obtain the mean activity coefficient. In the case of 1:1 salts, this is trivial and yields

$$\ln\gamma_\pm = -(500/M_1)^{1/2}A_\phi\{(2^{3/2}/\rho)\ln(1 + \rho 2^{-1/2}x_2^{1/2}) + (x_2^{1/2} - x_2^{3/2})/(1 + \rho 2^{-1/2}x_2^{1/2})\} \qquad (17)$$

For the complete expression for the activity coefficient of the solute, alternate standard states must be considered. If the pure fused salt exists at the temperature of interest, it is usually preferable to take it as the standard state, $\gamma_\pm = 1$, at $x_2 = 1$. Then for a 1:1 electrolyte

$$\ln\gamma_\pm = w_2 z_1^2 - (500/M_1)^{1/2}A_\phi\{(2^{3/2}/\rho)\ln[(1 + \rho 2^{-1/2}x_2^{1/2})/(1 + \rho 2^{-1/2})] + (x_2^{1/2} - x_2^{3/2})/(1 + \rho 2^{-1/2}x_2^{1/2})\}$$
$$\text{(pure salt standard state)} \qquad (18)$$

The standard state based on the infinitely dilute solute must be used if the pure solute is not liquid or may be desired for comparative purposes. This basis, $\gamma_\pm = 1$ at $x_2 = 0$, requires just a change in eq 18 by a constant and yields

$$\ln\gamma_\pm = w_2(z_1^2 - 1) - (500/M_1)^{1/2}A_\phi[(2^{3/2}/\rho)\ln(1 + \rho 2^{-1/2}x_2^{1/2}) + (x_2^{1/2} - x_2^{3/2})/(1 + \rho 2^{-1/2}x_2^{1/2})]$$
$$\text{(infinitely dilute standard state)} \qquad (19)$$

The activity-coefficient curves, including those for the electrostatic terms separately, are shown in Figures 4 and 5 for the two systems $(Li,K)NO_3-H_2O$ and $(Ag,Tl)NO_3-H_2O$, respectively. Experimental points for γ_1 are shown (except those at 119 °C, for which a separate curve would be needed). It is a numerical accident that the two standard states for $\gamma_\pm$ yield results so nearly the same as to be indistinguishable on Figure 4. The meaning of these results is discussed in the next section.

Discussion

Most apparent in Figures 4 and 5 is the large electrostatic effect for $\gamma_\pm$ as compared to the small effect for γ_1 in the region of dilute aqueous solution. Thus, without the guidance from Debye–Hückel theory, one would not obtain even approximately correct curves for the activity of the salt on the infinitely dilute standard state. However, on the fused salt standard state, where $\gamma_\pm = 1$ at $x_2 = 1$, the major portion of the curve for $\gamma_\pm$ is reasonably well defined by the experiments on water activity with the simple equations of the nonelectrolyte type.

The postulate mentioned above, that the departure from random distribution of ions is roughly constant from concentrated solutions through to the pure fused salt, is supported by the relative constancy of $\gamma_\pm{}^{el}$ over that range of composition. There is, of course, some ambiguity in the separation of the electrostatic effect; hence, this postulate cannot be proven exactly from data such as these.

It is interesting to compare volumetric data[20] with the ratio (b_1/b_2) from the activity equations. For $(Li,K)NO_3$–H_2O the ratio of molar volume of water to the average volume per ion is 0.87. The more exact equation with the Debye–Hückel term was fitted with $(b_1/b_2) = 1.0$, whereas the more approximate equation fitted rather well with either 1.2 or 1.0 for this ratio. For the system $(Ag,Tl)NO_3$–H_2O, the experimental volume ratio is 0.82 while the (b_1/b_2) values are 0.56 from eq 18 or 0.50 from eq 1. While the (b_1/b_2) ratios are in the general vicinity of the volume ratios in each case, they depart in opposite directions in the two examples. Thus other factors in addition to molecular volumes affect the (b_1/b_2) value for these systems as is the case for intermetallic solutions and to some extent even for organic molecular solutions.

Further study, both theoretical and experimental, of these fused salt–molecular liquid systems will be interesting.

Acknowledgment. This work was supported by the Division of Chemical Sciences, Office of Basic Energy Sciences, U.S. Department of Energy, under Contract No. W-7405-Eng-48.

References and Notes

(1) C. A. Kraus, *J. Phys. Chem.*, **58**, 673 (1954).
(2) A. N. Campbell and W. G. Paterson, *Can. J. Chem.*, **36**, 1004 (1958).
(3) A. N. Campbell, G. H. Debus, and E. M. Kartzmark, *Can. J. Chem.*, 33, 1508 (1955).
(4) P. Debye and E. Hückel, *Phys. Z.*, **24**, 185, 305 (1923).
(5) L. Onsager, *Phys. Z.*, **28**, 27 (1927).
(6) R. P. Seward, *J. Am. Chem. Soc.*, **73**, 515 (1951).
(7) C. W. Davies, "Ion Association", Butterworths, London 1962. Note especially Chapter 10.
(8) T. B. Tripp and J. Braunstein, *J. Am. Chem. Soc.*, **73**, 1984 (1954).
(9) H. Braunstein and J. Braunstein, *J. Chem. Thermodyn.*, **3**, 419 (1971).
(10) J. F. Gibbard and G. Scatchard, *J. Chem. Eng. Data*, **18**, 293 (1973).
(11) M-C Trudelle, M. Abraham, and J. Sangster, *Can. J. Chem.*, **55**, 1713 (1977).
(12) R. A. Robinson and R. H. Stokes, "Electrolyte Solutions", 2nd ed., Butterworths, London, 1965.
(13) J. J. van Laar, "Sechs Vortrage uber das Thermodynamische Potential", Vieweg-Verlag, 1906. See also *Z. Phys. Chem. (Leipzig)*, **72**, 723 (1919).
(14) J. M. Prausnitz, "Molecular Thermodynamics of Fluid-Phase Equilibria", Prentice-Hall, Englewood Cliffs, N.J., 1969.
(15) R. H. Stokes and R. A. Robinson, *J. Am. Chem. Soc.*, **70**, 1870 (1948).
(16) K. S. Pitzer, *J. Phys. Chem.*, **77**, 268 (1973).
(17) K. S. Pitzer, *Acc. Chem. Res.*, **10**, 371 (1977).
(18) D. J. Bradley and K. S. Pitzer, *J. Phys. Chem.*, **83**, 1599 (1979).
(19) K. S. Pitzer and L. Brewer, Lewis and Randall's "Thermodynamics", McGraw-Hill, New York, 1961, Chapter 22.
(20) "Gmelin Handbuch der Anorganischen Chemie", Springer-Verlag, West Berlin.

Comment: Electrolytes. From Dilute Solutions to Fused Salts

Most electrolytic solutions are limited in range of composition by the solubility of the solid salt. But there are exceptions, especially at high temperatures, where there is continuous miscibility to the pure liquid electrolyte. Since the molality becomes infinite for the fused salt, it must be abandoned as a measure of composition and the mole fraction used instead. This paper gives the general conceptual basis for successful treatment of such systems. Subsequently, refinements for higher accuracy were added, and the entire method generalized to mixed systems of an unlimited number of components. All of these features are included in a paper by Clegg, Pitzer, and Brimblecombe, *J. Phys. Chem.* **96**, 9470–9479 (1992).